STRUCTURAL STEEL DESIGNER'S HANDBOOK

Other McGraw-Hill Books Edited by Frederick S. Merritt

Merritt • STANDARD HANDBOOK FOR CIVIL ENGINEERS
Merritt & Ricketts • BUILDING DESIGN AND CONSTRUCTION HANDBOOK

Other McGraw-Hill Books of Interest

Beall • MASONRY DESIGN AND DETAILING
Breyer • DESIGN OF WOOD STRUCTURES
Brown • FOUNDATION BEHAVIOR AND REPAIR
Faherty & Williamson • WOOD ENGINEERING AND CONSTRUCTION HANDBOOK
Gaylord & Gaylord • STRUCTURAL ENGINEERING HANDBOOK
Harris • NOISE CONTROL IN BUILDINGS
Kubal • WATERPROOFING THE BUILDING ENVELOPE
Newman • STANDARD HANDBOOK OF STRUCTURAL DETAILS FOR BUILDING CONSTRUCTION
Sharp • BEHAVIOR AND DESIGN OF ALUMINUM STRUCTURES
Waddell & Dobrowolski • CONCRETE CONSTRUCTION HANDBOOK

STRUCTURAL STEEL DESIGNER'S HANDBOOK

Roger L. Brockenbrough Editor

R. L. Brockenbrough & Associates, Inc.
Pittsburgh, Pennsylvania

Frederick S. Merritt Editor

Consulting Engineer, West Palm Beach, Florida

Second Edition

McGRAW-HILL, INC.

New York   San Francisco   Washington, D.C.   Auckland   Bogotá
Caracas   Lisbon   London   Madrid   Mexico City   Milan
Montreal   New Delhi   San Juan   Singapore
Sydney   Tokyo   Toronto

Library of Congress Cataloging-in-Publication Data

Structural steel designer's handbook / Roger L. Brockenbrough, editor,
   Frederick S. Merritt, editor.—2nd ed.
      p.   cm.
   Includes index.
   ISBN 0-07-008776-8
   1. Building, Iron and steel.   2. Steel, Structural.
I. Brockenbrough, R. L.   II. Merritt, Frederick S.
TA684.S79   1994
624.1′821—dc20   93-38088
                                                              CIP

1 2 3 4 5 6 7 8 9 0   DOH/DOH   9 9 8 7 6 5 4 3

ISBN 0-07-008776-8

*The sponsoring editor for this book was Larry Hager, the editing supervisor
was Peggy Lamb, and the production supervisor was Suzanne W. Babeuf. It
was set in Times Roman by Techna Type, Inc.*

Printed and bound by R. R. Donnelley & Sons Company.

This book is printed on acid-free paper.

CONTENTS

Section 3. General Structural Theory   *Ronald D. Ziemian*   3.1

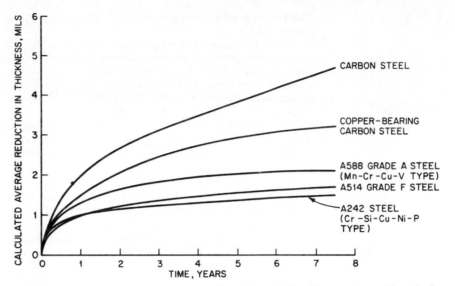

FIGURE 1.2 Corrosion curves for structural steels in an industrial atmosphere. *(From R. L. Brockenbrough and B. G. Johnston, USS Steel Design Manual, R. L. Brockenbrough & Associates, Inc., Pittsburgh, Pa., with permission.)*

with different chemical compositions; the minimum yield point ranges from 50 to 75 ksi depending on grade and thickness.

A852 is a quenched-and-tempered HSLA plate steel of the weathering type. It is intended for welded bridges and buildings and similar applications where weight savings, durability, and good notch toughness are important. It provides a minimum yield point of 70 ksi in thickness up to 4 in. The resistance to atmospheric corrosion is typically four times that of carbon steel.

1.1.4   Heat-Treated Constructional Alloy Steels

Steels that contain alloying elements in excess of the limits for carbon steel and are heat treated to obtain a combination of high strength and toughness are termed **constructional alloy steels.** Having a yield strength of 100 ksi, these are the strongest steels in general structural use.

A514 includes several grades of quenched and tempered steels, to permit use of various compositions developed by producers to obtain the specified strengths. Maximum thickness ranges from 1¼ to 6 in depending on the grade. Minimum yield strength for plate thicknesses over 2½ in is 90 ksi. Steels furnished to this specification can provide a resistance to atmospheric corrosion up to four times that of structural carbon steel depending on the grade.

Constructional alloy steels are also frequently selected because of their ability to resist abrasion. For many types of abrasion, this resistance is related to hardness or tensile strength. Therefore, constructional alloy steels may have nearly twice the resistance to abrasion provided by carbon steel. Also available are numerous grades that have been heat treated to increase the hardness even more.

1.1.5   Bridge Steels

Steels for application in bridges are covered by A709, which includes steel in several of the categories mentioned above. Under this specification, grades 36, 50, 70, and

100 are steels with yield strengths of 36, 50, 70, and 100 ksi, respectively. (See also Table 10.18.)

The grade designation is followed by the letter W, indicating whether ordinary or high atmospheric corrosion resistance is required. An additional letter, T or F, indicates that Charpy V-notch impact tests must be conducted on the steel. The T designation indicates that the material is to be used in a non-fracture-critical application as defined by AASHTO; the F indicates use in a fracture-critical application.

A trailing numeral, 1, 2, or 3, indicates the testing zone, which relates to the lowest ambient temperature expected at the bridge site. (See Table 1.2.) As indicated by the first footnote in the table, the service temperature for each zone is

TABLE 1.2 Charpy V-Notch Toughness for A709 Bridge Steels*

Grade	Maximum thickness, in, inclusive	Joining/ fastening method	Minimum average energy, ft·lb	Test temperature, °F		
				Zone 1	Zone 2	Zone 3
			Non-fracture-critical members			
36T	4	Mech./Weld.	15	70	40	10
50T,† 50WT†	2	Mech./Weld.	15			
	2 to 4	Mechanical	15	70	40	10
	2 to 4	Welded	20			
70WT‡	2½	Mech./Weld.	20			
	2½ to 4	Mechanical	20	50	20	−10
	2½ to 4	Welded	25			
100T, 100WT	2½	Mech./Weld.	25			
	2½ to 4	Mechanical	25	30	0	−30
	2½ to 4	Welded	35			
			Fracture-critical members			
36F	1½	Mech./Weld.	25	70	40	10
	1½ to 4	Mech./Weld.	25			−10
50F,† 50WF†	1½	Mech./Weld.	25			10
	1½ to 2	Mech./Weld.	25	70	40	−10
	2 to 4	Mechanical	25			−10
	2 to 4	Welded	30			−10
70WF‡	1½	Mech./Weld.	30			−10
	1½ to 2½	Mech./Weld.	30	20	20	−30
	2½ to 4	Mechanical	30			−30
	2½ to 4	Welded	35			−30
100F, 100WF	2½	Mech./Weld.	35			−30
	2½ to 4	Mechanical	35	0	0	−30
	2½ to 4	Welded	45			NA

*Minimum **service** temperatures:
    Zone 1, 0°F; Zone 2, below 0 to −30°F; Zone 3, below −30 to −60°F.
†If yield strength exceeds 65 ksi, reduce test temperature by 15°F for each 10 ksi above 65 ksi.
‡If yield strength exceeds 85 ksi, reduce test temperature by 15°F for each 10 ksi above 85 ksi.

Section 4. Analysis of Special Structures   *Frederick S. Merritt and Louis F. Geschwinder*   4.1

Section 5. Connections   *W. A. Thornton and T. Kane*   5.1

Section 6. Building Design Criteria   *R. A. LaBoube*   6.1

Section 11. Beam and Girder Bridges *Alfred Hedefine and John Swindlehurst*   11.1

Section 12. Truss Bridges *John M. Kulicki, Joseph E. Prickett, and David H. LeRoy*   12.1

Section 13. Arch Bridges   *Arthur W. Hedgren, Jr.*   13.1

Section 14. Cable-Suspended Bridges   *Walter Podolny, Jr.*   14.1

CONTRIBUTORS

Boring, Delbert F. *Regional Director, Construction Codes and Standards, American Iron and Steel Institute, Columbus, Ohio* (SECTION 6 BUILDING DESIGN CRITERIA)

Brockenbrough, Roger L. *R. L. Brockenbrough & Associates, Inc., Pittsburgh, Pennsylvania* (SECTION 1 PROPERTIES OF STRUCTURAL STEELS)

Cuoco, Daniel A. *Vice President, Lev Zetlin Associates, Inc., New York, New York* (SECTION 8 FLOOR AND ROOF SYSTEMS)

Geschwinder, Louis F. *Professor of Architectural Engineering, Pennsylvania State University, University Park, Pennsylvania* (SECTION 4 ANALYSIS OF SPECIAL STRUCTURES)

Haris, Ali A. K. *Vice President, Structural Engineering, Senior Project Manager, Ellerbe Becket, Inc., Kansas City, Missouri* (SECTION 7 DESIGN OF BUILDING MEMBERS)

Hedgren, Arthur W. Jr. *Senior Vice President, HDR Engineering, Inc., Pittsburgh, Pennsylvania* (SECTION 13 ARCH BRIDGES)

Hedefine, Alfred *President, Parsons, Brinckerhoff, Quade & Douglas, Inc., New York, New York* (SECTION 11 BEAM AND GIRDER BRIDGES)

Kane, T. *Cives Steel Company, Roswell, Georgia* (SECTION 5 CONNECTIONS)

Kulicki, John M. *Senior Vice President, Modjeski and Masters, Inc., Harrisburg, Pennsylvania* (SECTION 12 TRUSS BRIDGES)

LeRoy, David H. *Associate, Modjeski and Masters, Inc., Harrisburg, Pennsylvania* (SECTION 12 TRUSS BRIDGES)

LaBoube, Roger A. *Associate Professor of Civil Engineering, University of Missouri–Rolla, Rolla, Missouri* (SECTION 6 BUILDING DESIGN CRITERIA)

Marshall, Richard W. *Vice President, American Steel Erectors, Inc., Allentown, Pennsylvania* (SECTION 2 FABRICATION AND ERECTION)

Merritt, Frederick S. *Consulting Engineer, West Palm Beach, Florida* (SECTION 1 PROPERTIES OF STRUCTURAL STEELS; SECTION 4 ANALYSIS OF SPECIAL STRUCTURES)

Nickerson, Robert L., P.E. *Consultant—NBE, Ltd., Hempstead, Maryland* (SECTION 10 APPLICATION OF DESIGN CRITERIA FOR BRIDGES)

Peshek, Charles, Jr. *Consulting Engineer, Napierville, Illinois* (SECTION 2 FABRICATION AND ERECTION)

Podolny, Walter, Jr. *Bridge Division, Office of Engineering, Federal Highway Administration, U.S. Department of Transportation, Washington, D.C.* (SECTION 14 CABLE-SUSPENDED BRIDGES)

Prickett, Joseph E. *Associate, Modjeski and Masters, Inc., Harrisburg, Pennsylvania* (SECTION 12 TRUSS BRIDGES)

Roeder, Charles W. *Professor of Civil Engineering, University of Washington, Seattle, Washington* (SECTION 9 LATERAL-FORCE DESIGN)

Swindlehurst, John *Senior Professional Associate, Parsons Brinckerhoff, Inc., West Trenton, New Jersey* (SECTION 11 BEAM AND GIRDER BRIDGES)

Thornton, William A. *Chief Engineer, Cives Steel Company, Roswell, Georgia* (SECTION 5 CONNECTIONS)

Ziemian, Ronald D. *Assistant Professor of Civil Engineering, Bucknell University, Lewisburg, Pennsylvania* (SECTION 3 GENERAL STRUCTURAL THEORY)

ABOUT THE EDITORS

Roger L. Brockenbrough retired from U.S. Steel as a senior research consultant in 1991 after 30 years of service. He currently serves as a consultant to companies on the development and evaluation of new products. He is chairman of the American Iron and Steel Institute Committee on Specifications for the Design of Cold-Formed Steel Structural Members, and sits on committees of several other professional organizations. Mr. Brockenbrough is a registered Professional Engineer in four states and the author of more than 30 publications on steel design and related topics.

Frederick S. Merritt is a consulting engineer with vast experience in building and bridge design, structural analysis, and construction management. Formerly senior editor of *Engineering News Record* magazine, he is the author or editor of many books on civil engineering and construction-related topics. Mr. Merritt is a Fellow of the American Society of Civil Engineers and a Senior Member of ASTM.

PREFACE

This handbook has been developed to serve as a comprehensive reference source for designers of steel structures. Included is information on materials, fabrication, erection, structural theory, and connections, as well as the many facets of designing structural-steel systems and members for buildings and bridges. The information presented applies to a wide range of structures.

The handbook should be useful to consulting engineers; architects; construction contractors; fabricators and erectors; engineers employed by federal, state, and local governments; and educators. It will also be a good reference for engineering technicians and detailers. The material has been presented in easy-to-understand form to make it useful to professionals and those with more limited experience. Numerous examples, worked out in detail, illustrate design procedures.

The thrust is to provide practical techniques for cost-effective design as well as explanations of underlying theory and criteria. Design methods and equations from leading specifications are presented for ready reference. This includes those of the American Institute of Steel Construction (AISC), the American Association of State Highway and Transportation Officials (AASHTO), and the American Railway Engineering Association (AREA). Both the traditional allowable-stress design (ASD) approach and the load-and-resistance-factor design (LRFD) approach are presented. Nevertheless, users of this handbook would find it helpful to have the latest edition of these specifications on hand, because they are changed annually, as well as the AISC "Steel Construction Manual," ASD and LRFD.

Contributors to this book are leading experts in design, construction, materials, and structural theory. They offer know-how and techniques gleaned from vast experience. They include well-known consulting engineers, university professors, and engineers with an extensive fabrication and erection background. This blend of experiences contributes to a broad, well-rounded presentation.

The book begins with an informative section on the types of steel, their mechanical properties, and the basic behavior of steel under different conditions. Topics such as cold-work, strain-rate effects, temperature effects, fracture, and fatigue provide in-depth information. Aids are presented for estimating the relative weight and material cost of steels for various types of structural members to assist in selecting the most economical grade. A review of fundamental steel-making practices, including the now widely used continuous-casting method, is presented to give designers better knowledge of structural steels and alloys and how they are produced.

Because of their impact on total cost, a knowledge of fabrication and erection methods is a fundamental requirement for designing economical structures. Accordingly, the book presents description of various shop fabrication procedures, including cutting steel components to size, punching, drilling, and welding. Available erection equipment is reviewed, as well as specific methods used to erect bridges and buildings.

A broad treatment of structural theory follows to aid engineers in determining the forces and moments that must be accounted for in design. Basic mechanics, traditional tools for analysis of determinate and indeterminate structures, matrix methods, and other topics are discussed. Structural analysis tools are also presented for various special structures, such as arches, domes, cable systems, and orthotropic plates. This information is particularly useful in making preliminary designs and verifying computer models.

Connections have received renewed attention in current structural steel design, and improvements have been made in understanding their behavior in service and in design techniques. A comprehensive section on design of structural connections presents approved methods for all of the major types, bolted and welded. Information on materials for bolting and welding is included.

Successive sections cover design of buildings, beginning with basic design criteria and other code requirements, including minimum design dead, live, wind, seismic, and other loads. A state-of-the-art summary describes current fire-resistant construction, as well as available tools that allow engineers to design for fire protection and avoid costly tests. In addition, the book discusses the resistance of various types of structural steel to corrosion and describes corrosion-prevention methods.

A large part of the book is devoted to presentation of practical approaches to design of tension, compression, and flexural members, composite and noncomposite.

One section is devoted to selection of floor and roof systems for buildings. This involves decisions that have major impact on the economics of building construction. Alternative support systems for floors are reviewed, such as the stub-girder and staggered-truss systems. Also, framing systems for short and long-span roof systems are analyzed.

Another section is devoted to design of framing systems for lateral forces. Both traditional and newer-type bracing systems, such as eccentric bracing, are analyzed.

Over one-third of the handbook is dedicated to design of bridges. Discussions of design criteria cover loadings, fatigue, and the various facets of member design. Information is presented on use of weathering steel. Also, tips are offered on how to obtain economical designs for all types of bridges. In addition, numerous detailed calculations are presented for design of rolled-beam and plate-girder bridges, straight and curved, composite and noncomposite, box girders, orthotropic plates, and continuous and simple-span systems.

Notable examples of truss and arch designs, taken from current practice, make these sections valuable references in selecting the appropriate spatial form for each site, as well as executing the design.

The concluding section describes the various types of cable-supported bridges and the cable systems and fittings available. In addition, design of suspension bridges and cable-stayed bridges is covered in detail.

The authors and editors are indebted to numerous sources for the information presented. Space considerations preclude listing all, but credit is given wherever feasible, especially in bibliographies throughout the book.

The reader is cautioned that independent professional judgment must be exercised when information set forth in this handbook is applied. Anyone making use of this information assumes all liability arising from such use.

Roger L. Brockenbrough
Frederick S. Merritt

TABLE 1.1 Specified Minimum Properties for Structural Steel Shapes and Plates*

ASTM designation	Plate-thickness range, in	ASTM group for structural shapes†	Yield stress, ksi	Tensile strength, ksi‡	Elongation, %	
					In 2 in§	In 8 in
Carbon steels						
A36	8 maximum	1–5	36	58–80	23–21	20
	over 8	1–5	32	58–80	23	20
A573						
Grade 58	1½ maximum	¶	32	58–71	24	21
Grade 65	1½ maximum	¶	35	65–77	23	20
Grade 70	1½ maximum	¶	42	70–90	21	18
High-strength low-alloy steels						
A242	¾ maximum	1 and 2	50	70	21	18
	Over ¾ to 1½ max	3	46	67	21	18
	Over 1½ to 4 max	4 and 5	42	63	21	18
A588	4 maximum	1–5	50	70	21	18
	Over 4 to 5 max	1–5	46	67	21	18
	Over 5 to 8 max	1–5	42	63	21	—
A572						
Grade 42	6 maximum	1–5	42	60	24	20
Grade 50	4 maximum	1–5	50	65	21	18
Grade 60	1¼ maximum	1 and 2	60	75	18	16
Grade 65	1¼ maximum	1	65	80	17	15
Heat-treated carbon and HSLA steels						
A633						
Grade A	4 maximum	¶	42	63–83	23	18
Grade C	Over 2½ to 4 max	¶	50	70–90	23	18
Grade D	Over 2½ to 4 max	¶	50	70–90	23	18
Grade E	4 maximum	¶	60	80–100	23	18
	Over 4 to 6 max	¶	55	75–95	23	18
A678						
Grade A	1½ maximum	¶	50	70–90	22	—
Grade B	2½ maximum	¶	60	80–100	22	—
Grade C	¾ maximum	¶	75	95–115	19	—
	Over ¾ to 1½ max	¶	70	90–110	19	—
	Over 1½ to 2 max	¶	65	85–105	19	—
Grade D	3 maximum	¶	75	90–110	18	—
A852	4 maximum	¶	70	90–110	19	—

TABLE 1.1 Specified Minimum Properties for Structural Steel Shapes and Plates* (Continued)

ASTM designation	Plate-thickness range, in	ASTM group for structural shapes†	Yield stress, ksi	Tensile strength, ksi‡	Elongation, % In 2 in§	Elongation, % In 8 in
Heat-treated constructional alloy steels						
A514	2½ maximum	¶	100	110–130	18	—
	Over 2½ to 6 max.	¶	90	100–130	16	—

*The following are approximate values for all the steels:
  Modulus of elasticity—29×10^3 ksi.
  Shear modulus—11×10^3 ksi.
  Poisson's ratio—0.30.
  Yield stress in shear—0.57 times yield stress in tension.
  Ultimate strength in shear—⅔ to ¾ times tensile strength.
  Coefficient of thermal expansion—6.5×10^{-6} in per in per deg F for temperature range -50 to $+150°$F.
  Density—490 lb/ft³.
†See ASTM A6 for structural shape group classification.
‡Where two values are shown for tensile strength, the first is minimum and the second is maximum.
§The minimum elongation values are modified for some thicknesses in accordance with the specification for the steel. Where two values are shown for the elongation in 2 in, the first is for plates and the second for shapes.
¶Not applicable.

of carbon steel to atmospheric corrosion can be doubled by specifying a minimum copper content of 0.20%.) Typical corrosion curves for several steels exposed to industrial atmosphere are shown in Fig. 1.2.

A588 and A242 steels are called **weathering steels** because, when subjected to alternate wetting and drying in most bold atmospheric exposures, they develop a tight oxide layer that substantially inhibits further corrosion. They are often used bare (unpainted) where the oxide finish that develops is desired for aesthetic reasons or for economy in maintenance. Bridges and exposed building framing are typical examples of such applications. Designers should investigate potential applications thoroughly, however, to determine whether a weathering steel will be suitable. Information on bare-steel applications is available from steel producers.

A572 specifies columbium-vanadium HSLA steels in four grades with minimum yield points of 42, 50, 60, and 65 ksi. Grade 42 in thicknesses up to 6 in and grade 50 in thicknesses up to 4 in are used for welded bridges. All grades may be used for riveted or bolted construction and for welded construction in most applications other than bridges.

1.1.3 Heat-Treated Carbon and HSLA Steels

Both carbon and HSLA steels can be heat treated to provide yield points in the range of 50 to 75 ksi. This provides an intermediate strength level between the as-rolled HSLA steels and the heat-treated constructional alloy steels.

A633 is a normalized HSLA plate steel for applications where improved notch toughness is desired. Available in four grades with different chemical compositions, the minimum yield point ranges from 42 to 60 ksi depending on grade and thickness.

A678 includes quenched-and-tempered plate steels (both carbon and HSLA compositions) with excellent notch toughness. It is also available in four grades

SECTION 1
PROPERTIES OF STRUCTURAL STEELS

PART 1

INTRINSIC PROPERTIES

R. L. Brockenbrough
R. L. Brockenbrough & Associates, Inc.,
Pittsburgh, Pennsylvania

This section presents and discusses the properties of structural steels that are of importance in design and construction. Designers should be familiar with these properties so that they can select the most economical combination of suitable steels for each application and use the materials efficiently and safely.

In accordance with contemporary practice, the steels described in this section are given the names of the corresponding specifications of ASTM, 1916 Race St., Philadelphia, PA 19103. For example, all steels covered by ASTM A588, "Specification for High-strength Low-alloy Structural Steel," are called A588 steel.

1.1 STRUCTURAL STEEL SHAPES AND PLATES

Steels for structural uses may be classified by chemical composition, tensile properties, and method of manufacture as carbon steels, high-strength low-alloy steels (HSLA), heat-treated carbon steels, and heat-treated constructional alloy steels. A typical stress-strain curve for a steel in each classification is shown in Fig. 1.1 to illustrate the increasing strength levels provided by the four classifications of steel. The availability of this wide range of specified minimum strengths, as well as other material properties, enables the designer to select an economical material that will perform the required function for each application.

Some of the most widely used steels in each classification are listed in Table 1.1 with their specified strengths in shapes and plates. These steels are weldable, but the welding materials and procedures for each steel must be in accordance with approved methods. Welding information for each of the steels is available from most steel producers and in publications of the American Welding Society.

1.1.1 Carbon Steels

A steel may be classified as a carbon steel if (1) the maximum content specified for alloying elements does not exceed the following: manganese—1.65%, silicon—0.60%, copper—0.60%; (2) the specified minimum for copper does not exceed 0.40%; and (3) no minimum content is specified for other elements added to obtain a desired alloying effect.

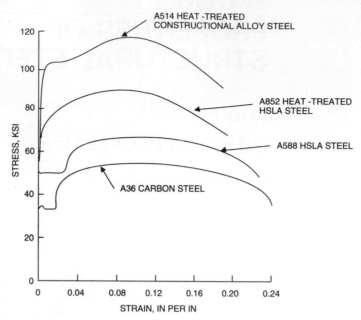

FIGURE 1.1   Typical stress-strain curves for structural steels. (Curves have been modified to reflect minimum specified properties.)

A36 steel is the principal carbon steel for bridges, buildings, and many other structural uses. This steel provides a minimum yield point of 36 ksi in all structural shapes and in plates up to 8 in thick.

A573, the other carbon steel listed in Table 1.1, is available in three strength grades for plate applications in which improved notch toughness is important.

1.1.2   High-Strength Low-Alloy Steels

Those steels which have specified minimum yield points greater than 40 ksi and achieve that strength in the hot-rolled condition, rather than by heat treatment, are known as HSLA steels. Because these steels offer increased strength at moderate increases in price over carbon steels, they are economical for a variety of applications.

A242 steel is a **weathering steel,** used where resistance to atmospheric corrosion is of primary importance. Steels meeting this specification usually provide a resistance to atmospheric corrosion at least four times that of structural carbon steel. However, when required, steels can be selected to provide a resistance to atmospheric corrosion of five to eight times that of structural carbon steels. A specified minimum yield point of 50 ksi can be furnished in plates up to ¾ in thick and the lighter structural shapes. It is available with a lower yield point in thicker sections, as indicated in Table 1.1.

A588 is the primary weathering steel for structural work. It provides a 50-ksi yield point in plates up to 4 in thick and in all structural sections; it is available with a lower yield point in thicker plates. Several grades are included in the specification to permit use of various compositions developed by steel producers to obtain the specified properties. This steel provides about four times the resistance to atmospheric corrosion of structural carbon steels.

These relative corrosion ratings are determined from the slopes of corrosion-time curves and are based on carbon steels not containing copper. (The resistance

Section 7. Design of Building Members   *Ali A. K. Haris*   **7.1**

Section 8. Floor and Roof Systems   *Daniel A. Cuoco*   **8.1**

considerably less than the Charpy V-notch impact-test temperature. This accounts for the fact that the dynamic loading rate in the impact test is more severe than that to which the structure is subjected. The toughness requirements depend on fracture criticality, grade, thickness, and method of connection.

(R. L. Brockenbrough, Sec. 9 in *Standard Handbook for Civil Engineers*, 4th ed., F. S. Merritt, ed., McGraw-Hill, Inc., New York.)

1.2   STEEL-QUALITY DESIGNATIONS

Steel plates, shapes, sheetpiling, and bars for structural uses—such as the load-carrying members in buildings, bridges, ships, and other structures—are usually ordered to the requirements of ASTM A6 and are referred to as **structural-quality steels.** (A6 does not indicate a specific steel.) This specification contains general requirements for delivery related to chemical analysis, permissible variations in dimensions and weight, permissible imperfections, conditioning, marking, and tension and bend tests of a large group of structural steels. (Specific requirements for the chemical composition and tensile properties of these steels are included in the specifications discussed in Art. 1.1.) All the steels included in Table 1.1 are structural-quality steels.

In addition to the usual die stamping or stenciling used for identification, plates and shapes of certain steels covered by A6 are marked in accordance with a color code, when specified by the purchaser, as indicated in Table 1.3.

Steel plates for pressure vessels are usually furnished to the general requirements of ASTM A20 and are referred to as **pressure-vessel-quality steels.** Generally, a greater number of mechanical-property tests and additional processing are required for pressure-vessel-quality steel.

1.3   RELATIVE COST OF STRUCTURAL STEELS

Because of the many strength levels and grades now available, designers usually must investigate several steels to determine the most economical one for each application. As a guide, relative material costs of several structural steels used as

TABLE 1.3 Identification Colors

Steels	Color
A36	None
A242	Blue
A514	Red
A572 grade 42	Green and white
A572 grade 50	Green and yellow
A572 grade 60	Green and gray
A572 grade 65	Green and blue
A588	Blue and yellow
A852	Blue and orange

tension members, beams, and columns are discussed below. The comparisons are based on cost of steel to fabricators (steel producer's price) because, in most applications, cost of a steel design is closely related to material costs. However, the total fabricated and erected cost of the structure should be considered in a final cost analysis. Thus the relationships shown should be considered as only a general guide.

Tension Members. Assume that two tension members of different-strength steels have the same length. Then, their material-cost ratio C_2/C_1 is

$$\frac{C_2}{C_1} = \frac{A_2}{A_1} \frac{p_2}{p_1} \tag{1.1}$$

where A_1 and A_2 are the cross-sectional areas and p_1 and p_2 are the material prices per unit weight. If the members are designed to carry the same load at a stress that is a fixed percentage of the yield point, the cross-sectional areas are inversely proportional to the yield stresses. Therefore, their relative material cost can be expressed as

$$\frac{C_2}{C_1} = \frac{F_{y1}}{F_{y2}} \frac{p_2}{p_1} \tag{1.2}$$

where F_{y1} and F_{y2} are the yield stresses of the two steels. The ratio p_2/p_1 is the relative price factor. Values of this factor for several steels are given in Table 1.4, with A36 steel as the base. The table indicates that the relative price factor is always less than the corresponding yield-stress ratio. Thus the relative cost of tension members calculated from Eq. (1.2) favors the use of high-strength steels.

Beams. The optimal section modulus for an elastically designed I-shaped beam results when the area of both flanges equals half the total cross-sectional area of the member. Assume now two members made of steels having different yield points and designed to carry the same bending moment, each beam being laterally braced and proportioned for optimal section modulus. Their relative weight W_2/W_1 and relative cost C_2/C_1 are influenced by the web depth-to-thickness ratio d/t. For

TABLE 1.4 Relative Price Factors*

Steel	Minimum yield stress, ksi	Relative price factor	Ratio of minimum yield stresses	Relative cost of tension members
A36	36	1.00	1.00	1.00
A572 grade 42	42	1.09	1.17	0.93
A572 grade 50	50	1.12	1.39	0.81
A588 grade A	50	1.23	1.39	0.88
A852	70	1.52	1.94	0.78
A514 grade B	100	2.07	2.78	0.75

*Based on plates ¾ × 96 × 240 in. Price factors for shapes tend to be lower. A852 and A514 steels are not available in shapes. Prices were those in effect Feb. 1, 1992.

example, if the two members have the same d/t values, such as a maximum value imposed by the manufacturing process for rolled beams, the relationships are

$$\frac{W_2}{W_1} = \left(\frac{F_{y1}}{F_{y2}}\right)^{2/3} \tag{1.3}$$

$$\frac{C_2}{C_1} = \frac{p_2}{p_1}\left(\frac{F_{y1}}{F_{y2}}\right)^{2/3} \tag{1.4}$$

If each of the two members has the maximum d/t value that precludes elastic web buckling, a condition of interest in designing fabricated plate girders, the relationships are

$$\frac{W_2}{W_1} = \left(\frac{F_{y1}}{F_{y2}}\right)^{1/2} \tag{1.5}$$

$$\frac{C_2}{C_1} = \frac{p_2}{p_1}\left(\frac{F_{y1}}{F_{y2}}\right)^{1/2} \tag{1.6}$$

Table 1.5 shows relative weights and relative material costs for several structural steels. These values were calculated from Eqs. (1.3) to (1.6) and the relative price factors given in Table 1.4, with A36 steel as the base. The table shows the decrease in relative weight with increase in yield stress. The relative material costs show that when bending members are thus compared for girders, the cost of A572 grade 50 steel is lower than that of A36 steel, and the cost of other steels is higher. For rolled beams, all the HSLA steels have marginally lower relative costs, and A572 grade 50 has the lowest cost.

Because the comparison is valid only for members subjected to the same bending moment, it does not indicate the relative costs for girders over long spans where the weight of the member may be a significant part of the loading. Under such conditions, the relative material costs of the stronger steels decrease from those shown in the table because of the reduction in girder weights. Also, significant economies can sometimes be realized by the use of hybrid girders, that is, girders having a lower-yield-stress material for the web than for the flange. HSLA steels, such as A572 grade 50, are often more economical for composite beams in the

TABLE 1.5   Relative Material Cost for Beams*

Steel	Plate girders		Rolled beams	
	Relative weight	Relative material cost	Relative weight	Relative material cost
A36	1.000	1.00	1.000	1.00
A572 grade 42	0.927	1.01	0.903	0.98
A572 grade 50	0.848	0.95	0.805	0.91
A588 grade A	0.848	1.04	0.805	0.99
A852	0.775	1.18		
A514 grade B	0.600	1.24		

*Comparisons are based on prices in effect Feb. 1, 1992.

floors of buildings. Also, A588 steel is often preferred for bridge members in view of its greater durability.

Columns.   The relative material cost for two columns of different steels designed to carry the same load may be expressed as

$$\frac{C_2}{C_1} = \frac{F_{c1}}{F_{c2}}\frac{p_2}{p_1} = \frac{F_{c1}/p_1}{F_{c2}/p_2} \tag{1.7}$$

where F_{c1} and F_{c2} are the column buckling stresses for the two members. This relationship is similar to that given for tension members, except that buckling stress is used instead of yield stress in computing the relative price-strength ratios. Buckling stresses can be calculated from basic column-strength criteria. (T. Y. Galambos, *Structural Stability Research Council Guide to Design Criteria for Metal Structures*, John Wiley & Sons, Inc., New York.) In general, the buckling stress is considered equal to the yield stress at a slenderness ratio L/r of zero and decreases to the classical Euler value with increasing L/r.

Relative price-strength ratios for A572 grade 50, the most popular high-strength low-alloy steel, at L/r values from zero to 120 are shown graphically in Fig. 1.3. As before, A36 steel is the base. Therefore, ratios less than 1.00 indicate a material

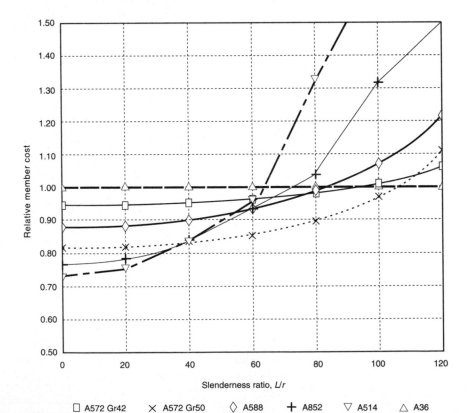

FIGURE 1.3 Curves show for several structural steels the variation of relative price-strength ratios, A36 steel being taken as unity, with slenderness ratios of compression members.

cost lower than that of A36 steel. The figure shows that for L/r from zero to about 100, A572 grade 50 steel is more economical than A36 steel. Thus the former is frequently used for columns in building construction, particularly in the lower stories, where slenderness ratios are smaller than in the upper stories.

1.4 STEEL SHEET AND STRIP FOR STRUCTURAL APPLICATIONS

Steel sheet and strip are used for many structural applications, including cold-formed members in building construction and the stressed skin of transportation equipment. Mechanical properties of several of the more frequently used sheet steels are presented in Table 1.6.

ASTM A570 covers seven strength grades of uncoated, hot-rolled, carbon-steel sheets and strip. (See ASTM A611 for cold-rolled carbon-steel sheet.) A446 covers several grades of galvanized, carbon-steel sheets. The various weights of zinc coat-

TABLE 1.6 Specified Minimum Mechanical Properties for Steel Sheet and Strip for Structural Applications

ASTM designation	Final condition	Yield point, ksi	Tensile strength, ksi	Elongation, % In 2 in*	In 8 in
A446	Galvanized				
Grade A		33	45	20	
Grade B		37	52	18	
Grade C		40	55	16	
Grade D		50	65	12	
Grade E		80	82	—	
Grade F		50	70	12	
A570	Hot-rolled				
Grade 30		30	49	25	19
Grade 33		33	52	23	18
Grade 36		36	53	22	17
Grade 40		40	55	21	16
Grade 45		45	60	19	14
Grade 50		50	65	17	12
Grade 55		55	70	15	10
A606	Hot-rolled, cut length	50	70	22	
	Hot-rolled, coils	45	65	22	
	Cold-rolled	45	65	22	
A607	Hot- or cold-rolled				
Grade 45		45	60†	25–23	
Grade 50		50	65†	22–20	
Grade 55		55	70†	20–18	
Grade 60		60	75†	18–16	
Grade 65		65	80†	16–14	
Grade 70		70	85†	14–12	

*Modified for some thicknesses in accordance with the specification. Where two values are given, the first is for hot-rolled, the second for cold-rolled steel.

†For class 1 product. Reduce tabulated strengths 5 ksi for class 2.

ing available for A446 sheets afford excellent corrosion protection in many applications.

A607, available in six strength levels, covers high-strength, low-alloy columbium or vanadium, or both, hot- and cold-rolled steel sheet and strip. The material may be in either cut lengths or coils. It is intended for structural or miscellaneous uses where greater strength and weight savings are important. A607 is available in two classes, each with six similar strength levels, but class 2 offers better formability and weldability than class 1. Without addition of copper, these steels are equivalent in resistance to atmospheric corrosion to plain carbon steel. With copper, however, resistance is twice that of carbon steel.

A606 covers high-strength, low-alloy, hot- and cold-rolled steel sheet and strip with enhanced corrosion resistance. This material is intended for structural or miscellaneous uses where weight savings or high durability are important. It is available, in cut lengths or coils, in either type 2 or type 4, with corrosion resistance two or four times, respectively, that of plain carbon steel.

1.5 TUBING FOR STRUCTURAL APPLICATIONS

Structural tubing is being used more frequently in modern construction. It is often preferred to other steel members when resistance to torsion is required and when a smooth, closed section is aesthetically desirable. In addition, structural tubing often may be the economical choice for compression members subjected to moderate to light loads. Square and rectangular tubing is manufactured either by cold or hot forming welded or seamless round tubing in a continuous process. A500 cold-formed carbon-steel tubing (Table 1.7) is produced in four strength grades in each of two product forms, shaped (square or rectangular) or round. A minimum

TABLE 1.7 Specified Minimum Mechanical Properties of Structural Tubing

ASTM designation	Product form	Yield point, ksi	Tensile strength, ksi	Elongation in 2 in, %
A500	Shaped			
Grade A		33	45	25
Grade B		42	58	23
Grade C		46	62	21
Grade D		36	58	23
A500	Round			
Grade A		39	45	25
Grade B		46	58	23
Grade C		50	62	21
Grade D		36	58	23
A501	Round or shaped	36	58	23
A618	Round or shaped			
Grades Ia, Ib, II				
Walls ≤¾ in		50	70	22
Walls >¾				
to 1½ in		46	67	22
Grade III		50	65	20

yield point of up to 46 ksi is available for shaped tubes and up to 50 ksi for round tubes.

A501 tubing is a hot-formed carbon-steel product. It provides a yield point equal to that of A36 steel in tubing having a wall thickness of 1 in or less.

A618 tubing is a hot-formed HSLA product. It provides a minimum yield point of 33 to 50 ksi depending on grade and wall thickness. The three grades all have enhanced resistance to atmospheric corrosion. Grades Ia and Ib can be used in the bare condition for many applications when properly exposed to the atmosphere.

1.6   STEEL CABLE FOR STRUCTURAL APPLICATIONS

Steel cables have been used for many years in bridge construction and are occasionally used in building construction for the support of roofs and floors. The types of cables used for these applications are referred to as **bridge strand** or **bridge rope.** In this use, **bridge** is a generic term that denotes a specific type of high-quality strand or rope.

A **strand** is an arrangement of wires laid helically about a center wire to produce a symmetrical section. A **rope** is a group of strands laid helically around a core composed of either a strand or another wire rope. The term **cable** is often used indiscriminately in referring to wires, strands, or ropes. Strand may be specified under ASTM A586; wire rope, under A603.

During manufacture, the individual wires in bridge strand and rope are generally galvanized to provide resistance to corrosion. Also, the finished cable is prestretched. In this process, the strand or rope is subjected to a predetermined load of not more than 55% of the breaking strength for a sufficient length of time to remove the "structural stretch" caused primarily by radial and axial adjustment of the wires or strands to the load. Thus, under normal design loadings, the elongation that occurs is essentially elastic and may be calculated from the elastic-modulus values given in Table 1.8.

TABLE 1.8 Mechanical Properties of Steel Cables

Minimum breaking strength, ksi,* of selected cable sizes			Minimum modulus of elasticity, ksi,* for indicated diameter range	
Nominal diameter, in	Zinc-coated strand	Zinc-coated rope	Nominal diameter range, in	Minimum modulus, ksi
½	30	23	Prestretched zinc-coated strand	
¾	68	52		
1	122	91.4	½ to 2⁹⁄₁₆	24,000
1½	276	208	2⅝ and over	23,000
2	490	372	Prestretched zinc-coated rope	
3	1076	824		
4	1850	1460	⅜ to 4	20,000

*Values are for cables with class A zinc coating on all wires. Class B or C can be specified where additional corrosion protection is required.

Strands and ropes are manufactured from cold-drawn wire and do not have a definite yield point. Therefore, a working load or design load is determined by dividing the specified minimum breaking strength for a specific size by a suitable safety factor. The breaking strengths for selected sizes of bridge strand and rope are listed in Table 1.8.

1.7   TENSILE PROPERTIES

The tensile properties of steel are generally determined from tension tests on small specimens or coupons in accordance with standard ASTM procedures. The behavior of steels in these tests is closely related to the behavior of structural-steel members under static loads. Because, for structural steels, the yield points and moduli of elasticity determined in tension and compression are nearly the same, compression tests are seldom necessary.

Typical tensile stress-strain curves for structural steels are shown in Fig. 1.1. The initial portion of these curves is shown at a magnified scale in Fig. 1.4. Both sets of curves may be referred to for the following discussion.

Strain Ranges.   When a steel specimen is subjected to load, an initial **elastic range** is observed in which there is no permanent deformation. Thus, if the load is removed, the specimen returns to its original dimensions. The ratio of stress to

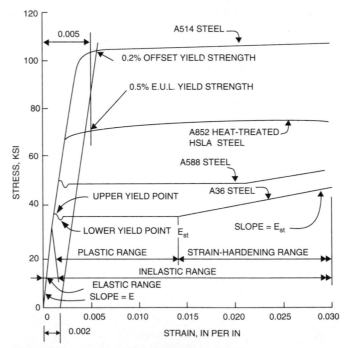

FIGURE 1.4 Partial stress-strain curves for structural steels strained through the plastic region into the strain-hardening range. *(From R. L. Brockenbrough and B. G. Johnston, USS Steel Design Manual, R. L. Brockenbrough & Associates, Inc., Pittsburgh, Pa., with permission.)*

strain within the elastic range is the **modulus of elasticity,** or **Young's modulus** E. Since this modulus is consistently about 29×10^3 ksi for all the structural steels, its value is not usually determined in tension tests, except in special instances.

The strains beyond the elastic range in the tension test are termed the **inelastic range.** For as-rolled and high-strength low-alloy (HSLA) steels, this range has two parts. First observed is a **plastic range,** in which strain increases with no appreciable increase in stress. This is followed by a **strain-hardening range,** in which strain increase is accompanied by a significant increase in stress. The curves for heat-treated steels, however, do not generally exhibit a distinct plastic range or a large amount of strain hardening.

The strain at which strain hardening begins (ϵ_{st}) and the rate at which stress increases with strain in the strain-hardening range (the strain-hardening modulus E_{st}) have been determined for carbon and HSLA steels. The average value of E_{st} is 600 ksi, and the length of the yield plateau is 5 to 15 times the yield strain. (T. V. Galambos, "Properties of Steel for Use in LRFD," *Journal of the Structural Division, American Society of Civil Engineers,* Vol. 104, No. ST9, 1978.)

Yield Point, Yield Strength, and Tensile Strength. As illustrated in Fig. 1.4, carbon and HSLA steels usually show an upper and lower yield point. The upper yield point is the value usually recorded in tension tests and thus is simply termed the **yield point.**

The heat-treated steels in Fig. 1.4, however, do not show a definite yield point in a tension test. For these steels it is necessary to define a **yield strength,** the stress corresponding to a specified deviation from perfectly elastic behavior. As illustrated in the figure, yield strength is usually specified in either of two ways: For steels with a specified value not exceeding 80 ksi, yield strength is considered as the stress at which the test specimen reaches a 0.5% extension under load (0.5% EUL) and may still be referred to as the yield point. For higher-strength steels, the yield strength is the stress at which the specimen reaches a strain 0.2% greater than that for perfectly elastic behavior.

Since the amount of inelastic strain that occurs before the yield strength is reached is quite small, yield strength has essentially the same significance in design as yield point. These two terms are sometimes referred to collectively as **yield stress.**

The maximum stress reached in a tension test is the tensile strength of the steel. After this stress is reached, increasing strains are accompanied by decreasing stresses. Fracture eventually occurs.

Proportional Limit. The proportional limit is the stress corresponding to the first visible departure from linear-elastic behavior. This value is determined graphically from the stress-strain curve. Since the departure from elastic action is gradual, the proportional limit depends greatly on individual judgment and on the accuracy and sensitivity of the strain-measuring devices used. The proportional limit has little practical significance and is not usually recorded in a tension test.

Ductility. This is an important property of structural steels. It allows redistribution of stresses in continuous members and at points of high local stresses, such as those at holes or other discontinuities.

In a tension test, ductility is measured by percent elongation over a given gage length or percent reduction of cross-sectional area. The percent elongation is determined by fitting the specimen together after fracture, noting the change in gage length and dividing the increase by the original gage length. Similarly, the percent reduction of area is determined from cross-sectional measurements made on the specimen before and after testing.

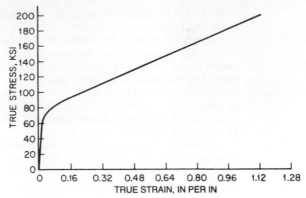

FIGURE 1.5 Curve shows the relationship between true stress and true strain for 50-ksi yield-point HSLA steel.

Both types of ductility measurements are an index of the ability of a material to deform in the inelastic range. There is, however, no generally accepted criterion of minimum ductility for various structures.

Poisson's Ratio. The ratio of transverse to longitudinal strain under load is known as **Poisson's ratio** ν. This ratio is about the same for all structural steels—0.30 in the elastic range and 0.50 in the plastic range.

True-Stress–True-Strain Curves. In the stress-strain curves shown previously, stress values were based on original cross-sectional area, and the strains were based on the original gauge length. Such curves are sometimes referred to as **engineering-type stress-strain curves.** However, since the original dimensions change significantly after the initiation of yielding, curves based on instantaneous values of area and gage length are often thought to be of more fundamental significance. Such curves are known as **true-stress–true-strain curves.** A typical curve of this type is shown in Fig. 1.5.

The curve shows that when the decreased area is considered, the true stress actually increases with increase in strain until fracture occurs instead of decreasing after the tensile strength is reached, as in the engineering stress-strain curve. Also, the value of true strain at fracture is much greater than the engineering strain at fracture (though until yielding begins true strain is less than engineering strain).

1.8 PROPERTIES IN SHEAR

The ratio of shear stress to shear strain during initial elastic behavior is the **shear modulus** G. According to the theory of elasticity, this quantity is related to the modulus of elasticity E and Poisson's ratio ν by

$$G = \frac{E}{2(1 + \nu)} \tag{1.8}$$

Thus a minimum value of G for structural steels is about 11×10^3 ksi. The yield stress in shear is about 0.57 times the yield stress in tension. The shear strength, or shear stress at failure in pure shear, varies from two-thirds to three-fourths the tensile strength for the various steels. Because of the generally consistent relationship of shear properties to tensile properties for the structural steels, and because of the difficulty of making accurate shear tests, shear tests are seldom performed.

1.9 HARDNESS TESTS

In the Brinell hardness test, a small spherical ball of specified size is forced into a flat steel specimen by a known static load. The diameter of the indentation made in the specimen can be measured by a micrometer microscope. The **Brinell hardness number** may then be calculated as the ratio of the applied load, in kilograms, to the surface area of the indentation, in square millimeters. In practice, the hardness number can be read directly from tables for given indentation measurements.

The Rockwell hardness test is similar in principle to the Brinell test. A spheroconical diamond penetrator is sometimes used to form the indentation, and the depth of the indentation is measured with a built-in, differential depth-measurement device. This measurement, which can be read directly from a dial on the testing device, becomes the **Rockwell hardness number.**

In either test, the hardness number depends on the load and type of penetrator used; therefore, these should be indicated when listing a hardness number. Other hardness tests, such as the Vickers tests, are also sometimes used. Tables are available that give approximate relationships between the different hardness numbers determined for a specific material.

Hardness numbers are considered to be related to the tensile strength of steel. Although there is no absolute criterion to convert from hardness numbers to tensile strength, charts are available that give approximate conversions (see ASTM A370). Because of its simplicity, the hardness test is widely used in manufacturing operations to estimate tensile strength and to check the uniformity of tensile strength in various products.

1.10 EFFECT OF COLD WORK ON TENSILE PROPERTIES

In the fabrication of structures, steel plates and shapes are often formed at room temperatures into desired shapes. These cold-forming operations cause inelastic deformation, since the steel retains its formed shape. To illustrate the general effects of such deformation on strength and ductility, the elemental behavior of a carbon-steel tension specimen subjected to plastic deformation and subsequent tensile reloadings will be discussed. However, the behavior of actual cold-formed structural members is more complex.

As illustrated in Fig. 1.6, if a steel specimen is unloaded after being stressed into either the plastic or strain-hardening range, the unloading curve follows a path parallel to the elastic portion of the stress-strain curve. Thus a residual strain, or **permanent set,** remains after the load is removed. If the specimen is promptly reloaded, it will follow the unloading curve to the stress-strain curve of the virgin (unstrained) material.

If the amount of plastic deformation is less than that required for the onset of strain hardening, the yield stress of the plastically deformed steel is about the same as that of the virgin material. However, if the amount of plastic deformation is sufficient to cause strain hardening, the yield stress of the steel is larger. In either instance, the tensile strength remains the same, but the ductility, measured from

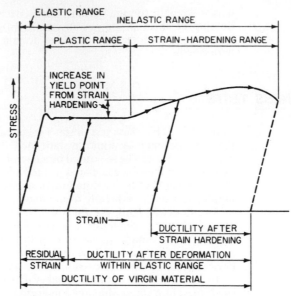

FIGURE 1.6   Stress-strain diagram (not to scale) illustrating the effects of strain-hardening steel. *(From R. L. Brockenbrough and B. G. Johnston, USS Steel Design Manual, R. L. Brockenbrough & Associates, Inc., Pittsburgh, Pa., with permission.)*

the point of reloading, is less. As indicated in Fig. 1.6, the decrease in ductility is nearly equal to the amount of inelastic prestrain.

A steel specimen that has been strained into the strain-hardening range, unloaded, and allowed to age for several days at room temperature (or for a much shorter time at a moderately elevated temperature) usually shows the behavior indicated in Fig. 1.7 during reloading. This phenomenon, known as **strain aging,** has the effect of increasing yield and tensile strength while decreasing ductility.

Most of the effects of cold work on the strength and ductility of structural steels can be eliminated by thermal treatment, such as stress relieving, normalizing, or annealing. However, such treatment is not often necessary.

(G. E. Dieter, Jr., *Mechanical Metallurgy*, 3rd ed., McGraw-Hill, Inc., New York.)

1.11   EFFECT OF STRAIN RATE ON TENSILE PROPERTIES

Tensile properties of structural steels are usually determined at relatively slow strain rates to obtain information appropriate for designing structures subjected to static loads. In the design of structures subjected to high loading rates, such as those caused by impact loads, however, it may be necessary to consider the variation in tensile properties with strain rate.

Figure 1.8 shows the results of rapid tension tests conducted on a carbon steel, two HSLA steels, and a constructional alloy steel. The tests were conducted at three strain rates and at three temperatures to evaluate the interrelated effect of these variables on the strength of the steels. The values shown for the slowest and the intermediate strain rates on the room-temperature curves reflect the usual room-

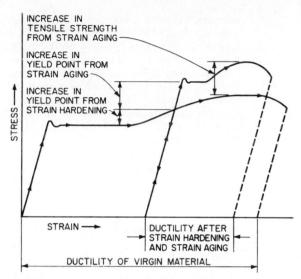

FIGURE 1.7 Effects of strain aging are shown by stress-strain diagram (not to scale). *(From R. L. Brockenbrough and B. G. Johnston, USS Steel Design Manual, R. L. Brockenbrough & Associates, Inc., Pittsburgh, Pa., with permission.)*

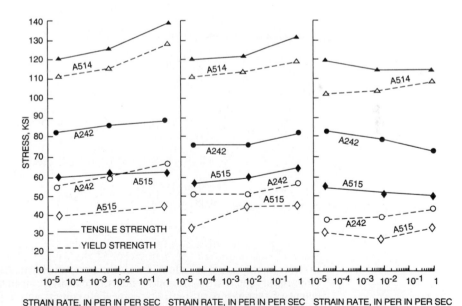

FIGURE 1.8 Effects of strain rate on yield and tensile strengths of structural steels at low, normal, and elevated temperatures. *(From R. L. Brockenbrough and B. G. Johnston, USS Steel Design Manual, R. L. Brockenbrough & Associates, Inc., Pittsburgh, Pa., with permission.)*

temperature yield stress and tensile strength, respectively. (In determination of yield stress, ASTM E8 allows a maximum strain rate of $\frac{1}{16}$ in per in per min, or 1.04×10^{-3} in per in per sec. In determination of tensile strength, E8 allows a maximum strain rate of 0.5 in per in per min, or 8.33×10^{-3} in per in per sec.)

The curves in Fig. 1.8a and b show that the tensile strength and 0.2% offset yield strength of all the steels increase as the strain rate increases at $-50°F$ and at room temperature. The greater increase in tensile strength is about 15%, for A514 steel, whereas the greatest increase in yield strength is about 48%, for A515 carbon steel. However, Fig. 1.8c shows that at 600°F, increasing the strain rate has a relatively small influence on the yield strength. But a faster strain rate causes a slight decrease in the tensile strength of most of the steels.

Ductility of structural steels, as measured by elongation or reduction of area, tends to decrease with strain rate. Other tests have shown that modulus of elasticity and Poisson's ratio do not vary significantly with strain rate.

1.12 EFFECT OF ELEVATED TEMPERATURES ON TENSILE PROPERTIES

The behavior of structural steels subjected to short-time loadings at elevated temperatures is usually determined from short-time tension tests. In general, the stress-strain curve becomes more rounded and the yield strength and tensile strength are reduced as temperatures are increased. The ratios of the elevated-temperature value to room-temperature value of yield and tensile strengths of several structural steels are shown in Fig. 1.9a and b, respectively.

Modulus of elasticity decreases with increasing temperature, as shown in Fig. 1.9c. The relationship shown is nearly the same for all structural steels. The variation in shear modulus with temperature is similar to that shown for the modulus of elasticity. But Poisson's ratio does not vary over this temperature range.

The following expressions for elevated-temperature property ratios, which were derived by fitting curves to short-time data, have proven useful in analytical modeling (R. L. Brockenbrough, "Theoretical Stresses and Strains from Heat Curving," *Journal of the Structural Division, American Society of Civil Engineers,* Vol. 96, No. ST7, 1970):

$$F_y/F_y' = 1 - \frac{T - 100}{5833} \qquad 100°F < T < 800°F \tag{1.9}$$

$$F_y/F_y' = (-720{,}000 + 4200 - 2.75T^2)10^{-6} \qquad 800°F < T < 1200°F \tag{1.10}$$

$$E/E' = 1 - \frac{T - 100}{5000} \qquad 100°F < T < 700°F \tag{1.11}$$

$$E/E' = (500{,}000 + 1333T - 1.111T^2)10^{-6} \qquad 700°F < T < 1200°F \tag{1.12}$$

$$\alpha = (6.1 + 0.0019T)10^{-6} \qquad 100°F < T < 1200°F \tag{1.13}$$

In these equations, F_y/F_y' and E/E' are the ratios of elevated-temperature to room-temperature yield strength and modulus of elasticity, respectively, α is the coefficient of thermal expansion per degree Fahrenheit, and T is the temperature in degrees Fahrenheit.

Ductility of structural steels, as indicated by elongation and reduction-of-area values, decreases with increasing temperature until a minimum value is reached. Thereafter, ductility increases to a value much greater than that at room temperature. The exact effect depends on the type and thickness of steel. The initial

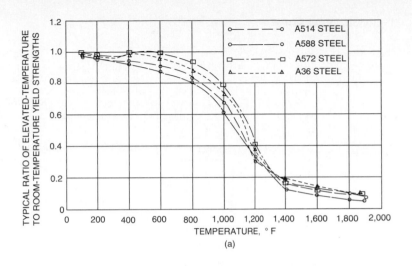

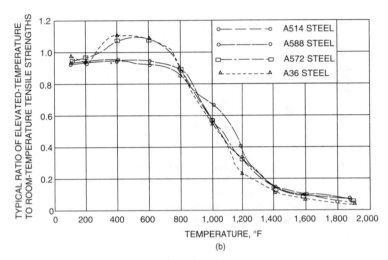

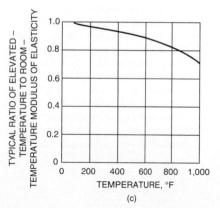

FIGURE 1.9 Effect of temperature on (*a*) yield strengths, (*b*) tensile strengths, and (*c*) modulus of elasticity of structural steels. *(From R. L. Brockenbrough and B. G. Johnston, USS Steel Design Manual, R. L. Brockenbrough & Associates, Inc., Pittsburgh, Pa., with permission.)*

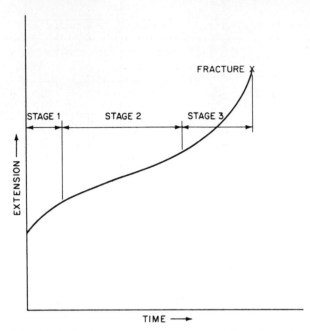

FIGURE 1.10 Creep curve for structural steel in tension (schematic). *(From R. L. Brockenbrough and B. G. Johnston, USS Steel Design Manual, R. L. Brockenbrough & Associates, Inc., Pittsburgh, Pa., with permission.)*

decrease in ductility is caused by strain aging and is most pronounced in the temperature range of 300 to 700°F. Strain aging also accounts for the increase in tensile strength in this temperature range shown for two of the steels in Fig. 1.9b.

Under long-time loadings at elevated temperatures, the effects of creep must be considered. When a load is applied to a specimen at an elevated temperature, the specimen deforms rapidly at first but then continues to deform, or creep, at a much slower rate. A schematic creep curve for a steel subjected to a constant tensile load and at a constant elevated temperature is shown in Fig. 1.10. The initial elongation occurs almost instantaneously and is followed by three stages. In stage 1, elongation increases at a decreasing rate. In stage 2, elongation increases at a nearly constant rate. And in stage 3, elongation increases at an increasing rate. The failure, or creep-rupture, load is less than the load that would cause failure at that temperature in a short-time loading test.

Table 1.9 indicates typical creep and rupture data for a carbon steel, an HSLA steel, and a constructional alloy steel. The table gives the stress that will cause a given amount of creep in a given time at a particular temperature.

For special elevated-temperature applications in which structural steels do not provide adequate properties, special alloy and stainless steels with excellent high-temperature properties are available.

1.13 FATIGUE

A structural member subjected to cyclic loadings may eventually fail through initiation and propagation of cracks. This phenomenon is called **fatigue** and can occur at stress levels considerably below the yield stress.

TABLE 1.9 Typical Creep Rates and Rupture Stresses for Structural Steels at Various Temperatures

Test Temperature, °F	Stress, ksi, for creep rate of		Stress, ksi for rupture in		
	0.0001% per hr*	0.00001% per hr†	1000 hours	10,000 hours	100,000 hours
A36 steel					
800	21.4	13.8	38.0	24.8	16.0
900	9.9	6.0	18.5	12.4	8.2
1000	4.6	2.6	9.5	6.3	4.2
A588 grade A steel‡					
800	34.6	29.2	44.1	35.7	28.9
900	20.3	16.3	28.6	22.2	17.3
1000	11.4	8.6	17.1	12.0	8.3
1200	1.7	1.0	3.8	2.0	1.0
A514 grade F steel‡					
700	—	—	101.0	99.0	97.0
800	81.0	74.0	86.0	81.0	77.0

*Equivalent to 1% in 10,000 hours.
†Equivalent to 1% in 100,000 hours.
‡Not recommended for use where temperatures exceed 800°F.

Extensive research programs conducted to determine the fatigue strength of structural members and connections have provided information on the factors affecting this property. These programs included studies of large-scale girder specimens with flange-to-web fillet welds, flange cover plates, stiffeners, and other attachments. The studies showed that the **stress range** (algebraic difference between maximum and minimum stress) and **notch severity** of details are the most important factors. Yield point of the steel had little effect. The knowledge developed from these programs has been incorporated into specifications of the American Institute of Steel Construction and the American Association of State Highway and Transportation Officials, which offer detailed provisions for fatigue design.

1.14 BRITTLE FRACTURE

Under sufficiently adverse combinations of tensile stress, temperature, loading rate, geometric discontinuity (notch), and restraint, a steel member may experience a brittle fracture. All these factors need not be present. In general, a **brittle fracture** is a failure that occurs by cleavage with little indication of plastic deformation. In contrast, a **ductile fracture** occurs mainly by shear, usually preceded by considerable plastic deformation.

Design against brittle fracture requires selection of the proper grade of steel for the application and avoiding notchlike defects in both design and fabrication. An awareness of the phenomenon is important so that steps can be taken to minimize the possibility of this undesirable, usually catastrophic failure mode.

An empirical approach and an analytical approach directed toward selection and evaluation of steels to resist brittle fracture are outlined below. These methods are actually complementary and are frequently used together in evaluating material and fabrication requirements.

Charpy V-Notch Test. Many tests have been developed to rate steels on their relative resistance to brittle fracture. One of the most commonly used tests is the Charpy V-notch test, which specifically evaluates notch toughness, that is, the resistance to fracture in the presence of a notch. In this test, a small square bar with a specified-size V-shaped notch at its midlength (type A impact-test specimen of ASTM A370) is simply supported at its ends as a beam and fractured by a blow from a swinging pendulum. The amount of energy required to fracture the specimen or the appearance of the fracture surface is determined over a range of temperatures. The appearance of the fracture surface is usually expressed as the percentage of the surface that appears to have fractured by shear.

A **shear fracture** is indicated by a dull or fibrous appearance. A shiny or crystalline appearance is associated with a **cleavage fracture.**

The data obtained from a Charpy test are used to plot curves, such as those in Fig. 1.11, of energy or percentage of shear fracture as a function of temperature. The temperature near the bottom of the energy-temperature curve, at which a selected low value of energy is absorbed, often 15 ft·lb, is called the **ductility transition temperature** or the **15-ft·lb transition temperature.** The temperature at which the percentage of shear fracture decreases to 50% is often called the **fracture-appearance transition temperature.** These transition temperatures serve as a rating of the resistance of different steels to brittle fracture. The lower the transition temperature, the greater is the notch toughness.

Of the steels in Table 1.1, A36 steel generally has about the highest transition temperature. Since this steel has an excellent service record in a variety of structural applications, it appears likely that any of the structural steels, when designed and fabricated in an appropriate manner, could be used for similar applications with little likelihood of brittle fracture. Nevertheless, it is important to avoid unusual temperature, notch, and stress conditions to minimize susceptibility to brittle fracture.

In applications where notch toughness is considered important, the minimum Charpy V-notch value and test temperature should be specified, because there may be considerable variation in toughness within any given product designation unless specifically produced to minimum requirements. The test temperature may be specified higher than the lowest operating temperature to compensate for a lower rate of loading in the anticipated application. (See Art. 1.1.5.)

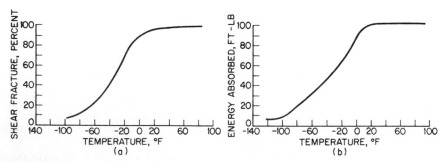

FIGURE 1.11 Transition curves from Charpy-V notch impact tests. (*a*) Variation of percent shear fracture with temperature. (*b*) Variation of absorbed energy with temperature.

It should be noted that as the thickness of members increases, the inherent restraint increases and tends to inhibit ductile behavior. Thus special precautions or greater toughness, or both, is required for tension or flexural members comprised of thick material. (See Art. 1.17.)

Fracture-Mechanics Analysis.   Fracture mechanics offers a more direct approach for prediction of crack propagation. For this analysis, it is assumed that a **crack,** which may be defined as a flat, internal defect, is always present in a stressed body. By linear-elastic stress analysis and laboratory tests on a precracked specimen, the defect size is related to the applied stress that will cause crack propagation and brittle fracture, as outlined below.

Near the tip of a crack, the stress component f perpendicular to the plane of the crack (Fig. 1.12a) can be expressed as

$$f = \frac{K_I}{\sqrt{2\pi r}} \tag{1.14}$$

where r is distance from tip of crack and K_I is a stress-intensity factor related to geometry of crack and to applied loading. The factor K_I can be determined from elastic theory for given crack geometries and loading conditions. For example, for a through-thickness crack of length $2a$ in an infinite plate under uniform stress (Fig. 1.12a),

$$K_I = f_a\sqrt{\pi a} \tag{1.15}$$

where f_a is the nominal applied stress. For a disk-shaped crack of diameter $2a$ embedded in an infinite body (Fig. 1.12b), the relationship is

$$K_I = 2f_a\sqrt{\frac{a}{\pi}} \tag{1.16}$$

If a specimen with a crack of known geometry is loaded until the crack propagates rapidly and causes failure, the value of K_I at that stress level can be calculated from the derived expression. This value is termed the **fracture toughness** K_c.

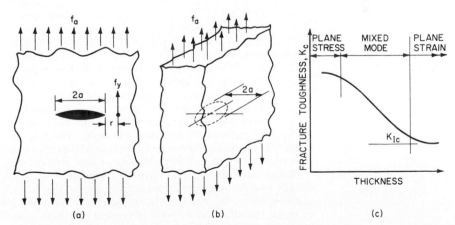

FIGURE 1.12   Fracture mechanics analysis for brittle fracture. (*a*) Sharp crack in a stressed infinite plate. (*b*) Disk-shaped crack in an infinite body. (*c*) Relation of fracture toughness to thickness.

A precracked tension or bend-type specimen is usually used for such tests. As the thickness of the specimen increases and the stress condition changes from plane stress to plane strain, the fracture toughness decreases to a minimum value, as illustrated in Fig. 1.12c. This value of plane-strain fracture toughness, designated K_{Ic}, may be regarded as a fundamental material property.

Thus, if K_{Ic} is substituted for K_I, for example, in Eq. (1.15) or (1.16), a numerical relationship is obtained between the crack geometry and the applied stress that will cause fracture. With this relationship established, brittle fracture may be avoided by determining the maximum-size crack present in the body and maintaining the applied stress below the corresponding level. The tests must be conducted at or correlated with temperatures and strain rates appropriate for the application, because fracture toughness decreases with temperature and loading rate. Correlations have been made to enable fracture toughness values to be estimated from the results of Charpy V-notch tests.

Fracture-mechanics analysis has proven quite useful, particularly in critical applications. Fracture-control plans can be established with suitable inspection intervals to ensure that imperfections, such as fatigue cracks, do not grow to critical size.

(J. M. Barsom and S. T. Rolfe, *Fracture and Fatigue Control in Structures; Applications of Fracture Mechanics,* Prentice-Hall, Inc., Englewood Cliffs, N.J.)

1.15 RESIDUAL STRESSES

Stresses that remain in structural members after rolling or fabrication are known as **residual stresses.** The magnitude of the stresses is usually determined by removing longitudinal sections and measuring the strain that results. Only the longitudinal stresses are usually measured. To meet equilibrium conditions, the axial force and moment obtained by integrating these residual stresses over any cross section of the member must be zero.

In a hot-rolled structural shape, the residual stresses result from unequal cooling rates after rolling. For example, in a wide-flange beam, the center of the flange cools more slowly and develops tensile residual stresses that are balanced by compressive stresses elsewhere on the cross section (Fig. 1.13a). In a welded member, tensile residual stresses develop near the weld and compressive stresses elsewhere provide equilibrium, as shown for the welded box section in Fig. 1.13b.

For plates with rolled edges (UM plates), the plate edges have compressive residual stresses (Fig. 1.13c). However, the edges of flame-cut plates have tensile residual stresses (Fig. 1.13d). In a welded I-shaped member, the stress condition in the edges of flanges before welding is reflected in the final residual stresses (Fig. 1.13e). Although not shown in Fig. 1.13, the residual stresses at the edges of sheared-edge plates vary through the plate thickness. Tensile stresses are present on one surface and compressive stresses on the opposite surface.

The residual-stress distributions mentioned above are usually relatively constant along the length of the member. However, residual stresses also may occur at particular locations in a member, because of localized plastic flow from fabrication operations, such as cold straightening or heat straightening.

When loads are applied to structural members, the presence of residual stresses usually causes some premature inelastic action; that is, yielding occurs in localized portions before the nominal stress reaches the yield point. Because of the ductility of steel, the effect on strength of tension members is not usually significant, but excessive tensile residual stresses, in combination with other conditions, can cause fracture. In compression members, residual stresses decrease the buckling load

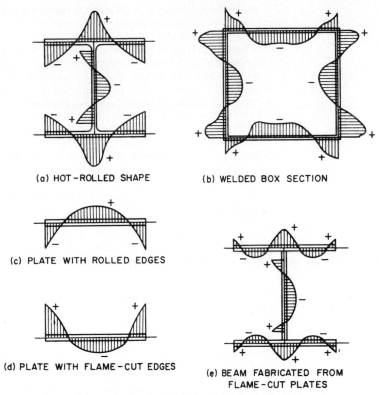

(a) HOT-ROLLED SHAPE

(b) WELDED BOX SECTION

(c) PLATE WITH ROLLED EDGES

(d) PLATE WITH FLAME-CUT EDGES

(e) BEAM FABRICATED FROM
FLAME-CUT PLATES

FIGURE 1.13 Typical residual-stress distributions (+ indicates tension and −
compression).

from that of an ideal or perfect member. However, current design criteria in general
use for compression members account for the influence of residual stress.

In bending members that have residual stresses, a small inelastic deflection of
insignificant magnitude may occur with the first application of load. However, under
subsequent loads of the same magnitude, the behavior is elastic. Furthermore, in
"compact" bending members, the presence of residual stresses has no effect on
the ultimate moment (plastic moment). Consequently, in the design of statically
loaded members, it is not usually necessary to consider residual stresses.

1.16 LAMELLAR TEARING

In a structural steel member subjected to tension, elongation and reduction of area
in sections normal to the stress are usually much lower in the through-thickness
direction than in the planar direction. This inherent directionality is of small consequence in many applications, but it does become important in design and fabrication of structures with highly restrained joints because of the possibility of
lamellar tearing. This is a cracking phenomenon that starts underneath the surface
of steel plates as a result of excessive through-thickness strain, usually associated
with shrinkage of weld metal in highly restrained joints. The tear has a steplike

appearance consisting of a series of terraces parallel to the surface. The cracking may remain completely below the surface or may emerge at the edges of plates or shapes or at weld toes.

Careful selection of weld details, filler metal, and welding procedure can restrict lamellar tearing in heavy welded constructions, particularly in joints with thick plates and heavy structural shapes. Also, when required, structural steels can be produced by special processes, generally with low sulfur content and inclusion control, to enhance through-thickness ductility.

The most widely accepted method of measuring the susceptibility of a material to lamellar tearing is the tension test on a round specimen, in which is observed the reduction in area of a section oriented perpendicular to the rolled surface. The reduction required for a given application depends on the specific details involved. The specifications to which a particular steel can be produced are subject to negotiations with steel producers.

(R. L. Brockenbrough, Chap. 1.2 in *Constructional Steel Design—An International Guide,* R. Bjorhovde et al., eds., Elsevier Science Publishers, Ltd., New York.)

1.17 WELDED SPLICES IN HEAVY SECTIONS

Shrinkage during solidification of large welds in structural steel members causes, in adjacent restrained metal, strains that can exceed the yield-point strain. In thick material, triaxial stresses may develop because there is restraint in the thickness direction as well as in planar directions. Such conditions inhibit the ability of a steel to act in a ductile manner and increase the possibility of brittle fracture. Therefore, for members subject to primary tensile stresses due to axial tension or flexure in buildings, the American Institute of Steel Construction (AISC) specifications for structural steel buildings impose special requirements for welded splicing of either group 4 or group 5 rolled shapes or of shapes built up by welding plates more than 2 in thick. The specifications include requirements for notch toughness, removal of weld tabs and backing bars (welds ground smooth), generous-sized weld-access holes, preheating for thermal cutting, and grinding and inspecting cut edges. Even for primary compression members, the same precautions should be taken for sizing weld access holes, preheating, grinding, and inspection.

Most heavy wide-flange shapes and tees cut from these shapes have regions where the steel has low toughness, particularly at flange-web intersections. These low-toughness regions occur because of the slower cooling there and, because of the geometry, the lower rolling pressure applied there during production. Hence, to ensure ductility and avoid brittle failure, bolted splices should be considered as an alternative to welding.

("AISC Specification for Structural Steel Buildings—Allowable Stress Design and Plastic Design" and "Load and Resistance Factor Design Specification for Structural Steel Buildings," American Institute of Steel Construction; R. L. Brockenbrough, Sec. 9, in *Standard Handbook for Civil Engineers,* 4th ed., McGraw-Hill, Inc., New York.)

1.18 VARIATIONS IN MECHANICAL PROPERTIES

Tensile properties of structural steel may vary from specified minimum values. Product specifications generally require that properties of the material "as represented by the test specimen" meet certain values. With some exceptions, ASTM specifications dictate a test frequency for structural-grade steels of only two tests per heat (in each strength level produced, if applicable) and more frequent testing for pressure-vessel grades. If the heats are very large, the test specimens qualify a considerable amount of product. As a result, there is a possibility that properties at locations other than those from which the specimens were taken will be different from those specified.

For plates, a test specimen is required by ASTM A6 to be taken from a corner. If the plates are wider than 24 in, the longitudinal axis of the specimen should be oriented transversely to the final direction in which the plates were rolled. For other products, however, the longitudinal axis of the specimen should be parallel to the final direction of rolling.

For structural shapes, the test specimen should be taken from the web. The flange, which is thicker, will usually have lower properties.

An extensive study commissioned by the American Iron and Steel Institute (AISI) compared yield points at various sample locations with the official product test. The studies indicated that the average difference at the check locations was -0.7 ksi. For the top and bottom flanges, at either end of beams, the average difference at check locations was -2.6 ksi.

Although the test value at a given location may be less than that obtained in the official test, the difference is offset to the extent that the value from the official test exceeds the specified minimum value. For example, a statistical study made to develop criteria for load and resistance factor design showed that the mean yield points exceeded the specified minimum yield point F_y as indicated below and with the indicated coefficient of variation (COV).

Flanges of rolled shapes $1.05 F_y$, COV = 0.10

Webs of rolled shapes $1.10 F_y$, COV = 0.11

Plates $1.10 F_y$, COV = 0.11

Also, these values incorporate an adjustment to the lower "static" yield points.

For similar reasons, the notch toughness can be expected to vary throughout a product.

(R. L. Brockenbrough, Chap. 1.2, in *Constructional Steel Design—An International Guide,* R. Bjorhovde, ed., Elsevier Science Publishers, Ltd., New York.)

SECTION 1
PROPERTIES OF STRUCTURAL STEELS

PART 2

EFFECTS OF STEELMAKING AND FABRICATION ON PROPERTIES

Frederick S. Merritt
Consulting Engineer, West Palm Beach, Florida

1.19 CHANGES IN CARBON STEELS ON HEATING AND COOLING

As pointed out in Art. 1.12, heating changes the tensile properties of steels. Actually, heating changes many steel properties. Often, the primary reason for such changes is a change in structure brought about by heat. Some of these structural changes can be explained with the aid of an iron-carbon equilibrium diagram (Fig. 1.14).

The diagram maps out the constituents of carbon steels at various temperatures as carbon content ranges from 0 to 5%. Other elements are assumed to be present only as impurities, in negligible amounts.

If a steel with less than 2% carbon is very slowly cooled from the liquid state, a solid solution of carbon in gamma iron will result. This is called **austenite.** (Gamma iron is a pure iron whose crystalline structure is face-centered cubic.)

If the carbon content is about 0.8%, the carbon remains in solution as the austenite slowly cools, until the A_1 temperature (1340°F) is reached. Below this temperature, the austenite transforms to the eutectoid **pearlite.** This is a mixture of ferrite and **cementite** (iron carbide, Fe_3C). Pearlite, under a microscope, has a characteristic platelike, or lamellar, structure with an iridescent appearance, from which it derives its name.

If the carbon content is less than 0.8%, as is the case with structural steels, cooling austenite below the A_3 temperature line causes transformation of some of the austenite to **ferrite.** (This is a pure iron, also called **alpha iron,** whose crystalline structure is body-centered cubic.) Still further cooling to below the A_1 line causes the remaining austenite to transform to pearlite. Thus, as indicated in Fig. 1.14, low-carbon steels are **hypoeutectoid steels,** mixtures of ferrite and pearlite.

Ferrite is very ductile but has low tensile strength. Hence carbon steels get their high strengths from the pearlite present or, more specifically, from the cementite in the pearlite.

The iron-carbon equilibrium diagram shows only the constituents produced by slow cooling. At high cooling rates, however, equilibrium cannot be maintained. Transformation temperatures are lowered, and steels with microstructures other than pearlitic may result. Properties of such steels differ from those of the pearlitic steels. Heat treatments of steels are based on these temperature effects.

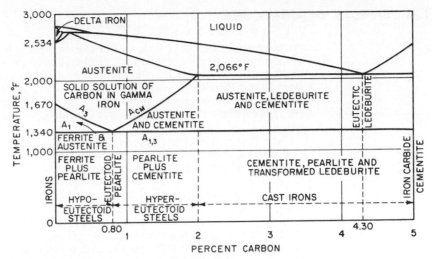

FIGURE 1.14 Iron-carbon equilibrium diagram.

If a low-carbon austenite is rapidly cooled below about 1300°F, the austenite will transform at constant temperature into steels with one of four general classes of microstructure:

Pearlite, or lamellar, microstructure results from transformations in the range 1300 to 1000°F. The lower the temperature, the closer is the spacing of the platelike elements. As the spacing becomes smaller, the harder and tougher the steels become. Steels such as A36, A572, and A588 have a mixture of a soft ferrite matrix and a hard pearlite.

Bainite forms in transformations below about 1000°F and above about 450°F. It has an acicular, or needlelike, microstructure. At the higher temperatures, bainite may be softer than the pearlitic steels. However, as the transformation temperature is decreased, hardness and toughness increase.

Martensite starts to form at a temperature below about 500°F, called the M_s temperature. The transformation differs from those for pearlitic and bainitic steels in that it is not time-dependent. Martensite occurs almost instantly during rapid cooling, and the percentage of austenite transformed to martensite depends only on the temperature to which the steel is cooled. For complete conversion to martensite, cooling must extend below the M_f temperature, which may be 200°F or less. Like bainite, martensite has an acicular microstructure, but martensite is harder and more brittle than pearlitic and bainitic steels. Its hardness varies with carbon content and to some extent with cooling rate. For some applications, such as those where wear resistance is important, the high hardness of martensite is desirable, despite brittleness. Generally, however, martensite is used to obtain tempered martensite, which has superior properties.

Tempered martensite is formed when martensite is reheated to a subcritical temperature after quenching. The tempering precipitates and coagulates carbides. Hence the microstructure consists of carbide particles, often spheroidal in shape, dispersed in a ferrite matrix. The result is a loss in hardness but a considerable improvement in ductility and toughness. The heat-treated carbon and HSLA steels and quenched and tempered constructional steels discussed in Art. 1.1 are low-carbon martensitic steels.

(Z. D. Jastrzebski, *Nature and Properties of Engineering Materials,* John Wiley & Sons, Inc., New York.)

1.20 *EFFECTS OF GRAIN SIZE*

As indicated in Fig. 1.14, when a low-carbon steel is heated above the A_1 temperature line, austenite, a solid solution of carbon in gamma iron, begins to appear in the ferrite matrix. Each island of austenite grows until it intersects its neighbor. With further increase in temperature, these grains grow larger. The final grain size depends on the temperature above the A_3 line to which the metal is heated. When the steel cools, the relative coarseness of the grains passes to the ferrite-plus-pearlite phase.

At rolling and forging temperatures, therefore, many steels grow coarse grains. Hot working, however, refines the grain size. The temperature at the final stage of the hot-working process determines the final grain size. When the finishing temperature is relatively high, the grains may be rather coarse when the steel is air-cooled. In that case, the grain size can be reduced if the steel is normalized (reheated to just above the A_3 line and again air-cooled). (See Art. 1.21.)

Fine grains improve many properties of steels. Other factors being the same, steels with finer grain size have better notch toughness because of lower transition temperatures (see Art. 1.14) than coarser-grained steels. Also, decreasing grain size improves bendability and ductility. Furthermore, fine grain size in quenched and tempered steel improves yield strength. And there is less distortion, less quench cracking, and lower internal stress in heat-treated products.

On the other hand, for some applications, coarse-grained steels are desirable. They permit deeper hardening. If the steels should be used in elevated-temperature service, they offer higher load-carrying capacity and higher creep strength than fine-grained steels.

Austenitic-grain growth may be inhibited by carbides that dissolve slowly or remain undissolved in the austenite or by a suitable dispersion of nonmetallic inclusions. Steels produced this way are called **fine-grained.** Steels not made with grain-growth inhibitors are called **coarse-grained.**

When heated above the critical temperature, 1340°F, grains in coarse-grained steels grow gradually. The grains in fine-grained steels grow only slightly, if at all, until a certain temperature, the coarsening temperature, is reached. Above this, abrupt coarsening occurs. The resulting grain size may be larger than that of coarse-grained steel at the same temperature. Note further that either fine-grained or coarse-grained steels can be heat-treated to be either fine-grained or coarse-grained (see Art. 1.21).

The usual method of making fine-grained steels involves controlled aluminum deoxidation (see also Art. 1.23). The inhibiting agent in such steels may be a submicroscopic dispersion of aluminum nitride or aluminum oxide.

(W. T. Lankford, Jr., ed., *The Making, Shaping and Treating of Steel,* Association of Iron and Steel Engineers, Pittsburgh, Pa.)

1.21 *ANNEALING AND NORMALIZING*

Structural steels may be annealed to relieve stresses induced by cold or hot working. Sometimes, also, annealing is used to soften metal to improve its formability or machinability.

Annealing involves austenitizing the steel by heating it above the A_3 temperature line in Fig. 1.14, then cooling it slowly, usually in a furnace. This treatment improves ductility but decreases tensile strength and yield point. As a result, further heat treatment may be necessary to improve these properties.

Structural steels may be normalized to refine grain size. As pointed out in Art. 1.20, grain size depends on the finishing temperature in hot rolling.

Normalizing consists of heating the steel above the A_3 temperature line, then cooling the metal in still air. Thus the rate of cooling is more rapid than in annealing. Usual practice is to normalize from 100 to 150°F above the critical temperature. Higher temperatures coarsen the grains.

Normalizing tends to improve notch toughness by lowering ductility and fracture transition temperatures. Thick plates benefit more from this treatment than thin plates. Requiring fewer roller passes, thick plates have a higher finishing temperature and cool slower than thin plates, thus have a more adverse grain structure. Hence the improvement from normalizing is greater for thick plates.

1.22   EFFECTS OF CHEMISTRY ON STEEL PROPERTIES

Chemical composition determines many characteristics of steels important in construction applications. Some of the chemicals present in commercial steels are a consequence of the steelmaking process. Other chemicals may be added deliberately by the producers to achieve specific objectives. Specifications therefore usually require producers to report the chemical composition of the steels.

During the pouring of a heat of steel, producers take samples of the molten steel for chemical analysis. These heat analyses are usually supplemented by product analyses taken from drillings or millings of blooms, billets, or finished products. ASTM specifications contain maximum and minimum limits on chemicals reported in the heat and product analyses, which may differ slightly.

Principal effects of the elements more commonly found in carbon and low-alloy steels are discussed below. Bear in mind, however, that the effects of two or more of these chemicals when used in combination may differ from those when each alone is present. Note also that variations in chemical composition to obtain specific combinations of properties in a steel usually increase cost, because it becomes more expensive to make, roll, and fabricate.

Carbon is the principal strengthening element in carbon and low-alloy steels. In general, each 0.01% increase in carbon content increases the yield point about 0.5 ksi. This, however, is accompanied by increase in hardness and reduction in ductility, notch toughness, and weldability, raising of the transition temperatures, and greater susceptibility to aging. Hence limits on carbon content of structural steels are desirable. Generally, the maximum permitted in structural steels is 0.30% or less, depending on the other chemicals present and the weldability and notch toughness desired.

Aluminum, when added to silicon-killed steel, lowers the transition temperature and increases notch toughness. If sufficient aluminum is used, up to about 0.20%, it reduces the transition temperature even when silicon is not present. However, the larger additions of aluminum make it difficult to obtain desired finishes on rolled plate. Drastic deoxidation of molten steels with aluminum or aluminum and titanium, in either the steelmaking furnace or the ladle, can prevent the spontaneous increase in hardness at room temperature called **aging.** Also, aluminum restricts grain growth during heat treatment and promotes surface hardness by nitriding.

Boron in small quantities increases hardenability of steels. It is used for this purpose in quenched and tempered low-carbon constructional alloy steels. However, more than 0.0005 to 0.004% boron produces no further increase in hardenability. Also, a trace of boron increases strength of low-carbon, plain molybdenum (0.40%) steel.

Chromium improves strength, hardenability, abrasion resistance, and resistance to atmospheric corrosion. However, it reduces weldability. With small amounts of chromium, low-alloy steels have higher creep strength than carbon steels and are used where higher strength is needed for elevated-temperature service. Also, chromium is an important constituent of stainless steels.

Columbium in very small amounts produces relatively larger increases in yield point but smaller increases in tensile strength of carbon steel. However, the notch toughness of thick sections is appreciably reduced.

Copper in amounts up to about 0.35% is very effective in improving the resistance of carbon steels to atmospheric corrosion. Improvement continues with increases in copper content up to about 1% but not so rapidly. Copper increases strength, with a proportionate increase in fatigue limit. Copper also increases hardenability, with only a slight decrease in ductility and little effect on notch toughness and weldability. However, steels with more than 0.60% copper are susceptible to precipitation hardening. And steels with more than about 0.5% copper often experience hot shortness during hot working, and surface cracks or roughness develop. Addition of nickel in an amount equal to about half the copper content is effective in maintaining surface quality.

Hydrogen, which may be absorbed during steelmaking, embrittles steels. Ductility will improve with aging at room temperature as the hydrogen diffuses out of the steel, faster from thin sections than from thick. When hydrogen content exceeds 0.0005%, flaking, internal cracks or bursts, may occur when the steel cools after rolling, especially in thick sections. In carbon steels, flaking may be prevented by slow cooling after rolling, to permit the hydrogen to diffuse out of the steel.

Manganese increases strength, hardenability, fatigue limit, notch toughness, and corrosion resistance. It lowers the ductility and fracture transition temperatures. It hinders aging. Also, it counteracts hot shortness due to sulfur. For this last purpose, the manganese content should be three to eight times the sulfur content, depending on the type of steel. However, manganese reduces weldability.

Molybdenum increases yield strength, hardenability, abrasion resistance, and corrosion resistance. It also improves weldability. However, it has an adverse effect on toughness and transition temperature. With small amounts of molybdenum, low-alloy steels have higher creep strength than carbon steels and are used where higher strength is needed for elevated-temperature service.

Nickel increases strength, hardenability, notch toughness, and corrosion resistance. It is an important constituent of stainless steels. It lowers the ductility and fracture transition temperatures, and it reduces weldability.

Nitrogen increases strength, but it may cause aging. It also raises the ductility and fracture transition temperatures.

Oxygen, like nitrogen, may be a cause of aging. Also, oxygen decreases ductility and notch toughness.

Phosphorus increases strength, fatigue limit, and hardenability, but it decreases ductility and weldability and raises the ductility transition temperature. Additions of aluminum, however, improve the notch toughness of phosphorus-bearing steels. Phosphorus improves the corrosion resistance of steel and works very effectively together with small amounts of copper toward this result.

Silicon increases strength, notch toughness, and hardenability. It lowers the ductility transition temperature, but it also reduces weldability. Silicon often is used as a deoxidizer in steelmaking (see Art. 1.23).

Sulfur, which enters during the steelmaking process, can cause hot shortness. This results from iron sulfide inclusions, which soften and may rupture when heated. Also, the inclusions may lead to brittle failure by providing stress raisers from which fractures can initiate. And high sulfur contents may cause porosity and hot cracking in welding unless special precautions are taken. Addition of manganese, however, can counteract hot shortness. It forms manganese sulfide, which is more refractory than iron sulfide. Nevertheless, it usually is desirable to keep sulfur content below 0.05%.

Titanium increases creep and rupture strength and abrasion resistance. It plays an important role in preventing aging. It sometimes is used as a deoxidizer in steelmaking (see Art. 1.23) and grain-growth inhibitor (see Art. 1.20).

Tungsten increases creep and rupture strength, hardenability, and abrasion resistance. It is used in steels for elevated-temperature service.

Vanadium, in amounts up to about 0.12%, increases rupture and creep strength without impairing weldability or notch toughness. It also increases hardenability and abrasion resistance. Vanadium sometimes is used as a deoxidizer in steelmaking (see Art. 1.23) and grain-growth inhibitor (see Art. 1.20).

In practice, carbon content is limited so as not to impair ductility, notch toughness, and weldability. To obtain high strength, therefore, resort is had to other strengthening agents that improve these desirable properties or at least do not impair them as much as carbon. Often, the better these properties are required to be at high strengths, the more costly the steels are likely to be.

Attempts have been made to relate chemical composition to weldability by expressing the relative influence of chemical content in terms of **carbon equivalent.** One widely used formula, for example, is

$$C_{eq} = C + \frac{(Mn + Si)}{6} + \frac{(Cr + Mo + V)}{5} + \frac{(Ni + Cu)}{15} \qquad (1.17)$$

where C = carbon content, %
   Mn = manganese content, %
   Si = silicon content, %
   Cr = chromium content, %
   Mo = molybdenum, %
   V = vanadium, %
   Ni = nickel content, %
   Cu = copper, %

Carbon equivalent is related to the maximum rate at which a weld and adjacent plate may be cooled after welding, without underbead cracking occurring. The higher the carbon equivalent, the lower will be the allowable cooling rate. Also, use of low-hydrogen welding electrodes and preheating becomes more important with increasing carbon equivalent. (*Structural Welding Code—Steel,* American Welding Society, Miami, Fla.)

Though carbon provides high strength in steels economically, it is not a necessary ingredient. Very high strength steels are available that contain so little carbon that they are considered carbon-free.

Maraging steels, carbon-free iron-nickel martensites, develop yield strengths from 150 to 300 ksi, depending on alloying composition. As pointed out in Art. 1.19, iron-carbon martensite is hard and brittle after quenching and becomes softer and more ductile when tempered. In contrast, maraging steels are relatively soft and ductile initially but become hard, strong, and tough when aged. They are fabricated while ductile and later strengthened by an aging treatment. These steels have high resistance to corrosion, including stress-corrosion cracking.

(W. T. Lankford, Jr., ed., *The Making, Shaping and Treating of Steel,* Association of Iron and Steel Engineers, Pittsburgh, Pa.)

1.23 EFFECTS OF STEELMAKING METHODS

Structural steel is usually produced today by one of two production processes. In the traditional process, iron or "hot metal" is produced in a blast furnace and then further processed in a basic oxygen furnace to make the steel for the desired products. Alternatively, steel can be made in an electric arc furnace that is charged mainly with steel scrap instead of hot metal. In either case, the steel must be produced so that undesirable elements are reduced to levels allowed by pertinent specifications to minimize adverse effects on properties.

The liquid steel produced by these two processes has traditionally been cast into ingot molds to solidify and await further processsing. As an alternative, a more efficient form of structural steel is produced by continuous casting. In this process, the liquid steel is poured through a water-cooled mold to produce a continuous slab or beam blank for beams. The slab is then reduced by rolling to produce the final product form, such as plate, and beams are rolled into structural shapes from the blanks.

In a **blast furnace,** iron ore, coke, and flux (limestone and dolomite) are charged into the top of a large refractory-lined furnace. Heated air is blown in at the bottom and passed up through the bed of raw materials. A supplemental fuel such as gas, oil, or powdered coal is also usually charged. The iron is reduced to metallic iron and melted; then it is drawn off periodically through tap holes into transfer ladles. At this point, the molten iron includes several other elements (manganese, sulfur, phosphorus, and silicon) in amounts greater than permitted for steel, and thus further processing is required.

In a **basic oxygen furnace,** the charge consists of hot metal from the blast furnace and steel scrap. Oxygen, introduced by a jet blown into the molten metal, reacts with the impurities present to facilitate the removal or reduction in level of unwanted elements, which are trapped in the slag or in the gases produced. Also, various fluxes are added to reduce the sulfur and phosphorus contents to desired levels. In this batch process, large heats of steel may be produced in less than an hour.

An **electric-arc furnace** does not require a hot metal charge but relies mainly on steel scrap. The metal is heated by an electric arc between large carbon electrodes that project through the furnace roof into the charge. Oxygen is injected to speed the process. This is a versatile batch process that can be adapted to producing small heats where various steel grades are required, but it also can be used to produce large heats.

Ladle treatment is an integral part of most steelmaking processes. The ladle receives the product of the steelmaking furnace so that it can be moved and poured into either ingot molds or a continuous casting machine. While in the ladle, the chemical composition of the steel is checked, and alloying elements are added as required. Also, deoxidizers are added to remove dissolved oxygen. Processing can be done at this stage to reduce further sulfur content, remove undesirable non-metallics, and change the shape of remaining inclusions. Thus significant improvements can be made in the toughness, transverse properties, and through-thickness ductility of the finished product. Vacuum degassing, argon bubbling, induction stirring, and the injection of rare earth metals are some of the many procedures that may be employed.

Ingot molds, in the traditional process, receive the steel poured from the ladle to cool and solidify. After sufficient solidification of the steel, the ingots are moved to a soaking pit, where they are held at the proper temperature. The ingots are subsequently cropped at their ends to remove undesirable material and then rolled into products, one by one.

Ideally, an ingot should be homogeneous, with a fine, equiaxial crystal structure. It should not contain nonmetallic inclusions or cavities and should be free of chemical segregation. In practice, however, because of uneven cooling and release of gases in the mold, an ingot may develop any of a number of internal and external defects. Some of these may be eliminated or minimized during the rolling operation. Prevention or elimination of the others often adds to the cost of steels.

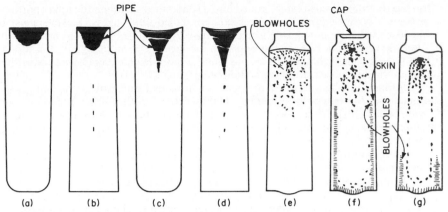

FIGURE 1.15   Types of ingot and molds. (*a*) to (*d*) indicate internal conditions in killed steel. (*a*) Big-end-up, hot-topped mold. (*b*) Big-end-down, hot-topped mold. (*c*) Big-end-up, not hot-topped. (*d*) Big-end-down, not hot-topped. (*e*) Semikilled steel. (*f*) Capped steel. (*g*) Rimmed steel.

Depending on taper, ingot molds are classified as big-end-down (Fig. 1.15*b* and *d*) and big-end-up (Fig. 1.15*a* and *c*). The big-end-down molds, in turn, are classified as open-top (Fig. 1.15*d*) and bottle-top (Fig. 1.15*b*). The mold shape can facilitate concentrating and removing some of the undesirable material from the ingot.

Steel cools unevenly in a mold because the liquid at the mold walls solidifies first and cools more rapidly than metal in the interior of the ingot. Gases, chiefly oxygen, dissolved in the liquid, are released as the liquid cools. Four types of ingot may result—killed, semikilled, capped, and rimmed—depending on the amount of gases dissolved in the liquid, the carbon content of the steel, and the amount of deoxidizers added to the steel.

Figure 1.15 shows ingot structures that differ in amount of gas developed in the mold and in type of mold. A **fully killed ingot** (Fig. 1.15*a* to *d*) develops no gas; the molten steel lies dead in the mold. The top surface solidifies relatively fast. Pipe, an intermittently bridged shrinkage cavity, forms below the top. Fully killed steels usually are poured in big-end-up molds with "hot tops" (Fig. 1.15*a*) to confine the pipe to the hot top, which is later discarded. A **semikilled ingot** (Fig. 1.15*e*) develops a slight amount of gas. The gas, trapped when the metal solidifies, forms blowholes in the upper portion of the ingot. A **capped ingot** (Fig. 1.15*f*) develops rimming action, a boiling caused by evolution of gas, forcing the steel to rise. The action is stopped by a metal cap secured to the mold. Strong upward currents along the sides of the mold sweep away bubbles that otherwise would form blowholes in the upper portion of the ingot. Blowholes do form, however, in the lower portion, separated by a thick solid skin from the mold walls. A **rimmed ingot** (Fig. 1.15*g*) develops a violent rimming action, confining blowholes to only the bottom quarter of the ingot.

Rimmed or capped steels cannot be produced if too much carbon is present (0.30% or more) because insufficient oxygen will be dissolved in the steels to cause the rimming action. Killed and semikilled steels require additional costs for deoxidizers if carbon content is low, and the deoxidation products form nonmetallic inclusions in the ingot. Hence it often is advantageous for steel producers to make low-carbon steels by rimmed or capped practice and high-carbon steels by killed or semikilled practice.

Pipe, or shrinkage cavities, generally is small enough in rimmed, capped, and semikilled steels to be eliminated by rolling. Big-end-down killed ingots without a hot top (Fig. 1.15*d*) also can be rolled into a sound product when the lower portion of the cavity is clean and not oxidized. For killed-steel ingots, however, when freedom from pipe is required, the steel should be cast in big-end-up molds with a hot top (Fig. 1.15*a*). A comparison with the big-end-down mold with hot top in Fig. 1.15*b* indicates that the big-end-up mold is more effective in concentrating the pipe in the hot top for cropping and discard.

Blowholes in the interior of an ingot, small voids formed by entrapped gases, usually are eliminated during rolling. If they extend to the surface, they may be oxidized and form seams when the ingot is rolled, because the oxidized metal cannot be welded together. Properly made ingots have a thick enough skin over blowholes to prevent oxidation.

Segregation in ingots depends on the chemical composition and on turbulence from gas evolution and convection currents in the molten metal. Killed steels have less segregation than semikilled steels, and these types of steels have less segregation than capped or rimmed steels. In rimmed steels, the effects of segregation are so marked that interior and outer regions differ enough in chemical composition to appear to be different steels. The boundary between these regions is sharp.

Rimmed steels are made without additions of deoxidizers to the furnace and with only small additions to the ladle to ensure sufficient evolution of gas. When properly made, rimmed ingots have little pipe and a good surface. Such steels are preferred where surface finish is important and the effects of segregation will not be harmful.

Capped steels are made much like rimmed steels but with less rimming action. Capped steels have less segregation. They are used to make sheet, strip, skelp, tinplate, wire, and bars.

Semikilled steel is deoxidized less than killed steel. Most deoxidation is accomplished with additions of a deoxidizer to the ladle. Semikilled steels are used in structural shapes and plates.

Killed steels usually are deoxidized by additions to both furnace and ladle. Generally, silicon compounds are added to the furnace to lower the oxygen content of the liquid metal and stop oxidation of carbon (block the heat). This also permits addition of alloying elements that are susceptible to oxidation. Silicon or other deoxidizers, such as aluminum, vanadium, and titanium, may be added to the ladle to complete deoxidation. Aluminum, vanadium, and titanium have the additional beneficial effect of inhibiting grain growth when the steel is normalized. (In the hot-rolled conditions, such steels have about the same ferrite grain size as semikilled steels.) Killed steels deoxidized with aluminum and silicon (**made to fine-grain practice**) often are specified for construction applications because of better notch toughness and lower transition temperatures than semikilled steels of the same composition.

(W. T. Lankford, Jr., ed., *The Making, Shaping and Treating of Steel*, Association of Iron and Steel Engineers, Pittsburgh, Pa.)

1.24 *EFFECTS OF HOT ROLLING*

For product produced from ingots, the first step in rolling is to reduce the cross section to the form of a billet, slab, or bloom. These forms permit correction of defects before finish rolling, shearing into convenient lengths for final rolling, reheating for further rolling, and transfer to other mills, if desired, for that processing.

Blooms and billets are distinguished principally by size. **Billets** are smaller, generally in the range 2 × 2 to 5 × 5 in. **Blooms** may be square or oblong and generally are in the range 6 × 6 to 12 × 12 in. Billets may be produced directly from small-cross-section ingots, but usually they are made from blooms. Final products obtained from billets include bars, wires, small angles, window sash, and fence posts.

Slabs and blooms are distinguished principally by shape. Generally, **slabs** have a width considerably greater than the thickness, whereas blooms usually are nearly square. Thickness of slabs often ranges from 2 to 9 in, and width from 24 to 60 in. Structural steel plates are rolled from slabs; shapes are made from blooms.

Ingots are rolled into blooms and slabs in primary mills. In general, these have three functions. The major function is to roll ingots into blooms and slabs. A secondary function is to produce pieces with desired dimensions and weights. And a third function is to carry out certain auxiliary operations: conditioning the products (removing defects), cropping portions with injurious defects, straightening bent pieces, correcting other physical conditions that exceed acceptable tolerances, and collecting crops, roll scale, and other byproducts for delivery to steelmaking facilities.

The major function requires that first the ingots be heated to between 2150 and 2450°F so that they will be sufficiently plastic for economic reduction by rolling to finished cross-sectional sizes. This heating is performed in soaking pits. Then, the ingots are gradually reduced in cross section, but increased in length, by passage through a series of rollers with decreasing spaces between rolls. This hot working welds voids in the ingots. It also breaks up the crystalline structure, which was made coarse by the heating. Next, portions of the product that are physically or chemically unsuited for final use are cropped and discarded. Also, specimens may be removed for testing. Finally, blooms and slabs not destined for immediate additional rolling are cooled. The cooling is controlled in furnaces to minimize development of internal stresses and prevent flakes, small internal ruptures, which may be caused by hydrogen in the steel (see Art. 1.22).

The **continuous casting** process is used to produce semifinished products directly from liquid steel, thus bypassing the ingot molds and primary mills. With continuous casting, the steel is poured from sequenced ladles to maintain a desired level in a tundish above an oscillating water-cooled copper mold (Fig. 1.16). The outer skin of the steel strand solidifies as it passes through the mold, and this action is further aided by water sprayed on the skin just after the strand exits the mold. The strand passes through sets of supporting rolls, curving rolls, and straightening rolls and is then rolled into slabs. The slabs are cut to length from the moving strand and held for subsequent rolling into finished product. Not only is the continuous casting process a more efficient method, but it also results in improved quality through more consistent chemical composition and better surfaces on the finished product.

Plates, produced from slabs or directly from ingots, are distinguished from sheet, strip, and flat bars by size limitations in ASTM A6. Generally, plates are heavier, per linear foot, than these other products. Plates are formed with straight horizontal rolls and later trimmed (sheared or gas cut) on all edges. Universal plates, or universal mill plates, rarely available, are formed between vertical and horizontal rolls and are trimmed on the ends only.

Slabs usually are reheated in a furnace and descaled with high-pressure water sprays before they are rolled into plates. The plastic slabs or ingots are gradually brought to desired dimensions by passage through a series of rollers. In the last rolling step, the plates pass through leveling, or flattening, rollers. Generally, the thinner the plate, the more flattening required. After passing through the leveler, plates are cooled uniformly, then sheared or gas cut to desired length, while still hot.

Some of the plates may be heat treated, depending on grade of steel and intended use. For carbon steel, the treatment may be annealing, normalizing, or stress

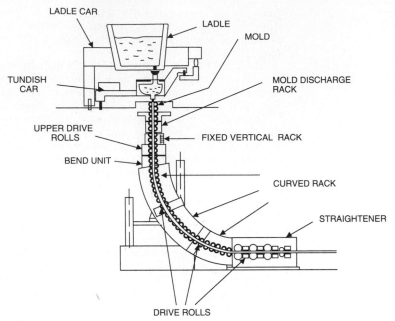

FIGURE 1.16 Schematic of slab caster.

relieving. Plates of HSLA or constructional alloy steels may be quenched and tempered. Some mills provide facilities for on-line heat treating or for thermo-mechanical processing (controlled rolling). Other mills heat treat off-line.

Shapes are rolled from continuously cast beam blanks or from blooms that first are reheated to 2250°F. Rolls gradually reduce the plastic blooms to the desired shapes and sizes. The shapes then are cut to length for convenient handling with a hot saw. After that, they are cooled uniformly. Next, they are straightened, in a roller straightener for smaller sizes or in a gag press for larger sizes. Finally, they are cut to desired length, usually by hot shearing, hot sawing, or cold sawing. Also, column ends may be milled to close tolerances.

ASTM A6 requires that material for delivery "shall be free from injurious defects and shall have a workmanlike finish." The specification permits manu-facturers to condition plates and shapes "for the removal of injurious surface imperfections or surface depressions by grinding, or chipping and grind-ing. . . ." Except in alloy steels, small surface imperfections may be corrected by chipping or grinding, then depositing weld metal with low-hydrogen electrodes. Conditioning also may be done on blooms, slabs, and billets before they are made into other products. In addition to chipping and grinding, they may be scarfed to remove surface defects.

Hand chipping is done with a cold chisel in a pneumatic hammer. Machine chipping may be done with a planer or a milling machine.

Scarfing, by hand or machine, removes defects with an oxygen torch. This can create problems that do not arise with other conditioning methods. When the heat source is removed from the conditioned area, a quenching effect is produced by rapid extraction of heat from the hot area by the surrounding relatively cold areas. The rapid cooling hardens the steel, the amount depending on carbon content and hardenability of the steel. In low-carbon steels, the effect may be insignificant. In high-carbon and alloy steels, however, the effect may be severe. If preventive measures are not taken, the hardened area will crack. To prevent scarfing cracks, the steel should be preheated before scarfing to between 300 and 500°F and, in some cases, postheated for stress relief. The hardened surface later can be removed by normalizing or annealing.

Internal structure and many properties of plates and shapes are determined largely by the chemistry of the steel, rolling practice, cooling conditions after rolling, and heat treatment, where used. Because the sections are rolled in a temperature range at which steel is austenitic (see Art. 1.19), internal structure is affected in several ways:

The interior of ingots consists of large crystals, called **dendrites,** characterized by a branching structure. Growth of individual dendrites occurs principally along their longitudinal axes perpendicular to the ingot surfaces. Heating for rolling tends to eliminate dendritic segregation, so that the rolled products are more homogeneous than ingots. Furthermore, during rolling, the dendritic structure is broken up. Also, recrystallization occurs. The final austenitic grain size is determined by the temperature of the steel during the last passes through the rolls (see Art. 1.20). In addition, dendrities and inclusions are reoriented in the direction of rolling. As a result, ductility and bendability are much better in the longitudinal direction than in the transverse, and these properties are poorest in the thickness direction.

The cooling rate after rolling determines the distribution of ferrite and the grain size of the ferrite. Since air cooling is the usual practice, the final internal structure and, therefore, the properties of plates and shapes depend principally on the chemistry of the steel, section size, and heat treatment. By normalizing the steel and by use of steels made to fine-grain practice (with grain-growth inhibitors, such as aluminum, vanadium, and titanium), grain size can be refined and properties consequently improved.

In addition to the preceding effects, rolling also may induce residual stresses in plates and shapes (see Art. 1.15). Still other effects are a consequence of the final thickness of the hot-rolled material.

Thicker material requires less rolling, the finish rolling temperature is higher, and the cooling rate is slower than for thin material. As a consequence, thin material has a superior microstructure. Furthermore, thicker material can have a more unfavorable state of stress because of stress raisers, such as tiny cracks and inclusions, and residual stresses.

Consequently, thin material develops higher tensile and yield strengths than thick material of the same steel. ASTM specifications for structural steels recognize this usually by setting lower yield points for thicker material. A36 steel, however, has the same yield point for all thicknesses. To achieve this, the chemistry is varied for plates and shapes and for thin and thick plates. Thicker plates contain more carbon and manganese to raise the yield point. This cannot be done for high-strength steels because of the adverse effect on notch toughness, ductility, and weldability.

Thin material has greater ductility and lower transition temperatures than thick material of the same steel. Since normalizing refines the grain structure, thick material improves relatively more with normalizing than does thin material. The improvement is even greater with silicon-aluminum-killed steels.

(W. T. Lankford, Jr., ed., *The Making, Shaping and Treating of Steel,* Association of Iron and Steel Engineers, Pittsburgh, Pa.)

1.25   *EFFECTS OF PUNCHING HOLES AND SHEARING*

Excessive cold working of exposed edges of structural-steel members can cause embrittlement and cracking and should be avoided. Punching holes and shearing during fabrication are cold-working operations that can cause brittle failure.

Bolt holes, for example, may be formed by drilling, punching, or punching followed by reaming. Drilling is preferable to punching, because punching drastically coldworks the material at the edge of a hole. This makes the steel less ductile and raises the transition temperature. The degree of embrittlement depends on type of steel and plate thickness. Furthermore, there is a possibility that punching can produce short cracks extending radially from the hole. Consequently, brittle failure can be initiated at the hole when the member is stressed.

Should the material around the hole become heated, an additional risk of failure is introduced. Heat, for example, may be supplied by an adjacent welding operation. If the temperature should rise to the 400 to 850°F range, strain aging will occur in material susceptible to it. The result will be a loss in ductility.

Reaming a hole after punching can eliminate the short, radial cracks and the risks of embrittlement. For that purpose, the hole diameter should be increased from $\frac{1}{16}$ to $\frac{1}{4}$ in by reaming, depending on material thickness and hole diameter.

Shearing has about the same effects as punching. If sheared edges are to be left exposed, $\frac{1}{16}$ in or more material, depending on thickness, should be trimmed, usually by grinding or machining. Note also that rough machining, for example, with edge planers making a deep cut, can produce the same effects as shearing or punching.

(M. E. Shank, *Control of Steel Construction to Avoid Brittle Failure,* Welding Research Council, New York.)

1.26 EFFECTS OF WELDING

Failures in service rarely, if ever, occur in properly made welds of adequate design.

If a fracture occurs, it is initiated at a notchlike defect. Notches occur for various reasons. The toe of a weld may form a natural notch. The weld may contain flaws that act as notches. A welding-arc strike in the base metal may have an embrittling effect, especially if weld metal is not deposited. A crack started at such notches will propagate along a path determined by local stresses and notch toughness of adjacent material.

Preheating before welding minimizes the risk of brittle failure. Its primary effect initially is to reduce the temperature gradient between the weld and adjoining base metal. Thus, there is less likelihood of cracking during cooling and there is an opportunity for entrapped hydrogen, a possible source of embrittlement, to escape. A consequent effect of preheating is improved ductility and notch toughness of base and weld metals, and lower transition temperature of weld.

Rapid cooling of a weld can have an adverse effect. One reason that arc strikes that do not deposit weld metal are dangerous is that the heated metal cools very fast. This causes severe embrittlement. Such arc strikes should be completely removed. The material should be preheated, to prevent local hardening, and weld metal should be deposited to fill the depression.

Welding processes that deposit weld metal low in hydrogen and have suitable moisture control often can eliminate the need for preheat. Such processes include use of low-hydrogen electrodes and inert-arc and submerged-arc welding.

Pronounced segregation in base metal may cause welds to crack under certain fabricating conditions. These include use of high-heat-input electrodes and deposition of large beads at slow speeds, as in automatic welding. Cracking due to segregation, however, is rare for the degree of segregation normally occurring in hot-rolled carbon-steel plates.

Welds sometimes are peened to prevent cracking or distortion, although special welding sequences and procedures may be more effective. Specifications often prohibit peening of the first and last weld passes. Peening of the first pass may crack or punch through the weld. Peening of the last pass makes inspection for cracks difficult. Peening considerably reduces toughness and impact properties of the weld metal. The adverse effects, however, are eliminated by the covering weld layer (last pass).

(M. E. Shank, *Control of Steel Construction to Avoid Brittle Failure*, Welding Research Council, New York; R. D. Stout and W. D. Doty, *Weldability of Steels*, Welding Research Council, New York.)

SECTION 2
FABRICATION AND ERECTION

Charles Peshek
Consulting Engineer, Naperville, Illinois
Formerly Director, Fabrication Operations & Standards, AISC

Richard W. Marshall
Vice President, American Steel Erectors, Inc., Allentown, Pennsylvania

Designers of steel-framed structures should be familiar not only with strength and serviceability requirements for the structures but also with fabrication and erection methods. These may determine whether a design is practical and cost-efficient. Furthermore, load capacity and stability of a structure may depend on design assumptions made as to type and magnitude of stresses and strains induced during fabrication and erection.

2.1 SHOP DETAIL DRAWINGS

Detail drawings are the means by which the intent of the designer is conveyed to the fabricating shop. They may be prepared by drafters (shop detailers) in the employ of the fabricator or by an independent detailing firm contracted by the steel fabricator. Detail drawings can be generated by computer with software developed for that purpose.

The detailer works from the engineering and architectural drawings and specifications to obtain member sizes, grades of steel, controlling dimensions, and all information pertinent to the fabrication process. After the detail drawings have been completed, they are meticulously checked by an experienced detailer, called a **checker,** before they are submitted for approval to the engineer or architect. After approval, the shop drawings are released to the shop for fabrication.

There are essentially two types of detail drawings, erection drawings and shop working drawings. Erection drawings are used by the erector in the field. They consist of line diagrams showing the location and orientation of each member or assembly, called **shipping pieces,** which will be shipped to the construction site. Each shipping piece is identified by a piece mark, which is painted on the member and shown in the erection drawings on the corresponding member.

Shop working drawings, simply called **details,** are prepared for every member of a steel structure. All information necessary for fabricating the piece is shown clearly on the detail. The size and location of all holes are shown, as well as the type, size, and length of welds.

While shop detail drawings are absolutely imperative in fabrication of structural steel, they are used also by inspectors to ascertain that members are being made as detailed. In addition, the details have lasting value to the owner of the structure in that he or she knows exactly what he or she has, should any alterations or additions be required at some later date.

To enable the detailer to do his or her job, the designer should provide the following information:

For simple-beam connections: Reactions of beams should be shown on design drawings, particularly when the fabricator must develop the connections. For unusual or complicated connections, it is good practice for the designer to consult with a fabricator during the design stages of a project to determine what information should be included in the design drawings.

For rigid beam-to-column connections: Some fabricators prefer to be furnished the moments and forces in such connections. With these data, fabricators can develop an efficient connection best suited to their practices.

For welding: Weld sizes and types of electrode should, in general, be shown on design drawings. Designers unfamiliar with welding can gain much by consulting with a fabricator, preferably while the project is being designed.

For fasteners: The type of fastening must be shown in design drawings. When specifying high-strength bolts, designers must indicate whether the bolts are to be used in slip-critical or bearing-type connections.

For tolerances: If unusual tolerances for dimensional accuracy exist, these must be clearly shown on the design drawings. Unusual tolerances are those which are more stringent than tolerances specified in the general specification of the type of structure under consideration, such as the AISC "Code of Standard Practice for Steel Buildings and Bridges" and "Specification for Structural Steel Buildings, Allowable Stress Design and Plastic Design," and "Load and Resistance Factor Design Specification for Structural Steel Buildings," AASHTO "Standard Specifications for Highway Bridges"; and ASTM A6, "General Requirements for Delivery of Rolled Steel Plates, Shapes, Sheet Piling, and Bars for Structural Use."

2.2 CUTTING, SHEARING, AND SAWING

Flame cutting steel with an oxygen-fed torch is one of the most useful methods in steel fabrication. The torch is used extensively to cut material to proper size, including stripping flange plates from a wider plate, or cutting beams to required lengths. The torch is also used to cut complex curves or forms, such as those encountered in finger-type expansion devices for bridge decks. In addition, two torches are sometimes used simultaneously to cut a member to size and bevel its edge in preparation for welding. Also, torches may be gang-mounted for simultaneous multiple cutting.

Flame-cutting torches may be manually held or mechanically guided. Mechanical guides may take the form of a track on which is mounted a small self-propelled unit that carries the torch. This type is used principally for making long cuts, such as those for stripping flange plates or trimming girder web plates to size. Another type of mechanically guided torch is used for cutting intricately detailed pieces. This machine has an arm that supports and moves the torch. The arm may be controlled by a device following the contour of a template or may be computer-controlled.

In the flame-cutting process, the torch burns a mixture of oxygen and gas to bring the steel at the point where the cut is to be started to preheat temperature of about 1600°F. At this temperature, the steel has a great affinity for oxygen. The torch then releases pure oxygen under pressure through the cutting tip. This oxygen combines immediately with the steel. As the torch moves along the cut line, the oxidation, coupled with the erosive force of the oxygen stream, produces a cut about ⅛ in wide. Once cutting begins, the heat of oxidation helps to heat the material.

Structural steel of certain grades and thicknesses may require additional preheat. In those cases, flame is played on the metal ahead of the cut.

In such operations as stripping plate-girder flange plates, it is desirable to flame-cut both edges of the plate simultaneously. This limits distortion by imposing shrinkage stresses of approximately equal magnitude in both edges of the plate. For this reason, plates to be supplied by a mill for multiple cutting are ordered with sufficient width to allow a flame cut adjacent to the mill edges.

Plasma-arc cutting is an alternative process for steel fabrication. A tungsten electrode may be used, but hafnium is preferred because it eliminates the need for expensive inert shielding gases. Advantages of this method include faster cutting, easy removal of dross, and lower operating cost. Disadvantages include higher equipment cost, limitation of thickness of cut to 1½ in, slightly beveled edges, and a wider kerf.

Shearing is used in the fabricating shop to cut certain classes of plain material to size. Several types of shears are available. **Guillotine-type shears** are used to cut plates of moderate thickness. Some plate shears, called **rotary-plate shears,** have a rotatable cutting head that allows cutting on a bevel. **Angle shears** are used to cut both legs of an angle with one stroke. **Rotary-angle shears** can produce beveled cuts.

Sawing with a high-speed friction saw is often employed in the shop on light beams and channels ordered to multiple lengths. Sawing is also used for relatively light columns, because the cut produced is suitable for bearing and sawing is faster and less expensive than milling. Some fabricators utilize cold sawing as a means of cutting beams to nearly exact length when accuracy is demanded by the type of end connection being used.

2.3 PUNCHING AND DRILLING

Bolt holes in structural steel are usually produced by punching (within thickness limitations). The American Institute of Steel Construction (AISC) limits the thickness for punching to the nominal diameter of the bolt plus ⅛ in. In thicker material, the holes may be made by subpunching and reaming or by drilling. Multiple punches are generally used for large groups of holes, such as for beam splices. Drilling is more time-consuming and therefore more costly than punching. Both drill presses and multiple-spindle drills are used, and the flanges and webs may be drilled simultaneously.

2.4 CNC MACHINES

Computer numerically controlled (CNC) machines that offer increased productivity are used increasingly for punching, cutting, and other operations. Their use can reduce the time required for material handling and layout, as well as for punching, cutting, or shearing. Such machines can handle plates up to 30 by 120 in by 1¼ in thick. CNC machines are also available for fabricating W shapes, including punching or drilling, flame-cutting copes, weld preparation (bevels and rat holes) for splices and moment connections, and similar items. CNC machines have the capacity to drill holes up to 1⁹⁄₁₆ in in diameter in either flanges or web. Production is of high quality and accuracy.

2.5 BOLTING

Most field connections are made by bolting, either with high-strength bolts (ASTM A325 or A490) or with ordinary machine bolts (A307 bolts), depending on strength requirements. Shop connections frequently are welded but may use these same types of bolts.

When high-strength bolts are used, the connections should satisfy the requirements of the "Specification for Structural Joints Using ASTM A325 or A490 Bolts," approved by the Research Council on Structural Connections (RCSC) of the Engineering Foundation. Joints with high-strength bolts are designed as either bearing-type or slip-critical connections (see Art. 5.3). Bearing-type connections have a higher allowable load or design strength. Slip-critical connections always must be fully tightened to specified minimum values. Bearing-type connections may be either "snug tight" or fully tightened depending on the type of connection and service conditions. AISC specifications for structural steel buildings require fully tensioned high-strength bolts (or welds) for certain connections (see Sec. 6.14.2). The AASHTO "Standard Specifications for Highway Bridges" require slip-critical joints where slippage would be detrimental to the serviceability of the structure, including joints subjected to fatigue loading or significant stress reversal. In all other cases, connections may be made with "snug tight" high-strength bolts or A307 bolts, as may be required to develop the necessary strength. For tightening requirements, see Art. 5.14.

2.6 WELDING

Use of welding in fabrication of structural steel for buildings and bridges is governed by one or more of the following: American Welding Society Specifications D1.1, "Structural Welding Code," and D1.5, "Bridge Welding Code," and the AISC "Specification for Structural Steel Buildings." In addition to these specifications, welding may be governed by individual project specifications or standard specifications of agencies or groups, such as state departments of transportation.

Steels to be welded should be of a "weldable grade," such as A36, A441, A572, A588, or A514. Such steels may be welded by any of several welding processes: shielded metal arc, submerged arc, gas metal arc, flux-cored arc, electroslag, electrogas, and stud welding. Some processes, however, are preferred for certain grades and some are excluded, as indicated in the following.

The AWS "Structural Welding Code" and other specifications exempt from tests and qualification most of the common welded joints used in steel structures. The details of such "prequalified joints" are shown in AWS D1.0 and in the AISC "Steel Construction Manuals." It is advantageous to use these joints where applicable to avoid costs for additional qualification tests.

Shielded metal arc welding (SMAW) produces coalescence, or fusion, by the heat of an electric arc struck between a coated metal electrode and the material being joined, or **base metal.** The electrode supplies filler metal for making the weld, gas for shielding the molten metal, and flux for refining this metal. This process is commonly known also as **manual, hand,** or **stick welding.** Pressure is not used on the parts to be joined.

When an arc is struck between the electrode and the base metal, the intense heat forms a small molten pool on the surface of the base metal. The arc also decomposes the electrode coating and melts the metal at the tip of the electrode. The electron stream carries this metal in the form of fine globules across the gap and deposits and mixes it into the molten pool on the surface of the base metal. (Since deposition of electrode material does not depend on gravity, arc welding is feasible in various positions, including overhead.) The decomposed coating of the electrode forms a gas shield around the molten metal that prevents contact with the air and absorption of impurities. In addition, the electrode coating promotes electrical conduction across the arc, helps stabilize the arc, adds flux, slag-forming materials, to the molten pool to refine the metal, and provides materials for controlling the shape of the weld. In some cases, the coating also adds alloying elements. As the arc moves along, the molten metal left behind solidifies in a homogeneous deposit, or weld.

The electric power used with shielded metal arc welding may be direct or alternating current. With direct current, either straight or reverse polarity may be used. For straight polarity, the base metal is the positive pole and the electrode is the negative pole of the welding arc. For reverse polarity, the base metal is the negative pole and the electrode is the positive pole. Electrical equipment with a welding-current rating of 400 to 500 A is usually used for structural steel fabrication. The power source may be portable, but the need for moving it is minimized by connecting it to the electrode holder with relatively long cables.

The size of electrode (core wire diameter) depends primarily on joint detail and welding position. Electrode sizes of $\frac{1}{8}$, $\frac{5}{32}$, $\frac{3}{16}$, $\frac{7}{32}$, $\frac{1}{4}$, and $\frac{5}{16}$ in are commonly used. Small-size electrodes are 14 in long, and the larger sizes are 18 in long. Deposition rate of the weld metal depends primarily on welding current. Hence use of the largest electrode and welding current consistent with good practice is advantageous.

About 57 to 68% of the gross weight of the welding electrodes results in weld metal. The remainder is attributed to spatter, coating, and stub-end losses.

Shielded metal arc welding is widely used for manual welding of low-carbon steels, such as A36, and HSLA steels, such as A572 and A588. Though stainless steels, high-alloy steels, and nonferrous metals can be welded with this process, they are more readily welded with the gas metal arc process.

Submerged-arc welding (SAW) produces coalescence by the heat of an electric arc struck between a bare metal electrode and the base metal. The weld is shielded by flux, a blanket of granular fusible material placed over the joint. Pressure is not used on the parts to be joined. Filler metal is obtained either from the electrode or from a supplementary welding rod.

The electrode is pushed through the flux to strike an arc. The heat produced by the arc melts adjoining base metal and flux. As welding progresses, the molten flux forms a protective shield above the molten metal. On cooling, this flux solidifies under the unfused flux as a brittle slag that can be removed easily. Unfused flux is recovered for future use. About 1.5 lb of flux is used for each pound of weld wire melted.

Submerged-arc welding requires high currents. The current for a given cross-sectional area of electrode often is as much as 10 times as great as that used for manual welding. Consequently, the deposition rate and welding speeds are greater than for manual welding. Also, deep weld penetration results. Consequently, less edge preparation of the material to be joined is required for submerged-arc welding than for manual welding. For example, material up to $\frac{3}{8}$ in thick can be groove-welded, without any preparation or root opening, with two passes, one from each side of the joint. Complete fusion of the joint results.

Submerged-arc welding may be done with direct or alternating current. Conventional welding power units are used but with larger capacity than those used for manual welding. Equipment with current ratings up to 4000 A is used.

The process may be completely automatic or semiautomatic. In the semiautomatic process, the arc is moved manually. One-, two-, or three-wire electrodes can be used in automatic operation, two being the most common. Only one electrode is used in semiautomatic operation.

Submerged-arc welding is widely used for welding low-carbon steels and HSLA steels. Though stainless steels, high-alloy steels, and nonferrous metals can be welded with this process, they are generally more readily welded with the gas-shielded metal-arc process.

Gas metal arc welding (GMAW) produces coalescence by the heat of an electric arc struck between a filler-metal electrode and base metal. Shielding is obtained from a gas or gas mixture (which may contain an inert gas) or a mixture of a gas and flux.

This process is used with direct or alternating current. Either straight or reverse polarity may be employed with direct current. Operation may be automatic or semiautomatic. In the semiautomatic process, the arc is moved manually.

As in the submerged-arc process, high current densities are used, and deep weld penetration results. Electrodes range from 0.020 to $\frac{1}{8}$ in in diameter, with corresponding welding currents of about 75 to 650 A.

Practically all metals can be welded with this process. It is superior to other presently available processes for welding stainless steels and nonferrous metals. For these metals, argon, helium, or a mixture of the two gases is generally used for the shielding gas. For welding of carbon steels, the shielding gas may be argon, argon with oxygen, or carbon dioxide. Gas flow is regulated by a flowmeter. A rate of 25 to 50 ft^3/hr of arc time is normally used.

Flux-cored arc welding (FCAW) is similar to the GMAW process except that a flux-containing tubular wire is used instead of a solid wire. Shielding is provided by decomposition of the flux materials in the wire. Additional shielding is often provided by an externally supplied shielding gas fed through the electrode gun. The flux performs functions similar to the electrode coatings used for SMAW.

Electroslag welding (ESW) produces fusion with a molten slag that melts filler metal and the surfaces of the base metal. The weld pool is shielded by this molten slag, which moves along the entire cross section of the joint as welding progresses. The electrically conductive slag is maintained in a molten condition by its resistance to an electric current that flows between the electrode and the base metal.

The process is started much like the submerged-arc process by striking an electric arc beneath a layer of granular flux. When a sufficiently thick layer of hot molten slag is formed, arc action stops. The current then passes from the electrode to the base metal through the conductive slag. At this point, the process ceases to be an arc welding process and becomes the electroslag process. Heat generated by resistance to flow of current through the molten slag and weld puddle is sufficient to melt the edges at the joint and the tip of the welding electrode. The temperature of the molten metal is in the range of 3500°F. The liquid metal coming from the filler wire and the molten base metal collect in a pool beneath the slag and slowly solidify to form the weld. During welding, since no arc exists, no spattering or intense arc flash occurs.

Because of the large volume of molten slag and weld metal produced in electroslag welding, the process is generally used for welding in the vertical position. The parts to be welded are assembled with a gap 1 to 1¼ in wide. Edges of the joint need only be cut squarely, by either machine or flame.

Water-cooled copper shoes are attached on each side of the joint to retain the molten metal and slag pool and to act as a mold to cool and shape the weld surfaces.

The copper shoes automatically slide upward on the base-metal surfaces as welding progresses.

Preheating of the base metal is usually not necessary in the ordinary sense. Since the major portion of the heat of welding is transferred into the joint base metal, preheating is accomplished without additional effort.

The electroslag process can be used to join plates from $1\frac{1}{4}$ to 18 in thick. The process cannot be used on heat-treated steels without subsequent heat treatment. AWS and other specifications prohibit the use of ESW for welding quenched-and-tempered steel or for welding dynamically loaded structural members subject to tensile stresses or to reversal of stress.

Electrogas welding (EGW) is similar to electroslag welding in that both are automatic processes suitable only for welding in the vertical position. Both utilize vertically traveling, water-cooled shoes to contain and shape the weld surface. The electrogas process differs in that once an arc is established between the electrode and the base metal, it is continuously maintained. The shielding function is performed by helium, argon, carbon dioxide, or mixtures of these gases continuously fed into the weld area. The flux core of the electrode provides deoxidizing and slagging materials for cleansing the weld metal. The surfaces to be joined, preheated by the shielding gas, are brought to the proper temperature for complete fusion by contact with the molten slag. The molten slag flows toward the copper shoes and forms a protective coating between the shoes and the faces of the weld. As weld metal is deposited, the copper shoes, forming a weld pocket of uniform depth, are carried continuously upward.

The electrogas process can be used for joining material from $\frac{1}{2}$ to more than 2 in thick. The process cannot be used on heat-treated material without subsequent heat treatment. AWS and other specifications prohibit the use of EGW for welding quenched-and-tempered steel or for welding dynamically loaded structural members subject to tensile stresses or to reversal of stress.

Stud welding produces coalescence by the heat of an electric arc drawn between a metal stud or similar part and another work part. When the surfaces to be joined are properly heated, they are brought together under pressure. Partial shielding of the weld may be obtained by surrounding the stud with a ceramic ferrule at the weld location.

Stud welding usually is done with a device, or gun, for establishing and controlling the arc. The operator places the stud in the chuck of the gun with the flux end protruding. Then the operator places the ceramic ferrule over this end of the stud. With timing and welding-current controls set, the operator holds the gun in the welding position, with the stud pressed firmly against the welding surface, and presses the trigger. This starts the welding cycle by closing the welding-current contactor. A coil is activated to lift the stud enough to establish an arc between the stud and the welding surface. The heat melts the end of the stud and the welding surface. After the desired arc time, a control releases a spring that plunges the stud into the molten pool.

Direct current is used for stud welding. A high current is required for a very short time. For example, welding currents up to 2500 A are used with arc time of less than 1 sec for studs up to 1 in in diameter.

(O. W. Blodgett, *Design of Welded Structures,* The James F. Lincoln Arc Welding Foundation, Cleveland, Ohio.) See also Arts. 5.15 to 5.23.

2.7 CAMBER

Camber is a curvature built into a member or structure so that when it is loaded, it deflects to a desired shape. Camber, when required, might be for dead load only, dead load and partial live load, or dead load and full live load. The decision to camber and how much to camber is one made by the designer.

Rolled beams are generally cambered in a large press, known as a **bulldozer** or **gag press,** or through the use of heat. In a gag press, the beam is inched along and given an incremental bend at many points.

When heat is used to camber a beam, the flange to be shortened is heated with an oxygen-fed torch. As the flange is heated, it tries to elongate. But because it is held to its original length by the unheated web, the flange is forced to upset, increase inelastically in thickness, to relieve its compressive stresses. Since the increase in thickness is inelastic, the flange will not return to its original thickness on cooling. When the flange is allowed to cool, therefore, it must shorten to return to its original volume. The heated flange therefore experiences a net shortening that produces the camber.

Experience has shown that the residual stresses remaining in a beam after cambering are little different from those due to differential cooling rates of the elements of the shape after it has been produced by hot rolling. Note that allowable design stresses are based to some extent on the fact that residual stresses virtually always exist.

Plate girders usually are cambered by cutting the web plate to the cambered shape before the flanges are attached.

Large bridge and roof trusses are cambered by fabricating the members to lengths that will yield the desired camber when the trusses are assembled. For example, each compression member is fabricated to its geometric (loaded) length plus the calculated axial deformation under load. Similarly, each tension member is fabricated to its geometric length minus the axial deformation.

2.8 SHOP PREASSEMBLY

When the principal operations on a main member, such as punching, drilling, and cutting, are completed, and when the detail pieces connecting to it are fabricated, all the components are brought together to be fitted up, i.e., temporarily assembled with fit-up bolts, clamps, or tack welds. At this time, the member is inspected for dimensional accuracy, squareness, and, in general, conformance with shop detail drawings. Misalignment in holes in mating parts should be detected then and holes reamed, if necessary, for insertion of bolts. When fit-up is completed, the member is bolted or welded with final shop connections.

The foregoing type of shop preassembly or fit-up is an ordinary shop practice, routinely performed on virtually all work. There is another class of fit-up, however, mainly associated with highway and railroad bridges, that may be required by project specifications. These may specify that the holes in bolted field connections and splices be reamed while the members are assembled in the shop. Such requirements should be reviewed carefully before they are specified. The steps of subpunching (or subdrilling), shop assembly, and reaming for field connections add significant costs. Modern CNC drilling equipment can provide full-size holes located with a high degree of accuracy. AASHTO specifications, for example, include provisions for reduced shop assembly procedures when CNC drilling operations are used.

Where assembly and reaming are required, the following guidelines apply:

Splices in bridge girders are commonly reamed assembled. Alternatively, the abutting ends and the splice material may be reamed to templates independently.

Ends of floorbeams and their mating holes in trusses or girders usually are reamed to templates separately.

For reaming truss connections, three methods are in use in fabricating shops. The particular method to be used on a job is dictated by the project specifications or the designer.

Associated with the reaming methods for trusses is the method of cambering trusses. Highway and railroad bridge trusses are cambered by increasing the geometric (loaded) length of each compression member and decreasing the geometric length of each tension member by the amount of axial deformation it will experience under load (see Art. 2.7).

Method 1 (RT, or Reamed-template, Method). All members are reamed to geometric angles (angles between members under load) and cambered (no-load) lengths. Each chord is shop-assembled and reamed. Web members are reamed to metal templates. The procedure is as follows:

With the bottom chord assembled in its loaded position (with a minimum length of three abutting sections), the field connection holes are reamed. (**Section,** as used here and in methods 2 and 3, means fabricated member. A chord section, or fabricated member, usually is two panels long.)

With the top chord assembled in its loaded position (with a minimum length of three abutting sections), the field connection holes are reamed.

The end posts of heavy trusses are normally assembled and the end connection holes reamed, first for one chord and then for the other. The angles between the end post and the chords will be the geometric angles. For light trusses, however, the end posts may be treated as web members and reamed to metal templates.

The ends of all web members and their field holes in gusset plates are reamed separately to metal templates. The templates are positioned on the gusset plates to geometric angles. Also, the templates are located on the web members and gusset plates so that when the unloaded member is connected, the length of the member will be its cambered length.

Method 2 (Gary or Chicago Method). All members are reamed to geometric angles and cambered lengths. Each chord is assembled and reamed. Web members are shop-assembled and reamed to each chord separately. The procedure is as follows:

With the bottom chord assembled in its geometric (loaded) alignment (with a minimum number of three abutting sections), the field holes are reamed.

With the top chord assembled in its geometric position (with a minimum length of three abutting sections), the holes in the field connections are reamed.

The end posts and all web members are assembled and reamed to each chord separately. All members, when assembled for reaming, are aligned to geometric angles.

Method 3 (Fully Assembled Method). The truss is fully assembled, then reamed. In this method, the bottom chord is assembled and blocked into its cambered (unloaded) alignment, and all the other members are assembled to it. The truss, when fully assembled to its cambered shape, is then reamed. Thus the members are positioned to cambered angles, not geometric angles.

When the extreme length of trusses prohibits laying out the entire truss, method 3 can be used sectionally. For example, at least three abutting complete sections (top and bottom chords and connecting web members) are fully assembled in their cambered position and reamed. Then complete sections are added to and removed from the assembled sections. The sections added are always in their cambered position. There should always be at least two previously assembled and reamed sections in the layout. Although reaming is accomplished sectionally, the procedure fundamentally is the same as for a full truss assembly.

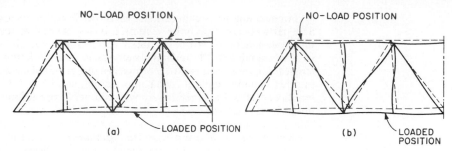

FIGURE 2.1   Effects of reaming methods on truss assembly. (*a*) Truss configurations produced in methods 1 and 2. (*b*) Truss shapes produced in method 3.

In methods 1 and 2, field connections are reamed to cambered lengths and geometric angles, whereas in method 3, field connections are reamed to cambered lengths and angles. To illustrate the effects of these methods on an erected and loaded truss, Fig. 2.1*a* shows by dotted lines the shape of a truss that has been reamed by either method 1 or 2 and then fully connected, but without load. As the members are fitted up (pinned and bolted), the truss is forced into its cambered position. Bending stresses are induced into the members because their ends are fixed at their geometric (not cambered) angles. This bending is indicated by exaggerated S curves in the dotted configuration. The configuration shown in solid lines in Fig. 2.1*a* represents the truss under the load for which the truss was cambered. Each member now is strained; the fabricated length has been increased or decreased to the geometric length. The angles that were set in geometric position remain geometric. Therefore, the S curves induced in the no-load assembly vanish. Secondary bending stresses, for practical purposes, have been eliminated. Further loading or a removal of load, however, will produce some secondary bending in the members.

Figure 2.1*b* illustrates the effects of method 3. Dotted lines represent the shape of a truss reamed by method 3 and then fully connected, but without load. As the members are fitted up (pinned and bolted), the truss takes its cambered position. In this position, as when they were reamed, members are straight and positioned to their cambered angles, hence have no induced bending. The solid lines in Fig. 2.1*b* represent the shape of the truss under the load for which the truss was cambered. Each member now is strained; the fabricated length has been increased or decreased to its geometric length. The angles that were set in the cambered (no-load) position are still in that position. As a result, S curves are induced in the members, as indicated in Fig. 2.1*b* by exaggerated S curves in solid lines. Secondary stresses due to bending, which do not occur under camber load in methods 1 and 2, are induced by this load in method 3. Further loading will increase this bending and further increase the secondary stresses.

Bridge engineers should be familiar with the reaming methods and see that design and fabrication are compatible.

2.9   ROLLED SECTIONS

Hot-rolled sections produced by rolling mills and delivered to the fabricator include the following designations: W shapes, wide-flange shapes with essentially parallel flange surfaces, S shapes, American Standard beams with slope of 16⅔% on inner flange surfaces; HP shapes, bearing-pile shapes (similar to W shapes but with flange and web thicknesses equal), M shapes (miscellaneous shapes that are similar to W, S, or HP but do not meet that classification), C shapes (American Standard channel shape with slope of 16⅔% on inner flange surfaces), MC shapes (miscellaneous channels similar to C), L shapes and angles, and ST (structural tees cut from W, M, or S shapes). Such material, as well as plates and bars, is referred to collectively as **plain material.**

To fulfill the needs of a particular contract, some of the plain material may be purchased from a local warehouse or may be taken from the fabricator's own stock. The major portion of plain material, however, is ordered directly from a mill to specific properties and dimensions. Each piece of steel on the order is given an identifying mark through which its origin can be traced. Mill test reports, when required, are furnished by the mill to the fabricator to certify that the requirements specified have been met.

Steel shapes, such as beams, columns, and truss chords, that constitute main material for a project are often ordered from the mill to approximately their final length. The exact length ordered depends on the type of end connection or end preparation and the extent to which the final length can be determined at the time of ordering. The length ordered must take into account the mill tolerances on length. These range for wide-flange shapes from $\pm\frac{3}{8}$ to $\pm\frac{1}{2}$ in or more, depending on size and length of section (see ASTM A6). Beams that are to have standard framed or seated end connections therefore are ordered to such lengths that they will not be delivered too long. When connection material is attached, it is positioned to produce the desired length. Beams that will frame directly to other members, as is often the case in welded construction, must be ordered to such lengths that they cannot be delivered too short. In addition, an allowance for trimming must be added.

Wide-flange shapes used as columns are ordered with an allowance for finishing the ends.

Items such as angles for bracing or truss-web members, detail material, and light members in general are ordered in long pieces from which several members can be cut.

Plate material such as that for use in plate-girder webs is generally ordered to required dimensions plus additional amounts for trim and camber.

Plate material such as that for use in plate-girder flanges or built-up column webs and flanges is generally ordered to the required length plus trim allowance but in multiple widths for flame cutting or stripping to required widths.

The dimensions in which standard sections are ordered, i.e., multiple widths, multiple lengths, etc., are given careful consideration by the fabricator because the mill unit prices for the material depend on dimensions as well as on physical properties and chemistry. Computers are often used to optimize ordering of material.

Fabrication of standard sections entails several or all of the following operations: template making, layout, punching and drilling, fitting up and reaming, bolting, welding, finishing, inspection, cleaning, painting, and shipping.

2.10   BUILT-UP SECTIONS

These are members made up by a fabricator from two or more standard sections. Examples of common built-up sections are shown in Fig. 2.2. Built-up members are specified by the designer when the desired properties or configuration cannot be obtained in a single hot-rolled section. Built-up sections can be bolted or welded. Welded members, in general, are less expensive because much less handling is

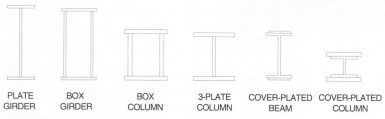

| PLATE GIRDER | BOX GIRDER | BOX COLUMN | 3-PLATE COLUMN | COVER-PLATED BEAM | COVER-PLATED COLUMN |

FIGURE 2.2 Typical built-up structural sections.

required in the shop and because of more efficient utilization of material. The clean lines of welded members also produce a better appearance.

Cover-plated rolled beams are used when the required bending capacity is not available in a rolled standard beam or when depth limitations preclude use of a deeper rolled beam or plate girder. Cover-plated beams are also used in composite construction to obtain the efficiency of a nonsymmetrical section.

Cover-plate material is ordered to multiple widths for flame cutting or stripping to the required width in the shop. For this reason, when several different design conditions exist in a project, it is good practice, as well as good economy, for the designer to specify as few different cover-plate thicknesses as possible and to vary the width of plate for the different members.

For bolted sections, cover plates and rolled-beam flanges are punched separately and are then brought together for fit-up. Sufficient temporary fitting bolts are installed to hold the cover plates in alignment, and minor mismatches of holes in mating parts are cleaned up by reaming. For welded sections, cover plates are held in position with small intermittent tack welds until final welding is done.

Plate girders are specified when the moment capacity, stiffness, or on occasion, web shear capacity cannot be obtained in a rolled beam. They usually are fabricated by welding.

Welded plate girders consist of a web plate, a top flange plate, a bottom flange plate, and stiffener plates. Web material is ordered from the mill to the width between flange plates plus an allowance for trim and camber, if required. Flange material is ordered to multiple widths for stripping to the desired widths in the shop.

When an order consists of several identical girders having shop flange splices, fabricators usually first lay the flange material end to end in the ordered widths and splice the abutting ends with the required groove welds. The long, wide plates thus produced are then stripped to the required widths. For this procedure, the flanges should be designed to a constant width over the length of the girder. This method is advantageous for several reasons: Flange widths permit groove welds sufficiently long to justify use of automatic welding equipment. Run-out tabs for starting and stopping the welds are required only at the edges of the wide, un-stripped plate. All plates can be stripped from one setup. And much less finishing is required on the welds.

After web and flange plates are cut to proper widths, they are brought together for fit-up and final welding. The web-to-flange welds, usually fillet welds, are positioned for welding with maximum efficiency. For relatively small welds, such as ¼- or ⁵⁄₁₆-in fillets, a girder may be positioned with web horizontal to allow welding of both flanges simultaneously. The girder is then turned over, and the corresponding welds are made on the other side. When relatively large fillet welds are required, the girder is held in a fixture with the web at an angle of about 45° to allow one weld at a time to be deposited in the flat position. In either method, the web-to-flange welds are made with automatic welding machines that produce welds of good quality at a high rate of deposition. For this reason, fabricators would prefer to use continuous fillet welds rather than intermittent welds, though an intermittent weld may otherwise satisfy design requirements.

After web-to-flange welds are made, the girder is trimmed to its detailed length. This is not done earlier because of the difficulty of predicting the exact amount of girder shortening due to shrinkage caused by the web-to-flange welds.

If holes are required in web or flange, the girder is drilled next. This step requires moving the whole girder to the drills. Hence, for economy, holes in main material should be avoided because of the additional amount of heavy-load handling required. Instead, holes should be located in detail material, such as stiffeners, which can be punched or drilled before they are welded to the girder.

The next operation applies the stiffeners to the web. Stiffener-to-web welds often are fillet welds. They are made with the web horizontal. The welds on each side of a stiffener may be deposited simultaneously with automatic welding equipment. For this equipment, many fabricators prefer continuous welds to intermittent welds. When welds are large, however, the girder may be positioned for flat, or downhand, welding of the stiffeners.

Variation in stress along the length of a girder permits reductions in flange material. For minimum weight, flange width and thickness might be decreased in numerous steps. But a design that optimizes material seldom produces an economical girder. Each change in width or thickness requires a splice. The cost of preparing a splice and making a weld may be greater than the cost of material saved to avoid the splice. Therefore, designers should hold to a minimum flange splices made solely to save material. Sometimes, however, the length of piece that can be handled may make splices necessary.

Welded crane girders differ from ordinary welded plate girders principally in that the upper surface of the top flange must be held at constant elevation over the span. A step at flange splices is undesirable. Since lengths of crane girders usually are such that flange splices are not made necessary by available lengths of material, the top flange should be continuous. In unusual cases where crane girders are long and splices are required, the flange should be held to a constant thickness. (It is not desirable to compensate for a thinner flange by deepening the web at the splice.) Depending on other elements that connect to the top flange of a crane girder, such as a lateral-support system or horizontal girder, holding the flange to a constant width also may be desirable.

Horizontally curved plate girders for bridges constitute a special case. Two general methods are used in fabricating them. In one method, the flanges are cut from a wide plate to the prescribed curve. Then the web is bent to this curve and welded to the flanges. In the second method, the girder is fabricated straight and then curved by application of heat to the flanges. This method, which is recognized by the AASHTO "Standard Specifications for Highway Bridges," is preferred by many fabricators because less scrap is generated in cutting flange plates, savings may accrue from multiple welding and stripping of flange plates, and the need for special jigs and fittings for assembling a girder to a curve is avoided.

("Fabrication Aids for Continuously Heat-Curved Girders" and "Fabrication Aids for Girders Curved with V-Heats," American Institute of Steel Construction, Chicago, Ill.)

Procedures used in fabricating other built-up sections, such as box girders and box columns, are similar to those for welded girders.

Columns generally require the additional operation of end finishing for bearing. For welded columns, all the welds connecting main material are made first, to eliminate uncertainties in length due to shrinkage caused by welding. After the ends are finished, detail material, such as connection plates for beams, is added.

For economy, holes in main material should be avoided, to eliminate the additional handling in moving the entire column to the drill. Holes preferably should be located in detail material.

2.11 CLEANING AND PAINTING

The AISC "Specification for Structural Steel Buildings" provides that, in general, steelwork to be concealed within the building need not be painted and that steel encased in concrete should not be painted. Inspection of old buildings has revealed that the steel withstands corrosion virtually the same whether painted or not.

When paint is required, a shop coat is applied as a primer for subsequent field coats. It is intended to protect the steel for only a short period of exposure.

Steel to be painted must be thoroughly cleaned of all loose mill scale, loose rust, dirt, and other foreign matter. Unless the fabricator is otherwise directed, cleaning of structural steel is ordinarily done with a wire brush.

Treatment of architecturally exposed structural steel varies somewhat from that for steel in unexposed situations. Since surface preparation is the most important factor affecting performance of paint on structural steel surfaces, it is common for blast cleaning to be specified for the steel to be exposed as a means of removing all mill scale.

Mill scale that forms on structural steel after hot rolling protects the steel from corrosion but only as long as this scale is intact and adheres firmly to the steel. Intact mill scale, however, is seldom encountered on fabricated steel because of weathering during storage and shipment and because of loosening caused by fabricating operations. Undercutting of mill scale, which leads to a mill scale lifting type of paint failure, is attributable to the broken or cracked condition of mill scale at the time of painting. When architecturally exposed structural steel is used, a little mill scale lifting and resulting rust staining may injure the appearance of a building. On industrial buildings, a little rust staining might not be objectionable. But where appearance is of paramount importance, descaling by blast cleaning is the preferred way of preparing the surface of architecturally exposed steel for painting.

Steel to be exposed to the weather and left unpainted, such as A588 or A242 steel, is generally treated in one of two ways, depending on the application.

For structures where appearance is not important and minimal maintenance is the prime consideration, the steel may be erected with no surface preparation at all. While it retains mill scale, the steel will not have a uniform color. But when the scale loses its adherence and flakes off, the exposed metal will form the tightly adherent oxide coating characteristic of this type of steel, and eventually, a uniform color will result.

Where uniform color of bare, unpainted steel is important, the steel must be freed of scale by blast cleaning to near-white metal. In such applications, extra precautions must be exercised to protect the blasted surfaces from scratches and staining.

(*Steel Structures Painting Manual*, vol. I, *Good Painting Practice*, vol. II, *Systems and Specifications*, Steel Structures Painting Council, 4400 Fifth Ave., Pittsburgh, PA 15213.)

2.12 FABRICATION TOLERANCES

Dimensional tolerances, permissible variations from theoretical dimensions, have been established for hot-rolled structural steel because of the speed with which it must be rolled, wear and deflection of the rolls, human differences between mill operators, and differential cooling rates of the elements of a section. Also, mills cut rolled sections to length while they are still hot. Tolerances that must be met before structural steel can be shipped from mill to fabricator are listed in ASTM A6, "General Requirements for Delivery of Rolled Steel Plates, Shapes, Sheet Piling and Bars for Structural Use."

Tolerances are specified for the dimensions and straightness of plates, hot-rolled shapes, and bars. For example, flanges of rolled beams need not be perfectly square with the web and need not be perfectly centered on the web.

Specifications covering fabrication of structural steel do not, in general, require closer tolerances than those in A6 but rather extend the definition of tolerances to fabricated members. Tolerances for the fabrication of structural steel, both hot-rolled and built-up members, will be found in standard codes, such as the AISC "Specification for Structural Steel Buildings," both the ASD and LRFD editions; AISC "Code of Standard Practice for Steel Buildings and Bridges"; AWS D1.1 "Structural Welding Code—Steel"; AWS D1.5 "Bridge Welding Code"; AASHTO "Standard Specifications for Highway Bridges"; and AASHTO "Guide Specification for Fracture-Critical Non-Redundant Steel Bridge Members."

The tolerance on length of material as delivered to the fabricator as defined in A6 is one case where the A6 tolerance may not be suitable for the final member. For example, A6 allows wide-flange beams 24 in or less deep to vary (plus or minus) from ordered length by $3/8$ in plus an additional $1/16$ in for each additional 5-ft increment over 30 ft. The AISC specification for length of fabricated steel, however, allows beams to vary from detailed length only $1/16$ in for members 30 ft or less long and $1/8$ in for members longer than 30 ft. For beams with framed or seated end connections, the fabricator can tolerate allowable variations in length by setting the end connections on the beam so as to not exceed the overall fabrication tolerance of $\pm 1/16$ or $\pm 1/8$ in. Members that must connect directly to other members, without framed or seated end connections, must be ordered from the mill with a little additional length to permit the fabricator to trim them to within $\pm 1/16$ or $\pm 1/8$ in of the desired length.

The AISC "Code of Standard Practice for Steel Buildings and Bridges" states that "permissible tolerances for out-of-square or out-of-parallel, depth, width and symmetry of rolled shapes are as specified in ASTM Specification A6. No attempt to match abutting cross-sectional configurations is made unless specifically required by the contract documents. The as-fabricated straightness tolerances of members are one-half of the standard camber and sweep tolerances in ASTM A6."

Designers should be familiar with the tolerances allowed by the specifications covering each job. If they require more restrictive tolerances, they must so specify on the drawings and must be prepared for possible higher costs of fabrication.

2.13 ERECTION EQUIPMENT

Steel buildings and bridges are generally erected with cranes, derricks, or specialized units. Mobile cranes include crawler cranes and truck cranes; stationary cranes include tower cranes and climbing cranes. Stiffleg derricks and guy derricks are stationary hoisting machines. A high line is an example of a specialized unit. These various types of erection equipment used for steel construction are also used for precast and cast-in-place concrete construction.

One of the most common machines for steel erection is the crawler crane (Fig. 2.3). Self-propelled, such cranes are mounted on a mobile base having endless tracks or crawlers for propulsion. The base of the crane contains a turntable that

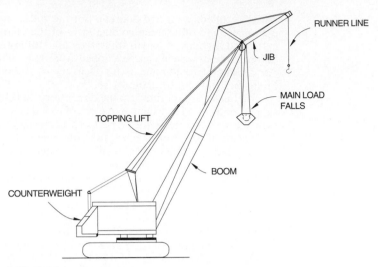

FIGURE 2.3 Crawler crane.

allows 360° rotation. Crawlers come with booms up to 450 ft high and capacities up to 350 tons. Self-contained counterweights prevent overturning.

Truck cranes (Fig. 2.4) are similar in many respects to crawler cranes. The principal difference is that truck cranes are mounted on rubber tires and are therefore much more mobile on hard surfaces. Truck cranes can be used with booms up to 350 ft long and have capacities up to 250 tons. Truck cranes have outriggers (that are always used) to provide stability.

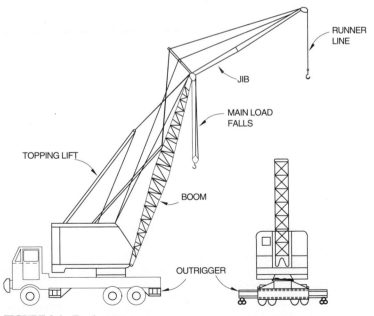

FIGURE 2.4 Truck crane.

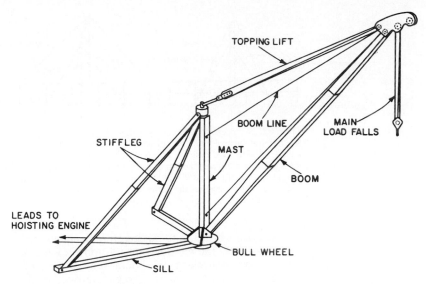

FIGURE 2.5   Stiffleg derrick.

A stiffleg derrick (Fig. 2.5) consists of a boom and a vertical mast rigidly supported by two legs. The two legs are capable of resisting either tensile or compressive forces, hence the name **stiffleg.** Stiffleg derricks are extremely versatile in that they can be used in a permanent location as yard derricks or can be mounted on a wheel-equipped frame for use as a traveler in bridge erection. A stiffleg derrick also can be mounted on a device known as a **creeper** and thereby lift itself vertically on a structure as it is being erected. Stiffleg derricks can range from small, 5-ton units to large, 250-ton units, with 80-ft masts and 180-ft booms.

A guy derrick (Fig. 2.6) is commonly associated with the erection of tall multistory buildings. It consists of a boom and a vertical mast supported by wire-rope

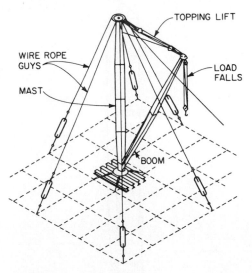

FIGURE 2.6   Guy derrick.

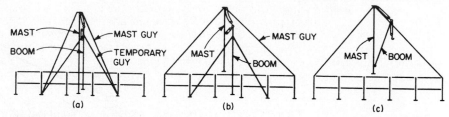

FIGURE 2.7 Steps in jumping a guy derrick. (*a*) Removed from its seat with the topping lift falls, the boom is revolved 180° and placed in a temporary jumping shoe. The boom top is temporarily guyed. (*b*) The load falls are attached to the mast above its center of gravity. Anchorages of the mast guys are shifted to the top of the next tier. The boom then raises the mast to its new position. The mast guys are adjusted and the load falls unhooked. (*c*) The temporary guys on the boom are removed. The mast raises the boom with the topping lift falls and places it in the boom seat, ready for operation.

guys which are attached to the structure being erected. Although a guy derrick can be rotated 360°, the rotation is handicapped by the presence of the guys. To clear the guys while swinging, the boom must be shorter than the mast and must be brought up against the mast. The guy derrick has the advantage of being able to climb vertically (jump) under its own power, such as illustrated for the construction of a building in Fig. 2.7. Guy derricks have been used up to 160 ft long and with capacities up to 250 tons.

Tower cranes in various forms are used extensively for erection of buildings and bridges. Several manufacturers offer accessories for converting conventional truck or crawler cranes into tower cranes. Such a tower crane (Fig. 2.8) is characterized

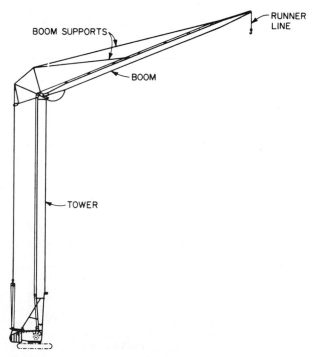

FIGURE 2.8 Tower crane on crawler-crane base.

by a vertical tower, which replaces the conventional boom, and a long boom at the top that can usually accommodate a jib as well. With the main load falls suspended from its end, the boom is raised or lowered to move the load toward or away from the tower. The cranes are counterweighted in the same manner as conventional truck or crawler cranes. Capacities of these tower cranes vary widely depending on the machine, tower height, and boom length and angle. Such cranes have been used with towers 250 ft high and booms 170 ft long. They can usually rotate 360°.

Other types of tower cranes with different types of support are shown in Fig. 2.9*a* through *c*. The type selected will vary with the type of structure erected and erection conditions. Each type of support shown may have either the kangaroo (topping lift) or the hammerhead (horizontal boom) configuration.

2.14 *ERECTION METHODS FOR BUILDINGS*

The determination of how to erect a building depends on many variables that must be studied by the erection engineer long before steel begins to arrive at the erection site. It is normal and prudent to have this erection planning developed on drawings and in written procedures. Such documents outline the equipment to be used, methods of supporting the equipment, conditions for use of the equipment, and sequence of erection. In many areas, such documents are required by law. The work plan that evolves from them is valuable because it can result in economies in the costly field work. Special types of structures may require extensive planning to ensure stability of the structure during erection.

Mill buildings, warehouses, shopping centers, and low-rise structures that cover large areas usually are erected with truck or crawler cranes. Selection of the equipment to be used is based on site conditions, weight and reach for the heavy lifts, and availability of equipment. Preferably, erection of such building frames starts at one end, and the crane backs away from the structure as erection progresses. The underlying consideration at all times is that an erected member should be stable before it is released from the crane. High-pitched roof trusses, for example, are often unstable under their own weight without top-chord bracing. If roof trusses are long and shipped to the site in several sections, they are often spliced on the ground and lifted into place with one or two cranes.

Multistory structures, or portions of multistory structures that lie within reach and capacity limitations of crawler cranes, are usually erected with crawler cranes. For tall structures, a crawler crane places steel it can reach and then erects the guy derrick (or derricks), which will continue erection. Alternatively, tower crawler cranes (see Fig. 2.8) and climbing tower cranes (Fig. 2.9) are used extensively for multistory structures. Depending on height, these cranes can erect a complete structure. They allow erection to proceed vertically, completing floors or levels for other trades to work on before the structure is topped out.

Use of any erecting equipment that loads a structure requires the erector to determine that such loads can be adequately withstood by the structure or to install additional bracing or temporary erection material that may be necessary. For example, guy derricks impart loads at guys, and at the base of the boom, a horizontal thrust that must be provided for. On occasion, floorbeams located between the base of the derrick and guy anchorages must be temporarily laterally supported to resist imposed compressive forces. Considerable temporary bracing is required in a multistory structure when a climbing crane is used. This type of crane imposes horizontal and vertical loads on the structure or its foundation.

The sequence of placing the members of a multistory structure is, in general, columns, girders, bracing, and beams. The exact order depends on the erection

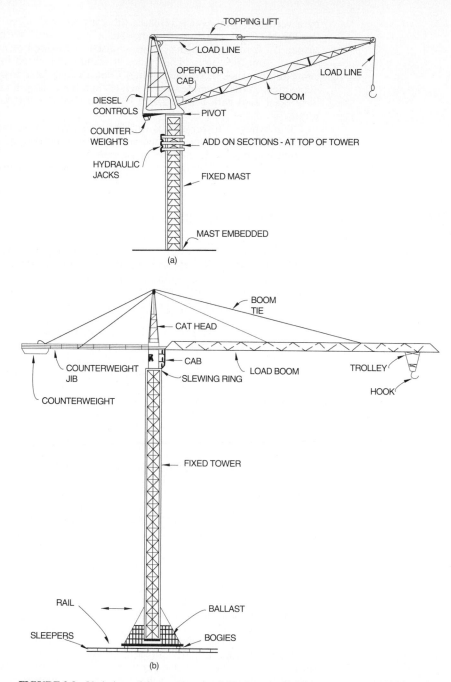

FIGURE 2.9 Variations of the tower crane: (*a*) kangaroo; (*b*) hammerhead; (*c*) climbing crane.

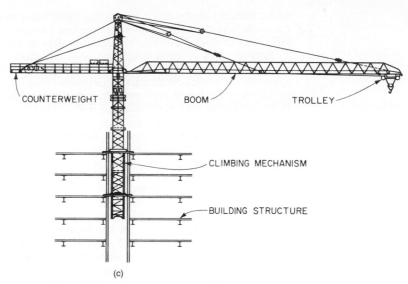

COUNTERWEIGHT   BOOM   TROLLEY

CLIMBING MECHANISM

BUILDING STRUCTURE

(c)

FIGURE 2.9 *Continued.*

equipment and type of framing. Planning must ensure that all members can be erected and that placement of one member does not prohibit erection of another.

Structural steel is erected by ironworkers who perform a multitude of tasks. The ground crew selects the proper members to hook onto the crane and directs crane movements in delivering the piece to the "connectors." The connectors direct the piece into its final location, place sufficient temporary bolts for stability, and unhitch the crane. A fitting-up crew, following the connectors, aligns the beams, plumbs the columns, and installs whatever temporary wire-rope bracing is necessary to maintain alignment. Following the fitting-up crew are the gangs who make the permanent connection. This work, which usually follows several stories behind member erection, may include tightening high-strength bolts or welding connections. An additional operation usually present is placing and welding metal deck to furnish a working floor surface for subsequent operations. Safety codes require permanent floor surfaces two floors below the erection work above.

In field-welded multistory buildings with continuous beam-to-column connections, the procedure is slightly different from that for bolted work. The difference is that the welded structure is not in its final alignment until beam-to-column connections are welded because of shrinkage caused by the welds. To accommodate the shrinkage, the joints must be opened up or the beams must be detailed long so that, after the welds are made, the columns are pulled into plumb. It is necessary, therefore, to erect from the more restrained portion of the framing to the less restrained. If a structure has a braced center core, that area will be erected first to serve as a reference point, and steel will be erected toward the perimeter of the structure. If the structure is totally unbraced, an area in the center will be plumbed and temporarily braced for reference. Welding of column splices generally is done several floors behind the erection of members.

2.15 ERECTION PROCEDURE FOR BRIDGES

Bridges are erected by a variety of methods. The choice of method in a particular case is influenced by type of structure, length of span, site conditions, manner in which material is delivered to the site, and equipment available. Bridges over navigable waterways are sometimes limited to erection procedures that will not inhibit traffic flow; for example, falsework may be prohibited.

Regardless of erection procedure selected, there are two considerations that override all others. The first is the security and stability of the structure under all conditions of partial construction, construction loading, and wind loading that will be encountered during erection. The second consideration is that the bridge must be erected in such a manner that it will perform as intended. For example, in continuous structures, this can mean that jacks must be used on the structure to effect the proper stress distribution. These considerations will be elaborated upon later as they relate to erection of particular types of bridges.

Simple-beam bridges are often erected with a crawler or truck crane. Bridges of this type generally require a minimal amount of engineering and are put up routinely by an experienced erector. One problem that does occur with beam spans, however, and especially composite beam spans, arises from lateral instability of the top flange during lifting or before placement of permanent bracing. Beams or girders that are too limber to lift unbraced require temporary compression-flange support, often in the form of a stiffening truss. Lateral support also may be provided by assembling two adjacent members on the ground with their bracing or cross members and erecting the assembly in one piece. Beams that can be lifted unbraced but are too limber to span alone also can be handled in pairs. Or it may be necessary to hold them with the crane until bracing connections can be made.

Continuous-beam bridges are erected in much the same way as simple-beam bridges. One or more field splices, however, will be present in the stringers of continuous beams. With bolted field splices, the holes in the members and connection material have been reamed in the shop to insure proper alignment of the member. With a welded field splice, it is generally necessary to provide temporary connection material to support the member and permit adjustment for alignment and proper positioning for welding. For economy, field splices should be located at points of relatively low bending moment. It is also economical to allow the erector some option regarding splice location, which may materially affect erection cost. The arrangement of splices in Fig. 2.10a, for example, will require, if falsework is to be avoided, that both end spans be erected first, then the center spans. The splice arrangement shown in Fig. 2.10b will allow erection to proceed from one

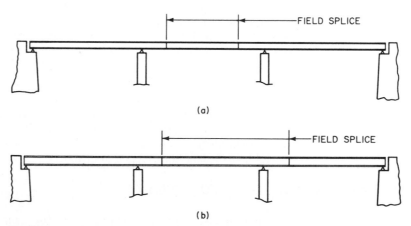

FIGURE 2.10 Field splices in girder bridges.

end to the other. While both arrangements are used, one may have advantages over the other in a particular situation.

Horizontally curved girder bridges are similar to straight-girder bridges, except for torsional effects. If use of falsework is to be avoided, it is necessary to resist the torques by assembling two adjacent girders with their diaphragms and temporary or permanent lateral bracing and erect the assembly as a stable unit. Diaphragms and their connections must be capable of withstanding end moments induced by girder torques.

Truss bridges require a vast amount of investigation to determine the practicability of a desired erection scheme or the limitations of a necessary erection scheme. The design of truss bridges, whether simple or continuous, generally assumes that the structure is complete and stable before it is loaded. The erector, however, has to impose dead loads, and often live loads, on the steel while the structure is partly erected. The structure must be erected safely and economically in a manner that does not overstress any member or connection.

Erection stresses may be of opposite sign and of greater magnitude than the design stresses. When designed as tension members but subjected to substantial compressive erection stresses, the members may be braced temporarily to reduce their effective length. If bracing is impractical, they may be made heavier. Members designed as compression members but subjected to tensile forces during erection are investigated for adequacy of area of net section where holes are provided for connections. If the net section is inadequate, the member must be made heavier.

Once an erection scheme has been developed, the erection engineer analyzes the structure under erection loads in each erection stage and compares the erection stresses with the design stresses. At this point, the engineer plans for reinforcing or bracing members, if required. The erection loads include the weights of all members in the structure in the particular erection stage and loads from whatever erection equipment may be on the structure. Wind loads are added to these loads.

In addition to determining member stresses, the erection engineer usually calculates reactions for each erection stage, whether they be reactions on abutments or piers or on falsework. Reactions on falsework are needed for design of the falsework. Reactions on abutments and piers may reveal a temporary uplift that must be provided for, by counterweighting or use of tie-downs. Often, the engineer also computes deflections, both vertical and horizontal, at critical locations for each erection stage to determine size and capacity of jacks that may be required on falsework or on the structure.

When all erection stresses have been calculated, the engineer prepares detailed drawings showing falsework, if needed, necessary erection bracing with its connections, alterations required for any permanent member or joint, installation of jacks and temporary jacking brackets, and bearing devices for temporary reactions on falsework. In addition, drawings are made showing the precise order in which individual members are to be erected.

Figure 2.11 shows the erection sequence for a through-truss cantilever bridge over a navigable river. For illustrative purpose, the scheme assumes that falsework is not permitted in the main channel between piers and that a barge-mounted crane will be used for steel erection. Because of the limitation on use of falsework, the erector adopts the cantilever method of erection. The plan is to erect the structure from both ends toward the center.

Note that top chord U13–U14, which is unstressed in the completed structure, is used as a principal member during erection. Note also that in the suspended span all erection stresses are opposite in sign to the design stresses.

As erection progresses toward the center, a negative reaction may develop at the abutments (panel point LO). The uplift may be counteracted by tie-downs to the abutment.

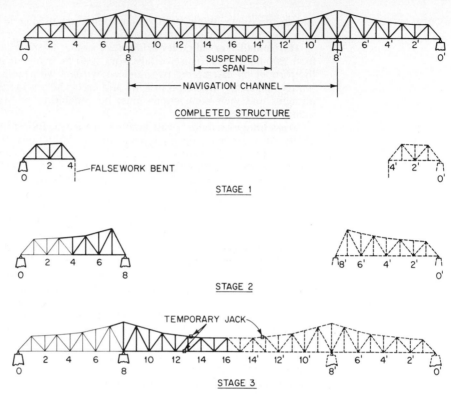

FIGURE 2.11 Erection stages for a continuous-truss bridge. In stage 1, with falsework at panel point 4, the portion of the truss from the abutment to that point is assembled on the ground and then erected on the abutment and the falsework. The operations are duplicated at the other end of the bridge. In stage 2, members are added by cantilevering over the falsework, until the piers are reached. Panel points 8 and 8' are landed on the piers by jacking down at the falsework, which then is removed. In stage 3, main-span members are added by cantilevering over the piers, until midspan is reached. Jacks are inserted at panel points L13, U13, and U13'. The main span is closed by jacking. The jacks then are unloaded to hang the suspended span and finally are removed.

Hydraulic jacks, which are removed after erection has been completed, are built into the chords at panel points U13, L13, and U13'. The jacks provide the necessary adjustent to allow closing of the span. The two jacks at U13 and L13 provide a means of both horizontal and vertical movement at the closing panel point, and the jack at U13' provides for vertical movement of the closing panel point only.

2.16 FIELD TOLERANCES

Permissible variations from theoretical dimensions of an erected structure are specified in the AISC "Code of Standard Practice for Steel Buildings and Bridges." It states that variations are within the limits of good practice or erected tolerance when they do not exceed the cumulative effect of permissible rolling and fabricating and erection tolerances. These tolerances are restricted in certain instances to total cumulative maximums.

The AISC "Code of Standard Practice" has a descriptive commentary that fully outlines and explains the application of the mill, fabrication, and erection tolerances for a building or bridge. Also see Art. 2.12 for a listing of specifications and codes that may require special or more restrictive tolerances for a particular type of structure.

An example of tolerances that govern the plumbness of a multistory building is the tolerance for columns. In multistory buildings, columns are considered to be plumb if the error does not exceed 1:500, except for columns adjacent to elevator shafts and exterior columns, for which additional limits are imposed. The tolerances governing the variation of columns, as erected, from their theoretical centerline are sometimes wrongfully construed to be lateral-deflection (drift) limitations on the completed structure when, in fact, the two considerations are unrelated.

(*Steel Construction Manual*, American Institute of Steel Construction.)

SECTION 3
GENERAL STRUCTURAL THEORY

Ronald D. Ziemian
*Assistant Professor of Civil Engineering, Bucknell University,
Lewisburg, Pennsylvania*

Safety and serviceability constitute the two primary requirements in structural design. For a structure to be safe, it must have adequate strength and ductility when resisting occasional extreme loads. To ensure that a structure will perform satisfactorily at working loads, functional or serviceability requirements also must be met. An accurate prediction of the behavior of a structure subjected to these loads is indispensable in designing new structures and evaluating existing ones.

The behavior of a structure is defined by the displacements and forces produced within the structure as a result of external influences. In general, structural theory consists of the essential concepts and methods for determining these effects. The process of determining them is known as **structural analysis.** If the assumptions inherent in the applied structural theory are in close agreement with actual conditions, such an analysis can often produce results that are in reasonable agreement with performance in service.

3.1 FUNDAMENTALS OF STRUCTURAL THEORY

Structural theory is based primarily on the following set of laws and properties. These principles often provide sufficient relations for analysis of structures.

Laws of mechanics. These consist of the rules for static equilibrium and dynamic behavior.

Properties of materials. The material used in a structure has a significant influence on its behavior. Strength and stiffness are two important material properties. These properties are obtained from experimental tests and may be used in the analysis either directly or in an idealized form.

Laws of deformation. These require that structure geometry and any incurred deformation be compatible; i.e., the deformations of contiguous structural components are in agreement such that all components fit together to define the deformed state of the entire structure.

STRUCTURAL MECHANICS—STATICS

An understanding of basic mechanics is essential for comprehending structural theory. Mechanics is a part of physics that deals with the state of rest and the motion of bodies under the action of forces. For convenience, mechanics is divided into two parts: statics and dynamics.

Statics is that branch of mechanics that deals with bodies at rest or in equilibrium under the action of forces. In elementary mechanics, bodies may be idealized as rigid when the actual changes in dimensions caused by forces are small in comparison with the dimensions of the body. In evaluating the deformation of a body under the action of loads, however, the body is considered deformable.

3.2 PRINCIPLES OF FORCES

The concept of force is an important part of mechanics. Created by the action of one body on another, force is a vector, consisting of magnitude and direction. In addition to these values, point of action or line of action is needed to determine the effect of a force on a structural system.

Forces may be concentrated or distributed. A **concentrated force** is a force applied at a point. A **distributed force** is spread over an area. It should be noted that a concentrated force is an idealization. Every force is in fact applied over some finite area. When the dimensions of the area are small compared with the dimensions of the member acted on, however, the force may be considered concentrated. For example, in computation of forces in the members of a bridge, truck wheel loads are usually idealized as concentrated loads. These same wheel loads, however, may be treated as distributed loads in design of a bridge deck.

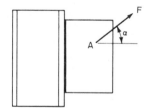

A set of forces is **concurrent** if the forces all act at the same point. Forces are **collinear** if they have the same line of action and are **coplanar** if they act in one plane.

Figure 3.1 shows a bracket that is subjected to a force **F** having magnitude F and direction defined by angle α. The force acts at point A. Changing any one of these designations changes the effect of the force on the bracket.

FIGURE 3.1 Arrow **F** represents force acting on a bracket.

Because of the additive properties of forces, force **F** may be resolved into two concurrent force components $\mathbf{F}_x$ and $\mathbf{F}_y$ in the perpendicular directions x and y, as shown in Fig. 3.2a. Adding these forces $\mathbf{F}_x$ and $\mathbf{F}_y$ will result in the original force **F** (Fig. 3.2b). In this case, the magnitudes and angle between these forces are defined as

$$F_x = F \cos \alpha \tag{3.1a}$$

$$F_y = F \sin \alpha \tag{3.1b}$$

$$F = \sqrt{F_x^2 + F_y^2} \tag{3.1c}$$

$$\alpha = \tan^{-1} \frac{F_y}{F_x} \tag{3.1d}$$

Similarly, a force **F** can be resolved into three force components F_x, F_y, and F_z aligned along three mutually perpendicular axes x, y, and z, respectively (Fig. 3.3). The magnitudes of these forces can be computed from

$$F_x = F \cos \alpha_x \tag{3.2a}$$

$$F_y = F \cos \alpha_y \tag{3.2b}$$

$$F_z = F \cos \alpha_z \tag{3.2c}$$

$$F = \sqrt{F_x^2 + F_y^2 + F_z^2} \tag{3.2d}$$

where α_x, α_y, and α_z are the angles between **F** and the axes and $\cos \alpha_x$, $\cos \alpha_y$, and $\cos \alpha_z$ are the **direction cosines** of **F**.

The resultant **R** of several concurrent forces F_1, F_2, and F_3 (Fig. 3.4a) may be determined by first using Eqs. (3.2) to resolve each of the forces into components parallel to the assumed x, y, and z axes (Fig. 3.4b). The magnitude of each of the perpendicular force components can then be summed to define the magnitude of the resultant's force components R_x, R_y, and R_z as follows:

$$R_x = \Sigma F_x = F_{1x} + F_{2x} + F_{3x} \tag{3.3a}$$

$$R_y = \Sigma F_y = F_{1y} + F_{2y} + F_{3y} \tag{3.3b}$$

$$R_z = \Sigma F_z = F_{1z} + F_{2z} + F_{3z} \tag{3.3c}$$

The magnitude of the resultant force **R** can then be determined from

$$R = \sqrt{R_x^2 + R_y^2 + R_z^2} \tag{3.4}$$

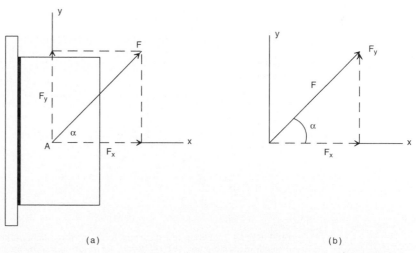

(a)   (b)

FIGURE 3.2   (a) Force **F** resolved into components, F_x along the x axis and F_y along the y axis. (b) Addition of forces F_x and F_y yields the original force **F**.

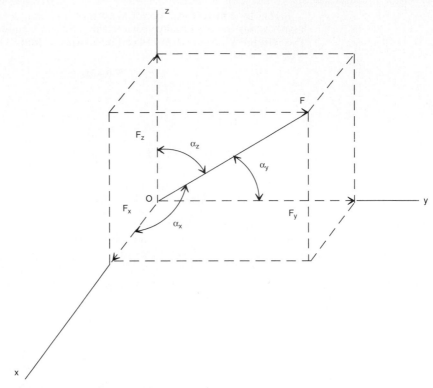

FIGURE 3.3   Resolution of a force in three dimensions.

The direction of **R** is determined by its direction cosines (Fig. 3.4c):

$$\cos \alpha_x = \frac{\Sigma F_x}{R} \qquad \cos \alpha_y = \frac{\Sigma F_y}{R} \qquad \cos \alpha_z = \frac{\Sigma F_z}{R} \qquad (3.5)$$

where α_x, α_y, and α_z are the angles between R and the x, y, and z axes, respectively.

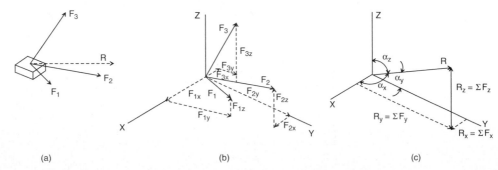

FIGURE 3.4   Addition of concurrent forces in three dimensions. (*a*) Forces $\mathbf{F}_1$, $\mathbf{F}_2$, and $\mathbf{F}_3$ act through the same point. (*b*) The forces are resolved into components along x, y, and z axes. (*c*) Addition of the components yields the components of the resultant force, which, in turn, are added to obtain the resultant.

If the forces acting on the body are nonconcurrent, they can be made concurrent by changing the point of application of the acting forces. This requires incorporating moments so that the external effect of the forces will remain the same (see Art. 3.3).

3.3 MOMENTS OF FORCES

A force acting on a body may have a tendency to rotate it. The measure of this tendency is the **moment** of the force about the axis of rotation. The moment of a force about a specific point equals the product of the magnitude of the force and the normal distance between the point and the line of action of the force. Moment is a vector.

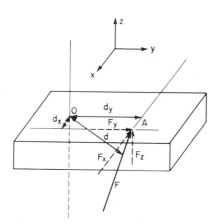

FIGURE 3.5 Moment of force **F** about an axis through point O equals the sum of the moments of the components of the force about the axis.

Suppose a force **F** acts at a point A on a rigid body (Fig. 3.5). For an axis through an arbitrary point O and parallel to the z axis, the magnitude of the moment **M** of **F** about this axis is the product of the magnitude F and the normal distance, or moment arm, d. The distance d between point O and the line of action of **F** can often be difficult to calculate. Computations may be simplified, however, with the use of **Varignon's theorem,** which states that the moment of the resultant of any force system about any axis equals the algebraic sum of the moments of the components of the force system about the same axis. The magnitude of the moment **M** may then be calculated as

$$M = F_x d_y + F_y d_x \tag{3.6}$$

where F_x = component of **F** parallel to the x axis
F_y = component of **F** parallel to the y axis
d_y = distance of F_x from axis through O
d_x = distance of F_y from axis through O

Because the component $\mathbf{F}_z$ is parallel to the axis through O, it has no tendency to rotate the body about this axis and hence does not produce any additional moment.

In general, any force system can be replaced by a single force and a moment. In some cases, the resultant may only be a moment, while for the special case of all forces being concurrent, the resultant will only be a force.

For example, the force system shown in Fig. 3.6a can be resolved into the equivalent force and moment system shown in Fig. 3.6b. The force **F** would have components $\mathbf{F}_x$ and $\mathbf{F}_y$ as follows:

$$F_x = F_{1x} + F_{2x} \tag{3.7a}$$

$$F_y = F_{1y} + F_{2y} \tag{3.7b}$$

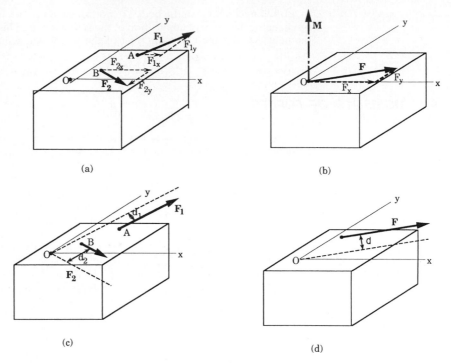

FIGURE 3.6 Resolution of nonconcurrent forces. (*a*) Nonconcurrent forces F_1 and F_2 resolved into force components parallel to *x* and *y* axes. (*b*) The forces are resolved into a moment **M** and a force **F**. (*c*) **M** is determined by adding moments of the force components. (*d*) The forces are resolved into a couple comprising **F** and a moment arm.

The magnitude of the resultant force **F** can then be determined from

$$F = \sqrt{F_x^2 + F_y^2} \tag{3.8}$$

With Varignon's theorem, the magnitude of moment **M** may then be calculated from

$$M = F_{1x}d_{1y} + F_{2x}d_{2y} + F_{1y}d_{1x} + F_{2y}d_{2x} \tag{3.9}$$

with d_1 and d_2 defined as the moment arms in Fig. 3.6*c*. Note that the direction of the moment would be determined by the sign of Eq. (3.9); with a right-hand convention, positive would be a counterclockwise and negative a clockwise rotation.

This force and moment could further be used to compute the line of action of the resultant of the forces $\mathbf{F}_1$ and $\mathbf{F}_2$ (Fig. 3.6*d*). The moment arm *d* could be calculated as

$$d = \frac{M}{F} \tag{3.10}$$

It should be noted that the four force systems shown in Fig. 3.6 are equivalent.

3.4 *EQUATIONS OF EQUILIBRIUM*

When a body is in **static equilibrium,** no translation or rotation occurs in any direction. Since there is no translation, the sum of the forces acting on the body must be zero. Since there is no rotation, the sum of the moments about any point must be zero.

In a two-dimensional space, these conditions can be written:

$$\Sigma F_x = 0 \tag{3.11a}$$

$$\Sigma F_y = 0 \tag{3.11b}$$

$$\Sigma M = 0 \tag{3.11c}$$

where ΣF_x and ΣF_y are the sum of the components of the forces in the direction of the perpendicular axes x and y, respectively, and ΣM is the sum of the moments of all forces about any point in the plane of the forces.

Figure 3.7a shows a truss that is in equilibrium under a 20-kip (20,000-lb) load. By Eq. (3.11), the sum of the reactions, or forces R_L and R_R, needed to support the truss, is 20 kips. (The process of determining these reactions is presented in Art. 3.29.) The sum of the moments of all external forces about any point is zero. For instance, the moment of the forces about the right support reaction R_R is

$$\Sigma M = (40 \times 15) - (30 \times 20) = 600 - 600 = 0$$

(Since only vertical forces are involved, the equilibrium equation for horizontal forces does not apply.)

A **free-body diagram** of a portion of the truss to the left of section AA is shown in (Fig. 3.7b). The internal forces in the truss members cut by the section must balance the external force and reaction on that part of the truss; i.e., all forces acting on the free body must satisfy the three equations of equilibrium [Eq. (3.11)].

For three-dimensional structures, the equations of equilibrium may be written

$$\Sigma F_x = 0 \qquad \Sigma F_y = 0 \qquad \Sigma F_z = 0 \tag{3.12a}$$

$$\Sigma M_x = 0 \qquad \Sigma M_y = 0 \qquad \Sigma M_z = 0 \tag{3.12b}$$

The three force equations [Eqs. (3.12a)] state that for a body in equilibrium there is no resultant force producing a translation in any of the three principal directions. The three moment equations [Eqs. (3.12b)] state that for a body in

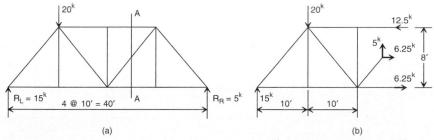

(a)   (b)

FIGURE 3.7 Forces acting on a truss. (*a*) Reactions R_L and R_R maintain equilibrium of the truss under 20-kip load. (*b*) Forces acting on truss members cut by section $A-A$ maintain equilibrium.

equilibrium there is no resultant moment producing rotation about any axes parallel to any of the three coordinate axes.

Furthermore, in statics, a structure is usually considered rigid or nondeformable, since the forces acting on it cause very small deformations. It is assumed that no appreciable changes in dimensions occur because of applied loading. For some structures, however, such changes in dimensions may not be negligible. In these cases, the equations of equilibrium would need to be defined according to the deformed geometry of the structure (Art. 3.46).

(J. L. Meriam, *Mechanics*, Part I: *Statics*, John Wiley & Sons, Inc., New York; F. P. Beer and E. R. Johnston, *Vector Mechanics for Engineers–Statics and Dynamics*, 5th ed., McGraw-Hill, Inc., New York.)

3.5 FRICTIONAL FORCES

Suppose a body A transmits a force $\mathbf{F}_{AB}$ onto a body B through a contact surface assumed to be flat (Fig. 3.8a). For the system to be in equilibrium, body B must react by applying an equal and opposite force $\mathbf{F}_{BA}$ on body A. $\mathbf{F}_{BA}$ may be resolved into a **normal force N** and a force $\mathbf{F}_f$ parallel to the plane of contact (Fig. 3.8b).

The force $\mathbf{F}_f$ is called a **frictional force.** When there is no lubrication, the resistance to sliding is referred to as **dry friction.** The primary cause of dry friction is the microscopic roughness of the surfaces.

For a system including frictional forces to remain static (sliding not to occur), $\mathbf{F}_f$ cannot exceed a limiting value that depends partly on the normal force transmitted across the surface of contact. Because this limiting value also depends on the nature of the contact surfaces, it must be determined experimentally. For example, the limiting value is increased considerably if the contact surfaces are rough.

The limiting value of a frictional force for a body at rest is larger than the frictional force when sliding is in progress. The frictional force between two bodies that are motionless is called **static friction,** and the frictional force between two sliding surfaces is called **sliding** or **kinetic friction.**

Experiments indicate that the limiting force for dry friction $\mathbf{F}_u$ is proportional to the normal force $\mathbf{N}$:

$$\mathbf{F}_u = \mu_s \mathbf{N} \tag{3.13a}$$

where μ_s is the coefficient of static friction. For sliding not to occur, the frictional force $\mathbf{F}_f$ must be less than or equal to $\mathbf{F}_u$. If $\mathbf{F}_f$ exceeds this value, sliding will occur.

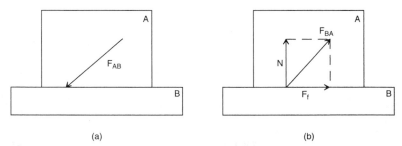

(a)   (b)

FIGURE 3.8   (a) Force $\mathbf{F}_{AB}$ tends to slide body A along the surface of body B. (b) Friction force $\mathbf{F}_f$ opposes motion.

In this case, the resulting frictional force is

$$\mathbf{F}_k = \mu_k \mathbf{N} \tag{3.13b}$$

where μ_k is the coefficient of kinetic friction.

Consider a block of negligible weight resting on a horizontal plane and subjected to a force $\mathbf{P}$ (Fig. 3.9a). From Eq. (3.1), the magnitudes of the components of $\mathbf{P}$ are

$$P_x = P \sin \alpha \tag{3.14a}$$

$$P_y = P \cos \alpha \tag{3.14b}$$

For the block to be in equilibrium, $\Sigma F_x = P_x - F_f = 0$ and $\Sigma F_y = P_y - N = 0$. Hence,

$$P_x = F_f \tag{3.15a}$$

$$P_y = N \tag{3.15b}$$

For sliding not to occur, the following inequality must be satisfied:

$$F_f \leq \mu_s N \tag{3.16}$$

Substitution of Eqs. (3.15) into Eq. (3.16) yields

$$P_x \leq \mu_s P_y \tag{3.17}$$

Substitution of Eqs. (3.14) into Eq. (3.17) gives

$$P \sin \alpha \leq \mu_s P \cos \alpha$$

which simplifies to

$$\tan \alpha \leq \mu_s \tag{3.18}$$

This indicates that the block will just begin to slide if the angle α is gradually decreased to the angle ϕ, where $\tan \phi = \mu_s$.

For the free-body diagram of the two-dimensional system down in Fig. 3.9b, the resultant force $\mathbf{R}$ of forces $\mathbf{F}_u$ and $\mathbf{N}$ is located within a plane sector of angle 2ϕ. In three-dimensional systems, if no motion occurs, $\mathbf{R}$ is located within a cone of angle 2ϕ, called the **cone of friction.**

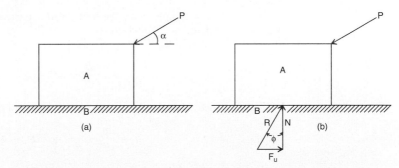

FIGURE 3.9 (a) Force $\mathbf{P}$ acting at an angle α tends to slide block A against friction with plane B. (b) When motion begins, the angle ϕ between the resultant $\mathbf{R}$ and the normal force $\mathbf{N}$ is the angle of static friction.

(F. P. Beer and E. R. Johnston, *Vector Mechanics for Engineers—Statics and Dynamics*, 5th ed., McGraw-Hill, Inc., New York.)

STRUCTURAL MECHANICS—DYNAMICS

Dynamics is that branch of mechanics which deals with bodies in motion. Dynamics is further divided into **kinematics,** the study of motion without regard to the forces causing the motion, and **kinetics,** the study of the relationship between forces and resulting motions.

3.6   KINEMATICS

Kinematics relates displacement, velocity, acceleration, and time. Most engineering problems in kinematics can be solved by assuming that the moving body is rigid and the motions occur in one plane.

Plane motion of a rigid body may be divided into four categories: **rectilinear translation,** in which all points of the rigid body move in straight lines; **curvilinear translation,** in which all points of the body move on congruent curves; **rotation,** in which all particles move in a circular path; and **plane motion,** a combination of translation and rotation in a plane.

Rectilinear translation is often of particular interest to designers. Let an arbitrary point P displace a distance Δs to P' during time interval Δt. The average velocity of the point during this interval is $\Delta s/\Delta t$. The **instantaneous velocity** is obtained by letting Δt approach zero:

$$v = \lim_{\Delta t \to 0} \frac{\Delta s}{\Delta t} = \frac{ds}{dt} \qquad (3.19)$$

Let Δv be the difference between the instantaneous velocities at points P and P' during the time interval Δt. The average acceleration is $\Delta v/\Delta t$. The **instantaneous acceleration** is

$$a = \lim_{\Delta t \to 0} \frac{\Delta v}{\Delta t} = \frac{dv}{dt} = \frac{d^2s}{dt^2} \qquad (3.20)$$

Suppose, for example, that the motion of a particle is described by the time-dependent displacement function $s(t) = t^4 - 2t^2 + 1$. By Eq. (3.19), the velocity of the particle would be

$$v = \frac{ds}{dt} = 4t^3 - 4t$$

By Eq. (3.20), the acceleration of the particle would be

$$a = \frac{dv}{dt} = \frac{d^2s}{dt^2} = 12t^2 - 4$$

With the same relationships, the displacement function $s(t)$ could be determined from a given acceleration function $a(t)$. This can be done by integrating the acceleration function twice with respect to time t. The first integration would yield the velocity function $v(t) = \int a(t)\, dt$, and the second would yield the displacement function $s(t) = \int\int a(t)\, dt\, dt$.

These concepts can be extended to incorporate the relative motion of two points A and B in a plane. In general, the displacement s_A of A equals the vector sum of the displacement s_B of B and the displacement s_{AB} of A relative to B:

$$s_A = s_B + s_{AB} \tag{3.21}$$

Differentiation of Eq. (3.21) with respect to time gives the velocity relation

$$v_A = v_B + v_{AB} \tag{3.22}$$

The acceleration of A is related to that of B by the vector sum

$$a_A = a_B + a_{AB} \tag{3.23}$$

These equations hold for any two points in a plane. They need not be points on a rigid body.

(J. L. Meriam, *Mechanics*, Part II: *Dynamics*, John Wiley & Sons, Inc., New York; F. P. Beer and E. R. Johnston, *Vector Mechanics for Engineers—Statics and Dynamics*, 5th ed., McGraw-Hill, Inc., New York.)

3.7   KINETICS

Kinetics is that part of dynamics that includes the relationship between forces and any resulting motion.

Newton's second law relates force and acceleration by

$$\mathbf{F} = m\mathbf{a} \tag{3.24}$$

where the force $\mathbf{F}$ and the acceleration $\mathbf{a}$ are vectors having the same direction, and the mass m is a scalar.

The acceleration, for example, of a particle of mass m subject to the action of concurrent forces, F_1, F_2, and F_3, can be determined from Eq. (3.24) by resolving each of the forces into three mutually perpendicular directions x, y, and z. The sums of the components in each direction are given by

$$\Sigma F_x = F_{1x} + F_{2x} + F_{3x} \tag{3.25a}$$

$$\Sigma F_y = F_{1y} + F_{2y} + F_{3y} \tag{3.25b}$$

$$\Sigma F_z = F_{1z} + F_{2z} + F_{3z} \tag{3.25c}$$

The magnitude of the resultant of the three concurrent forces is

$$\Sigma F = \sqrt{(\Sigma F_x)^2 + (\Sigma F_y)^2 + (\Sigma F_z)^2} \tag{3.26}$$

The acceleration of the particle is related to the force resultant by

$$\Sigma F = ma \tag{3.27}$$

The acceleration can then be determined from

$$a = \frac{\Sigma F}{m} \tag{3.28}$$

In a similar manner, the magnitudes of the components of the acceleration vector **a** are

$$a_x = \frac{d^2x}{dt^2} = \frac{\Sigma F_x}{m} \qquad (3.29a)$$

$$a_y = \frac{d^2y}{dt^2} = \frac{\Sigma F_y}{m} \qquad (3.29b)$$

$$a_z = \frac{d^2z}{dt^2} = \frac{\Sigma F_z}{m} \qquad (3.29c)$$

Transformation of Eq. (3.27) into the form

$$\Sigma F - ma = 0 \qquad (3.30)$$

provides a condition in dynamics that often can be treated as an instantaneous condition in statics; i.e., if a mass is suddenly accelerated in one direction by a force or a system of forces, an inertia force *ma* will be developed in the opposite direction so that the mass remains in a condition of dynamic equilibrium. This concept is known as **d'Alembert's principle.**

The principle of motion for a single particle can be extended to any number of particles in a system:

$$\Sigma F_x = \Sigma m_i a_{ix} = m\bar{a}_x \qquad (3.31a)$$

$$\Sigma F_y = \Sigma m_i a_{iy} = m\bar{a}_y \qquad (3.31b)$$

$$\Sigma F_z = \Sigma m_i a_{iz} = m\bar{a}_z \qquad (3.31c)$$

where, for example, ΣF_x = algebraic sum of all x-component forces acting on the system of particles

$\Sigma m_i a_{ix}$ = algebraic sum of the products of the mass of each particle and the x component of its acceleration

m = total mass of the system

$\bar{a}_x$ = acceleration of the center of the mass of the particles in the x direction

Extension of these relationships permits calculation of the location of the center of mass (centroid) of an object:

$$\bar{x} = \frac{\Sigma m_i x_i}{m} \qquad (3.32a)$$

$$\bar{y} = \frac{\Sigma m_i x_i}{m} \qquad (3.32b)$$

$$\bar{z} = \frac{\Sigma m_i x_i}{m} \qquad (3.32c)$$

where $\bar{x}, \bar{y}, \bar{z}$ = coordinates of center of mass of the system

m = total mass of the system

$\Sigma m_i x_i$ = algebraic sum of the products of the mass of each particle and its x coordinate

$\Sigma m_i y_i$ = algebraic sum of the products of the mass of each particle and its y coordinate

$\Sigma m_i z_i$ = algebraic sum of the products of the mass of each particle and its z coordinate

Concepts of impulse and momentum are useful in solving problems where forces are expressed as a function of time. These problems include both the kinematics and the kinetics parts of dynamics.

By Eqs. (3.29), the equations of motion of a particle with mass m are

$$\Sigma F_x = ma_x = m \frac{dv_x}{dt} \qquad (3.33a)$$

$$\Sigma F_y = ma_y = m \frac{dv_y}{dt} \qquad (3.33b)$$

$$\Sigma F_z = ma_z = m \frac{dv_z}{dt} \qquad (3.33c)$$

Since m for a single particle is constant, these equations also can be written as

$$\Sigma F_x \, dt = d(mv_x) \qquad (3.34a)$$

$$\Sigma F_y \, dt = d(mv_y) \qquad (3.34b)$$

$$\Sigma F_z \, dt = d(mv_z) \qquad (3.34c)$$

The product of mass and linear velocity is called **linear momentum.** The product of force and time is called **linear impulse.**

Equations (3.34) are an alternate way of stating Newton's second law. The action of ΣF_x, ΣF_y, and ΣF_z during a finite interval of time t can be found by integrating both sides of Eqs. (3.34):

$$\int_{t_0}^{t_1} \Sigma F_x \, dt = m(v_x)_{t_1} - m(v_x)_{t_0} \qquad (3.35a)$$

$$\int_{t_0}^{t_1} \Sigma F_y \, dt = m(v_y)_{t_1} - m(v_y)_{t_0} \qquad (3.35b)$$

$$\int_{t_0}^{t_1} \Sigma F_z \, dt = m(v_z)_{t_1} - m(v_z)_{t_0} \qquad (3.35c)$$

That is, **the sum of the impulses on a body equals its change in momentum.**
(J. L. Meriam, *Mechanics*, Part II: *Dynamics*, John Wiley & Sons, Inc., New York; F. P. Beer and E. R. Johnston, *Vector Mechanics for Engineers—Statics and Dynamics*, 5th ed., McGraw-Hill, Inc., New York.)

MECHANICS OF MATERIALS

Mechanics of materials, or **strength of materials,** incorporates the strength and stiffness properties of a material into the static and dynamic behavior of a structure.

3.8 STRESS-STRAIN DIAGRAMS

Suppose that a homogeneous steel bar with a constant cross-sectional area A is subjected to tension under axial loads P (Fig. 3.10a). A gage length L is selected away from the ends of the bar, to avoid disturbances by the end attachments that apply the load. The load P is increased in increments, and the corresponding elongation δ of the original gage length is measured. Figure 3.10b shows the plot of a typical load-deformation relationship resulting from this type of test.

Assuming that the load is applied concentrically, the **strain** at any point along the gage length will be $\varepsilon = \delta/L$, and the **stress** at any point in the cross section of the bar will be $f = P/A$. Under these conditions, it is convenient to plot the relation between stress and strain. Figure 3.11 shows the resulting plot of a typical stress-strain relationship resulting from this test.

3.9 COMPONENTS OF STRESS AND STRAIN

Suppose that a plane cut is made through a solid in equilibrium under the action of some forces (Fig. 3.12a). The distribution of force on the area A in the plane may be represented by an equivalent resultant force $\mathbf{R_A}$ through point O (also in the plane) and a couple producing moment $\mathbf{M_A}$ (Fig. 3.12b).

Three mutually perpendicular axes x, y, and z at point O are chosen such that axis x is normal to the plane and y and z are in the plane. $\mathbf{R_A}$ can be resolved into components $\mathbf{R_x}$, $\mathbf{R_y}$, and $\mathbf{R_z}$, and $\mathbf{M_A}$ can be resolved into $\mathbf{M_x}$, $\mathbf{M_y}$, and $\mathbf{M_z}$ (Fig. 3.12c). Component $\mathbf{R_x}$ is called **normal force.** $\mathbf{R_y}$ and $\mathbf{R_z}$ are called **shearing forces.** Over area A, these forces produce an average **normal stress** R_x/A and average **shear stresses** R_y/A and R_z/A, respectively. If the area of interest is shrunk to an infinitesimally small area around the point O, then the average stresses would approach limits, called **stress components,** f_x, v_{xy}, and v_{xz}, at point O. Thus, as indicated in Fig. 3.12d,

$$f_x = \lim_{A \to 0} \left(\frac{R_x}{A} \right) \qquad (3.36a)$$

$$v_{xy} = \lim_{A \to 0} \left(\frac{R_y}{A} \right) \qquad (3.36b)$$

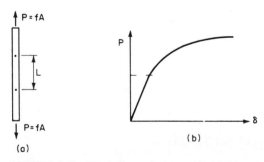

FIGURE 3.10 Elongations of test specimen (a) are measured from gage length L and plotted in (b) against load.

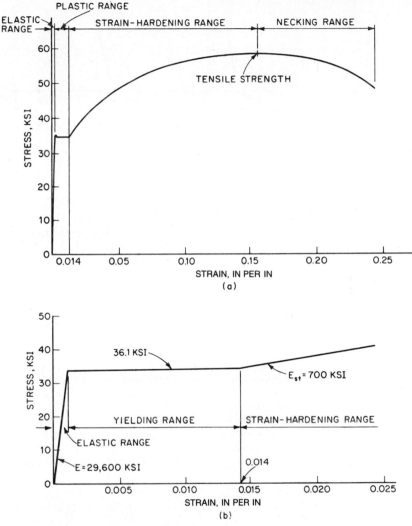

FIGURE 3.11 (*a*) Stress-strain diagram for A36 steel. (*b*) Portion of that diagram in the yielding range.

$$v_{xz} = \lim_{A \to 0} \left(\frac{R_z}{A} \right) \tag{3.36c}$$

Because the moment $\mathbf{M_A}$ and its corresponding components are all taken about point O, they are not producing any additional stress at this point.

If another plane is cut through O that is normal to the y axis, the area surrounding O in this plane will be subjected to a resultant force $\mathbf{R}_y$ through O and a moment $\mathbf{M}_y$. If the area is made to approach zero, the stress components f_y, v_{yx}, and v_{yz} are obtained. Similarly, if a third plane cut is made through O, normal to the z direction, the stress components are f_z, v_{zx}, and v_{zy}.

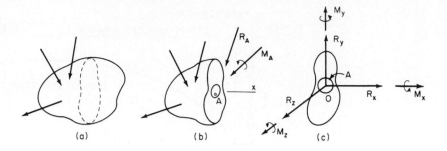

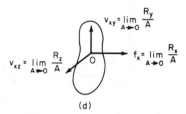

FIGURE 3.12 Stresses at a point in a body due to external loads. (*a*) Forces acting on the body. (*b*) Forces acting on a portion of the body. (*c*) Resolution of forces and moments about coordinate axes through point *O*. (*d*) Stresses at point *O*.

The normal-stress component is denoted by f and a single subscript, which indicates the direction of the axis normal to the plane. The shear-stress component is denoted by v and two subscripts. The first subscript indicates the direction of the normal to the plane, and the second subscript indicates the direction of the axis to which the component is parallel.

The state of stress at a point O is shown in Fig. 3.13 on a rectangular parallelepiped with length of sides Δx, Δy, and Δz. The parallelepiped is taken so small that the stresses can be considered uniform and equal on parallel faces. The stress

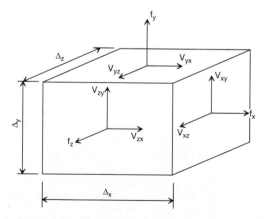

FIGURE 3.13 Components of stress at a point.

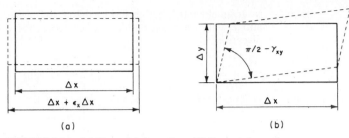

FIGURE 3.14 (*a*) Normal deformation. (*b*) Shear deformation.

at the point can be expressed by the nine components shown. Some of these components, however, are related by equilibrium conditions:

$$v_{xy} = v_{yx} \qquad v_{yz} = v_{zy} \qquad v_{zx} = v_{xz} \tag{3.37}$$

Therefore, the actual state of stress has only six independent components.

A component of strain corresponds to each component of stress. Normal strains ε_x, ε_y, and ε_z are the changes in unit length in the x, y, and z directions, respectively, when the deformations are small (for example, ε_x is shown in Fig. 3.14*a*). Shear strains γ_{xy}, γ_{zy}, and γ_{zx} are the decreases in the right angle between lines in the body at O parallel to the x and y, z and y, and z and x axes, respectively (for example, γ_{xy} is shown in Fig. 3.14*b*). Thus, similar to a state of stress, a state of strain has nine components, of which six are independent.

3.10 *STRESS-STRAIN RELATIONSHIPS*

Structural steels display linearly elastic properties when the load does not exceed a certain limit. Steels also are **isotropic;** i.e., the elastic properties are the same in all directions. The material also may be assumed **homogeneous,** so the smallest element of a steel member possesses the same physical property as the member. It is because of these properties that there is a linear relationship between components of stress and strain. Established experimentally (see Art. 3.8), this relationship is known as **Hooke's law.** For example, in a bar subjected to axial load, the normal strain in the axial direction is proportional to the normal stress in that direction, or

$$\varepsilon = \frac{f}{E} \tag{3.38}$$

where E is the **modulus of elasticity,** or **Young's modulus.**

If a steel bar is stretched, the width of the bar will be reduced to account for the increase in length (Fig. 3.14*a*). Thus the normal strain in the x direction is accompanied by lateral strains of opposite sign. If ε_x is a tensile strain, for example, the lateral strains in the y and z directions are contractions. These strains are related to the normal strain and, in turn, to the normal stress by

$$\varepsilon_y = -\nu\varepsilon_x = -\nu\frac{f_x}{E} \tag{3.39a}$$

$$\varepsilon_z = -\nu\varepsilon_x = -\nu\frac{f_x}{E} \tag{3.39b}$$

where ν is a constant called **Poisson's ratio.**

If an element is subjected to the action of simultaneous normal stresses f_x, f_y, and f_z uniformly distributed over its sides, the corresponding strains in the three directions are

$$\varepsilon_x = \frac{1}{E}[f_x - \nu(f_y + f_z)] \tag{3.40a}$$

$$\varepsilon_y = \frac{1}{E}[f_y - \nu(f_x + f_z)] \tag{3.40b}$$

$$\varepsilon_z = \frac{1}{E}[f_z - \nu(f_x + f_y)] \tag{3.40c}$$

Similarly, shear strain γ is linearly proportional to shear stress v

$$\gamma_{xy} = \frac{v_{xy}}{G} \qquad \gamma_{yz} = \frac{v_{yz}}{G} \qquad \gamma_{zx} = \frac{v_{zx}}{G} \tag{3.41}$$

where the constant G is the **shear modulus of elasticity,** or **modulus of rigidity.** For an isotropic material such as steel, G is directly proportional to E:

$$G = \frac{E}{2(1 + \nu)} \tag{3.42}$$

The analysis of many structures is simplified if the stresses are parallel to one plane. In some cases, such as a thin plate subject to forces along its edges that are parallel to its plane and uniformly distributed over its thickness, the stress distribution occurs all in one plane. In this case of **plane stress,** one normal stress, say f_z, is zero, and corresponding shear stresses are zero: $v_{zx} = 0$ and $v_{zy} = 0$.

In a similar manner, if all deformations or strains occur within a plane, this is a condition of **plane strain.** For example, $\varepsilon_z = 0$, $\gamma_{zx} = 0$, and $\gamma_{zy} = 0$.

3.11   PRINCIPAL STRESSES AND MAXIMUM SHEAR STRESS

When stress components relative to a defined set of axes are given at any point in a condition of plane stress or plane strain (see Art. 3.10), this state of stress may be expressed with respect to a different set of axes that lie in the same plane. For example, the state of stress at point O in Fig. 3.15a may be expressed in terms of either the x and y axes with stress components f_x, f_y, and v_{xy} or the x' and y' axes with stress components $f_{x'}$, $f_{y'}$, and $v_{x'y'}$ (Fig. 3.15b). If stress components f_x, f_y, and v_{xy} are given and the two orthogonal coordinate systems differ by an angle α with respect to the original x axis, the stress components $f_{x'}$, $f_{y'}$, and $v_{x'y'}$ can be determined by statics. The transformation equations for stress are

$$f_{x'} = \tfrac{1}{2}(f_x + f_y) + \tfrac{1}{2}(f_x - f_y)\cos 2\alpha + v_{xy}\sin 2\alpha \tag{3.43a}$$

$$f_{y'} = \tfrac{1}{2}(f_x + f_y) - \tfrac{1}{2}(f_x - f_y)\cos 2\alpha - v_{xy}\sin 2\alpha \tag{3.43b}$$

$$v_{x'y'} = -\tfrac{1}{2}(f_x - f_y)\sin 2\alpha + v_{xy}\cos 2\alpha \tag{3.43c}$$

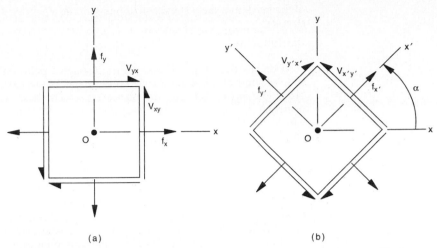

FIGURE 3.15 (*a*) Stresses at point O on planes perpendicular to x and y axes. (*b*) Stresses relative to rotated axes.

From these equations, an angle α_p can be chosen to make the shear stress $v_{x'y'}$ equal zero. From Eq. (3.43*c*), with $v_{x'y'} = 0$,

$$\tan 2\alpha_p = \frac{2v_{xy}}{f_x - f_y} \tag{3.44}$$

This equation indicates that two perpendicular directions, α_p and $\alpha_p + (\pi/2)$, may be found for which the shear stress is zero. These are called **principal directions.** On the plane for which the shear stress is zero, one of the normal stresses is the maximum stress f_1 and the other is the minimum stress f_2 for all possible states of stress at that point. Hence the normal stresses on the planes in these directions are called the **principal stresses.** The magnitude of the principal stresses may be determined from

$$f = \frac{f_x + f_y}{2} \pm \sqrt{\left(\frac{f_x - f_y}{2}\right)^2 + v_{xy}{}^2} \tag{3.45}$$

where the algebraically larger principal stress is given by f_1 and the minimum principal stress is given by f_2.

Suppose that the x and y directions are taken as the principal directions, that is, $v_{xy} = 0$. Then Eqs. (3.43) may be simplified to

$$f_{x'} = \tfrac{1}{2}(f_1 + f_2) + \tfrac{1}{2}(f_1 - f_2) \cos 2\alpha \tag{3.46a}$$

$$f_{y'} = \tfrac{1}{2}(f_1 + f_2) - \tfrac{1}{2}(f_1 - f_2) \cos 2\alpha \tag{3.46b}$$

$$v_{x'y'} = -\tfrac{1}{2}(f_1 - f_2) \sin 2\alpha \tag{3.46c}$$

By Eq. (3.46*c*), the maximum shear stress occurs when $\sin 2\alpha = \pi/2$, i.e., when $\alpha = 45°$. Hence the maximum shear stress occurs on each of two planes that bisect the angles between the planes on which the principal stresses act. The magnitude of the maximum shear stress equals one-half the algebraic difference of the principal stresses:

$$v_{max} = -\tfrac{1}{2}(f_1 - f_2) \qquad (3.47)$$

If on any two perpendicular planes through a point only shear stresses act, the state of stress at this point is called **pure shear.** In this case, the principal directions bisect the angles between the planes on which these shear stresses occur. The principal stresses are equal in magnitude to the unit shear stress in each plane on which only shears act.

3.12 *MOHR'S CIRCLE*

Equations (3.46) for stresses at a point O can be represented conveniently by **Mohr's circle** (Fig. 3.16). Normal stress f is taken as the abscissa, and shear stress v is taken as the ordinate. The center of the circle is located on the f axis at $(f_1 + f_2)/2$, where f_1 and f_2 are the maximum and minimum principal stresses at the point, respectively. The circle has a radius of $(f_1 - f_2)/2$. For each plane passing through the point O there are two diametrically opposite points on Mohr's circle that

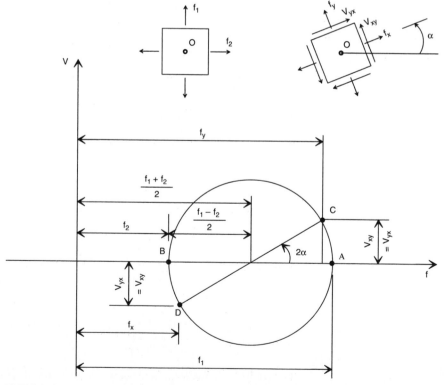

FIGURE 3.16 Mohr circle for obtaining, from principal stresses at a point, shear and normal stresses on any plane through the point.

correspond to the normal and shear stresses on the plane. Thus Mohr's circle can be used conveniently to find the normal and shear stresses on a plane when the magnitude and direction of the principal stresses at a point are known.

Use of Mohr's circle requires the principal stresses f_1 and f_2 to be marked off on the abscissa (points A and B in Fig. 3.16, respectively). Tensile stresses are plotted to the right of the v axis and compressive stresses to the left. (In Fig. 3.16, the principal stresses are indicated as tensile stresses.) A circle is then constructed that has radius $(f_1 + f_2)/2$ and passes through A and B. The normal and shear stresses f_x, f_y, and v_{xy} on a plane at an angle α with the principal directions are the coordinates of points C and D on the intersection of the circle and the diameter making an angle 2α with the abscissa. A counterclockwise angle change α in the stress plane represents a counterclockwise angle change of 2α on Mohr's circle. The stresses f_x, v_{xy}, and f_y, v_{yx} on two perpendicular planes are represented on Mohr's circle by points $(f_x, -v_{xy})$ and (f_y, v_{yx}), respectively.

Mohr's circle also can be used to obtain the principal stresses when the normal stresses on two perpendicular planes and the shearing stresses are known. Figure 3.17 shows construction of Mohr's circle from these conditions. Points C (f_x, v_{xy})

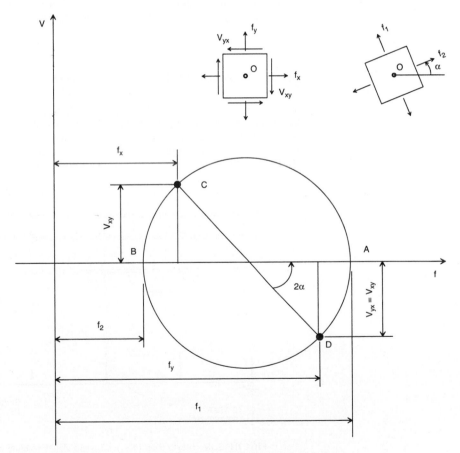

FIGURE 3.17 Mohr circle for determining principal stresses at a point.

and D (f_y, $-v_{xy}$) are plotted and a circle is constructed with CD as a diameter. Based on this geometry, the abscissas of points A and B that correspond to the principal stresses can be determined.

(I. S. Sokolnikoff, *Mathematical Theory of Elasticity*; S. P. Timoshenko and J. N. Goodier, *Theory of Elasticity*; and Chi-Teh Wang, *Applied Elasticity*; and F. P. Beer and E. R. Johnston, *Mechanics of Materials*, McGraw-Hill, Inc., New York; A. C. Ugural and S. K. Fenster, *Advanced Strength and Applied Elasticity*, Elsevier Science Publishing, New York.)

BASIC BEHAVIOR OF STRUCTURAL COMPONENTS

The combination of the concepts for statics (Arts. 3.2 to 3.5) with those of mechanics of materials (Arts. 3.8 to 3.12) provides the essentials for predicting the basic behavior of members in a structural system.

Structural members often behave in a complicated and uncertain way. To analyze the behavior of these members, i.e., to determine the relationships between the external loads and the resulting internal stresses and deformations, certain idealizations are necessary. Through this approach, structural members are converted to such a form that an analysis of their behavior in service becomes readily possible. These idealizations include mathematical models that represent the type of structural members being assumed and the structural support conditions (Fig. 3.18).

3.13   TYPES OF STRUCTURAL MEMBERS AND SUPPORTS

Structural members are usually classified according to the principal stresses induced by loads that the members are intended to support. **Axial-force members** (**ties** or **struts**) are those subjected to only tension or compression. A **column** is a member that may buckle under compressive loads due to its slenderness. **Torsion members,** or **shafts,** are those subjected to twisting moment, or torque. A **beam** supports

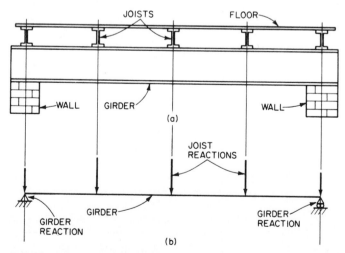

FIGURE 3.18   Idealization of (*a*) joist-and-girder framing by (*b*) concentrated loads on a simple beam.

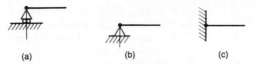

FIGURE 3.19 Representation of types of ideal supports: (*a*) roller, (*b*) hinged support, (*c*) fixed support.

loads that produce bending moments. A **beam-column** is a member in which both bending moment and compression are present.

In practice, it may not be possible to erect truly axially loaded members. Even if it were possible to apply the load at the centroid of a section, slight irregularities of the member may introduce some bending. For analysis purposes, however, these bending moments may often be ignored, and the member may be idealized as axially loaded.

There are three types of ideal supports (Fig. 3.19). In most practical situations, the support conditions of structures may be described by one of these three. Figure 3.19*a* represents a support at which horizontal movement and rotation are unrestricted, but vertical movement is restrained. This type of support is usually shown by **rollers.** Figure 3.19*b* represents a **hinged,** or **pinned, support,** at which vertical and horizontal movements are prevented, while only rotation is permitted. Figure 3.19*c* indicates a **fixed support,** at which no translation or rotation is possible.

3.14 *AXIAL-FORCE MEMBERS*

In an axial-force member, the stresses and strains are uniformly distributed over the cross section. Typical examples of this type of member are shown in Fig. 3.20.

Since the stress is constant across the section, the equation of equilibrium may be written as

$$P = Af \tag{3.48}$$

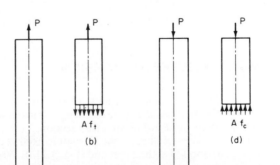

FIGURE 3.20 Stresses in axially loaded members: (*a*) bar in tension, (*b*) tensile stresses in bar, (*c*) strut in compression, (*d*) compressive stresses in strut.

where P = axial load
   f = tensile, compressive, or bearing stress
   A = cross-sectional area of the member

Similarly, if the strain is constant across the section, the strain ε corresponding to an axial tensile or compressive load is given by

$$\varepsilon = \frac{\Delta}{L} \qquad (3.49)$$

where L = length of member
   Δ = change in length of member

Assuming that the material is an isotropic linear elastic medium (see Art. 3.9), Eqs. (3.48) and (3.49) are related according to Hooke's law $\varepsilon = f/E$, where E is the modulus of elasticity of the material. The change in length Δ of a member subjected to an axial load P can than be expressed by

$$\Delta = \frac{PL}{AE} \qquad (3.50)$$

Equation (3.50) relates the load applied at the ends of a member to the displacement of one end of the member relative to the other end. The factor L/AE represents the **flexibility** of the member. It gives the displacement due to a unit load.

Solving Eq. (3.50) for P yields

$$P = \frac{AE}{L} \Delta \qquad (3.51)$$

The factor AE/L represents the **stiffness** of the member in resisting axial loads. It gives the magnitude of an axial load needed to produce a unit displacement.

Equations (3.50) to (3.51) hold for both tension and compression members. However, since compression members may buckle prematurely, these equations may apply only if the member is relatively short (Arts. 3.46 and 3.49).

3.15 *MEMBERS SUBJECTED TO TORSION*

Forces or moments that tend to twist a member are called **torsional loads.** In shafts, the stresses and corresponding strains induced by these loads depend on both the shape and size of the cross section.

Suppose that a circular shaft is fixed at one end and a twisting couple, or **torque,** is applied at the other end (Fig. 3.21a). When the angle of twist is small, the circular cross section remains circular during twist. Also, the distance between any two sections remains the same, indicating that there is no longitudinal stress along the length of the member.

Figure 3.21b shows a cylindrical section with length dx isolated from the shaft. The lower cross section has rotated with respect to its top section through an angle $d\theta$, where θ is the total rotation of the shaft with respect to the fixed end. With no stress normal to the cross section, the section is in a state of pure shear (Art.

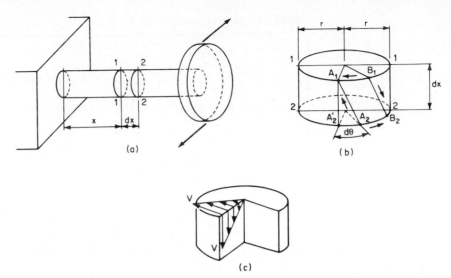

FIGURE 3.21 (*a*) Circular shaft in torsion. (*b*) Deformation of a portion of the shaft. (*c*) Shear in shaft.

3.9). The shear stresses act normal to the radii of the section. The magnitude of the shear strain γ at a given radius r is given by

$$\gamma = \frac{A_2 A_2'}{A_1 A_2'} = r\frac{d\theta}{dx} = \frac{r\theta}{L} \tag{3.52}$$

where L = total length of the shaft
$d\theta/dx = \theta/L$ = angle of twist per unit length of shaft

Incorporation of Hooke's law ($v = G\gamma$) into Eq. (3.52) gives the shear stress at a given radius r:

$$v = \frac{Gr\theta}{L} \tag{3.53}$$

where G is the shear modulus of elasticity. This equation indicates that the shear stress in a **circular** shaft varies directly with distance r from the axis of the shaft (Fig. 3.21*c*). The maximum shear stress occurs at the surface of the shaft.

From conditions of equilibrium, the twisting moment T and the shear stress v are related by

$$v = \frac{rT}{J} \tag{3.54}$$

where $J = \int r^2 \, dA = \pi r^4/2$ = **polar moment of inertia**
dA = differential area of the circular section

By Eqs. (3.53) and (3.54), the applied torque T is related to the relative rotation of one end of the member to the other end by

$$T = \frac{GJ}{L}\theta \tag{3.55}$$

TABLE 3.1 Torsional Constants and Shears

	Polar moment of inertia J	Maximum shear* v_{max}
(circle, $2r$)	$\dfrac{1}{2}\pi r^4$	$\dfrac{2T}{\pi r^3}$ at periphery
(square, a)	$0.141a^4$	$\dfrac{T}{208a^3}$ at midpoint of each side
(rectangle, a, b)	$ab^3\left[\dfrac{1}{3}-0.21\dfrac{b}{a}\left(1-\dfrac{b^4}{12a^4}\right)\right]$	$\dfrac{T(3a+1.8b)}{a^2b^2}$ at midpoint of longer sides
(triangle, a)	$0.0217a^4$	$\dfrac{20T}{a^3}$ at midpoint of each side
(annulus, $2r$, $2R$)	$\dfrac{1}{2}\pi\left(R^4-r^4\right)$	$\dfrac{2TR}{\pi\left(R^4-r^4\right)}$ at outer periphery

*T = twisting moment, or torque

The factor GJ/L represents the stiffness of the member in resisting twisting loads. It gives the magnitude of a torque needed to produce a unit rotation.

Noncircular shafts behave differently under torsion from the way circular shafts do. In noncircular shafts, cross sections do not remain plane, and radial lines through the centroid do not remain straight. Hence the direction of the shear stress is not normal to the radius, and the distribution of shear stress is not linear. If the end sections of the shaft are free to warp, however, Eq. (3.55) may be applied generally when relating an applied torque T to the corresponding member deformation θ. Table 3.1 lists values of J and maximum shear stress for various types of sections.

(*Torsional Analysis of Steel Members*, American Institute of Steel Construction; F. Arbabi, *Structural Analysis and Behavior*, McGraw-Hill, Inc., New York.)

3.16 BENDING STRESSES AND STRAINS IN BEAMS

Beams are structural members subjected to lateral forces that cause bending. There are distinct relationships between the load on a beam, the resulting internal forces and moments, and the corresponding deformations.

Consider the uniformly loaded beam with a symmetrical cross section in Fig. 3.22. Subjected to bending, the beam carries this load to the two supporting ends, one of which is hinged and the other of which is on rollers. Experiments have shown that strains developed along the depth of a cross section of the beam vary

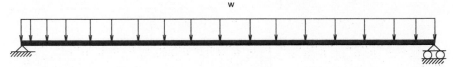

FIGURE 3.22 Uniformly loaded, simply supported beam.

linearly; i.e., a plane section before loading remains plane after loading. Based on this observation, the stresses at various points in a beam may be calculated if the stress-strain diagram for the beam material is known. From these stresses, the resulting internal forces at a cross section may be obtained.

Figure 3.23a shows the symmetrical cross section of the beam shown in Fig. 3.22. The strain varies linearly along the beam depth (Fig. 3.23b). The strain at the top of the section is compressive and decreases with depth, becoming zero at a certain distance below the top. The plane where the strain is zero is called the **neutral axis.** Below the neutral axis, tensile strains act, increasing in magnitude downward. With use of the stress-strain relationship of the material (e.g., see Fig. 3.11), the cross-sectional stresses may be computed from the strains (Fig. 3.23c).

If the entire beam is in equilibrium, then all its sections also must be in equilibrium. With no external horizontal forces applied to the beam, the net internal horizontal forces at any section must sum to zero:

$$\int_{c_b}^{c_t} f(y) \, dA = \int_{c_b}^{c_t} f(y)b(y) \, dy = 0 \tag{3.56}$$

where dA = differential unit of cross-sectional area located at a distance y from the neutral axis

$b(y)$ = width of beam at distance y from the neutral axis

$f(y)$ = normal stress at a distance y from the neutral axis

c_b = distance from neutral axis to beam bottom

c_t = distance from neutral axis to beam top

The moment M at this section due to internal forces may be computed from the stresses $f(y)$:

$$M = \int_{c_b}^{c_t} f(y)b(y)y \, dy \tag{3.57}$$

The moment M is usually considered positive when bending causes the bottom of the beam to be in tension and the top in compression. To satisfy equilibrium requirements, M must be equal in magnitude but opposite in direction to the moment at the section due to the loading.

3.16.1 Bending in the Elastic Range

If the stress-strain diagram is linear, then the stresses would be linearly distributed along the depth of the beam corresponding to the linear distribution of strains:

$$f(y) = \frac{f_t}{c_t} y \tag{3.58}$$

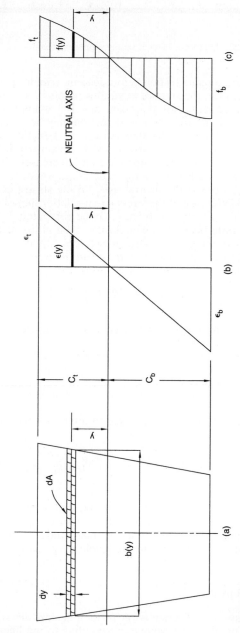

FIGURE 3.23 (*a*) Symmetrical section of a beam develops (*b*) linear strain distribution and (*c*) nonlinear stress distribution.

where f_t = stress at top of beam
$\quad\quad y$ = distance from the neutral axis

Substitution of Eq. (3.58) into Eq. (3.56) yields

$$\int_{c_b}^{c_t} \frac{f_t}{c_t} yb(y)\, dy = \frac{f_t}{c_t} \int_{c_b}^{c_t} yb(y)\, dy = 0 \tag{3.59}$$

Equation (3.59) provides a relationship that can be used to locate the neutral axis of the section. For the section shown in Fig. 3.23, Eq. (3.59) indicates that the neutral axis coincides with the centroidal axis.

Substitution of Eq. (3.58) into Eq. (3.57) gives

$$M = \int_{c_b}^{c_t} \frac{f_t}{c_t} b(y)y^2\, dy = \frac{f_t}{c_t} \int_{c_b}^{c_t} b(y)y^2\, dy = f_t \frac{I}{c_t} \tag{3.60}$$

where $\int_{c_b}^{c_t} b(y)y^2\, dy = I$ = **moment of inertia** of cross section about the neutral axis. The factor I/c_t is the **section modulus** S_t for the top surface.

Substitution of f_t/c_t from Eq. (3.58) into Eq. (3.60) gives the relation between moment and stress at any distance y from the neutral axis:

$$M = \frac{I}{y} f(y) \tag{3.61a}$$

$$f(y) = M \frac{y}{I} \tag{3.61b}$$

Hence, for the bottom of the beam,

$$M = f_b \frac{I}{c_b} \tag{3.62}$$

where I/c_b is the section modulus S_b for the bottom surface.

For a section symmetrical about the neutral axis,

$$c_t = c_b \quad\quad f_t = f_b \quad\quad S_t = S_b \tag{3.63}$$

For example, a rectangular section with width b and depth d would have a moment of inertia $I = bd^3/12$ and a section modulus for both compression and tension $S = I/c = bd^2/6$. Hence,

$$M = Sf = \frac{bd^2}{6} f \tag{3.64a}$$

$$f = \frac{M}{S} = M \frac{6}{bd^2} \tag{3.64b}$$

The geometric properties of several common types of cross sections are given in Table 3.2.

3.16.2   Bending in the Plastic Range

If a beam is heavily loaded, all the material at a cross section may reach the yield stress f_y [that is, $f(y) = \pm f_y$]. Although the strains would still vary linearly with

TABLE 3.2   Properties of Sections

	$A = \dfrac{\text{Area}}{bh}$	$c = \text{depth to centroid} \div h$	$I = \text{moment of inertia}$ about centroidal axis $\div bh^3$
	1.0	$\dfrac{1}{2}$	$\dfrac{1}{12}$
	1.0	$\dfrac{b}{2h}\sin\alpha + \dfrac{1}{2}\cos\alpha$	$\dfrac{1}{12}\left(\dfrac{b}{h}\sin\alpha\right)^2 + \dfrac{1}{12}\cos^2\alpha$
	$1 - \left(1 - \dfrac{b'}{b}\right)\left(1 - \dfrac{2t}{h}\right)$	$\dfrac{1}{2}$	$\dfrac{1}{12}\left[1 - \left(1 - \dfrac{b'}{b}\right)\left(1 - \dfrac{2t}{h}\right)^3\right]$
	$\dfrac{\pi}{4} = 0.785398$	$\dfrac{1}{2}$	$\dfrac{\pi}{64} = 0.049087$
	$\dfrac{\pi}{4}\left(1 - \dfrac{h_1^2}{h^2}\right)$	$\dfrac{1}{2}$	$\dfrac{\pi}{64}\left(1 - \dfrac{h_1^4}{h^4}\right)$
	$\dfrac{2}{3}$	$\dfrac{3}{5}$	$\dfrac{8}{175}$
	$\dfrac{2}{3}$	$\dfrac{3}{5}$	$\dfrac{8}{175}$
	$1 - \dfrac{h_1}{h}$	$\dfrac{1}{2}$	$\dfrac{1}{12}\left(1 - \dfrac{h_1^3}{h^3}\right)$
	$1 - \dfrac{b_1}{b}\left(\dfrac{h_1}{h}\right)$	$\dfrac{1}{2}$	$\dfrac{1}{12}\left[1 - \dfrac{b_1}{b}\left(\dfrac{h_1}{h}\right)^3\right]$

TABLE 3.2   Properties of Sections *(Continued)*

	$A = \dfrac{\text{Area}}{bh}$	$c = $ depth to centroid $\div h$	$I = $ moment of inertia about centroidal axis $\div bh^3$
	$1 - \left(1 - \dfrac{b'}{b}\right)\left(\dfrac{h_1}{h}\right)$	$\dfrac{1}{2}$	$\dfrac{1}{12}\left[1 - \left(1 - \dfrac{b'}{b}\right)\left(\dfrac{h_1}{h}\right)^3\right]$
	$1 - \dfrac{b_1}{b}\left(1 - \dfrac{t}{h}\right)$	$\dfrac{1}{2}\dfrac{1 - \dfrac{b_1}{b}\left(1 - \dfrac{t^2}{h^2}\right)}{1 - \dfrac{b_1}{b}\left(1 - \dfrac{t}{h}\right)}$	$\dfrac{1}{3}\left\{1 - a\left(1 - \dfrac{t^3}{h^3}\right) - \dfrac{3}{4}\dfrac{\left[1 - a\left(1 - \dfrac{t^2}{h^2}\right)\right]^2}{\left[1 - a\left(1 - \dfrac{t}{h}\right)\right]}\right\}$ $a = \dfrac{b_1}{b}$
	$\dfrac{t}{h} + \dfrac{b'}{b}\left(1 - \dfrac{t}{h}\right)$	$\dfrac{1}{2}\dfrac{\left(\dfrac{t}{h}\right)^2 + \dfrac{b'}{b}\left(1 - \dfrac{t^2}{h^2}\right)}{\dfrac{t}{h} + \dfrac{b'}{b}\left(1 - \dfrac{t}{h}\right)}$	$\dfrac{1}{3}\left\{1 - a\left(1 - \dfrac{t^3}{h^3}\right) - \dfrac{3}{4}\dfrac{\left[1 - a\left(1 - \dfrac{t^2}{h^2}\right)\right]^2}{\left[1 - a\left(1 - \dfrac{t}{h}\right)\right]}\right\}$ $a = \dfrac{b - b'}{b}$
	$\dfrac{1}{2}$	$\dfrac{2}{3}$	$\dfrac{1}{36}$
	$\dfrac{(1 + k)}{2}$	$\dfrac{(2 + k)}{3(1 + k)}$	$\dfrac{1}{36}\dfrac{(1 + 4k + k^2)}{(1 + k)}$

depth (Fig. 3.24b), the stress distribution would take the form shown in Fig. 3.24c. In this case, Eq. (3.57) becomes the **plastic moment:**

$$M_p = f_y \int_0^{c_t} b(y)y \, dy + f_y \int_0^{c_b} b(y)y \, dy = Z f_y \qquad (3.65)$$

where $\int_0^{c_t} b(y)y \, dy + f_y \int_0^{c_b} b(y)y \, dy = Z = $ **plastic section modulus.** For a rectangular section (Fig. 3.24a),

$$M_p = b f_y \int_0^{h/2} y \, dy + b f_y \int_0^{-h/2} y \, dy = \frac{bh^2}{4} f_y \qquad (3.66)$$

Hence the plastic modulus Z equals $bh^2/4$ for a rectangular section.

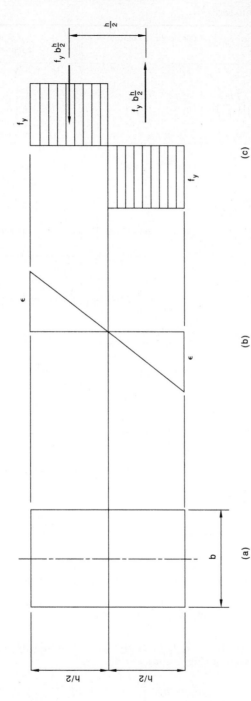

FIGURE 3.24 For a rectangular beam (*a*) in the plastic range, strain distribution (*b*) is linear, while stress distribution (*c*) is rectangular.

3.17   *SHEAR STRESSES IN BEAMS*

In addition to normal stresses (Art. 3.16), beams are subjected to shearing. Shear stresses vary over the cross section of a beam. At every point in the section, there are both a vertical and a horizontal unit shear stress, equal in magnitude [Eq. (3.37)].

To determine these stresses, consider the portion of a beam with length dx between vertical sections 1–1 and 2–2 (Fig. 3.25). At a horizontal section a distance y from the neutral axis, the horizontal shear force $\Delta H(y)$ equals the difference between the normal forces acting above the section on the two faces:

$$\Delta H(y) = \int_y^{c_t} f_2(y)b(y) \, dy - \int_y^{c_t} f_1(y)b(y) \, dy \qquad (3.67)$$

where $f_2(y)$ and $f_1(y)$ are the bending-stress distributions at sections 2–2 and 1–1, respectively.

If the bending stresses vary linearly with depth, then, according to Eq. (3.61),

$$f_2(y) = \frac{M_2 y}{I} \qquad (3.68a)$$

$$f_1(y) = \frac{M_1 y}{I} \qquad (3.68b)$$

where M_2 and M_1 are the internal bending moments at sections 2–2 and 1–1,

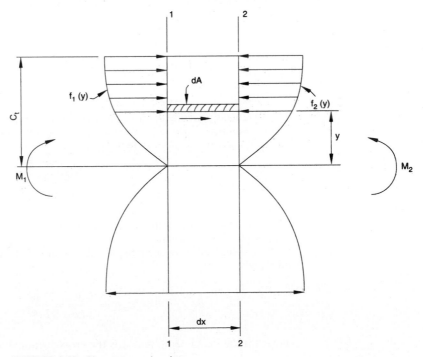

FIGURE 3.25   Shear stresses in a beam.

respectively, and I is the moment of inertia about the neutral axis of the beam cross section. Substitution in Eq. (3.67) gives

$$\Delta H(y) = \frac{M_2 - M_1}{I} \int_y^{c_t} yb(y)\,dy = \frac{Q(y)}{I}\,dM \qquad (3.69)$$

where $Q(y) = \int_y^{c_t} yb(y)\,dy =$ **static moment** about neutral axis of the area above the plane at a distance y from the neutral axis

$b(y) =$ width of beam

$dM = M_2 - M_1$

Division of $\Delta H(y)$ by the area $b(y)\,dx$ yields the shear stress at y:

$$v(y) = \frac{\Delta H(y)}{b(y)\,dx} = \frac{Q(y)}{Ib(y)}\frac{dM}{dx} \qquad (3.70)$$

Integration of $v(y)$ over the cross section provides the total internal vertical shear force V on the section:

$$V = \int_{c_b}^{c_t} v(y)b(y)\,dy \qquad (3.71)$$

To satisfy equilibrium requirements, V must be equal in magnitude but opposite in direction to the shear at the section due to the loading.

Substitution of Eq. 3.70 into Eq. 3.71 gives

$$V = \int_{c_b}^{c_t} \frac{Q(y)}{Ib(y)}\frac{dM}{dx}b(y)\,dy = \frac{dM}{dx}\frac{1}{I}\int_{c_b}^{c_t} Q(y)\,dy = \frac{dM}{dx} \qquad (3.72)$$

inasmuch as $I = \int_{c_b}^{c_t} Q(y)\,dy$. Equation (3.72) indicates that shear is the rate of change of bending moment along the span of the beam.

Substitution of Eq. (3.72) into Eq. (3.70) yields an expression for calculating the shear stress at any section depth:

$$v(y) = \frac{VQ(y)}{Ib(y)} \qquad (3.73)$$

According to Eq. (3.73), the maximum shear stress occurs at a depth y when the ratio $Q(y)/b(y)$ is maximum.

For rectangular cross sections, the maximum shear stress occurs at middepth and equals

$$v_{\max} = \frac{3}{2}\frac{V}{bh} = \frac{3}{2}\frac{V}{A} \qquad (3.74)$$

where h is the beam depth and A is the cross-sectional area.

3.18 SHEAR, MOMENT, AND DEFORMATION RELATIONSHIPS IN BEAMS

The relationship between shear and moment identified in Eq. (3.72), that is, $V = dM/dx$, indicates that the shear force at a section is the rate of change of the bending moment. A similar relationship exists between the load on a beam and the shear at a section. Figure 3.26b shows the resulting internal forces and moments for the portion of beam dx shown in Fig. 3.26a. Note that when the internal shear acts upward on the left of the section, the shear is positive, and when the shear acts upward on the right of the section, it is negative. For equilibrium of the vertical forces,

$$\Sigma F_y = V - (V + dV) + w(x)\, dx = 0 \qquad (3.75)$$

Solving for $w(x)$ gives

$$w(x) = \frac{dV}{dx} \qquad (3.76)$$

This equation indicates that the rate of change in shear at any section equals the load per unit length at that section. When concentrated loads act on a beam, Eqs. (3.72) and (3.76) apply to the region of the beam between the concentrated loads.

Beam Deflections.   To this point, only relationships between the load on a beam and the resulting internal forces and stresses have been established. To calculate the deflection at various points along a beam, it is necessary to know the relationship between load and the deformed curvature of the beam or between bending moment and this curvature.

When a beam is subjected to loads, it deflects. The deflected shape of the beam taken at the neutral axis may be represented by an elastic curve $\delta(x)$. If the slope of the deflected shape is such that $d\delta/dx \ll 1$, the radius of curvature R at a point x along the span is related to the derivatives of the ordinates of the elastic curve $\delta(x)$ by

$$\frac{1}{R} = \frac{d^2\delta}{dx^2} = \frac{d}{dx}\left(\frac{d\delta}{dx}\right) = \phi \qquad (3.77)$$

$1/R$ is referred to as the **curvature** ϕ of a beam. It represents the rate of change of the slope $\theta = d\delta/dx$ of the neutral axis.

Consider the deformation of the dx portion of a beam shown in Fig. 3.26b. Before the loads act, sections 1–1 and 2–2 are vertical (Fig. 3.27a). After the loads act, plane sections remain plane, and the portion becomes trapezoidal. The top of the beam shortens an amount $\varepsilon_t\, dx$ and the beam bottom an amount $\varepsilon_b\, dx$, where ε_t is the compressive unit strain at the beam top and ε_b is the tensile unit strain at the beam bottom. Each side rotates through a small angle. Let the angle of rotation of section 1–1 be $d\theta_1$ and that of section 2–2, $d\theta_2$ (Fig. 3.27b). Hence the angle between the two faces will be $d\theta_1 + d\theta_2 = d\theta$. Since $d\theta_1$ and $d\theta_2$ are small, the total shortening of the beam top between sections 1–1 and 2–2 is also given by $c_t\, d\theta = \varepsilon_t\, dx$, from which $d\theta/dx = \varepsilon_t/c_t$, where c_t is the distance from the neutral axis to the beam top. Similarly, the total lengthening of the beam bottom is given by $c_b\, d\theta = \varepsilon_b\, dx$, from which $d\theta/dx = \varepsilon_b/c_b$, where c_b is the distance from the

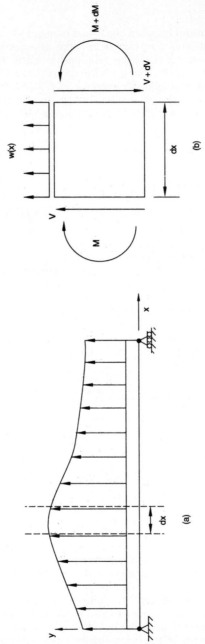

FIGURE 3.26 (a) Beam with distributed loading. (b) Internal forces and moments on a section of the beam.

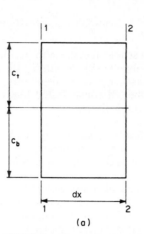
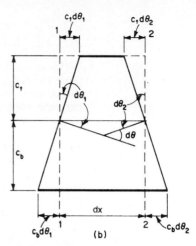

FIGURE 3.27   (a) Portion of an unloaded beam. (b) Deformed portion after beam is loaded.

neutral axis to the beam bottom. By definition, the beam curvature is therefore given by

$$\phi = \frac{d}{dx}\left(\frac{d\delta}{dx}\right) = \frac{d\theta}{dx} = \frac{\varepsilon_t}{c_t} = \frac{\varepsilon_b}{c_b} \qquad (3.78)$$

When the stress-strain diagram for the material is linear, $\varepsilon_t = f_t/E$ and $\varepsilon_b = f_b/E$, where f_b and f_t are the unit stresses at top and bottom surfaces and E is the modulus of elasticity. By Eq. (3.60), $f_t = M(x)c_t/I(x)$ and $f_b = M(x)c_b/I(x)$, where x is the distance along the beam span where the section dx is located and $M(x)$ is the moment at the section. Substitution for ε_t and f_t or ε_b and f_b in Eq. (3.78) gives

$$\phi = \frac{d^2\delta}{dx^2} = \frac{d}{dx}\left(\frac{d\delta}{dx}\right) = \frac{d\theta}{dx} = \frac{M(x)}{EI(x)} \qquad (3.79)$$

Equation (3.79) is of fundamental importance, for it relates the internal bending moment along the beam to the curvature or second derivative of the elastic curve $\delta(x)$, which represents the deflected shape. Equations (3.72) and (3.76) further relate the bending moment $M(x)$ and shear $V(x)$ to an applied distributed load $w(x)$. From these three equations, the following relationships between load on the beam, the resulting internal forces and moments, and the corresponding deformations can be shown:

$$\delta(x) = \text{elastic curve representing the deflected shape} \qquad (3.80a)$$

$$\frac{d\delta}{dx} = \theta(x) = \text{slope of the deflected shape} \qquad (3.80b)$$

$$\frac{d^2\delta}{dx^2} = \phi = \frac{M(x)}{EI(x)} = \text{curvature of the deflected shape and also the moment-curvature relationship} \qquad (3.80c)$$

$$\frac{d^3\delta}{dx^3} = \frac{d}{dx}\left[\frac{M(x)}{EI(x)}\right] = \frac{V(x)}{EI(x)} = \text{shear-deflection relationship} \qquad (3.80d)$$

$$\frac{d^4\delta}{dx^4} = \frac{d}{dx}\left[\frac{V(x)}{EI(x)}\right] = \frac{w(x)}{EI(x)} = \text{load-deflection relationship} \qquad (3.80e)$$

These relationships suggest that the shear force, bending moment, and beam slope and deflection may be obtained by integrating the load distribution. For some simple cases this approach can be used conveniently. However, it may be cumbersome when a large number of concentrated loads act on a structure. Other methods are suggested in Arts. 3.32 to 3.39.

Shear, Moment, and Deflection Diagrams. Figures 3.28 to 3.49 show some special cases in which shear, moment, and deformation distributions can be expressed in analytic form. The figures also include diagrams indicating the variation of shear, moment, and deformations along the span. A diagram in which shear is plotted along the span is called a **shear diagram.** Similarly, a diagram in which bending moment is plotted along the span is called a **bending-moment diagram.**

Consider the simply supported beam subjected to a downward-acting, uniformly distributed load w (units of load per unit length) in Fig. 3.31a. The support reactions R_1 and R_2 may be determined from equilibrium equations. Summing moments about the left end yields

$$\Sigma M = R_2 L - wL\frac{L}{2} = 0 \qquad R_2 = \frac{wL}{2}$$

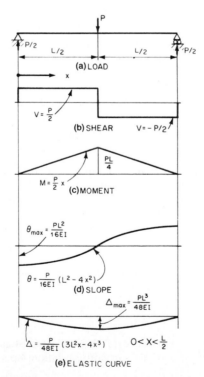

FIGURE 3.28 Shears, moments, and deformations for midspan load on a simple beam.

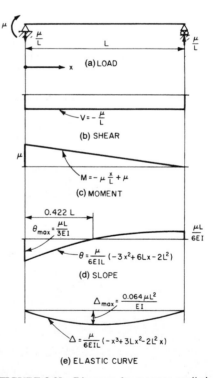

FIGURE 3.29 Diagrams for moment applied at one end of a simple beam.

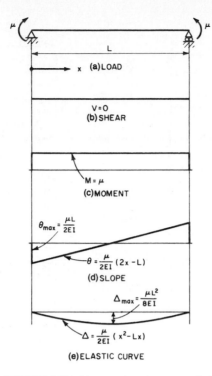

FIGURE 3.30 Diagrams for moments applied at both ends of a simple beam.

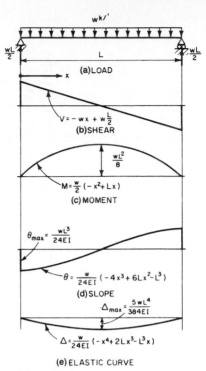

FIGURE 3.31 Shears, moments, and deformations for uniformly loaded simple beam.

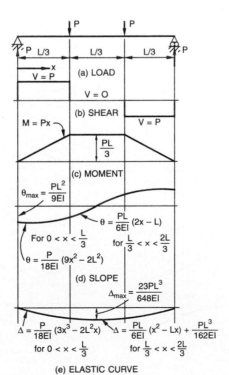

FIGURE 3.32 Simple beam with concentrated load at the third points.

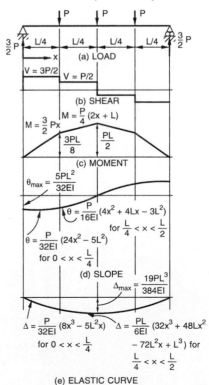

FIGURE 3.33 Diagrams for simple beam loaded at quarter points.

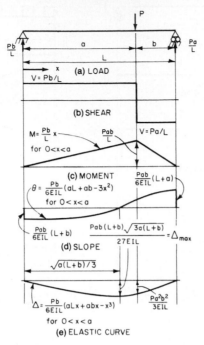

FIGURE 3.34 Diagrams for concentrated load on a simple beam.

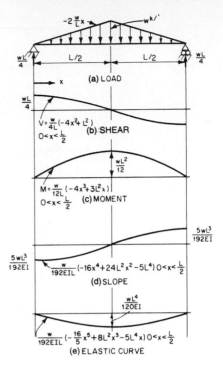

FIGURE 3.35 Symmetrical triangular load on a simple beam.

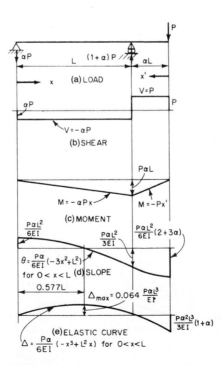

FIGURE 3.36 Concentrated load on a beam overhang.

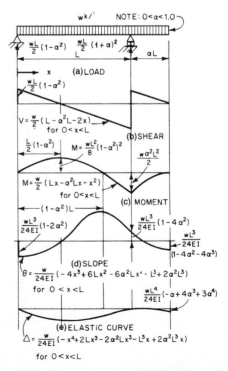

FIGURE 3.37 Uniformly loaded beam with overhang.

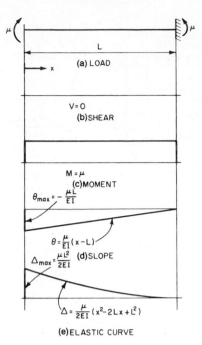

FIGURE 3.38 Shears, moments, and deformations for moment at one end of a cantilever.

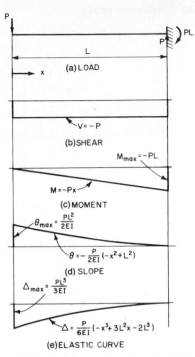

FIGURE 3.39 Diagrams for concentrated load on a cantilever.

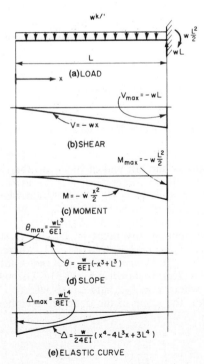

FIGURE 3.40 Shears, moments, and deformations for uniformly loaded cantilever.

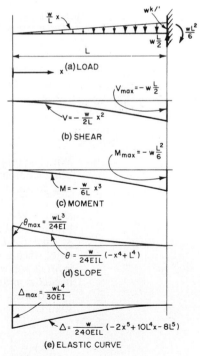

FIGURE 3.41 Triangular load on cantilever with maximum at support.

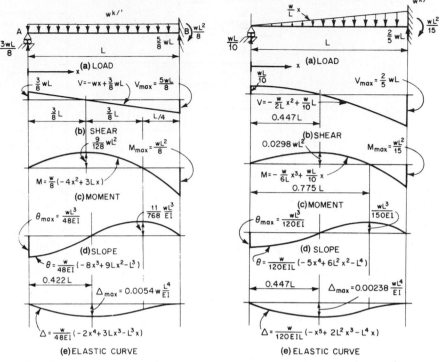

FIGURE 3.42 Uniform load on beam with one end fixed, one end on rollers.

FIGURE 3.43 Triangular load on beam with one end fixed, one end on rollers.

R_1 may then be found from equilibrium of vertical forces:

$$\Sigma F_y = R_1 + R_2 - wL = 0 \qquad R_1 = \frac{wL}{2}$$

With the origin taken at the left end of the span, the shear at any point can be obtained from Eq. (3.80e) by integration: $V = \int -w \, dx = -wx + C_1$, where C_1 is a constant. When $x = 0$, $V = R_1 = wL/2$, and when $x = L$, $V = -R_2 = -wL/2$. For these conditions to be satisfied, $C_1 = wL/2$. Hence the equation for shear is $V(x) = -wx + wL/2$ (Fig. 3.31b).

The bending moment at any point is, by Eq. (3.80d), $M(x) = \int V \, dx = \int (-wx + wL/2) \, dx = -wx^2/2 + wLx/2 + C_2$, where C_2 is a constant. In this case, when $x = 0$, $M = 0$. Hence $C_2 = 0$, and the equation for bending moment is $M(x) = \frac{1}{2}w(-x^2 + Lx)$, as shown in Fig. 3.31c. The maximum bending moment occurs at midspan, where $x = L/2$, and equals $wL^2/8$.

From Eq. (3.80c), the slope of the deflected member at any point along the span is

$$\theta(x) = \int \frac{M(x)}{EI} \, dx = \int \frac{w}{2EI} (-x^2 + Lx) \, dx = \frac{w}{2EI} \left(-\frac{x^3}{3} + \frac{wLx^2}{2} \right) + C_3$$

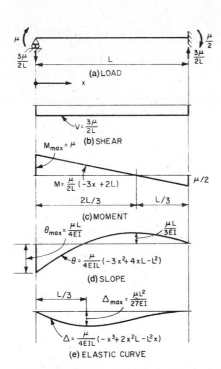

FIGURE 3.44 Moment applied at one end of a beam with a fixed end.

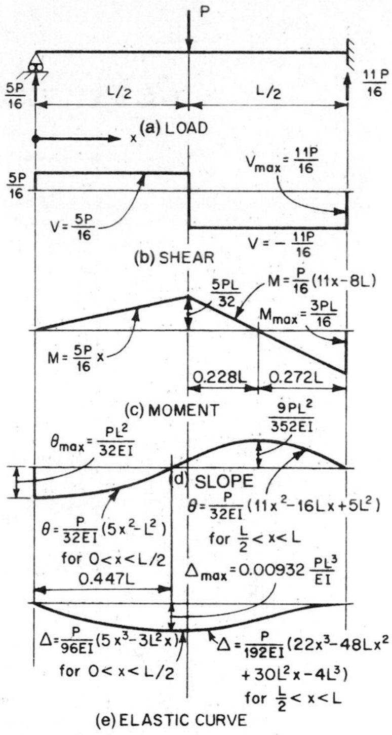

FIGURE 3.45 Load at midspan of beam with one fixed end, one end on rollers.

where C_3 is a constant. When $x = L/2$, $\theta = 0$. Hence $C_3 = -wL^3/24EI$, and the equation for slope is

$$\theta(x) = \frac{w}{24EI}(-4x^3 + 6Lx^2 - L^3)$$

(See Fig. 3.31d.)

The deflected-shape curve at any point is, by Eq. (3.80b),

$$\delta(x) = \frac{w}{24EI}\int(-4x^3 + 6Lx^2 - L^3)\,dx$$

$$= -wx^4/24EI + wLx^3/12EI - wL^3x/24EI + C_4$$

where C_4 is a constant. In this case, when $x = 0$, $\delta = 0$. Hence $C_4 = 0$, and the equation for deflected shape is

$$\delta(x) = \frac{w}{24EI}(-x^4 + 2Lx^3 - L^3x)$$

as shown in Fig. 3.31e. The maximum deflection occurs at midspan, where $x = L/2$, and equals $-5wL^4/384EI$.

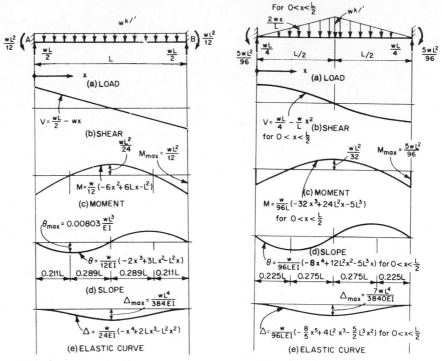

FIGURE 3.46 Shears, moments, and deformations for uniformly loaded fixed-end beam.

FIGURE 3.47 Diagrams for triangular load on a fixed-end beam.

For concentrated loads, the equations for shear and bending moment are derived in the region between the concentrated loads, where continuity of these diagrams exists. Consider the simply supported beam subjected to a concentrated load at midspan (Fig. 3.28a). From equilibrium equations, the reactions R_1 and R_2 equal $P/2$. With the origin taken at the left end of the span, $w(x) = 0$ when $x \neq L/2$. Integration of Eq. (3.80e) gives $V(x) = C_3$, a constant, for $x = 0$ to $L/2$, and $V(x) = C_4$, another constant, for $x = L/2$ to L. Since $V = R_1 = P/2$ at $x = 0$, $C_3 = P/2$. Since $V = -R_2 = -P/2$ at $x = L$, $C_4 = -P/2$. Thus, for $0 \leq x < L/2$, $V = P/2$, and for $L/2 < x \leq L$, $V = -P/2$ (Fig. 3.28b). Similarly, equations for bending moment, slope, and deflection can be expressed from $x = 0$ to $L/2$ and again for $x = L/2$ to L, as shown in Figs. 3.28c, 3.28d, and 3.28e, respectively.

In practice, it is usually not convenient to derive equations for shear and bending-moment diagrams for a particular loading. It is generally more convenient to use equations of equilibrium to plot the shears, moments, and deflections at critical points along the span. For example, the internal forces at the quarter span of the uniformly loaded beam in Fig. 3.31 may be determined from the free-body diagram in Fig. 3.50. From equilibrium conditions for moments about the right end,

$$\sum M = M + \left(\frac{wL}{4}\right)\left(\frac{L}{8}\right) - \left(\frac{wL}{2}\right)\left(\frac{L}{4}\right) = 0 \qquad (3.81a)$$

$$M = \frac{3wL^2}{32} \qquad (3.81b)$$

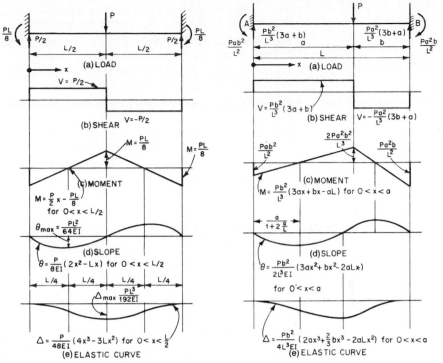

FIGURE 3.48 Shears, moments, and deformations for load at midspan of a fixed-end beam.

FIGURE 3.49 Diagrams for concentrated load on a fixed-end beam.

Also, the sum of the vertical forces must equal zero:

$$\sum F_y = \frac{wL}{2} - \frac{wL}{4} - V = 0 \tag{3.82a}$$

$$V = \frac{wL}{4} \tag{3.82b}$$

Several important concepts are demonstrated in the preceding examples:

• The shear at a section is the algebraic sum of all forces on either side of the section.

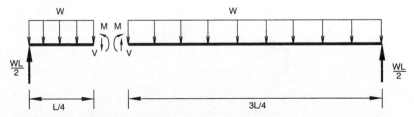

FIGURE 3.50 Bending moment and shear at quarter point of a uniformly loaded simple beam.

- The bending moment at a section is the algebraic sum of the moments about the section of all forces on either side of the section.

- A maximum bending moment occurs at an **inflection point,** i.e., where the shear or slope of the bending-moment diagram is zero.

- Where there is no distributed load along a span, the shear diagram is a horizontal line. (Shear is a constant, which may be zero.)

- The shear diagram changes sharply at the point of application of a concentrated load.

- The difference between the bending moments at two sections of a beam equals the area under the shear diagram between the two sections.

- The difference between the shears at two sections of a beam equals the area under the load diagram between those sections.

3.19 SHEAR DEFLECTIONS IN BEAMS

Shear deformations in a beam add to the deflections due to bending discussed in Art. 3.18. Deflections due to shear are generally small, but in some cases they should be taken into account.

When a cantilever is subjected to load P (Fig. 3.51a), a portion dx of the span undergoes a shear deformation (Fig. 3.51b). For an elastic material, the angle γ equals the ratio of the shear stress v to the shear modulus of elasticity G. Assuming that the shear on the element is distributed uniformly, which is an approximation, the deflection of the beam $d\delta_s$ caused by the deformation of the element is

$$d\delta_s = \gamma \, dx = \frac{v}{G} \, dx \approx \frac{V}{AG} \, dx \tag{3.83}$$

Figure 3.52c shows the corresponding shear deformation. The total shear deformation at the free end of a cantilever is

$$\delta_s \approx \int_0^L \frac{V}{AG} \, dx = \frac{PL}{AG} \tag{3.84}$$

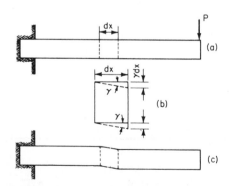

FIGURE 3.51 (a) Cantilever with a concentrated load. (b) Shear deformation of a small portion of the beam. (c) Shear deflection of the cantilever.

The shear deflection given by Eq. (3.84) is usually small compared with the flexural deflection for different materials and cross-sectional shapes. For example, the flexural deflection at the free end of a cantilever is $\delta_f = PL^3/3EI$. For a rectangular section made of steel with $G \approx 0.4E$, the ratio of shear deflection to flexural deflection is

$$\frac{\delta_s}{\delta_f} = \frac{PL/AG}{PL^3/3EI} = \frac{5}{8} \left(\frac{h}{L}\right)^2 \tag{3.85}$$

where h = depth of the beam. Thus, for a beam of rectangular section when $h/L = 0.1$, the shear deflection is less than 1% of the flexural deflection.

Shear deflections can be approximated for other types of beams in a similar way. For example, the midspan shear deflection for a simply supported beam loaded with a concentrated load at the center is $PL/4AG$.

3.20 MEMBERS SUBJECTED TO COMBINED FORCES

Most of the relationships presented in Arts. 3.16 to 3.19 hold only for symmetrical cross sections, e.g., rectangles, circles, and wide-flange beams, and only when the plane of the loads lies in one of the axes of symmetry. There are several instances where this is not the case, e.g., members subjected to axial load and bending and members subjected to torsional loads and bending.

Combined Axial Load and Bending. For short, stocky members subjected to both axial load and bending, stresses may be obtained by superposition if (1) the deflection due to bending is small and (2) all stresses remain in the elastic range. For these cases, the total stress normal to the section at a point equals the algebraic sum of the stress due to axial load and the stress due to bending about each axis:

$$f = \pm \frac{P}{A} \pm \frac{M_x}{S_x} \pm \frac{M_y}{S_y} \tag{3.86}$$

where P = axial load
A = cross-sectional area
M_x = bending moment about the centroidal x axis
S_x = elastic section modulus about the centroidal x axis
M_y = bending moment about the centroidal y axis
S_y = elastic section modulus about the centroidal y axis

If bending is only about one axis, the maximum stress occurs at the point of maximum moment. The two signs for the axial and bending stresses in Eq. (3.86) indicate that when the stresses due to the axial load and bending are all in tension or all in compression, the terms should be added. Otherwise, the signs should be obeyed when performing the arithmetic. For convenience, compressive stresses can be taken as negative and tensile stresses as positive.

Bending and axial stress are often caused by eccentrically applied axial loads. Figure 3.52 shows a column carrying a load P with eccentricity e_x and e_y. The stress in this case may be found by incorporating the resulting moments $M_x = Pe_x$ and $M_y = Pe_y$ into Eq. (3.86).

If the deflection due to bending is large, M_x and M_y should include the additional moment produced by second-order effects. Methods for incorporating these effects are presented in Arts. 3.46 to 3.48.

3.21 UNSYMMETRICAL BENDING

When the plane of loads acting transversely on a beam does not contain any of the beam's axes of symmetry, the loads may tend to produce twisting as well as bending. Figure 3.53 shows a horizontal channel twisting even though the vertical load H acts through the centroid of the section.

The **bending axis** of a beam is the longitudinal line through which transverse loads should pass to preclude twisting as the beam bends. The **shear center** for any

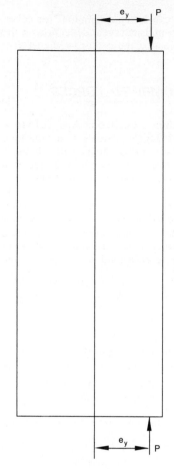

FRONT ELEVATION

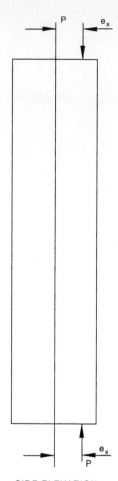

SIDE ELEVATION

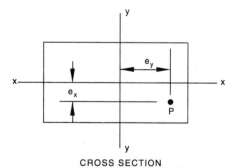

CROSS SECTION

FIGURE 3.52 Eccentrically loaded column.

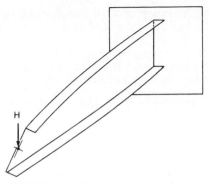

FIGURE 3.53   Twisting of a channel.

section of the beam is the point in the section through which the bending axis passes.

For sections having two axes of symmetry, the shear center is also the centroid of the section. If a section has an axis of symmetry, the shear center is located on that axis but may not be at the centroid of the section. Figure 3.54 shows a channel section in which the horizontal axis is the axis of symmetry. Point O represents the shear center. It lies on the horizontal axis but not at the centroid C. A load at the section must pass through the shear center if twisting of the member is not to occur. The location of the shear center relative to the center of the web can be obtained from

$$e = \frac{b/2}{1 + \frac{1}{6}(A_w/A_f)}$$ (3.87)

where b = width of flange overhang
$A_f = t_f b$ = area of flange overhang
$A_w = t_w h$ = web area

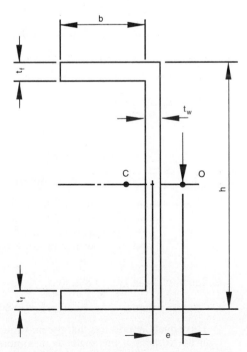

FIGURE 3.54   Relative position of shear center O and centroid C of a channel.

(F. Bleich, *Buckling Strength of Metal Structures*, McGraw-Hill, Inc., New York; F. B. Seely and J. O. Smith, *Advanced Mechanics of Materials*, John Wiley & Sons, Inc., New York.)

For a member with an unsymmetrical cross section subject to combined axial load and biaxial bending, Eq. (3.86) must be modified to include the effects of unsymmetrical bending. In this case, stress in the elastic range is given by

$$f = \frac{P}{A} + \frac{M_y - M_x(I_{xy}/I_x)}{I_y - (I_{xy}/I_x)I_{xy}} x + \frac{M_x - M_y(I_{xy}/I_y)}{I_x - (I_{xy}/I_y)I_{xy}} y \qquad (3.88)$$

where A = cross-sectional area
M_x, M_y = bending moment about x–x and y–y axes
I_x, I_y = moment of inertia about x–x and y–y axes
x, y = distance of stress point under consideration from y–y and x–x axes
I_{xy} = product of inertia

$$I_{xy} = \int xy \, dA \qquad (3.89)$$

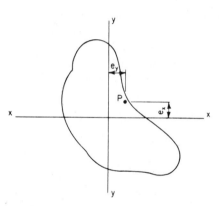

FIGURE 3.55 Eccentric load P on an unsymmetrical cross section.

Moments M_x and M_y may be caused by transverse loads or eccentricities of axial loads. An example of the latter case is shown in Fig. 3.55. For an axial load P, $M_x = Pe_x$ and $M_y = Pe_y$, where e_x and e_y are eccentricities with respect to the x–x and y–y axes, respectively.

To show an application of Eq. (3.88) to an unsymmetrical section, stresses in the lintel angle in Fig. 3.56 will be calculated for $M_x = 200$ in-kips, $M_y = 0$, and $P = 0$. The centroidal axes x–x and y–y are 2.6 and 1.1 in from the bottom and left side, respectively, as shown in Fig. 3.56. The moments of inertia are $I_x = 47.82$ in⁴ and $I_y = 11.23$ in⁴. The product of inertia can be calculated by dividing the angle into two rectangular parts and then applying Eq. (3.89):

$$I_{xy} = \int xy \, dA = A_1 x_1 y_1 + A_2 x_2 y_2$$

$$= 7(-0.6)(0.9) + 3(1.4)(-2.1) = -12.6 \qquad (3.90)$$

where A_1 and A_2 = cross-sectional areas of parts 1 and 2
x_1 and x_2 = horizontal distance from the angle's centroid to the centroid of parts 1 and 2
y_1 and y_2 = vertical distance from the angle's centroid to the centroid of parts 1 and 2

Substitution in Eq. (3.88) gives

$$f = 6.64x + 5.93y$$

This equation indicates that the maximum stresses normal to the cross section occur at the corners of the angle. A maximum compressive stress of 25.43 ksi occurs at the upper right corner, where $x = -0.1$ and $y = 4.4$. A maximum tensile stress of 22.72 ksi occurs at the lower left corner, where $x = -1.1$ and $y = -2.6$.

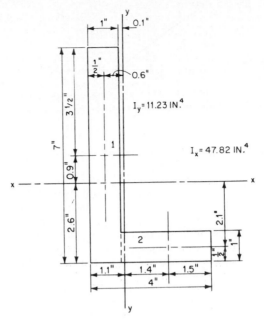

FIGURE 3.56   Steel lintel angle.

(F. B. Seely and J. O. Smith, *Advanced Mechanics of Materials*, John Wiley & Sons, Inc., New York; I. H. Shames, *Mechanics of Deformable Solids*, Prentice-Hall, Inc., Englewood Cliffs, N.J.; F. R. Shanley, *Strength of Materials*, McGraw-Hill, Inc., New York.)

CONCEPTS OF WORK AND ENERGY

The concepts of work and energy are often useful in structural analysis. These concepts provide a basis for some of the most important theorems of structural analysis.

3.22 WORK OF EXTERNAL FORCES

Whenever a force is displaced by a certain amount or a displacement is induced by a certain force, **work** is generated. The increment of work done on a body by a force F during an incremental displacement ds from its point of application is

$$dW = F \, ds \cos \alpha \qquad (3.91)$$

where α is the angle between F and ds (Fig. 3.57). Equation (3.91) implies that work is the product of force and the component of displacement in the line of action of the force, or the product of displacement and the component of force along the path of the displacement. If the component of the displacement is in the same direction as the force or the component of the force acts in the same direction

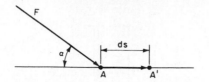

FIGURE 3.57 Force performs work in direction of displacement.

as the path of displacement, the work is positive; otherwise, the work is negative. When the line of action of the force is perpendicular to the direction of displacement ($\alpha = \pi/2$), no work is done.

When the displacement is a finite quantity, the total work can be expressed as

$$W = \int F \cos \alpha \, ds \qquad (3.92)$$

Integration is carried out over the path the force travels, which may not be a straight line.

The work done by the weight of a body, which is the force, when it is moved in a vertical direction is the product of the weight and vertical displacement. According to Eq. (3.91) and with α the angle between the downward direction of gravity and the imposed displacement, the weight does positive work when movement is down. It does negative work when movement is up.

In a similar fashion, the rotation of a body by a moment M through an incremental angle $d\theta$ also generates work. The increment of work done in this case is

$$dW = M \, d\theta \qquad (3.93)$$

The total work done during a finite angular displacement is

$$W = \int M \, d\theta \qquad (3.94)$$

3.23 VIRTUAL WORK AND STRAIN ENERGY

Consider a body of negligible dimensions subjected to a force F. Any displacement of the body from its original position will create work. Suppose a small displacement δs is assumed but does not actually take place. This displacement is called a **virtual displacement,** and the work δW done by force F during the displacement δs is called **virtual work.** Virtual work also is done when a virtual force δF acts over a displacement s.

Virtual Work on a Particle. Consider a particle at location A that is in equilibrium under the concurrent forces F_1, F_2, and F_3 (Fig. 3.58a). Hence equilibrium requires that the sum of the components of the forces along the x axis be zero:

$$\Sigma F_x = F_1 \cos \alpha_1 - F_2 \cos \alpha_2 + F_3 \cos \alpha_3 = 0 \qquad (3.95)$$

where α_1, α_2, α_3 = angle force makes with the x axis. If the particle is displaced a virtual amount δs along the x axis from A to A' (Fig. 3.58b), then the total virtual work done by the forces is the sum of the virtual work generated by displacing each of the forces F_1, F_2, and F_3. According to Eq. (3.91),

$$\delta W = F_1 \cos \alpha_1 \, \delta s - F_2 \cos \alpha_2 \, \delta s + F_3 \cos \alpha_3 \, \delta s \qquad (3.96)$$

Factoring δs from the right side of Eq. (3.96) and substituting the equilibrium relationship provided in Eq. (3.95) gives

$$\delta W = (F_1 \cos \alpha_1 - F_2 \cos \alpha_2 + F_3 \cos \alpha_3) \, \delta s = 0 \qquad (3.97)$$

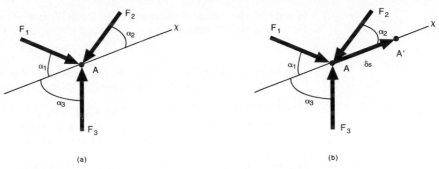

FIGURE 3.58 (*a*) Forces act on a particle *A*. (*b*) Forces perform virtual work over virtual displacement δ*s*.

Similarly, the virtual work is zero for the components along the *y* and *z* axes. In general, Eq. (3.97) requires

$$\delta W = 0 \qquad\qquad (3.98)$$

That is, virtual work must be equal to zero for a single particle in equilibrium under a set of forces.

In a rigid body, distances between particles remain constant, since no elongation or compression takes place under the action of forces. The virtual work done on each particle of the body when it is in equilibrium is zero. Hence the virtual work done by the entire rigid body is zero.

In general, then, for a rigid body in equilibrium, $\delta W = 0$.

Virtual Work on a Rigid Body.   This principle of virtual work can be applied to idealized systems consisting of rigid elements. As an example, Fig. 3.59 shows a horizontal lever, which can be idealized as a rigid body. If a virtual rotation of δθ

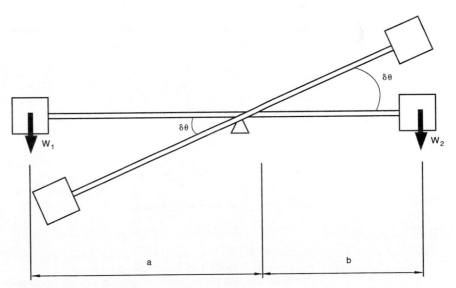

FIGURE 3.59   Virtual rotation of a lever.

is applied, the virtual displacement for force W_1 is $a\ \delta\theta$, and for force W_2, $b\ \delta\theta$. Hence the virtual work during this rotation is

$$\delta W = W_1 a\ \delta\theta - W_2 b\ \delta\theta \tag{3.99}$$

If the lever is in equilibrium, $\delta W = 0$. Hence $W_1 a = W_2 b$, which is the equilibrium condition that the sum of the moments of all forces about a support should be zero.

When the body is not rigid but can be distorted, the principle of virtual work as developed above cannot be applied directly. However, the principle can be modified to apply to bodies that undergo linear and nonlinear elastic deformations.

Strain Energy in a Bar.   The internal work U done on elastic members is called **elastic potential energy,** or **strain energy.** Suppose, for example, that a bar (Fig. 3.60a) made of an elastic material, such as steel, is gradually elongated an amount Δ_f by a force P_f. As the bar stretches with increases in force from 0 to P_f, each increment of internal work dU may be expressed by Eq. (3.91) with $\alpha = 0$:

$$dU = P\ d\Delta \tag{3.100}$$

where $d\Delta$ = the current increment in displacement in the direction of P
$\quad\ P$ = the current applied force, $0 \le P \le P_f$

Equation (3.100) also may be written as

$$\frac{dU}{d\Delta} = P \tag{3.101}$$

which indicates that the derivative of the internal work with respect to a displacement (or rotation) gives the corresponding force (or moment) at that location in the direction of the displacement (or rotation).

After the system comes to rest, a condition of equilibrium, the total internal work is

$$U = \int P\ d\Delta \tag{3.102}$$

The current displacement Δ is related to the applied force P by Eq. (3.51); that is, $P = EA\Delta/L$. Substitution into Eq. (3.102) yields

$$U = \int_0^{\Delta_f} \frac{EA}{L} \Delta\ d\Delta = \frac{EA\Delta_f^2}{2L} = \frac{LP_f^2}{2EA} = \frac{1}{2} P_f \Delta_f \tag{3.103}$$

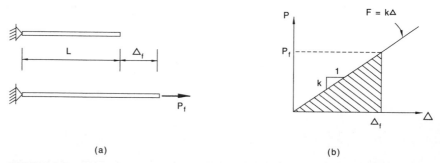

(a)   (b)

FIGURE 3.60   (a) Bar in tension elongates. (b) Energy stored in the bar is represented by the area under the load-displacement curve.

When the force is plotted against displacement (Fig. 3.60b), the internal work is the shaded area under the line with slope $k = EA/L$.

When the bar in Fig. 3.60a is loaded and in equilibrium, the internal virtual work done by P_f during an additional virtual displacement $\delta\Delta$ equals the change in the strain energy of the bar:

$$\Delta U = k\Delta_f\,\delta\Delta \tag{3.104}$$

where Δ_f is the original displacement produced by P_f.

Principle of Virtual Work.   This example illustrates the **principle of virtual work.** If an elastic body in equilibrium under the action of external loads is given a virtual deformation from its equilibrium condition, the work done by the external loads during this deformation equals the change in the internal work or strain energy, that is,

$$\delta W = \delta U \tag{3.105}$$

Hence, for the loaded bar in equilibrium (Fig. 3.60a), the external virtual work equals the internal virtual strain energy:

$$P_f\,\delta\Delta = k\Delta_f\,\delta\Delta \tag{3.106}$$

[For rigid bodies, no internal strain energy is generated, that is, $\delta U = k\Delta_f\,\delta\Delta = 0$, and Eq. (3.106) reduces the Eq. (3.98).] The example may be generalized to any constrained (supported) elastic body acted on by forces $P_1, P_2, P_3, \ldots$ for which the corresponding displacements are $\Delta_1, \Delta_2, \Delta_3, \ldots$. Equation (3.100) may then be expanded to

$$dU = \Sigma P_i\,d\Delta_i \tag{3.107}$$

Similarly, Eq. (3.101) may be generalized to

$$\frac{\partial U}{\partial \Delta_i} = P_i \tag{3.108}$$

The increase in strain energy due to the increments of the deformations is given by substitution of Eq. (3.108) into Eq. (3.107):

$$dU = \sum \frac{\partial U}{\partial \Delta_i}\,d\Delta_i = \frac{\partial U}{\partial \Delta_1}\,d\Delta_1 + \frac{\partial U}{\partial \Delta_2}\,d\Delta_2 + \frac{\partial U}{\partial \Delta_3}\,d\Delta_3 + \cdots \tag{3.109}$$

If specific deformations in Eq. (3.109) are represented by virtual displacements, load and deformation relationships for several structural systems may be obtained from the principle of virtual work.

Strain energy also can be generated when a member is subjected to other types of loads or deformations. The strain-energy equation can be written as a function of either load or deformation.

Strain Energy in Shear.   For a member subjected to pure shear, strain energy is given by

$$U = \frac{V^2L}{2AG} \tag{3.110a}$$

$$U = \frac{AG\Delta^2}{2L} \tag{3.110b}$$

where V = shear load
Δ = shear deformation
L = length over which the deformation takes place
A = shear area
G = shear modulus of elasticity

Strain Energy in Torsion. For a member subjected to torsion,

$$U = \frac{T^2L}{2JG} \tag{3.111a}$$

$$U = \frac{JG\theta^2}{2L} \tag{3.111b}$$

where T = torque
θ = angle of twist
L = length over which the deformation takes place
J = polar moment of inertia
G = shear modulus of elasticity

Strain Energy in Bending. For a member subjected to pure bending (constant moment),

$$U = \frac{M^2L}{2EI} \tag{3.112a}$$

$$U = \frac{EI\theta^2}{2L} \tag{3.112b}$$

where M = bending moment
θ = angle through which one end of beam rotates with respect to the other end
L = length over which the deformation takes place
I = moment of inertia
E = modulus of elasticity

For beams carrying transverse loads, the total strain energy is the sum of the energy for bending and that for shear.

Virtual Forces. Virtual work also may be created when a system of **virtual forces** is applied to a structure that is in equilibrium. In this case, the principle of virtual work requires that external virtual work, created by virtual forces acting over their induced displacements, equals the internal virtual work or strain energy. This concept is often used to determine deflections. For convenience, virtual forces are often represented by unit loads. Hence this method is frequently called the **unit-load method** or **dummy-unit-load method.**

Dummy-Unit-Load Method. A unit load is applied at the location and in the direction of an unknown displacement Δ produced by given loads on a structure. According to the principle of virtual work, the external work done by the unit load equals the change in strain energy in the structure:

$$1\Delta = \Sigma fd \tag{3.113}$$

where Δ = deflection in desired direction produced by given loads
f = force in each element of the structure due to the unit load
d = deformation in each element produced by the given loads

The summation extends over all elements of the structure.

For a vertical component of a deflection, a unit vertical load should be used. For a horizontal component of a deflection, a unit horizontal load should be used. And for a rotation, a unit moment should be used.

For example, the deflection in the axial-loaded member shown in Fig. 3.60a can be determined by substituting $f = 1$ and $d = P_f L / EA$ into Eq. (3.113). Thus $1\Delta_f = 1 P_f L / EA$ and $\Delta_f = P_f L / EA$.

For applications of the unit-load method for analysis of large structures, see Arts. 3.31 and 3.33.3.

(C. H. Norris et al., *Elementary Structural Analysis*, 4th ed.; and M. S. Naschi, *Stress, Stability,* and *Chaos in Structural Engineering: An Energy Approach,* McGraw-Hill, Inc., New York.)

3.24   *CASTIGLIANO'S THEOREMS*

If strain energy U, as defined in Art. 3.23, is expressed as a function of external forces, the partial derivative of the strain energy with respect to one of the external forces P_i gives the displacement Δ_i corresponding to that force:

$$\frac{\partial U}{\partial P_i} = \Delta_i \qquad (3.114)$$

This is known as **Castigliano's first theorem.**

If no displacement can occur at a support and Castigliano's theorem is applied to that support, Eq. (3.114) becomes

$$\frac{\partial U}{\partial P_i} = 0 \qquad (3.115)$$

Equation (3.115) is commonly called the **principle of least work,** or **Castigliano's second theorem.** It implies that any reaction components in a structure will take on loads that will result in a minimum strain energy for the system. Castigliano's second theorem is restricted to linear elastic structures. On the other hand, Castigliano's first theorem is only limited to elastic structures and hence can be applied to nonlinear structures.

As an example, the principle of least work will be applied to determine the force in the vertical member of the truss shown in Fig. 3.61. If S_a denotes the force in the vertical bar, then vertical equilibrium requires the force in each of the inclined bars to be $(P - S_a)/(2 \cos \alpha)$. According to Eq. (3.103), the total strain energy in the system is

$$U = \frac{S_a^2 L}{2EA} + \frac{(P - S_a)^2 L}{4EA \cos^3 \alpha} \qquad (3.116)$$

The internal work in the system will be minimum when

$$\frac{\partial U}{\partial S_a} = \frac{S_a L}{EA} - \frac{(P - S_a)L}{2EA \cos^3 \alpha} = 0 \qquad (3.117a)$$

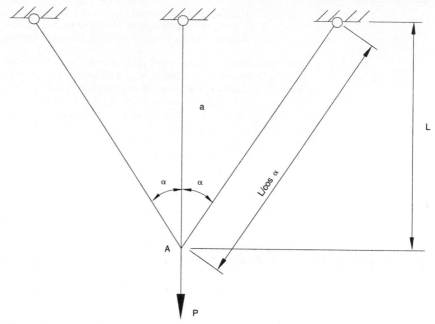

FIGURE 3.61 Statically indeterminate truss.

Solution of Eq. (3.117a) gives the force in the vertical bar as

$$S_a = \frac{P}{1 + 2\cos^3 \alpha} \tag{3.117b}$$

(N. J. Hoff, *Analysis of Structures*, John Wiley & Sons, Inc., New York.)

3.25 RECIPROCAL THEOREMS

If the bar shown in Fig. 3.62a, which has a stiffness $k = EA/L$, is subjected to an axial force P_1, it will deflect $\Delta_1 = P_1/k$. According to Eq. (3.103), the external work done is $P_1\Delta_1/2$. If an additional load P_2 is then applied, it will deflect an additional amount $\Delta_2 = P_2/k$ (Fig. 3.62b). The additional external work generated is the sum of the work done in displacing P_1, which equals $P_1\Delta_2$, and the work done in displacing P_2, which equals $P_2\Delta_2/2$. The total external work done is

$$W = \tfrac{1}{2}P_1\Delta_1 + \tfrac{1}{2}P_2\Delta_2 + P_1\Delta_2 \tag{3.118}$$

According to Eq. (3.103), the total internal work or strain energy is

$$U = \tfrac{1}{2}k\Delta_f^2 \tag{3.119}$$

where $\Delta_f = \Delta_1 + \Delta_2$. For the system to be in equilibrium, the total external work must equal the total internal work, that is,

$$\tfrac{1}{2}P_1\Delta_1 + \tfrac{1}{2}P_2\Delta_2 + P_1\Delta_2 = \tfrac{1}{2}k\Delta_f^2 \tag{3.120}$$

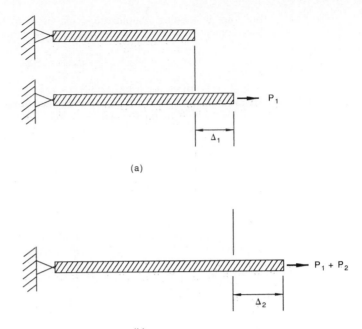

FIGURE 3.62 (*a*) Load on a bar performs work over displacement Δ_1. (*b*) Additional work is performed by both a second load and the original load.

If the bar is then unloaded and then reloaded by placing P_1 on the bar first and later applying P_1, the total external work done would be

$$W = \tfrac{1}{2}P_2\Delta_2 + \tfrac{1}{2}P_1\Delta_1 + P_2\Delta_1 \tag{3.121}$$

The total internal work would be the same as that for the first loading sequence because the total deflection of the system is still $\Delta_f = \Delta_1 + \Delta_2$. This implies that for a linear elastic system, the sequence of loading does not affect resulting deformations and corresponding internal forces. That is, in a **conservative system,** work is path-independent.

For the system to be in equilibrium under this loading, the total external work would again equal the total internal work:

$$\tfrac{1}{2}P_2\Delta_2 + \tfrac{1}{2}P_1\Delta_1 + P_2\Delta_1 = \tfrac{1}{2}k\Delta_f^2 \tag{3.122}$$

Equating the left sides of Eqs. (3.120) and (3.122) and simplifying give

$$P_1\Delta_2 = P_2\Delta_1 \tag{3.123}$$

This example, specifically Eq. (3.123), also demonstrates **Betti's theorem:** For a linearly elastic structure, the work done by a set of external forces P_1 acting through the set of displacements Δ_2 produced by another set of forces P_2 equals the work done by P_2 acting through the displacements Δ_1 produced by P_1.

Betti's theorem may be applied to a structure in which two loads P_i and P_j act at points i and j, respectively. P_i acting alone causes displacements Δ_{ii} and Δ_{ji}, where the first subscript indicates the point of displacement and the second indicates the point of loading. Application next of P_j to the system produces additional displacements Δ_{ij} and Δ_{jj}. According to Betti's theorem, for any P_i and P_j,

$$P_i \Delta_{ij} = P_j \Delta_{ji} \qquad (3.124)$$

If $P_i = P_j$, then, according to Eq. (3.124), $\Delta_{ij} = \Delta_{ji}$. This relationship is known as **Maxwell's theorem of reciprocal displacements:** For a linear elastic structure, the displacement at point i due to a load applied at another point j equals the displacement at point j due to the same load applied at point i.

ANALYSIS OF STRUCTURAL SYSTEMS

A **structural system** consists of the primary load-bearing structure, including its members and connections. An analysis of a structural system consists of determining the reactions, deflections, and internal forces and corresponding stresses caused by external loads. Methods for determining these depend on both the external loading and the type of structural system that is assumed to resist these loads.

3.26 TYPES OF LOADS

Loads are forces that act or may act on a structure. For the purpose of predicting the resulting behavior of the structure, the loads, or external influences, including forces, consequent displacements, and support settlements, are presumed to be known. These influences may be specified by law, e.g., building codes, codes of recommended practice, or owner specifications, or they may be determined by engineering judgment. Loads are typically divided into two general classes: **dead load,** which is the weight of a structure including all of its permanent components, and **live load,** which is comprised of all loads other than dead loads.

The type of load has an appreciable influence on the behavior of the structure on which it acts. In accordance with this influence, loads may be classified as static, dynamic, long time, or repetitive.

Static loads are those applied so slowly that the effect of time can be ignored. All structures are subject to some static loading, e.g., their own weight. There is, however, a large class of loads that usually is approximated by static loading for convenience. Occupancy loads and wind loads are often assumed static. All the analysis methods presented in the following articles, with the exception of Arts. 3.52 to 3.55, assume that static loads are applied to structures.

Dynamic loads are characterized by very short durations, and the response of the structure depends on time. Earthquake shocks, high-level wind gusts, and moving live loads belong in this category.

Long-duration loads are those which act on a structure for extended periods of time. For some materials and levels of stress, such loads cause structures to undergo deformations under constant load that may have serious effects. Creep and relaxation of structural materials may occur under long-duration loads. The weight of a structure and any superimposed dead load fall in this category.

Repetitive loads are those applied and removed many times. If repeated for a large number of times, they may cause the structure to fail in fatigue. Moving live load is in this category.

3.27 COMMONLY USED STRUCTURAL SYSTEMS

Structures are typically too complicated to analyze in their real form. To determine the response of a structure to external loads, it is convenient to convert the structural system to an idealized form. Stresses and displacements in trusses, for example, are analyzed based on the following assumptions.

3.27.1 Trusses

A **truss** is a structural system constructed of linear members forming triangular patterns. The members are assumed to be straight and connected to one another by frictionless hinges. All loading is assumed to be concentrated at these connections (joints). By virtue of these properties, truss members are subject only to axial load. In reality, these conditions may not be satisfied; for example, connections are never frictionless, and hence some moments may develop in adjoining members. In practice, however, assumption of the preceding conditions is reasonable.

If all the members are coplanar, then the system is called a **planar truss.** Otherwise, the structure is called a **space truss.** The exterior members of a truss are called **chords,** and the diagonals are called **web members.**

Trusses often act as beams. They may be constructed horizontally; examples include roof trusses and bridge trusses. They also may be constructed vertically; examples include transmission towers and internal lateral bracing systems for buildings or bridge towers and pylons. Trusses often can be built economically to span several hundred feet.

Roof trusses, in addition to their own weight, support the weight of roof sheathing, roof beams or purlins, wind loads, snow loads, suspended ceilings, and sometimes cranes and other mechanical equipment. Provisions for loads during construction and maintenance often need to be included. All applied loading should be distributed to the truss in such a way that the loads act at the joints. Figure 3.63 shows some common roof trusses.

Bridge trusses are typically constructed in pairs. If the roadway is at the level of the bottom chord, the truss is a **through truss.** If it is level with the top chord, it is a **deck truss.** The floor system consists of floor beams, which span in the transverse direction and connect to the truss joints; stringers, which span longitudinally and connect to the floor beams; and a roadway or deck, which is carried by the stringers. With this system, the dead load of the floor system and the bridge live loads it supports, including impact, are distributed to the truss joints. Figure 3.64 shows some common bridge trusses.

3.27.2 Rigid Frames

A **rigid frame** is a structural system constructed of members that resist bending moment, shear, and axial load and with connections that do not permit changes in the angles between the members under loads. Loading may be either distributed along the length of members, such as gravity loads, or entirely concentrated at the connections, such as wind loads.

If the axial load in a frame member is negligible, the member is commonly referred to as a **beam.** If moment and shear are negligible and the axial load is compressive, the member is referred to as a **column.** Members subjected to moments, shears, and compressive axial forces are typically called **beam-columns.** (Most vertical members are called **columns,** although technically they behave as beam-columns.)

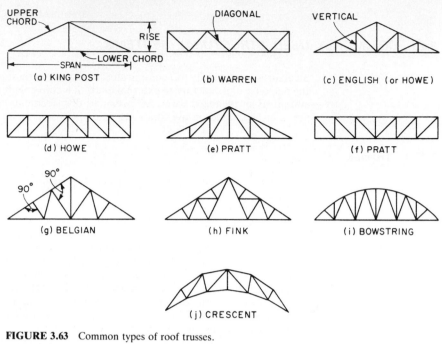

FIGURE 3.63 Common types of roof trusses.

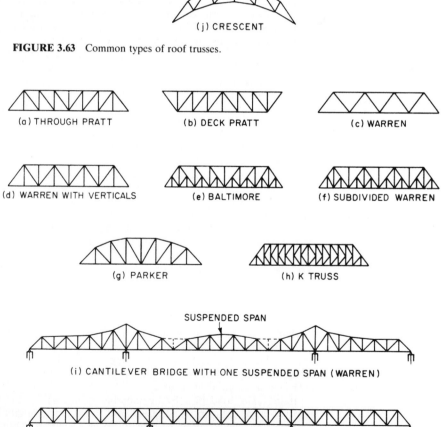

FIGURE 3.64 Common types of bridge trusses.

If all the members are coplanar, the frame is called a **planar frame.** Otherwise, it is called a **space frame.** One plane of a space frame is called a **bent.** The area spanning between neighboring columns on a specific level is called a **bay.**

3.27.3 Continuous Beams

A **continuous beam** is a structural system that carries load over several spans by a series of rigidly connected members that resist bending moment and shear. The loading may be either concentrated or distributed along the lengths of members. The underlying structural system for many bridges is often a set of continuous beams.

3.28 DETERMINACY AND GEOMETRIC STABILITY

In a **statically determinate system,** all reactions and internal member forces can be calculated solely from equations of equilibrium. However, if equations of equilibrium alone do not provide enough information to calculate these forces, the system is **statically indeterminate.** In this case, adequate information for analyzing the system will only be gained by also considering the resulting structural deformations. Static determinacy is never a function of loading. In a statically determinate system, the distribution of internal forces is not a function of member cross section or material properties.

In general, the degree of static determinacy n for a truss may be determined by

$$n = m - \alpha j + R \qquad (3.125)$$

where m = number of members
j = number of joints including supports
α = dimension of truss ($\alpha = 2$ for a planar truss and $\alpha = 3$ for a space truss)
R = number of reaction components

Similarly, the degree of static determinacy for a frame is given by

$$n = 3(\alpha - 1)(m - j) + R \qquad (3.126)$$

where $\alpha = 2$ for a planar frame and $\alpha = 3$ for a space frame.

If n is greater than zero, the system is geometrically stable and statically indeterminate; if n is equal to zero, it is statically determinate and may or may not be stable; if n is less than zero, it is always geometrically unstable. Geometric instability of a statically determinate truss ($n = 0$) may be determined by observing that multiple solutions to the internal forces exist when applying equations of equilibrium.

Figure 3.65 provides several examples of statically determinate and indeterminate systems. In some cases, such as the planar frame shown in Fig. 3.65e, the frame is statically indeterminate for computation of internal forces, but the reactions can be determined from equilibrium equations.

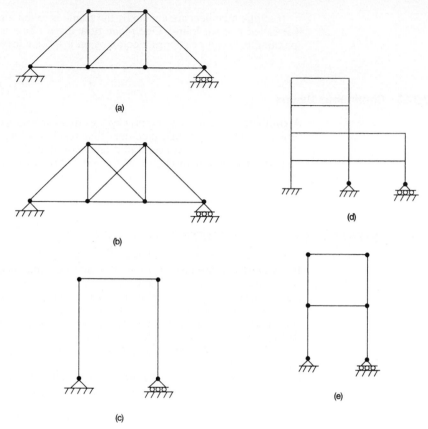

FIGURE 3.65   Examples of statically determinate and indeterminate systems: (*a*) Statically determinate truss ($n = 0$); (*b*) statically indeterminate truss ($n = 1$); (*c*) statically determinate frame ($n = 0$); (*d*) statically indeterminate frame ($n = 15$); (*e*) statically indeterminate frame ($n = 3$).

3.29   *CALCULATION OF REACTIONS IN STATICALLY DETERMINATE SYSTEMS*

For statically determinate systems, reactions can be determined from equilibrium equations [Eq. (3.11) or (3.12)]. For example, in the planar system shown in Fig. 3.66, reactions R_1, H_1, and R_2 can be calculated from the three equilibrium equations. The beam with overhang carries a uniform load of 3 kips/ft over its 40-ft horizontal length, a vertical 60-kip concentrated load at *C*, and a horizontal 10-kip concentrated load at *D*. Support *A* is hinged; it can resist vertical and horizontal forces. Support *B*, 30 ft away, is on rollers; it can resist only vertical force. Dimensions of the member cross sections are assumed to be small relative to the spans.

Support *A* but not support *B* can resist horizontal loads. Since the sum of the horizontal forces must equal zero and there is a 10-kip horizontal load at *D*, the horizontal component of the reaction at *A* is $H_1 = 10$ kips.

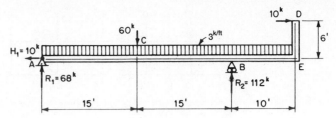

FIGURE 3.66 Beam with overhang with uniform and concentrated loads.

The vertical reaction at A can be computed by setting the sum of the moments of all forces about B equal to zero:

$$3 \times 40 \times 10 + 60 \times 15 - 10 \times 6 - 30R_1 = 0$$

from which $R_1 = 68$ kips. Similarly, the reaction at B can be found by setting the sum of the moments about A of all forces equal to zero:

$$3 \times 40 \times 20 + 60 \times 15 + 10 \times 6 - 30R_2 = 0$$

from which $R_2 = 112$ kips. Alternatively, equilibrium of vertical forces can be used to obtain R_2, given $R_1 = 68$:

$$R_2 + R_1 - 3 \times 40 - 60 = 0$$

Solution of this equation also yields $R_2 = 112$ kips.

3.30 FORCES IN STATICALLY DETERMINATE TRUSSES

A convenient method for determining the member forces in a truss is to isolate a portion of the truss. A section should be chosen such that it is possible to determine the forces in the cut members with the equations of equilibrium [Eq. (3.11) or (3.12)]. Compressive forces act toward the panel point, and tensile forces act away from the panel point.

3.30.1 Method of Sections

To calculate the force in member a of the truss in Fig. 3.67a, the portion of the truss in Fig. 3.67b is isolated by passing section x–x through members a, b, and c. Equilibrium of this part of the truss is maintained by the 10-kip loads at panel points U_1 and U_2, the 25-kip reaction, and the forces S_a, S_b, and S_c in members a, b, and c, respectively. S_a can be determined by equating to zero the sum of the moments of all the external forces about panel point L_3, because the other unknown forces S_b and S_c pass through L_3 and their moments therefore equal zero. The corresponding equilibrium equation is

$$-9S_a + 36 \times 25 - 24 \times 10 - 12 \times 10 = 0$$

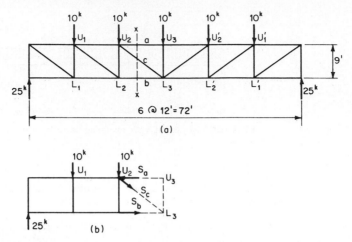

FIGURE 3.67 (*a*) Truss with loads at panel points. (*b*) Stresses in members cut by section *x—x* hold truss in equilibrium.

Solution of this equation yields $S_a = 60$ kips. Similarly, S_b can be calculated by equating to zero the sum of the moments of all external forces about panel point U_2:

$$-9S_b + 24 \times 25 - 12 \times 10 = 0$$

from which $S_b = 53.3$ kips.

Since members *a* and *b* are horizontal, they do not have a vertical component. Hence diagonal *c* must carry the entire vertical shear on section *x–x*: $25 - 10 - 10 = 5$ kips. With 5 kips as its vertical component and a length of 15 ft on a rise of 9 ft,

$$S_c = \text{¹⁵⁄₉} \times 5 = 8.3 \text{ kips}$$

When the chords are not horizontal, the vertical component of the diagonal may be found by subtracting from the shear in the section the vertical components of force in the chords.

3.30.2 Method of Joints

A special case of the method of sections is choice of sections that isolate the joints. With the forces in the cut members considered as external forces, the sum of the horizontal components and the sum of the vertical components of the external forces acting at each joint must equal zero.

Since only two equilibrium equations are available for each joint, the procedure is to start with a joint that has two or fewer unknowns (usually a support). When these unknowns have been found, the procedure is repeated at successive joints with no more than two unknowns.

For example, for the truss in Fig. 3.68*a*, at joint 1 there are three forces: the reaction of 12 kips, force S_a in member *a*, and force S_c in member *c*. Since *c* is horizontal, equilibrium of vertical forces requires that the vertical component of force in member *a* be 12 kips. From the geometry of the truss, $S_a = 12 \times \text{¹⁵⁄₉} =$

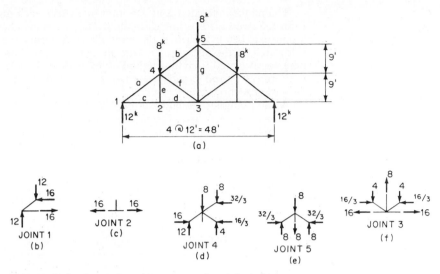

FIGURE 3.68 Calculation of truss stresses by method of joints.

20 kips. The horizontal component of S_a is $20 \times \frac{12}{15} = 16$ kips. Since the sum of the horizontal components of all forces acting at joint 1 must equal zero, $S_c = 16$ kips.

At joint 2, the force in member e is zero because no vertical forces are present there. Hence the force in member d equals the previously calculated 16-kip force in member c. Forces in the other members would be determined in the same way (see Fig. 3.68d, e, and f).

3.31 *DEFLECTIONS OF STATICALLY DETERMINATE TRUSSES*

In Art. 3.23, the basic concepts of virtual work and specifically the unit-load method are presented. Employing these concepts, this method may be adapted readily to computing the deflection at any panel point (joint) in a truss.

Specifically, Eq. (3.113), which equates external virtual work done by a virtual unit load to the corresponding internal virtual work, may be written for a truss as

$$1\Delta = \sum_{i=1}^{n} f_i \frac{P_i L_i}{E_i A_i} \tag{3.127}$$

where Δ = displacement component to be calculated (also the displacement at and in the direction of an applied unit load)

n = total number of members

f_i = axial force in member i due to unit load applied where Δ occurs—horizontal or vertical unit load for horizontal or vertical displacement, moment for rotation

P_i = axial force in member i due to the given loads

L_i = length of member i

E_i = modulus of elasticity for member i

A_i = cross-sectional area of member i

To find the deflection Δ at any joint in a truss, each member force P_i resulting from the given loads is first calculated. Then each member force f_i resulting from a unit load being applied at the joint where Δ occurs and in the direction of Δ is calculated. If the structure is statically determinate, both sets of member forces may be calculated from the method of joints (Sec. 3.30.2). Substituting each member's forces P_i and f_i and properties L_i, E_i, and A_i, into Eq. (3.127) yields the desired deflection Δ.

As an example, the midspan downward deflection for the truss shown in Fig. 3.68a will be calculated. The member forces due to the 8-kip loads are shown in Fig. 3.69a. A unit load acting downward is applied at midspan (Fig. 3.69b). The member forces due to the unit load are shown in Fig. 3.69b. On the assumption that all members have area $A_i = 2$ in^2 and modulus of elasticity $E_i = 29,000$ ksi, Table 3.3 presents the computations for the midspan deflection Δ. Members not stressed by either the given loads, $P_i = 0$, or the unit load, $f_i = 0$, are not included in the table. The resulting midspan deflection is calculated as 0.31 in.

TABLE 3.3 Calculation of Truss Deflections

Member	P_i, kips	f_i	$\dfrac{1000L_i}{E_iA_i}$	$f_i\dfrac{P_iL_i}{E_iA_i}$, in
a	−20.00	−0.83	3.103	0.052
b	−13.33	−0.83	3.103	0.034
c	16.00	0.67	2.483	0.026
d	16.00	0.67	2.483	0.026
g	8.00	1.00	3.724	0.030
h	−20.00	−0.83	3.103	0.052
i	−13.33	−0.83	3.103	0.034
j	16.00	0.67	2.483	0.026
k	16.00	0.67	2.483	0.026
				$\Delta = \Sigma = 0.306$

3.32 FORCES IN STATICALLY DETERMINATE BEAMS AND FRAMES

Similar to the method of sections for trusses discussed in Art. 3.30, internal forces in statically determinate beams and frames also may be found by isolating a portion of these systems. A section should be chosen so that it will be possible to determine the unknown internal forces from only equations of equilibrium [Eq. (3.11) or (3.12)].

As an example, suppose that the forces and moments at point A in the roof purlin of the **gable frame** shown in Fig. 3.70a are to be calculated. Support B is a hinge. Support C is on rollers. Support reactions R_1, H_1, and R_2 are determined from equations of equilibrium. For example, summing moments about B yields

$$\Sigma M = 30 \times R_2 + 12 \times 8 - 15 \times 12 - 30 \times 6 = 0$$

from which $R_2 = 8.8$ kips. $R_1 = 6 + 12 + 6 - 8.8 = 15.2$ kips.

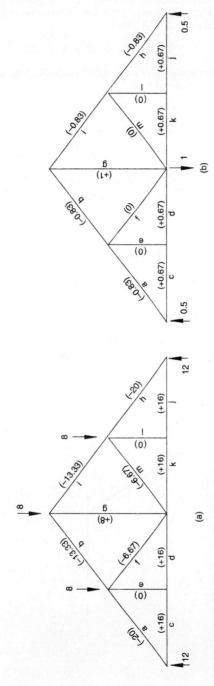

FIGURE 3.69 (a) Loaded truss with stresses in members shown in parentheses. (b) Stresses in truss due to a unit load applied for calculation of midspan deflection.

The portion of the frame shown in Fig. 3.70*b* is then isolated. The internal shear V_A is assumed normal to the longitudinal axis of the rafter and acting downward. The axial force P_A is assumed to cause tension in the rafter. Equilibrium of moments about point A yields

$$\Sigma M = M_A + 10 \times 6 + (12 + 10 \tan 30) \times 8 - 10 \times 15.2 = 0$$

from which $M_A = -50.19$ kips-ft. Vertical equilibrium of this part of the frame is maintained with

$$\Sigma F_y = 15.2 - 6 + P_A \sin 30 - V_A \cos 30 = 0 \qquad (3.128)$$

Horizontal equilibrium requires that

$$\Sigma F_x = 8 + P_A \cos 30 + V_A \sin 30 = 0 \qquad (3.129)$$

Simultaneous solution of Eqs. (3.128) and (3.129) gives $V_A = 3.96$ kips and $P_A = -11.53$ kips. The negative value indicates that the rafter is in compression.

3.33   DEFORMATIONS IN BEAMS

Article 3.18 presents relationships between a distributed load on a beam, the resulting internal forces and moments, and the corresponding deformations. These relationships provide the key expressions used in the **conjugate-beam method** and the **moment-area method** for computing beam deflections and slopes of the neutral axis under loads. The unit-load method used for this purpose is derived from the principle of virtual work (Art. 3.23).

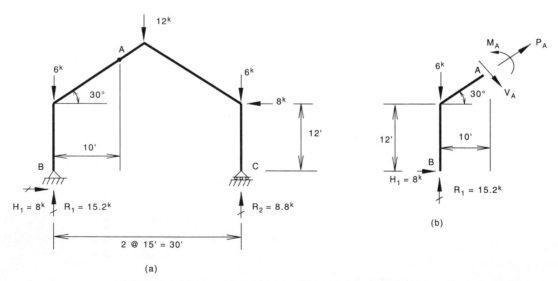

FIGURE 3.70   (*a*) Loaded gable frame. (*b*) Internal forces hold portion of frame in equilibrium.

3.33.1 Conjugate-Beam Method

For a beam subjected to a distributed load $w(x)$, the following integral relationships hold:

$$V(x) = \int w(x)\, dx \tag{3.130a}$$

$$M(x) = \int V(x)\, dx = \int\int w(x)\, dx\, dx \tag{3.130b}$$

$$\theta(x) = \int \frac{M(x)}{EI(x)}\, dx \tag{3.130c}$$

$$\delta(x) = \int \theta(x)\, dx = \int\int \frac{M(x)}{EI(x)}\, dx\, dx \tag{3.130d}$$

Comparison of Eqs. (3.130a) and (3.130b) with Eqs. (3.130c) and (3.130d) indicates that for a beam subjected to a distributed load $w(x)$, the resulting slope $\theta(x)$ and deflection $\delta(x)$ are equal, respectively, to the corresponding shear distribution $\overline{V}(x)$ and moment distribution $\overline{M}(x)$ generated in an associated or conjugate beam subjected to the distributed load $M(x)/EI(x)$. $M(x)$ is the moment at x due to the actual load $w(x)$ on the original beam.

In some cases, the supports of the real beam should be replaced by different supports for the conjugate beam to maintain the consistent θ-to-$\overline{V}$ and δ-to-$\overline{M}$ correspondence. For example, at the fixed end of a cantilevered beam, there is no rotation ($\theta = 0$) and no deflection ($\delta = 0$). Hence, at this location in the conjugate beam, $\overline{V} = 0$ and $\overline{M} = 0$. This can only be accomplished with a free-end support; i.e., a fixed end in a real beam is represented by a free end in its conjugate beam. A summary of the corresponding support conditions for several conjugate beams is provided in Fig. 3.71.

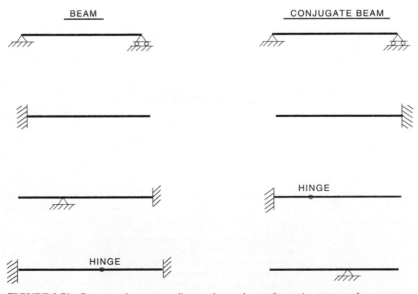

FIGURE 3.71 Beams and corresponding conjugate beams for various types of supports.

The sign convention to be employed for the conjugate-beam method is as follows:

A positive M/EI segment in the real beam should be placed as a downward (negative) distributed load $\overline{w}$ on the conjugate beam. A negative M/EI segment should be applied as an upward (positive) $\overline{w}$.

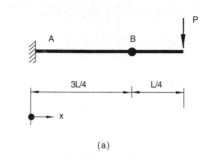

(a)

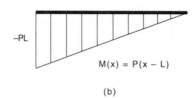

$M(x) = P(x - L)$

(b)

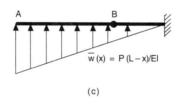

$\overline{w}(x) = P(L-x)/EI$

(c)

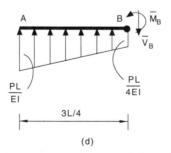

(d)

FIGURE 3.72 Deflection calculations for a cantilever by the conjugate beam method. (*a*) Cantilever beam with a load on the end. (*b*) Bending-moment diagram. (*c*) Conjugate beam loaded with M/EI distribution. (*d*) Deflection at B equals the bending moment at B due to the M/EI loading.

Positive shear $\overline{V}$ in the conjugate beam corresponds to a clockwise (negative) slope θ in the real beam. Negative $\overline{V}$ corresponds to a counterclockwise (positive) θ.

Positive moment $\overline{M}$ in the conjugate beam corresponds to a downward (negative) deflection δ in the real beam. Negative $\overline{M}$ corresponds to upward (positive) δ.

As an example, suppose the deflection at point B in the cantilevered beam shown in Fig. 3.72*a* is to be calculated. With no distributed load between the tip of the beam and its support, the bending moments on the beam are given by $M(x) = P(x - L)$ (Fig. 3.72*b*). The conjugate beam is shown in Fig. 3.72*c*. It has the same physical dimensions (E, I, and L) as the original beam but interchanged support conditions and is subject to a distributed load $w(x) = P(L - x)/EI$, as indicated in Fig. 3.72*c*. Equilibrium of the free-body diagram shown in Fig. 3.73*d* requires $\overline{V}_B = +15PL^2/32EI$ and $\overline{M}_B = +27PL^3/128EI$. The slope in the real beam at point B is then equal to the negative of the conjugate shear at this point, $\theta_B = -\overline{V}_B = -15PL^2/32EI$. Similarly, the deflection at point I' is the negative of the conjugate moment, $\delta_B = -\overline{M}_B = -27PL^3/128EI$. See also Sec. 3.33.2.

3.33.2 Moment-Area Method

Similar to the conjugate-beam method, the moment-area method is based on Eqs. (3.130*a*) to (3.130*d*). It expresses the deviation in the slope and tangential deflection between points A and B on a deflected beam:

$$\theta_B - \theta_A = \int_{x_A}^{x_B} \frac{M(x)}{EI(x)} \, dx \qquad (3.131a)$$

$$t_B - t_A = \int_{x_A}^{x_B} \frac{M(x)x}{EI(x)} \, dx \qquad (3.131b)$$

Equation (3.131a) indicates that the change in slope of the elastic curve of a beam between any two points equals the area under the M/EI diagram between these points. Similarly, Eq. (3.131b) indicates that the tangential deviation of any point on the elastic curve with respect to the tangent to the elastic curve at a second point equals the moment of the area under the M/EI diagram between the two points taken about the first point.

For example, deflection δ_B and rotation θ_B at point B in the cantilever shown in Fig. 3.72a are

$$\theta_B = \theta_A + \int_0^{3L/4} \frac{M(x)}{EI} \, dx$$

$$= 0 + \left(-\frac{PL}{4EI} \frac{3L}{4} - \frac{1}{2} \frac{3PL}{4EI} \frac{3L}{4} \right)$$

$$= -\frac{15PL^2}{32EI}$$

$$t_B = t_A + \int_0^{3L/4} \frac{M(x)x}{EI} \, dx$$

$$= 0 + \left(-\frac{PL}{4EI} \frac{3L}{4} \frac{1}{2} \frac{3L}{4} - \frac{1}{2} \frac{3PL}{4EI} \frac{3L}{4} \frac{2}{3} \frac{3L}{4} \right)$$

$$= -\frac{27PL^3}{128EI}$$

For this particular example $t_A = 0$, and hence $\delta_B = t_B$.

The moment-area method is particularly useful when a point of zero slope can be identified. In cases where a point of zero slope cannot be located, deformations may be more readily calculated with the conjugate-beam method. As long as the bending-moment diagram can be defined accurately, both methods can be used to calculate deformations in either statically determinate or indeterminate beams.

3.33.3 Unit-Load Method

Article 3.23 presents the basic concepts of the unit-load method. Article 3.34 employs this method to compute the deflections of a truss. The method also can be adapted to compute deflections in beams.

The deflection Δ at any point of a beam due to bending can be determined by transforming Eq. (3.113) to

$$1\Delta = \int_0^L \frac{M(x)}{EI(x)} m(x) \, dx \tag{3.132}$$

where $M(x)$ = moment distribution along the span due to the given loads
E = modulus of elasticity
I = cross-sectional moment of inertia
L = beam span
$m(x)$ = bending-moment distribution due to a unit load at the location and in the direction of deflection Δ

As an example of the use of Eq. (3.132), the midspan deflection will be determined for a prismatic, simply supported beam under a uniform load w (Fig. 3.73a). With support A as the origin, the equation for bending moment due to the uniform load is $M(x) = wLx/2 - wx^2/2$ (Fig. 3.73b). For a unit vertical load at midspan (Fig. 3.73c), the equation for bending moment in the left half of the beam is

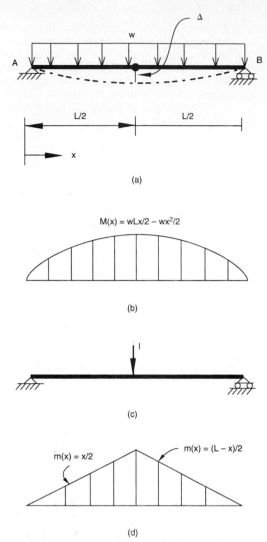

FIGURE 3.73 Deflection calculations for a simple beam by unit load method. (*a*) Uniformly loaded beam. (*b*) Bending-moment diagram for the uniform load. (*c*) Unit load at midspan. (*d*) Bending-moment diagram for the unit load.

$m(x) = x/2$ and in the right half $m(x) = (L - x)/2$ (Fig. 3.73*d*). By Eq. (3.132), the deflection is

$$\Delta = \frac{1}{EI} \int_0^{L/2} \left(\frac{wLx}{2} - \frac{wx^2}{2} \right) \frac{x}{2} \, dx + \frac{1}{EI} \int_{L/2}^{L} \left(\frac{wLx}{2} - \frac{wx^2}{2} \right) \frac{L - x}{2} \, dx$$

from which $\Delta = 5wL^4/384EI$. If the beam were not prismatic, EI would be a function of x and would be inside the integral.

Equation (3.113) also can be used to calculate the slope at any point along a beam span. Figure 3.74a shows a simply supported beam subjected to a moment M_A acting at support A. The resulting moment distribution is $M(x) = M_A (1 - x/L)$ (Fig. 3.74b). Suppose that the rotation θ_B at support B is to be determined. Application of a unit moment at B (Fig. 3.74c) results in the moment distribution $m(x) = x/L$ (Fig. 3.74d). By Eq. (3.132), on substitution of θ_B for Δ, the rotation at B is

$$\theta_B = \int_0^L \frac{M(x)}{EI(x)} m(x) \, dx = \frac{M_A}{EI} \int_0^L \left(1 - \frac{x}{L}\right) \frac{x}{L} \, dx = \frac{M_A L}{6EI}$$

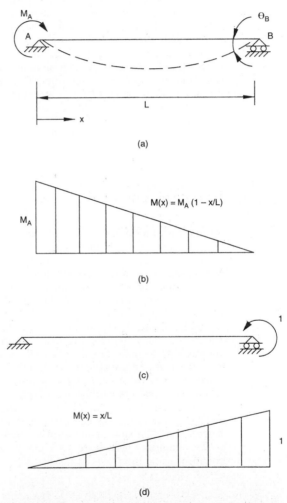

(a)

(b)

(c)

(d)

FIGURE 3.74   Calculation of end rotations of a simple beam by the unit-load method. (a) Moment applied at one end. (b) Bending-moment diagram for the applied moment. (c) Unit load applied at end where rotation is to be determined. (d) Bending-moment diagram for the unit load.

(C. H. Norris et al., *Elementary Structural Analysis*, 4th ed., McGraw-Hill, Inc., New York; J. McCormac and R. E. Elling, *Structural Analysis—A Classical and Matrix Approach*, Harper and Row Publishers, New York.)

3.34 METHODS FOR ANALYSIS OF STATICALLY INDETERMINATE SYSTEMS

For a statically indeterminate structure, equations of equilibrium alone are not sufficient to permit analysis (see Art. 3.28). For such systems, additional equations must be derived from requirements ensuring compatibility of deformations. The relationship between stress and strain affects compatibility requirements. In Arts. 3.35 to 3.39, linear elastic behavior is assumed; i.e., in all cases stress is assumed to be directly proportional to strain.

There are two basic approaches for analyzing statically indeterminate structures, force methods and displacement methods. In the **force methods,** forces are chosen as redundants to satisfy equilibrium. They are determined from compatibility conditions (see Art. 3.35). In the **displacement methods,** displacements are chosen as redundants to ensure geometric compatibility. They are also determined from equilibrium equations (see Art. 3.36). In both methods, once the unknown redundants are determined, the structure can be analyzed by statics.

3.35 FORCE METHOD (METHOD OF CONSISTENT DEFLECTIONS)

For analysis of a statically indeterminate structure by the force method, the degree of indeterminacy (number of redundants) n should first be determined (see Art. 3.28). Next, the structure should be reduced to a statically determinate structure by release of n constraints or redundant forces $(X_1, X_2, X_3, \ldots, X_n)$. Equations for determination of the redundants may then be derived from the requirements that equilibrium must be maintained in the reduced structure and deformations should be compatible with those of the original structure.

Displacements $\delta_1, \delta_2, \delta_3, \ldots, \delta_n$ in the reduced structure at the released constraints are calculated for the loads on the original structure. Next, a separate analysis is performed for each released constraint j to determine the displacements at all the released constraints for a unit load applied at j in the direction of the constraint. The displacement f_{ij} at constraint i due to a unit load at released constraint j is called a **flexibility coefficient.**

Next, displacement compatibility at each released constraint is enforced. For any constraint i, the displacement δ_i due to the given loading on the reduced structure and the sum of the displacements $f_{ij}X_j$ in the reduced structure caused by the redundant forces are set equal to known displacement Δ_i of the original structure:

$$\Delta_i = \delta_i + \sum_{j=1}^{n} f_{ij}X_j \qquad i = 1, 2, 3, \ldots, n \qquad (3.133)$$

If the redundant i is a support that has no displacement, then $\Delta_i = 0$. Otherwise, Δ_i will be a known support displacement. With n constraints, Eq. (3.133) provides n equations for solution of the n unknown redundant forces.

As an example, the continuous beam shown in Fig. 3.75a will be analyzed. If axial-force effects are neglected, the beam is indeterminate to the second degree ($n = 2$). Hence two redundants should be chosen for removal to obtain a statically determinate (reduced) structure. For this purpose, the reactions R_B at support B and R_C at support C are selected. Displacements of the reduced structure may then be determined by any of the methods presented in Art. 3.33. Under the loading shown in Fig. 3.75a, the deflections at the redundants are $\delta_B = -5.395$ in and $\delta_C = -20.933$ in (Fig. 3.75b). Application of an upward-acting unit load to the

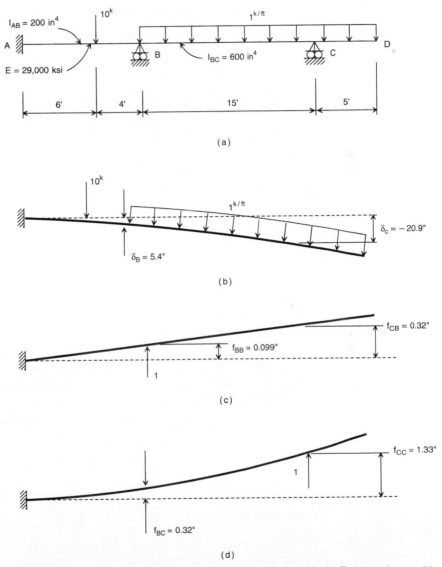

FIGURE 3.75   Analysis of a continuous beam by the force method. (*a*) Two-span beam with concentrated and uniform loads. (*b*) Displacements of beam when supports at *B* and *C* are removed. (*c*) Displacements for unit load at *B*. (*d*) Displacements for unit load at *C*.

reduced beam at B results in deflections $f_{BB} = 0.0993$ in at B and $f_{CB} = 0.3228$ in at C (Fig. 3.75c). Similarly, application of an upward-acting unit load at C results in $f_{BC} = 0.3228$ in at B and $f_{CC} = 1.3283$ in at C (Fig. 3.75d). Since deflections cannot occur at supports B and C, Eq. (3.133) provides two equations for displacement compatibility at these supports:

$$0 = -5.395 + 0.0993R_B + 0.3228R_C$$

$$0 = -20.933 + 0.3228R_B + 1.3283R_C$$

Solution of these simultaneous equations yields $R_B = 14.77$ kips and $R_C = 12.17$ kips. With these two redundants known, equilibrium equations may be used to determine the remaining reactions as well as to draw the shear and moment diagrams (see Art. 3.32).

In the preceding example, in accordance with the reciprocal theorem (Art. 3.22), the flexibility coefficients f_{CB} and f_{BC} are equal. In linear elastic structures, the displacement at constraint i due to a load at constraint j equals the displacement at constraint j when the same load is applied at constraint i; that is, $f_{ij} = f_{ji}$. Use of this relationship can significantly reduce the number of displacement calculations needed in the force method.

The force method also may be applied to statically indeterminate trusses and frames. In all cases, the general approach is the same.

(F. Arbabi, *Structural Analysis and Behavior*, McGraw-Hill, Inc., New York; and J. McCormac and R. E. Elling, *Structural Analysis—A Classical and Matrix Approach*, Harper and Row Publishers, New York.)

3.36 DISPLACEMENT METHODS

For analysis of a statically determinate or indeterminate structure by any of the displacement methods, independent displacements of the joints, or **nodes,** are chosen as the unknowns. If the structure is defined in a three-dimensional, orthogonal coordinate system, each of the three translational and three rotational displacement components for a specific node is called a **degree of freedom.** The displacement associated with each degree of freedom is related to corresponding deformations of members meeting at a node so as to ensure geometric compatibility.

Equilibrium equations relate the unknown displacements $\Delta_1, \Delta_2, \ldots, \Delta_n$ at degrees of freedom $1, 2, \ldots, n$, respectively, to the loads P_i on these degrees of freedom in the form

$$P_1 = k_{11}\Delta_1 + k_{12}\Delta_2 + \cdots + k_{1n}\Delta_n$$

$$P_2 = k_{21}\Delta_1 + k_{22}\Delta_2 + \cdots + k_{2n}\Delta_n$$

$$\vdots$$

$$P_n = k_{n1}\Delta_1 + k_{n2}\Delta_2 + \cdots + k_{nn}\Delta_n$$

or more compactly as

$$P_i = \sum_{j=1}^{n} k_{ij}\Delta_j \qquad \text{for } i = 1, 2, 3, \ldots, n \tag{3.134}$$

Member loads acting between degrees of freedom are converted to equivalent loads acting at these degrees of freedom.

The typical k_{ij} coefficient in Eq. (3.134) is a **stiffness coefficient.** It represents the resulting force (or moment) at point i in the direction of load P_i when a unit displacement at point j in the direction of Δ_j is imposed and all other degrees of freedom are restrained against displacement. P_i is the given concentrated load at degree of freedom i in the direction of Δ_i.

When loads, such as distributed loads, act between nodes, an equivalent force and moment should be determined for these nodes. For example, the nodal forces for one span of a continuous beam are the fixed-end moments and simple-beam reactions, both with signs reversed. **Fixed-end moments** for several beams under various loads are provided in Fig. 3.76. (See also Arts. 3.37, 3.38, and 3.39.)

(F. Arbabi, *Structural Analysis and Behavior*, McGraw-Hill, Inc., New York.)

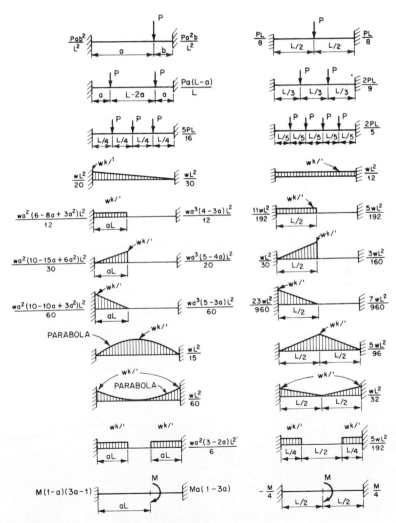

FIGURE 3.76 Fixed-end moments in beams.

3.37 SLOPE-DEFLECTION METHOD

One of several displacement methods for analyzing statically indeterminate structures that resist loads by bending involves use of slope-deflection equations. This method is convenient for analysis of continuous beams and rigid frames in which axial force effects may be neglected. It is not intended for analysis of trusses.

Consider a beam AB (Fig. 3.77a) that is part of a continuous structure. Under loading, the beam develops end moments M_{AB} at A and M_{BA} at B and end rotations θ_A and θ_B. The latter are the angles that the tangents to the deformed neutral axis at ends A and B, respectively, make with the original direction of the axis. (Counterclockwise rotations and moments are assumed positive.) Also, let Δ_{BA} be the displacement of B relative to A (Fig. 3.77b). For small deflections, the rotation of the chord joining A and B may be approximated by $\Phi_{BA} = \Delta_{BA}/L$. The end moments, end rotations, and relative deflection are related by the slope-deflection equations:

$$M_{AB} = \frac{2EI}{L}(2\theta_A + \theta_B - 3\Phi_{BA}) + M_{AB}^F \qquad (3.135a)$$

$$M_{BA} = \frac{2EI}{L}(\theta_A + 2\theta_B - 3\Phi_{BA}) + M_{BA}^F \qquad (3.135b)$$

where E = modulus of elasticity of the material
I = moment of inertia of the beam
L = span

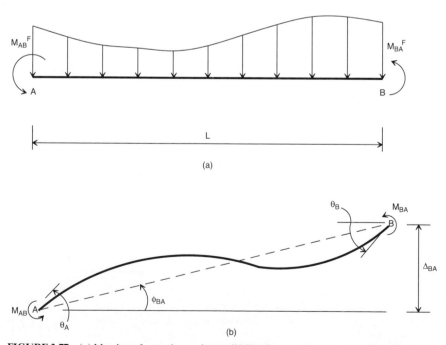

FIGURE 3.77   (a) Member of a continuous beam. (b) Elastic curve of the member for end moment and displacement of an end.

M_{AB}^F = fixed-end moment at A

M_{BA}^F = fixed-end moment at B

Use of these equations for each member in a structure plus equations for equilibrium at the member connections is adequate for determination of member displacements. These displacements can then be substituted into the equations to determine the end moments.

As an example, the beam in Fig. 3.75a will be analyzed by employing the slope-deflection equations [Eqs. (3.135a and b)]. From Fig. 3.76, the fixed-end moments in span AB are

$$M_{AB}^F = \frac{10 \times 6 \times 4^2}{10^2} = 9.60 \text{ ft-kips}$$

$$M_{BA}^F = -\frac{10 \times 4 \times 6^2}{10^2} = -14.40 \text{ ft-kips}$$

The fixed-end moments in BC are

$$M_{BC}^F = 1 \times \frac{15^2}{12} = 18.75 \text{ ft-kips}$$

$$M_{CB}^F = -1 \times \frac{15^2}{12} = -18.75 \text{ ft-kips}$$

The moment at C from the cantilever is $M_{CD} = 12.50$ ft-kips.

If $E = 29,000$ ksi, $I_{AB} = 200$ in^4, and $I_{BC} = 600$ in^4, then $2EI_{AB}/L_{AB} = 8055.6$ ft-kips and $2EI_{BC}/L_{BC} = 16,111.1$ ft-kips. With $\theta_A = 0$, $\Phi_{BA} = 0$, and $\Phi_{CB} = 0$, Eq. (3.135) yields

$$M_{AB} = 8,055.6\theta_B + 9.60 \tag{3.136}$$

$$M_{BA} = 2 \times 8,055.6\theta_B - 14.40 \tag{3.137}$$

$$M_{BC} = 2 \times 16,111.1\theta_B + 16,111.1\theta_C + 18.75 \tag{3.138}$$

$$M_{CB} = 16,111.1\theta_B + 2 \times 16,111.1\theta_C - 18.75 \tag{3.139}$$

Also, equilibrium of joints B and C requires that

$$M_{BA} = -M_{BC} \tag{3.140}$$

$$M_{CB} = -M_{CD} = -12.50 \tag{3.141}$$

Substitution of Eqs. (3.137) and (3.138) in Eq. (3.140) and Eq. (3.139) in Eq. (3.141) gives

$$48,333.4\theta_B + 16,111.1\theta_C = -4.35 \tag{3.142}$$

$$16,111.1\theta_B + 32,222.2\theta_C = 6.25 \tag{3.143}$$

Solution of these equations yields $\theta_B = -1.86 \times 10^{-4}$ and $\theta_C = 2.87 \times 10^{-4}$ radians. Substitution in Eqs. (3.136) to (3.139) gives the end moments: $M_{AB} = 8.1$, $M_{BA} = -17.4$, $M_{BC} = 17.4$, and $M_{CB} = -12.5$ ft-kips. With these moments and switching the signs of moments at the left end of members to be consistent with the sign convention in Art. 3.18, the shear and bending-moment diagrams

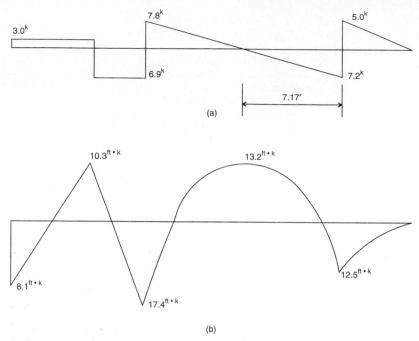

FIGURE 3.78 Shear diagram (*a*) and moment diagram (*b*) for the continuous beam in Fig. 3.75*a*.

shown in Fig. 3.78*a* and *b* can be obtained. This example also demonstrates that a valuable by-product of the displacement method is the calculation of several of the node displacements.

If axial force effects are neglected, the slope-deflection method also can be used to analyze rigid frames.

(J. McCormac and R. E. Elling, *Structural Analysis—A Classical and Matrix Approach*, Harper and Row Publishers, New York.)

3.38 *MOMENT-DISTRIBUTION METHOD*

The moment-distribution method is one of several displacement methods for analyzing continuous beams and rigid frames. Moment distribution, however, provides an alternative to solving the system of simultaneous equations that result with other methods, such as slope deflection. (See Arts. 3.36, 3.37, and 3.39.)

Moment distribution is based on the fact that the bending moment at each end of a member of a continuous frame equals the sum of the fixed-end moments due to the applied loads on the span and the moments produced by rotation of the member ends and of the chord between these ends. Given fixed-end moments, the moment-distribution method determines moments generated when the structure deforms.

Figure 3.79 shows a structure consisting of three members rigidly connected at joint *O* (ends of the members at *O* must rotate the same amount). Supports at *A*, *B*, and *C* are fixed (rotation not permitted). If joint *O* is locked temporarily to prevent rotation, applying a load on member *OA* induces fixed-end moments at

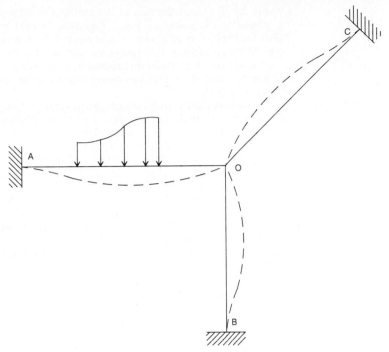

FIGURE 3.79   Straight members rigidly connected at joint O. Dash lines show deformed shape after loading.

A and O. Suppose fixed-end moment $M_{OA}{}^F$ induces a counterclockwise moment on locked joint O. Now, if the joint is released, $M_{OA}{}^F$ rotates it counterclockwise. Bending moments are developed in each member joined at O to balance $M_{OA}{}^F$. Bending moments are also developed at the fixed supports A, B, and C. These moments are said to be carried over from the moments in the ends of the members at O when the joint is released.

The total end moment in each member at O is the algebraic sum of the fixed-end moment before release and the moment in the member at O caused by rotation of the joint, which depends on the relative stiffness of the member. Stiffness of a prismatic fixed-end beam is proportional to EI/L, where E is the modulus of elasticity, I the moment of inertia, and L the span.

When a fixed joint is unlocked, it rotates if the algebraic sum of the bending moments at the joint does not equal zero. The moment that causes the joint to rotate is the **unbalanced moment.** The moments developed at the far ends of each member of the released joint when the joint rotates are **carry-over moments.**

In general, if all joints are locked and then one is released, the amount of unbalanced moment distributed to member i connected to the unlocked joint is determined by the **distribution factor** D_i the ratio of the moment distributed to i to the unbalanced moment. For a prismatic member,

$$D_i = \frac{E_i I_i / L_i}{\sum_{j=1}^{n} E_j I_j / L_j} \qquad (3.144)$$

where $\Sigma_{j=1}^{n} E_j I_j / L_j$ is the sum of the stiffness of all n members, including member i, joined at the unlocked joint. Equation (3.144) indicates that the sum of all distribution factors at a joint should equal 1.0. Members cantilevered from a joint contribute no stiffness and therefore have a distribution factor of zero.

The amount of moment distributed from an unlocked end of a prismatic member to a locked end is ½. This **carry-over factor** can be derived from Eqs. (3.135a and b) with $\theta_A = 0$.

Moments distributed to fixed supports remain at the support; i.e., fixed supports are never unlocked. At a pinned joint (non-moment-resisting support), all the unbalanced moment should be distributed to the pinned end on unlocking the joint. In this case, the distribution factor is 1.0.

To illustrate the method, member end moments will be calculated for the continuous beam shown in Fig. 3.75a. All joints are initially locked. The concentrated load on span AB induces fixed-end moments of 9.60 and -14.40 ft-kips at A and B, respectively (see Art 3.37). The uniform load on BC induces fixed-end moments of 18.75 and -18.75 ft-kips at B and C, respectively. The moment at C from the cantilever CD is 12.50 ft-kips. These values are shown in Fig. 3.80a.

The distribution factors at joints where two or more members are connected are then calculated from Eq. (3.144). With $EI_{AB}/L_{AB} = 200E/120 = 1.67E$ and $EI_{BC}/L_{BC} = 600E/180 = 3.33E$, the distribution factors are $D_{BA} = 1.67E/(1.67E + 3.33E) = 0.33$ and $D_{BC} = 3.33/5.00 = 0.67$. With $EI_{CD}/L_{CD} = 0$ for a cantilevered member, $D_{CB} = 10E/(0 + 10E) = 1.00$ and $D_{CD} = 0.00$.

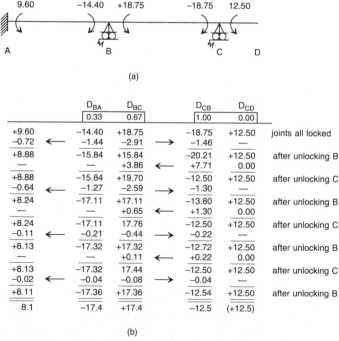

(a)

	D_{BA}	D_{BC}		D_{CB}	D_{CD}	
	0.33	0.67		1.00	0.00	
+9.60	−14.40	+18.75		−18.75	+12.50	joints all locked
−0.72 ←	−1.44	−2.91 →		−1.46	—	
+8.88	−15.84	+15.84		−20.21	+12.50	after unlocking B
—	—	+3.86 ←		+7.71	0.00	
+8.88	−15.84	+19.70		−12.50	+12.50	after unlocking C
−0.64 ←	−1.27	−2.59 →		−1.30	—	
+8.24	−17.11	+17.11		−13.80	+12.50	after unlocking B
—	—	+0.65 ←		+1.30	0.00	
+8.24	−17.11	17.76		−12.50	+12.50	after unlocking C
−0.11 ←	−0.21	−0.44 →		−0.22	—	
+8.13	−17.32	+17.32		−12.72	+12.50	after unlocking B
—	—	+0.11 ←		+0.22	0.00	
+8.13	−17.32	17.44		−12.50	+12.50	after unlocking C
−0.02 ←	−0.04	−0.08 →		−0.04	—	
+8.11	−17.36	+17.36		−12.54	+12.50	after unlocking B
8.1	−17.4	+17.4		−12.5	(+12.5)	

(b)

FIGURE 3.80 (a) Fixed-end moments for beam in Fig. 3.75a. (b) Steps in moment distribution. Fixed-end moments are given in the top line, final moments in the bottom line, in ft-kips.

Joints not at fixed supports are then unlocked one by one. In each case, the unbalanced moments are calculated and distributed to the ends of the members at the unlocked joint according to their distribution factors. The distributed end moments, in turn, are "carried over" to the other end of each member by multiplication of the distributed moment by a carry-over factor of $\frac{1}{2}$. For example, initially unlocking joint B results in an unbalanced moment of $-14.40 + 18.75 = 4.35$ ft-kips. To balance this moment, -4.35 ft-kips is distributed to members BA and BC according to their distribution factors: $M_{BA} = -4.35 D_{BA} = -4.35 \times 0.33 = -1.44$ ft-kips and $M_{BC} = -4.35 D_{BC} = -2.91$ ft-kips. The carry-over moments are $M_{AB} = M_{BA}/2 = -0.72$ and $M_{CB} = M_{BC}/2 = -1.46$. Joint B is then locked, and the resulting moments at each member end are summed: $M_{AB} = 9.60 - 0.72 = 8.88$, $M_{BA} = -14.40 - 1.44 = -15.84$, $M_{BC} = 18.75 - 2.91 = 15.84$, and $M_{CB} = -18.75 - 1.46 = -20.21$ ft-kips. When the step is complete, the moments at the unlocked joint balance, that is, $-M_{BA} = M_{BC}$.

The procedure is then continued by unlocking joint C. After distribution of the unbalanced moments at C and calculation of the carry-over moment to B, the joint is locked, and the process is repeated for joint B. As indicated in Fig. 3.80b, iterations continue until the final end moments at each joint are calculated to within the designer's required tolerance.

There are several variations of the moment-distribution method. This method may be extended to determine moments in rigid frames that are subject to drift, or sidesway.

(C. H. Norris et al., *Elementary Structural Analysis*, 4th ed., McGraw-Hill, Inc., New York; J. McCormac and R. E. Elling, *Structural Analysis—A Classical and Matrix Approach*, Harper and Row Publishers, New York.)

3.39 MATRIX STIFFNESS METHOD

As indicated in Art. 3.36, displacement methods for analyzing structures relate force components acting at the joints, or **nodes,** to the corresponding displacement components at these joints through a set of equilibrium equations. In matrix notation, this set of equations [Eq. (3.134)] is represented by

$$\mathbf{P} = \mathbf{K}\Delta \tag{3.145}$$

where $\mathbf{P}$ = column vector of nodal external load components $\{P_1, P_2, \ldots, P_n\}^T$
$\quad\quad \mathbf{K}$ = stiffness matrix for the structure
$\quad\quad \Delta$ = column vector of nodal displacement components: $\{\Delta_1, \Delta_2, \ldots, \Delta_n\}^T$
$\quad\quad n$ = total number of degrees of freedom
$\quad\quad T$ = transpose of a matrix (columns and rows interchanged)

A typical element k_{ij} of $\mathbf{K}$ gives the load at nodal component i in the direction of load component P_i that produces a unit displacement at nodal component j in the direction of displacement component Δ_j. Based on the reciprocal theorem (see Art. 3.25), the square matrix K is symmetrical, that is, $k_{ij} = k_{ji}$.

For a specific structure, Eq. (3.145) is generated by first writing equations of equilibrium at each node. Each force and moment component at a specific node must be balanced by the sum of member forces acting at that joint. For a two-dimensional frame defined in the xy plane, force and moment components per node include F_x, F_y, and M_z. In a three-dimensional frame, there are six force and moment components per node: F_x, F_y, F_z, M_x, M_y, and M_z.

From member force-displacement relationships similar to Eq. (3.135), member force components in the equations of equilibrium are replaced with equivalent displacement relationships. The resulting system of equilibrium equations can be put in the form of Eq. (3.145).

Nodal boundary conditions are then incorporated into Eq. (3.145). If, for example, there are a total of n degrees of freedom, of which m degrees of freedom are restrained from displacement, there would be $n - m$ unknown displacement components and m unknown restrained force components or reactions. Hence a total of $(n - m) + m = n$ unknown displacements and reactions could be determined by the system of n equations provided with Eq. (3.145).

Once all displacement components are known, member forces may be determined from the member force-displacement relationships.

For a prismatic member subjected to the end forces and moments shown in Fig. 3.81a, displacements at the ends of the member are related to these member forces by the matrix expression

$$
\begin{Bmatrix} F'_{xi} \\ F'_{yi} \\ M'_{zi} \\ F'_{xj} \\ F'_{yj} \\ M'_{zj} \end{Bmatrix} = \frac{E}{L^3} \begin{bmatrix} AL^2 & 0 & 0 & -AL^2 & 0 & 0 \\ 0 & 12I & 6IL & 0 & -12I & 6IL \\ 0 & 6IL & 4IL^2 & 0 & -6IL & 2IL^2 \\ -AL^2 & 0 & 0 & AL^2 & 0 & 0 \\ 0 & -12I & -6IL & 0 & 12I & -6IL \\ 0 & 6IL & 2IL^2 & 0 & -6IL & 4IL^2 \end{bmatrix} \begin{Bmatrix} \Delta'_{xi} \\ \Delta'_{yi} \\ \theta'_{zi} \\ \Delta'_{xj} \\ \Delta'_{yj} \\ \theta'_{zj} \end{Bmatrix} \qquad (3.146)
$$

where L = length of member (distance between i and j)
$\quad\quad\;\; E$ = modulus of elasticity
$\quad\quad\;\; A$ = cross-sectional area of member
$\quad\quad\;\; I$ = moment of inertia about neutral axis in bending

In matrix notation, Eq. (3.146) for the ith member of a structure can be written

$$\mathbf{S}'_i = \mathbf{k}'_i \boldsymbol{\delta}'_i \qquad (3.147)$$

where $\mathbf{S}'_i$ = vector forces and moments acting at the ends of member i
$\quad\quad\;\; \mathbf{k}'_i$ = stiffness matrix for member i
$\quad\quad\;\; \boldsymbol{\delta}'_i$ = vector of deformations at the ends of member i

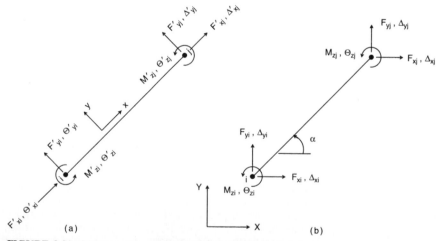

FIGURE 3.81 Member of a continuous structure. (a) Forces at the ends of the member and deformations are given with respect to the member local coordinate system; (b) with respect to the structure global coordinate system.

The force-displacement relationships provided by Eqs. (3.146) and (3.147) are based on the member's xy **local coordinate system** (Fig. 3.81a). If this coordinate system is not aligned with the structure's XY **global coordinate system,** these equations must be modified or transformed. After transformation of Eq. (3.147) to the global coordinate system, it would be given by

$$\mathbf{S}_i = \mathbf{k}_i \boldsymbol{\delta}_i \qquad (3.148)$$

where $\mathbf{S}_i = \boldsymbol{\Gamma}_i^T \mathbf{S}_i'$ = force factor for member i
$\mathbf{k}_i = \boldsymbol{\Gamma}_i^T \mathbf{k}_i' \boldsymbol{\Gamma}_i$ = member stiffness matrix
$\boldsymbol{\delta}_i = \boldsymbol{\Gamma}_i^T \boldsymbol{\delta}_i'$ = displacement vector for member i
$\boldsymbol{\Gamma}_i$ = transformation matrix for member i

For the member shown in Fig. 3.81b, which is defined in two-dimensional space, the **transformation matrix** is

$$\boldsymbol{\Gamma} = \begin{bmatrix} \cos\alpha & \sin\alpha & 0 & 0 & 0 & 0 \\ -\sin\alpha & \cos\alpha & 0 & 0 & 0 & 0 \\ 0 & 0 & 1 & 0 & 0 & 0 \\ 0 & 0 & 0 & \cos\alpha & \sin\alpha & 0 \\ 0 & 0 & 0 & -\sin\alpha & \cos\alpha & 0 \\ 0 & 0 & 0 & 0 & 0 & 1 \end{bmatrix} \qquad (3.149)$$

where α = angle between the structure's global X axis and the member's local x axis.

Example. To demonstrate the matrix displacement method, the rigid frame shown in Fig. 3.82a will be analyzed. The two-dimensional frame has three joints, or nodes, A, B, and C, and hence a total of nine possible degrees of freedom (Fig. 3.82b). The displacements at node A are not restrained. Nodes B and C have zero displacement. For both AB and AC, modulus of elasticity E = 29,000 ksi, area A = 1 in^2, and moment of inertia I = 10 in^4. Forces will be computed in kips; moments, in kip-in.

At each degree of freedom, the external forces must be balanced by the member forces. This requirement provides the following equations of equilibrium with reference to the global coordinate system:

At the free degrees of freedom at node A, $\Sigma F_{xA} = 0$, $\Sigma F_{yA} = 0$, and $\Sigma M_{zA} = 0$:

$$10 = F_{xAB} + F_{xAC} \qquad (3.150a)$$

$$-200 = F_{yAB} + F_{yAC} \qquad (3.150b)$$

$$0 = M_{zAB} + M_{zAC} \qquad (3.150c)$$

At the restrained degrees of freedom at node B, $\Sigma F_{xB} = 0$, $\Sigma F_{yB} = 0$, and $\Sigma M_{zB} = 0$:

$$R_{xB} - F_{xBA} = 0 \qquad (3.151a)$$

$$R_{yB} - F_{yBA} = 0 \qquad (3.151b)$$

$$M_{zB} - M_{zBA} = 0 \qquad (3.151c)$$

At the restrained degrees of freedom at node C, $\Sigma F_{xC} = 0$, $\Sigma F_{yC} = 0$, and $\Sigma M_{zC} = 0$:

$$R_{xC} - F_{xCA} = 0 \qquad (3.152a)$$

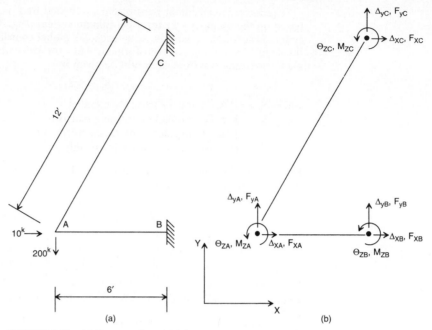

FIGURE 3.82 (*a*) Two-member rigid frame, with modulus of elasticity $E = 29,000$ ksi, area $A = 1$ in², and moment of inertia $I = 10$ in⁴. (*b*) Degrees of freedom at nodes.

$$R_{yC} - F_{yCA} = 0 \qquad (3.152b)$$

$$M_{zC} - M_{zCA} = 0 \qquad (3.152c)$$

where subscripts identify the direction, member, and degree of freedom.

Member force components in these equations are then replaced by equivalent displacement relationships with the use of Eq. (3.148). With reference to the global coordinates, these relationships are as follows:

For member AB with $\alpha = 0°$, $\mathbf{S}_{AB} = \mathbf{\Gamma}_{AB}^T \mathbf{k}'_{AB} \mathbf{\Gamma}_{AB} \mathbf{\delta}_{AB}$:

$$
\begin{Bmatrix} F_{xAB} \\ F_{yAB} \\ M_{zAB} \\ F_{xBA} \\ F_{yBA} \\ M_{zBA} \end{Bmatrix}
=
\begin{bmatrix}
402.8 & 0 & 0 & -402.8 & 0 & 0 \\
0 & 9.324 & 335.6 & 0 & -9.324 & 335.6 \\
0 & 335.6 & 16111 & 0 & -335.6 & 8056 \\
-402.8 & 0 & 0 & 402.8 & 0 & 0 \\
0 & -9.324 & -335.6 & 0 & 9.324 & -335.6 \\
0 & 335.6 & 8056 & 0 & -335.6 & 16111
\end{bmatrix}
\begin{Bmatrix} \Delta_{xA} \\ \Delta_{yA} \\ \Theta_{zA} \\ \Delta_{xB} \\ \Delta_{yB} \\ \Theta_{zB} \end{Bmatrix}
$$

$$(3.153)$$

For member AC with $\alpha = 60°$, $\mathbf{S}_{AC} = \mathbf{\Gamma}_{AC}^T \mathbf{k}'_{AC} \mathbf{\Gamma}_{AC} \mathbf{\delta}_{AC}$:

$$
\begin{Bmatrix} F_{xAC} \\ F_{yAC} \\ M_{zAC} \\ F_{xCA} \\ F_{yCA} \\ M_{zCA} \end{Bmatrix}
=
\begin{bmatrix}
51.22 & 86.70 & -72.67 & -51.22 & -86.70 & -72.67 \\
86.70 & 151.3 & 41.96 & -86.70 & -151.3 & 41.96 \\
-72.67 & 41.96 & 8056 & 72.67 & -41.96 & 4028 \\
-51.22 & -86.70 & 72.67 & 51.22 & 86.70 & 72.67 \\
-86.70 & -151.3 & -41.96 & 86.70 & 151.3 & -41.96 \\
-72.67 & 41.96 & 4028 & 72.67 & -41.96 & 8056
\end{bmatrix}
\begin{Bmatrix} \Delta_{xA} \\ \Delta_{yA} \\ \Theta_{zA} \\ \Delta_{xC} \\ \Delta_{yC} \\ \Theta_{zC} \end{Bmatrix}
$$

$$(3.154)$$

Incorporating the support conditions $\Delta_{xB} = \Delta_{yB} = \Theta_{zB} = \Delta_{xC} = \Delta_{yC} = \Theta_{zC} = 0$ into Eqs. (3.153) and (3.154) and then substituting the resulting displacement relationships for the member forces in Eqs. (3.150) to (3.152) yields

$$
\begin{Bmatrix} 10 \\ -200 \\ 0 \\ R_{xB} \\ R_{yB} \\ M_{zB} \\ R_{xC} \\ R_{yC} \\ M_{zC} \end{Bmatrix} = \begin{bmatrix} 402.8 + 51.22 & 0 + 86.70 & 0 - 72.67 \\ 0 + 86.70 & 9.324 + 151.3 & 335.6 + 41.96 \\ 0 - 72.67 & 335.6 + 41.96 & 16111 + 8056 \\ -402.8 & 0 & 0 \\ 0 & -9.324 & -335.6 \\ 0 & 335.6 & 8056 \\ -51.22 & -86.70 & 72.67 \\ -86.70 & -151.3 & -41.96 \\ -72.67 & 41.96 & 4028 \end{bmatrix} \begin{Bmatrix} \Delta_{xA} \\ \Delta_{yA} \\ \Theta_{zA} \end{Bmatrix}
$$

$$(3.155)$$

Equation (3.155) contains nine equations with nine unknowns. The first three equations may be used to solve the displacements at the free degrees of freedom $\Delta_f = \mathbf{K}_{\bar{f}f}^{-1}\mathbf{P}_f$:

$$
\begin{Bmatrix} \Delta_{xA} \\ \Delta_{yA} \\ \Theta_{zA} \end{Bmatrix} = \begin{bmatrix} 454.0 & 86.70 & -72.67 \\ 86.70 & 160.6 & 377.6 \\ -72.67 & 377.6 & 24167 \end{bmatrix}^{-1} \begin{Bmatrix} 10 \\ -200 \\ 0 \end{Bmatrix} = \begin{Bmatrix} 0.3058 \\ -1.466 \\ 0.0238 \end{Bmatrix} \quad (3.156a)
$$

These displacements may then be incorporated into the bottom six equations of Eq. (3.155) to solve for the unknown reactions at the restrained nodes, $\mathbf{P}_s = \mathbf{K}_{sf}\Delta_f$:

$$
\begin{Bmatrix} R_{xB} \\ R_{yB} \\ M_{zB} \\ R_{xC} \\ R_{yC} \\ M_{zC} \end{Bmatrix} = \begin{bmatrix} -402.8 & 0 & 0 \\ 0 & -9.324 & -335.6 \\ 0 & 335.6 & 8056 \\ -51.22 & -86.70 & 72.67 \\ -86.70 & -151.3 & -41.96 \\ -72.67 & 41.96 & 4028 \end{bmatrix} \begin{Bmatrix} 0.3058 \\ -1.466 \\ 0.0238 \end{Bmatrix} = \begin{Bmatrix} -123.2 \\ 5.67 \\ -300.1 \\ 113.2 \\ 194.3 \\ 12.2 \end{Bmatrix} \quad (3.156b)
$$

With all displacement components now known, member end forces may be calculated. Displacement components that correspond to the ends of a member should be transformed from the global coordinate system to the member's local coordinate system, $\boldsymbol{\delta}' = \boldsymbol{\Gamma}\boldsymbol{\delta}$.

For member AB with $\alpha = 0°$:

$$
\begin{Bmatrix} \Delta'_{xA} \\ \Delta'_{yA} \\ \Theta'_{zA} \\ \Delta'_{xB} \\ \Delta'_{yB} \\ \Theta'_{zB} \end{Bmatrix} = \begin{bmatrix} 1 & 0 & 0 & 0 & 0 & 0 \\ 0 & 1 & 0 & 0 & 0 & 0 \\ 0 & 0 & 1 & 0 & 0 & 0 \\ 0 & 0 & 0 & 1 & 0 & 0 \\ 0 & 0 & 0 & 0 & 1 & 0 \\ 0 & 0 & 0 & 0 & 0 & 1 \end{bmatrix} \begin{Bmatrix} 0.3058 \\ -1.466 \\ 0.0238 \\ 0 \\ 0 \\ 0 \end{Bmatrix} = \begin{Bmatrix} 0.3058 \\ -1.466 \\ 0.0238 \\ 0 \\ 0 \\ 0 \end{Bmatrix} \quad (3.157a)
$$

For member AC with $\alpha = 60°$:

$$
\begin{Bmatrix} \Delta'_{xA} \\ \Delta'_{yA} \\ \Theta'_{zA} \\ \Delta'_{xC} \\ \Delta'_{yC} \\ \Theta'_{zC} \end{Bmatrix} = \begin{bmatrix} 0.5 & 0.866 & 0 & 0 & 0 & 0 \\ -0.866 & 0.5 & 0 & 0 & 0 & 0 \\ 0 & 0 & 1 & 0 & 0 & 0 \\ 0 & 0 & 0 & 0.5 & 0.866 & 0 \\ 0 & 0 & 0 & -0.866 & 0.5 & 0 \\ 0 & 0 & 0 & 0 & 0 & 1 \end{bmatrix} \begin{Bmatrix} 0.3058 \\ -1.466 \\ 0.0238 \\ 0 \\ 0 \\ 0 \end{Bmatrix} = \begin{Bmatrix} -1.1117 \\ -0.9978 \\ 0.0238 \\ 0 \\ 0 \\ 0 \end{Bmatrix}
$$

$$(3.157b)$$

Member end forces are then obtained by multiplying the member stiffness matrix by the member end displacements, both with reference to the member local coordinate system, $S' = k'\delta'$.

For member AB in the local coordinate system:

$$
\begin{Bmatrix} F'_{xAB} \\ F'_{yAB} \\ M'_{zAB} \\ F'_{xBA} \\ F'_{yBA} \\ M'_{zBA} \end{Bmatrix} =
\begin{bmatrix}
402.8 & 0 & 0 & -402.8 & 0 & 0 \\
0 & 9.324 & 335.6 & 0 & -9.324 & 335.6 \\
0 & 335.6 & 16111 & 0 & -335.6 & 8056 \\
-402.8 & 0 & 0 & 402.8 & 0 & 0 \\
0 & -9.324 & -335.6 & 0 & 9.324 & -335.6 \\
0 & 335.6 & 8056 & 0 & -335.6 & 16111
\end{bmatrix}
$$

$$
\times \begin{Bmatrix} 0.3058 \\ -1.466 \\ 0.0238 \\ 0 \\ 0 \\ 0 \end{Bmatrix} = \begin{Bmatrix} 123.2 \\ -5.67 \\ -108.2 \\ -123.2 \\ 5.67 \\ -300.1 \end{Bmatrix} \quad (3.158)
$$

For member AC in the local coordinate system:

$$
\begin{Bmatrix} F'_{xAC} \\ F'_{yAC} \\ M'_{zAC} \\ F'_{xCA} \\ F'_{yCA} \\ M'_{zCA} \end{Bmatrix} =
\begin{bmatrix}
201.4 & 0 & 0 & -201.4 & 0 & 0 \\
0 & 1.165 & 83.91 & 0 & -1.165 & 83.91 \\
0 & 83.91 & 8056 & 0 & -83.91 & 4028 \\
-201.4 & 0 & 0 & 201.4 & 0 & 0 \\
0 & -1.165 & -83.91 & 0 & 1.165 & -83.91 \\
0 & 83.91 & 4028 & 0 & -83.91 & 8056
\end{bmatrix}
$$

$$
\times \begin{Bmatrix} -1.1117 \\ -0.9978 \\ 0.0238 \\ 0 \\ 0 \\ 0 \end{Bmatrix} = \begin{Bmatrix} -224.9 \\ 0.836 \\ 108.2 \\ 224.9 \\ -0.836 \\ 12.2 \end{Bmatrix} \quad (3.159)
$$

At this point all displacements, member forces, and reaction components have been determined.

The matrix displacement method can be used to analyze both determinate and indeterminate frames, trusses, and beams. Because the method is based primarily on manipulating matrices, it is employed in most structural-analysis computer programs. In the same context, these programs can handle substantial amounts of data, which enables analysis of large and often complex structures.

(W. McGuire and R. H. Gallagher, *Matrix Structural Analysis*, John Wiley & Sons, Inc., New York; D. L. Logan, *A First Course in the Finite Element Method*, PWS-Kent Publishing, Boston, Mass.)

3.40 INFLUENCE LINES

In studies of the variation of the effects of a moving load, such as a reaction, shear, bending moment, or stress, at a given point in a structure, use of diagrams called **influence lines** is helpful. An influence line is a diagram showing the variation of an effect as a unit load moves over a structure.

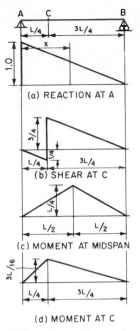

FIGURE 3.83 Influence diagrams for a simple beam.

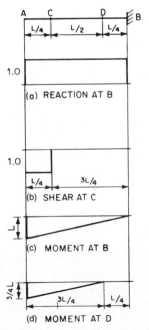

FIGURE 3.84 Influence diagrams for a cantilever.

An influence line is constructed by plotting the position of the unit load as the abscissa and as the ordinate at that position, to some scale, the value of the effect being studied. For example, Fig. 3.83a shows the influence line for reaction A in simple-beam AB. The sloping line indicates that when the unit load is at A, the reaction at A is 1.0. When the load is at B, the reaction at A is zero. When the unit load is at midspan, the reaction at A is 0.5. In general, when the load moves from B toward A, the reaction at A increases linearly: $R_A = (L - x)/L$, where x is the distance from A to the position of the unit load.

Figure 3.83b shows the influence line for shear at the quarter point C. The sloping lines indicate that when the unit load is at support A or B, the shear at C is zero. When the unit load is a small distance to the left of C, the shear at C is −0.25; when the unit load is a small distance to the right of C, the shear at C is 0.75. The influence line for shear is linear on each side of C.

Figures 3.83c and d show the influence lines for bending moment at midspan and quarter point, respectively. Figures 3.84 and 3.85 give influence lines for a cantilever and a simple beam with an overhang.

Influence lines can be used to calculate reactions, shears, bending moments, and other effects due to fixed and moving loads. For example, Fig. 3.86a shows a simply supported beam of 60-ft span subjected to a dead load $w = 1.0$ kip per ft and a live load consisting of three concentrated loads. The reaction at A due to the dead load equals the product of the area under the influence line for the reaction at A (Fig. 3.86b) and the uniform load w. The maximum reaction at A due to the live loads may be obtained by placing the concentrated loads as shown in Fig. 3.86b and equals the sum of the products of each concentrated load and the ordinate of the influence line at the location of the load. The sum of the dead-load reaction and the maximum live-load reaction there-

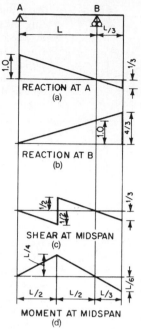

REACTION AT A
(a)

REACTION AT B
(b)

SHEAR AT MIDSPAN
(c)

MOMENT AT MIDSPAN
(d)

FIGURE 3.85 Influence diagrams for a beam with overhang.

fore is

$$R_A = \tfrac{1}{2} \times 1.0 \times 60 \times 1.0$$
$$+ 16 \times 1.0 + 16 \times 0.767$$
$$+ 4 \times 0.533 = 60.4 \text{ kips}$$

Figure 3.86c is the influence diagram for midspan bending moment with a maximum ordinate $L/4 = {}^{60}\!/_4 = 15$. Figure 3.86c also shows the influence diagram with the live loads positioned for maximum moment at midspan. The dead-load moment at midspan is the product of w and the area under the influence line. The midspan live-load moment equals the sum of the products of each live load and the ordinate at the location of each load. The sum of the dead-load moment and the maximum live-load moment equals

$$M = \tfrac{1}{2} \times 15 \times 60 \times 1.0 + 16 \times 15$$
$$+ 16 \times 8 + 4 \times 8 = 850 \text{ ft-kips}$$

An important consequence of the reciprocal theorem presented in Art. 3.25 is the **Mueller-Breslau principle:** The influence line of a certain effect is to some scale the deflected shape of the structure when that effect acts.

The effect, for example, may be a reaction, shear, moment, or deflection at a point. This principle is used extensively in obtaining influence lines for statically indeterminate structures (see Art. 3.28).

Figure 3.87a shows the influence line for reaction at support B for a two-span continuous beam. To obtain this influence line, the support at B is replaced by a

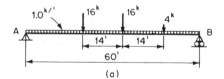

(a)

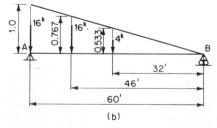

(b)

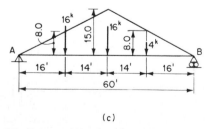

(c)

FIGURE 3.86 Determination for moving loads on a simple beam (*a*) of maximum end reaction (*b*) and maximum midspan moment (*c*) from influence diagrams.

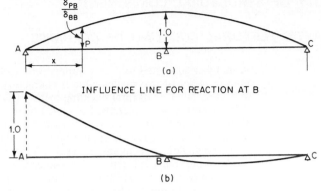

INFLUENCE LINE FOR REACTION AT B

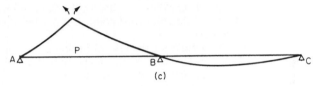

(b)

INFLUENCE LINE FOR REACTION AT A

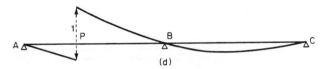

INFLUENCE LINE FOR MOMENT AT P

INFLUENCE LINE FOR SHEAR AT P

FIGURE 3.87   Influence lines for a two-span continuous beam.

unit upward-concentrated load. The deflected shape of the beam is the influence line of the reaction at point B to some scale. To show this, let δ_{BP} be the deflection at B due to a unit load at any point P when the support at B is removed, and let δ_{BB} be the deflection at B due to a unit load at B. Since, actually, reaction R_B prevents deflection at B, $R_B\delta_{BB} - \delta_{BP} = 0$. Thus $R_B = \delta_{BP}/\delta_{BB}$. By Eq. (3.124), however, $\delta_{BP} = \delta_{PB}$. Hence

$$R_B = \frac{\delta_{BP}}{\delta_{BB}} = \frac{\delta_{PB}}{\delta_{BB}} \tag{3.160}$$

Since δ_{BB} is constant, R_B is proportional to δ_{PB}, which depends on the position of the unit load. Hence the influence line for a reaction can be obtained from the deflection curve resulting from replacement of the support by a unit load. The magnitude of the reaction may be obtained by dividing each ordinate of the deflection curve by the displacement of the support due to a unit load applied there.

Similarly, influence lines may be obtained for reaction at A and moment and shear at P by the Mueller-Breslau principle, as shown in Figs. 3.87b, c, and d, respectively.

(C. H. Norris et al., *Elementary Structural Analysis*, 4th ed.; and F. Arbabi, *Structural Analysis and Behavior*, McGraw-Hill, Inc., New York.)

INSTABILITY OF STRUCTURAL COMPONENTS

3.41   ELASTIC FLEXURAL BUCKLING OF COLUMNS

A member subjected to pure compression, such as a column, can fail under axial load in either of two modes. One is characterized by excessive axial deformation and the second by **flexural buckling** or excessive lateral deformation.

For short, stocky columns, Eq. (3.48) relates the axial load P to the compressive stress f. After the stress exceeds the yield point of the material, the column begins to fail. Its load capacity is limited by the strength of the material.

In long, slender columns, however, failure may take place by buckling. This mode of instability is often sudden and can occur when the axial load in a column reaches a certain critical value. In many cases, the stress in the column may never reach the yield point. The load capacity of slender columns is not limited by the strength of the material but rather by the stiffness of the member.

Elastic buckling is a state of lateral instability that occurs while the material is stressed below the yield point. It is of special importance in structures with slender members.

A formula for the critical buckling load for pin-ended columns was derived by Euler in 1757 and is still in use. For the buckled shape under axial load P for a pin-ended column of constant cross section (Fig. 3.88a), Euler's column formula can be derived as follows:

With coordinate axes chosen as shown in Fig. 3.88b, moment equilibrium about one end of the column requires

$$M(x) + Py(x) = 0 \tag{3.161}$$

where $M(x)$ = bending moment at distance x from one end of the column
$y(x)$ = deflection of the column at distance x

Substitution of the moment-curvature relationship [Eq. (3.79)] into Eq. (3.161) gives

$$EI \frac{d^2y}{dx^2} + Py(x) = 0 \tag{3.162}$$

where E = modulus of elasticity of the material
I = moment of inertia of the cross section about the bending axis

The solution to this differential equation is

$$y(x) = A \cos \lambda x + B \sin \lambda x \tag{3.163}$$

where $\lambda = \sqrt{P/EI}$
A, B = unknown constants of integration

Substitution of the boundary condition $y(0) = 0$ into Eq. (3.163) indicates that $A = 0$. The additional boundary condition $y(L) = 0$ indicates that

$$B \sin \lambda L = 0 \tag{3.164}$$

where L is the length of the column. Equation (3.164) is often referred to as a **transcendental equation.** It indicates that either $B = 0$, which would be a trivial

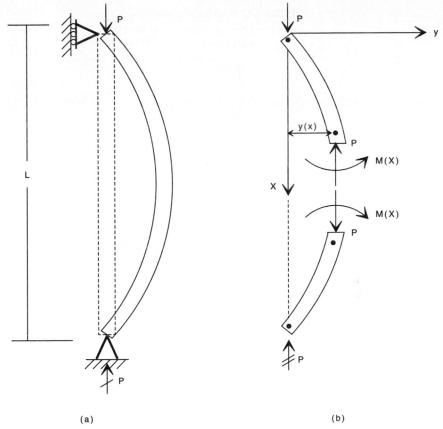

FIGURE 3.88 Buckling of a pin-ended column under axial load. (*b*) Internal forces hold the column in equilibrium.

solution, or that λL must equal some multiple of π. The latter relationship provides the minimum critical value of P:

$$\lambda L = \pi \qquad P = \frac{\pi^2 EI}{L^2} \tag{3.165}$$

This is the **Euler formula** for pin-ended columns. On substitution of Ar^2 for I, where A is the cross-sectional area and r the radius of gyration, Eq. (3.165) becomes

$$P = \frac{\pi^2 EA}{(L/r)^2} \tag{3.166}$$

L/r is called the **slenderness ratio** of the column.

Euler's formula applies only for columns that are perfectly straight, have a uniform cross section made of a linear elastic material, have end supports that are ideal pins, and are concentrically loaded.

Equations (3.165) and (3.166) may be modified to approximate the critical buckling load of columns that do not have ideal pins at the ends. Table 3.4 illustrates some ideal end conditions for slender columns and corresponding critical buckling

TABLE 3.4 Buckling Formulas for Columns

Type of column	Effective length	Critical buckling load
	L	$\dfrac{\pi^2 EI}{L^2}$
	$L/2$	$\dfrac{4\pi^2 EI}{L^2}$
	$\approx 0.7L$	$\approx \dfrac{2\pi^2 EI}{L^2}$
	$2L$	$\dfrac{\pi^2 EI}{4L^2}$

loads. It indicates that elastic critical buckling loads may be obtained for all cases by substituting an **effective length** KL for the length L of the pinned column assumed for the derivation of Eq. (3.166):

$$P = \frac{\pi^2 EA}{(KL/r)^2} \qquad (3.167)$$

Equation (3.167) also indicates that a column may buckle about either the section's major or minor axis depending on which has the greater slenderness ratio KL/r.

In some cases of columns with open sections, such as a cruciform section, the controlling buckling mode may be one of twisting instead of lateral deformation. If the warping rigidity of the section is negligible, **torsional buckling** in a pin-ended column will occur at an axial load of

$$P = \frac{GJA}{I_\rho} \qquad (3.168)$$

where G = shear modulus of elasticity
$\quad\quad\ J$ = torsional constant
$\quad\quad\ A$ = cross-sectional area
$\quad\quad\ I_\rho$ = polar moment of inertia = $I_x + I_y$

If the section possesses a significant amount of warping rigidity, the axial buckling load is increased to

$$P = \frac{A}{I_\rho}\left(GJ + \frac{\pi^2 EC_w}{L^2}\right) \qquad (3.169)$$

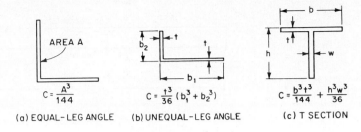

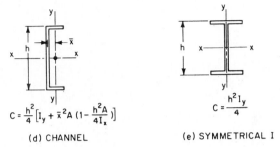

FIGURE 3.89 Torsion-bending constants for torsional buckling. $A =$ cross-sectional area; $I_x =$ moment of inertia about $x–x$ axis; $I_y =$ moment of inertia about $y–y$ axis. *(After F. Bleich, Buckling Strength of Metal Structures, McGraw-Hill Inc., New York.)*

where C_w is the warping constant, a function of cross-sectional shape and dimensions (see Fig. 3.89).

(S. P. Timoshenko and J. M. Gere, *Theory of Elastic Stability*, and F. Bleich, *Buckling Strength of Metal Structures*, McGraw-Hill, Inc., New York; T. V. Galambos, *Guide to Stability of Design of Metal Structures*, John Wiley & Sons, Inc, New York; W. McGuire, *Steel Structures*, Prentice-Hall, Inc., Englewood Cliffs, N.J.)

3.42 *ELASTIC LATERAL BUCKLING OF BEAMS*

Bending of the beam shown in Fig. 3.90*a* produces compressive stresses within the upper portion of the beam cross section and tensile stresses in the lower portion. Similar to the behavior of a column (Art. 3.41), a beam, although the compressive stresses may be well within the elastic range, can undergo lateral buckling failure. Unlike a column, however, the beam is also subjected to tension, which tends to restrain the member from lateral translation. Hence, when **lateral buckling** of the beam occurs, it is through a combination of twisting and out-of-plane bending (Fig. 3.90*b*).

For a simply supported beam of rectangular cross section subjected to uniform bending, buckling occurs at the critical bending moment

$$M_{cr} = \frac{\pi}{L} \sqrt{EI_y GJ} \qquad (3.170)$$

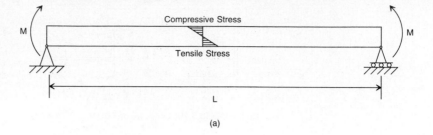

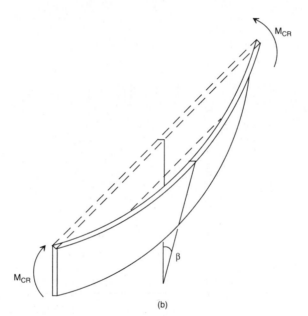

FIGURE 3.90 (*a*) Simple beam subjected to equal end moments. (*b*) Elastic lateral buckling of the beam.

where L = unbraced length of the member
$\quad\quad E$ = modulus of elasticity
$\quad\quad I_y$ = moment of inertial about minor axis
$\quad\quad G$ = shear modulus of elasticity
$\quad\quad J$ = torsional constant

As indicated in Eq. (3.170), the critical moment is proportional to both the lateral bending stiffness EI_y/L and the torsional stiffness of the member GJ/L.

For the case of an open section, such as a wide-flange or I-beam section, warping rigidity can provide additional torsional stiffness. Buckling of a simply supported beam of open cross section subjected to uniform bending occurs at the critical bending moment

$$M_{cr} = \frac{\pi}{L} \sqrt{EI_y \left(GJ + EC_w \frac{\pi^2}{L^2} \right)} \qquad (3.171)$$

where C_w is the warping constant, a function of cross-sectional shape and dimensions (see Fig. 3.89).

In Eq. (3.170) and (3.171), the distribution of bending moment is assumed to be uniform. For the case of a nonuniform bending-moment gradient, buckling often occurs at a larger critical moment. Approximations of this critical bending moment M'_{cr} may be obtained by multiplying M_{cr} given by Eq. (3.170) or (3.171) by an amplification factor:

$$M'_{cr} = C_b M_{cr} \qquad (3.172)$$

where $C_b = 1.75 + 1.05(M_1/M_2) + (M_1/M_2)^2 \leq 2.3$. M_1 is the smaller and M_2 is the larger end moment in the unbraced length of the beam. M_1/M_2 is positive when the moments cause reverse curvature and negative when they cause single curvature.

C_b equals 1.0 for unbraced cantilevers and for members where the moment within a significant portion of the unbraced segment is greater than or equal to the larger of the segment end moments.

(S. P. Timoshenko and J. M. Gere, *Theory of Elastic Stability*, and F. Bleich, *Buckling Strength of Metal Structures*, McGraw-Hill, Inc., New York; T. V. Galambos, *Guide to Stability of Design of Metal Structures*, John Wiley & Sons, Inc., New York; W. McGuire, *Steel Structures*, Prentice-Hall, Inc., Englewood Cliffs, N.J.; *Load and Resistance Factor Design Specification for Structural Steel Buildings*, American Institute of Steel Construction, Chicago, Ill.)

3.43 ELASTIC FLEXURAL BUCKLING OF FRAMES

In Arts. 3.41 and 3.42, elastic instabilities of isolated columns and beams are discussed. Most structural members, however, are part of a structural system where the ends of the members are restrained by other members. In these cases, the instability of the system governs the critical buckling loads on the members. It is therefore important that frame behavior be incorporated into stability analyses. For details of such analyses, see T. V. Galambos, *Guide to Stability of Design of Metal Structures*, John Wiley & Sons, Inc, New York; S. Timoshenko and J. M. Gere, *Theory of Elastic Stability*, and F. Bleich, *Buckling Strength of Metal Structures*, McGraw-Hill, Inc., New York.

3.44 LOCAL BUCKLING

Buckling may sometimes occur in the form of wrinkles in thin elements such as webs, flanges, cover plates, and other parts that make up a section. This phenomenon is called **local buckling.**

The critical buckling stress in rectangular plates with various types of edge support and edge loading in the plane of the plates is given by

$$f_{cr} = k \frac{\pi^2 E}{12(1 - \nu^2)(b/t)^2} \qquad (3.173)$$

TABLE 3.5 Values of k for Buckling Stress in Thin Plates

$\frac{a}{b}$	All edges clamped — Case 1	All edges simply supported — Case 2	Clamped edges — Case 3	Clamped edges — Case 4
0.4	28.3	8.4		9.4
0.6	15.2	5.1	13.4	7.1
0.8	11.3	4.2	8.7	7.3
1.0	10.1	4.0	6.7	7.7
1.2	9.4	4.1	5.8	7.1
1.4	8.7	4.5	5.5	7.0
1.6	8.2	4.2	5.3	7.3
1.8	8.1	4.0	5.2	7.2
2.0	7.9	4.0	4.9	7.0
2.5	7.6	4.1	4.5	7.1
3.0	7.4	4.0	4.4	7.1
3.5	7.3	4.1	4.3	7.0
4.0	7.2	4.0	4.2	7.0
∞	7.0	4.0	4.0	∞

where k = constant that depends on the nature of loading, length-to-width ratio of plate, and edge conditions

E = modulus of elasticity

ν = Poisson's ratio [Eq. (3.39)]

b = length of loaded edge of plate, or when the plate is subjected to shearing forces, the smaller lateral dimension

t = plate thickness

Table 3.5 lists values of k for various types of loads and edge support conditions. (From formulas, tables, and curves in F. Bleich, *Buckling Strength of Metal Structures*, S. P. Timoshenko and J. M. Gere, *Theory of Elastic Stability*, and G. Gernard, *Introduction to Structural Stability Theory*, McGraw-Hill, Inc., New York.)

NONLINEAR BEHAVIOR OF STRUCTURAL SYSTEMS

Contemporary methods of steel design require engineers to consider the behavior of a structure as it reaches its limit of resistance. Unless premature failure occurs due to local buckling, fatigue, or brittle fracture, the limit-state behavior will most likely include a nonlinear response. As a frame is being loaded, **nonlinear behavior** can be attributed primarily to second-order effects associated with changes in geometry and yielding of members and connections.

3.45   COMPARISONS OF ELASTIC AND INELASTIC ANALYSES

In Fig. 3.91, the empirical limit-state response of a frame is compared with response curves generated in four different types of analyses: **first-order elastic analysis, second-order elastic analysis, first-order inelastic analysis,** and **second-order inelastic analysis.** In a first-order analysis, **geometric nonlinearities** are not included. These effects are accounted for, however, in a second-order analysis. **Material nonlinear** behavior is not included in an elastic analysis but is incorporated in an inelastic analysis.

In most cases, second-order and inelastic effects have interdependent influences on frame stability; i.e., second-order effects can lead to more inelastic behavior, which can further amplify the second-order effects. Producing designs that account for these nonlinearities requires use of either conventional methods of linear elastic

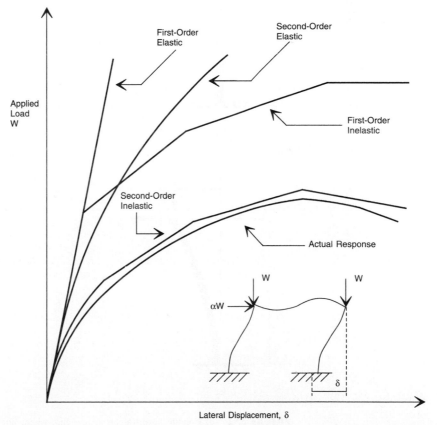

FIGURE 3.91   Load-displacement responses for a rigid frame determined by different methods of analysis.

analysis (Arts. 3.29 to 3.39) supplemented by semiempirical or judgmental allowances for nonlinearity or more advanced methods of nonlinear analysis.

3.46 GENERAL SECOND-ORDER EFFECTS

A column unrestrained at one end with length L and subjected to horizontal load H and vertical load P (Fig. 3.92a) can be used to illustrate the general concepts of second-order behavior. If E is the modulus of elasticity of the column material and I is the moment of inertia of the column, and the equations of equilibrium are formulated on the undeformed geometry, the first-order deflection at the top of

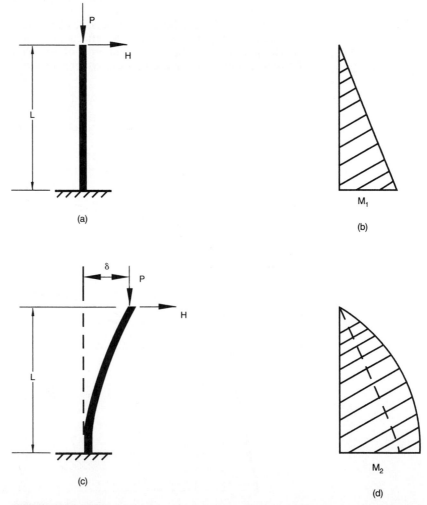

FIGURE 3.92   (a) Column unrestrained at one end, where horizontal and vertical loads act. (b) First-order maximum bending moment M_1 occurs at the base. (c) The column with top displaced by the forces. (d) Second-order maximum moment M_2 occurs at the base.

the column is $\Delta_1 = HL^3/3EI$, and the first-order moment at the base of the column is $M_1 = HL$ (Fig. 3.92b). As the column deforms, however, the applied loads move with the top of the column through a deflection δ. In this case, the actual second-order deflection $\delta = \Delta_2$ not only includes the lateral deflection due to the horizontal load H but also the deflection due to the eccentricity generated with respect to the neutral axis of the column when the vertical load P is displaced (Fig. 3.92c). From equations of equilibrium for the deformed geometry, the second-order base moment is $M_2 = HL + P\Delta_2$ (Fig. 3.92d). The additional deflection and moment generated are examples of second-order effects or geometric nonlinearities.

In a more complex structure, the same type of second-order effects can be present. They may be attributed primarily to two factors: the axial force in a member having a significant influence on the bending stiffness of the member and the relative lateral displacement at the ends of members. Where it is essential that these destabilizing effects are incorporated within a limit-state design procedure, general methods are presented in Arts. 3.47 and 3.48.

3.47 APPROXIMATE AMPLIFICATION FACTORS FOR SECOND-ORDER EFFECTS

One method for approximating the influences of second-order effects (Art. 3.46) is through the use of **amplification factors** that are applied to first-order moments. Two factors are typically used. The first factor accounts for the additional deflection and moment produced by a combination of compressive axial force and lateral deflection δ along the span of a member. It is assumed that there is no relative lateral translation between the two ends of the member. The additional moment is often termed the $P\delta$ **moment.** For a member subject to a uniform first-order bending moment M_{nt} and axial force P (Fig. 3.93) with no relative translation of the ends of the member, the amplification factor is

$$B_1 = \frac{1}{1 - P/P_e} \tag{3.174}$$

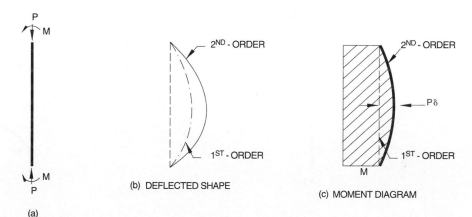

FIGURE 3.93 $P\delta$ effect for beam-column with uniform bending.

where P_e is the elastic critical buckling load about the axis of bending (see Art. 3.4[1]). Hence the moments from a second-order analysis when no relative translation of the ends of the member occurs may be approximated by

$$M_{2nt} = B_1 M_{nt} \tag{3.175}$$

where $B_1 \geq 1$.

The amplification factor in Eq. (3.174) may be modified to account for a nonuniform moment or moment gradient (Fig. 3.94) along the span of the member:

$$B_1 = \frac{C_m}{1 - P/P_e} \tag{3.176}$$

where C_m is a coefficient whose value is to be taken as follows:

1. For compression members with ends restrained from joint translation and not subject to transverse loading between supports, $C_m = 0.6 - 0.4(M_1/M_2)$. M_1 is the smaller and M_2 is the larger end moment in the unbraced length of the member. M_1/M_2 is positive when the moments cause reverse curvature and negative when they cause single curvature.

2. For compression members subject to transverse loading between supports, $C_m = 1.0$.

The second amplification factor accounts for the additional deflections and moments that are produced in a frame that is subject to sidesway, or drift. By combination of compressive axial forces and relative lateral translation of the ends of members, additional moments are developed. These moments are often termed the $P\Delta$ **moments.** In this case, the moments M_{lt} determined from a first-order analysis are amplified by the factor

$$B_2 = \frac{1}{1 - \dfrac{\Sigma P}{\Sigma P_e}} \tag{3.177}$$

where ΣP = total axial load of all columns in a story

ΣP_e = sum of the elastic critical buckling loads about the axis of bending for all columns in a story

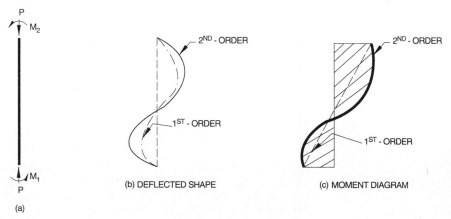

(a)

(b) DEFLECTED SHAPE

(c) MOMENT DIAGRAM

FIGURE 3.94 $P\delta$ effect for beam-column with nonuniform bending.

Hence the moments from a second-order analysis when lateral translation of the ends of the member occurs may be approximated by

$$M_{2lt} = B_2 M_{lt} \tag{3.178}$$

For an unbraced frame subjected to both horizontal and vertical loads, both $P\delta$ and $P\Delta$ second-order destabilizing effects may be present. To account for these effects with amplification factors, two first-order analyses are required. In the first analysis, *nt* (no translation) moments are obtained by applying only vertical loads while the frame is restrained from lateral translation. In the second analysis, *lt* (linear translation) moments are obtained for the given lateral loads and the restraining lateral forces resulting from the first analysis. The moments from an actual second-order analysis may then be approximated by

$$M = B_1 M_{nt} + B_2 M_{lt} \tag{3.179}$$

(T. V. Galambos, *Guide to Stability of Design of Metal Structures*, John Wiley & Sons, Inc, New York; W. McGuire, *Steel Structures*, Prentice-Hall, Inc., Englewood Cliffs, N.J.; *Load and Resistance Factor Design Specifications for Structural Steel Buildings*, American Institute of Steel Construction, Chicago, Ill.)

3.48 GEOMETRIC STIFFNESS MATRIX METHOD FOR SECOND-ORDER EFFECTS

The conventional matrix stiffness method of analysis (Art. 3.39) may be modified to include directly the influences of second-order effects described in Art. 3.46. When the response of the structure is nonlinear, however, the linear relationship in Eq. (3.145), $\mathbf{P} = \mathbf{K\Delta}$, can no longer be used. An alternative is a numerical solution obtained through a sequence of linear steps. In each step, a load increment is applied to the structure and the stiffness and geometry of the frame are modified to reflect its current loaded and deformed state. Hence Eq. (3.145) is modified to the incremental form

$$\delta\mathbf{P} = \mathbf{K_t}\delta\mathbf{\Delta} \tag{3.180}$$

where $\delta\mathbf{P}$ = the applied load increment

$\mathbf{K_t}$ = the modified or tangent stiffness matrix of the structure

$\delta\mathbf{\Delta}$ = the resulting increment in deflections

The tangent stiffness matrix $\mathbf{K_t}$ is generated from nonlinear member force-displacement relationships. They are reflected by the nonlinear member stiffness matrix

$$\mathbf{k}' = \mathbf{k}'_E + \mathbf{k}'_G \tag{3.181}$$

where $\mathbf{k}'_E$ = the conventional elastic stiffness matrix (Art. 3.39)

$\mathbf{k}'_G$ = a geometric stiffness matrix which depends not only on geometry but also on the existing internal member forces.

In this way, the analysis ensures that the equations of equilibrium are sequentially being formulated for the deformed geometry and that the bending stiffness of all members is modified to account for the presence of axial forces.

Inasmuch as a piecewise linear procedure is used to predict nonlinear behavior, accuracy of the analysis increases as the number of load increments increases. In many cases, however, good approximations of the true behavior may be found with relatively large load increments. The accuracy of the analysis may be confirmed by comparing results with an additional analysis that uses smaller load steps.

(J. S. Przemieniecki, *Theory of Matrix Structural Analysis*, McGraw-Hill, Inc., New York; W. F. Chen and E. M. Lui, *Stability Design of Steel Frames*, CRC Press, Inc., Boca Raton, Fla.; T. V. Galambos, *Guide to Stability Design Criteria for Metal Structures*, John Wiley & Sons, Inc., New York)

3.49 *GENERAL MATERIAL NONLINEAR EFFECTS*

Most structural steels can undergo large deformations before rupturing. For example, yielding in ASTM A36 steel begins at a strain of about 0.0012 in per in and continues until strain hardening occurs at a strain of about 0.014 in per in. At rupture, strains can be on the order of 0.25 in per in. These material characteristics affect the behavior of steel members strained into the yielding range and form the basis for the plastic theory of analysis and design.

The **plastic capacity** of members is defined by the amount of axial force and bending moment required to completely yield a member's cross section. In the absence of bending, the plastic capacity of a section is represented by the **axial yield load**

$$P_y = AF_y \tag{3.182}$$

where A = cross-sectional area
F_y = yield stress of the material

For the case of flexural and no axial force, the plastic capacity of the section is defined by the **plastic moment**

$$M_p = ZF_y \tag{3.183}$$

where Z is the plastic section modulus (Art. 3.16). The plastic moment of a section can be significantly greater than the moment required to develop first yielding in the section, defined as the **yield moment**

$$M_y = SF_y \tag{3.184}$$

where S is the elastic section modulus (Art. 3.16). The ratio of the plastic modulus to the elastic section modulus is defined as a section's **shape factor**

$$s = \frac{Z}{S} \tag{3.185}$$

The shape factor indicates the additional moment beyond initial yielding that a section can develop before becoming completely yielded. The shape factor ranges from about 1.1 for wide-flange sections to 1.5 for rectangular shapes and 1.7 for round sections.

For members subjected to a combination of axial force and bending, the plastic capacity of the section is a function of the section geometry. For example, one estimate of the plastic capacity of a wide-flange section subjected to an axial force P and a major-axis bending moment M_{xx} is defined by the interaction equation

$$\frac{P}{P_y} + 0.85 \frac{M_{xx}}{M_{px}} = 1.0 \tag{3.186}$$

where M_{px} = major-axis plastic moment capacity = $Z_{xx}F_y$. An estimate of the minor-axis plastic capacity of wide-flange section is

$$\left(\frac{P}{P_y}\right)^2 + 0.84 \frac{M_{yy}}{M_{py}} = 1.0 \tag{3.187}$$

where M_{yy} = minor-axis bending moment, and M_{py}^- = minor-axis plastic moment capacity = $Z_{yy}F_y$.

When one section of a member develops its plastic capacity, an increase in load can produce a large rotation or axial deformation or both, at this location. When a large rotation occurs, the fully yielded section forms a **plastic hinge.** It differs from a true hinge in that some deformation remains in a plastic hinge after it is unloaded.

The plastic capacity of a section may be equal to or greater than the ultimate strength of the member or the structure in which it exists. First, if the member is part of a redundant system (Art. 3.28), the structure can sustain additional load by distributing the corresponding effects away from the plastic hinge and to the remaining unyielded portions of the structure. Means for accounting for this behavior are incorporated into inelastic methods of analysis.

Secondly, there is a range of strain hardening beyond F_y that corresponds to large strains but in which a steel member can develop an increased resistance to additional loads. This assumes, however, that the section is adequately braced and sized so that local or lateral buckling does not occur.

Material nonlinear behavior can be demonstrated by considering a simply supported beam with span L = 400 in and subjected to a uniform load w (Fig. 3.95a). The maximum moment at midspan is M_{max} = $wL^2/8$ (Fig. 3.95b). If the beam is made of a W24 × 103 wide-flange section with a yield stress F_y = 36 ksi and a section modulus S_{xx} = 245 in^3, the beam will begin to yield at a bending moment of $M_y = F_yS_{xx}$ = 36 × 245 = 8820 in-kips. Hence, when beam weight is ignored, the beam is permitted to carry a uniform load $w = 8M_y/L^2$ = 8 × 8820/400^2 = 0.44 kips/in.

A W24 × 103 shape, however, has a plastic section modulus Z_{xx} = 280 in^3. Consequently, the plastic moment equals $M_p = F_yZ_{xx}$ = 36 × 280 = 10,080 in-kips. When beam weight is ignored, this moment is produced by a uniform load $w = 8M_p/L^2$ = 8 × 10,080/400^2 = 0.50 kips/in, an increase of 14% over the load at initiation of yield. The load developing the plastic moment is often called the **limit,** or **ultimate load.** It is under this load that the beam, with hinges at each of its supports, develops a plastic hinge at midspan (Fig. 3.95c) and becomes unstable. If strain-hardening effects are neglected, a kinematic mechanism has formed, and no further loading can be resisted.

If the ends of a beam are fixed as shown in Fig. 3.96a, the midspan moment is M_{mid} = $wL^2/24$. The maximum moment occur at the ends, M_{end} = $wL^2/12$ (Fig. 3.96b). If the beam has the same dimensions as the one in Fig. 3.95a, the beam begins to yield at uniform load $w = 12M_y/L^2$ = 12 × 8820/400^2 = 0.66 kips/in. If additional load is applied to the beam, plastic hinges eventually form at the ends of the beam at load $w = 12M_p/L^2$ = 12 × 10,080/400^2 = 0.76 kips/in. Although plastic hinges exist at the supports, the beam is still stable at this load. Under additional loading, it behaves as a simply supported beam with moments M_p at each end (Fig. 3.96c) and a maximum moment M_{mid} = $wL^2/8 - M_p$ at midspan (Fig. 3.96d). The limit load of the beam is reached when a plastic hinge forms at midspan, M_{mid} = M_p, thus creating a mechanism (Fig. 3.96e). The uniform load at which this occurs is $w = 2M_p × 8/L^2$ = 2 × 10,080 × 8/400^2 = 1.01 kips/in, a load that is 53% greater than the load at which initiation of yield occurs and 33% greater than the load that produces the first plastic hinges.

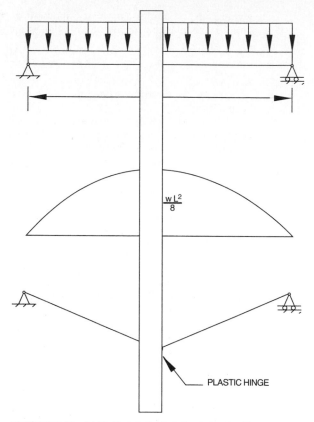

FIGURE 3.95 (*a*) Uniformly loaded simple beam. (*b*) Moment diagram. (*c*) Development of a plastic hinge at midspan.

3.50 CLASSICAL METHODS OF PLASTIC ANALYSIS

In continuous structural systems with many members, there are several ways that mechanisms can develop. The **limit load,** or **load creating a mechanism,** lies between the loads computed from upper-bound and lower-bound theorems. The **upper-bound theorem** states that a load computed on the basis of an assumed mechanism will be greater than, or at best equal to, the true limit load. The **lower-bound theorem** states that a load computed on the basis of an assumed bending-moment distribution satisfying equilibrium conditions, with bending moments nowhere exceeding the plastic moment M_p, is less than, or at best equal to, the true limit load. The plastic moment is $M_p = ZF_y$, where Z = plastic section modulus and F_y = yield stress. If both theorems yield the same load, it is the true ultimate load.

In the application of either theorem, the following conditions must be satisfied at the limit load: External forces must be in equilibrium with internal forces; there must be enough plastic hinges to form a mechanism; and the plastic moment must not be exceeded anywhere in the structure.

The process of investigating mechanism failure loads to determine the maximum load a continuous structure can sustain is called **plastic analysis.**

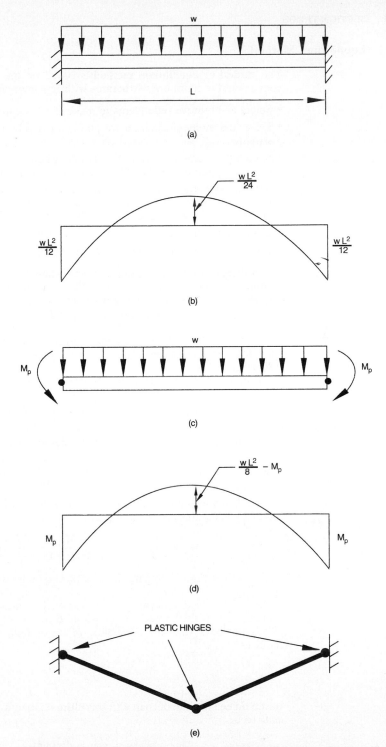

FIGURE 3.96 (*a*) Uniformly loaded fixed-end beam. (*b*) Moment diagram. (*c*) Beam with plastic hinges at both ends. (*d*) Moment diagram for the plastic condition. (*e*) Beam becomes unstable when plastic hinge also develops in the interior.

3.50.1 Equilibrium Method

The **statical** or **equilibrium method** is based on the lower-bound theorem. It is convenient for continuous structures with few members. The steps are

- Select and remove redundants to make the structure statically determinate.
- Draw the moment diagram for the given loads on the statically determinate structure.
- Sketch the moment diagram that results when an arbitrary value of each redundant is applied to the statically determinate structure.
- Superimpose the moment diagrams in such a way that the structure becomes a mechanism because there are a sufficient number of the peak moments that can be set equal to the plastic moment M_p.
- Compute the ultimate load from equilibrium equations.
- Check to see that M_p is not exceeded anywhere.

To demonstrate the method, a plastic analysis will be made for the two-span continuous beam shown in Fig. 3.97a. The moment at C is chosen as the redundant. Figure 3.97b shows the bending-moment diagram for a simple support at C and the moment diagram for an assumed redundant moment at C. Figure 3.97c shows the combined moment diagram. Since the moment at D appears to exceed the moment at B, the combined moment diagram may be adjusted so that the right span becomes a mechanism when the peak moments at C and D equal the plastic moment M_p (Fig. 3.97d).

If $M_C = M_D = M_p$, then equilibrium of span CE requires that at D,

$$M_p = \frac{2L}{3}\left(\frac{2P}{3} - \frac{M_p}{L}\right) = \frac{4PL}{9} - \frac{2M_p}{3}$$

from which the ultimate load P_u may be determined as

$$P_u = \frac{9}{4L}\left(M_p + \frac{2M_p}{3}\right) = \frac{15M_p}{4L}$$

The peak moment at B should be checked to ensure that $M_B \le M_p$. For the ultimate load P_u and equilibrium in span AC,

$$M_B = \frac{P_u L}{4} - \frac{M_p}{2} = \frac{15M_p}{4L}\frac{L}{4} - \frac{M_p}{2} = \frac{7M_p}{16} < M_p$$

This indicates that at the limit load, a plastic hinge will not form in the center of span AC.

If the combined moment diagram had been adjusted so that span AC becomes a mechanism with the peak moments at B and C equaling M_p (Fig. 3.97e), this would not be a statically admissible mode of failure. Equilibrium of span AC requires

$$P_u = \frac{6M_p}{L}$$

Based on equilibrium of span CE, this ultimate load would cause the peak moment at D to be

$$M_D = \frac{4P_u L}{9} - \frac{2M_p}{3} = \frac{6M_p}{L}\frac{4L}{9} - \frac{2M_p}{3} = 2M_p$$

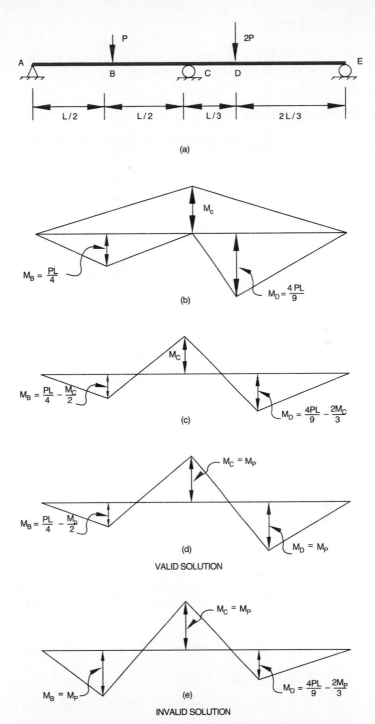

FIGURE 3.97 (*a*) Two-span continuous beam with concentrated loads. (*b*) Moment diagrams for positive and negative moments. (*c*) Combination of moment diagrams in (*b*). (*d*) Valid solution for ultimate load is obtained with plastic moments at peaks at *C* and *D*, and M_p is not exceeded anywhere. (*e*) Invalid solution results when plastic moment is assumed to occur at *B* and M_p is exceeded at *D*.

In this case, M_D violates the requirement that M_p cannot be exceeded. The moment diagram in Fig. 3.97e is not valid.

3.50.2 Mechanism Method

As an alternative, the **mechanism method** is based on the upper-bound theorem. It includes the following steps:

- Determine the locations of possible plastic hinges.
- Select plastic-hinge configurations that represent all possible mechanism modes of failure.
- Using the principle of virtual work, which equates internal work to external work, calculate the ultimate load for each mechanism.
- Assume that the mechanism with the lowest critical load is the most probable and hence represents the ultimate load.
- Check to see that M_p is not exceeded anywhere.

To illustrate the method, the ultimate load will be found for the continuous beam in Fig. 3.97a. Basically, the beam will become unstable when plastic hinges form at B and C (Fig. 3.98a) or C and D (Fig. 3.98b). The resulting constructions are called either **independent** or **fundamental mechanisms.** The beam is also unstable when hinges form at B, C, and D (Fig. 3.98c). This configuration is called a **composite** or **combination mechanism** and also will be discussed.

Applying the principle of virtual work (Art. 3.23) to the beam mechanism in span AC (Fig. 3.98d), external work equated to internal work for a virtual end rotation θ gives

$$\theta P \frac{L}{2} = 2\theta M_p + \theta M_p$$

from which $P = 6M_p/L$.

Similarly, by assuming a virtual end rotation α at E, a beam mechanism in span CE (Fig. 3.98e) yields

$$2P \frac{L}{3} 2\alpha = 2\alpha M_p + 3\alpha M_p$$

from which $P = 15M_p/4L$.

Of the two independent mechanisms, the latter has the lower critical load. This suggests that the ultimate load is $P_u = 15M_p/4L$.

For the combination mechanism (Fig. 3.98f), application of virtual work yields

$$\theta P \frac{L}{2} + 2P \frac{L}{3} 2\alpha = 2\theta M_p + \theta M_p + 2\alpha M_p + 3\alpha M_p$$

from which

$$P = (6M_p/L)[3(\theta/\alpha) + 5]/[3(\theta/\alpha) + 8]$$

In this case, the ultimate load is a function of the value assumed for the ratio θ/α. If θ/α equals zero, the ultimate load is $P = 15M_p/4L$ (the ultimate load for span CE as an independent mechanism). The limit load as θ/α approaches infinity is $P = 6M_p/L$ (the ultimate load for span AC as an independent mechanism). For

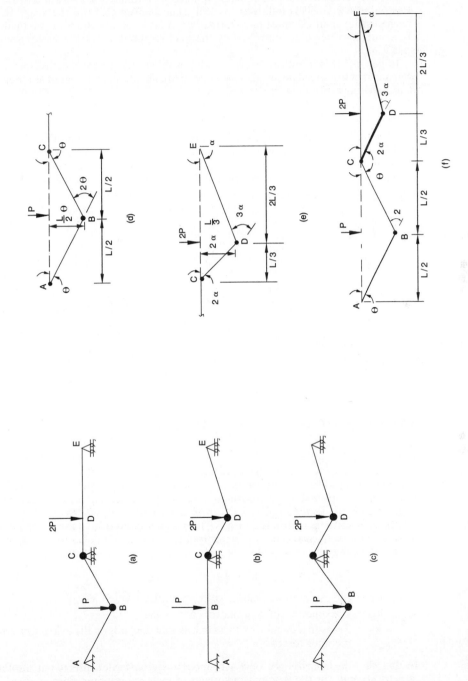

FIGURE 3.98 Plastic analysis of two-span continuous beam by the mechanism method. Beam mechanisms form when plastic hinges occur at (a) B and C, (b) C and D, and (c) B, C, and D. (d), (e), (f) show virtual displacements assumed for the mechanisms in (a), (b), and (c), respectively.

all positive values of θ/α, this equation predicts an ultimate load P such that $15M_p/4L \leq P \leq 6M_pL$. This indicates that for a mechanism to form the span AC, a mechanism in span CE must have formed previously. Hence the ultimate load for the continuous beam is controlled by the load required to form a mechanism in span CE.

In general, it is useful to determine all possible independent mechanisms from which composite mechanisms may be generated. The number of possible independent mechanisms m may be determined from

$$m = p - r \tag{3.188}$$

where p = the number of possible plastic hinges and r = the number of redundancies. Composite mechanisms are selected in such a way as to maximize the total external work or minimize the total internal work to obtain the lowest critical load. Composite mechanisms that include the displacement of several loads and elimination of plastic hinges usually provide the lowest critical loads.

3.50.3   Extension of Classical Plastic Analysis

The methods of plastic analysis presented in Secs. 3.50.1 and 3.50.2 can be extended to analysis of frames and trusses. However, such analyses can become complex, especially if they incorporate second-order effects (Art. 3.46) or reduction in plastic-moment capacity for members subjected to axial force and bending (Art. 3.49).

(E. H. Gaylord, Jr., et al., *Design of Steel Structures*, McGraw-Hill, Inc., New York; W. Prager, *An Introduction to Plasticity*, Addison-Wesley Publishing Company, Inc., Reading, Mass., L. S. Beedle, *Plastic Design of Steel Frames*, John Wiley & Sons, Inc., New York; *Plastic Design in Steel—A Guide and Commentary*, Manual and Report No. 41, American Society of Civil Engineers; R. O. Disque, *Applied Plastic Design in Steel*, Van Nostrand Reinhold Company, New York.)

3.51   *CONTEMPORARY METHODS OF INELASTIC ANALYSIS*

Just as the conventional matrix stiffness method of analysis (Art. 3.39) may be modified to directly include the influences of second-order effects (Art. 3.48), it also may be modified to incorporate nonlinear behavior of structural materials. Loads may be applied in increments to a structure and the stiffness and geometry of the frame changed to reflect its current deformed and possibly yielded state. The tangent stiffness matrix $\mathbf{K_t}$ in Eq. (3.180) is generated from nonlinear member force-displacement relationships. To incorporate material nonlinear behavior, these relationships may be represented by the nonlinear member stiffness matrix

$$\mathbf{k}' = \mathbf{k}'_E + \mathbf{k}'_G + \mathbf{k}'_P \tag{3.189}$$

where $\mathbf{k}'_E$ = the conventional elastic stiffness matrix (Art. 3.39)

$\quad\,\, \mathbf{k}'_G$ = a geometric stiffness matrix (Art. 3.48)

$\quad\,\, \mathbf{k}'_P$ = a plastic reduction stiffness matrix that depends on the existing internal member forces.

In this way, the analysis not only accounts for second-order effects but also can directly account for the destabilizing effects of material nonlinearities.

In general, there are two basic inelastic stiffness methods for investigating frames: the **plastic-zone** or **spread of plasticity method** and the **plastic-hinge** or

concentrated plasticity method. In the plastic-zone method, yielding is modeled throughout a member's volume, and residual stresses and material strain-hardening effects can be included directly in the analysis. In a plastic-hinge analysis, material nonlinear behavior is modeled by the formation of plastic hinges at member ends. Hinge formation and any corresponding plastic deformations are controlled by a **yield surface,** which may incorporate the interaction of axial force and biaxial bending.

(T. V. Galambos, *Guide to Stability Design Criteria for Metal Structures*, John Wiley & Sons, New York; W. F. Chen and E. M. Lui, *Stability Design of Steel Frames*, CRC Press, Inc., Boca Raton, Fla.; W. F. Chen and T. Atsuta, *Theory of Beam-Columns*, vol. 2: *Space Behavior and Design*, McGraw-Hill, Inc., New York.)

TRANSIENT LOADING

Dynamic loads are one of the types of loads to which structures may be subjected (Art. 3.26). When dynamic effects are insignificant, they usually are taken into account in design by application of an impact factor or an increased factor of safety. In many cases, however, an accurate analysis based on the principles of dynamics is necessary. Such an analysis is particularly desirable when a structure is acted on by unusually strong wind gusts, earthquake shocks, or impulsive loads, such as blasts.

3.52 *GENERAL CONCEPTS OF STRUCTURAL DYNAMICS*

There are many types of dynamic loads. **Periodic loads** vary cyclically with time. **Nonperiodic loads** do not have a specific pattern of variation with time. **Impulsive dynamic loading** is independent of the motion of the structure. **Impactive dynamic loading** includes the interaction of all external and internal forces and thus depends on the motions of the structure and of the applied load.

To define a loading within the context of a dynamic or transient analysis, one must specify the direction and magnitude of the loading at every instant of time. The loading may come from either time-dependent forces being applied directly to the structure or from time-dependent motion of the structure's supports, such as a steel frame subjected to earthquake loading.

The term **response** is often used to describe the effects of dynamic loads on structures. More specifically, a response to dynamic loads may represent the displacement, velocity, or acceleration at any point within a structure over a duration of time.

A reciprocating or oscillating motion of a body is called **vibration.** If vibration takes place in the absence of external forces but is accompanied by external or internal frictional forces, or both, it is **damped free vibration.** When frictional forces are also absent, the motion is **undamped free vibration.** If a disturbing force acts on a structure, the resulting motion is **forced vibration** (see also Art. 3.53).

In Art. 3.36, the concept of a degree of freedom is introduced. Similarly, in the context of dynamics, a structure will have n degrees of freedom if n displacement components are required to define the deformation of the structure at any time. For example, a mass M attached to a spring with a negligible mass compared with M represents a one-degree-of-freedom system (Fig. 3.99a). A two-mass system interconnected by weightless springs (Fig. 3.99b) represents a two-degree-of-free-

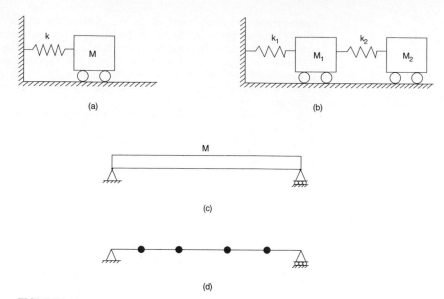

FIGURE 3.99 Idealization of dynamic systems. (*a*) Single-degree-of-freedom system. (*b*) Two-degree-of-freedom system. (*c*) Beam with uniformly distributed mass. (*d*) Equivalent lumped-mass system for beam in (*c*).

dom system. The beam with the uniformly distributed mass in Fig. 3.99*c* has an infinite number of degrees of freedom because an infinite number of displacement components are required to completely describe its deformation at any instant of time.

Because the behavior of a structure under dynamic loading is usually complex, corresponding analyses are generally performed on idealized representations of the structure. In such cases, it is often convenient to represent a structure by one or more dimensionless weights interconnected to each other and to fixed points by weightless springs. For example, the dynamic behavior of the beam shown in Fig. 3.99*c* may be approximated by lumping its distributed mass into several concentrated masses along the beam. These masses would then be joined by members that have bending stiffness but no mass. Such a representation is often called an **equivalent lumped-mass model.** Figure 3.99*d* shows an equivalent four-degree-of-freedom, lumped-mass model of the beam shown in Fig. 3.99*c* (see also Art. 3.53).

3.53 *VIBRATION OF SINGLE-DEGREE-OF-FREEDOM SYSTEMS*

Several dynamic characteristics of a structure can be illustrated by studying single-degree-of-freedom systems. Such a system may represent the motion of a beam with a weight at center span and subjected to a time-dependent concentrated load $P(t)$ (Fig. 3.100*a*). It also may approximate the lateral response of a vertically loaded portal frame constructed of flexible columns, fully restrained connections, and a rigid beam that is also subjected to a time-dependent force $P(t)$ (Fig. 3.100*b*).

In either case, the system may be modeled by a single mass that is connected to a weightless spring and subjected to time-dependent or dynamic force $P(t)$ (Fig. 3.100*c*). The magnitude of the mass m is equal to the given weight W divided by

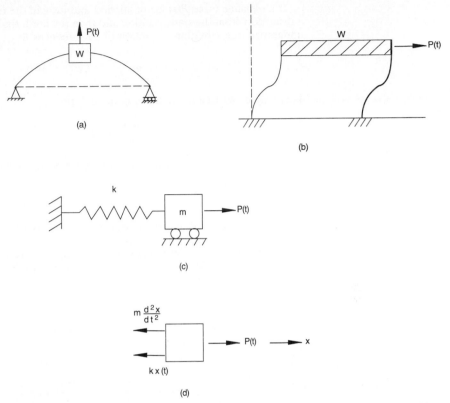

FIGURE 3.100 Dynamic response of single-degree-of-freedom systems. Beam (*a*) and rigid frame (*b*) are represented by a mass on a weightless spring (*c*). Motion of mass (*d*) under variable force is resisted by the spring and inertia of the mass.

the acceleration of gravity $g = 386.4 \, \text{in/sec}^2$. For this model, the weight of structural members is assumed negligible compared with the load W. By definition, the stiffness k of the spring is equal to the force required to produce a unit deflection of the mass. For the beam, a load of $48EI/L^3$ is required at center span to produce a vertical unit deflection; thus $k = 48EI/L^3$, where E is the modulus of elasticity, psi; I the moment of inertia, in^4; and L the span of the beam, in. For the frame, a load of $2 \times 12EI/h^3$ produces a horizontal unit deflection; thus $k = 24EI/h^3$, where I is the moment of inertia of each column, in^4, and h is the column height, in. In both cases, the system is presumed to be loaded within the elastic range. **Deflections are assumed to be relatively small.**

At any instant of time, the dynamic force $P(t)$ is resisted by both the spring force and the inertia force resisting acceleration of the mass (Fig. 3.100*d*). Hence, by d'Alembert's principle (Art. 3.7), dynamic equilibrium of the body requires

$$m \frac{d^2x}{dt^2} + kx(t) = P(t) \tag{3.190}$$

Equation (3.190) represents the controlling differential equation for modeling the motion of an undamped forced vibration of a single-degree-of-freedom system.

If a dynamic force $P(t)$ is not applied and instead the mass is initially displaced a distance x from the static position and then released, the motion would represent undamped free vibration. Equation (3.190) reduces to

$$m \frac{d^2x}{dt^2} + kx(t) = 0 \qquad (3.191)$$

This may be written in the more popular form

$$\frac{d^2x}{dt^2} + \omega^2 x(t) = 0 \qquad (3.192)$$

where $\omega = \sqrt{k/m}$ = **natural circular frequency,** radians per sec. Solution of Eq. (3.192) yields

$$x(t) = A \cos \omega t + B \sin \omega t \qquad (3.193)$$

where the constants A and B can be determined from the initial conditions of the system.

For example, if, before being released, the system is displaced x' and provided initial velocity v', the constants in Eq. (3.193) are found to be $A = x'$ and $B = v'/\omega$. Hence the equation of motion is

$$x(t) = x' \cos \omega t + v' \sin \omega t \qquad (3.194)$$

This motion is periodic, or harmonic. It repeats itself whenever $\omega t = 2\pi$. The time interval or **natural period of vibration** T is given by

$$T = \frac{2\pi}{\omega} = 2\pi \sqrt{\frac{m}{k}} \qquad (3.195)$$

The **natural frequency** f, which is the number of cycles per unit time, or hertz (Hz), is defined as

$$f = \frac{1}{T} = \frac{\omega}{2\pi} = \frac{1}{2\pi} \sqrt{\frac{k}{m}} \qquad (3.196)$$

For undamped free vibration, the natural frequency, period, and circular frequency depend only on the system stiffness and mass. They are independent of applied loads or other disturbances.

(J. M. Biggs, *Introduction to Structural Dynamics*; C. M. Harris and C. E. Crede, *Shock and Vibration Handbook*, 3rd ed.; L. Meirovitch, *Elements of Vibration Analysis*, McGraw-Hill, Inc., New York.)

3.54 *MATERIAL EFFECTS OF DYNAMIC LOADS*

Dynamic loading influences material properties as well as the behavior of structures. In dynamic tests on structural steels with different rates of strain, both yield stress and yield strain increase with an increase in strain rate. The increase in yield stress is significant for A36 steel in that the average dynamic yield stress reaches 41.6 ksi for a time range of loading between 0.01 and 0.1 sec. The strain at which strain hardening begins also increases, and in some cases the ultimate strength can increase slightly. In the elastic range, however, the modulus of elasticity typically remains constant. (See Art. 1.11.)

3.55 REPEATED LOADS

Some structures are subjected to repeated loads that vary in magnitude and direction. If the resulting stresses are sufficiently large and are repeated frequently, the members may fail because of fatigue at a stress smaller than the yield point of the material (Art. 3.8).

Test results on smooth, polished specimens of structural steel indicate that, with complete reversal, there is no strength reduction if the number of the repetitions of load is less than about 10,000 cycles. The strength, however, begins to decrease at 10,000 cycles and continues to decrease up to about 10 million cycles. Beyond this, strength remains constant. The stress at this stage is called the **endurance,** or **fatigue, limit.** For steel subjected to bending with complete stress reversal, the endurance limit is on the order of 50% of the tensile strength. The endurance limit for direct stress is somewhat lower than for bending stress.

The fatigue strength of actual structural members is typically much lower than that of test specimens because of the influences of surface roughness, connection details, and attachments (see Arts. 1.13 and 6.22).

SECTION 4
ANALYSIS OF SPECIAL STRUCTURES

Frederick S. Merritt
Consulting Engineer, West Palm Beach, Florida

and

Louis F. Geschwindner
Professor of Architectural Engineering, The Pennsylvania State University, University Park, Pennsylvania

The general structural theory presented in Sec. 3 can be used to analyze practically all types of structural steel framing. For some frequently used complex framing, however, a specific adaptation of the general theory often expedites the analysis. In some cases, for example, formulas for reactions can be derived from the general theory. Then the general theory is no longer needed for an analysis. In some other cases, where use of the general theory is required, specific methods can be developed to simplify analysis.

This section presents some of the more important specific formulas and methods for complex framing. Usually, several alternative methods are available, but space does not permit their inclusion. The methods given in the following were chosen for their general utility when analysis will not be carried out with a computer.

4.1 THREE-HINGED ARCHES

An **arch** is a beam curved in the plane of the loads to a radius that is very large relative to the depth of section. Loads induce both bending and direct compressive stress. Reactions have horizontal components, though all loads are vertical. Deflections, in general, have horizontal as well as vertical components. At supports, the horizontal components of the reactions must be resisted. For the purpose, tie rods, abutments, or buttresses may be used. With a series of arches, however, the reactions of an interior arch may be used to counteract those of adjoining arches.

A three-hinged arch is constructed by inserting a hinge at each support and at an internal point, usually the crown, or high point (Fig. 4.1). This construction is statically determinate. There are four unknowns—two horizontal and two vertical

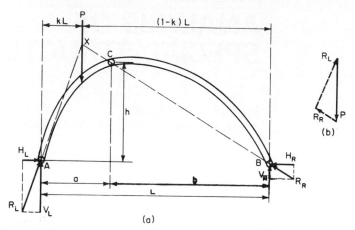

FIGURE 4.1 Three-hinged arch. (*a*) Determination of line of action of reactions. (*b*) Determination of reactions.

components of the reactions—but four equations based on the laws of equilibrium are available.

1. The sum of the horizontal forces acting on the arch must be zero. This relates the horizontal components of the reactions:

$$H_L = H_R = H \tag{4.1}$$

2. The sum of the moments about the left support must be zero. For the arch in Fig. 4.1, this determines the vertical component of the reaction at the right support:

$$V_R = Pk \tag{4.2}$$

where P = load at distance kL from left support
L = span

3. The sum of the moments about the right support must be zero. This gives the vertical component of the reaction at the left support:

$$V_L = P(1 - k) \tag{4.3}$$

4. The bending moment at the crown hinge must be zero. (The sum of the moments about the crown hinge also is zero but does not provide an independent equation for determination of the reactions.) For the right half of the arch in Fig. 4.1, $Hh - V_R b = 0$, from which

$$H = \frac{V_R b}{h} = \frac{Pkb}{h} \tag{4.4}$$

The influence line for H for this portion of the arch thus is a straight line, varying from zero for a unit load over the support to a maximum of ab/Lh for a unit load at C.

Reactions of three-hinge arches also can be determined graphically by taking advantage of the fact that the bending moment at the crown hinge is zero. This requires that the line of action of reaction R_R at the right support pass through C.

This line intersects the line of action of load P at X (Fig. 4.1). Because P and the two reactions are in equilibrium, the line of action of reaction R_L at the left support also must pass through X. As indicated in Fig. 4.1b, the magnitudes of the reactions can be found from a force triangle comprising P and the lines of action of the reactions.

After the reactions have been determined, the stresses at any section of the arch can be found by application of the equilibrium laws (Art. 4.4).

(T. Y. Lin and S. D. Stotesbury, *Structural Concepts and Systems for Architects and Engineers*, 2d Ed., Van Nostrand Reinhold Company, New York.)

4.2 TWO-HINGED ARCHES

A two-hinged arch has hinges only at the supports (Fig. 4.2a). Such an arch is statically indeterminate. Determination of the horizontal and vertical components of each reaction requires four equations, whereas the laws of equilibrium supply only three (Art. 4.1).

Another equation can be written from knowledge of the elastic behavior of the arch. One procedure is to assume that one of the supports is on rollers. The arch then becomes statically determinate. Reactions V_L and V_R and horizontal movement of the support δx can be computed for this condition with the laws of equilibrium (Fig. 4.2b). Next, with the support still on rollers, the horizontal force H required to return the movable support to its original position can be calculated (Fig. 4.2c). Finally, the reactions of the two-hinged arch of Fig. 4.2a are obtained by adding the first set of reactions to the second (Fig. 4.2d).

The structural theory of Sec. 3 can be used to derive a formula for the horizontal component H of the reactions. For example, for the arch of Fig. 4.2a, δx is the

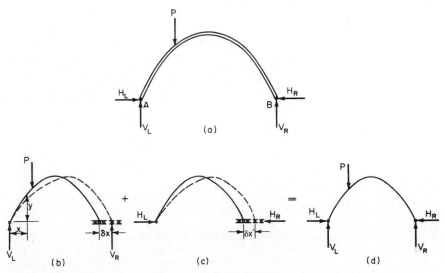

FIGURE 4.2 Two-hinged arch. Reactions of loaded arches (a) and (d) may be found as the sum of reactions in (b) and (c) with one support movable horizontally.

horizontal movement of the support due to loads on the arch. Application of virtual work gives

$$\delta x = \int_A^B \frac{My \, ds}{EI} - \int_A^B \frac{N \, dx}{AE} \tag{4.5}$$

where M = bending moment at any section due to loads on the arch
  y = vertical ordinate of section measured from immovable hinge
  I = moment of inertia of arch cross section
  A = cross-sectional area of arch at the section
  E = modulus of elasticity
  ds = differential length along arch axis
  dx = differential length along the horizontal
  N = normal thrust on the section due to loads

Unless the thrust is very large, the second term on the right of Eq. (4.5) can be ignored.

Let $\delta x'$ be the horizontal movement of the support due to a unit horizontal force applied to the hinge. Application of virtual work gives

$$\delta x' = -\int_A^B \frac{y^2 \, ds}{EI} - \int_A^B \frac{\cos^2 \alpha \, dx}{AE} \tag{4.6}$$

where α is the angle the tangent to axis at the section makes with horizontal. Neither this equation nor Eq. (4.5) includes the effect of shear deformation and curvature. These usually are negligible.

In most cases, integration is impracticable. The integrals generally must be evaluated by approximate methods. The arch axis is divided into a convenient number of elements of length Δs, and the functions under the integral sign are evaluated for each element. The sum of the results is approximately equal to the integral.

For the arch of Fig. 4.2,

$$\delta x + H \, \delta x' = 0 \tag{4.7}$$

When a tie rod is used to take the thrust, the right-hand side of the equation is not zero but the elongation of the rod $HL/A_s E$, where L is the length of the rod and A_s its cross-sectional area. The effect of an increase in temperature t can be accounted for by adding to the left-hand side of the equation $EctL$, where L is the arch span and c the coefficient of expansion.

For the usual two-hinged arch, solution of Eq. (4.7) yields

$$H = -\frac{\delta x}{\delta x'} = \frac{\displaystyle\sum_A^B (My \, \Delta s/EI) + \sum_A^B N \, \Delta s/AE}{\displaystyle\sum_A^B (y^2 \, \Delta s/EI) + \sum_A^B (\cos^2 \alpha \, \Delta x/AE)} \tag{4.8}$$

After the reactions have been determined, the stresses at any section of the arch can be found by application of the equilibrium laws (Art. 4.4).

Circular Two-Hinged Arch Example. A circular two-hinged arch of 175-ft radius with a rise of 29 ft must support a 10-kip load at the crown. The modulus of elasticity E is constant, as is I/A, which is taken as 40.0. The arch is divided into 12 equal segments, 6 on each symmetrical half. The elements of Eq. (4.8) are given in Table 4.1 for each half of the arch.

TABLE 4.1   Example of Two-Hinged Arch Analysis

α, radians	My, kip-ft	y^2, ft^2	$N \cos \alpha$, kips	$\cos^2 \alpha$
0.0487	12,665	829.0	0.24	1.00
0.1462	9,634	736.2	0.72	0.98
0.2436	6,469	568.0	1.17	0.94
0.3411	3,591	358.0	1.58	0.89
0.4385	1,381	154.8	1.92	0.82
0.5360	159	19.9	2.20	0.74
TOTAL	33,899	2,665.9	7.83	5.37

From Eq. (4.8) and with the values in Table 4.1 for one-half the arch, the horizontal reaction may be determined. The flexural contribution yields

$$H = \frac{2.0(33899)}{2.0(2665.9)} = 12.71 \text{ kips}$$

Addition of the axial contribution yields

$$H = \frac{2.0[33899 - 40.0(7.83)]}{2.0[2665.9 + 40.0(5.37)]} = 11.66 \text{ kips}$$

It may be convenient to ignore the contribution of the thrust in the arch under actual loads. If this is the case, $H = 11.77$ kips.

(F. Arbabi, *Structural Analysis and Behavior*, McGraw-Hill, Inc., New York.)

4.3 FIXED ARCHES

In a **fixed arch,** translation and rotation are prevented at the supports (Fig. 4.3). Such an arch is statically indeterminate. With each reaction comprising a horizontal and vertical component and a moment (Art. 4.1), there are a total of six reaction components to be determined. Equilibrium laws provide only three equations. Three more equations must be obtained from a knowledge of the elastic behavior of the arch.

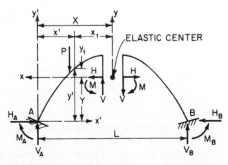

FIGURE 4.3   Fixed arch may be analyzed as two cantilevers.

One procedure is to consider the arch cut at the crown. Each half of the arch then becomes a cantilever. Loads along each cantilever cause the free ends to deflect and rotate. To permit the cantilevers to be joined at the free ends to restore the original fixed arch, forces must be applied at the free ends to equalize deflections and rotations. These conditions provide three equations.

Solution of the equations, however, can be simplified considerably if the center of coordinates is shifted to the elastic center of the arch and the coordinate axes are properly oriented. If the unknown forces and moments V, H, and M are determined at the elastic center (Fig. 4.3), each equation will contain only one unknown. When the unknowns at the elastic center have been determined, the shears, thrusts, and moments at any points on the arch can be found from the laws of equilibrium.

Determination of the location of the elastic center of an arch is equivalent to finding the center of gravity of an area. Instead of an increment of area dA, however, an increment of length ds multiplied by a width $1/EI$ must be used, where E is the modulus of elasticity and I the moment of inertia of the arch cross section.

In most cases, integration is impracticable. An approximate method is usually used, such as the one described in Art. 4.2.

Assume the origin of coordinates to be temporarily at A, the left support of the arch. Let x' be the horizontal distance from A to a point on the arch and y' the vertical distance from A to the point. Then the coordinates of the elastic center are

$$X = \frac{\sum_{A}^{B} (x' \, \Delta s/EI)}{\sum_{A}^{B} (\Delta s/EI)} \qquad Y = \frac{\sum_{A}^{B} (y' \, \Delta s/EI)}{\sum_{A}^{B} (\Delta s/EI)} \tag{4.9}$$

If the arch is symmetrical about the crown, the elastic center lies on a normal to the tangent at the crown. In this case, there is a savings in calculation by taking the origin of the temporary coordinate system at the crown and measuring coordinates parallel to the tangent and the normal. Furthermore, Y, the distance of the elastic center from the crown, can be determined from Eq. (4.9) with y' measured from the crown and the summations limited to the half arch between crown and either support. For a symmetrical arch also, the final coordinates should be chosen parallel to the tangent and normal to the crown.

For an unsymmetrical arch, the final coordinate system generally will not be parallel to the initial coordinate system. If the origin of the initial system is translated to the elastic center, to provide new temporary coordinates $x_1 = x' - X$ and $y_1 = y' - Y$, the final coordinate axes should be chosen so that the x axis makes an angle α, measured clockwise, with the x_1 axis such that

$$\tan 2\alpha = \frac{2 \sum_{A}^{B} (x_1 y_1 \, \Delta s/EI)}{\sum_{A}^{B} (x_1^2 \, \Delta s/EI) - \sum_{A}^{B} (y_1^2 \, \Delta s/EI)} \tag{4.10}$$

The unknown forces H and V at the elastic center should be taken parallel, respectively, to the final x and y axes.

The free end of each cantilever is assumed connected to the elastic center with a rigid arm. Forces H, V, and M act against this arm, to equalize the deflections produced at the elastic center by loads on each half of the arch. For a coordinate system with origin at the elastic center and axes oriented to satisfy Eq. (4.10), application of virtual work to determine deflections and rotations yields

$$H = \frac{\sum\limits_{A}^{B} (M'y \, \Delta s / EI)}{\sum\limits_{A}^{B} (y^2 \, \Delta s / EI)}$$

$$V = \frac{\sum\limits_{A}^{B} (M'x \, \Delta s / EI)}{\sum\limits_{A}^{B} (x^2 \, \Delta s / EI)} \qquad (4.11)$$

$$M = \frac{\sum\limits_{A}^{B} (M' \, \Delta s / EI)}{\sum\limits_{A}^{B} (\Delta s / EI)}$$

where M' is the average bending moment on each element of length Δs due to loads. To account for the effect of an increase in temperature t, add $EctL$ to the numerator of H, where c is the coefficient of expansion and L the distance between abutments. Equations (4.11) may be similarly modified to include deformations due to secondary stresses.

With H, V, and M known, the reactions at the supports can be determined by application of the equilibrium laws. In the same way, the stresses at any section of the arch can be computed (Art. 4.4).

(S. Timoshenko and D. H. Young, *Theory of Structures*, McGraw-Hill, Inc., New York; S. F. Borg and J. J. Gennaro, *Advanced Structural Analysis*, Van Nostrand Reinhold Company, New York; G. L. Rogers and M. L. Causey, *Mechanics of Engineering Structures*, John Wiley & Sons, Inc., New York; J. Michalos, *Theory of Structural Analysis and Design*, The Ronald Press Company, New York.)

4.4 STRESSES IN ARCH RIBS

When the reactions have been determined for an arch (Arts. 4.1 to 4.3), the principal forces acting on any cross section can be found by applying the equilibrium laws. Suppose, for example, the forces H, V, and M acting at the elastic center of a fixed arch have been computed, and the moment M_x, shear S_x, and axial thrust N_x normal to a section at X (Fig. 4.4) are to be determined. H, V, and the load P may be resolved into components parallel to the thrust and shear, as indicated in Fig. 4.4. Then, equating the sum of the forces in each direction to zero gives

$$N_x = V \sin \theta_x + H \cos \theta_x + P \sin(\theta_x - \theta)$$
$$S_x = V \cos \theta_x - H \sin \theta_x + P \cos(\theta_x - \theta) \qquad (4.12)$$

Equating moments about X to zero yields

$$M_x = Vx + Hy - M + Pa \cos \theta + Pb \sin \theta \qquad (4.13)$$

For structural steel members, the shearing force on a section usually is assumed to be carried only by the web. In built-up members, the shear determines the size

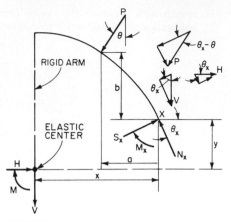

FIGURE 4.4 Arch stresses at any point may be determined from forces at the elastic center.

and spacing of fasteners or welds between web and flanges. The full (gross) section of the arch rib generally is assumed to resist the combination of axial thrust and moment.

4.5 PLATE DOMES

A **dome** is a three-dimensional structure generated by translation and rotation or only rotation of an arch rib. Thus a dome may be part of a sphere, ellipsoid, paraboloid, or similar curved surface.

Domes may be thin-shell or framed, or a combination. Thin-shell domes are constructed of sheet metal or plate, braced where necessary for stability, and are capable of transmitting loads in more than two directions to supports. The surface is substantially continuous from crown to supports. Framed domes, in contrast, consist of interconnected structural members lying on the dome surface or with points of intersection lying on the dome surface (Art. 4.6). In combination construction, covering material may be designed to participate with the framework in resisting dome stresses.

Plate domes are highly efficient structurally when shaped, proportioned, and supported to transmit loads without bending or twisting. Such domes should satisfy the following conditions:

The plate should not be so thin that deformations would be large compared with the thickness. Shearing stresses normal to the surface should be negligible. Points on a normal to the surface before it is deformed should lie on a straight line after deformation. And this line should be normal to the deformed surface.

Stress analysis usually is based on the membrane theory, which neglects bending and torsion. Despite the neglected stresses, the remaining stresses are in equilibrium, except possibly at boundaries, supports, and discontinuities. At any interior point of a thin-shell dome, the number of equilibrium conditions equals the number of unknowns. Thus, in the membrane theory, a plate dome is statically determinate.

The membrane theory, however, does not hold for certain conditions: concentrated loads normal to the surface and boundary arrangements not compatible with equilibrium or geometric requirements. Equilibrium or geometric incompatibility induces bending and torsion in the plate. These stresses are difficult to compute even for the simplest type of shell and loading, yet they may be considerably larger than the membrane stresses. Consequently, domes preferably should be designed to satisfy membrane theory as closely as possible.

Make necessary changes in dome thickness gradual. Avoid concentrated and abruptly changing loads. Change curvature gradually. Keep discontinuities to a minimum. Provide reactions that are tangent to the dome. Make certain that the reactions at boundaries are equal in magnitude and direction to the shell forces there. Also, at boundaries, ensure, to the extent possible, compatibility of shell deformations with deformations of adjoining members, or at least keep restraints to a minimum. A common procedure is to use as a support a husky ring girder and to thicken the shell gradually in the vicinity of this support. Similarly, where a circular opening is provided at the crown, the opening usually is reinforced with a ring girder, and the plate is made thicker than necessary for resisting membrane stresses.

Dome surfaces usually are generated by rotating a plane curve about a vertical axis, called the **shell axis.** A plane through the axis cuts the surface in a meridian, whereas a plane normal to the axis cuts the surface in a circle, called a **parallel** (Fig. 4.5a). For stress analysis, a coordinate system for each point is chosen with the x axis tangent to the meridian, y axis tangent to the parallel, and z axis normal to the surface. The membrane forces at the point are resolved into components in the directions of these axes (Fig. 4.5b).

Location of a given point P on the surface is determined by the angle θ between the shell axis and the normal through P and by the angle ϕ between the radius through P of the parallel on which P lies and a fixed reference direction. Let r_θ be the radius of curvature of the meridian. Also, let r_ϕ, the length of the shell normal between P and the shell axis, be the radius of curvature of the normal section at P. Then,

$$r_\phi = \frac{a}{\sin \theta} \tag{4.14}$$

where a is the radius of the parallel through P.

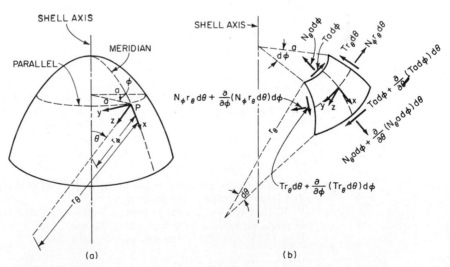

(a)   (b)

FIGURE 4.5   Thin-shell dome. (a) Coordinate system for analysis. (b) Forces acting on a small element.

Figure 4.5*b* shows a differential element of the dome surface at *P*. Normal and shear forces are distributed along each edge. They are assumed to be constant over the thickness of the plate. Thus, at *P*, the meridional unit force is N_θ, the unit hoop force N_ϕ, and the unit shear force *T*. They act in the direction of the *x* or *y* axis at *P*. Corresponding unit stresses at *P* are N_θ/t, N_ϕ/t, and T/t, where *t* is the plate thickness.

Assume that the loading on the element per unit of area is given by its *X*, *Y*, *Z* components in the direction of the corresponding coordinate axis at *P*. Then, the equations of equilibrium for a shell of revolution are

$$\frac{\partial}{\partial \theta}(N_\theta r_\phi \sin \theta) + \frac{\partial T}{\partial \phi} r_\theta - N_\phi r_\theta \cos \theta + X r_\theta r_\phi \sin \theta = 0$$

$$\frac{\partial N_\phi}{\partial \phi} r_\theta + \frac{\partial}{\partial \theta}(T r_\phi \sin \theta) + T r_\theta \cos \theta + Y r_\theta r_\phi \sin \theta = 0 \tag{4.15}$$

$$N_\theta r_\phi + N_\phi r_\theta + Z r_\theta r_\phi = 0$$

When the loads also are symmetrical about the shell axis, Eqs. (4.15) take a simpler form and are easily solved, to yield

$$N_\theta = -\frac{R}{2\pi a} \sin \theta = -\frac{R}{2\pi r_\phi} \sin^2 \theta \tag{4.16}$$

$$N_\phi = \frac{R}{2\pi r_\theta} \sin^2 \theta - Z r_\phi \tag{4.17}$$

$$T = 0 \tag{4.18}$$

where *R* is the resultant of total vertical load above parallel with radius *a* through point *P* at which stresses are being computed.

For a spherical shell, $r_\theta = r_\phi = r$. If a vertical load *p* is uniformly distributed over the horizontal projection of the shell, $R = \pi a^2 p$. Then the unit meridional thrust is

$$N_\theta = -\frac{pr}{2} \tag{4.19}$$

Thus there is a constant meridional compression throughout the shell. The unit hoop force is

$$N_\phi = -\frac{pr}{2} \cos 2\theta \tag{4.20}$$

The hoop forces are compressive in the upper half of the shell, vanish at $\theta = 45°$, and become tensile in the lower half.

If, for a spherical dome, a vertical load *w* is uniform over the area of the shell, as might be the case for the weight of the shell, then $R = 2\pi r^2(1 - \cos \theta)w$. From Eqs. (4.16) and (4.17), the unit meridional thrust is

$$N_\theta = -\frac{wr}{1 + \cos \theta} \tag{4.21}$$

In this case, the compression along the meridian increases with θ. The unit hoop force is

$$N_\phi = wr \left(\frac{1}{1 + \cos \theta} - \cos \theta \right) \qquad (4.22)$$

The hoop forces are compressive in the upper part of the shell, reduce to zero at $51°50'$, and become tensile in the lower part.

A ring girder usually is provided along the lower boundary of a dome to resist the tensile hoop forces. Under the membrane theory, however, shell and girder will have different strains. Consequently, bending stresses will be imposed on the shell. Usual practice is to thicken the shell to resist these stresses and provide a transition to the husky girder.

Similarly, when there is an opening around the crown of the dome, the upper edge may be thickened or reinforced with a ring girder to resist the compressive hoop forces. The meridional thrust may be computed from

$$N_\theta = -wr \frac{\cos \theta_0 - \cos \theta}{\sin^2 \theta} - P \frac{\sin \theta_0}{\sin^2 \theta} \qquad (4.23)$$

and the hoop forces from

$$N_\phi = wr \left(\frac{\cos \theta_0 - \cos \theta}{\sin^2 \theta} - \cos \theta \right) + P \frac{\sin \theta_0}{\sin^2 \theta} \qquad (4.24)$$

where $2\theta_0$ = angle of opening
P = vertical load per unit length of compression ring

4.6   RIBBED DOMES

As pointed out in Art. 4.5, domes may be thin-shell, framed, or a combination. One type of framed dome consists basically of arch ribs with axes intersecting at a common point at the crown and with skewbacks, or bases, uniformly spaced along a closed horizontal curve. Often, to avoid the complexity of a joint with numerous intersecting ribs at the crown, the arch ribs are terminated along a compression ring circumscribing the crown. This construction also has the advantage of making it easy to provide a circular opening at the crown should this be desired. Stress analysis is substantially the same whether or not a compression ring is used. In the following, the ribs will be assumed to extend to and be hinged at the crown. The bases also will be assumed hinged. Thrust at the bases may be resisted by abutments or a tension ring.

Despite these simplifying assumptions, such domes are statically indeterminate because of the interaction of the ribs at the crown. Degree of indeterminancy also is affected by deformations of tension and compression rings. In the following analysis, however, these deformations will be considered negligible.

It usually is convenient to choose as unknowns the horizontal component H and vertical component V of the reaction at the bases of each rib. In addition, an unknown force acts at the crown of each rib. Determination of these forces requires solution of a system of equations based on equilibrium conditions and common displacement of all rib crowns. Resistance of the ribs to torsion and bending about the vertical axis is considered negligible in setting up these equations.

As an example of the procedure, equations will be developed for analysis of a spherical dome under unsymmetrical loading. For simplicity, Fig. 4.6 shows only

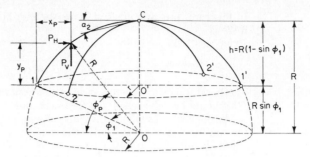

FIGURE 4.6   Arch ribs in a spherical dome with hinge at crown.

two ribs of such a dome. Each rib has the shape of a circular arc. Rib $1C1'$ is subjected to a load with horizontal component P_H and vertical component P_V. Coordinates of the load relative to point 1 are (x_P, y_P). Rib $2C2'$ intersects rib $1C1'$ at the crown at an angle $\alpha_2 \leq \pi/2$. A typical rib rCr' intersects rib $1C1'$ at the crown at an angle $\alpha_r \leq \pi/2$. The dome contains n identical ribs.

A general coordinate system is chosen with origin at the center of the sphere, which has radius R. The base of the dome is assigned a radius r. Then, from the geometry of the sphere,

$$\cos \phi_1 = \frac{r}{R} \tag{4.25}$$

For any point (x, y),

$$x = R(\cos \phi_1 - \cos \phi) \tag{4.26}$$

$$y = R(\sin \phi - \sin \phi_1) \tag{4.27}$$

And the height of the crown is

$$h = R(1 - \sin \phi_1) \tag{4.28}$$

where ϕ_1 = angle radius vector to point 1 makes with horizontal
       ϕ = angle radius vector to point (x, y) makes with horizontal

Assume temporarily that arch $1C1'$ is disconnected at the crown from all the other ribs. Apply a unit downward vertical load at the crown (Fig. 4.7a). This produces vertical reactions $V_1 = V_{1'} = \frac{1}{2}$ and horizontal reactions

$$H_1 = -H_{1'} = r/2h = \cos \phi_1/2(1 - \sin \phi_1)$$

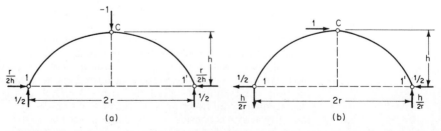

(a)   (b)

FIGURE 4.7   Reactions for a three-hinged rib (a) for a vertical downward load and (b) for a horizontal load at the crown.

Here and in the following discussion upward vertical loads and horizontal loads acting to the right are considered positive. At the crown, downward vertical displacements and horizontal displacements to the right will be considered positive.

For $\phi_1 \le \phi \le \pi/2$, the bending moment at any point (x, y) due to the unit vertical load at the crown is

$$m_V = \frac{x}{2} - \frac{ry}{2h} = \frac{r}{2}\left(1 - \frac{\cos \phi}{\cos \phi_1} - \frac{\sin \phi - \sin \phi_1}{1 - \sin \phi_1}\right) \tag{4.29}$$

For $\pi/2 \le \phi \le \pi$,

$$m_V = \frac{r}{2}\left(1 + \frac{\cos \phi}{\cos \phi_1} - \frac{\sin \phi - \sin \phi_1}{1 - \sin \phi_1}\right) \tag{4.30}$$

By application of virtual work, the downward vertical displacement d_V of the crown produced by the unit vertical load is obtained by dividing the rib into elements of length Δs and computing

$$d_V = \sum_1^{1'} \frac{m_V^2 \, \Delta s}{EI} \tag{4.31}$$

where E = modulus of elasticity of steel
I = moment of inertia of cross section about horizontal axis

The summation extends over the length of the rib.

Next, apply at the crown a unit horizontal load acting to the right (Fig. 4.7b). This produces vertical reactions $V_1 = -V_{1'} = -h/2r = -(1 - \sin \phi_1)/2 \cos \phi_1$ and $H_1 = H_{1'} = -\frac{1}{2}$.

For $\phi_1 \le \phi \le \pi/2$, the bending moment at any point (x, y) due to the unit horizontal load at the crown is

$$m_H = -\frac{hx}{2r} + \frac{y}{2} = \frac{h}{2}\left(\frac{\cos \phi}{\cos \phi_1} - 1 + \frac{\sin \phi - \sin \phi_1}{1 - \sin \phi_1}\right) \tag{4.32}$$

For $\pi/2 \le \phi \le \pi$,

$$m_H = \frac{h}{2}\left(\frac{\cos \phi}{\cos \phi_1} + 1 - \frac{\sin \phi - \sin \phi_1}{1 - \sin \phi_1}\right) \tag{4.33}$$

By application of virtual work, the displacement d_H of the crown to the right induced by the unit horizontal load is obtained from the summation over the arch rib

$$d_H = \sum_1^{1'} \frac{m_H^2 \, \Delta s}{EI} \tag{4.34}$$

Now, apply an upward vertical load P_V on rib $1C1'$ at (x_P, y_P), with the rib still disconnected from the other ribs. This produces the following reactions:

$$V_1 = -P_V \frac{2r - x}{2r} = -\frac{P_V}{2}\left(1 + \frac{\cos \phi_P}{\cos \phi_1}\right) \tag{4.35}$$

$$V_{1'} = -\frac{P_V}{2}\left(1 - \frac{\cos \phi_P}{\cos \phi_1}\right) \tag{4.36}$$

$$H_1 = -H_{1'} = V_{1'}\frac{r}{h} = -\frac{P_V}{2}\frac{\cos\phi_1 - \cos\phi_P}{1 - \sin\phi_1} \tag{4.37}$$

where ϕ_P is the angle that the radius vector to the load point (x_P, y_P) makes with the horizontal $\leq\pi/2$. By application of virtual work, the horizontal and vertical components of the crown displacement induced by P_V may be computed from

$$\delta_{HV} = \sum_1^{1'} \frac{M_V m_H \, \Delta s}{EI} \tag{4.38}$$

$$\delta_{VV} = \sum_1^{1'} \frac{M_V m_V \, \Delta s}{EI} \tag{4.39}$$

where M_V is the bending moment produced at any point (x, y) by P_V.

Finally, apply a horizontal load P_H acting to the right on rib $1C1'$ at (x_P, y_P), with the rib still disconnected from the other ribs. This produces the following reactions:

$$V_{1'} = -V_1 = -P_H\frac{y}{2r} = -\frac{P_H}{2}\frac{\sin\phi_P - \sin\phi_1}{\cos\phi_1} \tag{4.40}$$

$$H_{1'} = -V_{1'}\frac{r}{h} = -\frac{P_H}{2}\frac{\sin\phi_P - \sin\phi_1}{1 - \sin\phi_1} \tag{4.41}$$

$$H_1 = -\frac{P_H}{2}\frac{2 - \sin\phi_1 - \sin\phi_P}{1 - \sin\phi_1} \tag{4.42}$$

By application of virtual work, the horizontal and vertical components of the crown displacement induced by P_H may be computed from

$$\delta_{HH} = \sum_1^{1'} \frac{M_H m_H \, \Delta s}{EI} \tag{4.43}$$

$$\delta_{VH} = \sum_1^{1'} \frac{M_H m_V \, \Delta s}{EI} \tag{4.44}$$

Displacement of the crown of rib $1C1'$, however, is resisted by a force X exerted at the crown by all the other ribs. Assume that X consists of an upward vertical force X_V and a horizontal force X_H acting to the left in the plane of $1C1'$. Equal but oppositely directed forces act at the junction of the other ribs.

Then the actual vertical displacement at the crown of rib $1C1'$ is

$$\delta_V = \delta_{VV} + \delta_{VH} - X_V d_V \tag{4.45}$$

Now, if V_r is the downward vertical force exerted at the crown of any other rib r, then the vertical displacement of that crown is

$$\delta_V = V_r d_V \tag{4.46}$$

Since the vertical displacements of the crowns of all ribs must be the same, the right-hand side of Eqs. (4.45) and (4.46) can be equated. Thus,

$$\delta_{VV} + \delta_{VH} - X_V d_V = V_r d_V = V_s d_V \tag{4.47}$$

where V_s is the vertical force exerted at the crown of another rib s. Hence

$$V_r = V_s \tag{4.48}$$

And for equilibrium at the crown,

$$X_V = \sum_{r=2}^{n} V_r = (n - 1)V_r \tag{4.49}$$

Substituting in Eq. (4.47) and solving for V_r yields

$$V_r = \frac{\delta_{VV} + \delta_{VH}}{nd_V} \tag{4.50}$$

The actual horizontal displacement at the crown of rib $1C1'$ is

$$\delta_H = \delta_{HV} + \delta_{HH} - X_H d_H \tag{4.51}$$

Now, if H_r is the horizontal force acting to the left at the crown of any other rib r, not perpendicular to rib $1C1'$, then the horizontal displacement of that crown parallel to the plane of rib $1C1'$ is

$$\delta_H = \frac{H_r d_H}{\cos \alpha_r} \tag{4.52}$$

Since for all ribs the horizontal crown displacements parallel to the plane of $1C1'$ must be the same, the right-hand side of Eqs. (4.51) and (4.52) can be equated. Hence

$$\delta_{HV} + \delta_{HH} - X_H d_H = \frac{H_r d_H}{\cos \alpha_r} = \frac{H_s d_H}{\cos \alpha_s} \tag{4.53}$$

where H_s is the horizontal force exerted on the crown of any other rib s and α_s is the angle between rib s and rib $1C1'$. Consequently,

$$H_s = H_r \frac{\cos \alpha_s}{\cos \alpha_r} \tag{4.54}$$

For equilibrium at the crown,

$$X_H = \sum_{s=2}^{n} H_s \cos \alpha_s = H_r \cos \alpha_r + \sum_{s=3}^{n} H_s \cos \alpha_s \tag{4.55}$$

Substitution of H_s as given by Eq. (4.54) in this equation gives

$$X_H = H_r \cos \alpha_r + \frac{H_r}{\cos \alpha_r} \sum_{s=3}^{n} \cos^2 \alpha_s = \frac{H_r}{\cos \alpha_r} \sum_{s=2}^{n} \cos^2 \alpha_s \tag{4.56}$$

Substituting this result in Eq. (4.53) and solving for H_r yields

$$H_r = \frac{\cos \alpha_r}{1 + \displaystyle\sum_{s=2}^{n} \cos^2 \alpha_s} \frac{\delta_{H'} + \delta_{H''}}{d_H} \tag{4.57}$$

Then, from Eq. (4.56),

$$X_H = \frac{\displaystyle\sum_{s=2}^{n} \cos^2 \alpha_s}{1 + \displaystyle\sum_{s=2}^{n} \cos^2 \alpha_s} \frac{\delta_{HV} + \delta_{HH}}{d_H} \tag{4.58}$$

Since X_V, X_H, V_r, and H_r act at the crown of the ribs, the reactions they induce can be determined by multiplication by the reactions for a unit load at the crown. For the unloaded ribs, the reactions thus computed are the actual reactions. For the loaded rib, the reactions should be superimposed on those computed for P_V from Eqs. (4.35) to (4.37) and for P_H from Eqs. (4.40) to (4.42).

Superimposition can be used to determine the reactions when several loads are applied simultaneously to one or more ribs.

Hemispherical Domes.   For domes with ribs of constant moment of inertia and comprising a complete hemisphere, formulas for the reactions can be derived. These formulas may be useful in preliminary design of more complex domes.

If the radius of the hemisphere is R, the height h and radius r of the base of the dome also equal R. The coordinates of any point on rib $1C1'$ then are

$$x = R(1 - \cos \phi) \qquad y = R \sin \phi \qquad 0 \le \phi \le \frac{\pi}{2} \tag{4.59}$$

Assume temporarily that arch $1C1'$ is disconnected at the crown from all the other ribs. Apply a unit downward vertical load at the crown. This produces reactions

$$V_1 = V_{1'} = \tfrac{1}{2} \qquad H_1 = -H_{1'} = \tfrac{1}{2} \tag{4.60}$$

The bending moment at any point is

$$m_V = \frac{R}{2} (1 - \cos \phi - \sin \phi) \qquad 0 \le \phi \le \frac{\pi}{2} \tag{4.61a}$$

$$m_V = \frac{R}{2} (1 + \cos \phi - \sin \phi) \qquad \frac{\pi}{2} \le \phi \le \pi \tag{4.61b}$$

By application of virtual work, the downward vertical displacement d_V of the crown is

$$d_V = \int \frac{m_V^2 \, ds}{EI} = \frac{R^3}{EI} \left(\frac{\pi}{2} - \frac{3}{2} \right) \tag{4.62}$$

Next, apply at the crown a unit horizontal load acting to the right. This produces reactions

$$V_1 = -V_{1'} = -\tfrac{1}{2} \qquad H_1 = H_{1'} = -\tfrac{1}{2} \tag{4.63}$$

The bending moment at any point is

$$m_H = \frac{R}{2} (\cos \phi - 1 + \sin \phi) \qquad 0 \le \phi \le \frac{\pi}{2} \tag{4.64a}$$

$$m_H = \frac{R}{2} (\cos \phi + 1 - \sin \phi) \qquad \frac{\pi}{2} \le \phi \le \pi \tag{4.64b}$$

By application of virtual work, the displacement of the crown d_H to the right is

$$d_H = \int \frac{m_H^2 \, ds}{EI} = \frac{R^3}{EI} \left(\frac{\pi}{2} - \frac{3}{2} \right) \tag{4.65}$$

Now, apply an upward vertical load P_V on rib $1C1'$ at (x_P, y_P), with the rib still disconnected from the other ribs. This produces reactions

$$V_1 = -\frac{P_V}{2} (1 + \cos \phi_P) \tag{4.66}$$

$$V_{1'} = -\frac{P_V}{2} (1 - \cos \phi_P) \tag{4.67}$$

$$H_1 = H_{1'} = -\frac{P_V}{2} (1 - \cos \phi_P) \tag{4.68}$$

where $0 \leq \phi_P \leq \pi/2$. By application of virtual work, the vertical component of the crown displacement is

$$\delta_{VV} = \int \frac{M_V m_V \, ds}{EI} = \frac{P_V R^3}{EI} C_{VV} \tag{4.69}$$

$$C_{VV} = \frac{1}{4} \left(-\phi_P + 2 \sin \phi_P - 3 \cos \phi_P + \sin \phi_P \cos \phi_P - \sin^2 \phi_P \right.$$
$$\left. - 2\phi_P \cos \phi_P - 2 \cos^2 \phi_P + 5 - \frac{3\pi}{2} + \frac{3\pi}{2} \cos \phi_P \right) \tag{4.70}$$

For application to downward vertical loads, $-C_{VV}$ is plotted in Fig. 4.8. Similarly, the horizontal component of the crown displacement is

$$\delta_{HV} = \int \frac{M_V m_H \, ds}{EI} = -\frac{P_V R^3}{EI} C_{HV} \tag{4.71}$$

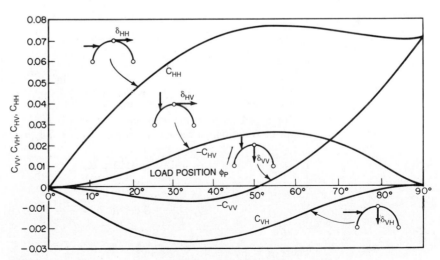

FIGURE 4.8 Coefficients for computing reactions of dome ribs.

$$C_{HV} = \frac{1}{4}\left(-\phi_P + 2\sin\phi_P + 3\cos\phi_P + \sin\phi_P\cos\phi_P - \sin^2\phi_P\right.$$

$$\left. - 2\phi_P\cos\phi_P - 2\cos^2\phi_P - 1 + \frac{\pi}{2} + \frac{\pi}{2}\cos\phi_P\right) \quad (4.72)$$

For application to downward vertical loads, $-C_{HV}$ is plotted in Fig. 4.8.

Finally, apply a horizontal load P_H acting to the right on rib $1C1'$ at (x_P, y_P), with the rib still disconnected from the other ribs. This produces reactions

$$V_1 = V_{1'} = -\frac{P_H}{2}\sin\phi_P \quad (4.73)$$

$$H_1 = -P_H(1 - \tfrac{1}{2}\sin\phi_P) \quad (4.74)$$

$$H_{1'} = -\frac{P_H}{2}\sin\phi_P \quad (4.75)$$

By application of virtual work, the vertical component of the crown displacement is

$$\delta_{VH} = \int \frac{M_H m_V\,ds}{EI} = \frac{P_H R^3}{EI}C_{VH} \quad (4.76)$$

$$C_{VH} = \frac{1}{4}\left[-\phi_P + 3\left(\frac{\pi}{2} - 1\right)\sin\phi_P - 2\cos\phi_P - \sin\phi_P\cos\phi_P\right.$$

$$\left. + \sin^2\phi_P - 2\phi_P\sin\phi_P + 2\right] \quad (4.77)$$

Values of C_{VH} are plotted in Fig. 4.8. The horizontal component of the displacement is

$$\delta_{HH} = \int \frac{M_H m_H\,ds}{EI} = \frac{P_H R^3}{EI}C_{HH} \quad (4.78)$$

$$C_{HH} = \frac{1}{4}\left[\phi_P + \left(\frac{\pi}{2} - 3\right)\sin\phi_P + 2\cos\phi_P + \sin\phi_P\cos\phi_P\right.$$

$$\left. - \sin^2\phi_P + 2\phi_P\sin\phi_P - 2\right] \quad (4.79)$$

Values of C_{HH} also are plotted in Fig. 4.8.

For a vertical load P_V acting upward on rib $1C1'$, the forces exerted on the crown of an unloaded rib are, from Eqs. (4.50) and (4.57),

$$V_r = \frac{\delta_{VH}}{nd_V} = \frac{2P_V C_{VH}}{n(\pi - 3)} \quad (4.80)$$

$$H_r = \frac{\delta_{HH}}{d_H}\beta\cos\alpha_r = -\frac{2P_V C_{HH}}{\pi - 3}\beta\cos\alpha_r \quad (4.81)$$

where $\beta = 1 \bigg/ \left(1 + \sum_{s=2}^{n}\cos^2\alpha_s\right)$

The reactions on the crown of the loaded rib are, from Eqs. (4.49) and (4.58),

$$X_V = (n - 1)V_r = \frac{n - 1}{n} \frac{2P_V C_{VV}}{\pi - 3} \tag{4.82}$$

$$X_H = \frac{\delta_{HV}}{d_H} \gamma = -\frac{2P_V C_{HV}}{\pi - 3} \tag{4.83}$$

where $\gamma = \beta \sum\limits_{s=2}^{n} \cos^2 \alpha_s$

For a horizontal load P_H acting to the right on rib $1C1'$, the forces exerted on the crown of an unloaded rib are, from Eqs. (4.50) and (4.57),

$$V_r = \frac{\delta_{VH}}{nd_V} = \frac{2P_H C_{VH}}{n(\pi - 3)} \tag{4.84}$$

$$H_r = \frac{\delta_{HH}}{d_H} \beta \cos \alpha_r = \frac{2P_H C_{HH}}{\pi - 3} \beta \cos \alpha_r \tag{4.85}$$

The reactions on the crown of the loaded rib are, from Eqs. (4.49) and (4.58),

$$X_V = (n - 1)V_r = \frac{n - 1}{n} \frac{2P_H C_{VH}}{\pi - 3} \tag{4.86}$$

$$X_H = \frac{\delta_{HV}}{d_H} \gamma = \frac{2P_H C_{HH}}{\pi - 3} \gamma \tag{4.87}$$

The reactions for each rib caused by the crown forces can be computed with Eqs. (4.60) and (4.63). For the unloaded ribs, the actual reactions are the sums of the reactions caused by V_r and H_r. For the loaded rib, the reactions due to the load must be added to the sum of the reactions caused by X_V and X_H. The results are summarized in Table 4.2 for a unit vertical load acting downward ($P_V = -1$) and a unit horizontal load acting to the right ($P_H = 1$).

4.7 RIBBED AND HOOPED DOMES

Article 4.5 noted that domes may be thin-shelled, framed, or a combination. It also showed how thin-shelled domes can be analyzed. Article 4.6 showed how one type of framed dome, ribbed domes, can be analyzed. This article shows how to analyze another type, ribbed and hooped domes.

This type also contains regularly spaced arch ribs around a closed horizontal curve. It also may have a tension ring around the base and a compression ring around the common crown. In addition, at regular intervals, the arch ribs are intersected by structural members comprising a ring, or hoop, around the dome in a horizontal plane (Fig. 4.9). The rings resist horizontal displacement of the ribs at the points of inter-

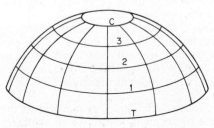

FIGURE 4.9 Ribbed and hooped dome.

TABLE 4.2   Reactions of Ribs of Hemispherical Ribbed Dome

Constant moment of inertia for each of n identical ribs

$$\beta = \frac{1}{1 + \sum\limits_{s=2}^{n} \cos^2 \alpha_s} \qquad \gamma = \beta \sum_{s=2}^{n} \cos^2 \alpha_s$$

ϕ_P = angle the radius vector to load from center of hemisphere makes with horizontal

α_r = angle between loaded and unloaded rib $\leq \pi/2$

Reactions of loaded rib Unit downward vertical load	Reactions of unloaded rib Unit downward vertical load
$V_1 = \dfrac{1}{2} + \dfrac{1}{2}\cos\phi_P - \dfrac{n-1}{n}\dfrac{C_{VV}}{\pi-3} + \dfrac{C_{HV}}{\pi-3}\gamma$	$V_r = \dfrac{C_{VV}}{n(\pi-3)} - \dfrac{C_{HV}}{\pi-3}\beta\cos\alpha_r$
$V_{1'} = \dfrac{1}{2} - \dfrac{1}{2}\cos\phi_P - \dfrac{n-1}{n}\dfrac{C_{VV}}{\pi-3} - \dfrac{C_{HV}}{\pi-3}\gamma$	$V_{r'} = \dfrac{C_{VV}}{n(\pi-3)} + \dfrac{C_{HV}}{\pi-3}\beta\cos\alpha_r$
$H_1 = \dfrac{1}{2} - \dfrac{1}{2}\cos\phi_P - \dfrac{n-1}{n}\dfrac{C_{VV}}{\pi-3} + \dfrac{C_{HV}}{\pi-3}\gamma$	$H_r = \dfrac{C_{VV}}{n(\pi-3)} - \dfrac{C_{HV}}{\pi-3}\beta\cos\alpha_r$
$H_{1'} = -\dfrac{1}{2} + \cos\phi_P + \dfrac{n-1}{n}\dfrac{C_{VV}}{\pi-3} + \dfrac{C_{HV}}{\pi-3}\gamma$	$H_{r'} = -\dfrac{C_{VV}}{n(\pi-3)} - \dfrac{C_{HV}}{\pi-3}\beta\cos\alpha_r$
Unit horizontal load acting to right	Unit horizontal load acting to right
$V_1 = -\dfrac{1}{2}\sin\phi_P - \dfrac{n-1}{n}\dfrac{C_{VH}}{\pi-3} + \dfrac{C_{HH}}{\pi-3}\gamma$	$V_r = \dfrac{C_{VH}}{n(\pi-3)} - \dfrac{C_{HH}}{\pi-3}\beta\cos\alpha_r$
$V_{1'} = \dfrac{1}{2}\sin\phi_P - \dfrac{n-1}{n}\dfrac{C_{VH}}{\pi-3} - \dfrac{C_{HH}}{\pi-3}\gamma$	$V_{r'} = \dfrac{C_{VH}}{n(\pi-3)} + \dfrac{C_{HH}}{\pi-3}\beta\cos\alpha_r$
$H_1 = -1 + \dfrac{1}{2}\sin\phi_P - \dfrac{n-1}{n}\dfrac{C_{VH}}{\pi-3} + \dfrac{C_{HH}}{\pi-3}\gamma$	$H_r = \dfrac{C_{VH}}{n(\pi-3)} - \dfrac{C_{HH}}{\pi-3}\beta\cos\alpha_r$
$H_{1'} = -\dfrac{1}{2}\sin\phi_P + \dfrac{n-1}{n}\dfrac{C_{VH}}{\pi-3} + \dfrac{C_{HH}}{\pi-3}\gamma$	$H_{r'} = -\dfrac{C_{VH}}{n(\pi-3)} - \dfrac{C_{HH}}{\pi-3}\beta\cos\alpha_r$

section. If the rings are made sufficiently stiff, they may be considered points of support for the ribs horizontally. Some engineers prefer to assume the ribs hinged at those points. Others assume the ribs hinged only at tension and compression rings and continuous between those hoops. In many cases, the curvature of rib segments between rings may be ignored.

Figure 4.10a shows a rib segment 1-2 assumed hinged at the rings at points 1 and 2. A distributed downward load W induces bending moments between points 1 and 2 and shears assumed to be $W/2$ at 1 and 2. The ring segment above, 2-3, applies a thrust at 2 of $\Sigma W/\sin\theta_2$, where ΣW is the sum of the vertical loads on the rib from 2 to the crown and θ_2 is the angle with the horizontal of the tangent to the rib at 2.

These forces are resisted by horizontal reactions at the rings and a tangential thrust, provided by a rib segment below 1 or an abutment at 1. For equilibrium, the vertical component of the thrust must equal $W + \Sigma W$. Hence the thrust equals

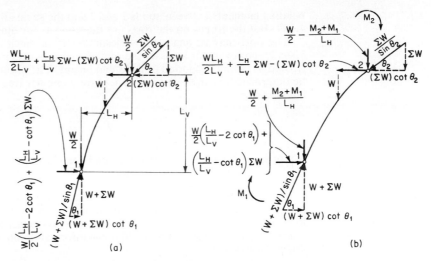

FIGURE 4.10 Forces acting on a segment of a dome rib between hoops. (*a*) Ends of segment assumed hinged. (*b*) Rib assumed continuous.

$(W + \Sigma W)/\sin \theta_1$, where θ_1 is the angle with the horizontal of the tangent to the rib at 1.

Setting the sum of the moments about 1 equal to zero yields the horizontal reaction supplied by the ring at 2:

$$H_2 = \frac{WL_H}{2L_V} + \frac{L_H}{L_V} \Sigma W - (\Sigma W) \cot \theta_2 \qquad (4.88)$$

where L_H = horizontal distance between 1 and 2
$\quad L_V$ = vertical distance between 1 and 2

Setting the sum of the moments about 2 equal to zero yields the horizontal reaction supplied by the ring at 1:

$$H_1 = \frac{W}{2} \left(\frac{L_H}{L_V} - 2 \cot \theta_1 \right) + \left(\frac{L_H}{L_V} - \cot \theta_1 \right) \Sigma W \qquad (4.89)$$

For the direction assumed for H_2, the ring at 2 will be in compression when the right-hand side of Eq. (4.88) is positive. Similarly, for the direction assumed for H_1, the ring at 1 will be in tension when the right-hand side of Eq. (4.89) is positive. Thus the type of stress in the rings depends on the relative values of L_H/L_V and $\cot \theta_1$ or $\cot \theta_2$. Alternatively, it depends on the difference in the slope of the thrust at 1 or 2 and the slope of the line from 1 to 2.

Generally, for maximum stress in the compression ring about the crown or tension ring around the base, a ribbed and hooped dome should be completely loaded with full dead and live loads. For an intermediate ring, maximum tension will be produced with live load extending from the ring to the crown. Maximum compression will result when the live load extends from the ring to the base.

When the rib is treated as continuous between crown and base, moments are introduced at the ends of each rib segment (Fig. 4.10b). These moments may be computed in the same way as for a continuous beam on immovable supports, neglecting the curvature of rib between supports. The end moments affect the bending moments between points 1 and 2 and the shears there, as indicated in Fig. 4.10b. But the forces on the rings are the same as for hinged rib segments.

The rings may be analyzed by elastic theory in much the same way as arches. Usually, however, for loads on the ring segments between ribs, these segments are treated as simply supported or fixed-end beams. The hoop tension or thrust T may be determined, as indicated in Fig. 4.11 for a circular ring, by the requirements of equilibrium:

$$T = \frac{Hn}{2\pi} \qquad (4.90)$$

where H = radial force exerted on ring by each rib
n = number of load points

The procedures outlined neglect the effects of torsion and of friction in joints, which could be substantial. In addition, deformations of such domes under overloads often tend to redistribute those loads to less highly loaded members. Hence more complex analyses without additional information on dome behavior generally are not warranted.

Many domes have been constructed as part of a hemisphere, such that the angle made with the horizontal by the radius vector from the center of the sphere to the base of the dome is about 60°. Thus the radius of the sphere is nearly equal to the diameter of the dome base, and the rise-to-span ratio is about $1 - \sqrt{3/2}$, or 0.13. Some engineers believe that high structural economy results with such proportions.

(Z. S. Makowski, *Analysis, Design, and Construction of Braced Domes*, Granada Technical Books, London, England.)

4.8 SIMPLE SUSPENSION CABLES

The objective of this and the following article is to present general procedures for analyzing simple cable suspension systems. The numerous types of cable systems available make it impractical to treat anything but the simplest types. Additional information may be found in Sec. 14, which covers suspension bridges and cable-stayed structures.

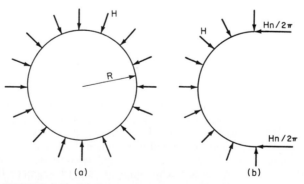

FIGURE 4.11 (a) Forces acting on a complete hoop of a dome. (b) Forces acting on half of a hoop.

Characteristics of Cables. A **suspension cable** is a linear structural member that adjusts its shape to carry loads. The primary assumptions in the analysis of cable systems are that the cables carry only tension and that the tension stresses are distributed uniformly over the cross section. Thus no bending moments can be resisted by the cables.

For a cable subjected to gravity loads, the equilibrium positions of all points on the cable may be completely defined, provided the positions of any three points on the cable are known. These points may be the locations of the cable supports and one other point, usually the position of a concentrated load or the point of maximum sag. For gravity loads, the shape of a cable follows the shape of the moment diagram that would result if the same loads were applied to a simple beam. The maximum sag occurs at the point of maximum moment and zero shear for the simple beam.

The tensile force in a cable is tangent to the cable curve and may be described by horizontal and vertical components. When the cable is loaded only with gravity loads, the horizontal component at every point along the cable remains constant. The maximum cable force will occur where the maximum vertical component occurs, usually at one of the supports, while the minimum cable force will occur at the point of maximum sag.

Since the geometry of a cable changes with the application of load, the common approaches to structural analysis, which are based on small-deflection theories, will not be valid, nor will superposition be valid for cable systems. In addition, the forces in a cable will change as the cable elongates under load, as a result of which equations of equilibrium are nonlinear. A common approximation is to use the linear portion of the exact equilibrium equations as a first trial and to converge on the correct solution with successive approximations.

A cable must satisfy the second-order linear differential equation

$$Hy'' = q \qquad\qquad (4.91)$$

where H = horizontal force in cable
$\quad y$ = rise of cable at distance x from low point (Fig. 4.12)
$\quad y'' = d^2y/dx^2$
$\quad q$ = gravity load per unit span

4.8.1 Catenary

Weight of a cable of constant cross section represents a vertical loading that is uniformly distributed along the length of cable. Under such a loading, a cable takes the shape of a catenary.

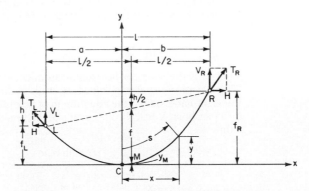

FIGURE 4.12 Cable with supports at different levels.

To determine the stresses in and deformations of a catenary, the origin of coordinates is taken at the low point C, and distance s is measured along the cable from C (Fig. 4.12). With q_o as the load per unit length of cable, Eq. (4.91) becomes

$$Hy'' = \frac{q_o \, ds}{dx} = q_o\sqrt{1 + y'^2} \tag{4.92}$$

where $y' = dy/dx$. Solving for y' gives the slope at any point of the cable:

$$y' = \sinh \frac{q_o x}{H} = \frac{q_o x}{H} + \frac{1}{3!}\left(\frac{q_o x}{H}\right)^3 + \cdots \tag{4.93}$$

A second integration then yields

$$y = \frac{H}{q_o}\left(\cosh \frac{q_o x}{H} - 1\right) = \frac{q_o}{H}\frac{x^2}{2!} + \left(\frac{q_o}{H}\right)^3 \frac{x^4}{4!} + \cdots \tag{4.94}$$

Equation (4.94) is the catenary equation. If only the first term of the series expansion is used, the cable equation represents a parabola. Because the parabolic equation usually is easier to handle, a catenary often is approximated by a parabola.

For a catenary, length of arc measured from the low point is

$$s = \frac{H}{q_o}\sinh \frac{q_o x}{H} = x + \frac{1}{3!}\left(\frac{q_o}{H}\right)^2 x^3 + \cdots \tag{4.95}$$

Tension at any point is

$$T = \sqrt{H^2 + q_o^2 s^2} = H + q_o y \tag{4.96}$$

The distance from the low point C to the left support L is

$$a = \frac{H}{q_o}\cosh^{-1}\left(\frac{q_o}{H} f_L + 1\right) \tag{4.97}$$

where f_L is the vertical distance from C to L. The distance from C to the right support R is

$$b = \frac{H}{q_o}\cosh^{-1}\left(\frac{q_o}{H} f_R + 1\right) \tag{4.98}$$

where f_R is the vertical distance from C to R.

Given the sags of a catenary f_L and f_R under a distributed vertical load q_o, the horizontal component of cable tension H may be computed from

$$\frac{q_o l}{H} = \cosh^{-1}\left(\frac{q_o f_L}{H} + 1\right) + \cosh^{-1}\left(\frac{q_o f_R}{H} + 1\right) \tag{4.99}$$

where l is the span, or horizontal distance, between supports L and $R = a + b$. This equation usually is solved by trial. A first estimate of H for substitution in the right-hand side of the equation may be obtained by approximating the catenary by a parabola. Vertical components of the reactions at the supports can be computed from

$$R_L = \frac{H \sinh q_o a}{H} \qquad R_R = \frac{H \sinh q_o b}{H} \tag{4.100}$$

See also Art. 14.6.

4.8.2 Parabola

Uniform vertical live loads and uniform vertical dead loads other than cable weight generally may be treated as distributed uniformly over the horizontal projection of the cable. Under such loadings, a cable takes the shape of a parabola.

To determine cable stresses and deformations, the origin of coordinates is taken at the low point C (Fig. 4.12). With w_o as the uniform load on the horizontal projection, Eq. (4.91) becomes

$$Hy'' = w_o \qquad (4.101)$$

Integration gives the slope at any point of the cable:

$$y' = \frac{w_o x}{H} \qquad (4.102)$$

A second integration then yields the parabolic equation

$$y = \frac{w_o x^2}{2H} \qquad (4.103)$$

The distance from the low point C to the left support L is

$$a = \frac{l}{2} - \frac{Hh}{w_o l} \qquad (4.104)$$

where l = span, or horizontal distance, between supports L and $R = a + b$
h = vertical distance between supports

The distance from the low point C to the right support R is

$$b = \frac{l}{2} + \frac{Hh}{w_o l} \qquad (4.105)$$

When supports are not at the same level, the horizontal component of cable tension H may be computed from

$$H = \frac{w_o l^2}{h^2}\left(f_R - \frac{h}{2} \pm \sqrt{f_L f_R}\right) = \frac{w_o l^2}{8f} \qquad (4.106)$$

where f_L = vertical distance from C to L
f_R = vertical distance from C to R
f = sag of cable measured vertically from chord LR midway between supports (at $x = Hh/w_o l$)

As indicated in Fig. 4.12,

$$f = f_L + \frac{h}{2} - y_M \qquad (4.107)$$

where $y_M = Hh^2/2w_o l^2$. The minus sign should be used in Eq. (4.106) when low point C is between supports. If the vertex of the parabola is not between L and R, the plus sign should be used.

The vertical components of the reactions at the supports can be computed from

$$V_L = w_o a = \frac{w_o l}{2} - \frac{Hh}{l} \qquad V_R = w_o b = \frac{w_o l}{2} + \frac{Hh}{l} \qquad (4.108)$$

Tension at any point is

$$T = \sqrt{H^2 + w_o^2 x^2}$$

Length of parabolic arc RC is

$$L_{RC} = \frac{b}{2} \sqrt{1 + \left(\frac{w_o b}{H}\right)^2} + \frac{H}{2w_o} \sinh \frac{w_o b}{H} = b + \frac{1}{6}\left(\frac{w_o}{H}\right)^2 b^3 + \cdots \quad (4.109)$$

Length of parabolic arc LC is

$$L_{LC} = \frac{a}{2} \sqrt{1 + \left(\frac{w_o a}{H}\right)^2} + \frac{H}{2w_o} \sinh \frac{w_o a}{H} = a + \frac{1}{6}\left(\frac{w_o}{H}\right)^2 a^3 + \cdots \quad (4.110)$$

When supports are at the same level, $f_L = f_R = f$, $h = 0$, and $a = b = l/2$. The horizontal component of cable tension H may be computed from

$$H = \frac{w_o l^2}{8f} \quad (4.111)$$

The vertical components of the reactions at the supports are

$$V_L = V_R = \frac{w_o l}{2} \quad (4.112)$$

Maximum tension occurs at the supports and equals

$$T_L = T_R = \frac{w_o l}{2} \sqrt{1 + \frac{l^2}{16f^2}} \quad (4.113)$$

Length of cable between supports is

$$L = \frac{l}{2} \sqrt{1 + \left(\frac{w_o l}{2H}\right)^2} + \frac{H}{w_o} \sinh \frac{w_o l}{2H}$$

$$= l\left(1 + \frac{8}{3}\frac{f^2}{l^2} - \frac{32}{5}\frac{f^4}{l^4} + \frac{256}{7}\frac{f^6}{l^6} + \cdots\right) \quad (4.114)$$

If additional uniformly distributed load is applied to a parabolic cable, the elastic elongation is

$$\Delta L = \frac{Hl}{AE}\left(1 + \frac{16}{3}\frac{f^2}{l^2}\right) \quad (4.115)$$

where A = cross-sectional area of cable
E = modulus of elasticity of cable steel
H = horizontal component of tension in cable

The change in sag is approximately

$$\Delta f = \frac{15}{16}\frac{l}{f}\left(\frac{\Delta L}{5 - 24f^2/l^2}\right) \quad (4.116)$$

If the change is small and the effect on H is negligible, this change may be computed from

$$\Delta f = \frac{15}{16} \frac{Hl^2}{AEf} \left(\frac{1 + 16f^2/3l^2}{5 - 24f^2/l^2} \right) \qquad (4.117)$$

For a rise in temperature t, the change in sag is about

$$\Delta f = \frac{15}{16} \frac{l^2 ct}{f(5 - 24f^2/l^2)} \left(1 + \frac{8}{3} \frac{f^2}{l^2} \right) \qquad (4.118)$$

where c is the coefficient of thermal expansion.

4.8.3   Example—Simple Cable

A cable spans 300 ft and supports a uniformly distributed load of 0.2 kips per ft. The unstressed equilibrium configuration is described by a parabola with a sag at the center of the span of 20 ft. $A = 1.47$ in^2 and $E = 24,000$ ksi. Successive application of Eqs. (4.111), (4.115), and (4.116) results in the values shown in Table 4.3. It can be seen that the process converges to a solution after five cycles.

(H. Max Irvine, *Cable Structures*, MIT Press, Cambridge, Mass.; Prem Krishna, *Cable-Suspended Roofs*, McGraw-Hill, Inc., New York; J. B. Scalzi et al., *Design Fundamentals of Cable Roof Structures*, U.S. Steel Corp., Pittsburgh, Pa.; J. Szabo and L. Kollar, *Structural Design of Cable-Suspended Roofs*, Ellis Horwood Limited, Chichester, England.)

4.9   CABLE SUSPENSION SYSTEMS

Single cables, such as those analyzed in Art. 4.8, have a limited usefulness when it comes to building applications. Since a cable is capable of resisting only tension, it is limited to transferring forces only along its length. The vast majority of structures require a more complex ability to transfer forces. Thus it is logical to combine cables and other load-carrying elements into systems. Cables and beams or trusses are found in combination most often in suspension bridges (see Sec. 14), while for

TABLE 4.3   Example Cable Problem

Cycle	Sag, ft	Horizontal force, kips, from Eq. (4.111)	Change in length, ft, from Eq. (4.115)	Change in sag, ft, from Eq. (4.116)	New sag, ft
1	20.00	112.5	0.979	2.81	22.81
2	22.81	98.6	0.864	2.19	22.19
3	22.19	101.4	0.887	2.31	22.31
4	22.31	100.8	0.883	2.29	22.29
5	22.29	100.9	0.884	2.29	22.29

buildings it is common to combine multiple cables into cable systems, such as three-dimensional networks or two-dimensional cable beams and trusses.

Like simple cables, cable systems behave nonlinearly. Thus accurate analysis is difficult, tedious, and time-consuming. As a result, many designers use approximate methods or preliminary designs that appear to have successfully withstood the test of time. Because of the numerous types of systems and the complexity of analysis, only general procedures will be outlined in this article, which deals with cable systems in which the loads are carried to supports only by cables.

Networks consist of two or three sets of parallel cables intersecting at an angle. The cables are fastened together at their intersections. **Cable trusses** consist of pairs of cables, generally in a vertical plane. One cable of each pair is concave downward, the other concave upward (Fig. 4.13). The two cables of a cable truss play different roles in carrying load. The sagging cable, whether it is the upper cable (Fig. 4.13*a* or *b*), the lower cable (Fig. 14.13*d*), or in both positions (Fig. 4.13*c*), carries the gravity load, while the rising cable resists upward load and provides damping. Both cables are initially tensioned, or prestressed, to a predetermined shape, usually parabolic. The prestress is made large enough that any compression that may be induced in a cable by superimposed loads only reduces the tension in the cable; thus compressive stresses cannot occur. The relative vertical position of the cables is maintained by vertical spreaders or by diagonals. Diagonals in the truss plane do not appear to increase significantly the stiffness of a cable truss.

Figure 4.13 shows four different arrangements of cables with spreaders to form a cable truss. The intersecting types (Fig. 4.13*b* and *c*) usually are stiffer than the others, for given size cables and given sag and rise.

For supporting roofs, cable trusses often are placed radially at regular intervals. Around the perimeter of the roof, the horizontal component of the tension usually is resisted by a circular or elliptical compression ring. To avoid a joint with a jumble of cables at the center, the cables usually are also connected to a tension ring circumscribing the center.

Cable trusses may be analyzed as discrete or continuous systems. For a discrete system, the spreaders are treated as individual members and the cables are treated as individual members between each spreader. For a continuous system, the spreaders are replaced by a continuous diaphragm that ensures that the changes in sag and rise of cables remain equal under changes in load.

To illustrate the procedure for a cable truss treated as a continuous system, the type shown in Fig. 4.13*d* and again in Fig. 4.14 will be analyzed. The bottom cable will be the load-carrying cable. Both cables are prestressed and are assumed to be parabolic. The horizontal component H_{iu} of the initial tension in the upper cable is given. The resulting rise is f_u, and the weight of cables and spreaders is taken as w_c. Span is l.

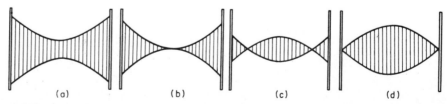

FIGURE 4.13   Planar cable systems. (*a*) Completely separated cables. (*b*) Cables intersecting at midspan. (*c*) Crossing cables. (*d*) Cables meeting at supports.

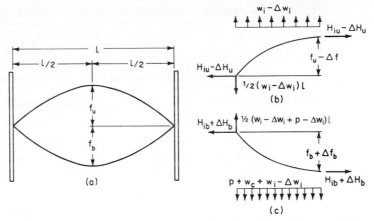

FIGURE 4.14 (a) Cable system with discrete spreaders replaced by an equivalent diaphragm. (b) Forces acting on the top cable. (c) Forces acting on the bottom cable.

The horizontal component of the prestress in the bottom cable H_{ib} can be determined by equating the bending moment in the system at midspan to zero:

$$H_{ib} = \frac{f_u}{f_b} H_{iu} + \frac{w_c l^2}{8 f_b} = \frac{(w_c + w_i)l^2}{8 f_b} \qquad (4.119)$$

where f_b = sag of lower cable
$\quad w_i$ = uniformly distributed load exerted by diaphragm on each cable when cables are parabolic

Setting the bending moment at the high point of the upper cable equal to zero yields

$$w_i = \frac{8 H_{iu} f_u}{l^2} \qquad (4.120)$$

Thus the lower cable carries a uniform downward load $w_c + w_i$, while the upper cable is subjected to a distributed upward force w_i.

Suppose a load p uniformly distributed horizontally is now applied to the system (Fig. 4.14a). This load may be dead load or dead load plus live load. It will decrease the tension in the upper cable by ΔH_u and the rise by Δf (Fig. 4.14b). Correspondingly, the tension in the lower cable will increase by ΔH_b and the sag by Δf (Fig. 4.14c). The force exerted by the diaphragm on each cable will decrease by Δw_i.

The changes in tension may be computed from Eq. (4.117). Also, application of this equation to the bending-moment equations for the midpoints of each cable and simultaneous solution of the resulting pair of equations yields the changes in sag and diaphragm force. The change in sag may be estimated from

$$\Delta f = \frac{1}{H_{iu} + H_{ib} + (A_u f_u^2 + A_b f_b^2)16E/3l^2} \frac{pl^2}{8} \qquad (4.121)$$

where A_u = cross-sectional area of upper cable
$\quad A_b$ = cross-sectional area of lower cable

The decrease in uniformly distributed diaphragm force is given approximately by

$$\Delta w_i = \frac{(H_{iu} + 16A_u E f_u^2/3l^2)p}{H_{iu} + H_{ib} + (A_u f_u^2 + A_b f_b^2)16E/3l^2} \tag{4.122}$$

And the change in load on the lower cable is nearly

$$p - \Delta w_i = \frac{(H_{ib} + 16A_b E f_b^2/3l^2)p}{H_{iu} + H_{ib} + (A_u f_u^2 + A_b f_b^2)16E/3l^2} \tag{4.123}$$

In Eqs. (4.121) to (4.123), the initial tensions H_{iu} and H_{ib} generally are relatively small compared with the other terms and can be neglected. If then $f_u = f_b$, as is often the case, Eq. (4.122) simplifies to

$$\Delta w_i = \frac{A_u}{A_u + A_b} p \tag{4.124}$$

and Eq. (4.123) becomes

$$p - \Delta w_i = \frac{A_b}{A_u + A_b} p \tag{4.125}$$

The horizontal component of tension in the upper cable for load p may be computed from

$$H_u = H_{iu} - \Delta H_u = \frac{w_i - \Delta w_i}{w_i} H_{iu} \tag{4.126}$$

The maximum vertical component of tension in the upper cable is

$$V_u = \frac{(w_i - \Delta w_i)l}{2} \tag{4.127}$$

The horizontal component of tension in the lower cable may be computed from

$$H_b = H_{ib} + \Delta H_b = \frac{w_c + w_i + p - \Delta w_i}{w_c + w_i} H_{ib} \tag{4.128}$$

The maximum vertical component of tension in the lower cable is

$$V_b = \frac{(w_c + w_i + p - \Delta w_i)l}{2} \tag{4.129}$$

In general, in analysis of cable systems, terms of second-order magnitude may be neglected, but changes in geometry should not be ignored.

Treatment of a cable truss as a discrete system may be much the same as that for a cable network considered a discrete system. For loads applied to the cables between joints, or nodes, the cable segments between nodes are assumed parabolic. The equations given in Art. 4.8 may be used to determine the forces in the segments and the forces applied at the nodes. Equilibrium equations then can be written for the forces at each joint.

These equations, however, generally are not sufficient for determination of the forces acting in the cable system. These forces also depend on the deformed shape of the network. They may be determined from equations for each joint that take into account both equilibrium and displacement conditions.

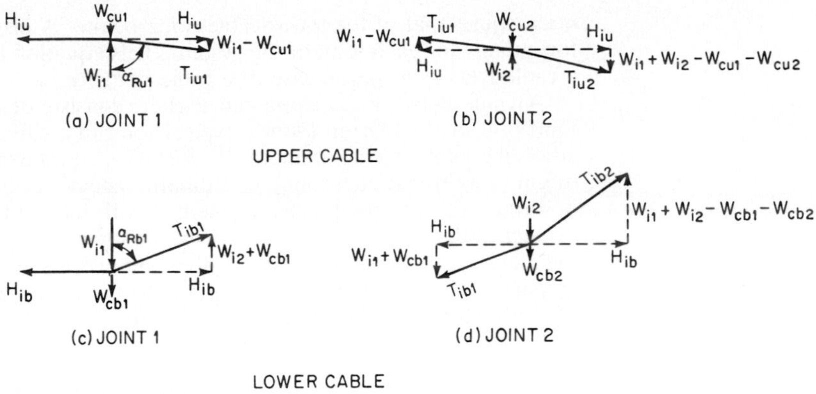

(a) JOINT 1

(b) JOINT 2

UPPER CABLE

(c) JOINT 1

(d) JOINT 2

LOWER CABLE

FIGURE 4.15   Forces acting at joints of a cable system with spreaders.

For a cable truss (Fig. 4.14a) prestressed initially into parabolic shapes, the forces in the cables and spreaders can be found from equilibrium conditions, as indicated in Fig. 4.15. With the horizontal component of the initial tension in the upper cable H_{iu} given, the prestress in the segment to the right of the high point of that cable (joint 1, Fig. 4.15a) is $T_{iu1} = H_{iu}/\cos \alpha_{Ru1}$. The vertical component of this tension equals $W_{i1} - W_{cu1}$, where W_{i1} is the force exerted by the spreader and W_{cu1} is the load on joint 1 due to the weight of the upper cable. [If the cable is symmetrical about the high point, the vertical component of tension in the cable segment is $(W_{i1} - W_{cu1})/2$.] The direction cosine of the cable segment $\cos \alpha_{Ru1}$ is determined by the geometry of the upper cable after it is prestressed.Hence W_{i1} can be computed readily when H_{iu} is known.

With W_{i1} determined, the initial tension in the lower cable at its low point (joint 1, Fig. 4.15c) can be found from equilibrium requirements in similar fashion and by setting the bending moment at the low point equal to zero. Similarly, the cable and spreader forces at adjoining joints (joint 2, Fig. 4.15b and d) can be determined.

Suppose now vertical loads are applied to the system. They can be resolved into concentrated vertical loads acting at the nodes, such as the load P at a typical joint O_b of the bottom cable, shown in Fig. 4.16b. The equations of Art. 4.8 can be used for the purpose. The loads will cause vertical displacements δ of all the joints. The spreaders, however, ensure that the vertical displacement of each upper-cable

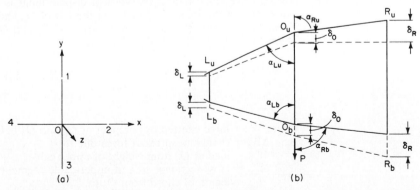

(a)

(b)

FIGURE 4.16   (a) Typical joint in a cable network. (b) Displacement of the cables in a network caused by a load acting at a joint.

node equals that of the lower-cable node below. A displacement equation can be formulated for each joint of the system. This equation can be obtained by treating a cable truss as a special case of a cable network.

A cable network, as explained earlier, consists of interconnected cables. Let joint 0 in Fig. 4.16a represent a typical joint in a cable network and 1, 2, 3, . . . adjoining joints. Cable segments 01, 02, 03, . . . intersect at 0. 0 is selected as the origin of a three-dimensional, coordinate system.

In general, a typical cable segment 0r will have direction cosines $\cos \alpha_{rx}$ with respect to the x axis, $\cos \alpha_{ry}$ with respect to the y axis, and $\cos \alpha_{rz}$ with respect to the z axis. A load P at 0 can be resolved into components P_x parallel to the x axis, P_y parallel to the y axis, and P_z parallel to the z axis. Similarly, the displacement of any joint r can be resolved into components δ_{rx}, δ_{ry}, and δ_{rz}. For convenience, let

$$\Delta_x = \delta_{rx} - \delta_{0x} \qquad \Delta_y = \delta_{ry} - \delta_{0y} \qquad \Delta_z = \delta_{rz} - \delta_{0z} \qquad (4.130)$$

For a cable-network joint, in general, then, where n cable segments interconnect, three equations can be established:

$$\sum_{r=1}^{n} \left[\frac{EA_r}{l_r} \cos \alpha_{rz} (\Delta_x \cos \alpha_{rx} + \Delta_y \cos \alpha_{ry} + \Delta_z \cos \alpha_{rz}) \right] + P_z = 0 \qquad (4.131a)$$

$$\sum_{r=1}^{n} \left[\frac{EA_r}{l_r} \cos \alpha_{ry} (\Delta_x \cos \alpha_{rx} + \Delta_y \cos \alpha_{ry} + \Delta_z \cos \alpha_{rz}) \right] + P_y = 0 \qquad (4.131b)$$

$$\sum_{r=1}^{n} \left[\frac{EA_r}{l_r} \cos \alpha_{rx} (\Delta_x \cos \alpha_{rx} + \Delta_y \cos \alpha_{ry} + \Delta_z \cos \alpha_{rz}) \right] + P_x = 0 \qquad (4.131c)$$

where E = modulus of elasticity of steel cable
A_r = cross-sectional area of cable segment 0r
l_r = length of chord from 0 to r

These equations are based on the assumption that deflections are small and that, for any cable segment, initial tension T_i can be considered negligible compared with EA.

For a cable truss, $n = 2$ for a typical joint. If only vertical loading is applied, only Eq. (4.131a) is needed. At typical joints O_u of the upper cable and O_b of the bottom cable (Fig. 4.16b), the vertical displacement is denoted by δ_O. The displacements of the joints L_u and L_b on the left of O_u and O_b are indicated by δ_L. Those of the joints R_u and R_b on the right of O_u and O_b are represented by δ_R. Then, for joint O_u, Eq. (4.131a) becomes

$$\frac{EA_{Lu}}{l_{Lu}} \cos^2 \alpha_{Lu} (\delta_L - \delta_O) + \frac{EA_{Ru}}{l_{Ru}} \cos^2 \alpha_{Ru} (\delta_R - \delta_O) = W_i - \Delta W_i - W_{cu}$$

$$(4.132)$$

where W_i = force exerted by spreader at O_u and O_b before application of P
ΔW_i = change in spreader force due to P
W_{cu} = load at O_u from weight of upper cable
A_{Lu} = cross-sectional area of upper-cable segment on the left of O_u
l_{Lu} = length of chord from O_u to L_u
A_{Ru} = cross-sectional area of upper-cable segment on the right of O_u
l_{Ru} = length of chord from O_u to R_u

For joint O_b, Eq. (4.131a) becomes, on replacement of subscript u by b,

$$\frac{EA_{Lb}}{l_{Lb}}\cos^2\alpha_{Lb}(\delta_L - \delta_O) + \frac{EA_{Rb}}{l_{Rb}}\cos^2\alpha_{Rb}(\delta_R - \delta_O) = -P - W_i + \Delta W_i - W_{cb}$$

(4.133)

where W_{cb} is the load at O_b due to weight of lower cable and spreader.

Thus, for a cable truss with m joints in each cable, there are m unknown vertical displacements δ and m unknown changes in spreader force ΔW_i. Equations (4.132) and (4.133), applied to upper and lower nodes, respectively, provide $2m$ equations. Simultaneous solution of these equations yields the displacements and forces needed to complete the analysis.

The direction cosines in Eqs. (4.131) to (4.133), however, should be those for the displaced cable segments. If the direction cosines of the original geometry of a cable network are used in these equations, the computed deflections will be larger than the true deflections, because cables become stiffer as sag increases. The computed displacements, however, may be used to obtain revised direction cosines. The equations may then be solved again to yield corrected displacements. The process can be repeated as many times as necessary for convergence, as was shown for a single cable in Art 4.8.

For cable networks in general, convergence can often be speeded by computing the direction cosines for the third cycle of solution with node displacements that are obtained by averaging the displacements at each node computed in the first two cycles.

(H. Mollman, "Analysis of Plane Prestressed Cable Structures," *Journal of the Structural Division*, ASCE, vol. 96, no. ST10, *Proceedings Paper* 7598, October 1970, pp. 2059–2082; D. P. Greenberg, "Inelastic Analysis of Suspension Roof Structures," *Journal of the Structural Division*, ASCE, vol. 96, no. ST5, *Proceedings Paper* 7284, May 1970, pp. 905–930; H. Tottenham and P. G. Williams, "Cable Net: Continuous System Analysis," *Journal of the Engineering Mechanics Division*, ASCE, vol. 96, no. EM3, *Proceedings Paper* 7347, June 1970, pp. 277–293; A. Siev, "A General Analysis of Prestressed Nets," *Publications, International Association for Bridge and Structural Engineering*, vol. 23, pp. 283–292, Zurich, Switzerland, 1963; A. Siev, "Stress Analysis of Prestressed Suspended Roofs," *Journal of the Structural Division*, ASCE, vol. 90, no. ST4, *Proceedings Paper* 4008, August 1964, pp. 103–121; C. H. Thornton and C. Birnstiel, "Three-Dimensional Suspension Structures," *Journal of the Structural Division*, ASCE, vol. 93, no. ST2, *Proceedings Paper* 5196, April 1967, pp. 247–270. *Cable Roof Structures*, Bethlehem Steel Corporation.)

4.10 PLANE-GRID FRAMEWORKS

A **plane grid** comprises a system of two or more members occurring in a single plane, interconnected at intersections, and carrying loads perpendicular to the plane. Grids comprised of beams, all occurring in a single plane, are referred to as **single-layer grids.** Grids comprised of trusses and those with bending members located in two planes with members maintaining a spacing between the planes are usually referred to as **double-layer grids.**

The connection between the grid members is such that all members framing into a particular joint will be forced to deflect the same amount. They are also connected so that bending moment is transferred across the joint. Although it is possible that torsion may be transferred into adjacent members, normally, torsion is not considered in grids comprised of steel beams because of their low torsional stiffness.

Methods of analyzing single- and double-layer framing generally are similar. This article therefore will illustrate the technique with the simpler plane framing and with girders instead of plane trusses. Loading will be taken as vertical. Girders will be assumed continuous at all nodes, except terminals.

Girders may be arranged in numerous ways for plane-grid framing. Figure 4.17 shows some ways of placing two sets of girders. The grid in Fig. 4.17a consists of orthogonal sets laid perpendicular to boundary girders. Columns may be placed at the corners, along the boundaries, or at interior nodes. In the following analysis, for illustrative purposes, columns will be assumed only at the corners, and interior girders will be assumed simply supported on the boundary girders. With wider spacing of interior girders, the arrangement shown in Fig. 4.17b may be preferable. With beams in alternate bays spanning in perpendicular directions, loads are uniformly distributed to the girders. Alternatively, the interior girders may be set parallel to the main diagonals, as indicated in Fig. 4.17c. The method of analysis for this case is much the same as for girders perpendicular to boundary members. The structure, however, need not be rectangular or square, nor need the interior members be limited to two sets of girders.

Many methods have been used successfully to obtain exact or nearly exact solutions for grid framing, which may be highly indeterminate. These include consistent deflections, finite differences, moment distribution or slope deflection, flat plate analogy, and model analysis. This article will be limited to illustrating the use of the method of consistent deflections.

In this method, each set of girders is temporarily separated from the other sets. Unknown loads satisfying equilibrium conditions then are applied to each set. Equations are obtained by expressing node deflections in terms of the loads and equating the deflection at each node of one set to the deflection of the same node in another set. Simultaneous solution of the equations yields the unknown loads on each set. With these now known, bending moments, shears, and deflections of all the girders can be computed by conventional methods.

For a simply supported grid, the unknowns generally can be selected and the equations formulated so that there is one unknown and one equation for each interior node. The number of equations required, however, can be drastically reduced if the framing is made symmetrical about perpendicular axes and the loading is symmetrical or antisymmetrical. For symmetrical grids subjected to unsymmetrical loading, the amount of work involved in analysis often can be decreased by resolving loads into symmetrical and antisymmetrical components. Figure 4.18 shows how this can be done for a single load unsymmetrically located on a grid. The analysis requires the solution of four sets of simultaneous equations and addition of the results, but there are fewer equations in each set than for unsymmetrical loading. The number of unknowns may be further decreased when the

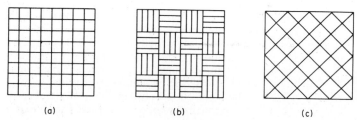

(a)   (b)   (c)

FIGURE 4.17 Orthogonal grids. (a) Girders on short spacing. (b) Girders on wide spacing with beams between them. (c) Girders set diagonally.

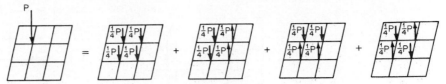

FIGURE 4.18 Resolution of a load into symmetrical and antisymmetrical components.

proportion of a load at a node to be assigned to a girder at that node can be determined by inspection or simple computation. For example, for a square orthogonal grid, each girder at the central node carries half the load there when the grid loading is symmetrical or antisymmetrical.

For analysis of simply supported grid girders, influence coefficients for deflection at any point induced by a unit load are useful. They may be computed from the following formulas.

The deflection at a distance xL from one support of a girder produced by a concentrated load P at a distance kL from that support (Fig. 4.19) is given by

$$\delta = \frac{PL^3}{6EI} x(1 - k)(2k - k^2 - x^2)$$

$$0 \le x \le k \quad (4.134)$$

$$\delta = \frac{PL^3}{6EI} k(1 - x)(2x - x^2 - k^2)$$

$$k \le x \le 1 \quad (4.135)$$

where L = span of simply supported girder

E = modulus of elasticity of the steel

I = moment of inertia of girder cross section

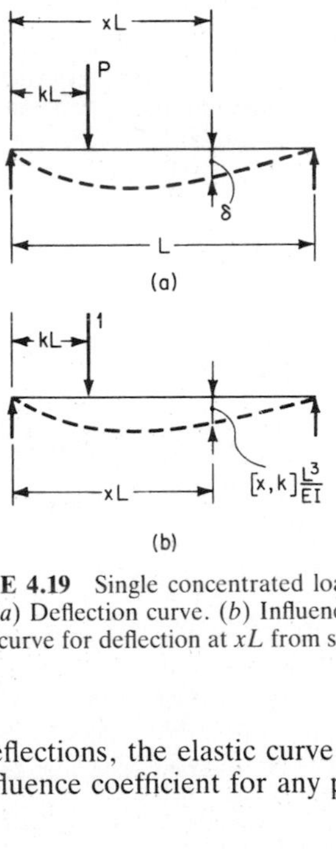

FIGURE 4.19 Single concentrated load on a beam. (*a*) Deflection curve. (*b*) Influence-coefficients curve for deflection at xL from support.

For deflections, the elastic curve is also the influence curve, when $P = 1$. Hence the influence coefficient for any point of the girder may be written

$$\delta' = \frac{L^3}{EI} [x, k] \qquad (4.136)$$

where
$$[x, k] = \begin{cases} \dfrac{x}{6} (1 - k)(2k - k^2 - x^2) & 0 \le x \le k \\[2mm] \dfrac{k}{6} (1 - x)(2x - x^2 - k^2) & k \le x \le 1 \end{cases} \qquad (4.137)$$

The deflection at a distance xL from one support of the girder produced by concentrated loads P at distances kL and $(1 - k)L$ from that support (Fig. 4.20) is given by

$$\delta = \frac{PL^3}{EI} (x, k) \qquad (4.138)$$

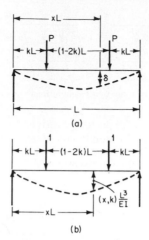

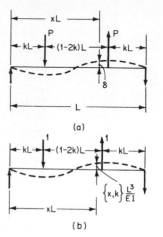

FIGURE 4.20 Two equal downward-acting loads symmetrically placed on a beam. (*a*) Deflection curve. (*b*) Influence-coefficients curve.

FIGURE 4.21 Equal upward and downward concentrated loads symmetrically placed on a beam. (*a*) Deflection curve. (*b*) Influence-coefficients curve.

where
$$(x, k) = \begin{cases} \dfrac{x}{6}(3k - 3k^2 - x^2) & 0 \le x \le k \\ \dfrac{k}{6}(3x - 3x^2 - k^2) & k \le x \le \dfrac{1}{2} \end{cases} \tag{4.139}$$

The deflection at a distance xL from one support of the girder produced by a downward concentrated load P at distance kL from the support and an upward concentrated load P at a distance $(1 - k)L$ from the support (antisymmetrical loading, Fig. 4.21) is given by

$$\delta = \frac{PL^3}{EI}\{x, k\} \tag{4.140}$$

where
$$[x, k] = \begin{cases} \dfrac{x}{6}(1 - 2k)(k - k^2 - x^2) & 0 \le x \le k \\ \dfrac{k}{6}(1 - 2x)(x - x^2 - k^2) & k \le x \le \dfrac{1}{2} \end{cases} \tag{4.141}$$

For convenience in analysis, the loading carried by the grid framing is converted into concentrated loads at the nodes. Suppose, for example, that a grid consists of two sets of parallel girders, as in Fig. 4.17, and the load at interior node r is P_r. Then it is convenient to assume that one girder at the node is subjected to an unknown force X_r there, and the other girder therefore carries a force $P_r - X_r$ at the node. With one set of girders detached from the other set, the deflections produced by these forces can be determined with the aid of Eqs. (4.134) to (4.141).

A simple example will be used to illustrate the application of the method of consistent deflections. Assume an orthogonal grid within a square boundary (Fig. 4.22*a*). There are $n = 4$ equal spaces of width h between girders. Columns are located at the corners A, B, C, and D. All girders have a span $nh = 4h$ and are simply supported at their terminals, though continuous at interior nodes. To simplify the example, all girders are assumed to have equal and constant moment of inertia

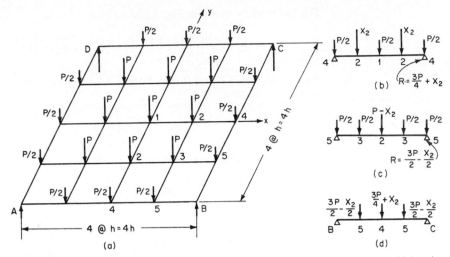

FIGURE 4.22 Square bay with orthogonal grid. (a) Loads distributed to joints. (b) Loads on midspan girder. (c) Loads on quarter-point girder. (d) Loads on boundary girder.

I. Interior nodes carry a concentrated load P. Exterior nodes, except corners, are subjected to a load $P/2$.

Because of symmetry, only five different nodes need be considered. These are numbered from 1 to 5 in Fig. 4.22a, for identification. By inspection, loads P at nodes 1 and 3 can be distributed equally to the girders spanning in the x and y directions. Thus, when the two sets of parallel girders are considered separated, girder 4-4 in the x direction carries a load of $P/2$ at midspan (Fig. 4.22b). Similarly, girder 5-5 in the y direction carries loads of $P/2$ at the quarter points (Fig. 4.22c).

Let X_2 be the load acting on girder 4-4 (x direction) at node 2 (Fig. 4.22b). Then $P - X_2$ acts on girder 5-5 (y direction) at midspan (Fig. 4.22c). The reactions R of girders 4-4 and 5-5 are loads on the boundary girders (Fig. 4.22d).

Because of symmetry, X_2 is the only unknown in this example. Only one equation is needed to determine it.

To obtain this equation, equate the vertical displacement of girder 4-4 (x direction) at node 2 to the vertical displacement of girder 5-5 (y direction) at node 2. The displacement of girder 4-4 equals its deflection plus the deflection of node 4 on BC. Similarly, the displacement of girder 5-5 equals its deflection plus the deflection of node 5 on AB or its equivalent BC.

When use is made of Eqs. (4.136) and (4.138), the deflection of girder 4-4 (x direction) at node 2 equals

$$\delta_2 = \frac{n^3 h^3}{EI}\left\{\left[\frac{1}{4}, \frac{1}{2}\right]\frac{P}{2} + \left(\frac{1}{4}, \frac{1}{4}\right)X_2\right\} + \delta_4 \qquad (4.142a)$$

where δ_4 is the deflection of BC at node 4. By Eq. (4.137), $[\frac{1}{4}, \frac{1}{2}] = (\frac{1}{48})(\frac{11}{16})$. By Eq. (4.139), $(\frac{1}{4}, \frac{1}{4}) = \frac{1}{48}$. Hence

$$\delta_2 = \frac{n^3 h^3}{48EI}\left(\frac{11}{32}P + X_2\right) + \delta_4 \qquad (4.142b)$$

For the loading shown in Fig. 4.22d,

$$\delta_4 = \frac{n^3 h^3}{EI} \left\{ \left[\frac{1}{2}, \frac{1}{2} \right] \left(\frac{3P}{4} + X_2 \right) + \left(\frac{1}{2}, \frac{1}{4} \right) \left(\frac{3P}{2} - \frac{X_2}{2} \right) \right\} \tag{4.143a}$$

By Eq. (4.137), $[\frac{1}{2}, \frac{1}{2}] = \frac{1}{48}$. By Eq. (4.139), $(\frac{1}{2}, \frac{1}{4}) = (\frac{1}{48})(\frac{11}{8})$. Hence Eq. (4.143a) becomes

$$\delta_4 = \frac{n^3 h^3}{48EI} \left(\frac{45}{16} P + \frac{5}{16} X_2 \right) \tag{4.143b}$$

Similarly, the deflection of girder 5-5 (y direction) at node 2 equals

$$\delta_2 = \frac{n^3 h^3}{EI} \left\{ \left[\frac{1}{2}, \frac{1}{2} \right] (P - X_2) + \left(\frac{1}{2}, \frac{1}{4} \right) \frac{P}{2} \right\} + \delta_5 = \frac{n^3 h^3}{48EI} \left(\frac{27}{16} P - X_2 \right) + \delta_5 \tag{4.144}$$

For the loading shown in Fig. 4.22d,

$$\delta_5 = \frac{n^3 h^3}{EI} \left\{ \left[\frac{1}{4}, \frac{1}{2} \right] \left(\frac{3P}{4} + X_2 \right) + \left(\frac{1}{4}, \frac{1}{4} \right) \left(\frac{3P}{2} - \frac{X_2}{2} \right) \right\}$$

$$= \frac{n^3 h^3}{48EI} \left(\frac{129}{64} P + \frac{3}{16} X_2 \right) \tag{4.145}$$

The needed equation for determining X_2 is obtained by equating the right-hand side of Eqs. (4.142b) and (4.144) and substituting δ_4 and δ_5 given by Eqs. (4.143b) and (4.145). The result, after division of both sides of the equation by $n^3 h^3 / 48EI$, is

$$\tfrac{11}{32}P + X_2 + \tfrac{45}{16}P + \tfrac{5}{16}X_2 = \tfrac{27}{16}P - X_2 + \tfrac{129}{64}P + \tfrac{3}{16}X_2 \tag{4.146}$$

Solution of the equation yields

$$X_2 = \frac{35P}{136} = 0.257P \quad \text{and} \quad P - X_2 = \frac{101P}{136} = 0.743P$$

With these forces known, the bending moments, shears, and deflections of the girders can be computed by conventional methods.

To examine a more general case of symmetrical framing, consider the orthogonal grid with rectangular boundaries in Fig. 4.23a. In the x direction, there are n spaces of width h. In the y direction, there are m spaces of width k. Only members symmetrically placed in the grid are the same size. Interior nodes carry a concentrated load P. Exterior nodes, except corners, carry $P/2$. Columns are located at the corners. For identification, nodes are numbered in one quadrant. Since the loading, as well as the framing, is symmetrical, corresponding nodes in the other quadrants may be given corresponding numbers.

At any interior node r, let X_r be the load carried by the girder spanning in the x direction. Then $P - X_r$ is the load at that node applied to the girder spanning in the y direction. For this example, therefore, there are six unknowns X_r, because r ranges from 1 to 6. Six equations are needed for determination of X_r. They may be obtained by the method of consistent deflections. At each interior node, the vertical displacement of the x-direction girder is equated to the vertical displacement of the y-direction girder, as in the case of the square grid.

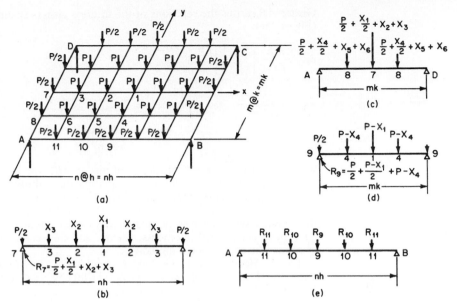

FIGURE 4.23 Rectangular bay with orthogonal girder grid. (*a*) Loads distributed to joints. (*b*) Loads on longer midspan girder. (*c*) Loads on shorter boundary girder AD. (*d*) Loads on shorter midspan girder. (*e*) Loads on longer boundary girder AB.

To indicate the procedure for obtaining these equations, the equation for node 1 in Fig. 4.23*a* will be developed. When use is made of Eqs. (4.136) and (4.138), the deflection of girder 7-7 at node 1 (Fig. 4.23*b*) equals

$$\delta_1 = \frac{n^3 h^3}{E I_7} \left\{ \left[\frac{1}{2}, \frac{1}{2} \right] X_1 + \left(\frac{1}{2}, \frac{1}{3} \right) X_2 + \left(\frac{1}{2}, \frac{1}{6} \right) X_3 \right\} + \delta_7 \qquad (4.147)$$

where I_7 = moment of inertia of girder 7-7
$\quad\quad \delta_7$ = deflection of girder AD at node 7

Girder AD carries the reactions of the interior girders spanning in the x direction (Fig. 4.23*c*):

$$\delta_7 = \frac{m^3 k^3}{E I_{AD}} \left\{ \left[\frac{1}{2}, \frac{1}{2} \right] \left(\frac{P}{2} + \frac{X_1}{2} + X_2 + X_3 \right) + \left(\frac{1}{2}, \frac{1}{4} \right) \left(\frac{P}{2} + \frac{X_4}{2} + X_5 + X_6 \right) \right\}$$

$$(4.148)$$

where I_{AD} is the moment of inertia of girder AD. Similarly, the deflection of girder 9-9 at node 1 (Fig. 4.23*d*) equals

$$\delta_1 = \frac{m^3 k^3}{E I_9} \left\{ \left[\frac{1}{2}, \frac{1}{2} \right] (P - X_1) + \left(\frac{1}{2}, \frac{1}{4} \right) (P - X_4) \right\} + \delta_9 \qquad (4.149)$$

where I_9 = moment of inertia of girder 9-9
$\quad\quad \delta_9$ = deflection of girder AB at node 9

Girder AB carries the reactions of the interior girders spanning in the y direction (Fig. 4.23e):

$$\delta_9 = \frac{n^3 h^3}{EI_{AB}} \left\{ \left[\frac{1}{2}, \frac{1}{2}\right] \left(\frac{P}{2} + \frac{P - X_1}{2} + P - X_4\right) \right.$$

$$+ \left(\frac{1}{2}, \frac{1}{3}\right) \left(\frac{P}{2} + \frac{P - X_2}{2} + P - X_5\right)$$

$$\left. + \left(\frac{1}{2}, \frac{1}{6}\right) \left(\frac{P}{2} + \frac{P - X_3}{2} + P - X_6\right) \right\} \quad (4.150)$$

where I_{AB} is the moment of inertia of girder AB. The equation for vertical displacement at node 1 is obtained by equating the right-hand side of Eqs. (4.147) and (4.149) and substituting δ_7 and δ_9 given by Eqs. (4.148) and (4.150).

After similar equations have been developed for the other five interior nodes, the six equations are solved simultaneously for the unknown forces X_r. When these have been determined, moments, shears, and deflections for the girders can be computed by conventional methods.

(A. W. Hendry and L. G. Jaeger, *Analysis of Grid Frameworks and Related Structures*, Prentice-Hall, Inc., Englewood Cliffs, N.J.; Z. S. Makowski, *Steel Space Structures*, Michael Joseph, London.)

4.11 FOLDED PLATES

Planar structural members inclined to each other and connected along their longitudinal edges comprise a **folded-plate structure** (Fig. 4.24). If the distance between supports in the longitudinal direction is considerably larger than that in the transverse direction, the structure acts much like a beam in the longitudinal direction. In general, however, conventional beam theory does not accurately predict the stresses and deflections of folded plates.

A folded-plate structure may be considered as a series of girders or trusses leaning against each other. At the outer sides, however, the plates have no other members to lean against. Hence the edges at boundaries and at other discontinuities should be reinforced with strong members to absorb the bending stresses there.

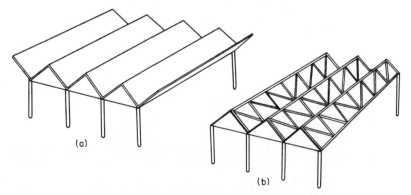

(a)

(b)

FIGURE 4.24   Folded plate roofs. (*a*) Solid plates. (*b*) Trussed plates.

At the supports also, strong members are needed to transmit stresses from the plates into the supports. The structure may be simply supported, or continuous, or may cantilever beyond the supports.

Another characteristic of folded plates that must be taken into account is the tendency of the inclined plates to spread. As with arches, provision must be made to resist this displacement. For the purpose, diaphragms or ties may be placed at supports and intermediate points.

The plates may be constructed in different ways. For example, each plate may be a stiffened steel sheet or hollow roof decking (Fig. 4.24a). Or it may be a plate girder with solid web. Or it may be a truss with sheet or roof decking to distribute loads transversely to the chords (Fig. 4.24b).

A folded-plate structure has a two-way action in transmitting loads to its supports. In the transverse direction, the plates act as slabs spanning between plates on either side. Each plate then serves as a girder in carrying the load received from the slabs longitudinally to the supports.

The method of analysis to be presented assumes the following: The material is elastic, isotropic, and homogeneous. Plates are simply supported but continuously connected to adjoining plates at fold lines. The longitudinal distribution of all loads on all plates is the same. The plates carry loads transversely only by bending normal to their planes and longitudinally only by bending within their planes. Longitudinal stresses vary linearly over the depth of each plate. Buckling is prevented by adjoining plates. Supporting members such as diaphragms, frames, and beams are infinitely stiff in their own planes and completely flexible normal to their planes. Plates have no torsional stiffness normal to their own planes. Displacements due to forces other than bending moments are negligible.

With these assumptions, the stresses in a steel folded-plate structure can be determined by developing and solving a set of simultaneous linear equations based on equilibrium conditions and compatibility at fold lines. The following method of analysis, however, eliminates the need for such equations.

Figure 4.25a shows a transverse section through part of a folded-plate structure. An interior element, plate 2, transmits the vertical loading on it to joints 1 and 2. Usual procedure is to design a 1-ft-wide strip of plate 2 at midspan to resist the transverse bending moment. (For continuous plates and cantilevers, a 1-ft-wide strip at supports also would be treated in the same way as the midspan strip.) If the load is w_2 kips per ft^2 on plate 2, the maximum bending moment in the transverse strip is $w_2 h_2 a_2 / 8$, where h_2 is the depth (feet) of the plate and a_2 is the horizontal projection of h_2.

The 1-ft-wide transverse strip also must be capable of resisting the maximum shear $w_2 h_2 / 2$ at joints 1 and 2. In addition, vertical reactions equal to the shear must be provided at the fold lines. Similarly, plate 1 applies a vertical reaction W_1 kips per ft at joint 1, and plate 3, a vertical reaction $w_3 h_3 / 2$ at joint 2. Thus the total vertical force from the 1-ft-wide strip at joint 2 is

$$R_2 = \tfrac{1}{2}(w_2 h_2 + w_3 h_3) \tag{4.151}$$

Similar transverse strips also load the fold line. It may be considered subject to a uniformly distributed load R_2 kips per ft. The inclined plates 2 and 3 then carry this load in the longitudinal direction to the supports (Fig. 4.25c). Thus each plate is subjected to bending in its inclined plane.

The load to be carried by plate 2 in its plane is determined by resolving R_1 at joint 1 and R_2 at joint 2 into components parallel to the plates at each fold line (Fig. 4.25b). In the general case, the load (positive downward) of the nth plate is

$$P_n = \frac{R_n}{k_n \cos \phi_n} - \frac{R_{n-1}}{k_{n-1} \cos \phi_n} \tag{4.152}$$

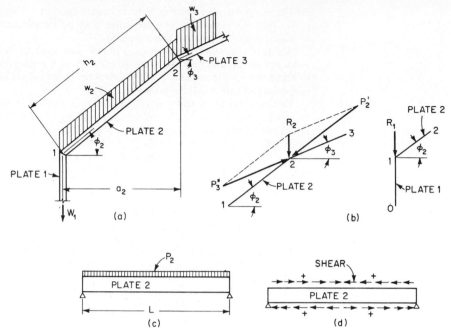

FIGURE 4.25 Forces on folded plates. (*a*) Transverse section. (*b*) Forces at joints 1 and 2. (*c*) Plate 2 acting as girder. (*d*) Shears on plate 2.

where R_n = vertical load, kips per ft, on joint at top of plate n
R_{n-1} = vertical load, kips per ft, on joint at bottom of plate n
ϕ_n = angle, deg, plate n makes with the horizontal
k_n = $\tan \phi_n - \tan \phi_{n+1}$

This formula, however, cannot be used directly for plate 2 in Fig. 4.25*a* because plate 1 is vertical. Hence the vertical load at joint 1 is carried only by plate 1. So plate 2 must carry

$$P_2 = \frac{R_2}{k_2 \cos \phi_2} \tag{4.153}$$

To avoid the use of simultaneous equations for determining the bending stresses in plate 2 in the longitudinal direction, assume temporarily that the plate is disconnected from plates 1 and 3. In this case, maximum bending moment, at midspan, is

$$M_2 = \frac{P_2 L^2}{8} \tag{4.154}$$

where L is the longitudinal span (ft). Maximum bending stresses then may be determined by the beam formula $f = \pm M/S$, where S is the section modulus. The positive sign indicates compression, and the negative sign tension.

For solid-web members, $S = I/c$, where I is the moment of inertia of the plate cross section and c is the distance from the neutral axis to the top or bottom of the plate. For trusses, the section modulus (in³) with respect to the top and bottom, respectively, is given by

$$S_t = A_t h \qquad S_b = A_b h \tag{4.155}$$

where A_t = cross-sectional area of top chord, in²
A_b = cross-sectional area of bottom chord, in²
h = depth of truss, in

In the general case of a folded-plate structure, the stress in plate n at joint n, computed on the assumption of a free edge, will not be equal to the stress in plate $n + 1$ at joint n, similarly computed. Yet, if the two plates are connected along the fold line n, the stresses at the joint should be equal. To restore continuity, shears are applied to the longitudinal edges of the plates (Fig. 4.25d). The unbalanced stresses at each joint then may be adjusted by converging approximations, similar to moment distribution.

If the plates at a joint were of constant section throughout, the unbalanced stress could be distributed in proportion to the reciprocal of their areas. For a symmetrical girder, the unbalance should be distributed in proportion to the factor

$$F = \frac{1}{A} \left(\frac{h^2}{2r^2} + 1 \right) \tag{4.156}$$

where A = cross-sectional area, in², of girder
h = depth, in, of girder
r = radius of gyration, in, of girder cross section

And for an unsymmetrical truss, the unbalanced stress at the top should be distributed in proportion to the factor

$$F_t = \frac{1}{A_t} + \frac{1}{A_b + A_t} \tag{4.157}$$

The unbalance at the bottom should be distributed in proportion to

$$F_b = \frac{1}{A_b} + \frac{1}{A_b + A_t} \tag{4.158}$$

A carry-over factor of $-\frac{1}{2}$ may be used for distribution to the adjoining edge of each plate. Thus the part of the unbalance assigned to one edge of a plate at a joint should be multiplied by $-\frac{1}{2}$, and the product should be added to the stress at the other edge.

After the bending stresses have been adjusted by distribution, if the shears are needed, they may be computed from

$$T_n = T_{n-1} - \frac{f_{n-1} + f_n}{2} A_n \tag{4.159}$$

for true plates, and for trusses, from

$$T_n = T_{n-1} - f_{n-1} A_b - f_n A_t \tag{4.160}$$

where T_n = shear, kips, at joint n
f_n = bending stress, ksi, at joint n
A_n = cross-sectional area, in², of plate n

Usually, at a boundary edge, joint 0, the shear is zero. With this known, the shear at joint 1 can be computed from the preceding equations. Similarly, the shear can be found at successive joints. For a simply supported, uniformly loaded, folded plate, the shear stress f_v (ksi) at any point on an edge n is approximately

$$f_v = \frac{T_{\max}}{18Lt}\left(\frac{1}{2} - \frac{x}{L}\right) \tag{4.161}$$

where x = distance, ft, from a support
t = web thickness of plate, in
L = longitudinal span, ft, of plate

As an illustration of the method, stresses will be computed for the folded-plate structure in Fig. 4.26a. It may be considered to consist of four inverted-V girders,

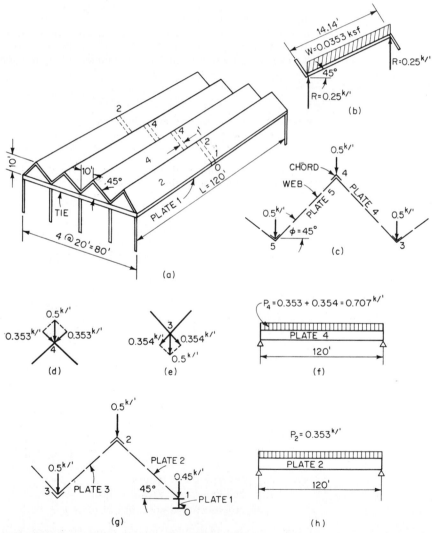

FIGURE 4.26 (*a*) Folded-plate roof. (*b*) Plate reactions for transverse span. (*c*) Loads at joints of typical interior transverse section. (*d*) Forces at joint 4. (*e*) Forces at joint 3. (*f*) Plate 4 acting as girder. (*g*) Loads at joints of outer transverse section. (*h*) Plate 2 acting as girder.

each simply supported with a span of 120 ft. The plates are inclined at an angle of 45° with the horizontal. With a rise of 10 ft and horizontal projection $a = 10$ ft, each plate has a depth $h = 14.14$ ft. The structure is subjected to a uniform load $w = 0.0353$ ksf over its surface. The inclined plates will be designed as trusses. The boundaries, however, will be reinforced with a vertical member, plate 1. The structure is symmetrical about joint 5.

As indicated in Fig. 4.26a, a 1-ft-wide strip is selected transversely across the structure at midspan. This strip is designed to transmit the uniform load w to the folds. It requires a vertical reaction of $0.0353 \times 14.14/2 = 0.25$ kip per ft along each joint (Fig. 4.26b). Because of symmetry, a typical joint then is subjected to a uniform load of $2 \times 0.25 = 0.5$ kip per ft (Fig. 4.26c). At joint 1, the top of the vertical plate, however, the uniform load is 0.25 plus a load of 0.20 on plate 1, or 0.45 kip per ft (Fig. 4.26g).

The analysis may be broken into two parts, to take advantage of simplification permitted by symmetry. First, the stresses may be determined for a typical interior inverted-V girder. Then, the stresses may be computed for the boundary girders, including plate 1.

The typical interior girder consists of plates 4 and 5, with load of 0.5 kip per ft at joints 3, 4, and 5 (Fig. 4.26c). This load may be resolved into loads in the plane of the plates, as indicated in Fig. 4.26d and e. Thus a typical plate, say plate 4, is subjected to a uniform load of 0.707 kip per ft (Fig. 4.26f). Hence the maximum bending moment in this truss is

$$M = \frac{0.707(120)^2}{8} = 1273 \text{ ft-kips}$$

Assume now that each chord is an angle $8 \times 8 \times \%_{16}$ in, with an area of 8.68 in². Then the chords, as part of plate 4, have a maximum bending stress of

$$f = \pm\frac{1273}{8.68 \times 14.14} = \pm 10.36 \text{ ksi}$$

Since the plate is typical, adjoining plates also impose an equal stress on the same chords. Hence the total stress in a typical chord is $\pm 10.36 \times 2 = \pm 20.72$ ksi, the stress being compressive along ridges and tensile along valleys.

To prevent the plates composing the inverted-V girder from spreading, a tie is needed at each support. This tie is subjected to a tensile force

$$P = R \cos \phi = 0.707(^{120}\!/_2)0.707 = 30 \text{ kips}$$

The boundary inverted-V girder consists of plates 1, 2, and 3, with a vertical load of 0.5 kip per ft at joints 2 and 3 and 0.45 kip per ft on joint 1. Assume that plate 1 is a W36 × 135. The following properties of this shape are needed: $A = 39.7$ in², $h = 35.55$ in, $A_f = 9.44$ in², $r = 14$ in, $S = 439$ in³. Assume also that the top flange of plate 1 serves as the bottom chord of plate 2. Thus this chord has an area of 9.44 in².

With plate 1 vertical, the load on joint 1 is carried only by plate 1. Hence, as indicated by the resolution of forces in Fig. 4.26d, plate 2 carries a load in its plane of 0.353 kip per ft (Fig. 4.26h). The maximum bending moment due to this load is

$$M = \frac{0.353(120)^2}{8} = 637 \text{ ft-kips}$$

Assume now that the plates are disconnected along their edges. Then the maximum bending stress in the top chord of plate 2, including the stress imposed by bending of plate 3, is

$$f_t = \frac{637}{8.68 \times 14.14} + 10.36 = 5.18 + 10.36 = 15.54 \text{ ksi}$$

and the maximum stress in the bottom chord is

$$f_b = \frac{-637}{9.44 \times 14.14} = -4.77 \text{ ksi}$$

For the load of 0.45 kip per ft, plate 1 has a maximum bending moment of

$$M = \frac{0.45(120)^2 12}{8} = 9730 \text{ in-kips}$$

The maximum stresses due to this load are

$$f = \frac{M}{S} = \pm\frac{9730}{439} = \pm 22.16 \text{ ksi}$$

Because the top flange of the girder has a compressive stress of 22.16 ksi, whereas acting as the bottom chord of the truss, the flange has a tensile stress of 4.77 ksi, the stresses at joint 1 must be adjusted. The unbalance is 22.16 + 4.77 = 26.93 ksi.

The distribution factor at joint 1 for plate 2 can be computed from Eq. (4.158):

$$F_2 = \frac{1}{9.44} + \frac{1}{9.44 + 8.68} = 0.1611$$

The distribution factor for plate 1 can be obtained from Eq. (4.156):

$$F_1 = \frac{1}{39.7}\left[\frac{(35.5)^2}{2(14)^2} + 1\right] = 0.1062$$

Hence the adjustment in the stress in the girder top flange is

$$\frac{-26.93 \times 0.1062}{0.1062 + 0.1611} = -10.70 \text{ ksi}$$

The adjusted stress in that flange then is 22.16 − 10.70 = 11.46 ksi. The carry-over to the bottom flange is $(-\frac{1}{2})(-10.70) = 5.35$ ksi. And the adjusted bottom flange stress is −22.16 + 5.35 = −16.87 ksi.

Plate 2 receives an adjustment of 26.93 − 10.70 = 16.23 ksi. As a check, its adjusted stress is −4.77 + 16.23 = 11.46 ksi, the same as that in the top flange of plate 1. The carry-over to the top chord is $(-\frac{1}{2})16.23 = -8.12$. The unbalanced stress now present at joint 2 may be distributed in a similar manner, the distributed stresses may be carried over to joints 1 and 3, and the unbalance at those joints may be further distributed. The adjustments beyond joint 2, however, will be small.

(V. S. Kelkar and R. T. Sewell, *Fundamentals of the Analysis and Design of Shell Structures*, Prentice-Hall, Englewood Cliffs, N.J.)

4.12 ORTHOTROPIC PLATES

Plate equations are applicable to steel plate used as a deck. Between reinforcements and supports, a constant-thickness deck, loaded within the elastic range, acts as an isotropic elastic plate. But when a deck is attached to reinforcing ribs or is continuous over relatively closely spaced supports, its properties change in those directions. The plate becomes **anistropic.** And if the ribs and floorbeams are perpendicular to each other, the plate is **orthogonal-anistropic,** or **orthotropic** for short.

An orthotropic-plate deck, such as the type used in bridges, resembles a plane-grid framework (Art. 4.10). But because the plate is part of the grid, an orthotropic-plate structure is even more complicated to analyze. In a bridge, the steel deck plate, protected against traffic and weathering by a wearing surface, serves as the top flange of transverse floorbeams and longitudinal girders and is reinforced longitudinally by ribs (Fig. 4.27). The combination of deck with beams and girders permits design of bridges with attractive long, shallow spans.

Ribs, usually of constant dimensions and closely spaced, are generally continuous at floorbeams. The transverse beams, however, may be simply supported at girders. The beams may be uniformly spaced at distances ranging from about 4 to 20 ft. Rib spacing ranges from 12 to 24 in.

Ribs may be either open (Fig. 4.28a) or closed (Fig. 4.28b). Open ribs are easier to fabricate and field splice. The underside of the deck is readily accessible for inspection and maintenance. Closed ribs, however, offer greater resistance to torsion. Load distribution consequently is more favorable. Also, less steel and less welding are required than for open-rib decks.

Because of the difference in torsional rigidity and load distribution with open and closed ribs, different equations are used for analyzing the two types of decks. But the general procedure is the same for both.

Stresses in an orthotropic plate are assumed to result from bending of four types of members:

Member I comprises the plate supported by the ribs (Fig. 4.29a). Loads between the ribs cause the plate to bend.

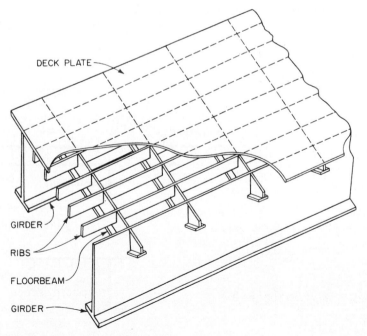

FIGURE 4.27 Orthotropic plate.

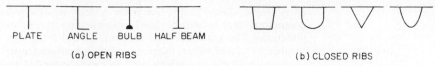

FIGURE 4.28   Types of ribs for orthotropic plates.

Member II consists of plate and longitudinal ribs. The ribs span between and are continuous at floorbeams (Fig. 4.29b). Orthotropic analysis furnishes distribution of loads to ribs and stresses in the member.

Member III consists of the reinforced plate and the transverse floorbeams spanning between girders (Fig. 4.29c). Orthotropic analysis gives stresses in beams and plate.

Member IV comprises girders and plate (Fig. 4.29d). Stresses are computed by conventional methods. Hence determination of girder and plate stresses for this member will not be discussed in this article.

The plate theoretically should be designed for the maximum principal stresses that result from superposition of all bending stresses. In practice, however, this is

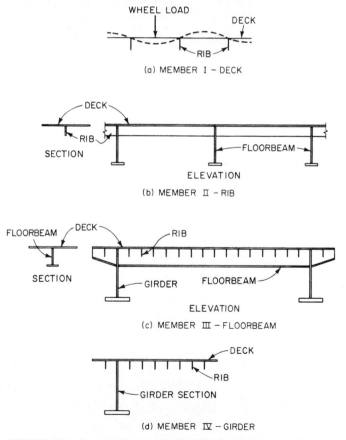

FIGURE 4.29   Four members treated in analysis of orthotropic plates.

not done because of the low probability of the maximum stress and the great reserve strength of the deck as a membrane (second-order stresses) and in the inelastic range.

Special attention, however, should be given to stability against buckling. Also, loading should take into account conditions that may exist at intermediate erection stages.

Despite many simplifying assumptions, orthotropic-plate theories that are available and reasonably in accord with experiments and observations of existing structures require long, tedious computations. (Some or all of the work, however, may be done with computers to speed up the analysis.) The following method, known as the **Pelikan-Esslinger method,** has been used in design of several orthotropic-plate bridges. Though complicated, it still is only an approximate method. Consequently, several variations of it also have been used.

In one variation, members II and III are analyzed in two stages. For the first stage, the floorbeams are assumed as rigid supports for the continuous ribs. Dead- and live-load shears, reactions, and bending moments in ribs and floorbeams then are computed for this condition. For the second stage, the changes in live-load shears, reactions, and bending moments are determined with the assumption that the floorbeams provide elastic support.

Analysis of Member I. Plate thickness generally is determined by a thickness criterion. If the allowable live-load deflection for the span between ribs is limited to $1/300$th of the rib spacing, and if the maximum deflection is assumed as one-sixth of the calculated deflection of a simply supported, uniformly loaded plate, the thickness (in) should be at least

$$t = 0.065a\sqrt[3]{p} \tag{4.162}$$

where a = spacing, in, of ribs
p = load, ksi

The calculated thickness may be increased, perhaps $1/16$ in, to allow for possible metal loss due to corrosion.

The ultimate bearing capacity of plates used in bridge decks may be checked with a formula proposed by K. Kloeppel:

$$p_u = \frac{6.1f_u t}{a}\sqrt{\epsilon_u} \tag{4.163}$$

where p_u = loading, ksi, at ultimate strength
ϵ_u = elongation of the steel, in per in, under stress f_u
f_u = ultimate tensile strength, ksi, of the steel
t = plate thickness, in

Open-Rib Deck—Member II, First Stage. Resistance of the orthotropic plate between the girders to bending in the transverse, or x, direction and torsion is relatively small when open ribs are used compared with flexural resistance in the y direction (Fig. 4.30a). A good approximation of the deflection w (in) at any point (x, y) may therefore be obtained by assuming the flexural rigidity in the x direction and torsional rigidity to be zero. In this case, w may be determined from

$$D_y\frac{\partial^4 w}{\partial y^4} = p(x, y) \tag{4.164}$$

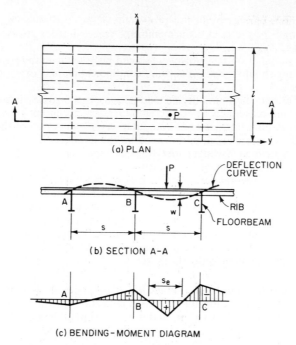

FIGURE 4.30 (*a*) For orthotropic-plate analysis, the *x* axis lies along a floorbeam, the *y* axis along a girder. (*b*) A rib deflects like a continuous beam. (*c*) Length of positive region of rib bending-moment diagram determines effective rib span s_e.

where D_y = flexural rigidity of orthotropic plate in longitudinal, or *y*, direction, in-kips

$p(x, y)$ = load expressed as function of coordinates *x* and *y*, ksi

For determination of flexural rigidity of the deck, the rigidity of ribs is assumed to be continuously distributed throughout the deck. Hence the flexural rigidity in the *y* direction is

$$D_y = \frac{EI_r}{a} \tag{4.165}$$

where E = modulus of elasticity of steel, ksi

I_r = moment of inertia of one rib and effective portion of plate, in^4

a = rib spacing, in

Equation (4.164) is analogous to the deflection equation for a beam. Thus strips of the plate extending in the *y* direction may be analyzed with beam formulas with acceptable accuracy.

In the first stage of the analysis, bending moments are determined for one rib and the effective portion of the plate as a continuous beam on rigid supports. (In this and other phases of the analysis, influence lines or coefficients are useful. See, for example, Table 4.4 and Fig. 4.31.) Distribution of live load to the rib is based on the assumption that the ribs act as rigid supports for the plate. For a distributed

TABLE 4.4 Influence Coefficients for Continuous Beam on Rigid Supports

Constant moment of inertia and equal spans

y/s	Midspan moments at C m_C/s	End moments at 0 m_0/s	Reactions at 0 r_0
0	0	0	1.000
0.1	0.0215	−0.0417	0.979
0.2	0.0493	−0.0683	0.922
0.3	0.0835	−0.0819	0.835
0.4	0.1239	−0.0849	0.725
0.5	0.1707	−0.0793	0.601
0.6	0.1239	−0.0673	0.468
0.7	0.0835	−0.0512	0.334
0.8	0.0493	−0.0331	0.207
0.9	0.0215	−0.0153	0.093
1.0	0	0	0
1.2	−0.0250	0.0183	−0.110
1.4	−0.0311	0.0228	−0.137
1.6	−0.0247	0.0180	−0.108
1.8	−0.0122	0.0089	−0.053
2.0	0	0	0
2.2	0.0067	−0.0049	0.029
2.4	0.0083	−0.0061	0.037
2.6	0.0066	−0.0048	0.029
2.8	0.0032	−0.0023	0.014
3.0	0	0	0

load with width B in, centered over the rib, the load carried by the rib is given in Table 4.5 for B/a ranging from 0 to 3. For B/a from 3 to 4, the table gives the load taken by one rib when the load is centered between two ribs. The value tabulated in this range is slightly larger than that for the load centered over a rib. Uniform dead load may be distributed equally to all the ribs.

The effective width of plate as the top flange of the rib is a function of the rib span and end restraints. In a loaded rib, the end moments cause two inflection points to form. In computation of the effective width, therefore, the effective span s_e (in) of the rib should be taken as the distance between those points, or length of positive-moment region of the bending-moment diagram (Fig. 4.30c). A good approximation is

$$s_e = 0.7s \qquad (4.166)$$

where s is the floorbeam spacing (in).

The ratio of effective plate width a_o (in) to rib spacing a (in) is given in Table 4.5 for a range of values of B/a and a/s_e. Multiplication of a_o/a by a gives the width of the top flange of the T-shaped rib (Fig. 4.32).

Open-Rib Deck—Member III, First Stage.   For the condition of floorbeams acting as rigid supports for the rib, dead-load and live-load moments for a beam are computed with the assumption that the girders provide rigid support. The effective

TABLE 4.5 Analysis Ratios for Open Ribs

Ratio of load width to rib spacing B/a	Ratio of load on one rib to total load R_o/P	Ratio of effective plate width to rib spacing a_o/a for the following ratios of rib spacing to effective rib span a/s_e							
		0	0.1	0.2	0.3	0.4	0.5	0.6	0.7
0	1.000	2.20	2.03	1.62	1.24	0.964			
0.5	0.959	2.16	1.98	1.61	1.24	0.970	0.777		
1.0	0.854	2.03	1.88	1.56	1.24	0.956	0.776		
1.5	0.714	1.83	1.73	1.47	1.19	0.938	0.776		
2.0	0.567	1.60	1.52	1.34	1.12	0.922	0.760	0.641	
2.5	0.440	1.34	1.30	1.18	1.04	0.877	0.749	0.636	0.550
3.0	0.354	1.15	1.13	0.950	0.936	0.827	0.722	0.626	0.543
3.5	0.296	0.963	0.951	0.902	0.832	0.762	0.675	0.604	0.535
4.0	0.253	0.853	0.843	0.812	0.760	0.699	0.637	0.577	0.527

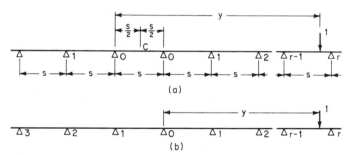

FIGURE 4.31 Continuous beam with constant moment of inertia and equal spans on rigid supports. (*a*) Coordinate *y* for load location for midspan moment at *C*. (*b*) Coordinate *y* for reaction and end moment at *O*.

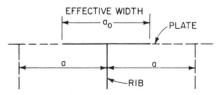

FIGURE 4.32 Effective width of open rib.

width s_o (in) of the plate acting as the top flange of the T-shaped floorbeam is a function of the span and end restraints. For a simply supported beam, in the computation of effective plate width, the effective span l_e (in) may be taken approximately as

$$l_e = l \qquad (4.167)$$

where l is the floorbeam span (in).

For determination of floorbeam shears, reactions, and moments, s_o may be taken as the floorbeam spacing. For stress computations, the ratio of effective plate width s_o to effective beam spacing s_f (in) may be obtained from Table 4.6. When all beams are equally loaded

$$s_f = s \qquad (4.168)$$

The effect of using this relationship for calculating stresses in unequally loaded floorbeams generally is small. Multiplication of s_o/s_f given by Table 4.6 by s yields, for practical purposes, the width of the top flange of the T-shaped floorbeam.

Open-Rib Deck—Member II, Second Stage. In the second stage, the floorbeams act as elastic supports for the ribs under live loads. Deflection of the beams, in proportion to the load they are subjected to, relieves end moments in the ribs but increases midspan moments.

Evaluation of these changes in moments may be made easier by replacing the actual live loads with equivalent sinusoidal loads. This permits use of a single mathematical equation for the deflection curve over the entire floorbeam span. For this purpose, the equivalent loading may be expressed as a Fourier series. Thus, for the coordinate system shown in Fig. 4.30a, a wheel load P kips distributed over a deck width B (in) may be represented by the Fourier series

$$Q_{nx} = \sum_{n=1}^{\infty} Q_n \sin \frac{n\pi x}{l} \qquad (4.169)$$

where n = integer
x = distance, in, from support
l = span, in, or distance, in, over which equivalent load is distributed

For symmetrical loading, only odd numbers need be used for n. For practical purposes, Q_n may be taken as

$$Q_n = \frac{2P}{l} \sin \frac{n\pi x_P}{l} \qquad (4.170)$$

where x_P is the distance (in) of P from the girder.

Thus, for two equal loads P centered over x_P and c in apart,

$$Q_n = \frac{4P}{l} \sin \frac{n\pi x_P}{l} \cos \frac{n\pi c}{2l} \qquad (4.171)$$

For m pairs of such loads centered, respectively, over $x_1, x_2, \ldots , x_m$,

$$Q_n = \frac{4P}{l} \cos \frac{n\pi c}{2l} \sum_{r=1}^{m} \sin \frac{n\pi x_r}{l} \qquad (4.172)$$

TABLE 4.6 Effective Width of Plate

a_e/s_e e_e/s_e s_f/l_e	a_o/a_e e_o/e_e s_o/s_f	a_e/s_e e_e/s_e s_f/l_e	a_o/a_e e_o/e_e s_o/s_f	a_e/s_e e_e/s_e s_f/l_e	a_o/a_e e_o/e_e s_o/s_f	a_e/s_e e_e/s_e s_f/l_e	a_o/a_e e_o/e_e s_o/s_f
0	1.10	0.20	1.01	0.40	0.809	0.60	0.622
0.05	1.09	0.25	0.961	0.45	0.757	0.65	0.590
0.10	1.08	0.30	0.921	0.50	0.722	0.70	0.540
0.15	1.05	0.35	0.870	0.55	0.671	0.75	0.512

TABLE 4.7 Influence Coefficients for Continuous Beam on Elastic Supports

Constant moment of inertia and equal spans

Flexibility coefficient γ	Midspan moments m_C/s, for unit load at support:			End moments m_0/s, for unit load at support:				Reactions r_0, for unit load at support:		
	0	1	2	0	1	2	3	0	1	2
0.05	0.027	−0.026	−0.002	0.100	−0.045	−0.006	0.001	0.758	0.146	−0.034
0.10	0.045	−0.037	−0.010	0.142	−0.053	−0.021	0.001	0.611	0.226	−0.010
0.50	0.115	−0.049	−0.049	0.260	−0.031	−0.066	−0.032	0.418	0.256	0.069
1.00	0.161	−0.040	−0.069	0.323	−0.001	−0.079	−0.059	0.353	0.245	0.098
1.50	0.193	−0.029	−0.079	0.363	0.023	−0.082	−0.076	0.319	0.236	0.111
2.00	0.219	−0.019	−0.083	0.395	0.043	−0.181	−0.087	0.297	0.228	0.118
4.00	0.291	0.019	−0.087	0.479	0.104	−0.066	−0.108	0.250	0.206	0.127
6.00	0.341	0.049	−0.081	0.534	0.147	−0.048	−0.115	0.226	0.192	0.128
8.00	0.379	0.076	−0.073	0.577	0.182	−0.031	−0.115	0.210	0.182	0.128
10.00	0.411	0.098	−0.064	0.612	0.211	−0.015	−0.113	0.199	0.175	0.127

For a load W distributed over a lane width B (in) and centered over x_W,

$$Q_n = \frac{4W}{n\pi} \sin \frac{n\pi x_W}{l} \sin \frac{n\pi B}{2l} \tag{4.173}$$

And for a load W distributed over the whole span,

$$Q_n = \frac{4W}{n\pi l} \tag{4.174}$$

Bending moments and reactions for the ribs on elastic supports may be conveniently evaluated with influence coefficients. Table 4.7 lists such coefficients for midspan moment, end moment, and reaction of a rib for a unit load over any support (Fig. 4.33).

The influence coefficients are given as a function of the flexibility coefficient of the floorbeam:

$$\gamma = \frac{l^4 I_r}{\pi^4 s^3 a I_f} \tag{4.175}$$

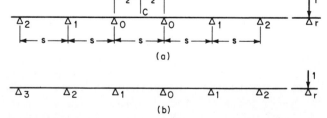

FIGURE 4.33 Continuous beam with constant moment of inertia and equal spans on elastic supports. Load over support for (*a*) midspan moment at *C*, and (*b*) reaction and support moment at *O*.

where I_r = moment of inertia, in, of rib, including effective width of plate
I_f = moment of inertia, in, of floorbeam, including effective width of plate
s = rib span, in
a = rib spacing, in

For calculation of change in moment in the rib, use of only the first term of the series for the equivalent load Q_{1x} yields sufficiently accurate results in calculating the change in moments due to elasticity of the floorbeams. The increase in moment at midspan of a rib may be computed from

$$\Delta M_C = Q_{1x} sa \sum \frac{r_m m_{Cm}}{s} \qquad (4.176)$$

where r_m = influence coefficient for reaction of rib at support m when floorbeam provides rigid support
m_{Cm} = influence coefficient for midspan moment at C for load at support m when floorbeams provide elastic support

The summation in Eq. (4.176) is with respect to subscript m; that is, the effects of reactions at all floorbeams are to be added. The decrease in end moment in the rib may be similarly computed.

The effective width a_o of the rib in this stage may be taken as $1.10a$.

Open-Rib Deck—Member III, Second Stage. Deflection of a floorbeam reduces the reactions of the ribs there. As a result, bending moments also are decreased. The more flexible the floorbeams, the longer the portion of deck over which the loads are distributed and the greater the decrease in beam moment.

The decrease may be computed from

$$\Delta M_f = Q_{1x} \left(\frac{l}{\pi}\right)^2 \left(r_0 - \sum r_m r_{0m}\right) \qquad (4.177)$$

where r_0 = influence coefficient for reaction of rib at support 0 (floorbeam for which moment reduction is being computed) when beam provides rigid support
r_m = influence coefficient for reaction of rib at support m when floorbeam provides rigid support
r_{0m} = influence coefficient for reaction of rib at support 0 for load over support m when floorbeams provide elastic support

The summation of Eq. (4.177) is with respect to m; that is, the effects of reactions at all floorbeams are to be added.

For computation of shears, reactions, and moments, the effective width of plate as the top flange of the floorbeam may be taken as the beam spacing s. For calculation of stresses, the effective width s_o may be obtained from Table 4.6 with $s_f = s$.

Open-Rib Deck—Member IV. The girders are analyzed by conventional methods. The effective width of plate as the top flange on one side of each girder may be taken as half the distance between girders on that side.

Closed-Rib Deck—Member II, First Stage. Resistance of closed ribs to torsion generally is large enough that it is advisable not to ignore torsion. Flexural rigidity in the transverse, or x, direction (Fig. 4.30a) may be considered negligible compared with torsional rigidity and flexural rigidity in the y direction. A good approximation of the deflection w (in) may therefore be obtained by assuming the flexural rigidity in the x direction to be zero. In that case, w may be determined from

$$D_y \frac{\partial^4 w}{\partial y^4} + 2H \frac{\partial w^4}{\partial x^2 \, \partial y^2} = p(x, y) \tag{4.178}$$

where D_y = flexural rigidity of the orthotropic plate in longitudinal, or y, direction, in-kips
H = torsional rigidity of orthotropic plate, in-kips
$p(x, y)$ = load expressed as function of coordinates x and y, ksi

In the computation of D_y and H, the contribution of the plate to these parameters must be included. For the purpose, the effective width of plate acting as the top flange of a rib is obtained as the sum of two components. One is related to the width a (in) of the rib at the plate. The second is related to the distance e (in) between ribs (Fig. 4.34). These components may be computed with the aid of Table

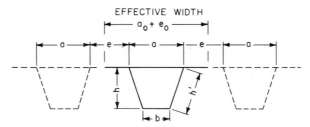

FIGURE 4.34 Effective width of a closed rib.

4.6. For use with this table, the effective rib span s_e (in) may be found from Eq. (4.166). For determination of shears, reactions, and moments, it is sufficiently accurate to take $a_e = a$ and $e_e = e$. (The error in using the resulting section for stress computations usually also will be small.) Then, in terms of the values given by Table 4.6, the effective plate width for a closed rib is

$$a'_o + e'_o = \frac{a_o}{a_e} a_e + \frac{e_o}{e_e} e_e \tag{4.179}$$

The flexural rigidity in the longitudinal direction usually is taken as the average for the orthotropic plate. Thus

$$D_y = \frac{EI_r}{a + e} \tag{4.180}$$

where E = modulus of elasticity of steel, ksi
I_r = moment of inertia, in^4, of rib, including effective plate width

Because of the flexibility of the orthotropic plate in the transverse direction, the full cross section is not completely effective in resisting torsion. Hence the formula for computing H includes a reduction factor ν:

$$H = \frac{1}{2} \frac{\nu GK}{a + e} \tag{4.181}$$

where G = shearing modulus of elasticity of steel = 11,200 ksi
K = torsional factor, a function of the cross section

In general, for hollow closed ribs, the torsional factor may be determined from

$$K = \frac{4A_r^2}{p_e/t_r + a/t_p} \tag{4.182}$$

where A_r = mean of area enclosed by inner and outer boundaries of rib, in^2
p_e = perimeter of rib, exclusive of top flange, in
t_r = rib thickness, in
t_p = plate thickness, in

For a trapezoidal rib, for example,

$$K = \frac{(a + b)^2 h^2}{(b + 2h')/t_r + a/t_p} \tag{4.183}$$

where b = width, in, of rib base
h = depth, in, of rib
h' = length, in, of rib side

The reduction factor ν may be determined analytically. The resulting formulas however, are lengthy, and their applicability to a specific construction is questionable. For a major structure, it is advisable to verify the torsional rigidity, and perhaps also the flexural rigidities, by model tests. For a trapezoidal rib, the reduction factor may be closely approximated by

$$\frac{1}{\nu} = 1 + \frac{GK}{EI_p} \frac{a^2}{12(a + e)^2} \left(\frac{\pi}{s_e}\right)^2 \left[\left(\frac{e}{a}\right)^3 + \left(\frac{e - b}{a + b} + \frac{b}{a}\right)^2\right] \tag{4.184}$$

where I_p = moment of inertia, in per in, of plate alone = $t_p^3/10.92$
s_e = effective rib span for torsion, in = $0.81s$
s = rib span, in

As for open ribs, analysis of an orthotropic plate with closed ribs is facilitated by use of influence coefficients. For computation of these coefficients, Eq. (4.178) reduces to the homogeneous equation

$$D_y \frac{\partial^4 w}{\partial y^4} + 2H \frac{\partial w^4}{\partial x^2 \, \partial y^2} = 0 \tag{4.185}$$

The solution can be given as an infinite series consisting of terms of the form

$$w_n = (C_{1n} \sinh \alpha_n y + C_{2n} \cosh \alpha_n y + C_{3n} \alpha_n y + C_{4n}) \sin \frac{n\pi x}{l} \tag{4.186}$$

where n = integer ranging from 1 to ∞ (odd numbers for symmetrical loads)
l = floorbeam span, in
x = distance, in, from girder
C_{rn} = integration constant, determined by boundary conditions

$$\alpha_n = \frac{n\pi}{l} \sqrt{\frac{2H}{D_y}} \tag{4.187}$$

$\sqrt{H/D_y}$ is called the **plate parameter.**
Because of the sinusoidal form of the deflection surface [Eq. (4.186)] in the x direction, it is convenient to represent loading by an equivalent expressed in a

Fourier series [Eqs. (4.169) to (4.174)]. Convergence of the series may be improved, however, by distributing the loading over a width smaller than l but larger than that of the actual loading.

Influence coefficients for the ribs may be computed with the use of a carry-over factor κ_n for a sinusoidal moment applied at a rigid support. If a moment M is applied at one support of a continuous closed rib, the moment induced at the other end is κM. The carry-over factor for a closed rib is given by

$$\kappa_n = \sqrt{k_n^2 - 1} - k_n \qquad (4.188)$$

where $k_n = (\alpha_n s \coth \alpha_n s - 1)/\beta_n$
$\beta_n = 1 - \alpha_n s/\sinh \alpha_n s$

Thus there is a carry-over factor for each value of n.

Next needed for the computation of influence coefficients are shears, reactions, support moments, and interior moments in a rib span as a sinusoidal load with unit amplitude moves over that span. Since an influence curve also is a deflection curve for the member, these values can be obtained from Eq. (4.186). Consider a longitudinal section through the deflection surface at $x = l/2n$.

Then the influence coefficient for the moment at the support from which y is measured may be computed from

$$m_{0n} = \frac{\kappa_n s}{\beta_n(1 - \kappa_n^2)} \left[-\frac{\kappa_n - \cosh \alpha_n s}{\sinh \alpha_n s} \sinh \alpha_n y - \cosh \alpha_n y + (\kappa_n - 1)\frac{y}{s} + 1 \right]$$

$$(4.189)$$

The coefficient should be determined for each value of n. (If the loading is symmetrical, only odd values of n are needed.) To obtain the influence coefficient when the load is in the next span, m_{0n} should be multiplied by κ_n. And when the load is in either of the following two spans, m_{0n} should be multiplied by κ_n^2 and κ_n^3, respectively.

The influence coefficient for the bending moment at midspan may be calculated from

$$m_{Cn} = \frac{\sinh \alpha_n y}{2\alpha_n \cosh (\alpha_n s/2)} + \frac{\kappa_n s}{2\beta(1 - \kappa_n) \cosh (\alpha_n s/2)}$$

$$\times \left(\tanh \frac{\alpha_n s}{2} \sinh \alpha_n y - \cosh \alpha_n y + 1 \right) \qquad y \leq \frac{s}{2} \quad (4.190)$$

With the influence coefficients known, the bending moment in the rib at $x = l/2n$ can be obtained from

$$M_0 = (a + e) \sum_{n=1}^{\infty} Q_{nx} m_{0n} \qquad \text{or} \qquad M_C = (a + e) \sum_{n=1}^{\infty} Q_{nx} m_{Cn} \quad (4.191)$$

for each equivalent load Q_{nx}. As before, a is the width (in) of the rib at the plate and e is the distance (in) between adjoining ribs (Fig. 4.34).

Closed-Rib Deck—Member III, First Stage. In this stage, the rib reactions on the floorbeams are computed on the assumption that the beams provide rigid support. The reactions may be calculated with the influence coefficients in Table 4.4. The effective width of plate acting as top flange of the floorbeam may be obtained from Table 4.6 with s_f equal to the floorbeam spacing and, for a simply supported beam, $l_e = l$.

Closed-Rib Deck—Members II and III, Second Stage. The analysis for the case of floorbeams providing elastic support is much the same for a closed-rib deck as for open ribs. Table 4.7 can supply the needed influence coefficients. The flexibility coefficient, however, should be computed from

$$\gamma = \frac{l^4 I_r}{\pi^4 s^3 (a + e) I_f} \qquad (4.192)$$

Similarly, the change in midspan moment in a rib can be found from Eq. (4.176) with $a + e$ substituted for a. And the change in floorbeam moment can be obtained from Eq. (4.177). The effective width of plate used for first-stage calculations generally can be used for the second stage with small error.

Closed-Rib Deck—Member IV. The girders are analyzed by conventional methods. The effective width of plate as the top flange on one side of each girder may be taken as half the distance between girders on that side.

(*Design Manual for Orthotropic Steel Plate Deck Bridges*, American Institute of Steel Construction, Chicago, Ill.; M. S. Troitsky, *Orthotropic Bridges Theory and Design*, The James F. Lincoln Arc Welding Foundation, P.O. Box 3035, Cleveland, Ohio 44117; S. P. Timoshenko and S. Woinowsky-Krieger, *Theory of Plates and Shells*, McGraw-Hill, Inc., New York.)

SECTION 5
CONNECTIONS

W. A. Thornton
Chief Engineer, Cives Steel Company, Roswell, Ga.

and

T. Kane
Technical Manager, Cives Steel Company, Roswell, Ga.

In this section, the term *connections* is used in a general sense to include all types of joints in structural steel made with fasteners or welds. Emphasis, however, is placed on the more commonly used connections, such as beam-column connections, main-member splices, and truss connections.

Recommendations apply to buildings and to both highway and railway bridges unless otherwise noted. This material is based on the specifications of the American Institute of Steel Construction (AISC), "Load and Resistance Factor Design Specification for Structural Steel Buildings," 1986, and "Specification for Structural Steel Buildings—Allowable Stress Design and Plastic Design," 1989; the American Association of State Highway and Transportation Officials (AASHTO), "Standard Specifications for Highway Bridges," 1992; and the American Railway Engineering Association (AREA), "Manual for Railway Engineering," 1992.

5.1 LIMITATIONS ON USE OF FASTENERS AND WELDS

Structural steel fabricators prefer that job specifications state that "shop connections shall be made with bolts or welds" rather than restricting the type of connection that can be used. This allows the fabricator to make the best use of available equipment and to offer a more competitive price. For bridges, however, standard specifications restrict fastener choice.

High-strength bolts may be used in either slip-critical or bearing-type connections (Art. 5.3), subject to various limitations. Bearing-type connections have higher allowable loads and should be used where permitted. Also, bearing-type connections may be either fully tensioned or snug-tight, subject to various limitations. Snug-tight bolts are much more economical to install and should be used where permitted.

Bolted slip-critical connections must be used for bridges where stress reversal may occur or slippage is undesirable. In bridges, connections subject to computed tension or combined shear and computed tension must be slip-critical. Bridge construction requires that bearing-type connections with high-strength bolts be limited to members in compression and secondary members.

Carbon-steel bolts should not be used in connections subject to fatigue.

In building construction, snug-tight bearing-type connections can be used for most cases, including connections subject to stress reversal due to wind or seismic loading. The American Institute of Steel Construction (AISC) requires that fully tensioned high-strength bolts or welds be used for connections indicated in Sec. 6.14.2.

The AISC imposes special requirements on use of welded splices and similar connections in heavy sections. This includes ASTM A6 group 4 and 5 shapes and splices in built-up members with plates over 2 in thick subject to tensile stresses due to tension or flexure. Charpy V-notch tests are required, as well as special fabrication and inspection procedures. Where feasible, bolted connections are preferred to welded connections for such sections (see Art. 1.17).

In highway bridges, fasteners or welds may be used in field connections wherever they would be permitted in shop connections. In railroad bridges, the American Railway Engineering Association (AREA) recommended practice requires that field connections be made with high-strength bolts. Welding may be used only for minor connections that are not stressed by live loads and for joining deck plates or other components that are not part of the load-carrying structure.

5.2   BOLTS IN COMBINATION WITH WELDS

In new work, ASTM A307 bolts or high-strength bolts used in bearing-type connections should not be considered as sharing the stress in combination with welds. Welds, if used, should be provided to carry the entire stress in the connection. High-strength bolts proportioned for slip-critical connections may be considered as sharing the stress with welds.

In welded alterations to structures, existing rivets and high-strength bolts tightened to the requirements for slip-critical connections are permitted for carrying stresses resulting from loads present at the time of alteration. The welding needs to be adequate to carry only the additional stress.

In bridges, welds and bolts in the same connection plane should not be considered as sharing the stress.

If two or more of the general types of welds (groove, fillet, plug, slot) are combined in a single joint, the effective capacity of each should be separately computed with reference to the axis of the group in order to determine the allowable capacity of the combination.

AREA does not permit the use of plug or slot welds but will accept fillet welds in holes and slots.

FASTENERS

In steel erection, fasteners commonly used include bolts, welded studs, and pins. Properties of these are discussed in the following articles.

5.3   HIGH-STRENGTH BOLTS, NUTS, AND WASHERS

For general purposes, A325 and A490 high-strength bolts may be specified. Each type of bolt can be identified by the ASTM designation and the manufacturer's

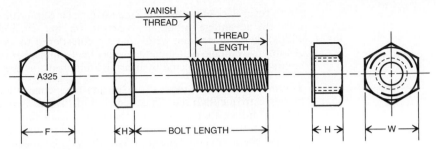

FIGURE 5.1   High-strength structural steel bolt and nut.

mark on the bolt head and nut (Fig. 5.1). The cost of A490 bolts is 15 to 20% greater than that of A325 bolts.

Job specifications often require that "main connections shall be made with bolts conforming to the Specification for Structural Joints Using ASTM A325 and A490 Bolts." This specification, approved by the Research Council on Structural Connections (RCSC) of the Engineering Foundation, establishes bolt, nut, and washer dimensions, minimum fastener tension, and requirements for design and installation.

As indicated in Table 5.1, many sizes of high-strength bolts are available. Most standard connection tables, however, apply primarily to ¾- and ⅞-in bolts. Shop and erection equipment is generally set up for these sizes, and workers are familiar with them.

Bearing versus Slip-Critical Joints.   Connections made with high-strength bolts may be slip-critical (material joined being clamped together by the tension induced in the bolts by tightening them) or bearing-type (material joined being restricted from moving primarily by the bolt shank). In bearing-type connections, bolt threads may be included in or excluded from the shear plane. Different stresses are allowed for each condition. The slip-critical connection is the most expensive, because it requires that the faying surfaces be free of paint, grease, and oil. Hence this type of connection should be used only where required by the governing design specification, e.g., where it is undesirable to have the bolts slip into bearing or where

TABLE 5.1   Thread Lengths for High-Strength Bolts

Bolt diameter, in	Nominal thread, in	Vanish thread, in	Total thread, in
½	1.00	0.19	1.19
⅝	1.25	0.22	1.47
¾	1.38	0.25	1.63
⅞	1.50	0.28	1.78
1	1.75	0.31	2.06
1⅛	2.00	0.34	2.34
1¼	2.00	0.38	2.38
1⅜	2.25	0.44	2.69
1½	2.25	0.44	2.69

stress reversal could cause slippage (Art. 5.1). Slip-critical connections, however, have the advantage in building construction that when used in combination with welds, the fasteners and welds may be considered to share the stress (Art. 5.2). Another advantage that sometimes may be useful is that the strength of slip-critical connections is not affected by bearing limitations, as are other types of fasteners.

Threads in Shear Planes. The bearing-type connection with threads in shear planes is frequently used. Since location of threads is not restricted, bolts can be inserted from either side of a connection. Either the head or the nut can be the element turned. Paint is permitted on the faying surfaces.

The bearing-type connection with threads excluded from shear planes is the most economical high-strength bolted connection, because fewer bolts generally are needed for a given capacity. But this type should be used only after careful consideration of the difficulties involved in excluding the threads from the shear planes. The location of the thread runout depends on which side of the connection the bolt is entered and whether a washer is placed under the head or the nut. This location is difficult to control in the shop but even more so in the field. The difficulty is increased by the fact that much of the published information on bolt characteristics does not agree with the basic specification used by bolt manufacturers (American National Standards Institute B18.2.1).

Total nominal thread lengths and vanish thread lengths for high-strength bolts are given in Table 5.1. It is common practice to allow the last $\frac{1}{8}$ in of vanish thread to extend across a single shear plane.

In order to determine the required bolt length, the value shown in Table 5.2 should be added to the grip (i.e., the total thickness of all connected material, exclusive of washers). For each hardened flat washer that is used, add $\frac{5}{32}$ in, and for each beveled washer, add $\frac{5}{16}$ in. The tabulated values provide appropriate allowances for manufacturing tolerances and also provide for full thread engagement with an installed heavy hex nut. The length determined by the use of Table 5.2 should be adjusted to the next longer $\frac{1}{4}$-in length.

Washer Requirements. The RCSC specification requires that design details provide for washers in connections with high-strength bolts as follows:

1. A hardened beveled washer should be used to compensate for the lack of parallelism where the outer face of the bolted parts has a greater slope than 1:20 with respect to a plane normal to the bolt axis.

TABLE 5.2 Lengths to be Added to Grip

Nominal bolt size, in	Addition to grip for determination of bolt length, in
$\frac{1}{2}$	$\frac{11}{16}$
$\frac{5}{8}$	$\frac{7}{8}$
$\frac{3}{4}$	1
$\frac{7}{8}$	$1\frac{1}{8}$
1	$1\frac{1}{4}$
$1\frac{1}{8}$	$1\frac{1}{2}$
$1\frac{1}{4}$	$1\frac{5}{8}$
$1\frac{3}{8}$	$1\frac{3}{4}$
$1\frac{1}{2}$	$1\frac{7}{8}$

2. For A325 and A490 bolts for slip-critical connections and connections subject to direct tension, hardened washers are required as specified in items 3 through 7 below. For bolts permitted to be tightened only snug-type, if a slotted hole occurs in an outer ply, a flat hardened washer or common plate washer shall be installed over the slot. For other connections with A325 and A490 bolts, hardened washers are not generally required.

3. When the calibrated wrench method is used for tightening the bolts, hardened washers shall be used under the element turned by the wrench.

4. For A490 bolts tensioned to the specified tension, hardened washers shall be used under the head and nut in steel with a specified yield point less than 40 ksi.

5. A hardened washer conforming to ASTM F436 shall be used for A325 or A490 bolts 1 in or less in diameter tightened in an oversized or short slotted hole in an outer ply.

6. Hardened washers conforming to F436 but at least $\frac{5}{16}$ in thick shall be used, instead of washers of standard thickness, under both the head and nut of A490 bolts more than 1 in in diameter tightened in oversized or short slotted holes in an outer ply. This requirement is not met by multiple washers even though the combined thickness equals or exceeds $\frac{5}{16}$ in.

7. A plate washer or continuous bar of structural-grade steel, but not necessarily hardened, at least $\frac{5}{16}$ in thick and with standard holes, shall be used for an A325 or A490 bolt 1 in or less in diameter when it is tightened in a long slotted hole in an outer ply. The washer or bar shall be large enough to cover the slot completely after installation of the tightened bolt. For an A490 bolt more than 1 in in diameter in a long slotted hole in an outer ply, a single hardened washer (not multiple washers) conforming to F436, but at least $\frac{5}{16}$ in thick, shall be used instead of a washer or bar of structural-grade steel.

The requirements for washers specified in items 4 and 5 above are satisfied by other types of fasteners meeting the requirements of A325 or A490 and with a geometry that provides a bearing circle on the head or nut with a diameter at least equal to that of hardened F436 washers. Such fasteners include "twist-off" bolts with a splined end that extends beyond the threaded portion of the bolt. During installation, this end is gripped by a special wrench chuck and is sheared off when the specified bolt tension is achieved.

The RCSC specification permits direct tension-indicating devices, such as washers incorporating small, formed arches designed to deform in a controlled manner when subjected to the tightening force. The specification also provides guidance on use of such devices to assure proper installation (Art. 5.14).

5.4 CARBON-STEEL OR UNFINISHED (MACHINE) BOLTS

"Secondary connections may be made with unfinished bolts conforming to the Specifications for Low-carbon Steel ASTM A307" is an often-used specification. (Unfinished bolts also may be referred to as **machine, common,** or **ordinary bolts**.) When this specification is used, secondary connections should be carefully defined to preclude selection by ironworkers of the wrong type of bolt for a connection (see also Art. 5.1). A307 bolts generally have no identification marks on their square, hexagonal, or countersunk heads (Fig. 5.2), as do high-strength bolts.

Use of high-strength bolts where A307 bolts provide the required strength merely adds to the cost of a structure. High-strength bolts cost at least 10% more than machine bolts.

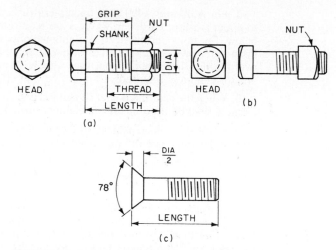

FIGURE 5.2   Unfinished (machine) or common bolts. (*a*) With hexagonal nut and bolt. (*b*) With square head and nut. (*c*) With countersunk head.

A disadvantage of A307 bolts is the possibility that the nuts may loosen. This may be eliminated by use of lock washers. Alternatively, locknuts can be used or threads can be jammed, but either is more expensive than lock washers.

5.5   WELDED STUDS

Fasteners with one end welded to a steel member frequently are used for connecting material. Shear connectors in composite construction are a common application. Welded studs also are used as anchors to attach wood, masonry, or concrete to steel. Types of studs and welding guns vary with manufacturers.

Table 5.3 lists approximate allowable loads for several sizes of threaded studs. Check manufacturer's data for studs to be used. Chemical composition and physical properties may differ from those assumed for this table.

TABLE 5.3   Allowable Loads (kips) on Threaded Welded Studs

(*ASTM A108, grade 1015, 1018, or 1020*)

Stud size, in	Tension	Single shear
⅝	6.9	4.1
¾	10.0	6.0
⅞	13.9	8.3
1	18.2	10.9

Use of threaded studs for steel-to-steel connections can cut costs. For example, fastening rail clips to crane girders with studs eliminates drilling of the top flange of the girders and may permit a reduction in flange size.

FIGURE 5.3 Welded stud.

In designs with threaded studs, clearance must be provided for stud welds. Usual sizes of these welds are indicated in Fig. 5.3 and Table 5.4. The dimension C given is the minimum required to prevent burn-through in stud welding. Other design considerations may require greater thicknesses.

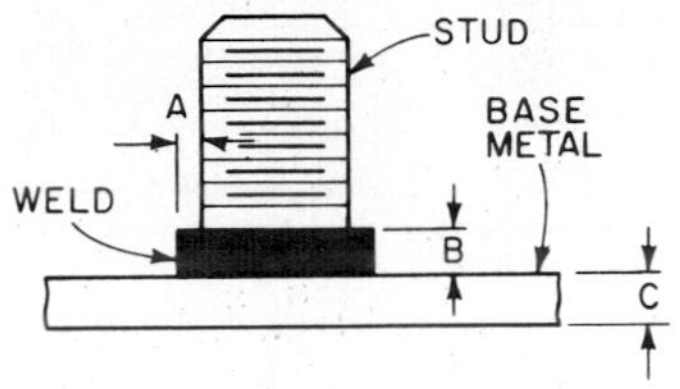

5.6 PINS

A pinned connection is used to permit rotation of the end of a connected member. Some aspects of the design of a pinned connection are the same as those of a bolted bearing connection. The pin serves the same purpose as the shank of a bolt. But since only one pin is present in a connection, forces acting on a pin are generally much greater than those on a bolt. Shear on a pin can be resisted by selecting a large enough pin diameter and an appropriate grade of steel. Bearing on thin webs or plates can be brought within allowable values by addition of reinforcing plates. Because a pin is relatively long, bending, ignored in bolts, must be investigated in choosing a pin diameter. Arrangements of plates on the pin affect bending stresses. Hence plates should be symmetrically placed and positioned to minimize stresses.

Finishing of the pin and its effect on bearing should be considered. Unless the pin is machined, the roundness tolerance may not permit full bearing, and a close fit of the pin may not be possible. The requirements of the pin should be taken into account before a fit is specified.

Pins may be made of any of the structural steels permitted by AISC, AASHTO, and AREA specifications, ASTM A108 grades 1016 through 1030, and A668 classes C, D, F, and G.

Pins must be forged and annealed when they are more than 7 in in diameter for railroad bridges. Smaller pins may be forged and annealed or cold-finished carbon-steel shafting. In pins larger than 9 in in diameter, a hole at least 2 in in diameter must be bored full length along the axis. This work should be done after the forging has been allowed to cool to a temperature below the critical range,

TABLE 5.4 Minimum Weld and Base-Metal Dimensions (in) for Threaded Welded Studs

Stud size, in	A	B and C
⅝	⅛	¼
¾	3/16	5/16
⅞	3/16	⅜
1	¼	7/16

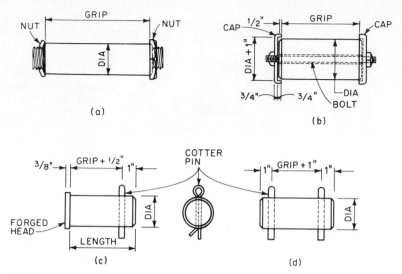

FIGURE 5.4   Pins. (*a*) With recessed nuts. (*b*) With caps and through bolt. (*c*) With forged head and cotter pin. (*d*) With cotter at each end (used in horizontal position).

with precautions taken to prevent injury by too rapid cooling, and before the metal is annealed. The hole permits passage of a bolt with threaded ends for attachment of nuts or caps at the pin ends.

When reinforcing plates are needed on connected material, the plates should be arranged to reduce eccentricity on the pin to a minimum. One plate on each side should be as wide as the outstanding flanges will permit. At least one full-width plate on each segment should extend to the far end of the stay plate. Other reinforcing plates should extend at least 6 in beyond the near edge. All plates should be connected with fasteners or welds arranged to transmit the bearing pressure uniformly over the full section.

In buildings, pinhole diameters should not exceed pin diameters by more than $\frac{1}{32}$ in. In bridges, this requirement holds for pins more than 5 in in diameter, but for smaller pins, the tolerance is reduced to $\frac{1}{50}$ in.

Length of pin should be sufficient to secure full bearing on the turned body of the pin of all connected parts. Pins should be secured in position and connected material restrained against lateral movement on the pins. For the purpose, ends of a pin may be threaded, and hexagonal recessed nuts or hexagonal solid nuts with washers may be screwed on them (Fig. 5.4*a*). Usually made of malleable castings or steel, the nuts should be secured by cotter pins in the screw ends or by burred threads. Bored pins may be held by a recessed cap at each end, secured by a nut on a bolt passing through the caps and the pin (Fig. 5.4*b*). In building work, a pin may be secured with cotter pins (Fig. 5.4*c* and *d*).

The most economical method is to drill a hole in each end for cotter pins. This, however, can be used only for horizontal pins. When a round must be turned down to obtain the required fit, a head can be formed to hold the pin at one end. The other end can be held by a cotter pin or threaded for a nut.

Example.   Determine the diameter of pin required to carry a 320-kip reaction of a deck-truss highway bridge (Fig. 5.5).

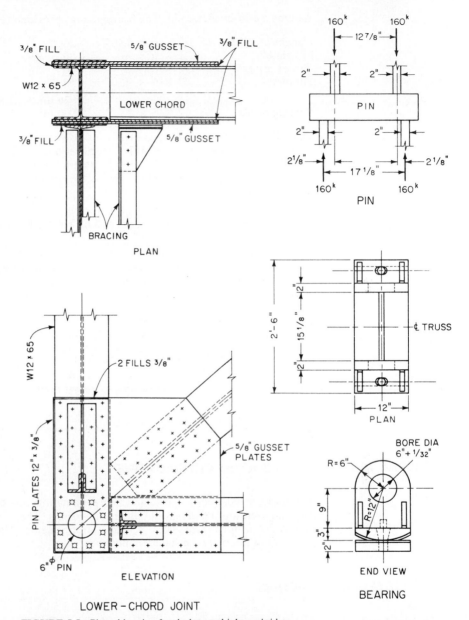

FIGURE 5.5 Pinned bearing for deck-truss highway bridge.

Bearing.   For A36 steel, American Association of State Highway and Transportation Officials (AASHTO) specifications permit a bearing stress of 14 ksi on pins subject to rotation, such as those used in rockers and hinges. Hence the minimum bearing area on the pin must equal

$$A = {}^{320}\!/_{14} = 22.8 \text{ in}^2$$

Assume a 6-in-diameter pin. The bearing areas provided (Fig. 5.5) are

Flanges of W12 × 65   2 × 6 × 0.605 = 7.26
Fill plates   2 × 6 × ⅜   = 4.50
Gusset plates   2 × 6 × ⅝   = 7.50
Pin plates   2 × 6 × ⅜   = 4.50
                                      ────────────────
                                      23.76 in² > 22.8
Bearing plates   2 × 6 × 2   = 24.00 in² > 22.8

The 6-in pin is adequate for bearing.

Shear.   For A36 steel, AASHTO specifications permit a shear stress on pins of 14 ksi. As indicated in the loading diagram for the pin in Fig. 5.5, the reaction is applied to the pin at two points. Hence the shearing area equals $2 \times \pi(6)^2/4 = 56.6$. Thus the shearing stress is

$$f_v = \frac{320}{56.6} = 5.65 \text{ ksi} < 14$$

The 6-in pin is adequate for shear.

Bending.   For A36 steel, AASHTO specifications permit a bending stress of 20 ksi. From the loading diagram for the pin (Fig. 5.5), the maximum bending moment is $M = 160 \times 2\frac{1}{8} = 340$ in-kips. The section modulus of the pin is

$$S = \frac{\pi d^3}{32} = \frac{\pi(6)^3}{32} = 21.2 \text{ in}^3$$

Thus the maximum bending stress in the pin is

$$f_b = \frac{340}{21.2} = 16 \text{ ksi} < 20$$

The 6-in pin also is satisfactory in bending.

GENERAL CRITERIA FOR BOLTED CONNECTIONS

Standard specifications for structural steel for buildings and bridges contain general criteria governing the design of bolted connections. They cover such essentials as permissible fastener size, sizes of holes, arrangements of fasteners, size and attachment of fillers, and installation methods.

5.7 FASTENER DIAMETERS

Minimum bolt diameters are ½ in for buildings and ¾ in for railroad bridges. Structural shapes that do not permit use of ⅝-in fasteners may be used only in handrails. In highway-bridge members carrying calculated stress, ¾-in fasteners are the smallest permitted, in general, but ⅝-in fasteners may be used in 2½-in stressed legs of angles and in flanges of sections requiring ⅝-in fasteners (controlled by required installation clearance to web and minimum edge distance).

In general, a connection with a few large-diameter fasteners costs less than one of the same capacity with many small-diameter fasteners. The fewer the fasteners, the fewer the number of holes to be formed and the less installation work. Larger-diameter fasteners are particularly favorable in connections where shear governs, because the load capacity of a fastener in shear varies with the square of the fastener diameter. For practical reasons, however, ¾- and ⅞-in-diameter fasteners are preferred.

Maximum Fastener Diameters in Angles. In bridges, the diameter of fasteners in angles carrying calculated stress may not exceed one-fourth the width of the leg in which they are placed. In angles where the size is not determined by calculated stress, 1-in fasteners may be used in 3½-in legs, ⅞-in fasteners in 3-in legs, and ¾-in fasteners in 2½-in legs. In addition, in highway bridges, ⅝-in fasteners may be used in 2-in legs.

5.8 FASTENER HOLES

Standard specifications require that holes for bolts be ¹⁄₁₆ in larger than the nominal fastener diameter. In computing net area of a tension member, the diameter of the hole should be taken ¹⁄₁₆ in larger than the hole diameter.

Standard specifications also require that the holes be punched or drilled. Punching usually is the most economical method. To prevent excessive damage to material around the hole, however, the specifications limit the maximum thickness of material in which holes may be punched full size. These limits are summarized in Table 5.5.

In buildings, holes for thicker material may be either drilled from the solid or subpunched and reamed. The die for all subpunched holes and the drill for all subdrilled holes should be at least ¹⁄₁₆ in smaller than the nominal fastener diameter.

In highway bridges, holes for material not within the limits given in Table 5.5 should be subdrilled or drilled full size. Holes in all field connections and field splices of main members of trusses, arches, continuous beams, bents, towers, plate girders, and rigid frames should be subpunched, or subdrilled when required by thickness limitations, and subsequently reamed while assembled or to a steel template. Holes for floorbeam and stringer field end connections should be similarly formed. The die for subpunched holes and the drill for subdrilled holes should be ³⁄₁₆ in smaller than the nominal fastener diameter.

A contractor has the option of forming, with parts for a connection assembled, subpunched holes and reaming or drilling full-size holes. The contractor also has the option of drilling or punching holes full size in unassembled pieces or connec-

TABLE 5.5 Maximum Material Thickness (in) for Punching Fastener Holes*

	AISC	AASHTO	AREA
A36 steel	$d + \frac{1}{8}$†	¾§	⅞
High-strength steels	$d + \frac{1}{8}$†	⅝§	¾
Quenched and tempered steels	½‡	½§	

*Unless subpunching or subdrilling and reaming are used.
†d = fastener diameter, in.
‡A514 steel.
§But not more than five thicknesses of metal.

tions with suitable numerically controlled drilling or punching equipment. In this case, the contractor may be required to demonstrate, by means of check assemblies, the accuracy of this drilling or punching procedure. Holes drilled or punched by numerically controlled equipment are formed to size through individual pieces, but they may instead be formed by drilling through any combination of pieces held tightly together.

In railway bridges, holes for shop and field bolts may be punched full size, within the limits of Table 5.5, in members that will not be stressed by vertical live loads. This provision applies to, but is not limited to, the following: stitch bolts, sway bracing or connecting material, lacing stay plates, diaphragms that do not transfer shear or other forces, inactive fillers, and stiffeners not at bearing points.

Shop-bolt holes to be reamed may be subpunched. Methods permitted for shop-bolt holes in rolled beams and plate girders, including stiffeners and active fillers at bearing points, depend on material thickness and, in some cases, on strength. In materials not thicker than the nominal bolt diameter less ⅛ in, the holes should be subpunched ⅛ in less in diameter than the finished holes and then reamed to size with parts assembled. In A36 material thicker than ⅞ in (¾ in for high-strength steels), the holes should be subdrilled ¼ in less in diameter than the finished holes and then reamed to size with parts assembled.

A special provision applies to the case where matching shop-bolt holes in two or more plies are required to be reamed with parts assembled. If the assembly consists of more than five plies with more than three plies of main material, the matching holes in the other plies also should be reamed with parts assembled. Holes in those plies should be subpunched ⅛ in less in diameter than the finished hole.

Other shop-bolt holes should be subpunched ¼ in less in diameter than the finished hole and then reamed to size with parts assembled.

Field splices in plate girders and in truss chords should be reamed or drilled full size with members assembled. Truss-chord assemblies should consist of at least three abutting sections. Milled ends of the compression chords should have full bearing.

Field-bolt holes may be subpunched or subdrilled ¼ in less in diameter than finished holes in individual pieces. The subsized holes should then be reamed to size through steel templates with hardened steel bushings. In A36 steel thicker than ⅞ in (¾ in for high-strength steels), field-bolt holes may be subdrilled ¼ in less in diameter than the finished holes and then reamed to size with parts assembled or drilled full size with parts assembled. Field-bolt holes for sway bracing should conform to the requirements for shop-bolt holes.

5.9 MINIMUM NUMBER OF FASTENERS

In buildings, connections carrying calculated stresses, except lacing, sag bars, and girts, should be designed to support at least 6 kips.

In highway bridges, connections, including angle bracing but not lacing bars and handrails, should contain at least two fasteners. Web shear splices should have at least two rows of fasteners on each side of the joint.

In railroad bridges, connections should have at least three fasteners per plane of connection.

Long Grips. In buildings, if A307 bolts in a connection carry calculated stress and have grips exceeding five diameters, the number of these fasteners used in the connection should be increased 1% for each additional 1/16 in in the grip.

5.10 CLEARANCES FOR FASTENERS

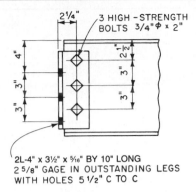

2L-4" x 3½" x ⁵⁄₁₆" BY 10" LONG
2 ⅝" GAGE IN OUTSTANDING LEGS
WITH HOLES 5 ½" C TO C

FIGURE 5.6 Staggered holes provide clearance for high-strength bolts.

Designs should provide ample clearance for tightening high-strength bolts. Detailers who prepare shop drawings for fabricators generally are aware of the necessity for this and can, with careful detailing, secure the necessary space. In tight situations, the solution may be staggering of holes (Fig. 5.6), variations from standard gages (Fig. 5.7), use of knife-type connections, or use of a combination of shop welds and field bolts.

Minimum clearances for tightening high-strength bolts are indicated in Fig. 5.8 and Table 5.6.

5.11 FASTENER SPACING

Pitch is the distance (in) along the line of principal stress between centers of adjacent fasteners. It may be measured along one or more lines of fasteners. For example, suppose bolts are staggered along two parallel lines. The pitch may be given as the distance between successive bolts in each line separately. Or it may be given as the distance, measured parallel to the fastener lines, between a bolt in one line and the nearest bolt in the other line.

Gage is the distance (in) between adjacent lines of fasteners along which pitch is measured or the distance (in) from the back of an angle or other shape to the first line of fasteners.

The minimum distance between centers of fasteners should be at least three times the fastener diameter. (The AISC specification, however, permits 2⅔ times the fastener diameter.)

Limitations also are set on maximum spacing of fasteners, for several reasons. In built-up members, **stitch fasteners,** with restricted spacings, are used between components to ensure uniform action. Also, in compression members, such fasteners are required to prevent local buckling. In bridges, **sealing fasteners** must be

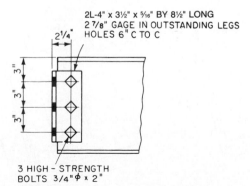

2L-4" x 3½" x ⁵⁄₁₆" BY 8½" LONG
2 ⅞" GAGE IN OUTSTANDING LEGS
HOLES 6" C TO C

3 HIGH – STRENGTH
BOLTS 3/4"∅ x 2"

FIGURE 5.7 Increasing the gage in framing angles provides clearance for high-strength bolts.

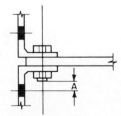

FIGURE 5.8 The usual minimum clearances *A* for high-strength bolts are given in Table 5.6.

TABLE 5.6 Clearances for High-Strength Bolts

Bolt dia, in	Nut height, in	Usual min clearance, in A	Min clearance for twist-off bolts, in A	
			Small tool	Large tool
$\frac{5}{8}$	$\frac{5}{8}$	1	$1\frac{5}{8}$	—
$\frac{3}{4}$	$\frac{3}{4}$	$1\frac{1}{4}$	$1\frac{5}{8}$	$1\frac{7}{8}$
$\frac{7}{8}$	$\frac{7}{8}$	$1\frac{3}{8}$	$1\frac{5}{8}$	$1\frac{7}{8}$
1	1	$1\frac{7}{16}$		$1\frac{7}{8}$
$1\frac{1}{8}$	$1\frac{1}{8}$	$1\frac{9}{16}$		—
$1\frac{1}{4}$	$1\frac{1}{4}$	$1\frac{11}{16}$		—

closely spaced to seal the edges of plates and shapes in contact to prevent penetration of moisture. Maximum spacing of fasteners is governed by the requirements for sealing or stitching, whichever requires the smaller spacing.

For sealing, the pitch of fasteners on a single line adjoining a free edge of an outside plate or shape should not exceed 7 in or $4 + 4t$ in, where t is the thickness (in) of the thinner outside plate or shape (Fig. 5.9a). (See also the maximum edge distance, Art. 5.12). If there is a second line of fasteners uniformly staggered with those in the line near the free edge, a smaller pitch for the two lines can be used if the gage g (in) for these lines is less than $1\frac{1}{2} + 4t$. In this case, the staggered pitch (in) should not exceed $4 + 4t - \frac{3}{4}g$ or 7 in but need not be less than half the requirement for a single line (Fig. 5.9b).

Bolted joints in unpainted weathering steel require special limitations on pitch: 14 times the thickness of the thinnest part, not to exceed 7 in (AISC specification).

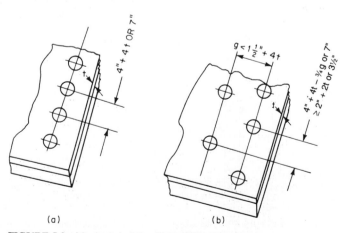

       (a)   (b)

FIGURE 5.9 Maximum pitch of bolts for sealing.

TABLE 5.7 Minimum Edge Distances (in) for Fastener Holes in Steel for Buildings

Fastener diameter, in	At sheared edges	At rolled edges of plates, shapes, or bars or gas-cut edges*
½	⅞	¾
⅝	1⅛	⅞
¾	1¼	1
⅞	1½†	1⅛
1	1¾†	1¼
1⅛	2	1½
1¼	2¼	1⅝
Over 1¼	1¾d‡	1¼d‡

*All edge distances in this column may be reduced ⅛ in when the hole is at a point where stress does not exceed 25% of the maximum allowed stress in the element.

†These may be 1¼ in at the ends of beam connection angles.

‡d = fastener diameter in.

From AISC "Specification for Structural Steel Buildings."

5.12 EDGE DISTANCE OF FASTENERS

Minimum distances from centers of fasteners to any edges are given in Tables 5.7 and 5.8.

The AISC specifications for structural steel for buildings have the following provisions for minimum edge distance: The distance from the center of a standard hole to an edge of a connected part should not be less than the applicable value from Table 5.7 nor the value from the equation

$$L_e \leq 2P/F_u t \tag{5.1}$$

TABLE 5.8 Minimum Edge Distances (in) for Fastener Holes in Steel for Bridges

Fastener diameter, in	At sheared or flame-cut edges		In flanges of beams or channels		At other rolled or planed edges	
	Highway	Railroad	Highway	Railroad	Highway	Railroad
½		⅞		⅝		¾
⅝	1⅛	1⅛	⅞	13/16	1	15/16
¾	1¼	1 5/16	1	15/16	1⅛	1⅛
⅞	1½	1½	1⅛	1⅛	1¼	1 5/16
1	1¾	1¾	1¼	1¼	1½	1½
Over 1		1¾d*		1¼d*		1½d*

*d = fastener diameter, in.

where L_e = the distance from the center of a standard hole to the edge of the connected part, in

P = force transmitted by one fastener to the critical connected part, kips

F_u = specified minimum tensile strength of the critical connected part, ksi

t = thickness of the critical connected part, in

Also, L_e should not be less than $1\frac{1}{2}d$ when $F_p = 1.2F_u$, where d is the diameter of the bolt (in) and F_p is the allowable bearing stress of the critical connected part (ksi).

The AASHTO specifications for highway bridges require the minimum distance from the center of any bolt in a standard hole to a sheared or flame-cut edge to be as shown in Table 5.8. When there is only a single transverse fastener in the direction of the line of force in a standard or short slotted hole, the distance from the center of the hole to the edge connected part should not be less than $1\frac{1}{2}d$ when $F_p = 1.1F_u$.

The AREA *Manual for Railway Engineering* minimum edge distance for a sheared edge is given in Table 5.8. The distance between the center of the nearest bolt and the end of the connected part toward which the pressure of the bolt is directed should be not less than $2df_p/F_u$, where d is the diameter of the bolt (in) and f_p is the computed bearing stress due to the service load (ksi).

Maximum edge distances are set for sealing and stitch purposes. AISC specifications limit the distance from center of fastener to nearest edge of parts in contact to 12 times the thickness of the connected part, with a maximum of 6 in. The AASHTO maximum is 5 in or 8 times the thickness of the thinnest outside plate. (AISC gives the same requirement for unpainted weathering steel.) The AREA maximum is 6 in or 4 times the plate thickness plus 1.5 in.

5.13 FILLERS

A **filler** is a plate inserted in a splice between a gusset or splice plate and stress-carrying members to fill a gap between them. Requirements for fillers included in the AISC specifications for structural steel for buildings are as follows.

In welded construction, a filler $\frac{1}{4}$ in or more thick should extend beyond the edge of the splice plate and be welded to the part on which it is fitted (Fig. 5.10).

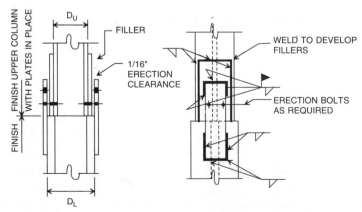

FIGURE 5.10 Typical welded splice of columns when depth D_u of the upper column is nominally 2 in less than depth D_L of the lower column.

The welds should be able to transmit the splice-plate stress, applied at the surface of the filler, as an eccentric load. The welds that join the splice plate to the filler should be able to transmit the splice plate stress and should have sufficient length to prevent overstress of the filler along the toe of the welds. A filler less than ¼ in thick should have edges flush with the splice-plate edges. The size of the welds should equal the sum of the filler thickness and the weld size necessary to resist the splice plate stress.

In bearing connections with bolts carrying computed stress passing through fillers thicker than ¼ in, the fillers should extend beyond the splice plate (Fig. 5.11). The filler extension should be secured by sufficient bolts to distribute the load on the member uniformly over the combined cross section of member and filler. Alternatively, an equivalent number of bolts should be included in the connection. Fillers ¼ to ¾ in thick need not be extended if the allowable shear stress in the bolts is reduced by the factor $0.4(t - 0.25)$, where t is the total thickness of the fillers but not more than ¾ in.

The AASHTO specifications for highway bridges require the following: Fillers through which stress-carrying fasteners pass should preferably be extended beyond the gusset or splice material. The extension should be secured by enough additional fasteners to carry the stress in the filler. This stress should be calculated as the total load on the member divided by the combined cross-sectional area of the member and filler. Alternatively, additional fasteners may be passed through the gusset or splice material without extending the filler. If a filler is less than ¼ in thick, it should not be extended beyond the splice material. Additional fasteners are not required. Fillers ¼ in or more thick should not consist of more than two plates, unless the engineer gives permission.

The American Railway Engineering Association (AREA) does not require additional bolts for development of fillers in high-strength bolted connections.

5.14 INSTALLATION OF FASTENERS

All parts of a connection should be held tightly together during installation of fasteners. Drifting done during assembling to align holes should not distort the metal or enlarge the holes. Holes that must be enlarged to admit fasteners should be reamed. Poor matching of holes is cause for rejection.

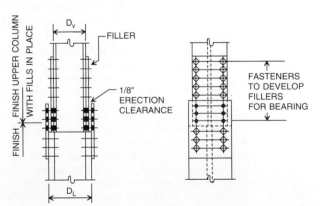

FIGURE 5.11 Typical bolted splice of columns when depth D_u of the upper column is nominally 2 in less than depth D_L of the lower column.

For connections with high-strength bolts, surfaces, when assembled, including those adjacent to bolt heads, nuts, and washers, should be free of scale, except tight mill scale. The surfaces also should be free of defects that would prevent solid seating of the parts, especially dirt, burrs, and other foreign material. Contact surfaces within slip-critical joints should be free of oil, paint, lacquer, and rust inhibitor.

Each high-strength bolt should be tightened so that when all fasteners in the connection are tight it will have the total tension (kips) given in Table 6.18, for its diameter. Tightening should be done by the turn-of-the-nut method or with properly calibrated wrenches.

High-strength bolts usually are tightened with an impact wrench. Only where clearance does not permit its use will bolts be hand-tightened.

Requirements for joint assembly and tightening of connections are given in the "Specification for Structural Joints Using ASTM A325 or A490 Bolts," Research Council on Structural Connections of the Engineering Foundation. The provisions applicable to connections requiring full pretensioning include the following.

Calibrated-wrench Method. When a calibrated wrench is used, it must be set to cut off tightening when the required tension (Table 6.18) has been exceeded by 5%. The wrench should be tested periodically (at least daily on a minimum of three bolts of each diameter being used). For the purpose, a calibrating device that gives the bolt tension directly should be used. In particular, the wrench should be calibrated when bolt size or length of air hose is changed.

When bolts are tightened, bolts previously tensioned may become loose because of compression of the connected parts. The calibrated wrench should be reapplied to bolts previously tightened to ensure that all bolts are tensioned to the prescribed values.

Turn-of-the-nut Method. When the turn-of-the-nut method is used, tightening may be done by impact or hand wrench. This method involves three steps:

1. **Fit-up of connection.** Enough bolts are tightened a sufficient amount to bring contact surfaces together. This can be done with fit-up bolts, but it is more economical to use some of the final high-strength bolts.

2. **Snug tightening of bolts.** All high-strength bolts are inserted and made snug-tight (tightness obtained with a few impacts of an impact wrench or the full effort of a person using an ordinary spud wrench). While the definition of snug-tight is rather indefinite, the condition can be observed or learned with a tension-testing device.

3. **Nut rotation from snug-tight position.** All bolts are tightened by the amount of nut rotation specified in Table 5.9. If required by bolt-entering and wrench-operation clearances, tightening, including by the calibrated-wrench method, may be done by turning the bolt while the nut is prevented from rotating.

Direct-Tension-Indicator Tightening. Two types of direct-tension-indicator devices are available: washers and twist-off bolts. The hardened-steel load-indicator washer has dimples on the surface of one face of the washer. When the bolt is torqued, the dimples depress to the manufacturer's specification requirements, and proper torque can be measured by the use of a feeler gage. Special attention should be given to proper installation of flat hardened washers when load-indicating washers are used with bolts installed in oversize or slotted holes and when the load-indicating washers are used under the turned element.

The twist-off bolt is a bolt with an extension to the actual length of the bolt. This extension will twist off when torqued to the required tension by a special

TABLE 5.9 Number of Nut or Bolt Turns from Snug-Tight Condition for High-Strength Bolts*

Bolt length (Fig. 5.1)	Slope of outer faces of bolted parts		
	Both faces normal to bolt axis	One face normal to bolt axis and the other sloped†	Both faces sloped†
Up to 4 diameters	⅓	½	⅔
Over 4 diameters but not more than 8 diameters	½	⅔	⅚
Over 8 diameters but not more than 12 diameters‡	⅔	⅚	1

*Nut rotation is relative to the bolt regardless of whether the nut or bolt is turned. For bolts installed by ½ turn and less, the tolerance should be ±30°. For bolts installed by ⅔ turn and more, the tolerance should be ±45°. This table is applicable only to connections in which all material within the grip of the bolt is steel.

†Slope is not more than 1:20 from the normal to the bolt axis, and a beveled washer is not used.

‡No research has been performed by RCSC to establish the turn-of-the-nut procedure for bolt lengths exceeding 12 diameters. Therefore, the required rotation should be determined by actual test in a suitable tension-measuring device that simulates conditions of solidly fitted steel.

torque gun. A representative sample of at least three bolts and nuts for each diameter and grade of fastener should be tested in a calibration device to demonstrate that the device can be torqued to 5% greater tension than that required in Table 6.18.

When the direct tension indicator involves an irreversible mechanism such as yielding or fracture of an element, bolts should be installed in all holes and brought to the snug-tight condition. All fasteners should then be tightened, progressing systematically from the most rigid part of the connection to the free edges in a manner that will minimize relaxation of previously tightened fasteners prior to final twist off or yielding of the control or indicator element of the individual devices. In some cases, proper tensioning of the bolts may require more than a single cycle of systematic tightening.

WELDS

Welded connections often are used because of simplicity of design, fewer parts, less material, and decrease in shop handling and fabrication operations. Frequently, a combination of shop welding and field bolting is advantageous. With connection angles shop welded to a beam, field connections can be made with high-strength bolts without the clearance problems that may arise in an all-bolted connection.

Welded connections have a rigidity that can be advantageous if properly accounted for in design. Welded trusses, for example, deflect less than bolted trusses, because the end of a welded member at a joint cannot rotate relative to the other members there. If the end of a beam is welded to a column, the rotation there is practically the same for column and beam.

A disadvantage of welding, however, is that shrinkage of large welds must be considered. It is particularly important in large structures where there will be an accumulative effect.

Properly made, a properly designed weld is stronger than the base metal. Improperly made, even a good-looking weld may be worthless. Properly made, a weld has the required penetration and is not brittle.

Prequalified joints, welding procedures, and procedures for qualifying welders are covered by AWS D1.1, "Structural Welding Code—Steel," and AWS D1.5, "Bridge Welding Code," American Welding Society. Common types of welds with structural steels intended for welding when made in accordance with AWS specifications can be specified by note or by symbol with assurance that a good connection will be obtained.

In making a welded design, designers should specify only the amount and size of weld actually required. Generally, a 5⁄16-in weld is considered the maximum size for a single pass.

The cost of fit-up for welding can range from about one-third to several times the cost of welding. In designing welded connections, therefore, designers should consider the work necessary for the fabricator and the erector in fitting members together so they can be welded.

5.15   WELDING MATERIALS

Weldable structural steels permissible in buildings and bridges are listed with required electrodes in Tables 5.10 and 5.11. Welding electrodes and fluxes should conform to AWS 5.1, 5.5, 5.17, 5.18, 5.20, 5.23, 5.25, 5.26, 5.28, or 5.29 or applicable provisions of AWS D1.1 or D1.5. Weld metal deposited by electroslag or electrogas welding processes should conform to the requirements of AWS D1.1 or D1.5 for these processes. For bridges, the impact requirements in D1.5 are mandatory. Welding processes are described in Art. 2.6.

For welded connections in buildings, the electrodes or fluxes given in Table 5.10 should be used in making complete-penetration groove welds. These welds can be designed with allowable stresses for base metal indicated in Table 6.19.

For welded connections in bridges, the electrodes or fluxes given in Table 5.11 should be used in making complete-penetration groove welds. These welds can be designed with allowable stresses for base metal indicated in Table 10.8. For allowable stresses in welds, see Art. 10.9.

5.16   TYPES OF WELDS

The main types of welds used for structural steel are fillet, groove, plug, and slot. The most commonly used weld is the fillet. For light loads, it is the most economical, because little preparation of material is required. For heavy loads, groove welds are the most efficient, because the full strength of the base metal can be obtained easily. Use of plug and slot welds generally is limited to special conditions where fillet or groove welds are not practical.

More than one type of weld may be used in a connection. If so, the allowable capacity of the connection is the sum of the effective capacities of each type of weld used, separately computed with respect to the axis of the group.

Tack welds may be used for assembly or shipping. They are not assigned any stress-carrying capacity in the final structure. In some cases, these welds must be removed after final assembly or erection.

Fillet welds have the general shape of an isosceles right triangle (Fig. 5.12). The size of the weld is given by the length of leg. The strength is determined by the

TABLE 5.10   Matching Filler-Metal Requirements for Complete-Penetration Groove Welds in Building Construction

Base metal*	Welding process			
	Shielded metal-arc	Submerged-arc	Gas metal-arc	Flux colored arc
A36†, A53 grade B A500 grades A and B A501, A529, and A570 grades 30 through 50	AWS A5.1 or A5.5§ E60XX E70XX E70XX-X	AWS A5.17 or A5.23§ F6XX-EXXX F7XX-EXXX or F7XX-EXX-XX	AWS A5.18 ER70S-X	AWS A5.20 E6XT-X E7XT-X (Except −2, −3, −10, −GS)
A242‡, A441, A572 grade 42 and 50, and A588‡ (4 in. and under)	AWS A5.1 or A5.5§ E7015, E7016, E7018, E7028 E7015-X, E7016-X, E7018-X	AWS A5.17 or A5.23§ F7XX-EXXX F7XX-EXX-XX	AWS A5.18 ER70S-X	AWS A5.20 E7XT-X (Except −2, −3, −10, −GS)
A572 grades 60 and 65	AWS A5.5§ E8015-X, E8016-X E8018-X	AWS A5.23§ F8XX-EXX-XX	AWS A5.28§ ER 80S-X	AWS A5.29§ E8XTX-X
A514 over 2½ in thick	AWS A5.5§ E10015-X, E10016-X, E10018-X	AWS A5.23§ F10XX-EXX-XX	AWS A5.28§ ER 100S-X	AWS A5.29§ E10XTX-X
A514 2½ in thick and under	AWS A5.5§ E11015-X E11016-X E11018-X	AWS A5.23§ F11XX-EXX-XX	AWS A5.28§ ER110S-X	AWS A5.29§ E11XTX-X

*In joints involving base metals of different groups, low-hydrogen filler-metal requirements applicable to the lower-strength group may be used. The low-hydrogen processes are subject to the technique requirements applicable to the higher-strength group.

†Only low-hydrogen electrodes may be used for welding A36 steel more than 1 in thick for dynamically loaded structures.

‡Special welding materials and procedures (e.g., E80XX-X low-alloy electrodes) may be required to match the notch toughness of base metal (for applications involving impact loading or low temperature) or for atmospheric corrosion and weathering characteristics.

§Deposited weld metal should have a minimum impact strength of 20 ft-lb at 0°F when Charpy V-notch specimens are required.

throat thickness, the shortest distance from the root (intersection of legs) to the face of the weld. If the two legs are unequal, the nominal size of the weld is given by the shorter of the legs. If welds are concave, the throat is diminished accordingly, and so is the strength.

Fillet welds are used to join two surfaces approximately at right angles to each other. The joints may be lap (Fig. 5.13) or tee or corner (Fig. 5.14). Fillet welds also may be used with groove welds to reinforce corner joints. In a skewed tee joint, the included angle of weld deposit may vary up to 30° from the perpendicular, and one corner of the edge to be connected may be raised, up to ³⁄₁₆ in. If the separation is greater than ¹⁄₁₆ in, the weld leg should be increased by the amount of the root opening.

Groove welds are made in a groove between the edges of two parts to be joined. These welds generally are used to connect two plates lying in the same plane (butt joint), but they also may be used for tee and corner joints.

Standard types of groove welds are named in accordance with the shape given the edges to be welded: square, single vee, double vee, single bevel, double bevel, single U, double U, single J, and double J (Fig. 5.15). Edges may be shaped by

TABLE 5.11   Matching Filler-Metal Requirements for Complete-Penetration Groove Welds in Bridge Construction
(*a*) *Qualifed in Accordance with AWS D1.5 and Paragraph 5.6*

Base metal*	Welding process†		
	Shielded metal-arc	Submerged-arc	Flux-cored arc with external shielding gas
A36/M270 grade 36	AWS A5.1 or A5.5 E7016, E7018, or E7028	AWS A5.17 F6A-EXXX F7A0-EXXX	AWS A5.20 E6XT-1,5 E7XT-1,5
A572 M270 grade 50 type 1, 2, or 3	AWS A5.1 or A5.5 E7016, E7018, or E7028		
A588/M270 grade 50W 4 in thick and under		AWS A5.17 F7A-EXXX	AWS A5.20 E7XT-1,5
A852/M270 grade 70	AWS A5.5 E9018-M		
A852/M270 grade 70‡		AWS A5.23 F9AX-EXXX-X	AWS A5.29 E9XT1-X E9XT5-X
A514/M270 grade 100 A517/M270 grade 100 Over 2½ in thick	AWS A5.5 E10018-M		

(*b*) *Qualified in accordance with AWS D1.5 and Paragraph 5.7*

Base metal*	Welding process†					
	Flux-cored arc, self-shielding	Gas metal-arc	Electroslag§	Electrogas§	Submerged-arc	Shielded metal-arc
A36/M270 grade 36	AWS A5.20 E6XT-6,8 E7XT-6,8 AWS A5.29 E6XT8-8 E7XT8-X	AWS A5.18 ER70S- 2,3,6,7	AWS A5.25 FES 60-XXXX FES 70-XXXX FES72-XXXX	AWS A5.26 EG60XXXX EG62XXXX EG70XXXX EG72XXXX		
A572/M270 grade 50 or A588/M270 grade 50W 4 in thick and under	AWS A5.20 E7XT-6,8 AWS A5.29 E7XT8-X	AWS A5.18 ER 70S- 2,3,6,7	AWS A5.25 FES 70-XXXX FES 72-XXXX	AWS A526 EG70XXXX EG72XXXX		
A852/M270 grade 70‡		As Approved by Engineer				
A514/M270 grade 100 or A517/M270 grade 100 over 2½ in thick	With external shielding gas AWS A5.29 E100 T5-K3 E101 T1-K7	AWS A5.28 ER110S-1 ER100S-2			AWS A5.23 F10A-EM2-M2	

(b) Qualified in accordance with AWS D1.5 and Paragraph 5.7 (Continued)

Base metal*	Welding process†					
	Flux-cored arc, self-shielding	Gas metal-arc	Electroslag§	Electrogas§	Submerged-arc	Shielded metal-arc
A514/M270 grade 100 or A517/M270 grade 100 2½ in thick or less	With external shielding gas AWS A5.29 E110T5-K3,K4 E111T1-K4	AWS A5.28 ER110S-1			AWS A5.23 F11A4-EM3-M3	AWS A5.5 E11018-M

*In joints involving base metals of two different yield strengths, filler metal applicable to the lower-strength base metal may be used.
†Electrode specifications with the same yield and tensile properties, but with lower impact-test temperature may be substituted (e.g., F7A2-EXXX may be substituted for F7A0-EXXX).
‡Special welding materials and procedures may be required to match atmospheric, corrosion and weathering characteristics. See AWS D1.5.
§Not authorized for tension and reversal members.

flame cutting, arc-air gouging, or edge planing. Material up to ⅝ in thick, however, may be groove-welded with square-cut edges, depending on the welding process used.

Groove welds should extend the full width of parts joined. Intermittent groove welds and butt joints not fully welded throughout the cross section are prohibited.

Groove welds also are classified as complete-penetration and partial-penetration welds.

In a **complete-penetration weld,** the weld material and the base metal are fused throughout the depth of the joint. This type of weld is made by welding from both sides of the joint or from one side to a backing bar or backing weld. When the joint is made by welding from both sides, the root of the first-pass weld is chipped or gouged to sound metal before the weld on the opposite side, or back pass, is made. The throat dimension of a complete-penetration groove weld, for stress

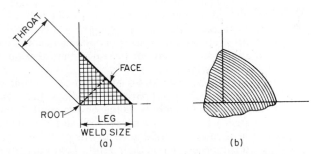

FIGURE 5.12 Fillet weld. (*a*) Theoretical cross section. (*b*) Actual cross section.

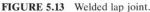

FIGURE 5.13 Welded lap joint.

FIGURE 5.14 (*a*) Tee joint. (*b*) Corner joint.

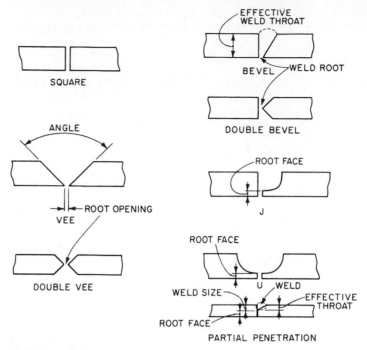

FIGURE 5.15   Groove welds.

computations, is the full thickness of the thinner part joined, exclusive of weld reinforcement.

Partial-penetration welds generally are used when forces to be transferred are small. The edges may not be shaped over the full joint thickness, and the depth of weld may be less than the joint thickness (Fig. 5.15). But even if edges are fully shaped, groove welds made from one side without a backing strip or made from both sides without back gouging are considered partial-penetration welds. They often are used for splices in building columns carrying axial loads only. In bridges, such welds should not be used where tension may be applied normal to the axis of the welds.

Plug and slot welds are used to transmit shear in lap joints and to prevent buckling of lapped parts. In buildings, they also may be used to join components of built-up members. (Plug or slot welds, however, are not permitted on A514 steel.) The welds are made, with lapped parts in contact, by depositing weld metal in circular or slotted holes in one part. The openings may be partly or completely filled, depending on their depth. Load capacity of a plug or slot completely welded equals the product of hole area and allowable stress. Unless appearance is a main consideration, a fillet weld in holes or slots is preferable.

Economy in Selection.   In selecting a weld, designers should consider not only the type of joint but also the type of weld that would require a minimum amount of metal. This would yield a saving in both material and time.

While strength of a fillet weld varies with size, volume of metal varies with the square of the size. For example, a ½-in fillet weld contains four times as much metal per inch of length as a ¼-in weld but is only twice as strong. In general, a

TABLE 5.12 Number of Passes for Welds

Weld size,* in	Fillet welds	Single-bevel groove welds (back-up weld not included)		Single-V groove welds (back-up weld not included)		
		30° bevel	45° bevel	30° open	60° open	90° open
3/16	1					
1/4	1	1	1	2	3	3
5/16	1					
3/8	3	2	2	3	4	6
7/16	4					
1/2	4	2	2	4	5	7
5/8	6	3	3	4	6	8
3/4	8	4	5	4	7	9
7/8		5	8	5	10	10
1		5	11	5	13	22
1 1/8		7	11	9	15	27
1 1/4		8	11	12	16	32
1 3/8		9	15	13	21	36
1 1/2		9	18	13	25	40
1 3/4		11	21			

*Plate thickness for groove welds.

smaller but longer fillet weld costs less than a larger but shorter weld of the same capacity.

Furthermore, small welds can be deposited in a single pass. Large welds require multiple passes. They take longer, absorb more weld metal, and cost more. As a guide in selecting welds, Table 5.12 lists the number of passes required for some frequently used types of welds.

Double-V and double-bevel groove welds contain about half as much weld metal as single-V and single-bevel groove welds, respectively (deducting effects of root spacing). Cost of edge preparation and added labor of gouging for the back pass, however, should be considered. Also, for thin material, for which a single weld pass may be sufficient, it is uneconomical to use smaller electrodes to weld from two sides. Furthermore, poor accessibility or less favorable welding position (Art. 5.18) may make an unsymmetrical groove weld more economical, because it can be welded from only one side.

When bevel or V grooves can be flame-cut, they cost less than J and U grooves, which require planning or arc-air gouging.

5.17 STANDARD WELDING SYMBOLS

These should be used on drawings to designate welds and provide pertinent information concerning them. The basic parts of a weld symbol are a horizontal line and an arrow:

Extending from either end of the line, the arrow should point to the joint in the same manner as the electrode would be held to do the welding.

Welding symbols should clearly convey the intent of the designer. For the purpose, sections or enlarged details may have to be drawn to show the symbols, or notes may be added. Notes may be given as part of welding symbols or separately. When part of a symbol, the note should be placed inside a tail at the opposite end of the line from the arrow:

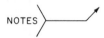

Type and length of weld are indicated above or below the line. If noted below the line, the symbol applies to a weld on the arrow side of the joint, the side to which the arrow points. If noted above the line, the symbol indicates that the other side, the side opposite the one to which the arrow points (not the far side of the assembly), is to be welded.

A fillet weld is represented by a right triangle extending above or below the line to indicate the side on which the weld is to be made. The vertical leg of the triangle is always on the left.

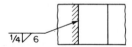

The preceding symbol indicates that a ¼-in fillet weld 6 in long is to be made on the arrow side of the assembly. The following symbol requires a ¼-in fillet weld 6 in long on both sides.

If a weld is required on the far side of an assembly, it may be assumed necessary from symmetry, shown in sections or details, or explained by a note in the tail of the welding symbol. For connection angles at the end of a beam, far-side welds generally are assumed:

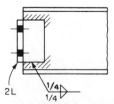

Length of weld is not shown on the symbol in this case because the connection requires a continuous weld the full length of each angle on both sides of the angle.

Care must be taken not to omit length unless a continuous full-length weld is wanted. "Continuous" should be written on the weld symbol to indicate length when such a weld is required. In general, a tail note is advisable to specify welds on the far side, even when the welds are the same size.

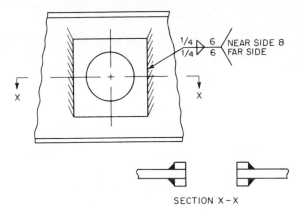

SECTION X–X

For many members, a stitch or intermittent weld is sufficient. It may be shown as

$$\sqrt{1/4 \, V \, 2\text{--}10}$$

This symbol calls for ¼-in fillet welds on the arrow side. Each weld is to be 2 in long. Spacing of welds is to be 10 in center to center. If the welds are to be staggered on the arrow and other sides, they can be shown as

$$\frac{1/4 \diagdown 2\text{--}10}{1/4 \, V \, 2\text{--}10}$$

Usually, intermittent welds are started and finished with a weld at least twice as long as the length of the stitch welds. This information is given in a tail note:

$$\frac{1/4 \diagdown 2\text{--}10}{1/4 \, V \, 2\text{--}10} \, \Big/ \, 4"$$

When the welding is to be done in the field rather than in the shop, a triangular flag should be placed at the intersection of arrow and line:

$$\frac{3/16 \diagdown 8}{1/4 \, V \, 3}$$

This is important in ensuring that the weld will be made as required. Often, a tail note is advisable for specifying field welds.

A continuous weld all around a joint is indicated by a small circle around the intersection of line and arrow:

Such a symbol would be used, for example, to specify a weld joining a pipe column to a base plate. The all-around symbol, however, should not be used as a substitute for computation of actual weld length required. Note that the type of weld is indicated below the line in the all-around symbol, regardless of shape or extent of joint.

The preceding devices for providing information with fillet welds also apply to groove welds. In addition, groove-weld symbols also must designate material preparation required. This often is best shown on a cross section of the joint.

A square-groove weld (made in thin material) without root opening is indicated by

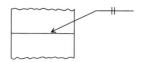

Length is not shown on the welding symbol for groove welds because these welds almost always extend the full length of the joint.

A short curved line below a square-groove symbol indicates weld contour. A short straight line in that position represents a flush weld surface. If the weld is not to be ground, however, that part of the symbol is usually omitted. When grinding is required, it must be indicated in the symbol.

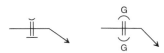

The root-opening size for a groove weld is written in within the symbol indicating the type of weld. For example, a ⅛-in root opening for a square-groove weld is specified by

And a ⅛-in root opening for a bevel weld, not to be ground, is indicated by

In this and other types of unsymmetrical welds, the arrow not only designates the arrow side of the joint but also points to the side to be shaped for the groove weld. When the arrow has this significance, the intention often is emphasized by an extra break in the arrow.

The angle at which the material is to be beveled should be indicated with the root opening:

A double-bevel weld is specified by

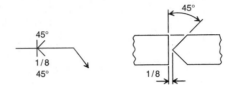

A single-V weld is represented by

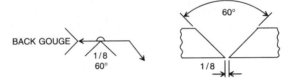

A double-V weld is indicated by

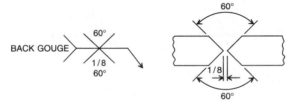

Summary.   Standard symbols for various types of welds are summarized in Fig. 5.16. The symbols do not indicate whether backing, spacer, or extension bars are required. These should be specified in general notes or shown in detail drawings. Preparation for J and U welds is best shown by an enlarged cross section. Radius and depth of preparation must be given.

In preparing a weld symbol, insert size, weld-type symbol, length of weld, and spacing, in that order from left to right. The perpendicular leg of the symbol for fillet, bevel, J, and flare-bevel welds should be on the left of the symbol. Bear in

	BACK WELDS	FILLET WELDS	PLUG OR SLOT WELDS	GROOVE WELDS							WELD ALL AROUND	FIELD WELD
				SQUARE	VEE	BEVEL	U	J	FLARE VEE	FLARE BEVEL		
ARROW SIDE	⌐	◸	◿	⊓	⌒	⌐	⌣	⊓	⌒	⌒	○	↗
OTHER SIDE	◹	◺	◺	⌣	⌣	⌣	⌣	⌣	⌣	⌣	○	↙
BOTH SIDES		◿		⧖	⧓	⧓	⧓	⧓	⧓	⧓		

FIGURE 5.16 Summary of welding symbols.

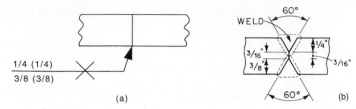

FIGURE 5.17 Penetration information is given on the welding symbol in (*a*) for the weld shown in (*b*). Penetration must be at least ³⁄₁₆ in. Second side must be back-gouged before the weld on that side is made.

mind also that arrow-side and other-side welds are the same size, unless otherwise noted. When billing of detail material discloses the identity of the far side with the near side, the welding shown for the near side also will be duplicated on the far side. Symbols apply between abrupt changes in direction of welding unless governed by the all-around symbol or dimensioning shown.

Where groove preparation is not symmetrical and complete, additional information should be given on the symbol. Also, it may be necessary to give weld-penetration information, as in Fig. 5.17. For the weld shown, penetration from either side must be a minimum of ³⁄₁₆ in. The second side should be back-gouged before the weld there is made.

Welds also may be a combination of different groove and fillet welds. While symbols can be developed for these, designers will save time by supplying a sketch or enlarged cross section. It is important to convey the required information accurately and completely to the workers who will do the job. Actually, it is common practice for designers to indicate what is required of the weld and for fabricators and erectors to submit proposed procedures.

5.18 WELDING POSITIONS

The position of the stick electrode relative to the joint when a weld is being made affects welding economy and quality. In addition, AWS specifications D1.0 and D1.5 prohibit use of some welding positions for some types of welds. Careful designing should eliminate the need for welds requiring prohibited welding positions and employ welds that can be efficiently made.

The basic welding positions are as follows:

Flat, with face of weld nearly horizontal. Electrode is nearly vertical, and welding is performed from above the joint.

Horizontal, with axis of weld horizontal. For groove welds, the face of weld is nearly vertical. For fillet welds, the face of weld usually is about 45° relative to horizontal and vertical surfaces.

Vertical, with axis of weld nearly vertical. (Welds are made upward.)

Overhead, with face of weld nearly horizontal. Electrode is nearly vertical, and welding is performed from below the joint.

Where possible, welds should be made in the flat position. Weld metal can be deposited faster and more easily. Generally, the best and most economical welds are obtained. In a shop, the work usually is positioned to allow flat or horizontal welding. With care in design, the expense of this positioning can be kept to a minimum. In the field, vertical and overhead welding sometimes may be necessary. The best assurance of good welds in these positions is use of proper electrodes by experienced welders.

The AWS specifications require that only the flat position be used for submerged-arc welding, except for certain sizes of fillet welds. Single-pass fillet welds may be made in the flat or the horizontal position in sizes up to $\frac{5}{16}$ in with a single electrode and up to $\frac{1}{2}$ in with multiple electrodes. Other positions are prohibited.

When groove-welded joints can be welded in the flat position, submerged-arc and gas metal-arc processes usually are more economical than the manual shielded metal-arc process.

GENERAL CRITERIA FOR WELDED CONNECTIONS

5.19 LIMITATIONS ON FILLET-WELD DIMENSIONS

For a given size of fillet weld, cooling rate is faster and restraint is greater with thick plates than with thin plates. To prevent cracking due to resulting internal stresses, specifications set minimum sizes for fillet welds, depending on plate thickness (Table 5.13).

In bridges, seal welds should be continuous. Size should be changed only when required for strength or by changes in plate thickness.

To prevent overstressing of base material at a fillet weld, standard specifications also limit the maximum weld size. They require that allowable stresses in adjacent base material not be exceeded when a fillet weld is stressed to its allowed capacity.

Example. Two angles transfer a load of 120 kips to a $\frac{3}{8}$-in-thick plate through four welds (Fig. 5.18). Assume that the welding process is shielded metal-arc using E70XX electrodes and steel is ASTM A36.

TABLE 5.13 Minimum Fillet-Weld Sizes and Plate-Thickness Limits

Sizes of fillet welds,* in			Minimum plate thickness for fillet welds on each side of the plate, in	
Buildings† AWS D1.1	Bridges‡ AWS D1.5	Maximum plate thickness, in§	36-ksi steel	50-ksi steel
$\frac{1}{8}$¶	—	$\frac{1}{4}$	—	—
$\frac{3}{16}$	—	$\frac{1}{2}$	0.38	0.28
$\frac{1}{4}$	$\frac{1}{4}$	$\frac{3}{4}$	0.51	0.37
$\frac{5}{16}$	$\frac{5}{16}$	Over $\frac{3}{4}$	0.64	0.46

*Weld size need not exceed the thickness of the thinner part joined, but AISC and AWS D1.5 require that care be taken to provide sufficient preheat to ensure weld soundness.

†When low-hydrogen welding is employed, AWS D1.1 permits the thinner part joined to be used to determine the minimum size of fillet weld.

‡Smaller fillet welds may be approved by the engineer based on applied stress and use of appropriate preheat.

§Plate thickness is the thickness of the thicker part joined.

¶Minimum weld size for structures subjected to dynamic loads is $\frac{3}{16}$ in.

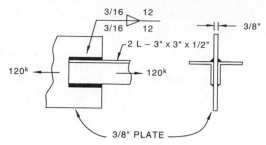

FIGURE 5.18 Welds on two sides of a plate induce stresses in it.

Allowable shear stress in fillet weld $F_v = 0.3 \times$ nominal tensile strength of weld metal $= 0.3 \times 70$ ksi $= 21.0$ ksi. The capacity of 1 in of $\frac{5}{16}$-in fillet weld $= 0.707(\frac{5}{16})21.0 = 4.64$ kips. Since there are welds on both sides of plate, the effective thickness required for a $\frac{5}{16}$-in fillet weld is 0.64 in (Table 5.13). Therefore, the effective capacity of a $\frac{5}{16}$-in fillet weld is $4.64 \times 0.375/0.64 = 2.7$ kips. For four welds,

$$\text{Length of weld required} = \frac{120}{2.7 \times 4} = 11.1 \text{ in}$$

The minimum size of fillet weld (Table 5.13) should be used, a $\frac{3}{16}$-in fillet weld with a capacity of $0.707(\frac{3}{16})21.0 = 2.78$ kips per in. Required minimum plate thickness is 0.38 in (Table 5.13). This weld satisfies the AISC requirement that the length of welds should be at least twice the distance between welds to be fully effective.

The capacity of a fillet weld per inch of length, with 21.0-ksi allowable stress, can be computed conveniently by multiplying the weld size in sixteenths of an inch by 0.928, since $0.707 \times \frac{2}{16} = 0.928$. For example, the capacity of 1 in of $\frac{5}{16}$-in fillet weld is $0.928 \times 5 = 4.64$, as in the preceding example.

A limitation also is placed on the maximum size of fillet welds along edges. One reason is that edges of rolled shapes are rounded, and weld thickness consequently is less than the nominal thickness of the part. Another reason is that if weld size and plate thickness are nearly equal, the plate corner may melt into the weld, reducing the length of weld leg and the throat. Hence standard specifications require the following: **Along edges of material less than ¼ in thick, maximum size of fillet weld may equal material thickness. But along edges of material ¼ in or more thick, the maximum size should be $\frac{1}{16}$ in less than the material thickness.**

Weld size may exceed this, however, if drawings definitely show that the weld is to be built out to obtain full throat thickness. AWS D1.1 requires that the minimum effective length of a fillet weld be at least four times the nominal size, or else the weld must be considered not to exceed 25% of the effective length. AWS D1.5 requires that the minimum effective length of a fillet weld be at least four times the nominal size or 1½ in, whichever is greater.

Suppose, for example, a ½-in weld is only 1½ in long. Its effective size is $1\frac{1}{2}/4 = \frac{3}{8}$ in.

Subject to the preceding requirements, intermittent fillet welds may be used in buildings to transfer calculated stress across a joint or faying surfaces when the strength required is less than that developed by a continuous fillet weld of the smallest permitted size. Intermittent fillet welds also may be used to join components of built-up members in buildings. But such welds are prohibited in bridges, in general, because of the requirements for sealing edges to prevent penetration of moisture and to avoid fatigue failures.

Intermittent welds are advantageous with light members, where excessive welding can result in straightening costs greater than the cost of welding. Intermittent welds often are sufficient and less costly than continuous welds (except girder fillet welds made with automatic welding equipment).

Weld lengths specified on drawings are effective weld lengths. They do not include distances needed for start and stop of welding.

To avoid the adverse effects of starting or stopping a fillet weld at a corner, welds extending to corners should be returned continuously around the corners in the same plane for a distance of at least twice the weld size. This applies to side and top fillet welds connecting brackets, beam seats, and similar connections, on the plane about which bending moments are computed. End returns should be indicated on design and detail drawings.

Fillet welds deposited on opposite sides of a common plane of contact between two parts must be interrupted at a corner common to both welds.

If longitudinal fillet welds are used alone in end connections of flat-bar tension members, the length of each fillet weld should at least equal the perpendicular distance between the welds. The transverse spacing of longitudinal fillet welds in end connections should not exceed 8 in unless the design otherwise prevents excessive transverse bending in the connections.

5.20 LIMITATIONS ON PLUG AND SLOT WELD DIMENSIONS

In material ⅝ in or less thick, the thickness of plug or slot welds should be the same as the material thickness. In material more than ⅝ in thick, the weld thickness should be at least half the material thickness but not less than ⅝ in.

Diameter of hole for a plug weld should be at least the depth of hole plus ⅜ in, but the diameter should not exceed 2¼ times the thickness of the weld metal.

Thus the hole diameter in ¾-in plate could be a minimum of ¾ + ⅜ = 1⅛ in. Depth of metal would be at least ⅝ in > (1.125/2.25 = 0.05 in) > (½ × ¾ = ⅜ in).

Plug welds may not be spaced closer center to center than four times the hole diameter.

Length of slot for a slot weld should not exceed 10 times the weld thickness. Width of slot should be at least depth of hole plus ⅜ in, but the width should not exceed 2¼ times the weld thickness.

Thus, width of slot in ¾-in plate could be a minimum of ¾ + ⅜ = 1⅛ in. Weld metal depth would be at least ⅝ in > (1.125/2.25 = 0.5 in) > (½ × ¾ = ⅜ in). If the minimum depth is used, the slot could be up to 10 × ⅝ = 6¼ in long.

Slot welds may be spaced no closer than four times their width in a direction transverse to the slot length. In the longitudinal direction, center-to-center spacing should be at least twice the slot length.

5.21 WELDING PROCEDURES

Welds should be qualified and should be made only by welders, welding operators, and tackers qualified as required in AWS D1.1 for buildings and AWS D1.5 for bridges.

Welding should not be permitted under any of the following conditions:

When the ambient temperature is below 0°F.

When surfaces are wet or exposed to rain, snow, or high wind.

When welders are exposed to inclement conditions.

Surfaces and edges to be welded should be free from fins, tears, cracks, and other defects. Also, surfaces at and near welds should be free from loose scale, slag, rust, grease, moisture, and other material that may prevent proper welding. AWS specifications, however, permit mill scale that withstands vigorous wire brushing, a light film of drying oil, or antispatter compound to remain. But the specifications require all mill scale to be removed from surfaces on which flange-to-web welds are to be made by submerged-arc welding or shielded metal-arc welding with low-hydrogen electrodes.

Parts to be fillet-welded should be in close contact. The gap between parts should not exceed ³⁄₁₆ in. If it is ¹⁄₁₆ in or more, fillet-weld size should be increased by the amount of separation. The separation between faying surfaces for plug and slot welds, and for butt joints landing on a backing, should not exceed ¹⁄₁₆ in. Parts to be joined at butt joints should be carefully aligned. Where the parts are effectively restrained against bending due to eccentricity in alignment, an offset not exceeding 10% of the thickness of the thinner part joined, but in no case more than ¹⁄₈ in, is permitted as a departure from theoretical alignment. When correcting misalignment in such cases, the parts should not be drawn in to a greater slope than ¹⁄₂ in in 12 in.

For permissible welding positions, see Art. 5.18. Work should be positioned for flat welding, whenever practicable.

In general, welding procedures and sequences should avoid needless distortion and should minimize shrinkage stresses. As welding progresses, welds should be deposited so as to balance the applied heat. Welding of a member should progress from points where parts are relatively fixed in position toward points where parts have greater relative freedom of movement. Where it is impossible to avoid high residual stresses in the closing welds of a rigid assembly, these welds should be made in compression elements. Joints expected to have significant shrinkage should be welded before joints expected to have lesser shrinkage, and restraint should be kept to a minimum. If severe external restraint against shrinkage is present, welding should be carried continuously to completion or to a point that will ensure freedom from cracking before the joint is allowed to cool below the minimum specified preheat and interpass temperature.

In shop fabrication of cover-plated beams and built-up members, each component requiring splices should be spliced before it is welded to other parts of the member. Up to three subsections may be spliced to form a long girder or girder section.

With too rapid cooling, cracks might form in a weld. Possible causes are shrinkage of weld and heat-affected zone, austenite-martensite transformation, and entrapped hydrogen. Preheating the base metal can eliminate the first two causes. Preheating reduces the temperature gradient between weld and adjacent base metal, thus decreasing the cooling rate and resulting stresses. Also, if hydrogen is present, preheating allows more time for this gas to escape. Use of low-hydrogen electrodes, with suitable moisture control, also is advantageous in controlling hydrogen content.

High cooling rates occur at arc strikes that do not deposit weld metal. Hence arc strikes outside the area of permanent welds should be avoided. Cracks or blemishes resulting from arc strikes should be ground to a smooth contour and checked for soundness.

TABLE 5.14 Requirements of AWS D1.1 for Minimum Preheat and Interpass Temperature (°F) for Welds in Buildings*

Thickness of thickest part at point of welding, in	Shielded metal-arc with other than low-hydrogen electrodes ASTM A36†, A53 grade B, A501, A529 A570 all grades	Shielded metal-arc with low-hydrogen electrodes; submerged-arc, gas, metal-arc or flux-cored arc ASTM A36†, A53 grade B, A242 A441, A501, A529 A570 all grades, A572 grades 42, 50, A588	Shielded metal-arc with low-hydrogen electrodes; submerged-arc, gas metal-arc or flux-cored arc ASTM A572 grades 60 and 65	Shielded metal-arc with low-hydrogen electrodes; submerged-arc, with carbon or alloy steel wire neutral flux, gas metal arc or flux cored arc ASTM A514
To ¾	0‡	0‡	50	50
Over ¾ to 1½	150	50	150	125
Over 1½ to 2½	225	150	225	175
Over 2½	300	225	300	225

*In joints involving combinations of base metals, preheat as specified for the higher-strength steel being welded.
†Use only low-hydrogen electrodes when welding A36 steel more than 1 in thick for dynamically loaded structures.
‡When the base-metal temperature is below 32°F, the base metal should be preheated to at least 70°F and the minimum temperature maintained during welding.

To avoid cracks and for other reasons, standard specifications require that under certain conditions, before a weld is made the base metal must be preheated. Tables 5.14 and 5.15 list typical preheat and interpass temperatures. The tables recognize that as plate thickness, carbon content, or alloy content increases, higher preheats are necessary to lower cooling rates and to avoid microcracks or brittle heat-affected zones.

Preheating should bring to the specified preheat temperature the surface of the base metal within a distance equal to the thickness of the part being welded, but not less than 3 in, of the point of welding. This temperature should be maintained as a minimum interpass temperature while welding progresses.

Preheat and interpass temperatures should be sufficient to prevent crack formation. Temperatures above the minimums in Tables 5.14 and 5.15 may be required for highly restrained welds.

For A514, A517, and A852 steels, the maximum preheat and interpass temperature should not exceed 400°F for thicknesses up to 1½ in, inclusive, and 450°F for greater thicknesses. Heat input during the welding of these quenched and tempered steels should not exceed the steel producer's recommendation. Use of stringer beads to avoid overheating is advisable.

Peening sometimes is used on intermediate weld layers for control of shrinkage stresses in thick welds to prevent cracking. It should be done with a round-nose tool and light blows from a power hammer after the weld has cooled to a temperature warm to the hand. The root or surface layer of the weld or the base metal at the edges of the weld should not be peened. Care should be taken to prevent scaling or flaking of weld and base metal from overpeening.

When required by plans and specifications, welded assemblies should be stress-relieved by heat treating. (See AWS D1.1 and D1.5 for temperatures and holding times required.) Finish machining should be done after stress relieving.

TABLE 5.15 Requirements of AWS D1.5 for Minimum Preheat and Interpass Temperatures (°F) for Welds in Bridges*

Thickness of thickest part at point of welding, in	Shielded metal-arc, submerged-arc, gas metal-arc or flux-cored arc			Shielded metal-arc, submerged-arc, gas metal-arc or flux-cored arc
	ASTM A36/M270 grade 36, A572 grade 50/M270 grade 50, A588/M270 grade 50W			ASTM A852/M270 grade 70, A514/M270 grade 100, A517/M270 grade 100
	General	AREA fracture critical		
		A36, A572	A588	
To ¾	50	100	100	50
Over ¾ to 1½	70	150	200	125
Over 1½ to 2½	150	200	300	175
Over 2½	225	300	350	225

*In joints involving combinations of base metals, preheat as specified for the higher-strength steel being welded. When the base-metal temperature is below 32°F, preheat the base metal to at least 70°F, and maintain this minimum temperature during welding.

Tack and other temporary welds are subject to the same quality requirements as final welds. For tack welds, however, preheat is not mandatory for single-pass welds that are remelted and incorporated into continuous submerged-arc welds. Also, defects such as undercut, unfilled craters, and porosity need not be removed before final submerged-arc welding. Welds not incorporated into final welds should be removed after they have served their purpose, and the surface should be made flush with the original surface.

Before a weld is made over previously deposited weld metal, all slag should be removed, and the weld and adjacent material should be brushed clean.

Groove welds should be terminated at the ends of a joint in a manner that will ensure sound welds. Where possible, this should be done with the aid of weld tabs or runoff plates. AWS D1.5 requires removal of weld tabs after completion of the weld in bridge construction. AWS D1.1 does not require removal of weld tabs for statically loaded structures but does require it for dynamically loaded structures. The ends of the welds then should be made smooth and flush with the edges of the abutting parts.

After welds have been completed, slag should be removed from them. The metal should not be painted until all welded joints have been completed, inspected, and accepted. Before paint is applied, spatter, rust, loose scale, oil, and dirt should be removed.

AWS D1.1 and D1.5 present details of techniques acceptable for welding buildings and bridges, respectively. These techniques include handling of electrodes and fluxes and maximum welding currents.

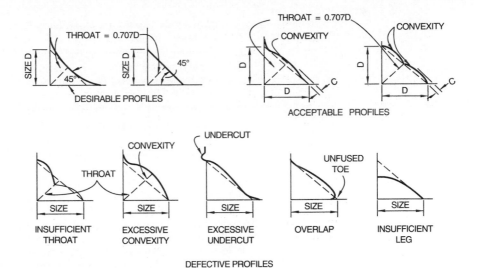

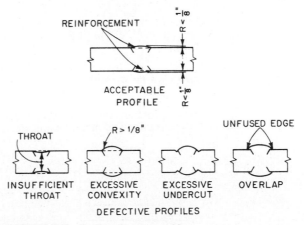

FIGURE 5.19 Profiles of fillet welds.

5.22 WELD QUALITY

A basic requirement of all welds is thorough fusion of weld and base metal and of successive layers of weld metal. In addition, welds should not be handicapped by craters, undercutting, overlap, porosity, or cracks. (AWS D1.1 and D1.5 give acceptable tolerances for these defects.) If craters, excessive concavity, or undersized welds occur in the effective length of a weld, they should be cleaned and filled to the full cross section of the weld. Generally, all undercutting (removal of base metal at the toe of a weld) should be repaired by depositing weld metal to restore the original surface. Overlap (a rolling over of the weld surface with lack of fusion at an edge), which may cause stress concentrations, and excessive convexity should be reduced by grinding away of excess material (see Figs. 5.19 and 5.20). If excessive porosity,

FIGURE 5.20 Profiles of groove welds.

TABLE 5.16 AWS D1.1 Limits on Convexity of Fillet Welds

Measured leg size or width of surface bead, in	Maximum convexity, in
5/16 or less	1/16
Over 5/16 but less than 1	1/8
1 or more	3/16

excessive slag inclusions, or incomplete fusion occur, the defective portions should be removed and rewelded. If cracks are present, their extent should be determined by acid etching, magnetic-particle inspection, or other equally positive means. Not only the cracks but also sound metal 2 in beyond their ends should be removed and replaced with the weld metal. Use of a small electrode for this purpose reduces the chances of further defects due to shrinkage. An electrode not more than 5/32 in in diameter is desirable for depositing weld metal to compensate for size deficiencies.

AWS D1.1 limits convexity C to the valves in Table 5.16. AWS D1.5 limits C to 0.06 in plus 7% of the measured face of the weld.

Weld-quality requirements should depend on the job the welds are to do. Excessive requirements are uneconomical. Size, length, and penetration are always important for a stress-carrying weld and should completely meet design requirements. Undercutting, on the other hand, should not be permitted in main connections, such as those in trusses and bracing, but small amounts might be permitted in less important connections, such as those in platform framing for an industrial building. Type of electrode, similarly, is important for stress-carrying welds but not so critical for many miscellaneous welds. Again, poor appearance of a weld is objectionable if it indicates a bad weld or if the weld will be exposed where aesthetics is a design consideration, but for many types of structures, such as factories, warehouses, and incinerators, the appearance of a good weld is not critical. A sound weld is important. But a weld entirely free of porosity or small slag inclusions should be required only when the type of loading actually requires this perfection.

Welds may be inspected by one or more methods: visual inspection; nondestructive tests, such as ultrasonic, x-ray, dye penetration, and magnetic particle; and cutting of samples from finished welds. Designers should specify which welds are to be examined, extent of the examination, and methods to be used.

5.23 WELDING CLEARANCE AND SPACE

Designers and detailers should detail connections to ensure that welders have ample space for positioning and manipulating electrodes and for observing the operation with a protective hood in place. Electrodes may be up to 18 in long and 3/8 in in diameter.

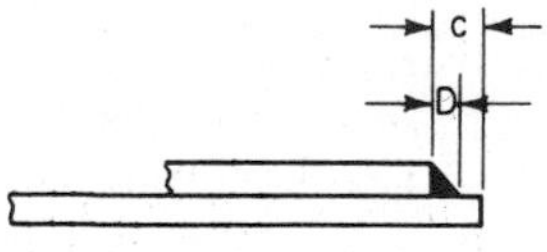

In addition, adequate space must be provided for deposition of the required size of fillet weld. For example, to provide an adequate landing c (in) for the fillet weld of size D (in) in Fig. 5.21, c should be at least $D + 5/16$. In building-

FIGURE 5.21 Minimum landing for a fillet weld.

column splices, however, $c = D + \frac{3}{16}$ often is used for welding splice plates to fillers.

DESIGN OF CONNECTIONS

The general procedure for designing a connection is as follows: Calculate the loads and locate their lines of actions. Make a preliminary layout. The connection should be as compact as practicable, to conserve material, yet satisfy the applicable criteria of governing specifications and should minimize eccentricity of loading. Select a type of fastener or weld, and estimate the size. Compute the allowable capacity per unit. Determine the number of units needed, and arrange them for maximum efficiency of the connection and for compactness. Provide adequate cross-sectional areas for the parts to be connected. Make a final layout. Check clearances to be sure that the connection can be fabricated and erected as designed. Check spacings to ensure that requirements of governing specifications are satisfied.

Axially stressed members meeting at a joint should have their gravity axes intersect at a point, if possible. Otherwise, the joint must be designed for bending stresses due to the eccentricity of the loading. Also, connections should be made symmetrical about the axis of the members, if possible.

5.24 *MINIMUM CONNECTIONS*

In buildings, connections carrying calculated stresses, except for lacing, sag bars, and girts, should be designed to support at least 6 kips.

In highway bridges, connections, except for lacing bars and handrails, should contain at least two fasteners or equivalent weld. The smallest angle that may be used in bracing is $3 \times 2\frac{1}{2}$ in. In railroad bridges, the minimum number of fasteners per plane of connection is three.

5.25 *HANGER CONNECTIONS*

In buildings, end connections for hangers should be designed for the full loads on the hangers. In trusses, however, the AISC specification requires that end connections should develop not only the design load but also at least 50% of the effective strength of the members. This does not apply if a smaller amount is justified by an engineering analysis that considers other factors, including loads from handling, shipping, and erection. This requirement is intended only for shop-assembled trusses where such loads may be significantly different from the loads for which the trusses were designed.

In highway bridges, connections should be designed for the average of the calculated stress and the strength of the members. But the connections should be capable of developing at least 75% of the strength of the members.

In railroad bridges, end connections of main tension members should have a strength at least equal to that of the members. Connections for secondary and bracing members should develop at least the average of the calculated stress and the strength of the members. But bracing members used as ties or struts to reduce the unsupported length of a member need not be connected for more than the flexural strength of that member.

When a connection is made with fasteners, the end fasteners carry a greater load than those at the center of the connection. Because of this, AISC and AASHTO reduce fastener strength when the length of a connection exceeds 50 in.

5.25.1 Bolted Lap Joints

Tension members serving as hangers may be connected to supports in any of several ways. One of the most common is use of a lap joint, with fasteners or welds.

Example.   A pair of A36 steel angles in a building carry a 60-kip vertically suspended load (Fig. 5.22). Size the angles and gusset plate, and determine the number of $\frac{7}{8}$-in-diameter A325N (threads included) bolts required.

Bolts.   Capacity of one bolt in double shear (i.e., two shear planes) is $r_v = 2 \times 21 \times 0.4418 = 18.6$ kips. Thus $60/18.6 = 3.23$ or 4 bolts are required.

Angles.   Gross area required $= 60/21.6 = 2.78$ in². Try two angles $3 \times 3 \times \frac{1}{4}$ with area $= 2.88$ in². Then net area $A_n = 2.88 - 1 \times 0.25 \times 2 = 2.38$ in², and effective net area $A_e = UA_n = 0.85 \times 2.38 = 2.02$ in². (The U factor accounts for shear lag, because only one leg of an angle is bolted.) The allowable load based on fracture of the net area is $2.02 \times 0.5 \times 58 = 58.6$ kips < 60 kips. The angles are not adequate. Try two angles $3 \times 3 \times \frac{5}{16}$ with area $= 3.55$ in². The net area is $A_n = 3.55 - 1 \times 0.3125 \times 2 = 2.92$ in², $A_e = 0.85 \times 2.92 = 2.48$ in², and the allowable load based on net area fracture is $2.48 \times 0.5 \times 58 = 71.9$ kips > 60 kips—OK.

Gusset.   Assume that the gusset is 10 in wide at the top bolt. Thus the net width is $10 - 1 = 9$ in, and since, for $U = 0.85$, this is greater than $0.85 \times 10 = 8.5$, use 8.5 in for the effective net width. For the gross area (yield limit state), thickness required $= 60/(10 \times 21.6) = 0.28$ in. For the effective net area (fracture limit state), thickness required $= 60/(8.5 \times 29) = 0.24$ in. Yield controls at 0.28 in. Use a $\frac{5}{16}$-in gusset.

Additional checks required are for bearing and angle-leg tearout (block shear fracture).

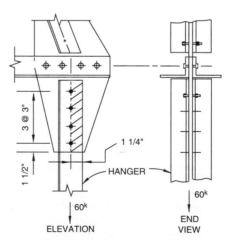

FIGURE 5.22 Hanger supported by gusset-plate connection with bolts.

Bearing.   For a bolt spacing of 3 in and an edge distance of 1½ in, the allowable bearing stress is $1.2F_u = 1.2 \times 58 = 69.6$ ksi, and the allowable bearing load is $69.6 \times 0.3125 \times 0.875 \times 4 = 76.1$ kips > 60 kips—OK.

Tearout.   The cross-hatched region of the angle leg in Fig. 5.22 can tear out as a block. To investigate this, determine the tearout capacity of the angles: The shear area is

$$A_v = (10.5 - 3.5 \times 0.9375) \times 0.3125 \times 2 = 4.51 \text{ in}^2$$

The tension area is

$$A_t = (1.25 - 0.5 \times 0.9375) \times 0.3125 \times 2 = 0.49 \text{ in}^2$$

The tearout resistance is

$$P_{to} = 4.51 \times 0.3 \times 58 + 0.49 \times 0.5 \times 58 = 92.7 \text{ kips} > 60 \text{ kips—OK}$$

5.25.2   Welded Lap Joints

For welded lap joints, standard specifications require that the amount of lap be five times the thickness of the thinner part joined, but not less than 1 in. Lap joints with plates or bars subjected to axial stress should be fillet-welded along the end of both lapped parts, unless the deflection of the lapped parts is sufficiently restrained to prevent the joint from opening under maximum loading.

Welded end connections have the advantage of avoiding deductions of hole areas in determining net section of tension members.

If the tension member consists of a pair of angles required to be stitched together, ring fills and fully tensioned bolts can be used (Fig. 5.23a). With welded connections, welded stitch bars (Fig. 5.23b) should be used. Care should be taken to avoid undercutting at the toes of the angles at end-connection welds and stitch welds. The end-connection welds may be placed equally on the toe and heel of the angles, ignoring the small eccentricity. The welds should be returned around the end of each angle.

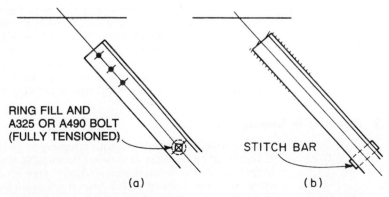

RING FILL AND
A325 OR A490 BOLT
(FULLY TENSIONED)

STITCH BAR

(a)

(b)

FIGURE 5.23   Pair of steel angles stitched together (*a*) with ring fill and high-strength bolts, (*b*) with stitch bar and welds.

Welded connections have the disadvantage of requiring fitting. Where there are several identical pieces, however, jigs can be used to reduce fit-up time.

Example.   Suppose the hanger in the preceding example is to be connected to the gusset plate by a welded lap joint with fillet welds (Fig. 5.24).

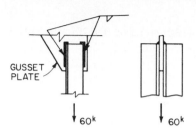

FIGURE 5.24 Hanger supported by gusset-plate connection with welds.

Since the two angles will now have no holes, no net-area fracture-limit-state checks need be made. As before, the gross area required is $60/21.6 = 2.78$ in^2. Try two angles $3 \times 3 \times \frac{1}{4}$, with gross area $= 2.88$ in^2. Because only one leg of the angle is connected, shear lag is a consideration here, just as it was in the bolted case. Thus the effective area is $0.85 \times 2.88 = 2.45$ in^2, and the capacity is $2.45 \times 29.0 = 71.0$ kips > 60 kips—OK. Note that the fracture allowable stress of $0.5F_u$ is used here. The limit state within the confines of the connection is fracture, not yield. Yield is the limit state in the angles outside the connection where the stress distribution in the angles becomes uniform.

The maximum-size fillet weld that can be used along the edge of the $\frac{1}{4}$-in material is $\frac{3}{16}$ in. To provide space for landing the fillet welds along the back and edge of each angle, the minimum width of gusset plate should be $3 + 2(\frac{3}{16} + \frac{5}{16}) = 4$ in, if $\frac{3}{16}$-in welds are used. A preliminary sketch of the joint (Fig. 5.24) indicates that with a minimum width of 4 in, the gusset plate will be about 6 in wide at the ends of the angles, where the load will reach 60 kips on transfer from the welds. For a 6-in width, the required thickness of plate is $60/(22 \times 6) = 0.46$ in. Use a $\frac{1}{2}$-in plate.

For a $\frac{1}{2}$-in plate, the minimum-size fillet weld is $\frac{3}{16}$ in. Since the maximum size permitted also is $\frac{3}{16}$ in, use a $\frac{3}{16}$-in fillet weld. If an E70XX electrode is used to make the welds, the allowable shear is 21 ksi. The capacity of the welds then is $0.707 \times \frac{3}{16} \times 21 = 2.78$ kips per in. Length of weld required equals $60(2 \times 2.78) = 10.8$ in. Hence supply a total length of fillet welds of at least 11 in, with at least $5\frac{1}{2}$ in along the toe and $5\frac{1}{2}$ in along the heel of each angle.

To check the $\frac{1}{2}$-in-thick gusset plate, divide the capacity of the two opposite welds by the allowable shear stress: Gusset thickness required is $2 \times 2.78/14.5 = 0.38$ in < 0.50 in—OK. One rule of thumb for fillet welds on both faces opposite each other is to make the gusset thickness twice the weld size. However, this rule is too conservative in the present case because the gusset cannot fail through one section under each weld. A better way is to check for gusset tearout. The shear tearout area is $A_v = 6 \times 0.5 \times 2 = 6.0$ in^2, and the tension tearout area is $A_t = 3 \times 0.5 \times 1 = 1.5$ in^2. Thus the tearout capacity is $P_{to} = 6 \times 0.3 \times 58 + 1.5 \times 0.5 \times 58 = 148$ kips > 60 kips—OK. This method recognizes a true limit state, whereas matching fillet-weld size to gusset-plate thickness does not.

5.25.3 Fasteners in Tension

As an alternative to lap joints, with fasteners or welds in shear, hangers also may be supported by fasteners in tension. Permissible tension in such fasteners equals the product of the reduced cross-sectional area at threads and allowable tensile stress. The stress is based on full-head rivets or on bolts with hexagonal or square heads and nuts. Flattened- or countersunk-head fasteners, therefore, should not be used in joints where they will be stressed in tension.

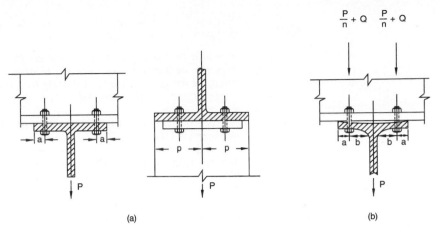

FIGURE 5.25 Two connections with bolts in tension (*a*) develop prying action (*b*), imposing forces *Q* on the bolts.

Bolts have a deliberately applied pretension. The tension is maintained by compression in the connected parts. A tensile force applied to a fastener relieves the compression in the connected parts without increasing the tension in the fastener. Unless the tensile force is large enough to permit the connected parts to separate, the tension in the fastener will not exceed the pretension.

Generally, the total force on a fastener in tension equals the average force on all the fasteners in the joint plus force due to eccentricity, if present. Sometimes, however, the configuration of the joint produces a prying effect on the fasteners that may be serious and should be investigated.

Figure 5.25*a* shows a connection between the flange of a supporting member and the flange of a T shape (tee, half wide-flange beam, or pair of angles with plate between), with bolts in tension. The load *P* is concentrically applied. If the prying force is ignored, the average force on any fastener is P/n, where *n* is the total number of fasteners in the joint. But, as indicated in Fig. 5.25*b*, distortion of the T flange induces an additional prying force *Q* in the fastener. This force is negligible when the connected flanges are thick relative to the fastener gages or when the flanges are thin enough to be flexible.

Note that in Fig. 5.25*b* though only the T is shown distorted, either flange may distort enough to induce prying forces in the fasteners. Hence theoretical determination of *Q* is extremely complex. Research has shown, however, that the following approach from the AISC "Manual of Steel Construction—ASD" gives reliable results: Let α = ratio of moment M_2 at bolt line to moment δM_1, at stem line where δ = ratio of net area (along the line of bolts) to gross area (at face of stem or angle leg), and α' = value of α for which the required thickness is a minimum or allowable tension per bolt is a maximum.

$$\alpha' = \frac{1}{\delta(1 + \rho)} \left[\left(\frac{t_c}{t} \right)^2 - 1 \right] \tag{5.2}$$

The allowable tensile load (kips) per fastener, including the effect of prying action, is given by

$$T_a = B \left(\frac{t}{t_c} \right)^2 (1 + \delta\alpha') \tag{5.3}$$

where $t_c = \sqrt{\dfrac{8Bb'}{pF_y}}$

B = allowable bolt tension, including the effect of shear, if any, kips

t = thickness of thinnest connected flange or angle leg, in

p = length of flange or angle leg tributary to a bolt, in

F_y = flange or angle yield stress, ksi

a' = $a + d/2$

a = distance from center of fastener to edge of flange or angle leg, in (not to exceed $1.25b$)

b' = $b - d/2$

b = distance from center of fastener to face of tee stem or angle leg, in

ρ = b'/a'

δ = $1 - d'/p$

d = bolt diameter, in

d' = bolt-hole diameter, in

If $\alpha' > 1$, $\alpha' = 1$ should be used to calculate T_a. If $\alpha' < 0$, $T_a = B$.

Example.   A tee stub hanger is to transfer a 60-kip load to the flange of a wide-flange beam through four 1-in-diameter A325 bolts in tension (Fig. 5.26). The tee stub is half of a W18 $\times$ 60 made of A572 grade 50 steel, with F_y = 50 ksi, b = 1.792 in, a = 1.777 in $< (1.25 \times 1.792 = 2.24$ in). Other needed dimensions are given in Fig. 5.26. Check the adequacy of this connection.

$$B = 44 \times 0.7854 = 34.6 \text{ kips}$$

$$t = 0.695 \text{ in}$$

$$p = 4.5 \text{ in}$$

$$a' = 1.775 + 0.5 = 2.277 \text{ in}$$

$$b' = 1.792 - 0.5 = 1.292 \text{ in}$$

$$d' = 1.0625 \text{ in}$$

$$\rho = 1.292/2.277 = 0.567$$

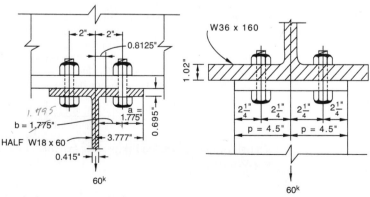

FIGURE 5.26   Example of hanger with bolts in tension.

Substitution in Eq. (5.2) yields

$$t_c = \sqrt{\frac{8 \times 34.6 \times 1.292}{4.5 \times 50}} = 1.261 \text{ in}$$

With $\delta = 1 - 1.0625/4.5 = 0.7639$ and $t_c = 1.261$,

$$\alpha' = \frac{1}{0.7639 \times 1.568} \left[\left(\frac{1.261}{0.695}\right)^2 - 1 \right] = 1.915$$

Since $\alpha' > 1$, use $\alpha' = 1$ to compute the allowable tensile load per fastener, including prying action:

$$T_a = 34.6 \left(\frac{0.695}{1.261}\right)^2 1.7639 = 18.5 \text{ kips}$$

Since the load per bolt $T = 60/4 = 15$ kips is less than the allowable load per bolt of 18.5 kips, the connection is satisfactory.

This method checks both the bolts and the tee flange or angle leg. No further checks are required. The prying force Q is not explicitly calculated but instead is built into the formulas. The prying force can be calculated as follows:

$$\alpha = \frac{1}{\delta} \left[\frac{T}{B} \left(\frac{t_c}{t}\right)^2 - 1 \right] \tag{5.4}$$

where T is the load per bolt (kips). And

$$Q = B\alpha\rho\delta \left(\frac{t}{t_c}\right)^2 \tag{5.5}$$

Substitution of the connection dimensions and $T = {}^{60}/_4 = 15$ kips gives

$$\alpha = \frac{1}{0.7639} \left[\frac{15}{34.6} \left(\frac{1.261}{0.695}\right)^2 - 1 \right] = 0.559$$

$$Q = 34.6 \times 0.559 \times 0.567 \times 0.7639 \left(\frac{0.695}{1.261}\right)^2 = 2.55 \text{ kips}$$

and the total load per bolt is $15 + 2.55 = 17.55$ kips. Note that the previous conclusion that the connection is satsifactory is confirmed by this calculation; that is,

$$T_a = 18.5 \text{ kips} > T + Q = 17.55 \text{ kips}$$

AASHTO requires that Q be estimated by the empirical formula

$$Q = \left[\frac{3b}{8a} - \frac{t^3}{20} \right] T \tag{5.6}$$

where all quantities are as defined above, except that b is measured from the center of the bolt to the toe of the fillet of the flange or angle leg; there is no restriction on a, and $B = 39.5 \times 0.7854 = 31.0$ kips. Hence

$$Q = \left[\frac{3 \times 1.1875}{8 \times 1.777} - \frac{0.695^3}{20} \right] 15 = 3.51 \text{ kips}$$

Since $T + Q = 15 + 3.51 = 18.5$ kips < 31.0 kips, the bolts are OK. The tee flange, however, should be checked independently. This can be done as follows.

Assume that the prying force $Q = 3.51$ acts at a distance a from the bolt line. The moment at the bolt line then is

$$M_b = 3.51 \times 1.777 = 6.24 \text{ kip-in}$$

and that at the toe of the fillet, with $T + Q = 18.5$ kips, is

$$M_f = 3.51 \times (1.777 + 1.1875) - 18.5 \times 1.1875 = -11.6 \text{ kip-in}$$

The maximum moment in the flange is thus 11.6 kip-in, and the bending stress in the flange is

$$f_b = \frac{11.6 \times 6}{4.5 \times 0.695^2} = 32.0 < 0.75 \times 50 = 37.5 \text{ ksi—OK}$$

5.25.4   Welded Butt Joints

A hanger also may be connected with a simple welded butt joint (Fig. 5.27). The allowable stress for the complete-penetration groove weld is the same as for the base metal used.

Examples.   A bar hanger carries a 60-kip load and is supported through a complete-penetration groove weld at the edge of a tee stub.

For A36 steel with allowable tensile stress of 22 ksi, the bar should have an area of $^{60}\!/_{22} = 2.73$ in^2. A bar $5\frac{1}{2} \times \frac{1}{2}$ with an area of 2.75 in^2 could be used. The weld strength is equal to that of the base metal. Hence no allowance need be made for its presence.

Suppose, however, that the hanger is to be used in a highway bridge and the load will range from 30 kips in compression (minus) to 60 kips in tension (plus). The design then will be governed by the allowable stresses in fatigue. These depend on the stress range and the number of cycles of load the structure will be subjected to. Design is to be based on 100,000 cycles. Assuming a redundant load path and the detail of Fig. 5.27, load category C applies with an allowable stress range $SR = 35.5$ ksi, according to AASHTO. For A36 steel, the maximum tensile stress is 20 ksi. At 60-kip tension, the area required is $^{60}\!/_{20} = 3$ in^2. A bar $5 \times \frac{5}{8}$ could be tried. It has an area of 3.125 in^2. Then the maximum tensile stress is $60/3.125 = 19.2$ ksi, and the maximum compressive stress is $30/3.125 = 9.6$ ksi. Thus the stress range is $19.2 - (-9.6) = 28.8$ ksi < 35.5 ksi—OK. The $5 \times \frac{5}{8}$ bar is satisfactory.

As another example, consider the preceding case subjected to 2 million cycles. The allowable stress range is now 13 ksi. The required area is $[60 - (-30)]/13 = 6.92$ in^2. Try a bar $5 \times 1\frac{1}{2}$ with area 7.5 in^2. The maximum tensile stress is $60/7.5 = 8.0$ ksi (<20 ksi), and the maximum compressive stress is $30/7.5 = 4.0$ ksi. Thus the actual stress range is $8.0 - (-4.0) = 12.0$ ksi < 13.0 ksi—OK. The $5 \times 1\frac{1}{2}$ bar is satisfactory.

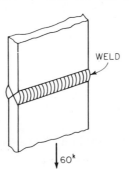

FIGURE 5.27   Groove weld for hanger connection.

5.26 TENSION SPLICES

Design rules for tension splices are substantially the same as those for hanger connections. In buildings, splices should develop the strength required by the stresses at point of splice. For groove welds, however, the full strength of the smaller spliced member should be developed.

In highway bridges, splices should be designed for the larger of the following: 75% of the strength of the member or the average of the calculated stress at point of splice and the strength of the member there. Where a section changes size at a splice, the strength of the smaller section may be used in sizing the splice. In tension splices, the strength of the member should be calculated for the net section.

In railroad bridges, tension splices in main members should have the same strength as the members. Splices in secondary and bracing members should develop the average of the strength of the members and the calculated stresses at the splices.

When fillers are used, the requirements discussed in Art. 5.13 should be satisfied.

In groove-welded tension splices between parts of different widths or thicknesses, a smooth transition should be provided between offset surfaces or edges. The slope with respect to the surface or edge of either part should not exceed 1:2.5 (equivalent to about 5:12, or 22°). Thickness transition may be accomplished with sloping weld faces, or by chamfering the thicker part, or by a combination of the two methods.

Splices may be made with complete-penetration groove welds, preferably without splice plates. The basic allowable unit stress for such welds is the same as for the base metal joined. For fatigue, however, the allowable stress range F_{sr} for base metal adjacent to continuous flange-web fillet welds may be used for groove-welded splices only if

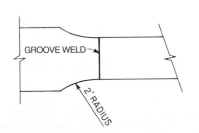

FIGURE 5.28 Recommended taper for unequal-width plates at groove-welded splice.

1. The parts joined are of equal thickness.

2. The parts joined are of equal widths or, if of unequal widths, the parts are tapered as indicated in Fig. 5.28, or, except for A514 and A517 steels, tapered with a uniform slope not exceeding 1:2.5.

3. Weld soundness is established by radiographic or ultrasonic testing.

4. The weld is finished smooth and flush with the base metal on all surfaces by grinding in the direction of applied stress, leaving surfaces free from depressions. Chipping may be used if it is followed by such grinding. The grinding should not reduce the thickness of the base metal by more than $\frac{1}{32}$ in or 5% of the thickness, whichever is smaller.

Groove-welded splices that do not conform to all these conditions must be designed for reduced stress range assigned to base metal adjacent to groove welds.

For riveted and bolted flexural members, splices in flange parts between field splices should be avoided. In any one flange, not more than one part should be spliced at the same cross section.

Fatigue need not be considered when calculating bolt stresses but must be taken into account in design of splice plates.

Example. A plate girder in a highway bridge is to be spliced at a location where a 12-in-wide flange is changed to a 16-in-wide flange (Fig. 5.29). Maximum bending moments at the splice are $+700$ kip-ft (tension) and -200 kip-ft (compression). Steel is A36. The girder is redundant and is subjected to not more than 2 million cycles of stress. The connection is slip-critical; bolts are A325SC, $\frac{7}{8}$ in in diameter, in standard holes. The web has nine holes for $\frac{7}{8}$-in bolts. The tension-flange splice in Fig. 5.29 is to be checked.

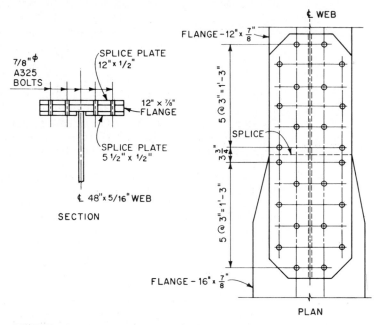

FIGURE 5.29 Tension-flange splice for highway-bridge plate girder.

According to AASHTO specifications for highway bridges, members loaded primarily in bending should be designed for stresses computed for the gross section. If, however, the areas of flange holes exceed 15% of the flange area, the excess should be deducted from the gross area. With two bolt holes, net width of the flange is $12 - 2 = 10$ in. With four holes, the net width along a zigzag section through the holes is $12 - 4 + 2 \times 3^2/(4 \times 3) = 9.5$ in, with the addition of $s^2/4g$ for two chains, where $s =$ bolt pitch $= 3$ in and $g =$ gage $= 3$ in. The four-hole section, with less width, governs. In this case, the ratio of hole area to flange area is $(12 - 9.5)/12 = 0.21 > 0.15$. The area reduction for the flange is $12(0.21 - 0.15)0.875 = 0.63$ in^2, and reduced flange area is $12 \times 0.875 - 0.63 = 9.87$ in^2. With moment of inertia I_{gw} of the web equal to $0.3125(48)^3/12 = 2880$ in^4, the effective gross moment inertia of the girder is

$$I_g = 9.87 \left(\frac{48.875}{2}\right)^2 \times 2 + 2880 = 14{,}670 \text{ in}^4$$

AASHTO requires that the design tensile stress on the net section not exceed $0.5F_u = 29$ ksi. The net moment of inertia with flange area $= 9.5 \times 0.875 =$

8.3125 in² and nine web holes is

$$I_n = 2880 + 2 \times 8.3125 \left(\frac{48.875}{2}\right)^2 - (2 \times 1 \times 0.3125(5^2 + 10^2 + 15^2 + 20^2)$$

$$= 2880 + 9928 - 469 = 12,339 \text{ in}^4$$

The net moment of inertia of the web is

$$I_{nw} = 2880 - 469 = 2411 \text{ in}^4$$

Bending stresses in the girder are computed as follows:

Gross section: $\qquad f_b = \dfrac{700 \times 12 \times 24.875}{14,670} = 14.2 \text{ ksi} < 20 \text{ ksi—OK}$

Net section: $\qquad f_b = 14.2 \times 9.87/8.3125 = 16.9 \text{ ksi} < 29 \text{ ksi—OK}$

Stress range: $\quad f_{sr} = \dfrac{[700 - (-200)]12 \times 24.875}{14,670} = 18.3 \text{ ksi} \approx 18 \text{ ksi—OK}$

The flange splice must have a capacity not less than that based on

1. The average of the calculated design stress at the point of splice and the allowable stress of the member at the same point
2. 75% of the allowable stress in the member

From criterion 1, the design stress $F_{d_1} = (14.2 + 20)/2 = 17.1$ ksi, and from criterion 2, the design stress $F_{d_2} = 0.75 \times 20 = 15$ ksi. Therefore, design for $F_d = 17.1$ ksi on the gross section.

The required splice-plate sizes are determined as follows:

$$\text{Design moment} = \frac{700 \times 17.1}{14.2} = 843 \text{ kip-ft}$$

$$\text{Moment in flange} = 843 \left(\frac{14,670 - 2880}{14,670}\right) = 678 \text{ kip-ft}$$

$$\text{Flange force} = \frac{678 \times 12}{48.875} = 166 \text{ kips}$$

$$\text{Area required} = \frac{166}{20} = 8.3 \text{ in}^2$$

This area should be split approximately equally to inside and outside plates. Try an outside plate ½ × 12 with area of 6 in² and two inside plates ½ × 5½ each with area of 5.5 in². Total plate area is 11.5 in² > 8.3 in²—OK. Net width of the outside plate with two holes deducted is 12 − 2 = 10 in and of the inside plates with two holes deducted is 11 − 2 = 9 in. The net width of the outside plate along a zigzag section through four holes is 12 − 4 + 2 × 3²/(4 × 3) = 9.5 in. The net width of the inside plate with four holes is 11 − 4 + 2 × 3²/(4 × 3) = 8.5 in. The zigzag section with four holes controls. The net area is 0.5(9.5 + 8.5) = 9.0 in². Hence the stress in the plates is 166/9.0 = 18.4 ksi < 29 ksi—OK.

Since the gross area of the splice plates exceeds that of the flange, there is no need to check fatigue of the plates.

Capacity of the bolts in this slip-critical connection, with class A surfaces and 12 A325 7/8-in-diameter bolts in standard holes, is computed as follows: Allowable shear stress is $F_v = 15.5$ ksi. The capacity of one bolt thus is $r_v = 15.5 \times 3.14(0.875)^2/4 = 9.32$ kips. Shear capacity of 12 bolts in double shear is $2 \times 12 \times 9.32 = 224$ kips > 166 kips—OK. The bearing capacity of the bolts, with an allowable stress of $1.1F_u = 1.1 \times 65 = 71.5$ ksi, is $71.5 \times 0.875 \times 0.875 \times 12 = 657$ kips > 166 kips—OK.

5.27 COMPRESSION SPLICES

The requirements for strength of splice given in Art. 5.26 for tension splices apply also to compression splices.

Compression members may be spliced with complete-penetration groove welds. As for tension splices, with such welds, it is desirable that splice plates not be used.

Groove-welded compression splices may be designed for the basic allowable stresses for base metal. Fatigue does not control if the splice will always be in compression.

Groove-welded compression splices should be made with a smooth transition when the offset between surfaces at either side of the joint is greater than the thickness of the thinner part connected. The slope relative to the surface or edge of either part should not exceed 1:2½ (5:12, or 22°). For smaller offsets, the face of the weld should be sloped 1:2½ from the surface of the thinner part or sloped to the surface of the thicker part if this requires a lesser slope.

At riveted or bolted splices, compression members may transmit the load through the splice plates or through bearing.

Example. Investigate the bolted flange splice in the highway-bridge girder of the example in Art. 5.26 for use with the compression flange.

The computations for the bolts are the same for the compression splice as for the tension splice. Use 12 bolts, for sealing, in a staggered pattern (Fig. 5.29).

The splice-plate net area for the tension flange is OK for the compression flange because it is subject to the smaller moment 200 kip-ft rather than 700 kip-ft. The gross area, however, must carry the force due to the 700-kip-ft moment that causes compression in the compression splice. By calculations in Art. 5.26, the splice plates are OK for the maximum compressive stress of 20 ksi and for stress range. Therefore, the splice plates for the compression splice can be the same as those required for tension.

To avoid buckling of the splice plates, the ends of compression members in bolted splices should be in close contact, whether or not the load is transmitted in bearing, unless the splice plates are checked for buckling. For the compression splice, the unsupported length of plate L is 3.75 in, and the effective length is $0.65L = 2.44$ in. The slenderness ratio of the plate then is $2.44\sqrt{12}/0.5 = 16.90$. Hence the allowable compression stress is

$$F_a = 16,980 - 0.53(16.90)^2 = 16,829 \text{ psi} = 16.8 \text{ ksi}$$

The compressive stress in the inside splice plates due to the loads is $166/11.5 = 14.4$ ksi < 16.8 ksi—OK.

Columns in Buildings. Ends of compression members may be milled to ensure full bearing at a splice. When such splices are fabricated and erected under close inspection, the designer may assume that stress transfer is achieved entirely through bearing. In this case, splice material and fasteners serve principally to hold the connected parts in place. But they also must withstand substantial stresses during erection and before floor framing is placed. Consequently, standard specifications generally require that splice material and fasteners not only be arranged to hold all parts in place but also be proportioned for 50% of the computed stress. The AISC specification, however, exempts building columns from this requirement. But it also requires that all joints with stress transfer through bearing be proportioned to resist any tension that would be developed by specified lateral forces acting in conjunction with 75% of the calculated dead-load stress and no live load.

When fillers are used, the requirements discussed in Art. 5.13 should be satisfied.

In multistory buildings, changes in column sizes divide the framing vertically into tiers. Joints usually are field bolted or welded as successive lengths are erected. For convenience in connecting beam and girder framing, column splices generally are located 2 to 3 ft above floor level. Also, for convenience in handling and erection, columns usually are erected in two- or three-story lengths.

To simplify splicing details while securing full bearing at a change in column size, wide-flange shapes of adjoining tiers should be selected with the distance between the inner faces of the flanges constant; for example, note that the T distances given in the AISC "Steel Construction Manuals" for W14 $\times$ 43 and heavier W14 sections are all 11 or 11¼ in. If such sections are not used, or if upper and lower members are not centered, full bearing will not be obtained. In such cases, stress transfer may be obtained with filler plates attached to the flanges of the smaller member and finished to bear on the larger member. Enough fasteners must be used to develop in single shear the load transmitted in bearing. Instead of fillers, however, a butt plate may be interposed in the joint to provide bearing for the upper and lower sections (Fig. 5.30). The plate may be attached to either shaft with tack welds or clip angles. Usually, it is connected to the upper member, because a plate atop the lower one may interfere with beam and girder erection.

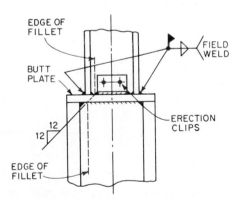

FIGURE 5.30   Column splice with butt plate.

The butt plate should be thick enough to resist the bending and shear stresses imposed by the eccentric loading. Generally, a 1½-in-thick plate can be used when a W8 section is centrally seated over a W10 and a 2-in-thick plate between other sections. Plates of these thicknesses need not be planed as long as they provide satisfactory bearing. Types and sizes of welds are determined by the loads.

For direct load only, on column sections that can be spliced without plates, partial-penetration bevel or J welds may be used. Figure 5.31 illustrates a typical splice with a partial-penetration single-bevel groove weld (AWS prequalified, man-

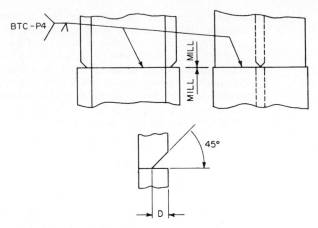

FIGURE 5.31 Column splice with partial-penetration groove welds.

ual, shielded metal-arc welded joint BTC-P4). Weld sizes used depend on thickness of the column flange. AISC recommends the partial-penetration groove welds in Table 5.17.

TABLE 5.17 Recommended Sizes of Partial-Penetration Groove Welds for Column Splices

Thickness of thinner flange, in	Effective weld size, in
Over ½ to ¾ inclusive*	¼
Over ¾ to 1½ inclusive	5/16
Over 1½ to 2¼ inclusive	3/8
Over 2¼ to 6 inclusive	½
Over 6	5/8

*Up to ½ in, use splice plates.

A similar detail may be used for splices subjected to bending moments. A full moment splice can be made with complete-penetration groove welds (Fig. 5.32). When the bevel is formed, a shoulder of at least ⅛ in should remain for landing and plumbing the column.

When flange plates are not used, column alignment and stability may be achieved with temporary field-bolted lugs. These usually are removed after the splice is welded, to meet architectural requirements.

To facilitate column erection, holes often are needed to receive the pin of a lifting hitch. When splice plates are used, the holes can be provided in those plates (Fig. 5.34). When columns are spliced by groove welding, without splice plates, however, a hole must be provided in the column web, or auxiliary plates, usually temporary, must be attached for lifting. The lifting-lug detail in Fig. 5.33 takes advantage of the constant distance between inside faces of flanges of W14 sections.

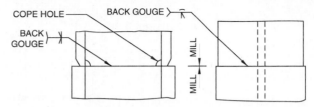

FIGURE 5.32 Column splice with complete-penetration groove welds to resist bending moments.

Bolted column splices generally are made with flange plates. Fillers are inserted (Fig. 5.34) when the differences in column sizes are greater than can be absorbed by erection clearance.

To provide erection clearance when columns of the same nominal depth are spliced, a $\frac{1}{16}$-in filler often is inserted under each splice plate on the lower column. Or with the splice plates shop fastened to the lower column, the fastener holes may be left open on the top line below the milled joint until the upper member is connected. This detail permits the erector to spring the splice plates apart enough for placement of the upper shaft.

For the detail in Fig. 5.34, the usual maximum gage for the flanges should be used for G_1 and G_2 (4, 5½, or 11½ in). The widths of fillers and splice plates are then determined by minimum edge distances (1¼ in for ¾-in fasteners, 1½ in for ⅞-in fasteners, and 1¾-in for 1-in fasteners).

Fill plates not in bearing, as shown in Fig. 5.34, may be connected for shipping, with two fasteners or with welds, to the upper column. Fillers milled to carry load in bearing or thick fills that will carry load should be attached with sufficient fasteners or welds to transmit that load (Art. 5.13).

When fasteners are used for a column splice, it is good practice to space the holes so that the shafts are pulled into bearing during erection. If this is not done, it is possible for fasteners in the upper shaft to carry the entire load until the fasteners deform and the joint slips into bearing.

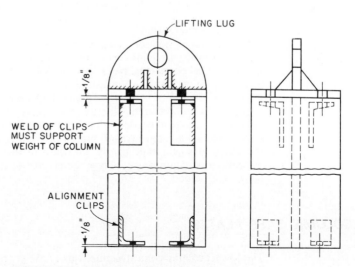

FIGURE 5.33 Lifting lug facilitates column erection.

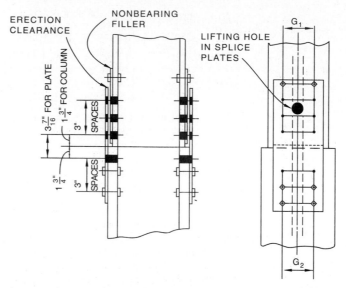

FIGURE 5.34 Column splice with flange plates. A hole in each plate is used for erection of the lower column.

Usually, for direct load only, thicknesses used for splice plates are as follows:

Column weight lb/ft	Splice-plate thickness, in
Up to 132	⅜
Up to 233	½
Over 233	¾

Four fasteners are used in each half of a ⅜-in splice plate and six fasteners in each half of a ½-in or heavier splice plate.

If, however, the joint carries tension or bending moments, the plate size and number of fasteners must be designed to carry the load. An equivalent amount of weld may be used instead of fasteners.

A combination of shop-welded and field-welded connection material is usually required for moment connections. Splice plates are shop welded to the top of the lower shaft and field welded to the bottom of the upper shaft. For field alignment, a web splice plate can be shop bolted or welded to one shaft and field bolted to the other. A single plate nearly full width of the web with two to four fasteners, depending on column size, allows the erector to align the shafts before the flange plates are welded.

5.28 COLUMN BASE PLATES

The lowest columns of a structure usually are supported on a concrete foundation. To prevent crushing of the concrete, base plates are inserted between the steel and concrete to distribute the load. For very heavy loads, a grillage, often encased in concrete, may be required. It consists of one or more layers of steel beams with pipe separators between them and tie rods through the pipe to prevent separation.

The area (in^2) of base plate required may be computed from

$$A = \frac{P}{F_p} \tag{5.7}$$

where P = load, kips
F_p = allowable bearing pressure on support, ksi

The allowable pressure depends on strength of concrete in the foundation and relative sizes of base plate and concrete support area. If the base plate occupies the full area of the support, $F_p = 0.35f'_c$ where f'_c is the 28-day compressive strength of the concrete. If the base plate covers less than the full area, $F_p = 0.35f'_c \sqrt{A_2/A_1} \leq 0.70f'_c$, where A_1 is the base-plate area ($B \times N$), and A_2 is the full area of the concrete support.

Eccentricity of loading or presence of bending moment at the column base increases the pressure on some parts of the base plate and decreases it on other parts. To compute these effects, the base plate may be assumed completely rigid so that the pressure variation on the concrete is linear.

Plate thickness may be determined by treating projections m and n of the base plate beyond the column as cantilevers. The cantilever dimensions m and n are usually defined as shown in Fig. 5.35. (If the base plate is small, the area of the base plate inside the column profile should be treated as a beam.) Yield-line analysis shows that an equivalent cantilever dimension n' can be defined as $n' = \frac{1}{4}\sqrt{db_f}$, and the required base plate thickness t_p can be calculated from

$$t_p = 2l\sqrt{\frac{f_p}{F_y}} \tag{5.8}$$

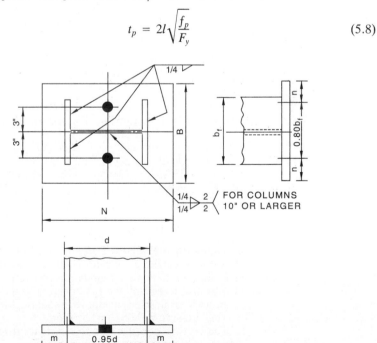

FIGURE 5.35 Column welded to a base plate.

where l = max (m, n, n'), in
    $f_p = P/(BN) \le F_p$, ksi
    F_y = yield strength of base plate, ksi
    P = column axial load, kips

For columns subjected only to direct load, the welds of column to base plate, as shown in Fig. 5.35, are required principally for withstanding erection stresses. For columns subjected to uplift, the welds must be proportioned to resist the forces.

Base plates are tied to the concrete foundation with hooked anchor bolts embedded in the concrete. When there is no uplift, anchor bolts commonly used are ¾ in in diameter, about 1 ft 6 in long plus a 3-in hook.

Instead of welds, anchor bolts may be used to tie the column and base plate to the concrete foundation. With anchor bolts up to about 1¼ in in diameter, heavy clip angles may be fastened to the columns to transfer overturning or uplift forces from the column to the anchor bolts. For large uplift forces, stiffeners may be required with the anchor bolts (Fig. 5.36). The load is transferred from column to bolts through the stiffener-plate welds. The cap plate should be designed for bending.

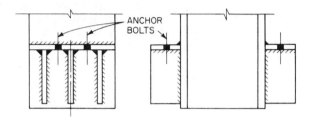

FIGURE 5.36   Column base with stiffeners for anchor bolts.

Example—Base Plate with Concentric Column Load.   A W10 × 54 column of A36 steel carries a concentric load of 120 kips. The foundation concrete has a 28-day specified strength f'_c = 3000 psi. Design a base plate of A36 steel to occupy the full concrete area.

Since $A_2/A_1 = 1$, the allowable bearing stress on the concrete is $F_p = 0.35 \times 3000 = 1050$ psi = 1.05 ksi. The required base-plate area is 120/1.05 = 114 in². Since the column size is $d \times b_f = 10.125 \times 10$, try a base plate 12 × 12; area = 144 in². Then, from Fig. 5.35, $m = (12 - 0.95 \times 10.125)/2 = 1.19$ in, $n = (12 - 0.80 \times 10)/2 = 2.00$ in, and $n' = \sqrt{10.125 \times 10}/4 = 2.52$ in. Hence, $f_p = 120/144 = 0.83$ ksi < 1.05 ksi—OK. The plate thickness required, from Eq. (5.8), with $F_y = 36$ ksi, is

$$t_p = 2 \times 2.52\sqrt{0.83/36} = 0.77 \text{ in}$$

Use a ⅞-in-thick plate.

Example—Base Plate for Column with Bending Moment.   A W10 × 54 column of A36 steel carries a concentric load of 120 kips and a bending moment of 29 ft-kips (Fig. 5.37). For the foundation concrete, f'_c = 3000 psi. Design a base plate of A36 steel to occupy the full concrete support area.

As a first trial, select a plate 18 in square. Area provided is 324 in². The allowable bearing pressure is $0.35f'_c = 1050$ psi.

The bearing pressure due to the concentric load is 120/324 = 0.370 ksi. The maximum pressures due to the moment are

$$\pm\frac{M}{BN^2/6} = \pm\frac{29 \times 12}{18 \times 18^2/6} = \pm 0.358 \text{ ksi}$$

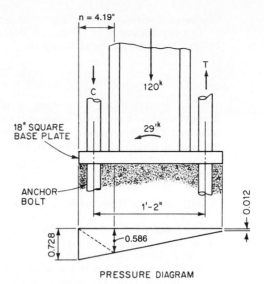

FIGURE 5.37   Example of column base-plate design.

Hence the maximum total pressure is $0.370 + 0.358 = 0.728$ ksi < 1.050. The minimum pressure is $0.370 - 0.358 = 0.012$ ksi. Thus the plate area is satisfactory.

The projection of the plate beyond the ends of the flanges is

$$n = (18 - 0.80 \times 10)/2 = 5 \text{ in}$$

By Eq. (5.8), the thickness required for this cantilever is

$$t_p = 2 \times 5 \sqrt{0.370/36} = 1.01 \text{ in}$$

The projection of the plate in the perpendicular direction (Fig. 5.37) is

$$m = (18 - 0.95 \times 10.125)/2 = 4.19 \text{ in}$$

Over the distance m, the decrease in bearing pressure is $4.19 (0.728 - 0.012)/$ $18 = 0.142$ ksi. Therefore, the pressure under the flange is $0.728 - 0.142 = 0.561$ ksi. The projection acts as a cantilever with a trapezoidal load varying from 0.728 to 0.561 ksi. The bending moment in this cantilever at the flange is

$$M = 18 \frac{(4.19)^2 0.561}{6} + \frac{(4.19)^2 0.728}{3} = 106 \text{ in-kips}$$

The thickness required to resist this moment is

$$t_p = \sqrt{\frac{6 \times 106}{18 \times 27}} = 1.14 \text{ in}$$

Use a 1¼-in-thick plate.

The overturning moment of 29 ft-kips induces forces T and C in the anchor bolts equal to $29/1.167 = 24.9$ kips. If two bolts are used, the force per bolt is $24.9/2 = 12.5$ kips. Then, with an allowable stress of 20 ksi, each bolt should have a nominal area of

$$12.5/20 = 0.625 \text{ in}^2$$

Use two 1-in-diameter bolts, each with a nominal area of 0.785 in².

The circumference of a 1-in bolt is 3.14 in. With an allowable bond stress of 160 psi, length of the bolts should be

$$L = \frac{12,500}{160 \times 3.14} = 25 \text{ in}$$

Also, because of the overturning moment, the weld of the base plate to each flange should be capable of resisting a force of $29 \times {}^{12}\!/_{10} = 34.8$ kips. Assume a minimum E70XX weld of ⁵⁄₁₆-in capacity equal to $0.707 \times 21 \times {}^5\!/_{16} = 4.63$ kips per in. The length of weld required then is

$$L = \frac{34.8}{4.63} = 7.5 \text{ in}$$

A weld along the flange width would be more than adequate.

Pressure under Part of Base Plate.   The equivalent eccentricity of the moment acting at the base of a column equals the moment divided by the concentric load: $e = M/P$. When the eccentricity exceeds one-sixth the length of base plate (P lies outside the middle third of the plate), part of the plate no longer exerts pressure against the concrete. The pressure diagram thus extends only part of the length of the base plate. If the pressure variation is assumed linear, the pressure diagram is triangular. For moderate eccentricities (one anchor bolt required on each side of the column), the design procedure is similar to that for full pressure over the base plate. For large eccentricities (several anchor bolts required on each side of the column), design may be treated like that of a reinforced-concrete member subjected to bending and axial load.

Example.   A W10 × 54 column of A36 steel carries a concentric load of 120 kips and a bending moment of 45 ft-kips (Fig. 5.38). For the foundation concrete, $f'_c = 3000$ psi. Design a base plate of A36 steel to occupy the full concrete support area.

As a first trial, select a plate 20 in square. The eccentricity of the load then is $45 \times {}^{12}\!/_{120} = 4.50$ in $> ({}^{20}\!/_6 = 3.33$ in). Therefore, there will be pressure over only part of the plate. Assume this length to be d.

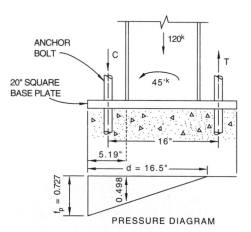

FIGURE 5.38 Loading produces pressure over part of a column base.

For equilibrium, the area of the triangular pressure diagram must equal the 120-kip vertical load:

$$120 = \tfrac{1}{2}Bdf_p = 10df_p \qquad (5.9)$$

where B = width of plate = 20 in and f_p = maximum bearing pressure (ksi). The allowable bearing pressure is $0.35f_c' = 0.35 \times 3000 = 1050$ psi = 1.050 ksi.

For equilibrium also, the sum of the moments about the edge of the base plate at the point of maximum pressure must be zero:

$$120 \times 10 - 45 \times 12 - \tfrac{1}{6}Bd^2f_p = 0$$

Substitution of $B = 20$ and rearrangement of terms yields

$$660 = 3.33d^2f_p \qquad (5.10)$$

Dividing Eq. (5.9) by (5.10) eliminates f_p, and solution of the result gives $d = 16.5$ in. For this value of d, Eq. (5.9) gives $f_p = 0.727$ ksi < 1.050. The size of the base plate is satisfactory.

The projection of the plate beyond the flange is $(20 - 0.95 \times 10.125)/2 = 5.19$ in. The pressure under the flange is $0.727(16.5 - 5.19)/16.5 = 0.498$ ksi. The projection acts as a cantilever with a trapezoidal load varying from 0.727 to 0.498 ksi. The bending moment in this cantilever at the flange is

$$M = 20\left[\frac{(5.19)^2 0.498}{6} + \frac{(5.19)^2 0.727}{3}\right] = 175.3 \text{ in } = \text{ kips}$$

The thickness required to resist this moment is

$$t = \sqrt{\frac{6 \times 175.3}{20 \times 27}} = 1.40 \text{ in}$$

Use a 1½-in-thick plate.

The overturning moment induces forces T and C in the anchor bolts equal to $45/1.33 = 33.7$ kips. If two bolts are used, the force per bolt is $33.7/2 = 16.85$ kips. Then, with an allowable stress of 20 ksi, each bolt should have a nominal area of $16.85/20 = 0.843$ in^2. Use two 1⅛-in-diameter bolts, each with a nominal area of 0.994 in^2.

The circumference of a 1⅛-in bolt is 3.53 in. With an allowable bond stress of 160 psi, length of the bolts should be

$$L = \frac{16,850}{160 \times 3.53} = 30 \text{ in}$$

Assume a minimum weld of 5⁄16 in for connecting the base plate to each flange. The weld has a capacity of 4.63 kips per in. Welds 10 in long on each flange provide a resisting moment of $4.63 \times 10 \times 10 = 463$ in-kips. Since the moment to be resisted is $45 \times 12 = 540$ in-kips, welds along the inside faces of the flanges must resist the 77-in-kip difference. With a smaller moment arm than the outer welds, because of the 5⁄8-in flange thickness, the length of weld per flange required is

$$L = \frac{77}{4.63(10 - 2 \times 0.625)} = 1.9 \text{ in}$$

The outer-face welds should be returned at least 1 in along the inside faces of the flanges on opposite sides of the web.

Setting of Base Plates.   The welds of flanges to base plate in the preceding examples usually are made in the shop for small plates. Large base plates often are shipped separately, to be set before the columns are erected. When a column is set in place, the base must be leveled and then grouted. For this purpose, the footing must be finished to the proper elevation. The required smooth bearing area may be obtained with a steel leveling plate about ¼ in thick. It is easy to handle, set to the proper elevation, and level. Oversized holes punched in this plate serve as templates for setting anchor bolts. Alternatively, columns may be set with wedges or shims instead of a leveling plate.

Large base plates may be set to elevation and leveled with shims or with leveling screws (Fig. 5.39). In those cases, the top of concrete should be set about 2 in below the base plate to permit adjustments to be made and grout to be placed under the plate, to ensure full bearing. One or more large holes may be provided in the plate for grouting.

Planing of Base Plates.   Base plates, except thin rolled-steel bearing plates and surfaces embedded in grout, should be planed on all bearing surfaces. Also, all bearing surfaces of rolled-steel bearing plates over 4 in thick should be planed. Rolled-steel bearing plates over 2 in but not more than 4 in thick may be planed or straightened by pressing. Rolled-steel bearing plates 2 in or less in thickness, when used for column bases, and the bottom surfaces of grouted plates need not be planed. Design thickness of a plate is that after milling. So finishing must be taken into account in ordering the plate.

5.29 BEAM BEARING PLATES

Beams may be supported directly on concrete or masonry if the bearing pressure is within the allowable. The flanges, however, will act as cantilevers, loaded by the bearing pressure f_p (ksi). The maximum bending stress (ksi) may be computed from

$$f_b = \frac{3f_p(B/2 - k)^2}{t^2} \tag{5.11}$$

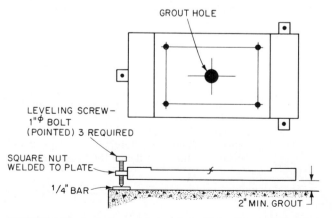

FIGURE 5.39   Large base plate installed with leveling screws.

where B = flange width, in
k = distance from bottom of beam to web toe of fillet, in
t = flange thickness, in

The allowable bending stress in this case is $0.75F_y$, where F_y is the steel yield stress (ksi).

In the absence of building code regulations, the allowable bearing stress F_p (psi) may be taken as 400 for sandstone and limestone, 250 for brick, $0.35f'_c$ for concrete when the _full_ area of the support is covered by the bearing steel, and $0.35f'_c \sqrt{A_2/A_1} \leq 0.70f'_c$ if the base plate covers less than the full area, where f'_c is the 28-day compressive strength of the concrete.

When the bearing pressure under a beam flange exceeds the allowable, a bearing plate should be inserted under the flange to distribute the beam load over the concrete or masonry. The beam load may be assumed to be uniformly distributed to the bearing plate over an area of $2kN$, where k is the distance (in) from bottom of beam to web toe of fillet and N is the length of plate (Fig. 5.40).

Example. A W12 × 26 beam of A36 steel with an end reaction of 19 kips rests on a brick wall (Fig. 5.40). Length of bearing is limited to 6 in. Design a bearing plate of A36 steel.

With allowable bearing pressure of 0.250 ksi, plate area required is $19/0.250 = 76$ in^2. Since the plate length is limited to 6 in, the plate width should be at least $\frac{76}{6} = 12.7$ in. Use a 6 × 13 in plate, area 78 in^2. The actual bearing pressure then will be $f_p = \frac{19}{78} = 0.243$ ksi.

For a W12 × 26 the distance $k = 0.875$ in (see beam dimensions in AISC manual). The plate projection acting as a cantilever then is $l = B/2 - k = 13/2 - 0.875 = 5.62$ in. As for a column base (Art. 5.28), the required thickness can be computed for a bearing pressure of 0.243 ksi from Eq. (5.8):

$$t_p = 2 \times 5.62 \sqrt{0.243/36} = 0.923 \text{ in}$$

Use a 1-in-thick bearing plate.

The beam web must be checked for web yielding over the bearing plate, over a length of $N + 2.5k = 8.188$ in. The allowable bearing stress is $0.66F_y = 23.6$ ksi for A36 steel. For a web thickness of 0.230 in, the web stress due to the 19-kip reaction of the beam is

$$f = \frac{19}{8.188 \times 0.230} = 10.1 \text{ ksi} < 23.6\text{—OK}$$

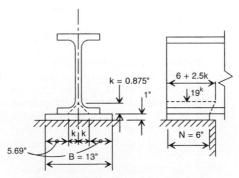

FIGURE 5.40 Beam seated on bearing plate on concrete or masonry support.

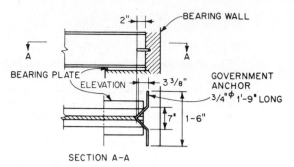

FIGURE 5.41   Wall-bearing beam anchored with government anchor.

In addition to web yielding, web crippling must be checked. The allowable web crippling load is given by

$$R_{cp} = 34t_w^2 \left[1 + 3 \left(\frac{N}{d}\right)\left(\frac{t_w}{t_f}\right)^{1.5} \right] \sqrt{\frac{F_y t_f}{t_w}} \qquad (5.12)$$

Substitution of plate and beam dimensions, including beam depth $d = 12.22$ in and flange thickness t_f of 0.380 in yields

$$R_{cp} = 34 \times 0.230^2 \left[1 + 3 \left(\frac{6}{12.22}\right)\left(\frac{0.230}{0.380}\right)^{1.5} \right] \sqrt{\frac{36 \times 0.380}{0.230}} = 23.5 \text{ kips}$$

Since 23.5 kips > 19 kips, the beam is OK for avoiding web crippling.

Beams usually are not attached to bearing plates. The plates are shipped separately and grouted in place before the beams are erected. Wall-bearing beams usually are anchored to the masonry or concrete. Government anchors (Fig. 5.41) generally are preferred for this purpose.

5.30   SHEAR SPLICES

In buildings, splices of members subjected principally to shear should develop the strength required by the stresses at point of splice. For groove welds, however, the full strength of the smaller spliced member should be developed.

In highway bridges, shear splices should be designed for the larger of the following: 75% of the strength of the member or the average of the calculated stress at point of splice and the strength of the member there. Splices of rolled flexural members, however, may be designed for the calculated maximum shear multiplied by the ratio of splice design moment to actual moment at the splice.

In railroad bridges, shear splices in main members should have the same strength as the members. Splices in secondary members should develop the average of the strength of the members and the calculated stresses at the splices.

When fillers are used, the requirements discussed in Art. 5.13 should be satisfied.

Shear splices may be made with complete-penetration groove welds, preferably without splice plates. Design rules for groove welds are practically the same as for compression splices (Art. 5.27), though values of allowable stresses are different.

Shear splices are most often used for splicing girder webs. In such applications, splice plates should be symmetrically arranged on opposite sides of the web. For bridges, they should extend the full depth of the girder between flanges. Bolted splices should have at least two rows of fasteners on each side of the joint.

Generally, it is desirable to locate flange and web splices at different sections of a girder. But this is not always practical. Long continuous girders, for example, often require a field splice, which preferably is placed at a section with low bending stresses, such as the dead-load inflection points. In such cases it could be troublesome and costly to separate flange and web splices.

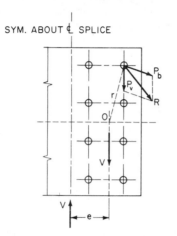

SYM. ABOUT ₵ SPLICE

FIGURE 5.42   Bolted splice for a girder web.

Sometimes, a web splice can be placed where it is required to transmit only shear. Usually, however, a web splice must be designed for both shear and moment. Even in a web splice subjected to pure shear, moment is present if splice plates are used, because of the eccentricity of the shear. For example, as indicated in Fig. 5.42, a group of fasteners on one side of the joint transmits the shear V (kips) from the web to the splice plates. This shear acts through the center of gravity O of the fastener group. On the other side of the joint, a similar group of fasteners transmits the shear from the splice plates to the web on that side. This shear acts through the center of gravity of that fastener group. The two transmitted shears, then, form a couple $2Ve$, where e is half the distance between the shears. This moment must be taken by the fasteners in both groups. In the design of a splice, however, it generally is simpler to work with only one side of the joint. Hence, when the forces on one fastener group are computed, the fasteners should be required to carry the shear V plus a moment Ve. (In a symmetrical splice, e is the distance from O to the splice.)

The classical elastic method assumes that in resisting the moment, each fastener in the group tends to rotate about O. As a consequence:

The reaction P_b of each fastener acts normal to the radius vector from O to the fastener (Fig. 5.42).

The magnitude of P_b is proportional to the distance r from O.

The resisting moment provided by each fastener is proportional to its distance from O.

The applied moment equals the total resisting moment, the sum of the resisting moments of the fasteners in the group.

These consequences result in the relationship

$$M = \frac{P_m J}{c} \qquad P_m = \frac{Mc}{J} \qquad (5.13)$$

where M = applied moment, in-kips
P_m = load due to M on outermost fastener in group, kips
J = sum of squares of distances of fasteners in group from center of gravity of group, in^2 (analogous to polar moment of inertia)
c = distance of outermost fastener from center of gravity, in

Hence, when the moment applied to a fastener group is known, the maximum stress in the fasteners can be computed from Eq. (5.13).

This stress has to be added vectorially to the shear on the fastener (Fig. 5.42). The shear (kips) is given by

$$P_v = \frac{V}{n} \qquad (5.14)$$

where n is the number of fasteners in group. The resultant stress must be less than the allowable capacity of the fastener.

Depending on the fastener pattern, the largest resultant stress does not necessarily occur in the outermost fastener. Vectorial addition of shear and bending stresses may have to be performed for the most critical fasteners in a group to determine the maximum.

In computing the fastener stresses, designers generally find that the computations are simpler if the forces and distances are resolved into their horizontal and vertical components. Advantage can be taken of the fact that

$$J = I_x + I_y \qquad (5.15)$$

where I_y = sum of squares of distances measured horizontally from center of gravity to fasteners, in^2

I_x = sum of squares of distances measured vertically from center of gravity to fasteners, in^2

I_x and I_y are analogous to moment of inertia.

Fatigue need not be considered when calculating bolt stresses but should be taken into account in designing splice plates subjected to bending moments.

Example. The 48 × 5/16 in A36 steel web of a plate girder in a highway bridge is to be field spliced with 7/8-in-diameter A325 bolts (Fig. 5.43). Maximum shear is 95 kips, and maximum bending moments are 700 and −200 ft-kips. The moment of inertia I_g of the gross section of the girder is 14,670 in^4. Design the web splice for A36 steel.

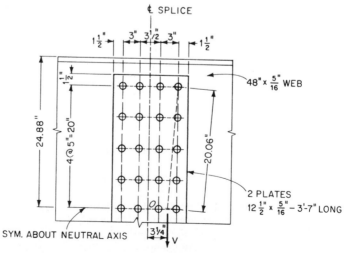

FIGURE 5.43 Example of design of a girder-web splice.

With an allowable stress of 12 ksi, the shear capacity of the web is

$$V_c = 12 \times 48 \times \tfrac{5}{16} = 180 \text{ kips}$$

One design criterion, then, is $0.75V_c = 0.75 \times 180 = 135$ kips. A second design criterion is the average shear:

$$V_{\text{av}} = \tfrac{1}{2}(95 + 180) = 137.5 \text{ kips} > 135$$

This governs for the bolts in the web splice.

The gross moment of inertia of the web is 2,880 in⁴. From this should be subtracted the moment of inertia of the holes. With the bolts arranged as shown in Fig. 5.43 and area per hole $= 1 \times \tfrac{5}{16} = 0.313$ in^2,

$$I_x = 2[(5)^2 + (10)^2 + (15)^2 + (20)^2]0.313 = 1500 \times 0.313 = 470 \text{ in}^4$$

Hence the net moment of inertia of the web is $2880 - 470 = 2410$ in⁴. Allowable bending stress for the girder is 20 ksi.

The maximum bending stress for a moment of 700 ft-kips and $I_g = 14,670$ in⁴ is, for the gross section,

$$f_b = \frac{700 \times 12 \times 24,875}{14,670} = 14.2 \text{ ksi} < 20 \text{ ksi—OK}$$

and the average stress for design of the splice is $F_b = (14.2 + 20)/2 = 17.1$ ksi $> 0.75 \times 20$. Hence the girder design moment is $700 \times 17.1/14.2 = 843$ ft-kips. The net moment carried by the web then is $843 \times 2880/14,670 = 166$ ft-kips, to which must be added the moment due to the 3.25-in eccentricity of the shear. The total web moment thus is $166 \times 12 + 137.5 \times 3.25 = 2439$ in-kips.

For determination of the maximum stress in the bolts due to this moment, J is needed for use in Eq. (5.13). It is obtained from Eq. (5.15) and bolt-hole calculations:

$$I_x = 4(5^2 + 10^2 + 15^2 + 20^2) = 3000$$

$$I_y = 18(1.5)^2 \qquad\qquad = \underline{41}$$

$$J = 3,041 \text{ in}^2$$

By Eq. (5.13) then, the load on the outermost bolt due to moment is

$$P_m = \frac{2,439 \times 20.06}{3041} = 1609 \text{ kips}$$

The vertical component of this load is

$$P_v = \frac{1609 \times 1.5}{20.06} = 1.20 \text{ kips}$$

And the horizontal component is

$$P_h = \frac{1609 \times 20}{20.06} = 16.04 \text{ kips}$$

By Eq. (5.14), the load per bolt due to shear is $P_v = 137.5/18 = 7.63$ kips. Consequently, the total load on the outermost bolt is the resultant

$$R = \sqrt{(1.20 + 7.63)^2 + (16.04)^2} = 18.31 \text{ kips}$$

The allowable capacity of a $\frac{7}{8}$-in A325 bolt in double shear is

$$2 \times 0.601 \times 15.5 = 18.6 \text{ kips} > 18.31 \text{ kips}$$

The bolts and bolt arrangement are satisfactory.

For determination of the maximum stress in the splice plates, note that they are 43 in deep and must carry the moment $M_w = 2439$ in-kips. Try two plates, each $\frac{5}{16}$ in thick. For the gross section, the moment of inertia is

$$I_{gp} = \frac{1}{12} \times 0.3125 \times 43^3 \times 2 = 4141 \text{ in}^4$$

and for the net section,

$$I_{np} = 4141 - 0.3125 \times 1(5^2 + 10^2 + 15^2 + 20^2)4 = 3203 \text{ in}^4$$

The stress on the gross section is

$$f_b = \frac{2439}{4141} \times 21.5 = 12.7 \text{ ksi} < 20 \text{ ksi—OK}$$

and that on the net is

$$f_b = \frac{2439}{3203} \times 21.5 = 16.4 \text{ ksi} < 29 \text{ ksi—OK}$$

For the stress range, the moment in the web is $[700 - (-200)]2880/14,670 = 177$ ft-kips. The eccentric moment is $[95 - (-2/7 \times 95)]1.75 = 214$ in-kips. Hence the stress range is

$$f_{sr} = \frac{(177 \times 12 + 214)21.5}{4141} = 12.1 \text{ ksi} < 18 \text{ ksi—OK}$$

The assumed web plates are satisfactory.

Alternative Splices.   Bolted splices may be made in several different ways when moment is not to be transmitted across the joint. Figure 5.44 shows two examples.

Field-bolted splices generally are more economical than field-welded splices. Groove welds may be more economical, however, if they are made in the shop. Field welding is complicated by the need to control the welding sequence to minimize residual stresses. This often requires that flange-to-web welds on each side of the joint be omitted during fabrication and made after the splice welds. Also, the web interferes with the groove welding of the flange. To avoid the risk of a poor weld, the web should be coped (Fig. 5.45). This provides space for welding, backup bars, cleaning, and other necessary operations.

5.31 *BRACKET CONNECTIONS*

Brackets are projections that carry loads. The connection of a bracket to a support has to transmit both shear and moment. Fasteners or welds may be used for the

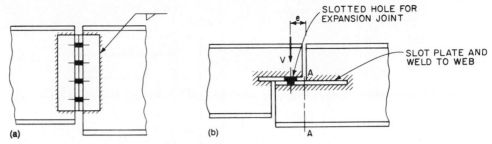

FIGURE 5.44 Girder splice. (*a*) With web plates, shop welds, and field bolts. (*b*) Designed to transmit only shear. Nevertheless, section *A–A* should be designed to resist bending moment *Ve* due to shear.

purpose. Connections may be made with fasteners or welds subjected only to shear or to combined shear and tension.

Figure 5.46 shows a plate bracket connected with bolts in single shear. Design of this type of connection is similar to that of the web splice in Fig. 5.42. Each fastener is subjected to a shear load and a load due to moment. The load due to moment on any fastener acts normal to the radius vector from the center of gravity *O* of the fastener group to the fastener. The maximum load due to moment is given by Eq. (5.13). The resultant of this load and the shear load must be less than the allowable capacity of the fastener. Vectorial addition of the loads due to bending and shear must be performed for the critical fasteners to determine the maximum resultant.

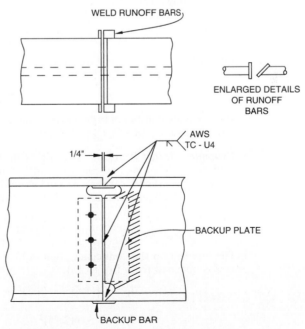

FIGURE 5.45 Welded girder splice. Web plate serves both as an erection connection and as backup for the web groove weld.

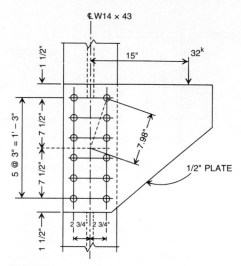

FIGURE 5.46   Bracket with bolts in single shear.

As for a web splice, computations usually are simpler if distances and forces are resolved into horizontal and vertical components, and if J for Eq. (5.13) is obtained from Eq. (5.15).

Tests have indicated that use of the actual moment arm l in computing stresses in a bracket connection is too conservative. Consequently, a smaller moment arm L often is used. It may be computed from the following empirical formulas:

For a single row of fasteners:

$$L = l - \frac{1 + 2n}{4} \tag{5.16}$$

For two or more rows of fasteners:

$$L = l - \frac{1 + n}{2} \tag{5.17}$$

where n is the number of fasteners per row. These equations hold only for symmetric fastener groups not subject to tension.

Example—Bolted Bracket.   Investigate the bracket connection in Fig. 5-46. The A36 steel bracket is to be connected with $\frac{7}{8}$-in-diameter A325N bolts to a building column. The bracket carries a 32-kip load 15 in from the center of the column web.

The effective moment arm is given by Eq. (5.17):

$$L = 15 - \frac{1 + 6}{2} = 11.5 \text{ in}$$

The moment on the connection then is $32 \times 11.5 = 368$ in-kips.

J is obtained from Eq. (5.15):

$$I_x = 4[(1.5)^2 + (4.5)^2 + (7.5)^2] = 315$$

$$I_y = 2 \times 6(2.75)^2 \qquad\qquad = \underline{91}$$
$$J = \overline{406} \text{ in}^2$$

By Eq. (5.13), the load on the outermost bolts due to moment is

$$P_m = \frac{368 \times 7.98}{406} = 7.24 \text{ kips}$$

The vertical component of this load is

$$P_v = \frac{7.24 \times 2.75}{7.98} = 2.50 \text{ kips}$$

And the horizontal component is

$$P_h = \frac{7.24 \times 7.50}{7.98} = 6.80 \text{ kips}$$

By Eq. (5.14), the load per bolt due to shear is $^{32}/_{12} = 2.65$ kips. Consequently, the total load on the outermost bolt is the resultant:

$$R = \sqrt{(2.50 + 2.65)^2 + (6.80)^2} = 8.53 \text{ kips}$$

The allowable capacity of a $\frac{7}{8}$-in bolt is $0.601 \times 21 = 12.6$ kips > 8.53 kips. The connection is unsatisfactory.

An ultimate-strength method may be used instead of the preceding procedure and gives an accurate estimate of the strength of eccentrically loaded bolt groups. The method assumes that fastener groups rotate about an instantaneous center. (This point coincides with the centroid of a group only when the moment arm l becomes very large.) Tables in the AISC ASD and LRFD manuals are based on this method.

Example—Ultimate-Strength Method for Bolted Brackets. Investigate the bracket connection in Fig. 5.46. The bracket of A36 steel is to be connected to the building column with $\frac{7}{8}$-in-diameter A325N (noncritical) bolts. The bracket carries a 10-kip dead load and a 22-kip live load due to a traveling crane. Use the tables for eccentrically load fastener groups in the AISC LRFD manual to check the bolts and the plate.

The factored load (required strength) is

$$P = 1.2 \times 10 + 1.6 \times 22 = 47 \text{ kips}$$

The design strength of the bolts (bearing type, threads included in shear plane, 54-ksi shear strength) is

$$r_v = 0.6013 \times 54 \times 0.65 = 21.1 \text{ kips}$$

where 0.65 is a reduction factor for ultimate strength. From the AISC LRFD manual Table XII for moment arm $X_o = 15$ in and number of bolts per row $n = 6$, the coefficient $C = 3.77$. Therefore, the design strength of the bolt group is

$$P_u = Cr_v = 3.77 \times 21.1 = 79.5 \text{ kips} > 47 \text{ kips—OK} \qquad (5.18)$$

The plate should be checked for bending and shear. For the gross section, required bending strength is

$$f_b = \frac{47(15 - 2.75 - 0.5)}{0.5(18)^2/6} = 20.5 \text{ ksi}$$

The strength of the gross section is

$$F_b = 0.9 \times 36 = 32.4 \text{ ksi} > 20.5 \text{ ksi}$$

The plate is satisfactory for bending (yielding) of the gross section. For bending of the net section, from the AISC LRFD manual, the section modulus is $S_{net} = 18 \text{ in}^3$ and the required strength is

$$f_b = \frac{47(15 - 2.75)}{18} = 31.9 \text{ ksi}$$

Fracture, rather than yielding, however, is the limit state for the net section. The fracture strength of a net section in tension is $0.75 F_u A_e$, where F_u is the specified tensile strength and A_e is the effective net area. Since bending induces a tensile stress over the top half of the bracket, and because the yield limit on the gross section has already been checked, it is reasonable to assume that the fracture (rupture) design strength of the bracket net section is $0.75 \times 58 = 43.5 \text{ ksi} > 31.9$ ksi. The bracket is satisfactory for fracture of the net section.

The shear on the gross section is

$$f_v = \frac{47}{0.5 \times 18} = 5.2 < (0.8 \times 0.7 \times 36 = 20.2 \text{ ksi}) \text{—OK}$$

The shear on the net section is

$$f_v = \frac{47}{0.5(18 - 6 \times 0.9375)} = 7.6 < (0.75 \times 0.6 \times 58 = 26.1 \text{ ksi})$$

The plate is satisfactory for shear.

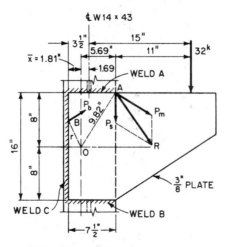

FIGURE 5.47 Bracket with fillet welds in shear.

A plate bracket such as the one in Fig. 5.46 also can be connected to a support with fillet welds in shear. Design of the welds for such a connection can be performed by the classical elastic method, which is analogous to that for fasteners. For example, for the bracket in Fig. 5.47, the shear due to the 32-kip load induces a shear (kips per in) equal to the load divided by the total length of weld (weld A + weld B + weld C). The moment due to the load tends to rotate the welds about their center of gravity O. As a consequence:

The force acting at any point of a weld acts normal to the radius vector from O to the point. In Fig. 5.47, P_b, the force due to the moment of the 32-kip load about O, is normal to the radius vector OB.

The magnitude of P_b (kips per in) is proportional to the distance r from O.

The resisting moment per inch of weld is proportional to the square of the distance from O.

The applied moment equals the total resisting moment, the sum of the resisting moments of all the welds in the group.

These consequences result in the relationship

$$M = \frac{P_m J}{c} \qquad P_m = \frac{Mc}{J} \tag{5.19}$$

where M = applied moment, in-kips
$\quad\quad P_m$ = load due to M on the point of a weld most distant from the center of gravity of the weld group, kips per in
$\quad\quad J$ = sum of squares of distances of unit weld lengths from center of gravity of group, in^3 (analogous to polar moment of inertia)
$\quad\quad c$ = distance of outermost point from center of gravity, in

Hence, when the moment applied to a weld group is known, the maximum stress in the welds can be computed from Eq. (5.19). This stress has to be added vectorially to the shear on the weld. The resultant stress must be less than the allowable capacity of the weld.

Depending on the weld pattern, the largest resultant stress does not necessarily occur at the outermost point of the weld group. Vectorial addition of shear and bending stresses may have to be performed for the most critical points in a group to determine the maximum.

Computation of weld stresses generally is simplified if the forces and distances are resolved into their horizontal and vertical components. Advantage can be taken of the fact that

$$J = I_x + I_y \tag{5.20}$$

where I_y = sum of squares of distances measured horizontally from center of gravity of weld group to unit lengths of welds, in^3
$\quad\quad I_x$ = sum of squares of distances measured vertically from center of gravity of weld group to unit lengths of welds, in^3

I_x and I_y are analogous to moment of inertia.

Example—Welded Bracket Connection. Investigate the bracket connection in Fig. 5.47. The A36 steel bracket is to be connected with fillet welds made with E70XX electrodes to a building column. The bracket carries a 32-kip load 15 in from the center of the column web.

Because of symmetry, the center of gravity O of the weld group is located vertically halfway between top and bottom of the 16-in-deep plate. The horizontal location of O relative to the vertical weld is obtained by dividing the moments of the weld lengths about the vertical weld by the total length of welds:

$$\bar{x} = \frac{2 \times 7.5 \times 7.5/2}{2 \times 7.5 + 16} = \frac{56.2}{31} = 1.81 \text{ in}$$

J is obtained from Eq. (5.20):

Welds A and B: $2 \times 7.5(8)^2 = \quad 961$ $\qquad$ Welds A and B: $\dfrac{2(7.5)^3}{12} = \quad 70$

Weld C: $\qquad \dfrac{(16)^3}{12} = \underline{\quad 341\quad}$ $\qquad$ $2 \times 7.5 \left(\dfrac{7.5}{2} - 1.81\right)^2 = \quad 56$

$\qquad\qquad\qquad\qquad\quad I_x = 1302 \text{ in}^3$ $\qquad\qquad$ Weld C: $16(1.81)^2 = \underline{\quad 53\quad}$

$\qquad\qquad\qquad\qquad\qquad\qquad\qquad\qquad\qquad\qquad\qquad\qquad\qquad\quad I_y = 179 \text{ in}^3$

$$J = 1301 + 179 = 1480 \text{ in}^3$$

By Eq. (5.19), the stress on the most distant point A in the weld group due to moment is

$$P_m = \frac{32(15 + 1.69)9.82}{1480} = 3.54 \text{ kips per in}$$

The vertical component of this load is

$$P_v = \frac{3.54 \times 5.69}{9.82} = 2.05 \text{ kips per in}$$

And the horizontal component is

$$P_h = \frac{3.54 \times 8}{9.82} = 2.88 \text{ kips per in}$$

The shear load on the welds is $32/(2 \times 7.5 + 16) = 1.03$ kips per in. Consequently, the total load on the outermost point is the resultant

$$R = \sqrt{(1.03 + 2.05)^2 + (2.88)^2} = 4.22 \text{ kips per in}$$

For an allowable stress of 21 ksi, the weld size required is

$$D = \frac{4.22}{0.707 \times 21} = 0.284 \text{ in}$$

Use a ⁵⁄₁₆-in weld.

Instead of the elastic method used in the preceding example, an ultimate-strength method based on instantaneous center of rotation can be used. The tables for eccentrically loaded weld groups in the AISC manuals—ASD and LRFD—are based on this method.

Example—Ultimate-Strength Method for Welded Brackets. Investigate the bracket connection of the preceding example using the ultimate-strength method.

From Fig. 5.47 and the AISC ASD manual Table XXIII, $l = 16$ in, $kl = 7.5$ in, and $al = 11 + 7.5 - 1.81 = 16.69$ in, from which $a = 16.69/16 = 1.04$. By interpolation in Table XXIII, coefficient $C = 0.585$. Hence the required weld size in number of sixteenths of an inch is

$$D = \frac{32}{0.585 \times 16} = 3.41$$

Use a ¼-in fillet weld. (In comparison, the elastic analysis requires a ⁵⁄₁₆-in fillet weld.)

For a ⁵⁄₁₆-in fillet weld, the minimum plate thickness is ⅜ in, to permit inspection of the weld size. Try a ⅜-in-thick plate and check for shear and bending. For shear,

$$f_v = 32/(0.375 \times 16) = 5.33 \text{ ksi} < 14.4 \text{ ksi—OK}$$

For bending,

$$f_b = \frac{32 \times 11}{0.375(16)^2/6} = 22.0 \text{ ksi} \simeq 21.6 \text{ ksi—OK}$$

Fasteners in Shear and Tension. Bolted brackets also may be attached to supports with the fasteners subjected to combined shear and tension. For the bracket in Fig. 5.48, for example, each bolt is subjected to a shear (kips) of

$$P_v = \frac{P}{n} \qquad (5.21)$$

where P = load, kips, on the bracket
n = total number of fasteners

In addition, the moment is resisted by the upper fasteners in tension and the pressure of the lower part of the bracket against the support. Usual practice with prestressed, high-strength bolts assumes the neutral axis at the center of gravity of the fastener group, the load on each fastener proportional to distance from the center of gravity, and the resisting moment of each fastener proportional to the square of this distance. The load (kips) on the most heavily stressed fastener may be computed from

$$P_m = \frac{Mc}{I} \qquad (5.22)$$

where M = moment on fastener group, in-kips
I = moment of inertia of fasteners about netural axis, in⁴
c = distance of outermost fastener in tension from neutral axis, in

In bearing-type connections, the allowable stresses are determined from an interaction equation for tension and shear. However, in slip-critical connections, since the shear load is carried by friction at the faying surface, the reduction in friction resistance above the neutral axis of the bolt group (due to the tensile force from bending) is compensated for by an increase in friction resistance below the neutral axis (due to the compressive force from bending). Thus an interaction equation is not required in this case, but both the shear and tensile stresses must be less than those allowable. Also, since slip is a serviceability limit state, the strength-limit state of bearing also must be checked. In addition, the tension forces on the fasteners and the bending of the flanges must be checked for prying (Sec. 5.25.3).

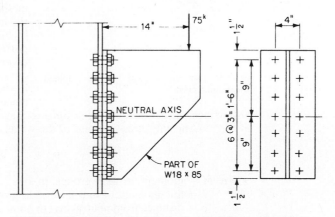

FIGURE 5.48 Bracket with bolts in combined shear and tension.

Example—Bracket with Bolts in Tension and Shear.   Investigate the slip-critical connection in Fig. 5.48. The A36 steel bracket is to be connected with ⅞-in-diameter A325 bolts to a flange of a building column. The bracket carries a 75-kip load 14 in from the flange.

The allowable stress in shear for a slip-critical connection is 17 ksi. Actual shearing stress is, by Eq. (5.21),

$$f_v = \frac{75}{14 \times 0.601} = 8.91 \text{ ksi} < 17 \text{ ksi}$$

Tensile stress in the bolts, induced by the moment, must be less than the 44 ksi allowable. The moment of inertia of the bolt group is

$$I = 2 \times 2 \times 0.601[(3)^2 + (6)^2 + (9)^2] = 303 \text{ in}^4$$

The tensile stress in the top row of bolts is, by Eq. (5.22),

$$f_t = \frac{75 \times 14 \times 9}{303} = 31.1 \text{ ksi} < 44 \text{ ksi}$$

For the check for bearing, the interaction equation is

$$F_t = \sqrt{44^2 - 4.39 f_v^2} = \sqrt{44^2 - 4.39 \times 8.91^2} = 39.8 \text{ ksi}$$

Since $f_t = 31.1$ ksi < 39.8 ksi, the connection is satisfactory.

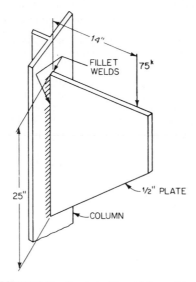

FIGURE 5.49   Bracket with welds in combined shear and tension.

Welds in Tension and Shear.   Brackets also can be connected to supports with fillet welds in combined shear and tension. For the bracket in Fig. 5.49, for example, the welds carry a shear (kips per in) of

$$P_v = \frac{P}{L} \qquad (5.23)$$

where P = load on bracket, kips
L = total length of welds, in

In addition, the moment is resisted by the upper portion of the welds in tension and the pressure of the lower part of the bracket against the support. The elastic method assumes that the weld carries all the load (bearing pressure between the bracket and the lower part of the support is neglected). The neutral axis is taken at the center of gravity of the weld group. The intensity of load at any point in the weld is proportional to the distance from the neutral axis, and the resisting moment per inch is proportional to that distance. The tension (kips per in) at the most heavily stressed point of the welds may be computed from

$$P_m = \frac{Mc}{I} \qquad (5.24)$$

where M = moment on weld groups, in-kips
I = line moment of inertia of weld group about neutral axis, in³
c = distance from neutral axis to point of welds farthest from neutral axis, in

Example—Brackets with Welds in Tension and Shear.   An A36 steel plate bracket is to be connected with fillet welds on two sides of the plate to a building column (Fig. 5.49). The bracket carries a 75-kip load 14 in from the column flange. The welds are to be made with E70XX electrodes.

Assume a plate ½ in thick. With an allowable stress of 21.6 ksi, the length of plate at the support should be at least

$$L = \sqrt{\frac{75 \times 14 \times 6}{0.5 \times 21.6}} = 24.2 \text{ in}$$

To keep down the size of welds, use a plate 25 × ½ in.

By Eq. (5.23), the shear on the two welds is

$$P_v = \frac{75}{2 \times 25} = 1.50 \text{ kips per in}$$

By Eq. (5.24), the maximum tensile stress in the welds is

$$P_m = \frac{75 \times 14 \times 25/2}{2(25)^3/12} = 5.03 \text{ kips per in}$$

The resultant of the shear and tensile stresses is

$$R = \sqrt{(1.50)^2 + (5.03)^2} = 5.26 \text{ kips per in}$$

With an allowable stress of 21 ksi, the weld size required is

$$D = \frac{5.26}{0.707 \times 21} = 0.344 \text{ in}$$

Use ⅜-in fillet welds.

The plate should be checked for shear and bending. For shear, the stress is

$$f_v = 75/(0.5 \times 25) = 6.0 \text{ ksi} < 14.4 \text{ ksi—OK}$$

For bending, the stress is

$$f_b = \frac{75 \times 14}{0.5(25)^2/6} = 20.2 \text{ ksi} < 21.6 \text{ ksi—OK}$$

The ultimate-strength method also can be used for the bracket weld of Fig. 5.49.

Example—Ultimate-Strength Method for Welded Bracket.   Determine the required weld size for the bracket in the preceding example by the ultimate-strength method.

From the AISC ASD manual Table XIX (special case), with $k = 0$, $l = 25$, and $al = 14$, from which $a = 0.56$, the coefficient C is determined by interpolation to be 0.719. Thus the weld size required, in number of sixteenths of an inch, is $D = 75/(0.719 \times 25) = 4.17$. Use 5/16-in fillet welds.

5.32 CONNECTIONS FOR SIMPLE BEAMS

End connections of beams to their supports are classified as simple-beam, rigid-frame, and semirigid connections.

Simple, or **conventional, connections** are assumed free to rotate under loads. They are designed to carry shear only. The AISC specifications for structural steel buildings require that connections of this type have adequate inelastic rotation capacity to avoid overstressing the fasteners or welds.

Rigid-frame connections, transmitting bending moment as well as shear, are used to provide complete continuity in a frame (Art. 5.33).

Semirigid connections provide end restraint intermediate between the rigid and flexible types (Art. 5.33).

For simple connections, design drawings should give the end reactions for each beam. If no information is provided, the detailer may design the connections for one-half the maximum allowable total uniform load on each beam.

Simple connections are of two basic types: framed (Fig. 5.50*a*) and seated (Fig. 5.50*b*). A **framed connection** transfers the load from a beam to a support through one or two connection angles, or a shear plate attached to the supporting member, or a tee attached to either the supporting or supported member. A **seated connection** transfers the load through a seat under the beam bottom flange. A top, or cap, angle should be used with seated connections to provide lateral support. It may be attached to the beam top flange, as shown in Fig. 5.50*b*, or to the top portion of the web. With both framed and seated connections, the beam end is stopped ½ in short of the face of the supporting member, to allow for inaccuracies in beam length.

5.32.1 Framed Connections

These generally are more economical of material than seated connections. For example, in a symmetrical, bolted, framed connection, the fasteners through the web are in double shear. In a seated connection, the fasteners are in single shear. Hence framed connections are used where erection clearances permit, e.g., for connections to column flanges or to girders with flanges at the same level as the beam flanges. Seated connections, however, usually are more advantageous for connections to column webs because placement of beams between column flanges is easier. Seats also are useful in erection because they provide support for beams while field holes are aligned and fasteners are installed. Furthermore, seated connections may be more economical for deep beams. They require fewer field bolts, though the total number of shop and field fasteners may be larger than those required for a framed connection of the same capacity.

The AISC manual lists capacities and required design checks for beam connections for buildings. Design is facilitated when this information can be used. For cases where such connections are not suitable, beam connections can be designed by the principles and methods given for brackets in Art. 5.31.

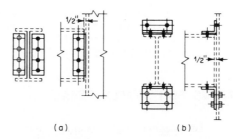

(a)   (b)

FIGURE 5.50 Simple-beam connections. (*a*) Framed. (*b*) Seated.

Vertical fastener spacing in framed connections is standardized at 3 in. The top gage line also is set 3 in below the beam top, when practicable. Closer spacing may be used, however, as long as AISC specification restrictions on minimum spacing are met.

To ensure adequate stiffness and stability, the length of the connection material in a framed connection should be at least half the distance T between flange-web fillets of the beam.

Distance between inner gage lines of outstanding legs or flange of connection material is standardized at 5½ in, but sometimes a shorter spacing is required to meet AISC specification reqirements for minimum edge distance.

Thickness of connection material may be determined by shear on a vertical section, availability of material of needed thicknesses, or the bearing value for the nominal fastener diameter.

When a beam frames into a girder with tops of both at the same level, the top of the beam generally is notched, or coped, to remove enough of the flange and web to clear the girder flange. Depth of cut should be sufficient to clear the web fillet (k distance for a rolled section). Length of cope should be sufficient to clear the girder flange by ½ to ¾ in. A fillet with smooth transition should be provided at the intersection of the horizontal and vertical cuts forming the cope.

For beams framing into column flanges, most fabricators prefer connections attached to the columns in the shop. Then the beams require punching only. Thus less handling and fewer operations are required in the shop. Furthermore, with connection material attached to the columns, erectors have more flexibility in plumbing the steel before field bolts are tightened or field welds made.

Some of the standardized framed connections in the AISC manual are arranged to permit substitution of welds for fasteners. For example, welds A in Fig. 5.51a

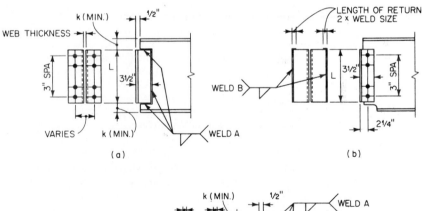

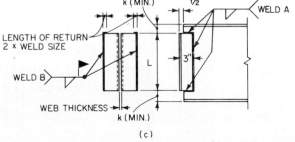

FIGURE 5.51 Welded framed connections. (*a*) Welds replace bolts of the standardized connection at the web. (*b*) Welds replace bolts of the standardized connection for outstanding legs of framing angles. (*c*) All-welded standardized connection.

replace fasteners for the web connections. Welds *B* replace fasteners in the outstanding legs (Fig. 5.51*b*). Angle thickness must be at least the weld size plus ¹⁄₁₆ in and a minimum of ⁵⁄₁₆. Holes may be provided for erection bolts in legs that are to be field welded. When fasteners are used in outstanding legs, the bearing capacity of supporting material should be investigated.

Welds *A* are eccentrically located. They receive the load from the beam web and the connection transmits the load to the support at the back of the outstanding legs. Hence there is an eccentricity of load equal to the distance from the back of the outstanding legs to the center of gravity of welds *A*. Therefore, when the connections in Fig. 5.51 are used, the combination of vertical shear and moment on welds *A* should be taken into account in design, unless the tables in the AISC manual are used.

For the connection in Fig. 5.51*b*, the welds usually are made in the shop. Consequently, the beam bottom must be coped to permit the beam to be inserted between the angles in the field.

Welds *B* also are eccentrically loaded. The beam reaction is transmitted from the center of web to the welds along the toes of the outstanding legs. This moment, too, should be taken into account in design. To prevent cracking, the vertical welds at the top of the angles should be returned horizontally for a distance of twice the weld size.

Standardized framed connections of the type shown in Fig. 5.51*c* were developed especially for welding of both the web legs and outstanding legs of the connection angles. Use of this type for an all-welded connection is more economical than a combination of welds *A* and *B* in Fig. 5.51*a* and *b*.

5.32.2 Seated Connections

These may be unstiffened, as shown in Fig. 5.50*b* and Fig. 5.53*a*, or stiffened, as shown in Fig. 5.52 and Fig. 5.53*b*. A stiffened seat usually is used when loads to be carried exceed the capacities of the outstanding leg of standardized unstiffened seats. Tables in the AISC manual facilitate the design of both types of connections.

Common applications of unstiffened seats include beam connections to column webs and to deep girders. Occasionally, such connections are used at column flanges. Stiffened seats are rarely used for connections other than to column webs, because the stiffeners may protrude through the fireproofing or architectural finish and thus are esthetically undesirable.

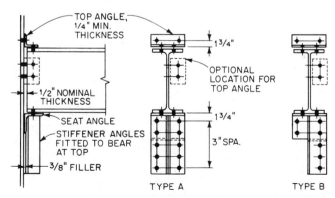

FIGURE 5.52 Standardized stiffened-seat connections with bolts.

Design of a seated connection generally is based on the assummption that the seat carries the full beam reaction. The top, or cap, angle only provides lateral support. Even for large beams, this angle can be small and can be attached with only two fasteners in each leg (Fig. 5.52) or a toe weld along each leg (Fig. 5.53).

With the nominal tolerance of ½ in between beam end and face of support, the length of support provided a beam end by a seat angle equals the width of the outstanding leg less ½ in. Thus a typical 4-in-wide angle leg provides 3½ in of bearing. Because of the short bearing, the capacity of a seated connection may be determined by the thickness of the beam web, for resisting web yielding and crippling. Beam-load tables in the AISC manual list for each size of beam the maximum beam reaction R permitted for a 3½-in bearing.

Also, the tables give four values, R_1, R_2, R_3, and R_4, which allow seats of other lengths to be checked for yielding and crippling. For yielding, the required bearing length N is

$$N = (R - R_1)/R_2 \qquad (5.25)$$

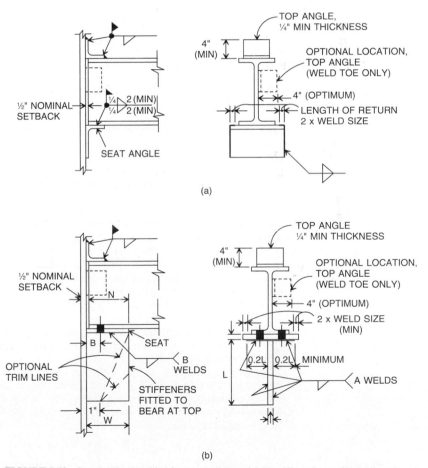

FIGURE 5.53 Standardized welded-seat connections. (*a*) Unstiffened seat. (*b*) Stiffened seat.

and for crippling,

$$N = (R - R_3)/R_4 \qquad\qquad (5.26)$$

The larger of these two bearing lengths must be used, but N may not be less than k, the distance from the outer face of flange to the toe of web fillet. In Eqs. (5.25) and (5.26),

$$R_1 = 0.66F_y(2.5k)t_w, \text{ kips}$$

$$R_2 = 0.66F_yt_w, \text{ kips}$$

$$R_3 = 34F_y^{0.5}t_w^{1.5}t_f^{0.5}, \text{ kips}$$

$$R_4 = 102F_y^{0.5}t_w^3/(t_fd), \text{ kips}$$

$$F_y = \text{beam yield stress, ksi}$$

$$t_w = \text{beam web thickness, in}$$

$$t_f = \text{beam flange thickness, in}$$

$$d = \text{beam depth, in}$$

For unstiffened seats, the bearing length is assumed to extend from the beam end toward midspan. The beam end is taken as ¾ in from the face of support, to account for the ½-in nominal setback and to allow for possible underrun in the beam length. The reaction is assumed centered on the bearing length N. For stiffened seats, the bearing length is assumed to extend from the end of the seat toward the beam end. Again, the reaction is assumed centered on the bearing length. In design of the seat, however, an eccentricity from the face of the support of 80% of the beam-seat width is used if it is larger than the eccentricity based on the reaction position at the center of N.

The AISC manual has tables to facilitate the design of both unstiffened and stiffened seats.

Unstiffened Seats.   The capacity of the outstanding leg of an unstiffened seat is determined by its resistance to bending. The critical section for bending is assumed to be located at the toe of the fillet of the outstanding leg. When reactions are so large that more than 3½ in of bearing is required, stiffened seats usually are used.

In addition to the capacity of the oustanding leg, the capacity of an unstiffened seat depends also on the fasteners or welds used. The small eccentricity of the beam reaction generally is neglected in determining fastener capacities.

Example—Bolted Unstiffened Seat.   For A36 steel, design an unstiffened seat to support the 10-kip reaction of a W8 × 18 beam on a column web. Use A307 bolts and 3½-in column gage.

Assume a seat angle 4 × 3 in with 4-in leg outstanding. Bearing length is 3½ in. Maximum reaction of a W8 × 18 for this bearing is 23 kips. With web thickness $t = 0.23$ in and $k = ¾$ in for this beam, the 10-kip reaction may be assumed distributed over a length N calculated from data for the W8 in the AISC tables for allowable uniform loads:

$$N = (R - R_1)/R_2 = (10 - 10.2)/5.46 = -0.04$$

$$N = (R - R_3)R_4 = (10 - 12.9)/2.77 = -1.05$$

Since N may not be less than $k = 0.75$ for the W8 and the preceding calculations yield a negative value for N, use $N = 0.75$ in. The reaction then acts at a distance of $N/2 = 0.75/2 = 0.375$ in from the end of the beam and $0.375 + 0.75 = 1.125$ in from the face of the support.

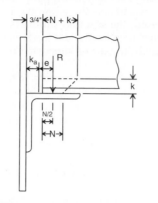

FIGURE 5.54 Example of design of an un-stiffened-seat connection.

For the 3½-in column gage, the seat angle need be only 6 in long. Assume an angle thickness of ½ in. For a $4 \times 3 \times \frac{1}{2}$ angle, the distance k_a from the back of the angle to the toe of the fillet of the outstanding leg (Fig. 5.54) is $^{15}/_{16}$ in. Hence, for this angle, the moment arm of the reaction is $1.125 - 0.9375 = 0.1875$ in. Consequently, the bending moment, at a section through the toe of the fillet of the outstanding leg, is $10 \times 0.1875 = 1.875$ in-kips. For an allowable stress of 27 ksi, the required section modulus for the 6-in-long leg is $1.875/27 = 0.0694$ in³. A $6 \times \frac{1}{2}$-in rectangular section has a section modulus of $6(\frac{1}{2})^2/6 = 0.25$ in³. Hence the ½-in thickness is satisfactory.

For an allowable shear stress of 10 ksi on A307 bolts, the bolts must have an area of $10/10 = 1$ in². Two $^7/_8$-in-diameter bolts supply $2 \times 0.601 = 1.202$ in². Use two such bolts to attach the seat angle to the column web.

Unstiffened-Seat Connections.   When a seated connection is used for a beam at a girder web, a side angle, attached to the beam web, usually is used instead of a top angle. The beam need not be fastened to the seat, since the side angle provides lateral support for both beam and girder. But for seated connections on columns, the beam bottom flange should always be connected to the seat with welds or fasteners. This aids alignment of the column during erection and braces the column. The top angle should be located above the seat a distance equal to actual beam depth plus a small tolerance, say ⅛ in, to compensate for mill variations in beam depth.

The top row of fasteners attaching a seat angle to the support should be placed as close as permissible to the top of the seat. This will reduce the deflection of the seat angle caused by the eccentricity of the beam reaction.

For welded unstiffened seats, a weld is placed along each end of the vertical leg of the seat angle for the width of the leg. To prevent the angle from tearing away from the support at the top, the welds are returned along the top of the seat at the heel of the angle for a distance equal to twice the weld size (Fig. 5.53a). (While horizontal welds at heel and toe of the seat angle would be more effective in resisting the eccentric loading, the heel weld might interfere with seating of the beam if actual beam length exceeded the specified length.) The welds are subjected by the eccentric loading to both shear and tension. They may be designed in the same way as the welds for the bracket in Fig. 5.49 (Art. 5.31).

Sometimes, available space on a support limits the length of seat angle that can be used. In such cases, it may be advisable to cut the bottom flange of the beam to reduce its width sufficiently to provide space on the seat for weld returns and beam-to-seat welds.

For channels, the outstanding leg of an unstiffened seat usually is 6 in wide, instead of the typical 4 in. Because channel flanges have only one gage line, the wider leg is used to obtain space for two erection bolts.

Stiffened Seats.   These require that stiffeners be fitted to bear against the underside of the seat. The stiffeners must be sized to provide adequate length of bearing for the beam, to prevent web yielding and crippling. Area of stiffeners must be adequate to carry the beam reaction at the allowable bearing stress.

When fasteners are used, the seat and stiffeners usually are angles (Fig. 5.52). A filler with the same thickness as the seat angle is inserted below the seat angle, between the stiffeners and the face of support. For light loads, a single stiffener angle may be used (type B, Fig. 5.52). For heavier loads, two stiffener angles may be required (type A, Fig. 5.52). Outstanding legs of these angles need not be stitched together. To accommodate the gage of fasteners in the supporting member, paired stiffeners may be separated. But the separation must be at least 1 in wide and not more than twice $k - t_s$, where k is the distance from outer surface of beam flange to web toe of fillet (in), and t_s is the stiffener thickness (in).

For standardized stiffened-seat connections, ⅜-in-thick seat angles are specified. The outstanding leg is made wide enough to extend beyond the outstanding leg of the stiffener angle. The width of the vertical leg of the seat angle is determined by the type of connection.

In determination of the bearing capacity of a stiffener, the effective width of the outstanding leg of the stiffener generally is taken as ½ in less than the actual width.

When stiffened seats are to be welded, they can be fabricated by welding two plates to form a tee (Fig. 5.53b) or by cutting a T shape from a wide-flange or I beam. When two plates are used, the stiffener should be fitted to bear against the underside of the seat. Thickness of the seat plate usually equals that of the stiffener but should not be less than ⅜ in.

The stiffener usually is attached to the face of the support with two fillet welds over the full length L of the stiffener. The welds should be returned a distance of at least $0.2L$ along the underside of the seat on each side of the stiffener. The welds are subjected to both shear and tension because of the eccentricity of the loading on the seat. Design is much the same as for the bracket in Fig. 5.49 (Art. 5.31).

Size and length of welds between a seat plate and stiffener should be equal to or greater than the corresponding dimensions of the horizontal returns.

Stiffener and seat should be made as narrow as possible while providing required bearing. This will minimize the eccentricity of the load on the welds. For a channel, however, the seat plate, but not the stiffener, usually is made 6 in wide, to provide space for two erection bolts. In this case, the seat projects beyond the stiffener, and length of welds between seat and stiffener cannot exceed the stiffener width.

Determination of stiffener thickness may be influenced by the thickness of the web of the beam to be supported and by the size of weld between seat and support. One recommendation is that stiffener thickness be at least the product of beam-web thickness and the ratio of yield strength of web to yield strength of seat material. A second rule is based on Table 5.13. To prevent an A36 plate from being overstressed in shear by the pair of vertical fillet welds (E70XX electrodes), thickness of plate should be at least

$$t = \frac{2 \times 0.707D \times 21}{14.4} \simeq 2D \qquad (5.27)$$

where D is weld size (in).

Similarly, overstressing in shear should be avoided in the web of a supporting member of A36 steel where stiffened seats are placed on opposite sides of the web. If the vertical welds are also on opposite sides of a part of the web, the maximum weld size is half the web thickness.

Example—Welded Stiffened Seat.   For A36 steel, design a welded stiffened seat of the type shown in Fig. 5.53*b* to carry the 85-kip reaction of a W27 × 94 beam on a column web.

The tables for allowable uniform loads in the AISC ASD manual give for a W27 × 94, $R_1 = 41.8$, $R_2 = 11.6$, $R_3 = 60.4$, and $R_4 = 3.59$. Also, $k = 1\frac{7}{16}$ in. Thus the required bearing length (in) is 6.85 in, the largest value of k,

$$N = (85 - 41.8)/11.6 = 3.72 \quad \text{and} \quad N = (85 - 60.4)/3.59 = 6.85$$

The required width of seat is $6.85 + 0.5 = 7.35$ in. Use an 8-in-wide seat. From the AISC ASD manual, Table VIII, the required length L of stiffener plate (Fig. 5.53*b*) is 16 in when $\frac{5}{16}$-in fillet welds (welds A) are used between the stiffener plate and the column web. According to Table VIII, this connection is good for 94.4 kips > 85 kips—OK.

The thickness of the stiffener plate is taken as twice the weld size when A36 plate and E70 electrodes are used. Therefore, the stiffener plate should be $\frac{5}{8}$ in thick (dimension t of Fig. 5.53*b*). The seat-plate thickness is usually taken to be the same as the stiffener plate but should not be less than $\frac{3}{8}$ in.

The weld between the stiffener and the seat plate (welds B of Fig. 5.53*b*) is usually taken to be the same size as welds A, but an AISC minimum weld can be used as long as it develops the strength of the $0.2L$ returns of welds A to the seat plate. In the calculation of the capacity of the fillet weld, it is convenient to use the capacity per inch per sixteenths of an inch of weld size, which is equal to $21.0 \times 0.707/16 = 0.928$ kips for a 21.0-ksi allowable stress. Thus the design load for welds B is $0.2 \times 16 \times 5 \times 0.928 \times 2 = 29.7$ kips. The length of welds B is 8 in. The required weld size in number of sixteenths of an inch is

$$D = 29.7/(8 \times 2 \times 0.928) = 2.0$$

Thus a $\frac{3}{16}$-in fillet weld will work, but the AISC minimum weld for the $\frac{5}{8}$-in stiffener plate is $\frac{1}{4}$ in. Use a $\frac{1}{4}$-in fillet weld for welds B.

5.33   MOMENT CONNECTIONS

End connections of beams to supports are classified in Art. 5.32 as simple-beam, rigid-frame, or semirigid. Various types of simple connections are described in that article. Rigid and semirigid connections are discussed in this article. These connections transmit both shear and bending moment from beam to support, whereas simple connections transmit only shear. Moment-transmitting, or moment, connections often are used in buildings between girders and columns to provide resistance to lateral forces.

Rigid connections provide complete continuity in a frame. Semirigid connections provide end restraint intermediate between the rigid and flexible types. Design drawings should clearly state requirements for moment connections and list the end shears and moments for which the connections are to be designed and detailed. Otherwise, the fabricator may assume that simple connections are to be used. Also, the designer should supply detail sketches of the connections.

Design of moment connections usually is based on the assumption that vertical shear is carried by a web connection to the support, and moment, by flange connections to the support. Or vertical shear may be transmitted through stiffened or unstiffened seats, as for simple connections (Art. 5.32).

For moment connections with fasteners, flange connections often are made with tees cut from I beams (Fig. 5.55) to take advantage of the thick webs and relatively narrow, thick flanges of these beams. Part of an S20 or an S24 is commonly used. The stem of a tee is fastened to each beam flange, and the flange of each tee to the support. For connections with relatively small moments, an angle or part of a channel may be used instead of a tee. Vertical shear is transmitted to the support by angle or tee connections on the beam web or by a stiffened or unstiffened seat.

Clearance, generally of about ¼ in, should be provided between the top flange and top connection material to allow for mill variations in beam depth. To compensate for depth deficiencies, shims should be used between the flange and connection material to maintain the specified spacing of the connection material for top and bottom flanges.

This connection material must be sized to transmit from beam flange to support a force equal to the moment divided by the beam depth. The number of fasteners required between the flanges and the top or bottom connection material is determined by dividing this force by the allowable load per fastener for single shear.

The fasteners between the tension-flange connection and the support are subjected to both tension and a prying action. Forces due to the prying action may be computed from Eq. (5.5) or (5.6). The direct tension per fastener equals the transmitted flange force divided by the number of fasteners in the connection at the support.

Unless the support material is sufficiently thick or is stiffened, the flange forces may cause severe distortions. Thicknesses of column web and flanges should be checked to see if stiffeners are required. (See criteria in Art. 6.23.)

Welding can simplify a moment connection considerably. But care must be taken in designing such a connection that fabrication does not become too exacting and therefore more costly. Figure 5.56 shows an economical rigid connection with welds and high-strength field bolts. Sufficient bolts are provided to carry the shear. The bolts also hold the beam in place during erection until the flange welds are made. An angle welded to the column serves as an erection seat and a backup for the bottom-flange groove weld. A backup bar is provided for the top-flange weld.

Figure 5.57 shows an economical semirigid connection with welds and high-strength field bolts. The connection plate on the top flange is narrower than the flange to provide space for fillet welds between connection material and flange. These welds start at a distance from the support of 1.5 times the width of the connection plate and extend to the end of the plate farthest from the support. Less than full-length welds are used to limit the amount of moment transmitted, by

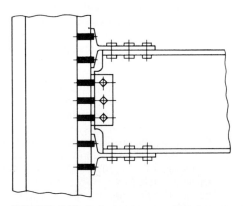

FIGURE 5.55 Bolted moment connection of a beam to a column.

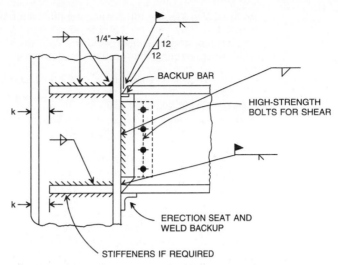

FIGURE 5.56 Rigid-frame connection with welds and bolts.

permitting the connection material to yield when excessive moment is imposed. The connection plate for the bottom flange is wider than that flange to allow down-hand fillet welding. High-strength bolts transmit the vertical shear from the beam web to a plate welded to the support.

Example. Design a semirigid connection of the type shown in Fig. 5.57 for a W18 × 50 framed into one flange of a W12 × 53 column. The connection is to transmit a moment of 110 ft-kips and a vertical shear of 27 kips. All material other than fasteners and welds is A36 steel. Use A325N (noncritical) bolts (threads included in shear plane) and E70XX electrodes.

Top Connection Plate. For a moment of 110 ft-kips, the connection plates, 18 in apart, are subjected to a force $T = 110 \times {}^{12}\!/_{18} = 73.3$ kips. With an allowable

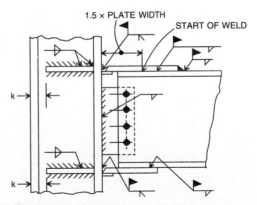

FIGURE 5.57 Semirigid connection with welds and high-strength bolts.

stress of 21.6 ksi, the top connection plate should have a cross-sectional area of at least $A_p = 73.3/21.6 = 3.39$ in^2. Use a $6 \times \frac{5}{8}$ in plate ($A_p = 3.75$ in^2).

The minimum-size fillet weld for $\frac{5}{8}$-in-thick material is $\frac{1}{4}$ in. Assume a $\frac{1}{4}$-in weld for connecting the top plate to the flange:

$$\text{Capacity} = 0.928 \times 4 = 3.71 \text{ kips per in}$$

Length of weld required is $73.3/3.71 = 19.75$ in. Use a 6-in weld across the end of the plate and a 7-in weld along each side (total 20 in). The side welds start at a distance from the column flange of 1.5 times the width of the connection plate, or 9 in.

Bottom Connection Plate.   A $9 \times \frac{1}{2}$ in plate supplies an area of 4.50 in^2 > 3.39. With a $\frac{1}{4}$-in fillet weld for connecting the plate to the bottom flange, weld length required is 19.75 in, as for the top flange. A 10-in-long weld along each side of the connection plate is the minimum required. (To avoid overhead welding, an end weld is not specified for this plate.)

Bolts.   Try $\frac{3}{4}$-in-diameter bolts for the shear plate. Each bolt can carry $0.4418 \times 21 = 9.3$ kips, so the number of bolts required is $27/9.3 = 2.9$, or 3 bolts. For three bolts at 3-in pitch and $1\frac{1}{4}$-in edge distance, the web connection plate is $8\frac{1}{2}$ in long. Try a $\frac{1}{4}$-in-thick plate. The following capacity checks are then made:

Gross shear:

$$R_g = 8.5 \times 0.25 \times 14.4 = 30.6 \text{ kips} > 27 \text{ kips—OK}$$

Net shear:

$$R_n = (8.5 - 3 \times 0.9375)0.25 \times 0.3 \times 58 = 24.7 \text{ kips} < 27 \text{ kips—NG}$$

Try a $\frac{3}{8}$-in-thick plate.

Net shear:

$$R_n = 24.7 \times 0.375/0.25 = 37.1 \text{ kips} > 27 \text{ kips—OK}$$

Bearing on shear plate:

$$\text{Top bolt edge distance} = 1\frac{1}{4} \text{ in} < (1.5 \times 0.875 = 1.3125 \text{ in})$$

Bearing allowable for the top bolt is

$$r_p = 0.5 \times 58 \times 0.375 \times 1.25 = 13.6 \text{ kips}$$

For the other two bolts, the 3-in pitch $> 3 \times 0.875 = 2.625$ in, so the bearing allowable for these bolts is

$$r_p = 1.2 \times 58 \times 0.875 \times 0.375 = 22.8 \text{ kips}$$

Therefore, the allowable bearing load on the web shear plate is $13.6 + 2 \times 22.8 = 59.2$ kips > 27 kips—OK.

Bearing on beam web:

$$R_p = 3 \times 1.2 \times 58 \times 0.355 \times 0.875 = 64.8 \text{ kips} > 27 \text{ kips—OK}$$

Weld of shear plate to column flange: The W12 $\times$ 53 column flange is $\frac{9}{16}$ in thick, which requires a $\frac{1}{4}$-in minimum fillet weld. The capacity of these welds applied to each side of the ⅜-in shear plate is

$$R_w = 0.928 \times 4 \times 2 \times 8.5 \times 0.375/0.51 = 46.4 \text{ kips} > 27 \text{ kips—OK}$$

The factor 0.375/0.51 recognizes that a plate about ½-in thick is required to support two ¼-in fillet welds.

Column Stiffeners.   The force delivered to the column by the top and bottom flange connection plates is 73.3 kips. Assuming that this is due to gravity loads, the AISC specification requires that a flange force $P_{bf} = \frac{5}{3} \times 73.3 = 122$ kips be used to check the column. There are three limit states that must be checked: flange bending, web yielding, and web buckling. A fourth limit state, web crippling, does not control for hot-rolled shapes. The following calculations use data from the AISC ASD manual tables for allowable axial loads on columns.

Flange bending:

$$P_{fb} = 74 \text{ kips} < 122 \text{ kips}$$

Need for a stiffener is indicated.

Web yielding:

$$P_{wy} = P_{wo} + tP_{wi} = 78 + 0.5 \times 12 = 84 \text{ kips} < 122 \text{ kips}$$

Again, need for a stiffener is indicated.

Web buckling:

$$P_{wb} = 106 \text{ kips} < 122 \text{ kips}$$

In this case also, need for a stiffener is indicated.

The three limit states do not all apply to both flanges. Flange bending applies only to the column adjacent to the beam-tension flange. Web yielding applies to both tension and compression flanges; hence the thinner of the two flange plates is used for simplification. Web buckling applies only to the compression flange. Because moments are often reversible, and for simplicity of calculation and fabrication, it is common practice to use the same stiffeners in both locations. Thus it is also common to apply all three limit states to both stiffener locations. With this philosophy, the column capacity is the smallest of P_{fb}, P_{wy}, and P_{wb}, that is, 74 kips. The stiffeners and their welds must be sized to carry what the column cannot; that is, $P_s = 122 - 74 = 48$ kips. Thus each stiffener carries ⁴⁸⁄₂ = 24 kips. For a yield stress of 36 ksi, the stiffener area required is 24/36 = 0.67 in².

In addition to the size required by load, the AISC specification gives the following minimum criteria for stiffeners:

1. The width of each stiffener plus one-half the thickness of the column web shall not be less than one-third the width of the flange or moment connection plate delivering the concentrated force.

2. The thickness of stiffeners shall not be less than one-half the thickness of the flange or plate delivering the concentrated load. Also, the width/thickness ratio shall not exceed $95/\sqrt{F_y}$, where F_y is the specified yield stress of the steel.

3. The welds joining stiffeners to the column web shall be sized to carry the force in the stiffener caused by unbalanced moments on opposite sides of the column.

From these criteria, the minimum stiffener size is as follows: Stiffener width, from criterion 1, should be at least ⅓ × 9 + ³⁄₁₆ = 3³⁄₁₆, say 4 in. Stiffener thickness, from criterion 2, should not be less than $\frac{1}{2} \times \frac{5}{8} = \frac{5}{16}$ in, but the width/thickness ratio may not be more than $95/\sqrt{36} = 15.8$. Since $4/0.3125 = 12.8 < 15.8$, a $\frac{5}{8}$-in stiffener is OK, but use a $\frac{3}{8}$-in-minimum thickness. Thus try a stiffener $\frac{3}{8} \times 4$ in. Area $= 0.375 \times 4 = 1.5$ in$^2 > 0.67$ in^2—OK.

Weld of stiffener to column flange: To clear the fillet between the web and flange of the column, usual practice takes a $\frac{3}{4}$-in snip off the corner of the stiffener. The effective stiffener area is thus $0.375 \times 3.25 = 1.22$ in$^2 > 0.67$ in^2—still OK. The length of the weld at the top and bottom of the stiffener is 3.25 in. The weld is required to carry $24 \times \frac{3}{5} = 14.4$ kips. The size needed is

$$D = 14.4/(2 \times 3.25 \times 0.928) = 2.39 \text{ sixteenths of an inch}$$

Since the column flange thickness is $\frac{9}{16}$ in, the AISC minimum fillet weld is $\frac{1}{4}$ in. Use $\frac{1}{4}$-in fillet welds.

Weld of stiffener to column web: Since the column has a beam framing to one flange only in this example, the stiffeners need not be extended to the other flange but can be kept back the k distance from the other flange, as shown in Fig. 5.57. If a full-depth stiffener is required, it extends to the k distance of the "unattached" flange. Depending on the controlling limit state, either a partial-depth (at least one-half the column depth) stiffener or a full-depth stiffener is required. The limit states of flange bending and web yielding allow the use of partial-depth stiffeners, but web buckling requires a full-depth stiffener. Since the check for web buckling indicated inadequate resistance, a full-depth stiffener is required, which, in this case, means extending the stiffener to the k distance, as shown in Fig. 5.57. Thus the required length of the stiffener is $12 - 1\frac{1}{4} - \frac{9}{16} = 10\frac{3}{16}$ in. Use a stiffener 10 in long. Then the size of fillet weld required is

$$D = 14.4/(10 \times 2 \times 0.928) = 0.84 \text{ sixteenths of an inch}$$

Use a $\frac{3}{16}$-in fillet weld, the AISC minimum.

Since there is a similar stiffener on the opposite side of the column web, check that the column web can support the fillet welds on both sides. In this case, the shear force can disperse both up and down, so there are two shear planes in the web to be compared with the fillet strength. Thus, for the weld,

$$r_{\text{weld}} = 0.84 + 0.928 \times 4 = 3.1 \text{ kips per in}$$

For the web,

$$r_{\text{web}} = 0.345 \times 14.4 \times 2 = 9.9 \text{ kips per in} > 3.1 \text{ kips per in}$$

The web can support the fillet welds.

Panel shear in column web: The panel zone is the area of the column between the beam flanges or connection plates. The panel-zone shear capacity is the shear capacity of the column, $V_c = 0.4F_y d_c t_w$, where d_c is the depth of the column and t_w is its web thickness. For the W12 $\times$ 53 of A36 steel, $V_c = 0.4 \times 36 \times 12.06 \times 0.345 = 59.9$ kips. This capacity does not, however, exceed the applied force from the beam flange, $T = 73.3$ kips. The panel zone appears to fail in shear. However, when a moment is applied to a column by a beam at a moment connection, column moments equal and opposite to the applied beam moment develop. If the story height above and below the beam in question is H ft, and inflection points are assumed at the usual half-story heights, the story shear induced in the column due to M, the applied beam moment plus the associated column moments, is $V_s = M/H$. If the story height is 10 ft, $V_s = \frac{110}{10} = 11$ kips. Story shear always acts counter to the applied force T from the beam flange. Thus the net panel shear is 73.3 − 11 = 62.3 kips, which is about 4% greater than the column capacity of 59.9 kips. At an overstress of 4%, the column may be considered satisfactory without further reinforcement, because material properties will generally be more than 4% over the minimums.

If the column must be reinforced, a web doubler plate can be used to increase the web thickness to the required value. However, it is usually less expensive to choose a heavier column and avoid the fabrication expense of the doubler plate. A W12 × 58 column (web thickness 4.35% greater than the W12 × 53) eliminates the marginal doubler requirement. Also, it may be more economical to choose a still heavier column and eliminate the stiffeners as well. For example, a W12 × 79 column can carry the 110-ft-kip beam moment and 27-kip shear without stiffeners.

5.34 BEAMS SEATED ATOP SUPPORTS

There are many cases where beams are supported only at the bottom flange. In one-story buildings, for example, beams often are seated atop the columns. In bridges, beams generally are supported on bearings under the bottom flange. In all these cases, precautions should be taken to ensure lateral stability of the beams (see also Art. 5.29).

Regardless of the location of support, the compression flange of the beams should be given adequate lateral support. Such support may be provided by a deck or by bracing. In addition, beams supported at the lower flange should be braced against forces normal to the plane of the web and against eccentric vertical loading. The eccentricity may be caused by a column not being perfectly straight or perfectly plumb. Or the eccentricity may be produced by rolling imperfections in the beam; for example, the web may not be perfectly perpendicular to the flanges.

To resist such loadings, AASHTO and AREA require that deck spans be provided with cross frames or diaphragms at each support. AREA specifies that cross frames be used for members deeper than 3 ft 6 in and spaced more than 4 ft on centers. Members not braced with cross frames should have I-shaped diaphragms as deep as depth of beams permit. AASHTO prefers diaphragms at least half the depth of the members. Cross frames or diaphragms should be placed in all bays. This bracing should be proportioned to transmit all lateral forces, including centrifugal and seismic forces and cross winds on vehicles, to the bearings.

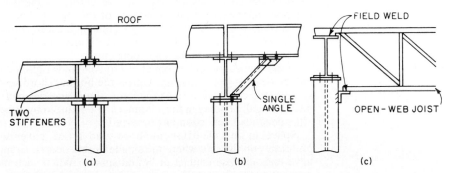

FIGURE 5.58 Bracing for girders. (*a*) Stiffeners brace a girder that is continuous over the top of a column. (*b*) Knee brace restrains a girder seated atop a column. (*c*) Open-web joist provides lateral support for a girder atop a column.

In through bridge spans, girders should be stiffened against lateral deformation by gusset plates or knee braces with solid webs. These stiffeners should be connected to girder stiffeners and to floorbeams.

In buildings, lateral support may be provided a girder seated atop a column by fastening cross members to it and stiffening the web at the support (Fig. 5.58*a*). Or a knee brace may be inserted at the support between the bottom flange of the girder and the bottom flange of a cross member (Fig. 5.58*b*). Where an open-web joist frames into a girder seated atop a column, lateral support can be provided by connecting the top chord of the joist to the top flange of the girder and attaching the bottom chord as shown in Fig. 5.58*c*.

5.35 *TRUSS CONNECTIONS*

Truss members generally are designed to carry only axial forces. At panel points, where members intersect, it is desirable that the forces be concurrent, to avoid bending moments. Hence the gravity axes of the members should be made to intersect at a point if practicable. When this cannot be done, the connections should be designed for eccentricity present. Also, groups of fasteners or welds at the ends of each member should have their centers of gravity on or near the gravity axis of the member. Otherwise, the member must be designed for eccentric loading. In building trusses not subject to repeated variation in stress, however, eccentricity of welds and fasteners may be neglected at the ends of single-angle, double-angle, and similar members.

At panel points, several types of connections may be used. Members may be pin-connected to each other (Art. 5.6), or they may be connected through gusset plates, or they may be welded directly to each other. In bridges, fasteners should be symmetrical with the axis of each member as far as practicable. If possible, the design should fully develop the elements of the member.

Design of connections at gusset plates is similar to that for gusset-plate connections illustrated in Art. 5.25. In all cases, gusset plates should be trimmed to minimize material required and for good appearance. When cuts are made, minimum edge distances for fasteners and seats for welds should be maintained. Re-entrant cuts should be avoided.

At columns, if the reactions of trusses on opposite sides of a support differ by only a small amount (up to 20%), the gravity axes of end diagonal and chord may be allowed to intersect at or near the column face (Fig. 5.59). The connections need be designed only for the truss reactions. If there is a large difference in reactions of trusses on opposite sides of the support, or if only one truss frames into a column, the gravity axes of diagonal and chord should be made to intersect on the column gravity axis. Shear and moment in the column flange should be considered in design of the end connection.

If the connection of bottom chord to column is rigid, deflection of the truss will induce bending in the column. Unless the bottom chord serves as part of a bracing system, the connection to the column should be flexible. When the connection must be rigid, bending in the column can be reduced by attaching the bottom chord after dead-load deflection has occurred.

Splices in truss chords should be located as near panel points as practicable and preferably on the side where the smaller stress occurs. Compression chords should have ends in close contact at bolted splices. When such members are fabricated and erected with close inspection and detailed with milled ends in full-contact bearing, the splices may be held in place with high-strength bolts proportioned for at least 50% of the lower allowable stress of the sections spliced.

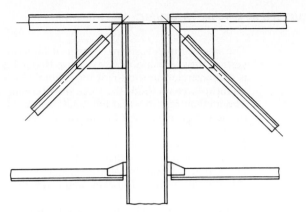

FIGURE 5.59   Truss connection to a column. The gravity axes of the top chords and the diagonals meet near the faces of the support.

In other cases, compression and tension chords should be designed for not less than the average of the calculated stress at the point of splice and the strength of the member at that point but not less than 75% of the strength of the member (Arts. 5.26 and 5.27). Tension and compression chords may be spliced with complete-penetration groove welds without splice plates.

Gusset Plates.   These should be sized to resist all loads imposed. The design of gussets and the connection of members to the gusset is based on statics and yield strength, which are the basic ingredients of the lower-bound theorem of limit analysis. Basically, this theorem states that if equilibrium is satisfied in the structure (or connection) and yield strength is nowhere exceeded, the applied load will, at most, be equal to the load required to fail the connection. In other words, the connection will be safe. In addition to yield strength, gussets and their associated connections also must be checked to ensure that they do not fail by fracture (lack of ductility) and by instability (buckling).

AASHTO has a stability criterion for the free edges of gusset plates that requires that an edge be stiffened when its length exceeds plate thickness times $347/\sqrt{F_y}$, where F_y is the specified yield stress (ksi). This criterion is intended to prevent a gusset from flexing when the structure deforms and the angles between the members change. Because of repeated loading, this flexing can cause fatigue cracks in gussets. For statically loaded structures, which include structures subjected to wind and seismic loads, this flexing is not detrimental to structural safety and may even be desirable in seismic design because it allows more energy absorption in the members of the structure and it reduces premature fracture in the connections.

For buildings, the AISC seismic specification has detailing requirements to allow gussets to flex in certain situations. Although the AISC ASD and LRFD manuals have no specific requirements regarding gusset stability, it is important that gusset buckling be controlled to prevent changes in structure geometry that could render the structure unserviceable or cause catastrophic collapse (see also Art. 5.36).

Fracture in gusset-plate connections also must be prevented not only when connections are made with fillet welds, which have limited ductility in their transverse direction, but also with bolts, because of the possibility of tearout fracture and nonuniform distribution of tension and shear to the bolts.

5.36 *CONNECTIONS FOR BRACING*

The lateral force–resisting system of large buildings is sometimes provided by a vertical truss with connections such as that in Fig. 5.60. The design of this connection is demonstrated in the following example by a method adopted by the AISC, the **uniform-force method,** the force distributions of which are indicated in Fig. 5.61. The method requires that only uniform forces, no moments, exist along the edges of the gusset plate used for the connection.

Example.   Design the bracing connection of Fig. 5.60. The column is to be made of grade 50 steel and other steel components of A36 steel. This connection comprises four connection interfaces: diagonal brace to gusset plate, gusset plate to column, gusset plate to beam, and beam to column.

Brace to Gusset.   The diagonal brace is a tube, slotted to accept the gusset plate, and is field welded to the gusset with four $\frac{1}{4}$-in fillet welds. For an allowable stress of 21.0 ksi, a $\frac{1}{16}$-in fillet weld has a capacity of 0.928 kips per in of length. Hence the weld length required for the 200-kip load is $200/(0.928 \times 4 \times 4) = 13.9$ in. Use 16 in of $\frac{1}{4}$-in fillet weld. The $\frac{1}{2}$-in tube wall is more than sufficient to support this weld.

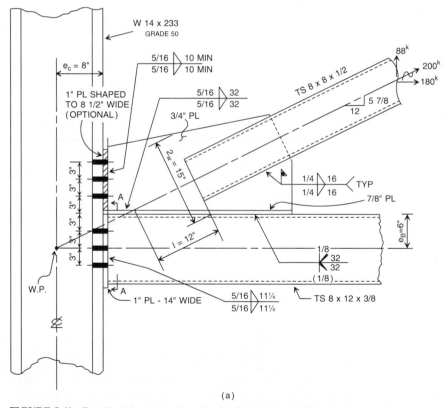

(a)

FIGURE 5.60 Details of the connection of a tubular brace to a tubular beam at a column. (*a*) A gusset plate is welded to the brace, to a plate welded to the top of the beam, and to end plates, which are field bolted to the column. (*b*) Section *A–A* through the tubular beam.

Tearout.   Next, the gusset is checked for tearout. The shear tearout area is $A_v = 2 \times 16t$, where t is the gusset thickness. The tension tearout area is $A_t = 8t$. The load required to produce tearout therefore is

$$P_{to} = 0.3 \times 58 \times 32t + 0.5 \times 58 \times 8t \geq 200 \text{ kips}$$

Solution of this inequality gives $t \geq 0.253$ in. Try a gusset thickness of 5/16 in.

Buckling.   For a check for gusset buckling, the critical or "Whitmore section" has a width $l_w = 15$ in, and the gusset column length $l = 12$ in (Fig. 5.60a). The slenderness ratio of the gusset column is

$$kl/r = 0.5 \times 12\sqrt{12}/0.3125 = 66.4$$

For the yield stress $F_y = 36$ ksi, the allowable compressive stress on the Whitmore section (from the AISC ASD manual, Table C36) is $F_a = 16.80$. The actual stress is

$$f_a = 200/(15 \times 0.3125) = 42.7 \text{ ksi} > 16.80 \text{ ksi—NG}$$

Recalculation indicates that an 11/16-in plate has sufficient strength. Most fabricators, however, stock 3/4-in plate and will prefer to use it. For 3/4-in plate, $F_a = 20.1$ ksi and

$$kl/r = 0.5 \times 12\sqrt{12}/0.75 = 27.7$$

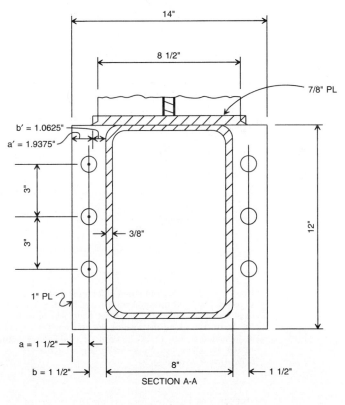

SECTION A-A

(b)

FIGURE 5.60   *Continued.*

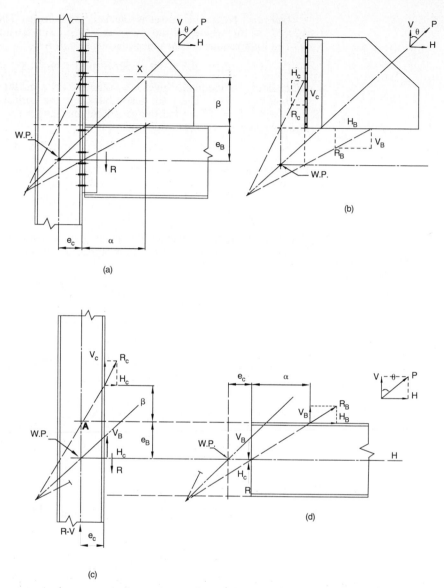

FIGURE 5.61   Determination of forces in a connection of a brace to a beam at a column through a gusset plate, by the uniform force method. (*a*) Lines of action of forces when no moments exist along the edges of the gusset plate. (*b*) Horizontal and vertical forces act at the top of the beam and along the face of the column. (*c*) Location of connection interfaces are related as indicated by Eq. (5.28). (*d*) For an axial force P acting on the brace, the force components in (*b*) are given by: $H_B = \alpha P/r$, $V_B = e_B P/r$, $H_C = e_C P/r$, and $V_C = \beta P/r$, where $r = \sqrt{(\alpha + e_C)^2 + (\beta + e_B)^2}$.

$$f_a = 200/(15 \times 0.75) = 17.8 \text{ ksi} < 20.1 \text{ ksi—OK}$$

Stress Components. The forces at the gusset-to-column and gusset-to-beam interfaces are determined from the geometry of the connection. As shown in Figs. 5.60a and 5.61, for the beam, $e_b = 6$ in; for the column, $e_c = 8$ in, and tan $\theta = 12/5.875 = 2.0426$. The parameters α and β determine the location of the centroids of the horizontal and vertical edge connections of the gusset plate:

$$\alpha - \beta \tan \theta = e_B \tan \theta - e_c \tag{5.28}$$

This constraint must be satisfied for no moments to exist along the edges of the gusset, only uniform forces. Try $\alpha = 16$ in, estimated, based on the 32-in length of the horizontal connection plate B (Fig. 5.62a), which is dictated by geometry.

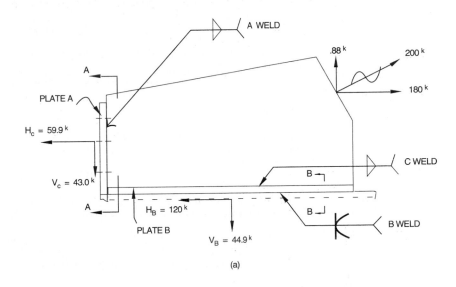

(a)

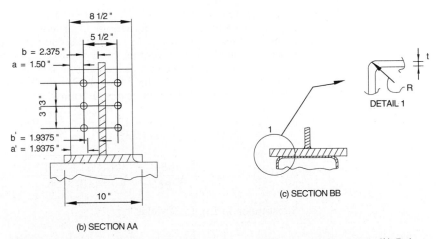

FIGURE 5.62 Gusset plate in Fig. 5.60a. (a) Force distribution on the gusset. (b) Bolt arrangement in plate A. (c) Attachment of gusset to plate B.

By Eq. (5.28),

$$\beta = \frac{16 - 6 \times 2.0426 + 8}{2.0426} = 5.75 \simeq 6$$

For $\beta = 6$, three rows of bolts can be used for the connection of the gusset to the column. The distance from the working point WP, the intersection of the axes of the brace, beam, and column, to X (Fig. 5.61a) is

$$r = \sqrt{(16 + 8)^2 + (5.75 + 6)^2} = 26.72 \text{ in}$$

The stress components are

$$H_B = 200 \times 16/26.72 = 120 \text{ kips}$$

$$V_B = 200 \times 6/26.72 = 44.9 \text{ kips}$$

$$V_C = 200 \times 5.75/26.72 = 43.0 \text{ kips}$$

$$H_C = 200 \times 8/26.72 = 59.9 \text{ kips}$$

These forces are shown in Fig. 5.62.

Gusset to Column. Try six A325N (noncritical) $\frac{7}{8}$-in-diameter bolts. For an allowable stress of 21 ksi, the shear capacity per bolt is

$$r_v = 21 \times 0.6013 = 12.6 \text{ kips}$$

These bolts are subjected to a shear $V_C = 43$ kips and a tension (or compression if the brace force reverses) $H_C = 59.9$ kips. The required shear capacity per bolt is $^{43}\!/_6 = 7.2$ kips < 12.6 kips—OK.

The allowable bolt tension when combined with shear is

$$B = 0.6013\sqrt{44^2 - 4.39(7.2/0.6013)^2} = 21.8 \text{ kips}$$

The applied tension per bolt is $59.9/6 = 9.98$ kips < 21.8 kips—OK.

Prying Action on Plate A. Check the end plate (Plate A in Fig. 5.62) for an assumed thickness of 1 in. Equations (5.2) and (5.3) are used to compute the allowable stress in the plate. As shown in section A–A (Fig. 5.62b), the bolts are positioned on $5\frac{1}{2}$-in cross centers. Since the gusset is $\frac{3}{4}$ in thick, the distance from the center of a bolt to the face of the gusset is $b = (5.5 - 0.75)/2 = 2.375$ in and $b' = b - d/2 = 2.375 - 0.875/2 = 1.9375$ in. The distance from the center of a bolt to the edge of the plate is $a = 1.50$ in $< (1.25b = 2.9969)$. For $a = 1.50$, $a' = a + d/2 = 1.50 + 0.875/2 = 1.9375$ in; $\rho = b'/a' = 1.9375/1.9375 = 1.0$; $p = 3$ in; and $\delta = i - d'/p = 1 - 0.9375/3 = 0.6875$.

For use in Eqs. (5.2) and (5.3),

$$t_c = \sqrt{\frac{8Bb'}{pF_y}} = \sqrt{\frac{8 \times 21.8 \times 1.9375}{3 \times 36}} = 1.769 \text{ in}$$

Substitution in Eq. (5.2) yields

$$\alpha' = \frac{1}{\delta(1 + \rho)}\left[\left(\frac{t_c}{t}\right)^2 - 1\right] = \frac{1}{0.6875 \times 2.00}\left[\left(\frac{1.769}{1}\right)^2 - 1\right] = 1.549$$

Since $\alpha' > 1$, use $\alpha' = 1$. Then, from Eq. (5.3), the allowable stress is

$$T_a = B \left(\frac{t}{t_c}\right)^2 (1 + \delta\alpha') = 21.8 \left(\frac{1}{1.769}\right)^2 (1 + 0.6875)$$

$$= 11.76 \text{ kips} > 9.98 \text{ kips—OK}$$

The column flange can be checked for bending (prying) in the same way, but since the flange thickness $t_f = 1.720$ in $\gg 1$, the flange is satisfactory.

Weld A.   This weld (Fig. 5.62a) has a length of about 10 in and the load it must carry is

$$P_A = \sqrt{H_C^2 + V_C^2} = \sqrt{59.9^2 + 43^2} = 73.7 \text{ kips}$$

The weld size required in sixteenths of an inch is

$$D = \frac{73.7}{2 \times 10 \times 0.928} = 3.97$$

A ¼-in fillet weld is sufficient, but the minimum fillet weld permitted by the AISC specifications for structural steel buildings for 1-in-thick steel is ⁵⁄₁₆ in (unless low-hydrogen electrodes are used). Use a ⁵⁄₁₆-in weld.

Gusset to Beam.   Weld B (Fig. 5.62a) is 32 in long and carries a load

$$P_B = \sqrt{H_B^2 + V_B^2} = \sqrt{120^2 + 44.9^2} = 128 \text{ kips}$$

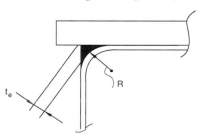

FIGURE 5.63   Effective throat of a flare bevel-groove weld.

This weld is a flare bevel-groove weld that attaches plate B to the top corners of the beam (Fig. 5.63). Satisfying ASTM tube specification A500, this shape has a radius $R = 3t$, where $t =$ tube thickness = ⅜ in (Fig. 5.62c). The AISC specifications assume that $R = 2t$. Use of this radius to compute weld throat t_e (Fig. 5.63) is conservative for the following reason: The capacity of the flare groove weld, when it is made with E70 electrodes, is given by the product of t_e and the weld length and the allowable stress, 21 ksi, and the AWS defines t_e as ⁵⁄₁₆R when the groove is filled flush (Fig. 5.63).

With $R = 2t$, $t_e = 2 \times 0.375 \times$ ⁵⁄₁₆ = 0.234 in. The throat required is

$$T_{\text{req}} = \frac{128}{2 \times 32 \times 21} = 0.095 \text{ in} < 0.234 \text{ in—OK}$$

Thus use a flush flare bevel-groove weld.

Experience shows that tube corner radii vary significantly from producer to producer, so verify in the shop with a radius gage that the tube radius is $2 \times 0.375 = 0.75$ in. If $R < 0.75$, verify that ⁵⁄₁₆$R \geq$ (0.095 in ≈⅛ in). If ⁵⁄₁₆$R <$ ⅛ in, supplement the flare groove weld with a fillet weld (Fig. 5.64) to obtain an effective throat of ⅛ in.

Plate B.   This plate is used to transfer the vertical load $V_B = 44.9$ kips from the "soft" center of the tube top wall to the stiff side walls of the tube. For determination of thickness t_B, plate B can be treated as a simply supported beam subjected to a midspan concentrated load from the gusset plate (Fig. 5.65). The

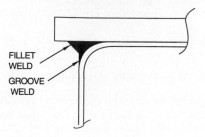

FIGURE 5.64   Fillet weld reinforces a flare bevel-groove weld.

bending moment in the beam is 44.9 × ¾ = 89.8 in-kips. The section modulus is $32t_B^2/6$. The allowable bending stress for the plate is $0.75F_y = 0.75 \times 36 = 27$ ksi. Substitution in the flexure formula yields the equation

$$27 = \frac{89.8 \times 6}{32t_B^2}$$

from which $t_B = 0.79$ in. This indicates that a ⅞-in-thick plate will suffice. As a check of this 32-in-long plate for shear, the stress is computed:

$$f_v = \frac{120}{2 \times 32 \times 0.875} = 2.14 \text{ ksi} < 14.4 \text{ ksi—OK}$$

Weld C.   The required weld size in number of sixteenths of an inch is

$$D = \frac{120}{2 \times 32 \times 0.928} = 2.02$$

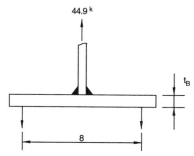

FIGURE 5.65   Plate *B* of Fig. 5.62*a* acts as a simple beam under load from the gusset plate.

But use a ⁵⁄₁₆-in fillet weld, the AISC minimum when low-hydrogen electrodes are not used.

Beam to Column.   Plate *A*, 1 in thick, extends on the column flange outward from the gusset-to-column connection but is widened to 14 in to accommodate the bolts to the column flange. Figures 5.60*b* and 5.66 show the arrangement.

Weld D.   This weld (Fig. 5.66) has a length of $2(12 - 2 \times 0.375) = 22.5$ in, clear of the corner radii, and carries a

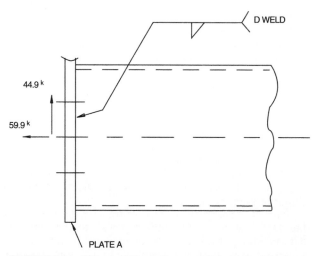

FIGURE 5.66   Plate *A* of Fig. 5.62*a* transmits loads from the beam to the column.

load

$$P_B = \sqrt{H_C^2 + V_B^2} = \sqrt{59.9^2 + 44.9^2} = 74.9 \text{ kips}$$

The fillet-weld size required in number of sixteenths of an inch is

$$D = \frac{74.9}{22.5 \times 0.928} = 3.59$$

Use the AISC minimum $5/16$-in fillet weld.

Bolts. Shear per bolt is $44.9/6 = 7.48$ kips < 12.6 kips—OK. The tensile load per bolt is $59.9/6 = 9.98$ kips. Allowable tension per bolt for combined shear and tension is

$$B = 0.6013 \sqrt{44^2 - 4.39(7.48/0.6013)^2} = 21.3 \text{ kips} > 9.98 \text{ kips—OK}$$

Equations (5.2) and (5.3) are used to check prying (bending) of plate A. From Fig. 5.60b, the distance from the center of a bolt to a wall of the tube is $b = 1.5$ in and $b' = b - d/2 = 1.5 - 0.875/2 = 1.0625$ in. The distance from the center of a bolt to an edge of the plate is

$$a = (14 - 11)/2 = 1.5 \text{ in} < 1.25b$$

Also for use with Eqs. (5.2) and (5.3),

$$a' = a + d/2 = 1.5 + 0.875/2 = 1.9375 \text{ in}$$

$$\rho = b'/a' = 1.0625/1.9375 = 0.5484$$

$$p = 3 \text{ in}$$

$$\delta = 1 - d'/p = 1 - 0.9375/3 = 0.6875$$

For use in Eqs. (5.2) and (5.3),

$$t_c = \sqrt{\frac{8 \times 21.3 \times 1.0625}{3 \times 36}} = 1.2948$$

Substitution in Eq. (5.2) yields

$$\alpha' = \frac{1}{\delta(1 + \rho)} \left[\left(\frac{t_c}{t}\right)^2 - 1 \right] = \frac{1}{0.6875 \times 1.5484} \left[\left(\frac{1.2948}{1}\right)^2 - 1 \right] = 0.6354$$

From Eq. (5.3), the allowable stress is

$$T_a = B \left(\frac{t}{t_c}\right)^2 (1 + \delta\alpha') = 21.3 \left(\frac{1}{1.2948}\right)^2 (1 + 0.6354 \times 0.6875)$$

$$= 18.3 \text{ kips} > 9.98 \text{ kips—OK}$$

Column Flange. This must be checked for bending caused by the axial load of 59.9 kips. Figure 5.67 shows a yield surface that can be used for this purpose. Based on this yield surface, the effective length of column flange tributary to each pair of bolts, for use in Eqs. (5.2) and (5.3) to determine the effect of prying, may be taken as

$$p_{\text{eff}} = \frac{p(n - 1) + \pi b/2 + 2c}{n} \tag{5.29}$$

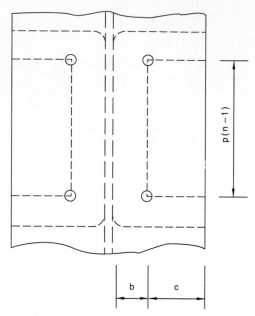

FIGURE 5.67   Yield lines for the flange of the column in Fig. 5.60a.

where n = number of rows of bolts

$$b = (8 + 3 - 1.070)/2 = 4.965$$
$$c = (15.890 - 8 - 3)/2 = 2.445$$
$$p_{\text{eff}} = \frac{3(3 - 1) + 4.965\,\pi/2 + 2 \times 2.445}{3} = 6.23 \text{ in}$$

Also for use with Eqs. (5.2) and (5.3).

$$b' = 4.965 - 0.875/2 = 4.528 \text{ in}$$

$$a = (14 - 11)/2 = 1.5 < 1.25b$$

$$a' = 1.5 + 0.875/2 = 1.9375 \text{ in}$$

$$\rho = 4.528/1.9375 = 2.337$$

$$\delta = 1 - 0.9375/6.23 = 0.850$$

$$t_c = \sqrt{8 \times 21.3 \times 4.528/(6.23 \times 50)} = 1.574 \text{ in}$$

$$t = \text{column flange thickness} = 1.720 \text{ in}$$

Substitution in Eq. (5.2) gives

$$\alpha' = \frac{1}{0.850 \times 3.337}\left[\left(\frac{1.574}{1.720}\right)^2 - 1\right] = -0.0573$$

Since $\alpha' < 0$, use $\alpha' = 0$, and the allowable stress is

$$T_a = B = 21.3 \text{ kips} > 9.98 \text{ kips—OK}$$

This completes the design (see Fig. 5.60).

5.37 CRANE-GIRDER CONNECTIONS

Supports of crane girders must be capable of resisting static and dynamic horizontal and vertical forces and stress reversal. Consequently, heavily loaded crane girders (for cranes with about 75-ton capacity or more) usually are supported on columns carrying no other loads. Less heavily loaded crane girders may be supported on building columns, which usually extend above the girders, to carry the roof.

It is not advisable to connect a crane girder directly to a column. End rotations and contraction and expansion of the girder flanges as the crane moves along the girder induce severe stresses and deformations at the connections that could cause failure. Hence, vertical support should be provided by a seat, and horizontal support by flexible connections. These connections should offer little restraint to end rotation of the girders in the plane of the web but should prevent the girders from tipping over and should provide lateral support to the compression flange.

When supported on building columns, the crane girders may be seated on brackets (very light loads) or on a setback in those columns. (These arrangements are unsatisfactory for heavy loads, because of the eccentricity of the loading.) When a setback is used, the splice of the deep section and the shallow section extending upward must be made strong. Often, it is desirable to reinforce the flange that supports the girder.

Whether seated on building columns or separate columns, crane girders usually are braced laterally at the supports against the building columns. In addition, when separate columns are used, they too should be braced at frequent intervals against the building columns to prevent relative horizontal movement and buckling. The bracing should be flexible in the vertical direction, to avoid transfer of axial stress between the columns.

Figure 5.68 shows the connection of a crane girder supported on a separate column. At the top flange, horizontal thrust is transmitted from the channel through a clip angle and a tee to the building column. Holes for the fasteners in the vertical leg of the clip angle are horizontal slots, to permit sliding of the girders longitudinally on the seat. The seat under the bottom flange extends to and is attached to the building column. This plate serves as a flexible diaphragm. Stiffeners may be placed on the building column opposite the top and bottom connections, if needed. Stiffeners also may be required for the crane-girder web. If so, they should be located over the flanges of the crane-girder column.

5.38 RIGID-FRAME KNEES

Rigid-frame bridges and one-story rigid frames in buildings generally are used for long spans. Consequently, the knee, where beam and column intersect, is subjected to large bending moments, in addition to shear and thrust. The knee must be designed to transmit these forces from beam end to column.

For lightly loaded or short-span rigid frames, the intersection may be a simple rectangle (Fig. 5.69a) or trapezoid. The beam flanges and outer column flange extend uninterrupted to their terminals. The inner column flange ends at the bottom flange of the beam, and stiffeners are placed on the beam web directly above it. If required, diagonal stiffeners may be used to reinforce the web of the knee (Figs. 5.70c and 5.71).

For more heavily loaded or longer-span rigid frames, a haunched connection may be used (Fig. 5.69b). Stiffeners are placed on the web wherever there are changes in direction of the flange. An alternative design (Fig. 5.69c) replaces the

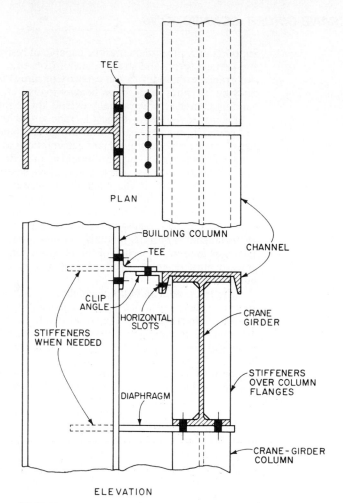

FIGURE 5.68   Crane girder connection at building column.

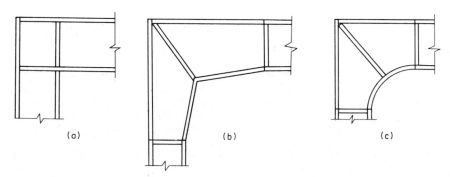

FIGURE 5.69   Rigid-frame knees. (*a*) Rectangular. (*b*) Haunched. (*c*) Curved.

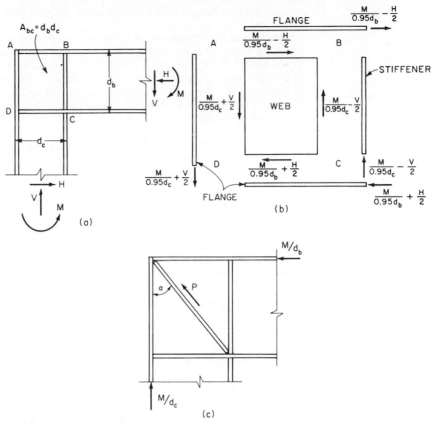

FIGURE 5.70 Rectangular knee. (*a*) Forces acting on the knee. (*b*) Forces acting on components of the knee. (*c*) Knee with diagonal stiffener.

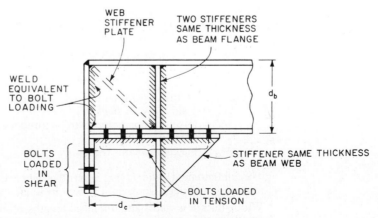

FIGURE 5.71 Welded and bolted rectangular knee.

straight bottom flanges with a smooth curve. In buildings, the top flange also may be curved.

Because of the complexity of the stress distribution in a knee, design usually is based on empirical rules resulting from tests and experience and on plastic theory. When plastic theory is used, a plastic hinge is assumed to form at the knee. At that stage, bending moment transmitted by the knee remains constant with increase in load, and relative rotation of beam and column ends can occur. The following rules for plastic design in the AISC "Specification for Structural Steel Buildings— ASD" are pertinent:

- Strength, determined by rational analysis, shall not be less than that required to support a factored load equal to 1.7 times the given live and dead loads, or 1.3 times these loads acting in conjunction with 1.3 times specified wind or earthquake forces.

- Web stiffeners are required on a member at a point where a plastic hinge would form.

- All connections, the rigidity of which is essential to the continuity assumed as the basis of the analysis, shall be capable of resisting the moments, shears, and axial loads to which they would be subjected by the full factored loading, or any partial distribution thereof.

- Tapered or curved corner connections shall be so proportioned that the full plastic bending strength of the section adjacent to the connection can be developed, if required.

    High-strength bolts, A307 bolts and welds shall be proportioned to resist the forces produced at factored loads, using 1.7 times allowable working stresses. In general, groove welds are preferable to fillet welds, but their use is not mandatory.

- Members shall be adequately braced to resist lateral and torsional displacements at the plastic-hinge locations associated with the failure mechanism. The laterally unsupported distance (in) l_{cr} from such braced hinge locations to similarly braced adjacent points on the frame shall not exceed the value determined from Eq. (5.30a) or (5.30b) as applicable.

When $1.0 > M/M_p > -0.5$,

$$\frac{l_{cr}}{r_y} = \frac{1375}{F_y} + 25 \qquad (5.30a)$$

When $-0.5 > M/M_p > -1.0$,

$$\frac{l_{cr}}{r_y} = \frac{1375}{F_y} \qquad (5.30b)$$

where r_y = radius of gyration of cross section about weak axis, in
F_y = yield stress of steel, ksi
M = lesser of moments due to factored loads at the ends of the unbraced segment, ft-kips
M_p = plastic moment, ft-kips

The end-moment ratio M/M_p is positive when the segment is bent in reverse curvature and negative when it is bent in single curvature.

- Width/thickness ratios and unbraced lengths of all parts of the connection that would be subject to compression must meet AISC specification requirements applicable to plastic design. Sheared edges and punched holes should not be used in portions of the knee subject to tension.

- When the knee is proportioned elastically for the moments that would exist within its length, the continuous frame may be analyzed as a mechanism with a hinge at the small end of the haunch, instead of at the point of intersection of connected members. This should require less steel.

Knee design usually is based on the assumption that flanges carry the bending moment, and the web the shear.

Figure 5.70*b* shows the assumed forces acting on the rectangular knee in Fig. 5.70*a*. Flange forces produced by bending moment are assumed to act at the centroid of the flanges. The lever arms are assumed to be 95% of the depth of the members. The beam flanges impose on the horizontal edges of the web horizontal shears equal to the approximate flange forces $M/0.95d_b + H/2$, where H is the horizontal thrust (kips) and M is the moment (in-kips). The column flanges impose on the vertical edges of the web vertical shears equal to the approximate flange forces $M/0.95d_c \pm V/2$, where V is the vertical shear (kips).

If the beam develops a plastic hinge at the knee, the horizontal forces acting on the web equal $M_p/0.95d_b \pm H/2$, where M_p is the plastic moment (in-kips) for the beam. Thus the web must transmit a maximum shear equal to $M_p/0.95d_b + H/2$. The maximum shear resistance of the web is $td_cF_y/\sqrt{3}$, where t = web thickness (in) and $F_y/\sqrt{3}$ is the yield strength (ksi) of the web steel in shear. Hence web thickness should be at least

$$ t = \frac{1.92M_p}{A_{bc}F_y} + \frac{0.866H}{d_cF_y} \tag{5.31} $$

where A_{bc} is the area of knee web (in^2) = d_bd_c.

If the beam web extends into the knee and the web thickness is less than that required by Eq. (5.31), the web should be reinforced with diagonal stiffeners or a doubler plate. Stiffeners are preferable. They form the diagonals of a truss with the flanges as chords and verticals (Fig. 5.70*c*). The stiffeners then have to resist a maximum horizontal force (kips) equal to the flange force minus the web shear resistance:

$$ P = \frac{M_p}{0.95d_b} + \frac{H}{2} - \frac{td_cF_y}{\sqrt{3}} \tag{5.32} $$

where t is the actual web thickness (in). Hence the total area of diagonal stiffeners required (in^2) is

$$ A_{st} = \frac{P}{F_y \sin \alpha} \tag{5.33} $$

where α is the angle between these stiffeners and the vertical.

Beam flanges may be extended unchanged into the knee. If adequate for maximum beam stresses and resistance to local buckling, they will be satisfactory for the knee. Member *AD* should be a continuation of the outer column flange or a plate of equal cross-sectional area groove-welded to it. Stiffener *BC* transmits the forces in the other column flange into the beam web. A half-depth stiffener normally is adequate.

Figure 5.71 shows a rigid-frame knee with shop welds and field bolts. Bolts and welds at the beam seat must resist the horizontal force $M/d_b + H/2$. Bolts and welds at the vertical plate on the outer column flange must resist the force

$M/d_c + V/2$. For the seat, use a plate with cross-sectional dimensions at least equal to those of the beam flange. Determine the size of the plate on the outer column flange from the net section required to resist the tension $M/d_c + V/2$.

Haunched knees, such as the one in Fig. 5.69b, should be designed to resist at every section applied moments M_x by elastic theory, with a plastic hinge at the beam end of the knee, or by plastic theory. For plastic design, each vertical section from the beam end to the diagonal stiffener and each horizontal section from the diagonal stiffener to the top of the column should provide a plastic moment M_{px} equal to or greater than M_x at that section. The slope of the bottom flange can be adjusted to keep component sizes within desirable limits. Width/thickness ratios and unbraced length of components should be checked against recommended criteria for plastic design. The influence of axial thrust and shear on the plastic moment may be neglected. Stiffeners can be designed for truss action with the flanges.

Knees with curved haunch can be designed in a similar manner. Radial stiffeners should be provided at points of tangency, at the midpoint of the curve, and at intermediate points if necessary to assist the web in resisting the shear and to restrain cross bending of the curved flange. To resist cross bending, the width-thickness ratio of the curved flange should not exceed

$$\frac{b}{t} = \frac{2R}{b} \tag{5.34}$$

where R = radius of curvature of flange
b = flange width

SECTION 6
BUILDING DESIGN CRITERIA

R. A. LaBoube
Associate Professor of Civil Engineering, University of Missouri–Rolla, Rolla, Missouri

Building designs generally are controlled by local or state building codes. In addition, designs must satisfy owner requirements and specifications. For buildings on sites not covered by building codes, or for conditions not included in building codes or owner specifications, designers must use their own judgment in selecting design criteria. This section has been prepared to provide information that will be helpful for this purpose. It summarizes the requirements of model building codes and standard specifications and calls attention to recommended practices.

The American Institute of Steel Construction (AISC) promulgates several standard specifications, but two are of special importance in building design. One is the "Specification for Structural Steel Buildings—Allowable Stress Design (ASD) and Plastic Design." The second is the "Load and Resistance Factor Design (LRFD) Specification for Structural Steel Buildings," which takes into account the strength of steel in the plastic range and utilizes the concepts of first-order theory of probability and reliability. The standards for both ASD and LRFD are reviewed in this section.

Steels used in structural applications are specified in accordance with the applicable specification of ASTM. Where heavy sections are to be spliced by welding, special material notch-toughness requirements may be applicable, as well as special fabrication details (see Arts. 1.13, 1.14, and 1.20).

6.1 BUILDING CODES

A **building code** is a legal ordinance enacted by public bodies, such as city councils, regional planning commissions, states, or federal agencies, establishing regulations governing building design and construction. Building codes are enacted to protect public health, safety, and welfare.

A building code presents minimum requirements to protect the public from harm. It does not necessarily indicate the most efficient or most economical practice.

Building codes specify design techniques in accordance with generally accepted theory. They present rules and procedures that represent the current generally accepted engineering practices.

A building code is a consensus document that relies on information contained in other recognized codes or standard specifications, e.g., the model building codes

6.1

promulgated by building officials associations and standards of AISC, ASTM, and the American National Standards Institute (ANSI). Information generally contained in a building code addresses all aspects of building design and construction, e.g., fire protection, mechanical and electrical installations, plumbing installations, design loads and member strengths, types of construction and materials, and safeguards during construction. For its purposes, a building code adopts provisions of other codes or specifications either by direct reference or with modifications.

6.2 APPROVAL OF SPECIAL CONSTRUCTION

Increasing use of specialized types of construction not covered by building codes has stimulated preparation of special-use permits or approvals. Model codes individually and collectively have established formal review procedures that enable manufacturers to attain approval of building products. These code-approval procedures entail a rigorous engineering review of all aspects of product design.

6.3 STANDARD SPECIFICATIONS

Standard specifications are consensus documents sponsored by professional or trade associations to protect the public and to avoid, as much as possible, misuse of a product or method and thus promote the responsible use of the product. Examples of such specifications are the American Institute of Steel Construction (AISC) allowable stress design (ASD) and load and resistance factor design (LRFD) specifications; the American Iron and Steel Institute's (AISI's) "Specification for the Design of Cold-Formed Steel Structural Members," the Steel Joist Institute's "Standard Specifications Load Tables and Weight Tables for Steel Joists and Joist Girders," and the American Welding Society's (AWS's) "Structural Welding Code—Steel" (AWS D1.1).

Another important class of standard specifications defines acceptable standards of quality of building materials, standard methods of testing, and required workmanship in fabrication and erection. Many of these widely used specifications are developed by ASTM. As need arises, ASTM specifications are revised to incorporate the latest technological advances. The complete ASTM designation for a specification includes the year in which the latest revision was approved. For example, A588/A588M-91 refers to specification A588, adopted in 1991. The *M* indicates that it includes alternative metric units.

In addition to standards for product design and building materials, there are standard specifications for minimum design loads, e.g., "Minimum Design Loads for Buildings and Other Structures" (ASCE 7-88), American Society of Civil Engineers, and "Low-Rise Building Systems Manual," Metal Building Manufacturers Association.

It is advisable to use the latest editions of standards, recommended practices, and building codes.

6.4 BUILDING OCCUPANCY LOADS

Safe yet economical building designs necessitate application of reasonable and prudent design loads. Computation of design loads can require a complex analysis involving such considerations as building end use, location, and geometry.

6.4.1 Building Code–Specified Loads

Before initiating a design, engineers must become familiar with the load requirements of the local building code. All building codes specify minimum design loads. These include, when applicable, dead, live, wind, earthquake, and impact loads, as well as earth pressures.

Dead, floor live, and roof live loads are considered vertical loads and generally are specified as force per unit area, e.g., lb per ft^2 or kPa. These loads are often referred to as **gravity loads.** In some cases, concentrated dead or live loads also must be considered.

Wind loads are assumed to act normal to building surfaces and are expressed as pressures, e.g., psf or kPa. Depending on the direction of the wind and the geometry of the structure, wind loads may exert either a positive or negative pressure on a building surface.

All building codes and project specifications require that a building have sufficient strength to resist imposed loads without exceeding the design strength in any element of the structure. Of equal importance to design strength is the design requirement that a building be functional as stipulated by serviceability considerations. Serviceability requirements are generally given as allowable or permissible maximum deflections, either vertical or horizontal, or both.

6.4.2 Dead Loads

The dead load of a building includes weights of walls, permanent partitions, floors, roofs, framing, fixed service equipment, and all other permanent construction (Table 6.1). The American Society of Civil Engineers (ASCE) standard, "Minimum Design Loads for Buildings and Other Structures" (ASCE 7-88), gives detailed information regarding computation of dead loads for both normal and special considerations.

6.4.3 Floor Live Loads

Typical requirements for live loads on floors for different occupancies are summarized in Table 6.2. These minimum design loads may differ from requirements of local or state building codes or project specifications. The engineer of record for the building to be constructed is responsible for determining the appropriate load requirements.

Temporary or movable partitions should be considered a floor live load. For structures designed for live loads exceeding 80 lb per ft^2, however, the effect of partitions may be ignored, if permitted by the local building code.

Live Load Reduction.   Because of the small probability that a member supporting a large floor area will be subjected to full live loading over the entire area, building codes permit a reduced live load based on the areas contributing loads to the member (influence area). **Influence area** is defined as the floor area over which the influence surface for structural effects on a member is significantly different from zero. Thus the influence area for an interior column comprises the four surrounding bays (four times the conventional tributary area), and the influence area for a corner column is the adjoining corner bay (also four times the tributary area, or area next to the column and enclosed by the bay center lines). Similarly, the influence area for a girder is two times the tributary area and equals the panel area for a two-way slab.

TABLE 6.1 Minimum Design Dead Loads

Component	Load, lb/ft²
Ceilings	
Acoustical fiber tile	1
Gypsum board (per ⅛-in thickness)	0.55
Mechanical duct allowance	4
Plaster on tile or concrete	5
Plaster on wood lath	8
Suspended steel channel system	2
Suspended metal lath and cement plaster	15
Suspended metal lath and gypsum plaster	10
Wood furring suspension system	2.5
Coverings, roof, and wall	
Asbestos-cement shingles	4
Asphalt shingles	2
Cement tile	16
Clay tile (for mortar add 10 lb):	
Book tile, 2-in	12
Book tile, 3-in	20
Ludowici	10

Component	Load, lb/ft²
Waterproofing membranes:	
Bituminous, gravel-covered	5.5
Bituminous, smooth surface	1.5
Liquid applied	1.0
Single-ply, sheet	0.7
Wood sheathing (per inch thickness)	3
Wood shingles	3
Floor fill	
Cinder concrete, per inch	9
Lightweight concrete, per inch	8
Sand, per inch	8
Stone concrete, per inch	12
Floors and floor finishes	
Asphalt block (2-in), ½-in mortar	30
Cement finish (1-in) on stone-concrete fill	32
Ceramic or quarry tile (¾-in) on ½-in mortar bed	16
Ceramic or quarry tile (¾-in) on 1-in mortar bed	23

Component	Load, lb/ft²
Frame partitions	
Movable steel partitions	4
Wood or steel studs, ½-in gypsum board each side	8
Wood studs, 2 × 4, unplastered	4
Wood studs, 2 × 4, plastered one side	12
Wood studs, 2 × 4, plastered two sides	20
Frame walls	
Exterior stud walls:	
2 × 4 @ 16 in, ⅝-in gypsum, insulated, ⅜-in siding	11
2 × 6 @ 16 in, ⅝-in gypsum, insulated, ⅜-in siding	12
Exterior stud walls with brick veneer	48
Windows, glass, frame and sash	8
Masonry walls*	
Clay brick wythes:	
4 in	39
8 in	79

Material	lb/ft²
Roman	12
Spanish	19
Composition:	
Three-ply ready roofing	1
Four-ply felt and gravel	5.5
Five-ply felt and gravel	6
Copper or tin	1
Deck, metal, 20 ga	2.5
Deck, metal, 18 ga	3
Decking, 2-in wood (Douglas fir)	5
Decking, 3-in wood (Douglas fir)	8
Fiberboard, 1/2-in	0.75
Gypsum sheathing, 1/2-in	2
Insulation, roof boards (per inch thickness):	
Cellular glass	0.7
Fibrous glass	1.1
Fiberboard	1.5
Perlite	0.8
Polystyrene foam	0.2
Urethane foam with skin	0.5
Plywood (per 1/8-in thickness)	0.4
Rigid insulation, 1/2-in	0.75
Skylight, metal frame, 3/8-in wire glass	8
Slate, 3/16-in	7
Slate, 1/4-in	10

Material	lb/ft²
Concrete fill finish (per inch thickness)	12
Hardwood flooring, 7/8-in	4
Linoleum or asphalt tile, 1/4-in	1
Marble and mortar on stone-concrete fill	33
Slate (per inch thickness)	15
Solid flat tile on 1-in mortar base	23
Subflooring, 3/4-in	3
Terrazzo (1 1/2-in) directly on slab	19
Terrazzo (1-in) on stone-concrete fill	32
Terrazzo (1-in), 2-in stone concrete	32
Wood block (3-in) on mastic, no fill	10
Wood block (3-in) on 1/2-in mortar base	16

Floors, wood-joist (no plaster) double wood floor

Joist sizes, in	12-in spacing, lb/ft²	16-in spacing, lb/ft²	24-in spacing, lb/ft²
2 × 6	6	5	5
2 × 8	6	6	5
2 × 10	7	6	6
2 × 12	8	7	6

	lb/ft²
12 in	115
16 in	155

Hollow concrete masonry unit wythes:

	4 in	6 in	8 in	10 in	12 in
Wythe thickness (in)	4	6	8	10	12
Unit percent solid	70	55	52	50	48
Light weight units (105 pcf):					
No grout	22	27	35	42	49
48 o.c.		31	40	49	58
40 o.c. (Grout spacing)		33	43	53	63
32 o.c. (Grout spacing)		34	45	56	66
24 o.c.		37	49	61	72
16 o.c.		42	56	70	84
Full grout		57	77	98	119
Normal Weight Units (135 pcf):					
No grout	29	35	45	54	63
48 o.c.		33	50	61	72
40 o.c. (Grout spacing)		36	53	65	77
32 o.c. (Grout spacing)		38	55	68	80
24 o.c.		41	59	73	86
16 o.c.		47	66	82	98
Full grout		64	87	110	133

Solid concrete masonry unit wythes (incl. concrete brick):

	4	6	8	10	12
Wythe thickness, in	4	6	8	10	12
Lightweight units (105 pcf)	32	49	67	84	102
Normal weight units (135 pcf)	41	63	86	108	131

*Weights of masonry include mortar but not plaster. For plaster, add 5 lb/ft² for each face plastered. Values given represent averages. In some cases there is a considerable range of weight for the same construction.

TABLE 6.2 Minimum Design Live Loads

a. Uniformly distributed design live loads

Occupancy or use	Live load, lb/ft²	Occupancy or use	Live load, lb/ft²
Armories and drill rooms	150	Manufacturing	
Assembly areas and theaters		Light	125
Fixed seats (fastened to floor)	60	Heavy	250
Lobbies	100	Marquees and canopies	75
Movable seats	100	Office buildings[b]	
Platforms (assembly)	100	Lobbies	100
Stage floors	150	Offices	50
Balconies (exterior)	100	Penal institutions	
On one- and two-family residences only, and not exceeding 100 ft²	60	Cell blocks	40
		Corridors	100
Bowling alleys, poolrooms, and similar recreational areas	75	Residential	
		Dwellings (one- and two-family)	
Corridors		Uninhabitable attics without storage	10
First floor	100	Uninhabitable attics with storage	20
Other floors, same as occupancy served except as indicated		Habitable attics and sleeping areas	30
		All other areas	40
Dance halls and ballrooms	100	Hotels and multifamily buildings	
Decks (patio and roof)		Private rooms and corridors serving them	40
Same as area served, or for the type of occupancy accommodated		Public rooms, corridors, and lobbies serving them	100
Dining rooms and restaurants	100	Schools	
Fire escapes	100	Classrooms	40
On single-family dwellings only	40	Corridors above first floor	80
Garages (passenger cars only)	50	Sidewalks, vehicular driveways, and yards, subject to trucking[a]	250
For trucks and buses use AASHTO[a] lane loads (see Table 6.2b for concentrated-load requirements)		Stadium and arena bleachers[c]	100
Grandstands[c]	100	Stairs and exitways	100
Gymnasiums, main floors and balconies	100	Storage warehouses	
Hospitals		Light	125
Operating rooms, laboratories	60	Heavy	250
Private rooms	40	Stores	
Wards	40	Retail	
Corridors above first floor	80	First floor	100
Libraries		Upper floors	75
Reading rooms	60	Wholesale, all floors	125
Stack rooms[d]	150	Walkways and elevated platforms (other than exitways)	60
Corridors above first floor	80	Yards and terraces (pedestrians)	100

TABLE 6.2 Minimum Design Live Loads (*Continued*)

b. Concentrated live loads[e]

Location	Load, lb
Elevator machine room grating (on 4-in^2 area)	300
Finish, light floor-plate construction (on 1-in^2 area)	200
Garages:	
Passenger cars:	
Manual parking (on 20-in^2 area)	2,000
Mechanical parking (no slab), per wheel	1,500
Trucks, buses (on 20-in^2 area) per wheel	16,000
Office floors (on area 2.5 ft square)	2,000
Roof-truss panel point over garage, manufacturing, or storage floors	2,000
Scuttles, skylight ribs, and accessible ceilings (on area 2.5 ft square)	200
Sidewalks (on area 2.5 ft square)	8,000
Stair treads (on 4-in^2 area at center of tread)	300

c. Minimum design loads for materials

Material	Load, lb/ft^3	Material	Load, lb/ft^2
Aluminum, cast	165	Earth (not submerged) (*Continued*):	
Bituminous products:		Sand and gravel, dry, loose	100
Asphalt	81	Sand and gravel, dry, packed	120
Petroleum, gasoline	42	Sand and gravel, wet	120
Pitch	69	Gold, solid	1205
Tar	75	Gravel, dry	104
Brass, cast	534	Gypsum, loose	70
Bronze, 8 to 14% tin	509	Ice	57.2
Cement, portland, loose	90	Iron, cast	450
Cement, portland, set	183	Lead	710
Cinders, dry, in bulk	45	Lime, hydrated, loose	32
Coal, anthracite, piled	52	Lime, hydrated, compacted	45
Coal, bituminous or lignite, piled	47	Magnesium alloys	112
Coal, peat, dry, piled	23	Mortar, hardened:	
Charcoal	12	Cement	130
Copper	556	Lime	110
Earth (not submerged):		Riprap (not submerged):	
Clay, dry	63	Limestone	83
Clay, damp	110	Sandstone	90
Clay and gravel, dry	100	Sand, clean and dry	90
Silt, moist, loose	78		
Silt, moist, packed	96		

TABLE 6.2   Minimum Design Live Loads (*Continued*)

c. Minimum design loads for materials (Continued)

Material	Load, lb/ft^3	Material	Load, lb/ft^2
Sand, river, dry	106	Stone, ashlar (*Continued*):	
Silver	656	Sandstone	140
Steel	490	Shale, slate	155
Stone, ashlar:		Tin, cast	459
Basalt, granite, gneiss	165	Water, fresh	62.4
Limestone, marble, quartz	160	Water, sea	64

[a]American Association of State Highway and Transportation Officials lane loads should also be considered where appropriate.

[b]File and computer rooms should be designed for heavier loads; depending on anticipated installations. See also Corridors.

[c]For detailed recommendations, see American National Standard for Assembly Seating, Tents, and Air-Supported Structures, ANSI/NFPA 102.

[d]For the weight of books and shelves, assume a density of 65 pcf, convert it to a uniformly distributed area load, and use the result if it exceeds 150 lb/ft^2.

[e]Use instead of uniformly distributed live load, except for roof trusses, if concentrated loads produce greater stresses or deflections. Add impact factor for machinery and moving loads: 100% for elevators, 20% for light machines, 50% for reciprocating machines, 33% for floor or balcony hangers. For craneways, add a vertical force equal to 25% of the maximum wheel load; a lateral force equal to 10% of the weight of trolley·and lifted load, at the top of each rail; and a longitudinal force equal to 10% of maximum wheel loads, acting at top of rail.

The standard, "Minimum Design Loads for Buildings and Other Structures" (ASCE 7-88), American Society of Civil Engineers, permits a reduced live load L (lb per ft^2) computed from Eq. (6.1) for design of members with an influence area of 400 ft^2 or more:

$$L = L_o(0.25 + 15/\sqrt{A_I}) \tag{6.1}$$

where L_o = unreduced live load, lb per ft^2
A_I = influence area, ft^2

The reduced live load should not be less than $0.5L_o$ for members supporting one floor nor $0.4L_o$ for all other loading situations. If live loads exceed 100 lb per ft^2, and for garages for passenger cars only, design live loads may be reduced 20% for members supporting more than one floor. For members supporting garage floors, one-way slabs, roofs, or areas used for public assembly, no reduction is permitted if the design live load is 100 lb per ft^2 or less.

6.4.4   Concentrated Loads

Some building codes require that members be designed to support a specified concentrated live load in addition to the uniform live load. The concentrated live load may be assumed to be uniformly distributed over an area of 2.5 ft^2 and located to produce the maximum stresses in the members. Table 6.2b lists some typical loads that may be specified in building codes.

6.4.5   Pattern Loading

This is an arrangement of live loads that produces maximum possible stresses at a point in a continuous beam. The member carries full dead and live loads, but full live load may occur only in alternating spans or some combination of spans. In a high-rise building frame, maximum positive moments are produced by a checkerboard pattern of live load, i.e., by full live load on alternate spans horizontally and alternate bays vertically. Maximum negative moments at a joint occur, for most practical purposes, with full live loads only on the spans adjoining the joint. Thus pattern loading may produce critical moments in certain members and should be investigated.

6.5 ROOF LOADS

In northern areas, roof loads are determined by the expected maximum snow loads. However, in southern areas, where snow accumulation is not a problem, minimum roof live loads are specified to accommodate the weight of workers, equipment, and materials during maintenance and repair.

6.5.1 Roof Live Loads

Some building codes specify that design of flat, curved, or pitched roofs should take into account the effects of occupancy and rain loads and be designed for minimum live loads, such as those given in Table 6.3. Other codes require that structural members in flat, pitched, or curved roofs be designed for a live load L_r (lb per ft^2 of horizontal projection) computed from

$$L_r = 20R_1R_2 \geq 12 \tag{6.2}$$

where R_1 = reduction factor for size of tributary area
   = 1 for $A_t \leq 200$
   = $1.2 - 0.001A_t$ for $200 < A_t < 600$
   = 0.6 for $A_t \geq 600$

TABLE 6.3 Roof Live Loads (lb per ft^2) of Horizontal Projection*

Roof slope	Tributary loaded area, ft^2, for any structural member		
	0 to 200	201 to 600	Over 600
Flat or rise less than 4:12 Arch or dome with rise less than ⅛ of span	20	16	12
Rise 4:12 to less than 12:12 Arch or dome with rise ⅛ span to less than ⅜ span	16	14	12
Rise 12:12 or greater Arch or dome with rise ⅜ of span or greater	12	12	12

*As specified in "Low-Rise Building Systems Manual," Metal Building Manufacturers Association, Cleveland, Ohio.

A_t = tributary area, or area contributing load to the structural member, ft^2 (Sec. 6.4.3)

R_2 = reduction factor for slope of roof

= 1 for $F \le 4$

= $1.2 - 0.05F$ for $4 < F < 12$

= 0.6 for $F \ge 12$

F = rate of rise for a pitched roof, in/ft

= rise-to-span ratio multiplied by 32 for an arch or dome

6.5.2 Snow Loads

Determination of design snow loads for roofs is often based on the maximum ground snow load in a 50-year mean recurrence period (2% probability of being exceeded in any year). This load or data for computing it from an extreme-value statistical analysis of weather records of snow on the ground may be obtained from the local building code or the National Weather Service. Maps showing ground snow loads for various regions are presented in model building codes and standards, such as "Minimum Design Loads for Buildings and Other Structures" (ASCE 7-88), American Society of Civil Engineers. The map scales, however, may be too small for use for some regions, especially where the amount of local variation is extreme or high country is involved.

The "Low-Rise Building Systems Manual," Metal Building Manufacturers Association, Cleveland, Ohio, recommends that the design roof snow load be determined from

$$p_f = 0.7 p_g \tag{6.3}$$

where p_f = roof snow load, lb per ft^2

p_g = ground snow load for 50-year recurrence period, lb per ft^2

Some building codes and ASCE 7-88 specify a formula that takes into account the consequences of a structural failure in view of the end use of the building to be constructed and the wind exposure of the roof:

$$p_f = 0.7 C_e C_t I p_g \tag{6.4}$$

where C_e = wind exposure factor (Table 6.4)

C_t = thermal effects factor (Table 6.5)

I = importance factor for end use (Table 6.6)

In their provisions for roof design, codes and standards also allow for the effect of roof slopes, snow drifts, and unbalanced snow loads. The structural members should be investigated for the maximum possible stress that the loads might induce.

6.6   WIND LOADS

Wind loads are randomly applied dynamic loads. The intensity of the wind pressure on the surface of a structure depends on wind velocity, air density, orientation of the structure, area of contact surface, and shape of the structure. Because of the complexity involved in defining both the dynamic wind load and the behavior of an indeterminate steel structure when subjected to wind loads, the design criteria adopted by building codes and standards have been based on the application of an

TABLE 6.4 Exposure Factor for Eq. (6.4)

Nature of site*	Exposure factor C_e
A. Windy area with roof exposed on all sides with no shelter afforded by terrain, higher structures, or trees†	0.8
B. Windy areas with little shelter available†	0.9
C. Locations in which snow removal by wind cannot be relied on to reduce roof loads because of terrain, higher structures, or several trees nearby	1.0
D. Areas that do not experience much wind and where terrain, higher structures, or several trees shelter the roof†	1.1
E. Densely forested areas that experience little wind, with roof located tight in among conifers	1.2

*The conditions discussed should be representative of those which are likely to exist during the life of the structure. Roofs that contain several large pieces of mechanical equipment or other obstructions do not qualify for siting category A.

†Obstructions within a distance of $10h_o$ provide shelter, where h_o is the height of the obstruction above the roof level. If the obstruction is created by deciduous trees, which are leafless in winter, C_e may be reduced by 0.1.

TABLE 6.5 Thermal Factor for Eq. (6.4)

Thermal condition*	Thermal factor C_t
Heated structure	1.0
Structure kept just above freezing	1.1
Unheated structure	1.2

*These conditions should be representative of those which are likely to exist during the life of the structure.

TABLE 6.6 Importance Factor for Eq. (6.4)

Category	Description of buildings and structures in category	Importance factor I
I	Those not included below	1.0
II	Primary occupancy permits 300 persons or more to congregate in one area	1.1
III	Essential facilities, such as hospitals, fire and police stations, communication centers, utilities, military, shelters	1.2
IV	Low hazard to human life, such as minor storage facilities, agricultural buildings, temporary structures	0.8

equivalent static wind pressure. This equivalent static design wind pressure p (psf) is defined in a general sense by

$$p = qGC_p \tag{6.5}$$

where q = velocity pressure, psf
  G = gust response factor to account for fluctuations in wind speed
  C_p = pressure coefficient or shape factor that reflects the influence of the wind on the various parts of a structure

Velocity pressure is computed from

$$q = K_z(IV)^2 \tag{6.6}$$

where K_z = velocity exposure coefficient that accounts for variation of velocity with height and ground roughness
  I = importance factor associated with the type of occupancy (Table 6.7)
  V = basic wind speed (Fig. 6.1), mi/h, but not less than 70 mi/h

Velocity pressures due to wind to be used in building design vary with type of terrain, distance above ground level, importance of building, likelihood of hurricanes, and basic wind speed recorded near the building site. The wind pressures are assumed to act horizontally on the building area projected on a vertical plane normal to the wind direction.

The basic wind speed used in design is the fastest-mile wind speed recorded at a height of 10 m (32.8 ft) above open, level terrain with a 50-year mean recurrence interval.

Unusual wind conditions often occur over rough terrain and around ocean promontories. Basic wind speeds applicable to such regions should be selected with the aid of meteorologists and the application of extreme-value statistical analysis to anemometer readings taken at or near the site of the proposed building. Generally, however, minimum basic wind velocities are specified in local building codes and in national model building codes but should be used with discretion, because actual velocities at a specific site and on a specific building may be significantly larger. In the absence of code specifications and reliable data, basic wind speed at a height of 10 m above grade may be estimated from Fig. 6.1.

TABLE 6.7 Importance Factor for Eq. (6.6)

Category	Descriptions of structures in category	Importance factor I	
		100 mi from hurricane oceanline* and in other areas	At hurricane oceanline
I	Those not included below	1.00	1.05
II	Primary occupancy permits 300 or more persons in one area	1.07	1.11
III	Hospital, fire, and police stations and other essential facilities	1.07	1.11
IV	Low hazard to human life	0.95	1.00

*I may be determined by linear interpolation between a hurricane oceanline, such as the Atlantic Ocean and the Gulf of Mexico, and 100 mi inland.

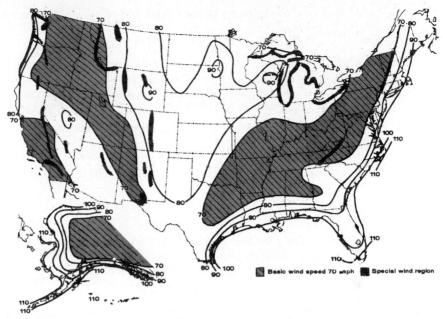

FIGURE 6.1 Contours on map of the United States show basic wind speeds (fastest-mile speeds recorded 10 m above ground) for open terrain and grasslands with 50-year mean recurrence interval.

For design purposes, wind pressures should be determined in accordance with the degree to which terrain surrounding the proposed building exposes it to the wind. Exposures may be classified as follows:

Exposure A applies to centers of large cities, where for at least one-half mile upwind from the building the majority of structures are over 70 ft high and lower buildings extend at least one more mile upwind.

Exposure B applies to wooded or suburban terrain or to urban areas with closely spaced buildings mostly less than 70 ft high, where such conditions prevail upwind for a distance from the building of at least 1500 ft or 10 times the building height.

Exposure C exists for flat, open country or exposed terrain with obstructions less than 30 ft high.

Exposure D applies to flat, unobstructed areas exposed to wind blowing over a large expanse of water with a shoreline at a distance from the building of not more than 1500 ft or 10 times the building height.

For design purposes also, the following formulas may be used to determine, for heights z (in feet) greater than 15 ft above ground, a pressure coefficient K for converting wind speeds to pressures.

For exposure A, for heights up to 1500 ft above ground level,

$$K = 0.000517 \left(\frac{z}{32.8}\right)^{2/3}$$
(6.7)

For z less than 15 ft, $K = 0.00031$.

For exposure B, for heights up to 1200 ft above ground level,

$$K = 0.00133 \left(\frac{z}{32.8}\right)^{4/9} \tag{6.8}$$

For z less than 15 ft, $K = 0.00095$.
   For exposure C, for heights up to 900 ft above ground level,

$$K = 0.00256 \left(\frac{z}{32.8}\right)^{2/7} \tag{6.9}$$

For z less than 15 ft, $K = 0.0020$.
   For exposure D, for heights up to 700 ft above ground level,

$$K = 0.00357 \left(\frac{z}{32.8}\right)^{1/5} \tag{6.10}$$

For z less than 15 ft, $K = 0.0031$.
   Wind pressures on low buildings are different at a specific elevation from those on tall buildings. Hence building codes may give different formulas for pressures for the two types of construction. In any case, however, design wind pressure should be a minimum of 10 psf.

Multistory Buildings.   For design of the main wind force–resisting system of ordinary, rectangular, multistory buildings, the design pressure at any height z (ft) above ground may be computed from

$$p_{zw} = G_o C_{pw} q_z \tag{6.11}$$

where p_{zw} = design wind pressure on windward wall, psf
   G_o = gust response factor
   C_{pw} = external pressure coefficient
   q_z = velocity pressure computed from Eq. (6.6)

For windward walls, C_{pw} may be taken as 0.8. For side walls, C_{pw} may be assumed as -0.7 (suction). For roofs and leeward walls, the design pressure at elevation z is

$$p_{zl} = G_o C_p q_h \tag{6.12}$$

where p_{zl} = design pressure on roof or leeward wall, psf
   C_p = external pressure coefficient for roof or leeward wall
   q_h = velocity pressure at mean roof height h

In these equations, the gust response factor may be taken approximately as

$$G_o = 0.65 + \frac{8.58D}{(h/30)^n} \geq 1 \tag{6.13}$$

where D = 0.16 for exposure A, 0.10 for exposure B, 0.07 for exposure C, and 0.05 for exposure D
   n = $\frac{1}{3}$ for exposure A, $\frac{2}{9}$ for exposure B, $\frac{1}{7}$ for exposure C, and 0.1 for exposure D
   h = mean roof height, ft

For leeward walls subjected to suction, C_p depends on the ratio of the depth d to width b of the building and may be assumed as follows:

$$d/b = 1 \text{ or less} \qquad 2 \qquad 4 \text{ or more}$$

$$C_p = -0.5 \qquad -0.3 \qquad -0.2$$

The negative sign indicates suction. Table 6.8 lists values of C_p for pressures on roofs.

Flexible Buildings. These are structures with a fundamental natural frequency less than 1 Hz or with a ratio of height to least horizontal dimension (measured at midheight for buildings with tapers or setbacks) exceeding 5. For such buildings, the main wind force–resisting system should be designed for a pressure on windward walls at any height z (ft) above ground computed from

$$p_{zw} = G_f C_{pw} q_z \tag{6.14}$$

where G_f is the gust response factor determined by analysis of the system taking into account its dynamic properties. For leeward walls of flexible buildings,

$$p_{zl} = G_f C_p q_h \tag{6.15}$$

(In ASCE 7-88, G_f is represented by $\overline{G}$. Requiring a knowledge of the fundamental frequency, structural damping characteristics, and type of exposure of the building, the formula for G_f is complicated, but computations may be simplified somewhat by use of tables and charts in the standard.)

One-Story Buildings. For design of the main wind force–resisting system of rectangular, one-story buildings, the design pressure at any height z (ft) above ground

TABLE 6.8 External Pressure Coefficients C_p for roofs*

Flat roofs						-0.7
Wind parallel to ridge of sloping roof						
$\quad h/b$ or $h/d \leq 2.5$						-0.7
$\quad h/b$ or $h/d > 2.5$						-0.8
Wind perpendicular to ridge of sloping roof, at angle θ with horizontal						
$\quad$ Leeward side						-0.7
$\quad$ Windward side						

	Slope of roof θ, deg					
h/d	10	20	30	40	50	60 or more
0.3 or less	0.2	0.2	0.3	0.4	0.5	
0.5	-0.9	-0.75	-0.2	0.3	0.5	0.01θ
1.0	-0.9	-0.75	-0.2	0.3	0.5	
1.5 or more	-0.9	-0.9	-0.9	0.35	0.21	

*h = height of building, ft; d = depth, ft, of building in direction of wind; b = width, ft, of building transverse to wind.
$\quad$ Based on data in ASCE 7-88.

may be computed for windward walls from

$$p_{zw} = (G_oC_p + C_{p1})q_z \qquad (6.16)$$

where C_{p1} = 0.75 if the percentage of openings in one wall exceeds that of other walls by 10% or more
= 0.25 for all other cases

For roofs and leeward walls, the design pressure at elevation z is

$$p_{zl} = G_oC_pq_h - C_{p2}q_z \qquad (6.17)$$

where C_{p2} = +0.75 or −0.25 if the percentage of openings in one wall exceeds that of other walls by 10% or more
= ±0.25 for all other cases

(Positive signs indicate pressures acting toward a wall; negative signs indicate pressures acting away from the wall.)

Codes and standards may present the gust factors and pressure coefficients in different formats. Coefficients from different codes and standards should not be mixed.

Designers should exercise judgment in selecting wind loads for a building with unusual shape, response-to-load characteristics, or site exposure where channeling of wind currents or buffeting in the wake of upwind obstructions should be considered in design. Wind-tunnel tests on a model of the structure and its neighborhood may be helpful in supplying design data. (See also Sec. 9.)

("Minimum Design Loads for Buildings and Other Structures," ASCE 7-88; and K. C. Mehta et al., *Guide to the Use of the Wind Load Provisions*," American Society of Civil Engineers.)

6.7   SEISMIC LOADS

Earthquakes have occurred in many states. Figure 6.2 shows a map of the United States that has been partitioned into zones that reflect the relative severity of seismic activity, as indicated in "Minimum Design Loads for Buildings and Other Structures" (ASCE 7-88), American Society of Civil Engineers.

The engineering approach to seismic design differs from that for other load types. For live, wind, or snow loads, the intent of a structural design is to preclude structural damage. However, to achieve an economical seismic design, codes and standards permit local yielding of a structure during a major earthquake. **Local yielding** absorbs energy but results in permanent deformations of structures. Thus seismic design incorporates not only application of anticipated seismic forces but also use of structural details that ensure adequate ductility to absorb the seismic forces without compromising the stability of structures. Provisions for this are included in the AISC specifications for structural steel for buildings.

The forces transmitted by an earthquake to a structure result from vibratory excitation of the ground. The vibration has both vertical and horizontal components. However, it is customary for building design to neglect the vertical component because most structures have reserve strength in the vertical direction due to gravity-load design requirements.

Seismic requirements in building codes and standards attempt to translate the complicated dynamic phenomenon of earthquake force into a simplified equivalent static force to be applied to a structure for design purposes. For example, ASCE

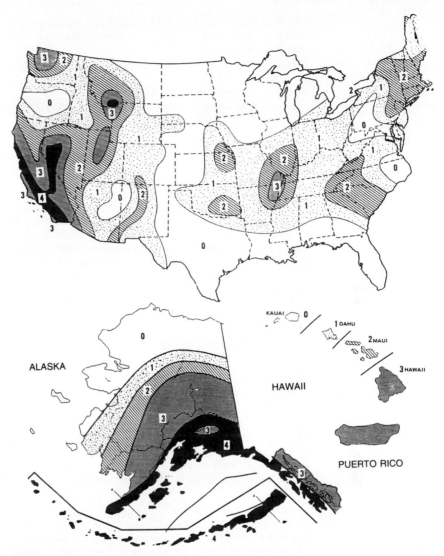

FIGURE 6.2 Map of the United States partitioned into zones that indicate the relative severity of seismic activity.

Seismic Coefficients

Seismic Zone	Coefficient Z
4	1
3	$3/4$
2	$3/8$
1	$3/16$
0	$1/8$

7-88 stipulates that the total lateral force, or base shear, V (kips) acting in the direction of each of the principal axes of the main structural system should be computed from

$$V = ZIKCSW \qquad (6.18)$$

where Z = seismic zone coefficient (Fig. 6.2)
I = occupancy importance factor (Table 6.9)
K = horizontal shear force factor (Table 6.10)
C = seismic coefficient = $\frac{1}{15}T^{0.5} < 0.12$
S = soil profile coefficient (Table 6.11)
T = fundamental elastic period of vibration, s
W = total dead load, kips

The total dead load W includes the weight of all permanent features of the structure. In storage and warehouse structures, the dead load also should include 25% of the floor live load. Where the ground snow load is greater than 30 psf, the snow load also should be included in the definition of W.

A rigorous evaluation of the fundamental elastic period T requires consideration of the intensity of loading and the response of the structure to the loading. To expedite design computations, T, seconds, may be determined from Eq. (6.19) or (6.20).

For isolated shear walls or for braced frames,

$$T = 0.05h_n/D_s \qquad (6.19)$$

For moment-resisting frames,

$$T = C_T h_n^{3/4} \qquad (6.20)$$

where C_T = 0.035 for steel frames
h_n = height from the base to the uppermost level in the main structure, ft
D_s = longest dimension of a shear wall or braced frame in a direction parallel to the applied forces, ft

The lateral force V should be distributed over the height of the structure as concentrated loads at each floor level or story. The concentrated load at the top of the building F_t (kips) is determined from

$$F_t = 0.07TV \qquad (6.21)$$

TABLE 6.9   Occupancy Importance Factor for Eq. (6.18)*

Category	Description of buildings and structures in category	Importance factor I
I	Those not included below	1.00
II	Primary occupancy permits 300 persons or more to congregate in one area	1.25
III	Essential facilities, such as hospitals, fire and police stations, communication centers, utilities, military, shelters	1.50
IV	Low hazard to human life, such as minor storage facilities, agricultural buildings, temporary structures	—

*As specified in "Minimum Design Loads for Buildings and Other Structures" (ASCE 7-88), American Society of Civil Engineers.

TABLE 6.10 Horizontal Force Factor K for Eq. (6.18)

Lateral force–resisting construction	K
Bearing-wall system providing support for a major portion of the gravity loads and with:	
Reinforced concrete shear walls*	1.33
Braced frames*	1.33
Masonry shear walls*	1.33
One-, two-, or three-story light wood or metal frame-wall system	1.00
Space-frame system† providing support for gravity loads with seismic force resistance provided by shear walls* or braced frames*	1.00
Moment-resisting-frame system providing support for gravity loads with seismic force resistance provided by moment-resisting joints and with:	
Ordinary steel frames	1.00
Special frames‡	0.67
Intermediate concrete frames§	1.25
Dual system with an essentially complete space frame providing support for gravity loads and with seismic force resistance provided by a combination of a special moment-resisting-frame system and shear walls or braced frames	0.80
With intermediate concrete frames§	1.00
Elevated tanks plus full content, where tanks are supported on four or more cross-braced legs and are not supported on a building	2.50
Structures other than buildings	2.00

*A shear wall is a wall capable of resisting lateral forces parallel to itself. A braced frame is a truss system or equivalent capable of resisting lateral forces and in which members carry primarily axial stresses.

†A space frame is a three-dimensional structural system without bearing walls and composed of interconnected members that are laterally supported so as to function as a complete self-contained unit.

‡A special moment-resisting space frame supports gravity loads and is composed of ductile members and joints capable of resisting seismic loads primarily by flexure. The AISC specifications for structural steel for buildings contain requirements for design of such members and joints. In dual systems, the moment-resisting frame should be capable of resisting at least 25% of the design seismic forces.

§An intermediate concrete frame resists seismic loads in the same manner as a special frame but has semiductile joints. This type of frame is not permitted in Zones 3 and 4.

TABLE 6.11 Soil Profile Coefficient S^* for Eq. (6.18)

Soil profile type	S
S_1—rock, material with a shear-wave velocity exceeding 2500 ft/s, or stiff, stable deposits of sand, gravel, or stiff clay overlying rock at a depth less than 200 ft	1.0
S_2—stable deposits of sand, gravel, or stiff clay overlying rock at a depth exceeding 200 ft	1.2
S_3—sand or weak to medium-stiff clays 30 ft or more deep (the clay layer may include layers of sand or gravel)	1.5

*For sites where soil characteristics are not known in detail or do not fit the preceding descriptions, profile S_2 or S_3, whichever gives the larger value for CS should be used.

F_t need not exceed $0.25V$ and may be taken as zero when $T \leq 0.7s$. A concentrated force F_x (kips) applied at each level x, including the top, is computed from

$$F_x = (V - F_t) \frac{w_x h_x}{\Sigma w_i h_i} \tag{6.22}$$

where w_x, w_i = portion of W assigned to level x, i
h_x, h_i = height of level x or i above the base, ft

For parts of structures and nonstructural components, as described in Table 6.12, including their anchorage to the main structural system, the design lateral force F_p (kips) is given by

$$F_p = ZIC_p W_p \tag{6.23}$$

where C_p = horizontal force factor (Table 6.12)
W_p = weight of component, kips

Provision also should be made in design of structural framing for horizontal torsion, overturning effects, and building drift.

Provisions for seismic design are continually evolving; hence designers should be alert for changes in criteria.

TABLE 6.12   Horizontal Force Factor C_p^*

Part or portion of building	Direction of horizontal force	C_p
Exterior bearing and nonbearing walls; interior bearing walls and partitions; interior nonbearing walls and partitions; masonry or concrete fences over 6 ft high	Normal to flat surface	0.3
For elements laterally self-supported only at ground level	Normal to flat surface	0.2
Cantilever elements: Parapets	Normal to flat surface $\}$	0.8
Chimneys or stacks	Any direction	
Exterior and interior ornamentations and appendages	Any direction	0.8
When connected to part of or housed within a building: Penthouses, anchorage and supports for chimneys, stacks, and tanks, including contents Storage racks with upper storage level at more than 8 feet in height, plus contents All equipment or machinery Fire sprinkler system	Any direction	0.3
Suspended ceiling framing systems (applies to Seismic Zones 2, 3, and 4 only)	Any direction	0.3
Connections for prefabricated structural elements other than walls, with positive anchorage and with force applied at center of gravity of assembly	Any direction	0.3
Access floor systems	Any direction	0.3

*As specified in "Minimum Design Loads for Buildings and Other Structures" (ASCE 7-88), American Society of Civil Engineers.

(N. M. Newmark and E. Rosenblueth, *Fundamentals of Earthquake Engineering*, Prentice-Hall, Englewood Cliffs, N.J.: S. Okamoto, *Introduction to Earthquake Engineering*, John Wiley & Sons, Inc., New York; F. S. Merritt and J. Ricketts, *Building Design and Construction Handbook*, 5th Ed., McGraw-Hill, Inc., New York.)

6.8 IMPACT LOADS

The live loads specified in building codes and standards include an allowance for ordinary impact loads. Where structural members will be subjected to unusual vibrations or impact loads, such as those described in Table 6.13, provision should be made for them in design of the members. Most building codes specify a percentage increase in live loads to account for impact loads. Impact loads for cranes are given in Art. 6.9.

6.9 CRANE-RUNWAY LOADS

Design of structures to support cranes involves a number of important considerations, such as determination of maximum wheel loads, allowance for impact, effects due to multiple cranes operating in single or multiple aisles, and traction and braking forces, application of crane stops, and cyclic loading and fatigue. The following suggested design criteria extracted from the "Low-Rise Building Systems Manual," Metal Building Manufacturers Association, Cleveland, Ohio, apply to cranes classified as infrequent service, light service, moderate service, and heavy service. For criteria for cranes classified as severe and continuous severe service, consult "Specification for Design and Construction of Mill Buildings" (AISE Standard No. 13), Association of Iron and Steel Engineers, Pittsburgh, Pa.

Computation of crane design loads encompass the following three auxiliary loads:

R_C = rated capacity of the crane
W_h = weight of hoist with trolley
W_t = total weight of the crane including bridge with end trucks, hoist with trolley, and cab with walkway for cab-operated cranes

The **wheel load** is the sum of the vertical auxiliary and collateral crane loads without impact. For a bridge crane, the maximum wheel load is attained with the trolley loaded at the rated capacity and positioned close to the crane girder to be designed.

TABLE 6.13 Minimum Percentage Increase in Live Load on Structural Members for Impact

Type of member	Source of impact	Percent
Supporting	Elevators and elevator machinery	100
Supporting	Light machines, shaft or motor-driven	20
Supporting	Reciprocating machines or power-driven units	50
Hangers	Floors or balconies	33

When the maximum wheel load W_L is not specified, it may be conservatively approximated by

$$W_L = R + 0.5(W_h + W_t)/N_b \qquad (6.24)$$

where N_b is the number of end-truck wheels at one end of the bridge.

For an underhung monorail crane, the maximum wheel load may be calculated from

$$W_L = R + W_h/N_m \qquad (6.25)$$

where N_m is the number of trolley wheels.

A crane induces impact on its supporting structure. Therefore, except for manually powered cranes, for which impact forces may be neglected, the maximum live loads should be increased by percentages as follows to allow for impact:

For vertical impact on runway girders, connections, and support brackets, 25%, but it may be ignored for building frames, runway columns, and building foundations.

For lateral forces on runway girders, 20% of $(R + W_h)$, half of which should be applied at the top of each rail.

For longitudinal forces caused by movement and braking parallel to the runway girders, to be resisted by a longitudinal bracing system, 10% of the maximum wheel loads applied at the top of the rail.

Runway girders, their connections, support brackets, building frames and runway columns should be designed for the positions of cranes that produce maximum stresses. The crane runway also should be designed to resist forces that may be exerted by the crane against runway stops. For a crane aisle inside a building, the runway stop should be designed for a force equal to 10% of $(W_t + W_h)$ acting longitudinally at the center of the end-truck bumper. For stops not inside a building, the design loading should include wind on the surface of the crane. For moderate-to high-use cranes, fatigue should be considered (Art. 6-22).

6.10 RESTRAINT LOADS

This type of loading is caused by changes in dimensions or geometry of structures or members due to the behavior of material, type of framing, or details of construction used. Stresses induced must be considered where they may increase design requirements. They may occur as the result of foundation settlement or temperature or shrinkage effects that are restrained by adjoining construction or installations.

6.11 COMBINED LOADS

The types of loads described in Arts. 6.4 to 6.10 may act simultaneously. Maximum stresses or deformations, therefore, may result from some combination of the loads. Building codes specify various combinations that should be investigated, depending on whether allowable stress design (ASD) or load and resistance factor design (LRFD) is used.

For ASD, the following are typical combinations that should be investigated:

1. D

2. $D + L + (L_r \text{ or } S \text{ or } R)$

3. $0.75[D + L + (L_r \text{ or } S \text{ or } R) + T]$

4. $D + A$

5. $0.75[D + (W \text{ or } E)]$

6. $0.75[D + (W \text{ or } E) + T]$

7. $D + A + (S \text{ or } 0.5W \text{ or } E)$

8. $0.75[D + L + (L_r \text{ or } S \text{ or } R) + (W \text{ or } E)]$

9. $0.75(D + L + W + 0.5S)$

10. $0.75(D + L + 0.5W \text{ or } S)$

11. $0.66[D + L + (L_r \text{ or } S \text{ or } R) + (W \text{ or } E) + T]$

where D = dead load
L = floor live load, including impact
L_r = roof live load
A = loads from cranes and materials handling systems
S = roof snow load
R = rain load
W = wind load
E = earthquake load
T = restraint loads

Instead of the factors 0.75 and 0.66, allowable stresses may be increased one-third and one-half, respectively.

S in load combination 7 may be taken as zero for snowloads ≤ 13 psf, as $0.5S$ for $13 <$ snowload ≤ 31 psf, and as $0.75S$ for snowloads > 31 psf. For the case of $D + E + A$ for load combination 7, the auxiliary crane loads should include only the weight of the crane, including the bridge with end trucks and hoist and trolley.

For LRFD, the "Load and Resistance Factor Design Specification," American Institute of Steel Construction, prescribes the following factored loads:

1. $1.4D$

2. $1.2D + 1.6L + 0.5(L_r \text{ or } S \text{ or } R)$

3. $1.2D + 1.6(L_r \text{ or } S \text{ or } R) + (0.5L \text{ or } 0.8W)$

4. $1.2D + 1.3W + 0.5L + 0.5(L_r \text{ or } S \text{ or } R)$

5. $1.2D + 1.5E + (0.5L \text{ or } 0.2S)$

6. $0.9D - (1.3W \text{ or } 1.5E)$

In the above combinations, R is the load due to initial rainwater or ice, exclusive of ponding. As with the ASD load combinations, the most critical load combination may occur when one or more of the loads are not acting.

6.12 ASD AND LRFD SPECIFICATIONS

The American Institute of Steel Construction (AISC) has developed design specifications for structural steel with two different design approaches: "Specification for Structural Steel Buildings—Allowable Stress Design (ASD) and Plastic Design" and "Load and Resistance Factor Design (LRFD) Specification for Structural Steel Buildings." Building codes either adopt by reference or incorporate both these approaches. It is the prerogative of the designer to select the approach to employ; this decision is generally based on economics. The approaches should not be mixed.

ASD.   The AISC specification for ASD establishes allowable unit stresses that, under service loads on a structure, may not be exceeded in structural members or their connections. Allowable stresses incorporate a safety factor to compensate for uncertainties in design and construction. Common allowable unit stresses of the AISC ASD specification are summarized in Table 6.14.

LRFD.   The AISC specification for LRFD requires that factors be applied to both service loads and the nominal resistance (strength) of members and connections. To account for uncertainties in estimating the service loads, load factors generally greater than unity are applied to them (Art. 6.11). To reflect the variability inherent in predictions of the strength of a member or connection, the nominal resistance R_n is multiplied by a resistance factor ϕ less than unity. To ensure that a member or connection has sufficient strength to support the service loads, the service loads multiplied by the appropriate load factors (factored loads) should not exceed the design strength ϕR_n. Table 6.14 summarizes the formulas for design strength specified in the AISC LRFD specification.

6.13 AXIAL TENSION

The AISC LRFD specification gives the design strength P_n (kips) of a tension member as

$$\phi_t P_n = 0.9 F_y A_g \le 0.75 F_u A_e \qquad (6.26)$$

where A_e = effective net area, in^2
A_g = gross area of member, in^2
F_y = specified minimum yield strength, ksi
F_u = specified minimum tensile strength, ksi
ϕ_t = resistance factor for tension

For ASD, the allowable unit tension stresses are $0.60 F_y$ on the gross area and $0.50 F_u$ on the effective net area.

The effective net area A_e of a tension member for both LRFD and ASD is defined as follows, with A_n = net area (in^2) of the member:

When the load is directly introduced by connectors into each of the elements of a cross section,

$$A_e = A_n \qquad (6.27)$$

For a bolted connection, when the load is introduced into some but not all of the elements of a cross section,

$$A_e = U A_n \qquad (6.28)$$

TABLE 6.14 Design Strength and Allowable Stresses for W-Shape Structural Steel Members for Buildings*

Type of stress or failure	LRFD design strength	ASD allowable unit stresses
Tension (Art. 6.13):		
Fracture in net section	$0.75F_u A_e$	$0.5F_u A_e$
Yielding in gross section	$0.9F_y A_g$	$0.6F_y A_g$
Shear (Art. 6.14.1):		
Shear in beam web	$0.54F_y A_w$	$0.4F_y A_w$
For fasteners and welds (Arts. 6.14.2 and 6.14.3)		
Compression:		
Axial load (Sec. 6.16.2)	$0.85F_{cr}A_g$	$F_a A_g$

Compression axial load, LRFD:

$$\text{For } \lambda_C > 1.5:$$
$$F_{cr} = 0.658^{\lambda_c^2}F_y$$
$$\lambda_c = \frac{KL}{r\pi}\sqrt{\frac{F_y}{E}}$$
$$\text{For } \lambda_C > 1.5:$$
$$F_{cr} = [0.877/\lambda^2]F_y$$

Compression axial load, ASD:

$$\text{For } KL/r \le C_c:$$
$$F_a = \frac{\left[1 - \dfrac{(KL/r)^2}{2C_c^2}\right]F_y}{\dfrac{5}{3} + \dfrac{3(KL/r)}{8C_c} - \dfrac{(KL/r)^3}{8C_c^3}}$$
$$\text{For } KL/r > C_c:$$
$$F_a = \frac{12\pi^2 E}{23(KL/r)^2}$$

	LRFD design strength	ASD allowable unit stresses
Bending:		
Minor axis bending (Art. 6.17)	$0.9F_y Z_y$	$0.75F_y S_y$
Major axis bending for compact shape (Art. 6.17)	$0.9M_n$	$F_b S_x$

Bending, LRFD:

$$\text{For } L_b \le L_p:$$
$$M_n = M_p$$
$$\text{For } L_p < L_b < L_r:$$
$$M_n = C_b\left[M_p - (M_p - M_r)\left(\frac{L_b - L_p}{L_r - L_p}\right)\right] \le M_p$$
$$\text{For } L_b > L_r:$$
$$M_n = M_{cr} \le C_b M_r$$
$$M_{cr} = C_b\frac{\pi}{L_b}$$
$$\times \sqrt{EI_y GJ + \left(\frac{\pi E}{L_b}\right)^2 I_y C_w}$$

Bending, ASD:

$$\text{For } L_b < L_c:$$
$$F_b = 0.66F_y$$
$$\text{For } L_b > L_c:$$
$$F_b = \frac{12 \times 10^3 C_b}{Ld/A_f} \le 0.60\ F_y$$
$$\text{For } \sqrt{\frac{102 \times 10^3 C_b}{F_y}} \le \frac{L}{r_T}$$
$$\le \sqrt{\frac{510 \times 10^3 C_b}{F_y}}:$$
$$F_b = \left[\frac{2}{3} - \frac{F_y(L/r_T)^2}{1530 \times 10^3 C_b}\right]F_y$$
$$\le 0.60F_y$$
$$\text{For } \frac{L}{r_T} \ge \sqrt{\frac{510 \times 10^3 C_b}{F_y}}:$$
$$F_b = \frac{170 \times 10^3 C_b}{(L/r_T)^2} \le 0.60F_y$$

Bearing (Art. 6.18)

*For definitions of symbols, see articles referenced in the table.

For a welded connection, when the load is introduced into some but not all of the elements of a cross section,

$$A_e = UA_g \tag{6.29}$$

U is defined by the following:

$U = 0.90$ for W, M, or S shapes with width of flanges at least two-thirds the depth of section and for structural tees cut from these shapes if connection is to the flange.

$U = 0.85$ for W, M, or S shapes not meeting the preceding conditions, for structural tees cut from these shapes, and for all other shapes and built-up sections. Bolted or riveted connections should have at least three fasteners per line in the direction of applied force.

$U = 0.75$ for all members with bolted or riveted connections with only two fasteners per line in the direction of applied force.

When load is transmitted through welds transverse to the load to some but not all of the cross-sectional elements of W, M, or S shapes or structural tees cut from them, A_e = the area of the directly connected elements.

When load is transmitted to a plate by welds along both edges at its end, the length of the welds should be at least equal to the width of the plate. A_e is given by Eq. (6.29) with $U = 1.00$ when $l > 2w$, 0.87 for $2w > l > 1.5w$, and 0.75 for $1.5w > l > w$, where l = length (in) of weld and w = plate width (distance between edge welds, in).

For design of built-up tension members, see Art. 6.29.

Because of stress concentrations around holes, the AISC specifications establish stringent requirements for design of eyebars and pin-connected members. The tensile strength of pin-connected members for LRFD is given by

$$\phi P_n = 0.75 P_n = 1.50 t b_{eff} F_u \tag{6.30}$$

where t = thickness of pin-connected plate, in
$b_{eff} = 2t + 0.63 \le d'$
d' = distance from the edge of the pin hole to the edge of the plate in the direction normal to the applied force, in

See the AISC LRFD specification for other applicable requirements.

Equations (6.26) and (6.30) assume static loads. Design strength may have to be decreased for alternating or cyclic loading (Art. 6.24).

For allowable tension in welds, see Art. 6.14.3.

The design strength of rivets, bolts, and threaded parts is given in Table 6.15. High-strength bolts required to support loads in direct tension should have a large enough cross section that their average tensile stress, computed for the nominal bolt area and independent of any initial tightening force, will not exceed the appropriate design stress in Table 6.15. In determining the loads, tension resulting from prying action produced by deformation of the connected parts should be added to other external loads.

(See also Arts. 6.15 and 6.20 to 6.24. For built-up tension members, see Art. 6.29.)

6.14 SHEAR

For beams and plate girders, the web area for shear calculations A_w (in²) is the product of the overall depth, d (in) and thickness, t (in) of the web. The AISC LRFD and ASD specifications for structural steel for buildings specify the same nominal equations but present them in different formats.

TABLE 6.15 Design Tensile Strength of Fasteners*

	LRFD	ASD
Description of Fasteners	Nominal strength, ksi	Allowable tension, F_t, ksi
A307 bolts	45.0	20.0
A325 bolts, when threads are *not* excluded from shear planes	90.0	44.0
A325 bolts, when threads *are* excluded from shear planes	90.0	44.0
A490 bolts, when threads are *not* excluded from shear planes	112.5	54.0
A490 bolts, when threads *are* excluded from the shear planes	112.5	54.0
Threaded parts, when threads are *not* excluded from the shear planes	$0.75F_u$	$0.33F_u$
Threaded parts, when threads *are* excluded from the shear planes	$0.75F_u$	$0.33F_u$
A502, grade 1, hot-driven rivets	45.0	23.0
A502, grades 2 and 3, hot-driven rivets	60.0	29.0

*Resistance factor $\phi = 0.75$.

6.14.1 Shear in Webs

For LRFD, the design shear strength ϕV_n (kips) is given by Eqs. (6.31) to (6.33) with $\phi = 0.90$. For $h/t \leq 187\sqrt{k/F_y}$,

$$\phi V_n = 0.54 F_y A_w \tag{6.31}$$

For $187\sqrt{k/F_y} < h/t \leq 234\sqrt{k/F_y}$,

$$\phi V_n = 0.54 F_y A_w \frac{187\sqrt{k/F_y}}{h/t} \tag{6.32}$$

For $h/t > 234\sqrt{k/F_y}$,

$$\phi V_n = A_w \frac{23,760}{(h/t)^2} \tag{6.33}$$

where h = clear distance between flanges less the fillet or corner radius at each flange for a rolled shape and the clear distance between flanges for a built-up section, in

t = web thickness, in

k = web buckling coefficient

$= 5 + 5/(a/h)^2$ if $a/h \leq 3.0$

$= 5$ if $a/h > 3.0$ or $[260/(h/t)]^2$

a = clear distance between transverse stiffeners, in

F_y = specified minimum yield stress of the web, ksi

A higher design shear strength is permitted for plate girders if the spacing of intermediate stiffeners is such that $a/h < [260/(h/t)]^2 < 3.0$.

The design shear stress F_v (ksi) in ASD is given by Eqs. (6.34) or (6.35). For $h/t \leq 380/\sqrt{F_y}$,

$$F_v = 0.40F_y \tag{6.34}$$

For $h/t > 380/\sqrt{F_y}$,

$$F_v = F_yC_v/2.89 \leq 0.4F_y \tag{6.35}$$

where $C_v = \dfrac{45,000k_v}{F_y(h/t)^2}$ when $C_v < 0.8$

$ = \dfrac{190}{h/t}\sqrt{\dfrac{k_v}{F_y}}$ when $C_v > 0.8$

$ k_v = 4.00 + \dfrac{5.34}{(a/h)^2}$ when $a/h < 1.0$

$ = 5.34 + \dfrac{4.00}{(a/h)^2}$ when $a/h > 1.0$

6.14.2 Shear in Bolts

For rivets, bolts, and threaded parts, design shear strengths are specified in Tables 6.16 and 6.17. The design strengths permitted for high-strength bolts depend on whether a connection is slip-critical or bearing type.

TABLE 6.16   Design Shear Strength of Fasteners in Bearing-Type Connections

	Shear strength, ksi	
	LRFD	ASD
Description of fasteners	Nominal strength	Allowable shear F_v
A307 bolts*	27.0	10.0
A325 bolts, when threads are *not* excluded from shear planes†	54.0	21.0
A325 bolts, when threads *are* excluded from shear planes†	72.0	30.0
A490 bolts, when threads are *not* excluded from shear planes†	67.5	28.0
A490 bolts, when threads *are* excluded from the shear planes†	90.0	40.0
Threaded parts, when threads are *not* excluded from the shear planes†	$0.45F_u$	$0.17F_u$
Threaded parts, when threads *are* excluded from the shear planes†	$0.60F_u$	$0.22F_u$
A502, grade 1, hot-driven rivets†	36.0	17.5
A502, grades 2 and 3, hot-driven rivets†	48.0	22.0

*Resistance factor $\phi = 0.60$.
†Resistance factor $\phi = 0.65$.

TABLE 6.17   Allowable Loads F_s (ksi) for Slip-Critical Connections*

Type of bolt	Standard-size holes	Oversized and short-slotted holes	Long-slotted holes	
			Transverse loading	Parallel loading
A325	17	15	12	10
A490	21	18	15	13

*Applies to both ASD and LRFD. LRFD design for slip-critical connections is made for service loads. For LRFD, $\phi = 1.0$ except for long-slotted holes when the load is parallel to the slot for which $\phi = 0.85$. For LRFD, when the loading combination includes wind or seismic loads, the combined load effects at service loads may be multiplied by 0.75.

Slip-critical joints are connections in which slip would be detrimental to the serviceability of the structure in which the joints are components. These include connections subject to fatigue loading or significant load reversal or in which bolts are installed in oversized holes or share loads with welds at a common faying surface. In slip-critical joints, the fasteners are high-strength bolts tightened to a specified tension. This applies a clamping force to the connected plies, and the resulting frictional resistance to slip P_s (kips) is required to equal or exceed the load on the connection:

$$P_s = F_s A_b N_b N_s \tag{6.36}$$

where F_s = allowable slip load per unit area of bolt, ksi
A_b = area corresponding to nominal bolt area, in^2
N_b = number of bolts in the connection
N_s = number of slip planes

Table 6.17 lists allowable values of F_s for connections with ASTM A325 and A490 high-strength bolts and with surfaces with clean mill scale or blast-cleaned surfaces with coatings that provide a mean slip coefficient of at least 0.33. Higher values of F_s may be used for joints with slip coefficients of 0.50 or more (see "Specification for Structural Joints Using ASTM A325 or A490 Bolts," Research Council on Structural Connections of the Engineering Foundation).

Bolts in slip-critical joints should be pretensioned to the values given in Table 6.18 to provide at least the minimum clamping force needed to prevent slip under service loads. Design checks for slip-critical joints should be made for service loading for both ASD and LRFD.

In bearing-type connections, load is resisted by shear in and bearing on the bolts. Design strength is influenced by the presence of threads; i.e., a bolt with threads excluded from the shear plane is assigned a higher design strength than a bolt with threads included in the shear plane (see Table 6.16). Design stresses are assumed to act on the nominal body area of bolts in both ASD and LRFD.

Bearing-type connections are assigned higher design strengths than slip-critical joints and hence are more economical. Also, erection is faster with bearing-type joints because the bolts need not be highly tensioned.

In connections where slip can be tolerated, bolts not subject to tension loads nor loosening or fatigue due to vibration or load fluctuations need only be made snug-tight. This can be accomplished with a few impacts of an impact wrench or by full manual effort with a spud wrench sufficient to bring connected plies into firm contact. Slip-critical joints and connections subject to direct tension should be indicated on construction drawings.

TABLE 6.18 Minimum Pretension (kips) for Bolts*

Bolt size, in	A325 bolts	A490 bolts
½	12	15
⅝	19	24
¾	28	35
⅞	39	49
1	51	64
1⅛	56	80
1¼	71	102
1⅜	85	121
1½	103	148
Over 1½	—	0.7 T.S.

* Equal to 70% of minimum tensile strengths (T.S.) of bolts, rounded off to the nearest kip.

Where permitted by building codes, ASTM A307 bolts or snug-tight high-strength bolts may be used for connections that are not slip critical. The AISC specifications for structural steel for buildings require that fully tensioned, high-strength bolts (Table 6.18) or welds be used for the following joints:

Column splices in multistory framing, if it is more than 200 ft high, or when it is between 100 and 200 ft high and the smaller horizontal dimension of the framing is less than 40% of the height, or when it is less than 100 ft high and the smaller horizontal dimension is less than 25% of the height.

Connections, in framing more than 125 ft high, on which column bracing is dependent and connections of all beams or girders to columns.

Crane supports, as well as roof-truss splices, truss-to-column joints, column splices and bracing, and crane supports, in framing supporting cranes with capacity exceeding 5 tons.

Connections for supports for impact loads, loads causing stress reversal, or running machinery.

The height of framing should be measured from curb level (or mean level of adjoining land when there is no curb) to the highest point of roof beams for flat roofs or to mean height of gable for roofs with a rise of more than 2⅔ in 12. Penthouses may be excluded.

6.14.3 Shear in Welds

Welds subject to static loads should be proportioned for the design strengths in Table 6.19.

The effective area of groove and fillet welds for computation of design strength is the effective length times the effective throat thickness. The effective area for a plug or slot weld is taken as the nominal cross-sectional area of the hole or slot in the plane of the faying surface.

TABLE 6.19 Design Shear Strength for Welds, ksi

| Types of weld and stress | Material | LRFD | | ASD |
		Resistance factor ϕ	Nominal strength* F_{BM} or F_w	Allowable stress
Complete penetration groove weld				
Tension normal to effective area	Base	0.90	F_y	
Compression normal to effective area	Base	0.90	F_y	Same as base metal
Tension or compression parallel to axis of weld				
Shear on effective area	Base	0.90	$0.60F_y$	$0.30 \times$ nominal tensile strength of weld metal
	Weld electrode	0.80	$0.60F_{EXX}$	
Partial penetration groove welds				
Compression normal to effective area	Base	0.90	F_y	Same as base metal
Tension or compression parallel to axis of weld				
Shear parallel to axis of weld	Base	0.75	$0.60F_{EXX}$	$0.30 \times$ nominal tensile strength of weld metal
	Weld electrode			
Tension normal to effective area	Base	0.90	F_y	$0.30 \times$ nominal tensile strength of weld metal
	Weld Electrode	0.80	$0.60F_{EXX}$	
Fillet welds				
Shear on effective area	Base	0.75	$0.60F_{EXX}$	$0.30 \times$ nominal tensile strength of weld metal
	Weld electrode			
Tension or compression parallel to axis of weld	Base	0.90	F_y	Same as base metal
Plug or slot welds				
Shear parallel to faying surfaces (on effective area)	Base	0.75	$0.60F_{EXX}$	$0.30 \times$ nominal tensile strength of weld metal
	Weld electrode			

*Design strength is the smaller of F_{BM} and F_w:
 F_{BM} = nominal strength of base metal to be welded, ksi
 F_w = nominal strength of weld electrode material, ksi
 F_y = specified minimum yield stress of base metal, ksi
F_{EXX} = classification strength of weld metal, as specified in appropriate AWS specifications, ksi

Effective length of fillet welds, except fillet welds in holes or slots, is the overall length of the weld, including returns. For a groove weld, the effective length is taken as the width of the part joined.

The effective throat thickness of a fillet weld is the shortest distance from the root of the joint to the nominal face of the weld. However, for fillet welds made by the submerged-arc process, the effective throat thickness is taken as the leg size for ⅜-in and smaller welds and equal to the theoretical throat plus 0.11 in for fillet welds larger than ⅜ in.

The effective throat thickness of a complete-penetration groove weld equals the thickness of the thinner part joined. Table 6.20 shows the effective throat thickness for partial-penetration groove welds. Flare bevel and flare V-groove welds when flush to the surface of a bar or 90° bend in a formed section should have effective throat thicknesses of ⁵⁄₁₆ and ½ times the radius of the bar or bend, respectively, and when the radius is 1 in or more, for gas-metal arc welding, ¾ of the radius.

To provide adequate resistance to fatigue, design stresses should be reduced for welds and base metal adjacent to welds in connections subject to stress fluctuations (see Art. 6.22). To ensure adequate placement of the welds to avoid stress concentrations and cold joints, the AISC specifications set maximum and minimum limits on the size and spacing of the welds. These are discussed in Art. 5.19.

6.15 COMBINED TENSION AND SHEAR

Combined tension and shear stresses are of concern principally for fasteners, plate-girder webs, and ends of coped beams, gusset plates, and similar locations.

6.15.1 Tension and Shear in Bolts

The AISC "Load and Resistance Factor Design (LRFD) Specification for Structural Steel Buildings" contains interaction formulas for design of bolts subject to combined tension and shear in bearing-type connections. The specification stipulates that the tension stress applied by factored loads must not exceed the design tension stress F_t (ksi) computed from the appropriate formula (Table 6.21) when the applied shear stress f_v (ksi) is caused by the same factored loads. This shear stress must not exceed the design shear strength.

For bolts in a slip-critical connection, the LRFD specification indicates that the design strength for shear alone as given in Table 6.17, is to be reduced by the

TABLE 6.20 Effective Throat Thickness of Partial-Penetration Groove Welds

Welding process	Welding position	Included angle at root of groove	Effective throat thickness
Shielded metal arc Submerged arc Gas metal arc Flux-cored arc	All	J or U joint Bevel or V joint ≥60°	Depth of chamfer
		Bevel or V joint <60° but ≥45°	Depth of chamfer minus ⅛-in

TABLE 6.21 Tension Stress Limit F_t (ksi) for Fasteners in Bearing-Type Connections

Type of bolt	Type of design	Threads in the shear plane	
		Included	Excluded
A307	LRFD	$39 - 1.8f_v \leq 30$	
	ASD	$26 - 1.8f_v \leq 20$	
A325	LRFD	$85 - 1.8f_v \leq 68$	$85 - 1.4f_v \leq 68$
	ASD	$\sqrt{44^2 - 4.39f_v^2}$	$\sqrt{44^2 - 2.15f_v^2}$
A490	LRFD	$106 - 1.8f_v \leq 84$	$106 - 1.4f_v \leq 84$
	ASD	$\sqrt{54^2 - 3.75f_v^2}$	$\sqrt{54^2 - 1.82f_v^2}$

factor $(1 - T/T_b)$, where T_b is the minimum pretension force (kips; see Table 6.18), and T is the service tensile force applied to the bolts.

According to the AISC "Specification for Structural Steel Buildings—Allowable Stress Design," the applied tension stress must not exceed the allowable tension stress F_t as given by Table 6.21. The applied shear stress must not exceed the allowable shear stress. When the allowable stresses are increased for wind or seismic loads, the constants, except for the coefficients of f_v, in the equations may be increased one-third. To account for combined loading for a slip-critical connection, allowable shear stress is to be reduced by the factor $(1 - f_t A_b/T_b)$, where T_b is the minimum pretension force (ksi; see Table 6.18), and f_t is the average tensile stress (ksi) applied to the bolts.

6.15.2 Tension and Shear in Girder Webs

Large shear forces may be developed in plate girders designed for tension-field action. Therefore, at locations of high shear and bending (i.e., when $0.6V_n/M_n \leq V_u/M_u \leq V_n/0.75M_n$), according to the AISC LRFD specification, the member should be reviewed for the following interaction of flexure and shear:

$$M_u/M_n + 0.625V_u/V_n \leq 1.375\phi \qquad (6.37)$$

where M_n = nominal flexural strength computed as indicated in Art. 6.17.2
M_u = required flexural strength to resist the factored loads
ϕ = reduction factor = 0.90
V_u = nominal shear strength computed as indicated in Art. 6.14.1
V_n = required shear strength to resist the factored loads

For ASD, the plate-girder web should be so proportioned that bending tensile stresses F_b due to moment in the plane of the web do not exceed

$$F_b = (0.825 - 0.375f_v/F_v)F_y \leq 0.6F_y \qquad (6.38)$$

where f_v = applied shear stress
F_v = allowable shear stress with tension-field action
F_y = minimum specified yield stress of the web

The AISC ASD specification also imposes additional requirements on webs with yield point greater than 65 ksi.

6.15.3 Block Shear

This is a failure mode that may occur at the ends of coped beams, in gusset plates, and in similar locations. It is a tearing failure mode involving shear rupture along one path, such as through a line of bolt holes, and tensile rupture along a perpendicular line.

The AISC LRFD specification for structural steel buildings specifies that the design strength for the limit state of rupture along a shear failure path in main members should be taken as $\phi F_n A_{ns}$, where ϕ = reduction factor = 0.75, F_n = 60% of the minimum specified tensile strength F_u (ksi), and A_{ns} = net area subject to shear.

The AISC ASD specification specifies allowable shear and tensile stresses for the end connections of beams where the top flange is coped and in similar situations where failure might occur by shear along a plane through fasteners or by a combination of shear along a plane through fasteners and tension along a perpendicular plane. The shear stress F_v should not exceed $0.30 F_u$ acting on the net shear area. Also, the tensile stress F_t should not exceed $0.50 F_u$ acting on the net tension area (Art. 6.25).

6.16 COMPRESSION

Compressive forces can produce local or overall buckling failures in a steel member. **Overall buckling** is the out-of-plane bending exhibited by an axially loaded column or beam (Art. 6.17). **Local buckling** may manifest itself as a web failure beneath a concentrated load or over a reaction or as buckling of a flange or web along the length of a beam or column.

6.16.1 Local Buckling

Local buckling characteristics of the cross section of a member subjected to compression may affect its strength. With respect to potential for local buckling, sections may be classified as compact, noncompact, or slender-element (Art. 6.23).

6.16.2 Axial Compression

Design of members that are subjected to compression applied through the centroidal axis (axial compression) is based on the assumption of uniform stress over the gross area. This concept is applicable to both load and resistance factor design (LRFD) and allowable stress design (ASD).

Design of an axially loaded compression member or column for both LRFD and ASD utilizes the concept of **effective** column length KL. The buckling coefficient K is the ratio of the effective column length to the unbraced length L. Values of K depend on the support conditions of the column to be designed. The AISC specifications for LRFD and ASD indicate that K should be taken as unity for columns in braced frames unless analysis indicates that a smaller value is justified. Analysis is required for determination of K for unbraced frames, but K should not be less than unity. Design values for K recommended by the Structural Stability Research Council for use with six idealized conditions of rotation and translation at column supports are illustrated in Fig. 6.3 (see also Arts. 7.4 and 7.9).

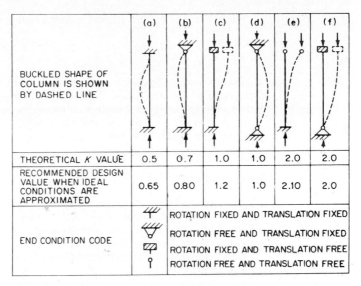

	(a)	(b)	(c)	(d)	(e)	(f)
BUCKLED SHAPE OF COLUMN IS SHOWN BY DASHED LINE						
THEORETICAL K VALUE	0.5	0.7	1.0	1.0	2.0	2.0
RECOMMENDED DESIGN VALUE WHEN IDEAL CONDITIONS ARE APPROXIMATED	0.65	0.80	1.2	1.0	2.10	2.0
END CONDITION CODE		ROTATION FIXED AND TRANSLATION FIXED				
		ROTATION FREE AND TRANSLATION FIXED				
		ROTATION FIXED AND TRANSLATION FREE				
		ROTATION FREE AND TRANSLATION FREE				

FIGURE 6.3 Effective length factor K for columns.

The axially compression strength of a column depends on its stiffness measured by the slenderness ratio KL/r, where r is the radius of gyration about the plane of buckling. For serviceability considerations, AISC recommends that KL/r not exceed 200.

LRFD strength for a compression member ϕP_n (kips) is given by

$$\phi P_n = 0.85 A_g F_{cr} \tag{6.39}$$

with $\phi = 0.85$. For $\lambda_c \leq 1.5$,

$$F_{cr} = 0.658^{\lambda_c^2} F_y \tag{6.40a}$$

for $\lambda_c > 1.5$,

$$F_{cr} = \frac{0.877}{\lambda_c^2} F_y \tag{6.40b}$$

where $\lambda_c = (KL/r\pi) \sqrt{F_y/E}$

F_y = minimum specified yield stress of steel, ksi

A_g = gross area of member, in^2

E = elastic modulus of the steel = 29,000 ksi

For the strength of composite columns, see Art. 6.26.4; for built-up columns, see Art. 6.28.

For ASD, the allowable compression stress depends on whether buckling will be elastic or inelastic, as indicated by the slenderness ratio

$$C_c = \sqrt{2\pi^2 E/F_y} \tag{6.41}$$

When $KL/r < C_c$, the allowable compression stress F_a (kips) on the gross section should be computed from

$$F_a = \frac{1 - (KL/r)^2/2C_c^2}{5/3 + 3(KL/r)/8C_c - (KL/r)^3/8C_c^3} F_y \tag{6.42}$$

When $KL/r > C_c$, the allowable compression stress is

$$F_a = \frac{12\pi^2 E}{23(KL/r)^2} \tag{6.43}$$

Tables of allowable loads for columns are contained in the AISC "Manual of Steel Construction" for ASD and for LRFD. For composite compression members, see Art. 6.26.4; for built-up compression members, see Art. 6.28.

(T. V. Galambos, *Guide to Stability Design Criteria for Metal Structures*, John Wiley & Sons, Inc., New York.)

6.16.3   Concentrated Loads on Beams

Large concentrated loads or reactions on flexural members may cause their webs to fail by yielding or crippling unless the webs are made sufficiently thick to preclude this or are assisted by bearing stiffeners. Also, adequate bearing length should be provided on the flange of the member.

Web yielding manifests as a stress concentration in a web beneath a concentrated load. The AISC LRFD specification for structural steel buildings limits the design strength of the web at the toe of the fillet under a concentrated load to ϕR_n (kips), where $\phi = 1.0$ and R_n is determined from Eq. (6.44) or (6.45).

When the concentrated loads is applied at a distance from the end of the member greater than the member depth,

$$R_n = (5k + N)F_y t_w \tag{6.44}$$

When the load acts at or near the end of the member,

$$R_n = (2.5k + N)F_y t_w \tag{6.45}$$

where N = length of bearing on the flange of the member, in
  k = distance from outer face of flange to web toe of fillet, in
  t_w = web thickness, in
  F_y = specified minimum yield stress of the web steels, ksi

Web crippling is a buckling of a slender web beneath a concentrated load. For unstiffened webs of beams under concentrated loads, the AISC LRFD specification sets the crippling load at ϕR_n (kips), where $\phi = 0.75$ and R_n is determined from Eq. (6.46) or (6.47).

When the concentrated load is applied at a distance of at least $d/2$ from the member end,

$$R_n = 135 t_w^2 \left[1 + 3(N/d)(t_w/t_f)^{1.5}\right]\sqrt{F_y t_f/t_w} \tag{6.46}$$

When the concentrated load is applied less than $d/2$ from the member end,

$$R_n = 68 t_w^2 \left[1 + 3(N/d)(t_w/t_f)^{1.5}\right]\sqrt{F_y t_f/t_w} \tag{6.47}$$

where d = overall depth of the member, in
  t_f = flange thickness, in

If stiffeners are provided and extended at least one-half the web depth, web crippling need not be checked.

The AISC ASD specification contains similar criteria for web yielding and crippling. For yielding, the following allowable stress limits should be met.

When the concentrated load is applied at a distance from the member end that is greater than the member depth,

$$\frac{R}{t_w(N + 5\,k)} \leq 0.66F_y \tag{6.48}$$

When the concentrated load is applied at or near the member end,

$$\frac{R}{t_w(N + 2.5k)} \leq 0.66F_y \tag{6.49}$$

where R is the applied concentrated load or reaction (kip).

Bearing stiffeners should be provided on the web when the applied concentrated load or reaction exceeds the following web crippling limits.

When the concentrated load is applied at a distance of at least $d/2$ from the member end,

$$R = 67.5t_w^2\,[1 + 3\,(N/d)\,(t_w/t_f)^{1.5}]\sqrt{F_y t_f/t_w} \tag{6.50}$$

When the concentrated load is applied less than $d/2$ from the member end,

$$R = 34t_w^2\,[1 + 3\,(N/d)\,(t_w/t_f)^{1.5}]\sqrt{F_y t_f/t_w} \tag{6.51}$$

When an unreinforced web is unable to achieve the required strength, bearing stiffeners should be provided. See Art. 6.25 for design of bearing stiffeners.

6.17 BENDING STRENGTH

For a member subjected to flexure, the bending strength depends on the shape of the member, width/thickness or depth/thickness ratios of its elements, location and direction of loading, and the support given to the compression flange.

Higher strengths are assigned to symmetrical and compact shapes. Flexural strength may be reduced, however, based on the spacing of lateral supports that prevent displacement of the compression flange and twist of the cross section.

6.17.1 Compact Shapes

The AISC LRFD and ASD specifications define compact sections similarly: **Compact sections** are sections capable of developing a fully plastic stress distribution and possess a rotation capacity of about 3 before the onset of local buckling. **Rotational capacity** is the incremental angular rotation that a section can accept before local failure occurs, defined as $R = (\theta_u/\theta_p) - 1$, where θ_u is the overall rotation attained at the factored-load state and θ_p is the idealized rotation corresponding to elastic-theory deformations for the case where the moment equals M_p, the plastic bending moment (Sec. 6.17.2).

A section is considered compact if its flanges are continuously connected to its web or webs and the width/thickness or depth/thickness ratios of its compression elements do not exceed the following: for the flanges of beams, rolled or welded, and channels, $65/\sqrt{F_y}$, and for the flanges of box and hollow structural sections of uniform thickness, $190/\sqrt{F_y}$, where F_y is the minimum specified yield stress of the flange steel. The limiting depth/thickness ratio of webs is $640/\sqrt{F_y}$. (See also Art. 6.23 and Table 6.24.)

For flanges of I-shaped members and tees, the width is half the full nominal width for rolled shapes and the distance from the free edge to the first line of fasteners or welds for built-up sections. For webs, the depth is the full nominal depth.

6.17.2 LRFD Bending Strength

According to the AISC LRFD specification, the flexural design strength for a compact shape is determined by the limit state of lateral-torsional buckling with an upper limit of yielding of the cross section.

The flexural design strength ϕM_n for a doubly or singly symmetrical I-shape member with $\phi = 0.90$ is determined from Eqs. (6.52) to (6.54), depending on the relationship between the laterally unbraced length of the compression flange L_b and the limiting unbraced lengths for full plastic bending capacity L_p or inelastic-torsional buckling L_r.

When $L_b \leq L_p$,

$$M_n = M_p \tag{6.52}$$

When $L_b \leq L_r$,

$$M_n = C_b \left[M_p - (M_p - M_r) \left(\frac{L_b - L_p}{L_r - L_p} \right) \right] \leq M_p \tag{6.53}$$

When $L_b > L_r$,

$$M_n = M_{cr} \leq C_b M_r \tag{6.54}$$

where M_p = plastic bending moment = $F_y Z$
Z = plastic section modulus (computed for complete yielding of the beam cross section)
$M_r = (F_y - F_r)S_x$
S_x = section modulus about major axis
$M_{cr} = C_b \dfrac{\pi}{L_b} \sqrt{EI_y GJ + \left(\dfrac{\pi E}{L_b} \right)^2 I_y C_w}$
F_y = minimum specified yield stress of the compression flange, ksi
$C_b = 1.75 + 1.05\,(M_1/M_2) + 0.3\,(M_1/M_2)^2 \leq 2.3$, where M_1 is the smaller and M_2 the larger end moment in the unbraced segment of the beam; M_1/M_2 is positive when the moments cause reverse curvature and negative when bent in single curvature
   = 1.0 for unbraced cantilevers and for members where the moment within a significant portion of the unbraced segment is greater than or equal to the larger of the segment end moments
L_b = distance between points braced against lateral displacement of the compression flange, or between points braced to prevent twist of the cross section
$L_p = 300\,r_y/\sqrt{F_y}$ for I-shape members and channels bent about their major axis
$L_r = \dfrac{r_y X_1}{(F_{yw} - F_r)} \sqrt{1 + \sqrt{1 + X_2(F_{yw} - F_r)^2}}$
$X_1 = \dfrac{\pi}{S_x} \sqrt{\dfrac{EGJA}{2}}$
$X_2 = 4 \dfrac{C_w}{I_y} \left(\dfrac{S_x}{GJ} \right)^2$

r_y = radius of gyration about the minor axis
E = modulus of elasticity of steel = 29,000 ksi
A = area of the cross section of the member
G = shear modulus of elasticity of steel = 11,000 ksi
J = torsional constant for the section
F_{yw} = yield stress of web, ksi
I_y = moment of inertia about minor axis
C_w = warping constant
F_r = compressive residual stress in flange; 10 ksi for rolled shapes, 16.5 ksi for welded shapes

For singly symmetrical I-shaped members with the compression flange larger than the tension flange, use S_{xc} (section modulus referred to compression flange) instead of S_x in Eqs. (6.53) and (6.54).

For noncompact shapes, consideration should be given to the reduction in flexural strength because of local buckling of either the compression flange or the compression portion of the web. Appendix F1.7 and Appendix G of the AISC LRFD specification provide design guidance for evaluating the strength of such members.

Because of the enhanced lateral stability of circular or square shapes and shapes bending about their minor axis, the nominal moment capacity is defined by $M_n = M_p$, where M_p is evaluated for the minor axis and $\phi = 0.90$.

6.17.3 ASD Bending Stresses

The ASD requirements for bending strength follow, in concept, the LRFD provisions in that allowable stresses are defined based on the member cross section, the width/thickness and depth/thickness ratios of its elements, the direction of loading, and the extent of lateral support provided to the compression flange.

The allowable bending stress for a compact shape depends on the laterally unsupported length L of the compression flange. The allowable stress also depends on the stiffness of the compression part of the cross section as measured by L/r_T, where r_T is the radius of gyration of a section comprising the compression flange and one-third of the web area in compression, taken about an axis in the plane of the web.

The largest bending stress permitted for a compact section symmetrical about and loaded in the plane of its minor axis is

$$F_b = 0.66F_y \tag{6.55}$$

This stress can be used, however, only if L does not exceed the smaller of the values of L_c computed from Eqs. (6.56) and (6.57):

$$L_c = 76b_f/\sqrt{F_y} \tag{6.56}$$

$$L_c = \frac{20{,}000}{(d/A_f)F_y} \tag{6.57}$$

where b_f = width of flange, in
d = nominal depth of the beam, in
A_f = area of the compression flange, in
F_y = minimum specified yield stress, ksi

When $L > L_c$, the allowable bending stress for compact or noncompact sections is the larger of F_b (ksi) computed from Eq. (6.58) and Eq. (6.59), (6.60), or (6.61):

$$F_b = 12,000C_b/(Ld/A_f) \leq 0.60F_y \tag{6.58}$$

where C_b is the bending coefficient defined in Sec. 6.17.2. When $L/r_T \leq \sqrt{102,000C_b/F_y}$,

$$F_b = 0.60F_y \tag{6.59}$$

When $\sqrt{102,000C_b/F_y} \leq L/r_T \leq \sqrt{510,000C_b/F_y}$,

$$F_b = \left[\frac{2}{3} - \frac{F_y(L/r_T)^2}{1,530,000C_b}\right] F_y \leq 0.60F_y \tag{6.60}$$

When $L/r_T \leq \sqrt{510,000C_b/F_y}$,

$$F_b = 170,000C_b/(L/r_T)^2 \leq 0.60F_y \tag{6.61}$$

The AISC specifications for structural steel buildings do not require lateral bracing for members having equal strength about both major and minor axes, nor for bending about the weak axis when loads pass through the shear center.

For I- and H-shape members symmetrical about both axes, with compact flanges continuously connected to the web and solid rectangular sections subjected to bending about the minor axis, the allowable bending stress is

$$F_b = 0.75F_y \tag{6.62}$$

This stress is also permitted for solid round and square bars.

For shapes not covered in the preceding, refer to the AISC specifications for structural steel buildings.

6.18   BEARING

For bearing on the contact area of milled surfaces, pins in reamed, drilled, or bored holes, and ends of a fitted bearing stiffener, the LRFD design strength is ϕR_n (kips), where $\phi = 0.75$ and $R_n = 2.0F_yA_{pb}$. The projected bearing area is represented by A_{pb}, and F_y is the specified minimum yield stress (ksi) for the part in bearing with the lower yield stress.

Expansion rollers and rockers are limited to an LRFD bearing strength ϕR_n, with $\phi = 0.75$:

$$R_n = 1.5(F_y - 13)Ld/20 \tag{6.63}$$

where d = diameter of roller or rocker, in
    L = length of bearing, in

The ASD allowable stress for bearing F_p (ksi) on the contact surface of milled surfaces and pins is $0.9F_y$. On the contact area of expansion rollers and rockers,

$$F_p = \left(\frac{F_y - 13}{20}\right) 0.66d \tag{6.64}$$

6.19 COMBINED BENDING AND COMPRESSION

Design of a structural member for loading that induces both bending and axial compression should take into account not only the primary stresses due to the combined loading but also secondary effects. Commonly called **P-delta effects,** these result from two sources: (1) Incremental bending moments caused by buckling of the member that create eccentricity δ of the axial compression load with respect to the neutral axis, and (2) secondary moments produced in a member of a rigid frame due to sidesway of the frame that creates eccentricity Δ of the axial compression load with respect to the neutral axis. Both the AISC LRFD and ASD specifications for structural steel buildings contain provisions for the influence of second-order effects; each specification, however, treats P-delta differently.

6.19.1 LRFD Strength in Bending and Compression

The LRFD specification presents two interaction equations for determining the strength of a member under combined bending and axial compression. The equation to use depends on the ratio of the required compressive strength P_u (kips) to resist the factored load to the nominal compressive strength ϕP_n (kips) computed from Eq. (6.39), where $\phi = \phi_c =$ resistance factor for compression $= 0.85$.

For $P_u/(\phi_c P_n) \geq 0.2$,

$$\frac{P_u}{\phi P_n} + \frac{8}{9}\left(\frac{M_{ux}}{\phi_b M_{nx}} + \frac{M_{uy}}{\phi_b M_{ny}}\right) \leq 1.0 \tag{6.65}$$

For $P_u/(\phi P_n) < 0.2$,

$$\frac{P_u}{2\phi P_n} + \left(\frac{M_{ux}}{\phi_b M_{nx}} + \frac{M_{uy}}{\phi_b M_{ny}}\right) \leq 1.0 \tag{6.66}$$

where x, y = indexes representing axis of bending about which a moment is applied

M_u = required flexural strength to resist the factored load

M_n = nominal flexural strength determined as indicated in Art. 6.17.2

ϕ_b = resistance factor for flexure = 0.90

The required flexural strength M_u should be evaluated with due consideration given to second-order moments. The moments may be determined for a member in a rigid frame by a second-order analysis or from

$$M_u = B_1 M_{nt} + B_2 M_{lt} \tag{6.67}$$

where M_{nt} = required flexural strength in member if there is no lateral translation of the frame

M_{lt} = required flexural strength in member as a result of lateral translation of the frame only (first-order analysis)

$B_1 = \dfrac{C_m}{(1 - P_u/P_e)} \geq 1$

$P_e = A_g F_y/\lambda_c^2$, where λ_c is defined as in Art. 6.16.2 with $K \leq 1.0$ in the plane of bending

$C_m = 0.6 - 0.4M_1/M_2$ for restrained compression members in frames braced against joint translation and not subject to transverse loading between their supports in the plane of bending, where M_1/M_2 is the ratio of the smaller to larger moments at the ends of that portion of the member unbraced in the plane of bending under consideration; M_1/M_2 is positive when the member is bent in reverse curvature, negative when bent in single curvature

$= 0.85$ for members whose ends are restrained or 1.0 for members whose ends are unrestrained in frames braced against joint translation in the plane of loading and subjected to transverse loading between their supports, unless the value of C_m is determined by rational analysis

$$B_2 = \frac{1}{1 - \Delta_{oh}\Sigma P_u/\Sigma HL} \quad \text{or} \quad \frac{1}{1 - \Sigma P_u/\Sigma P'_e}$$

ΣP_u = required axial load strength of all columns in a story, kips

Δ_{oh} = translation deflection of the story under consideration, in

ΣH = sum of all story horizontal forces producing Δ_{oh}, kips

L = story height, in

$P'_e = A_g F_y/\lambda_c^2$ (kips), where λ_c is the slenderness parameter defined in Art. 6.16.2 with $K \geq 1.0$ in the plane of bending

6.19.2   ASD for Bending and Compression

In ASD, the interaction of bending and axial compression is governed by Eqs. (6.68) and (6.69) or (6.70):

$$\frac{f_a}{F_a} + \frac{C_{mx}f_{bx}}{\left(1 - \dfrac{f_a}{F'_{ex}}\right)F_{bx}} + \frac{C_{my}f_{by}}{\left(1 - \dfrac{f_a}{F'_{ey}}\right)F_{by}} \leq 1.0 \tag{6.68}$$

$$\frac{f_a}{0.60\,F_y} + \frac{f_{bx}}{F_{bx}} + \frac{f_{by}}{F_{by}} \leq 1.0 \tag{6.69}$$

When $f_a/F_a \leq 0.15$, Eq.(6.70) is permitted in lieu of Eqs. (6.68) and (6.69):

$$\frac{f_a}{F_a} + \frac{f_{bx}}{F_{bx}} + \frac{f_{by}}{F_{by}} \leq 1.0 \tag{6.70}$$

where x, y = indexes representing the bending axis to which the applicable stress applies

F_a = allowable stress for axial force alone, ksi

F_b = allowable compression stress for bending moment alone, ksi

F'_e = Euler stress divided by a safety factor, ksi

$= 12\,\pi^2 E/23(KL/r_b)^2$

L = unbraced length in plane of bending, in

r_b = radius of gyration about bending plane, in

K = effective length factor in plane of bending

f_a = axial compression stress due to loads, ksi

f_b = compression bending stress at design section due to loads, ksi

C_m = coefficient defined in Art. 6.19.1

F_y = yield stress of the steel, ksi

F'_e as well as allowable stresses may be increased one-third for wind and seismic loads combined with other loads (Art. 6.21).

6.20 COMBINED BENDING AND TENSION

For combined axial tension and bending, the AISC LRFD specification stipulates that members should be proportioned to satisfy the same interaction equations as for axial compression and bending, Eqs. (6.65) and (6.66), Art. 6.19.1, but with

$$P_u = \text{required tensile strength, kips}$$
$$P_n = \text{nominal tensile strength, kips}$$
$$M_u = \text{required flexural strength}$$
$$M_n = \text{nominal flexural strength}$$
$$\phi_t = \text{resistance factor for tension} = 0.90$$
$$\phi_b = \text{resistance factor for flexure} = 0.90$$

The ASD interaction equation for combined axial tension and bending is similar to Eq. (6.70):

$$\frac{f_a}{F_t} + \frac{f_{bx}}{F_{bx}} + \frac{f_{by}}{F_{by}} \le 1.0 \qquad (6.71)$$

but with f_b = computed bending tensile stress, ksi
$\quad\quad\quad f_a$ = computed axial tensile stress, ksi
$\quad\quad\quad F_b$ = allowable bending stress, ksi
$\quad\quad\quad F_t$ = allowable tensile stress, ksi

6.21 WIND AND SEISMIC STRESSES

The AISC ASD specification permits allowable stresses due to wind or earthquake forces, acting alone or in combination with dead and live load, to be increased by one-third (Art. 6.11). The required section computed on this basis, however, may not be less than that required for the design dead, live, and impact loads computed without the one-third stress increase. This stress increase cannot be applied in combination with load-reduction factors in load combinations. Also, the one-third stress increase does not apply to stresses resulting from fatigue loading.

In LRFD design, generally equivalent allowances are made by the prescribed load factors.

6.22 FATIGUE LOADING

Fatigue damage may occur to members supporting machinery, cranes, vehicles, and other mobile equipment. Such damage is not likely in members subject to few load changes or small stress fluctuations. For example, full design wind or seismic loads occur too infrequently to justify stress reductions for fatigue. For members susceptible to fatigue damage, the AISC specifications for structural steel buildings limit the range of stress fluctuation. Because the cyclic stresses that induce fatigue

TABLE 6.22 Allowable Stress Range (ksi) for Repeated Loads*

Stress category	Number of loading cycles			
	20,000† to 100,000‡	100,000‡ to 500,000§	500,000§ to 2 million¶	More than 2 million¶
A	63	37	24	24
B	49	29	18	16
B′	39	23	15	12
C	35	21	13	10
D	28	16	10	7
E	22	13	8	5
E′	16	9	6	3
F	15	12	9	8

*As specified in the AISC ASD and LRFD specifications for structural steel buildings.
†Equivalent to about two applications every day for 25 years.
‡Equivalent to about 10 applications every day for 25 years.
§Equivalent to about 50 applications every day for 25 years.
¶Equivalent to about 200 applications every day for 25 years.

damage occur due to service loads, both the AISC LRFD and ASD specifications contain the same design stress ranges (Table 6.22).

Stress range is the algebraic difference between minimum and maximum stresses. For stress reversal at any point, for example, stress range equals the sum of maximum repeated tensile and compressive stresses (ignoring sign) or the sum of maximum shearing stresses of opposite direction that result from different arrangements of live load.

The probability of fatigue damage depends on the number of load applications, number of loading cycles, stress range, and presence of stress raisers. Members that will be subjected to less than 20,000 cycles will not sustain damage unless the stress range is large.

For increasing repetitions of load, damage occurs at decreasing stress ranges. This behavior is reflected in Table 6.22, which lists allowable stress ranges in accordance with the number of cycles of repeated loads and stress categories for structural details. These details are described in Table 6.23, in which the diagrams are provided as illustrative examples and are not intended to exclude other similar construction.

Design of members to resist fatigue cannot be executed with the certainty with which members can be designed to resist static loading. However, it is often possible to reduce the magnitude of a stress concentration below the minimum value that will cause fatigue failure.

In general, avoid design details that cause severe stress concentrations or poor stress distribution. Provide gradual changes in section. Eliminate sharp corners and notches. Do not use details that create high localized constraint. Locate unavoidable stress raisers at points where fatigue conditions are the least severe. Place connections at points where stress is low and fatigue conditions are not severe. Provide structures with multiple load paths or redundant members, so that a fatigue crack in any one of the several primary members is not likely to cause collapse of the entire structure.

TABLE 6.23 Stress Categories for Determination of Allowable Stress Ranges* (for tensile stresses or for stress reversal, except as noted)

Structural detail	Stress category	Diagram number
Plain material		
Base metal with rolled or cleaned surface. Flame-cut edges with ANSI smoothness of 1,000 or less	A	1,2
Built-up members		
Base metal in members without attachments, built-up plates or shapes connected by continuous full-penetration groove welds or by continuous fillet welds parallel to the direction of applied stress	B	3,4,5,6
Base metal in members without attachments, built-up plates, or shapes connected by full-penetration groove welds with backing bars not removed, or by partial-penetration groove welds parallel to the direction of applied stress	B′	3,4,5,6
Base metal at toe welds on girder webs or flanges adjacent to welded transverse stiffeners	C	7
Base metal at ends of partial-length, welded cover plates that are narrower than the flange and have square or tapered ends, with or without welds across the ends, or wider than the flange with welds across the ends		
Flange thickness ≤ 0.8 in	E	5
Flange thickness > 0.8 in	E′	5
Base metal at end of partial-length, welded cover plates wider than the flange without welds across the ends	E′	5
Mechanically fastened connections		
Base metal at gross section of high-strength-bolted, slip-critical connections, except axially loaded joints that induce out-of-plane bending in connected material	B	8
Base metal at net section of other mechanically fastened joints	D	8,9

TABLE 6.23   (*Continued*)

Structural detail	Stress category	Diagram number
Mechanically fastened connections		
Base metal at net section of fully tensioned high-strength-bolted, bearing connections	B	8,9
Groove welds		
Base metal and weld metal at full-penetration, groove-welded splices of parts of similar cross section; welds ground flush, with grinding in the direction of applied stress and with weld soundness established by radiographic or ultrasonic inspection	B	10,11
Base metal and weld metal at full-penetration, groove-welded splices at transitions in width or thickness; welds ground to provide slopes no steeper than 1 to 2½, with grinding in the direction of applied stress, and with weld soundness established by radiographic or ultrasonic inspection		
A514 base metal	B′	12,13
Other base metals	B	12,13
Base metal and weld metal at full-penetration, groove-welded splices, with or without transitions having slopes no greater than 1 to 2½ when reinforcement is not removed but weld soundness is established by radiographic or ultrasonic inspection	C	10,11,12,13
Base metal at details attached by full-penetration groove welds, their terminations ground smooth, subject to longitudinal or transverse loading, or both, when the detail embodies a transition radius R, and for transverse loading; weld soundness should be established by radiographic or ultrasonic inspection		
Longitudinal loading		
$R > 24$ in	B	14
24 in $> R > 6$ in	C	14
6 in $> R > 2$ in	D	14
2 in $> R$	E	14

TABLE 6.23 (*Continued*)

Structural detail	Stress category	Diagram number
Groove welds		
Detail base metal for transverse loading: equal thickness; reinforcement removed		
$R > 14$ in	B	14
24 in $> R > 6$ in	C	14
6 in $> R > 2$ in	D	14
2 in $> R$	E	14,15
Detail base metal for transverse loading: equal thickness; reinforcement not removed		
$R > 24$ in	C	14
24 in $> R > 6$ in	C	14
6 in $> R > 2$ in	D	14
2 in $> R$	E	14,15
Detail base metal for transverse loading: unequal thickness; reinforcement removed		
$R > 2$ in	D	14
2 in $> R$	E	14,15
Detail base metal for transverse loading: unequal thickness; reinforcement not removed		
all R	E	14,15
Detail base metal for transverse loading		
$R > 6$ in	C	19
6 in $> R > 2$ in	D	19
2 in $> R$	E	19
Base metal at detail attached by full-penetration groove welds subject to longitudinal loading		
$2 < a < 12b$ or 4 in	D	15
$a > 12b$ or 4 in when $b \leq 1$ in	E	15
$a > 12b$ or 4 in when $b > 1$ in	E'	15
Partial-penetration groove welds		
Weld metal of partial-penetration, transverse groove welds, based on effective throat area of the welds	F	16

13

Groove weld

R

14

a

15

16

b = thickness

17

b

18

TABLE 6.23 *(Continued)*

Structural detail	Stress category	Diagram number
Fillet welds		
Weld metal of continuous or intermittent longitudinal or transverse fillet welds—shear-stress range, including stress reversal	F	15,17,18,20,21
Fillet-welded connections		
Base metal at intermittent fillet welds	E	
Base metal at junction of axially loaded members with fillet-welded end connections. Welds should be dispersed about the axis of the member to balance weld stresses		
$b \le 1$ in	E	17,18
$b > 1$ in	E′	17,18
Base metal at members connected with transverse fillet welds		
$b \le \frac{1}{2}$ in	C	20,21
$b > \frac{1}{2}$ in		
Base metal in fillet-welded attachments, subject to transverse loading, where the weld termination embodies a transition radius, weld termination ground smooth, and main material subject to longitudinal loading:		
$R > 2$ in	D	19
$R < 2$ in	E	19
Base metal at detail attached by fillet welds or partial-penetration groove welds subject to longitudinal loading		
$a < 2$ in	C	15,23,24,25,26
2 in $< a < 12b$ or 4 in	D	15,23,24,26
$a > 12b$ or 4 in when $b \le 1$ in	E	15,23,24,26
$a > 12b$ or 4 in when $b > 1$ in	E′	15,23,24,26
Base metal attached by fillet welds or partial-penetration groove welds subjected to longitudinal loading when the weld termination embodies a transition radius with the weld termination ground smooth:		
$R > 2$ in	D	19
$R \le 2$ in	E	19
Base metal at stud-type shear connector attached by fillet weld or automatic end weld	C	22

19

20

21

22

23

24

TABLE 6.23   (*Continued*)

Structural detail	Stress category	Diagram number
Fillet-welded connections (*Continued*)		
Shear stress including stress reversal, on nominal area of stud-type shear connectors	F	25
Plug or Slot welds		
Base metal at plug or slot welds	E	27
Shear, including stress reversal, on plug or slot welds	F	27

*Based on provisions in the AISC ASD and LRFD specifications for structural steel buildings.

(J. H. Faupel, *Engineering Design*, John Wiley & Sons, Inc., New York; C. H. Norris et al., *Structural Design for Dynamic Loads*, McGraw-Hill, Inc., New York; W. H. Munse, *Fatigue of Welded Steel Structures*, Welding Research Council, 345 East 47th Street, New York, N.Y. 10017.)

6.23 LOCAL PLATE BUCKLING

When compression is induced in an element of a cross section, e.g., a beam or column flange or web, that element may buckle. Such behavior is called **local buckling.** Provision to prevent it should be made in design because, if it should occur, it can impair the ability of a member to carry additional load. Both the AISC ASD and LRFD specifications for structural steel buildings recognize the influence of local buckling by classifying steel sections as compact, noncompact, or slender-element.

A **compact** section has compression elements with width/thickness ratios less than λ_p given in Table 6.24. If one of the compression elements of a cross section exceeds λ_p but does not exceed λ_r given in Table 6.24, the section is **noncompact.** If the width/thickness ratio of an element exceeds λ_r, the section is classified as **slender-element.**

When the width/thickness ratios exceed the limiting value λ_r, the AISC specifications require a reduction in the allowable strength of the member.

The limits on width/thickness elements of compression elements as summarized in Table 6.24 for LRFD and ASD depend on the type of member and whether the element is supported, normal to the direction of the compressive stress, on either one or two edges parallel to the stress. See note *a* in Table 6.24.

6.24 DESIGN PARAMETERS FOR TENSION MEMBERS

To prevent undue vibration of tension members, AISC specifications for ASD and LRFD suggest that the slenderness ratio L/r be limited to 300. This limit does not

TABLE 6.24   Maximum Width/Thickness Ratios b/t^a for Compression Elements for Buildings[b]

Description of element	ASD and LRFD[c] Compact, λ_p	ASD[c] Noncompact[d]	LRFD[c] Noncompact, λ_r
Projecting flange element of I-shaped rolled beams and channels in flexure	$65/\sqrt{F_y}$	$95/\sqrt{F_y}$	$\dfrac{141}{\sqrt{F_y - 10}}$
Projecting flange element of I-shaped hybrid or welded beams in flexure	$65/\sqrt{F_y}$	$95/\sqrt{F_y/k_c}$[e]	$\dfrac{106}{\sqrt{F_{yw} - 16.5}}$
Projecting flange element of I-shaped sections in pure compression; plates projecting from compression elements; outstanding legs of pairs of angles in continuous contact; flanges of channels in pure compression	Not specified	$95/\sqrt{F_y}$	$95/\sqrt{F_y}$
Flanges of square and rectangular box and hollow structural sections of uniform thickness subject to bending or compression; flange cover plates and diaphragm plates between lines of fasteners or welds	$190/\sqrt{F_y}$	$238/\sqrt{F_y}$	$\dfrac{238}{\sqrt{F_y - F_r}}$
Unsupported width of cover plates perforated with a succession of access holes	Not specified	$317/\sqrt{F_y}$	$317/\sqrt{F_y - F_r}$
Legs of single angle struts; legs of double angle struts with separators; unstiffened elements, i.e., supported along one edge	Not specified	$76/\sqrt{F_y}$	$76/\sqrt{F_y}$
Stems of tees	Not specified	$127/\sqrt{F_y}$	$127/\sqrt{F_w}$
All other uniformly compressed stiffened elements, i.e., supported along two edges	Not specified	$253/\sqrt{F_y}$	$253/\sqrt{F_y}$
Webs in flexural compression[a]	$640/\sqrt{F_y}$	$760/\sqrt{F_b}$	$970/\sqrt{F_y}$
D/t for circular hollow sections[f] In axial compression for ASD In flexure for ASD In axial compression for LRFD In flexure for LRFD In plastic design for LRFD	$3300/F_y$ $3300/F_y$ $2070/F_y$ $2070/F_y$ $1300/F_y$	Not specified	$3300/F_y$ $8970/F_y$

[a] t = element thickness. For unstiffened elements supported along only one edge, parallel to the direction of the compression force, the width b should be taken as follows: For flanges of I-shaped members and tees, half the full nominal width; for legs of angles and flanges of channels and zees, the full nominal dimension; for plates, the distance from the free edge to the first row of fasteners or line of welds; and for stems of tees, the full nominal depth.

For stiffened elements (supported along two edges parallel to the direction of the compression force) the width should be taken as follows: For webs of rolled or formed sections, the clear distance between flanges less the fillet or corner radius at each flange; for webs of built-up sections, the distance between adjacent lines of fasteners or the clear distance between flanges when welds are used; for flange or diaphragm plates in built-up sections, the distance between adjacent lines of fasteners or lines of welds; and for flanges of rectangular hollow structural sections, the clear distance between webs less the inside corner radius on each side. If the corner radius is not known, the flat width may be taken as the total section width minus three times the thickness.

[b] As required in AISC specifications for ASD and LRFD. These specifications also set specific limitations on plate-girder components.

[c] F_y = specified minimum yield stress of the steel (ksi), but for hybrid beams, use F_{yf}, the yield strength (ksi) of the flanges.

F_b = allowable bending stress (ksi) in the absence of axial force.

F_r = compressive residual stress in flange (ksi; 10 ksi for rolled shapes, 16.5 ksi for welded shapes).

[d] Elements with width/thickness ratios that exceed the noncompact limits should be designed as slender sections.

[e] $k_c = 4.05/(h/t)^{0.46}$ for $h/t > 70$; otherwise, $k_c = 1$.

[f] D = outside diameter; t = section thickness.

apply to rods. Tension members have three strength-limit states, yielding in the gross section, fracture in the net section, and block shear (see Table 6.14).

For fracture in the net section, as defined by the AISC specifications, the critical net section is the critical cross section over which failure is likely to occur through a chain of holes. The critical section may be normal to the tensile force, on a diagonal, or along a zigzag line, depending on which is associated with the smallest area.

A net section is determined by the net width and the thickness of the joined part. **Net width** is defined as the gross width less the sum of the diameters of all holes in the chain plus $s^2/4g$ for each gage space in the chain, where s is the spacing center-to-center in the direction of the tensile force (pitch) of consecutive holes and g is the transverse spacing center-to-center (gage) of the same consecutive holes. The **critical net section** is defined by the chain of holes with the smallest net width.

For angles, the **gross width** is the sum of the width of the legs less the thickness. In determining the net section, the gage should be taken as the sum of the gages, measured from the backs of the angles, less the thickness of leg.

In the computation of net section for any tension member, the width of the bolt hole should be taken as $\frac{1}{16}$ in greater than the nominal dimension of the hole normal to the direction of the applied stress. Nominal hole dimensions are summarized in Table 6.25.

For design of splice and gusset plates in bolted connections, the net area should be evaluated, as indicated above, except that the actual net area may not exceed 85% of the gross area.

6.25 DESIGN PARAMETERS FOR ROLLED BEAMS AND PLATE GIRDERS

In the design of beams, girders, and trusses, the span should be taken as the distance between the center of gravity of supports. At supports, a flexural member should be restrained against torsion or rotation about its longitudinal axis. Usually, this requires that the top and bottom flanges of the beam be laterally braced. A slender flexural member seated atop a column may become unstable because of the flexibility of the columns if only the top flange is laterally restrained. Therefore, the bottom flange also must be restrained, by bracing or continuity at column connections, to prevent relative rotation between the beam and the column.

TABLE 6.25 Nominal Bolt Hole Dimensions, in

Bolt diameter	Round-hole diameter		Slotted holes—width × length	
	Standard	Oversize	Short-slot	Long-slot
$\frac{1}{2}$	$\frac{9}{16}$	$\frac{5}{8}$	$\frac{9}{16} \times \frac{11}{16}$	$\frac{9}{16} \times 1\frac{1}{4}$
$\frac{5}{8}$	$\frac{11}{16}$	$\frac{13}{16}$	$\frac{11}{16} \times \frac{7}{8}$	$\frac{11}{16} \times 1\frac{9}{16}$
$\frac{3}{4}$	$\frac{13}{16}$	$\frac{15}{16}$	$\frac{13}{16} \times 1$	$\frac{13}{16} \times 1\frac{7}{8}$
$\frac{7}{8}$	$\frac{15}{16}$	$1\frac{1}{16}$	$\frac{15}{16} \times 1\frac{1}{8}$	$\frac{15}{16} \times 2\frac{3}{16}$
1	$1\frac{1}{16}$	$1\frac{1}{4}$	$1\frac{1}{16} \times 1\frac{5}{16}$	$1\frac{1}{16} \times 2\frac{1}{2}$
$\geq 1\frac{1}{8}$	$d + \frac{1}{16}$	$d + \frac{5}{16}$	$(d + \frac{1}{16}) \times (d + \frac{3}{8})$	$(d + \frac{1}{16}) \times 2.5d$

6.25.1 Flange Area

Plate girders, cover-plated beams, and rolled or welded beams should be proportioned on the basis of the gross section. The AISC LRFD specification does not require a deduction for shop or field bolt holes in either flange unless the reduction in flange area is greater than 15% of the gross flange area. In such cases, the area in excess of 15% should be deducted.

The AISC ASD specification for structural steel buildings does not require reduction in flange area for bolt holes when

$$0.5F_u A_{fn} \geq 0.6F_y A_{fg} \tag{6.72}$$

where A_{fg} = gross flange area
A_{fn} = net flange area
F_u = specified minimum tensile strength of the flange steel, ksi
F_y = specified minimum yield stress of the flange steel, ksi

If $0.5F_u A_{fn} < 0.6F_y A_{fg}$, the member flexural properties should be computed using an effective tension flange area:

$$A_{fe} = \frac{5}{6}\frac{F_u}{F_y} A_{fn} \tag{6.73}$$

Welds and bolts connecting the flange to the web or joining a cover plate to a flange should be of adequate size to resist the total horizontal shear due to bending of the member. In addition, flange-to-web connections should be adequate to transmit to the web any direct tension loads applied to the flange.

Cover plates used to increase the flexural strength of a beam should, according to the AISC ASD specification, extend beyond the theoretical cutoff a distance sufficient to develop the capacity of the plate. The AISC LRFD specification gives no specific requirements for the cutoff point of a cover plate.

6.25.2 Web Area

Webs are the shear-carrying elements of beams and girders. At any point along the length of a flexural member, the applied shear must be less than the strength of the gross web area, the product of the overall depth d and thickness t_w of the web.

Generally, shear is not a controlling limit state for the design of a rolled shape. Webs of rolled shapes are thick enough that shear buckling does not occur. However, the shear strength of the web is a major design issue when proportioning a plate girder.

To prevent buckling of the compression flange of a plate girder into the web before the flange yields, both the ASD and LRFD specifications limit the web depth/thickness ratio to

$$h/t_w = 14,000/\sqrt{F_y(F_y + 16.5)} \tag{6.74}$$

where h is the distance between adjacent lines of fasteners or clear distance between flanges for welded flange-to-web connections. However, when transverse stiffeners are utilized and they are spaced not more than $1.5h$ apart, the maximum depth/thickness ratio is

$$h/t_w = 2000/\sqrt{F_y} \tag{6.75}$$

6.25.3 Stiffener Requirements

Stiffeners are often a less costly alternative to increasing the thickness of the web to prevent buckling. Angles or plates normal to a web may be used as bearing stiffeners or transverse or longitudinal stiffeners to prevent buckling of the web.

Bearing stiffeners, when required at locations of concentrated loads or end reactions, should be placed in pairs, normal to and on opposite sides of the web. They should be designed as columns. The column section is assumed to consist of two stiffeners and a strip of the web. The web strip is taken as $25t_w$ for interior stiffeners and $12t_w$ for stiffeners at the end of the web. For computing the design strength of the assumed column section, the effective length may be taken as three-fourths of the stiffener length.

When the load normal to the flange is tensile, the stiffeners should be welded to the loaded flange. When the load normal to the flange is compressive, the stiffeners should either bear on or be welded to the loaded flange. It is essential that a load path exist for the stiffeners to contribute effectively to the member strength.

Transverse intermediate stiffeners, which may be attached to the web either singly or in pairs, increase the shear buckling strength of the web. The stiffeners also increase the carrying capacity of the web through tension-field action.

Intermediate stiffeners are required when the web depth/thickness ratio h/t_w exceeds 260 or when the required shear strength cannot be achieved by an un-reinforced web. The AISC LRFD specification indicates that when $h/t_w \leq 418/\sqrt{F_y}$, transverse stiffeners are not necessary.

When bearing is not needed for transmission of a load or reaction, intermediate stiffeners may be stopped short of the tension flange a distance up to six times the web thickness. The stiffener-to-web weld should be stopped between four and six times the web thickness from the near toe of the web-to-flange weld.

A rectangular-plate compression flange, however, may twist unless prevented by stiffeners. Hence single stiffeners should be attached to the compression flange. When lateral bracing is connected to stiffeners, the stiffeners should be attached to the compression flange with fasteners or welds capable of transmitting 1% of the total flange stress, unless the flange is composed only of angles.

For design of a transverse stiffener, the AISC ASD and LRFD specifications establish minimum requirements for stiffener dimensions. For LRFD, the moment of inertia for a transverse stiffener used to develop the web design shear strength should be at least

$$I_{st} = at_w^3 j \tag{6.76}$$

where a = clear distance between stiffeners, in
$$j = \frac{2.5}{(a/h)^2} - 2 \geq 0.5$$

The moment of inertia should be taken about an axis in the center of the web for stiffener pairs or about the face in contact with the web plate for single stiffeners.

For ASD, the moment of inertia of a single stiffener or pair of stiffeners, with reference to an axis in the plane of the web, should be at least

$$I_{st} = (h/50)^4 \tag{6.77}$$

The gross area (in^2), or total area when stiffeners are attached to the web in pairs, should be at least

$$A_{st} = \frac{1 - C_v}{2} \left[\frac{a}{h} - \frac{(a/h)^2}{1 + (a/h)^2} \right] YDht_w \tag{6.78}$$

where Y = ratio of yield stress of web to that of stiffener
D = 1 for stiffener pairs
= 1.8 for single-angle stiffeners
= 2.4 for single-plate stiffeners
C_v = coefficient defined for Eq. (6.35)

For connecting stiffeners to girder webs, maximum spacing for bolts is 12 in center-to-center. Clear distance between intermittent fillet welds should not exceed 16 times the web thickness nor 10 in.

6.26 *CRITERIA FOR COMPOSITE CONSTRUCTION*

In composite construction, a steel beam and a concrete slab act together to resist bending. The slab, in effect, serves as a cover plate and allows use of a lighter steel section. The AISC ASD and LRFD specifications for structural steel buildings treat two cases of composite members: (1) totally encased members that depend on the natural bond between the steel and concrete, and (2) steel members with shear connectors, mechanical anchorages between a concrete slab and the steel.

6.26.1 Composite Beams with Shear Connectors

The most common application of composite construction is a simple or continuous beam with shear connectors. For such applications, the design may be based on an effective concrete-steel T-beam, where the width of the concrete slab on each side of the beam centerline may not be taken more than

1. One-eighth of the beam span, center-to-center of supports
2. One-half the distance of the centerline of the adjacent beam
3. The distance from beam centerline to the edge of the slab

LRFD for Composite Beams.   The AISC LRFD specification requires that the design positive-moment capacity ϕM_n of a composite beam be computed as follows:
For $h_c/t_w \leq 640/\sqrt{F_y}$, $\phi = 0.85$ and M_n is to be determined based on the plastic stress distribution (Fig. 6.4). For $h_c/t_w > 640/\sqrt{F_y}$, $\phi = 0.90$ and M_n is to be determined from superposition of elastic stresses on the transformed section, with

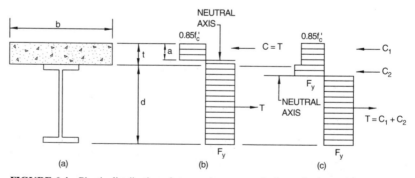

FIGURE 6.4   Plastic distribution of stresses in a composite beam in the positive-moment region. (*a*) Structural steel beam connected to concrete slab for composite action. (*b*) Stress distribution when neutral axis is in the slab. (*c*) Stress distribution when neutral axis is in the web.

effects of shoring taken into account, where F_y is the specified yield stress of tension flange (ksi), t_w is the web thickness (in), and h_c is twice the distance (in) from the neutral axis of the steel beam alone to the inside face of the compression flange (less the fillet or corner radius for rolled beams) or to the nearest line of mechanical fasteners at the compression flange.

In the negative-moment region of a composite beam, the design flexural strength should be determined for the steel section alone unless provision is made to utilize composite action. When the steel beam is an adequately braced, compact shape, shear connectors connect the slab to the steel in the negative-moment region, and the slab reinforcement parallel to the steel beam, within the effective width of the slab, is adequately developed, the negative-moment capacity may be taken as $\phi_b M_n$, with $\phi_b = 0.85$. Reinforcement parallel to the beam may be included in computations of properties of the composite section.

When a composite beam will be shored during construction until the concrete has developed sufficient strength, composite action may be assumed in design to be available to carry all loads. When shores are not used during construction, the steel section acting alone should be designed to support all loads until the concrete attains 75% of its specified compressive strength.

Composite construction often incorporates steel deck that serves as a form for the concrete deck, is connected to the steel beam, and remains in place after the concrete attains its design strength. The preceding moment-capacity computations may be used for composite systems with metal deck if the deck meets the following criteria:

The steel-deck rib height should not exceed 3 in.

Average width of concrete rib or haunch and slab thickness above the steel deck should be at least 2 in.

Welded stud shear connectors should be ¾ in in diameter or less and extend at least 1½ in above the steel deck.

The AISC LRFD specification indicates that for ribs perpendicular to the steel beam, concrete below the top of the steel deck should be neglected in determining section properties, whereas for ribs parallel to the steel beam, that concrete may be included (see also Art. 6.26.2).

ASD for Composite Beams.   The AISC ASD specification requires that flexural strength of a composite beam be determined by elastic analysis of the transformed section with an allowable stress of $0.66F_y$, where F_y is the minimum specified yield stress (ksi) of the tension flange. This applies whether the beam is temporarily shored or unshored during construction.

The transformed section comprises the steel beam and an equivalent area of steel for the compression area of the concrete slab. The equivalent area is calculated by dividing the concrete area by the modular ratio n, where $n = E/E_c$, E is the elastic modulus of the steel beam, and E_c is the elastic modulus of the concrete.

If a composite beam will be constructed without shoring, the steel section should be assumed to act alone in carrying loads until the concrete has attained 75% of its specified compressive strength. The maximum allowable stress in the steel beam, in this case, is $0.9F_y$. After that time, the transformed section may be assumed to support all loads. The maximum allowable stress in the concrete is $0.45f_c'$, where f_c' is the specified concrete compressive strength.

When shear connectors are used, full composite action is obtained only when sufficient connectors are installed between points of maximum moment and points of zero moment to carry the horizontal shear between those points. When fewer connectors are provided, the increase in bending strength over that of the steel beam alone is directly proportional to the number of steel connectors. Thus, when adequate connectors for full composite action are not provided, the section modulus of the transformed composite section must be reduced accordingly.

For ASD, an effective section modulus may be computed from

$$S_{eff} = S_s + (S_{tr} - S_s) \sqrt{\frac{V_{h'}}{V_h}} \tag{6.79}$$

where V_h = smaller of total horizontal shears computed from Eqs. (6.82) and (6.81) or the shear from Eq. (6.83)

$V_{h'}$ = allowable horizontal shear load on all the connectors between point of maximum moment and nearest point of zero moment [see Eq. (6.86)]

S_s = section modulus of steel beam referred to the bottom flange

S_{tr} = section modulus of transformed composite section referred to its bottom flange, based on maximum permitted effective width of concrete flange

6.26.2 Shear Connectors

The purpose of shear connectors is to ensure composite action between a concrete slab and a steel beam by preventing the slab from slipping relative to or lifting off the flange to which the connectors are welded. Headed-stud or channel shear connectors are generally used. The studs should extend at least four stud diameters above the flange. The welds between the connectors and the steel flange should be designed to resist the shear carried by the connectors. When the welds are not directly over the beam web, they tend to tear out of a thin flange before their full shear-resisting capacity is attained. Consequently, the AISC ASD and LRFD specifications require that the diameter of studs not set directly over the web be $2\frac{1}{2}$ times the flange thickness or less. The specifications also limit the spacing center-to-center of shear connectors to a maximum of eight times the total slab thickness. Minimum spacings permitted are six stud diameters along the longitudinal axis of the beam and four stud diameters perpendicular to that axis.

Shear connectors, except those installed in the ribs of formed steel deck, should have at least 1 in of concrete cover in all directions.

The AISC LRFD specification requires that the total horizontal shear V_h necessary to develop full composite action between the point of maximum positive moment and the point of zero moment be the smaller of V_h (kips) computed from Eqs. (6.80) and (6.81):

$$V_h = A_s F_y \tag{6.80}$$

$$V_h = 0.85 f'_c A_c \tag{6.81}$$

where A_s = area of the steel cross section, in^2

A_c = area of concrete slab, in^2

F_y = specified minimum yield stress of steel tension flange, ksi

f'_c = specified compressive strength of concrete, ksi

The AISC ASD specification requires V_h to be the smaller of V_h computed from Eqs. (6.82) and (6.83):

$$V_h = A_s F_y / 2 \tag{6.82}$$

$$V_h = 0.85 f'_c A_c / 2 \tag{6.83}$$

If longitudinal reinforcing steel with area A'_s within the effective width of the concrete slab is included in the properties of the concrete, $\frac{1}{2} F_{yr} A'_s$ should be added to the right-hand side of Eq. (6.82).

In negative-moment regions of continuous composite beams, longitudinal rein-forcing steel may be placed within the effective width of the slab to aid the steel beam in carrying tension due to bending, when shear connectors are installed on the tension flange. The total horizontal shear (kips) to be resisted by the connectors between an interior support and each adjacent inflection point should be taken as

$$V_h = \frac{A_{sr}F_{yr}}{2} \tag{6.84}$$

where A_{sr} = total area of longitudinal reinforcing steel within the effective flange width at the interior support, in^2

F_{yr} = specified minimum yield stress of reinforcing steel, ksi

For full composite action, the number of shear connectors resisting V_h each side of a point of maximum moment should be at least

$$N = \frac{V_h}{q} \tag{6.85}$$

where q is the allowable shear load for one connector. If N_1 connectors are provided, where $N_1 < N$, they may be assumed capable of carrying a total horizontal shear (kips) of

$$V_{h'} = N_1 q \tag{6.86}$$

This shear is used in Eq. (6.79) to determine the effective section modulus.

Shear connectors generally can be spaced uniformly between points of maximum and zero moment. Some loading patterns, however, require closer spacing near inflection points or supports. The AISC specifications consequently require that when a concentrated load occurs in a region of positive bending moment, the number of connectors required between that load and the nearest point of zero moment should be at least

$$N_2 = N_1 \frac{M\beta/M_{max} - 1}{\beta - 1} \tag{6.88}$$

where M_{max} = maximum positive bending moment

M = bending moment at a concentrated load ($M < M_{max}$)

$\beta = S_{tr}/S_s$ or S_{eff}/S_s, whichever is applicable

S_s = section modulus of steel beam, referred to bottom flange

S_{tr} = section modulus of transformed composite section, referred to bottom flange, based on maximum permitted effective width of concrete flange

S_{eff} = effective section modulus [Eq. (6.79)]

The AISC LRFD specification indicates that the nominal strength q (kips) of a stud shear connector embedded in a solid concrete slab may be computed from

$$q = 0.5A_{sc}\sqrt{f_c' E_c} \le A_{sc}F_u \tag{6.88}$$

where A_{sc} = cross-sectional area of a stud connector, in^2

E_c = elastic modulus of the concrete = $w^{1.5}\sqrt{f_c'}$

w = weight of the concrete, lb/ft^3

F_u = tensile strength of a stud connector, ksi

For a channel shear connector embedded in a solid slab,

$$q = 0.3(t_f + 0.5t_w)L_c\sqrt{f_c' E_c} \tag{6.89}$$

TABLE 6.26   Allowable Shear Loads (kips) for Shear Connectors*

Connector sizes	Specified concrete compressive strength, ksi		
	3.0	3.5	$f'_c \geq 4.0$
Hooked or headed studs:			
½-in diameter × 2 in or more	5.1	5.5	5.9
⅝-in diameter × 2½ in or more	8.0	8.6	9.2
¾-in diameter × 3 in or more	11.5	12.5	13.3
⅞-in diameter × 3½ in or more	15.6	16.8	18.0
Channels:†			
C3 × 4.1	4.3w	4.7w	5.0w
C4 × 5.4	4.6w	5.0w	5.3w
C5 × 6.7	4.9w	5.3w	5.6w

*For concrete made with ASTM C33 aggregates.
†w = length of channel, in.
Based on the AISC ASD specification for structural steel buildings.

where t_f = thickness of channel flange, in
t_w = thickness of channel web, in
L = length of channel, in

For ASD, the allowable shear q (kips) for a connector is given in Table 6.26 for flat-soffit concrete slabs made with C33 aggregate. For concrete made with C330 lightweight aggregate, the factors in Table 6.27 should be applied.

Working values q for shear connectors in Table 6.26 incorporate a safety factor of 2.5 applied to ultimate strength. For use with concrete not conforming to ASTM C33 or C330 and for types of connectors not listed in the table, values of q should be established by tests.

6.26.3 Encased Beams

When shear connectors are not used, composite action may be attained by encasing a steel beam in concrete. To be considered composite construction, concrete and

TABLE 6.27   Correction Factors for Connector Shear Loads when Concrete Is Made with ASTM C330 Aggregates*

f'_c, ksi	Air-dry weight of concrete, lb/ft³						
	90	95	100	105	110	115	120
4 or less	0.73	0.76	0.78	0.81	0.83	0.86	0.88
5 or more	0.82	0.85	0.87	0.91	0.93	0.96	0.99

*Multiply the horizontal shear obtained from Table 6.26 for concrete made with C33 aggregate by the applicable correction factor to obtain the allowable shear load for one connector in a flat-soffit concrete slab made with rotary kiln–produced aggregate conforming to ASTM 330.

steel must satisfy the following requirements: Steel beams should be encased 2 in or more on their sides and bottom in concrete cast integrally with the slab (Fig. 6.5). The top of the steel beam should be at least 1½ in below the top of the slab and 2 in above its bottom. The encasement should be reinforced throughout its depth and across its bottom to prevent spalling of the concrete.

For such encased beams, composite action may be assumed produced by bond between the steel member and the concrete.

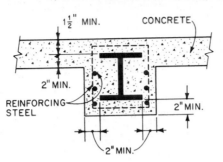

FIGURE 6.5 Concrete cover required for encased beams.

Design of an encased beam depends on whether the steel beam is shored temporarily when the concrete is cast. If the shoring remains in place until the concrete attains 75% of required strength, the composite section can be assumed to carry all loads. Without such shoring, the steel beam carries the dead load unassisted. Only loads applied after the concrete reaches 75% of its required strength can be assumed taken by the composite beam.

Since the beam then is completely braced laterally, the allowable bending stress in the steel flanges for ASD is $0.66F_y$, where F_y is the steel yield stress (ksi). Compressive stress in the concrete should not exceed $0.45f_c'$, where f_c' is the specified 28-day strength of the concrete (ksi). The concrete should be assumed unable to carry tension. Reinforcement to resist tension, however, may be provided in the concrete for negative moment in continuous beams and cantilevers. This reinforcement should be placed within the effective width.

For LRFD, the design flexural strength $\phi_b M_n$ should be computed with $\phi_b = 0.90$ and M_n determined from superposition of elastic stresses on the transformed composite section or from the plastic distribution on the steel section alone (Art. 6.17.2).

The **transformed composite section** is obtained by treating the concrete on the compression side of the neutral axis as an equivalent steel area. This is done by dividing that concrete area by n, the ratio of the modulus of elasticity of steel to that of the concrete.

6.26.4 Composite Columns

The AISC LRFD specification for structural steel buildings contains provisions for design of concrete-encased compression members. It sets the following requirements for qualification as a composite column: The cross-sectional area of the steel core—shapes, pipe, or tubing—should be at least 4% of the total composite area. The concrete should be reinforced with longitudinal load-carrying bars, continuous at framed levels, and lateral ties and other longitudinal bars to restrain the concrete, all with at least 1½ in of clear concrete cover. The cross-sectional area of transverse and longitudinal reinforcement should be at least 0.007 in² per in of bar spacing. Spacing of ties should not exceed two-thirds of the smallest dimension of the composite section. Strength of the concrete f_c' should be between 3 and 8 ksi for normal-weight concrete and at least 4 ksi for lightweight concrete. Specified minimum yield stress F_y of steel core and reinforcement should not exceed 55 ksi. Wall thickness of steel pipe or tubing filled with concrete should be at least $b\sqrt{F_y/3E}$ or $D\sqrt{F_y/8E}$, where b is the width of the face of a rectangular section, D is the outside diameter of a circular section, and E is the elastic modulus of the steel.

The AISC LRFD specification gives the design strength of an axially loaded composite column as ϕP_n, where $\phi = 0.85$ and P_n is determined from Eqs. (6.90) to (6.92):

$$\phi P_n = 0.85 A_s F_{cr} \tag{6.90}$$

For $\lambda_c \leq 1.5$,

$$F_{cr} = 0.758^{\lambda_c^2} F_{my} \tag{6.91}$$

For $\lambda_c > 1.5$,

$$F_{cr} = \frac{0.877}{\lambda_c^2} F_{my} \tag{6.92}$$

where $\lambda_c = (KL/r_m\pi)\sqrt{F_{my}/E_m}$
$\quad KL$ = effective length of column, in (Art. 6.16.2)
$\quad A_s$ = gross area of steel core, in^2
$\quad F_{my} = F_y + c_1 F_{yr}(A_r/A_s) + c_2 f_c'(A_c/A_s)$
$\quad E_m = E + c_3 E_c(A_c/A_s)$
$\quad r_m$ = radius of gyration of steel core, in ≤ 0.3 of the overall thickness of the composite cross section in the plane of buckling for steel shapes
$\quad A_c$ = cross-sectional area of concrete, in^2
$\quad A_r$ = area of longitudinal reinforcement, in^2
$\quad E_c$ = elastic modulus of concrete, ksi
$\quad F_{yr}$ = specified minimum yield stress of longitudinal reinforcement, ksi

For concrete-filled pipe and tubing, $c_1 = 1.0$, $c_2 = 0.85$, and $c_3 = 0.4$. For concrete-encased shapes, $c_1 = 0.7$, $c_2 = 0.6$ and $c_3 = 0.2$.

When the steel core consists of two or more steel shapes, they should be tied together with lacing, tie plates, or batten plates to prevent buckling of individual shapes before the concrete attains $0.75f_c'$.

Design strength of the concrete is $1.7\phi_c f_c' A_b$, where $\phi_c = 0.60$ and A_b = loaded area (in^2).

6.27   SERVICEABILITY

This is a state in which the function, appearance, maintainability, and durability of a building and comfort and safety of its occupants are preserved under normal usage. Generally, serviceability limits are based on engineering judgment, taking into account the purpose to be served by the structure, provisions for occupant safety and comfort, and characteristics of collateral building materials. Serviceability checks should be conducted for service loads, not factored loads.

Vertical Deflections.   Limits on vertical deflections of a flexural member are not prescribed by the AISC LRFD specification for structural steel buildings. However, the ASD specification limits deflection to $\frac{1}{360}$ of the span L for beams supporting plastered ceilings. Because the deflection limits are application-dependent, the specifications do not give comprehensive criteria. (See J. M. Fisher and M. A. West, *Serviceability Design Considerations for Low-Rise Buildings*, American Institute of Steel Construction.) The "Commentary" on the ASD specification suggests that the depth of fully stressed flexural members be at least $F_y L/800$ for floors and $F_y L/1000$ for roof purlins, except for flat roofs.

Ponding. Special consideration should be given to the vertical deflection of roof beams due to **ponding,** the retention of water caused solely by the deflections of flat-roof framing. The amount of retained water depends on the flexibility of the framing; as the beams deflect, the depth of ponding increases and increases the deflections even more. Failure of the beams is a possibility. Therefore, according to both the AISC LRFD and ASD specifications, roof systems should be investigated to ensure stability under ponding. Preferably, roofs should be provided with sufficient slope for drainage of rainwater. A flat roof with a steel deck supported on steel purlins may be considered stable, and no further investigation is required if Eqs. (6.93) and (6.94) are satisfied:

$$C_p + 0.9C_s \leq 0.25 \tag{6.93}$$

$$I_d \geq 25S^4 \times 10^{-6} \tag{6.94}$$

where I_d = moment of inertia of the steel deck, in^4

$$C_p = \frac{32L_sL_p{}^4}{10^7I_p}$$

$$C_s = \frac{32SL_s{}^4}{10^7I_s}$$

L_p = column spacing in direction of girders (length of primary members), ft

L_s = column spacing perpendicular to direction of girders (length of purlins), ft

S = spacing of purlins, ft

I_p = moment of inertia of a girder, in^4

I_s = moment of inertia of a purlin, in^4

For trusses and steel joists, the moment of inertia I_s should be reduced by 15% when used in Eq. (6.94).

For ponding analysis, the ASD specification requires, in addition, that the total bending stress due to dead loads, gravity loads, and ponding should not exceed $0.80F_y$ for primary and secondary members. Stresses resulting from wind or seismic forces need not be included in a ponding analysis.

Drift. Lateral deflections (sidesway or drift) under service loads of a steel frame should be limited so as not to injure building occupants or damage walls, partitions, or attached cladding. In computation of drift, a 10-year recurrence-interval wind is generally used. The 10-year wind pressure can be estimated at 75% of the 50-year wind pressure. (For design guidance on wind-drift limits, see J. M. Fisher and M. A. West, *Serviceability Design Considerations for Low-Rise Buildings*, American Institute of Steel Construction.

Vibrations. In design of beams and girders that will support large areas free of partitions or other sources of damping, the members should be proportioned to limit vibrations to levels below those perceptible to humans. The "Commentary" to the AISC ASD specification suggests that the depth of steel beams for such areas be at least $L/20$. (See also T. M. Murray, "Acceptability Criterion for Occupant-Induced Floor Vibrations," *Engineering Journal*, second quarter, 1981, American Institute of Steel Construction.

Camber. **Camber** is a curvature of a flexural member to offset some or all deflections due to service loads. The objective generally is to eliminate the appearance of sagging or to match the elevation of the member to adjacent building components when the member is loaded. Camber requirements should be specified in the design documents.

The ASD specification requires that trusses spanning 80 ft or more be cambered for about the dead-load deflection. Crane girders spanning 75 ft or more should be cambered for about the dead-load deflection plus one-half the live-load deflection.

6.28 BUILT-UP COMPRESSION MEMBERS

Design of built-up compression members should comply with the basic requirement for prevention of local and overall buckling of compression members as summarized in Arts. 6.16 and 6.23. To ensure, however, that individual components, such as plates and shapes, of a built-up member act together, the AISC ASD and LRFD specifications for structural steel buildings emphasize proper interconnection of the components. Many of the AISC requirements are the same for ASD and LRFD. The ASD specification, however, requires that all connections be welded or made with fully tightened, high-strength bolts.

Connections at Ends. For built-up columns bearing on base plates or milled surfaces, components in contact at or near the ends should be connected with rivets, bolts, or welds. Bolts should be placed parallel to the axis of the member not more than four diameters apart for a distance of at least $1\frac{1}{2}$ times the maximum width of the member. The weld should be continuous and at least as long as the maximum width of the member.

Intermediate Connections.   Between the end connections of built-up compression members, the longitudinal spacing of welds or bolts should be adequate to transfer applied forces. Along an outside plate, when welds are used along the plate edges or when bolts are provided at all adjacent gage lines (not staggered) at each section, the maximum spacing should not exceed 12 in or $127t/\sqrt{F_y}$, where t is the thickness of the thinner outside plate and F_y is the specified minimum yield stress of the steel. When bolts are staggered, the maximum spacing along each gage line should not exceed 18 in or $190t\sqrt{F_y}$.

For two rolled shapes in contact, the maximum longitudinal spacing of bolts or intermittent welds should not exceed 24 in. If two or more rolled shapes in compression members are separated by intermittent fillers, the shapes should be connected at the fillers by at least two intermediate connectors at intervals that limit local buckling. Accordingly, the AISC specifications restrict the slenderness ratio of each shape KL/r, where L is the connector spacing, to a maximum of three-fourths the governing slenderness ratio of the built-up member. The least radius of gyration r should be used in computation of the slenderness ratio of each component.

Components of a built-up compression member that has open sides may be tied together with lacing or perforated cover plates. Lacing may consist of flat bars, angles, channels, or other shapes inclined at an angle to the axis of the member of at least 60° for single lacing and 45° for double lacing. The lacing should terminate at tie plates that are connected across the ends of the member and that are at least as long as the distance s between the lines of connectors between tie plates and member components. Tie plates also should be installed at intermediate points where lacing is interrupted and should be at least $s/2$ long. Thickness of tie plates should be at least $s/50$. A tie plate should be connected to each component by at least three bolts or by welds with a length of at least one-third that of the tie plate. Spacing of bolts in tie plates should not exceed six diameters.

Lacing should be capable of resisting a shear force normal to the axis of the built-up member equal to 2% of the total compressive stress in the member in ASD (compressive design strength in LRFD). Spacing of lacing should limit the L/r of the flange, where L is the distance along the flange between its connections to the lacing, to a maximum of three-fourths the governing slenderness ratio of the built-up member. Maximum permissible slenderness ratio for single lacing is 140, and for double lacing, 200. For lacing in compression, the unsupported length should be taken as the distance between lacing connections to the built-up member for single lacing and 70% of that for double lacing. When the distance between the lines of connectors across the open sides of the built-up member exceeds 15 in, angles or double lacing, joined at intersections, should be used.

Continuous cover plates perforated with access holes are an alternative to lacing. The plates should be designed to resist axial stress and local buckling (Art. 6.23). Length of holes in the direction of stress should not exceed twice the width, and the holes should have a minimum radius of 1½ in. The clear distance between holes in the direction of stress should be at least the transverse distance between the nearest lines of connecting bolts or welds.

For built-up members made of weathering steels that will be exposed unpainted to atmospheric corrosion, designers should take precautions to ensure that corrosion is not accelerated by moisture entrapped between faying surfaces. Hence components should be held in tight contact. Spacing of fasteners between a plate and a shape or between two plate components in contact should not exceed 14 times the thickness of the thinnest part nor 7 in. Maximum edge distance should not exceed 8 times the thickness of the thinnest part nor 5 in.

Strength of Built-Up Compression Members.   For ASD, the equations for allowable stresses for axially loaded compression members given in Art. 6.16.2 also may be used for built-up compression members.

If the buckling mode of a built-up member involves relative deformation or slip between the components that induces shear in the connections between components, resistance to buckling may be lowered. The AISC LRFD specification requires that the design strength of a built-up compression member composed of two or more shapes be computed from Eqs. (6.39) to (6.40b) with the slenderness ratio KL/r in λ_c replaced by a modified slenderness ratio $(KL/r)_m$ given by Eqs. (6.95) to (6.97).

When snug-tight bolted connections are used,

$$\left(\frac{KL}{r}\right)_m = \sqrt{\left(\frac{KL}{r}\right)_o^2 + \left(\frac{a}{r_i}\right)^2} \tag{6.95}$$

When welds or fully tightened bolts, as for slip-critical joints, are used and $a/r_i > 50$,

$$\left(\frac{KL}{r}\right)_m = \sqrt{\left(\frac{KL}{r}\right)_o^2 + \left(\frac{a}{r_i} - 50\right)^2} \tag{6.96}$$

With the aforementioned connectors and $a/r_i \leq 50$,

$$\left(\frac{KL}{r}\right)_m = \left(\frac{KL}{r}\right)_o \tag{6.97}$$

where $(KL/r)_o$ = slenderness ratio of built-up member acting as a unit
a = longitudinal spacing between connectors
r_i = minimum radius of gyration of a component

6.29 BUILT-UP TENSION MEMBERS

The design strength and allowable stresses for prismatic built-up members subjected to axial tension by static forces are the same as for tension members given in Art. 6.13.

Components of a built-up tension member should be connected at frequent intervals to ensure that they act together, that faying surfaces intended to be in contact stay in contact, that excessive vibration of relatively thin parts does not occur, and that moisture will not penetrate between faying surfaces and cause corrosion.

The slenderness ratio L/r of any component, where L is the longitudinal spacing of fasteners, should preferably not exceed 300. Provisions for lacing and tie plates, perforated cover plates, and fasteners, except those provisions intended specifically for compression members, are the same as for built-up compression members (Art. 6.28).

6.30 PLASTIC DESIGN

Structural steel members often have considerable reserve load-carrying capacity after yielding occurs, e.g., at the outer surfaces of a beam. For a flexural member, this reserve capacity is quantified by the shape factor Z/S of the cross section, where Z is the plastic section modulus and S is the corresponding elastic section modulus. For W shapes, Z/S ranges from 1.10 to 1.15.

The AISC ASD and LRFD specifications recognize that for structures with continuous framing, additional reserve strength is available above that predicted by elastic theory. When loads induce yielding at points of maximum negative bending moment, plastic hinges start to form there, and rotation occurs without increase in stresses beyond the yield stress. Positive moments and corresponding stresses, however, increase. The result is a redistribution of moments, creating reserve load-carrying capacity.

The AISC ASD and LRFD specifications also recognize the reserve strength of properly braced, compact shapes when stresses are determined by elastic theory. In LRFD, the maximum moment capacity is defined as the plastic moment M_p instead of the yield moment M_y. In ASD, the allowable stress is $0.66F_y$ instead of $0.60F_y$.

The LRFD specification permits plastic analysis only for steels with yield stress not exceeding 65 ksi. Compression flanges involving plastic-hinge rotation and all webs should have a width/thickness ratio not exceeding λ_p in Table 6.24. The vertical bracing system of a multistory frame should be capable of preventing buckling of the structure and maintaining lateral stability under factored loads. In frames where lateral stability depends on the bending stiffness of rigidly connected beams and columns (rigid frames), the axial force in the columns under factored loads should not exceed $0.75A_gF_y$, where A_g is the column area and F_y is the yield stress of the steel. The slenderness parameter λ_c for columns should not exceed 150% of the effective length factor K (Art. 6.16.2).

The laterally unbraced length L_b of the compression flange of a flexural member at plastic hinges associated with the failure mechanism, for a compact section bent about the major axis, should not exceed L_{pd}, as given by Eq. (6.98). The equation applies to I-shape members, loaded in the plane of the web, that are doubly symmetrical or are singly symmetrical with the compression flange larger than the tension flange:

$$L_{pd} = \frac{3600 + 2200(M_l/M_p)}{F_y} r_y \qquad (6.98)$$

where F_y = specified minimum yield stress of compression flange, ksi

r_y = radius of gyration about minor axis, in

M_l = smaller moment at an end of the unbraced length of the member, in-kips

M_p = plastic moment computed for the fully plastic distribution for hybrid girders, in-kips

= $F_y Z$ for homogeneous sections

M_l/M_p is positive when moments cause reverse curvature. L_b is not limited for members with square or circular cross sections nor for beams bent about the minor axis. In the region of the last hinge to form before failure, the flexural design strength is ϕM_n, as given in Art. 6.17.2.

Moment redistribution is permitted for beams and girders, composite or non-composite, except for hybrid girders and members of A514 steel, that meet the preceding requirements. The qualifying members also should be continuous over supports or rigidly framed to columns with rivets, high-strength bolts, or welds. Such members, except for cantilevers, may be proportioned for 90% of negative moments that are induced by gravity loading and are maximum at points of support. The maximum positive moment, however, should be increased by 10% of the average negative moments. When beams or girders are rigidly framed to columns, the columns may be proportioned for the 10% reduction for the combined axial and bending loading, but the stress f_a due to concurrent axial loads on the columns should not exceed $0.15F_a$, where F_a is the axial design strength.

6.31 CABLE CONSTRUCTION

Cables, more commonly referred to as **wire rope,** are sometimes used in buildings to support long-span roofs, to suspend floorbeams from upper levels, and as bracing members to resist wind loads. Wire rope is made of three basic components: wires, strands, and a core. Each rope end usually is equipped with a fitting, an accessory for attaching the cable to an anchorage or part of a structure.

A **wire** is a single, continuous length of metal cold-drawn from a rod. A **strand** consists of multiple wires placed helically around a central wire to produce a symmetrical section. A **rope** is comprised of strands laid helically around a core composed of a strand or another wire. Strand, in general, is stronger and has a higher modulus of elasticity than rope of the same size but is less flexible.

Design Strength.   The **breaking strength** is the ultimate load registered for a wire rope during a tension test. For design, the breaking strength is divided by a design factor, which is the ratio of the breaking strength to the expected design load. This factor plays an important role in determining the life expectancy of a rope. (Excessive loading will impair a rope's serviceability.) The American Iron and Steel Institute "Wire Rope Users Manual" suggests a common design factor of 5.

Breaking strength is a function of rope classification, coating, and diameter. The AISI "Wire Rope Users Manual" lists industry-accepted strengths. ASTM A603, "Standard Specification for Zinc-Coated Steel Wire Rope," also gives minimum breaking strengths.

Design Considerations.   Cables should be assumed to have no resistance to bending. Design should take in to account stretch from both tensioning of the rope and the applied loads. Also, design should consider deflections and resulting stresses caused by changes in magnitude and position of loads. In addition, effects of temperature variations should be accounted for in the design of both the cables and the supporting structure. Design provisions are also necessary for expansion and contraction of wire rope.

Wire rope supporting floors or roofs should be proportioned with due consideration for limiting deflections produced by the design loads.

(See also Secs. 1 and 4.)

6.32   FIRE PROTECTION*

Building codes play a dominant role in defining the level of fire protection that is expected by society. Typically, fire protection is implemented in design through code compliance. As a consequence, a working knowledge of building codes is an important prerequisite for contemporary design.

In the past, keeping abreast of building codes was difficult, even for the largest design offices, since most major cities and a number of states maintained locally developed codes. Today, this impediment is less relevant. In the United States and Canada, with a few notable exceptions, the vast majority of cities, states, and provinces now enforce one of the following model codes:

- *National Building Code*, Building Officials and Code Administrators International, Homewood, Ill.
- *Standard Building Code*, Southern Building Code Congress International, Birmingham, Ala.
- *Uniform Building Code*, International Conference of Building Officials, Whittier, Calif.
- *National Building Code of Canada*, National Research Council of Canada, Ottawa, Ontario

Two fire-related characteristics of materials influence selection and design of structural systems: combustibility and fire resistance.

6.32.1   Combustible and Noncombustible Materials

Most fires are either accidental or caused by carelessness. Fires are usually small when they start and require fuel to grow in intensity and magnitude. In fact, many fires either self-extinguish due to a lack of readily available fuel or are extinguished by building occupants. Furthermore, even though most fires involve building contents, a combustible building itself may be the greatest potential source of fuel.

By definition, noncombustible materials such as stone, concrete, brick, and steel do not burn and therefore do not serve as sources of fuel. Although the physical properties of noncombustible materials may be adversely affected by elevated temperature exposures, these materials do not contribute to either the intensity or duration of fires. Wood, paper, and plastics are examples of combustible materials.

Tests conducted by the National Institute for Standards and Technology (formerly the National Bureau of Standards) indicate that an approximate relationship exists between the amount of available combustible material (**fire loading,** pounds of wood equivalent per square foot of floor area) and **fire severity,** hours of equivalent fire exposure (Fig. 6.6). Subsequent field surveys measured the fire loads typically found in buildings with different occupancies (Table 6.28).

---

*Article 6.32 was written by Delbert F. Boring, Regional Director, Construction Codes and Standards, American Iron and Steel Institute, Columbus, Ohio.

TABLE 6.28 Typical Occupancy Fire Loads and Fire Severity*

Type of occupancy	Occupancy fire load, psf	Equivalent fire severity, h
Assembly	5 to 10	½ to 1
Business	5 to 10	½ to 1
Educational	5 to 10	½ to 1
Hazardous	Variable	Variable
Industrial		
Low hazard	0 to 10	0 to 1
Moderate hazard	10 to 25	1 to 2½
Institutional	5 to 10	½ to 1
Mercantile	10 to 20	1 to 2
Residential	5 to 10	½ to 1
Storage		
Low hazard	0 to 10	0 to 1
Moderate hazard	10 to 30	1 to 3

*Based on data in *Fire Protection through Modern Building Codes*, American Iron and Steel Institute, Washington, D.C.

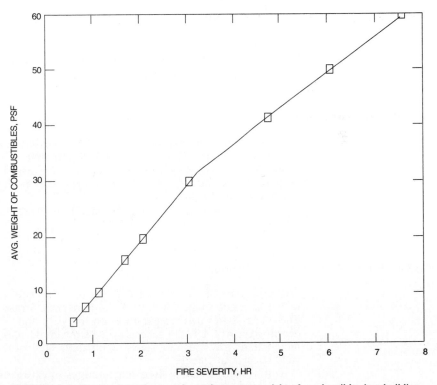

FIGURE 6.6 Curve relates fire severity to the average weight of combustibles in a building.

A reasonable estimate of the structural fire loading for conventional wood-frame construction is 7½ to 10 psf. For heavy-timber construction, the corresponding structural fire load may be on the order of 12½ to 17½ psf. As a consequence, building codes generally limit the permitted size (allowable height and area) of combustible buildings to a much greater degree than for noncombustible buildings.

6.32.2 Fire Resistance

In addition to regulating building construction based on the combustibility or non-combustibility of structures, building codes also specify fire-resistance requirements as a function of building occupancy and size, i.e., height and area. In general, **fire resistance** is defined as the relative ability of construction assemblies, such as, floors, walls, partitions, beams, girders, and columns, to prevent spread of fire to adjacent spaces or perform structurally when exposed to fire. Fire-resistance requirements are based on tests conducted in accordance with "Standard Methods of Fire Tests of Building Construction and Materials" (ASTM E119).

The ASTM E119 test method specifies a "standard" fire exposure that is used to evaluate the fire resistance of construction assemblies (Fig. 6.7). Fire-resistance requirements are specified in terms of the time during which an assembly continues to prevent the spread of fire or perform structurally when exposed to the "standard fire." Thus fire-resistance requirements are expressed in terms of hours or fractions thereof. The design of fire-resistant buildings is typically accomplished in a pre-scriptive fashion by selecting tested "designs" that meet specific building code requirements. Listings of fire-resistance ratings for construction assemblies are available from a number of sources:

- *Fire-Resistance Directory*, Underwriters Laboratories, Northbrook, Ill.
- *Fire-Resistance Ratings*, American Insurance Services Group, New York, N.Y.
- *Fire-Resistance Design Manual*, Gypsum Association, Washington, DC.

6.32.3 Fireproof Buildings

In the past, the term **fireproof** was frequently used to describe fire-resistant buildings. The use of this and terms such as **fireproofing** is unjustified and should be avoided. Experience has clearly demonstrated that large-loss fires (in terms of both property losses and loss of life) can and do occur in fire-resistant buildings. No building is truly fireproof.

(*Fire Protection Handbook*, National Fire Protection Association, Quincy, Mass.).

6.32.4 Effect of Temperature on Steel

The properties of virtually all building materials are adversely affected by the temperatures developed during standard fire tests. Structural steel is no exception. The effect of elevated temperatures on the yield and tensile strengths of steel is described in Art. 1.12. In general, yield strength decreases with large increases in temperature, but structural steels retain about 60% of their ambient-temperature yield strength at 1000°F.

During many building fires, temperatures in excess of 1000°F develop for relatively brief periods of time, but failures do not occur in structural steel members

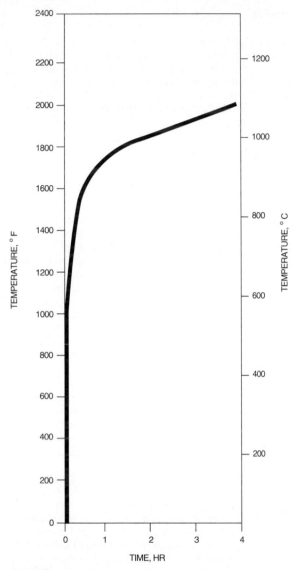

FIGURE 6.7 Variation of temperature with time in the standard fire test specified in ASTM E119.

inasmuch as they are rarely loaded enough to develop full design stresses. As a consequence, in many instances, bare structural steel has sufficient load-carrying capacity to withstand the effects of fire. This is not recognized in the standard fire tests, however, inasmuch as, in the tests, the temperatures are continuously increased and structural members are loaded to design capacities. Based on these tests, when building codes specify fire-resistant construction, they require fire-protection materials to "insulate" structural steel elements.

6.32.5 Fire-Protection Materials

A variety of different materials or systems are used to protect structural steel. The performance of these are directly determined during standard fire tests. In addition to the insulation characteristics evaluated in the tests, the physical integrity of fire-protection materials is extremely important and should be preserved during installation. Required fire-protection assemblies should be carefully inspected during and after construction to ensure that they are installed according to the manufacturers' recommendations and the appropriate fire-resistant designs.

Gypsum.   This material, in several forms, is widely used for fire protection (Fig. 6.8). As a plaster, it is applied over metal lath or gypsum lath. In the form of wallboard, gypsum is typically installed over cold-formed steel framing or furring.

The effectiveness of gypsum-based fire protection can be increased significantly by addition of lightweight mineral aggregates, such as vermiculite and perlite, to gypsum plaster. It is important that the mix be properly proportioned and applied in the required thickness and that the lath be correctly installed.

Three general types of gypsum wallboard are readily available: regular, type X, and proprietary. **Type X** wallboards have specially formulated cores that provide greater fire resistance than conventional wallboard of the same thickness. **Proprietary** wallboards also are available with even greater fire-resistant characteristics. It is therefore important to verify that the wallboard used is that specified for the desired fire-resistant design. In addition, the type and spacing of fasteners and, when appropriate, the type and support of furring channels should be in accordance with specifications.

("Design Data—Gypsum Products," Gypsum Association, Washington, D.C.)

Spray-Applied Materials.   The most widely used fire-protection materials for structural steel are mineral fiber and cementitious materials that are spray applied directly to the contours of beams, girders, columns, and floor and roof decks (Fig. 6.9). The spray-applied materials are based on proprietary formulations. Hence it is imperative that the manufacturer's recommendations for mixing and application be followed closely. Fire-resistant designs are published by Underwriters Laboratories.

Adhesion is an important characteristic of spray-applied materials. To ensure that it is attained, the structural steel should be free of dirt, oil, and loose scale; generally, the presence of light rust will not adversely affect adhesion. When the steel has been painted, however, field experience and testing have demonstrated that adhesion problems can arise. (Paint and primers are not generally required for corrosion protection when structural steel will be enclosed within a building or otherwise protected from the elements.) If paint is specified for structural steel that will subsequently be protected with spray-applied materials, the specifier should contact the paint and fire-protection material suppliers in advance to ensure that the two materials are compatible. Otherwise, bonding agents or other costly field modifications may be required to ensure adequate adhesion.

Suspended Ceiling Systems.   A wide variety of proprietary suspended ceiling systems are also available for protecting floors and beams and girders (Fig. 6.10). Fire-resistance ratings for such systems are published by Underwriters Laboratories. These systems are specifically designed for fire protection and require careful integration of ceiling tile, grid, and suspension systems. Also, openings for light fixtures, air diffusers, and similar accessories must be adequately protected. As a consequence, manufacturer's installation instructions should be closely followed.

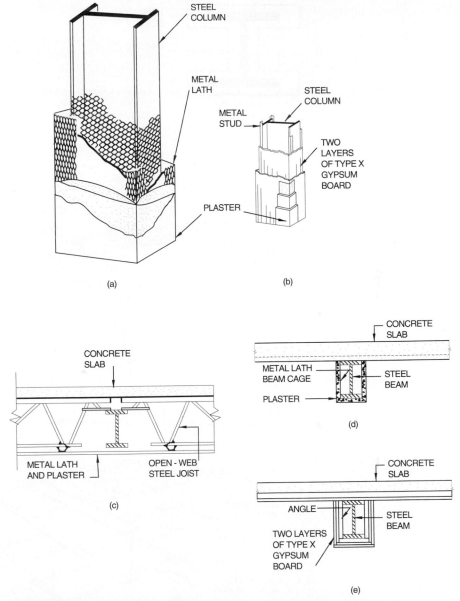

FIGURE 6.8   Some methods for applying gypsum as fire protection for structural steel: (*a*) Column enclosed in plaster on metal lath; (*b*) column boxed in with wallboard and plaster; (*c*) open-web joist with plaster ceiling; (*d*) beam enclosed in a plaster cage; (*e*) beam boxed in with wallboard.

In the case of load-transfer trusses or girders that support loads from more than one floor, building codes may require individual protection. As a consequence, suspended ceiling systems may not be permitted for this application.

Concrete and Masonry.   Concrete, once widely used for fire protecting structural steel, is not particularly efficient for this application because of its weight and

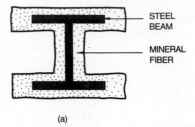

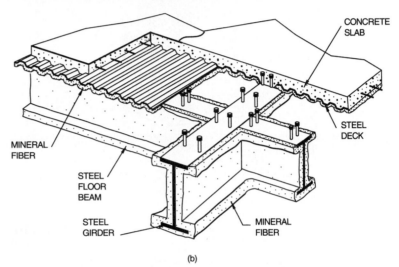

FIGURE 6.9   Mineral fiber spray applied to (a) steel beam; (b) beam-and-girder floor system, with steel floor deck supporting a concrete slab.

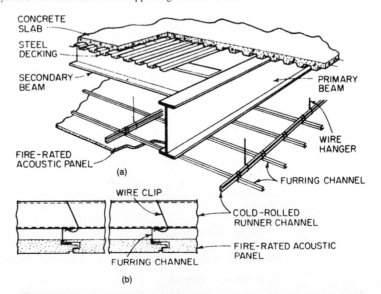

FIGURE 6.10   Steel floor system fire-protected on the underside by a suspended ceiling.

relatively high thermal conductivity. As a result, concrete is rarely used when the purpose is fire protection only.

Concrete floor slabs are acceptable as fire protection for the tops of flexural members. Concrete or masonry is also sometimes used to encase steel columns for architectural or structural purposes or when substantial resistance to physical damage is required (Fig. 6.11).

Design information on the fire resistance of steel columns encased in concrete or protected with precast-concrete column covers is available from the American Iron and Steel Institute, Washington, D.C. Information on the use of concrete masonry and brick to protect steel columns may be obtained from the National Concrete Masonry Association, Herndon, Va., and the Brick Institute of America, Reston, Va., respectively.

6.32.6 Architecturally Exposed Steel

This concept involves the architectural expression of structural systems on building exteriors in contrast to the general practice of concealing them behind decorative facades. Design of architecturally exposed steel is strongly influenced by building code requirements for fire-resistant construction.

One approach for meeting code requirements for structural fire protection when the appearance of architecturally exposed steel is desired is illustrated in Fig. 6.12. As shown, flanges of a steel column are fire protected with a spray-applied material, insulation is placed against the web between the flanges, and the assembly is enclosed in a metal cover with the shape of the column.

Another approach is to use tubular columns filled with water (Fig. 6.13). Originally patented in 1884, this system was neglected until the late 1960s, when it was adopted for the 64-story U.S. Steel Building in Pittsburgh, Pa. Since then, several other buildings have been designed using this concept. In a fire, the entrapped water in a tubular column is expected to limit the temperature rise in the steel. Generally, corrosion inhibitors should be added to the water, and in cold climates, antifreeze solution should be used for exterior columns.

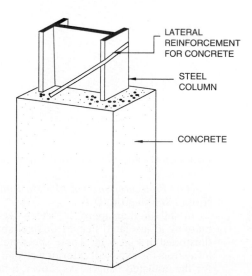

FIGURE 6.11 Concrete-encased steel column.

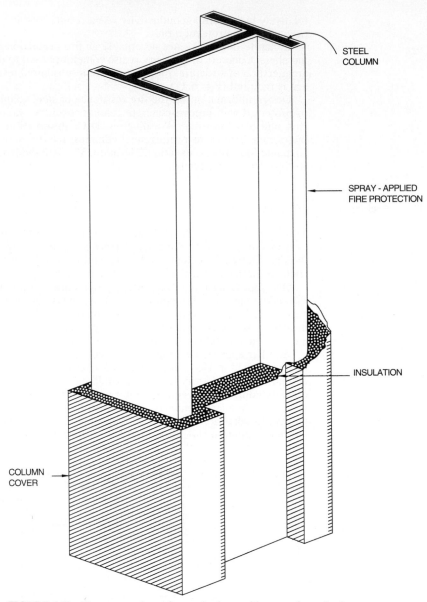

STEEL
COLUMN

SPRAY - APPLIED
FIRE PROTECTION

INSULATION

COLUMN
COVER

FIGURE 6.12   Fire-protected exterior steel column with exposed metal column covers.

(*Fire Protection through Modern Building Codes*, American Iron and Steel Institute, Washington, D.C.)

In still another approach, sheet-steel covers are applied on the outside of a
building to insulated flanges of steel spandrel girders to act as flame shields, as
illustrated in Fig. 6.14. These sheet-steel covers not only serve to deflect flames
away from the exposed, exterior web of a girder but also provide weather protection
for the insulated flanges. As shown, a flame-shielded spandrel girder is protected
in the interior of the building in a conventional manner.

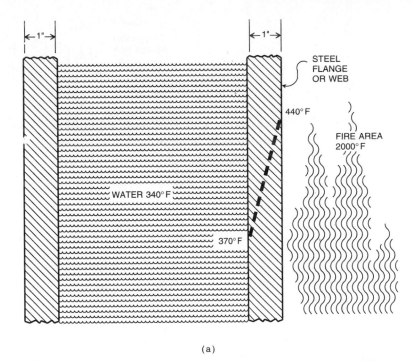

(a)

FIGURE 6.13 Tubular steel columns filled with water for fire resistance. (*a*) Temperature variation during exposure to fire. (*b*) Schematic arrangement of fire-protection system.

As illustrated by full-scale fire tests on flame-shielded spandrel girders, the standard fire test is not representative of the exposure that will be experienced by exterior columns and girders. Research on fire exposure conditions for exterior structural elements has led to development of a comprehensive design method for fire-safe exterior structural steel which has been adopted by some building codes.

(*Design Guide for Fire-Safe Structural Steel*, American Iron and Steel Institute, Washington, D.C.)

6.32.7 Restrained and Unrestrained Construction

One of the major sources of confusion with respect to design of fire-resistant buildings is the concept of restrained and unrestrained ratings. Fire-resistant design is based on the use of tested assemblies and is predicated on the assumption that test assemblies are "representative" of actual construction. In reality, this assumption is extremely difficult to implement in laboratory-scale fire tests. The primary difficulty arises from the size of available test furnaces, which typically can only accommodate floor specimens in the range of 15 by 18 ft in area. As a result, a typical test assembly actually represents a relatively small portion of a floor or roof structure. Thus, even though the standard fire test is frequently described as "large scale," it clearly is not "full scale."

In the attempt to model real floor systems in a representative manner, several problems arise. For example, since most floor slabs and roof decks are physically, if not structurally, continuous over beams and girders, real beams and girders are usually much larger than can be accommodated in available furnaces. Also, beams

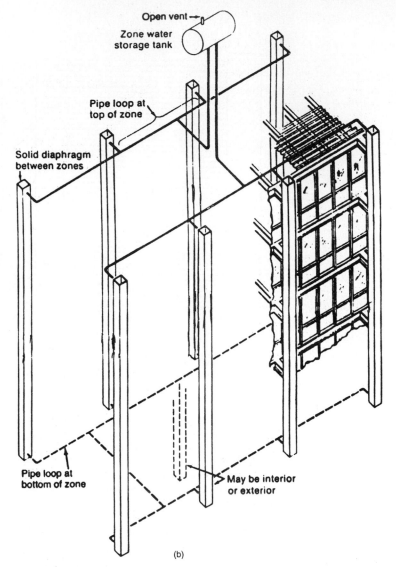

Open vent →

Zone water
storage tank

Pipe loop at
top of zone

Solid diaphragm
between zones

Pipe loop at
bottom of zone

May be interior
or exterior

(b)

FIGURE 6.13 *Continued.*

frame into columns and girders in a number of different ways. In some cases, connections are designed to resist only shear forces. In other cases, full- or partial-moment connections are provided. In short, given the cost of testing, the complexity of modern structural systems, and the size of available test facilities, it is unrealistic to assume that test assemblies can accurately model real construction systems.

In recognition of the practical difficulties associated with testing, ASTM E119 includes two test conditions, restrained and unrestrained. The restraint that is contemplated in fire testing is restraint against thermal expansion, not structural restraint in the traditional sense. When an assembly is supported or surrounded

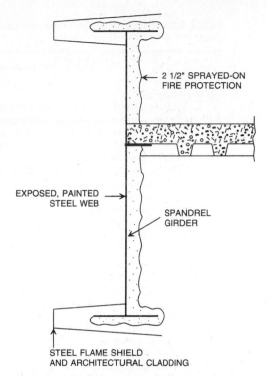

2 1/2" SPRAYED-ON
FIRE PROTECTION

EXPOSED, PAINTED
STEEL WEB

SPANDREL
GIRDER

STEEL FLAME SHIELD
AND ARCHITECTURAL CLADDING

FIGURE 6.14   Flame shields placed on flanges of a
spandrel girder to protect the web against flames.

by construction that is capable of resisting expansion, to some degree, thermal
stresses will be induced in the assembly in addition to those due to dead and live
loads. Originally, it was thought that thermal stresses would reduce the fire resistance of many assemblies. However, extensive research indicated that restraint
actually improved the fire resistance of many common types of floor systems. The
two test conditions in E119 recognize the complexity of this issue.

   The restrained condition applies when the assembly is supported or surrounded
by construction that is capable of resisting substantial thermal expansion throughout
the range of anticipated elevated temperatures. Otherwise, the assembly should
be considered free to rotate and expand at the supports and should be considered
unrestrained. Thus a floor system that is simply supported from a structural standpoint may often be restrained from a fire-resistance standpoint. To provide guidance
in the use of restrained and unrestrained ratings, ASTM E119 includes examples
in an explanatory appendix (Table 6.29) which indicate that most common types
of steel framing systems can be considered to be restrained from a fire-resistance
standpoint.

6.32.8   Temperatures of Fire-Exposed Structural Steel Elements

Basic heat-transfer principles indicate that the rate of temperature change of a
beam or column varies inversely with mass and directly with the surface area
through which heat is transferred to the member. Thus the weight-to-heated-
perimeter ratio W/D of a structural steel member significantly influences the tem-

TABLE 6.29 Examples of Restrained and Unrestrained Construction for Use in Fire Tests*

Type of construction	Condition
I. Wall bearing:	
a. Single-span and simply supported end spans of multiple bays.†	
(*1*) Open-web steel joists or steel beams, supporting concrete slab, precast units, or metal decking	Unrestrained
(*2*) Concrete slabs, precast units, or metal decking	Unrestrained
b. Interior spans of multiple bays:	
(*1*) Open-web steel joists, steel beams or metal decking, supporting continuous concrete slab	Restrained
(*2*) Open-web steel joists or steel beams, supporting precast units or metal decking	Unrestrained
(*3*) Cast-in-place concrete slab systems	Restrained
(*4*) Precast concrete where the potential thermal expansion is resisted by adjacent construction‡	Restrained
II. Steel framing:	
(*1*) Steel beams welded, riveted, or bolted to the framing members	Restrained
(*2*) All types of cast-in-place floor and roof systems (such as beam-and-slabs, flat slabs, pan joists, and waffle slabs) where the floor or roof system is secured to the framing members	Restrained
(*3*) All types of prefabricated floor or roof systems where the structural members are secured to the framing members and the potential thermal expansion of the floor or roof system is resisted by the framing system or the adjoining floor or roof construction‡	Restrained
III. Concrete framing:	
(*1*) Beams securely fastened to the framing members	Restrained
(*2*) All types of cast-in-place floor or roof systems (such as beam-and-slabs, flat slabs, pan joists, and waffle slabs) where the floor system is cast with the framing members	Restrained
(*3*) Interior and exterior spans of precast systems with cast-in-place joints resulting in restraint equivalent to that which would exist in condition III (*1*)	Restrained
(*4*) All types of prefabricated floor or roof systems where the structural members are secured to such systems and the potential thermal expansion of the floor or roof systems is resisted by the framing system or the adjoining floor or roof construction‡	Restrained
IV. Wood construction:	
All types	Unrestrained

*As recommended by ASTM Committee E-5 in the appendix to E119-88.

†Floor and roof systems can be considered restrained when they are tied into walls with or without tie beams, the walls being designed and detailed to resist thermal thrust from the floor or roof system.

‡For example, resistance to potential thermal expansion is considered to be achieved when:

(*1*) Continuous structural concrete topping is used,

(*2*) The space between the ends of precast units or between the ends of units and the vertical face of supports is filled with concrete or mortar, or

(*3*) The space between the ends of precast units and the vertical faces of supports, or between the ends of solid or hollow-slab units does not exceed 0.25% of the length for normal-weight concrete members or 0.1% of the length for structural lightweight concrete members.

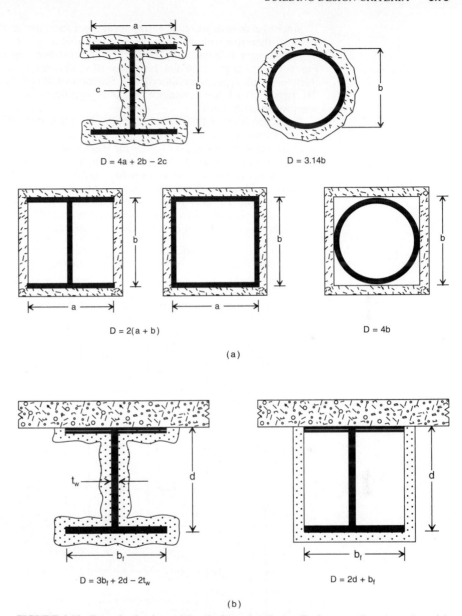

$D = 4a + 2b - 2c$ $D = 3.14b$

$D = 2(a + b)$ $D = 4b$

(a)

$D = 3b_f + 2d - 2t_w$ $D = 2d + b_f$

(b)

FIGURE 6.15 Formulas for determining the heated perimeter D of structural steel members (a) columns and (b) beams.

perature that the member will experience when exposed to fire. W is the weight per unit length of the member (lb/ft), and D is the inside perimeter of the fire protection material (in). Expressions for calculating D are illustrated in Fig. 6.15 for columns and beams with either contour or box protection. In short, the weight-to-heated-perimeter ratio defines the **thermal size** of a structural member.

Since the temperature of a structural steel member is strongly influenced by W/D, it therefore follows that the required thickness of fire-protection material is also strongly influenced by W/D. This interrelationship is clearly illustrated in Fig. 6.16, which gives the fire resistance of steel columns protected with different thicknesses of gypsum wallboard as a function of W/D. The curves show that in determination of fire resistance, W/D is significant, as is the thickness of the fire-protection material.

In recognition of this basic principle, several semiempirical design equations have been developed for determining the thicknesses of fire protection for structural steel elements as a function of W/D for specific fire-resistance ratings. These equations have been incorporated into the Underwriters Laboratories *Fire-Resistance Directory* and are described in the following publications available from the American Iron and Steel Institute: *Designing Fire Protection for Steel Columns*, *Designing Fire Protection for Steel Beams*, and *Designing Fire Protection for Steel Trusses*. These calculation methods also have been recognized by model building codes and are widely used in design of cost-effective, fire-resistant steel buildings.

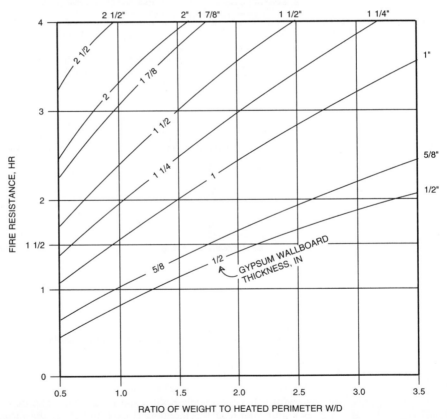

FIGURE 6.16 Variation in fire resistance of structural steel column with weight-to-heated-perimeter ratios and gypsum wallboard thickness.

6.32.9 Rational Fire Design

Building code requirements for structural fire protection generally are prescriptive and are based on standard fire tests. This approach suffers from the following significant deficiencies:

- The fire exposure is arbitrary and does not necessarily represent real building fires. In many cases, real fires result in high temperatures of short duration.
- In effect, the standard fire test presumes that structural members will be fully loaded at the time of a fire. In reality, fires occur randomly, and design requirements should be probability-based. Rarely will design structural loads occur simultaneously with fire.
- Given the scale of available laboratory facilities, structural interaction cannot be directly evaluated. In effect, ASTM E119 unrestrained ratings presume virtually no structural interaction, i.e., simple supports without continuity. To some degree, structural interaction is indirectly considered in establishing restrained ratings. The boundary conditions are arbitrary, however, and the extension to real buildings is largely based on judgment.

As a consequence of these shortcomings, a rational engineering design standard for structural fire protection is desirable. Standards of this type have been developed and are now routinely used in Japan, Australia, and throughout much of Europe (*International Fire Engineering Design for Steel Structures*: *State-of-the-Art*, International Iron and Steel Institute, Brussels, Belgium) and are being developed in the United States by the American Society of Civil Engineers in cooperation with the Society of Fire Protection Engineers.

(*The SFPE Handbook of Fire Protection Engineering*, National Fire Protection Association, Quincy, Mass.)

SECTION 7
DESIGN OF BUILDING MEMBERS

Ali A. K. Haris
*Vice President, Structural Engineering, Senior Project Manager,
Ellerbe Becket, Kansas City, Missouri*

Steel members in building structures can be part of the floor framing system to carry gravity loads, the vertical framing system, the lateral framing system to provide lateral stability to the building and resist lateral loads, or two or more of these systems. Floor members are normally called **joists, purlins, beams,** or **girders.** Roof members are also known as **rafters.**

Purlins, which support floors, roofs, and decks, are relatively close in spacing. Beams are floor members supporting the floor deck. Girders are steel members spanning between columns and usually supporting other beams. Transfer girders are members that support columns and transfer loads to other columns. The primary stresses in joists, purlins, beams, and girders are due to flexural moments and shear forces.

Vertical members supporting floors in buildings are designated **columns.** The most common steel shapes used for columns are wide-flange sections, pipes, and tubes. Columns are subject to axial compression and also often to bending moments. Slenderness in columns is a concern that must be addressed in the design.

Lateral framing systems may consist of the floor girders and columns that support the gravity floor loads but with rigid connections. These enable the flexural members to serve the dual function of supporting floor loads and resisting lateral loads. Columns, in this case, are subject to combined axial loads and moments. The lateral framing system also can consist of vertical diagonal braces or shear walls whose primary function is to resist lateral loads. Mixed bracing systems and rigid steel frames are also common in tall buildings.

Most steel floor framing members are considered simply supported. Most steel columns supporting floor loads only are considered as pinned at both ends. Other continuous members, such as those in rigid frames, must be analyzed as plane or space frames to determine the members' forces and moments.

Other main building components are steel trusses used for roofs or floors to span greater lengths between columns or other supports, built-up plate girders and stub girders for long spans or heavy loads, and open-web steel joists. See also Sec. 8.

This section addresses the design of these elements, which are common to most steel buildings, based on allowable stress design (ASD) and load and resistance factor design (LRFD). Design criteria for these methods are summarized in Sec. 6.

7.1   TENSION MEMBERS

Members subject to tension loads only include hangers, diagonal braces, truss members, and columns that are part of the lateral bracing system with significant uplift loads.

The AISC "LRFD Specification for Structural Steel Buildings." American Institute of Steel Construction (AISC) gives the nominal strength P_n (kips) of a cross section subject to tension only as the smaller of the capacity of yielding in the gross section,

$$P_n = F_y A_g \tag{7.1}$$

or the capacity at fracture in the net section,

$$P_n = F_u A_e \tag{7.2}$$

The factored load may not exceed either of the following:

$$P_u = \phi F_y A_g \qquad \phi = 0.9 \tag{7.3}$$

$$P_u = \phi F_u A_e \qquad \phi = 0.75 \tag{7.4}$$

where F_y and F_u are, respectively, the yield strength and the tensile strength (ksi) of the member. A_g is the gross area (in^2) of the member, and A_e is the effective cross-sectional area at the connection.

The effective area A_e is given by

$$A_e = U A_n \tag{7.5}$$

where A_n = net area of the member, in^2
     U = reduction coefficient
       = 1 when load is transmitted directly to each cross-sectional element by connectors
       = 0.9 for flange connections of W, M, or S shapes with flange widths at least two-thirds the depth and for structural tees cut from these shapes (Bolted connections should have at least three bolts per line in the direction of the loads.)
       = 0.85 for W, M, or S shapes with flange widths less than two-thirds the depth for structural tees cut from these shapes, and for all other shapes (Bolted connections should have at least three bolts per line in the direction of the loads.)
       = 0.75 for bolted connections of all members with only two bolts per line in the direction of the loads

If transverse welding is used to connect W, M, or S shapes or tees cut from these shapes when only a portion of the shape is welded, the effective area A_e is that area of the portion of the section that is welded. When load is transmitted by longitudinal welds along both edges at the end of a plate,

$$A_e = U A_g \tag{7.6}$$

where U = 1 when $l > 2w$
     = 0.87 when $2w \geq l > 1.5w$
     = 0.75 when $1.5w \geq l > w$
  l = weld length, in $> w$
  w = plate width (distance between welds), in

7.2 COMPARATIVE DESIGNS OF DOUBLE-ANGLE HANGER

A composite floor framing system is to be designed for sky boxes of a sports arena structure. The sky boxes are located about 15 ft below the bottom chord of the roof trusses. The sky-box framing is supported by an exterior column at the exterior edge of the floor and by steel hangers 5 ft from the inside edge of the floor. The hangers are connected to either the bottom chord of the trusses or to the steel beams spanning between trusses at roof level. The reactions due to service dead and live loads at the hanger locations are P_{DL} = 55 kips and P_{LL} = 45 kips. Hangers supporting floors and balconies should be designed for additional impact factors representing 33% of the live loads.

7.2.1 LRFD for Double-Angle Hanger

The factored axial tension load is the larger of

$$P_{UT} = 55 \times 1.2 + 45 \times 1.6 \times 1.33 = 162 \text{ kips (governs)}$$

$$P_{UT} = 55 \times 1.4 = 77 \text{ kips}$$

Double angles of A36 steel with one row of bolts will be used (F_y = 36 ksi and F_u = 58 ksi). The required area of the section is determined as follows: From Eq. (7.3), with P_U = 162 kips,

$$A_g = 162/(0.9 \times 36) = 5.00 \text{ in}^2$$

From Eq. (7.4),

$$A_e = 162/(0.75 \times 58) = 3.72 \text{ in}^2$$

Try two angles, 5 $\times$ 3 $\times$ ⅜ in, with A_g = 5.72 in². For 1-in-diameter A325 bolts with hole size $1\frac{1}{16}$ in, the net area of the angles is

$$A_n = 5.72 - 2 \times \tfrac{3}{8} \times \tfrac{17}{16} = 4.92 \text{ in}^2$$

and the effective area is

$$A_e = UA_n = 0.85 \times 4.92 = 4.18 \text{ in}^2 > 3.72 \text{ in}^2\text{—OK}$$

7.2.2 ASD for Double-Angle Hanger

The dead load on the hanger is 55 kips, and the live load plus impact is 45 $\times$ 1.33 = 60 kips (Art. 7.2.1). The total axial tension then is 55 + 60 = 115 kips. With the allowable tensile stress on the gross area of the hanger $F_t = 0.6F_y$ =

$0.6 \times 36 = 21.6$ ksi, the gross area A_g required for the hanger is

$$A_g = 115/21.6 = 5.32 \text{ in}^2$$

With the allowable tensile stress on the effective net area $F_t = 0.5F_u = 0.5 \times 58 = 29$ ksi,

$$A_e = 115/29 = 3.97 \text{ in}^2$$

Two angles $5 \times 3 \times \frac{3}{8}$ in provide $A_g = 5.72$ in$^2 > 5.32$ in^2—OK. For 1-in-diameter bolts in holes $1\frac{1}{16}$ in in diameter, the net area of the angles is

$$A_n = 5.72 - 2 \times \tfrac{3}{8} \times {}^{17}\!/_{16} = 4.92 \text{ in}^2$$

and the effective net area is

$$A_e = UA_n = 0.85 \times 4.92 = 4.18 \text{ in}^2 > 3.97 \text{ in}^2\text{—OK}$$

7.3 EXAMPLE—LRFD FOR WIDE-FLANGE TRUSS MEMBERS

One-way, long-span trusses are to be used to frame the roof of a sports facility. The truss span is 300 ft. All members are wide-flange sections. (See Fig. 7.1 for the typical detail of the bottom-chord splice of the truss).

Connections of the truss diagonals and verticals to the bottom chord are bolted. Slip-critical, the connections serve also as splices, with $1\frac{1}{8}$-in-diameter A325 bolts, in oversized holes to facilitate truss assembly in the field. The holes are $1\frac{7}{16}$ in in diameter. The bolts are placed in two rows in each flange. The number of bolts per row is more than two. The web of each member is also spliced with a plate with two rows of $1\frac{1}{8}$-in-diameter A325 bolts.

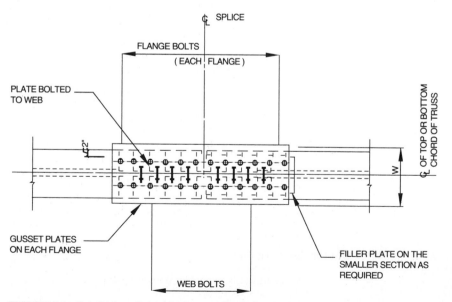

FIGURE 7.1 Detail of a splice in the bottom chord of a truss.

The structural engineer analyzes the trusses as pin-ended members. Therefore, all members are considered to be subject to axial forces only. Members of long-span trusses with significant deflections and large, bolted, slip-critical connections, however, may have significant bending moments. (See Art. 7.15 for an example of a design for combined axial load and bending moments.)

The factored axial tension in the bottom chord at midspan due to combined dead, live, theatrical, and hanger loads supporting sky boxes is $P_u = 2280$ kips.

With a wide-flange section of grade 50 steel ($F_y = 50$ ksi and $F_u = 65$ ksi), the required minimum gross area, from Eq. (7.3), is

$$A_g = P_u/\phi F_y = 2280/(0.9 \times 50) = 50.67 \text{ in}^2$$

Try a W14 $\times$ 176 section with $A_g = 51.8$ in^2, flange thickness $t_f = 1.31$ in, and web thickness $t_w = 0.83$ in. The net area is

$$A_n = 51.8 - (2 \times 1.31 \times 1.4375 \times 2 + 2 \times 0.83 \times 1.4375)$$

$$= 41.88 \text{ in}^2$$

Since all parts of the wide-flange section are connected at the splice connection, $U = 1$ for determination of the effective area from Eq. (7.5). Thus $A_e = A_n = 41.88$ in^2. From Eq. (7.4), the design strength is

$$\phi P_n = 0.75 \times 65 \times 41.88 = 2042 \text{ kips} < 2280 \text{ kips—NG}$$

Try a W14 $\times$ 193 with $A_g = 56.8$ in^2, $t_f = 1.44$ in, and $t_w = 0.89$ in. The net area is

$$A_n = 56.8 - (2 \times 1.44 \times 1.4375 \times 2 + 2 \times 0.89 \times 1.4375)$$

$$= 45.96 \text{ in}^2$$

From Eq. (7.4), the design strength is

$$\phi P_n = 0.75 \times 65 \times 45.96 = 2241 \text{ kips} < 2280 \text{ ksi—NG}$$

Use the next size, W14 $\times$ 211.

7.4 COMPRESSION MEMBERS

Steel members in buildings subject to compressive axial loads include columns, truss members, struts, and diagonal braces. Slenderness is a major factor in design of compression members. The slenderness ratio L/r is preferably limited to 200. Most suitable steel shapes are pipes, tubes, or wide-flange sections, as designated for columns in the AISC "Steel Construction Manual." Double angles, however, are commonly used for diagonal braces and truss members. Double angles can be easily connected to other members with gusset plates and bolts or welds.

The AISC "LRFD Specification for Structural Steel Buildings," American Institute of Steel Construction, gives the nominal strength P_n (kips) of a steel section in compression as

$$P_n = A_g F_{cr} \tag{7.7}$$

The factored load P_u (kips) may not exceed

$$P_u = \phi P_n \qquad \phi = 0.85 \tag{7.8}$$

The critical compressive stress F_{cr} (kips) is a function of material strength and slenderness. For determination of this stress, a column slenderness parameter λ_c is defined as

$$\lambda_c = \frac{KL}{r\pi}\sqrt{\frac{F_y}{E}} = \frac{KL}{r}\sqrt{\frac{F_y}{286,220}} \qquad (7.9)$$

where A_g = gross area of the member, in^2
K = effective length factor (Art. 6.16.2)
L = unbraced length of member, in
F_y = yield strength of steel, ksi
E = modulus of elasticity of steel material, ksi
r = radius of gyration corresponding to plane of buckling, in

When $\lambda_c \leq 1.5$, the critical stress is given by

$$F_{cr} = (0.658^{\lambda_c^2})F_y \qquad (7.10)$$

When $\lambda_c > 1.5$,

$$F_{cr} = \left(\frac{0.877}{\lambda_c^2}\right)F_y \qquad (7.11)$$

7.5   EXAMPLE—LRFD FOR STEEL PIPE IN AXIAL COMPRESSION

Pipe sections of A36 steel are to be used to support framing for the flat roof of a one-story factory building. The roof height is 18 ft from the tops of the steel roof beams to the finish of the floor. The steel roof beams are 16 in deep, and the bases of the steel-pipe columns are 1.5 ft below the finished floor. A square joint is provided in the slab at the steel column. Therefore, the concrete slab does not provide lateral bracing. The effective height of the column, from the base of the column to the center line of the steel roof beam, is

$$h = 18 + 1.5 - \frac{16}{2 \times 12} = 18.83 \text{ ft}$$

The dead load on the column is 30 kips. The live load due to snow at the roof is 36 kips. The factored axial load is the larger of the following:

$$P_u = 30 \times 1.4 = 42 \text{ kips}$$

$$P_u = 30 \times 1.2 + 36 \times 1.6 = 93.6 \text{ kips (governs)}$$

With the factored load known, the required pipe size may be obtained from a table in the AISC "Manual of Steel Construction—LRFD." For $KL = 19$ ft, a standard 6-in pipe (weight 18.97 lb per linear ft) offers the least weight for a pipe with a compression-load capacity of at least 93.6 kips. For verification of this selection, the following computations for the column capacity were made based on a radius of gyration $r = 2.25$ in. From Eq. (7.9),

$$\lambda_c = \frac{18.83 \times 12}{2.25}\sqrt{\frac{36}{286,220}} = 1.126 < 1.5$$

and $\lambda_c{}^2 = 1.269$. For $\lambda_c < 1.5$, Eq. (7.10) yields the critical stress

$$F_{cr} = 0.658^{1.269} \times 36 = 21.17 \text{ ksi}$$

The design strength of the 6-in pipe, then, from Eqs. (7.7) and (7.8), is

$$\phi P_n = 0.85 \times 5.58 \times 21.17 = 100.4 \text{ kips} > 93.6 \text{ kips—OK}$$

7.6 COMPARATIVE DESIGNS OF WIDE-FLANGE SECTION WITH AXIAL COMPRESSION

A wide-flange section is to be used for columns in a five-story steel building. A typical interior column in the lowest story will be designed to support gravity loads. (In this example, no eccentricity will be assumed for the load.) The effective height of the column is 18 ft. The axial loads on the column from the column above and from the steel girders supporting the second level are dead load 420 kips and live load (reduced according to the applicable building code) 120 kips.

7.6.1 LRFD for W Section with Axial Compression

The factored axial load is the larger of the following:

$$P_u = 420 \times 1.4 = 588 \text{ kips}$$

$$P_u = 420 \times 1.2 + 120 \times 1.6 = 696 \text{ kips (governs)}$$

To select the most economical section and material, assume that grade 36 steel costs $0.24 per pound and grade 50 steel costs $0.26 per pound at the mill. These costs do not include the cost of fabrication, shipping, or erection, which will be considered the same for both grades.

Use of the column design tables of the AISC "Manual of Steel Construction—LRFD" presents the following options:

For the column of grade 36 steel, select a W14 × 99, with a design strength $\phi P_n = 745$ kips.

$$\text{Cost} = 99 \times 18 \times 0.24 = \$428$$

For the column of grade 50 steel, select a W12 × 87, with a design strength $\phi P_n = 758$ kips.

$$\text{Cost} = 87 \times 18 \times 0.26 = \$407$$

Therefore, the W12 × 87 of grade 50 steel is the most economical wide-flange section.

7.6.2 ASD for W Section with Axial Compression

The dead- plus live-load axial compression totals $420 + 120 = 540$ kips (Art. 7.6.1).

Column design tables in the AISC "Steel Construction Manual—ASD" facilitate selection of wide-flange sections for various loads for columns of grades 36 and 50 steels.

For the column of grade 36 steel, with the slenderness ratio $KL = 18$ ft, the manual tables indicate that the least-weight section with a capacity exceeding 540 kips is a W14 × 109. It has an axial load capacity of 564 kips. Estimated cost of the W14 × 109 is $0.24 × 109 × 18 = $471.

LRFD requires a W14 × 99 of grade 36 steel, with an estimated cost of $428. Thus the cost savings by use of LRFD is $100(471 - 428)/428 = 9.1\%$.

For the column of grade 50 steel, with $KL = 18$ ft, the manual tables indicate that the least-weight section with a capacity exceeding 540 kips is a W14 × 90. It has an axial compression capacity of 609 kips. Estimated cost of the W14 × 90 is $0.26 × 90 × 18 = $421. Thus the grade 50 column costs less than the grade 36 column.

LRFD requires a W12 × 87 of grade 50 steel, with an estimated cost of $407. The cost savings by use of LRFD is $100(421 - 407)/421 = 3.33\%$.

This example indicates that when slenderness is significant in design of compression members, the savings with LRFD are not as large for slender members as for stiffer members, such as short columns or columns with a large radius of gyration about the x and y axes.

7.7   EXAMPLE—LRFD FOR DOUBLE ANGLES WITH AXIAL COMPRESSION

Double angles are the preferred steel shape for a diagonal in the vertical bracing part of the lateral framing system in a multistory building (Fig. 7.2). Lateral load on the diagonal in this example is due to wind only and equals 65 kips. The diagonals also support the steel beam at midspan. As a result, the compressive force on each brace due to dead loads is 15 kips, and that due to live loads is 10 kips. The maximum combined factored load is $P_u = 1.2 × 15 + 1.3 × 65 + 0.5 × 10 = 107.5$ kips.

The length of the brace is 19.85 ft, neglecting the size of the joint. Grade 36 steel is selected because slenderness is a major factor in determining the nominal capacity of the section. Selection of the size of double angles is based on trial and error, which can be assisted by load tables in the AISC "Manual of Steel Construction—LRFD" for columns of various shapes and sizes. For the purpose of illustration of the step-by-step design, double angles 6 × 4 × ⅝ in with ⅜-in spacing between the angles are chosen. Section properties are as follows: gross area $A_g = 11.7$ in² and the radii of gyration are $r_x = 1.90$ in and $r_y = 1.67$ in.

First, the slenderness effect must be evaluated to determine the corresponding critical compressive stresses. The effect of the distance between the spacer plates connecting the two angles is a new design criterion in LRFD. Assuming that the connectors are fully tightened bolts, the system slenderness is calculated as follows:

If maximum spacing between connectors $a = 80$ in, and minimum radius of gyration of an individual angle $r_i = 0.864$ in,

$$\frac{a}{r_i} = \frac{80}{0.864} = 92.59 > 50$$

The AISC "LRFD Specification for Structural Steel Buildings" defines the slenderness ratio for this condition as

$$\left(\frac{KL}{r}\right)_m = \sqrt{\left(\frac{KL}{r}\right)_o^2 + \left(\frac{a}{r_i} - 50\right)^2} \qquad (7.12)$$

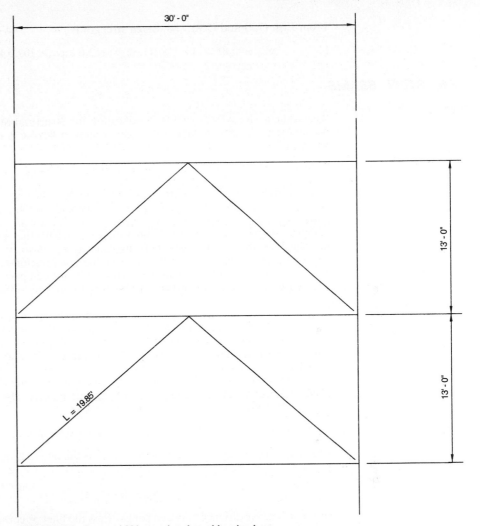

FIGURE 7.2 Inverted V-braces in a lateral bracing bent.

With $K = 1$, $r_y = 1.67$, and $L = 19.85$, substitution in Eq. (7.12) yields

$$\left(\frac{KL}{r}\right)_m = \sqrt{\left(\frac{19.85 \times 12}{1.67}\right)^2 + (92.59 - 50)^2} = 149$$

From Eq. 7.9, for determination of the critical stress F_{cr},

$$\lambda_c = 149 \sqrt{\frac{36}{286,220}} = 1.67 > 1.5$$

The critical stress, from Eq. (7.11), then is

$$F_{cr} = \left(\frac{0.877}{1.67^2}\right) 36 = 11.32 \text{ ksi}$$

From Eqs. (7.7) and (7.8), the design strength is

$$\phi P_n = 0.85 \times 11.7 \times 11.32 = 112.6 \text{ kips} > 107.5 \text{ kips—OK}$$

7.8 STEEL BEAMS

According to the AISC "LRFD Specification for Structural Steel Buildings," the nominal capacity M_p (in-kips) of a steel section in flexure is equal to the plastic moment:

$$M_p = ZF_y \tag{7.13}$$

where Z is the plastic section modulus (in^3), and F_y is the steel yield strength (ksi). But this applies only when local or lateral torsional buckling of the compression flange is not a governing criterion. The nominal capacity M_p is reduced when the compression flange is not braced laterally for a length that exceeds the limiting unbraced length for full plastic bending capacity L_p. Also, the nominal moment capacity is less than M_p when the ratio of the compression-element width to its thickness exceeds limiting slenderness parameters for compact sections. The same is true for the effect of the ratio of web depth to thickness. (See Arts. 6.17.1 and 6.17.2.)

In addition to strength requirements for design of beams, serviceability is important. Deflection limitations defined by local codes or standards of practice must be maintained in selecting member sizes. Dynamic properties of the beams are also important design parameters in determining the vibration behavior of floor systems for various uses.

The shear forces in the web of wide-flange sections should be calculated, especially if large concentrated loads occur near the supports. The AISC specification requires that the factored shear V_v (kips) not exceed

$$V_u = \phi_v V_n \qquad \phi_v = 0.90 \tag{7.14}$$

where ϕ_v is a capacity reduction factor and V_n is the nominal shear strength (kips). For $h/t_w \le 187\sqrt{k/F_{yw}}$,

$$V_n = 0.6F_{yw}A_w \tag{7.15}$$

where h = clear distance between flanges (less the fillet or corner radius for rolled shapes), in
$\quad k$ = web-plate buckling coefficient (Art. 6.14.1)
$\quad t_w$ = web thickness, in
$\quad F_{yw}$ = yield strength of the web, ksi
$\quad A_w$ = web area, in^2

For $187\sqrt{k/F_{yw}} < h/t_w \le 234\sqrt{k/F_{yw}}$,

$$V_n = 0.6F_{yw}A_w \frac{187\sqrt{k/F_{yw}}}{h/t_w} \tag{7.16}$$

For $h/t_w > 234\sqrt{k/F_{yw}}$,

$$V_n = A_w \frac{26{,}400k}{(h/t_w)^2} \tag{7.17}$$

k may be taken as 5 if no stiffeners are used.

7.9 COMPARATIVE DESIGNS OF SIMPLE-SPAN FLOORBEAM

Floor framing for an office building is to consist of open-web steel joists with a standard corrugated metal deck and 3-in-thick normal-weight concrete fill. The joists are to be spaced 3 ft center to center. Steel beams spanning 30 ft between columns support the joists. A bay across the building floor is shown in Fig. 7.3.

Floorbeam AB in Fig. 7.3 will be designed for this example. The loads are listed in Table 7.1. The live load is reduced in Table 7.1, as permitted by the *Uniform Building Code*. The reduction factor R is given by the smaller of

$$R = 0.0008(A - 150) \tag{7.18}$$

$$R = 0.231(1 + D/L) \tag{7.19}$$

$$R = 0.4 \text{ for beams} \tag{7.20}$$

where D = dead load
L = live load
A = area supported = $30(40 + 25)/2 = 975 \text{ ft}^2$

From Eq. (7.18), $R = 0.0008(975 - 150) = 0.66$.
From Eq. (7.19), $R = 0.231(1 + ^{73}/_{50}) = 0.568$.
From Eq. (7.20), $R = 0.4$ (governs), and the reduced live load is 50(1
0.4) = 30 lb per ft², as shown in Table 7.1.

7.9.1 LRFD for Simple-Span Floorbeam

If the beam's self-weight is assumed to be 45 lb/ft, the factored uniform load is the larger of the following:

$$W_u = 1.4[73(40 + 25)/2 + 45] = 3384.5 \text{ lb per ft}$$

$$W_u = 1.2[73(40 + 25)/2 + 45] + 1.6 \times 30(40 + 25)/2$$

$$= 4461 \text{ lb per ft (governs)}$$

The factored moment then is

$$M_u = 4.461(30)^2/8 = 501.9 \text{ kip-ft}$$

TABLE 7.1 Loads on Floorbeam AB in Fig. 7.3

Dead loads, lb per ft²	
Floor deck	45
Ceiling and mechanical ductwork	5
Open-web joists	3
Partitions	20
Total dead load (exclusive of beam weight)	73
Live loads, lb per ft²	
Full live load	50
Reduced live load: 50(1 − 0.4)	30

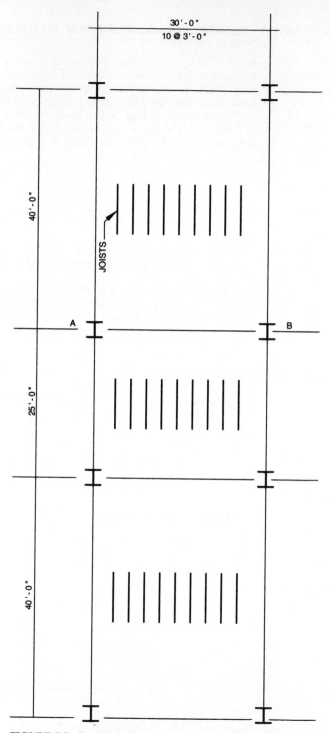

FIGURE 7.3 Part of the floor framing for an office building.

To select for beam *AB* a wide-flange section with F_y = 50 ksi, the top flange being braced by joists, the required plastic modulus Z_x is determined as follows: The factored moment M_u may not exceed the design strength of ϕM_r, and

$$\phi M_r = \phi Z_x F_y \qquad (7.21)$$

Therefore, from Eq. (7.21),

$$Z_x = \frac{501.9 \times 12}{0.9 \times 50} = 133.8 \text{ in}^3$$

A wide-flange section W24 × 55 with Z = 134 in³ is adequate.

Next, criteria are used to determine if deflections are acceptable. For the live-load deflection, the span L is 30 ft, the moment of inertia of the W24 × 55 is I = 1350 in⁴, and the modulus of elasticity E = 29,000 ksi. The live load is W_L = 30(40 + 25)/2 = 975 lb per ft. Hence the live-load deflection is

$$\Delta_L = \frac{5W_L L^4}{384EI} = \frac{5 \times 0.975 \times 30^4 \times 12^3}{384 \times 29,000 \times 1,350} = 0.454 \text{ in}$$

This value is less than $L/360$ = 30 × ¹²/₃₆₀ = 1 in, as specified in the *Uniform Building Code* (*UBC*). The *UBC* requires that deflections due to live load plus a factor K times deadload not exceed $L/240$. The K value, however, is specified as zero for steel. [The intent of this requirement is to include the long-term effect (creep) due to dead loads in the deflection criteria.] Hence the live-load deflection satisfies this criterion.

The immediate deflection due to the weight of the concrete and the floor framing is also commonly determined. If excessive deflections due to such dead loads are found, it is recommended that steel members be cambered to produce level floors and to avoid excessive concrete thickness during finishing the wet concrete.

In this example, the load due to the weight of the floor system is from Table 7.1 with the weight of the beam added,

$$W_{wt} = (45 + 3)(40 + 25)/2 + 55 = 1615 \text{ lb per ft}$$

The deflection due to this load is

$$\Delta_{wt} = \frac{5 \times 1.615 \times 30^4 \times 12}{384 \times 29,000 \times 1,350} = 0.752 \text{ in}$$

Therefore, cambering the beam ¾ in at midspan is recommended.

For review of the shear capacity of the section, the depth/thickness ratio of the web is

$$h/t_w = 54.6 < (187\sqrt{5\!/\!50} = 59.13)$$

From Eq. (7.15), the design shear strength is

$$\phi V_n = 0.9 \times 0.6 \times 50 \times 23.57 \times 0.395 = 251 \text{ kips}$$

The factored shear force near the support is

$$V_u = 4.461 \times 30/2 = 66.92 \text{ kips} < 251 \text{ kips—OK}$$

As illustrated in this example, it usually is not necessary to review the design of each simple beam with uniform load for shear capacity.

7.9.2 ASD for Simple-Span Floorbeam

The maximum moment due to the dead and live loads provided for Art. 7.9.1 is calculated as follows.

The total service load, after allowing a reduction in live load for size of area supported, is $73 + 30 = 103$ lb per ft^2. Assume that the beam weighs 60 lb per ft. Then, the total uniform load on the beam is

$$W_t = 103 \times 0.5(40 + 25) + 60 = 3408 \text{ lb per ft} = 3.408 \text{ kips per ft}$$

For this load, the maximum moment is

$$M = 3.408 \times 30^2/8 = 383.4 \text{ kip-ft}$$

For an allowable stress $F_b = 0.66F_y = 0.66 \times 50 = 33$ ksi, the required section modulus for the floorbeam is

$$S_r = M/F_b = 383.4 \times 12/33 = 139.4 \text{ in}^3$$

The least-weight wide-flange section with S exceeding 139.4 is a W21 × 68 or a W24 × 68. (If depth is not important, choose the latter because it will deflect less.)

LRFD requires a W24 × 55. The weight saving with LRFD is $100(68 - 55)/68 = 19.1\%$.

The percentage savings in weight with LRFD differs significantly from that in this example for a different ratio of live load to dead load. When live loads are relatively large, such as 100 psf for occupancy load in public areas, the savings in steel tonnage with LRFD is not as large as this example indicates.

Deflection calculation for ASD of the floorbeam is similar to that performed in Art. 7.9.1.

For review of shear stresses, the depth/thickness ratio of the web of the W24 × 68 is $h/t_w = 21/0.415 = 50.6$. Since this is less than $380/\sqrt{F_y} = 380/\sqrt{50} = 53.7$, the allowable shear stress is $F_v = 0.4 \times 50 = 20$ ksi. The vertical shear at the support is $V = 3.408 \times {}^{30}\!/_2 = 51.12$ kips. Hence the shear stress there is

$$f_v = 51.12/23.73 \times 0.415 = 5.19 \text{ ksi} < 20 \text{ ksi—OK}$$

7.10 EXAMPLE—LRFD FOR FLOORBEAM WITH UNBRACED TOP FLANGE

A beam of grade 50 steel with a span of 20 ft is to support the concentrated load of a stub pipe column at midspan. The factored concentrated load is 55 kips. No floor deck is present on either side of the beam to brace the top flange, and the pipe column is not capable of bracing the top flange laterally. The weight of the beam is assumed to be 50 lb/ft.

The factored moment at midspan is

$$M_u = 55 \times 20/4 + 0.050 \times 20^2/8 = 277.5 \text{ kip-ft}$$

A beam size for a first trial can be selected from a load-factor design table for steel with $F_y = 50$ ksi in the AISC "Steel Construction Manual—LRFD." The table lists several properties of wide-flange shapes, including plastic moment capacities ϕM_p. For example, an examination of the table indicates that the lightest beam with ϕM_p exceeding 277.5 kip-ft is a W18 × 40 with $\phi M_p = 294$ kip-ft. Whether this beam can be used, however, depends on the resistance of its top flange to

TABLE 7.2   Properties of Selected W Shapes for LRFD

Property	W18 × 35 grade 36	W18 × 40 grade 50	W21 × 50 grade 50	W21 × 62 grade 50
ϕM_p, kip-ft	180	294	413	540
L_p, ft	5.1	4.5	4.6	6.3
L_r, ft	14.8	12.1	12.5	16.6
ϕM_r, kip-ft	112	205	283	381
S_x, in^3	57.6	68.4	94.5	127
X_1, ksi	1590	1810	1730	1820
X_2, 1/ksi^2	0.0303	0.0172	0.0226	0.0159
r_y, in	1.22	1.27	1.30	1.77

buckling. The manual table also lists the limiting laterally unbraced lengths for full plastic bending capacity L_p and inelastic torsional buckling L_r. For the W18 × 40, $L_p = 4.5$ ft and $L_r = 12.1$ ft (Table 7.2).

In this example, then, the 20-ft unbraced beam length exceeds L_r. For this condition, the nominal bending capacity M_n is given by Eq. (6.54): $M_n = M_{cr} \leq C_b M_r$. For a simple beam with a concentrated load, the moment gradient C_b is unity. From the table in the manual for the W18 × 40 (grade 50), design strength $\phi M_r = 205$ kip-ft < 277.5 kip-ft. Therefore, a larger size is necessary.

The next step is to find a section that if its L_r is less than 20 ft, its M_r exceeds 277.5 kip-ft. The manual table indicated that a W21 × 50 has the required properties (Table 7.2). With the aid of Table 7.2, the critical elastic moment capacity ϕM_{cr} can be computed from

$$\phi M_{cr} = 0.90 \, \frac{C_b S_x X_1}{L_b / r_y} \sqrt{2 + \frac{X_1^2 X_2}{(L_b / r_y)^2}} \tag{7.22}$$

The beam slenderness ratio with respect to the y axis is

$$L_b / r_y = 20 \times 12 / 1.30 = 184.6$$

Thus the critical elastic moment capacity is

$$\phi M_{cr} = 0.90 \, \frac{1 \times 94.5 \times 1,730}{184.6} \sqrt{2 + \frac{1,730^2 \times 0.0226}{184.6^2}}$$

$$= 1,591 \text{ kip-in} = 132.6 \text{ kip-ft} < 277.5 \text{ kip-ft}$$

The W21 × 50 does not have adequate capacity. Therefore, trials to find the lowest-weight larger size must be continued. This trial-and-error process can be eliminated by using beam-selector charts in the AISC manual. These charts give the beam design moment corresponding to unbraced length for various rolled sections. Thus for $\phi M_r > 277.5$ kip-ft and $L = 20$ ft, the charts indicate that a W21 × 62 of grade 50 steel satisfies the criteria (Table 7.2). As a check, the following calculation is made with the properties of the W21 × 62 given in Table 7.2.

For use in Eq. (7.22), the beam slenderness ratio is

$$L_b / r_y = 20 \times 12 / 1.77 = 135.6$$

From Eq. (7.22), the critical elastic moment capacity is

$$\phi M_{cr} = 0.9 \frac{1 \times 127 \times 1820}{135.6} \sqrt{2 + \frac{1820^2 \times 0.0159}{135.6^2}}$$

$$= 3384 \text{ kip-in} = 282 \text{ kip-ft} > 277.5 \text{ kip-ft—OK}$$

7.11 EXAMPLE—LRFD FOR FLOORBEAM WITH OVERHANG

A floorbeam of A36 steel carrying uniform loads is to span 30 ft and cantilever over a girder for 7.5 ft (Fig. 7.4). The beam is to carry a dead load due to the weight of the floor plus assumed weight of beam of 1.5 kips per ft and due to partitions, ceiling, and ductwork of 0.75 kips per ft. The live load is 1.5 kips per ft.

Negative Moment. The cantilever is assumed to carry full live and dead loads, while the back span is subjected to the minimum dead load. This loading produces maximum negative moment and maximum unbraced length of compression (bottom) flange between the support and points of zero moment. The maximum factored load on the cantilever (Fig. 7.4a) is

$$W_{uc} = 1.2(1.5 + 0.75) + 1.6 \times 1.5 = 5.1 \text{ kips per ft}$$

The factored load on the backspan from dead load only is

$$W_{ub} = 1.2 \times 1.5 = 1.8 \text{ kips per ft}$$

Hence the maximum factored moment (at the support) is

$$-M_u = 5.1 \times 7.5^2/2 = 143.4 \text{ kip-ft}$$

From the bending moment diagram in Fig. 7.4b, the maximum factored moment in the backspan is 137.1 kip-ft, and the distance between the support of the cantilever and the point of inflection in the backspan is 5.3 ft. The compression flange is unbraced over this distance. The beam will be constrained against torsion at the support. Therefore, since the 7.5-ft cantilever has a longer unbraced length and its end will be laterally braced, design of the section should be based on $L_b = 7.5$ ft.

A beam size for a first trial can be selected from a load-factor design table in the AISC "Steel Construction Manual—LRFD." The table indicates that the lightest-weight section with ϕM_p exceeding 143.4 kip-ft and with potential capacity to sustain the large positive moment in the backspan is a W18 × 35. Table 7.2 lists section properties needed for computation of the design strength. The table indicates that the limiting unbraced length L_r for inelastic torsional buckling is 14.8 ft $> L_b$. The design strength should be computed from Eq. (6.53):

$$\phi M_n = C_b[\phi M_p - (\phi M_p - \phi M_r)(L_b - L_p)/(L_r - L_p)] \qquad (7.23)$$

For an unbraced cantilever, the moment gradient C_b is unity. Therefore, the design strength at the support is

$$\phi M_n = 1[180 - (180 - 112)(7.5 - 5.1)/(14.8 - 5.1)]$$

$$= 163.2 \text{ kip-ft} > 143.4 \text{ kip ft—OK}$$

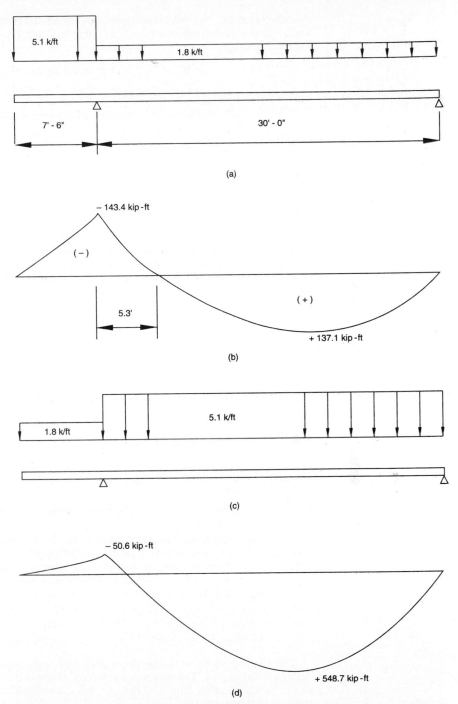

FIGURE 7.4 Loads and moments for a floorbeam with an overhang. (*a*) Placement of factored loads for maximum negative moment. (*b*) Factored moments for the loading in (*a*). (*c*) Placement of factored loads for maximum positive moment. (*d*) Factored moments for the loading in (*c*).

Positive Moment.   For maximum positive moment, the cantilever carries minimum load, whereas the backspan carries full load (Fig. 7.4c). Dead load is the minimum for the cantilever:

$$W_{uc} = 1.2 \times 1.5 = 1.8 \text{ kips per ft}$$

Maximum factored load on the backspan is

$$W_{ub} = 1.2(1.5 + 0.75) + 1.6 \times 1.5 = 5.1 \text{ kips per ft}$$

Corresponding factored moments are (Fig. 7.4d)

$$-M_u = 1.8 \times 7.5^2/2 = 50.6 \text{ kip-ft}$$

$$+M_u = 5.1 \times 30^2/8 - 50.6/2 + \frac{1}{2 \times 5.1}\left(\frac{50.6}{30}\right)^2 = 548.7 \text{ kip-ft}$$

Since the top flange of the beam is braced by the floor deck, the nominal capacity of the section is the plastic moment capacity ϕM_p. For the W18 × 35 selected for negative moment, Table 7.2 shows $\phi M_p = 180 < 548.7$ kip-ft. Hence this section is not adequate for the maximum positive moment. The least-weight beam with $\phi M_p > 548.7$ kip-ft is a W24 × 84 ($\phi M_p = 605$ kip-ft). If, however, the clearance between the beam and the ceiling does not limit the depth of the beam to 24 in, a W27 × 84 may be preferred; it has greater moment capacity and stiffness.

7.12 COMPOSITE BEAMS

Composite steel beam construction is common in multistory commercial buildings. Utilizing the concrete deck as the top (compression) flange of a steel beam to resist maximum positive moments produces an economical design. In general, composite floorbeam construction consists of the following:

- Concrete over a metal deck, the two acting as one composite unit to resist the total loads. The concrete is normally reinforced with welded wire mesh to control shrinkage cracks.
- A metal deck, usually 1½, 2, or 3 in deep, spanning between steel beams to carry the weight of the concrete until it hardens, plus additional construction loads.
- Steel beams supporting the metal deck, concrete, construction, and total loads. When unshored construction is specified, the steel beams are designed as·non-composite to carry the weight of the concrete until it hardens, plus additional construction loads. The steel section must be adequate to resist the total loads acting as a composite system integral with the floor slab.
- Shear connectors, studs, or other types of mechanical shear elements welded to the top flange of the steel beam to ensure composite action and to resist the horizontal shear forces between the steel beam and the concrete deck.

The effective width of the concrete deck as a flange of the composite beam is defined in Art. 6.26.1. The compression force C (kips) in the concrete is the smallest of the values given by Eqs. (7.24) to (7.26). Equation (7.24) denotes the design strength of the concrete:

$$C_c = 0.85 f_c' A_c \tag{7.24}$$

where f'_c = concrete compressive strength, ksi

A_c = area of the concrete within the effective slab width, in^2 (If the metal deck ribs are perpendicular to the beam, the area consists only of the concrete above the metal deck. If, however, the ribs are parallel to the beam, all the concrete, including the concrete in the ribs, comprises the area.)

Equation (7.25) gives the yield strength of the steel beam:

$$C_t = A_s F_y \tag{7.25}$$

where A_s = area of the steel section (not applicable to hybrid sections), in^2

F_y = yield strength of the steel, ksi

Equation (7.26) expresses the strength of the shear connectors:

$$C_s = \Sigma Q_n \tag{7.26}$$

where ΣQ_n is the sum of the nominal strength of the shear connectors between the point of maximum positive moment and zero moment on either side.

For full composite design, three locations of the plastic neutral axis are possible. The location depends on the relationship of C_c to the yield strength of the web, $P_{yw} = A_w F_y$, and C_t. The three cases are as follows (Fig. 7.5):

Case 1. The plastic neutral axis is located in the web of the steel section. This case occurs when the concrete compressive force is less than the web force $C_c \leq P_{yw}$.

Case 2. The plastic neutral axis is located within the thickness of the top flange of the steel section. This case occurs when $P_{yw} < C_c < C_t$.

Case 3. The plastic neutral axis is located in the concrete slab. This case occurs when $C_c \geq C_t$. (When the plastic axis occurs in the concrete slab, the tension in the concrete below the plastic neutral axis is neglected.)

The AISC ASD and LRFD "Specification for Structural Steel Buildings" restricts the number of studs in one rib of metal deck perpendicular to the axis of beam to three. Maximum spacing along the beam is 36 in $\leq 8t$, where t = total

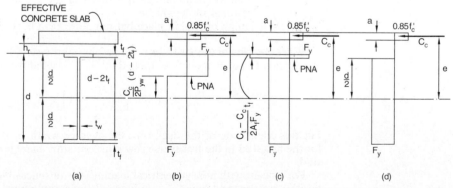

FIGURE 7.5 Stress distributions assumed for plastic design of a composite beam. (*a*) Cross section of composite beam. (*b*) Plastic neutral axis (PNA) in the web. (*c*) PNA in the steel flange. (*d*) PNA in the slab.

slab thickness (in). When the metal deck ribs are parallel to the axis of the beam, the number of rows of studs depends on the flange width of the beam.

The minimum spacing of studs is six diameters along the longitudinal axis of the beam (4 in for ¾-in-diameter studs) and four diameters transverse to the beam (3 in for ¾-in-diameter studs).

The total horizontal shear force C at the interface between the steel beam and the concrete slab is assumed to be transmitted by shear connectors. Hence the number of shear connectors required for composite action is

$$N_s = C/Q_n \qquad (7.27)$$

where Q_n = nominal strength of one shear connector, kips
N_s = number of shear studs between maximum positive moment and zero moment on each side of the maximum positive moment.

The nominal strength of a shear stud connector embedded in a solid concrete slab may be computed from

$$Q_n = 0.5A_{sc}\sqrt{f_c'E_c} \qquad (7.28)$$

where A_{sc} = cross-sectional area of stud, in²
f_c' = specified compressive strength of concrete, ksi
E_c = modulus of elasticity of the concrete, ksi
    = $w^{1.5}\sqrt{f_c'}$
w = unit weight of the concrete, lb/ft³

The strength Q_n, however, may not exceed $A_{sc}F_u$, where F_u is minimum tensile strength of a stud (ksi).

When the shear connectors are embedded in concrete on a metal deck, a reduction factor R should be applied to Q_n computed from Eq. (7.28).

When the ribs of the metal deck are perpendicular to the beam,

$$R = \frac{0.85}{\sqrt{N_r}}\frac{w_r}{h_r}\left(\frac{H_s}{h_r} - 1\right) \leq 1 \qquad (7.29)$$

where N_r = number of studs in one rib at a beam, not to exceed 3 in computations
w_r = average width of concrete rib or haunch, at least 2 in but not more than the minimum clear width near the top of the steel deck, in
h_r = nominal rib height, in
H_s = length of stud in place but not more than $h_r + 3$ in computations, in

When the ribs of the steel deck are parallel to the steel beam and $w_r/h_r < 1.5$, a reduction factor R should be applied to Q_n computed from Eq. (7.28):

$$R = 0.6\frac{w_r}{h_r}\left(\frac{H_s}{h_r} - 1\right) \leq 1 \qquad (7.30)$$

For this orientation of the deck ribs, the average width w_r should be at least 2 in for the first stud in the transverse row plus four stud diameters for each additional stud.

For a beam with nonsymmetrical loading, the distances between the maximum positive moment and point of zero moment (inflection point) on either side of the point of maximum moment will not be equal. Or, if one end of a beam has negative moment, then the inflection point will not be at that end.

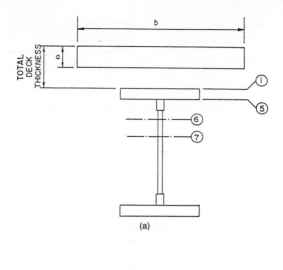

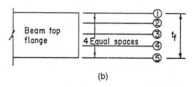

FIGURE 7.6 Seven locations of the plastic neutral axis used for determining the strength of a composite beam. (*a*) For cases 6 and 7, the PNA lies in the web. (*b*) For cases 1 through 5, the PNA lies in the steel flange.

When a concentrated load occurs on a beam, the number of shear connectors between the concentrated load and the inflection point should be adequate to develop the maximum moment at the concentrated load.

When the moment capacity of a fully composite beam is much greater than the applied moment, a partially composite beam may be utilized. It requires fewer shear connectors and thus has a lower construction cost. A partially composite design also may be used advantageously when the number of shear connectors required for a fully composite section cannot be provided because of limited flange width and length.

Figure 7.6 shows seven possible locations of the plastic neutral axis (PNA) in a steel section. The horizontal shear between the steel section and the concrete slab, which is equal to the compressive force in the concrete C, can be determined as illustrated in Table 7.3.

7.13 LRFD FOR COMPOSITE BEAM WITH UNIFORM LOADS

The typical floor construction of a multistory building is to have composite framing. The floor consists of 3¼-in-thick lightweight concrete over a 2-in-deep steel deck. The concrete weighs 115 lb/ft³ and has a compressive strength of 3.0 ksi. An additional 30% of the dead load is assumed for equipment load during construction. The deck is to be supported on steel beams with stud shear connectors on the top

TABLE 7.3 Q_n for Partial Composite Design (kips)

Location of PNA	Q_n and concrete compression
(1)	$A_s F_y$
(2) to (5)	$A_s F_y - 2\Delta A_f F_y^*$
(6)	$0.5[C(5) + C(7)]$†
(7)	$0.25 A_s F_y$

$^*\Delta A_f$ = area of the segment of the steel flange above the plastic neutral axis (PNA).
†$C(n)$ = compressive force at location (n).

flange for composite action (Art. 7.12). Unshored construction is assumed. Therefore, the beams must be capable of carrying their own weight, the weight of the concrete before it hardens, deck weight, and construction loads. Shear connectors will be ¾ in in diameter and 3½ in long. The floor system should be investigated for vibration, assuming a damping ratio of 5%.

A typical beam supporting the deck is 30 ft long. The distance to adjacent beams is 10 ft. Ribs of the deck are perpendicular to the beam. Uniform dead loads on the beam are construction, 0.50 kips per ft, plus 30% for equipment loads, and superimposed load, 0.25 kips per ft. Uniform live load is 0.50 kips per ft.

Beam Selection. Initially, a beam of A36 steel that can support the construction loads is selected. It is assumed to weigh 26 lb/ft. Thus the beam is to be designed for a service dead load of $0.5 \times 1.3 + 0.026 = 0.676$ kips per ft.

$$\text{Factored load} = 0.676 \times 1.4 = 0.946 \text{ kips per ft}$$

$$\text{Factored moment} = M_u = 0.946 \times 30^2/8 = 106.5 \text{ kip-ft}$$

The plastic section modulus required therefore is

$$Z = \frac{M_u}{\phi F_y} = \frac{106.5 \times 12}{0.9 \times 36} = 39.4 \text{ in}^3$$

Use a W16 × 26 ($Z = 44.2$ in^3 and moment of inertia $I = 301$ in^4).

The beam should be cambered to offset the deflection due to a dead load of $0.50 + 0.026 = 0.526$ kips per ft.

$$\text{Camber} = \frac{5 \times 0.526 \times 30^4 \times 12^3}{384 \times 29,000 \times 301} = 1.1 \text{ in}$$

Camber can be specified on the drawings as 1 in.

Strength of Fully Composite Section. Next, the composite steel section is designed to support the total loads. The live load may be reduced in accordance with area supported (Art. 7.9). The reduction factor is $R = 0.0008(300 - 150) = 0.12$. Hence the reduced live load is $0.5(1 - 0.12) = 0.44$ kips per ft. The factored load is the larger of the following:

$$1.2(0.50 + 0.25 + 0.026) + 1.6 \times 0.44 = 1.635 \text{ kips per ft}$$

$$1.4(0.5 + 0.25 + 0.026) = 1.086 \text{ kips per ft}$$

Hence the factored moment is

$$M_u = 1.635 \times 30^2/8 = 183.9 \text{ kip-ft}$$

The concrete-flange width is the smaller of $b = 10 \times 12 = 120$ in or $b = 2(30 \times ^{12}\!/_8) = 90$ in (governs).

The compressive force in the concrete C is the smaller of the values computed from Eqs. (7.24) and (7.25).

$$C_c = 0.85 f'_c A_c = 0.85 \times 3 \times 90 \times 3.25 = 745.9 \text{ kips}$$

$$C_t = A_s F_y = 7.68 \times 36 = 276.5 \text{ kips (governs)}$$

The depth of the concrete compressive-stress block (Fig. 7.5) is

$$a = \frac{C}{0.85 f'_c b} = \frac{276.5}{0.85 \times 3.0 \times 90} = 1.205 \text{ in}$$

Since $C_c > C_t$, the plastic neutral axis will line in the concrete slab (case 3, Art. 7.12). The distance between the compression and tension forces on the W16 × 26 (Fig. 7.5d) is

$$e = 0.5d + 5.25 - 0.5a$$

$$= 0.5 \times 15.69 + 5.25 - 0.5 \times 1.205 = 12.493 \text{ in}$$

The design strength of the W16 × 26 is

$$\phi M_n = 0.85 C_t e = 0.85 \times 276.5 \times 12.493/12 = 244.7 \text{ kip-ft} > 183.9 \text{ kip-ft—OK}$$

Partial Composite Design. Since the capacity of the full composite section is more than required, a partial composite section may be satisfactory. Seven values of the composite section (Fig. 7.6) are calculated as follows, with the flange area $A_f = 5.5 \times 0.345 = 1.898 \text{ in}^2$.

1. Full composite:

$$\Sigma Q_n = A_s F_y = 276.5 \text{ kips}$$

$$\phi M_n = 244.7 \text{ kip-ft}$$

2. Plastic neutral axis $\Delta A_f = A_f/4 = 0.4745$ in below the top of the top flange. From Table 7.3, $\Sigma Q_n = A_s F_y - 2\Delta A_f F_y$.

$$\Sigma Q_n = 276.5 - 2 \times 0.4745 \times 36 = 242.3$$

$$a = 242.3/(0.85 \times 3.0 \times 90) = 1.0558 \text{ in}$$

$$e = 15.69/2 + 5.25 - 1.0558/2 = 12.567 \text{ in}$$

$$M_n = 242.3 \times 12.567 + 0.5(276.5 - 242.3)$$
$$\times \left(15.69 - 0.345 \frac{276.5 - 242.3}{2 \times 1.898 \times 36} \right)$$

$$= 3{,}312 \text{ kip-in}$$

$$\phi M_n = 0.85 \times 3312/12 = 234.6 \text{ kip-ft}$$

3. PNA $\Delta A_f = A_f/2 = 0.949$ in below the top of the top flange:

$$\Sigma Q_n = 208.2 \text{ kips}$$

$$\phi M_n = 224.0 \text{ kip-ft}$$

4. PNA $\Delta A_f = 3A_f/4 = 1.4235$ in below the top of the top flange:

$$\Sigma Q_n = 174.0 \text{ kips}$$

$$\phi M_n = 212.8 \text{ kip-ft}$$

5. PNA at the bottom of the top flange ($\Delta A_f = A_f$):

$$\Sigma Q_n = 139.9 \text{ kips}$$

$$\phi M_n = 201.0 \text{ kip-ft}$$

6. Plastic neutral axis within the web. ΣQ_n is the average of items 5 and 7. (See Table 7.3.)

$$\Sigma Q_n = (139.9 + 69.1)/2 = 104.5 \text{ kips}$$

$$\phi M_n = 186.4 \text{ kip-ft}$$

7.
$$\Sigma Q_n = 0.25 \times 276.5 = 69.1 \text{ kips}$$

$$\phi M_n = 166.7 \text{ kip-ft}$$

From the partial composite values 2 to 7, value 6 is just greater than $M_u = 183.9$ kip-ft.

The AISC "Manual of Steel Construction" includes design tables for composite beams that greatly simplify the calculations. For example, the table for the W16 $\times$ 26, grade 36, composite beam gives ϕM_n for the seven positions of the PNA and for several values of the distance Y_2 (in) from the concrete compressive force C to the top of the steel beam. For the preceding example,

$$Y_2 = Y_{con} - a/2 \tag{7.31}$$

where Y_{con} = total thickness of floor slab, in
a = depth of the concrete compressive-stress block, in

From the table for case 6, $\Sigma Q_n = 104$ kips.

$$a = \frac{104}{0.85 \times 3.0 \times 90} = 0.453 \text{ in}$$

Substitution of a and $Y_{con} = 5.25$ in in Eq. (7.28) gives

$$Y_2 = 5.25 - 0.453/2 = 5.02 \text{ in}$$

The manual table gives the corresponding moment capacity for case 6 and $Y_2 = 5.02$ in as

$$\phi M_n = 186 \text{ kip-ft} > 183.9 \text{ kip-ft—OK}$$

The number of shear studs is based on $C = 104.5$ kips. The nominal strength

Q_n of one stud is given by Eq. (7.28). For a ¾-in stud, with shearing area A_{sc} = 0.442 in² and tensile strength F_u = 60 ksi, the limiting strength is $A_{sc}F_u$ = 0.442 × 60 = 26.5 kips. With concrete unit weight w = 115 lb/ft³ and compressive strength f_c' = 3.0 ksi, and modulus of elasticity E_c = 2136 ksi, the nominal strength given by Eq. (7.28) is

$$Q_n = 0.5 \times 0.442 \sqrt{3.0 \times 2136} = 17.7 \text{ kips} < 26.5 \text{ kips}$$

The number of shear studs required is 2 × 104.5/17.7 = 11.8. Use 12. The total number of metal deck ribs supported on the steel beam is 30. Therefore, only one row of shear studs is required, and no reduction factor is needed.

Deflection Calculations.   Deflections are calculated based on the partial composite properties of the beam. First, the properties of the transformed full composite section (Fig. 7.7) are determined.

The modular ratio E_s/E_n is n = 29,000/2136 = 13.6. This is used to determine the transformed concrete area A_1 = 3.25 × 90/13.6 = 21.52 in². The area of the W16 × 26 is 7.68 in², and its moment of inertia I_s = 301 in⁴. The location of the elastic neutral axis is determined by taking moments of the transformed concrete area and the steel area about the top of the concrete slab:

$$X = \frac{21.52 \times 3.25/2 + 7.68(0.5 \times 15.69 + 5.25)}{21.52 + 7.68} = 4.64 \text{ in}$$

The elastic transformed moment of inertia for full composite action is

$$I_{tr} = \frac{90 \times 3.25^3}{13.6 \times 12} + 21.52 \left(4.64 - \frac{3.25}{2}\right)^2$$

$$+ 7.68 \left(\frac{15.69}{2} + 5.25 - 4.64\right)^2 + 301 = 1065 \text{ in}^4$$

Since partial composite construction is used, the effective moment of inertia is determined from

$$I_{eff} = I_s + (I_{tr} - I_s) \sqrt{\Sigma Q_n/C_f} \qquad (7.32)$$

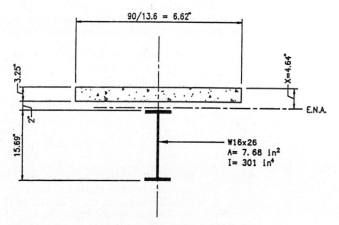

FIGURE 7.7 Transformed section of a composite beam.

where C_f = concrete compression force based on full composite action

$$I_{eff} = 301 + (1065 - 301) \sqrt{104.5/276.5} = 770.7 \text{ in}^4$$

I_{eff} is used to calculate the immediate deflection under service loads (without long-term effects).

For long-term effect on deflections due to creep of the concrete, the moment of inertia is reduced to correspond to a 50% reduction in E_c. Accordingly, the transformed moment of inertia with full composite action and 50% reduction in E_c is I_{tr} = 900.3 in^4 and is based on a modular ratio $2n$ = 27.2. The corresponding transformed concrete area is A_1 = 10.76 in^2.

The reduced effective moment of inertia for partial composite construction with long-term effect is determined from Eq. (7.32):

$$I_{eff} = 301 + (900.3 - 301) \sqrt{104.5/276.5} = 669.4 \text{ in}^4$$

Since unshored construction is specified, the deflection under the weight of concrete when placed and the steel weight is compensated for by the camber specified. Long-term effect due to these weights need not be considered because the concrete is not stressed by them.

Deflection due to long-term superimposed dead loads is

$$D_1 = \frac{5 \times 0.25 \times 30^4 \times 12^3}{384 \times 29,000 \times 669.4} = 0.235 \text{ in}$$

Deflection due to short-term (reduced) live load is

$$D_2 = \frac{5 \times 0.44 \times 30^4 \times 12^3}{384 \times 29,000 \times 770.7} = 0.358 \text{ in}$$

Total deflection is

$$D = D_1 + D_2 = 0.235 + 0.358 = 0.593 \text{ in} = L/607 \text{---OK}$$

Vibration Investigation.   The vibration study of composite beams is based on Wiss and Parmlee's "drop-of-the-heel" method and Murray's empirical equation. (T. M. Murray, "Design to Prevent Floor Vibrations," *AISC Engineering Journal,* third quarter, 1979, and "Acceptability Criterion for Occupant-Induced Floor Vibrations," *AISC Engineering Journal,* second quarter, 1989.)

The total dead load W_D considered in the vibration equations consists of the weight of the concrete and steel beam plus a percentage of the superimposed dead load. The percentage of superimposed dead load is 30% in this example:

$$W_D = 0.50 \times 30 + 0.026 \times 30 + 0.30 \times 0.25 \times 30$$

$$= 18.0 \text{ kips}$$

The frequency f (Hz) of a composite simple-span beam is given by

$$f = 1.57 \sqrt{\frac{gEI_t}{W_D L^3}} \tag{7.33}$$

where g = gravitational acceleration = 386.4 in/s^2
E = steel modulus of elasticity, ksi
I_t = transformed moment of inertia of the composite section, in^4
W_D = total weight on the beam, kips
L = span, in

Substitution of previously determined values into Eq. (7.33) yields

$$f = 1.57 \sqrt{\frac{386.4 \times 29{,}000 \times 770.7}{18.0(30 \times 12)^3}} = 5.03 \text{ Hz}$$

The amplitude A_o of a single beam is calculated by dividing the total floor amplitude A_{ot} by the number N_{eff} of effective beams:

$$A_o = A_{ot}/N_{eff} \tag{7.34}$$

For a constant $t_o = (1/\pi f) \tan^{-1} a \le 0.05$,

$$A_{ot} = 0.246L^3(0.10 - t_o)/EI_t \tag{7.35}$$

For $t_o > 0.05$,

$$A_{ot} = \frac{0.246L^3}{EI_t} \times \frac{1}{2\pi f} \sqrt{2(1 - a \sin a - \cos a) + a^2} \tag{7.36}$$

where $a = 0.1\pi f = 0.1\pi \times 5.03 = 1.58$ radians.

The number of effective beams can be determined from

$$N_{eff} = 2.967 - 0.05776(S/d_e) + 2.556 \times 10^{-8}L^4/I_t + 0.0001(L/S)^3 \ge 1.0 \tag{7.37}$$

where S = spacing of beams in the floor, in
d_e = effective depth of the slab, in
  = average slab thickness when the metal deck ribs are perpendicular to the beam
  = concrete thickness above the metal deck when the deck ribs are parallel to the beam

With $S = 120$ in and $d_e = 4.25$ in, the number of effective beams is

$$N_{eff} = 2.967 - 0.05776 \times \frac{120}{4.25} + 2.556 \times 10^{-8} \times \frac{360^4}{770.7}$$
$$+ 0.0001 \left(\frac{360}{120}\right)^3 = 1.90$$

For $t_o = (1/1.58\pi) \tan^{-1} 1.58 > 0.05$, the total floor amplitude is, from Eq. (7.36),

$$A_{ot} = \frac{0.246 \times 360^3}{29{,}000 \times 770.7} \times \frac{1}{2\pi \times 5.04}$$
$$\times \sqrt{2(1 - 1.58 \sin 1.58 - \cos 1.58) + 1.58^2}$$

$$= 0.188 \text{ in}$$

The amplitude of one beam then is, by Eq. (7.34),

$$A_o = A_{ot}/N_{eff} = 0.0188/1.9 = 0.0099 \text{ in}$$

The mean response rating is given by

$$R = 5.08 \left(\frac{fA_o}{D^{0.217}}\right)^{0.265} \tag{7.38}$$

where f = frequency of the composite beam, Hz
A_o = maximum amplitude of one beam, in
D = damping ratio

For the following values of R, the rating denotes

1. Imperceptible vibration
2. Barely perceptible vibration
3. Distinctly perceptible vibration
4. Strongly perceptible vibration
5. Severe vibration

For the assumed 5% damping ratio,

$$R = 5.08 \left(\frac{5.03 \times 0.0099}{0.05^{0.217}} \right)^{0.265} = 2.7$$

For $2.5 < 3.5$, the vibration may be considered distinctly perceptible.
Murray's equation gives the minimum acceptable damping (percent) as

$$D = 35A_o f + 2.5 = 35 \times 0.0099 \times 5.03 + 2.5 = 4.2 < 5.0 \text{ (acceptable)}$$

7.14 EXAMPLE—LRFD FOR COMPOSITE BEAM WITH CONCENTRATED LOADS AND END MOMENTS

The general information for design of a floor system is the same as that given in Art. 7.14. In this example, a girder of grade 50 steel is to support the floorbeams. (Deck ribs are parallel to the girder.) The girder loads and span are shown in Fig. 7.8 and Table 7.4. The spacing to the left adjacent girder is 30 ft and to the right girder 20 ft.

Dead-Load Moment for Unshored Beam.   The steel girder is to support construction dead loads, nonshored, with 30% additional dead load assumed applied during construction. The girder is assumed to weigh 44 lb/ft. The negative end moments

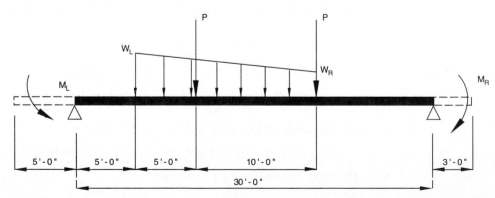

FIGURE 7.8 Composite beam with overhang carries two concentrated loads and a uniformly decreasing load over part of the span. Cantilever carries uniform loads.

TABLE 7.4 Concentrated and Partial Loads on Composite Beam

Type of load	Construction dead load	Superimposed dead load	Live load
Concentrated load P, kips	14.85	7.5	15.0
Negative moment M_L, kip-ft	22.5	7.5	20.0
Negative moment M_R, kip-ft	7.5	2.5	7.0
Partial-load start w_L, kips per ft	0.50	0.75	0.50
Partial-load end w_R, kips per ft	0.20	0.30	0.20

are neglected for this phase of the design since the concrete may be placed over the entire span between the supports but not over the cantilever.

The factored dead loads are

$$P_u = 14.85 \times 1.30 \times 1.4 = 27.03 \text{ kips}$$

$$W_{Lu} = 0.5 \times 1.30 \times 1.4 = 0.910 \text{ kips per ft}$$

$$W_{Ru} = 0.2 \times 1.30 \times 1.4 = 0.364 \text{ kips per ft}$$

$$W_{Gu} = 0.044 \times 1.4 = 0.062 \text{ kips per ft}$$

For the girder acting as a simple beam with a 30-ft span, the factored dead-load moment is $M_u = 328.0$ ft-kips, and the plastic modulus required is $Z = M_u/0.9F_y$ $= 328 \times 12/(0.9 \times 50) = 87.5$ in^3. The least-weight section with larger modulus is a W21 × 44, with $Z = 95.4$ in^3.

Camber.   This is computed for maximum deflection attributable to full construction dead loads. For this computation, the dead-load portion of the end moments is included. The loads are listed under construction dead load in Table 7.4.

The corresponding deflection is 1.09 in. A camber of 1 in may be specified.

Design for Maximum End Moment.   This takes into account the unbraced length of the girder. For the maximum possible unbraced length of the bottom (compression) flange of the steel section, only the dead loads act between supports. The factored dead loads are

$$P_u = 1.2 \times 14.85 = 17.82 \text{ kips}$$

$$W_{Lu} = 1.2 \times 0.5 = 0.60 \text{ kips per ft}$$

$$W_{Ru} = 1.2 \times 0.2 = 0.24 \text{ kips per ft}$$

$$W_{Gu} = 1.2 \times 0.044 = 0.053 \text{ kips per ft}$$

$$M_{Lu} = 1.2(22.5 + 7.5) + 1.6 \times 20 = 68.0 \text{ kip-ft}$$

$$M_{Ru} = 1.2(7.5 + 2.50) + 1.6 \times 7 = 23.2 \text{ kip-ft}$$

The unbraced length of the bottom flange is 2.9 ft. The cantilever length is 5 ft (governs). The design strength ϕM_n for a wide-flange section of grade 50 steel may be obtained from curves in the AISC "Steel Construction Manual—LRFD." A curve indicates that the W21 × 44 with an unbraced length of 5 ft has a design strength $\phi M_n = 356$ kip-ft.

Design for Positive Moment. For this computation, the load factor used for the negative dead-load moments is 1.2, with only dead load on the cantilevers. The load factor for live loads is 1.6.

The factored loads, with live loads reduced 40% for the size of areas supported, are

$$P_u = 1.2(14.85 + 7.5) + 1.6 \times 9.0 = 41.22 \text{ kips}$$

$$W_{Lu} = 1.2(0.5 + 0.75) + 1.6 \times 0.30 = 1.98 \text{ kips per ft}$$

$$W_{Ru} = 1.2(0.20 + 0.30) + 1.6 \times 0.12 = 0.792 \text{ kips per ft}$$

$$W_{Gu} = 1.2 \times 0.044 = 0.053 \text{ kips per ft}$$

$$M_{Lu} = 1.2 \times 22.5 = 27.0 \text{ kip-ft}$$

$$M_{Ru} = 1.2 \times 7.5 = 9.0 \text{ kip-ft}$$

For these loads, the factored maximum positive moment is $M_u = 509.6$ kip-ft.

For determination of the capacity of the composite beam, the effective concrete-flange width is the smaller of

$$b = 12(30 + 20)/2 = 300 \text{ in}$$

$$b = 12 \times 30/4 = 90 \text{ in (governs)}$$

Design tables for composite beams in the AISC manual greatly simplify calculation of design strength. For example, the table for the W21 × 44 grade 50 beam gives ϕM_n for seven positions of the plastic neutral axis (PNA) and for several values of the distance Y_2 from the top of the steel beam to the centroid of the effective concrete-flange force (ΣQ_n) (see Art. 7.13). Try $\Sigma Q_n = 260$ kips. The corresponding depth of the concrete compression block is

$$a = \frac{260}{0.85 \times 3.0 \times 90} = 1.133 \text{ in}$$

From Eq. (7.31), $Y_2 = 5.25 - 1.133/2 = 4.68$ in. The manual table gives the corresponding design strength for case 6 and $Y_2 = 4.68$ in, by interpolation, as

$$\phi M_n = 546 \text{ kip-ft} > (M_u = 509.6 \text{ kip-ft})$$

The maximum positive moment M_u occurs 13.25 ft from the left support (Fig. 7.8). The inflection points occur 0.49 and 0.19 ft from the left and right supports, respectively.

Shear Connectors. Next, the studs required to develop the maximum positive moment and the moments at the concentrated loads are determined. Welded studs ¾ in in diameter are to be used. As in Art. 7.13, the nominal strength of a stud is $Q_n = 17.7$ kips.

For development of the maximum positive moment on both sides of the point of maximum moment, with $\Sigma Q_n = 260$ kips, at least $260/17.7 = 14.69$ studs are required. Since the negative-moment region is small, it is not practical to limit the stud placement to the positive-moment region only. Therefore, additional studs are required for placement of connectors over the entire 30-ft span.

Stud spacing on the left of the point of maximum moment should not exceed

$$S_L = 12(13.25 - 0.49)/14.69 = 10.42 \text{ in}$$

Stud spacing on the right of the point of maximum moment should not exceed

$$S_R = 12(30 - 13.25 - 0.19)/14.69 = 13.53 \text{ in}$$

For determination of the number of studs and spacing required between the concentrated load P 10 ft from the left support (Fig. 7.8) and the left inflection point, the maximum load at that load is calculated to be $M_{Lu} = 502.1$ kip-ft. For the W21 $\times$ 44 grade 50 beam, the manual table indicates that for $\Sigma Q_n = 260$ kips and $Y_2 = 4.68$ in, as calculated previously, the design strength is $\phi M_n = 546$ kip-ft. For $\frac{3}{4}$-in studs and $\Sigma Q_n = 260$ kips, the required number of studs is 14.69. Spacing of these studs, which may not exceed 10.42 in, is also limited to

$$S_{PL} = 12(10 - 0.49)/14.69 = 7.77 \text{ in}$$

Hence the number of studs to be placed in the 10 ft between P and the left support is $10 \times 12/7.77 = 15.4$ studs. Use 16 studs.

For determination of the number of studs and spacing required between the concentrated load P 10 ft from the right support (Fig. 7.8) and the right inflection point, the maximum moment at that load is calculated to be $M_{Ru} = 481.2$ kip-ft. For the W21 $\times$ 44, the manual table indicates that, for case 7, $\Sigma Q_n = 163$ kips and $\phi M_n = 486$ kip-ft. The required number of studs for $\Sigma Q_n = 163$ kips is $163/17.7 = 9.21$ studs. Spacing of these studs, which may not exceed 13.53 in, is also limited to

$$S_{PR} = 12(10 - 0.19)/9.21 = 12.78 \text{ in}$$

The number of studs to be placed in the 10 ft between P and the right support is $10 \times 12/12.78$. Use 10 studs.

The number of studs required between the two concentrated loads equals the sum of the number required between the point of maximum moment and P on the left and right. On the left, the required number of studs is $13.25 \times 12/10.42 - 16 = -0.74$. Since the result is negative, use on the left the maximum permissible stud spacing of 36 in. On the right, the required number of studs is $16.75 \times 12/13.53 - 10 = 4.85$. Use 5 studs. The spacing should not exceed $12(16.75 - 10)/5 = 16.2$ in. Specification of one spacing for the middle segment, however, is more practical. Accordingly, the number of studs between the two concentrated loads would be based on the smallest spacing on either side of the point of maximum moment: $10 \times 12/16.2 = 7.4$. Use 8 studs spaced 15 in center to center.

It may be preferable to specify the total number of studs placed on the beam based on one uniform spacing. The spacing required to develop the maximum moment on either side of its location and between each concentrated load and a support is 7.77 in, as calculated previously. For this spacing over the 30-ft span, the total number of studs required is $30 \times 12/7.77 = 46.3$. Use 48 studs (the next even number).

Deflection Computations. The elastic properties of the composite beam, which consists of a W21 $\times$ 44 and a concrete slab 5.25 in deep (an average of 4.25 in deep) and 90 in wide, are as follows:

$$E_c = 115^{1.5}\sqrt{3.0} = 2136 \text{ ksi}$$

$$n = E_s/E_c = 29,000/2136 = 13.58$$

$$b/n = 90/13.58 = 6.63 \text{ in}$$

$$I_{tr} = 2496 \text{ in}^4$$

For determination of the effective moment of inertia I_{eff} at the location of the maximum moment, a reduced value of the transformed moment of inertia I_{tr} is used based on the partial-composite construction assumed in the computation of shear-connector requirements. For use in Eq. (7.32), the moment of inertia of the W21 × 44 is $I_s = 843$ in^4, $Q_n = 260$ kips, and C_f is the smaller of

$$C_f = 0.85 f_c' A_c = 0.85 \times 3.0 \times 4.25 \times 90 = 975.4 \text{ kips}$$

$$C_f = A_s F_y = 13.0 \times 50 = 650 \text{ kips (governs)}$$

$$I_{eff} = 843 + (2496 - 843)\sqrt{260/650} = 1888 \text{ in}^4$$

A reduced moment of inertia I_r due to long-time effect (creep of the concrete) is determined based on a modular ratio $2n = 2 \times 13.58 = 27.16$ and effective slab width $b/n = 90/27.16 = 3.31$ in. The reduced transformed moment of inertia is 2088 in^4 and the reduced effective moment of inertia is

$$I_r = 843 + (2088 - 843)\sqrt{260/650} = 1630 \text{ in}^4$$

The deflection computations for unshored construction exclude the weight of the concrete slab and steel beam. Whether or not the steel beam is adequately cambered, the assumption is made that the concrete will be finished as a level surface. Hence the concrete slab is likely to be thicker at midspan of the beams and deck.

For computation of the midspan deflections, the cantilevers are assumed to carry only dead load. From Table 7.4, the superimposed dead loads are $P_s = 7.5$ kips, $w_{Ls} = 0.75$ kips per ft, and $w_{Rs} = 0.30$ kips per ft. The dead-load end moments are $M_L = 22.5$ kip-ft and $M_R = 7.5$ kip-ft. For $I_r = 1630$ in^4, the maximum deflection due to these loads is

$$D = \frac{15,865,000}{29,000 \times 1630} = 0.336 \text{ in}$$

The deflection at the left concentrated load P is 0.296 in and at the second load, 0.288 in.

From Table 7.4, the live loads with a 40% reduction for size of area supported are $P_L = 9.0$ kips, $w_{LL} = 0.30$ kips per ft, and $w_{RL} = 0.12$ kips per ft. The maximum deflection due to these loads and with an effective moment of inertia of 1888 in^4 is 0.319 in. The deflection at the left load is 0.282 in and at the second load, 0.275 in.

Total deflections due to superimposed dead loads and live loads are

$$\text{Maximum deflection} = 0.336 + 0.319 = 0.655 \text{ in}$$

$$\text{Deflection at left load } P = 0.295 + 0.282 = 0.577 \text{ in}$$

$$\text{Deflection at right load } P = 0.288 + 0.275 = 0.563 \text{ in}$$

7.15 COMBINED AXIAL LOAD AND BIAXIAL BENDING

Members subject to axial compression or tension and bending about one or two axes, such as columns that are part of rigid frames in two directions, are designed to satisfy the following interaction equations. For symmetrical shapes when $P_u/\phi P_n \geq 0.2$,

$$\frac{P_u}{\phi P_n} + \frac{8}{9}\left(\frac{M_{ux}}{\phi_b M_{nx}} + \frac{M_{uy}}{\phi_b M_{ny}}\right) \leq 1.0 \tag{7.39a}$$

For $P_u/\phi P_n < 0.2$,

$$\frac{P_u}{2\phi P_n} + \left(\frac{M_{ux}}{\phi_b M_{nx}} + \frac{M_{uy}}{\phi_b M_{ny}}\right) \leq 1.0 \tag{7.39b}$$

where P_u = factored axial load, kips
P_n = nominal compressive or tensile strength, kips
M_u = factored bending moment, kip-in
M_n = nominal flexural strength, kip-in
ϕ_t = resistance factor for tension = 0.90
ϕ_b = resistance factor for flexure = 0.90

The factored moments M_{ux} and M_{uy} should include second-order effects, such as $P - \Delta$, for the factored loads. If second-order analysis is not performed, the factored moments can be calculated with magnifiers as follows:

$$M_u = B_1 M_{nt} + B_2 M_{lt} \tag{7.40}$$

where M_{nt} = factored bending moment based on the assumption that there is no lateral translation of the frame, kip-in
M_{lt} = factored bending moment as a result only of lateral translation of the frame, kip-in
$B_1 = \dfrac{C_m}{(1 - P_u/P_e)} \geq 1$
$P_e = A_g F_y/\lambda_c^2$
λ_c = slenderness parameter (Art. 7.4) with effective length factor $K \leq 1.0$ in the plane of bending
A_g = gross area of the member, in^2
F_y = specified minimum yield point of the member, ksi
C_m = coefficient defined for Eq. (6.67)
$B_2 = \dfrac{1}{1 - \Sigma P_u \dfrac{\Delta_{oh}}{\Sigma HL}}$ or $\dfrac{1}{1 - \dfrac{\Sigma P_u}{\Sigma P_e}}$
ΣP_u = sum of factored axial loads of all columns in a story, kips
Δ_{oh} = translation deflection of the story under consideration, in
ΣH = sum of all horizontal forces in a story that produce Δ_{oh}, kips
L = story height, in
$P_e = A_g F_y/\lambda_c^2$, where λ_c is the slenderness parameter with the effective length factor $K \geq 1.0$ in the plane of bending determined for the member when sway is permitted, kips

Since several computer analysis programs are available with the $P - \Delta$ feature included, it is advisable to determine the $P - \Delta$ effects by second-order analysis of framing subject to lateral loads. If the $P - \Delta$ effect is evaluated for frames subject to lateral as well as to vertical loads, the moment magnifier B_2 can be considered to be unity.

7.16 EXAMPLE—LRFD FOR WIDE-FLANGE COLUMN IN A MULTISTORY RIGID FRAME

Columns at the ninth level of a multistory building are to be part of a rigid frame that resists wind loads. Typical floor-to-floor height is 13 ft.

In the ninth story, a wide-flange column of grade 50 steel is to carry loads from a transfer girder, which supports an offset column carrying the upper levels. Therefore, the lower column discontinues at the ninth level. The loads on that column are as follows: dead load, 750 kips; superimposed dead load, 325 kips; and live load, 250 kips. The moments due to gravity loads at the beam-column connection are

$$\text{Dead-load major-axis moment} = 180 \text{ kip-ft}$$

$$\text{Live-load major-axis moment} = 75 \text{ kip-ft}$$

$$\text{Dead-load minor-axis moment} = 75 \text{ kip-ft}$$

$$\text{Live-load minor-axis moment} = 40 \text{ kip-ft}$$

The column axial loads and moments due to service lateral loads with $P - \Delta$ effect included are

$$\text{Axial load} = 600 \text{ kips}$$

$$\text{Major-axis moment} = 1050 \text{ kip-ft}$$

$$\text{Minor-axis moment} = 0.0$$

The beams attached to the flanges of the column with rigid welded connections are part of the rigid frame and have spans of 30 ft. The following beam sizes and corresponding stiffnesses, at top and bottom ends of the column apply.

The beams at both sides of the column at the floor above and the floor below are W36 × 300. The sum of the stiffnesses I_b/L_b of the beams is

$$\Sigma(I_b/L_b) = 20{,}300 \times 2/(30 \times 12) = 112.8 \text{ in}^3$$

where I_b is the beam moment of inertia (in⁴).

The effective length factor K_x corresponding to the case of frame with sidesway permitted is used in determining the axial-load capacity and the moment magnifier B_1. The moment magnifier B_2 is considered unity inasmuch as the $P - \Delta$ effect is included in the analysis.

Axial-Load Capacity.   Since the column is part of a wind-framing system, the K values should be computed based on column and beam stiffnesses. To determine the major-axis K_x, assume that a W14 × 426 with $I_{cx} = 6600$ in⁴ will be selected for the column. At the top of the column, where there is no column above the floor, the relative column-beam stiffness is

$$G_A = \frac{\Sigma(I_c/L_c)}{\Sigma(I_b/L_b)} = \frac{6600/12(13 - 3)}{112.8} = 0.49$$

At the column bottom, with a W14 × 426 column below,

$$G_B = \frac{\Sigma(I_c/L_c)}{\Sigma(I_b/L_b)} = \frac{2 \times 6600/12(13 - 3)}{112.8} = 0.98$$

From a nomograph for the case when sidesway is permitted (Fig. 7.9b), $K_x = 1.23$ (at the intersection with the K axis of a straight line connecting 0.49 on the G_A axis with 0.98 on the G_B axis).

Since the connection of beams to the column web is a simple connection with inhibited sidesway, $K_y = 1.0$.

The effective lengths to be used for determination of axial-load capacity are

$$K_x L_x = 1.23(13 - 3) = 12.3 \text{ ft}$$

$$K_y L_y = 1.0 \times 13 = 13 \text{ ft}$$

The W14 × 426 has radii of gyration $r_x = 7.26$ in and $r_y = 4.34$ in. Therefore, the slenderness ratios for the column are

$$K_x L_x/r_x = 12.3 \times 12/7.26 = 20.3$$

$$K_y L_y/r_y = 13 \times 12/4.34 = 35.9 \text{ (governs)}$$

Use of the AISC "Manual of Steel Construction—LRFD" tables for design axial strength of compression members simplifies evaluation of the trial column size. For the W14 × 426, grade 50 section, a table indicates that for $K_y L_y = 13$ ft, $\phi P_n = 4830$ kips.

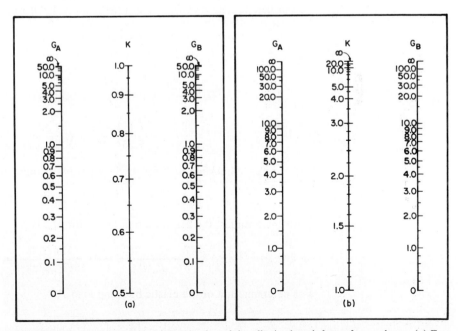

FIGURE 7.9 Nomographs for determination of the effective length factor for a column. (a) For use when sidesway is prevented. (b) For use when sidesway may occur.

Moment Capacity.   Next, the nominal bending-moment capacities are calculated. For strong-axis bending moment, $K_y L_y = 13$ ft is assumed for the flange lateral-buckling state. The limiting lateral unbraced length L_p (in) for plastic behavior for the W14 × 426 is

$$L_p = 300 r_y / \sqrt{F_y} = 300 \times 4.34 / \sqrt{50} = 184 \text{ in } 15.3 \text{ ft} > 13 \text{ ft}$$

Since the unbraced length is less than L_p,

$$\phi M_{nx} = 0.9 \times 869 \times 50/12 = 3259 \text{ kip-ft}$$

$$\phi M_{ny} = 0.9 Z_y F_y = 0.9 \times 434 \times 50/12 = 1628 \text{ kip-ft}$$

Interaction Equation for Dead Load.   For use in the interaction equation for axial load and bending [Eq. (7.39a) or (7.39b)], the factored dead load is

$$P_u = 1.4(750 + 325 + 0.426 \times 13) = 1513 \text{ kips}$$

The factored moments applied to columns due to any general loading conditions should include the second-order magnification. When the frame analysis does not include second-order effects, the factored column moment can be determined from Eq. (7.40).

Computer analysis programs usually include the second-order analysis ($P - \Delta$ effects). Therefore, the values of B_2 for moments about both column axes can be assumed to be unity. However, B_1 should be determined for evaluation of the nonsway magnifications. For a braced column (drift prevented), the slenderness coefficient K_x is determined from Fig. 7.9a with $G_A = 0.49$ and $G_B = 0.98$, previously calculated. The nomograph indicates that $K_x = 0.73$.

For determination of B_1 in Eq. (7.40), the column when loaded is assumed to have single curvature with end moments $M_1 = M_2$. Hence $C_m = 1$.

For determination of the elastic buckling load P_{ex}, the slenderness parameter is

$$\lambda_{cx} = \frac{KL_x}{r_x \pi} \sqrt{\frac{F_y}{E}} = \frac{KL_x}{r_x} \sqrt{\frac{F_y}{286,220}}$$

$$= \frac{0.73 \times 12(13 - 10)}{7.26} \sqrt{\frac{50}{286,220}} = 0.159$$

and the elastic buckling load for the beam cross-sectional area $A_g = 125$ in^2 is

$$P_{ex} = A_g F_y / \lambda_{cx}^2 = 125 \times 50/0.159^2 = 247,000 \text{ kips}$$

With these values, the magnification factor for M_{ux} is

$$B_{1x} = \frac{C_m}{1 - P_u/P_{ex}} = \frac{1.0}{1 - 1513/247,000} = 1.006$$

For determination of the elastic buckling load P_{ey},

$$\lambda_{cy} = \frac{1 \times 13 \times 12}{4.34} \sqrt{\frac{50}{286,220}} = 0.475$$

The elastic buckling load with respect to the y axis is

$$P_{ey} = A_g F_y / \lambda_{cy}^2 = 125 \times 50/0.475^2 = 27,700 \text{ kips}$$

With these values, the magnification factor for M_{uy} is

$$B_{1y} = \frac{C_m}{1 - P_u/P_{ey}} = \frac{1}{1 - 1513/27,700} = 1.058$$

Application of the magnification factor to the dead-load moments due to gravity loads yields

$$M_{ux} = 1.006 \times 1.4 \times 180 = 253.5 \text{ kip-ft}$$

$$M_{uy} = 1.058 \times 1.4 \times 75 = 111.1 \text{ kip-ft}$$

The interaction result, which may be considered a section efficiency ratio, is, from Eq. (7.39a) for $P_u/\phi P_n = 1513/4830 = 0.313 > 0.2$,

$$R = 0.313 + \frac{8}{9}\left(\frac{253.5}{3259} + \frac{111.1}{1628}\right)$$

$$= 0.313 + \frac{8}{9}(0.0778 + 0.682) = 0.443 < 1.0$$

Interaction Equation for Full Gravity Loading.   For use in the interaction equation based on factored loads and moments due to 1.2 times the dead load plus 1.6 times the live load,

$$P_u = 1.2(750 + 325 + 0.426 \times 13) + 1.6 \times 250 = 1697 \text{ kips.}$$

Determined in the same way as for the dead load, the magnification factors are

$$B_{1x} = \frac{1.0}{1 - 1697/247,000} = 1.007$$

$$B_{1y} = \frac{1.0}{1 - 1697/27,700} = 1.065$$

Application of the magnification factors to the factored moments yields

$$M_{ux} = 1.007(1.2 \times 180 + 1.6 \times 75) = 338.4 \text{ kip-ft}$$

$$M_{uy} = 1.065(1.2 \times 75 + 1.6 \times 40) = 164.0 \text{ kip-ft}$$

With $P_u/\phi P_n = 1697/4830 = 0.351 > 0.2$, substitution of the preceding values in Eq. (7.39a) yields

$$R = 0.351 + \frac{8}{9}\left(\frac{338.4}{3259} + \frac{164.0}{1628}\right)$$

$$= 0.351 + \frac{8}{9}(0.1038 + 0.1008) = 0.533 < 1$$

Interaction Equation with Wind Load.   For use in the interaction equation based on factored loads and moments due to 1.2 times the dead load plus 0.5 times the live load plus 1.3 times the wind load of 600 kips, including the $P - \Delta$ effect,

$$P_u = 1.2(750 + 325 + 0.426 \times 13) + 0.5 \times 250 + 1.3 \times 600$$

$$= 2202 \text{ kips}$$

Under wind action, double curvature may occur for strong-axis bending. For this condition, with $M_1 = M_2$,

$$C_{mx} = 0.6 - 0.4 \times 1 = 0.2$$

In this case, the magnification factor for strong-axis bending is

$$B_{1x} = \frac{0.2}{1 - 2202/247{,}000} = 0.202 < 1$$

Use $B_{1x} = 1.0$. The magnification factor for minor axis bending is, with $C_m = 1$ for single-curvature bending,

$$B_{1y} = \frac{1.0}{1 - 2202/27{,}700} = 1.0864$$

Application of the magnification factors to the factored moments yields

$$M_{ux} = 1.0(1.2 \times 180 + 0.5 \times 75 + 1.3 \times 1050)$$

$$= 1618 \text{ kip-ft}$$

$$M_{uy} = 1.0864(1.2 \times 75 + 0.5 \times 40) = 119.5 \text{ kip-ft}$$

With $P_u/\phi P_n = 2202/4830 = 0.456 > 0.2$, substitution of the preceding values in Eq. (7.39a) yields

$$R = 0.456 + \frac{8}{9}\left(\frac{1618}{3259} + \frac{119.5}{1628}\right)$$

$$= 0.456 + \frac{8}{9}(0.496 + 0.0734) = 0.96 < 1$$

This is the governing R value, and since it is less than unity, the column selected, W14 × 426, is adequate.

7.17 BASE PLATE DESIGN

Base plates are usually used to distribute column loads over a large enough area of supporting concrete construction that the design bearing strength of the concrete will not be exceeded. The factored load P_u is considered to be uniformly distributed under a base plate.

The nominal bearing strength f_p (ksi) of the concrete is given by

$$f_p = 0.85f_c'\sqrt{A_2/A_1} \qquad A_2/A_1 \le 2 \qquad (7.41)$$

where f_c' = specified compressive strength of concrete, ksi
A_1 = area of the base plate, in^2
A_2 = area of the supporting concrete that is geometrically similar to and concentric with the loaded area, in^2

In most cases, the bearing strength f_p is $0.85f'_c$ when the concrete support is slightly larger than the base plate or $1.7f'_c$ when the support is a spread footing, pile cap, or mat foundation. Therefore, the required area of a base plate for a factored load P_u is

$$A_1 = P_u/\phi_c f'_c \tag{7.42}$$

where ϕ_c is the strength reduction factor = 0.6. For a wide-flange column, A_1 should not be less than $b_f d$, where b_f is the flange width (in) and d is the depth of column (in).

The length N (in) of a rectangular base plate for a wide-flange column may be taken in the direction of d as

$$N = \sqrt{A_1} + \Delta > d \tag{7.43}$$

For use in Eq. (7.43),

$$\Delta = 0.5(0.95d - 0.80b_f) \tag{7.44}$$

The width B (in) parallel to the flanges, then, is

$$B = A_1/N \tag{7.45}$$

The thickness of the base plate t_p (in) is the largest of the values given by Eqs. (7.46) to (7.48):

$$t_p = m \sqrt{\frac{2P_u}{0.9F_y BN}} \tag{7.46}$$

$$t_p = n \sqrt{\frac{2P_u}{0.9F_y BN}} \tag{7.47}$$

$$t_p = c \sqrt{\frac{2P_o}{0.9F_y A_H}} \tag{7.48}$$

where m = projection of base plate beyond the flange and parallel to the web, in
      = $(N - 0.95d)/2$
    n = projection of base plate beyond the edges of the flange and perpendicular to the web, in
      = $(B - 0.80b_f)/2$
    P_o = $P_u b_f d/NB$
    A_H = $P_o/0.6f_p$
    c = $0.25[d + b_f - t_f - \sqrt{(d + b_f - t_f)^2 - 4(A_H - t_f b_f)}]$
    t_f = flange thickness, in

7.18 EXAMPLE—LRFD OF COLUMN BASE PLATE

A base plate of A36 steel is to distribute the load from a W14 × 233 column to a concrete pedestal whose size is slightly larger than that of the base plate. The pedestal concrete strength f'_c is 4.0 ksi. The factored load on the column is 1731 kips.

From Eq. (7.41), the nominal bearing strength of the concrete is

$$f_p = 0.85 \times 4.0 = 3.4 \text{ ksi}$$

The required area of the base plate is computed from Eq. (7.42):

$$A_1 = 1,731/(0.6 \times 3.4) = 848.5 \text{ in}^2$$

From Eq. (7.44) for determination of the length N of the base plate,

$$\Delta = 0.5(0.95 \times 16.04 - 0.80 \times 15.89) = 1.26 \text{ in}$$

From Eq. (7.43),

$$N = \sqrt{848.5} + 1.26 = 30.4 \text{ in}$$

Use a 31-in length. The width required then is

$$B = A_1/N = 848.5/31 = 27.4 \text{ in}$$

Use a 28-in-wide plate. The area of the plate is $A_1 = 868 \text{ in}^2$.
For determination of the base plate thickness, the projections beyond the column are

$$m = (31 - 0.95 \times 16.04)/2 = 7.88 \text{ in (governs)}$$

$$n = (28 - 0.8 \times 15.89)/2 = 7.64 \text{ in}$$

Equation (7.46) yields a larger thickness than Eq. (7.47) because $m > n$. From Eq. (7.46),

$$t_p = 7.88 \sqrt{\frac{2 \times 1731}{0.9 \times 36 \times 28 \times 31}} = 2.76 \text{ in}$$

For use in Eq. (7.48),

$$P_o = 1731 \times 15.89 \times 16.04/(31 \times 28) = 508.3 \text{ kips}$$

$$A_H = 508.3/(0.6 \times 3.4) = 249.2 \text{ in}^2$$

For calculation of c for use in Eq. (7.48),

$$d + b_f - t_f = 16.04 + 15.89 - 1.72 = 30.21 \text{ in}$$

$$c = 0.25[30.21 - \sqrt{30.21^2 - 4(249.2 - 1.72 \times 15.89)}] = 6.29$$

Then, the thickness given by Eq. (7.48) is

$$t_p = 6.29 \sqrt{\frac{2 \times 508.3}{0.9 \times 36 \times 249.2}} = 2.23 \text{ in} < 2.76 \text{ in}$$

Use a base plate $28 \times 2\frac{3}{4} \times 31$ in.

SECTION 8
FLOOR AND ROOF SYSTEMS

Daniel A. Cuoco
Principal, LZA/Thornton-Tomasetti Engineers,
New York, New York

Structural-steel framing provides designers with a wide selection of economical systems for floor and roof construction. Steel framing can achieve longer spans compared to other types of construction. This minimizes the number of columns and footings, thereby increasing speed of erection. Longer spans also provide more flexibility for interior-space planning.

Another advantage of steel construction is its ability to accommodate readily future structural modifications, such as openings for tenants' stairs and changes to heavier floor loadings. When reinforcement of existing steel structures is required, it can be accomplished by such measures as addition of framing members connected to existing members and field welding of additional steel plates to strengthen existing members.

FLOOR DECKS

The most common types of floor-deck systems currently used with structural steel construction are concrete fill on metal decks, precast-concrete planks, and cast-in-place concrete slabs.

8.1 CONCRETE FILL ON METAL DECK

The most prevalent type of floor deck used with steel frames is concrete fill on metal deck. The metal deck consists of cold-formed profiles made from steel sheet, usually having a yield strength of at least 33 ksi. Design requirements for metal deck are contained in the American Iron and Steel Institute's "Specification for the Design of Cold-Formed Steel Structural Members."

The concrete fill is usually specified to have a 28-day compressive strength of at least 3000 psi. Requirements for concrete design are contained in the American Concrete Institute Standard ACI 318, "Building Code Requirements for Reinforced Concrete."

Sheet thicknesses of metal deck usually range between 24 and 18 ga, although thicknesses outside this range are sometimes used. The design thicknesses corresponding to typical gage designations are shown in Table 8.1.

TABLE 8.1 Equivalent Thicknesses for Cold-Formed Steel

Gage designation	Design thickness, in
28	0.0149
26	0.0179
24	0.0239
22	0.0299
20	0.0359
18	0.0478
16	0.0598

Metal deck is commonly available in depths of 1½, 2, and 3 in. Generally, it is preferable to use a deeper deck that can span longer distances between supports and thereby reduce the number of beams required. For example, a beam spacing of about 15 ft can be achieved with 3-in deck. However, each project must be evaluated on an individual basis to determine the most efficient combination of deck depth and beam spacing.

For special long-span applications, metal deck is available with depths of 4½, 6, and 7½ in from some manufacturers.

Composite versus Noncomposite Construction. Ordinarily, composite construction with metal deck and structural-steel framing is used. In this case, the deck acts not only as a permanent form for the concrete slab but also, after the concrete hardens, as the positive bending reinforcement for the slab. To achieve this composite action, deformations are formed in the deck to provide a mechanical interlock with the concrete (Fig. 8.1). Although not serving a primary structural purpose,

FIGURE 8.1 Cold-formed steel decking used in composite construction with concrete fill.

welded wire fabric is usually placed within the concrete slab about 1 in below the top surface to minimize cracking due to concrete shrinkage and thermal effects. This welded wire fabric also provides, to a limited degree, some amount of crack control in negative-moment regions of the slab over supporting members.

Noncomposite metal deck is used as a form for concrete and is considered to be ineffective in resisting superimposed loadings. In cases where the deck is shored, or where the deck is unshored but the long-term reliability of the deck will be questionable, the deck is also considered to be ineffective in supporting the dead load of the concrete slab. For example, unless special precautions are taken to prevent deterioration, metal deck used in exposed parking structures is susceptible to corrosion and will eventually be ineffective. In such cases, the metal deck should be used solely as a form to support the concrete until it hardens. Reinforcement should be placed within the slab to resist all design loadings.

Noncellular versus Cellular Deck.   It is often desirable to distribute a building's electrical wiring within the floor deck system, in which case cellular metal deck can be used in lieu of noncellular deck. Cellular deck is essentially noncellular deck, such as that shown in Fig. 8.1, with a flat sheet added to the bottom of the deck to create cells (Fig. 8.2). Electrical, power, and telephone wiring is placed within the cells for distribution over the entire floor area. In many cases, a sufficient number of cells is obtained by combining alternate panels of cellular deck and noncellular deck, which is called a blended system (Fig. 8.3). When cellular deck is used, the 3-in depth is the minimum preferred because it provides convenient space for wiring. The 1½-in depth is rarely used.

For feeding wiring into the cells, a trench header is placed within the concrete above the metal deck, in a direction perpendicular to the cells (Fig. 8.4). Special attention should be given to the design of the structural components adjacent to the trench header, since composite action for both the floor deck and beams is lost

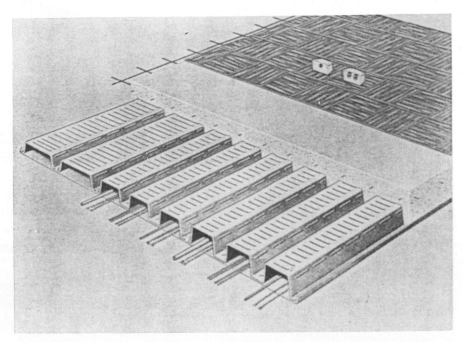

FIGURE 8.2   Cellular steel deck with concrete slab.

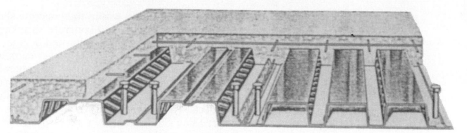

FIGURE 8.3 Blended deck, alternating cellular and noncellular panels, in composite construction.

in these areas. Where possible, the direction of the cells should be selected to minimize the total length of trench header required. Generally, by running the cells in the longitudinal direction of the building, the total length of trench header is significantly less than if the cells were run in the transverse direction (Fig. 8.5).

If a uniform grid of power outlets is desired, such as 5 ft by 5 ft on centers, preset outlets can be positioned above the cells and cast into the concrete fill. However, in many cases the outlet locations will be dictated by subsequent tenant layouts. In such cases, the concrete fill can be cored and afterset outlets can be installed at any desired location.

Shored versus Unshored Construction. To support the weight of newly placed concrete and the construction live loads applied to the metal deck, the deck can

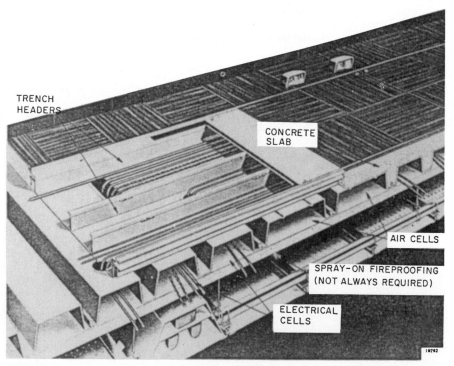

FIGURE 8.4 Cellular steel deck with trench header placed within the concrete slab to feed wiring to cells.

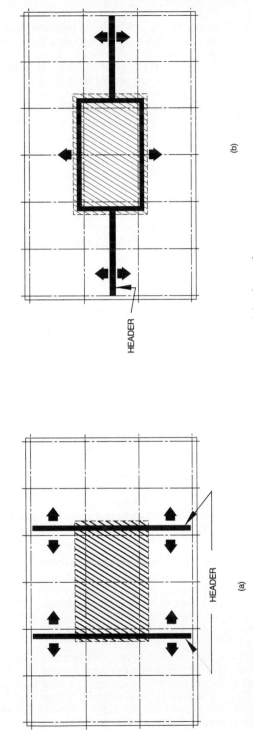

FIGURE 8.5 Floor layout for cellular deck with cells in different directions. Length of trench header serving them is less for (*a*) cells in the longitudinal direction than for (*b*) cells in the transverse direction.

8.5

either be shored or be designed to span between supporting members. If the deck is shored, a shallower-depth or thinner-gage deck can be used. The economy of shoring, however, should be investigated, inasmuch as the savings in deck cost may be more than offset by the cost of the shoring. Also, deflections that will occur after the shoring is removed should be evaluated. Another consideration is that use of shoring can sometimes affect the construction schedule, since the shoring is usually kept in place until the concrete fill has reached at least 75% of its specified 28-day compressive strength. In addition, when shoring is used, the concrete must be designed to resist the stresses resulting from the total dead load combined with all superimposed loadings.

When concrete is cast on unshored metal deck, the weight of the concrete causes the deck to deflect between supports. This deflection is usually limited to the lesser of $\frac{1}{180}$ the deck span or $\frac{3}{4}$ in. Since the top surface of the concrete slab will normally be finished level, there will be an additional amount of concrete placed as a result of the deck deflection. The added weight of this additional concrete must be taken into account in design of the metal deck to ensure adequate strength. The concrete fill, however, need only be designed for the stresses resulting from superimposed loadings.

Unshored metal-deck construction is the system most commonly used. The additional cost of the deeper or thicker deck is generally much less than the cost of shoring. To increase the efficiency of the unshored deck in supporting the weight of the unhardened concrete and construction live loads, from both a strength and deflection standpoint, the deck is normally extended continuously over supporting members for two or three spans, in lieu of single-span construction. For superimposed loadings, however, the composite slab is designed as a simple span, unless negative-moment reinforcement is provided over supports in accordance with conventional reinforced-concrete-slab design (disregarding the metal deck as compressive reinforcement).

Lightweight versus Normal-Weight Concrete. Either lightweight or normal-weight concrete can serve the structural function of the concrete fill placed on the metal deck. Although there is a cost premium associated with lightweight concrete, sometimes the savings in steel framing and foundation costs can outweigh the premium. Also, lightweight concrete in sufficient thickness can provide the necessary fire rating for the floor system and thus eliminate the need for additional fire protection (see "Fire Protection" below).

The tradeoffs in use of lightweight concrete versus normal-weight concrete plus fire protection should be evaluated on a project-by-project basis.

Fire Protection. Most applications of concrete fill on metal deck in buildings require that the floor-deck assembly have a fire rating. For noncellular metal deck, the fire rating is usually obtained either by providing sufficient concrete thickness above the metal deck or by applying spray-on fire protection to the underside of the metal deck. The fire rating for cellular metal deck is usually obtained by the latter method. As an alternative, a fire-rated ceiling system can be installed below the cellular or noncellular deck.

When the required fire rating is obtained by providing sufficient concrete-fill thickness, lightweight concrete requires a lesser thickness than normal-weight concrete for the same rating. For example, a 2-hour rating can be obtained by using either $3\frac{1}{4}$ in of lightweight concrete or $4\frac{1}{2}$ in of normal-weight concrete above the metal deck. The latter option is rarely used, since the additional thickness of heavier concrete penalizes the steel tonnage (i.e., heavier beams, girders, and columns) and the foundations.

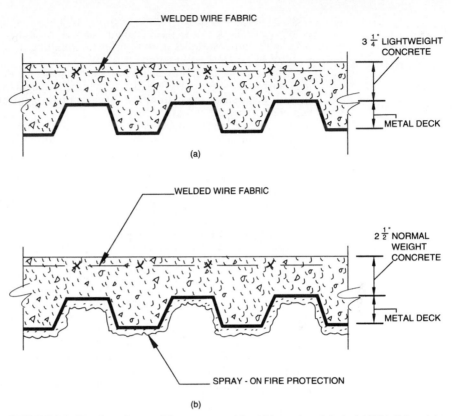

FIGURE 8.6 Two-hour fire-rated floor systems, with cold-formed steel deck. (*a*) With lightweight concrete fill; (*b*) with normal-weight concrete fill.

If spray-on fire protection is used on the underside of the metal deck, the thickness of concrete above the deck can be the minimum required to resist the applied floor loads. This minimum thickness is usually 2½ in, and the less expensive normal-weight concrete may be used instead of lightweight concrete. Therefore, the two options that are frequently considered for a 2-hour-rated, noncellular floor-deck system are 3¼-in lightweight concrete above the metal deck without spray-on fire protection and 2½-in normal-weight concrete above the metal deck with spray-on fire protection (Fig. 8.6). Since the dead load of the floor deck for the two options is essentially the same, the steel framing and foundations will also be the same. Thus, the comparison reduces to the cost of the more expensive lightweight concrete versus the cost of the normal-weight concrete plus the spray-on fire protection. Since the costs, and contractor preferences, vary with geographical location, the evaluation must be made on an individual project basis. (See also Art. 6.32.)

Diaphragm Action of Metal-Deck Systems. Concrete fill on metal deck readily serves as a relatively stiff diaphragm that transfers lateral loads, such as wind and seismic forces, at each floor level through in-plane shear to the lateral load-resisting elements of the structure, such as shear walls and braced frames. The resulting shear stresses can usually be accommodated by the combined strength of the concrete fill and metal deck, without need for additional reinforcement. Attachment of the metal deck to the steel framing, as well as attachment between adjacent deck units, must be sufficient to transfer the resulting shear stresses (see "Attachment of Metal Deck to Framing" below).

Additional shear reinforcement may be required in floor decks with large openings, such as those for stairs or shafts, with trench headers for electrical distribution, or with other shear discontinuities. Also, floors in multistory buildings in which cumulative lateral loads are transferred from one lateral load-resisting system to another, for example, from perimeter frames to interior shear walls, may be subjected to unusually large shear stresses that require a diaphragm strength significantly greater than that for a typical floor.

Attachment of Metal Deck to Framing.   Metal deck can be attached to the steel framing with puddle (arc spot) welds, screws, or power-driven fasteners. These attachments provide lateral bracing for the steel framing and, when applicable, transfer shear stresses resulting from diaphragm action. The maximum spacing of attachments to steel framing is generally 12 in.

Attachment of adjacent deck units to each other, that is, sidelap connection, can be made with welds, screws, or button punches. Generally, the maximum spacing of sidelap attachments is 36 in. In addition to diaphragm or other loading requirements, the type, size, and spacing of attachments is sometimes dictated by insurance (Factory Mutual or Underwriters' Laboratories) requirements.

Weld sizes generally range between ½-in and ¾-in minimum visible diameter. When metal deck is welded to steel framing, welding washers should be used if the deck thickness is less than 22 ga to minimize the possibility of burning through the deck. Sidelap welding is not recommended for deck thicknesses of 22 ga and thinner.

Screws can be either self-drilling or self-tapping. Self-drilling screws have drill threads formed at the screw end. This enables direct installation without the need for predrilling of holes in the steel framing or metal deck. Self-tapping screws require that a hole be drilled prior to installation. Typical screw sizes are No. 12 and No. 14 (with 0.216-in and 0.242-in shank diameter, respectively) for attachment of metal deck to steel framing. No. 8 and No. 10 screws (with 0.164-in and 0.190-in shank diameter, respectively) are frequently used for sidelap connections.

Power-driven fasteners are installed through the metal deck into the steel framing with pneumatic or powder-actuated equipment. Predrilled holes are not required. These types of fasteners are not used for sidelap connections.

Button punches can be used for sidelap connections of certain types of metal deck that utilize upstanding seams at the sidelaps. However, since uniformity of installation is difficult to control, button punches are not usually considered to contribute significantly to diaphragm strength.

The diaphragm capacity of various types and arrangements of metal deck and attachments are given in the Steel Deck Institute *Diaphragm Design Manual*.

8.2   *PRECAST-CONCRETE PLANK*

This is another type of floor deck that is used with steel-framed construction (Fig. 8.7). The plank is prefabricated in standard widths, usually ranging between 4 and 8 ft, and is normally prestressed with high-strength steel tendons. Shear keys formed at the edges of the plank are subsequently grouted, to allow loads to be distributed between adjacent planks. Voids are usually placed within the thickness of the plank to reduce the deadweight without causing significant reduction in plank strength.

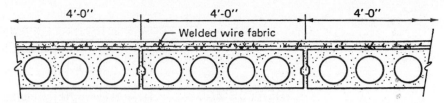

FIGURE 8.7 Precast-concrete plank floor with concrete topping.

The inherent fire resistance of the precast concrete plank obviates the need for supplementary fire protection.

Topped versus Untopped Planks. Precast planks can be structurally designed to sustain required loadings without need for a cast-in-place concrete topping. However, in many cases, it is advantageous to utilize a topping to eliminate differences in camber and elevation between adjacent planks at the joints and thus provide a smooth slab top surface. When a topping is used, the top surface of the plank may be intentionally roughened to achieve composite action between topping and plank. Thereby, the topping also serves as a structural component of the floor-deck system.

A cast-in-place concrete topping can also be used for embedment of conduits and outlets that supply electricity and communication services. Voids within the planks can also be used as part of the distribution system. When the topping is designed to act compositely with the plank, however, careful consideration must be given to the effects of these embedded items.

Dead-Load Deflection of Concrete Plank. In design of prestressed-concrete planks, the prestressing load balances a substantial portion of the dead load. As a result, relatively small dead-load deflections occur. For planks subjected to significant superimposed dead-load conditions of a sustained nature, for example, perimeter plank supporting an exterior masonry wall, additional prestressing to compensate for the added dead load, or some other stiffening method, is required to prevent large initial and creep deflections of the plank.

Diaphragm Action of Concrete-Plank Systems. The diaphragm action of a floor deck composed of precast-concrete planks can be enhanced by making field-welded connections between steel embedments located intermittently along the shear keys of adjacent planks. (See also Art. 8.1.)

Attachments of Concrete Plank to Framing. Precast-concrete planks are attached to and provide lateral bracing for supporting steel framing. A typical method of attachment is a field-welded connection between the supporting steel and steel embedments in the precast planks.

8.3 CAST-IN-PLACE CONCRETE SLABS

Use of cast-in-place concrete for floor decks in steel-framed construction is a traditional approach that was much more prevalent prior to the advent of metal deck and spray-on fire protection. For one of the more common types of cast-in-place concrete floors, the formwork is configured to encase the steel framing, to provide

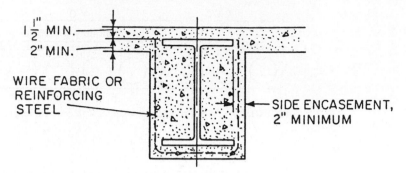

FIGURE 8.8 Minimum requirements for composite action with concrete-encased steel framing.

fire protection and lateral bracing for the steel (see Fig. 8.8). If the proper confinement details are provided, this encasement can also serve to achieve composite action between the steel framing and the floor deck.

Dead-load deflections should be calculated and, for long spans with large deflections, the formwork should be cambered to provide a level deck surface after removal of the formwork shoring. Diaphragm action is readily attainable with cast-in-place concrete floor decks. (See also Art. 8.1.)

ROOF DECKS

The systems used for floor decks (Arts. 8.1 to 8.3) can also be used for roof decks. When used as roof decks, these systems are overlaid by roofing materials, to provide a weathertight enclosure. Other roof deck systems are described in Arts. 8.4 to 8.7.

8.4 METAL ROOF DECK

Steel-framed buildings often utilize a roof deck composed simply of metal deck. When properly sloped for drainage, the metal deck itself can serve as a watertight enclosure. Alternatively, roofing materials can be placed on top of the deck. In either case, diaphragm action can be achieved by proper sizing and attachment of the metal deck. A fire rating can be provided by applying spray-on fire protection to the underside of the roof deck, or by installing a fire-rated ceiling system below the deck.

Metal roof deck usually is used for noncomposite construction. It is commonly available in depths of 1½, 2, and 3 in. Long-span roof deck is available with depths of 4½, 6, and 7½ in from some manufacturers. Cellular roof deck is sometimes used to provide a smooth soffit. When a lightweight insulating concrete fill is placed over the roof deck, the deck should be galvanized and also vented (perforated) to accelerate the drying time of the insulating fill.

Standing-Seam System.   When the metal roof deck is to serve as a weathertight enclosure, connection of deck units with standing seams offers the advantage of

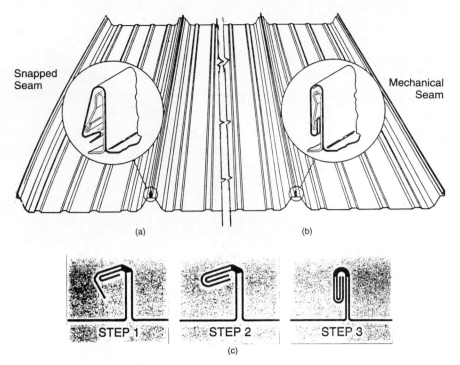

FIGURE 8.9 Standing-seam roof deck. (*a*) With snapped seam; (*b*) with mechanical seam; (*c*) steps in forming a seam.

placing the deck seam above the drainage surface of the roof, thereby minimizing the potential for water leakage (Fig. 8.9). The seams can simply be snapped together or, to enhance their weathertightness, can be continuously seamed by mechanical means with a field-operated seaming machine provided by the deck manufacturer. Some deck types utilize an additional cap piece over the seam, which is mechanically seamed in the field (see Fig. 8.10). Frequently, the seams contain a factory-applied sealant for added weather protection.

Thicknesses of standing-seam roof decks usually range between 26 and 20 ga. Typical spans range between 3 and 8 ft. A roof slope of at least ¼ in per ft should be provided for drainage of rainwater.

Standing-seam systems are typically attached to the supporting members with concealed anchor clips (Fig. 8.11) that allow unimpeded longitudinal thermal movement of the deck relative to the supporting structure. This eliminates buildup of stresses within the system and possible leakage at connections. Since these free-floating connections do not laterally brace the supporting members, however, supplementary bracing of roof framing is required.

8.5 LIGHTWEIGHT PRECAST-CONCRETE ROOF PANELS

Roof decks of lightweight precast-concrete panels typically span 5 to 10 ft between supports. Panel thicknesses range from 2 to 4 in, and widths are usually 16 to 24 in. Depending on the product, concrete density can vary from 50 to 115 lb per ft^3.

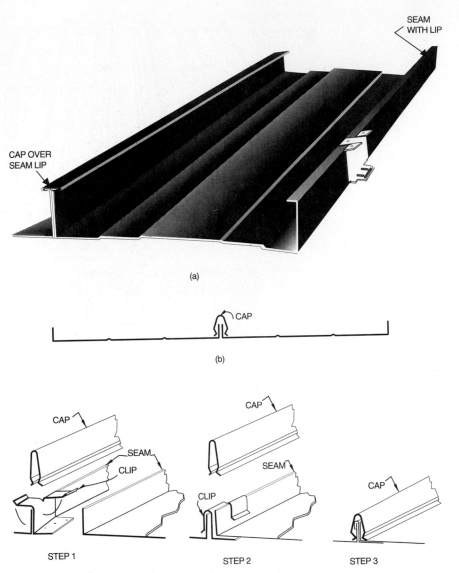

FIGURE 8.10   Standing-seam roof deck with cap installed over the seams. (*a*) Channel cap with flanges folded over lip of seam. (*b*) U-shaped cap clamps over clips on seam. (*c*) Steps in forming a seam with clamped cap.

Certain types of panels have diaphragm capacities. Many panels can achieve a fire rating when used as part of an approved ceiling assembly.

The panels are typically attached to steel framing with cold-formed-steel clips (see Fig. 8.12). The joints between panels are cemented on the upper side, usually with an asphaltic mastic compound. Insulation and roofing materials are normally placed on top of the panels. Some panels are nailable for application of certain types of roof finishes, such as slate, tile, and copper.

ANCHOR CLIP

FIGURE 8.11 Typical anchor clip for standing-seam roof deck.

8.6 WOOD-FIBER PLANKS

Planks formed of wood fibers bonded with portland cement provide a lightweight roof deck with insulating and acoustical properties. The typical density of this material ranges between 30 and 40 lb per ft^3. Some plank types have diaphragm capacities. When used as part of an approved ceiling assembly, many planks can achieve a fire rating.

The planks are usually supported by steel bulb tees (Fig. 8.13), which are nominally spaced 32 to 48 in on centers. The joint over the bulb tee is typically grouted with a gypsum-concrete grout, and roofing materials are applied to the top surface of the planks.

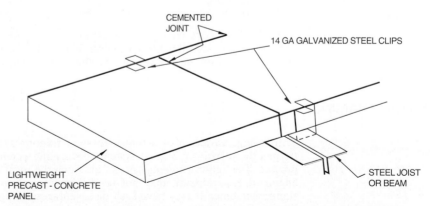

CEMENTED JOINT

14 GA GALVANIZED STEEL CLIPS

LIGHTWEIGHT PRECAST - CONCRETE PANEL

STEEL JOIST OR BEAM

FIGURE 8.12 Typical clips for attachment of precast-concrete panels to steel framing. The clips are driven into place for a wedge fit at diagonal corners of the panels. Minimum flange width for supporting member is preferably 4 in.

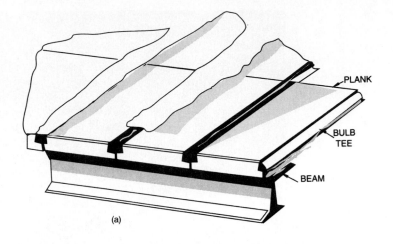

(a)

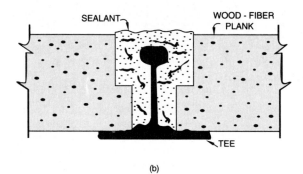

(b)

FIGURE 8.13 (*a*) Wood-fiber planks form roof deck. (*b*) Plank is supported by a steel bulb tee.

8.7 GYPSUM-CONCRETE DECKS

Poured gypsum concrete is typically used in conjunction with steel bulb tees, formboards, and galvanized reinforcing mesh (Fig. 8.14). Drainage slopes can be readily built into the roof deck by varying the thickness of gypsum.

FLOOR FRAMING

With a large variety of structural steel floor-framing systems available, designers frequently investigate several systems during the preliminary design stage of a project. The lightest framing system, although the most efficient from a structural engineering standpoint, may not be the best selection from an overall project standpoint, since it may have such disadvantages as high fabrication costs, large floor-to-floor heights, and difficulties in interfacing with mechanical ductwork.

Spandrel members are frequently subjected to torsional loadings induced by facade elements and thus require special consideration. In addition, design of these

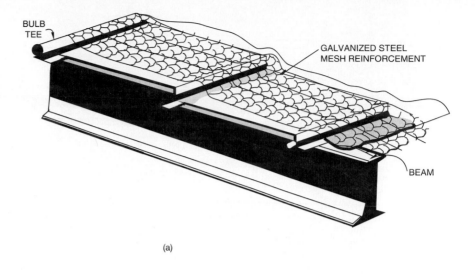

(a)

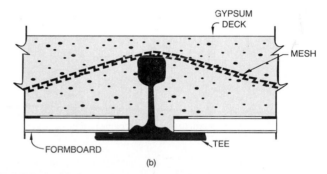

(b)

FIGURE 8.14   (*a*) Gypsum-concrete roof deck. (*b*) Cast on formboard, the deck is supported by a steel bulb tee.

members is frequently governed by deflection criteria established to avoid damage to, or to permit proper functioning of, the facade construction.

8.8   ROLLED SHAPES

Hot-rolled, wide-flange steel shapes are the most commonly used members for multistory steel-framed construction. These shapes, which are relatively simple to fabricate, are economical for beams and girders with short to moderate spans. In general, wide-flange shapes are readily available in several grades of steel, including ASTM A36 and the higher-strength ASTM A572.

Interfacing with mechanical ductwork is usually accomplished in one of two ways. First, the steel framing can be designed to incorporate the shallowest members that provide the required strength and stiffness, and the mechanical ductwork can be routed beneath the floor framing. As an alternative, deeper beams and girders than would otherwise be necessary can be used, and these members can be fab-

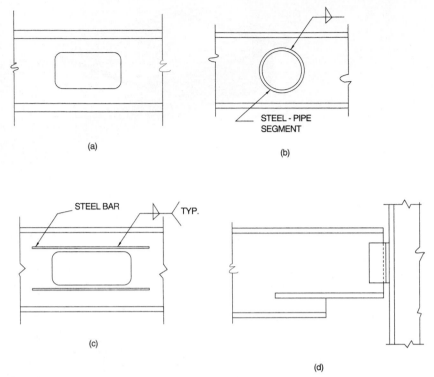

FIGURE 8.15 Penetrations for ducts and pipes in beam or girder webs. (*a*) Rectangular opening, unreinforced. (*b*) Circular opening reinforced with a steel-pipe segment. (*c*) Rectangular penetration reinforced with steel bars welded to the web. (*d*) Reinforced penetration at a column.

ricated with penetrations, or openings, that allow passage of ductwork and pipes. Openings can be either unreinforced, when located in zones subjected to low stress levels, or reinforced with localized steel plates, pipes, or angles (Fig. 8.15).

("Steel and Composite Beams with Openings," Steel Design Guide Series, no. 2, American Institute of Steel Construction.)

Composite versus Noncomposite Construction. Wide-flange beams and girders are frequently designed to act compositely with the floor deck. This enables the use of lighter or shallower members. Composite action is readily achieved through the use of shear connectors welded to the top flange of the beam or girder (Fig. 8.16). When the floor deck is composed of concrete fill on metal deck, the shear connectors are field-welded through the metal deck and onto the top flange of the beam or girder, prior to concrete placement.

In cases where increased future loadings are likely, such as file storage loading in office areas, additional shear connectors can be provided in the original design at minimal additional cost. When the increased loadings must be accommodated, reinforcement plates need only be welded to the easily accessible bottom flange of the beams and girders, since the added shear connectors have already been installed.

Noncomposite design is generally found to be more economical for relatively short spans, inasmuch as the added cost of shear connectors tends not to justify the savings in steel framing.

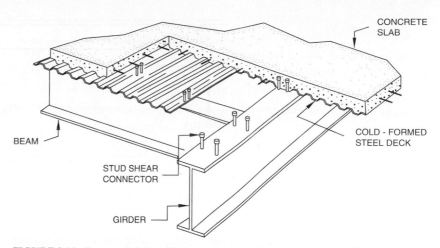

FIGURE 8.16 Beam and girder with shear connectors for composite action with concrete slab.

Shored versus Unshored Construction.   Composite floor framing can be designed as being either shored or unshored during construction. In most cases, unshored construction is used. This allows dead-load deflections to occur during the concrete placement, and the floors to be finished with a level surface. In such cases, the additional concrete dead load must be taken into account when designing the beams and girders.

When unshored construction is used for moderate spans with relatively large dead-load deflections, the beams and girders can be cambered for the dead-load deflection, thereby resulting in a level floor surface after placement of the concrete. When camber is specified, however, careful consideration should be given to the end restraint of the beam, for example, whether the beam frames into girders or into columns, even if simple connections are used throughout. End restraint reduces deflections, and camber that exceeds the actual dead-load deflection can sometimes be troublesome, since it may affect the fire rating (because of insufficient concrete-fill thickness over metal deck) the elevation of preset inserts in an electrified floor system, or installation of interior finishes.

Shored construction will result in lighter or shallower beams and girders than unshored construction, since the flexural members will act compositely with the floor deck in resisting the weight of the concrete when the shores are removed. However, consideration must be given to the deflections that will occur after shore removal, and whether the resulting floor levelness will be acceptable.

8.9 OPEN-WEB JOISTS

Although more frequently used for moderate- to long-span roof framing, open-web steel joists (Fig. 8.17) are sometimes used for floor framing in multistory buildings. Joists as floor members subjected to gravity loadings represent an efficient use of material, particularly since net uplift loadings that are sometimes applicable for roof joist design are not applicable for floor joist design. Also, the open webs of joists provide an effective means of routing mechanical ductwork throughout the floor.

Joists can be designed to act compositely with the floor deck by adding shear connectors to the top chord. In cases where increased future loadings are likely,

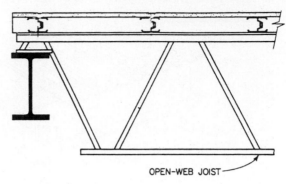

FIGURE 8.17 Open-web steel joist supports gypsum deck.

such as file storage loading in office areas, the web members can be oversized and additional shear connectors can be provided in the original design at minimal additional cost. At the time when the increased loadings must be accommodated, reinforcement plates need only be welded to the easily accessible bottom chord of the joists, since the added shear connectors and increased web sizes have already been provided.

8.10 LIGHTWEIGHT STEEL FRAMING

Cold-formed steel structural members can provide an extremely lightweight floor framing system. These members, usually C or Z shapes, are normally spaced 24 in center to center (c to c) and can span up to about 30 ft between supports. Because of their light weight, these members can be handled and installed easily and quickly. Connections of cold-formed members are usually accomplished by welding or by the use of self-drilling screws.

This type of floor-framing system is frequently used in conjunction with cold-formed steel load-bearing wall studs for low-rise construction. Spans are preferred short to keep depth of floor system small. This depth has a direct bearing on the overall height of structure, to which costs of several building components are proportional.

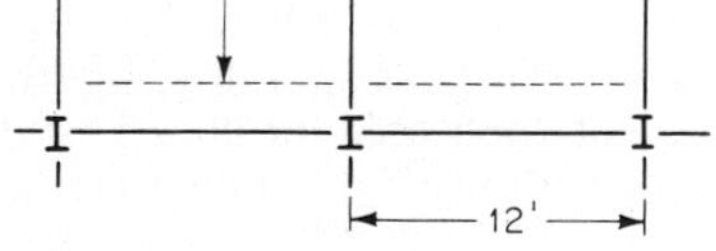

Space in apartment buildings often is so arranged that beams and columns can be confined, hidden from view, within walls and partitions. Since parallel walls or partitions usually are spaced about 12 ft apart, joists that span between beams located in those dividers can be short-span.

In Fig. 8.18, the joists span in the short direction of the panel, to obtain the least floor depth. They are supported on beams of greater depth hidden from view in the walls. With mo-

FIGURE 8.18 Typical short-span floor framing for a high-rise apartment building.

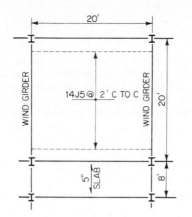

FIGURE 8.19   For economical framing, joists are supported on wind girders.

ment connections to the columns, these beams are designed to resist lateral forces on the building as well as vertical loading. (Depth of the beams may be dictated by lateral-force design criteria.) As part of moment-resisting frames, the beams usually are oriented to span parallel to the narrow dimension of the structure. In that case, the joists are set parallel to the long axis of the building. When beam and joist spans are nearly equal, framing costs generally will be lower if the joists are oriented to span between wind girders, regardless of their orientation (Fig. 8.19). This arrangement takes advantage of the substantial members required for lateral-force resistance without appreciably increasing their sizes to carry the joists.

The service core of a high-rise residential building, containing stairs, elevators, and shafts for ducts and pipes, usually is framed with lightweight, shallow beams. These are placed around openings to provide substantial support for point loading.

Because of lighter dead and live loads, columns in apartment buildings are much smaller than columns in office buildings and usually are less visible. Orientation of columns usually is determined by wind criteria and often is oriented as indicated in Fig. 8.20.

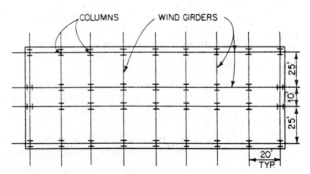

FIGURE 8.20   Typical framing plan for narrow, high-rise building orients columns for strong-axis resistance to lateral forces in the narrow direction.

8.11   TRUSSES

When relatively long spans are involved, trusses are frequently selected for the floor-framing system. As for open-web joists, mechanical ductwork can be easily routed through the web openings. Shear connectors can be added for composite action with the floor deck. Increased future loadings can be accommodated at a minimal cost premium by oversizing the web members and providing additional shear connectors in the original design.

8.12 STUB GIRDERS

The primary advantage of the stub-girder system is that it provides ample space for routing mechanical ductwork throughout a floor while achieving a reduced floor construction depth as compared to conventional steel framing.

This system utilizes floorbeams that are supported on top of, rather than framed into, stub girders. Thus, the floorbeams are designed as continuous members, which results in steel savings and reduced deflections. A stub girder consists of a shallow wide-flange member directly beneath the floorbeams, and intermittent wide-flange stubs, having the same depth as the floorbeams. The stubs are placed perpendicular to and between the floorbeams, leaving space for the passage of mechanical duct-work (Fig. 8.21). The stubs are welded to the top of the stub girder and connect to the floor deck, which is typically concrete fill on metal deck, thereby enabling the stub girder to act compositely with the floor deck.

8.13 STAGGERED TRUSSES

In an effort to provide a structural-steel framing system with a minimum floor-to-floor height for multistory residential construction, the staggered truss system was developed. This system consists of story-high trusses spanning the full width of a building. They are placed at alternate column lines in alternate stories, thus resulting in a staggered arrangement of trusses (Fig. 8.22). The trusses span about 60 ft between exterior columns, resulting in a column-free interior space. In addition to the simple checkerboard pattern, alternative stacking patterns are possible in order to accommodate varied interior layouts (Fig. 8.23).

At a typical floor, the deck spans between the top chord of one truss and the bottom chord of the adjacent truss. Since the staggered trusses are typically spaced 20 to 30 ft on centers, a long-span floor deck system is required. Precast-concrete plank with topping is frequently used, since, in addition to accommodating the span, the plank underside can be finished to provide an acceptable ceiling. An alternative system consists of long-span composite metal deck, having a depth of up to 7½ in, with concrete fill. The top and bottom chords of the trusses are usually wide-flange shapes to efficiently resist the bending stresses induced by the floor loadings.

Diagonal web members of the trusses are deleted at corridor openings. This results in bending stresses in the truss chords due to Vierendeel action. Consequently, corridors are typically located near the building centerline, that is, near midspan of the trusses, at points of minimum truss shear, thereby minimizing the chord bending stresses.

Lateral loads in the transverse direction are transferred to the truss top chords via diaphragm action of the floor deck. These loads are transmitted through the depth of the trusses to the bottom chords and are then transferred through the floor deck at that level to the adjacent-truss top chords. The overturning couple produced by the transfer of lateral load from the top chord to the bottom chord is resisted by a vertical couple at the ends of the truss. Only axial forces are induced in the exterior columns. Therefore, transverse lateral loads are transmitted down through the structure without creating bending stresses in the trusses or columns, except at truss openings.

In the longitudinal direction, lateral loads are transferred via floor diaphragm action to the exterior columns. These resist the loads by conventional means, such as rigid frames or braced bents. To provide added strength and stiffness, the exterior

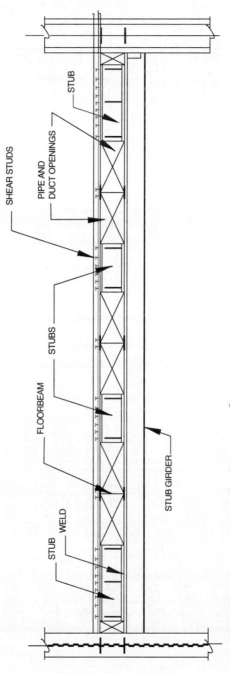

FIGURE 8.21 Stub girder supports floorbeams on top flange.

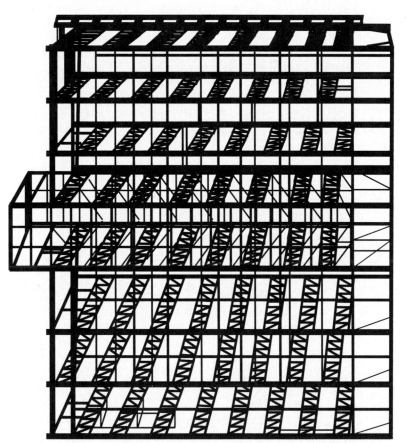

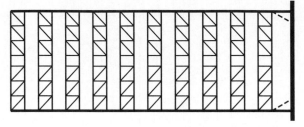

(b)

(a)

FIGURE 8.22 Staggered-truss system. (*a*) Story-high trusses are erected in alternate stories along alternate column lines; (*b*) typical vertical section through building.

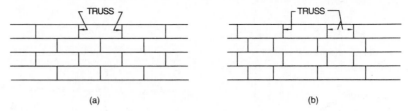

FIGURE 8.23 Stacking of trusses in staggered-truss systems. (*a*) Checkerboard pattern; (*b*) an alternative arrangement.

columns are usually oriented so that the strong axis assists in resisting lateral loads in the longitudinal direction.

To achieve the necessary structural interaction between the trusses and the floor deck and to provide the necessary continuity of the floor diaphragm, adequate connection by such means as weld plates or shear connectors must be provided between the various structural elements. Floor decks with large openings or other shear discontinuities may require additional reinforcement.

Although the staggered-truss system resists gravity and lateral loads primarily by axial stresses, consideration must be given to the bending stresses in the exterior columns that result from the truss deformations under gravity loads (Fig. 8.24). These bending stresses can be significantly reduced by cambering the trusses, thereby preloading the columns. An alternative is to provide slotted bottom-chord connections that are torqued or welded after dead load is applied.

8.14 CASTELLATED BEAMS

A special fabrication technique is applied to wide-flange shapes to produce castellated beams. This technique consists of cutting the web of a wide-flange shape along a corrugated pattern, separating and shifting the upper and lower pieces, and rewelding the two pieces along the middepth of the newly created beam (Fig. 8.25). The result is a beam with depth, strength, and stiffness greater than the original wide-flange shape, but that maintains the same weight per foot as the original wide-flange shape. In addition, the numerous hexagonal openings, or castellations, that are formed in the beam web can accommodate mechanical ductwork, thereby reducing the overall floor depth.

Castellated beams can be designed to act compositely with the floor deck. Economical spans range up to about 70 ft. For composite design, it is structurally more efficient to fabricate the beam from a heavier wide-flange shape for the lower portion than for the upper portion. As a rule of thumb, the deflection of a castellated beam is about 25% greater than the deflection of an equivalent beam with the same depth but without web openings.

The load capacity of a castellated beam is frequently dictated by the local strength of the web posts and the tee portions above and below the openings. Therefore, these beams are more efficient for supporting uniform loadings than for concentrated loadings. The latter produce web-shear distributions that tend to be less favorable because the perforated web has less capacity than the solid web.

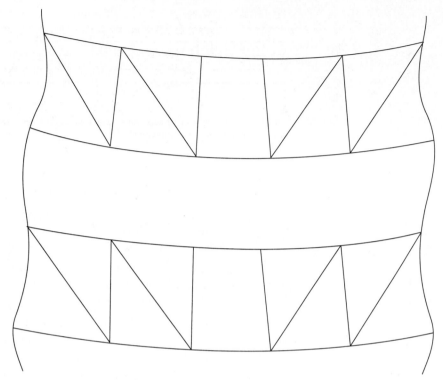

FIGURE 8.24   Deformations of staggered trusses induce bending in exterior columns.

8.15   ASD VERSUS LRFD

If member sizes computed in ASD and LRFD (Art. 6.12) are compared, the latter usually results in lighter or shallower members. For example, LRFD will typically result in material savings in the range of 10 to 15% when used for strength design of composite beams. In some cases, the material savings for certain components can be more than 30%. However, when the governing criterion is serviceability, such as deflection or vibration, ASD and LRFD will typically produce the same member sizes.

A comparison of composite beam and girder sizes obtained from ASD and LRFD for a typical 30-ft by 30-ft interior bay of an office building is shown in Fig. 8.26. A similar comparison for a 30-ft by 45-ft bay is shown in Fig. 8.27. These comparisons were based on the following assumptions:

- Beams and girders are ASTM A572, Grade 50, steel.
- Floor is 3-in, 20-ga composite metal deck with lightweight concrete fill with a total weight of 47 lb/ft^2.
- Total dead load (floor slab, partitions, ceiling, and mechanical) is 77 lb/ft^2.
- Live load is 80 lb/ft^2, with live-load reductions in accordance with size of loaded areas supported (Art. 6.4.3).

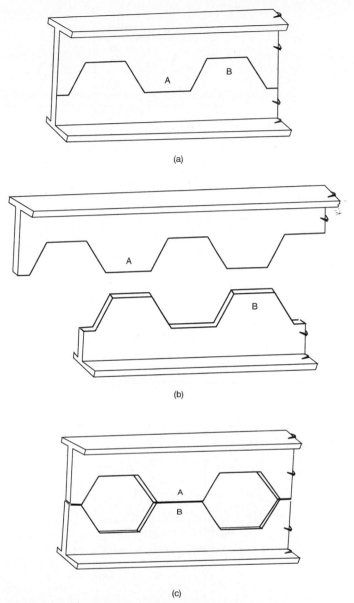

FIGURE 8.25 Steps in formation of a castellated beam. (*a*) Corrugated cut is made longitudinally in a wide-flange beam. (*b*) Half of the beam is moved longitudinally with respect to the other half and (*c*) welded to it.

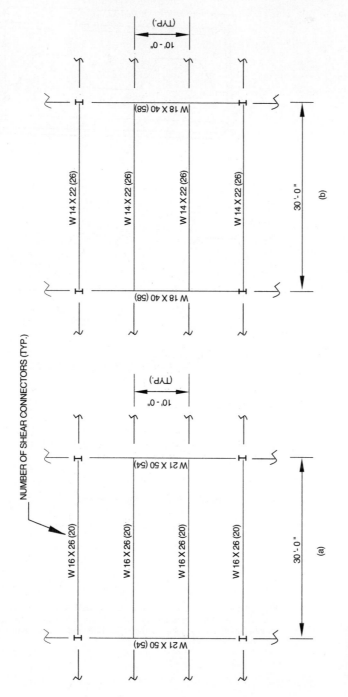

FIGURE 8.26 Sizes computed for beams and girders for a 30 × 30-ft interior bay of an office building (*a*) for ASD and (*b*) for LRFD.

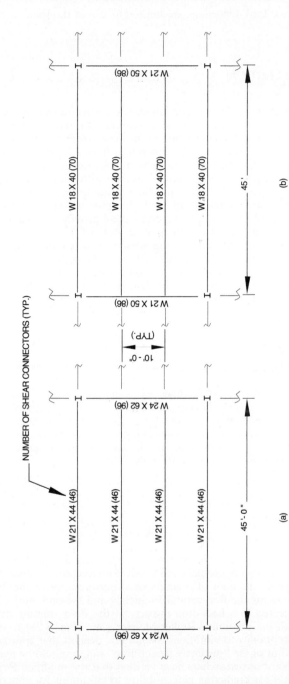

FIGURE 8.27 Sizes computed for floor framing for a 30 × 45-ft interior bay of an office building (*a*) for ASD and (*b*) for LRFD.

8.27

- Dead-load deflections are minimized by temporary shoring or cambering.
- Live-load deflections are limited to $\frac{1}{360}$ of the span.

8.16   DEAD-LOAD DEFLECTION

Although, in general, building codes restrict the magnitude of live-load deflections, they do not contain criteria or limitations relating to dead-load deflections.

The dead-load deflection of the floor framing system will not affect the levelness of the floor surface if the concrete is finished level despite the deflection or if the floor framing members are cambered for deflection due to the concrete dead load. In cases where the concrete is finished with a level surface, the slab will be thicker at midspan and, hence, the floor framing members should be designed for the additional concrete dead load. In cases where the floor framing members are cambered, care must be taken to avoid providing too much camber. (See "Shored versus Unshored Construction" in Art. 8.8.)

When shored construction is used, or when the concrete floor thickness is kept constant, that is, the top surface follows the deflected shape of the framing members to avoid the placement of additional concrete, the dead-load deflection of the floor-framing system should be evaluated to determine whether the resulting floor levelness will be acceptable.

8.17   FIRE PROTECTION

There are several methods by which fire ratings can be readily achieved for structural-steel floor framing systems. These methods include application of spray-on fire protection, encasement of the framing members in a fire-rated assembly, or installation of a fire-rated ceiling system below the framing. For open-web joists and lightweight steel framing, the last two options are usually more practical than spray-on fire protection. (See also Art. 6.32.)

8.18   VIBRATIONS

Although a floor system may be adequately designed from a strength standpoint, a serviceability problem will result if unacceptable vibrations occur during normal usage of the floor. The anticipated performance of the floor can be analyzed by computing the first natural frequency and the amplitude, that is, deflection when subjected to a heel-drop impact, of the floor framing member and plotting the result on a modified Reiher-Meister scale (Fig. 8.28) to determine the degree of perceptibility to vibrations. Generally, designs that approach or exceed the upper portion of the "distinctly perceptible" range should be avoided.

Various researchers have verified that the modified Reiher-Meister scale is accurate for predicting perceptibility to vibrations for concrete slab (including concrete fill on metal deck) floor systems framed with steel joists or steel beams.

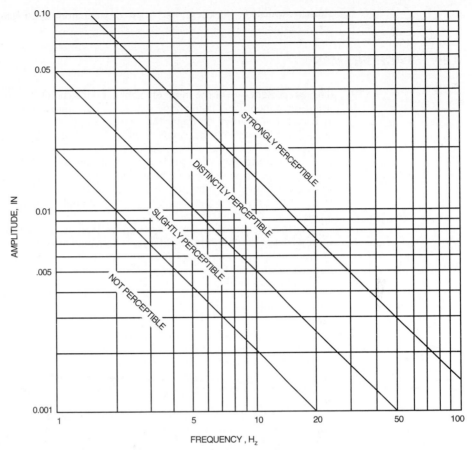

FIGURE 8.28 Modified Reiher-Meister scale relates perception of vibrations to amplitude and frequency.

(T. M. Murray, "Acceptability Criterion for Occupant-Induced Floor Vibrations," *AISC Engineering Journal*, vol. 18, no. 2.)

ROOF FRAMING

The systems used for floor framing (Arts. 8.8 to 8.14) can also be used for roof framing. Other roof framing systems are described below.

8.19 PLATE GIRDERS

For long spans or heavy loadings that exceed the capacity of standard rolled shapes, plate girders can be used. Plate girders are composed of individual steel plates that can vary in width, thickness, and grade of steel along their length to optimize the cross section.

8.20 SPACE FRAMES

These represent one of the more efficient uses of structural materials. Space frames are three-dimensional lattice-type structures that span in more than one direction. It is common practice to apply the "space frame" designation to structures that would more accurately be categorized as "space trusses," that is, assemblies of members pin-connected at the joints, or nodes.

In addition to providing great rigidity and inherent redundancy, space frames can span large areas economically, providing exceptional flexibility of usage within the structure by eliminating interior columns. Space frames possess a versatility of shape and form. They can utilize a standard module to generate flat grids, barrel vaults, domes, and free-form shapes.

The most common example of a space truss is the double-layer grid, which consists of top- and bottom-chord layers connected by web members. Various types of grid orientations can be utilized. Top- and bottom-chord members can be either parallel or skewed to the edges of the structure, and can be either parallel or skewed to one another (see Fig. 8.29). One of the advantages of having top and bottom chords skewed relative to one another is that the top-chord members have shorter lengths, thereby resulting in a more economical design for compressive forces.

Space frames spanning over large column-free areas are generally supported along the perimeter or at the corners. Overhangs are employed where possible to provide some amount of stress counteraction to relieve the interior chord forces and to provide a greater number of "active" diagonal web members to distribute the reactions at supports into the space frame. In cases where the reactions are very large, space-frame members near the supports are sometimes extended beneath the bottom chord, in the form of inverted pyramids, to the top of the columns. This effectively produces a column capital, which facilitates distribution of forces into the space frame.

The depth of a space frame is generally 4 to 8% of its span. To effectively utilize the two-way spanning capability of a space frame, the aspect (length-to-width) ratio

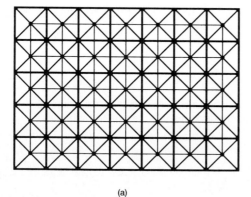

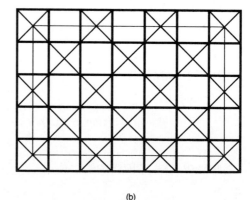

(a)   (b)

FIGURE 8.29 Types of space-frame grids. (*a*) Top and bottom chords parallel to edges of the structure. (*b*) Top and bottom chords skewed to each other.

should generally not exceed 1.5:1.0. For a 1.5:1.0 ratio, about 70% of the gravity loads are carried by the short span.

Types of members used for space frames may be structural steel hot-rolled shapes, or round or rectangular tubes, or cold-formed steel sections. Many space frames are capable of utilizing two or more different member types.

For some space-frame roof structures, the top chords also act as purlins to directly support the roofing system. In these cases, the top chords must be designed for a combination of axial and bending stresses. For other roof structures, a separate subframing system is utilized for the roofing system, and an interface connection to the space frame is provided at the top chord nodes. In these cases, the roofing system does not transmit bending stresses to the top chord members.

Regardless of the type of space frame, the essence of any such system is its node. Most space frame systems have concentric nodes; that is, the centroidal axes of all members framing into a node project to a common working point at the center of the node. Some systems, however, have eccentric joints. For these, local bending of the members must be considered in addition to the basic joint and member stresses.

Most space frames are assembled either in-place on a piece-by-piece basis, or in portions on the ground and then lifted into place. In some cases, where construction sequencing permits, the entire space frame can be preassembled on the ground and then lifted into place.

8.21 ARCHED ROOFS

These are advantageous for long bays, especially if large clearances are desirable along the center. Such braced barrel vaults have been used for hangars, gymnasiums, and churches. While these roofs can be supported on columns, they also can be extended to the ground, thus eliminating the need for walls (Fig. 8.30). The roofs usually are relatively lightweight, though spans are large, because they can be shaped so that load is transmitted to the foundations almost entirely by axial compressive stresses.

Designers have a choice of a wide variety of structural systems for cylindrical arches. Basically, they may be formed with structural framing of various types and a roof deck, or they may be stressed-skin construction.

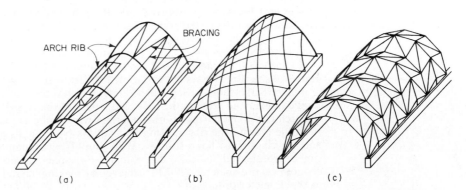

FIGURE 8.30 Cylindrical arches. (*a*) Ribbed; (*b*) diagonal grid (lamella); (*c*) pleated barrel.

Framing may consist of braced arch ribs (Fig. 8.30*a*), curved grids, or space frames. Depending on foundation and other conditions, arch ribs may be fixed-end, single-hinged, double-hinged (pinned), or triple-hinged (statically determinate). Much lighter members can be employed for a diagonal grid, or lamella, system (Fig. 8.30*b*), but many more members must be handled.

With stressed-skin construction, the roof deck acts integrally with the framing in carrying the load.

As in folded-plate construction, the stiffness can be increased by pleating or undulating the surface (Fig. 8.30*c*).

Regardless of the type of structural system selected, provision must be made for resisting the arch thrust. If ground conditions permit, the thrust may be resisted entirely by the foundations. Otherwise, ties must be used. Arches supported above grade may be buttressed or tied.

8.22   DOME ROOFS

These are preferable to arches where the large column-free area to be covered is circular, elliptical, or approximately an equal-sided polygon. They often have been used to roof exhibition buildings, arenas, planetariums, water reservoirs, and gas tanks. Also, the feasibility of covering large stadiums with domes has been demonstrated. Domes are relatively lightweight, despite long spans, because they can be shaped so that loads induce mainly axial stresses.

Domes may be readily supported on columns, without ties or buttresses, because they can be shaped to produce little or no thrust. For a shallow dome, a tension ring usually is provided around the base to resist thrusts. If desired, however, domes may be extended to grade, thus eliminating the need for walls (Fig. 8.31). If an opening is left at the crown, for example, for a lantern (Fig. 8.31*b*), a compression ring is installed around the opening to resist the thrusts. Also, if desired, portions of a dome may be made movable, to expose the building interior.

Designers also have a choice of a wide variety of structural systems for domes. In general, dome construction may be categorized as single-layer framing; double-layer (truss) framing, or space frame, for greater resistance to buckling; and stressed skin, with the roof deck acting integrally with structural framing. Greater stiffness can be obtained by dimpling, pleating (Fig. 8.31*c*), or undulating the surface.

In addition, each category offers a wide variety of designs. For example, more than one dozen types of braced, single-layer systems are in use. Two of these, which are frequently used, are illustrated in Fig. 8.31.

Figure 8.31*a* shows a ribbed dome. Its principal components are half arches. They are shown connected at the crown, but usually, to avoid a cramped joint with numerous members converging there, the ribs are terminated at a small-diameter compression ring circumscribing the crown. The opening may be used for light and ventilation. If the connections at top and bottom of the ribs permit rotation in the planes of the ribs, the system is statically determinate for all loads.

Figure 8.31*b* shows a Schwedler dome, which offers more even distribution of the dead load and reduces the unbraced length of the ribs. Principal members are the arch ribs and a series of horizontal rings with diameter increasing with distance from the crown. The ribs transmit loads to the base mainly by axial compression, and the rings resist hoop stresses. With simplifying assumptions, this system too can be considered statically determinate. For spherical domes of this type, an economical rise-span ratio is 0.13, achieved by making the radius of the dome equal to the diameter of its base.

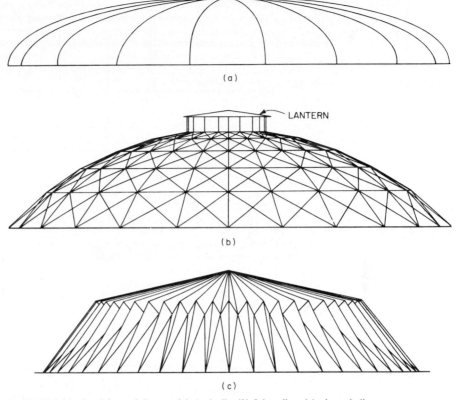

FIGURE 8.31   Steel-framed domes. (*a*) Arch rib; (*b*) Schwedler; (*c*) pleated rib.

8.23 CABLE STRUCTURES

High-strength steel cables are very efficient for long-span roof construction. They resist loads solely by axial tension. While the cables are relatively low cost for the load-carrying capacity provided, other necessary components of the system must be considered in making cost comparisons. Costs of these components increase slowly with increasing span. Consequently, the larger the column-free area required, the greater the likelihood that a cable roof will be the lowest-cost system for spanning the area.

Components other than cables that are needed are vertical supports and anchorages. Vertical supports are needed to provide required vertical clearances within the structure, because cables sag below their supports. Usually, the cables are supported on posts, or towers, or on walls.

Anchorages are required to resist the tension in the cables. Means employed for the purpose include heavy foundations, pile foundations, part of the building (Fig. 8.32*a*), perimeter compression rings and interior tension rings (Fig. 8.32*b*). For attachment to the anchorages, each cable usually comes equipped with end fittings, often threaded to permit a jack to grip and tension the cable and to allow

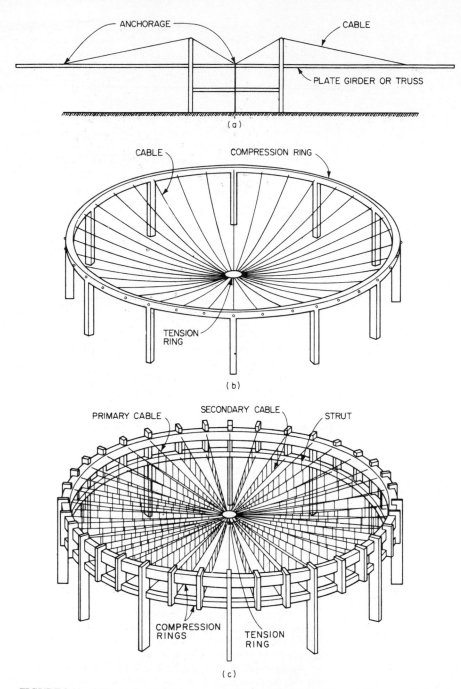

FIGURE 8.32   Cable roofs. (*a*) Cable-stayed cantilever roof; (*b*) single-layer cable-suspended roof; (*c*) double-layer cable-suspended roof.

use of a nut for holding the tensioned cable in place. In addition, bearing plates generally are needed for distributing the cable reaction.

Cable roofs may be classified as cable-stayed or cable-suspended. In a cable-stayed roof, the deck is carried by girders or trusses, which, in turn, are supported at one or more points by cables. This type of construction is advantageous where long-span cantilevers are needed, for example, for hangars (Fig. 8.32a). In a cable-suspended roof, the roof deck and other loads are carried directly by the cables (Fig. 8.32b).

The single-layer cable roof structure in Fig. 8.32b is composed of radial cables, a central tension ring, and a perimeter compression ring. Since this system is extremely lightweight, it is susceptible to wind uplift and wind-induced oscillations unless a heavy roof deck, such as precast-concrete panels, is utilized. Uplift and oscillation can be eliminated with the use of a double-layer cable roof (Fig. 8.32c) in which the primary and secondary cables are pretensioned during erection.

For a double-layer system with diagonal struts between the primary and secondary cables, truss action can be developed. If pretension is sufficiently high in the compression chord, compression induced by increasing load only decreases the tension in that chord but cannot cause stress reversal.

For both single- and double-layer systems, circular or elliptical layouts minimize bending in the perimeter compression ring and are thus more efficient than square or rectangular layouts.

Since the number of anchorages and connections does not increase linearly with increasing span, cable structures with longer spans can cost less per square foot of enclosed area than those with shorter spans. This is contrary to the economics of most other structural systems, which increase in cost per square foot of enclosed area as the span increases.

Another type of cable structure is the cable-truss dome, or "tensegrity" dome. It consists of a series of radial cable trusses, concentric cable hoops, a central tension ring, and a perimeter compression ring. The dome is prestressed during erection and is typically covered with fabric roofing.

Cable spacing depends on type of roof deck. Close spacing generally is economical, a maximum of 10 ft.

For watertightness and to avoid potential trouble due to roof movements at points where cables penetrate a roof, it is desirable to place cables either completely below or completely above the roof surface. If cables must penetrate a roof, the joints should be calked and sealed with a metal-protected, rubber-like collar.

In design of cable roofs, special consideration should be given to roof movements, especially if the roof deck does not offer a significant contribution to rigidity. Care should be taken that joints in a flexible roof do not open or that a concrete deck does not develop serious cracks, destroying the watertightness of the roof. Insulation may be necessary to prevent large thermal movements. Consideration should be given also to fire resistance. Sprinklers may be required or desirable. If the cables are galvanized, corrosion usually is unlikely, but the possibility should be investigated especially for chemically polluted atmospheres.

SECTION 9
LATERAL-FORCE DESIGN

Charles W. Roeder
*Professor of Civil Engineering, University of Washington,
Seattle, Washington*

Design of buildings for lateral forces requires a greater understanding of the load mechanism than many other aspects of structural design. To fulfill this need, this section provides a basic overview of current practice in seismic and wind design. It also discusses recent changes in design provisions and possible trends for the future.

There are fundamental differences between design methods for wind and earthquake loading. Wind-loading design is concerned with safety, but occupant comfort and serviceability is a dominant concern. Wind loading does not require any greater understanding of structural behavior beyond that required for gravity and other loading. As a result, the primary emphasis of the treatment of wind loading in this section is on the loading and the distribution of loading. Design for seismic loading is primarily concerned with structural safety during major earthquakes, but serviceability and the potential for economic loss are also of concern. Earthquake loading requires an understanding of the behavior of structural systems under large, inelastic, cyclic deformations. Much more detailed analysis of structural behavior is needed for application of earthquake design provisions, because structural behavior is fundamentally different for seismic loading, and there are a number of detailed requirements and provisions needed to assure acceptable seismic performance. Because of these different concerns, the two types of loading are discussed separately in the following.

9.1 DESCRIPTION OF WIND FORCES

The magnitude and distribution of wind velocity are the key elements in determining wind design forces. Mountainous or highly developed urban areas provide a rough surface, which slows wind velocity near the surface of the earth and causes wind velocity to increase rapidly with height above the earth's surface. Large, level open areas and bodies of water provide little resistance to the surface wind speed, and wind velocity increases more slowly with height. Wind velocity increases with height in all cases but does not increase appreciably above the critical heights of about 950 ft for open terrain to 1500 ft for rough terrain. This variation of wind speed over height has been modeled as a power law:

$$V_z = V \left(\frac{z}{z_g}\right)^n \tag{9.1}$$

where V is the basic wind velocity, or velocity measured at a height z_g above ground and V_z is the velocity at height z above ground. The coefficient n varies with the surface roughness. It generally ranges from 0.33 for open terrain to 0.14 for rough terrain. The wind speeds V_z and V are the fastest-mile wind speeds, which are approximately the fastest average wind speeds maintained over a distance of 1 mile. Basic wind speeds are measured at an elevation z_g above the surface of the earth at an open site. Design wind loads are based on a statistical analysis of the maximum fastest-mile wind speed expected within a given recurrence interval, such as 50 years. Statistical maps of wind speeds have been developed and are the basis of present design methods. However, the maps consider only regional variations in wind speed and do not consider tornadoes, tropical storms, or local wind currents. The wind speed data are maintained for open sites and must be corrected for other site conditions. (Wind speeds for elevations higher than the critical elevations mentioned previously are not affected by surface conditions.)

Wind speeds V_w are translated into pressure q by the equation

$$q = C_D \frac{\rho}{2} V_w^2 \tag{9.2}$$

where C_D is a drag coefficient and ρ is the density of air at standard atmospheric pressure. The drag coefficient C_D depends on the shape of the body or structure and is less than 1 if the wind flows around the body. The pressure q is the stagnation pressure q_s if $C_D = 1.0$, since the structure effectively stops the forward movement of the wind. Thus, on substitution in Eq. (9.2) of $C_D = 1.0$ and air density at standard atmospheric pressure,

$$q_s = 0.00256 V_w^2 \tag{9.3}$$

where the wind speed is in miles per hour and pressure, in psf.

The shape and geometry of the building have other effects on the wind pressure and pressure distribution. Large inward pressures develop on the windward walls of enclosed buildings and outward pressures develop on leeward walls, as illustrated in Fig. 9.1a. Buildings with openings on the windward side will allow air to flow into the building, and internal pressures may develop as depicted in Fig. 9.1b. These internal pressures cause loads on the over-all structure and structural frame. More important, these pressures place great demands on the attachment of roofing and external cladding. Openings in a side wall or leeward wall may cause an internal pressure in the building as illustrated in Fig. 9.1c and d. This buildup of internal pressure depends on the size of the openings for all walls and the geometry of the structure. Slopes of roofs may affect the pressure distribution, as illustrated in Fig. 9.1e. Projections and overhangs (Fig. 9.2) may also restrict the airflow and accumulate pressure. These effects must be considered in design.

The velocity used in the pressure calculation is the velocity of the wind relative to the structure. Thus, vibrations or movements of the structure occasionally may affect the magnitude of the relative velocity and the pressure. Some structures are sensitive to aerodynamic effects. They may be susceptible to dynamic instability, such as vortex shedding and flutter. These may occur where local airflow around the structure causes dynamic amplification of the structural response because of the interaction of the structural response with the airflow. These undesirable conditions require special analysis that takes into account the shape of the body, airflow

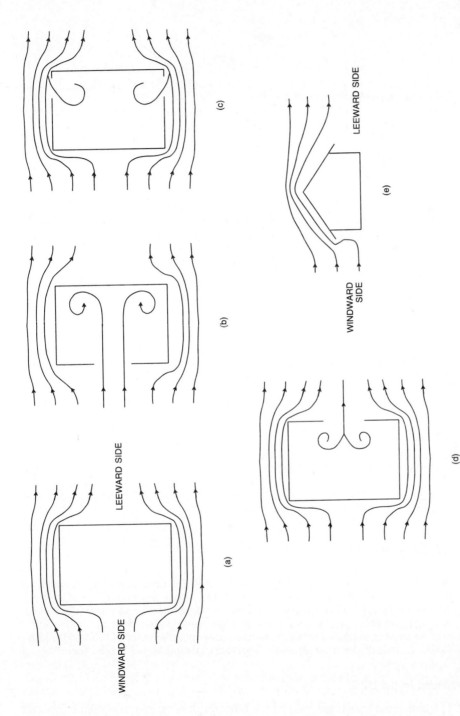

FIGURE 9.1 Plan view of a building indicating the wind loading on it with changes in velocity and direction of wind. (*a*) High pressure on a solid wall on the windward side but outward or reduced inward pressure on the leeward side. (*b*) Wind entering through an opening in the windward wall induces outward pressure on the interior of the walls. (*c*) and (*d*) Wind entering through openings in a side wall or a leeward wall produce internal pressures in the building. (*e*) On a sloping roof, high inward pressure develops on the windward side, outward or reduced inward pressure on the leeward side.

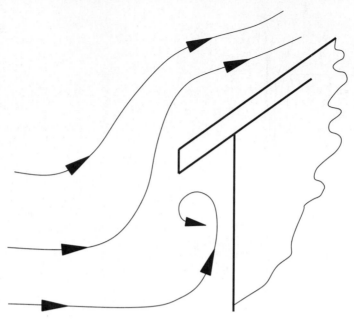

FIGURE 9.2 Roof overhang restricts airflow, creates large local forces on the structure.

around the body, dynamic characteristics of the structure, wind speed, and other related factors.

The fastest-mile wind speed is smaller than the short-duration wind speed due to gusting. Corrections are made in design calculations for the effect of gusting through use of gust factors, which increase design wind pressure to account for short-duration increases in wind speed. The gust factors are largely affected by the roughness of the surface of the earth. They decrease with increasing height, reduced surface roughness, and duration of gusting.

Although gusting provides only a short-duration dynamic loading to the structure, a major concern may be the vibration, rocking, or buffeting caused by the dynamic effect. The pressure distribution caused by these combined effects must be applied to the building as a wind load.

9.2 DETERMINATION OF WIND LOADS

Wind loading as described in Art. 9.1 is the basis for design wind loads specified in "Minimum Design Loads for Buildings and Other Structures," ASCE 7-88, American Society of Civil Engineers. Model building codes specify simplified methods based on these provisions for determining wind loads. These methods can be used for most structures. One such method is incorporated in the "Uniform Building Code" (UBC) of the International Conference of Building Officials, Inc.

9.2.1 Wind-Load Provisions in the UBC

The basic wind speeds specified by the UBC for the continental United States and Alaska are shown in Fig. 9.3. The contours on the map indicate wind speeds that

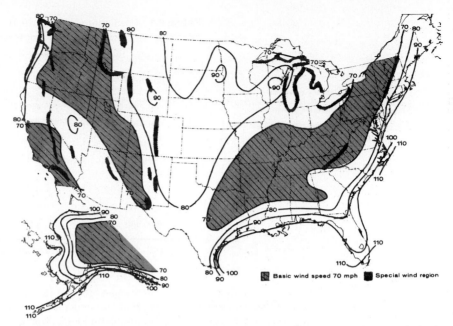

FIGURE 9.3 Contours indicate for regions of the continental United States and Alaska the basic wind speeds, mph, the fastest-mile speeds 10 m above ground in open terrain with a 2% annual probability of occurrence. *(Based on data in "Minimum Design Loads for Buildings and Other Structures," ASCE 7-88, American Society of Civil Engineers, and the "Uniform Building Code," 1991, International Conference of Building Officials.)*

have a 2% probability of being exceeded in a year at a height 10 m above ground on open sites. (These are wind speeds that are expected to occur once in 50 years.) The effects of extreme conditions, such as tornadoes, hurricanes, mountainous regions, or local wind currents are not included, but the possibility of occurrence should be taken into account in design.

The stagnation pressures q_s [Eq. (9.3)] at a height of 10 m above ground are provided in tabular form in the UBC:

Basic wind speeds, mph	70	80	90	100	110
Pressure q_s, psf	13	17	21	26	31

The UBC integrates the combined effects of gusting, changes of wind velocity with height above ground, and the local terrain or surface roughness of the earth in a coefficient, C_e. Values of C_e are given in the UBC for specific exposure conditions as a stepwise function of height (Table 9.1). The UBC defines three exposure conditions, B to D. Exposure C represents open terrain (assumed in Fig. 9.3). Exposure B applies to protected sites. Exposure D is an extreme exposure primarily intended for open shorelines and coastal regions. Coefficient C_e as well as stagnation pressure q_s are factors used in determination of design wind pressures.

The UBC also specifies an importance factor I to be assigned to a building so that more important structures are designed for larger forces to assure their serviceability after an extreme windstorm. For most buildings, $I = 1.0$. For such buildings as hospitals, fire and police stations, and communications centers, and where the primary occupancy is for assembly of 300 or more persons, $I = 1.15$.

TABLE 9.1 Coefficient C_e for Eq. (9.4)

	Exposure	
Height, ft*	C	D
0–20	1.2	0.7
20–40	1.3	0.8
40–60	1.5	1.0
60–100	1.6	1.1
100–150	1.8	1.3
150–200	1.9	1.4
200–300	2.1	1.6
300–400	2.2	1.8

*Height above average level of adjoining ground.

A final factor C_q depends on the geometry of the structure and its appendages and on the component or portion of the structure to be loaded. It is intended to account for the pressure distribution on buildings, which may affect the major load elements.

The design pressure p, psf, is then given by

$$p = C_e C_q q_s I \qquad (9.4)$$

The UBC presents two methods of distributing the pressures to the primary load-resisting system. Method 1 (Fig. 9.4b) is a normal-force method, which distributes pressures normal to the various parts of the building. The pressures act simultaneously in a direction normal to the plane of roofs or walls. In this method, $C_q = 0.8$ inward for all windward walls and 0.5 outward for all leeward walls. For winds parallel to the ridge line of sloped roofs and for flat roofs, $C_q = 0.7$ outward. For winds perpendicular to the ridge line, $C_q = 0.7$ outward on the leeward side. On the windward side:

$C_q = 0.7$ outward with roof slope less than 2:12

    $= 0.9$ outward and 0.3 inward with roof slope between 2:12 and 9:12

    $= 0.4$ inward with roof slope between 9:12 and 12:12

    $= 0.7$ inward with roof slope greater than 12:12.

Method 2 (Fig. 9.4c) uses a projected-area approach with horizontal and vertical pressures applied simultaneously to the vertical and horizontal projections of the building, respectively. For this case, $C_q = 1.4$ on the vertical projected area of any structure over 40 ft tall, 1.3 on the vertical projected area of any shorter structure, and 0.7 upward (uplift) on any horizontal projection.

Individual components and local areas may have local pressure concentrations due to local disturbance of the airflow (Fig. 9.2). These normally do not affect the design of load frames and major load-carrying elements, but they may require increased resistance for architectural elements, local structural members supporting these elements, and attachment details. The UBC also contains values of C_q for

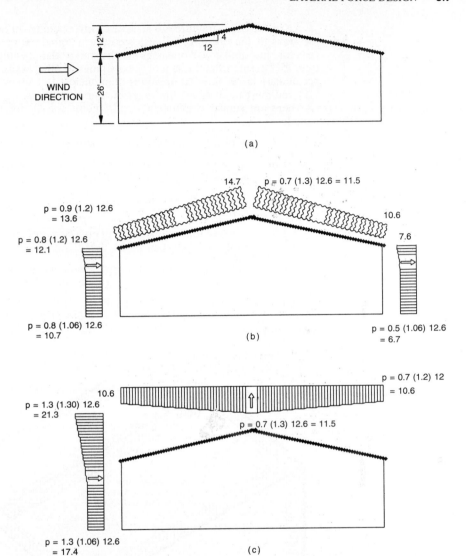

FIGURE 9.4 Distribution of wind pressure on a single-story building with sloping roof. (*a*) Building in open terrain subjected to a 70-mph wind; (*b*) pressures computed by the normal-force method; (*c*) pressures computed by the projected-area method.

these local conditions. Some of these component requirements for C_q for wall elements include:

1.2 inward for all wall elements

1.2 outward for wall elements of enclosed and unenclosed structures

1.6 outward for wall elements of open structures

1.3 inward and outward for all parapet walls

An unenclosed structure is a structure with openings in one or more walls, but the sums of the openings on each side are within 15% of each other. An open structure has similar wall openings but the sum of the openings on one wall is more than 15% greater that the sum of the openings of other walls. Open structures may accumulate larger internal pressures than enclosed or unenclosed structures (Fig. 9.1) and must be designed for larger outward pressures.

There are similar component requirements for C_q for roof elements. These include:

C_q = 1.7   outward for roof elements of open structures with slope less than 2:12

     = 1.6   outward or 0.8 inward for roof elements of open structures with slope greater than 2:12 but less than 7:12

     = 1.7   inward and outward for roof elements of open structures with slope greater than 7:12

     = 1.3   outward for roof elements of enclosed and unenclosed structures with roof slope less than 7:12

     = 1.3   outward or inward for roof elements of enclosed and unenclosed structures with roof slope greater than 7:12

Corners of wall elements must also be subjected to C_q = 1.5 outward or 1.2 inward for the lesser of 10 ft or 10% of the least width of the structure. Roof eaves and other projections are also collectors of concentrated wind pressure (Fig. 9.2).

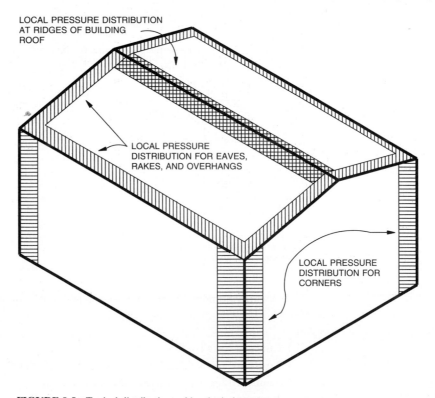

LOCAL PRESSURE DISTRIBUTION AT RIDGES OF BUILDING ROOF

LOCAL PRESSURE DISTRIBUTION FOR EAVES, RAKES, AND OVERHANGS

LOCAL PRESSURE DISTRIBUTION FOR CORNERS

FIGURE 9.5 Typical distributions of local wind pressures.

Building codes require considerations of these local pressure distributions with

C_q = 2.3   upward of roof rakes, ridges, and eaves without overhang and slope less than 2:12

    = 2.6   upward of roof rakes, ridges, and eaves without overhang and slope greater than 2:12 but less than 7:12

    = 1.6   upward of roof rakes, ridges, and eaves without overhang and slope greater than 7:12

    = 0.5   greater coefficient for overhanging elements and canopies.

These factors combine to produce a complex distribution of design pressures. Some of the distributions are illustrated in Fig. 9.5.

These localized distributions affect the strength of local elements and the strength of attachment details of local elements, but they do not affect the global strength requirements of the structure.

9.2.2 Other Provisions for Wind Loads

Alternative methods for determining wind loads, such as that in ASCE Standard 7-88, are available, and give more detailed provisions than those in the UBC (Art. 9.2.1) for defining and distributing wind loads. Tabulated data may be more detailed in these other methods, and more equations may be required. However, the pressure distributions are similar to that provided by the UBC.

These methods provide basic wind loads for buildings, but they do not specify how to estimate or control aerodynamic effects. Aerodynamic effects may result in interaction between the dynamic response of a structure and the wind flow around it. This interaction may amplify the dynamic response and cause considerable occupant discomfort during some windstorms.

Furthermore, local variations in wind velocity can be caused by adjacent buildings. The wind may be funneled onto the structure, or the structure may be protected by surrounding structures. Wind tunnel testing is often required for designing for these effects. Local wind variations are most likely to be significant for tall, slender structures. As a general rule, buildings with unusual geometry or a height more than 5 times the base dimension are logical candidates for a wind tunnel test. Such a test can reveal the predominant wind speeds and directions for the site, local effects such as channeling of the wind by surrounding buildings, effects of the new building on existing surrounding structures, the dynamic response of the building, and the interaction of the response with the wind velocity. The model used for the test can include the stiffness of the building, and wind pressures can be measured at critical locations. Major structures often are based on wind-tunnel-test results, since greater economy and more predictable structural performance are possible.

Special structures, such as antennas, transmission lines, and supports for signs and lighting, may also be susceptible to aerodynamic effects and require special analysis. Aerodynamic effects are beyond the scope of this section, but analytical methods of dealing with these are available. Wind tunnel testing may also be required for these systems.

(E. Simu and R. H. Scanlan, *Wind Effects on Structures*, Wiley-Interscience, New York.)

9.3 SEISMIC LOADS IN MODEL CODES

The "Uniform Building Code" of the International Conference of Building Officials has been the primary source of seismic design provisions for the United States. It adopts recommendations of the Structural Engineers Association of California (SEAOC). The UBC and SEAOC define design forces and establish detailed requirements for seismic design of many structural types. "Minimum Design Loads for Buildings and Other Structures," ASCE 7-88, American Society of Civil Engineers, specifies similar requirements. Another model code is the "National Earthquake Hazard Reduction Program (NEHRP) Recommended Provisions for the Development of Seismic Regulations for New Buildings," of the Building Seismic Safety Council, Federal Emergency Management Agency (FEMA), Washington, D.C. It specifies design forces but provides fewer requirements for design details than the UBC.

The American Institute of Steel Construction (AISC) promulgates "Seismic Design Provisions for Structural Steel Buildings—Load and Resistance Factor Design." This document does not establish design forces, but it provides detailed design requirements for steel structures in a load-and-resistance-factor design (LRFD) format. These provisions are similar to those of the UBC and SEAOC, except that they are based on LRFD rather than on allowable stress design (ASD).

Many of the engineers who participate in writing the UBC also help develop the NEHRP, AISC, ASCE, and other seismic design provisions. As a result, the codes are similar. The UBC provisions are emphasized in the following.

9.4 EQUIVALENT STATIC FORCES FOR SEISMIC DESIGN

The "Uniform Building Code" offers two methods for determining and distributing seismic design loads. One is a dynamic method, which is required to be used for a structure that is irregular or of unusual proportions (Art. 9.5). The other specifies equivalent static forces and is the most widely used, because of its relative simplicity.

The equivalent-static-force method defines the static shear at the base of a building as

$$V = \frac{ZICW}{R_w} \tag{9.5}$$

where W = total dead load, including permanent equipment, plus 10% for partition loads, snow loads exceeding 30 psf, and at least 25% of floor live loads in storage and warehouse occupancies. The base is the level at which seismic motions are imparted to the building or the level at which the structure, acting as a vibrator, is supported.

The coefficient C is determined by

$$C = \frac{1.25S}{T^{2/3}} \le 2.75 \tag{9.6}$$

This formula approximates the effects of a general design response spectrum. It results in larger seismic design forces for short-period structures than for longer-period ones.

S is a coefficient that accounts for the effects of soil conditions at the site. It should be determined from properly substantiated geotechnical data derived from site explorations and analysis. Where soil properties are not known in sufficient detail to determine the soil profile, $S = 1.5$ should be used. $S = 1.0$ may be used for rocklike material (for example, soil with shear-wave velocity exceeding 2500 fps) or where depth of soil is less than 200 ft and it is stiff or dense. For greater depths of such soil, $S = 1.2$. For soil extending to a depth of 40 ft and consisting of more than 20 ft of soft to medium stiff clay but not more than 40 ft of soft clay, $S = 1.5$. For greater depths of soft clay, $S = 2.0$.

T is the fundamental period of the structure. It may be computed by dynamic analysis or by an approximate equation such as

$$T = C_t h_n^{3/4} \tag{9.7}$$

where h_n is the height, ft, from the base of the building to level n, which is the uppermost level in the main portion of the structure. The factor $C_t = 0.035$ for steel moment-resisting frames, which are relatively flexible structures. $C_t = 0.030$ for eccentric-braced frames and reinforced concrete moment-resisting frames. $C_t = 0.02$ for braced frames and other relatively stiff structures.

Equation (9.7) yields periods that are shorter than those computed for some steel structures. Hence, when T is computed from the structural properties and deformational characteristics of the resisting elements, the UBC requires a minimum seismic design shear V that is at least 80% of the force computed with T from Eq. (9.7). The 80% limitation is particularly important for steel moment-resisting frames because it frequently controls their design.

The seismic design shear V depends on regional seismicity (Fig. 9.6), which is quantified by a zone factor Z, which approximates an effective peak ground acceleration (on firm soil for the region). $Z = 0.075$ for zone 1 in Fig. 9.6, 0.15 for zone 2, 0.20 for zone 3, 0.30 for zone 4, and 0.40 for zone 5. Although no damage may have been recorded for zone 0, it is advisable to design for at least a small value of Z, say $Z = 0.05$.

The importance factor I in Eq. (9.5) depends on the importance of the building. $I = 1.25$ for essential or hazardous facilities and 1.0 for standard or special-occupancy structures.

The coefficient R_w in Eq. (9.5) reduces the seismic design forces in recognition of the energy dissipation in ductile structures during a major earthquake. A measure of the ductility and inelastic behavior of a structure, R_w ranges from 4 to 12. The largest values of R_w are used for ductile structural systems. Thus, special moment-resisting frames, which are required to be detailed to be very ductile and are characterized by reliable seismic behavior, are assigned $R_w = 12$. (See also Art. 9.7.1.) Moment-resisting frames are three-dimensional structural systems in which the members and joints are capable of resisting lateral forces on a structure primarily by flexure.

For steel eccentric braced frames (Fig. 9.12), in which at least one end of each diagonal brace intersects a beam at a point away from the column-girder joint and which are not as ductile, $R_w = 10$. In an eccentric braced frame, a link beam must be inserted at least at one end of each brace. The link beam must be designed to yield in shear or bending to prevent buckling of the bracing. See also Art. 9.7.3.

Smaller values of R_w are required for ordinary moment-resisting frames ($R_w = 6$) and concentrically braced frames (Fig. 9.11), in which members are subjected primarily to axial forces, ($R_w = 8$), because they are less ductile. See also Art. 9.7.2.

For dual systems, such as special steel moment-resisting frames, capable of resisting at least $0.25V$, combined with steel eccentric braced frames or concrete shear walls, $R_w = 12$, but for the combination with steel concentrically braced frames, $R_w = 10$.

The UBC employs $3R_w/8$ as the ratio of the predicted maximum deflection expected during a design earthquake to the yield deflection. This suggests that the

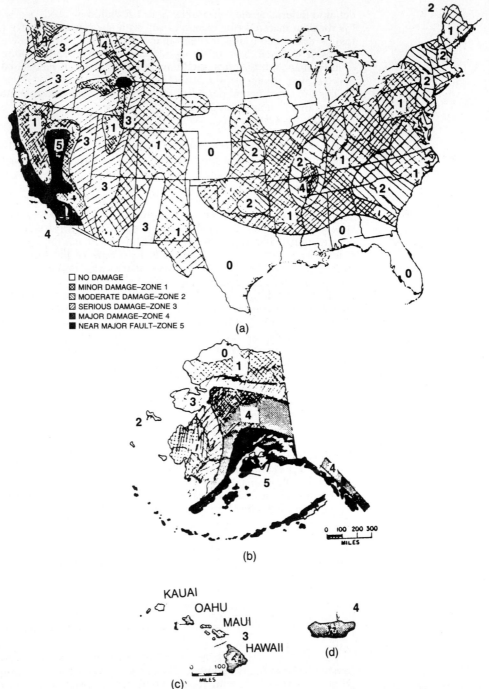

□ NO DAMAGE
⊠ MINOR DAMAGE–ZONE 1
◩ MODERATE DAMAGE–ZONE 2
▨ SERIOUS DAMAGE–ZONE 3
▩ MAJOR DAMAGE–ZONE 4
■ NEAR MAJOR FAULT–ZONE 5

(a)

(b)

(c)

(d)

FIGURE 9.6 Zones of probable seismic intensity in (*a*) mainland United States; (*b*) Alaska; (*c*) Hawaii; (*d*) Puerto Rico. (*Based on data in "Minimum Design Loads for Buildings and Other Structures," ASCE 7-88, American Society of Civil Engineers, and the "Uniform Building Code," 1991, International Conference of Building Officials.*)

expected ductility demand for an R_w of 12 is about $12 \times 3R_w/8 = 4.5$ times the yield deformation during a severe earthquake. There is evidence, however, that this predicted ductility demand is unrealistically small for some conditions and acceleration records.

The UBC also establishes limits on the story drift of the structure to the smaller of $0.03/R_w$ or 0.004 times the story height (may be $0.04/R_w$ or 0.005 for structures with an approximate period less than 0.7 second). These drift limits are particularly important for steel frames and often control the design for taller structures. However, the UBC permits use of forces associated with the true calculated fundamental vibration period rather than the 80% minimum forces associated with the strength-design requirements. This latter possibility often relaxes the drift limits, since steel frames may be quite flexible, with periods well above those predicted by Eq. (9.7).

See also Art. 9.6.

Force Distribution. The seismic base shear V [Eq. (9.5)] is distributed throughout the structure in accordance with its mass and stiffness. A concentrated force F_t, however, should be applied to the top of the structure if the period T is greater than 0.7 second.

$$F_t = 0.07VT \tag{9.8}$$

F_t provides a constant shear over the height of the structure. The static-force method is a single-mode design method, even though long-period structures are influenced by higher-mode response, and F_t simulates the effect of higher-mode response on the earthquake-load distribution.

The remainder of the base shear is distributed over the height of the structure in accordance with a first-mode approximation,

$$F_i = \frac{(V - F_t)\, w_i x_i}{\sum\limits_{j=1}^{n} w_j x_j} \tag{9.9}$$

where F_i and w_i are the seismic force and mass at the ith level and x_i is the height from the base to the ith floor. The force F_i at each floor is distributed horizontally in proportion to the distribution of the mass of the floor, and the forces are distributed to the vertical frames in proportion to their relative stiffness, including the effects of torsion. Floor slabs and the attachments between floor diaphragms and lateral load frames must have adequate strength and stiffness to distribute these inertial forces. The frames must be designed for a minimum torsion, which is produced by a mass eccentricity of 5% of the normal maximum base dimension plus the computed eccentricity between the centers of mass and rigidity.

9.5 DYNAMIC METHOD OF SEISMIC LOAD DISTRIBUTION

The "Uniform Building Code" static-force method (Art. 9.4) is based on a single-mode response with approximate load distributions and corrections for higher-mode response. These simplifications are appropriate for simple, regular structures. However, they do not consider the full range of seismic behavior in complex structures. The dynamic method of seismic analysis is required for many structures with unusual or irregular geometry, since it results in distributions of seismic design forces that are consistent with the distribution of mass and stiffness of the frames, rather than arbitrary and empirical rules. Irregular structures include frames with any of the following characteristics:

The lateral stiffness of any story is less than 70% of that of the story above or less than 80% of the average stiffness for the three stories above

The mass of any story is more than 150% of the effective mass for an adjacent story, except for a light roof above

The horizontal dimension of the lateral-force-resisting system in any story is more than 130% of that of an adjacent story

The story strength is less than 80% of the story above

Frames with a story strength that is less than 80% of that of the story above must be designed with consideration of the P-Δ effects caused by gravity loading combined with the seismic loading.

Frames with horizontal irregularities place great demands on floors acting as diaphragms and the horizontal load-distribution system. Special care is required in their design when any of the following conditions exist:

The maximum story drift due to torsional irregularity is more than 1.2 times the average story drift for the two ends of the structure.

There are reentrant corners in the plan of the structure with projections more than 15% of the plan dimension

The diaphragms are discontinuous or have cutouts or openings totaling more than 50% of the enclosed area or changes of stiffness of more than 50%

There are discontinuities in the lateral-force load path

Irregular structures commonly require use of a variation of the dynamic method of seismic analysis, since it provides a more appropriate distribution of design loads. Many of these structures should also be subjected to a step-by-step dynamic analysis for specific accelerations to check the design further.

The dynamic method is based on equations of motion for linear-elastic seismic response. The equation of motion for a single-degree-of-freedom system subjected to a seismic ground acceleration a_g may be expressed as

$$m \frac{d^2x}{dt^2} + c \frac{dx}{dt} + kx = -ma_g \tag{9.10}$$

where d^2x/dt^2 is the acceleration of the structure, dx/dt is the velocity relative to the ground motion, and x is the displacement from an equilibrium position. The coefficients m, c, and k are the mass, damping, and stiffness of the system, respectively. Equation (9.10) can be solved by a number of methods.

The maximum acceleration is often expressed as a function of the fundamental period of vibration of the structure in a response spectrum. The response spectrum depends on the acceleration record. Since response varies considerably with acceleration records and structural period, smoothed response spectra are commonly used in design to account for the many uncertainties in future earthquakes and actual structural characteristics.

Most structures are multidegree-of-freedom systems. The n equations of motion for a system with n degrees of freedom are commonly written in matrix form as

$$[\mathbf{M}]\{\ddot{\mathbf{x}}\} + [\mathbf{C}]\{\dot{\mathbf{x}}\} + [\mathbf{K}]\{\mathbf{x}\} = -[\mathbf{M}]\{\mathbf{B}\}\, a_g \tag{9.11}$$

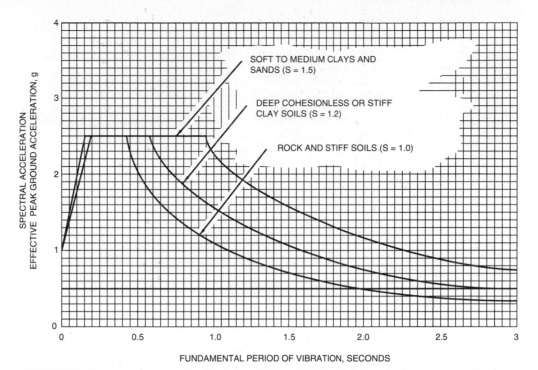

FIGURE 9.7 Response spectra for the 1979 Imperial Valley (California) College record, recommended for the dynamic method for determining seismic behavior of structures. *(After the "Uniform Building Code," 1991, International Conference of Building Officials.)*

where $[\mathbf{M}]$, $[\mathbf{C}]$, and $[\mathbf{K}]$ are $n \times n$ square matrices of the mass, damping, and stiffness, and $\{\ddot{\mathbf{x}}\}$, $\{\dot{\mathbf{x}}\}$, and $\{\mathbf{x}\}$ are column vectors of the acceleration, relative velocity, and relative displacement. The column vector $\{\mathbf{B}\}$ defines the direction of the ground acceleration relative to the orientation of the mass matrix. The multi-degree-of-freedom equations are coupled. They can be solved simultaneously by a number of methods. However, the single-degree-of-freedom response spectrum method is also commonly used for multidegree-of-freedom systems. The solution is assumed to be separable and the n eigenvalues (natural frequencies) ω_i and eigenvectors (mode shapes) $\{\mathbf{\Phi_i}\}$ are found. The solutions for the relative displacements, relative velocities and accelerations are then for i equal 1 to n:

$$\{\mathbf{x}_i\} = \sum_{j=1}^{n} \{\mathbf{\Phi}_j\}\, f_j(t) \tag{9.12}$$

$$\{\dot{\mathbf{x}}_i\} = \sum_{j=1}^{n} \{\mathbf{\Phi}_j\}\, \dot{f}_j(t) \tag{9.13}$$

$$\{\ddot{\mathbf{x}}_i\} = \sum_{j=1}^{n} \{\mathbf{\Phi}_j\}\, \ddot{f}_j(t) \tag{9.14}$$

The mode shapes are orthogonal with respect to the mass and the stiffness matrix. This orthogonality uncouples the equations of motion if the damping matrix is a diagonal matrix or proportional to a combination of the mass and stiffness matrix; that is, $\{\Phi_j\}^T[M]\{\Phi_i\}$ and $\{\Phi_j\}^T[K]\{\Phi_i\}$ are zero if $i \neq j$ and scalar numbers if $i = j$.

The response-spectrum technique can then be used to find the maximum values of $f_j(t)$ for each mode of vibration. Figure 9.7 shows the design response spectra recommended by the UBC unless site-specific spectra are employed. The response is based on calculations of the single-degree-of-freedom elastic response for a range of earthquake acceleration records and is normalized by the zone factor Z used in the static-force method. Given the modes of vibration for a multidegree-of-freedom system, a spectral acceleration for each mode, S_{ai}, can be determined from the response spectra. The base shear V_i acting in each mode can then be determined from

$$V_i = \frac{(\{\Phi_j\}^T[\mathbf{M}]\{\mathbf{B}\})^2}{\{\Phi_i\}^T[\mathbf{M}]\{\Phi_i\}} S_{ai} \qquad (9.15)$$

The distribution of this maximum base shear over the structure is

$$\{\mathbf{F}_i\} = \frac{(\{\Phi_i\}^T[\mathbf{M}]\{\mathbf{B}\})}{\{\Phi_i\}^T[\mathbf{M}]\{\Phi_i\}} [\mathbf{M}]\{\mathbf{B}\} S_{ai} \qquad (9.16)$$

Other response characteristics for each mode can be calculated from similar equations.

The maximum response in each mode does not occur at the same time for all modes. So some form of modal combination technique is used. The complete quadratic combination (CQC) method is one commonly used method for rationally combining these modal contributions. (E. L. Wilson et al., "A Replacement for the SRSS Method in Seismic Analysis," *Earthquake Engineering and Structural Dynamics*, vol. 9, pp. 187–194, 1981.) The method degenerates into a variation of the square root of the sum of the squares (SRSS) method when the modes of vibration are well-separated. The summation must include an adequate number of modes to assure that at least 90% of the mass of the structure is participating in the seismic loading.

The total seismic design force and the force distribution over the height and width of the structure for each mode can be determined by this method. The combined force distribution takes into account the variation of mass and stiffness of the structure, unusual aspects of the structure, and the dynamic response in the full range of modes of vibration, rather than the single mode used in the static-force method. The combined forces are used to design the structure, often reduced by R_w in accordance with the ductility of the structural system. In many respects, the dynamic method is much more rational than the static-force method, which involves many more assumptions for computing and distributing design forces. The dynamic method sometimes permits smaller seismic design forces than the static-force method. However, while it offers many rational advantages, the dynamic method is still a linear-elastic approximation to an inelastic-design method. As a result, it assumes that the inelastic response is distributed throughout the structure in the same manner as predicted by the elastic-mode shapes. This assumption may be inadequate if there is a brittle link in the system.

9.6   STRUCTURAL STEEL SYSTEMS FOR SEISMIC DESIGN

Since seismic loading is an inertial loading, the horizontal forces are dependent on the dynamic characteristics of the acceleration record and the structure. Seismic design codes use a response spectrum to model these dynamic characteristics. These forces are reduced in accordance with the ductility of the structure. This reduction is accomplished by the R_w factor in the static-force method, and the reduction may be quite large (Art. 9.5). The designer must assure that the structure is capable of developing the required ductility, and it is well-known that the available ductility varies with different structural systems. Figure 9.8 shows the inelastic dynamic response of two steel moment-resisting frames, which had identical mass and geometry but were designed for different seismic loads. The stiffness and natural periods of the frames are nearly identical, and they are all subjected to the same seismic acceleration record (1979 Imperial Valley College, Fig. 9.7). The story drift and inelastic deformation cycles are much larger for the frame with the smaller seismic design force. This shows that the smaller design force (larger R_w) requires a structure that maintains its integrity through larger inelastic cyclic deformations than if the structure had been designed for smaller R_w and larger seismic design forces.

Evaluation of Ductility.   Two major factors may affect evaluation of the ductility of structural systems. First, the ductility is often measured by the hysteretic behavior of the critical components. The hysteretic behavior is usually examined by observing the cyclic force-deflection (or moment-rotation) behavior as shown in Fig. 9.9. The slope of the curves represents the stiffness of the structure. The enclosed areas represent the energy that is dissipated, and this can be very large, because of the repeated cycles of vibration. These enclosed areas are sometimes full and fat (Fig. 9.9*a*), or they may be pinched or distorted (Fig. 9.9*b*). Structural framing with curves enclosing a large area, representing large dissipated energy, are regarded as superior systems for resisting seismic loading.

Special steel moment-resisting frames and eccentric braced frames, defined in Art. 9.4, have large hysteretic areas. As a result, they are designed for larger values of R_w, thus, smaller seismic forces and greater inelastic deformation (Fig. 9.8).

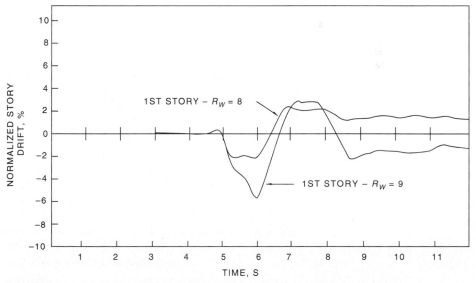

FIGURE 9.8 Curves show inelastic dynamic response of steel frames designed for different values of R_w, plotted for eight-story, weak-column, strong-beam framing with 2% damping (Imperial Valley College record).

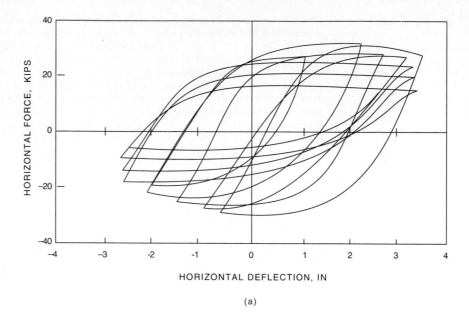

(a)

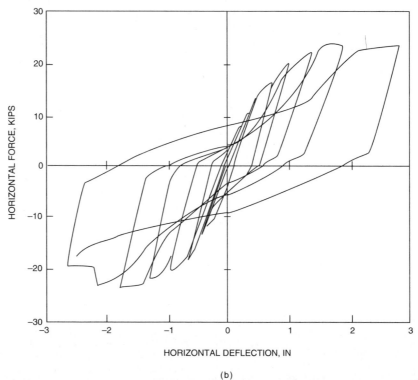

(b)

FIGURE 9.9 Hysteretic behavior of three steel frames. (*a*) Moment-resisting frame; (*b*) concentric braced frame; (*c*) eccentric braced frame.

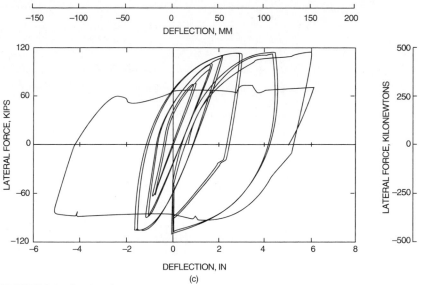

FIGURE 9.9 *Continued.*

While some steel structures are very ductile, not all structures have this great ductility. In particular, local and global buckling can easily change the hysteretic behavior from that of Fig. 9.9*a* to Fig. 9.9*b*. This change can dramatically decrease the available ductility of the structural system. As a result, braced frames, which are dominated by brace buckling during severe earthquakes, are assigned a smaller value of R_w and, hence, a larger seismic design force.

Effects of Inelastic Deformations. The distribution of inelastic deformation is a second factor that can effect the inelastic seismic performance of a structural system. Some structural systems concentrate the inelastic deformation (ductility demand) into a small portion of the structure. This can dramatically increase the ductility demand for that portion of the structure. This concentration of damage is sometimes related to factors that cause pinched hysteretic behavior, since buckling may change the stiffness distribution as well as affect the energy dissipation.

Ductility demand, however, can also be related to other factors. Figure 9.10 shows the computed inelastic response of two steel moment-resisting frames that have identical mass and nearly identical strength and stiffness and are subject to the same acceleration record as that in Fig. 9.8. The frames differ, however, in that one is designed to yield in the beams while the other is designed to yield in the columns. This difference in design concept results in a significant difference in seismic response and ductility demand. Design codes attempt to assure greater ductility from structures designed for smaller seismic forces, but attaining this objective is complicated by the fact that ductility and ductility demand are not fully understood.

Steel moment-resisting frames are somewhat flexible. This leads to relatively low seismic design forces from design response spectra, such as that of Fig. 9.7 and Eq. (9.6). While such frames may be ductile, the ductility can be lost if certain requirements, which are summarized in Art. 9.7.1, are not satisfied in design and construction of the frame. Figure 9.9*a* illustrates the cyclic force-deflection behavior of a ductile steel moment-resisting frame. It has stable strength and stiffness through large, repeated, inelastic deformations, which provide large energy dissipation.

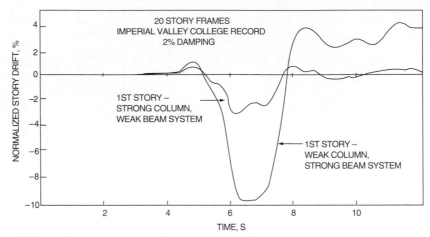

FIGURE 9.10   Curves show inelastic dynamic response of two steel frames with identical mass and nearly identical strength and stiffness but designed with two different strategies for determining inelastic deformations.

This behavior is important, since it dampens the inelastic response and improves the seismic performance of the structure without requiring excessive strength or deformation in the structure.

Concentric braced frames, defined in Art 9.4, economically provide much larger strength and stiffness than a moment-resisting frame with the same amount of steel. There are a wide range of bracing configurations, and considerable variations in structural performance may result from these different configurations. Figure 9.11 shows some concentric bracing configurations. The braces, which provide the bulk of the stiffness in concentrically braced frames, attract very large compressive and tensile forces during an earthquake. As a result, compressive buckling of the braces often dominates the behavior of these frames. The pinched cyclic force-deflection behavior shown in Fig. 9.9b commonly results, and failure of braces may be quite dramatic. Therefore, concentric braced frames are regarded as stiffer, stronger but less ductile than steel moment-resisting frames. Different design provisions are required for concentric braced frames than for moment-resisting steel frames. These are summarized in Art. 9.7.2.

Eccentric braced frames, defined in Art. 9.4, can combine the strength and stiffness of concentrically braced frames with the good ductility of moment-resisting frames. Eccentric braced frames incorporate a deliberately controlled eccentricity in the brace connections (Fig. 9.12). The eccentricity and the link beams are carefully chosen to prevent buckling of the brace, and provide a ductile mechanism for energy dissipation. If they are properly designed, eccentric braced frames lead to good inelastic performance as depicted in Fig. 9.9c, but they require yet another set of design provisions, which are summarized in Art. 9.7.3.

Dual systems, defined in Art 9.4, may combine the strength and stiffness of a braced frame and shear wall with the good inelastic performance of special steel moment-resisting frames. Dual systems are frequently assigned an R_w value and seismic design force that are intermediate to those required for either system acting alone. Design provisions provide limits and recommendations regarding the relative stiffness and distribution of resistance of the two components. Dual systems have led to a wide range of structural combinations for seismic design. Many of these are composite or hybrid structural systems. However, steel frames with composite

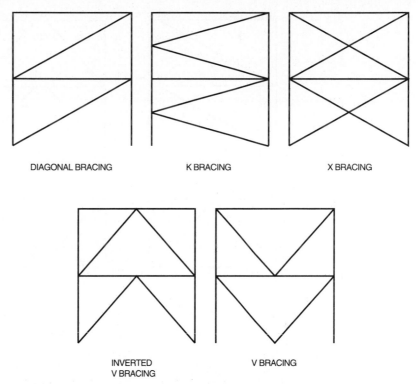

DIAGONAL BRACING   K BRACING   X BRACING

INVERTED
V BRACING   V BRACING

FIGURE 9.11 Typical configurations of concentric braced frames.

concrete floor slabs are not commonly used for developing seismic resistance, even though composite floors are commonly used for gravity-load design throughout the United States.

9.7 SEISMIC-DESIGN LIMITATIONS ON STEEL FRAMES

The "Uniform Building Code" contains a wide range of special seismic design requirements for steel frames. Similar provisions are contained in the "Seismic Provisions for Structural Steel Buildings" of the American Institute of Steel Construction. These requirements are intended to assure that steel frames actually achieve the ductility and behavior required for the assigned seismic design force. Use of systems with poor or uncertain seismic performance is restricted or prohibited for some applications. There are a range of specific requirements for moment-resisting frames and concentric and eccentric braced frames.

9.7.1 Limitations on Moment-Resisting Frames

Structural tests have shown that steel moment-resisting frames may provide excellent ductility and inelastic behavior under severe seismic loading. Because these frames are frequently quite flexible, drift limits often control the design. The UBC recognizes this ductility and assigns $R_w = 12$ to special moment-resisting frames

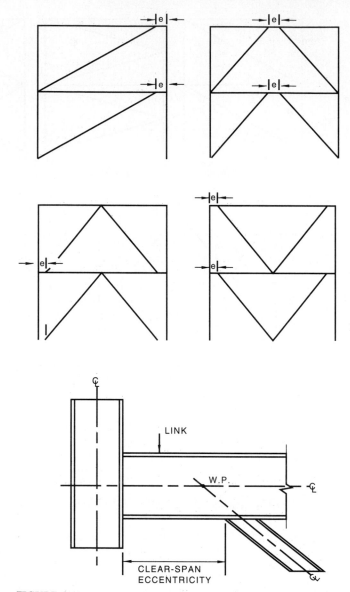

FIGURE 9.12 Typical configurations of eccentric braced frames. See also Fig. 9.15.

(Art. 9.4). These frames must satisfy a range of slenderness requirements to control local buckling during plastic deformation in a severe earthquake.

The unsupported length of bending members must be less than $96r_y$, where r_y is the radius of gyration about the weak axis. The objective is to control lateral torsional buckling during plastic deformation under cyclic loading. The flanges of beams must have a slenderness less than

$$\frac{b_f}{2t_f} \leq \frac{52}{\sqrt{F_y}} \tag{9.17}$$

where F_y = specified minimum yield stress, ksi, of the steel and b_f and t_f are the flange width, in, and flange thickness, in, respectively. The purpose of this requirement is to control flange buckling during the plastic deformation expected in a severe earthquake. The webs of beams must satisfy

$$\frac{d}{t_w} \le \frac{640 \left(1 - 3.74 \frac{f_a}{F_y}\right)}{\sqrt{F_y}} \qquad \frac{f_a}{F_y} \le 0.16 \qquad (9.18)$$

$$\frac{d}{t_w} \le \frac{257}{\sqrt{F_y}} \qquad \frac{f_a}{F_y} > 0.16 \qquad (9.19)$$

where f_a = computed axial compression, ksi; F_y is the yield stress, ksi; and d and t_w are the web depth, in, and web thickness, in, respectively. This is required to control web buckling during plastic deformation. These limitations are not specifically required for columns by the UBC, although it is logical to use them whenever yielding is expected in the columns. These limits are somewhat more conservative than the normal compactness requirements for steel design, because of the great ductility demand of seismic loading.

Seismic Loads for Columns. The UBC provisions require that columns and column splices be designed for the possibility of uplift and extreme compressive load combinations. The UBC specifies two special factored-load combinations for this purpose. For axial compression, columns should have the strength to resist

$$1.0P_{DL} + 0.7P_{LL} + 3\left(\frac{R_w}{8}\right) P_E \qquad (9.20)$$

For axial tension, design strength should equal or exceed

$$0.85P_{DL} + 3\left(\frac{R_w}{8}\right) P_E \qquad (9.21)$$

where P_{DL} and P_{LL} are the axial dead and live loads, kips; P_E is the axial seismic load; and R_w is the coefficient in Eq. (9.5). Similar but slightly different load combinations are required by the AISC "Provisions."

Beam-to-Column Connections. In special moment-resisting frames, beam-to-column connections must develop the plastic capacity of the beams. This is commonly achieved with bolted webs and full-penetration groove welds at the beam flanges (Fig. 9.13a). Many commonly used wide-flange sections, however, require welding of the web (Fig. 9.13b) if more than 30% of their plastic capacity is developed in the web; that is, web welding is required if

$$F_{yf}t_f b_f \le \frac{0.70Z_b F_{yb}}{d_b - t_f} \qquad (9.22)$$

where F_{yf} and F_{yb} are the yield stress, ksi, of the flange and beam; t_f and b_f, the flange thickness and width, in; d_b, the beam depth, in; and Z_b, the plastic section modulus, in^3, of the beam. Continuity plates or stiffeners (Fig. 9.13) are often required for the columns of special moment-resisting frames, because the beams transfer large bending moments to the columns during seismic loading. The bending moments induce large concentrated forces at the connections of the beam flanges to the columns (Fig. 9.14). Stiffeners are required opposite the compression flange

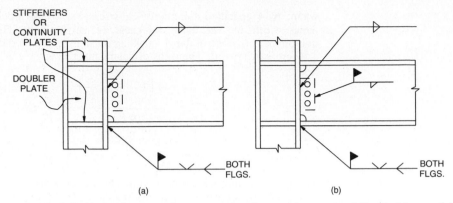

FIGURE 9.13 Typical beam-to-column connections for special moment-resisting steel frames. (*a*) Normal connection with angle shop welded to the column flange, but field bolted to the beam web, and the beam flanges field welded to the column; (*b*) welded beam-to-column connection except for erection bolts in the beam web.

of the beam when

$$d_c > \frac{4100t^3\sqrt{F_{yc}}}{P_{bf}} \tag{9.23}$$

where t is the web thickness, in, of the column; d_c, the column-web depth clear of fillets, in; F_{yc}, the column yield stress, ksi; and P_{bf}, the factored compression force in the beam flange, kips. Stiffeners are required opposite the tensile flange of the beam when

$$t_f < 0.4\ \sqrt{\frac{1.8b_f t_f F_{yb}}{F_{yc}}} \tag{9.24}$$

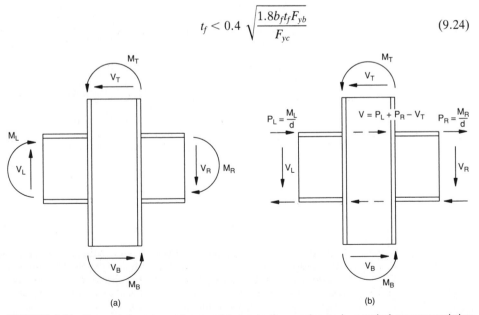

FIGURE 9.14 Forces acting on a column and beam in the panel zone in a typical moment-resisting connection during seismic loading. Forces in (*a*) are equivalent to those in (*b*).

These limitations prevent deformation of the column web under cyclic loading and assure a reasonably uniform distribution of stress in the flange welds.

Seismic bending moments in the beams cause large shear stresses in the column web in the panel zone of the connection (Fig. 9.14). The panel-zone shear strength, kips, may be computed from

$$V = 0.55F_{yc}d_c t \left[1 + \frac{3b_c t_{cf}^2}{d_b d_c t} \right] \tag{9.25}$$

where b_c is the width, in, of the column flange and t_{cf} the thickness, in, of the flange. The panel-zone shear strength need not exceed 80% of the plastic bending capacity of the beams intersecting the column panel zone, but the panel zone should be capable of resisting the bending moments due to gravity loads plus 1.85 times the bending moments due to the prescribed seismic design forces.

Equation (9.25) takes into account the fact that the strength of the panel zone is enhanced by the strength and stiffness of the column flanges, and that panel-zone yielding provides good, stable energy dissipation and inelastic performance. Also, this equation encourages panel-zone yielding over many other types of plastic deformation. Equation (9.25), however, permits some plastic deformation of the panel zone at loads well below the design load, since V is sometimes considerably larger than the panel-zone yield capacity,

$$V_s = 0.55F_{yc}d_c t_w \tag{9.26}$$

If the panel zone does not have the capacity required by Eq. (9.25), a doubler plate (Fig. 9.13a) or thicker column web is required. A minimum thickness t_z for the combined doubler plate and column web is prescribed:

$$t_z \geq \frac{d_z + w_z}{90} \tag{9.27}$$

where d_z and w_z are the depth between continuity plates and width between column flanges in the panel zone. Doubler plates must be stitched to the web of the column with plug welds to prevent local buckling of the plate, otherwise d_z cannot be included in Eq. (9.27).

Panel-zone requirements often control the lateral resistance of steel moment-resisting frames. However, this may cause some difficulties for structural designers. The UBC requires computation of the story drift due to panel-zone deformation, and there is no clear, simple method for calculating story drift in frames dominated by panel-zone yielding.

Special moment-resisting frames provide superior performance when yielding due to severe seismic loading occurs in the beams rather than the columns. This strong-column, weak-beam behavior is required, except in special cases. To ensure this behavior, the following relationship must be satisfied, except as indicated below.

$$\frac{\Sigma Z_c(F_{yc} - f_a)}{\Sigma Z_b F_{yb}} > 1.0 \qquad f_a \geq 0 \tag{9.28}$$

where Z_b and Z_c are the plastic section modulus, in^3, of the beam and the column. This requirement need not be met when $f_a < 0.4F_{yc}$ for all load combinations, except for those specified by Eqs. (9.20) and (9.21), and any of the following conditions hold:

1. The joint is at the top story of a multistory frame with fundamental period greater than 0.7 second.

2. The joint is in a single-story frame.

3. The sum of the resistances of the weak-column joints is less than 20% of the resistances for a specific story in the total frame and the sum of the resistances of the weak-column joints in a specific frame is less than 33% of the resistances for the frame.

4. The frame is designed without the 33% allowable-stress increase for seismic loads.

These provisions are somewhat controversial. Different specifications have different criteria for permitting weak-column joints. The AISC provisions allow use of weak-column joints if any of the following conditions are met:

1. The column has an axial force less than 30% of the yield force.

2. The columns in any story have a ratio of shear strength to design shear strength greater than the story above.

3. The column lateral shear strength is not included in the design to resist design seismic shears.

Research suggests that yielding of the columns results in concentration of damage in the structural frames (Fig. 9.10) and reduces the available ductility in the structure while increasing the ductility demand. However, many structural configurations quite naturally lead to weak-column, strong-beam behavior. In addition, the issue is further complicated by concern that panel-zone yielding may lead to an equivalent of weak-column, strong-beam behavior even though Eq. (9.28) is satisfied.

Some steel moment-resisting frames are not designed to satisfy the preceding conditions. In many cases, these frames are used in less seismically active zones. Sometimes, however, they are used in seismically active zones with larger seismic design forces; that is, they are designed with an R_w of 5 or 6. As a result, design forces would be 2.0 to 2.4 times those for the special frames. These ordinary moment-resisting frames must satisfy some but not all of the preceding conditions, depending on the seismic zone and design forces in the structure. In addition, some of the preceding restrictions on special moment-resisting frames may be eased in seismic zones other than 4 or 5.

9.7.2 Limitations on Concentric Braced Frames

Concentric braced steel frames are much stiffer and stronger than moment-resisting frames, and they frequently lead to economical structures. However, their inelastic behavior is usually inferior to that of special moment-resisting steel frames (Art. 9.6). One reason is that the behavior of concentric braced frames under large seismic forces is dominated by buckling. Furthermore, the columns must be designed for tensile loads and foundation uplift as well as for compression.

Figure 9.11 shows some of the common bracing configurations for concentric braced frames. Seismic design requirements vary with bracing configuration.

X bracing, for example, usually is very slender and has large tensile capacity and little compressive buckling capacity. It may be an economical design for lateral loads, but it permits concentration of inelastic deformations, and energy dissipation during major earthquakes is poor. As a result, X bracing is restricted to use in less seismically active zones or very short structures in more active zones.

K bracing causes yielding in the columns during severe seismic loading. One diagonal is in compression while the other is in tension, and the compression diagonal buckles well before the tensile brace yields. The buckling introduces large shears and bending moments in the columns. As a result, K bracing is prohibited in the more seismically active regions.

Because of these considerations, diagonal and chevron bracing are the primary systems for major structures in seismically active regions of the United States.

Chevron bracing (V or inverted V, shown in Fig. 9.11) causes beam yielding during severe seismic excitation, whereas K bracing causes column yielding. Beam flexure with chevron bracing induces deformations of floors during a major earthquake but provides additional energy dissipation, which may improve the seismic response during major earthquakes.

Diagonal bracing acts in tension for lateral loads in one direction and in compression for lateral loads in the other direction. The "Uniform Building Code" requires that the direction of the inclination of bracing with the diagonal bracing system be balanced, since braces have much larger capacity in tension than in compression.

Buckling of Bracing. In general, the energy dissipation of concentric braced frames is strongly influenced by postbuckling brace behavior. This is quite different for slender braces than for stocky braces. For example, the compressive strength of a slender brace is much smaller in later cycles of loading than it is in the first cycle. In addition, very slender braces offer less energy dissipation but are able to sustain more loading cycles and larger inelastic deformation than stocky braces. In view of this, the slenderness ratio of bracing is limited, in zones 4 and 5, to

$$\frac{L}{r} \leq \frac{720}{\sqrt{F_y}} \tag{9.29}$$

where L is the unsupported length, in; r is the least radius of gyration, in; and F_y is the yield stress, ksi. In addition, the allowable compressive stress in slender bracing is reduced to

$$F_{as} = \beta F_a \tag{9.30}$$

where F_a is the usual allowable compressive stress.

$$\beta = \frac{1}{1 + \dfrac{KL/r}{2C_c}} \tag{9.31}$$

where KL is the effective length and $C_c = \sqrt{2\pi^2 E/F_y}$, with E the modulus of elasticity, ksi. The reduction factor β need not be less than 0.8 for seismic zones 2 and 3 or lower.

Bracing, contributing most of the lateral strength and stiffness to frames, resists most of the seismic load. It is tempting, for economy, to design bracing as tension members only, since steel is very efficient in tension. However, this results in poor inelastic behavior under severe earthquake loading, a major reason for excluding X bracing from seismically active regions. On the other hand, more energy is dissipated in a brace yielding in tension than in a brace buckling in compression. As result, all bracing systems for seismic zones 4 and 5 must be designed so that at least 30%, but no more than 70%, of the base shear [Eq. (9.5)] is carried by bracing acting in tension, while the balance is carried by bracing acting in compression.

The overall and local slenderness of bracing is important. The ratio b/t of width to thickness of single-angle struts or double-angle braces that are separated by stitching elements is limited to

$$\frac{b}{t} \leq \frac{76}{\sqrt{F_y}} \tag{9.32}$$

For the stems of T sections, the ratio of depth to web thickness may not exceed

$$\frac{d}{t_w} = \frac{127}{\sqrt{F_y}} \tag{9.33}$$

Strength of Connections.   The UBC requires that the connections of the braces in concentric braced frames be stronger than the bracing members themselves. This assures that energy dissipation occurs in the members rather than the connections. The connection need not be designed for forces larger than $3R_w/8$ times the prescribed seismic forces. This product is regarded as the maximum force required to assure that a brace and connection remain elastic during a major earthquake. The net section of the brace must be considered in this connection evaluation. The ratio of net area A_e to gross area A_g must satisfy

$$\frac{A_e}{A_g} = \frac{1.2\alpha F^*}{F_u} \tag{9.34}$$

where α = fraction of the member force transferred across the net section
F^* = smaller of the strength of the bracing member or $3R_w/8$ times the prescribed seismic design force
F_u = minimum specified tensile strength of the brace steel, ksi

Selection of $\mathbf{R_w}$.   Once concentric bracing is selected for seismic design, R_w for the bent can be chosen. It varies between 6 and 10. The value of 6 is used if the bracing carries gravity loads in addition to seismic loading. Concentric braced frames dissipate energy by brace buckling, and buckling is less tolerable when the brace carries gravity load. A value of 8 is used if the bracing carries the entire lateral load but no gravity load. The value of 10 is used for a dual system where at least 25% of the base shear [Eq. (9.5)] is carried by a special moment-resisting frame. These coefficients result in seismic design loads that are 1.2 to 2.0 times the design loads required for special moment-resisting steel frames.

Effect of Brace Cross Sections.   The cross section of braces is not addressed in most building codes. Research has shown, however, that the cross section may have a considerable impact on seismic performance. For example, hollow steel tubes appear to be attractive for bracing, because the large radius of gyration permits a large buckling load. However, tubes sustain a concentration of damage during inelastic seismic deformation, which hastens their failure and makes them a less desirable alternative than angles, channels, or other open sections.

Concentric braced frames require care in design. However, the weight of steel required to support the gravity loads and develop required stiffness within the structure may be smaller with this type of braced frame than that required for many other framing systems. In addition, concentric braced frames often utilize simple, more economical connections, and this sometimes results in considerable savings in construction cost.

9.7.3   Eccentric Braced Frames

These combine the strength and stiffness of a concentric braced frame with the inelastic performance of a special moment-resisting frame (Fig. 9.9*c*). The UBC permits use of an R_w of 10 or 12 for an eccentric braced frame. This results in seismic design forces comparable to those required for special moment-resisting frames if the fundamental period of vibration is the same. However, braced frames are invariably stiffer than moment-resisting frames of similar geometry and have a shorter period. This results in a somewhat larger design load than for special moment-resisting frames under comparable conditions. (C. W. Roeder, and E. P. Popov, "Eccentrically Braced Steel Frames for Earthquakes," *Journal of Structural Division*, March 1978, American Society of Civil Engineers.)

There are a number of special design provisions that must be satisfied by eccentric braced frames. As defined in Art 9.4, a link must be provided at least at one end of each brace. The link beam should be designed so that it is the weak link of the structure under severe seismic loading. This is done by selecting the size of the steel section and the length of the link beam to match seismic-load design requirements. The weak link is assured by the requirement that the brace be designed for at least 1.5 times the axial force present when the strength of the link is developed. The column must remain elastic at 1.25 times the strength of the eccentric braced bay, a strength equivalent to the smaller of the shear strength or the reduced flexural strength of the brace. The link beam is expected to undergo a large amount of plastic deformation during a major earthquake. As a result, local slenderness requirements also must be met for the webs and flanges of beams. These provisions are essentially the same as those in Eqs. (9.17) to (9.19) for special moment-resisting frames.

Beams in eccentric braced frames must be restrained at both top and bottom flanges against lateral torsional buckling. For the purpose, lateral bracing should be installed at the ends of the link beams. Also, the unsupported length L_u, in, of a beam must not exceed

$$L_u = \frac{76b_f}{\sqrt{F_y}} \tag{9.35}$$

where b_f is the width, in, of the beam flange and F_y is the yield strength, ksi, of the beam steel. The lateral bracing at each end of a link beam must be able to resist a lateral force equal to $0.015F_y b_f t_f$, where t_f is the thickness, in, of the beam flange. Intermediate lateral bracing should have a strength equal to 1% of the force in the link-beam flange at the brace point.

The web of a link beam is subject to very high shear stress. Stiffeners therefore are needed to control shear buckling. The link beam must have full-depth stiffeners on both sides of the web at the brace end. Also, full-depth stiffeners should be installed at a distance b_f from each end of the link when it has a clear length between $1.6M_s/V_s$ and $2.6M_s/V_s$, where $M_s = ZF_y$, $V_s = 0.55F_y\, dt$, d = depth, in, of the web, and Z is the plastic section modulus. The stiffeners are required to have a minimum combined width of at least $b_f - t_w$. Thickness should be at least $0.75t_w$, but not less than ⅜ in.

The plastic capacity of long link beams is controlled by flexural yielding. Long link beams are those with clear lengths greater than $2.6M_s/V_s$ (Fig. 9.15). Such beams normally are less ductile than the shorter links, but the geometry of frames is such that less deformation is required of the link beams during an earthquake.

Shear yielding dominates the behavior of short link beams. These have a clear length less than $1.6M_s/V_s$. These beams can sustain much larger deformation without failure during cyclic loading, but the geometry of frames requires that the link beams sustain larger deformations during a major earthquake. Designers should balance the required and the available ductility. The "Uniform Building Code" presents guidelines for accomplishing this balance:

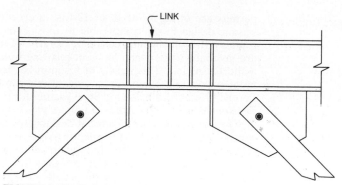

FIGURE 9.15 Typical connection details and stiffener arrangement for an eccentric braced frame.

When the braced frame sways a total of $3R_w/8$ times the drift caused by the design earthquake, the rotation of a short link relative to the rest of the beam should not be more than 0.06 radians. The relative rotation of a long link should not exceed 0.03 radians. The limiting rotation for link beams with intermediate lengths is determined by interpolation.

Intermediate stiffeners are required when the shear capacity V_s of the web controls the design or when the shear determined by the flexural strength exceeds $0.45F_y\, dt_w$ (Fig. 9.15). Spacing s for intermediate stiffeners for link beams with a relative rotation of 0.06 radians or less should not exceed

$$s = 38t_w - d/5 \qquad (9.36)$$

For a link-beam rotation of 0.03 radians or less,

$$s \le 56t_w - d/5 \qquad (9.37)$$

Spacing for in-between rotations is determined by interpolation.

Doubler plates and holes or penetrations are not permitted in the link beams. The connections must be strong enough to develop fully the plastic capacity of the link beams. This often requires that the web yield in beam-column connections to develop the full shear capacity of the link-beam section.

Eccentric braced frames are a rational attempt to design steel structures that fully develop the ductility of the steel without loss of strength and stiffness due to buckling. The design of these frames is somewhat more complicated than that of some other steel frames, but eccentric braced frames offer advantages in economical use of steel and seismic performance of structures.

9.8 FORCES IN FRAMES SUBJECTED TO LATERAL LOADS

The design loads for wind and seismic effects are applied to structures in accordance with the guidelines in Arts. 9.2 to 9.5. Next, the structure must be analyzed to determine forces and moments for design of the members and connections. Member and connection design proceeds quite normally for wind load design after these internal forces are determined, but seismic design is also subject to the detailed ductility considerations described in Arts. 9.6 and 9.7.

9.8.1 Approximate Analysis of Structural Frames

This is required for preliminary design and for interpretation and evaluation of computer results. Approximate methods are based on physical observations of the response of structures to applied loads. Two such methods are the portal and cantilever methods, often used for analyzing moment-resisting frames under lateral loads.

The portal method is used for buildings of intermediate or shorter height. In this method, a bent is treated as if it were composed of a series of two-column rigid frames, or portals. Each portal shares one column with an adjoining portal. Thus, an interior column serves as both the windward column of one portal and the leeward column of the adjoining portal. Horizontal shear in each story is distributed in equal amounts to interior columns, while each exterior column is assigned half the shear for an interior column, since exterior columns do not share the loads of adjacent portals. If the bays are unequal, shear may be apportioned to each column in proportion to the lengths of the girders it supports. When bays are equal, the axial load in interior columns due to lateral load is zero.

Inflection points (points of zero moment) are placed at midheight of the columns and midspan of beams. This approximates the deflected shapes and moment diagrams of those members under lateral loads. The location of the inflection points may be adjusted for special cases, such as fixed or pinned base columns, or roof beams and top-story columns, or other special situations. On the basis of the preceding assumptions, member forces and bending moments can be determined entirely from the equations of equilibrium. As an example, Fig. 9.16 indicates the geometry and loading of an eight-story moment-resisting frame, and Fig. 9.17 illustrates the use of the portal method on the upper stories of the frame. The frame has two interior columns. So one-third of the shear in each story is distributed to the interior columns and half of this, or one-sixth, is distributed to the exterior columns (Fig. 9.17). The other member forces are computed by equations of equilibrium on each subassemblage. For example, for the subassemblage at the top of the frame in Fig. 9.17, setting the sum of the moments equal to zero yields

$$\frac{l}{2} S_1 = 4.17 \frac{h}{2} \quad \text{or} \quad S_1 = 4.17 \frac{h}{l} \tag{9.38}$$

Setting the sum of the vertical forces equal to zero gives

$$A_4 = -S_1 = -4.17 \frac{h}{l} \tag{9.39}$$

Setting the sum of the horizontal forces equal to zero results in

$$A_1 = 25 - 4.17 = 20.83 \tag{9.40}$$

For the central top subassemblage:

$$\frac{l}{2}(S_1 + S_2) = 8.33 \frac{h}{2} \quad \text{or} \quad S_2 = \frac{h}{l}(8.33 - 4.17) = 4.16 \frac{h}{l} \tag{9.41}$$

The remaining axial and shear forces can be determined by this procedure, and bending moments can be determined directly from these forces from equilibrium equations.

The cantilever method is used for tall buildings. It is based on the recognition that axial shortening of the columns contributes to much of the lateral deflections of such buildings (Fig. 9.18). In this method, the floors are assumed to remain

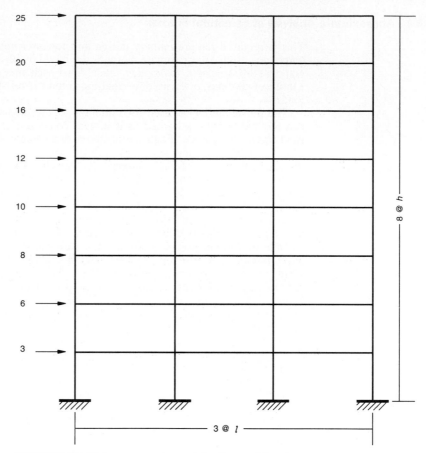

FIGURE 9.16   Eight-story moment-resisting frame subjected to static lateral loading.

plane, and the axial force in each column is assumed to be proportional to the distance of the column from the centroid of the columns. Inflection points are assumed to occur at midheight of the columns and at midspan of the beams. The internal moments and forces are determined from equations of equilibrium, as with the portal method. The determination of the forces and moments in the members at the top floors of the frame in Fig. 9.16 is illustrated in Fig. 9.19. The lateral forces cause overturning moments, which induce axial tensile and compressive forces in the columns in the columns. Therefore,

$$A_4 = -A_7 \quad \text{and} \quad A_5 = -A_6 \tag{9.42}$$

Since the exterior columns are 3 times as far from the centroid of the columns as the interior columns,

$$A_4 = 3A_5 \tag{9.43}$$

Setting the sum of the moments equal to zero gives

$$3lA_4 + lA_5 = 25\frac{h}{2} \quad \text{and} \quad A_5 = 1.25\frac{h}{l} \tag{9.44}$$

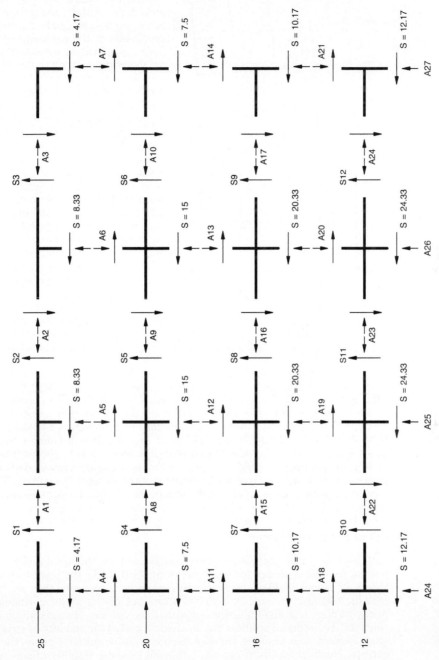

FIGURE 9.17 Forces at midspan of beams and midheight of columns in the frame of Fig. 9.16 as determined by the portal method.

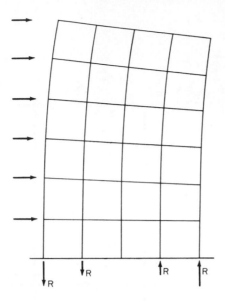

FIGURE 9.18 Drift of a moment-resisting frame assumed for analysis by the cantilever method.

Axial forces in the columns can be determined at other levels by the same procedure. Other shear forces and bending moments can be determined by application of the equations of equilibrium for individual subassemblages, as for the portal method.

Analysis of Dual Systems. Approximate analysis of braced frames can be performed as if the bracing were a truss. However, many braced structures are dual systems that combine moment-resisting-frame behavior with braced-frame behavior. Under these conditions, an approximate analysis can be performed by first distributing the lateral forces between the braced-frame and moment-resisting-frame portions of the structure in proportion to the relative stiffness of the components. Braced frames are commonly very stiff and normally would carry the largest portion of the lateral loads.

Once the initial distribution of member and connection forces and moments is completed, a preliminary design of the members can be performed. At this time, it is possible to reanalyze the structure by any of a number of linear-elastic, finite-element methods, for which computer programs are available.

While many major, existing structures were designed without benefit of computer analysis techniques, it is not advantageous to design modern buildings for wind and earthquake loading without this capability. It is needed to predict realistic structural response to wind loading and to evaluate occupant comfort, as well as for dynamic design for seismic loading, especially for buildings of unusual geometry. Both the seismic and wind load provisions in the "Uniform Building Code" result in local anomalies in the distribution of design forces due to the distribution of mass, stiffness, or local wind pressure, and many elements such as slabs and diaphragms may distribute large forces from one load element to another. The combination of these factors results in the requirement for finite-element analysis.

9.8.2 Nonlinear Analysis of Structural Frames

Although nonlinear analysis is not commonly used for structural design, it is important for seismic design for several reasons. First, while the seismic-design provisions of various building codes rely on linear-elastic concepts, they are based on inelastic response. Seismic behavior of structures during major earthquakes depends on nonlinear material behavior caused by yielding of steel and cracking of concrete. The reduced stiffness due to yielding makes the stability of structures of great concern, and ensuring stability requires consideration of geometric nonlinearities. Nonlinear analysis permits treatment of these stability effects with P-Δ moments (Fig. 9.20).

Second, design methods such as load-and-resistance-factor design encourage use of flexible, partly restrained (PR) connections. Such connections are inherently nonlinear in their response. Hence, it is necessary to analyze structures with at-

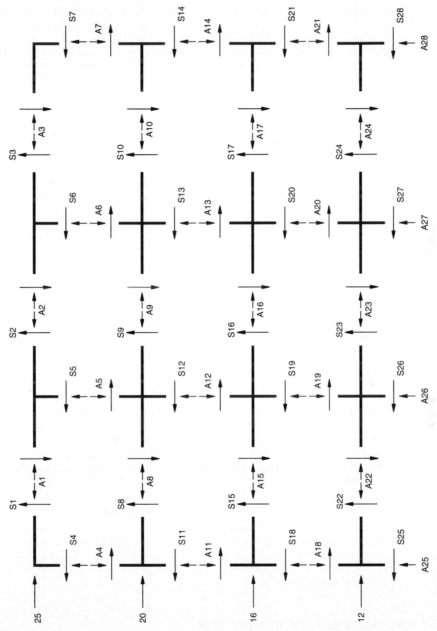

FIGURE 9.19 Cantilever method of determining the forces at midspan of beams and midheight of columns in the frame of Fig. 9.16.

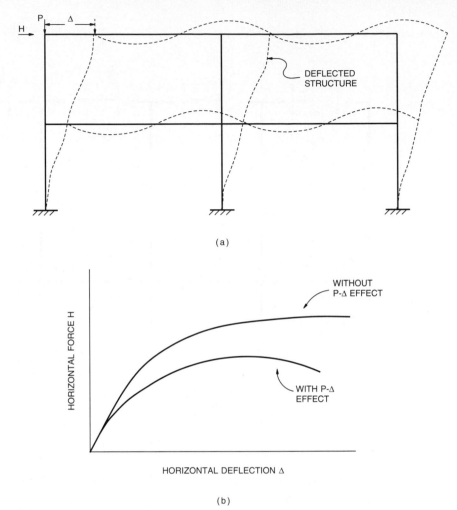

FIGURE 9.20 Sidesway of a two-story frame subjected to horizontal and vertical loads. (*a*) Position of deflected structure for drift Δ; (*b*) curves show relationship of horizontal force and drift with and without the P-Δ effect.

tention to the contribution of connection flexibility. Further nonlinearity may occur due to the effects of connection flexibility on frame stability and P-Δ moments. These nonlinear effects are not commonly considered in design at present. However, computer programs are available to model nonlinear frame behavior, and, as a result, it is highly probable that some buildings will be analyzed by these methods in the future.

9.9 MEMBER AND CONNECTION DESIGN FOR LATERAL LOADS

Wind loads on steel structures are determined by first establishing the pressure distributions on structures after considering the appropriate design wind velocity,

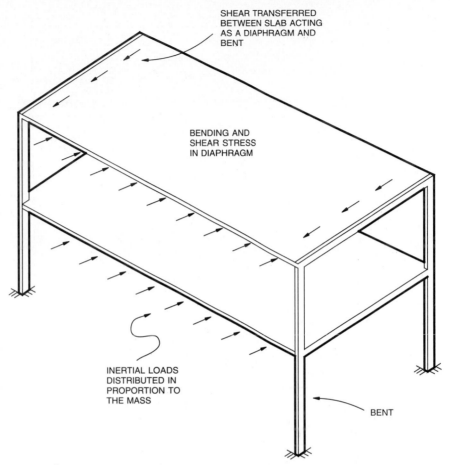

SHEAR TRANSFERRED
BETWEEN SLAB ACTING
AS A DIAPHRAGM AND
BENT

BENDING AND
SHEAR STRESS
IN DIAPHRAGM

INERTIAL LOADS
DISTRIBUTED IN
PROPORTION TO
THE MASS

BENT

FIGURE 9.21   Slab acting as a diaphragm distributes seismic loads to bents. Bending and shear stresses occur in the diaphragm.

the exposure condition, and the local variation of wind pressure on the structure (Art. 9.2). Then, the wind loads on frames and structural elements are determined by distributing the wind pressure in accordance with the tributary areas and relative stiffness of the various components.

Seismic design loads are determined by the static-force or dynamic methods. With the static-force method, the total base shear is determined by Eq. (9.5). It is distributed to bents and structural elements by simple rules combined with considerations of the distribution of mass and stiffness (Art. 9.4). With the dynamic method, the total range of dynamic modes of vibration are considered in determination of the base shear. This is distributed to the bents and components in accordance with the mode shapes. For both wind and seismic loading, forces and moments in members and connections can be first estimated by approximate analysis techniques (Art. 9.8.1). They may be ultimately computed by finite-element or other structural analysis techniques.

Once member and connection forces and moments are determined, design for lateral loads is similar to design for other loadings.

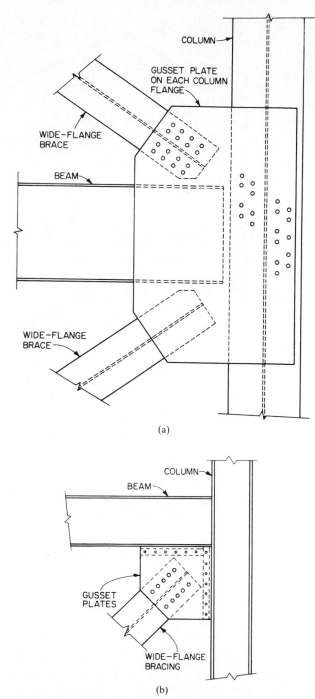

FIGURE 9.22 Typical gusset-plate connections at a column. (*a*) With braces above and below the joint; (*b*) with a brace only below the joint.

Connections used for wind loading run the full range of unrestrained (pinned), fully restrained (FR), or partly restrained (PR) connections. PR connections are frequently used for wind loading design, because they are economical and easily fabricated and constructed.

Connections used in seismic design are normally unrestrained or FR connections. PR connections have less seismic resistance than the members they are connecting, and therefore inelastic deformations during severe earthquakes are concentrated in the connections. PR connections have limited energy-dissipation capacity. Furthermore, the total ductility and deformation capacity of a structural frame under cyclic loading is uncertain, and the energy dissipation is concentrated in the connections. These combined effects have limited the use of PR connections in seismic design. Nevertheless, they offer many advantages and may be economical for use in less seismically active regions, rehabilitation projects, and perhaps, in the future, major seismic regions.

Wind loading design is based on elastic behavior of structures, and strength considerations are adequate for design of wind connections. Seismic design, in contrast, utilizes inelastic behavior and ductility of structures, and many design factors must be taken into account beyond the strength of the members and connections. These requirements are intended to assure adequate ductility of structures. Code provisions attempt to assure that inelastic deformations occur in members rather than connection (Art. 9.7).

Designs for wind and seismic loading often use floor slabs and other elements to distribute loads from one part of a structure to another (Fig. 9.21). Under these conditions, the slabs and other elements act as diaphragms. These may be considered deep beams and are subject to loadings and behavior quite different from that encountered in gravity-load design. It is important that this behavior be considered, and it is particularly important that the connection between the diaphragms and the structural elements be carefully designed. These connections often involve a composite connection between a steel structural member and a concrete slab, wall, or other component. The design rules for these composite connections are not as well-defined as those for most steel connections. However, there is general agreement that the connections should be designed for the largest forces to be transferred at the interface. Also, the design should recognize that large groups of shear connectors or other transfer elements do not necessarily behave as the sum of the individual elements.

Braced frames, which are economical systems for resisting both wind and earthquake loadings, frequently require large gusset plates in connections (Fig. 9.22). Different models for predicting the resistance of these connections may produce very different results and further research is needed to define their behavior. Hence, it is likely that there will be continuing changes in the design models for connections for lateral-load design. (See Sec. 5.)

Models used for design of connections should satisfy the equations of equilibrium, ensure support for maximum loads and deformations that are possible for connections, and recognize that large groups of connectors do not always behave as the sum of the individual connectors.

SECTION 10
APPLICATION OF DESIGN CRITERIA FOR BRIDGES

Robert L. Nickerson*

Consultant—NBE, Ltd., Hampstead, Maryland

The purpose of this section is to provide guidance to bridge designers for application of standard design specifications to the more common types of bridges and to provide rules of thumb to assist in obtaining cost-effective and safe structures. Because of the complexity of modern specifications for bridge design and construction and the large number of standards and guides with which designers must be familiar to ensure adequate designs, this section does not provide comprehensive treatment of all types of bridges. Because specifications are continually being revised, readers are cautioned to use the latest edition in practical applications.

10.1 STANDARD SPECIFICATIONS FOR BRIDGES

Designs for most highway bridges in the United States are governed by the "Standard Specifications for Highway Bridges" of the American Association of State Highway and Transportation Officials (AASHTO), 444 N. Capitol St., NW, Washington, DC 20001. AASHTO updates these specifications annually. Necessary revisions are published as "Interim Specifications." A new edition of the Standard Specifications is published about every fourth year and incorporates intervening "Interim Specifications." The design criteria for highway bridges in this section are based on the 15th (1992) edition.

For complex design-related items or modifications involving new technology, AASHTO issues tentative "Guide Specifications" to allow further assessment and refinement of the new criteria. AASHTO may adopt a "Guide Specification," after a trial period of use, as part of the Standard Specifications.

State highway departments usually adopt the AASHTO bridge specifications as their minimum standards for highway bridge design. Because conditions vary from state to state, however, many bridge owners modify the standard specifications to

---

*Revised Sec. 10, originally written by Frank D. Sears, Bridge Division, Federal Highway Administration, Washington, D.C. Information on railway bridges was added by Frederick S. Merritt.

meet specific needs. For example, California has specific requirements for earthquake resistance that may not be appropriate for many east-coast structures.

Recommended practices for design of railway bridges are promulgated by the American Railway Engineering Association (AREA), 2000 L St., NW, Washington, DC 20036. These practices, along with principles, data, specifications, plans, and economics pertaining to design and construction of the fixed plant of railways, are incorporated in the "AREA Manual for Railway Engineering." Issued in looseleaf from, the manual is kept current by issuance of annual supplements.

To ensure safe, cost-effective, and durable structures, designers should meet the requirements of the latest specifications and guides available. For unusual types of structures, or for bridges with spans longer than about 500 ft, designers should make a more detailed application of theory and performance than is possible with standard criteria or the practices described in this book. Use of much of the standard specifications, however, is appropriate for unusual structures, inasmuch as these generally are composed of components to which the specifications are applicable.

10.2 DESIGN METHODS FOR HIGHWAY BRIDGES

AASHTO "Standard Specifications for Highway Bridges" present two design methods for steel bridges: service-load, or allowable-stress, design (ASD) and strength, or load-factor, design (LFD). Both are being replaced by load-and-resistance-factor design (LRFD). This method utilizes factors based on the theory of reliability and statistical knowledge of load and material characteristics. (See also Sec. 6.) It identifies methods of modeling and analysis. It incorporates many of the existing AASHTO "Guide Specifications." Also, it includes features that are equally applicable to ASD and LFD that are not in past editions of the Standard Specifications. For example, the LRFD specifications include serviceability requirements for durability of bridge materials, inspectability of bridge components, maintenance that includes deck-replacement considerations in adverse environments, constructability, ridability, economy, and esthetics. Although procedures for ASD are presented in following articles, LFD or LRFD may often yield more economical results.

10.3 PRIMARY DESIGN CONSIDERATIONS

The primary purpose of a highway bridge is to carry safely (geometrically and structurally) the necessary traffic volumes and loads. Normally, traffic volumes, present and future, determine the number and width of traffic lanes, establish the need for, and width of, shoulders, and the minimum design truck weight. These requirements are usually established by the owner's planning and highway design section using the roadway design criteria contained in "A Policy on Geometric Design of Highways and Streets," American Association of State Highway and Transportation Officials. If lane widths, shoulders, and other pertinent dimensions are not established by the owner, this AASHTO Policy should be used for guidance. Ideally, bridge designers will be part of the highway design team to ensure that unduly complex bridge geometric requirements, or excessive bridge lengths are not generated during the highway-location approval process.

Traffic considerations for bridges are not necessarily limited to overland vehicles. In many cases, ships and construction equipment must be considered. Requirements for safe passage of extraordinary traffic over *and* under the structure may impose additional restrictions on the design that could be quite severe.

Past AASHTO "Standard Specifications for Highway Bridges" did not contain requirements for design for a specified service life for bridges. It has been assumed that, if the design provisions are followed, proper materials are specified, a quality assurance procedure is in place during construction, and adequate maintenance is performed, an acceptable service life will be achieved. An examination of the existing inventory of steel bridges throughout the United States indicates this to be generally true, although there are examples where service life is not acceptable. The predominant causes for reduced service life are geometric deficiencies because of increases in traffic that exceed the original design-traffic capacity. The LRFD specification addresses service life by requiring design and material considerations that will achieve a 75-year design life.

10.3.1 Deflection Limitations

In general, highway bridges consisting of simple or continuous spans should be designed so that deflection due to live load plus impact should not exceed $\frac{1}{800}$ the span. For bridges available to pedestrians in urban areas, this deflection should be limited to $\frac{1}{1000}$ the span. For cantilevers, the deflection should generally not exceed $\frac{1}{300}$ the cantilever arm, or $\frac{1}{375}$ where pedestrian traffic may be carried. (See also Art. 10.18.) In LRFD, these limits are optional.

Live-load deflection computations for beams and girders should be based on gross moment of inertia of cross section, or of transformed section for composite girders. For a truss, deflection computations should be based on gross area of each member, except for sections with perforated cover plates. For such sections, the effective area (net volume divided by length center to center of perforations) should be used.

Railroad-Bridge Deflections. Deflection requirements for railroad bridges are similar to those for highway bridges except that simple-span deflections are limited to $\frac{1}{640}$ the span under live load plus impact. Computations should be based on the loading that produces maximum stress at midspan.

10.3.2 Stringers and Floorbeams

Stringers are beams generally placed parallel to the longitudinal axis of the bridge, or direction of traffic, in highway bridges, such as truss bridges. Usually, they should be framed into floorbeams. But if they are supported on the top flanges of the floorbeams, it is desirable that the stringers be continuous over two or more panels. In bridges with wood floors, intermediate cross frames or diaphragms should be placed between stringers more than 20 ft long.

In skew bridges without end floorbeams, the stringers, at the end bearings, should be held in correct position by end struts also connected to the main trusses or girders. Lateral bracing in the end panels should be connected to the end struts and main trusses or girders.

Floorbeams preferably should be perpendicular to main trusses or girders. Also, connections to those members should be positioned to permit attachment of lateral bracing to both floorbeam and main truss or girder.

Main material of floorbeam hangers should not be coped or notched. Built-up hangers should have solid or perforated web plates or lacing.

All truss and girder spans should have end floorbeams. They should be designed to permit lifting the superstructure, for example, with jacks, without producing stresses exceeding the basic allowable stresses by more than 50%. Sufficient clearance should be provided at each end floorbeam to permit painting its side adjacent to the abutment back wall.

End connection angles of floorbeams and stringers should be at least ⅜ in thick in highway bridges. Each end connection for floorbeams and stringers generally should be made with two angles. These angles should be as long as the flanges will permit. If bracket or shelf angles are used during erection to furnish support, their load-carrying capacity should be ignored in determining the number of fasteners required to transmit end shear. End connections should be welded or made with high-strength bolts.

10.4 HIGHWAY DESIGN LOADINGS

The AASHTO "Standard Specifications for Highway Bridges" require bridges to be designed to carry dead and live loads and impact, or the dynamic effect of the live load. Structures should also be capable of sustaining other loads to which they may be subjected, such as longitudinal, centrifugal, thermal, seismic, and erection forces.

Dead Loads.   Designers should use the actual deadweights of materials specified for the structure. For the more commonly used materials, the Standard Specifications provide the weights to be used. For other materials, designers must determine the proper design loads. It is important that the dead loads used in design be noted on the contract plans for analysis purposes during possible future rehabilitations.

Live Loads.   There are four standard classes of highway vehicle loadings included in the Standard Specifications: H15, H20, HS15, and HS20. The AASHTO "Geometric Guide" states that the minimum design loading for new bridges should be HS20 (Fig. 10.1a) for all functional classes (local roads through freeways) of highways. Therefore, most bridge owners require design for HS20 truck loadings or greater. AASHTO also specifies an alternative loading of two 24-kip axles spaced 4 ft c to c.

The difference in truck gross weights is a direct ratio of the HS number; e.g., HS15 is 75% of HS20. (The difference between the H and HS trucks is the use of a third axle on an HS truck.) Many bridge owners, recognizing the trucking industries' use of heavier vehicles, are specifying design loadings greater than HS20.

For longer-span bridges, lane loadings are used to simulate multiple vehicles in a given lane. For example, for HS20 loading on a simple span, the lane load is 0.64 kips per ft plus an 18-kip concentrated load for moment and a 26-kip load for shear. A simple-span girder bridge with a span longer than about 140 ft would be subjected to a greater live-load design moment for the lane loading than for the truck loading (Table 10.5). (For end shear and reaction, the breakpoint is about 120 ft). Truck and lane loadings are not applied concurrently for ASD or LFD.

In ASD and LFD, if maximum stresses are induced in a member by loading of more than two lanes, the live load for three lanes should be reduced by 10%, and for four or more lanes, 25%. For LRFD, a reduction or increase depends on the method for live-load distribution.

For LRFD, the design vehicle design load is a combination of truck or tandem or lane loads and differs for positive and negative moment. Figure 10.1b and c shows the governing live loads for LRFD to produce maximum positive moment in a beam.

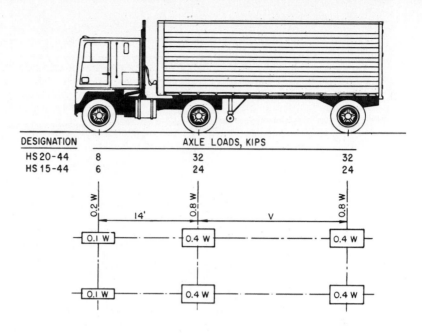

DESIGNATION	AXLE LOADS, KIPS		
HS 20-44	8	32	32
HS 15-44	6	24	24

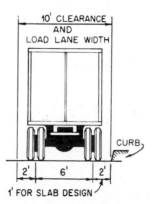

(a) STANDARD HS TRUCKS

FIGURE 10.1 Standard HS loadings for design of highway bridges. (*a*) Truck loading for ASD and LFD. *W* is the combined weight of the first two axles. *V* is the spacing of the axles, between 14 and 30 ft, inclusive, that produces maximum stresses.

Impact.   A factor is applied to vehicular live loads to represent increases in loading due to impact caused by a rough roadway surface or other disturbance. In the AASHTO Standard Specifications, the impact factor *I* is a function of span and is determined from

$$I = \frac{50}{L + 125} \le 0.30 \qquad (10.1)$$

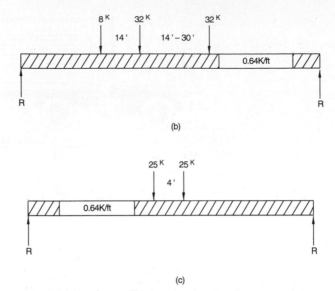

FIGURE 10.1 (*Continued*) (*b*) Truck and lane loading for load-and-resistance-factor design (LRFD). (*c*) Alternative loading for LRFD to be used if it produces greater stresses.

In this formula, L, ft, should be taken as follows:

	For moment	For shear
For simple spans	L = design span length for roadway decks, floorbeams, and longitudinal stringers	L = length of loaded portion from point of consideration to reaction
For cantilevers	L = length from point of consideration to farthermost axle	Use I = 0.30
For continuous spans	L = design length of span under consideration for positive moment; average of two adjacent loaded spand for negative moment	L = length as for simple spans

For LRFD, the impact factor is modified in recognition of the concept that the factor should be based on the type of bridge component, rather than the span. Table 10.1 gives the impact factors for LRFD. The impact factor is applied only to the truck portion of the live load.

Live Loads on Bridge Railings. Beginning in the 1960s, AASHTO specifications increased minimum design loadings for railings to a 10-kip load applied horizontally, intended to simulate the force of a 4000-lb automobile traveling at 60 mph and impacting the rail at a 25° angle. In 1989, AASHTO published AASHTO "Guide

TABLE 10.1 Impact Factors for Highway Bridges for LRFD

Component	Limit state	Impact factor, %
Deck joints	All	75
Components of bridge deck	All	50
All other components	Fatigue and fracture	15
	All	33

Specifications for Bridge Railings" requirements more representative of current vehicle impact loads and dependent on the class of highway crossing bridges. Since the effect of impact-type loadings are difficult to predict, the AASHTO Guide requires that railings be subjected to full-scale impact tests to a performance level *PL* that is a function of the highway type, design speed, percent of trucks in traffic, and bridge-rail offset. Generally, only low-volume, rural roads may utilize a rail tested to the PL-1 level, and high-volume interstate routes require a PL-3 rail. The full-scale tests apply the forces that must be resisted by the rail and its attachment details to the bridge deck.

PL-1 represents the forces delivered by an 1800-lb automobile traveling at 50 mph, or a 5400-lb pickup truck at 45 mph, and impacting the rail system at an angle of 20°. PL-2 represents the forces delivered from an automobile or pickup as in PL-1, but traveling at a speed of 60 mph, in addition to an 18,000-lb truck at 50 mph at an angle of 15°. PL-3 represents forces from an automobile or pickup as in PL-2, in addition to a 50,000-lb van-type tractor-trailer traveling at 50 mph and impacting at an angle of 15°.

The performance criteria require not only resistance to the vehicle loads but also acceptable performance of the vehicle after the impact. The vehicle may not penetrate or hurdle the railing, must remain upright during and after the collision, and be smoothly redirected by the railing. Thus, a rail system that can withstand the impact of a tractor-trailer truck, may not be acceptable if redirection of a small automobile is not satisfactory.

Because of the time and expense involved in full-scale testing, it is advantageous to specify previously tested and approved rails. State highway departments may provide these designs on request.

Earthquake Loads.   Seismic design is governed by the AASHTO "Standard Specifications for Seismic Design of Highway Bridges." Engineers should be familiar with the total content of these complex specifications to design adequate earthquake-resistant structures. These specifications are also the basis for the earthquake "extreme-event" limit state of the LRFD specifications, where the intent is to allow the structure to suffer damage but have a low probability of collapse during seismically induced ground shaking.

The purpose of the seismic design specifications is to ". . . establish design and construction provisions for bridges to minimize their susceptibility to damage from earthquakes." Each structure is assigned to a seismic performance category (SPC), which is a function of location relative to anticipated design ground accelerations and to the importance classification of the highway routing. The SPC assigned, in conjunction with factors based on the site soil profile and response modification factor for the type of structure, establishes the minimum design parameters that must be satisfied.

A loading that should be considered by designers for bridges that cross navigable waters is that induced by impact of large ships. Guidance for consideration of vessel impacts on a bridge is included in the AASHTO "Guide Specification and Commentary for Vessel Collision Design of Highway Bridges." This Guide Specification is based on probabilistic theories, accounting for differences in size and frequency of ships that will be using a waterway. The Guide is also the basis for the LRFD extreme-event limit state for vessel collision.

Thermal Loads.   Provisions must be included in bridge design for stresses *and* movements resulting from temperature variations to which the structure will be subjected. For steel structures, anticipated temperature extremes are as follows:

Moderate climate: 0 to 120°F

Cold climate: −30°F to +120°F

With a coefficient of expansion of 65×10^{-7} in/in/°F, the resulting change in length of a 100-ft-long bridge member is

Moderate climate: $120 \times 65 \times 10^{-7} \times 100 \times 12 = 0.936$ in

Cold climate: $150 \times 65 \times 10^{-7} \times 100 \times 12 = 1.170$ in

If a bridge is erected at the average of high and low temperatures, the resulting change in length will be one-half of the above.

For complex structures such as trusses and arches, length changes of individual members may induce secondary stresses that must be taken into account.

Longitudinal Forces.   Roadway decks are subjected to braking forces, which they transmit to supporting members. AASHTO Standard Specifications specify a longitudinal design force of 5% of the live load in all lanes carrying traffic in the same direction, without impact. The force should be assumed to act 6 ft above the deck.

Centrifugal Force on Highway Bridges.   Curved structures will be subjected to centrifugal forces by the live load. The force *CF*, as a percentage of the live load, without impact, should be applied 6 ft above the roadway surface, measured at centerline of the roadway.

$$CF = \frac{6.68S^2}{R} = 0.00117S^2D \qquad (10.2)$$

where S = design speed, mph
$\quad D$ = degree of curve = $5{,}729.65/R$
$\quad R$ = radius of curve, ft

Sidewalk Loadings.   In the interest of safety, many highway structures in non-urban areas are designed so that the full shoulder width of the approach roadway is carried across the structure. Thus, the practical necessity for a sidewalk or a refuge walk is eliminated.

In urban areas, however, structures should conform to the configuration of the approach roadways. Consequently, bridges normally require curbs or sidewalks, or both.

In these instances, sidewalks and supporting members should be designed for a live load of 85 psf. Girders and trusses should be designed for the following sidewalk live loads, lb per sq ft of sidewalk area:

Spans 0 to 25 ft 85
Spans 26 to 100 ft 60

$$\text{Spans over 100 ft} \ldots \ldots \ldots P = \left(30 + \frac{3{,}000}{L}\right)\frac{55 - W}{50} \le 60$$

where L = loaded length, ft and W = sidewalk width, ft.

There is no practical necessity that refuge walks on highway structures exceed 2 ft in width. Consequently, no live load need be applied. Current safety standards eliminate refuge walks on full-shoulder-width structures.

Structures designed for exclusive use of pedestrians should be designed for 85 psf.

Curb Loading.   Curbs should be designed to resist a lateral force of at least 0.50 kip per lin ft of curb. This force should be applied at the top of the curb or 10 in above the bridge deck if the curb is higher than 10 in.

Where sidewalk, curb, and traffic rail form an integral system, the traffic railing loading applies. Stresses in curbs should be computed accordingly.

Wind Loading on Highway Bridges.   The wind forces prescribed below are considered a uniformly distributed, moving live load. They act on the exposed vertical surfaces of all members, including the floor system and railing as seen in elevation, at an angle of 90° with the longitudinal axis of the structure. These forces are presumed for a wind velocity of 100 mph. They may be modified in proportion to the square of the wind velocity if conditions warrant change.

Superstructure.   For trusses and arches: 75 psf but not less than 0.30 kip per lin ft in the plane of loaded chord, nor 0.15 kip per lin ft in the plane of unloaded chord.

For girders and beams: 50 psf but not less than 0.30 kip per lin ft on girder spans.

Wind on Live Load.   A force of 0.10 kip per lin ft shall be applied to the live load, acting 6 ft above the roadway deck.

Substructure.   To allow for the effect of varying angles of wind in design of the substructure, the following longitudinal and lateral wind loads for the skew angles indicated should be assumed acting on the superstructure at the center of gravity of the exposed area.

When acting in combination with live load, the wind forces given in Table 10.2 may be reduced 70%. But they should be combined with the wind load on the live load, as given in Table 10.3.

TABLE 10.2   Skewed Superstructure Wind Forces for Substructure Design*

Skew angle of wind, deg	Trusses		Girders	
	Lateral load, psf	Longitudinal load, psf	Lateral load, psf	Longitudinal load, psf
0	75	0	50	0
15	70	12	44	6
30	65	28	41	12
45	47	41	33	16
60	25	50	17	19

*"Standard Specifications for Highway Bridges," American Association of State Highway and Transportation Officials.

TABLE 10.3 Wind Forces on Live Loads for
Substructure Design*

Skew angle of wind, deg	Lateral load, lb per lin ft	Longitudinal load, lb per lin ft
0	100	0
15	88	12
30	82	24
45	66	32
60	34	38

*"Standard Specifications for Highway Bridges," American
Association of State Highway and Transportation Officials.

For usual girder and slab bridges with spans not exceeding about 125 ft, the
following wind loads on the superstructure may be used for substructure design in
lieu of the more elaborate loading specified in Tables 10.2 and 10.3:

Wind on structure
  50 psf transverse
  12 psf longitudinal

Wind on live load
  100 psf transverse
  40 psf longitudinal

Transverse and longitudinal loads should be applied simultaneously.

Wind forces applied directly to the substructure should be assumed at 40 psf
for 100-mph wind velocity. For wind directions skewed to the substructure, this
force may be resolved into components perpendicular to end and side elevations,
acting at the center of gravity of the exposed areas. This wind force may be reduced
70% when acting in combination with live load.

Overturning Forces.   In conjunction with forces tending to overturn the structure, there should be added an upward wind force, applied at the windward quarter
point of the transverse superstructure width, of 20 psf, assumed acting on the deck
and sidewalk plan area. For this load also, a 70% reduction may be applied when
it acts in conjunction with live load.

Uplift on Highway Bridges.   Provision should be made to resist uplift by adequately attaching the superstructure to the substructure. AASHTO specifications
recommend engaging a mass of masonry equal to:

1. 100% of the calculated uplift caused by any loading or combination of loading
in which the live-plus-impact loading is increased 100%.
2. 150% of the calculated uplift at working-load level.

Anchor bolts under the above conditions should be designed at 150% of the
basic allowable stress.

Forces of Stream Current, Ice, and Drift on Highway Bridges.   All piers and
other portions of structures should be designed to resist the maximum stresses
induced by the forces of flowing water, floating ice, or drift.

The pressure of ice on piers should be calculated at 400 psi. The thickness of ice and height at which it applies should be determined by an investigation of the structure site.

The pressure P, psf, of flowing water on piers should be calculated from

$$P = KV^2 \qquad (10.3)$$

where V = velocity of water, fps, and K = constant ($1\frac{3}{8}$ for square ends, $\frac{1}{2}$ for angle ends where the angle is 30° or less, and $\frac{2}{3}$ for circular piers).

Buoyancy should be taken into account in the design of substructures, including piling, and of superstructures, where necessary.

Earth Pressure on Highway Bridges.   Structures that retain fills should be proportioned to withstand pressure as given by Rankine's formula. These structures, however, should be designed for an equivalent fluid pressure of at least 30 lb per cu ft.

For rigid frames, a maximum of one-half of the bending moments caused by lateral earth pressure may be used to reduce the positive moment in the beams, in the top slab, or in the top and bottom slab, as the case may be.

When highway traffic can come within a horizontal distance from the top of the structure equal to one-half its height, the earth pressure should have added to it a live-load surcharge pressure equal to at least 2 ft of earth. Where an adequately designed reinforced-concrete approach slab, supported at one end by the bridge, is provided, no live-load surcharge need be considered.

All designs should provide for the thorough drainage of backfilling material. Acceptable means include weep holes and crushed rock, pipe drains or gravel drains, and perforated drains.

10.5   LOAD COMBINATIONS AND EFFECTS FOR HIGHWAY BRIDGES

The following groups represent various combinations of service loads and forces to which a structure may be subjected. Every component of substructure and superstructure should be proportioned to resist all combinations of forces applicable to the type of bridge and its site.

For working-stress design, allowable unit stresses depend on the loading group, as indicated in Table 10.4. These stresses, however, do not govern for members subject to repeated stresses when allowable fatigue stresses are smaller. Note that no increase is permitted in allowable stresses for members carrying only wind loads. When the section required for each loading combination has been determined, the largest should be selected for the member being designed.

The "Standard Specifications for Highway Bridges" of the American Association of State Highway and Transportation Officials specifies for load-factor design factors to be applied to the various types of loads in loading combinations. These load factors are based on statistical analysis of loading histories. In addition, in LRFD, reduction factors are applied to the nominal resistance of materials in members and to compensate for various uncertainties in behavior.

To compare the effects of the design philosophies of ASD, LFD, and LRFD, the group loading requirements of the three methods will be examined. For simplification, only D, L, and I of Group I loading will be considered. All three

TABLE 10.4 Loading Combinations for Allowable-Stress Design

Group loading combination		Percentage of basic unit stress
I	$D + L + I + CF + E + B + SF$	100
IA	$D + 2(L + I)$	150
IB	$D + (L + I)^* + CF + E + B + SF$	†
II	$D + E + B + SF + W$	125
III	$D + L + I + CF + E + B + SF + 0.3W + WL + LF$	125
IV	$D + L + I + E + B + SF + T$	125
V	$D + E + B + SF + W + T$	140
VI	$D + I + CF + E + B + SF + 0.3W + WL + LF + T$	140
VII	$D + E + B + SF + EQ$	133
VIII	$D + L + I + CF + E + B + SF + ICE$	140
IX	$D + E + B + SF + W + ICE$	150
X‡	$D + L + I + E$	100

where D = dead load
L = live load
I = live-load impact
E = earth pressure (factored for some types of loadings)
B = buoyancy
W = wind load on structure
WL = wind load on live load of 0.10 kip per lin ft
LF = longitudinal force from live load
CF = centrifugal force
T = temperature
EQ = earthquake
SF = stream-flow pressure
ICE = ice pressure

*For overload live load plus impact as specified by the operating agency.
†Percentage $= \dfrac{\text{maximum unit stress (operating rating)}}{\text{allowable basic unit stress}} \times 100$
‡For culverts.

methods use the same general equation for determining the effects of the combination of loads:

$$N\Sigma(F \times \text{load}) \leq RF \times \text{nominal resistance} \qquad (10.4)$$

where
N = factor used in LRFD for ductility, redundancy, and operational importance of the bridge
    = 1.0 for ASD and LFD
$\Sigma(F \times \text{load})$ = sum of the factored loads for a combination of loads
F = load factor that is applied to a specific load
    = 1.0 for ASD; D, L, and I
load = one or more service loads that must be considered in the design
RF = resistance factor (safety factor for ASD) that is applied to a specific material
Nominal resistance = the strength of a material, based on the type of loading; e.g., tension, compression, or shear

For a flexural member subjected to bending by dead load, live load, and impact forces, let D, L, I represent the maximum tensile stress in the extreme surface due to dead load, live load, and impact, respectively. Then, for each of the design methods, the following must be satisfied:

ASD: $$D + L + I \leq 0.55F_y \qquad (10.5)$$

LFD: $$1.3D + 2.17(L + I) \leq F_y \qquad (10.6)$$

For strength limit state I,

LRFD: $$1.25D + 1.70(L + I) \leq F_y \qquad (10.7)$$

For LFD and LRFD, if the section is compact, the full plastic moment can be developed. Otherwise, the capacity is limited to the yield stress in the extreme surface.

The effect of the applied loads appears to be less for LRFD, but many other factors apply to LRFD designs that are not applicable to the other design methods. One of these is a difference in the design live-load model. Another major difference is that the LRFD specifications require checking of connections *and* components for minimum and maximum loadings. (Dead loads are to be varied by using a load factor of 0.9 to 1.25.) LRFD also requires checking for five different *strength* limit states, three *service* limit states, a *fatigue-and-fracture* limit state, and an *extreme-event* limit state. Although each structure may not have to be checked for all these limit states, the basic philosophy of the LRFD specifications is to assure serviceability over the design service life, safety of the bridge through redundancy and ductility of all components and connections, and survival (prevention of collapse) of the bridge when subjected to an extreme event; e.g., a 500-year flood. The limit states are described as follows:

Service limit states restrict stress, deformation, and crack width under regular service conditions.

Fatigue-and-fracture limit states limit the stress range under service conditions in accordance with the number of expected stress-range cycles.

Fracture limit states are taken as the material toughness requirements of the AASHTO Material Specifications.

Strength limit states ensure that stability and sufficient strength, both local and global, are provided to resist the statistically significant load combinations that a bridge will experience during its design life.

Extreme-event limit states ensure the structural survival of a bridge during a major earthquake or flood or when the structure is struck by a vessel, vehicle, or ice flow.

Example.   To compare the results of a design by ASD, LFD, and LRFD, a 100-ft, simple-span girder bridge is selected as a simple example. It has an 8-in-thick, noncomposite concrete deck, and longitudinal girders, made of grade 50 steel, spaced 12 ft c to c. It will carry HS20 live load. The section modulus S, in^3, will be determined for a laterally braced interior girder with a distribution factor of 1.0.

The bending moment due to dead loads is estimated to be about 2,200 ft-kips. The maximum moment due to the HS20 truck loading is 1,524 ft-kips (Table 10.5).

$$\text{Lane-load live-load moment} = \frac{wL^2}{8} = \frac{0.64(100)^2}{8} = 800 \text{ ft-kips}$$

TABLE 10.5 Maximum Moments, Shears, and Reactions for Truck or Lane Loads on One Lane, Simple Spans*

Span, ft	H15		H20		HS15		HS20	
	Moment†	End shear and end reaction‡	Moment†	End shear and end reaction‡	Moment†	End shear and end reaction‡	Moment†	End shear and end reaction‡
10	60.0§	24.0§	80.0§	32.0§	60.0§	24.0§	80.0§	32.0§
20	120.0§	25.8§	160.0§	34.4§	120.0§	31.2§	160.0§	41.6§
30	185.0§	27.2§	246.6§	36.3§	211.6§	37.2§	282.1§	49.6§
40	259.5§	29.1	346.0§	38.8	337.4§	41.4§	449.8§	55.2§
50	334.2§	31.5	445.6§	42.0	470.9§	43.9§	627.9§	58.5§
60	418.5	33.9	558.0	45.2	604.9§	45.6§	806.5§	60.8§
70	530.3	36.3	707.0	48.4	739.2§	46.8§	985.6§	62.4§
80	654.0	38.7	872.0	51.6	873.7§	47.7§	1,164.9§	63.6§
90	789.8	41.1	1,053.0	54.8	1,008.3§	48.4§	1,344.4§	64.5§
100	937.5	43.5	1,250.0	58.0	1,143.0§	49.0§	1,524.0§	65.3§
110	1,097.3	45.9	1,463.0	61.2	1,277.7§	49.4§	1,703.6§	65.9§
120	1,269.0	48.3	1,692.0	64.4	1,412.5§	49.8§	1,883.3§	66.4§
130	1,452.8	50.7	1,937.0	67.6	1,547.3§	50.7	2,063.1§	67.6
140	1,648.5	53.1	2,198.0	70.8	1,682.1§	53.1	2,242.8§	70.8
150	1,856.3	55.5	2,475.0	74.0	1,856.3	55.5	2,475.1	74.0
160	2,076.0	57.9	2,768.0	77.2	2,076.0	57.9	2,768.0	77.2
170	2,307.8	60.3	3,077.0	80.4	2,307.8	60.3	3,077.1	80.4
180	2,551.5	62.7	3,402.0	83.6	2,551.5	62.7	3,402.1	83.6
190	2,807.3	65.1	3,743.0	86.8	2,807.3	65.1	3,743.1	86.8
200	3,075.0	67.5	4,100.0	90.0	3,075.0	67.5	4,100.0	90.0
220	3,646.5	72.3	4,862.0	96.4	3,646.5	72.3	4,862.0	96.4
240	4,266.0	77.1	5,688.0	102.8	4,266.0	77.1	5,688.0	102.8
260	4,933.5	81.9	6,578.0	109.2	4,933.5	81.9	6,578.0	109.2
280	5,649.0	86.7	7,532.0	115.6	5,649.0	86.7	7,532.0	115.6
300	6,412.5	91.5	8,550.0	122.0	6,412.5	91.5	8,550.0	122.0

*Based on "Standard Specifications for Highway Bridges," American Association of State Highway and Transportation Officials. Impact not included.
†Moments in thousands of ft-lb (ft-kips).
‡Shear and reaction in kips. Concentrated load is considered placed at the support. Loads used are those stipulated for shear.
§Maximum value determined by standard truck loading. Otherwise, standard lane loading governs.

For both ASD and LFD, the impact factor is

$$I = \frac{50}{100 + 125} = 0.22$$

For LRFD, $I = 0.33$.

Allowable-Stress Design. The section modulus S for the girder for allowable-stress design is computed as follows: The design moment is

$$M = M_D + (1 + I)M_L = 2,200 + 1.22 \times 1,524 = 4,059 \text{ ft-kips}$$

For $F_y = 50$ ksi, the allowable stress is $F_b = 0.55 \times 50 = 27$ ksi. The section modulus required is then

$$S = \frac{M}{F_b} = \frac{4,059 \times 12}{27} = 1,804 \text{ in}^3$$

The section in Fig. 10.2, weighing 280.5 lb per ft, supplies a section modulus within 1% of required S—O.K.

Load-Factor Design. The design moment for LFD is

$$M_u = 1.3M_D + 2.17(1 + I)M_L$$

$$= 1.3 \times 2,200 + 2.17 \times 1.22 \times 1,524 = 6,895 \text{ ft-kips}$$

For $F_y = 50$ ksi, the section modulus required for LFD is

$$S = \frac{M_u}{F_y} = \frac{6,895 \times 12}{50} = 1,655 \text{ in}^3$$

If a noncompact section is chosen, this value of S is the required elastic section modulus. For a compact section, it is the plastic section modulus Z. Figure 10.3 shows a noncompact section supplying the required section modulus, with a ⅜-in-thick web and 1⅝-in-thick flanges. For a compact section, a ⅝-in-thick web is

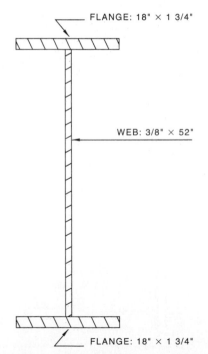

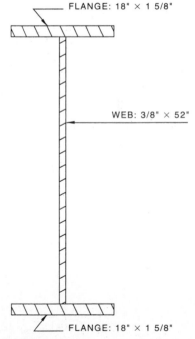

FIGURE 10.2 Girder with transverse stiffeners determined by ASD and LRFD for a 100-ft span: $S = 1799$ in³; $w = 280.5$ lb per ft.

FIGURE 10.3 Girder with transverse stiffeners determined by load-factor design for a 100-ft span: $S = 1681$ in³; $w = 265$ lb per ft.

required and 1¼-in-thick flanges are satisfactory. In this case, the noncompact girder is more efficient and will weigh 265 lb per ft.

Load-and-Resistance-Factor Design.   The live-load moment M_L is produced by a combination of truck and land loads, with impact applied only to the truck moment:

$$M_L = 1.33 \times 1524 + 800 = 2827 \text{ ft-kips}$$

The load factor N is a combination of factors applied to the loadings. Assume that the bridge has ductility (0.95), redundancy (0.95), and to is of operational importance (1.05). Thus, $N = 0.95 \times 0.95 \times 1.05 = 0.95$. The design moment for limit state I is

$$M_u = N[F_D M_D + F_L M_L]$$

$$= 0.95[1.25 \times 2200 + 1.75 \times 2827] = 7312 \text{ ft-kips}$$

Hence, the section modulus required for LRFD is

$$S = \frac{7312 \times 12}{50} = 1755 \text{ in}^3$$

The section selected for ASD (Fig. 10.2) is satisfactory for LRFD.
    For this example, the weight of the girder for LFD is 94% of that required for ASD and 90% of that needed for LRFD. The heavier girder required for LRFD is primarily due to the larger live load specified. For both LFD and LRFD, a compact section is advantageous, because it reduces the need for transverse stiffeners for the same basic weight of girder.

10.6   RAILROAD DESIGN LOADINGS

Figure 10.4 shows a part cross section of a typical through-girder railway bridge. Train wheel loads are transmitted from rails through ties to stringers, which carry the loads to transverse floorbeams and thence to girders.
    For structures with spans up to 400 ft, the American Railway Engineering Association (AREA) recommends that design be based on Cooper E80 loading (Fig. 10.5).
    Like highway loadings, railroad loadings have been increased over the years to accommodate new trends in transportation, and it is likely that this trend will

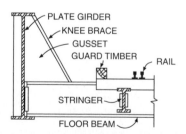

FIGURE 10.4   Part section through railway bridge.

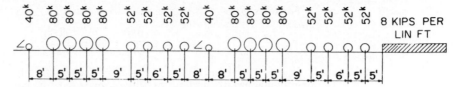

FIGURE 10.5 Cooper E80 loading for design of railway bridges.

continue in the future. Heavier Cooper E loadings will result in directly proportional increases in the concentrated and uniform loadings shown in Fig. 10.5.

To account for the effect of multiple tracks on a structure, the proportions of full live load on the tracks may be taken as

Two tracks .... Full live load
Three tracks...... Full live load on two tracks, one-half full live load on the third track
Four tracks .... Full live load on two tracks, one-half on one track, and one-quarter on the remaining track

Stability.   For spans and towers, stability should be investigated with live load on only one track, the leeward one for bridges with more than one track. The live load should be 1.2 kips per lin ft, without impact.

Impact on Railroad Bridges.   Impact loads should be applied at top of rail. Impact I, expressed as a percentage of axle loads, is given for open-deck bridges by Eqs. (10.8a and b) to (10.11) and modified by a factor determined by the number of tracks to be supported. For ballasted decks, a 10% reduction is permitted.

For rolling equipment without hammer blow (diesels, electric locomotives, tenders alone, etc.):

For $L < 80$ ft:

$$I = \frac{100}{S} + 40 - \frac{3L^2}{1600} \qquad (10.8a)$$

For $L \geq 80$ ft:

$$I = \frac{100}{S} + 16 + \frac{600}{L - 30} \qquad (10.8b)$$

For steam locomotives with hammer blow: *For beam spans, stringers, girders, floorbeams, posts of deck-truss spans carrying load from floorbeams only, and floorbeam hangers:*

For $L < 100$ ft:

$$I = \frac{100}{S} + 60 - \frac{L^2}{500} \qquad (10.9)$$

For $L \geq 100$ ft:

$$I = \frac{100}{S} + 10 + \frac{1800}{L - 40} \qquad (10.10)$$

For truss spans:

$$I = \frac{100}{S} + 15 + \frac{4000}{L + 25} \qquad (10.11)$$

In Eqs. (10.8*a* and *b*) to (10.11), S = distance, ft, c to c of single or groups of longitudinal beams, girders, or trusses, or length, ft, between supports of floorbeams or transverse girders; L = length, ft, center to center of supports for stringers, transverse floorbeams without stringers, longitudinal girders and trusses (main members), or length, ft, of the longer adjacent supported stringers, longitudinal beam, girder or truss for impact in floorbeams, floorbeam hangers, subdiagonals of trusses, transverse girders, supports for longitudinal and transverse girders and viaduct columns.

For members receiving load from more than one track, the impact percentage to be applied to the static live load is determined by the number of tracks:

When load is received from two tracks:

For $L < 175$ ft:

Full impact on two tracks

For 175 ft $\le L \le 225$ ft:

Full impact on one track and a percentage of full
impact on the other as given by $450 - 2L$

For $L > 225$ ft:

Full impact on one track and none on the other

When load is received from more than two tracks:

For all values of L:

Full impact on any two tracks

Longitudinal Forces on Railroad Bridges.   The longitudinal force from trains should be taken as 15% of the live load without impact.

Where rails are continuous (either welded or bolted joints) across the entire bridge from embankment to embankment, the effective longitudinal force should be taken as $L/1200$ times the force specified above, where L is the length of the bridge, ft, and $L/1200 \le 0.80$.

Where rails are not continuous, but are interrupted by a movable span, sliding rail expansion joints, or other devices, across the entire bridge from embankment to embankment, the effective longitudinal force should be taken as 15% of live load.

The effective longitudinal force should be taken on one track only. It should be distributed to the various components of the supporting structure, taking into account their relative stiffnesses, where appropriate, and the type of bearings.

The effective longitudinal force should be assumed to be applied at base of rail.

Centrifugal Forces on Railroad Bridges.   On curves, a centrifugal force corresponding to each axle load should be applied horizontally through a point 6 ft above the top of rail. This distance should be measured in a vertical plane along a line that is perpendicular to and at the midpoint of a radial line joining the tops of the rails. This force should be taken as a percentage C of the specified axle load without impact.

$$C = 0.00117S^2D \tag{10.12}$$

where S = train speed, mph
$\quad\quad D$ = degree of curve = $5729.65/R$
$\quad\quad R$ = radius of curve, ft

When the superelevation is 3 in less than that at which the resultant flange pressure between wheel and rail is zero,

$$C = 0.00117S^2D = 1.755(E + 3) \tag{10.13}$$

In that case, the actual superelevation, in, is given by

$$E = \frac{S^2D}{1500} - 3 = \frac{C - 5.265}{1.755} \tag{10.14}$$

and the permissible speed, mph, by

$$S = \sqrt{\frac{1500(E + 3)}{D}} \tag{10.15}$$

On curves, each axle load on each track should be applied vertically through the point defined above, 6 ft above top of rail. Impact should be computed and applied as indicated previously.

Preferably, the section of the stringer, girder, or truss on the high side of the superelevated track should be used also for the member on the low side, if the required section of the low-side member is smaller than that of the high-side member.

If the member on the low side is computed for the live load acting through the point of application defined above, impact forces need not be increased. Impact forces may, however, be applied at a value consistent with the selected speed, in which case the relief from centrifugal force acting at this speed should also be taken into account.

Lateral Forces from Equipment.   For bracing systems or for longitudinal members entirely without a bracing system, the lateral force to provide for the effect of the nosing of equipment, such as locomotives (in addition to the other lateral forces specified), should be a single moving force equal to 25% of the heaviest axle load. It should be applied at top of rail. This force may act in either lateral direction, at any point of the span.

On spans supporting multiple tracks, the lateral force from only one track should be used. Resulting vertical forces should be disregarded.

Wind Loading on Railroad Bridges.   AREA specifications require a wind load of 50 psf on unloaded bridges acting on the following surfaces:

Girder spans: 1½ times vertical projection

Truss spans: vertical projection of span plus any portion of leeward truss not shielded by the floor system

Viaduct towers and bents: vertical projection of all columns and tower bracing

On loaded bridges, a wind load of 30 psf, acting as described above, should be applied with a wind load of 0.30 kip per lin ft acting on the live load of one track at a distance 8 ft above top of rail. On girder and truss spans, the wind force should be at least 0.20 kip per lin ft for the loaded chord or flange, and 0.15 kip per lin ft for the unloaded chord or flange.

Loading Combinations on Railroad Bridges.   Every component of substructure and superstructure should be proportioned to resist all combinations of forces applicable to the type of bridge and its site. Members subjected to stresses from dead, live, impact, and centrifugal forces should be designed for the smaller of the basic allowable unit stress or the allowable fatigue stress.

With the exception of floorbeam hangers, members subject to stresses from other lateral or longitudinal forces, as well as to the dead, live, impact, and centrifugal forces, may be proportioned for 125% of the basic allowable unit stresses, without regard for fatigue. But the section should not be smaller than that required with basic unit stresses or allowable fatigue stresses when those lateral or longitudinal forces are not present. Note that there are two loading cases for wind: 50-psf wind on the unloaded bridge, or 0.30 kip per lin ft on one track and 30 psf on the bridge.

10.7 DISTRIBUTION OF HIGHWAY LOADS THROUGH DECKS

The "Standard Specifications for Highway Bridges" of the American Association of State Highway and Transportation Officials (AASHTO) requires that the width of a bridge roadway between curbs be divided into design traffic lanes 12 ft wide and loads located to produce maximum stress in supporting members. (Fractional parts of design lanes are not used.) Roadway widths from 20 to 24 ft, however, should have two design lanes, each equal to one-half the roadway width. Truck and lane loadings are assumed to occupy a width of 10 ft placed anywhere within the design lane to produce maximum effect. The following provisions are applicable to types of steel highway bridges commonly used.

If curbs, railings, and wearing surfaces are placed after the concrete deck has gained sufficient strength, their weight may be distributed equally to all stringers or beams. Otherwise, the dead load on the outside stringer or beam is the portion of the slab it carries.

The strength and stiffness of the deck determine, to some extent, the distribution of the live load to the supporting framing.

Shear.   For determining end shears and reactions, the deck may be assumed to act as a simple span between beams for lateral distribution of the wheel load. For shear elsewhere, the wheel load should be distributed by the method required for bending moment.

Moments in Longitudinal Beams.   To each interior longitudinal beam, the fraction of a wheel load listed in Table 10.6 should be applied for computation of live-load bending moments.

For an outer longitudinal beam, the live-load bending moments should be determined with the reaction of the wheel load when the deck is assumed to act as a simple span between beams. When four or more longitudinal beams carry a concrete deck, the fraction of a wheel load carried by an outer beam should be at least $S/5.5$ when the distance between that beam and the adjacent interior beam S, ft, is 6 or less. For $6 < S < 14$, the fraction should be at least $S/(4 + 0.25S)$. For $S > 14$, no minimum need be observed.

Moments in Transverse Beams. When a deck is supported directly on floorbeams, without stringers, each beam should receive the fraction of a wheel load listed in Table 10.7, as a concentrated load, for computation of live-load bending moments.

Distribution for LRFD. Research has led to recommendations for changes in the distribution factors DF in Tables 10.6 and 10.7. AASHTO has adopted these recommendations in a "Guide Specification" and as an approximate method in the LRFD Specifications, when a bridge meets specified requirements. As an alternative, a more refined method such as finite-element analysis is permitted.

As a simple example of results obtained with the older and newer methods, consider a bridge with a concrete deck and steel beams spaced 12 ft c to c. From Table 10.6, DF = 12/5.5 = 2.18 wheels per interior stringer. From the LRFD specification,

$$DF = 0.15 + (S/3)^{0.6}(S/L)^{0.2}(K_g/12Lt_s^3)^{0.1} \tag{10.16}$$

where S = beam spacing, ft
$\quad L$ = span, ft
$\quad t_s$ = thickness of concrete slab, in
$\quad K_g = n(I + Ae_g^2)$
$\quad n$ = modular ratio = ratio of steel modulus of elasticity E_s to the modulus of elasticity E_c of the concrete slab
$\quad I$ = moment of inertia, in^4, of the beam
$\quad A$ = area, in^2, of the beam
$\quad e_g$ = distance, in, from neutral axis of beam to center of gravity of concrete slab

TABLE 10.6 Fraction of Wheel Load DF Distributed to Longitudinal Beams*

Deck	Bridge with one traffic lane	Bridge with two or more traffic lanes
Concrete:		
On I-shaped steel beams	$S/7, S \le 10$†	$S/5.5, S \le 14$†
On steel box girders	$W_L = 0.1 + 1.7R + 0.85/N_w$‡	
Steel grid:		
Less than 4 in thick.	$S/4.5$	$S/4$
4 in or more thick	$S/6, S \le 6$†	$S/5, S \le 10.5$†
Timber:		
Plank	$S/4$	$S/3.75$
Strip 4 in thick or multiple-layer floors over 5 in thick.	$S/4.5$	$S/4$
Strip 6 in or more thick	$S/5, S \le 5$†	$S/4.25, S \le 6.5$†

*Based on "Standard Specifications for Highway Bridges," American Association of State Highway and Transportation Officials.
†For larger values of S, average beam spacing, ft, the load on each beam should be the reaction of the wheel loads with the deck assumed to act as a simple span between beams.
‡Provisions for reduction of live load do not apply to design of steel box girders with W_L, fraction of a wheel (both front and rear).

$\quad R$ = number of design traffic lanes N_w divided by number of box girders ($0.5 \le R \le 1.5$)
$\quad N_w = W_c/12$, reduced to nearest whole number
$\quad W_c$ = roadway width, ft, between curbs or barriers if curbs are not used.

TABLE 10.7 Fraction of Wheel Load Distributed to Transverse Beams*

Deck	Fraction per beam
Concrete	$S/6$†
Steel grid:	
Less than 4 in thick	$S/4.5$
4 in or more thick	$S/6$†
Timber:	
Plank	$S4$
Strip 4 in thick, wood block on 4-in plank subfloor, or multiple- layer floors more than 5 in thick............................	$S/4.5$
Strip 6 in or more thick	$S/5$†

*Based on "Standard Specifications for Highway Bridges," American Association of State Highway and Transportation Officials.
†When the spacing of beams S, ft, exceeds the denominator, the load on the beam should be the reaction of the wheel loads when the deck is assumed to act as a simple span between beams.

Assume the beams have a span L of 100 ft and consist of a $\frac{3}{8} \times 52$-in web and $1\frac{3}{4} \times 18$-in flanges, area $A = 82.5$ in² and moment of inertia $I = 49,897$ in⁴. Thickness of concrete slab t_s is 8 in and $n = 10$, so that $K_g = 1,317,726$. Hence, for $S = 12$ ft,

$$DF = 0.15 + \left(\frac{12}{3}\right)^{0.6}\left(\frac{12}{100}\right)^{0.2}\left(\frac{1,317,726}{1,200 \times 8^3}\right)^{0.1}$$

$$= 1.773 \text{ wheels per beam}$$

Thus, in this example, use of Eq. (10.16) to compute the distribution factor leads to a smaller live load on the beam. As a result, the maximum moments and shears will be smaller, and a lighter beam probably can be used.

This simple example demonstrates that the change in DF from the older method will change the required stringer cross section whether ASD or LRFD is used.

10.8   DISTRIBUTION OF RAILWAY LOADS THROUGH DECKS

The "AREA Manual for Railway Engineering" of the American Railway Engineering Association contains recommended practices for distribution of the live loads described in Art. 10.6 to support framing.

On open-deck bridges, ties within a length of 4 ft, but not more than three ties, may be assumed to support a wheel load. The live load should be considered a series of concentrated loads, however, for the design of beams and girders.

For ballasted-deck structures, live-load distribution is based on the assumption of standard crossties at least 8 ft long, about 8 in wide, and spaced not more than 2 ft c to c, with at least 6 in of ballast under the ties. For deck design, each axle load should be uniformly distributed over a length of 3 ft plus the minimum distance from bottom of tie to top of beams or girders, but not more than 5 ft or the minimum axle spacing of the loading. In the lateral direction, the axle load should be uniformly distributed over a width equal to the length of tie plus the minimum distance from bottom of tie to top of beams or girders. Deck thickness should be at least ½ in for steel plate, 3 in for timber, and 6 in for reinforced concrete.

For ballasted concrete decks supported by transverse steel beams without stringers, the portion of the maximum axle load to be carried by each beam is given by

$$P = \frac{1.15AD}{S} \qquad S \geq d \qquad (10.17)$$

where A = axle load
  S = axle spacing, ft
  D = effective beam spacing, ft
  d = beam spacing, ft

For bending moment, within the limitation that D may not exceed either axle or beam spacing, the effective beam spacing may be computed from

$$D = d \frac{1}{1 + d/aH} \left(0.4 + \frac{1}{d} + \frac{\sqrt{H}}{12} \right) \qquad (10.18)$$

where a = beam span, ft
  $H = nI_b/ah^3$
  n = ratio of modulus of elasticity of steel to that of concrete
  I_b = moment of inertia of beam, in^4
  h = thickness of concrete deck, in

For end shear, $D = d$. At each rail, a concentrated load of $P/2$ should be assumed acting on each beam.

D should be taken equal to d for bridges without a concrete deck or for bridges where the concrete slab extends over less than 75% of the floorbeam.

If $d > S$, P should be the maximum reaction of the axle loads with the deck between beams acting as a simple span.

For ballasted decks supported on longitudinal girders, axle loads should be distributed equally to all girders whose centroids lie within a lateral width equal to length of tie plus twice the minimum distance from bottom of tie to top of girders.

10.9 BASIC ALLOWABLE STRESSES FOR BRIDGES

Table 10.8 lists the basic allowable stresses for highway and railroad bridges recommended in AASHTO "Standard Specifications for Highway Bridges," and AREA "Manual of Railway Engineering." The stresses, ksi, are related to the minimum yield strength F_y, ksi, or minimum tensile strength F_u, ksi, of the material in all cases except those for which stresses are independent of the grade of steel being used.

The basic stresses may be increased for loading combinations (Art. 10.5). They may be superseded by allowable fatigue stresses (Art. 10.11).

Allowable Stresses in Welds. Standard specifications require that weld metal used in bridges conform to the "Bridge Welding Code," AWS D1.5, American Welding Society.

Yield and tensile strengths of weld metal usually are specified to be equal to or greater than the corresponding strengths of the base metal. The allowable stresses for welds in bridges generally are as follows:

Groove welds are permitted the same stress as the base metal joined. When base metals of different yield strengths are groove-welded, the lower yield strength governs.

TABLE 10.8 Basic Allowable Stresses, ksi, for Railroad and Highway Bridges[a]

Loading condition	Railroad bridges	Highway bridges
Tension:		
Axial, gross section without bolt holes		$0.55F_y$
Axial, net section	$0.55F_y$	$0.55F_y$[b]
Floorbeam hangers, including bending, net section with:		
Rivets in end connections	14	
High-strength bolts in end connections	20	
Bending, extreme fiber of rolled shapes, girders, and built-up sections, gross section[c]	$0.55F_y$	$0.55F_y$
Compression:		
Axial, gross section in:		
Stiffeners of plate girders	$0.55F_y$	$0.55F_y$
Splice material	$0.55F_y$	$0.55F_y$
Compression members:[d]		
$KL/r \leq 107/\sqrt{F_y}$	$0.55F_y$	
$107/\sqrt{F_y} \leq KL/r \leq 858/\sqrt{F_y}$	$0.6F_y - (F_y/166.2)^{3/2} KL/r$	
$KL/r \geq 858/\sqrt{F_y}$	$\dfrac{147{,}000}{(KL/r)^2}$	
$KL/r \leq C_c$		$\dfrac{F_y}{F.S.}\left[1 - \dfrac{(KL/r)^2 F_y}{4\pi^2 E}\right]$
$KL/r \geq C_c$		$\dfrac{\pi^2 E}{F.S.(KL/r)^2}$
Bending, extreme fiber of:		
Built-up[e] or rolled beam members (other than box type)		
Rolled channels[e]	$0.55F_y\left[1 - \dfrac{(l/r_y)^2 F_y}{1{,}800{,}000}\right]$	
Box-type members[f]		
I-type members loaded perpendicular to web	$0.55F_y$	
Rolled shapes, girders, and built-up sections with:		
Compression flange continuously supported		$0.55F_y$
Compression flange intermittently supported[g]		$\dfrac{50 \times 10^6 C_b}{S_{xc}}\left(\dfrac{I_{yc}}{L}\right)$ $\times \sqrt{0.772\dfrac{J}{I_{yc}} + 9.87\left(\dfrac{d}{l}\right)^2}$
Pins	$0.83F_y$	$0.80F_y$
Diagonal tension:		
Webs of girders and rolled beams at section where shear and bending occur simultaneously	$0.55F_y$	
Shear:		
Webs of rolled beams and plate girders, gross section	$0.35F_y$	$0.33F_y$
Pins	$0.42F_y$	$0.40F_y$
Bearing:		
Milled stiffeners and other steel parts in contact (rivets and bolts excluded)	$0.83F_y$	$0.80F_y$

TABLE 10.8 Basic Allowable Stresses, ksi, for Railroad and Highway Bridges[a] (*Continued*)

Loading condition	Railroad bridges	Highway bridges
Bearing (*Continued*)		
Pins:		
Not subject to rotation[h]	$0.75F_y$	$0.80F_y$
Subject to rotation (in rockers and hinges)	$0.375F_y$	$0.40F_y$

[a]F_y = minimum yield strength, ksi, and F_u = minimum tensile strength, ksi. Modulus of elasticity E = 29,000 ksi.

[b]Use $0.46F_u$ for ASTM A709, Grades 100/100W (M270) steels. Use net section if member has holes more than 1¼ in in diameter.

[c]When the area of holes deducted for high-strength bolts or rivets is more than 15% of the gross area, that area in excess of 15% should be deducted from the gross area in determining stress on the gross section.

[d]K = effective length factor. See Art. 6.16.2.

$C_c = \sqrt{2\pi^2 E/F_y}$

E = modulus of elasticity of steel, ksi

r = governing radius of gyration, in

L = actual unbraced length, in

F.S. = factor of safety = 2.12

[e]Applicable to built-up members and rolled beams symmetrical about major axis in plane of web. Welded members and channels may be designed with a higher stress equal to $10,500\,A_f/ld$, but not exceeding $0.55F_y$, if they have solid, rectangular flanges.

l = distance, in, between points of lateral support of compression flange

r_y = radius of gyration, in, of compression flange and that portion of web on compression side of axis of bending, about axis in plane of web

A_f = area, in², of smaller flange, excluding any portion of web

d = overall depth of member, in

[f]l/r_y is determined by: $\sqrt{\dfrac{3.95lS_x\,\sqrt{\Sigma S/t}}{A\,\sqrt{I_y}}}$

S_x = section modulus, in³, of box-type member about its major axis

A = total cross-sectional area, in², enclosed within the centerlines of the webs and flanges of the box-type member

S/t = ratio of length of any flange or web component to its thickness (neglect projections)

I_y = moment of inertia, in⁴, of box-type member about its minor axis

[g]Not to exceed $0.55F_y$.

l = length, in, of unsupported flange between lateral connections, knee braces, or other points of support

I_{yc} = moment of inertia of compression flange about the vertical axis in the plane of the web, in⁴

d = depth of girder, in

$J = \dfrac{[(bt^3)_c + (bt^3)_t + Dt_w^3]}{3}$, where b and t are the flange width and thickness, in, of the compression and tension flange, respectively, and t_w and D are the web thickness and depth, in, respectively

S_{xc} = section modulus with respect to compression flange, in³

$C_b = 1.75 + 1.05\,(M_1/M_2) + 0.3\,(M_1/M_2)^2 \le 2.3$ where M_1 is the smaller and M_2 the larger end moment in the unbraced segment of the beam; M_1/M_2 is positive when the moments cause reverse curvature and negative when bent in single curvature.

C_b = 1.0 for unbraced cantilevers and for members where the moment within a significant portion of the unbraced segment is greater than or equal to the larger of the segment end moments.

For the use of larger C_b values, see Structural Stability Research Council *Guide to Stability Design Criteria for Metal Structures*, 3d ed., p. 135. If cover plates are used, the allowable static stress at the point of theoretical cutoff should be determined by the formula.

[h]Applicable to pins used primarily in axially loaded members, such as truss members and cable adjusting links, and not applicable to pins used in members subject to rotation by expansion or deflection.

Fillet welds are allowed a shear stress of $0.27F_u$, where F_u is the tensile strength of the electrode classification or the tensile strength of the connected part, whichever is less. When quenched and tempered steels are joined, an electrode classification with strength less than that of the base metal may be used for fillet welds, but this should be clearly specified in the design drawings.

Plug welds are permitted a shear stress of 12.4 ksi.

These stresses may be superseded by fatigue requirements (Art. 10.11). The basic stresses may be increased for loading combinations as noted in Art. 10.5.

Effective area of groove and fillet welds for computation of stresses equals the effective length times effective throat thickness. The effective shearing area of plug welds equals the nominal cross-sectional area of the hole in the plane of the faying surface.

Effective length of a groove weld is the width of the parts joined, perpendicular to the direction of stress. The effective length of a straight fillet weld is the overall length of the full-sized fillet, including end returns. For a curved fillet weld, the effective length is the length of line generated by the center point of the effective throat thickness. For a fillet weld in a hole or slot, if the weld area computed from this length is greater than the area of the hole in the plane of the faying surface, the latter area should be used as the effective area.

Effective throat thickness of a groove weld is the thickness of the thinner piece of base metal joined. (No increase is permitted for weld reinforcement. It should be removed by grinding to improve fatigue strength.) The effective throat thickness of a fillet weld is the shortest distance from the root to the face, computed as the length of the altitude on the hypotenuse of a right triangle. For a combination partial-penetration groove weld and a fillet weld, the effective throat is the shortest distance from the root to the face minus $\frac{1}{8}$ in for any groove with an included angle less than 60° at the root of the groove.

In some cases, strength may not govern the design. Standard specifications set maximum and minimum limits on size and spacing of welds. These are discussed in Art. 5.19.

Rollers and Expansion Rockers.   Bearing, kips per lin in, on the diameter of a roller or rocker should not exceed

$$P = \frac{0.600d(F_y - 13)}{20} \qquad d \le 25 \tag{10.19}$$

$$P = \frac{3\sqrt{d}(F_y - 13)}{20} \qquad 25 \le d \le 125 \tag{10.20}$$

where d = diameter, in., of rocker or roller
   F_y = yield strength, ksi, in tension of steel in roller, rocker, or bearing plate, whichever is least

Expansion rollers should be at least 4 in in diameter for highway bridges and 6 in in diameter for railroad bridges.

Allowable Stresses for Bolts.   Bolted shear connections are classified as either bearing-type or slip-critical. The latter are required for connections subject to stress reversal, heavy impact, large vibrations, or where joint slippage would be detrimental to the serviceability of the bridge. These connections are discussed in Sec. 5. Bolted bearing-type connections are restricted to members in compression and secondary members.

Fasteners for bearing-type connections may be ASTM A307 carbon-steel bolts or A325 or A490 high-strength bolts. High-strength bolts are required for slip-critical connections and where fasteners are subjected to tension or combined tension and shear.

Bolts for highway bridges are generally $\frac{3}{4}$ or $\frac{7}{8}$ in in diameter. Holes for high-strength bolts may be standard, oversize, short-slotted, or long-slotted. Standard holes may be up to $\frac{1}{16}$ in larger in diameter than the nominal diameters of the bolts. Oversize holes may have a maximum diameter of $\frac{15}{16}$ in for $\frac{3}{4}$-in bolts and $1\frac{1}{16}$ in for $\frac{7}{8}$-in bolts. Minimum diameter of a slotted hole is the same as that of a standard hole. For $\frac{3}{4}$-in and $\frac{7}{8}$-in bolts, short-slotted holes may be up to 1 in and $1\frac{1}{8}$ in long, respectively, and long-slotted holes, a maximum of $1\frac{7}{8}$ and $2\frac{3}{16}$ in long, respectively.

In the computation of allowable loads for shear or tension on bolts, the cross-sectional area should be based on the nominal diameter of the bolts. For bearing, the area should be taken as the product of the nominal diameter of the bolt and the thickness of the metal on which it bears.

Allowable stresses for bolts specified in "Standard Specifications for Highway Bridges" of the American Association of State Highway and Transportation Officials (AASHTO) are summarized in Tables 10.9 and 10.10. The percentages of stress increase specified for load combinations in Art. 10.5 also apply to high-strength bolts in slip-critical joints, but the percentage may not exceed 133%.

In addition to satisfying these allowable-stress requirements, connections with high-strength bolts should also meet the requirements for combined tension and shear and for fatigue resistance.

Furthermore, the load P_s, kips, on a slip-critical connection should be less than

$$P_s = F_s A_b N_b N_s \qquad (10.21)$$

where F_s = allowable stress, ksi, given in Table 10.9 for a high-strength bolt in a slip-critical joint

A_b = area, in^2, based on the nominal bolt diameter

N_b = number of bolts in the connection

N_s = number of slip planes in the connection

Surfaces in slip-critical joints should be Class A, B, or C, as described in Table 10.9, but coatings providing a slip coefficient less than 0.33 may be used if the mean slip coefficient is determined by test. In that case, F_s for use in Eq. (10.21)

TABLE 10.9 Allowable Stresses, ksi, on Bolts in Highway Bridges

		Allowable shear, F_v				
		Slip-critical connections				
				Long-slotted holes		
ASTM designation	Allowable tension, F_t	Standard-size holes	Oversize and short-slotted holes	Transverse load	Parallel load	Bearing-type joints
A307	18					11
A325	39.5					19§
		15.5*	13.5	11*	9*	
		25†	21.5†	18†	15.5†	
		20‡	17‡	14.5‡	12.5‡	
A490	48.5					25¶
		19*	16*	13.5*	11.5*	
		30.5†	26†	21.5†	18†	
		24.5‡	20.5‡	17‡	14.5‡	

*Class A: When contact surfaces have a slip coefficient of 0.33, such as clean mill scale and blast-cleaned surfaces, with Class A coating.

†Class B: When contact surfaces have a slip coefficient of 0.50, such as blast-cleaned surfaces and such surfaces with Class B coating.

‡Class C: When contact surfaces have a slip coefficient of 0.40, such as hot-dipped galvanized and roughened surfaces.

Class A and B coatings include those with a mean slip coefficient of at least 0.33 or 0.50, respectively. See Appendix A, "Specification for Structural Joints Using ASTM A325 or A490 Bolts," Research Council on Structural Connections of the Engineering Foundation.

§Use 26.6 ksi when threads are excluded from the shear plane.

¶Use 35.0 ksi when threads are excluded from the shear plane.

TABLE 10.10 Allowable Bearing Stresses, ksi, on Bolted Joints in Highway Bridges

Conditions for connection material	A307 bolts	A325 bolts	A490 bolts
Threads permitted in shear planes	20		
Single bolt in line of force in a standard or short-slotted hole		$0.9F_u$*†	$0.9F_u$*†
Two or more bolts in line of force in standard or short-slotted holes		$1.1F_u$*†	$1.1F_u$*
Bolts in long-slotted holes		$0.9F_u$*†	$0.9F_u$*

*F_u = specified minimum tensile strength of connected parts. Connections with bolts in oversize holes or in slotted holes with the load applied less than about 80° or more than about 100° to the axis of the slot should be designed for a slip resistance less than that computed from Eq. (10.21).

†Not applicable when the distance, parallel to the load, from the center of a bolt to the edge of the connected part is less than $1\frac{1}{2}d$, where d is the nominal diameter of the bolt, or the distance to an adjacent bolt is less than $3d$.

should be taken as for Class A coatings but reduced in the ratio of the actual slip coefficient to 0.33.

Tension on high-strength bolts may result in prying action on the connected parts. See Art. 5.25.3.

Combined shear and tension on a slip-critical joint with high-strength bolts is limited by the interaction formulas in Eqs. (10.22) and (10.23). The shear f_v, ksi (slip load per unit area of bolt), for A325 bolts may not exceed

$$f_v = F_s(1 - 0.0159f_t) \tag{10.22}$$

and for A490 bolts,

$$f_v = F_s(1 - 0.0127f_t) \tag{10.23}$$

where f_t = tensile stress, ksi, due to loads, including stresses due to prying action. Combined shear and tension in a bearing-type connection is limited by the interaction equation,

$$f_v^2 + 0.36f_t^2 = F_v^2 \tag{10.24}$$

where f_v = computed shear stress, ksi, in bolt, and F_v = allowable shear, ksi, in bolt (Table 10.9). Equation (10.24) is based on the assumption that bolt threads are excluded from the shear plane.

Fatigue may control design of a bolted connection. To limit fatigue, service-load tensile stress on the area of a bolt based on the nominal diameter, including the effects of prying action, may not exceed the stress in Table 10.11. The prying force may not exceed 80% of the load.

10.10 FRACTURE CONTROL

As defined in the "Guide Specifications for Fracture Critical Non Redundant Steel Bridge Members" of the American Association of State Highway and Transpor-

TABLE 10.11 Allowable Tensile Fatigue Stresses for Bolts in Highway Bridges*

Number of cycles	A325 bolts	A490 bolts
20,000 or less	39.5	48.5
20,000 to 500,000	35.5	44.0
More than 500,000	27.5	34.0

*As specified in "Standard Specifications for Highway Bridges," American Association of State Highway and Transportation Officials.

tation Officials (AASHTO), a fracture-critical member (FCM) or member component is a tension member or component whose failure may result in collapse of the bridge of which it is a part. Although the definition is limited to tension members, failure of any member or component due to any type of stress or strain can also result in catastrophic failure.

The AASHTO "Standard Specifications for Highway Bridges" contains provisions for structural integrity. These recommend that, for new bridges, designers specify designs and details that employ continuity and redundancy to provide one or more alternate load paths. Also, external systems should be provided to minimize effects of probable severe loads.

The AASHTO LRFD specification, in particular, requires that multi-load-path structures be used unless "there are compelling reasons to the contrary." Also, main tension members and components whose failure may cause collapse of the bridge must be designated as FCM and the structural system must be designated nonredundant. Furthermore, the LRFD specification includes fracture control in the fatigue and fracture limit state.

Design of structures can be modified to eliminate the need for special measures to prevent catastrophe from a fracture, and when this is cost-effective, it should be done. Where use of an FCM is unavoidable, for example, the tie of a tied arch, as much redundancy as possible should be provided via continuity, internal redundancy through use of multiple plates, and similar measures.

Steels used in FCM must have the supplemental impact properties specified in the FCM guide specification. FCM should be so designated on the plans with the appropriate temperature zone (Table 1.2) based on the anticipated minimum service temperature.

Table 10.12 gives impact requirements for railway bridges.

TABLE 10.12 Charpy V-Notch Impact Requirements for Structural Steel for Railway Bridges*

Minimum service temperature, °F	Temperature zone designation	Minimum average energy, ft-lb
0 and above	1	15 at 70°F
−1 to −30	2	15 at 40°F
−31 to −60	3	15 at 10°F

*Based on data in the "Manual for Railway Engineering," American Railway Engineering Association. Impact test should be in accordance with the Charpy V-notch test.

10.11 REPETITIVE LOADINGS

Most structural damage to steel bridges is the result of repetitive loading from trucks or wind. Often, the damage is caused by secondary effects, for example, when live loads are distributed transversely through cross frames and induce large out-of-plane distortions that were not taken into account in design of the structure. Such strains may initiate small fatigue cracks. Under repetitive loads, the cracks grow. Unless the cracks are discovered early and remedial action taken, they may create instability under a combination of stress, loading rate, and temperature, and brittle fracture could occur. Proper detailing of steel bridges can prevent such fatigue cracking.

To reduce the probability of fracture, the structural steels included in the AASHTO specifications for M270 steels are required to have minimum impact properties (Art. 10.10). The higher the impact resistance of the steel, the larger a crack has to be before it is susceptible to unstable growth. With the minimum impact properties required for bridge steels, the crack should be large enough to allow discovery during the biannual bridge inspection before fracture occurs. The M270 specification requires average energy in a Charpy V-notch test of 15 ft-lb for grade 36 steels and ranging up to 35 ft-lb for grade 100 steels, at specified test temperatures. These depend on the lowest ambient service temperature (LAST) to which the structure may be subjected. Test temperatures are 70°F higher than the LAST to take into account the difference between the loading rate as applied by highway trucks and the Charpy V-notch impact tests.

Allowable Fatigue Stresses.   Members, connections, welds, and fasteners should be designed so that maximum stresses do not exceed the basic allowable stresses (Art. 10.9) and the range in stress due to loads does not exceed the allowable fatigue stress range. Table 10.13 lists allowable fatigue stress ranges in accordance with the number of cycles to which a member or component will be subjected and several stress categories for structural details. The details described in Table 6.23 for structural steel for buildings are generally applicable also to highway bridges. The diagrams are provided as illustrative examples and are not intended to exclude other similar construction. (See also Art. 6.22.) The allowable stresses apply to load combinations that include live loads and wind. For dead plus wind loads, use the stress range for 100,000 cycles.

Stress range is the algebraic difference between the maximum stress and the minimum stress. Tension stress is considered to have the opposite algebraic sign from compression stress.

Table 10.13a is applicable to redundant load-path structures. These provide multiple load paths so that a single fracture in a member or component cannot cause the bridge to collapse. The AASHTO standard specifications list as examples a simply supported, single-span bridge with several longitudinal beams and a multi-element eye bar in a truss. Table 10.13b is applicable to nonredundant load-path structures. The AASHTO specifications give as examples flange and web plates in bridges with only one or two longitudinal girders, one-element main members in trusses, hanger plates, and caps of single- or two-column bents.

Improved Provisions for Fatigue Design.   AASHTO has published "Guide Specifications for Fatigue Design of Steel Bridges," These indicate that the fatigue provisions in the "Standard Specifications for Highway Bridges" do not accurately reflect the actual fatigue conditions in such bridges; instead, they combine an artificially high stress range with an artificially low number of cycles to get a rea-

TABLE 10.13   Allowable Stress Range, ksi, for Repeated Loads on Highway Bridges[a]

(a) For redundant load-path structures

Stress category	Number of loading cycles			
	100,000[b]	500,000[c]	2,000,000[d]	More than 2,000,000[d]
A	63 (49)[e]	37 (29)[e]	24 (18)[e]	24 (16)[e]
B	49	29	18	16
B′	39	23	14.5	12
C	35.5	21	13	10 12[g]
D	28	16	10	7
E	22	13	8	4.5
E′	16	9.2	5.8	2.6
F	15	12	9	8

(b) For non-redundant load-path structures

Stress category	100,000[b]	500,000[c]	2,000,000[d]	More than 2,000,000[d]
A	50 (39)[e]	29 (23)[e]	24 (16)[e]	24 (16)[e]
B	39	23	16	16
B′	31	18	11	11
C	28	16	10 12[f]	9 11[f]
D	22	13	8	5
E[g]	17	10	6	2.3
E′	12	7	4	1.3
F	12	9	7	6

[a]Based on data in the "Standard Specifications for Highway Bridges," American Association of State Highway and Transportation Officials.

[b]Equivalent to about 10 applications every day for 25 years.

[c]Equivalent to about 50 applications every day for 25 years.

[d]Equivalent to about 200 applications every day for 25 years.

[e]Values in parentheses apply to unpainted weathering steel A709, all grades, when used in conformance with Federal Highway Administration "Technical Advisory on Uncoated Weathering Steel in Structures," Oct. 3, 1989.

[f]For welds of transverse stiffeners to webs or flanges of girders.

[g]AASHTO prohibits use of partial-length welded cover plates on flanges more than 0.8 in thick in nonredundant load-path structures.

sonable result. The actual effective stress ranges rarely exceed 5 ksi, whereas the number of truck passages in the design life of a bridge can exceed many million.

For this reason, these guide specifications give alternative fatigue-design procedures to those in the standard specifications. These procedures accurately reflect the actual conditions in bridges subjected to traffic loadings and provide the following additional advantages: (1) They permit more flexibility in accounting for differing traffic conditions at various sites. (2) They permit design for any desired design life. (3) They provide reasonable and consistent levels of safety over a broad range of design conditions. (4) They are based on extensive recent research and can be conveniently modified in the future if needed to reflect new research results. (5) They are consistent with recently developed fatigue-evaluation procedures for existing bridges.

The guide specifications use the same detail categories and corresponding fatigue strength data as the standard specifications. They also use methods of calculating stress ranges that are similar to those used with the standard specifications.

Thus, it is important that designers possess both the standard specifications and the guide specifications to design fatigue-resistant details properly. However, there is a prevailing misconception in the interpretation of the term "fatigue life." For example, the guide specifications state, "The safe fatigue life of each detail shall exceed the desired design life of the bridge." The implication is that the initiation of a fatigue crack is the end of the service life of the structure. In fact, the initiation of a fatigue crack does *not* mean the end of the life of an existing bridge, or even of the particular member, as documented by the many bridges that have experienced fatigue cracking and even full-depth fracture of main load-carrying members. These cracks and fractures have been successfully repaired by welding, drilling a hole at the crack tip, or placing bolted cover plates over a fracture. These bridges continue to function without reduction in load-carrying capacity or remaining service life.

The AASHTO load-and-resistance factor design specifications include a design limit state for fatigue and fracture. The specifications also incorporate the fatigue-design provisions of the existing guides into the standard specifications, but with some modifications. For example, the LRFD specifications specify a special fatigue design truck (Fig. 10.6) to be used in calculating the stress range resulting from repetitive loadings. The specifications also contain a fatigue-limit-state formula,

$$0.75\Delta f \le \Delta F_n \qquad (10.25)$$

where Δf = stress range, ksi, for the fatigue-truck load and ΔF_n = nominal fatigue resistance, ksi.

An understanding of what is or is not a fatigue-prone detail is necessary for designers to ensure that new designs do not include fatigue-sensitive details that will have to be repaired or replaced in the future. It is important that designers be familiar with current state of the art.

(John Fisher, "Fatigue Cracking of Steel Bridge Structures," FHWA Research Reports, Federal Highway Administration, Turner-Fairbank Highway Research Center, 6300 Georgetown Pike, McLean, VA 22101.)

10.12 DETAILING FOR EARTHQUAKES

Bridges must be designed so that catastrophic collapse cannot occur from seismic forces. Damage to a structure, even to the extent that it becomes unstable, is acceptable, but collapse is not!

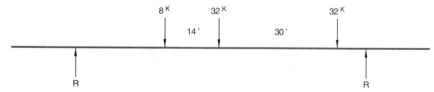

FIGURE 10.6 Design truck for calculation of fatigue stresses. Impact is taken as 15% of live load.

The "Standard Specifications for Seismic Design of Highway Bridges" of the American Association of State Highway and Transportation Officials contain standards for seismic design that are comprehensive in nature and embody several concepts that are significant departures from previous design provisions. They are based on the observed performance of bridges during past earthquakes and on recent research. The specifications include an extensive commentary that documents the basis for the standards and an example illustrating their use.

Although the specifications establish design seismic-force guidelines, of equal importance is the emphasis placed on proper detailing of bridge components. For instance, one of the leading causes of collapse when bridges are subjected to earthquakes is the displacement that occurs at bridge seats. If beam seats are not properly sized, the superstructure will fall off the substructure during an earthquake. Minimum support lengths to be provided at beam ends, based on seismic performance category, is a part of the specifications. Thus, to ensure earthquake-resistant structures, both displacements and loads must be taken into account in bridge design.

Retrofitting existing structures to provide earthquake resistance is also an important consideration for critical bridges. Guidance is provided in "Seismic Retrofitting Guidelines for Highway Bridges," Federal Highway Administration (FHWA) Report No. RD-83/007, and "Seismic Design and Retrofit Manual for Highway Bridges," FHWA Report No. IP-87-6, Federal Highway Administration, McLean, VA 22101.

10.13 DETAILING FOR BUCKLING

Prevention of buckling is a serious concern in bridge design, because of the potential for collapse. Three forms of buckling must be considered in bridge design.

10.13.1 Types of Buckling

The first, and most serious, is primary or Euler-type buckling. This form of buckling occurs when the capacity of an axially loaded compression member is exceeded and failure occurs with little or no warning. This is a rare occurrence with highway bridges, attesting to the adequacy of the current design provisions.

A second form of buckling is local plate buckling. This form of buckling usually manifests itself in the form of excessive distortion of plate elements. This may not be acceptable from a visual perspective, even though the member capacity may be sufficient. When very thin plates are specified, in the desire to achieve minimum weight and supposedly minimum cost, distortions due to welding may induce initial out-of-plane deformations that then develop into local buckling when the member is loaded. Proper welding techniques and use of transverse or longitudinal stiffeners, while maintaining recommended width-thickness limitations on plates and stiffeners, minimize the probability of local buckling.

The third, and perhaps the most likely form of buckling to occur in steel bridges, is lateral buckling. It develops when compression causes a flexural member to become unstable. Such buckling can be prevented by use of lateral bracing, members capable of preventing deformation normal to the direction of the compressive stress at the point of attachment.

Usually, lateral buckling is construction-related. For example, it can occur when a member is fabricated with very narrow compression flanges without adequate provision for transportation and erection stresses. It also can occur when adequate bracing is not provided during deck-placing sequences. Consequently, designers should ensure that compression flanges are proportioned to provide stability during all phases of the service life of bridges, including construction stages, when temporary lateral bracing may be required.

10.13.2   Maximum Slenderness Ratios of Bridge Members

Ratios of effective length to least radius of gyration of columns should not exceed the values listed in Table 10.14.

The length of top chords of half-through trusses should be taken as the distance between laterally supported panel points. The length of other truss members should be taken as the distance between panel-point intersections, or centers of braced points, or centers of end connections.

10.13.3   Plate-Buckling Criteria for Compression Elements

The "Standard Specifications for Highway Bridges" of the American Association of State Highway and Transportation Officials set a maximum width-thickness ratio b/t or D/t for compression members as given in Table 10.15.

10.13.4   Stiffening of Girder Webs

Bending of girders tends to buckle thin webs. This buckling may be prevented by making the web relatively thick compared with its depth (Table 10.15) or by stiffening the web with plates. With welded girders, plates attached normal to the web usually are used. The stiffeners may be set longitudinally or transversely (vertically), or both ways.

Bearing stiffeners are required for plate girders at concentrated loads, including all points of support. Rolled beams should have web stiffeners at bearings when the unit shear stress in the web exceeds 75% of the allowable shear. Bearing stiffeners should be placed in pairs, one stiffener on each side of the web. Plate stiffeners or the outstanding legs of angle stiffeners should extend as close as practicable to the outer edges of the flanges. The stiffeners should be ground to fit against the flange through which the concentrated load, or reaction, is transmitted, or they should be attached to that flange with full-penetration groove welds.

TABLE 10.14   Maximum Slenderness Ratios for Bridge Members

Member	Railroad	Highway
Main compression members	100	120
Wind and sway bracing in compression	120	140
Single lacing	140	
Double lacing	200	
Tension members	200	
Main		200
Main subject to stress reversal		140
Bracing		240

TABLE 10.15 Maximum Width-Thickness Ratios for Compression Elements of Highway Bridge Members

(a) Plates supported on only one side

Components	Limiting stress, ksi[a]	b/t for calculated stress less than the limiting stress[b]	b/t for calculated stress equal to the limiting stress[a]
Compression members[c]	$0.44F_y$	$51.4/\sqrt{f_a} \le 12$	$75/\sqrt{F_y}$
Welded-girder flange[d] Composite girder[d]	$0.55F_y$	$103/\sqrt{f_b} \le 24$ $122/\sqrt{f_{dl}}$	$140/\sqrt{F_y}$
Bolted-girder flange[e] Composite girder[e]	$0.55F_y$	$51.4/\sqrt{f_b} \le 12$ $61/\sqrt{f_{dl}}$	$70/\sqrt{F_y}$

(b) Plates supported on two sides

Component	Limiting stress, ksi[a]	b/t for calculated stress less than the limiting stress[b]	b/t for calculated stress equal to the limiting stress[a]
Girder web without stiffeners[f]	F_v	$270/\sqrt{f_v} \le 150$	$470/\sqrt{F_y}$
Girder web with transverse stiffeners[f]	F_b	$730/\sqrt{f_b} \le 170$	$990/\sqrt{F_y}$
Girder web with longitudinal stiffeners[f]	F_b	$1460/\sqrt{f_b} \le 340$	
Girder web with transverse stiffeners and one longitudinal stiffener[f]	F_b		$1980/\sqrt{F_y}$
Box-shapes—main plates or webs[g]	$0.44F_y$	$126/\sqrt{f_a} \le 45$	$190/\sqrt{F_y}$
Box or H shapes—solid cover plates or webs between main elements[g]	$0.44F_y$	$158/\sqrt{f_a} \le 50$	$240/\sqrt{F_y}$
Box shapes—perforated cover plates[g]	$0.44F_y$	$190/\sqrt{f_a} \le 55$	$285/\sqrt{F_y}$

[a]F_y = specified minimum yield strength of the steel, ksi
$\quad F_b$ = allowable bending stress, ksi
$\quad F_v$ = allowable shear stress, ksi
[b]f_a = computed compressive stress, ksi
$\quad f_b$ = computed compressive bending stress, ksi
$\quad f_v$ = computed shear stress, ksi
$\quad f_{dl}$ = top-flange compressive stress due to noncomposite dead load.
[c]For outstanding plates, outstanding legs of angles, and perforated plates at the perforations. Width b is the distance from the edge of plate or edge of perforation to the point of support. t is the thickness.
[d]b is the width of the compression flange and t is the thickness.
[e]b is the width of flange angles in compression, except those reinforced by plates. t is the thickness.
[f]b represents the depth of the web D, clear unsupported distance between flanges.
[g]When used as compression members. b is the distance between points of support for the plate and between roots of flanges for webs of rolled elements. t is the thickness.

They should be designed for bearing over the area actually in contact with the flange. No allowance should be made for the portions of the stiffeners fitted to fillets of flange angles or flange-web welds. (AREA recommends clipping plate stiffeners at 45° at upper and lower ends to clear such fillets or welds.) Connections of bearing stiffeners to the web should be designed to transmit the concentrated load, or reaction, to the web.

Bearing stiffeners should be designed as columns. For ordinary welded girders, the column section consists of the plate stiffeners and a strip of web. (At interior supports of continuous hybrid girders, however, when the ratio of web yield strength to tension-flange yield strength is less than 0.7, no part of the web should be considered effective.) For stiffeners consisting of two plates, the effective portion of the web is a centrally located strip $18t$ wide, where t is the web thickness, in (Fig. 10.7a). For stiffeners consisting of four or more plates, the effective portion of the web is a centrally located strip included between the stiffeners and extending beyond them a total distance of $18t$ (Fig. 10.7b). The radius of gyration should be computed about the axis through the center of the web. The width-thickness ratio of a stiffener plate or the outstanding leg of a stiffener angle should not exceed

$$\frac{b}{t} = \frac{69}{\sqrt{F_y}} \tag{10.26}$$

where F_y = yield strength, ksi, for stiffener steel

For highway bridges, no stiffeners, other than bearing stiffeners, are required, in general, if the depth-thickness ratio of the web does not exceed the value for girder webs without stiffeners in Table 10.15. But stiffeners may be required for attachment of cross frames.

Transverse stiffeners should be used for highway girders where D/t exceeds the aforementioned values, where D is the depth of the web, the clear unsupported distance between flanges. When transverse stiffeners are used, the web depth-thickness ratio should not exceed the values given in Table 10.15 for webs without longitudinal stiffeners and with one longitudinal stiffener. Intermediate stiffeners may be A36 steel, whereas web and flanges may be a higher grade.

Where required, transverse stiffeners may be attached to the highway-girder web singly or in pairs. Where stiffeners are placed on opposite sides of the web, they should be fitted tightly against the compression flange. Where a stiffener is placed on only one side of the web, it must be in bearing against, but need not be attached to the compression flange. Intermediate stiffeners need not bear against the tension flange. However, the distance between the end of the stiffener weld and the near edge of the web-to-flange fillet welds must not be less than $4t$ or more than $6t$.

Transverse stiffeners may be used, where not otherwise required, to serve as connection plates for diaphragms or cross frames. In such cases, the stiffeners must be rigidly connected to both the tension and compression flanges to prevent web

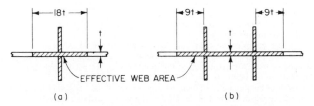

FIGURE 10.7   Effective column areas for design of stiffeners: (a) for one pair of stiffeners; (b) for two pairs.

fatigue cracks due to out-of-plane movements. The stiffener may be welded to both flanges, or a special bolted detail may be used to connect to the tension flange. The appropriate fatigue category must be used for the tension flange to reflect the detail used (see Art. 10.11).

Transverse stiffeners should be proportioned so that

$$I \geq d_o t^3 J \tag{10.27}$$

$$J = 2.5 \left(\frac{D}{d_o}\right)^2 - 2 \geq 0.5 \tag{10.28}$$

where I = moment of inertia, in^4, of transverse intermediate stiffener
J = ratio of rigidity of stiffener to web
d_o = actual distance, in, between transverse stiffeners
t = web thickness, in
D = depth of the web

For stiffener pairs, I should be taken about the center of the web. For single stiffeners, I should be taken about the web face in contact with the stiffeners. In either case, transverse stiffeners should project a distance, in, from the web of at least $b_f/4$, where b_f is the flange width, in, and at least $D'/30 + 2$, where D' is the girder depth, in. Thickness should be at least $\frac{1}{16}$ of this width.

Intermediate transverse stiffeners should have a gross cross-sectional area A, in^2, of at least

$$A = Y[0.15BDt_w(1 - C)(f_v/F_v) - 18t_w^2] \tag{10.29}$$

where Y = ratio of the yield strength of the web steel to the yield strength of the stiffener steel
t_w = web thickness, in
f_v = computed shear stress, ksi, in the web
F_v = allowable shear stress, ksi, in the web
B = 1.0 for pairs of stiffeners
= 1.8 for single angles
= 2.4 for single plates $\quad$ *shear yield*
C = ratio of buckling shear stress to shield year stress
= 1.0 when $D/t_w < 190\sqrt{k/F_y}$ $\tag{10.30a}$

$$= \frac{6000}{D/t_w}\sqrt{\frac{k}{F_y}} \quad \text{when} \quad 190\sqrt{k/F_y} \leq D/t_w \leq 237\sqrt{k/F_y} \tag{10.30b}$$

$$= \frac{45,000k}{(D/t_w)^2 F_y} \quad \text{when} \quad D/t_w > 237\sqrt{k/F_y} \tag{10.30c}$$

$$k = 5[1 + (D/d_o)^2] \tag{10.30d}$$

When A computed from Eq. (10.29) is very small or negative, transverse stiffeners need only satisfy Eq. (10.27) and the width-thickness limitations given previously.

Intermediate transverse stiffeners, with or without longitudinal stiffeners, should be spaced close enough that the computed shear stress f_v' does not exceed

$$F_v' = F_v\left[C + \frac{0.87(1 - C)}{\sqrt{1 + (d_o/D)^2}}\right] \tag{10.31a}$$

where C is defined by Eqs. (10.30a) to (10.30d). Spacing is limited to a maximum of $3D$, or for panels without longitudinal stiffeners, to ensure efficient fabrication, handling, and erection of the girders, to $67,600D(t_w/D)^2$. At a simple support, the first intermediate stiffener should be close enough to the support that the shear stress in the end panel does not exceed

$$f'_v = CF_y/3 \le F_y/3 \qquad (10.31b)$$

but not farther than $1.5D$.

If the shear stress is larger than $0.6F_v$ in a girder panel subjected to combined shear and bending moment, the bending stress F_s with live loads positioned for maximum moment at the section should not exceed

$$F_s = (0.754 - 0.34f_v/f'_v)F_y \qquad (10.32)$$

Fabricators should be given leeway to vary stiffener spacing and web thickness to optimize costs. Girder webs often compose 40 to 50% of the girder weight but only about 10% of girder bending strength. Hence, least girder weight may be achieved with minimum web thickness and many stiffeners but not necessarily at the lowest cost. Thus, the contract drawings should allow fabricators the option of choosing stiffener spacing. The contract drawings should also note the thickness requirements for a web with a minimum number of stiffeners. (A stiffener is required at every cross frame.) This allows fabricators to choose the most economical fabrication process. If desired, flange thicknesses can be reduced slightly if the thicker-web option is selected. In some cases, the most economical results may be obtained with a stiffened web having a thickness $\frac{1}{16}$ in less than that of an unstiffened web (Art. 10.18).

Preferably, the drawings should show the details for a range from unstiffened to fully stiffened webs. During the design stage, this is a relatively simple task. In contrast, after a construction contract has been awarded, the contractor cannot be expected to submit alternative girder designs, with or without value engineering, because it is often more trouble than the effort is worth. Contractors generally bid on what is shown on the plans, risking the possibility of losing the contract to a concrete alternative or to another contractor. On the other hand, by providing contract documents with sufficient flexibility, owners can profit from the fact that different fabricators have different methods of cost-effective fabrication that can be utilized on behalf of owners.

Longitudinal stiffeners should be used where D/t exceeds the values given in Table 10.15. They are required, even if the girder has transverse stiffeners, if the values of D/t for a web with transverse stiffeners is exceeded.

When required, a longitudinal stiffener should be attached to the web at a distance $D/5$ from the inner surface or leg of the compression flange, measured to the center of the stiffener plate. The stiffener should be proportioned so that

$$I \ge Dt^3 \left[2.4 \left(\frac{d_o}{D} \right)^2 - 0.13 \right] \qquad (10.33)$$

where I = moment of inertia, in^4, of longitudinal stiffener about edge in contact with web and d_o = actual distance, in, between transverse stiffeners. Width-thickness ratio of the longitudinal stiffener should not exceed

$$\frac{b_s}{t_s} = \frac{71.2}{\sqrt{f_b}} \qquad (10.34)$$

Bending stress in the stiffener should not exceed the allowable for the stiffener steel. The stiffener may be placed on only one side of the web. Not required to be continuous, it may be interrupted at transverse stiffeners.

Spacing of transverse stiffeners used with longitudinal stiffeners should satisfy Eq. (10.31a) but should not exceed 1.5 times the subpanel depth in the panel adjacent to a simple support as well as in interior panels. The limit on stiffener spacing given previously to ensure efficient handling of girders does not apply when longitudinal stiffeners are used. Also, in computation of required moment of inertia and area of transverse stiffeners from Eqs. (10.27) to (10.29), the maximum subpanel depth should be substituted for D.

Longitudinal stiffeners become economical for girder spans over 300 ft. Often, however, they are placed on fascia girders for esthetic reasons and may be used on portions of girders subject to tensile stresses or stress reversals. If this happens, designers should ensure that butt splices used by the fabricators for the longitudinal stiffeners are made with complete-penetration groove welds of top quality. (Plates of the sizes used for stiffeners are called *bar stock* and are available in limited lengths, which almost always make groove-welded splices necessary.) Many adverse in-service conditions have resulted from use of partial-penetration groove welds instead of complete-penetration.

10.13.5 Lateral Bracing

In highway girder bridges, AASHTO requires that the need for lateral bracing be investigated. The stresses induced in the flanges by the specified wind pressure must be within specified limits. In many cases lateral bracing will not be required, and a better structure can be achieved by eliminating fatigue prone details. Flanges attached to concrete decks or other decks of comparable rigidity will not require lateral bracing. When lateral bracing is required, it should be placed in the exterior bays between diaphragms or cross-frames, in or near the plane of the flange being braced.

Bracing consists of members capable of preventing rotation or lateral deformation of other members. This function may be served in some cases by main members, such as floorbeams where they frame into girders; in other cases by secondary members especially incorporated in the steel framing for the purpose; and in still other cases by other construction, such as a concrete deck. Preferably, bracing should transmit forces received to foundations or bearings, or to other members that will do so.

AASHTO specifications state that the smallest angle used in bracing should be $3 \times 2\frac{1}{2}$ in. Size of bracing often is governed by the maximum permissible slenderness ratio (Table 10.14) or width-thickness ratio of components (Table 10.15). Some designers prefer to design bracing for a percentage, often 2%, of the axial force in the member.

Bracing of Through Girders. Where plate girders comprise the main members of through spans, the compression flange must be braced to prevent buckling. Since that flange is above the level of the deck, neither cross bracing nor lateral bracing can be used in positive-moment areas. In such regions, the compression flange should be stiffened against lateral deformation with gusset plates or knee braces with solid webs. If floorbeams are used in the floor system, the brackets should be attached securely to the top flanges of those beams and to stiffeners on the main members. The brackets should extend to the top flange of the main girder. They should be as wide as clearance permits. If the unsupported length of the edge of the gusset plate or solid web exceeds 60 times its thickness, a stiffening plate or angles should be attached to the unsupported edge.

Through-truss, deck-truss, and spandrel-braced-arch highway bridges should have top and bottom lateral bracing (Fig. 10.8). For compression chords, lateral

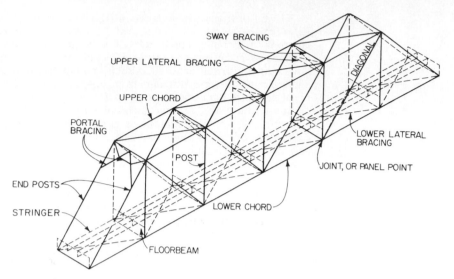

FIGURE 10.8 Components of a through-truss bridge.

bracing preferably should be as deep as the chords and connected to top and bottom flanges.

In railroad bridges, bottom lateral bracing should be placed in all spans, except deck types less than 50 ft long. Top lateral bracing should be incorporated in all deck spans and in through spans that have enough headroom. (A rigid deck may be considered lateral bracing in its plane.) Lateral bracing of the compression chords of trusses, compression flanges of deck girders, and posts of viaduct towers should be designed for a transverse shear in any panel equal to 2.5% of the total axial stress in the members in that panel, in addition to the shear from specified lateral forces.

In both highway and railroad bridges, if a double system of bracing is used (top and bottom laterals), both systems may be considered effective simultaneously if the members meet the requirements as both tension and compression members. The members should be connected at their intersections.

AASHTO specifications require that a horizontal wind force of 50 lb/ft^2 on the area of the superstructure exposed in elevation be included in determining the need for, or in designing, bracing. Half of the force should be applied in the plane of each flange. The maximum induced stresses F, ksi, in the bottom flange from the lateral forces can be computed from

$$F = RF_{cb} \tag{10.35a}$$

where $R = (0.2272L - 11)/S_d^{3/2}$ without bottom lateral bracing
$ = (0.059L - 0.640)/\sqrt{S_d}$ with bottom lateral bracing
$ L = $ span, ft
$ S_d = $ diaphragm or cross frame spacing, ft
$ F_{cb} = 72M_{cb}/t_f b_f^2$
$ M_{cb} = 0.08WS_d^2$
$ W = $ wind loading, kips per ft, along exterior flange
$ t_f = $ flange thickness, in
$ b_f = $ flange width, in

10.13.6 Cross Frames and Diaphragms for Deck Spans

In highway bridges, rolled beams and plate girders should be braced with cross frames or diaphragms at each end. Also, AASHTO specifications require that intermediate cross frames or diaphragms be spaced at intervals of 25 ft or less. They should be placed in all bays. Cross frames should be as deep as practicable. Diaphragms should be at least one-third and preferably one-half the girder depth. Cross frames and diaphragms should be designed for wind forces as described above for lateral bracing. The maximum horizontal force in the cross frames or diaphragms may be computed from

$$F_c = 1.14WS_d \qquad\qquad (10.35b)$$

End cross frames or diaphragms should be designed to transmit all lateral forces to the bearings. Cross frames between horizontally curved girders should be designed as main members capable of transferring lateral forces from the girder flanges.

In railroad bridges, cross frames or diaphragms also should be placed at ends and intermediate points of beam and girder spans. In open-deck construction, spacing should not exceed 18 ft. Where steel-plate, timber, or precast-concrete decking is used with ballasted-deck construction, cross frames or diaphragms without top lateral bracing should be spaced at intervals not exceeding 12 ft; or with top lateral bracing, at intervals not exceeding 18 ft. Where cast-in-place concrete decking is used with ballasted-deck construction, cross frames or diaphragms should be spaced at intervals not exceeding 24 ft. For girders or beams up to 54 in deep, concrete diaphragms with reinforcement extending through the main members may be used instead of steel diaphragms. Where ballast and track are carried on transverse beams without stringers, the beams should be connected with at least one line of longitudinal diaphragms per track. A cross frame or diaphragm should be placed at midspan when the length of longitudinal girders or beams exceeds the allowable spacing of cross frames or diaphragms.

Longitudinal railroad girders or beams with depth greater than 42 in and spaced more than 4 ft apart should be braced with cross frames. The angle of cross-frame diagonals with the vertical should not exceed 60°.

Longitudinal railroad girders or beams not requiring cross frames should be braced with I-shaped diaphragms. Their depth should be as large as the depth of the main members will permit.

Although AASHTO specifications require cross frames or diaphragms at intervals of 25 ft or less, it is questionable whether spacing that close is necessary for bridges in service. Often, a three-dimensional finite-element analysis will show that few, if any, cross frames or diaphragms are necessary. Inasmuch as most fatigue-related damage to steel bridge is a direct result of out-of-plane forces induced through cross frames, the possibility of eliminating them should be investigated for all new bridges, However, although cross frames may not be needed for service loads, they may be necessary to ensure stability during girder erection and deck placement.

The AASHTO LRFD specifications do not require cross frames or diaphragms but specify that the need for diaphragms or cross frames should be investigated for all stages of assumed construction procedures and the final condition. Diaphragms or cross frames required for conditions other than the final condition may be specified to be temporary bracing. If permanent cross frames or diaphragms are included in the structural model used to determine force effects, they should be designed for all applicable limit states for the calculated member loads. For plate girders, transverse stiffeners used as cross-frame connection stiffeners should be

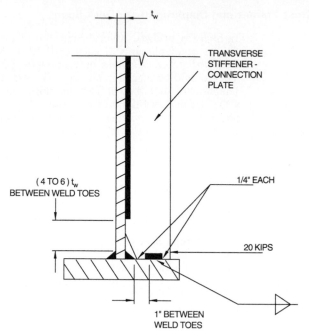

FIGURE 10.9   Girder connects to a cross frame through a transverse stiffener.

connected to both flanges to prevent distortion-induced fatigue cracking. Although many designers believe welding stiffeners to the tension flange is worse than leaving the connection stiffener unattached, experience has proven otherwise. Virtually no cracks result from the attachment weld, but a proliferation of cracks develop when connection stiffeners are not connected to the tension flange. The LRFD specifications also recommend that, where cross frames are used, the attachment be designed for a transverse force of 20 kips (Fig. 10.9). This applies to straight, nonskewed bridges when better information is not available.

10.13.7   Portal and Sway Bracing

End panels of simply supported, through-truss bridges have compression chords that slope to meet the bottom chords just above the bearings. Bracing between corresponding sloping chords of a pair of main trusses is called portal bracing (Fig. 10.8). Bracing between corresponding vertical posts of a pair of main trusses is called sway bracing (Fig. 10.8).

All through-truss bridges should have portal bracing, made as deep as clearance permits. (In railroad bridges, it should include knee braces.) Portal bracing preferably should be of the two-plane or box type, rigidly connected to the flanges of the end posts (sloping chords). If single-plane portal bracing is used, it should be set in the central transverse plane of the end posts. Diaphragms then should be placed between the webs of the end posts, to distribute the portal stresses.

Portal bracing should be designed to carry the end reaction of the top lateral system. End posts should be designed to transfer this reaction to the truss bearings.

Through trusses should have sway bracing at least 5 ft deep in highway bridges and 6 ft deep in railroad bridges at each intermediate panel point. Top lateral struts should be at least as deep as the top chord. When clearance does not permit the required depth of sway bracing in railroad bridges, knee braces should be placed at the ends of the top lateral struts and made as deep as practicable.

Deck trusses should have sway bracing between all corresponding panel points. This bracing should extend the full depth of the trusses below the floor system. End sway bracing should be designed to carry the top lateral forces to the supports through the truss end posts.

10.13.8 Bracing of Towers

Towers should be braced with double systems of diagonals and with horizontal struts at caps, bases, and intermediate panel points. Sections of members of longitudinal bracing in each panel should not be less than those of members in corresponding panels of the transverse bracing.

Column splices should be at or just above panel points. Bracing of a long column should fix the column about both axes at or near the same point.

Horizontal diagonal bracing should be placed, at alternate intermediate panel points, in all towers with more than two vertical panels. In double-track towers, horizontal bracing should be installed at the top to transmit horizontal forces.

Bottom struts of towers should be strong enough to slide the movable shoes with the structure unloaded, when the coefficient of friction is 0.25. Column bearings should be designed for expansion and contraction of the tower bracing.

10.14. CRITERIA FOR BUILT-UP TENSION MEMBERS

A tension member and all its components must be proportioned to meet the requirements for maximum slenderness ratio given in Table 10.14. The member also must be designed to ensure that the allowable tensile stress on the net section is not exceeded.

The **net section** of a high-strength-bolted tension member is the sum of the net sections of its components. The net section of a component is the product of its thickness and net width.

Net width is the minimum width normal to the stress minus an allowance for holes. The diameter of a hole for a fastener should be taken as $\frac{1}{8}$ in greater than the nominal fastener diameter. The chain of holes that is critical is the one that requires the largest deduction for holes and may lie on a straight line or in a zigzag pattern. The deduction for any chain of holes equals the sum of the diameters of all the holes in the chain less, for each gage space in the chain, $s^2/4g$, where s is the pitch, in, of any two successive holes and g is the gage, in, of those holes. The AREA recommends, however, that the net width not exceed 85% of the corresponding gross width.

For angles, the gross width should be taken as the sum of the widths of the legs less the thickness. The gage for holes in opposite legs is the sum of the gages from back of angle less the thickness. If a double angle or tee is connected with the angles or flanges back to back on opposite sides of a gusset plate, the full net section may be considered effective. But if double angles, or a single angle or tee, are connected on the same side of a gusset plate, the effective area should be taken as the net section of the connected leg or flange plus one-half the area of the outstanding leg. When angles connect to separate gusset plates, as in a double-webbed truss, and the angles are interconnected close to the gussets, for example, with stay plates, the full net area may be considered effective. Without such interconnection, only 80% of the net area may be taken as effective.

For built-up tension members with perforated plates, the net section of the plate through the perforation may be considered the effective area.

In pin-connected tension members other than eyebars, the net section across the pinhole should be at least 140%, and the net section back of the pinhole at least 100% of the required net section of the body of the member. The ratio of the net width, through the pinhole normal to the axis of the member, to thickness should be 8 or less. Flanges not bearing on the pin should not be considered in the net section across the pin.

To meet stress requirements, the section at pinholes may have to be reinforced with plates. These should be arranged to keep eccentricity to a minimum. One plate on each side should be as wide as the outstanding flanges will allow. At least one full-width plate on each segment should extend to the far side of the stay plate and the others at least 6 in beyond the near edge. These plates should be connected with fasteners or welds arranged to distribute the bearing pressure uniformly over the full section.

Eyebars should have constant thickness, no reinforcement at pinholes. Thickness should be between ½ and 2 in, but not less than ⅛ the width. The section across the center of the pinhole should be at least 135%, and the net section back of the pinhole at least 75% of the required net section of the body of the bar. The width of the body should not exceed the pin diameter divided by $\frac{3}{4} + F_y/400$, where F_y is the steel yield strength, ksi. The radius of transition between head and body of eyebar should be equal to or greater than the width of the head through the center of the pinhole.

Eyebars of a set should be symmetrical about the central plane of the truss and as nearly parallel and close together as practicable. But adjacent bars in the same panel should be at least ½ in apart. The bars should be held against lateral movement.

Stitching. In built-up members, welds connecting plates in contact should be continuous. Spacing of fasteners should be the smaller of that required for sealing, to prevent penetration of moisture (Art. 5.11), or stitching, to ensure uniform action. The pitch of stitch fasteners on any single line in the direction of stress should not exceed $24t$, where t = thickness, in, of the thinner outside plate or shape. If there are two or more lines of fasteners with staggered pattern, and the gage g, in, between the line under consideration and the farther adjacent line is less than $24t$, the staggered pitch in the two lines, considered together, should not exceed $24t$ or $30t - 3g/4$. The gage between adjacent lines of stitch fasteners should not exceed $24t$.

Cover Plates. When main components of a tension member are tied together with cover plates, the shear normal to the member in the planes of the plates should be assumed equally divided between the parallel plates. The shearing force should include that due to the weight of the member plus other external forces.

When perforated cover plates are used, the openings should be ovaloid or elliptical (minimum radius of periphery 1½ in). Length of perforation should not exceed twice its width. Clear distance between perforations in the direction of stress should not be less than the distance l between the nearer lines of connections of the plate to the member. The clear distance between the end perforation and end of the cover plate should be at least $1.25l$. For plates groove-welded to the flange edge of rolled components, l may be taken as the distance between welds when the width-thickness ratio of the flange projection is less than 7; otherwise, the distance i should be taken between the roots of the flanges. Thickness of a perforated plate should be at least $\frac{1}{50}$ of the distance between nearer lines of connection.

When stay plates are used to tie components together, the clear distance between them should be 3 ft or less. Length of end stay plates between end fasteners should be at least $1.25l$, and length of intermediate stay plates at least $0.563l$. Thickness of stay plates should not be less than $l/50$ in main members and $l/60$ in bracing. They should be connected by at least three fasteners on each side to the other components. If a continuous fillet weld is used, it should be at least $\frac{5}{16}$ in.

Tension-member components also may be tied together with end stay plates and lacing bars like compression members. The last fastener in the stay plates preferably should also pass through the end of the adjacent bar.

10.15 CRITERIA FOR BUILT-UP COMPRESSION MEMBERS

Compression members should be designed so that main components are connected directly to gusset plates, pins, or other members. Stresses should not exceed the allowable for the gross section. The radius of gyration and the effective area of a member with perforated cover plates should be computed for a transverse section through the maximum width of perforation. When perforations are staggered in opposite cover plates, the effective area should be considered the same as for a section with perforations in the same transverse plane.

Solid-Rib Arches. A compression member and all its components must be proportioned to meet the requirements for maximum slenderness ratio in Table 10.14. The member also must satisfy width-thickness requirements (Table 10.15). In addition, for solid-rib arches, longitudinal stiffeners are required when the depth-thickness ratio of each web exceeds

$$\frac{D}{t} = \frac{158}{\sqrt{f_a}} \le 60 \tag{10.36}$$

where D = unsupported distance, in, between flange components
   t = web thickness, in
   f_a = maximum compressive stress in web, ksi

If one longitudinal stiffener is used, it should have a moment of inertia I_s, in^4, of at least

$$I_s = 0.75Dt_w^{\ 3} \tag{10.37}$$

where D = clear unsupported depth of web, in, and t_w = web thickness, in. If the stiffener is placed at middepth of the web, the width-thickness ratio should not exceed

$$D/t_w = 237/\sqrt{f_a} \tag{10.38}$$

If two longitudinal stiffeners are used, each should have a moment of inertia of at least

$$I_s = 2.2Dt_w^{\ 3} \tag{10.39}$$

If the stiffeners are placed at the third points of the web depth, the width-thickness ratio should not exceed

$$D/t_w = 316/\sqrt{f_a} \qquad (10.40)$$

Maximum width-thickness ratio for an outstanding element of a stiffener is given by

$$\frac{b'}{t_s} = \frac{51.4}{\sqrt{f_a + f_b/3}} \leq 12 \qquad (10.41)$$

where b' = width of outstanding element, in
t_s = thickness of the element, in
f_b = maximum compressive bending stress, ksi

The preceding relationships for webs applies when

$$0.2 \leq f_b/(f_b + f_a) \leq 0.7 \qquad (10.42)$$

For flange plates between the webs of a solid-rib arch, the width-thickness ratio should not exceed

$$\frac{b_f}{t_f} = \frac{134}{\sqrt{f_a + f_b}} \leq 47 \qquad (10.43)$$

Maximum width-thickness ratio for the overhang of flange plates is given by

$$\frac{b'_f}{t_f} = \frac{51.4}{\sqrt{f_a + f_b}} \leq 12 \qquad (10.44)$$

Stitching. In built-up members, welds connecting plates in contact should be continuous. Spacing of fasteners should be the smaller of that required for sealing, to prevent penetration of moisture (Art. 5.11), or stitching, to ensure uniform action and prevent local buckling. The pitch of stitch fasteners on any single line in the direction of stress should not exceed $12t$, where t = thickness, in, of the thinner outside plate or shape. If there are two or more lines of fasteners with staggered pattern, and the gage g, in, between the line under consideration and the farther adjacent line is less than $24t$, the staggered pitch in the two lines, considered together, should not exceed $12t$ or $15t - 3g/8$. The gage between adjacent lines of stitch fasteners should not exceed $24t$.

Fastener Pitch at Ends. Pitch of fasteners connecting components of a compression member over a length equal to 1.5 times the maximum width of member should not exceed 4 times the fastener diameter. The pitch should be increased gradually over an equal distance farther from the end.

Shear. On the open sides of compression members, components should be connected with perforated plates or by lacing bars and end stay plates. The shear normal to the member in the planes of the plates or bars should be assumed equally divided between the parallel planes. The shearing force should include that due to the weight of the member, other external forces, and a normal shearing force, kips, given by

$$V = \frac{P}{100}\left(\frac{100}{L/r + 10} + \frac{L/r}{3,300/F_y}\right) \qquad (10.45)$$

where P = allowable compressive axial load on member, kips
   L = length of member, in
   r = radius of gyration, in, of section about axis normal to plane of lacing or perforated plate

Perforated Plates.   When perforated cover plates are used, the openings should be ovaloid or elliptical (minimum radius of periphery 1½ in). Length of perforation should not exceed twice its width. Clear distance between perforations in the direction of stress should not be less than the distance l between the nearer lines of connections of the plate to the member. The clear distance between the end perforation and end of the cover plate should be at least 1.25l. For plates groove-welded to the flange edge of rolled components, l may be taken as the distance between welds when the width-thickness ratio of the flange projection is less than 7; otherwise, the distance l should be taken between the roots of the flanges. Thickness should meet the requirements for perforated plates given in Table 10.15.

10.16   PLATE GIRDERS AND COVER-PLATED ROLLED BEAMS

Where longitudinal beams or girders support through bridges, the spans preferably should have two main members. They should be placed sufficiently far apart to prevent overturning by lateral forces.

   In railroad bridges, the distance between centers of outside girders should be at least ¹⁄₂₀ of the span for through bridges and ¹⁄₁₅ of the span for deck bridges. Where a track is supported on a pair of deck girders, they should be spaced at least 6½ ft c to c. If multiple girders are used, they should be arranged to distribute the track load as uniformly as possible to all members.

Spans.   For calculation of stresses, span is the distance between center of bearings or other points of support. For computing span-depth ratio for continuous beams, span should be taken as the distance between dead-load points of inflection.

Allowable-Stress Design.   Beams and plate girders should be proportioned by the moment-of-inertia method; that is, for pure bending, to satisfy the flexure formula:

$$\frac{I}{c} \geq \frac{M}{F_b} \tag{10.46}$$

where I = moment of inertia, in⁴, of gross section for compressive stress and of net section for tensile stress
   c = distance, in, from neutral axis to outermost surface
   M = bending moment at section, in kips
   F_b = allowable bending stress, ksi

The neutral axis should be taken along the center of gravity of the gross section. For computing the moment of inertia of the net section, the area of holes for high-strength bolts in excess of 15% of the flange area should be deducted from the gross area.

Span-Depth Ratio.   Depth of steel beams or girders for highway bridges should preferably be at least ¹⁄₂₅ of the span.
   For bracing requirements, see Art. 10.13.

Cover-Plated Rolled Beams.   Welds connecting a cover plate to a flange should be continuous and capable of transmitting the horizontal shear at any point. When the unit shear in the web of a rolled beam at a bearing exceeds 75% of the allowable shear for girder webs, bearing stiffeners should be provided to reinforce the web. They should be designed to satisfy the same requirements as bearing stiffeners for girders in Art. 10.13.

The theoretical end of a cover plate is the section at which the stress in the flange without that cover plate equals the allowable stress, exclusive of fatigue considerations. **Terminal distance,** or extension of cover plate beyond the theoretical end, is twice the nominal cover-plate width for plates not welded across their ends and 1.5 times the width for plates welded across their ends. Length of a cover plate should be at least twice the beam depth plus 3 ft. Thickness should not exceed twice the flange thickness.

Partial-length welded cover plates should extend beyond the theoretical end at least the terminal distance or a sufficient distance so that the stress range in the flange equals the allowable fatigue stress range for base metal at fillet welds, whichever is greater. Ends of tapered cover plates should be at least 3 in wide. Welds connecting a cover plate to a flange within the terminal distance should be of sufficient size to develop the computed stress in the cover plate at its theoretical end.

Because of their low fatigue strength, cover-plated beams are seldom cost-effective.

Girder Flanges.   Width-thickness ratios of compression flanges of plate girders should meet the requirements given in Art. 10.13. For other girders, see Arts. 10.17 and 10.19 to 10.20.

Each flange of a welded plate girder should consist of only one plate. To change size, plates of different thicknesses and widths may be joined end to end with complete-penetration groove welds and appropriate transitions (Art. 5.26).

Plate girders composed of flange angles, web plate, and cover plates attached with bolts or rivets are no longer used. In existing bolted girders, flange angles formed as large a part of the flange area as practicable. Side plates were used only where flange angles more than ⅞ in thick would otherwise be required. Except in composite design, the gross area of the compression flange could not be less than the gross area of the tension flange.

When cover plates were needed, at least one cover plate of the top flange extended full length of the girder unless the flange was covered with concrete. (AREA required one cover plate to be full length on each flange.) If more than one cover plate was desirable, the plates on each flange were made about the same thickness. When of unequal thickness, they were arranged so that they decreased in thickness from flange angles outward. No plate could be thicker than the flange angles. Fasteners connecting cover plates and flange were required to be adequate to transmit the horizontal shear at any point. Cover plates over 14 in wide should have four lines of fasteners.

Partial-length cover plates extended beyond the theoretical end far enough to develop the plate capacity or to reach a section where the stress in the remainder of the flange and cover plates equals the allowable fatigue stress range, whichever distance is greater.

Flange-to-Web Connections.   Welds or fasteners for connecting the flange of a plate girder to the web should be adequate to transmit the horizontal shear at any point plus any load applied directly to the flange. AASHTO permits the web to be connected to each flange with a pair of fillet welds.

For flange splices, see Arts. 5.26 and 5.27.

Girder Web and Stiffeners.   The web should be proportioned so that the average shear stress over the gross section does not exceed the allowable. In addition, depth-thickness ratio should meet the requirements of Art. 10.13. Also, stiffeners should be provided, where needed, in accordance with those requirements. For web splices, see Arts. 5.26, 5.27, and 5.30.

Camber.   Girders should be cambered to compensate for dead-load deflection. Also, on vertical curves, camber preferably should be increased or decreased to keep the flanges parallel to the profile grade line.
See also Art. 10.18.

10.17   COMPOSITE CONSTRUCTION WITH I GIRDERS

With shear connectors welded to the top flange of a beam or girder, a concrete slab may be made to work with that member in carrying bending stresses. In effect, a portion of the slab, called the **effective width,** functions much like a steel cover plate. In fact, the effective slab area may be transformed into an equivalent steel area for computation of composite-girder stresses and deflection. This is done by dividing the effective concrete area by the modular ratio n, the ratio of modulus of elasticity of steel, 29,000 ksi, to modulus of elasticity of the concrete. The equivalent area is assumed to act at the center of gravity of the effective slab. The equivalent steel section is called the **transformed section.**

Allowable-Stress Design.   Composite girders, in general, should meet the requirements of plate girders (Art. 10.16). Bending stresses in the steel girder alone and in the transformed section may be computed by the moment-of-inertia method, as indicated in Art. 10.16, or by load-factor design, and should not exceed the allowable for the material. The stress range at the shear connector must not exceed the allowable for a Category C detail.

The allowable concrete stress may be taken as $0.4f'_c$, where f'_c = unit ultimate compressive strength of concrete, psi, as determined by tests of 28-day-old cylinders. The allowable tensile stress of steel reinforcement for concrete should be taken as 20 ksi for A615 Grade 40 steel bars and 24 ksi for A615 Grade 60 steel bars. The modular ratio n may be assumed as follows:

f'_c	n
2,000–2,300	11
2,400–2,800	10
2,900–3,500	9
3,600–4,500	8
4,600–5,000	7
6,000 or more	6

To account for creep of the concrete under dead load, design of the composite section should include the larger of the dead-load stresses when the transformed section is determined with n or $3n$.

The neutral axis of the composite section preferably should lie below the top flange of the steel section. Concrete on the tension side should be ignored in stress computations.

Effective Slab Width.   The assumed effective width of slab should be equal to or less than one-quarter the span, distance center to center of girders, and 12 times the least slab thickness (Fig. 10.10). For exterior girders, the effective width on the exterior side should not exceed the actual overhang. When an exterior girder has a slab on one side only, the assumed effective width should be equal to or less than one-twelfth the span, half the distance to the next girder, and 6 times the least slab thickness (Fig. 10.10).

Span-Depth Ratios.   For composite highway girders, depth of steel girder alone should preferably be at least ⅓₀ of the span. Depth from top of concrete slab to bottom of bottom flange should preferably be at least ½₅ of the span. For continuous girders, spans for this purpose should be taken as the distance between dead-load inflection points.

Girder Web and Stiffeners.   The steel web should be proportioned so that the average shear stress over the gross section does not exceed the allowable. The effects of the steel flanges and concrete slab should be ignored. In addition, depth-thickness ratio should meet the requirements of Art. 10.13. Also, stiffeners should be provided, where needed, in accordance with those requirements. For web splices, see Arts. 5.26, 5.27, and 5.30.

Bending Stresses.   If, during erection, the steel girder is supported at intermediate points until the concrete slab has attained 75% of its required 28-day strength, the composite section may be assumed to carry the full dead load and all subsequent loads. When such shoring is not used, the steel girder alone must carry the steel and concrete dead loads. The composite section will support all loads subsequently applied. Thus, maximum bending stress in the steel of an unshored girder equals the sum of the dead-load stress in the girder alone plus stresses produced by loads on the composite section. Maximum bending stress in the concrete equals the stresses produced by those loads on the composite section at its top surface.

The positive-moment portion of continuous composite-girder spans should be designed in the same way as for simple spans. The negative-moment region need not be designed for composite action, in which case shear connectors need not be installed there. But additional connectors should be placed in the region of the dead-load inflection point as indicated later. If composite action is desired in the negative-moment portion, shear connectors should be installed. Then, longitudinal steel reinforcement in the concrete should be provided to carry the full tensile force. The concrete should be assumed to carry no tension.

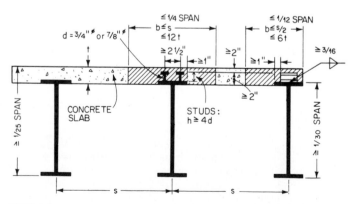

FIGURE 10.10   Effective width of concrete slab for composite construction.

Shear Connectors.   To ensure composite action, shear connectors must be capable of resisting both horizontal and vertical movements between concrete and steel. They should permit thorough compaction of the concrete so that their entire surfaces are in contact with the concrete. Usually, headed steel studs or channels, welded to the top flange of the girder, are used.

Channels should be attached transverse to the girder axis, with fillet welds at least along heel and toe. Minimum weld size permitted for this purpose is ³⁄₁₆ in.

Studs should be ¾- or ⅞-in nominal diameter. Overall length after welding should be at least 4 times the diameter. Steel should be A108, Grades 1015, 1018, or 1020, either fully or semikilled. The studs should be end-welded to the flange with automatically timed equipment. If a 360° weld is not obtained, the interrupted area may be repaired with a ³⁄₁₆-in fillet weld made by low-hydrogen electrodes in the shielded metal-arc process. Usually, two or more studs are installed at specific sections of a composite girder, at least four stud diameters c to c.

Clear depth of concrete cover over the tops of shear connectors should be at least 2 in. In addition, connectors should penetrate at least 2 in above the bottom of the slab. Clear distance between a flange edge and a shear-connector edge should not be less than 1 in in highway bridges, 1½ in in railroad bridges.

Pitch of Shear Connectors.   In general, shear connectors should not be spaced more than 24 in c to c along the span. Over interior supports of continuous beams, however, wider spacing may be used to avoid installation of connectors at points of high tensile stress.

Pitch may be determined by fatigue shear stresses due to change in horizontal shear or by ultimate-strength requirements for resisting total horizontal shear, whichever requires the smaller spacing. (Also, see the following method for stress design.)

Fatigue.   As live loads move across a bridge, the vertical shear at any point in a girder changes. For some position of the loading, vertical shear at the point due to live load plus impact reaches a maximum. For another position, shear there due to live load plus impact becomes a minimum, which may be opposite in sign to the maximum. The algebraic difference between maximum and minimum shear, kips, is the range of shear V_r.

The range of horizontal shear, kips per lin in, at the junction of a slab and girder at the point may be computed from

$$S_r = \frac{V_r Q}{I} \qquad (10.47)$$

where Q = statical moment, in³, about the neutral axis of the composite section, of the transformed compressive concrete area, or for negative bending moment, of the area of steel reinforcement in the concrete

I = moment of inertia, in⁴, about the neutral axis, of the transformed composite girder in positive-moment regions, and in negative-moment regions, the moment of inertia, in⁴, about the neutral axis, of the girder and concrete reinforcement if the girder is designed for composite action there, or without the reinforcement if the girder is noncomposite there

The allowable range of shear, kips per connector, is

FOR CHANNELS: $$Z_r = Bw \qquad (10.48)$$

FOR WELDED STUDS: $$Z_r = \alpha d^2 \qquad \left(\frac{h}{d} \geq 4\right) \qquad (10.49)$$

where w = transverse length of channel, in
d = stud diameter, in
h = overall stud height, in
B = 4 for 100,000 cycles of maximum stress
= 3 for 500,000 cycles
= 2.4 for 2,000,000 cycles (use this for railroad girders)
= 2.1 for more than 2,000,000 cycles
α = 13 for 100,000 cycles of maximum stress
= 10.6 for 500,000 cycles
= 7.85 for 2,000,000 cycles (use this for railroad girders)
= 5.50 for more than 2,000,000 cycles

The required pitch p_r, in, of shear connectors for fatigue is obtained from

$$p_r = \frac{\Sigma Z_r}{S_r} \tag{10.50}$$

where ΣZ_r is the allowable range of horizontal shear of all connectors at a cross section. Over interior supports of continuous beams, the pitch may be modified to avoid installation of connectors at points of high tensile stress. But the total number of connectors should not be decreased.

Ultimate Strength. The total number of connectors provided for fatigue, in accordance with Eq. (10.50), should be checked for adequacy at ultimate strength under dead load plus live load and impact. The connectors must be capable of resisting the horizontal forces H, kips, in positive-moment regions and in negative-moment regions. Thus, at points of maximum moment, H may be taken as the smaller of the values given by Eqs. (10.51) and (10.52).

$$H_1 = A_s F_y \tag{10.51}$$

$$H_2 = 0.85 f'_c bt \tag{10.52}$$

where A_s = cross-sectional area of steel girder, in^2
F_y = steel yield strength, ksi
f'_c = 28-day compressive strength of concrete, ksi
b = effective width of concrete slab, in
t = slab thickness, in

At points of maximum negative moment, H should be taken as

$$H_3 = A_{rs} F_{ry} \tag{10.53}$$

where A_{rs} = total area of longitudinal reinforcing steel at interior support within effective slab width, in^2, and F_{ry} = yield strength, ksi, of reinforcing steel. The total number of shear connectors required in any region then is

$$N = \frac{1,000H}{\phi Q_u} \tag{10.54}$$

where Q_u = ultimate strength of shear connector, lb, and ϕ = reduction factor, 0.85. In Eq. (10.54), the smaller of H_1 or H_2 should be used for H for determining the number of connectors required between a point of maximum positive moment and an end support in simple beams, and between a point of maximum positive moment and a dead-load inflection point in continuous beams. H_3 should be used for H for determining the total number of shear connectors required between a point of maximum negative moment and a dead-load inflection point in continuous beams. $H_3 = 0$ if slab reinforcement is not used in the computation of section properties for negative moment.

FOR CHANNELS:

$$Q_u = 550 \left(t_f + \frac{t_w}{2} \right) l \sqrt{f_c'} \qquad (10.55)$$

FOR WELDED STUDS:

$$Q_u = 0.4 d^2 \sqrt{f_c' E_c} \qquad \left(\frac{h}{d} > 4 \right) \qquad (10.56)$$

where E_c = modulus of the concrete, psi = $33 w^{3/2} \sqrt{f_c'}$
t_f = average thickness of channel flange, in
t_w = thickness of channel web, in
l = length of channel, in
f_c' = 28-day strength of concrete, psi
w = weight of the concrete, lb/ft³
d = stud diameter, in
h = stud height, in

Additional Connectors at Inflection Points. In continuous beams, the positive-moment region under live loads may extend beyond the dead-load inflection points, and additional shear connectors are required in the vicinity of those points when longitudinal reinforcing steel in the concrete slab is not used in computing section properties. The number needed is given by

$$N_c = A_{rs} \frac{f_r}{Z_r} \qquad (10.57)$$

where A_{rs} = total area, in², of longitudinal reinforcement at interior support within effective slab width
f_r = range of stress, ksi, due to live load plus impact in slab reinforcement over support (10 ksi may be used in the absence of accurate computations)
Z_r = allowable range, kips, of shear per connector, as given by Eqs. (10.48) and (10.49)

This number should be placed on either side of or centered about the inflection point for which it is computed, within a distance of one-third the effective slab width.

10.18 COST-EFFECTIVE PLATE-GIRDER DESIGNS

To get cost-effective results from the many different designs of fabricated girders that can satisfy the requirements of the "Standard Specifications for Highway Bridges" of the American Association of State Highway and Transportation Officials, designers should obtain advice from fabricators and contractors whenever possible. Also useful are steel-industry-developed rules-of-thumb intended to help designers. The following recommendations should be considered for all designs.

1. Load-factor design (LFD) yields more economical girder designs than does allowable-stress design (ASD). Modifying LFD by imposing higher loads than those for ASD, however, nullifies the usual cost advantage of LFD.

2. Properly designed for their environment, unpainted weathering-steel bridges are more economical in the long run than those requiring painting. Unpainted ASTM A588 weathering steel is the most economical.

3. The most economical painted design is that for hybrid girders. Painted homogenous girders of 50-ksi steel are a close second.

4. The fewer the girders, the greater the economy. Girder spacing must be compatible with deck design, but sometimes other factors govern selection of girder spacing. For economy, girder spacing should be 10 ft or more.

5. Transverse web stiffeners, except those serving as diaphragm or cross-frame connections, should be placed on only one side of a web.

6. Web depth may be several inches larger or smaller than the optimum without significant cost penalty.

7. A plate girder with a nominally stiffened web—$\frac{1}{16}$ in thinner than an unstiffened web—will be the least costly or very close to it. (Unstiffened webs are generally the most cost-effective for web depths less than 52 in. Nominally stiffened webs are most economical in the 52- to 72-in range. For greater depths, fully stiffened webs may be the most cost-effective.)

8. Web thickness should be changed only where splices occur. (Use standard-plate-thickness increments of $\frac{1}{16}$ in for plates up to 2 in thick and $\frac{1}{8}$-in increments for plates over 2 in thick.)

9. Longitudinal stiffeners should be considered for plate girders only for spans over 300 ft.

10. Not more than three plates should be butt-spliced to form the flanges of field sections up to 130 ft long. In some cases, it is advisable to extend a single flange-plate size the full length of a field section.

11. To justify a welded flange splice, about 700 lb of flange steel would have to be eliminated.

12. A constant flange width should be used between flange field splices. (Flange widths should be selected in 1-in increments.)

13. For most conventional cross sections, haunched girders are not advantageous for spans under 400 ft.

14. Bottom lateral bracing should be omitted where permitted by AASHTO specifications.

15. Elastomeric bearings are preferable to custom-fabricated steel bearings.

16. Composite construction may be advantageous in negative-moment regions of composite girders.

While these rules of thumb are generally valid, designers should bear in mind that such techniques as finite-element analysis, use of high-strength steels, and load-and-resistance-factor design make possible even better designs.

Consideration should be given to use of 40-in-deep rolled sections (meter beams). These may be cost-effective alternatives to welded girders for spans up to 100 ft or longer. Economy with these beams may be improved with end-bolted cover-plate details that allow use of category B stress ranges (Art. 10.11). Contract documents that allow either rolled beams or welded girders will be cost-effective for owners.

With fabricated girders, designers should ensure that flanges are wide enough to provide lateral stability for the girders during fabrication and erection. Flange width should be at least 12 in, but possibly even greater widths should be a minimum for deeper girders. The AISC recommends that, for shipping, handling, and erection, the ratio of length to width of compression flanges should be about 85 or less.

Designers also should avoid specifying thin flanges that make fabrication difficult. A thin flange is subject to excessive warping during welding of a web to the flange. To reduce warping, a flange should be at least ¾ in thick.

To minimize fabrication and deck forming costs when changes in the area of the top flange are required, the width should be held constant and required changes made by thickness transitions.

10.19 BOX GIRDERS

Closed-section members, such as box girders, often are used in highway bridges because of their rigidity, economy, appearance, and resistance to corrosion. Box girders have high torsional rigidity. With their wide bottom flanges (Fig. 10.11), relatively shallow depths can be used economically. And for continuous box girders, intermediate supports often can be individual, slender columns simply connected to concealed cross frames.

While box girders may be multicell (with three or more webs), single-cell girders, as illustrated in Fig. 10.11, are generally preferred. For short spans, such girders can be entirely shop-fabricated, permitting assembly by welding under closely controlled and economical conditions. Longer spans often can be prefabricated to the extent that only one field splice is necessary. One single-cell girder can be used to support bridges with one or two traffic lanes. But usually, multiple boxes are used to carry two or more lanes to keep box width small enough to meet shipping-clearance requirements.

Through the use of shear connectors welded to the top flanges, a concrete deck can be made to work with the box girders in carrying bending stresses. In such cases the concrete may be considered part of the top flange, and the steel top flange need be only wide enough for erection and handling stability, load distribution to the web, and placement of required shear connectors (Fig. 10.11).

Composite box girders are designed much like plate girders (Arts. 10.16 and 10.17). Criteria that are different are summarized in the following. For distribution of live loads to box girders, see Table 10.6.

Additional criteria apply to curved box-girder bridges.

Girder Spacing. The criteria are applicable to bridges with multiple single-cell box girders. Width center to center of top steel flanges in each girder should nearly equal the distance center to center between adjacent top steel flanges of adjacent boxes. (Width of boxes should nearly equal distance between boxes.) Cantilever overhang of deck, including curbs and parapets, should not exceed 6 ft or 60% of the distance between centers of adjacent top steel flanges of adjacent box girders.

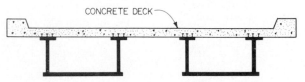

FIGURE 10.11 Composite construction with box girders.

Bracing. Diaphragms, cross frames, or other bracing should be provided within box girders at each support to resist transverse rotation, displacement, and distortion. Intermediate internal bracing for these purposes is not required if stability during concrete placement and curing has been otherwise ensured.

Lateral systems generally are not required between composite box girders. Need for a lateral system should be determined as follows: A horizontal load of 25 psf on the area of the girder exposed in elevation should be applied in the plane of the bottom flange. The resisting section should comprise the bottom flange serving as web, while portions of the box-girder webs, with width equal to 12 times their thickness, serve as flanges. A lateral system should be provided between bottom flanges if the combined stresses due to the 25-psf load and dead load of steel and deck exceed 150% of the allowable stress.

Access and Drainage. Manholes or other openings to the box interior should be provided for form removal, inspection, maintenance, drainage, or access to utilities.

Box-Girder Webs. Web plates may be vertical or inclined. A trapezoidal box generally requires a heavier bottom plate, and sometimes also a heavier concrete slab, but it may reduce the number of girders needed to support a deck. Design shear for an inclined web, kips, may be calculated from

$$V_w = \frac{V_v}{\cos \theta} \tag{10.58}$$

where V_v = vertical shear, kips, on web and θ = angle web makes with the vertical.

Transverse bending stresses due to distortion of the cross section and bottom-flange vibrations need not be considered if the web slope relative to the plane of the bottom flange is 4:1 or more and the bottom-flange width does not exceed 20% of the span. Furthermore, transverse bending stresses due to supplementary loadings, such as utilities, should not exceed 5 ksi. When any of the preceding limits are exceeded, transverse bending stresses due to all causes should be restricted to a maximum stress or stress range of 20 ksi.

Bottom Flange in Tension. Bending stress cannot be assumed uniformly distributed horizontally over very wide flanges. To simplify design, only a portion of such a flange should be considered effective, and the horizontal distribution of the bending stresses may be assumed uniform over that portion.

For simply supported girders, and between inflection points of continuous spans, the bottom flange may be considered completely effective if its width does not exceed one-fifth the span. For wider flanges, effective width equals one-fifth the span.

Unstiffened Compression Flanges. Compression flanges designed for the basic allowable stress of $0.55F_y$ need not be stiffened if the width-thickness ratio does not exceed

$$\frac{b}{t} = \frac{194}{\sqrt{F_y}} \tag{10.59}$$

where b = flange width between webs, in
$\quad\quad t$ = flange thickness, in
$\quad\quad F_y$ = steel yield strength for flange, ksi

When $194/\sqrt{F_y} < b/t \le 420/\sqrt{F_y}$, but not more than 60, the stress in an unstiffened bottom flange, ksi, should not exceed

$$F_b = F_y \left(0.326 + 0.224 \sin \frac{c\pi}{2} \right) \tag{10.60}$$

$$c = \frac{420 - (b/t) \sqrt{F_y}}{226} \tag{10.61}$$

When $b/t > 420/\sqrt{F_y}$, the stress, ksi, in the flange should not exceed

$$F_b = \frac{57{,}600}{(b/t)^2} \tag{10.62}$$

b/t preferably should not exceed 60, except in areas of low stress near inflection points.

Longitudinally Stiffened Compression Flanges.   When $b/t > 45$, use of longitudinal stiffeners should be considered. When used, they should be equally spaced across the compression flange. The number required depends heavily on the ratio of spacing to flange thickness.

For the flange, including the longitudinal stiffeners, to be designed for the basic allowable stress $0.55F_y$, this ratio should not exceed

$$\frac{w}{t} = \frac{97}{\sqrt{F_y/k}} \tag{10.63}$$

where w = width of flange, in, between longitudinal stiffeners or distance, in, from a web to nearest stiffener and k = buckling coefficient, which may be assumed to be between 2 and 4. For larger values of w/t, but not more than 60 or $210/\sqrt{F_y/k}$, the stress, ksi, in the flange should not exceed

$$F_b = F_y \left(0.326 + 0.224 \sin \frac{c'\pi}{2} \right) \tag{10.64}$$

$$c' = \frac{210 - (w/t) \sqrt{F_y/k}}{113} \tag{10.65}$$

When $210/\sqrt{F_y/k} < w/t \le 60$, the stress, ksi, should not exceed

$$F_b = \frac{14{,}400k}{(w/t)^2} \tag{10.66}$$

Stiffeners should be proportioned so that the depth-thickness ratio of any outstanding element does not exceed

$$\frac{d_s}{t_s} = \frac{82.2}{\sqrt{F_y}} \tag{10.67}$$

where d_s = depth, in, of outstanding element and t_s = thickness, in, of element. The moment of inertia, in^4, of each longitudinal stiffener about an axis through the base of the stiffener and parallel to the flange should be at least

$$I_s = \phi w t^3 \tag{10.68}$$

where $\phi = 0.07k^3n^4$ for $n > 1$
        $= 0.125k^3$ for $n = 1$
    n = number of longitudinal stiffeners

Longitudinal stiffeners should be extended to locations where the maximum stress in the flange does not exceed that allowed for base metal adjacent to or connected by fillet welds. At least one transverse stiffener should be installed near dead-load inflection points. It should be the same size as the longitudinal stiffeners.

Compression Flanges Stiffened Longitudinally and Transversely.   When $w/t > 97/\sqrt{F_y/k}$ and the number of longitudinal stiffeners exceeds two, addition of transverse stiffeners should be considered. They are not necessary, however, if the ratio of their spacing to flange width b exceeds 3. For the flange, including stiffeners, to be designed for the basic allowable stress $0.55F_y$, w/t for the longitudinal stiffeners should not exceed

$$\frac{w}{t} = \frac{97}{\sqrt{F_y/k_1}} \tag{10.69}$$

$$k_1 = \frac{[1 + (a/b)^2]^2 + 87.3}{(n + 1)^2(a/b)^2[1 + 0.1(n + 1)]} \leq 4 \tag{10.70}$$

where a = spacing, in, of transverse stiffeners. For larger values of w/t but not more than 60 or $210\sqrt{F_y/k_1}$, the stress, ksi, in the flange should not exceed

$$F_b = F_y\left(0.326 + 0.224 \sin \frac{c''\pi}{2}\right) \tag{10.71}$$

$$c'' = \frac{210 - (w/t) \sqrt{F_y/k_1}}{113} \tag{10.72}$$

When $210\sqrt{F_y/k_1} < w/t < 60$, the stress, ksi, should not exceed

$$F_b = \frac{14{,}400k_1}{(w/t)^2} \tag{10.73}$$

Spacing of transverse stiffeners should not exceed $4w$ when k_1 has its maximum value of 4.

When transverse stiffeners are used, each longitudinal stiffener should have a moment of inertia I_s as given by Eq. (10.68) with $\phi = 8$. Each transverse stiffener should have a moment of inertia, in^4, about an axis through its centroid parallel to its bottom edge of at least

$$I_t = \frac{0.10(n + 1)^3w^3f_bA_f}{Ea} \tag{10.74}$$

where f_b = maximum longitudinal bending stress, ksi, in flange in panels on either
        side of transverse stiffener
    A_f = area, in^2, of bottom flange, including stiffeners
    E = modulus of elasticity of flange steel, ksi

Depth-thickness ratio of outstanding elements should not exceed the value determined by Eq. (10.67).

Transverse stiffeners need not be connected to the flange. But they should be attached to the girder webs and longitudinal stiffeners. Each of these web connections should be capable of resisting a vertical force, kips,

$$R_w = \frac{F_y S_t}{2b} \tag{10.75a}$$

where S_t = section modulus, in³, of transverse stiffener and F_y = yield strength, ksi, of stiffener. Each connection of a transverse and longitudinal stiffener should be capable of resisting a vertical force, kips,

$$R_s = \frac{F_y S_t}{nb} \tag{10.75b}$$

Flange-to-Web Welds. Total effective thickness of welds connecting a flange to a web should at least equal the web thickness, except that when two or more diaphragms per span are provided, minimum size fillet welds may be used (see Art. 10.23). If fillet welds are used, they should be placed on both sides of the flange or web.

10.20 HYBRID GIRDERS

When plate girders are to be used for a bridge, costs generally can be cut by using flanges with higher yield strength than that of the web. Such construction is permitted for highway bridges under AASHTO specifications if the girders qualify as hybrid girders. Such girders are cost effective because the web of a plate girder contributes relatively little to the girder bending strength and the web shear strength depends on the depth/thickness ratio.

Hybrid girders, in general, may be designed for fatigue as if they were homogeneous plate girders of the flange steel. Composite and noncomposite I-shaped girders may qualify as hybrid.

Noncomposite girders must have both flanges of steel with the same yield strength. Yield strength of web steel should be lower, but not more than 35% less. Different areas may be used at the same cross section for top and bottom flanges. If, however, the bending stress in either flange exceeds $0.55F_{yw}$, where F_{yw} is the specified minimum yield stress of the web, ksi, the tension-flange area should be larger than the compression-flange area. In composite construction, the transformed area of the effective concrete slab or reinforcing steel should be included in the top-flange area.

Composite girders, in contrast, may have a compression flange of steel with yield strength less than that of the tension flange but not less than that of the web. Yield strength of web steel should be lower, but not by more than 35%, than the yield strength of the tension flange.

Criteria governing design of hybrid girders generally are the same as for homogeneous plate girders (Arts. 10.16 and 10.17). Those that differ follow.

Web. Average shear stress in the web should not exceed the allowable for the web steel.

The bending stress in the web may exceed the allowable for the web steel if the stress in each flange does not exceed the allowable for the flange steel multiplied by a reduction factor R.

$$R = 1 - \frac{\beta\psi(1-\alpha)^2(3-\psi+\psi\alpha)}{6+\beta\psi(3-\psi)} \tag{10.76}$$

where $\alpha = F_{yw}/F_{yf}$
F_{yw} = minimum specified yield strength of web, ksi
F_{yf} = minimum specified yield strength of flange, ksi
β = ratio of web area to tension-flange area
ψ = ratio of distance, in, between outer edge of tension flange and neutral axis (of the transformed section for composite girders) to depth, in, of steel section

In computation of maximum permissible depth-thickness ratios for a web, f_b should be taken as the calculated bending stress, ksi, in the compression flange divided by R.

In design of bearing stiffeners at interior supports of continuous hybrid girders for which $\alpha < 0.7$, no part of the web should be assumed to act in bearing.

Flanges. In composite girders, the bending stress in the concrete slab should not exceed the allowable stress for the concrete multiplied by R.

In computation of maximum permissible width-thickness ratios of a compression flange, f_b should be taken as the calculated bending stress, ksi, in the flange divided by R.

10.21 ORTHOTROPIC-DECK BRIDGES

In orthotropic-deck construction, the deck is a steel plate overlaid with a wearing surface and stiffened and supported by a rectangular grid. The steel deck assists its supports in carrying bending stresses. Main components usually are the steel deck plate, longitudinal girders, transverse floorbeams, and longitudinal ribs. Ribs may be open-type (Fig. 10.12a) or closed (Fig. 10.12b).

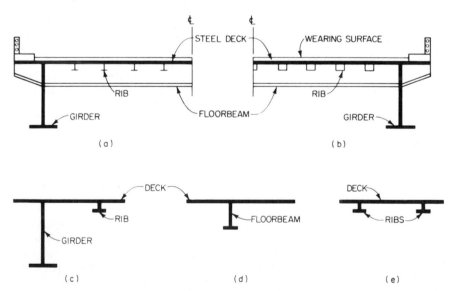

FIGURE 10.12 Orthotropic-plate construction. (*a*) With open ribs. (*b*) With closed ribs. (*c*) Deck and ribs act as the top flange of the main girder. (*d*) Deck acts as the top flange of the floorbeam. (*e*) Deck distributes loads to the ribs.

The steel deck acts as the top flange of the girders (system I, Fig. 10.12c). Also, the steel deck serves as the top flange of the ribs (Fig. 10.12e) and floorbeams (system II, Fig. 10.12d). In addition, the deck serves as an independent structural member that transmits loads to the ribs (system III, Fig. 10.12e).

Load Distribution. In determining direct effects of wheel loads on the deck plate, in design of system III for H20 or HS20 loadings, single-axle loads of 24 kips, or double-axle loads of 16 kips each spaced 4 ft apart, should be used. The contact area of one 12- or 8-kip wheel may be taken as 20 in wide (perpendicular to traffic) and 8 in long at the roadway surface. The loaded area of the deck may be taken larger by the thickness of the wearing surface on all sides, by assuming a 45° distribution of load through the pavement.

Deck Thickness. Usually, the deck plate is made of low-alloy steel with a yield point of 50 ksi. Thickness should be at least ⅜ in and is determined by allowable deflection under a wheel, unless greater thickness is required by design of system I or II. Deflection due to wheel load plus 30% impact should not exceed ⅟₃₀₀ of spacing of deck supports. Deflection computations should not include the stiffness of the wearing surface. When support spacing is 24 in or less, the deck thickness, in, that meets the deflection limitation is

$$t = 0.07ap^{1/3} \tag{10.77}$$

where a = spacing, in, of open ribs, or maximum spacing, in, of walls of closed ribs and p = pressure at top of steel deck under 12-kip wheel, ksi.

Allowable Stresses. Stresses in ribs and deck acting as the top flange of the girders and in the ribs due to local bending under wheel loads should be within the basic allowable tensile stress. But when the girder-flange stresses and local bending stresses are combined, they may total up to 125% of the basic allowable tensile stress. Local bending stresses are those in the deck plate due to distribution of wheel loads to ribs and beams. AASHTO standard specifications limit local transverse bending stresses for the wheel load plus 30% impact to a maximum of 30 ksi unless fatigue analysis or tests justify a higher allowable stress. If the spacing of transverse beams is at least 3 times that of the webs of the longitudinal ribs, local longitudinal and transverse bending stresses need not be combined with other bending stresses, as indicated in the following.

Elements of the longitudinal ribs and the portion of the deck plate between rib webs should meet the minimum thickness requirements given in Table 10-15. The stress f_a may be taken as the compressive bending stress due to bending of the rib, bending of the girder, or 75% of the sum of those stresses, whichever is largest. Unless analysis shows that compressive stresses in the deck induced by bending of the girders will not cause overall buckling of the deck, the slenderness ratio L/r of any rib should not exceed

$$\frac{L}{r} = 1000\sqrt{\frac{1.5}{F_y} - 2700\frac{F}{F_y^2}} \tag{10.78}$$

where L = distance, in, between transverse beams
r = radius of gyration, in³, about the horizontal centroidal axis of the rib plus effective area of deck plate
F_y = yield strength, ksi, of rib steel
F = maximum compressive stress, ksi (taken positive) of the deck plate acting as the top flange of the girders

The effective width, and hence the effective area, of the deck plate acting as the top flange of a longitudinal rib or a transverse beam should be determined by analysis of the orthotropic-plate system. Approximate methods may be used. (See, for example, Art. 4.12 or "Design Manual for Orthotropic Steel Plate Deck Bridges," American Institute of Steel Construction.) For the girders, the full width of the deck plate may be considered effective as the top flange if the girder span is at least 5 times the maximum girder spacing and 10 times the maximum distance from the web to the nearest edge of the deck. (For continuous beams, the span should be taken as the distance between inflection points.) If these conditions are not met, the effective width should be determined by analysis.

The elements of the girders and beams should meet requirements for minimum thickness given in Table 10.15 and for stiffeners (Arts. 10.31.4 and 10.19).

When connections between ribs and webs of beams, or holes in beam webs for passage of the ribs, or rib splices occur in tensile regions, they may affect the fatigue life of the bridge adversely. Consequently, these details should be designed to resist fatigue as described in Art. 10.11. Similarly, connections between the ribs and the deck plate should be designed for fatigue stresses in the webs due to transverse bending induced by wheel loads.

At the supports, some provision, such as diaphragms or cross frames, should be made to transmit lateral forces to the bearings and to prevent transverse rotation and other deformations.

The same method of analysis used to compute stresses in the orthotropic-plate construction should be used to calculate deflections. Maximum deflections of ribs, beams, and girders due to live load plus impact should not be more than $1/500$ of the span. See also Art. 11.10.

10.22 SPAN LENGTHS AND DEFLECTIONS

Many designers believe that steel girders, because of their lower weight per foot, should have longer spans than concrete beams for a bridge at the same location. This is not necessarily the case. The AISC has conducted studies that show that there are substantial economies for the steel alternative when the spans are kept the same, including the cost of extra substructure units. However, as with any preliminary study, site-specific considerations may indicate otherwise. For example, where the foundation or substructure costs, or both, are extremely high, it is probable that longer steel girders, with fewer substructure units, would be more cost-effective than shorter spans.

Deflection of steel bridges has always been important in design. If a bridge is too flexible, the public often complains about bridge vibrations, especially if sidewalks are present. There is also a concern that bridge vibrations may accelerate fatigue damage or cause premature deck deterioration. In an attempt to satisfy all these concerns, the AASHTO standard specifications include limitations on deflection and depth-span ratios as a means of ensuring sufficient stiffness of bridge members (Art. 10.3.1).

There is some doubt about the need for these limitations, especially relative to the potential for increased deck cracking. Many studies indicate flexibility of the superstructure is not a cause of increased deck cracking. The AISC notes that most European countries do not have live-load deflection limits in their design specifications.

The AASHTO LRFD specifications require that deflections be checked as part of the service limit state and include in the "Commentary" the statement: "Service limit states are intended to allow the bridge to perform acceptably for its service life. . . . Bridges should be designed to avoid undesirable structural or psychological effects from their deflection and vibrations. While no specific deflection, depth, or frequency limitations are specified herein, except for orthotropic decks, any large deviation from past successful practice regarding slenderness and deflections should be cause for review of the design to determine that it will perform satisfactorily.''

The LRFD specifications provide optional criteria for deflections that are essentially the same as those in the standard specifications. These provisions apply to all structures, not just steel, as was the case in the past. The LRFD specifications also require checking I-section members for permanent deflections.

10.23 *BEARINGS*

Bridges should be designed so that a total movement due to temperature change of 1¼ in per 100 ft can take place. Also, provisions should be made for changes in length of span resulting from live-load stresses. In spans over 300 ft long, allowance should be made for expansion and contraction in the floor system.

Expansion bearings are needed to permit such movements. (See also Art. 10.27.) In addition, to control the movements, at least one fixed bearing is required in each simple or continuous span. A fixed bearing should be firmly anchored against horizontal and vertical movement, but it may permit the end of the member supported to rotate in a vertical plane. An expansion bearing should permit only end rotation and movement parallel to the longitudinal axis of the supported member, unless provisions for transverse expansion are necessary.

Allowable bearing on granite is 800 psi and on sandstone or limestone, 400 psi, when the masonry projects 3 in or more beyond the edge of the bearing plate. For smaller projections, only 75% of these stresses is allowed. For reinforced concrete, the basic allowable stress f_c is 30% of the 28-day compressive strength. When the supporting surface is wider on all sides than the loaded area A_1, the allowable stress may be multiplied by $\sqrt{A_2/A_1} \leq 2$, where A_2 is the area of the supporting surface.

Bearings for spans of 50 ft or more should be designed to permit end rotation. For the purpose, curved bearing plates, elastomeric pads, or pin arrangements may be used. Elastomeric bearings are generally preferred. At expansion bearings, such spans may be provided with rollers, rockers, or sliding plates. Shorter spans may slide on metal plates with smooth surfaces.

In all cases, design of supports should ensure against accumulation of dirt, which could obstruct free movement of the span, and against trapping of water, which could accelerate corrosion. Beams, girders, or trusses should be supported so that bottom chords or flanges are above the bridge seat.

Self-lubricating bronze or copper-alloy sliding plates, with a coefficient of friction of 0.10 or less, may be used in expansion bearings instead of elastomeric pads, rollers, or rockers. These plates should be at least ½ in thick and chamfered at the ends.

Rockers generally are preferred to rollers because of the smaller probability of becoming frozen by dirt or corrosion. The upper surface of a rocker should have a pin or cylindrical bearing. The lower surface should be cylindrical with center of rotation at the center of rotation of the upper bearing surface. At the nominal centerline of bearing, the lower portion should be at least 1½ in thick. The effective length of rocker for computing line bearing stress should not exceed the length of the upper bearing surface plus the distance from lower to upper bearing surface. Adequate web material should be provided and arranged to ensure uniform load distribution over the effective length. The rocker should be doweled to the base plate.

Rollers are the alternative when the pressure on a rocker would require it to have too large a radius to keep bearing stress within the allowable. Rollers may be cylindrical or segmental. They should be at least 6 in diameter. They should be connected by substantial side bars and guided by gearing or other means to prevent lateral movement, skewing, and creeping. The roller nest should be designed so that the parts may be easily cleaned.

Effective Bearing Area.   Effective length of bearing area may be taken equal to effective length of rocker, or to roller length plus twice the thickness of the base plate. Effective width of bearing area may be taken as 4 times the base-plate thickness for rockers, or the distance between end rollers plus 4 times the base-plate thickness for rollers. The vertical load may be assumed uniformly distributed over the effective bearing area, except for eccentricity from rocker travel.

Sole plates and masonry plates should be at least ¾ in thick. For bearings with sliding plates but without hinges, the distance from centerline of bearing to edge of masonry plate, measured parallel to the longitudinal axis of the supported member, should not exceed 4 in plus twice the plate thickness. For spans on inclines exceeding 1% without hinged bearings, the bottom of the sole plate should be radially curved or beveled to be level.

Elastomeric pads are bearings made partly or completely of elastomer. They are used to transmit loads from a structural member to a support while allowing movements between the bridge and the support. Pads that are not all elastomer (reinforced pads) generally consist of alternate layers of steel or fabric reinforcement bonded to the elastomer. In addition to the reinforcement, the bearings may have external steel plates bonded to the elastomeric bearings. AASHTO prohibits tapered elastomeric layers in reinforced bearings.

The AASHTO "Standard Specifications for Highway Bridges" contain specifications for the materials, fabrication, and installation of the bearings. The specifications also present two methods for their design, both based on service loads without impact and the shear modulus at 73°F. The grade of elastomer permitted depends on the temperature zone in which the bridge is located. The specifications also require that either (1) a positive slip apparatus be installed and bridge components be able to withstand forces arising from a bearing force equal to twice the design shear force or (2) bridge components be able to sustain the forces arising from a bearing force equal to four times the design shear force. If the shear force exceeds one-fifth the dead-load compressive force, the bearing should be fixed against horizontal movement.

Design should allow for misalignment of girders because of fabrication or erection tolerances, camber, or other sources. It should also provide for subsequent replacement of bearings, when necessary. Also, it should ensure that bearings are not subjected to uplift when in service.

A beam or girder flange seated on an elastomeric bearing should be stiff enough to avoid damaging it. Stiffening may be achieved with a sole plate or bearing stiffeners. I beams and girders symmetrically placed on a bearing do not require such stiffening if the width-thickness ratio b_f/t_f of the bottom flange does not exceed

$$\frac{b_f}{t_f} = 2\sqrt{\frac{F_y}{3.4 f_c}} \qquad (10.79)$$

where b_f = total width, in, of the flange
    t_f = thickness, in, of flange or flange plus sole plate
    F_y = minimum yield strength, ksi, of girder steel
    f_c = average compressive stress P/A, ksi, due to dead plus live load, without impact

TFE pads are bearings with sliding surfaces made of polytetrafluoroethylene (TFE), which may consist of filled or unfilled sheet, fabric with TFE fibers, interlocked bronze and filled TFE structures, TFE-perforated metal composites and adhesives, or stainless-steel mating surfaces. The AASHTO standard specifications contain specifications for the materials, fabrication, and installation of the bearings.

The sliding surfaces of the pads permit translation or rotation by sliding of the TFE surfaces over a smooth, hard mating surface. This should preferably be made of stainless steel or other corrosion-resistant material. To prevent local stresses on the sliding surface, an expansion bearing should permit rotation of at least 1° due to live load, changes in camber during construction, and misalignment of the bearing. This may be achieved with such devices as hinges, curved sliding surfaces, elastomeric pads, or preformed fabric pads.

TFE sliding surfaces should be factory-bonded or mechanically fastened to a rigid backup material capable of resisting bending stresses to which the surfaces may be subjected. The surface mating to the TFE should be an accurate mate, flat, cylindrical, or spherical, as required, and should cover the TFE completely in all operating positions of the bearing. Preferably, the mating surface should be oriented so that sliding will cause dirt and dust to fall off it.

Pot Bearings.   Used mainly for long-span bridges, pot bearings are available as fixed, guided expansion, and nonguided expansion bearings, designed to provide for thermal expansion and contraction, rotation, camber changes, and creep and shrinkage of structural members. They consist of an elastomeric rotational element, confined and sealed by a steel piston and steel base pot. In effect, a structure supported on a pot bearing floats on a low-profile hydraulic cylinder, or pot, in which the liquid medium is an elastomer.

To facilitate rotation of the elastomeric rotational element, either TFE sheets are attached to the top and bottom of the elastomeric disk or the element is lubricated with a material compatible with the elastomer. To permit longitudinal or transverse movements, the upper surface of the steel piston is faced with a TFE sheet and supports a steel sliding-top bearing plate. The mating surface of that plate is faced with polished stainless steel.

Pot bearings have low resistance to bending in their plane. Consequently, a sole plate, beveled if necessary, should be provided on top of the bearing and a masonry plate should be installed on the bottom. A member should not be supported on both a pot bearing and a bearing with different properties.

To ensure contact between the piston and the elastomer, minimum load should be at least 20% of the design vertical load capacity.

Pedestals and shoes, if required, usually are made of cast steel or structural steel. Design should be based on the assumption that the vertical load is uniformly distributed over the entire bearing surface. The difference in width or length between top and bottom bearing surfaces should not exceed twice the vertical distance between them. For hinged bearings, this distance should be measured from the center of the pin.

AREA recommends that all components of cast-steel shoes and load-carrying parts of welded shoes be at least 1 in thick. AASHTO recommends that the web plates and angles connecting built-up pedestals and shoes to the base plate should be at least ⅝ in thick.

If pedestal size permits, webs should be rigidly connected transversely to ensure stability of the components. Webs and pinholes in them should be arranged to keep eccentricity to a minimum. The net section through a pinhole should provide at least 140% of the net area required for the stress transmitted through the pedestal or shoe. All parts of pedestals and shoes should be prevented from lateral movement on the pins.

Nuts with washers should be used to hold pins in place. Length of pins should be adequate for full bearing.

TABLE 10.16 Minimum Number of Anchor Bolts per Bearing

Span, ft	No. of bolts	Diameter, in	Embedment, in
(*a*) Trusses and girders			
50 or less............	2	1	10
51–100..............	2	1¼	12
101–150.............	2	1½	15
150 or more	4	1½	15
(*b*) Rolled beams			
All outer spans	2	1	10

Anchor bolts subject to tension should be designed to engage a mass of masonry that will provide resistance to uplift equal to 150% of the calculated uplift due to service loads or 100% of loading combinations for which live load plus impact is increased 100%, whichever is larger. The bolts, however, may be designed for 150% of the basic allowable stress. Resistance to pullout of anchor bolts may be obtained by use of swage bolts or by placing on each embedded end of a bolt a nut and washer or plate. Minimum requirements for number of bolts for each bearing, diameter, and embedment are given in Table 10.16.

10.24 DETAILING FOR WELDABILITY

Overdetailing of weld sizes and joint configurations can cause unnecessary fabrication and in-service problems and higher costs. Some designers believe "more weld metal is better" and "complete-penetration groove welds are better than fillet welds." But oversizing welds or specifying joint figurations that are not practical can cause inclusion of weld defects that are otherwise avoidable.

Whenever possible, designers should allow fabricators to select the type of joint to be used and the size of weld (Fig. 10.13). The "Standard Specifications for Highway Bridges" of the American Association of State Highway and Transportation Officials include maximum and minimum sizes for fillet welds as follows:

Limitations on Fillet-Weld Size. Size of a fillet weld may be the same as the material thickness up to ¼ in. For material ¼ in thick or more, size is limited to ¹⁄₁₆ in less than the material thickness, unless the drawings indicate that the weld should be built up to get full throat thickness.

Minimum size of fillet weld is based on the base-metal thickness of the thinner part joined, and single-pass welds must be used. For material ¾ in thick or less, weld size should be at least ¼ in. For thicker material, weld size may not be less than ⁵⁄₁₆ in. Only if the strength requirement exceeds that provided by the minimum size of fillet weld is it necessary to indicate the size of a fillet weld on the drawings. The "Bridge Welding Code," AWS D1.5, American Welding Society, provides adequate assurance of proper weld strength and quality. Letting fabricators select joint details for efficient utilization of their plant setup ensures the most cost-effective fabrication.

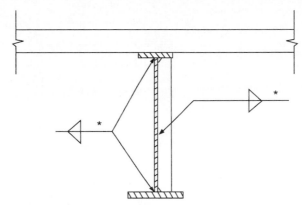

FIGURE 10.13 Symbols indicate welds to be made to a girder. Asterisks indicate that the weld sizes are to be selected by the fabricator. A note should be placed on the drawing to that effect. This does not apply when stress levels control.

The AASHTO specifications also require that the minimum length of a fillet weld be 4 times its size but at least 1½ in. If a fillet weld is subjected to repeated stress or to a tensile force not parallel to its axis, it should not end at a corner of a part or a member. Instead, it should be turned continuously around the corner for a distance equal to twice the weld size (if the return can be made in the same plane). Seal welds should be continuous.

Welding of Box Girders.   Poor detailing of a box girder or other type of enclosed member has been another source of fabrication problems and has contributed to adverse in-service performance when designs have not provided properly for fabrication. For example, designers often specify a complete-penetration groove weld for a corner, and the backing bar needed to ensure integrity of the weld is not always installed properly. Backing bars are sometimes left discontinuous, and this soon causes a fatigue crack to initiate. Also, when internal stiffeners are required for a box girder, which is frequently the case for large sections, assembly problems are encountered where welds or backing bars are interrupted at the stiffeners. Figure 10.14 shows a detail with backing bar that is not recommended for a box girder and a preferred arrangement that eliminates both the need for a backing bar and for welding to be done inside the box for attachment of the web to the top plate.

The assembly procedure requires first welding of the two webs to the bottom flange. For the purpose, continuous fillet welds are placed on one or both sides. Then, the stiffeners are welded to the webs (also to the compression flange if the member will be subjected to bending). Finally, the top flange is connected to the webs with fillet welds. The advantage of this procedure lies in the fact that it is usually practicable to get a fillet weld of better quality, easier to inspect with a nondestructive test, and less expensive than a complete-penetration weld.

10.25 STRINGER OR GIRDER SPACING

One of the major factors affecting the economy of highway bridges with a concrete deck on stringers or longitudinal girders is spacing of the main members. Older bridges typically had spacing of 8 ft or less. Now, however, longer concrete-deck

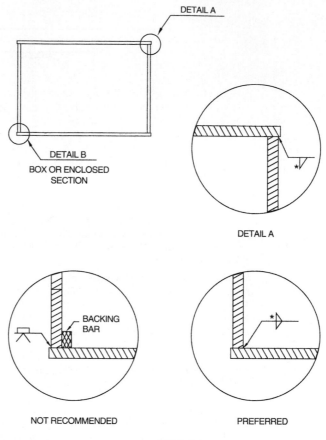

DETAIL A

DETAIL B

BOX OR ENCLOSED
SECTION

DETAIL A

BACKING
BAR

NOT RECOMMENDED

PREFERRED

DETAIL B

FIGURE 10.14   Corner joints for a box-shape member. Detail *A* requires a fillet weld between web and top flange. Asterisk indicates that
the size of the weld is to be selected by the fabricator. This does not
apply when stress levels control. Detail *B* shows two schemes for welding
of the web to the bottom flange, one not recommended and the other
preferred.

spans (up to 15 ft) are practicable through use of such devices as stay-in-place
metal or precast-concrete forms. This makes possible fewer girders. (To eliminate
the potential for fracture criticality when I-shape girders are used, there should be
at least three.) Although the steel weight per square foot of bridge may be higher
with fewer girders, the reduced costs of fabrication, handling, transportation, erecting, and painting, if required, usually provides substantial overall savings.

10.26   BRIDGE DECKS

Highway-bridge decks usually are constructed of reinforced concrete. Often, this
concrete is made with conventional aggregate and weighs about 150 lb per cu ft.
Sometimes, it is made with lightweight aggregate, resulting in 100 to 110 lb per cu

ft concrete. Lightweight aggregate normally consists of slag, expanded shale, or expanded clay.

In some concrete decks, the wearing surface is cast integrally with the structural slab. In others, a separate wearing surface, consisting of asphaltic concrete or conventional concrete, is added after the structural slab has been placed.

In instances where weight saving is important, particularly in movable spans, or in spans where aerodynamic stability is of concern, an open, steel-grid floor is specified. Where compromise is necessary, this grid is partly or completely filled with asphaltic or lightweight concrete to provide protection under the structure or to provide a more suitable riding surface.

For orthotropic-plate structures, is is necessary to provide over the steel deck a wearing surface on which traffic rides. These wearing surfaces are generally of three types: a layered system, stabilized mastic system, or thin combination coatings.

The layered system consists of a steel-deck prime coat, such as zinc metallizing, bituminous-base materials, or epoxy coatings. Over this coat is applied a copper or aluminum foil, or an asphalt mastic, followed by a leveling course of asphalt binder or stabilized mastic, and a surface course of stone-filled mastic asphalt or asphaltic concrete.

The stabilized mastic system consists of a prime coat on the steel, as in the layered system, followed by a layer of mastic, which is choked with rolled-in crushed rock.

Combination coatings contain filled epoxies or alkyd-resin binders in a single coating with silica sand.

A bridge deck serves as a beam on elastic foundations to transfer wheel loads to the supporting structural steel. In orthotropic bridges, the deck also contributes to the load-carrying capacity of longitudinal and transverse structural framing. In composite construction, the concrete deck contributes to the load-carrying capacities of girders. In fulfilling these functions, decks are subject to widely varying stresses and strains, due not only to load but also to temperature changes and strains of the main structure.

In general, bridge decks are designed as flexural members spanning between longitudinal or transverse beams and supporting wheel loads. A wheel usually is considered a concentrated load on the span but uniformly distributed in the direction normal to the span.

Concrete Slabs. The **effective span** S, ft, for a concrete slab supported on steel beams should be taken as the distance between edges of flanges plus half the width of a beam flange.

Allowable Stresses. The allowable compressive stress for concrete in design of slabs is $0.4f'_c$, where f'_c = 28-day compressive strength of concrete, ksi. The allowable tensile stress for reinforcing bars for grade 40 is 20 ksi and for grade 60, 24 ksi. Slabs designed for bending moment in accordance with the following provisions may be considered satisfactory for bond and shear.

Bending Moment. Because of the complexity of determining the exact load distribution, AASHTO specifications permit use of a simple empirical method. The method requires use of formulas for maximum bending moment due to live load (impact not included). Two principal cases are treated depending on the direction in which main reinforcement is placed. The formulas are summarized in Table 10.17. In these formulas, S is the effective span, ft, of the slab, as previously defined.

For rectangular slabs supported along all edges and reinforced in two directions perpendicular to the edges, the proportion of the load carried by the short span

TABLE 10.17 Live-Load Bending Moments, ft-kips per ft of Width, in Concrete Slabs*

Direction of main reinforcement and type of span	Loading	
	HS20	HS15
Perpendicular to traffic ($2 \leq S \leq 24$):		
Simple spans	$0.5(S + 2)$	$0.375(S + 2)$
Continuous spans	$\pm 0.4(S + 2)$	$\pm 0.3(S + 2)$
Cantilevers, $E = 0.8x + 3.75$†	$16x/E$†	$12x/E$†
Parallel to traffic:		
Simple spans:		
$S \leq 50$.............................	$0.900S$	$0.675S$
$50 < S \leq 100$........................	$1.3S - 20$	$0.750(1.3S - 20)$
Continuous spans	By analysis‡	By analysis‡
Cantilevers, $E = 0.35x + 3.2 \leq 7$†	$16x/E$†	$12x/E$†

*Based on "Standard Specifications for Highway Bridges," American Association of State Highway and Transportation Officials.

†x = distance, ft, from load to support.

‡Moments in continuous spans with main reinforcement parallel to traffic should be determined by analysis for the truck or appropriate lane loading. Distribution of wheel loads $E = 4 + 0.06S \leq 7$ ft. Lane loads should be distributed over a width of $2E$.

may be assumed for uniformly distributed loads as

$$p = \frac{b^4}{a^4 + b^4} \tag{10.80}$$

For a load concentrated at the center,

$$p = \frac{b^3}{a^3 + b^3} \tag{10.81}$$

where a = length of short span of slab, ft, and b = length of long span of slab, ft. If the length of slab exceeds 1.5 times the width, the entire load should be assumed carried by the reinforcement of the short span. The distribution width E, ft, for the load taken by either span should be determined as provided for other slabs in Table 10.17. Reinforcement determined for bending moments computed with these assumptions should be used in the center half of the short and long spans. Only 50% of this reinforcement need be used in the outer quarters. Supporting beams should be designed taking into account the nonuniform load distribution along their spans.

All slabs with main reinforcement parallel to traffic should be provided with edge beams. They may consist of a slab section with additional reinforcement, a beam integral with but deeper than the slab, or an integral, reinforced section of slab and curb. Simply supported edge beams should be designed for a live-load moment, ft-kips, of $1.6S$ for HS20 loading and $1.2S$ for HS15 loading, where S is the beam span, ft. For positive and negative moments in continuous beams, these values may be reduced 20%.

Distribution reinforcement is required in the bottom of all slabs transverse to the main reinforcement, for distribution of concentrated wheel loads. The minimum amounts to use are the following percentages of the main reinforcement steel required for positive moment:

For main reinforcement parallel to traffic, $\qquad \dfrac{100}{\sqrt{S}} \leq 50\%$

For main reinforcement perpendicular to traffic, $\qquad \dfrac{220}{\sqrt{S}} \leq 67\%$

where S = effective span of slab, ft. When main reinforcing steel is perpendicular to traffic, the distribution reinforcement in the outer quarters of the slab span need be only 50% of the required distribution reinforcement.

Transverse unsupported edges of the slab, such as at ends of a bridge or expansion joints, should be supported by diaphragms, edge beams, or other means, designed to resist moments and shears produced by wheel loads.

The effective length, ft, of slab resisting post loadings may be taken as

$$E = 0.8x + 3.75$$

where no parapet is used, with x = distance, ft, from center of post to point considered. If a parapet is used, $E = 0.8x + 5$.

Steel Grid Floors.   For grid floors filled with concrete, the load distribution and bending moments should be determined as for concrete slabs. The strength of the composite steel and concrete slab should be computed by the transformed-area method (Art. 10.17). If necessary to ensure adequate load transference normal to the main grid elements, reinforcement should be welded transverse to the main steel.

For open-grid floors, a wheel load should be distributed normal to the main bars over a distance equal to twice the center-to-center spacing of main bars plus 20 in for H20 loading, or 15 in for H15 loading. The portion of the load assigned to each bar should be uniformly distributed over a length equal to the rear-tire width (20 in for H20 loading and 15 in for H15). The strength of the section should be determined by the moment-of inertia method (Art. 10.16). Supports should be provided for all edges of open-grid floors.

10.27 EXPANSION JOINTS IN HIGHWAY BRIDGES

At expansion bearings and at other points where necessary, expansion joints should be installed in the floor system to permit it to move when the span deflects or changes length. If apron plates are used, they should be designed to bridge the joint and prevent accumulation of dirt on the bridge seats. Preferably, the apron plates should be connected to the end floorbeam. For amount of movement to provide for, see Art. 10.22.

Short-span bridges usually have expansion joints at one or both abutments. Longer-span structures usually have such joints at pier or off-pier hinges. Although these joints may relieve some forces caused by restraint of thermal movements, the joints have been a major source of bridge deterioration and poor ridability. The LRFD specifications of the American Association of State Highway and Transportation Officials (AASHTO) acknowledges that "Completely effective joint seals have yet to be developed for some situations . . . ." To provide more durable bridges, the goal in design should be to minimize the number of joints. One way to do this for multiple-span bridges is to use continuous beams or girders. Another, more general, alternative is to eliminate joints completely.

Some states permit jointless, or integral, steel-girder bridges with spans up to about 400 ft or longer. With this type of construction, restriction of the change in bridge length due to maximum temperature change induces longitudinal forces at fixed piers and abutments. This must be taken into account in design of substructures. Experience has shown, however, that the effect of these forces on superstructure design is negligible and that, with proper detailing, substructure design is relatively unaffected.

Tennessee is a major user of jointless steel-girder bridges for spans of 400 ft or more. Through experience, the highway department has developed details that are able to resist thermal forces and movements (Fig. 10.15), thus eliminating leaking bridge joints.

The AASHTO "Standard Specifications for Highway Bridges" specifies that movement calculations for integral abutments take into account not only temperature changes but also creep of the concrete deck and pavements. The abutments

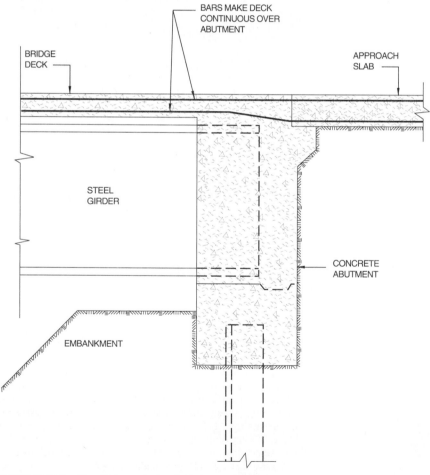

FIGURE 10.15   Details for an integral abutment.

should be designed to sustain the forces generated by restraint to thermal movements developed by the pressures of fills behind the abutments. (The specifications prohibit use of integral abutments constructed on spread footings keyed into rock.) Approach slabs should be connected directly to abutments and wingwalls, to prevent intrusion of water behind the abutments. Nevertheless, means should be provided for draining away water that may get entrapped.

The AASHTO specifications also require that details comply with recommendations in Technical Advisory T5140.13, Federal Highway Administration. These recommendations include the following:

Steel bridges with an overall length less than 300 ft should be constructed continuously and, if unrestained, have integral abutments. ("An unrestrained abutment is one that is free to rotate, such as a stub abutment on one row of piles or an abutment hinged at the footing."—"Structure Memorandum," State of Tennessee.) Greater lengths may be used when experience dictates such designs are satisfactory.

In the area immediately behind integral abutments, traffic will compact the fill where it is partly disturbed by abutment movement, if not prevented from doing so. For the purpose, approach slabs should be provided to span this area. The span length should be at least equal to a minimum of 4 ft for bearing on the soil plus the depth of the abutment (based on the assumption of a 1:1 slope from the bottom of the rear face of the abutment.) The Advisory suggests that a practical slab length is 14 ft.

The Advisory recommends that approach slabs be designed for live-load bending moments as indicated for the case of main reinforcement parallel to traffic in Table 10.17, with S = slab length minus 2 ft.

The Advisory also recommends that the slabs be anchored by steel reinforcement to the superstructure. In addition, positive anchorage should be provided between integral abutments and the superstructure. Figure 10.15 is an example of such construction.

The Advisory calls attention to a detail used by North Dakota that it considers desirable. To accommodate pavement growth and bridge movement, the state inserts a roadway expansion joint 50 ft away from the bridge.

Properly detailed and constructed, jointless bridges eliminate the maintenance that would be required if expansion joints were used, especially corrosion and deterioration of substructure and superstructure because of leakage. Also, jointless bridges provide better ridability. As a bonus, the cost of joints is eliminated.

10.28   BRIDGE STEELS AND CORROSION PROTECTION

One of the most important decisions designers have to make is selection of the proper grade of steel and corrosion-protection system. These should not only meet structural needs but also provide an economical structure capable of long-term, low-maintenance performance.

The "Standard Specifications for Highway Bridges" of the American Association of State Highway and Transportation Officials recognize structural steels designated M270 with a specified grade. These are equivalent to ASTM A709 steels of the same grades except for AASHTO-specified mandatory notch toughness. Material properties of M270 steels and other equivalent ASTM steels are listed in Table 10.18. (See also Table 1.1 and Art. 1.1.5). Steels that meet AASHTO M270 requirements are prequalified for use in welded bridges.

Designers should have available AASHTO "Standard Specifications for Transportation Materials and Methods of Sampling and Testing," Part 1, "Specifica-

TABLE 10.18 Highway-Bridge Structural Steels*

Type	Structural steel	High-strength low-alloy steel		Quenched and tempered low-alloy steel	High-yield-strength, quenched and tempered alloy steel	
AASHTO designation	M 270 grade 36	M 270 grade 50	M 270 grade 50W	M 270 grade 70W	M 270 grades 100/100W	
Equivalent ASTM designations	A 709 grade 36 A36	A 709 grade 50 A572 grade 50	A 709 grade 50W A588	A 709 grade 70W A852	A 709 grades 100/100W A514	
					Plates 2½ in over 2½ in thick or less thick to 4 in, incl.	
Minimum tensile strength, F_u	58	65	70	90	110	100
Minimum yield point or Minimum yield strength, F_y	36	50	50	70	100	90

*Based on the "Standard Specifications for Highway Bridges," American Association of State Highway and Transportation Officials. See also Table 1.1 and Art. 1.1.5.

tions," and Part 2, "Tests," to ensure that appropriate material properties are specified for their designs.

10.28.1 Minimum Steel Thickness

Because structural steel in bridges is exposed to the weather, minimum-thickness requirements are imposed on components to obtain a long life despite corrosion. Where steel will be exposed to unusual corrosive influences, the component should be increased in thickness beyond required thickness or specially protected against corrosion.

In highway bridges, structural steel components, except railings, fillers, and webs of certain rolled shapes, should be at least 5⁄16 in thick. Web thickness of rolled beams or channels should be at least 0.23 in. Closed ribs in orthotropic-plate decks should be at least 3⁄16 in thick. Fillers less than ¼ in thick should not be extended beyond splicing material.

In railroad bridges, steel, except for fillers, should not be less than 0.335 in thick. Gusset plates connecting chords and web members of trusses should be at least ½ in thick.

In addition, minimum thickness may be governed in highway and railroad bridges by slenderness ratios (Table 10.14) or maximum width-thickness or depth-thickness ratios (Table 10.15).

10.28.2 Weathering Steels

One way to achieve economy of structure is to use steel of a "weathering" grade when conditions permit. This is a type of steel that has enhanced atmospheric corrosion resistance *when properly used* and does not require painting, but it costs slightly more per pound than other steels of equivalent grade. The weathering grades are available only with yield points of 50 ksi and higher. Before selecting a weathering steel, designers should determine the corrosivity of the environment in which the bridge will be located as a first step. This will determine whether the use of an unpainted steel of grade 50W, 70W, or 100W (Table 10.18 and Arts. 1.1.4 and 1.1.5) is appropriate. These steels provide the most cost-effective grade that can be used in most situations and have proven to be capable of excellent performance even in areas where deicing salts are used. But use of good detailing practices, such as jointless bridges, is imperative to assure adequate performance (Art. 10.26).

The Federal Highway Administration "Guidelines for the Use of Unpainted Weathering Steel," to ensure a long-term and adequate performance of unpainted steels, recommends the following:

If the proposed structure is to be located at a site with any of the environmental or location characteristics noted below, use of uncoated weathering-grade steels should be considered with caution. A study of both the macroenvironment and microenvironment by a corrosion consultant may be required. In all environments, designers must pay careful attention to detailing, specifically as noted in the following recommendations for design details. Also, owners should implement, as a minimum, the maintenance actions as noted in the following.

Environments to be treated with caution include marine coastal areas; regions with frequent high rainfall, high humidity, or persistent fog; and industrial areas where concentrated chemical fumes may drift directly onto structures.

Locations to be treated with caution include grade separations in tunnel-like conditions, where concentration of vehicle exhausts may be highly corrosive; also, low-level water crossings, with clearance of 10 ft or less over stagnant, sheltered water or 8 ft or less over moving water.

Design details for uncoated steel in bridges and other highway structures require careful consideration of the following:

1. Elimination of bridge joints where possible.
2. If expansion joints are used, they must be able to control water that comes on the deck. A trough under the deck joint may serve to divert water away from vulnerable elements.
3. Painting all superstructure steel within a distance of 1½ times the depth of girder from bridge joints.
4. Avoiding use of welded drip bars where fatigue stresses may be critical.
5. Minimizing the number of bridge-deck scuppers.
6. Eliminating details that serve as water and debris "traps."
7. If box girders are used, they should be hermetically sealed, when possible, or provided with weep holes to allow proper drainage and circulation of air. All openings in boxes that are not sealed should be covered or screened.
8. Protecting pier caps and abutment walls to minimize staining.
9. Sealing overlapping surfaces exposed to water, to prevent capillary penetration of moisture.

Maintenance actions advisable include the following:

1. Implementing procedures designed to detect and minimize corrosion.
2. Controlling roadway drainage by diverting roadway drainage away from the bridge structure, cleaning troughs or resealing deck joints, maintaining deck drainage systems, and periodically cleaning and, when needed, repainting all

steel within a minimum distance of 1½ times the depth of the girder from bridge joints.

3. Regularly removing all dirt, debris, and other deposits that trap moisture.

4. Regularly removing all vegetation and other matter that can prevent the natural drying of wet steel surfaces.

5. Maintaining covers and screens over access holes.

The preceding recommendations are applicable to all structures, painted or unpainted, to ensure satisfactory performance. Unpainted structures that have been in existence for 20 or more years in environments consistent with these recommendations have provided excellent service, testifying to the adequacy of the weathering grades of steel.

10.28.3 Paint Systems

Where weathering grades of steel are not appropriate, only high-performance paint systems should be specified for corrosion protection. Designers should be aware, however, that recommendations for paint systems change periodically, primarily due to the need for consideration of environmental impacts. Lead-based paints, for example, are no longer acceptable due to their health hazard. Also, concern for the effect of volatile organic compounds on the ozone in the atmosphere has caused a change from mineral-based to water-based paints. Consequently, designers should ensure that only current technology is specified in contract documents.

Paint systems can be generally classed as organic or inorganic, depending on whether or not the vehicle or binder contains organic compounds. All systems contain three major ingredients in general: pigments; film-formers, or binders; and solvents.

The pigments are finely divided solids that give color, consistency, buildup, and durability.

The film-formers, or binders, are oils, resins, inorganic compounds, and other materials that make up the protective film of the coating. They are also called vehicle solids or nonvolatile vehicles. The vehicle contains all the liquid portion of the surface coating.

The solvents, also called thinners and volatiles, are liquids added to the paint to make them fluid enough for proper application. They evaporate, leaving the residue of pigment and binder to form the paint film by various drying and hardening processes.

Film formation of paint systems for bridge structures usually occurs by evaporation of volatile liquids. It may be the only method with oils or inert resins in suitable solvents. Other paint systems containing resins usually harden by a chemical process of polymerization by oxygen. Relatively new methods of film formation occur by chemical reactions between a catalyst and resinous material or between resins. Epoxies are an example of this process.

Compatibility of pigment and vehicle has an important effect on the durability of the paint coat.

Some binders, such as the hard resins, contribute hardness to the paint coat, whereas others, such as soft drying oils, provide flexibility, durability, and adhesion. Often, several binders occur in a paint formulation.

Penetration can be developed by proper binder selection. Highly polymerized oils and resins with large molecules penetrate less than those of lower polymerization.

Coating systems and paints recommended for use in highway bridges are given in Table 10.19.

("Steel Structures Painting Manual," Steel Structures Painting Council, Pittsburgh.)

Surface Preparation. Mill scale is formed on steel during the rolling process as a result of evaporation of water used in removal of production scale. It consists of an outer layer of iron oxide, an intermediate layer of magnetic oxide, and an inner layer of ferrous oxide. This mill scale acts as a cathode. At breaks in the scale, the steel acts as an anode in the presence of moisture. The resulting galvanic current corrodes the steel.

Mill scale is most commonly removed in the shop by sandblasting or centrifugal shotblasting. Acid pickling, hand- or power-tool cleaning, and flame cleaning are also used.

Research by the Steel Structures Painting Council indicates that pickling is better than, or equivalent to, many grades of shot- and sandblasting. Sandblasting is as good as, or better than, equivalent degrees of shotblasting. Wire brushing and solvent cleaning are poorer surface preparations than either blast cleaning or pickling. There is little difference in coating performance between that on white metal and commercial blast-cleaned surfaces. Tests also indicate vinyl paints perform much better than phenolics on all test surfaces.

TABLE 10.19 Coatings Systems and Paints for Highway Bridges[a]

Layer	Environment		
	High-pollution[b] and coastal[c]	Mild climate[d]	Mild climate[d] or maintenance repainting
Primer	Inorganic zinc[e]—3 mils	Organic zinc[e]—3 mils	Oil/alkyd[f]—2 mils
Intermediate coat	Epoxy—2 mils Vinyl wash primer[e]—0.3–0.5 mils	Epoxy—2 mils Vinyl wash primer[e]—0.3–0.5 mils	Oil/alkyd[f]—2 mils
Top coat	Epoxy, vinyl, or urethane[f]—2 mils	Epoxy, vinyl, or urethane[f]—2 mils	Oil/alkyd[f]—2 mils
Total system	5.3–7 mils	5.3–7 mils	6 mils

[a]Based on recommendations in AASHTO "Standard Specifications for Highway Bridges," 15th ed., 1992.

Coating and system thicknesses given in the table are the minimum recommended, except for vinyl wash primer. Coating systems listed for a specific environment may be used in a less severe environment. As a minimum, handling of paint should meet the provisions of the Steel Structures Painting Council (SSPC) PA Guide 3, "A Guide to Safety in Paint Application."

[b]High pollution—polluted environment such as industrial areas.

[c]Coastal zones—within 1000 ft of ocean or tidal water.

[d]Mild climate—other than a coastal region or high pollution environment.

[e]Requirement of Military Specifications should be met as follows: inorganic zinc paint, DOD-P-23236A (SH); organic zinc paint, DOD-P-21035A; vinyl wash primer, DOD-P-15328D.

[f]SSPC requirements should be met as follows: vinyl top-coat paint, SSPC-Paint 9; epoxy paint, SSPC-22; oil/alkyd primer and intermediate coat paint, SSPC-Paint 25; oil/alkyd top-coat paint, SSPC-Paint 104; urethane top-coat paint, SSPC-PS Guide 1700.

10.29 CONSTRUCTABILITY

Sometimes, unnecessary problems develop during construction of a bridge that could have easily been prevented with an appropriate design. Also, the construction procedures used by a contractor may lock in stresses unaccounted for in design that will adversely influence the service life of the bridge. Two specific areas where difficulties have occurred have been in construction of curved girder bridges and in deck-concrete placing sequences, especially when the bridge has a large skew.

As part of bridge design, the designers should assume an erection and concrete-placing sequence and check for construction stresses. The assumed methods should be included on the contract plans for the contractor's information, with the understanding that deviations will be accepted subject to the ability of the contractor to demonstrate that no adverse stresses will result from the proposed method.

The AASHTO LRFD specifications, to ensure that designers properly consider constructibility, specify that bridges be designed so that fabrication and erection can be performed without undue difficulty or distress and that effects of locked-in construction forces are within tolerable limits. When the method of construction of a bridge is not self-evident, or could induce unacceptable locked-in stresses, the designer should propose at least one feasible method on the plans. If the design requires some strengthening or temporary bracing or support during erection by the selected method, the plans should indicate the need thereof.

To provide for the above, designers should check for what is essentially a construction limit state. For the purpose, the following factored load should be used:

$$1.25[D + 1.5L + 1.25W + 1.0\Sigma \text{ (other forces as appropriate)}]$$

where D is the dead load, L is the live load, and W is the wind load. This should be applied to all designs, regardless of which specification is used.

10.30 INSPECTABILITY

Inspectability of all bridge members and connections is an essential design-stage consideration. This is especially apparent when the structure includes enclosed sections, such as box girders. Bridge service life has been impaired in the past when designers, concerned with stress distribution, either did not include access holes or made them so small it was impossible for an inspector to perform an adequate inspection. To ensure inspectability, experienced bridge inspectors should review the bridge design at an early stage of development.

Another consideration is safety of inspectors and traffic using the bridge during the inspection. A preferred method of inspection has been use of a "snooper" crane that allows easy access to underbridge members. But, on routes with very high traffic volumes, the presence of an inspection vehicle on the bridge creates a safety hazard to both inspection personnel and the traveling public. Other means of inspection should be provided in these instances, such as inspection ladders, walkways, catwalks, covered access holes, and provision for lighting, if necessary.

10.31 REFERENCE MATERIALS

Besides the "Standard Specifications for Highway Bridges" referred to frequently in preceding articles, the American Association of State Highway and Transportation Officials publishes numerous reference books, guide specifications, manuals, interim specifications, periodicals and other reference materials useful for design, fabrication, and erection of steel highway bridges. Obtain the latest catalog listing these from AASHTO, 444 N. Capitol St., NW, Washington, DC 20001.

The "Manual for Railway Engineering," American Railway Engineering Association, 2000 L St., NW, Washington, DC 20036, contains recommended practices for railway bridges.

SECTION 11
BEAM AND GIRDER BRIDGES

Alfred Hedefine*
President, Parsons, Brinckerhoff, Quade & Douglas, Inc.,
New York, NY

and

John Swindlehurst
Senior Professional Associate, Parsons, Brinckerhoff,
Quade & Douglas Inc., Newark, N.J.

Beams generally are the most economical type of framing for short-span bridges. Contemporary capabilities for extending beam construction to longer and longer spans safely and economically stem principally from the introduction of steel and the availability, in the early part of the twentieth century, of standardized rolled beams. By the late thirties, after wide-flange shapes became generally available, highway stringer bridges were erected with simply supported, wide-flange beams on spans up to about 110 ft. Riveted plate girders were used for highway-bridge spans up to about 150 ft. In the fifties, spans were extended to 300 ft by taking advantage of welding, continuity, and composite construction. And in the sixties, spans two and three times as long became economically feasible with the use of high-strength steels and box girders, or orthotropic-plate construction, or stayed girders. Thus, now, engineers, as a matter of common practice, design girder bridges for medium and long spans as well as for short spans.

11.1 CHARACTERISTICS OF BEAM BRIDGES

Rolled wide-flange shapes generally are the most economical type of construction for short-span bridges. The beams usually are used as stringers, set, at regular intervals, parallel to the direction of traffic, between piers or abutments (Fig. 11.1). The deck, placed on the top flange, often provides lateral support against buckling.

---

*With the assistance for the second edition of Mahir Sen, Bridge Engineer. Mr. Hedefine was assisted in preparation of Sec. 11 for the first edition by Mr. Swindlehurst, Robert Warshaw, Assistant Vice President, and Louis G. Silano, Associate, Parsons, Brinckerhoff, Quade & Douglas, Inc.

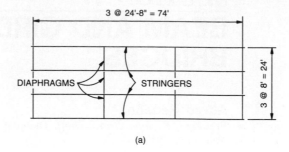

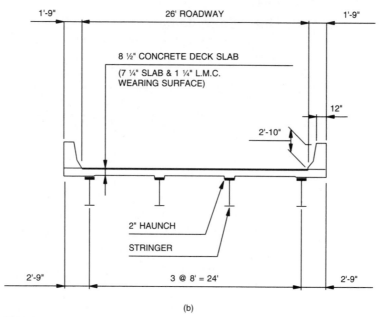

FIGURE 11.1 Two-lane highway bridge with rolled-beam stringers. (*a*) Framing plan. (*b*) Typical cross section.

Diaphragms between the beams offer additional bracing and also distribute loads laterally to the beams before the concrete deck has cured.

Spacing.   For railroad bridges, two stringers generally carry each track. They may, however, be more widely spaced than the rails, for stability reasons. If a bridge contains only two stringers, the distance between their centers should be at least 6 ft 6 in. When more stringers are used, they should be placed to distribute the track load uniformly to all beams.

For highway bridges, one factor to be considered in selection of stringer spacing is the minimum thickness of concrete deck permitted. For the deck to serve at maximum efficiency, its span between stringers should be at least that requiring the minimum thickness. But when stringer spacing requires greater than minimum thickness, the dead load is increased, cutting into the savings from use of fewer stringers. For example, if the minimum thickness of concrete slab is about 8 in, the stringer spacing requiring this thickness is about 8 ft for 4,000-psi concrete.

Thus, a 29-ft 6-in-wide bridge, with 26-ft roadway, could be carried on four girders with this spacing. The outer stringers then would be located 1 ft from the curb into the roadway, and the outer portion of the deck, with parapet, would cantilever 2 ft 9 in beyond the stringers.

If an outer stringer is placed under the roadway, the distance from the center of the stringer to the curb preferably should not exceed about 1 ft.

Stringer spacing usually lies in the range 6 to 15 ft. The smaller spacing generally is desirable near the upper limits of rolled-beam spans.

The larger spacing is economical for the longer spans where deep, fabricated, plate girders are utilized. Wider spacing of girders has resulted in development of long-span stay-in-place forms. This improvement in concrete-deck forming has made steel girders with a concrete deck more competitive.

Regarding deck construction, while conventional cast-in-place concrete decks are commonplace, precast-concrete deck slab bridges are often used and may prove practical and economical if stage construction and maintenance of traffic are required. Additionally, use of lightweight concrete, a durable and economical product, may be considered if dead weight is a problem.

Other types of deck are available such as steel orthotropic plates (Arts. 11.14 and 11.15). Also, steel grating decks may be utilized, whether unfilled, half-filled, or fully filled with concrete. The latter two deck-grating construction methods make it possible to provide composite action with the steel girder.

Short-Span Stringers.   For spans up to about 40 ft, noncomposite construction, where beams act independently of the concrete slab, and stringers of AASHTO M270 (ASTM A709), Grade 36 steel often are economical. If a bridge contains more than two such spans in succession, making the stringers continuous could improve the economy of the structure. Savings result primarily from reduction in number of bearings and expansion joints. A three-span continuous beam, for example, requires four bearings, whereas three simple spans need six bearings.

For such short spans, with relatively low weight of structural steel, fabrication should be kept to a minimum. Each fabrication item becomes a relatively large percentage of material cost. Thus, cover plates should be avoided. Also, diaphragms and their connections to the stringers should be kept simple. For example, they may be light channels field bolted or welded to plates welded to the beam webs (Fig. 11.2).

For spans 40 ft and less, each beam reaction should be transferred to a bearing plate through a thin sole plate welded to the beam flange. The bearing may be a flat steel plate or an elastomeric pad. At interior supports of continuous beams, sole plates should be wider than the flange. Then, holes needed for anchor bolts can be placed in the parts of the plates extending beyond the flange. This not only reduces fabrication costs by avoiding holes in the stringers but also permits use of lighter stringers, because the full cross section is available for moment resistance.

At each expansion joint, the concrete slab should be thickened to form a transverse beam, to protect the end of the deck. Continuous reinforcement is required for this beam. For the purpose, slotted holes should be provided in the ends of the steel beams to permit the reinforcement to pass through.

Live Loads.   Although AASHTO "Standard Specifications for Highway Bridges" specify for design H15-44, HS15-44, H20-44, and HS20-44 truck and lane loadings (Art. 10.4), many state departments of transportation are utilizing larger live loadings. The most common is HS20-44 plus 25% (HS25). An alternative military loading of two axles 4 ft apart, each axle weighing 24 kips, is usually also required and should be used if it causes higher stresses.

Dead Loads.   Superstructure design for bridges with a one-course deck slab should include a 25-psf additional dead load to provide for a future 2-in-thick

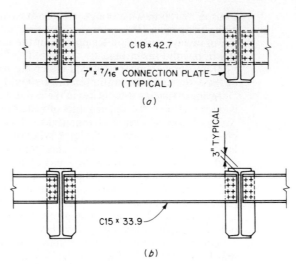

FIGURE 11.2 Diaphragms for rolled-beam stringers. (*a*) Intermediate diaphragm. (*b*) End diaphragm.

overlay wearing surface. Bridges with a two-course deck slab should not include this additional dead load. The assumption is that during repaving of the adjoining roadway, the 1¼-in wearing course (possibly latex modified concrete) will be removed and replaced only if necessary.

If metal stay-in-place forms are permitted for deck construction, consideration should be given to providing for an additional 8 psf to be included for the weight of the permanent steel form plus 5 psf for the additional thickness of deck concrete required. The specific additional dead load should be determined for the form to be utilized. The additional dead load is considered secondary and may be included in the superimposed dead load supported by composite construction, when used.

Long-Span Stringers. Composite construction with rolled beams (Art. 10.17) may become economical when simple spans exceed about 40 ft, or the end span of a continuous stringer exceeds 50 ft, or the interior span of a continuous stringer exceeds 65 ft. W36 rolled wide-flange beams of Grade 36 steel designed for composite action with the concrete slab are economical for spans up to about 85 ft, though such beams can be used for longer spans. When spans exceed 85 ft, consideration should be given to rolled beams made of high-strength steels, W40 rolled wide-flange beams, or to plate-girder stringers. In addition to greater economy than with noncomposite construction, composite construction offers smaller deflections or permits use of shallower stringers, and the safety factor is larger.

For long-span, simply supported, composite, rolled beams, costs often can be cut by using a smaller rolled section than required for maximum moment and welding a cover plate to the bottom flange in the region of maximum moment (partial-length cover plate). For the purpose, one plate of constant width and thickness should be used. It also is desirable for cover plates on continuous beams. The cover plate thickness should generally be limited to about 1 in and be either 2 in narrower or 2 in maximum wider than the flange. Longitudinal fillet welds attach the plate to the flange. Cover plates may be terminated and end-welded within the span at the theoretical cutoff point. American Association of State Highway and Transportation Officials (AASHTO) specifications provide for a Category E' allowable fatigue-stress range that must be utilized in the design of girders at this theoretic point.

Problems with fatigue cracking of the end weld and flange plate of older girders has caused designers to avoid terminating the cover plate within the span. Some state departments of transportation specify that cover plates be full length or terminated within 2 ft of the end bearings. The end attachments may be either special end welds or bolted connections.

Similarly, for continuous, noncomposite, rolled beams, costs often can be cut by welding cover plates to flanges in the regions of negative moment. Savings, however, usually will not be achieved by addition of a cover plate to the bottom flange in positive-moment areas. For composite construction, though, partial-length cover plates in both negative-moment and positive-moment regions can save money. In this case, the bottom cover plate is effective because the tensile forces applied to it are balanced by compressive forces acting on the concrete slab serving as a top cover plate.

For continuous stringers, composite construction can be used throughout or only in positive-moment areas. Costs of either procedure are likely to be nearly equal.

Design of composite stringers usually is based on the assumption that the forms for the concrete deck are supported on the stringers. Thus, these beams have to carry the weight of the uncured concrete. Alternatively, they can be shored, so that the concrete weight is transmitted directly to the ground. The shores are removed after the concrete has attained sufficient strength to participate in composite action. In that case, the full dead load may be assumed applied to the composite section. Hence, a slightly smaller section can be used for the stringers than with unshored erection. The savings in steel, however, may be more than offset by the additional cost of shoring, especially when provision has to be made for traffic.

Diaphragms for long-span rolled beams, as for short-span, should be of minimum permitted size. Also, connections should be kept simple (Fig. 11.2). At span ends, diaphragms should be capable of supporting the concrete edge beam provided to protect the end of the concrete slab.

For simply supported, long-span stringers, one end usually is fixed, whereas arrangements are made for expansion at the other end. Bearings may be built up of steel or they may be elastomeric pads. A single-thickness pad may be adequate for spans under 85 ft. For longer spans, laminated pads will be needed. Expansion joints in the deck may be made economically with extruded or preformed plastics.

Cambering of rolled-beam stringers is expensive. It often can be avoided by use of different slab-haunch depths over the beams.

11.2 EXAMPLE—ALLOWABLE-STRESS DESIGN OF COMPOSITE, ROLLED-BEAM STRINGER BRIDGE

To illustrate the design procedure, a two-lane highway bridge with simply supported, composite, rolled-beam stringers will be designed. As indicated in the framing plan in Fig. 11.1a, the stringers span 74 ft center to center (c to c) of bearings. The typical cross section in Fig. 11.1b shows a 26-ft-wide roadway flanked by 1-ft 9-in parapets. Structural steel to be used is Grade 36. Loading is HS25. Appropriate design criteria given in Sec. 10 will be used for this structure. Concrete to be used for the deck is Class A, with 28-day compressive strength $f'_c = 4,000$

psi and allowable compressive strength $f_c = 1,400$ psi. Modulus of elasticity $E_c = 33w^{1.5}\sqrt{f'_c} = 33(140)^{1.5}\sqrt{4,000} = 3,457,300$ psi.

Assume that the deck will be supported on four rolled-beam stringers, spaced 8 ft c to c, as shown in Fig. 11.1.

Concrete Slab.   The slab is designed to span transversely between stringers, as in noncomposite design. The effective span S is the distance between flange edges plus half the flange width, ft. In this case, if the flange width is assumed as 1 ft, $S = 8 - 1 + \frac{1}{2} = 7.5$ ft. For computation of dead load, assume a 9-in-thick slab, weight 112 lb/ft² plus 5 lb/ft² for stay-in-place forms. The 9-in-thick slab consists of a 7¾-in base slab plus a 1¼-in latex-modified concrete (LMC) wearing course. Total dead load then is 117 lb/ft². With a factor of 0.8 applied to account for continuity of the slab over the stringers, the maximum dead-load bending moment is

$$M_D = \frac{w_D S^2}{10} = \frac{117(7.5)^2}{10} = 660 \text{ ft-lb per ft}$$

From Table 10.17, the maximum live-load moment, with reinforcement perpendicular to traffic, plus a 25% increase for conversion to HS25 loading, equals

$$M_L = 1.25 \times 400(S + 2) = 500(7.5 + 2) = 4,750 \text{ ft-lb/ft}$$

Allowance for impact is 30% of this, or 1,425 ft-lb/ft. The total maximum moment then is

$$M = 660 + 4,750 + 1,425 = 6,835 \text{ ft-lb/ft}$$

For balanced design of the concrete slab, the depth $k_b d_b$ of the compression zone is determined from

$$k_b = \frac{1}{1 + f_s/nf_c} = \frac{1}{1 + 24,000/8(1,400)} = 0.318$$

where d_b = effective depth of slab, in, for balanced design
$\quad\quad f_s$ = allowable tensile stress for reinforcement, psi = 24,000 psi
$\quad\quad n$ = modular ratio = $E_s/E_c = 8$
$\quad\quad E_s$ = modulus of elasticity of the reinforcement, psi = 29,000,000 psi
$\quad\quad E_c$ = modulus of elasticity of the concrete, psi = 3,600,000 psi

For determination of the moment arm $j_b d_b$ of the tensile and compressive forces on the cross section,

$$j_b = 1 - k_b/3 = 1 - 0.318/3 = 0.894$$

Then the required depth for balanced design, with width of slab b taken as 1 ft, is

$$d_b = \sqrt{2M/f_c bjk} = 5.86 \text{ in}$$

For the assumed dimensions of the concrete slab, the depth from the top of slab to the bottom reinforcement is

$$d = 9 - 0.5 - 1 - 0.38 = 7.12 \text{ in}$$

The depth from bottom of slab to top reinforcement is

$$d = 7.75 + 1.25 - 2.75 - 0.38 = 5.88 \text{ in}$$

Since $d > d_b$, this will be an underreinforced section. Use $d = 5.88$ in. Then, the maximum compressive stress on a slab of the assumed dimensions is

$$f_c = \frac{M}{(kd)(jd)b/2} = \frac{6,835 \times 12}{1.87 \times 5.27 \times 12/2} = 1,390 < 1,400 \text{ psi}$$

Hence, a 9-in-thick concrete slab is satisfactory.

Required reinforcement area transverse to traffic is

$$A_s = \frac{12M}{f_s jd} = \frac{12 \times 6,835}{24,000 \times 5.27} = 0.65 \text{ in}^2/\text{ft}$$

Use No. 6 bars at 8-in intervals. These supply 0.66 in²/ft. For distribution steel parallel to traffic, use No. 5 bars at 9 in, providing an area about two-thirds of 0.65 in²/ft.

Stringer Design Procedure. A composite stringer bridge may be considered to consist of a set of T beams set side by side. Each T beam comprises a steel stringer and a portion of the concrete slab (Art. 10.17). The usual design procedure requires that a section be assumed for the steel stringer. The concrete is transformed into an equivalent area of steel. This is done for a short-duration load by dividing the effective area of the concrete flange by the ratio n of the modulus of elasticity of steel to the modulus of elasticity of the concrete, and for a long-duration load, under which the concrete may creep, by dividing by $3n$. Then, the properties of the transformed section are computed. Next, bending stresses are checked at top and bottom of the steel section and top of concrete slab. After that, cover-plate lengths are determined, web shear is investigated, and shear connectors are provided to bond the concrete slab to the steel section. Finally, other design details are taken care of, as in noncomposite design.

Fabrication costs often will be lower if all the stringers are identical. The outer stringers, however, carry different loads from those on interior stringers. Sometimes girder spacing can be adjusted to equalize the loads. If not, and the load difference is large, it may be necessary to provide different designs for inner and outer stringers. Exterior stringers, however, should have at least the same load capacity as interior stringers. Since the design procedure is the same in either case, only a typical interior stringer will be designed in this example.

Loads, Moments, and Shears. Assume that the stringers will not be shored during casting of the concrete slab. Hence, the dead load on each stringer includes the weight of an 8-ft-wide strip of concrete slab as well as the weights of steel shape, cover plate, and framing details. This dead load will be referred to as DL.

DEAD LOAD CARRIED BY STEEL BEAM, KIPS PER FT:

Slab: $0.150 \times 8 \times 7.75 \times 12$	=	0.775
Haunch—12×1 in: $0.150 \times 1 \times 1/12$	=	0.013
Stay-in-place forms: 0.013×7	=	0.091
Rolled beam and details—assume		0.296
DL per stringer		1.175

Maximum moment occurs at the center of the 74-ft span:

$$M_{DL} = 1.175(74)^2/8 = 804 \text{ ft-kips}$$

Maximum shear occurs at the supports and equals

$$V_{DL} = 1.175 \times 74/2 = 43.5 \text{ kips}$$

The safety-shaped parapets will be placed after the concrete has cured. Their weights may be equally distributed to all stringers. No allowance will be made for a future wearing surface, but provision will be made for the weight of the 1¼-in LMC wearing course. The total superimposed dead load will be designated SDL.

DEAD LOAD CARRIED BY COMPOSITE SECTION, KIPS PER FT

Two parapets: 1.060/4	0.265
LMC wearing course:	0.125
0.150 × 8 × 1.25/12	
SDL per stringer:	0.390

Maximum moment occurs at midspan and equals

$$M_{SDL} = 0.390(74)^2/8 = 267 \text{ ft-kips}$$

Maximum shear occurs at the supports and equals

$$V_{SDL} = 0.390 \times 74/2 = 14.4 \text{ kips}$$

The HS25 live load imposed may be a truck load or a lane load. For maximum effect with the truck load, the two 40-kip axle loads, with variable spacing V, should be placed 14 ft apart, the minimum permitted (Fig. 11.3a). Then, the distance of

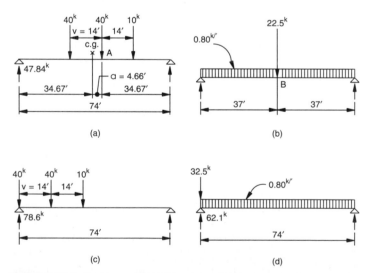

(a)

(b)

(c)

(d)

FIGURE 11.3 Positions of load for maximum stress in a simply supported stringer. (*a*) Maximum moment in the span with truck loads. (*b*) Maximum moment in the span with lane loading. (*c*) Maximum shear in the span with truck loads. (*d*) Maximum shear in the span with lane loading.

the center of gravity of the three axle loads from the center load is found by taking moments about the center load.

$$a = \frac{40 \times 14 - 10 \times 14}{40 + 40 + 10} = 4.67 \text{ ft}$$

Maximum moment occurs under the center axle load when its distance from midspan is the same as the distance of the center of gravity of the loads from midspan, or $4.67/2 = 2.33$ ft. Thus, the center load should be placed $7\frac{1}{2} - 2.33 = 34.67$ ft from a support (Fig. 11.3a). Then, the maximum moment due to the 90-kip truck load is

$$M_T = \frac{90(7\frac{1}{2} + 2.33)^2}{74} - 40 \times 14 = 1{,}321 \text{ ft-kips}$$

This loading governs, because the maximum moment due to lane loading (Fig. 11.3b) is smaller:

$$M_L = 0.80(74)^2/8 + 22.5 \times 7\frac{3}{4} = 962 < 1{,}321 \text{ ft-kips}$$

The distribution of the live load to a stringer may be obtained from Table 10.6, for a bridge with two traffic lanes.

$$\frac{S}{5.5} = \frac{8}{5.5} = 1.454 \text{ wheels} = 0.727 \text{ axle}$$

Hence, the maximum live-load moment is

$$M_{LL} = 0.727 \times 1{,}321 = 960 \text{ ft-kips}$$

While this moment does not occur at midspan as do the maximum dead-load moments, stresses due to M_{LL} may be combined with those from M_{DL} and M_{SDL} to produce the maximum stress, for all practical purposes.

For maximum shear with the truck load, the outer 32-kip load should be placed at the support (Fig. 11.3c). Then, the shear is

$$V_T = \frac{90(74 - 14 + 4.66)}{74} = 78.6 \text{ kips}$$

This loading governs, because the shear due to lane loading (Fig. 11.3d) is smaller:

$$V_L = 32.5 + 0.80 \times 7\frac{1}{2} = 62.1 < 78.6 \text{ kips}$$

Since the stringer receives 0.727 axle loads, the maximum shear on the stringer is

$$V_{LL} = 0.727 \times 78.6 = 57.1 \text{ kips}$$

Impact is the following fraction of live-load stress:

$$I = \frac{50}{L + 125} = \frac{50}{74 + 125} = 0.251$$

Hence, the maximum moment due to impact is

$$M_I = 0.251 \times 960 = 241 \text{ ft-kips}$$

and the maximum shear due to impact is

$$V_I = 0.251 \times 57.1 = 14.3 \text{ kips}$$

MIDSPAN BENDING MOMENTS, FT-KIPS:

M_{DL}	M_{SDL}	$M_{LL} + M_I$
804	267	1,201

END SHEAR, KIPS:

V_{DL}	V_{SDL}	$V_{LL} + V_I$	Total V
43.4	14.4	71.4	129.2

Properties of Composite Section.   The 9-in-thick roadway slab includes an allowance of 0.5 in for a wearing surface. Hence, the effective thickness of the concrete slab for composite action is 8.5 in.

The effective width of the slab as part of the top flange of the T beam is the smaller of the following:

¼ span = ¼ × 74 = 222 in
Stringer spacing, c to c = 8 × 12 = 96 in
12 × slab thickness = 12 × 8.5 = 102 in

Hence, the effective width is 102 in (Fig. 11.4).

To complete the T beam, a trial steel section must be selected. As a guide in doing this, formulas for estimated required flange area given in J. C. Hacker, "A Simplified Design of Composite Bridge Structures," *Journal of the Structural Division, ASCE, Proceedings Paper* 1432, November, 1957, may be used. To start, assume the rolled beam will be a 36-in-deep wide-flange shape, and take the allowable bending stress F_b as 20 ksi. The required bottom-flange area, in², then may be estimated from

$$A_{sb} = \frac{12}{F_b} \left(\frac{M_{DL}}{d_{cg}} + \frac{M_{SDL} + M_{LL} + M_I}{d_{cg} + t} \right) \tag{11.1a}$$

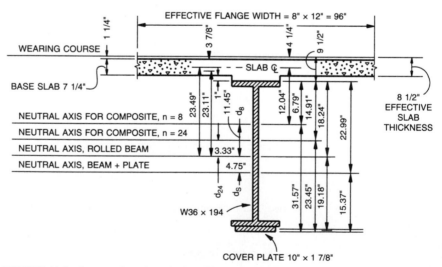

FIGURE 11.4   Cross section of composite stringer at midspan.

where d_{cg} = distance, in, between center of gravity of flanges of steel shape and t = thickness, in, of concrete slab. With d_{cg} assumed as 36 in, the estimated required bottom-flange area is

$$A_{sb} = \frac{12}{20}\left(\frac{804}{36} + \frac{267 + 1201}{36 + 8.5}\right) = 33.2 \text{ in}^2$$

The ratio $R = A_{st}/A_{sb}$, where A_{st} is the area, in², of the top flange of the steel beam, may be estimated to be

$$R = 50/(190 - L) = 50/(190 - 74) = 0.43 \qquad (11.1b)$$

Then, the estimated required area of the top flange is

$$A_{st} = RA_{sb} = 0.43 \times 33.2 = 14.3 \text{ in}^2$$

A W36 × 194 provides a flange with width 12.117 in, thickness 1.26 in, and area

$$A_{st} = 12.117 \times 1.26 = 15.27 > 14.3 \text{ in}^2\text{—OK}$$

With this shape, a bottom cover plate with an area of at least $33.2 - 15.27 = 17.9$ in². Maximum thickness permitted for a cover plate on a rolled beam is 1.5 times the flange thickness. In this case, therefore, plate thickness should not exceed $1.5 \times 1.26 = 1.89$ in. These requirements are met by a 10 × 1⅞-in plate, with an area of 18.75 in².

The trial section chosen consequently is a W36 × 194 with a partial-length cover plate 10 × 1⅞ in on the bottom flange (Fig. 11.4). Its neutral axis can be located by taking moments about the neutral axis of the rolled beam. This computation and that for the section moduli S_{st} and S_{sb} of the steel section are conveniently tabulated in Table 11.1.

TABLE 11.1   Steel Section for Maximum Moment

Material	A	d	Ad	Ad^2	I_o	I
W36 × 194	57.00				12,100	12,100
Cover plate 10 × 1⅞	18.75	−19.18	−359.6	6,898		6,898
	75.75		−359.6			18,998
$d_s = -359.6/75.75 = -4.75$ in				$-4.75 \times 359.6 =$		−1,708
					$I_{NA} =$	17,290

Distance from the neutral axis of the steel section to:

$$\text{Top of steel} = 18.24 + 4.75 = 22.99 \text{ in}$$

$$\text{Bottom of steel} = 18.24 - 4.75 + 1.88 = 15.37 \text{ in}$$

Section moduli	
Top of steel	Bottom of steel
$S_{st} = 17{,}290/22.99 = 752$ in³	$S_{sb} = 17{,}290/15.37 = 1{,}125$ in³

In computation of the properties of the composite section, the concrete slab, ignoring the haunch area, is transformed into an equivalent steel area. For the purpose, for this bridge, the concrete area is divided by the modular ratio $n = 8$ for short-time loading, such as live loads and impact. For long-time loading, such as superimposed dead loads, the divisor is $3n = 24$, to account for the effects of creep. The computations of neutral-axis location and section moduli for the composite section are tabulated in Table 11.2. To locate the neutral axis, moments are taken about the neutral axis of the rolled beam.

Stresses in Composite Section.   Since the stringers will not be shored when the concrete is cast and cured, the stresses in the steel section for load *DL* are determined with the section moduli of the steel section alone (Table 11.1). Stresses for load *SDL* are computed with section moduli of the composite section when $n = 24$ from Table 11.2*a*. And stresses in the steel for live loads and impact are calculated with section moduli of the composite section when $n = 8$ from Table 11.2*b* (Table 11.3*a*).

Stresses in the concrete are determined with the section moduli of the composite section with $n = 24$ for *SDL* from Table 11.2*a* and $n = 8$ for *LL + I* from Table 11.2*b* (Table 11.3*b*).

Since the bending stresses in steel and concrete are less than the allowable, the assumed steel section is satisfactory. Use the W36 × 194 with 10 × 1⅞-in bottom cover plate. Total weight of steel will be about 0.274 kip per ft, including 0.016 kip per ft for diaphragms, whereas 0.297 kip per ft was assumed in the dead-load calculations.

Maximum Shear Stress.   Though shear rarely is critical in wide-flange shapes adequate in bending, the maximum shear in the web should be checked. The total shear at the support has been calculated to be 129.3 kips. The web of the steel beam is about 36 in deep and the thickness is 0.770 in. Thus, the web area is

$$36 \times 0.770 = 27.7 \text{ in}^2$$

and the average shear stress is

$$f_v = \frac{129.3}{27.7} = 4.7 < 12 \text{ ksi}$$

This indicates that the beam has ample shear capacity.

End bearing stiffeners are not required for a rolled beam if the web shear does not exceed 75% of the allowable shear for girder webs, 12 ksi. The ratio of actual to allowable shears is

$$\frac{f_v}{F_v} = \frac{4.7}{12} = 0.39 < 0.75$$

Hence, bearing stiffeners are not required.

Cover-Plate Cutoff.   Bending moments decrease almost parabolically with distance from midspan, to zero at the supports. At some point on either side of the center, therefore, the cover plate is not needed for carrying bending moment. For locating this cutoff point, the properties of the composite section without the cover plate are needed, with $n = 24$ and $n = 8$ (Fig. 11.5). The computations are tabulated in Table 11.4.

TABLE 11.2 Composite Section for Maximum Moment

(a) For dead loads, $n = 24$

Material	A	d	Ad	Ad^2	I_o	I
Steel section	75.75		-360			18,998
Concrete 96 × 7.75/24*	31.00	23.11	716	16,556	155	16,711
	106.75		356			35,709

$d_{24} = 356/106.75 = 3.33$ in
$$-3.33 \times 356 = -1,185$$
$$I_{NA} = \quad 34,534$$

Distance from the neutral axis of the composite section to:

$$\text{Top of steel} = 18.24 - 3.33 = 14.91 \text{ in}$$

$$\text{Bottom of steel} = 18.24 + 3.33 + 1.88 = 23.45 \text{ in}$$

$$\text{Top of concrete} = 14.91 + 1 + 7.75 = 23.66 \text{ in}$$

Section moduli

Top of steel	Bottom of steel	Top of concrete
$S_{st} = 34,534/14.91$ $= 2,316 \text{ in}^3$	$S_{sb} = 34,534/23.45$ $= 1,473 \text{ in}^3$	$S_c = 34,534/23.66$ $= 1,460 \text{ in}^3$

(b) For live loads, $n = 8$

Material	A	d	Ad	Ad^2	I_o	I
Steel section	75.75		-360			18,998
Concrete 96 × 8.5/8	102.00	23.49	2,396	56,280	615	56,895
	177.75		2,036			75,893

$d_8 = 2,036/177.75 = 11.45$ in
$$-11.45 \times 2,036 = -23,312$$
$$I_{NA} = \quad 52,581$$

Distance from the neutral axis of the composite section to:

$$\text{Top of steel} = 18.24 - 11.45 = 6.79 \text{ in}$$

$$\text{Bottom of steel} = 18.24 + 11.45 + 1.88 = 31.57 \text{ in}$$

$$\text{Top of concrete} = 6.79 + 1 + 8.5 = 16.29 \text{ in}$$

Section moduli

Top of steel	Bottom of steel	Top of concrete
$S_{st} = 52,580/6.79$ $= 7,744 \text{ in}^3$	$S_{sb} = 52,580/31.57$ $= 1,666 \text{ in}^3$	$S_c = 52,580/16.29$ $= 3,228 \text{ in}^3$

*Depth of the top slab is taken as 7.75 in, inasmuch as the 1¼-in wearing course is included in the superimposed load.

TABLE 11.3 Stresses in the Composite Section, ksi, at Section of Maximum Moment

(a) Steel stresses	
Top of steel (compression)	Bottom of steel (tension)
DL: $f_b = \ \ \ 804 \times 12/752 \ \ = 12.83$	$f_b = 804 \times 12/1{,}125 \ \ \ = 8.58$
SDL: $f_b = \ \ \ 267 \times 12/2{,}316 = \ \ 1.38$	$f_b = 267 \times 12/1{,}473 \ \ \ = 2.18$
$LL + I$: $f_b = 1{,}202 \times 12/7{,}744 = \ \ \underline{1.86}$	$f_b = 1{,}202 \times 12/1{,}666 = \underline{8.66}$
Total: $\quad\quad\quad\quad\quad\quad\quad\quad\quad 16.07 < 20$	$19.42 < 20$

(b) Stresses at top of concrete
SDL: $f_c = \ \ \ 267 \times 12/(1{,}460 \times 24) = 0.09$
$LL + I$: $f_c = 1{,}202 \times 12/(3{,}228 \times \ \ 8) = \underline{0.56}$
$0.65 < 1.6$

The length L_{cp}, ft, required for the cover plate may be estimated by assuming that the curve of maximum moments is a parabola. Approximately,

$$L_{cp} = L \sqrt{1 - \frac{S'_{sb}}{S_{sb}}} \tag{11.2}$$

where L = span, ft
S'_{sb} = section modulus with respect to bottom of steel shape with lighter flange (without cover plate), in^3
S_{sb} = section modulus with respect to bottom of steel shape with heavier flange (with cover plate), in^3

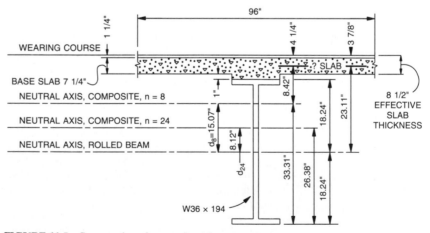

FIGURE 11.5 Cross section of composite stringer near supports.

TABLE 11.4   Composite Section Near Supports

(*a*) For dead loads, $n = 24$

Material	A	d	Ad	Ad^2	I_o	I
W36 × 194	57.0				12,100	12,100
Concrete 96 × 7.75/24	31.0	23.11	716	16,556	155	16,711
	88.0		716			28,811

$d_{24} = 716/88.0 = 8.14$ in $\qquad -8.14 \times 716 = -5,826$

Half-beam depth $= 18.24 \qquad\qquad\qquad\qquad I_{NA} = 22,985$

$\qquad\qquad 26.38$ in

$$S_{sb} = 22,985/26.38 = 871 \text{ in}^3$$

(*b*) For live loads, $n = 8$

Material	A	d	Ad	Ad^2	I_o	I
W36 × 194	57.0				12,100	12,100
Concrete 96 × 8.5/8	102.0	23.49	2,396	56,282	615	56,900
	159.0		2,396			69,000

$d_8 = 2,396/159 = 15.07$ in $\qquad -15.07 \times 2,396 = -36,110$

Half beam depth $= 18.24 \qquad\qquad\qquad\qquad I_{NA} = 32,890$

$\qquad\qquad 33.31$ in

$$S_{sb} = 32,890/33.31 = 987 \text{ in}^3$$

For the W36 × 194, $S'_{sb} = 665$. Hence,

$$L_{cp} = 74 \sqrt{1 - \frac{665}{1,125}} = 48 \text{ ft}$$

If the cover plate is welded along its ends, the terminal distance that the plate must be extended beyond its theoretical cutoff point is about 1.5 times the plate width. For the 10-in plate, therefore, the terminal distance is $1.5 \times 10 = 15$ in. Use 1.5 ft. Thus, L_{cp} must be increased by 2×1.5, to 51 ft.

Assume a 51-ft-long cover plate. It would then terminate 11.5 ft from each support (Fig. 11.6). The theoretical cutoff point is therefore $11.5 + 1.5 = 13.0$ ft from each support. The stresses at that point should be checked to ensure that allowable bending stresses in the composite section without the cover plate are not exceeded. Table 11.5*a* presents the calculations for maximum flexural tensile stress at the theoretical cutoff points, 13-ft from the supports, and Table 11.5*b*, calculations for stresses at the actual terminations of the cover plate, 11.5 ft from the supports. The composite section without the cover plate is adequate at the the-

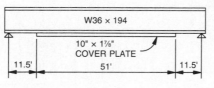

FIGURE 11.6   Elevation view of stringer.

TABLE 11.5 Stresses in Composite Steel Beam without Cover Plate

(*a*) At theoretical cutoff point, 13 ft from supports

Bending moments, ft-kips

M_{DL}	M_{SDL}	$M_{LL} + M_I$
466	155	744 (Fig. 11.7)

Stresses at bottom of steel (tension), ksi

DL: $f_b = 466 \times 12/665 = $ 8.41 (S_{sb} for W36 × 194)
SDL: $f_b = 155 \times 12/871 = $ 2.14 (S_{sb} from Table 11.4*a*)
$LL + I$: $f_b = 744 \times 12/987 = $ 9.04 (S_{sb} from Table 11.4*b*)

Total: 19.59 < 20

(*b*) At cover-plate terminal, 11.5 ft from support

Bending moments, ft-kips

M_{DL}	M_{SDL}	$M_{LL} + M_I$
422	140	677 (Fig. 11.8)

Stresses at bottom of steel (tension), ksi

DL: $f_b = 422 \times 12/665 = $ 7.62 (S_{sb} for W36 × 194)
SDL: $f_b = 140 \times 12/871 = $ 1.93 (S_{sb} from Table 11.4*a*)
$LL + I$: $f_b = 677 \times 12/987 = $ 8.23 (S_{sb} from Table 11.4*b*)

Total: 17.78

oretical cutoff point. But fatigue stresses in the beam should be checked at the actual termination of the plate, 11.5 ft from each support.

From Table 11.5*b*, the stress range equals the stress due to live load plus impact, 8.23 ksi. On the assumption that the bridge is a redundant-load-path structure, for base metal adjacent to a fillet weld (Category E′) subjected to 500,000 loading cycles, the allowable fatigue stress range permitted by AASHTO standard specifications is $F_{sr} = 9.2$ ksi > 8.23. The cover plate is satisfactory. (Because of past experience with fatigue cracking at termination welds for cover plates, however, the usual practice, when a cover plate is specified, is to extend it the full length of the beam.)

Cover-Plate Weld. The fillet weld connecting the cover plate to the bottom flange must be capable of resisting the shear at the bottom of the flange. The shear is a maximum at the end of the cover plate, 11.5 ft from the supports. The position of the truck load to produce maximum shear there is the same as that for maximum moment at those points (Fig. 11.8). Maximum shears and resulting shear stresses are given in Table 11.6.

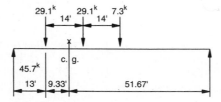

FIGURE 11.7 Position of truck load for maximum moment 13 ft from the support.

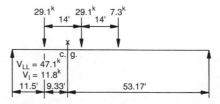

FIGURE 11.8 Position of truck load for maximum moment 11.5 ft from the support.

TABLE 11.6 Shear Stress 11.5 ft from Support

Shear, kips		
V_{DL}	V_{SDL}	$V_{LL} + V_I$
30.0	9.9	58.9

Shear stress, kips per in
DL: v = $30.0 \times 18.75 \times 14.43/17{,}290 = 0.47$ (*I* from Table 11.1)
SDL: v = $9.9 \times 18.75 \times 22.51/34{,}530 = 0.12$ (*I* from Table 11.2*a*)
LL + I: v = $58.9 \times 18.75 \times 30.63/52{,}580 = 0.64$ (*I* from Table 11.2*b*)
Total: 1.23

The shear stress at the section is computed from

$$v = \frac{VQ}{I} \tag{11.3}$$

where v = horizontal shear stress, kips per in
 V = vertical shear on cross section, kips
 Q = statical moment about neutral axis of area of cross section on one side of axis and not included between neutral axis and horizontal line through given point, in³
 I = moment of inertia, in⁴, of cross section about neutral axis

AASHTO specifications permit a stress $F_v = 0.27F_u = 15.7$ ksi in fillet welds when the base metal is Grade 36 steel. The minimum size of fillet weld permitted with the $1\frac{7}{8}$-in-thick cover plate is $\frac{5}{16}$ in. If a $\frac{5}{16}$-in weld is used on opposite sides of the plate, the two welds would be allowed to resist a shear stress of

$$v_a = 2 \times 0.313 \times 0.707 \times 15.7 = 6.9 > 1.23 \text{ kips per in}$$

Therefore, use $\frac{5}{16}$-in welds.

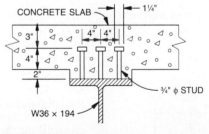

FIGURE 11.9 Welded studs on beam flange.

Shear Connectors. To ensure composite action of concrete deck and steel stringer, shear connectors welded to the top flange of the stringer must be embedded in the   concrete (Art. 10.17). For this structure, $\frac{3}{4}$-in dia. welded studs are selected. They are to be installed in groups of three at specified locations to resist the horizontal shear at the top of the steel stringer (Fig. 11.9). With height $h = 6$ in, they satisfy the requirement $h/d \geq 4$, where d = stud diameter, in.

With $f'_c = 4000$ psi for the concrete, the ultimate strength of a $\frac{3}{4}$-in-dia. welded stud is

$$S_u = 0.4d^2\sqrt{f'_c E_c} = 0.4(0.75)^2\sqrt{4{,}000 \times 3{,}600{,}000} = 27 \text{ kips}$$

This value is needed for determining the number of shear connectors required to develop the strength of the steel stringer or the concrete slab, whichever is smaller. At midspan, the strength of the rolled beam and cover plate, with area $A_s = 75.75$ in^2 from Table 11.1, is

$$P_1 = A_s F_y = 75.75 \times 36 = 2{,}727 \text{ kips}$$

The compressive strength of the concrete slab is

$$P_2 = 0.85 f_c' bt = 0.85 \times 4.0 \times 96 \times 8.5 = 2{,}774 < 2{,}727 \text{ kips}$$

Concrete strength governs. Hence, the number of studs provided between midspan and each support must be at least

$$N_1 = \frac{P_1}{\phi S_u} = \frac{2{,}727}{0.85 \times 27} = 119$$

With the studs placed in groups of three, therefore, there should be at least 40 groups on each half of the stringer.

Between the end of the cover plate and the support, the strength of the rolled beam alone, with $A_s = 57.0$, is

$$P_1 = A_s F_y = 57.0 \times 36 = 2{,}052 > 2{,}727 \text{ kips}$$

Steel strength still governs.

Pitch is determined by fatigue requirements. The allowable load range, kips per stud, may be computed from

$$Z_r = \alpha d^2 \tag{11.4}$$

With $\alpha = 10.6$ for 500,000 cycles of load (AASHTO specifications),

$$Z_r = 10.6(0.75)^2 = 5.97 \text{ kips per stud}$$

At the supports, the shear range $V_r = 71.4$ kips, the shear produced by live load plus impact. Consequently, with $n = 8$ for the concrete, and the transformed concrete area equal to 102 in^2 and $I = 32{,}980$ in^4 from Table 11.4b, the range of horizontal shear stress is

$$S_r = \frac{V_r Q}{I} = \frac{71.4 \times 102.0 \times 8.42}{32{,}980} = 1.859 \text{ kips per in}$$

Hence, the pitch required for stud groups near the supports is

$$p = \frac{3Z_r}{S_r} = \frac{3 \times 5.97}{1.8} = 9.63 \text{ in}$$

At 5 ft from the supports, the shear range $V_r = 66.1$ kips, produced by live load plus impact. Since the cross section is the same as at the support, the pitch required for the studs is

$$p = \frac{9.63 \times 71.4}{66.1} = 10.40 \text{ in}$$

At 25 ft from the supports, $V_r = 46.1$ kips (Fig. 11.10). With $I = 52{,}580$ in^4 from Table 11.2b, the range of horizon-

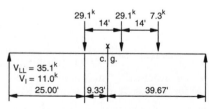

FIGURE 11.10   Position of loads for maximum shear 25 ft from the support.

tal shear stress is

$$S_r = \frac{V_r Q}{I} = \frac{46.1 \times 102.0 \times 12.00}{52,580} = 1.077 \text{ kips per in}$$

Hence, the pitch required is

$$p = \frac{3 \times 5.97}{1.077} = 16.6 \text{ in}$$

The shear-connector spacing selected to meet the preceding requirements is shown in Fig. 11.11.

Deflections. Dead-load deflections may be needed so that concrete for the deck may be finished to specified elevations. Cambering of rolled beams to offset dead-load deflections usually is undesirable because of the cost. The beams may, however, be delivered from the mill with a slight mill camber. If so, advantage should be taken of this, by fabricating and erecting the stringers with the camber upward.

The dead-load deflection has two components, one corresponding to *DL* and one to *SDL*. For computation for *DL*, the moment of inertia *I* of the steel section alone should be used. For *SDL*, *I* should apply to the composite section with *n* = 24 (Table 11.2*a*). Both components can be computed from

$$\delta = \frac{22.5wL^4}{EI} \tag{11.5}$$

where *w* = uniform load, kips per ft
  L = span, ft
  E = modulus of elasticity of steel, ksi
  I = moment of inertia of section about neutral axis

For *DL*, *w* = 1.175 kips per ft, and for *SDL*, *w* = 0.390 kip per ft.

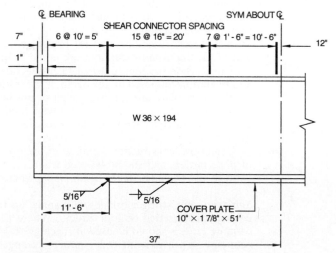

FIGURE 11.11 Shear connector spacing along the top flange of a stringer.

DEAD-LOAD DEFLECTION

$$DL: \delta = 22.5 \times 1.175(74)^2/(29{,}000 \times 17{,}290) = 1.60 \text{ in}$$

$$SDL: \delta = 22.5 \times 0.390(74)^4/(29{,}000 \times 34{,}530) = \underline{0.27}$$

Total: 1.87 in

Maximum live-load deflection should be checked and compared with $12L/800$. If desired, this deflection can be calculated accurately by the methods given in Sec. 3, including the effects of changes in moments of inertia. Or the midspan deflection of a simply supported stringer may be obtained with acceptable accuracy from the approximate formula:

$$\delta = \frac{324}{EI} P_T(L^3 - 555L + 4780) \tag{11.6}$$

where P_T = weight, kips, of one front truck wheel multiplied by the live-load distribution factor, plus impact, kips. In this case, $P_T = 10 \times 0.727 + 0.251 \times 10 \times 0.727 = 9.1$ kips. From Table 11.2b, for $n = 8$, $I = 52{,}580$. Hence,

$$\delta = \frac{324 \times 9.1}{29{,}000 \times 52{,}580}(74^3 - 555 \times 74 + 4{,}780) = 0.70 \text{ in}$$

And the deflection-span ratio is

$$\frac{0.74}{74 \times 12} = \frac{1}{1{,}200} < \frac{1}{800}$$

Thus, the live-load deflection is acceptable.

11.3 CHARACTERISTICS OF PLATE-GIRDER STRINGER BRIDGES

For simple or continuous spans exceeding about 85 ft, plate girders may be the most economical type of construction. Used as stringers instead of rolled beams, they may be economical even for long spans (350 ft or more). Design of such bridges closely resembles that for bridges with rolled-beam stringers (Arts. 11.1 and 11.2). Important exceptions are noted in this and following articles.

The decision whether to use plate girders often hinges on local fabrication costs and limitations imposed on the depth of the bridge. For shorter spans, unrestricted depth favors plate girders over rolled beams. For long spans, unrestricted depth favors deck trusses or arches. But even then, cable-supported girders may be competitive in cost. Stringent depth restrictions, however, favor through trusses or arches.

Composite construction significantly improves the economy and performance of plate girders and should be used wherever feasible. (See also Art. 11.1.) Advantage also should be taken of continuity wherever possible, for the same reasons.

Spacing.   For stringer bridges with spans up to about 175 ft, two lanes may be economically carried on four girders. Where there are more than two lanes, five or more girders should be used at spacings of 7 ft or more. With increase in span, economy improves with wider girder spacing, because of the increase in load-carrying capacity with increase in depth for the same total girder area.

For stringer bridges with spans exceeding 175 ft, girders should be spaced about

14 ft apart. Consequently, this type of construction is more advantageous where roadway widths exceed about 40 ft. For two-lane bridges in this span range, box girders or plate girders may be less costly.

Steel Grades.   In spans under about 100 ft, Grade 36 steel often will be more economical than higher-strength steels. For longer spans, however, designers should consider use of stronger steels, because some offer maintenance benefits as well as a favorable strength-cost ratio. But in small quantities, these steels may be expensive or unavailable. So where only a few girders are required, it may be uneconomical to use a high-strength steel for a light flange plate extending only part of the length of a girder.

In spans between 100 and 175 ft, hybrid girders, with stronger steels in the flanges than in the web (Art. 10.20), often will be more economical than girders completely of Grade 36 steel. For longer spans, economy usually is improved by making the web of higher-strength steels than Grade 36. In such cases, the cost of a thin web with stiffeners should be compared with that of a thicker web with fewer stiffeners and thus lower fabrication costs. Though high-strength steels may be used in flanges and web, other components, such as stiffeners, bracing, and connection details, should be of Grade 36 steel, because size is not determined by strength.

Haunches.   In continuous spans, bending moments over interior supports are considerably larger than maximum positive bending moments. Hence, theoretically, it is advantageous to make continuous girders deeper at interior supports than at midspan. This usually is done by providing a haunch, usually a deepening of the girders along a pleasing curve in the vicinity of those supports.

For spans under about 175 ft, however, girders with straight soffits may be less costly than with haunches. The expense of fabricating the haunches may more than offset savings in steel obtained with greater depth. With long spans, the cost of haunching may be further increased by the necessity of providing horizontal splices, which may not be needed with straight soffits. So before specifying a haunch, designers should make cost estimates to determine whether its use will reduce costs.

Web.   In spans up to about 100 ft, designers may have the option of specifying a web with stiffeners or a thicker web without stiffeners. For example, a $\frac{5}{16}$-in-thick stiffened plate or a $\frac{7}{16}$-in-thick unstiffened plate often will satisfy shear and buckling requirements in that span range. A girder with the thinner web, however, may cost more than with the thicker web, because fabrication costs may more than offset savings in steel. But if the unstiffened plate had to be thicker than $\frac{7}{16}$ in, the girder with stiffeners probably would cost less.

For spans over 100 ft, transverse stiffeners are necessary. Longitudinal stiffeners, with the thinner webs they permit, may be economical for Grade 36 as well as for high-strength steels.

Flanges.   In composite construction, plate girders offer greater flexibility than rolled beams, and thus can yield considerable savings in steel. Flange sizes of plate girders, for example, can be more closely adjusted to variations in bending stress along the span. Also, the grade of steel used in the flanges can be changed to improve economy. Furthermore, changes may be made where stresses theoretically permit a weaker flange, whereas with cover-plated rolled beams, the cover plate must be extended beyond the theoretical cutoff location.

Adjoining flange plates are spliced with a groove weld. It is capable of developing the full strength of the weaker plate when a gradual transition is provided between groove-welded material of different width or thickness. AASHTO specifies transition details that must be followed.

Designers should avoid making an excessive number of changes in sizes and grades of flange material. Although steel weight may be reduced to a minimum in that manner, fabrication costs may more than offset the savings in steel.

For simply supported, composite girders in spans under 100 ft, it may be uneconomical to make changes in the top flange. For spans between 100 and 175 ft, a single reduction in thickness of the top flange on either side of midspan may be economical. Over 175 ft, a reduction in width as well as thickness may prove worthwhile. More frequent changes are economically justified in the bottom flange, however, because it is more sensitive to stress changes along the span. In simply supported spans up to about 175 ft, the bottom flange may consist of three plates of two sizes—a center plate extending over about the middle 60% of the span and two thinner plates extending to the supports. (See Art. 10.18.)

Note that even though high-strength steels may be specified for the bottom flange of a composite girder, the steel in the top flange need not be of higher strength than that in the web. In a continuous girder, however, if the section is not composite in negative-moment regions, the section should be symmetrical about the neutral axis.

In continuous spans, sizes of top and bottom flanges may be changed economically once or twice in a negative-moment region, depending on whether only thickness need be changed or both width and thickness have to be decreased. Some designers prefer to decrease thickness first and then narrow the flange at another location. But a constant-width flange should be used between flange splices. In positive-moment regions, the flanges may be treated in the same way as flanges of simply supported spans.

Welding of stiffeners or other attachments to a tension flange usually should be avoided. Transverse stiffeners used as cross-frame connections, should be connected to both girder flanges (Art. 10.13.6). The flange stress should not exceed the allowable fatigue stress for base metal adjacent to or connected by fillet welds. Stiffeners, however, should be welded to the compression flange. Though not required for structural reasons, these welded connections increase lateral rigidity of a girder, which is a desirable property for transportation and erection.

Bracing.   Intermediate cross frames usually are placed in all bays and at intervals as close to 25 ft as practical, but no farther apart than 25 ft. Consisting of minimum-size angles, these frames provide a horizontal angle near the bottom flange and V bracing (Fig. 11.12) or X bracing. The angles usually are field-bolted to connection plates welded to each girder web. Eliminating gusset plates and bolting directly to stiffeners is often economical.

Cross frames also are required at supports. Those at interior supports of continuous girders usually are about the same as the intermediate cross frames. At end supports, however, provision must be made to support the end of the concrete deck. For the purpose, a horizontal channel of minimum weight, consistent with concrete edge-beam requirements, often is used near the top flange, with V or X bracing, and a horizontal angle near the bottom flange.

Lateral bracing in a horizontal plane near the bottom flange is sometimes required. The need for such bracing must be investigated, based on a wind pressure of 50 psf. (Spans with nonrigid decks may also require a top lateral system.) This bracing usually consists of crossing diagonal angles and the bottom angles of the cross frames.

Bearings.   Laminated elastomeric pads may be used economically as bearings for girder spans up to about 175 ft. Welded steel rockers or rollers are an alternative for all spans but may not meet seismic requirements. Seismic attenuation bearings, pot bearings, or spherical bearings with teflon guided surfaces for expansion are other alternatives.

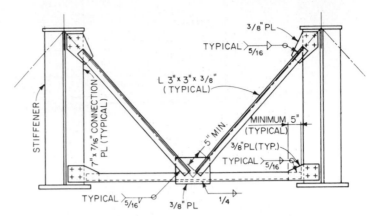

FIGURE 11.12   Intermediate cross frame for a stringer bridge.

Camber.   Plate girders should be cambered to compensate for dead-load deflections. When the roadway is on a grade, the camber should be adjusted so that the girder flanges will parallel the profile grade line. For the purpose, designers should calculate dead-load deflections at sufficient points along each span to indicate to the fabricator the desired shape for the unloaded stringer.

11.4 EXAMPLE—ALLOWABLE-STRESS DESIGN OF COMPOSITE, PLATE-GIRDER BRIDGE

To illustrate the design procedure, a two-lane highway bridge with simply supported, composite, plate-girder stringers will be designed. As indicated in the framing plan in Fig. 11.13a, the stringers span 100 ft c to c of bearings. The typical cross section in Fig. 11.13b shows a 26-ft roadway flanked by 1-ft 9-in-wide barrier curbs. Structural steel to be used is Grade 36. Loading is HS25. Appropriate design criteria given in Sec. 10 will be used for this structure. Concrete to be used for the deck is class A, with 28-day strength $f'_c = 4,000$ psi and allowable compressive stress $0.4f'_c = 1600$ psi. Modulus of elasticity $E_c = 3,457,300$ psi.

Assume that the deck will be supported on four plate-girder stringers, spaced 8 ft 4 in c to c, as indicated in Fig. 11.13.

Concrete Slab.   The slab is designed, to span transversely between stringers, in the same way as for rolled-beam stringers (Art. 11.2). A 9-in-thick one-course, concrete slab will be used with the plate-girder stringers.

Stringer Design Procedure.   The general design procedure outlined in Art. 11.2 for rolled beams also holds for plate girders. In this example, too, only a typical interior stringer will be designed.

Loads, Moments, and Shears.   Assume that the girders will not be shored during casting of the concrete slab. Hence, the dead load on each steel stringer includes the weight of an 8.33-ft-wide strip of slab as well as the weights of steel girder and framing details. This dead load will be referred to as *DL*.

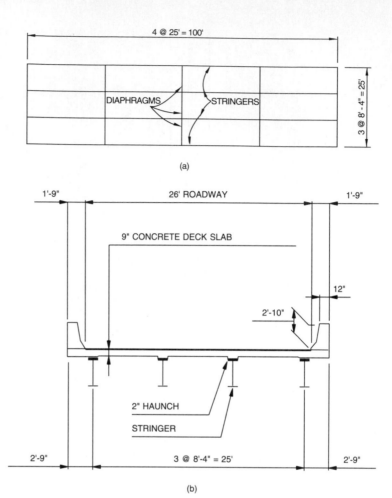

FIGURE 11.13 Two-lane highway bridge with plate-girder stringers. (*a*) Framing plan. (*b*) Typical cross section.

DEAD LOAD CARRIED BY STEEL BEAM, KIPS PER FT

Slab: $0.150 \times 8.33 \times \%_{12}$	$= 0.938$
Haunch—16×2 in: $0.150 \times 1.33 \times 0.167$	$= 0.034$
Steel stringer and framing details—assume:	0.327
Stay-in-place forms and additional concrete forms:	$\underline{0.091}$
DL per stringer:	1.390

Maximum moment occurs at the center of the 100-ft span and equals

$$M_{DL} = \frac{1.39(100)^2}{8} = 1,738 \text{ ft-kips}$$

Maximum shear occurs at the supports and equals

$$V_{DL} = \frac{1.39 \times 100}{2} = 69.5 \text{ kips}$$

Barrier curbs will be placed after the concrete slab has cured. Their weights may be equally distributed to all stringers. In addition, provision will be made for a future wearing surface, weight 25 psf. The total superimposed dead load will be designated *SDL*.

DEAD LOAD CARRIED BY COMPOSITE SECTION, KIPS PER FT

Two barrier curbs: 2 × 0.530/4	= 0.265
Future wearing surface: 0.025 × 8.33	= 0.208
SDL per stringer:	0.473

Maximum moment occurs at midspan and equals

$$M_{SDL} = \frac{0.473(100)^2}{8} = 592 \text{ ft-kips}$$

Maximum shear occurs at supports and equals

$$V_{SDL} = \frac{0.473 \times 100}{2} = 23.7 \text{ kips}$$

The HS25 live load imposed may be a truck load or lane load. But for this span, the truck load shown in Fig. 11.14a governs. The center of gravity of the three axles lies between the two heavier loads and is 4.66 ft from the center load. Maximum moment occurs under the center-axle load when its distance from midspan is the same as the distance of the center of gravity of the loads from midspan, or 4.66/2 = 2.33 ft. Thus, the center load should be placed 100/2 − 2.33 = 47.67 ft from a support (Fig. 11.14a). Then, the maximum moment is

$$M_T = \frac{90(100/2 + 2.33)^2}{100} - 40 \times 14 = 1,905 \text{ ft-kips}$$

The distribution of the live load to a stringer may be obtained from Table 10.6, for a bridge with two traffic lanes.

$$\frac{S}{5.5} = \frac{8.33}{5.5} = 1.516 \text{ wheels} = 0.758 \text{ axle}$$

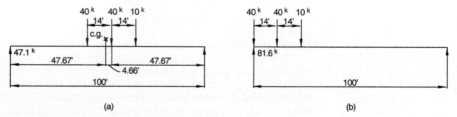

(a)   (b)

FIGURE 11.14   Positions of loads on a plate girder for maximum stress. (*a*) For maximum moment in the span. (*b*) For maximum shear in the span.

Hence, the maximum live-load moment is

$$M_{LL} = 0.758 \times 1,905 = 1,444 \text{ ft-kips}$$

While this moment does not occur at midspan as do the maximum dead-load moments, stresses due to M_{LL} may be combined with those from M_{DL} and M_{SDL} to produce the maximum stress, for all practical purposes.

For maximum shear with the truck load, the outer 40-kip load should be placed at the support (Fig. 11.14b). Then, the shear is

$$V_T = \frac{90(100 - 14 + 4.66)}{100} = 81.6 \text{ kips}$$

Since the stringer receives 0.758 axle load, the maximum shear on the stringer is

$$V_{LL} = 0.758 \times 81.6 = 61.9 \text{ kips}$$

Impact is taken as the following fraction of live-load stress:

$$I = \frac{50}{L + 125} = \frac{50}{100 + 125} = 0.222$$

Hence, the maximum moment due to impact is

$$M_I = 0.222 \times 1,444 = 321 \text{ ft-kips}$$

and the maximum shear due to impact is

$$V_I = 0.222 \times 61.9 = 13.8 \text{ kips}$$

MIDSPAN BENDING MOMENTS, FT-KIPS

M_{DL}	M_{SDL}	$M_{LL} + M_I$
1,738	592	1,765

END SHEAR, KIPS

V_{DL}	V_{SDL}	$V_{LL} + V_I$	Total V
69.5	23.7	75.7	168.9

Properties of Composite Section.   The 9-in thick roadway slab includes an allowance of 0.5 in for a wearing surface. Hence, the effective thickness of the concrete slab for composite action is 8.5 in.

The effective width of the slab as part of the top flange of the T beam is the smaller of the following:

¼ span = ¼ × 100 × 12 = 300 in
Stringer spacing, c to c = 8.33 × 12 = 100 in
12 × slab thickness = 12 × 8.5 = 102 in

Hence, the effective width is 100 in (Fig. 11.15).

To complete the T beam, a trial section must be selected for the plate girder. As a guide in doing this, formulas for estimating required flange area given in J. C. Hacker, "A Simplified Design of Composite Bridge Structures," *Journal of the Structural Division, ASCE, Proceedings Paper* 1432, November, 1957, may be used. To start, assume that the girder web will be 60 in deep. This satisfies the requirements that the depth-span ratio for girder plus slab exceed ¹⁄₂₅ and for girder alone, ¹⁄₃₀. With stiffeners, the web thickness is required to be at least ¹⁄₁₆₅ of the depth, or 0.364 in.

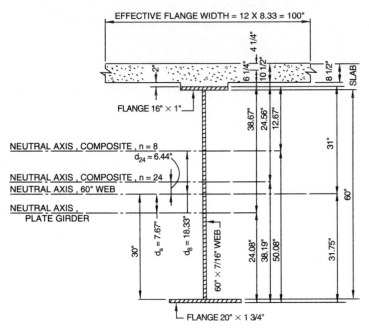

FIGURE 11.15   Cross section of composite plate girder at midspan.

Use a web plate $60 \times \frac{7}{16}$ in. With a cross-sectional area $A_w = 26.25$ in², the web will be subjected to maximum shearing stress considerably below the 12 ksi permitted.

$$f_v = \frac{168.9}{26.25} = 6.4 \text{ ksi} < 12$$

The required bottom-flange area may be estimated from Eq. (11.1a) with allowable bending stress $F_b = 20$ ksi and distance between centers of gravity of steel flanges taken as $d_{cg} = 63$ in.

$$A_{sb} = \frac{12}{20}\left(\frac{1{,}738}{63} + \frac{2{,}357}{63 + 9}\right) = 36.2 \text{ in}^2$$

Try a $20 \times 1\frac{3}{4}$-in bottom flange, area = 35 in².

The ratio of flange areas $R = A_{st}/A_{sb}$ may be estimated from Eq. (11.1b) as

$$R = \frac{50}{190 - L} = \frac{50}{190 - 100} = 0.55$$

Then, the estimated required area of the steel top flange is

$$A_{st} = RA_{sb} = 0.55 \times 35 = 19.3 \text{ in}^2$$

If the flange will be fully stressed in compression, the maximum permissible width-thickness ratio for the flange plate is 23 for Grade 36 steel. On the assumption of a 16-in-wide plate, the minimum thickness permitted is 16/23, or about ¾ in. Try a 16×1-in top flange, area = 16 in².

The trial section is shown in Fig. 11.15. Its neutral axis can be located by taking moments of web and flange areas about middepth of the web. This computation and that for the section moduli S_{st} and S_{sb} of the plate girder alone are conveniently tabulated in Table 11.7.

TABLE 11.7 Steel Section for Maximum Moment

Material	A	d	Ad	Ad^2	I_o	I
Top flange 16 × 1	16.0	30.50	488	14,880		14,880
Web 60 × 7⁄16	26.3				7,880	7,880
Bottom flange 20 × 1¾	35.0	−30.88	−1,081	33,380		33,380
	77.3		−593			56,140

$d_s = -593/77.3 = -7.67$ in $\qquad\qquad\qquad -7.67 \times 593 = -4,550$

$$I_{NA} = 51,590$$

Distance from neutral axis of steel section to:

$$\text{Top of steel} = 30 + 1 + 7.67 = 38.67 \text{ in}$$

$$\text{Bottom of steel} = 30 + 1.75 - 7.67 = 24.08 \text{ in}$$

Section moduli	
Top of steel	Bottom of steel
$S_{st} = 51,590/38.67 = 1,334$ in^3	$S_{sb} = 51,590/24.08 = 2,142$ in^3

In computation of the properties of the composite section, the concrete slab, ignoring the haunch area, is transformed into an equivalent steel area. For the purpose, for this bridge, the concrete area is divided by the modular ratio $n = 8$ for short-time loading, such as live loads and impact. For long-time loading, such as superimposed dead loads, the divisor is $3n = 24$, to account for the effects of creep. The computations of neutral-axis location and section moduli for the composite section are tabulated in Table 11.8. To locate the neutral axis, moments are taken about middepth of the girder web.

Stresses in Composite Section. Since the girders will not be shored when the concrete is cast and cured, the stresses in the steel section for load *DL* are determined with the section moduli of the steel section alone (Table 11.7). Stresses for load *SDL* are computed with section moduli of the composite section when $n = 24$ from Table 11.8*a*, and stresses in the steel for live loads and impact are calculated with section moduli of the composite section when $n = 8$ from Table 11.8*b* (Table 11.9*a*).

The width-thickness ratio of the compression flange now can be checked by the general formula applicable for any stress level:

$$\frac{b}{t} = \frac{103}{\sqrt{f_b}} = \frac{103}{\sqrt{19.24}} = 23.5 < \frac{24}{1}$$

Hence, the trial steel section is satisfactory.

Stresses in the concrete are determined with the section modulus of the composite section with $n = 24$ for *SDL* from Table 11.8*a* and $n = 8$ for *LL + I* from Table 11.8*b* (Table 11.9*b*). Therefore, the composite section is satisfactory. Use the section shown in Fig. 11.15 in the region of maximum moment.

Changes in Flange Sizes. One change in size of each flange will be made on both sides of midspan. (Though steel weight can be cut by reducing the area of the bottom flange in two steps, say from 1¾ to 1¼ in and then to ¾ in, higher fabrication

TABLE 11.8   Composite Section for Maximum Moment

(a) For dead loads, $n = 24$

Material	A	d	Ad	Ad^2	I_o	I
Steel section	77.3		-593			56,140
Concrete $100 \times 8.5/24$	35.4	37.25	1,319	49,120	210	49,330
	112.7		726			105,470

$d_{24} = 726/112.7 = 6.44$ in

$$-6.44 \times 726 = -4,680$$
$$I_{NA} = 100,790$$

Distance from neutral axis of composite section to:

Top of steel $= 31.00 - 6.44 = 24.56$ in

Bottom of steel $= 31.75 + 6.44 = 38.19$ in

Top of concrete $= 24.56 + 2 + 8.5 = 35.06$ in

Section moduli

Top of steel	Bottom of steel	Top of concrete
$S_{st} = 100,790/24.56$	$S_{sb} = 100,790/38.19$	$S_c = 100,790/35.06$
$= 4,104$ in^3	$= 2,639$ in^3	$= 2,875$ in^3

(b) For live loads, $n = 8$

Material	A	d	Ad	Ad^2	I_o	I
Steel section	77.3		-593			56,140
Concrete $100 \times 8.5/8$	106.3	37.25	3,958	147,500	640	148,140
	183.6		3,365			204,280

$d_{10} = 3,365/183.6 = 18.33$ in

$$-18.33 \times 3,365 = -61,680$$
$$I_{NA} = 142,600$$

Distance from neutral axis of composite section to:

Top of steel $= 31.00 - 18.33 = 12.67$ in

Bottom of steel $= 31.75 + 18.33 = 50.08$ in

Top of concrete $= 12.67 + 2 + 8.5 = 23.17$ in

Section moduli

Top of steel	Bottom of steel	Top of concrete
$S_{st} = 142,600/12.67$	$S_{sb} = 142,600/50.08$	$S_c = 142,600/23.17$
$= 11,254$ in^3	$= 2,847$ in^3	$= 6,154$ in^3

TABLE 11.9 Stresses, ksi, in Composite Plate Girder at Section of Maximum Moment

(a) Steel stresses	
Top of steel (compression)	Bottom of steel (tension)

Top of steel (compression)		Bottom of steel (tension)	
DL: $f_b = 1{,}738 \times 12/1{,}334 = 15.63$		$f_b = 1{,}738 \times 12/2{,}142 = 9.74$	
SDL: $f_b = 592 \times 12/4{,}104 = 1.73$		$f_b = 592 \times 12/2{,}639 = 2.69$	
$LL + I$: $f_b = 1{,}765 \times 12/11{,}254 = \underline{1.88}$		$f_b = 1{,}765 \times 12/2{,}847 = \underline{7.44}$	
Total:	$19.24 < 20$		$19.87 < 20$

(b) Stresses at top of concrete
SDL: $f_c = 592 \times 12/(2{,}875 \times 8) = 0.31$
$LL + I$: $f_c = 1{,}765 \times 12/(6{,}154 \times 8) = \underline{0.43}$
$0.74 < 1.6$

costs probably would offset the savings in steel costs.) Each flange will maintain its width throughout the span, but thickness will be reduced at locations to be determined.

For the top flange near the supports, assume a plate $16 \times \frac{3}{4}$ in, area = 12 in². Its width-thickness ratio is $16/\frac{3}{4} = 21.3 < 23$ and therefore is satisfactory. The cross section of the girder after the reduction in size of the top flange is shown in Fig. 11.16. The neutral axes of the steel section and of the composite section are

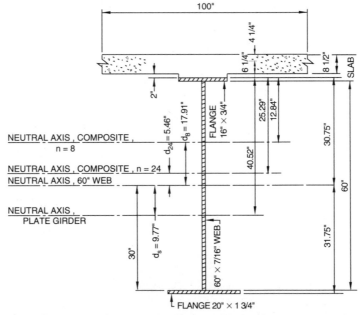

FIGURE 11.16 Cross section of composite plate girder about 30 ft from the supports.

located by taking moments of the areas about middepth of the girder web. These computations and those for the section moduli are given in Tables 11.10 and 11.11.

The location of the transition from the thicker plate to the thinner one can be estimated from Eq. (11.7), which gives the approximate length L_p, ft, of the thicker plate on the assumption of a parabolic bending-moment diagram.

$$L_p = L \sqrt{1 - \frac{S'_{st}}{S_{st}}} \tag{11.7}$$

where L = span, ft

S'_{st} = section modulus with respect to top of steel girder, with lighter flange, in^3

S_{st} = section modulus with respect to top of steel girder, with heavier flange, in^3

From Table 11.10, $S'_{st} = 1119$, and from Table 11.7, $S_{st} = 1334$. Hence,

$$L_p = 100 \sqrt{1 - \frac{1,119}{1,334}} = 40.1 \text{ ft}$$

Assume a 40-ft length for the 1-in top flange. It would then terminate 30 ft from the supports. Stresses should be checked at that location to ensure that they are within the allowable (Table 11.12).

Fatigue seldom governs at flange transitions in simply supported spans, but it should be checked for the final design.

For the bottom flange near the supports, assume a plate $20 \times \frac{7}{8}$ in, area = 17.5 in^2. The cross section of the girder after this reduction in size of the bottom flange is shown in Fig. 11.18. The neutral axes of the steel section and of the composite section are located by taking moments of the areas about middepth of the girder web. These computations and those for the section moduli are given in Tables 11.13 and 11.14.

The approximate location of the transition from the thicker bottom plate to the thinner one can be determined from Eq. (11.2), by setting the length L_p, ft, of the

TABLE 11.10 Steel Section about 30 Ft from Supports

Material	A	d	Ad	Ad^2	I_o	I
Top flange $16 \times \frac{3}{4}$	12.0	30.38	365	11,070		11,070
Web $60 \times \frac{7}{16}$	26.3				7,880	7,880
Bottom flange $20 \times 1\frac{3}{4}$	35.0	-30.88	$-1,081$	33,380		33,380
	73.3		-716			52,330
$d_s = -716/73.3 = -9.77$ in					$-7.77 \times 716 =$	$-7,000$
					$I_{NA} =$	45,330

Distance from neutral axis of steel section to:

$$\text{Top of steel} = 30 + 0.75 + 9.77 = 40.52 \text{ in}$$

Section modulus, top of steel

$$S_{st} = 45,330/40.52 = 1,119 \text{ in}^3$$

TABLE 11.11 Composite Section about 30 ft from Supports

(a) For dead loads, $n = 24$

Material	A	d	Ad	Ad^2	I_o	I
Steel section	73.3		-716			52,330
Concrete $100 \times 8\frac{1}{2}/24$	35.4	37.0	1,310	48,470	210	48,680
	108.7		594			101,010
$d_{24} = 594/108.7 = 5.46$ in					$-5.46 \times 594 =$	$-3,240$
					$I_{NA} =$	97,770

Distance from neutral axis of steel section to:

$$\text{Top of steel} = 30.75 - 5.46 = 25.29 \text{ in}$$

Section modulus, top of steel

$$S_{st} = 97,770/25.29 = 3,866 \text{ in}^3$$

(b) For live loads, $n = 8$

Material	A	d	Ad	Ad^2	I_o	I
Steel section	73.3		-716			52,330
Concrete $100 \times 8\frac{1}{2}/8$	106.3	37.0	3,933	145,520	640	146,160
	179.6		3,217			198,490
$d_8 = 3217/179.6 = 17.91$ in					$-17.91 \times 3,217 =$	$-57,620$
					$I_{NA} =$	140,870

Distance from neutral axis of steel section to:

$$\text{Top of steel} = 30.75 - 17.91 = 12.84 \text{ in}$$

Section modulus, top of steel

$$S_{st} = 140,870/12.84 = 10,971 \text{ in}^3$$

TABLE 11.12 Stresses in Composite Plate Girder 30 ft from Supports

	Bending moments, ft-kips	
M_{DL}	M_{SDL}	$M_{LL} + M_I$
1,460	498	1,518 (Fig. 11.17)

Stresses at top of steel (compression), ksi

$$
\begin{aligned}
DL: 1,460 \times 12/1,119 &= 15.66 \ (S_{st} \text{ from Table 11.10}) \\
SDL: \quad 498 \times 12/3,866 &= 1.55 \ (S_{st} \text{ from Table 11.11}a) \\
LL + I: 1,518 \times 12/10,971 &= 1.66 \ (S_{st} \text{ from Table 11.11}b) \\
\text{Total:} \quad & \underline{} \\
& 18.87 < 20
\end{aligned}
$$

thicker plate equal to L_{cp}. From Table 11.13, S'_{sb} = 1244, and from Table 11.7, S_{sb} = 2142. Hence,

$$L_p = 100 \sqrt{1 - \frac{1,244}{2,142}} = 64.7 \text{ ft}$$

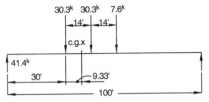

FIGURE 11.17 Positions of loads for maximum moment 30 ft from a support.

Assume a 66-ft length for the 1½-in bottom flange. It will then terminate 17 ft from the supports. Stresses are checked at that point to ensure that they are within the allowable (Table 11.15). Since the stresses are within the allowable, the bottom flange can be reduced in thickness to ⅞ in 17 ft from the supports.

Flange-to-Web Welds.   Fillet welds placed on opposite sides of the girder web to connect it to each flange must resist the horizontal shear between flange and web. The minimum size of weld permissible for the thickest plate at the connection usually determines the size of weld. In some cases, however, the size of weld may be governed by the maximum shear. In this example, shear does not govern, but the calculations are presented to illustrate the procedure.

The maximum shears, which occur at the supports, have been calculated previously but are included in Table 11.16*b*. Moments of inertia may be obtained from Table 11.13 for *DL* and from Table 11.14*b* for *SDL* and *LL + I*. Computations for the static moments *Q* of the flange areas are presented in Table 11.16*a*. Then,

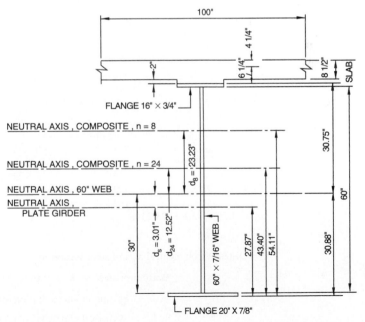

FIGURE 11.18 Cross section of composite plate girder near the supports.

TABLE 11.13   Steel Section Near Supports

Material	A	d	Ad	Ad^2	I_o	I
Top flange 16 × ¾	12.0	30.38	365	11,070		11,070
Web 60 × ⁷⁄₁₆	26.3				7,880	7,880
Bottom flange 20 × ⅞	17.5	−30.44	−533	16,220		16,220
	55.8		−168			35,170

$d_s = -168/55.8 = -3.01$ in

$$-3.01 \times 168 = -510$$
$$I_{NA} = 34,660$$

Distance from neutral axis of steel section to:

$$\text{Bottom of steel} = 30 + 0.88 - 3.01 = 27.87 \text{ in}$$

Section modulus, top of steel

$$S_{sb} = 34,660/27.87 = 1,244 \text{ in}^3$$

TABLE 11.14   Composite Section Near Supports

(a) For dead loads, $n = 24$

Material	A	d	Ad	Ad^2	I_o	I
Steel section	55.8		−168			35,170
Concrete 100 × 8.5/24	35.4	37.0	1,310	48,470	210	48,680
	91.2		1,142			83,850

$d_{24} = 1,142/91.2 = 12.52$ in

$$-1,252 \times 1,142 = -14,300$$
$$I_{NA} = 69,550$$

Distance from neutral axis of steel section to:

$$\text{Bottom of steel} = 30 + 0.88 + 12.52 = 43.40 \text{ in}$$

Section modulus, bottom of steel

$$S_{sb} = 69,550/43.40 = 1,602 \text{ in}^3$$

(b) For live loads, $n = 8$

Material	A	d	Ad	Ad^2	I_o	I
Steel section	55.8		−168			35,170
Concrete 100 × 8.5/8	106.3	37.0	3,933	145,520	640	146,160
	162.1		3,765			181,330

$d_8 = 3,765/162.1 = 23.23$ in

$$-23.23 \times 3,765 = -87,460$$
$$I_{NA} = 93,870$$

Distance from neutral axis of steel section to:

$$\text{Bottom of steel} = 30 + 0.88 + 23.23 = 54.11 \text{ in}$$

Section modulus, bottom of steel

$$S_{sb} = 93,870/54.11 = 1,735 \text{ in}^3$$

TABLE 11.15 Stresses in Composite Plate Girder 17 ft
from Supports

Bending moments, ft-kips		
M_{DL}	M_{SDL}	$M_{LL} + M_I$
981	335	1,044 (Fig. 11.19)

Stresses at top of steel (compression), ksi	
DL: 981 × 12/1,244 =	8.46 (S_{st} from Table 11.13)
SDL: 335 × 12/3,866 =	2.51 (S_{st} from Table 11.14a)
LL + I: 1,044 × 12/1,735 =	7.22 (S_{st} from Table 11.14b)
Total:	19.19 < 20

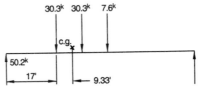

FIGURE 11.19 Positions of loads for maximum moment 17 ft from a support.

TABLE 11.16 Shear Stresses in Composite Plate Girder, ksi, at Supports

(a) Static moment Q, in³, of flange

Steel section only

Top flange: $Q_t = 12.0 × 33.39 = 401$
Bottom flange: $Q_b = 17.5 × 27.43 = 480$

Composite section, $n = 8$

Steel top flange: $Q_{st} = $ 12.0 × 7.15   =   86
Concrete slab: $Q_c = 106.3 × 13.77 = 1,464$
Total: $Q_t = $ 1,550
Steel bottom flange: $Q_b = $ 17.5 × 53.6   =   939

(b) Maximum shears, kips, at supports

V_{DL}	V_{SDL}	$V_{LL} + V_I$
69.5	23.7	75.7

(c) Shear stresses, kips per in

Top-flange welds	Bottom-flange welds
DL: 69.5 × 401/34,660 = 0.804	DL: 69.5 × 480/34,660 = 0.962
SDL: 23.7 × 1,550/93,870 = 0.391	SDL: 23.7 × 939/93,870 = 0.237
LL + I: 75.7 × 1,550/93,870 = 1.250	LL + I: 75.7 × 939/93,870 = 0.757
$v = 2.445$	$v = 1.956$

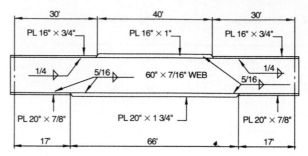

FIGURE 11.20 Plate girder with splices in top and bottom flanges.

shear, kips per in, on the two fillet welds can be computed from $v = VQ/I$ (Table 11.16c). The allowable stress on the weld is the smaller of the allowable shear stress, 12.4 ksi, for static loads and the allowable fatigue stress for 500,000 cycles of load. Fatigue does not govern in this example. Hence, the allowable load per weld is $12.4 \times 0.707 = 8.76$ kips per in. With a weld on each side of the web, the shear per weld is $2,445/2 = 1.223$. So the weld size required to resist the shear is $1.223/8.76 = 0.14$ in. (See *Flange-to-Web Welds*, p. 11.80.)

The minimum sizes of welds permitted with the flange material are all larger than this. Use two ¼-in fillets with the $16 \times$ ¾-in top flange, and two ⁵⁄₁₆-in fillets with the other flange plates (Fig. 11.20).

Stiffeners. See Arts. 11.9.4 to 11.9.6.

Bearings. See Arts. 11.9.8 and 11.9.9

Shear Connectors. See Art. 11.2.

Deflections. See Art. 11.2.

Load-Factor Design. See Art. 11.5.

11.5 EXAMPLE—LOAD-FACTOR DESIGN OF COMPOSITE PLATE-GIRDER BRIDGE

The "Standard Specifications for Highway Bridges" of the American Association for State Highway and Transportation Officials (AASHTO) allow load-factor design as an alternative method to allowable-stress design for design of simple and continuous beam and girder structures of moderate length and it is widely used for highway bridges.

Load-factor design (LFD) is a method of proportioning structural members for multiples of the design loads. The moments, shears, and other forces are determined by assuming elastic behavior of the structure. To ensure serviceability and durability, consideration is given to control of permanent deformations under overloads, to fatigue characteristics under service loadings, and to control of live-load deflections under service loadings. To illustrate load-factor design, a simply supported, composite, plate-girder stringer of the two-lane highway bridge in Art. 11.4 will be designed. The framing plan and the typical cross section are the same as for that bridge (Fig. 11.13). Structural steel is Grade 36, with yield strength $f_y = 36$ ksi and concrete for the deck slab is Class A, with 28-day strength $f'_c = 4,000$ psi. Loading is HS25.

11.5.1 Stringer Design Procedure

In the usual design procedure, the concrete deck slab is designed to span between the girders. A section is assumed for the steel stringer and classified as either symmetrical or unsymmetrical, compact or noncompact, braced or unbraced, and transversely or longitudinally stiffened. Section properties of a steel girder alone, and composite section properties of the steel girder and concrete slab are then determined, in a similar way as for allowable-stress design, for long- and short-duration loads. Next, flange local buckling is checked for the composite section. Fatigue stress checks are made for the most common connections found in a welded plate girder, such as those for transverse stiffeners, flange plate splices, gusset plates for lateral bracing, and flanges to webs.

The trial section is checked for compactness. The allowable stresses may have to be reduced if the section is noncompact and unbraced. Next, bending strength and shear capacity of the section are checked, and the section is adjusted as necessary. Then, transverse and longitudinal stiffeners are designed, if required. In addition, for a complete design, flange-web welds and shear connectors (fatigue to be included), bearing stiffeners (as concentrically loaded columns), lateral bracing (for wind loading) are designed and a deflection check is made.

11.5.2 Concrete Slab

The slab is designed to span transversely between stringers in the same way as for the allowable-stress method (Art. 11.4). A 9-in-thick, one-course concrete slab is used, as in Art. 11.4. The effective span S, the distance, ft, between flange edges plus half the flange width, is, for an assumed flange width of 16 in (1.33 ft),

$$S = 8.33 - 1.33 + 1.33/2 = 7.67 \text{ ft}$$

For computation of dead load,

Weight of concrete slab: $0.150 \times \%_{12}$ = 0.113
⅜-in extra concrete in stay-in-place   = 0.005
  forms: $0.150(\%)/12$
Future wearing surface   = 0.025
Total dead load w_D: 0.143 kips per ft

With a factor of 0.8 applied to account for continuity of slab over more than three stringers, the maximum dead-load bending moment is

$$M_D = \frac{w_D S^2}{10} = \frac{0.143(7.67)^2}{10} = 0.84 \text{ ft-kips per ft}$$

Maximum live-load moment, with reinforcement perpendicular to traffic, equals

$$M_L = \frac{(S + 2)}{32} P \qquad (11.8)$$

where P is the load on one rear wheel of a truck. Since $P = 16 \times 1.25 = 20$ kips for an HS25 truck,

$$M_L = \frac{(7.67 + 2)}{32} \, 20 = 6.04 \text{ ft-kips per ft}$$

Allowance for impact is 30% of this, or 1.81 ft-kips per ft. The total live load moment then is

$$M_L = 6.04 + 1.81 = 7.85 \text{ ft-kips per ft}$$

The factored total moment for AASHTO Group I loading on a straight bridge is

$$M_T = 1.3[DL + 1.67(LL + I) = 1.3(0.84 + 1.67 \times 7.85)$$

$$= 18.13 \text{ ft-kips per ft}$$

For a strip of slab $b = 12$ in wide, the effective depth d of the steel reinforcement is determined based on the assumption that No. 6 bars with 2.5 in of concrete cover will be used:

$$d = 9 - 2.5 - (\%)/2 = 6.13 \text{ in}$$

For determination of the moment capacity of the concrete slab, the depth of the equivalent rectangular compressive-stress block is given by

$$a = \frac{A_s f_y}{0.85 f'_c b} \qquad (11.9)$$

where A_s = the area, in^2, of the reinforcing steel.
For $f'_c = 4$ ksi and the yield strength of the reinforcing steel $F_y = 60$ ksi,

$$a = \frac{60 A_s}{0.85 \times 4 \times 12} = 1.47 A_s$$

Design moment strength ϕM_n is given by

$$\phi M_n = \phi A_s f_y (d - a/2) \qquad (11.10)$$

where the strength reduction factor $\phi = 0.90$ for flexure. If the nominal moment capacity ϕM_n is equated to the total factored moment M_T, the required area of reinforcement steel A_s can be obtained with Eq. (11.10) by solving a quadratic equation:

$$18.13 \times 12 = 0.9 \times 60 A_s (6.13 - 1.47 A_s/2)$$

from which $A_s = 0.72$ in^2 per ft. Number 6 bars at 7-in intervals supply 0.76 in^2 per ft and will be specified. The provided area should be checked to ensure that its ratio ρ to the concrete area does not exceed 75% of the balanced reinforcement ratio ρ_b.

$$\rho_b = \frac{0.85 \beta_1 f'_c}{f_y} \left(\frac{87}{87 + f_y} \right) \qquad (11.11)$$

where the factor $\beta_1 = 0.85$ for $f'_c = 4$-ksi concrete.

$$\rho_b = \frac{0.85 \times 0.85 \times 4}{60}\left(\frac{87}{87 + 60}\right) = 0.0285$$

For the provided steel area,

$$\rho = 0.76/(12 \times 6.13) = 0.0103 < (0.75\rho_b = 0.0214)\text{—OK}$$

AASHTO standard specifications state that, at any section of a flexural member where tension reinforcement is required by analysis, the reinforcement provided shall be adequate to develop a moment at least 1.2 times the cracking moment M_u calculated on the basis of modulus of rupture f_r for normal-weight concrete.

$$f_r = 7.5\sqrt{f_c'} \tag{11.12}$$

For $f_c' = 5$ ksi, $f_r = 7.5\sqrt{4000} = 474$ psi. The cracking moment is obtained from

$$M_u = f_r S \tag{11.13}$$

where the section modulus $S = bh^2/6 = 12 \times 8.5^2/6 = 144.5$ in^3. (One-half inch is deducted from the section for the wearing course.)

$$M_u = 474 \times 144.5/12,000 = 5.71 \text{ ft-kips per ft}$$

From Eq. (11.9), the depth of the equivalent rectangular stress block is

$$a = \frac{60 \times 0.76}{0.85 \times 4 \times 12} = 1.12 \text{ in}$$

Substitution of the preceding values in Eq. (11.10) yields the moment capacity

$$\phi M_n = 0.90 \times 0.76 \times 60(6.13 - 1.12/2)/12$$

$$= 19.05 > (1.2M_u = 6.85 \text{ ft-kips per ft})$$

Therefore, the minimum reinforcement requirement is satisfied.

For a complete slab design, serviceability requirements in the AASHTO standard specifications for fatigue and distribution of reinforcement in flexural members also need to be satisfied. Only a typical interior stringer will be designed in this example.

11.5.3 Loads, Moment and Shears

As in Art. 11.4, it is assumed that the girders will not be shored during casting of the concrete slab. The factored moments and shears will be obtained from the combination of dead load (DL) plus live load and impact ($LI + I$).

For the AASHTO Group I loading combination, the factored moment is

$$M_f = \gamma[\beta_D M_{DL} + 1.67(M_L + M_I)] \tag{11.14}$$

and the factored shear is

$$V_f = \gamma[\beta_D V_{DL} + 1.67(V_L + V_I)] \tag{11.15}$$

where γ is the load factor ($\gamma = 1.30$ for moment and 1.41 for shear) and β is 1.0.

Thus, the unfactored loads M_{DL}, M_{SDL} and V_{DL} in Art. 11.4 are still valid. Then,

$$M_{fT} = 1.30[1.0 \times 1{,}738 + 1.0 \times 592 + 1.67 \times (1{,}444 + 321)] = 6{,}862$$

$$V_{fT} = 1.41[1.0 \times 69.5 + 1.0 \times 23.7 + 1.67 \times (61.9 + 13.8)] = 309.7$$

FACTORED BENDING MOMENTS AT MIDSPAN, FT-KIPS

M_{fDL}	M_{fSDL}	$M_{fLL} + M_{fI}$	M_{fT}
2,260	770	3,832	6,862

FACTORED END SHEAR, KIPS

V_{fDL}	V_{fSDL}	$V_{fLL} + V_{fI}$	V_{fT}
98.0	33.4	178.3	309.7

11.5.4 Trial Girder Section

A trial section with a web plate $60 \times \frac{7}{16}$ in is assumed as in Art. 11.4. Bottom flange area can be estimated from

$$A_{sb} = \frac{12(M_{DL} + M_{LL} + M_I)}{F_y d} \tag{11.16}$$

For the preceding bending moments,

$$A_{sb} = \frac{12(2260 + 770 + 3832)}{36 \times 60} = 38.1 \text{ in}^2$$

Since the part of the web below the neutral axis will also carry some force, a bottom flange $20 \times 1\frac{1}{2}$ in ($A_{sb} = 30.0$ in^2) will be tried first. For the top flange plate, $A_{sb}/2 = 15.0$ in^2, a top flange of 16×1 in will be tried.

The concrete section for an interior stringer, not including the concrete haunch, is 8 ft 4 in wide (c to c of stringers) and $8\frac{1}{2}$ in deep ($\frac{1}{2}$ in of slab is deducted from the concrete depth for the wearing course). The concrete area $A_c = 8.33 \times 12 \times 8.50 = 850$ in^2. Thus, this is an unsymmetrical composite section.

Check for Local Buckling.   The trial section is assumed to be braced and non-compact. The width-thickness ratio b'/t of the projecting compression-flange element may not exceed

$$\frac{b'}{t} = \frac{69.6}{\sqrt{F_y}} \tag{11.17}$$

where b' is the width of the projecting element, t, the flange thickness, and F_y, the specified yield stress, ksi. For flange width $b = 16$ in and $F_y = 36$ ksi, the thickness should be at least

$$t = \frac{\sqrt{36}}{69.6} \times \frac{16}{2} = 0.69 \text{ in}$$

The 1-in-thick top flange is satisfactory.

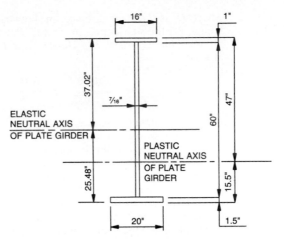

FIGURE 11.21 Cross section assumed for plate girder for load-factor design.

Properties of Trial Section.   The trial section is shown in Fig. 11.21. The computations for the location of the neutral axis and for the section moduli S_{st} and S_{sb} of the trial plate-girder section are tabulated in Table 11.17.

For unsymmetrical girders with transverse stiffeners but without longitudinal stiffeners, the minimum thickness of the web is obtained from:

$$\frac{D_c}{t_w} \le \frac{577}{\sqrt{F_y}} \qquad D_c > D/2 \tag{11.18}$$

TABLE 11.17   Steel Section for Maximum Factored Moment

Material	A	d	Ad	Ad^2	I_o	I
Top flange 16 × 1	16.0	30.50	488	14,880		14,880
Web 60 × ⁷⁄₁₆	26.3				7,880	7,880
Bottom flange 20 × 1½	30.0	−30.75	−923	28,370		28,370
	72.3		−435			51,130
$d_s = -435/72.3 = -6.02$ in					$-6.02 \times 435 =$	−2,620
					$I_{NA} =$	48,510

Distance from neutral axis of steel section to:

$$\text{Top of steel} = 30 + 1 + 6.02 = 37.02 \text{ in}$$

$$\text{Bottom of steel} = 30 + 1.50 - 6.02 = 25.48 \text{ in}$$

Section moduli	
Top of steel	Bottom of steel
$S_{st} = 48,510/37.02$	$S_{sb} = 48,510/25.48$
$= 1,310 \text{ in}^3$	$= 1,904 \text{ in}^3$

where D_c is the clear distance, in, between the neutral axis and the compression flange; D, the web depth, in; and t_w, the web thickness, in. For the trial section, $D_c = 37.02 > (D/2 = 30)$. Hence, from Eq. (11.18), the web thickness should be at least

$$t_w = \frac{D_c\sqrt{F_y}}{577} = \frac{37.02\sqrt{36}}{577} = 0.38 \text{ in, or } \tfrac{3}{8} \text{ in}$$

Since the assumed $\tfrac{7}{16}$-in web thickness exceeds $\tfrac{3}{8}$ in, the requirement for minimum web thickness without longitudinal stiffeners is met.

The computations for the location of the neutral axis and for the section moduli are given in Table 11.18 for the composite section, with $n = 8$ for short-time loading, such as live load and impact, and $n = 3 \times 8 = 24$ for long-time loading, such as superimposed dead loads. To locate the neutral axis, moments are taken about middepth of the girder web. Depth of the concrete haunch atop the girder is assumed to be 2 in. In addition, since the girder is composite, for prevention of flange buckling, the width-thickness ratio of the projecting element of the compression flange may not exceed

$$\frac{b'}{t} = \frac{69.6}{\sqrt{1.3f_{dl1}}} \tag{11.19}$$

where f_{dl1} is the compression stress ksi, in the top flange due to noncomposite dead load.

$$f_{dl1} = \frac{2,260 \times 12}{1,310} = 20.70 \text{ ksi}$$

From Eq. (11.19), for a flange width of 16 in, the flange thickness t_1 should be at least

$$t_1 = \frac{\sqrt{1.3 \times 20.7}}{69.6} \times \frac{16}{2} = 0.6 \text{ in}$$

The 1-in-thick top flange is satisfactory.

11.5.5 Fatigue Stresses

In the next step of the design procedure, fatigue stresses will be investigated. The four-stringer system is considered to have multiple load paths. A single fracture in a member cannot lead to collapse of the bridge. Hence, the structure is not fracture-critical.

Determination of the allowable stress range F_{sr} for fatigue is based on the stress category for the connection under consideration, the type of load path (redundant or nonredundant), and the stress cycle.

The bridge is located on a major highway (Case II) with an average daily truck traffic in one direction (ADTT) less than 2,500. The plate girders incorporate the four connection types tabulated in Table 11.19 with corresponding stress types and categories. For main (longitudinal) load-carrying members, the number of stress cycles of the maximum stress range for Case II, with ADTT < 2,500, the AASHTO standard specifications specify 500,000 loading cycles for truck loading and 100,000 for lane loading. Table 11.20 also lists for the four types of connections the allowable stress ranges F_{sr} for the redundant-load-path structure based on the connection

TABLE 11.18 Composite Section for Maximum Factored Moment

(a) For superimposed dead loads, $n = 24$

Material	A	d	Ad	Ad^2	I_o	I
Steel section	72.3		−435			51,130
Concrete $100 \times 8.5/24$	35.4	37.25	1319	49,120	210	49,330
	107.7		884			100,460
$d_{24} = 884/107.7 = 8.21$ in				$-8.21 \times 884 =$		−7,260
					$I_{NA} =$	93,200

Distance from neutral axis of composite section to:

$$\text{Top of steel} = 31.00 - 8.21 = 22.79 \text{ in}$$

$$\text{Bottom of steel} = 31.50 + 8.21 = 39.71 \text{ in}$$

$$\text{Top of concrete} = 22.79 + 2 + 8.5 = 33.29 \text{ in}$$

Top of steel	Bottom of steel	Top of concrete
$S_{st} = 93,200/22.79$ $= 4,089 \text{ in}^3$	$S_{sb} = 93,200/39.71$ $= 2,347 \text{ in}^3$	$S_c = 93,200/33.29$ $= 2,800 \text{ in}^3$

(b) For live loads, $n = 8$

Material	A	d	Ad	Ad^2	I_o	I
Steel section	72.3		−435			51,130
Concrete $100 \times 8.5/8$	106.3	37.25	3,958	147,440	640	148,080
	178.6		3,523			199,210
$d_8 = 3523/178.6 = 19.73$ in				$-19.73 \times 3,523 =$		−69,510
					$I_{NA} =$	129,700

Distance from neutral axis of composite section to:

$$\text{Top of steel} = 31.00 - 19.73 = 11.27 \text{ in}$$

$$\text{Bottom of steel} = 31.00 + 19.73 = 50.73 \text{ in}$$

$$\text{Top of concrete} = 11.27 + 2 + 8.5 = 21.77 \text{ in}$$

Top of steel	Bottom of steel	Top of concrete
$S_{st} = 129,700/11.27$ $= 11,510 \text{ in}^3$	$S_{sb} = 129,700/50.73$ $= 2,557 \text{ in}^3$	$S_c = 129,700/21.77$ $= 5,958 \text{ in}^3$

stress category and the number of stress cycles. The fatigue stress in the bottom flange is checked for unfactored HS25 loading on the composite section. For live-load moment plus impact, $M = 1,765$ ft-kips, and the corresponding stress is

$$f_b = \frac{1,765 \times 12}{2,557} = 8.3 \text{ ksi}$$

TABLE 11.19 Categories and Allowable Fatigue Stress Ranges for Connections*

Connection type	Stress type	Category	Allowable stress range F_{sr}, ksi	
			500,000 cycles	100,000 cycles
Toe of transverse stiffener	Tension or reversal	C	21	35.5
Groove weld at flanges	Tension or reversal	B	29	49
Gusset plate for lateral bracing	Tension or reversal	B		
Flange-to-web weld	Shear	F	12	15

*See AASHTO Specifications for full requirements that apply.

Since the plate girder under consideration is simply supported, the minimum live-load moment would be zero and the live-load stress range becomes $f_{sr} = 8.3$ ksi $< F_{sr} = 21$ ksi. The section is OK for fatigue.

11.5.6 Check for Compactness

The allowable stresses may have to be reduced if the section is noncompact and unbraced. Composite beams in positive bending qualify as compact when the web depth-thickness ratio D/t_w of the steel section meets the following requirement:

$$\frac{D}{t_w} \leq \frac{608}{\sqrt{F_y}} \tag{11.20}$$

where t_w is the web thickness, in, and D is the clear distance, in, between the flanges. For composite beams used in simple spans, D may be replaced by $2D_{cp}$, the distance, in, from the compression flange to the neutral axis in plastic bending. The compression depth of the composite section in plastic bending, including the slab, may not exceed

$$d_c = \frac{d + t_s}{7.5} \tag{11.21}$$

where d is the depth of the steel girder, in, and t_s is the thickness, in, of slab.

$$d_c = \frac{d + t_s}{7.5} = \frac{62.5 + 8.5}{7.5} = 9.47 \text{ in}$$

Therefore, the maximum allowed $D_{cp} = 9.47 - 8.5 = 0.97$ in. From Eq. (11.20), with D replaced by $2D_{cp} = 2 \times 0.97 = 1.94$,

$$t_w = \frac{\sqrt{36}}{608} \times 1.94 = 0.02 < \frac{7}{16} \text{ in}$$

The section meets the requirement for compactness.

Check of Unbraced Length of Top Flange. For live loads, the top flange of the girder is continuously supported by the concrete deck slab. But it is necessary to check the unbraced length L_b of the top flange for dead loads on the noncomposite section. For compact sections, spacing of lateral bracing of the compression flange may not exceed

$$\frac{L_b}{r_y} = \frac{[3.6 - 2.2 \, (M_1/M_u)]10^3}{F_y} \tag{11.22}$$

where r_y is the radius of gyration with respect to the y-y axis, in, M_1 is the smaller moment at the end of the unbraced length of the member, M_u is the maximum bending strength $= F_y Z$, and Z is the plastic section modulus, in³. For the 16 × 1-in top flange,

$$r_y = \sqrt{\frac{I}{A}} = \sqrt{\frac{1 \times 16^3/12}{1 \times 16}} = 4.62 \text{ in}$$

To determine the plastic modulus Z, the location of the axis that divides the section into two equal areas has to be found. The total area A of the girder is 72.25 in² (Table 11.17), and $A/2 = 36.13$ in². If $\bar{y}$ is the distance from top of steel to the axis, then $\bar{y} - 1$ is the web length from the axis to the flange. Since the web thickness is ⁷⁄₁₆ in and the flange area $A_f = 16$ in², $16 + (\bar{y} - 1)(⁷⁄₁₆) = 36.83$, and $\bar{y} = 47.0$ in (Fig. 11.21). Z is computed by taking moments about the axis:

$$Z = (16 \times 46.5 + 20.13 \times 23) + (80 \times 13.75 + 6.13 \times 7)$$

$$= 1,662 \text{ in}^3$$

The bending strength then is

$$M_u = F_y Z = 36 \times 1,662/12 = 4,986 \text{ ft-kips}$$

From Eq. (11.22) with $M_1 = M_{DL} = 1,738$ ft-kips, the maximum allowable unbraced length is

$$L_b = \frac{4.62[3.6 - 2.2(1,738/4,986)]10^3}{36} = 364 \text{ in}$$

Since the spacing of bracing cross frames is 25 ft $= 300$ in and L_b is larger, the section may be treated as braced and compact.

11.5.7 Bending Strength of Girder

The flexural stresses in the composite section of the interior girder are checked for all the factored loads to ensure that the maximum stresses do not exceed $F_y = 36$ ksi. The computations in Table 11.20 indicate that the composite section is OK.

11.5.8 Shear Capacity of Girder

For girders with transverse stiffeners, shear capacity V_u, ksi, is given by

$$V_u = V_p \left[C + \frac{0.87(1 - C)}{\sqrt{1 + (d_o/D)^2}} \right] \qquad \frac{d_o}{D} \leq 3 \qquad \frac{d_o}{D} \leq \frac{67,600}{(D/t_w)^2} \tag{11.23}$$

where V_p = shear yielding strength of web, ksi, $= 0.58 D t_w F_y$
d_o = spacing, in, of intermediate stiffeners
D = clear distance, in, between flanges

TABLE 11.20   Stresses, ksi, in Composite Girder at Section of Maximum Moment

Top of steel (compression)	Bottom of steel (tension)
DL: $f_b = 2,260 \times 12/1,310 = 20.70$	DL: $f_b = 2,260 \times 12/1,904 = 14.24$
SDL: $f_b = \quad 770 \times 12/4,089 = 2.26$	SDL: $f_b = \quad 770 \times 12/2,347 = 3.94$
LL + I: $f_b = 3,832 \times 12/11,510 = \underline{3.99}$	LL + I: $f_b = 3,832 \times 12/2,557 = \underline{17.98}$
Total: $\qquad\qquad\qquad\qquad 26.95 < 36$	Total: $\qquad\qquad\qquad\qquad 36.16 \approx 36$

Top of Concrete

$$SDL: f_c = \quad 770 \times 12/(2,800 \times 8) = 0.41$$

$$LL + I: f_c = 3,832 \times 12/(5,958 \times 8) = \underline{0.96}$$

$$1.37 < 4.0$$

t_w = web thickness, in
C = web buckling coefficient (Art. 10.13.4)

Stiffeners are usually equally spaced between cross frames. Spacing ranges up to the maximum of $1.5D$ for the first stiffener. The cross frames in the example are spaced about 25 ft apart. For a first trial, $d_o = 25/2 = 12.50$ ft = 150 in.

$$\frac{d_o}{D} = \frac{150}{60} = 2.5 < 3—\text{OK}$$

$$\frac{67,000}{[60/(\text{7/16})]^2} = 3.6 > \frac{d_o}{D}—\text{OK}$$

The plastic shear force is

$$V_p = 0.58 \times 36 \times 60 \times \text{7/16} = 548 \text{ kips}$$

The coefficient C, the buckling shear stress divided by the shear yield stress, is computed from

$$C = \frac{45,000k}{(D/t_w)^2 F_y} \qquad \frac{D}{t_w} > 237\sqrt{k/F_y} \qquad (11.24)$$

where $k = 5[1 + (d_o/D)^2] = 5[1 + (2.5/60)^2] = 5.8$ and $D/t_w = 60/(\text{7/16}) = 137$.

$$C = \frac{45,000 \times 5.8}{137^2 \times 36} = 0.39$$

From Eq. (11.23), the shear capacity of the girder is

$$V_u = 548 \left[0.39 + \frac{0.87(1 - 0.39)}{\sqrt{1 + (2.5)^2}} \right] = 322 \text{ kips}$$

The total factored end shear $V_{max} = 309.7$ kips < 322 kips. Thus, the section is adequate for shear.

11.5.9 Transverse Stiffener Design

For girders not meeting the shear capacity requirement, $V_u = CV_p$, transverse stiffeners are required. The trial section in this example meets this requirement. Girders designed to meet the shear requirement without transverse stiffeners normally have thicker webs but usually cost less than girders designed with thinner webs and transverse stiffeners, because of high welding costs for attaching stiffeners to webs. Another added advantage of a design without transverse stiffeners is elimination of the fatigue-prone welds between webs and stiffeners.

11.5.10 Shear Connectors

The horizontal shears at the interface of the concrete slab and steel girder are resisted by shear connectors throughout the simply supported span to develop composite action. Shear connectors are mechanical devices, such as welded studs or channels, placed in transverse rows across the top flange of the girder and embedded in the slab. The shear connectors for the girder will be designed for ultimate strength and the number of connectors provided for that purpose will be checked for fatigue.

For ultimate strength, the number N_1 of shear connectors required between a section of maximum positive moment and an adjacent end support should be at least

$$N_1 = P/\phi S_u \tag{11.25}$$

where S_u = ultimate strength of a shear connector, kips
ϕ = reduction factor = 0.85
P = force, kips in the concrete slab taken as the smaller of P_1 and P_2
$P_1 = A_s F_y$
$P_2 = 0.85 f'_c bt_s$
A_s = area, in^2, of the composite section
b = effective width, in, of slab for composite action
t_s = slab thickness, in
$P_1 = 72.3 \times 36 = 2{,}603$ kips
$P_2 = 0.85 \times 4 \times 100 \times 8.5 = 2{,}890 > 2{,}603$ kips

Hence, P for Eq. (11.25) = 2,603 kips.

For shear connectors, welded studs $\frac{7}{8}$ in in diameter and 6 in long will be used. According to AASHTO standard specifications, the ultimate strength S_u, kips, of welded studs for $h/d > 4$, where h is stud height, in, and d = stud diameter, in, may be determined from

$$S_u = 0.4d^2\sqrt{f'_c E_c} \tag{11.26}$$

where E_c = modulus of elasticity of the concrete, ksi = $1{,}800\sqrt{f'_c}$ = 3,600 ksi. For the $\frac{7}{8}$-in-dia. welded studs,

$$S_u = 0.4(\tfrac{7}{8})^2\sqrt{4 \times 3{,}600} = 36.75 \text{ kips}$$

from which the number of studs required is

$$N_1 = \frac{2{,}603}{0.85 \times 36.75} = 83$$

With the studs placed in groups of three, there should be at least 26 groups on each half of the girder.

Pitch is determined by fatigue requirements. The allowable load range, kips per stud, is given by Eq. (11.4), with $\alpha = 10.6$ for 500,000 cycles of loading. Hence, the allowable load range is

$$Z_r = 10.6(\tfrac{7}{8})^2 = 8.12 \text{ kips}$$

At the supports, the shear range $V_r = 75.7$ kips, the shear produced by live load plus impact service loads. Consequently, with $n = 8$ for the concrete, the transformed concrete area equal to 106.3 in^2, and $I = 129,700$ in^4 from Table 11.18b, the range of horizontal shear is

$$S_r = \frac{V_r O}{I} = \frac{75.7 \times 106.3 \times 17.52}{129,700} = 1.087 \text{ kips per in}$$

The pitch required for stud groups near a support is

$$p = \frac{3Z_r}{S_r} = \frac{3 \times 8.12}{1.087} = 22.41 \text{ in}$$

The average pitch required for ultimate strength for 26 groups between midspan and a support is $\tfrac{1}{2} \times 100 \times \tfrac{12}{26} = 23$ in. Use three $\tfrac{7}{8}$-in-dia. by 6-in-long studs per row, spaced at 18 in.

See also Arts. 11.9.4 to 11.9.6.

11.6 CHARACTERISTICS OF CURVED GIRDER BRIDGES

Past practice in design of new highways often located bridges first, then aligned the roadway with them. Current practice, in contrast, usually fits bridges into the desired highway alignment. Since curved crossings are sometimes unavoidable, and curved ramps at interchanges often must span other highways, railroads, or structures, bridges in those cases must be curved. Plate or box girders usually are the most suitable type of framing for such bridges.

Though the deck may be curved in accordance with the highway alignment, the girders may be straight or curved between skewed supports. Straight girders require less steel and have lower fabrication costs. But curved girders offer better appearance, and often the overall cost of a bridge with such girders may not be greater than that of a structure with straight members. Curved girders may reduce the number of foundations required because longer spans may be used; deck design and construction is simpler, because girder spacing and deck overhangs may be kept constant throughout the span; and cost savings may accrue from use of continuous girders, which may not be feasible with straight, skewed girders. Consequently, curved girders are generally used in curved bridges.

Curved girders introduce a new dimension in bridge design. The practice used for straight stringers of distributing loads to an individual stringer, as indicated in a standard specification, and then analyzing and designing the stringer by itself, cannot be used for curved bridges. For these structures, the entire superstructure must be designed as a unit. Diaphragms or cross frames as well as the stringers serve as main load-carrying members, because of the torsion induced by the curvature.

Analyses of such grids are very complicated, because they are statically indeterminate to a high degree. Computer programs, however, have been developed for performing the analyses. In addition, experience with rigorous analyses indicates that under certain conditions approximate methods give sufficiently accurate results.

The approximate methods described in this article are suitable for manual computations. They appear to be applicable to concentric, circular stringers where the arc between supports subtends an angle not much larger than about 0.5 radian, or about 30°. Also, where the spans are continuous, the methods may be used if the sum of the central angles subtended by each span does not exceed 90°. Accuracy of these methods, however, also seems to depend on the flexural rigidity of the deck in the radial direction and of the diaphragms.

The limitation of central angle indicates that the maximum span, along the arc, for a radius of curvature of 300 ft is about $300 \times 0.5 = 150$ ft for the approximate analysis. If the curved span is 200 ft, the approximate method should not be used unless the radius is at least $200/0.5 = 400$ ft.

Each simply supported or continuous girder should have at least one torsionally fixed support.

For box girders, in addition, accuracy depends on the ratio of bending stiffness to torsional rigidity EI/GK, where E is the modulus of elasticity, G the shearing modulus, I the moment of inertia for longitudinal bending, and K the torsional constant for the radial cross section. (For a hollow, rectangular tube,

$$K = \frac{4A^2}{\Sigma(l/t)} \tag{11.27}$$

where A = area, in^2, enclosed within the mean perimeter of tube
  l = length of a side, in
  t = thickness of that side, in

For inclusion in the summation in the denominator, a concrete slab in composite construction should be transformed into an equivalent steel plate by dividing the concrete cross-sectional area by the modular ratio n.) If the central angle of a curved span is about 0.5 radian, the approximate method should give satisfactory results if the weighted average of EI/GK in the span does not exceed 2.5.

A curved-girder bridge may have open framing, closed framing, or a combination of the two types. In open framing, curved plate girders are assisted in resisting torsion only by cross frames, diaphragms, or floorbeams at intervals along the span. In closed framing, the curved members may be box girders or plate girders assisted in resisting torsion by horizontal lateral bracing as well as by cross frames, diaphragms, or floorbeams.

11.6.1 Approximate Analysis of Open Framing

The approximate method for open framing derives from a rigorous method based on consistent deformations. Various components of the structure when distorted by loads must retain geometric compatibility with each other and simultaneously stay in equilibrium. The equations developed for these conditions can be satisfied only by a unique set of internal forces. In the rigorous method, a large number of such equations must be solved simultaneously. In the approximate method, considerable simplification is achieved by neglecting the stiffness of the plate girders in St. Venant (pure) torsion.

In the following, girders between the bridge centerline and the center of curvature are called **inner girders.** The rest are called **outer girders.**

The method will be described for a bridge with concentric circular stringers, equally spaced. Thus, for the four girders shown in Fig. 11.22a, if the distance

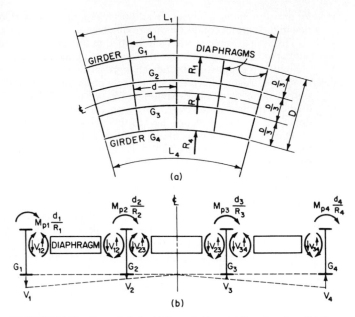

FIGURE 11.22 Curved-girder highway bridge. (*a*) Framing plan. (*b*) Cross section through the bridge at a diaphram.

from outer girder G_1 to inner girder G_4 is D, the girder spacing is $D/3$. The radius of the bridge centerline is R and of any girder G_n, R_n. Diaphragms are equally spaced at distance d apart along the centerline and placed radially between the girders.

Initially, the girders are assumed to be straight, and the span of each girder is taken as its developed length between supports. Preliminary moments M_p and shears V_p are computed as for straight girders.

These values must be corrected for the effects of curvature. The primary effect is a torque acting on every radial cross section of each girder. The torque per unit length at any section of a girder G_n is given approximately by

$$T_n = \frac{M_{pn}}{R_n} \tag{11.28}$$

where M_{pn} = preliminary bending moment at section and R_n = radius of G_n. If the diaphragm spacing along G_n is d_n and M_{pn} is taken as the preliminary moment at a diaphragm, then the total torque between diaphragms is

$$M_{tn} = \frac{M_{pn}d_n}{R_n} = \frac{M_{pn}d}{R} \tag{11.29}$$

This torque must be resisted by end moments in the diaphragms (Fig. 11.22*b*).

For equilibrium, the end moments on a diaphragm must be balanced by end shears forming an oppositely directed couple. For example, the diaphragm between G_2 and G_3 in Fig. 11.22*b* is subjected to end shears V_{23}. Also, the diaphragm between G_1 and G_2 is subjected to shears V_{12}. Consequently, G_2 is acted on by a net downward force V_2, called a V load, at the diaphragm

$$V_2 = V_{12} + V_{23} \tag{11.30}$$

where upward forces are taken as positive and downward forces as negative.

The V loads applied by the diaphragms are treated as additional loads on the girders.

For a bridge with two girders, the V load on the inner girder equals that on the outer girder, at a specific diaphragm, but is oppositely directed. Determined by equilibrium conditions at the diaphragm, this V load may be computed from

$$V = \frac{M_{p1} + M_{p2}}{K} \tag{11.31}$$

where $K = RD/d$ and $D =$ girder spacing in two-girder bridges and distance between inner and outer girders in bridges with more than two girders.

For a bridge with more than two girders, the method assumes that the V load on a girder at a diaphragm is proportional to the distance of the girder from the centerline of the bridge. Then, equilibrium conditions require that the V load on the outer girder of a multigirder bridge be computed from

$$V = \frac{\Sigma M_{pn}}{CK} \tag{11.32}$$

where $C =$ constant given in Table 11.21. The numerator in Eq. (11.13) consists of the sum of the preliminary moments in the girders at the line of diaphragms. Thus, for the four-girder bridge in Fig. 11.22b, the V load on G_1 and G_4 equals

$$-V_1 = V_4 = \frac{M_{p1} + M_{p2} + M_{p3} + M_{p4}}{1.11K} \tag{11.33}$$

By proportion, $-V_2 = V_3 = V_4/3$.

The bending moment produced by the V loads at any section of a girder G_n must be added to the preliminary moment at that section to produce the final bending moment M_n there. Thus,

$$M_n = M_{pn} + M_{vn} \tag{11.34}$$

where $M_{vn} =$ bending moment produced by V loads. Similarly, the shear due to the V loads must be added to the preliminary shears to yield the final shears. Stresses are computed in the same way as for straight girders.

Between diaphragms, the girder flanges resist the torsion. At any section, the stresses in the top and bottom flanges of a girder provide a couple equal to the torque but oppositely directed. The forces comprising this couple induce lateral

TABLE 11.21   Values of C for Eq. (11.32)

No. of girders	2	3	4	5	6	7	8	9	10
C	1.00	1.00	1.11	1.25	1.40	1.56	1.72	1.88	2.04

bending in the flanges. If q_n is the force per unit length of flange in girder G_n resisting torque,

$$q_n = \frac{M_n}{R_n h_n} \tag{11.35}$$

where h_n = distance between centroids of flanges. Each flange may be considered to act under this loading as a continuous beam spanning between diaphragms. The maximum negative moment for design purposes may be taken as

$$M_{Ln} = -\frac{0.1 M_n d_n^2}{R_n h_n} \tag{11.36}$$

The stress due to lateral bending should be added to that due to M_n to obtain the maximum stress in each flange. Where provision is made for composite action, however, the lateral bending stress in that flange may be neglected.

For preliminary design purposes, a rough approximation of the effects of curvature may be obtained by use of

$$p = 5.25 \frac{(1 + r)mL_c^2}{CRD} \tag{11.37}$$

where p = percent increase in moment in outer girder due to curvature
$r = \dfrac{\text{loading on inner girder}}{\text{loading on outer girder}} \left(\dfrac{R'}{R}\right)^2$
R' = radius of curvature of inner girder
R = radius of curvature of outer girder
m = number of girders
L_c = developed length of outer girder between supports when simply supported or between inflection points when continuous
C = constant given by Table 11.21
D = distance between inner and outer girders

11.6.2 Approximate Analysis of Closed Framing

Analysis of bridges with box girders or similar boxlike framing must take into account the torsional stiffness of these members. The method to be described is based on the following assumptions:

Girder cross sections are symmetrical about the vertical axis. Supports are radial. Curvature may vary so long as it does not change direction within a span. Diaphragms prevent distortion of the cross sections. Secondary stresses due to torsional warping are negligible.

Differential equations for determining the internal forces acting on a curved girder can be obtained from the equilibrium conditions for a differential segment (Fig. 11.23). Because upward and downward vertical forces must balance, the shear V is related to the loading w by

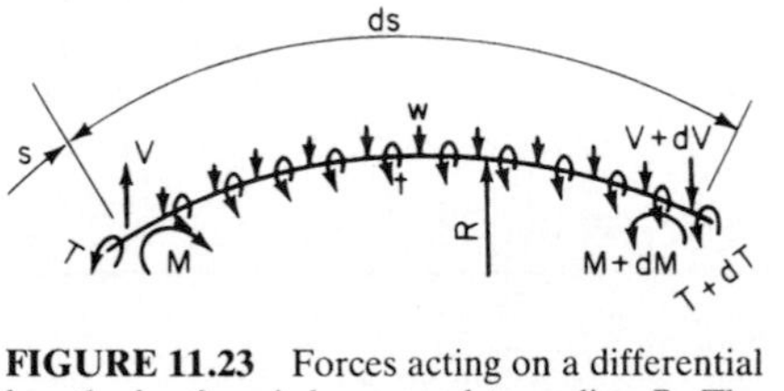

FIGURE 11.23 Forces acting on a differential length ds of a girder curved to radius R. The distributed vertical load w and torque t cause vertical end shears V, end moments M, and end torques T.

$$\frac{dV}{ds} = -w \tag{11.38}$$

Thus, as for a straight beam, the change in shear between any two sections of the girder equals the area of the load diagram between those sections. Because the sum of the moments about a radial plane must equal zero, bending moments M, torques T, and shears V are related by

$$\frac{dM}{ds} = -\frac{T}{R} + V \qquad (11.39)$$

where R = radius of curvature of girder. In the approximate method, with the limitations on central angle subtended by the span and on the ratio of bending stiffness to torsional stiffness, the T/R term can be ignored. Thus, the equation becomes

$$\frac{dM}{ds} = V \qquad (11.40)$$

As for straight beams, the change in bending moments between any two sections of the girder equals the area of the shear diagram between those sections.

Hence, bending moments in a curved girder of the closed-framing type may be computed approximately by treating it as a straight beam with span equal to the developed length of the curve.

A third equation is obtained by taking moments about a tangential plane:

$$\frac{dT}{ds} = \frac{M}{R} - t \qquad (11.41)$$

where t = applied torque. With the bending moments throughout the girder known, the torque at any section can be found from Eq. (11.41). [For a more rigorous solution, Eqs. (11.38), (11.39), and (11.41) may be solved simultaneously. This can be done by differentiating Eq. (11.39), solving for dT/ds, substituting the result in Eq. (11.41), and then solving the resulting second-order differential equation.]

Equation (11.41) indicates that the change in torque between any two sections of the girder equals the area of the $M/R - t$ diagram between those sections. Consequently, the torque on a curved girder of the closed-framing type can be determined by a method similar to the conjugate-beam method for determining deflections. In the approximate method, however, the moments M determined from Eq. (11.40) are used instead of those from the more complex rigorous solution.

Thus, first the bending-moment diagram (Fig. 11.24b) is obtained for the vertical loading on the developed length of the girder (Fig. 11.24a). Then, all ordinates are divided by the radius R. Next, the applied-torque diagram (Fig. 11.24d) is plotted for the twisting moments applied by the loading to the girder (Fig. 11.24c). The ordinates of this diagram are subtracted from the corresponding ordinates of the M/R diagram. The resulting $M/R - t$ diagram then is used as a loading diagram on the developed length of the girder (Fig. 11.24e). The resulting shears (Fig. 11.24f) equal the torques T in the curved girder. Note that positive $M/R - t$ is equivalent to an upward load on the conjugate beam.

The conjugate beam shown in Fig. 11.24c is simply supported. This requires that the angle of twist at the supports be zero. Hence, for this case, the curved girder is torsionally fixed at the supports. This condition is attained with a line of diaphragms at each support and a bearing under each web capable of resisting uplift, a common practice. Sometimes, interior supports of a continuous box girder are not fixed against torsion, for example, where a single bearing is placed under a diaphragm. In such cases, the span of the conjugate beam should be taken as the developed length of girder between supports that are fixed against torsion.

11.6.3 Loading

Dead loads may be distributed to curved girders in the same way as for straight girders. For live loads, the designer also may use any method commonly used for

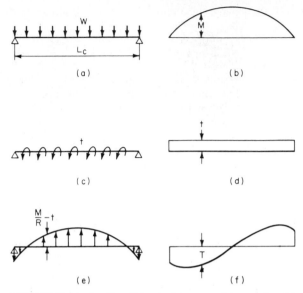

FIGURE 11.24 Loading diagrams for a curved box girder. (*a*) Uniform load *w* on the developed length. (*b*) Bending moment diagram for the uniform load. (*c*) Applied torque *t*. (*d*) Torsion diagram. (*e*) $M/R - t$ diagram applied as a load to the conjugate beam. (*f*) Shear diagram for the loading in (*e*).

straight girders. If the distribution procedure of the AASHTO "Standard Specifications for Highway Bridges" is used, however, a correction factor should be applied. The sum of the AASHTO live-load distribution factors for all the girders in a curved grid will usually exceed the number of wheel loads required for the roadway width. Hence, if these factors were used to compute the live-load moments in the girders at a line of diaphragms, the *V* loads there would be too large, because they are proportional to the sum of those moments. One way to correct the *V* loads determined with the AASHTO factors is to multiply the *V* loads by the ratio of the number of wheel loads required for the roadway width to the number of wheel loads determined by the sum of the AASHTO factors.

Impact may be taken into account, in the same way as for straight girders, as a percent increase in live load.

Centrifugal forces comprise a horizontal, radial loading on curved structures that does not apply to straight bridges. These forces are determined as a percentage of the live load, without impact (Art. 10.4). But the live load is restricted to one standard truck placed for maximum loading in each design lane.

Assumed to act 6 ft above the roadway surface, measured from the roadway centerline, centrifugal forces induce torques and horizontal shears in the superstructure. The shears may be assumed to be resisted by the concrete deck within its plane. The torques, however, must be resisted by the girders. In open-framing systems, the primary effect is on the preliminary bending moments. Resisting couples comprise upward and downward vertical forces in the girders. These forces increase bending moments in the outer girders (those farthest from the center of curvature) and decrease moments in the inner girders. The effect of centrifugal forces on *V* loads, however, is small, because *V* loads are determined by the sum of girder moments at a line of diaphragms and this sum is not significantly changed by centrifugal forces.

11.6.4 Sizing of Girders

Design rules for proportioning straight girders generally are applicable to curved girders, depth-span ratios, for example. But curvature does produce effects that should be considered for maximum economy. For instance, girder flanges in open-framing systems should be made as wide as practical to minimize lateral bending stresses. In some cases, where these stresses become too large, a reduction in spacing of diaphragms or cross frames may be desirable.

If curvature causes large adjustments to the preliminary moments in open framing, deepening of girders farthest from the center of curvature may be advantageous. This may be done without overall increase of the floor system, because of the superelevation of the deck.

Girder webs, in some cases, may have to be thicker than for straight girders with corresponding span, spacing, and loadings, because of the effects of curvature on shear. Reactions, too, may be significantly changed, and the effects on substructure design should be taken into account. For some sharply curved bridges, tie-downs may be required to prevent uplift at supports of girders closest to the center of curvature.

If horizontal lateral bracing is placed in an open-framing system, the effects of curvature should be examined more closely. Connections at frequent intervals can convert the system into the closed-framing type.

11.6.5 Fabrication

Curved plate girders usually are produced in one of two ways. One way is to mechanically bend the web to the desired curvature and then weld to it flange plates that have been flame-cut to the required shape. The procedure differs from fabrication of plate girders in handling procedures, layout for fabrication, and web-to-flange welding methods.

Alternatively, girders may be curved by selectively heating the flanges of members initially fabricated straight. In this method, less steel is required. The heating and cooling induce residual stresses, but research indicates that they do not affect fatigue strength.

Mechanical bending has been successfully used for curving rolled beams. This method induces bending stresses in the flanges that should be added to the stresses caused by loads.

(L. C. Bell and C. P. Heins, "Analysis of Curved Bridges," *Journal of the Structural Division, ASCE*, vol. 96, no. ST8, pp. 1657–1673, August, 1970.

P. P. Christiano and C. G. Culver, "Horizontally Curved Bridges Subject to Moving Load," *Journal of the Structural Division, ASCE*, vol. 95, no. ST8, pp. 1615–1643, August, 1969.

"Guide Specifications for Horizontally Curved Highway Bridges," and "Standard Specifications for Highway Bridges," American Association of State Highway and Transportation Officials.

R. L. Brockenbrough, "Distribution Factors for Curved I-Girder Bridges," *Journal of Structural Engineering, ASCE*, 1986.

M. A. Grubb, "Horizontally Curved I-Girder Bridge Analysis: V-Load Method," Transportation Research Board, 1984.)

11.7 EXAMPLE—ALLOWABLE-STRESS DESIGN OF CURVED STRINGER BRIDGE

The basic design procedures that apply to bridges with straight stringers apply also to bridges with curved stringers (Arts. 11.1 to 11.5). In determination of stresses, however, the effects of curvature must be taken into account (Art. 11.6).

To illustrate the design procedure, a curved, two-lane highway bridge with simply supported, composite, plate-girder stringers will be designed. As indicated in the framing plan in Fig. 11.25a, the stringers are concentric and the supports and diaphragms are placed radially. Outer girder G_1 spans 90 ft and has a radius of curvature R_1 of 300 ft. Spacing of diaphragms along this span is $d_1 = 15$ ft. Distance between inner and outer grids G_1 and G_3 is $D = 22$ ft c to c, and G_2 is midway between them. The typical cross section in Fig. 11.25b shows a 22-ft-wide roadway flanked by two 3-ft 3-in-wide safety walks.

Structural steel to be used is Grade 36. Concrete to be used for the deck is Class A, with 28-day strength $f'_c = 3000$ psi and allowable compressive stress $f_c = 1200$ psi. Appropriate design criteria given in Sec. 10 will be used for this structure. The approximate analysis described in Art. 11.6 for open framing will be applied to the design of the girders.

Concrete Slab.   The slab is designed, to span transversely between stringers, in the same way as for straight stringers (Art. 11.2). A 7.5-in-thick slab will be used with the curved plate-girder stringers.

Loads, Moments, and Shears for Stringers.   Assume that the girders will not be shored during casting of the concrete slab. Hence, the dead load on each steel

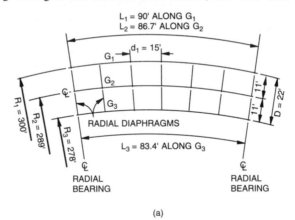

(a)

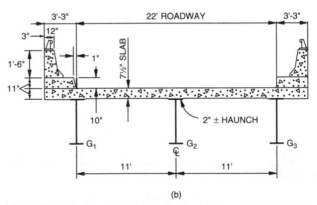

(b)

FIGURE 11.25   Two-lane highway bridge with curved stringers. (*a*) Framing plan. (*b*) Typical cross section.

TABLE 11.22 Dead Load Carried by Steel Beams, kips per ft

Stringers G_1 and G_3	
Slab: 0.150(1½ + 3.25)7.5/12	= 0.82
Haunch and extra concrete: 0.150(3.25 + 0.75)¾₁₂	= 0.15
Steel stringer and framing details—assume:	0.30
DL per stringer:	1.27

Stringer G_2	
Slab: 0.150 × 11 × 7.5/12	= 1.03
Haunch—18 × 2 in: 0.150 × 1.5 × ²⁄₁₂	= 0.04
Steel stringer and framing details—assume:	0.30
DL per stringer:	1.37

stringer includes the weight of the concrete slab as well as the weight of stringer and framing details. This dead load will be referred to as *DL* (see Table 11.22).

For design of the grid composed of the stringers and diaphragms not only is the maximum bending moment needed for each stringer but also the bending moment at each line of diaphragms (Table 11.23).

For computation of *V* loads,

$$K = \frac{RD}{d} = \frac{300 \times 22}{15} = 440$$

From Table 11.21, $C = 1.00$. *V* loads now can be computed from Eq. (11.32) and are listed in Table 11.24. They act at the diaphragms, downward on G_1, upward on G_3, to resist the torque due to curvature. The *V* load on G_2, the central girder in a three-girder grid, is assumed to be zero. The reaction due to these loads is

$$R_v = 4.64 + 7.43 + \frac{8.36}{2} = 16.25 \text{ kips}$$

The resulting bending moments are given in Table 11.25.

TABLE 11.23 Initial-Dead-Load Preliminary Moments, ft-kips

		Distance from support, ft		
	Span, ft	15, 75	30, 60	45
M_{p1} in G_1	90	714	1,143	1,286
M_{p2} in G_2	86.7	715	1,144	1,287
M_{p3} in G_3	83.4	613	981	1,104
ΣM_{pn}		2,042	3,268	3,677

TABLE 11.24 V Loads from $\Sigma M_{pn}/440$

Distance from support, ft	15, 75	30, 60	45
V loads on G_1 and G_3, kips	4.64	7.43	8.36

Final moments are the sum of the preliminary bending moments and the moments due to the V loads (Table 11.26).

Maximum shear occurs at the supports. For G_1, the maximum dead-load shear is the sum of the preliminary shear and V-load shear:

$$V_{DL} = \frac{1.27 \times 90}{2} + 16.25 = 73.4 \text{ kips}$$

Parapets, railings, and safety walks will be placed after the concrete slab has cured. Their weights may be equally distributed to all stringers. In addition, provision will be made for a future wearing surface, weight 20 psf. The total superimposed dead load will be designated *SDL*. Table 11.27 lists for the superimposed load on the composite section the dead loads, the preliminary bending moments, and the V loads. Table 11.28 gives the bending moments due to the V loads, and Table 11.29, the final superimposed dead-load moments. For G_1, the maximum shear due to the superimposed dead load is

$$V_{SDL} = \frac{0.607 \times 90}{2} + 7.57 = 34.9 \text{ kips}$$

The HS20-44 live load imposed may be a truck load or lane load. For these girder spans, however, truck loading governs. A standard truck should be placed within each 11-ft lane to produce maximum stresses in the stringers. The extreme left and right positions of the loading are shown in Fig. 11.26. The loads are distributed to the girders on the assumption that the concrete slab is simply supported on them (Table 11.30).

The trucks also should be positioned to produce maximum bending moment in each stringer for design of its central portion. Maximum dead-load moments, however, have been computed at midspan. Since moments are needed at diaphragm locations for computation of V loads and maximum moment occurs near midspan, it is convenient to place the trucks for maximum moment at midspan, where a diaphragm is located, and the error in so doing will be small. Hence, the central 32-kip axle of each truck is placed at midspan, with the other axles, 32 kips and 8 kips, 14 ft on either side.

TABLE 11.25   Initial-Dead-Load M_{pn}, ft-kips

	Distance from support, ft		
	15, 75	30, 60	45
M_{v1} in G_1	244	418	481
M_{v3} in G_3	−226	−387	−445

TABLE 11.26 Initial-Dead-Load Final Moments, ft-kips

	Distance from support, ft		
	15, 75	30, 60	45
M_1 in G_1	958	1,561	1,767
M_2 in G_2	715	1,144	1,287
M_3 in G_3	387	594	659

TABLE 11.27 Dead Load Carried by Composite Section, kips per ft

Two parapets: (⅓)2 × 0.150 × 1.5(1 + 1.25)/2	= 0.169
Two railings: (⅓)2 × 0.015	= 0.010
Two safety walks: (⅓)2 × 0.150[3.25(0.917 + 0.833)/2 − 0.833 × 0.083/2]	= 0.281
Future wearing surface: (⅓)0.020 × 22	= 0.147
SDL per stringer:	= 0.607

M_{pn} for superimposed dead load, ft-kips

		Distance from support, ft		
	Span, ft	15, 75	30, 60	45
M_{p1} in G_1	90	342	546	615
M_{p2} in G_2	86.7	317	507	571
M_{p3} in G_3	83.4	293	469	528
ΣM_{pn}		952	1,522	1,714

V Loads from $\Sigma M_{pn}/440$

Distance from support, ft	15, 75	30, 60	45
V loads on G_1 and G_3, kips	2.16	3.46	3.90

$$R_v = 2.16 + 3.46 + 3.90/2 = 7.57 \text{ kips}$$

TABLE 11.28 Superimposed Dead Load M_{vn}, ft-kips

	Distance from support, ft		
	15, 75	30, 60	45
M_{v1} in G_1	114	194	223
M_{v3} in G_3	−104	−180	−205

TABLE 11.29 Superimposed-Dead-Load Final Moments, ft-kips

	Distance from support, ft		
	15, 75	30, 60	45
M_1 in G_1	456	740	838
M_2 in G_2	317	507	571
M_3 in G_3	189	289	323

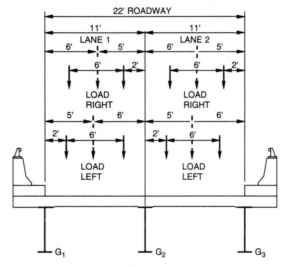

FIGURE 11.26 Position of truck wheel loads in design lanes.

TABLE 11.30 Fraction of Axle Load on Girders

	G_1	G_2	G_3
Load in lane 1 only, at extreme left	0.55	0.45	0
Load in lane 1 only, at extreme right	0.45	0.55	0
Load in lane 2 only, at extreme left	0	0.55	0.45
Load in lane 2 only, at extreme right	0	0.45	0.55

The midspan moments M_n for a truck in one lane only are used to produce the maximum midspan moments in each girder for a truck in each of the two lanes. The calculations are given in Tables 11.31 to 11.34.

For maximum live-load moments at other sections of each girder, the trucks should be placed in each lane as for the midspan moments and positioned to produce maximum preliminary moments at the sections. Then, V loads and the moments they cause should be calculated and added to the preliminary moments to yield M_{LL} at each section.

Maximum preliminary shear occurs at a support with a 32-kip axle at the support and the center of gravity of the loading $14 - 4.66 = 9.34$ ft from the support. For girder G_1, a truck should be placed at the left in both lanes 1 and 2 for maximum shear due to curvature. The truck in lane 1 should be located at a support for maximum shear, while the truck in lane 2 should be positioned for maximum midspan moments in G_2 and G_3. The maximum shear in G_1 caused by the truck in lane 2 equals the reaction $R_v = 5.07$, previously computed in determining maximum midspan moments (Table 11.33). The truck in lane 1 produces a maximum preliminary shear in G_1 of

$$V_{p1} = \frac{72(90 - 9.34)}{90} \times 0.55 = 35.5 \text{ kips}$$

This truck also induces the preliminary bending moments given in Table 11.35 in G_1 and G_2 at the diaphragms.

TABLE 11.31 M_{pn} for Maximum Live-Load Moment at Midspan, ft-kips

	Distance from support, ft				
	15	30	45	60	75
Girder G_1, span 90 ft					
Full axle load	596	1,192	1,340	968	484
0.55 axle load	328	656	737	532	266
0.45 axle load	268	536	603	436	218
Girder G_2, span 86.7 ft					
Full axle load	576	1,152	1,281	928	464
0.55 axle load	317	634	704	510	255
0.45 axle load	259	518	577	418	209
Girder G_3, span 83.4 ft					
Full axle load	556	1,109	1,221	889	444
0.55 axle load	306	610	671	489	244
0.45 axle load	250	499	550	400	200

TABLE 11.32 M_{pn} for Truck in One Lane Only, ft-kips and V Load, kips

	Distance from support, ft				
	15	30	45	60	75
At left in lane 1					
M_{p1} for G_1 (0.55 axle load)	328	656	737	532	266
M_{p2} for G_2 (0.45 axle load)	259	518	577	418	209
M_{p3} for G_3 (no load)	0	0	0	0	0
ΣM_{pn}	587	1,174	1,314	950	475
$V = \Sigma M_{pn}/440$	1.33	2.67	2.99	2.16	1.08
At right in lane 1					
M_{p1} for G_1 (0.45 axle load)	268	536	603	436	218
M_{p2} for G_2 (0.55 axle load)	317	634	704	510	255
M_{p3} for G_3 (no load)	0	0	0	0	0
ΣM_{pn}	585	1,170	1,307	946	473
$V = \Sigma M_{pn}/440$	1.33	2.66	2.97	2.15	1.08
At left in lane 2					
M_{p1} for G_1 (no load)	0	0	0	0	0
M_{p2} for G_2 (0.55 axle load)	317	634	704	510	255
M_{p3} for G_3 (0.45 axle load)	250	499	550	400	200
ΣM_{pn}	567	1133	1254	910	455
$V = \Sigma M_{pn}/440$	1.29	2.57	2.85	2.07	1.03
At right in lane 2					
M_{p1} for G_1 (no load)	0	0	0	0	0
M_{p2} for G_2 (0.45 axle load)	259	518	577	418	209
M_{p3} for G_3 (0.55 axle load)	306	610	671	489	244
ΣM_{pn}	565	1,128	1,248	907	453
$V = \Sigma M_{pn}/440$	1.28	2.56	2.84	2.06	1.03

The reaction due to the V loads is

$$R_v = 1.03 \times \tfrac{5}{6} + 1.02 \times \tfrac{4}{6} + 0.76 \times \tfrac{3}{6} + 0.51 \times \tfrac{2}{6} + 0.25 \times \tfrac{1}{6} = 2.12 \text{ kips}$$

Hence, the final shear in G_1 due to the trucks in both lanes is

$$V_{LL} = 35.5 + 2.12 + 5.07 = 42.7 \text{ kips}$$

Impact is taken as the following fraction of live-load stress:

$$I = \frac{50}{L_3 + 125} = \frac{50}{83.4 + 125} = 0.24$$

TABLE 11.33 *V*-Load Reactions, kips, and Final Midspan Moments M_n for Truck in One Lane Only, ft-kips

---

At left in lane 1

$R_v = 1.33 \times \frac{5}{6} + 2.67 \times \frac{4}{6} + 2.99 \times \frac{3}{6} + 2.16 \times \frac{2}{6} + 1.08 \times \frac{1}{6} = 5.28$

Midspan $M_{v1} = 5.28 \times 45 - 1.33 \times 30 - 2.67 \times 15 = 158$

Midspan $M_{v2} = 0$

Midspan $M_{v3} = -158 \times 83.4/90 = -146$

Midspan $M_1 = M_{p1} + M_{v1} = 737 + 158 = 895$

Midspan $M_2 = M_{p2} + M_{v2} = 577$

Midspan $M_3 = M_{p3} + M_{v3} = -146$

At right in lane 1

$R_v = 1.33 \times \frac{5}{6} + 2.66 \times \frac{4}{6} + 2.97 \times \frac{3}{6} + 2.15 \times \frac{2}{6} + 1.08 \times \frac{1}{6} = 5.26$

Midspan $M_{v1} = 5.26 \times 45 - 1.33 \times 30 - 2.66 \times 15 = 157$

Midspan $M_{v2} = 0$

Midspan $M_{v3} = -157 \times 83.4/90 = -145$

Midspan $M_1 = M_{p1} + M_{v1} = 603 + 157 = 760$

Midspan $M_2 = M_{p2} + M_{v2} = 704$

Midspan $M_3 = M_{p3} + M_{v3} = -145$

At left in lane 2

$R_v = 1.29 \times \frac{5}{6} + 2.57 \times \frac{4}{6} + 2.85 \times \frac{3}{6} + 2.07 \times \frac{2}{6} + 1.03 \times \frac{1}{6} = 5.07$

Midspan $M_{v1} = 5.07 \times 45 - 1.29 \times 30 - 2.57 \times 15 = 151$

Midspan $M_{v2} = 0$

Midspan $M_{v3} = -151 \times 83.4/90 = -140$

Midspan $M_1 = M_{p1} + M_{v1} = 151$

Midspan $M_2 = M_{p2} + M_{v2} = 704$

Midspan $M_3 = M_{p3} + M_{v3} = 550 - 140 = 410$

At right in lane 2

$R_v = 1.28 \times \frac{5}{6} + 2.56 \times \frac{4}{6} + 2.84 \times \frac{3}{6} + 2.06 \times \frac{2}{6} + 1.03 \times \frac{1}{6} = 5.05$

Midspan $M_{v1} = 5.05 \times 45 - 1.28 \times 30 - 2.56 \times 15 = 150$

Midspan $M_{v2} = 0$

Midspan $M_{v3} = -150 \times 83.4/90 = -140$

Midspan $M_1 = M_{p1} + M_{v1} = 150$

Midspan $M_2 = M_{p2} + M_{v2} = 577$

Midspan $M_3 = M_{p3} + M_{v3} = 671 - 140 = 531$

TABLE 11.34 Midspan Live-Load Moments M_{LL}, ft-kips

Girder	Truck position	M_{LL}
G_1	At left in lanes 1 and 2	$895 + 151 = 1{,}046$
G_2	At right in lane 1, left in lane 2	$704 + 704 = 1{,}408$
G_3	At right in lane 2	531

Thus, the maximum moments due to impact are

$$G_1:M_I = 0.24 \times 1{,}046 = 251 \text{ ft-kips}$$

$$G_2:M_I = 0.24 \times 1{,}408 = 338 \text{ ft-kips}$$

$$G_3:M_I = 0.24 \times 531 \quad = 127 \text{ ft-kips}$$

And the maximum shear in G_1 due to impact is

$$V_I = 0.24 \times 42.7 = 10.2 \text{ kips}$$

For centrifugal forces and radial wind forces on live load, both of which induce torques in the superstructure because they are assumed to act 6 ft above the roadway surface, allowable stresses may be increased 25%. Therefore, if the sum of the moments and shears produced by those forces does not exceed 25% of the moments and shears, respectively, without those forces, they may be ignored. For this structure, the effects of the wind and centrifugal forces are small enough to be neglected. But for illustrative purposes, they will be calculated.

Because of the sharp curvature, design speed is taken as 30 mph. Then, the centrifugal forces equal the following percentages of a truck load in lanes 1 and 2:

$$C = \frac{6.68S^2}{R} = \frac{6.68(30)^2}{295} = 20.4\% \qquad C = \frac{6.68(30)^2}{284} = 21.2\%$$

Application of these percentages to the axle load per lane permits use of the results of previous calculations for moments and shears. Thus, the horizontal force per axle is

$$H = 32 \times 0.204 + 32 \times 0.212 = 13.3 \text{ kips}$$

TABLE 11.35 M_p for Truck in Lane 1 Placed for Maximum Shear, ft-kips

		Distance from support, ft				
	Span, ft	15	30	45	60	75
M_{p1} for G_1	90	251	246	185	123	62
M_{p2} for G_2	86.7	203	201	151	100	50
ΣM_{pn}		454	447	336	223	112
$V = M_{pn}/440$		1.03	1.02	0.76	0.51	0.25

This force is assumed to act 6 ft above the roadway surface or about 8 ft above the centroidal axis of the girders. Thus, it causes a torque

$$T = 8 \times 13.3 = 106.4 \text{ ft-kips}$$

This is resisted by a couple compromising a downward vertical force on G_1 and an upward vertical force on G_3

$$P = \frac{106.4}{22} = 4.83 \text{ kips}$$

By proportion, the maximum moment M_C in G_1 due to the centrifugal forces can be obtained from the maximum moment M_{p1} previously computed for a truck load in lane 1, 1,340 ft-kips.

$$M_C = \frac{1{,}340 \times 4.83}{32} = 202 \text{ ft-kips}$$

Similarly, the maximum shear in G_1 due to the centrifugal forces is

$$V_C = \frac{35.5 \times 4.83}{32 \times 0.55} = 9.7 \text{ kips}$$

AASHTO specifications require a wind load on the live load of at least 0.1 kip per ft. This would cause a torque of $0.1 \times 8 = 0.8$ ft-kip per ft and a downward vertical force on G_1 of $0.8/22 = 0.0364$ kip per ft. Hence, the maximum shear in G_1 due to wind on live load is

$$V_{WL} = \frac{1}{2} \times 0.0364 \times 90 = 1.6 \text{ kips}$$

The maximum moment in G_1 due to this load is

$$M_{WL} = \frac{0.0364(90)^2}{8} = 36 \text{ ft-kips}$$

Combined, centrifugal forces and wind induce in G_1 a maximum shear

$$V_C + V_{WL} = 9.7 + 1.6 = 11.3 \text{ kips}$$

This is less than 25% of the total shear without these forces and may be ignored. Similarly, the combined maximum moment in G_1 is

$$M_C + M_{WL} = 202 + 36 = 238 \text{ ft-kips}$$

This is less than 25% of the total moment without these forces and may be ignored.

The maximum moments and shears for design therefore are as given in Tables 11.36 and 11.37.

Properties of Composite Section.   Design of the girders follows the procedures indicated for plate-girder stringers in Art. 11.4, except that the lateral bending stress in the flanges due to curvature must be taken into account.

For illustrative purposes, girder G_1 will be designed. The effective width of the concrete slab, governed by its 7-in effective thickness, is 84 in. A trial section for the plate girder is selected with the aid of the Hacker formulas, Eqs. (11.1a) and (11.1b), with the allowable stress reduced about 20% because of lateral bending.

Assume that the girder web will be 60 in deep. This satisfies the requirements that the depth-span ratio for girder plus slab exceed 1:25 and for girder alone 1:30.

TABLE 11.36 Midspan Bending Moments, ft-kips

	M_{DL}	M_{SDL}	$M_{LL} + M_I$
Girder G_1	1,767	838	1,297
Girder G_2	1,287	571	1,746
Girder G_3	659	323	658

With stiffeners, the web thickness is required to be at least $\frac{1}{165}$ of the depth, or 0.364 in.

Use a web plate $60 \times \frac{3}{8}$ in. With a cross-sectional area $A_w = 22.5$ in^2, the web will be subjected to a maximum shearing stress considerably below the 12 ksi permitted.

$$f_v = \frac{161.2}{22.5} = 7.2 < 12 \text{ ksi}$$

The required bottom-flange area may be estimated from Eq. (11.1a) with allowable bending stress $F_b = 20 - 4 = 16$ ksi. Distance between centers of gravity of steel flanges is assumed as $d_{cg} = 63$ in.

$$A_{sb} = \frac{12}{16}\left(\frac{1,767}{63} + \frac{2,135}{63 + 7}\right) = 43.8 \text{ in}^2$$

Try a 22×2-in bottom flange, area $= 44$ in^2.

The ratio of flange areas $R = A_{st}/A_{sb}$ may be estimated from Eq. (11.1b) as

$$R = \frac{50}{190 - L} = \frac{50}{190 - 90} = 0.50$$

Then, the estimated required area of the steel top flange is

$$A_{st} = RA_{sb} = 0.50 \times 43.8 = 21.9 \text{ in}^2$$

Try a $18 \times 1\frac{1}{2}$-in top flange, area $= 27$ in^2.

The trial section is shown in Fig. 11.27. Its neutral axis can be located by taking moments of web and flange areas about middepth of the web. This computation and that for the section moduli S_{st} and S_{sb} of the plate girder alone are conveniently tabulated in Table 11.38.

In computation of the properties of the composite section, the concrete slab, ignoring the haunch area, is transformed into an equivalent steel area, with $n = 30$ for superimposed dead load and $n = 10$ for live loads. The computations of

TABLE 11.37 End Shear, kips, in Girder G_1

V_{DL}	V_{SDL}	$V_{LL} + V_1$	Total V
73.4	34.9	52.9	161.2

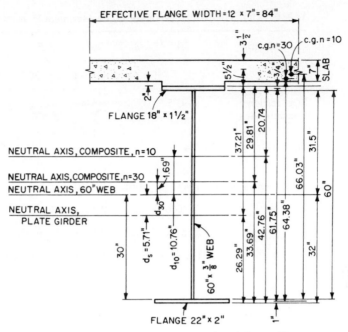

FIGURE 11.27   Cross section of composite plate girder at midspan.

TABLE 11.38   Steel Section for G_1 for Maximum Moment

Material	A	d	Ad	Ad^2	I_o	I
Top flange 18 × 1½	27.0	30.75	830	25,500		25,500
Web 60 × ⅜	22.5				6,750	6,750
Bottom flange 22 × 2	44.0	−31.0	−1.364	42,300		42,300
	93.5		−534			74,550
$d_s = -534/93.5 = -571$ in					$-5.71 \times 534 =$	−3,050
					$I_{NA} =$	71,500

Distance from neutral axis of steel section to:

$$\text{Top of steel} = 30 + 1.50 + 5.71 = 37.21 \text{ in}$$

$$\text{Bottom of steel} = 30 + 2.00 - 5.71 = 26.29 \text{ in}$$

Section moduli	
Top of steel	Bottom of steel
$S_{st} = 71,500/37.21 = 1,922$ in^3	$S_{sb} = 71,500/26.29 = 2,720$ in^3

TABLE 11.39 Composite Section for G_1 for Maximum Moment

Material	A	d	Ad	Ad^2	I_o	I
(a) For dead loads, $n = 30$						
Steel section	93.5		−534			74,550
Concrete 84 × 7/30	19.6	37.0	725	26,830	80	26,900
	113.1		191			101,450
$d_{s0} = 191/113.1 = 1.69$ in					$−1.69 × 191 =$	−320
					$I_{NA} =$	101,130

Distance from neutral axis of composite section to:

$$\text{Top of steel} = 31.50 - 1.69 = 29.81 \text{ in}$$

$$\text{Bottom of steel} = 32.00 + 1.69 = 33.69 \text{ in}$$

$$\text{Top of concrete} = 29.81 + 2 + 7 = 38.81 \text{ in}$$

Section moduli		
Top of steel	Bottom of steel	Top of concrete
$S_{st} = 101,130/29.81$	$S_{sb} = 101,130/33.69$	$S_c = 101,130/38.81$
$= 3,400 \text{ in}^3$	$= 3,010 \text{ in}^3$	$= 2,615 \text{ in}^3$

Material	A	d	Ad	Ad^2	I_o	I
(b) For live loads, $n = 10$						
Steel section	93.5		−534			74,550
Concrete 84 × 7/10	58.8	37.0	2,175	80,510	240	80,750
	152.3		1,641			155,300
$d_{10} = 1,641/152.3 = 10.76$ in					$−10.76 × 1,641 =$	−17,700
					$I_{NA} =$	137,600

Distance from neutral axis of composite section to:

$$\text{Top of steel} = 31.50 - 10.76 = 20.74 \text{ in}$$

$$\text{Bottom of steel} = 32.00 + 10.76 = 42.76 \text{ in}$$

$$\text{Top of concrete} = 20.74 + 2 + 7 = 29.74 \text{ in}$$

Section moduli		
Top of steel	Bottom of steel	Top of concrete
$S_{st} = 137,600/20.74$	$S_{sb} = 137,600/42.76$	$S_c = 137,600/29.74$
$= 6,630 \text{ in}^3$	$= 3,210 \text{ in}^3$	$= 4,630 \text{ in}^3$

neutral-axis location and section moduli for the composite section are tabulated in Table 11.39. To locate the neutral axis, moments of the areas are taken about middepth of the girder web.

Stresses in Composite Section.   Since the girders will not be shored when the concrete is cast and cured, the stresses in the steel section for load *DL* are determined with the section moduli of the steel section alone (Table 11.38). Stresses for load *SDL* are computed with the section moduli of the composite section when $n = 30$ (Table 11.39*a*). And stresses in the steel for live loads and impact are calculated with section moduli of the composite section when $n = 10$ (Table 11.39*b*). Lateral bending stresses in the bottom flange are superimposed on other stresses. Lateral bending moment in the top flange should be computed for load *DL*. For composite beams, lateral bending in the top flange under *SDL* and live load can be ignored.

The moments causing the lateral bending stresses can be computed from Eq. (11.36). For use in this calculation (Table 11.40), the centroid of the compression flange is located by taking the moment of the transformed area of the concrete slab about the centroid of the steel top flange. Thus, for the steel section alone, the distance between flange centroids is

$$h = 60 + \frac{2}{2} + \frac{1.5}{2} = 61.75 \text{ in}$$

For the composite section, $n = 30$

$$h = 61.75 + \frac{6.25 \times 19.6}{27.0 + 19.6} = 61.75 + 2.63 = 64.38 \text{ in}$$

For the composite section, $n = 10$,

$$h = 61.75 + \frac{6.25 \times 58.8}{27.0 + 58.8} = 61.75 + 4.28 = 66.03 \text{ in}$$

The section moduli of the top and bottom flanges about their central vertical axes are

$$S_{ft} = \frac{1.5(18)^2}{6} = 81 \text{ in}^3 \qquad S_{fb} = \frac{2(22)^2}{6} = 161.5 \text{ in}^3$$

Calculations of the steel stresses in G_1 are given in Table 11.41. The trial section is satisfactory.

Stresses in the concrete slab are determined with the section moduli of the composite section with $n = 30$ for *SDL* (Table 11.39*a*) and $n = 10$ for *LL* + *I* (Table 11.39*b*). The calculation is given in Table 11.42.

Therefore, the composite section for G_1 is satisfactory. Use for G_1 in the region of maximum moment the section shown in Fig. 11.27.

TABLE 11.40   Maximum Lateral Bending Moments, ft-kips

DL: $M_L = -0.1 \times 12 \times 1{,}767(15)^2/(300 \times 61.75)$	$= -26$
SDL: $M_L = -0.1 \times 12 \times 838(15)^2/(300 \times 64.38)$	$= -12$
LL + *I*: $M_L = -0.1 \times 12 \times 1{,}297(15)^2/(300 \times 66.03)$	$= -18$
Total:	-56

TABLE 11.41   Steel stresses in G_1, ksi

Top of steel (compression)		Bottom of steel (tension)	
DL: $f_b = 1.767 \times 12/1,992 =$	11.03	$f_b = 1,767 \times 12/2,720 =$	7.79
SDL: $f_b = 838 \times 12/3,400 \quad=$	2.95	$f_b = 838 \times 12/3,010 \quad=$	3.34
LL + I: $f_b = 1,297 \times 12/6,630 =$	2.34	$f_b = 1,297 \times 12/3,210 =$	4.85
L: $f_b = 26 \times {}^{12}\!/_{81} \quad=$	3.85	$f_b = 56 \times 12/161.5 \quad=$	4.16
Total:	$20.17 \approx 20$		$20.14 \approx 20$

The procedure is the same for design of other sections and for the other stringers. For design of other elements, see Arts. 11.2 and 11.4. Fatigue design is similar to that for straight girders.

11.8 DECK PLATE-GIRDER BRIDGES WITH FLOORBEAMS

For long spans, use of fewer but deeper girders to span the long distance between supports becomes more efficient. With appropriately spaced stringers between the main girders of highway bridges, depth of concrete roadway slab can be kept to the minimum permitted, thus avoiding increase in dead load from the deck. Spans of the longitudinal stringers are kept short by supporting them on transverse floorbeams spanning between the girders. If spacing of the floorbeams is 25 ft or less, additional diaphragms or cross frames between the girders are not required.

This type of construction can be used with deck or through girders. Through girders carry the roadway between them. Their use generally is limited to locations where vertical clearances below the bridge are critical. Deck girders carry the roadway on the top flange. They generally are preferred for highway bridges where vertical clearances are not severely restricted, because the girders, being below the deck, do not obstruct the view from the deck. Structurally, deck girders have the advantage that the concrete deck is available for bracing the top flange of the girders and for composite action. Bracing of the bottom flange is accomplished with horizontal lateral bracing.

The design procedure for through plate girders is described in Art. 11.10. Article 11.9 presents an example to indicate the design procedure for a deck girder bridge with floorbeams and stringers. In general, design of the stringers is much like that for a stringer bridge (Art. 11.2), and design of the girders is much like that for the girders of a multigirder bridge (Art. 11.4). In the following example, however, the stringers and girders are not designed for composite action. See also Art. 11.3.

TABLE 11.42   Stresses in G_1 at Top of Concrete, ksi

SDL: $f_c = 838 \times 12/(2,615 \times 30) \quad= 0.13$	
LL + I: $f_c = 1,297 \times 12/(4,630 \times 10) = 0.34$	
Total:	$0.47 < 1.2$

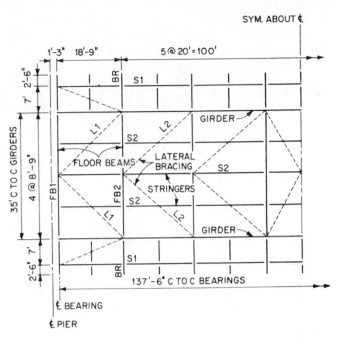

FIGURE 11.28 Framing plan for four-lane highway bridge with deck plate girders.

11.9 EXAMPLE—ALLOWABLE-STRESS DESIGN OF DECK PLATE-GIRDER BRIDGE WITH FLOORBEAMS

Two simply supported, welded, deck plate girders carry the four lanes of a highway bridge on a 137.5-ft span. The girders are spaced 35 ft c to c. Loads are distributed to the girders by longitudinal stringers and floorbeams (Fig. 11.28). The typical cross section in Fig. 11.29 shows a 48-ft roadway flanked by 3-ft-wide safety walks. Grade 50 steel is to be used for the girders and Grade 36 for stringers, floorbeams, and other components. Concrete to be used for the deck is class A, with 28-day strength $f'_c = 3,000$ psi and allowable compressive stress $f_c = 1,200$ psi. Appropriate design criteria given in Sec. 10 will be used for this structure.

11.9.1 Design of Concrete Slab

The slab is designed, to span transversely between stringers, in the same way as for rolled-beam stringers (Art. 11.2). A 7.5-in thick concrete slab will be used.

11.9.2 Design of Interior Stringer

Spacing of interior stringers c to c is 8.75 ft. Simply supported, a typical stringer S2 spans 20 ft. Table 11.43 lists the dead loads on S2. Maximum dead-load moment

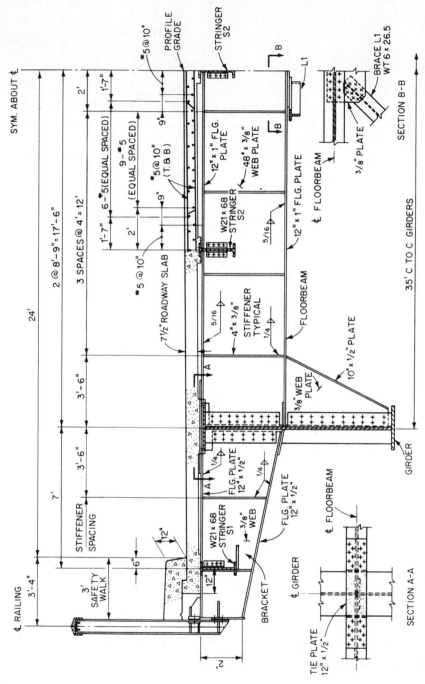

FIGURE 11.29 Typical cross section of deck-girder bridge at a floorbeam.

TABLE 11.43 Dead Load on S2, kips per ft

Slab: $0.150 \times 8.75 \times 7.5/12 =$	0.82
Haunch—assume:	0.033
Stringer—assume:	0.068
DL per stringer:	0.923

occurs at midspan and equals

$$M_{DL} = \frac{0.923(20)^2}{8} = 46.1 \text{ ft-kips}$$

Maximum dead-load shear occurs at the supports and equals

$$V_{DL} = \frac{0.923 \times 20}{2} = 9.2 \text{ kips}$$

The live load distributed to the stringer with spacing $S = 8.75$ ft is

$$\frac{S}{5.5} = \frac{8.75}{5.5} = 1.59 \text{ wheel loads} = 0.795 \text{ axle loads}$$

Maximum moment induced in a 20-ft span by a standard truck is 160 ft-kips. Hence, the maximum live-load moment in a stringer is

$$M_{LL} = 0.795 \times 160 = 127.2 \text{ ft-kips}$$

Maximum shear caused by the truck is 41.6 kips. Consequently, maximum live-load shear in the stringer is

$$V_{LL} = 0.795 \times 41.6 = 33.0 \text{ kips}$$

Impact is taken as 30% of live-load stress, because

$$I = \frac{50}{L + 125} = \frac{50}{20 + 125} = 0.35 > 0.30$$

So the maximum moment due to impact is

$$M_I = 0.30 \times 127.2 = 38.1 \text{ ft-kips}$$

and the maximum shear due to impact is

$$V_I = 0.30 \times 33.0 = 9.9 \text{ kips}$$

Maximum moments and shears in S2 are summarized in Table 11.44.

TABLE 11.44 Maximum Moments and Shears in S2

	DL	LL	I	Total
Moments, ft-kips	46.1	127.2	38.1	211.4
Shears, kips	9.2	33.0	9.9	52.1

With an allowable bending stress F_b = 20 ksi for a stringer of Grade 36 steel, the section modulus required is

$$S = \frac{M}{F_b} = \frac{211.4 \times 12}{20} = 127 \text{ in}^3$$

With an allowable shear stress F_v = 12 ksi, the web area required is

$$A_w = \frac{52.1}{12} = 4.33 \text{ in}^2$$

Use a W21 × 68. It provides a section modulus of 139.9 in³ and a web area of 0.43 × 21.13 = 9.1 in².

11.9.3 Design of an Exterior Stringer

Simply supported, S1 spans 20 ft. It carries sidewalk as well as truck loads (Fig. 11.28). Dead loads are apportioned between S1 and the girder, 7 ft away, by treating the slab as simply supported at the girder. Table 11.45 lists the dead loads on S1.
Maximum dead-load moment occurs at midspan and equals

$$M_{DL} = \frac{1.15(20)^2}{8} = 57.5 \text{ ft-kips}$$

Maximum dead-load shear occurs at the supports and equals

$$V_{DL} = \frac{1.15 \times 20}{2} = 11.5 \text{ kips}$$

TABLE 11.45 Dead Load on S1, kips per ft

Railing: 0.070 × 9.83/7	= 0.098
Sidewalk: 0.150 × 1 × 3 × 8/7	= 0.514
Slab: 0.150 × 8 × 7.5/12 × 4/7	= 0.428
Stringers, brackets, framing details—assume:	0.110
DL per stringer:	1.150

The live load from the roadway distributed to the exterior stringer with spacing $S = 7$ ft from the girder is

$$\frac{S}{4.0 + 0.25S} = \frac{7}{4.0 + 0.25 \times 7} = 1.22 \text{ wheel loads} = 0.61 \text{ axle loads}$$

Maximum moment induced in a 20-ft span by a standard truck load is 160 ft-kips. Hence, the maximum live-load moment in S1 is

$$M_{LL} = 0.61 \times 160 = 97.7 \text{ ft-kips}$$

Maximum shear caused by the truck is 41.6 kips. Therefore, maximum live-load shear in S1 is

$$V_{LL} = 0.61 \times 41.6 = 25.4 \text{ kips}$$

Impact for a 20-ft span is 30% of live-load stress. Hence, maximum moment due to impact is

$$M_I = 97.7 \times 0.3 = 29.3 \text{ ft-kips}$$

and maximum shear due to impact is

$$V_I = 25.4 \times 0.3 = 7.6 \text{ kips}$$

Sidewalk loading at 85 psf on the 3-ft-wide sidewalk imposes a uniformly distributed load w_{SLL} on the stringer. With the slab assumed simply supported at the girder,

$$w_{SLL} = \frac{0.085 \times 3 \times 8}{7} = 0.29 \text{ kip per ft}$$

This causes a maximum moment of

$$M_{SLL} = \frac{0.29(20)^2}{8} = 14.5 \text{ ft-kips}$$

and a maximum shear of

$$V_{SLL} = \frac{0.29 \times 20}{2} = 2.9 \text{ kips}$$

Maximum moments and shears in S1 are summarized in Table 11.46.

TABLE 11.46 Maximum Moments and Shears in S1

	DL	LL	I	Total
Moments, ft-kips	57.5	97.7	29.3	184.5
Shears, kips	11.5	25.4	7.6	44.5

If the exterior stringer has at least the capacity of the interior stringers, the allowable stress may be increased 25% when the effects of sidewalk live load are combined with those from dead load, traffic live load, and impact. In this case, the moments and shears due to sidewalk live load are less than 25% of the moments and shears without that load. Hence, they may be ignored.

With an allowable bending stress $F_b = 20$ ksi for Grade 36 steel, the section modulus required for S1 is

$$S = \frac{M}{F_b} = \frac{184.5 \times 12}{20} = 111 \text{ in}^3$$

With an allowable shear stress $F_v = 12$ ksi, the web area required is

$$A_w = \frac{44.5}{12} = 3.7 \text{ in}^2$$

Use a W21 × 68, as for S2.

11.9.4 Design of an Interior Floorbeam

Floorbeam FB2 is considered to be a simply supported beam with 35-ft span and symmetrical 9.5-ft brackets, or overhangs (Fig. 11.29). It carries a uniformly distributed dead load due to its own weight and that of a concrete haunch, assumed at 0.21 kip per ft. Also, FB2 carries a concentrated load from S1 of $2 \times 11.5 = 23.0$ kips and a concentrated load from each of three interior stringers S2 of $2 \times 9.2 = 18.4$ kips (Fig. 11.30).

Moments and Shears in Main Span. Because of the brackets, negative moments occur and reach a maximum at the supports. The maximum negative dead-load moment is

$$M_{DL} = -0.21(9.5)^2 - 23.0 \times 7 = -171 \text{ ft-kips}$$

The reaction at either support under the symmetrical dead load is

$$R_{DL} = \frac{3 \times 18.4}{2} + 23 + \frac{0.21 \times 54}{2} = 56.3 \text{ kips}$$

Maximum dead-load shear in the overhang is

$$V_{DL} = 23 + 0.21 \times 9.5 = 25.0 \text{ kips}$$

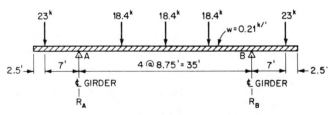

FIGURE 11.30 Dead loads on a floorbeam of the deck-girder bridge.

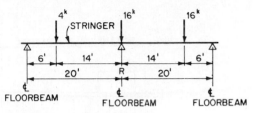

FIGURE 11.31 Positions of loads on a stringer for maximum live load on a floorbeam.

Hence, the maximum shear between girders is

$$V_{DL} = 56.3 - 25.0 = 31.3 \text{ kips}$$

Maximum positive dead-load moment occurs at midspan and equals

$$M_{DL} = 31.3 \times 17.5 - 18.4 \times 8.75 - \frac{0.21(17.5)^2}{2} - 171 = 184 \text{ ft-kips}$$

Maximum live-load stresses in the floorbeam occur when the center truck wheels pass over it (Fig. 11.31). In that position, the wheels impose on FB2 a load

$$W = 16 + \frac{16 \times 6}{20} + \frac{4 \times 6}{20} = 22 \text{ kips}$$

For maximum positive moment, trucks should be placed in the two central lanes, as close to midspan as permissible (Fig. 11.32). Then, the maximum moment is

$$M_{LL} = 2 \times 22 \times 15.5 - 22 \times 6 = 550 \text{ ft-kips}$$

Maximum negative moment occurs at a support with a truck in the outside lane with a wheel 2 ft from the curb (Fig. 11.33). This moment equals

$$M_{LL} = -22 \times 4.5 = -99 \text{ ft-kips}$$

Maximum live-load shear between girders occurs at support A with three lanes closest to that support loaded, as indicated in Fig. 11.34. Because three lanes are loaded, the floorbeam need to be designed for only 90% of the resulting shear. The reaction at A is

$$R_{LL} = \frac{0.90 \times 22(39.5 + 33.5 + 27.5 + 21.5 + 15.5 + 9.5)}{35} = 83.2 \text{ kips}$$

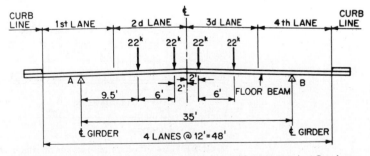

FIGURE 11.32 Positions of loads for maximum positive moment in a floorbeam.

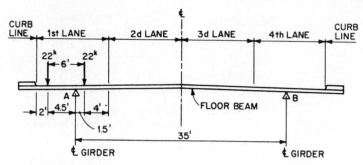

FIGURE 11.33   Positions of loads for maximum negative moment and maximum shear in the overhang of a floorbeam.

Subtraction of the shear in the bracket for this loading gives the maximum live-load shear between girders:

$$V_{LL} = 83.2 - 0.9 \times 22 = 63.4 \text{ kips}$$

The maximum live-load shear in the overhang is produced by the loading in Fig. 11.33 and is V_{LL} in 22 kips.

Impact is taken as 30% of live-load stress, because

$$I = \frac{50}{L + 125} = \frac{50}{35 + 125} = 0.31 > 0.30$$

Sidewalk loading transmitted by exterior stringers S1 to the floorbeam equals $2 \times 2.9 = 5.8$ kips. This induces a shear in the overhang $V_{SLL} = 5.8$ kips. Also, it causes a reaction

$$R_{SLL} = \frac{5.8 \times 42}{35} = 7.0 \text{ kips}$$

Subtraction of the overhang shear gives the maximum shear between girders

$$V_{SLL} = 7.0 - 5.8 = 1.2 \text{ kips}$$

Maximum negative moment due to the sidewalk live load is

$$M_{SLL} = -5.8 \times 7 = 41 \text{ ft-kips}$$

Results of preceding calculations are summarized in Table 11.47.

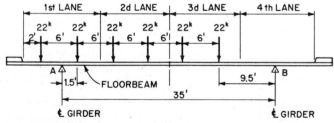

FIGURE 11.34   Position of loads for maximum shear at A in main span of floorbeam.

TABLE 11.47 Maximum Moments, Shears, and Reactions in Floorbeam FB2

	DL	LL	I	SLL	Total
Negative moments, ft-kips	−171	−99	−30	−41	−341
Positive moments, ft-kips	184	550	165	......	899
Shear in main span, kips	31.3	63.4	19.0	1.2	114.9
Shear in overhang, kips	25.0	22.0	6.6	5.8	59.4
Reaction, kips	56.3	83.2	25.0	7.0	171.5

Main-Span Section. FB2 will be designed as a plate girder of Grade 36 steel. Assume a 48-in-deep web. If the floorbeam is not stiffened longitudinally, web thickness must be at least $t = D/170 = 48/170 = 0.283$ in. To satisfy the allowable shear stress of 12 ksi, with a maximum shear from Table 11.47 of 114.9 kips, web thickness should be at least $t = 114.9/(12 \times 48) = 0.20$ in. These requirements could be met with a ⁵⁄₁₆-in web, the minimum thickness required. But fewer stiffeners will be needed if a slightly thicker plate is selected. So use a 48 × ⅜-in web.

Assume that the tension and compression flanges will be the same size and that each flange will have two holes for ⅞ in-dia. high-strength bolts. To satisfy the allowable bending stress of 20 ksi, with a maximum moment of 899 ft-kips from Table 11.47, flange area should be about

$$A_f = \frac{899 \times 12}{20(48 + 1)} = 11 \text{ in}^2$$

With an allowance for the bolt holes, assume for each flange a plate 12 × 1 in. Width-thickness ratio of 12:1 for the compression flange is less than the 24:1 maximum and is satisfactory.

The trial section assumed is shown in Fig. 11.35. Moment of inertia and section modulus of the net section are calculated as shown in Table 11.48. Distance from

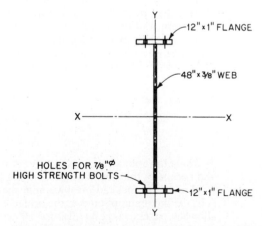

FIGURE 11.35 Cross section of floorbeam in main span.

TABLE 11.48 Moment of Inertia of Floorbeam FB2 at Midspan

Material	A	d	Ad^2 or I_o
2 flanges 12 × 1	24	24.5	14,400
Web 48 × ⅜	18		3,456
		$I_g =$	17,856
4 holes 1 × 1	4	24.5	−2,400
		$I_{net} =$	15,456 in⁴

neutral axis to top or bottom of the floorbeam is 25 in. Hence, the section modulus provided is

$$S_{net} = \frac{15,456}{25} = 616 \text{ in}^3$$

Maximum bending stress therefore is

$$f_b = \frac{899 \times 12}{616} = 17.6 < 20 \text{ ksi}$$

The section is satisfactory.

A check of the weight of the floorbeam is desirable to verify the assumptions made in dead-load calculations. Weight of slab haunch, beam, and details was assumed at 0.21 kip per ft. Average weight of haunch will be about 0.05 kip per ft. Thus, the assumed weight of floorbeam and details was about 0.16 kip per ft. If 8% of the weight is assumed in details, actual weight is 1.08(2 × 40.8 + 61.2) = 154 < 160 lb per ft assumed.

Flange-to-Web Welds. Each flange will be connected to the web by a fillet weld on opposite sides of the web. These welds must resist the horizontal shear between flange and web. The minimum size of weld permissible for the thickest plate at the connection usually determines the size of weld. In some cases, however, the size of weld may be determined by the maximum shear. In this example, shear does not govern, but the calculations are presented to illustrate the procedure.

The gross moment of inertia, 17,856 in⁴, is used in computing the shear v, kips per in, between flange and web. From Table 11.47, maximum shear is 114.9 kips. Still needed is the static moment Q of the flange area:

$$Q = 12 \times 1 \times 24.5 = 294 \text{ in}^3$$

Then, the shear is

$$v = \frac{VQ}{I} = \frac{114.9 \times 294}{17,856} = 1.89 \text{ kips per in}$$

The allowable stress on the weld is not determined by fatigue. It is sufficient and necessary that the base metal in the flange be investigated for fatigue and the weld metal be checked for maximum shear stress. For fatigue, the stress category is B. On the assumption that the bridge is a nonredundant-load-path structure, the allowable stress range in FB2 for 500,000 cycles of loading is 23 ksi. The stress range due to live-load plus impact moments is 12[550 − (−99) + 165 − (−30)]/ 616 = 16.4 ksi < 23 ksi. The base metal is satisfactory for fatigue.

The allowable shear stress is $F_v = 0.27F_u = 0.27 \times 58 = 15.6$ ksi. Hence, the allowable load per weld is $15.6 \times 0.707 = 11.03$ kips per in, and for two welds, 22.06 kips per in. So the weld size required to resist the shear is $1.89/22.06 = 0.09$ in. The minimum size of weld permitted with the 1-in thick flange plate, however is $\frac{5}{16}$ in. Use two $\frac{5}{16}$-in welds at each flange.

Connection to Girder.   Connection of the floorbeam to the girder is made with 18 A325 high-strength bolts. Each has a capacity in a slip-critical connection with Class A surface of 9.3 kips. For the maximum shear of 114.9 kips, the number of bolts required is $114.9/9.3 = 13$. The 18 provided are satisfactory.

Main-Span Stiffeners.   Bearing stiffeners are not needed, because the web is braced at the supports by the connections with the girders. Whether intermediate transverse stiffeners are needed can be determined from Table 10.15. The compressive bending stress at the support is

$$f_b = \frac{341 \times 12 \times 25}{17,856} = 5.73 \text{ ksi}$$

For a girder web with transverse stiffeners, the depth-thickness ratio should not exceed

$$\frac{D}{t} = \frac{730}{\sqrt{5.73}} = 304 > 170$$

Hence, web thickness should be at least $48/170 = 0.28 < 0.375$ in. Actual $D/t_w = 48/0.375 = 126 < 150$. The average shear stress at the support is

$$f_v = \frac{114.9}{18} = 6.38 \text{ ksi}$$

The limiting shear stress for the girder web without stiffeners is, from Table 10.15,

$$F_v = \left(\frac{270}{126}\right)^2 = 4.59 \text{ ksi} < 6.38 \text{ ksi}$$

Hence, web thickness for shear should be at least

$$t_w = \frac{48}{270}\sqrt{6.38} = 0.45 \text{ in}$$

This is larger than the $\frac{3}{8}$-in web thickness assumed. Therefore, intermediate transverse stiffeners are required. (A change in web thickness from $\frac{3}{8}$ to $\frac{7}{16}$ in would eliminate the need for the stiffeners.)

Stiffener spacing is determined by the shear stress computed from Eq. (10.31a). Assume that the stiffener spacing $d_o = 48$ in $=$ the web depth D. Hence, $d_o/D = 1$. From Eq. (10.30d), for use in Eq. (10.31a), $k = 5(1 + 1^2) = 10$ and $\sqrt{k/F_y} = \sqrt{10/36} = 0.527$. Since $D/t_w = 126$, C in Eq. (10.30a) is determined by the parameter $126/0.527 = 239 > 237$. Hence, C is given by Eq. (10.30c):

$$C = \frac{45,000k}{(D/t_w)^2 F_y} = \frac{45,000 \times 10}{126^2 \times 36} = 0.787$$

From Eq. (10.31a), the maximum allowable shear for $d_o = 48$ in is

$$F_v' = F_v \left[C + \frac{0.87(1 - C)}{\sqrt{1 + (d_o/D)^2}} \right]$$

$$= 12 \left[0.787 + \frac{0.87(1 - 0.787)}{\sqrt{1 + 1^2}} \right] = 11.02 > 6.38 \text{ ksi}$$

Since the allowable stress is larger than the computed stress, the stiffeners can be spaced 48 in apart. (Because of the brackets, the floorbeam can be considered continuous at the supports. Thus, the stiffener spacing need not be half the calculated spacing, as would be required for the first two stiffeners at simple supports.) Stiffener locations are shown in Fig. 11.29.

Stiffeners may be placed in pairs and welded on each side of the web with two ¼-in welds. Moment of inertia provided by each pair must satisfy Eq. (10.27), with J as given by Eq. (10.28).

$$J = 2.5 \left(\frac{48}{48} \right)^2 - 2 = 0.5$$

$$I = 48(⅜)^3 0.5 = 1.27 \text{ in}^4$$

Use $4 \times$ ⅜-in stiffeners. They satisfy minimum requirements for thickness of projection from the web and provide a moment of inertia

$$I = 0.375(2 \times 4 + 0.375)^3/12 = 18.36 > 1.27 \text{ in}^4$$

11.9.5 Design of Floorbeam Bracket

The floorbeam brackets are designed next. They can be tapered, because the rapid decrease in bending stress from the girder outward permits a corresponding reduction in web depth. To ensure adequate section throughout, the brackets are tapered from the 48-in depth of the floorbeam main span to 2 ft at the outer end (Fig. 11.29).

Splice at Girder. Bracket flanges are made the same size as the plate required for the moment splice to the main span. This plate is assumed to carry the full maximum negative moment of -341 ft-kips. With an allowable bending stress of 20 ksi, the splice plates then should have an area of at least

$$A_f = \frac{341 \times 12}{48.5 \times 20} = 4.2 \text{ in}^2$$

Use a $12 \times$ ½-in plate, with gross area of 6 in². After deduction of two holes for ⅞-in dia bolts, it provides a net area of $6 - 2 \times 1 \times ½ = 5$ in². Hence, the bracket flanges also are $12 \times$ ½-in plates. Use minimum-size ¼-in flange-to-web fillet welds.

The number of bolts required in the splice is determined by whichever is larger, 75% of the strength of the splice plate or the average of the calculated stress and strength of plate. The calculated stress is $20 \times 4.2/5 = 16.8$ ksi. The average stress is $(20 + 16.8)/2 = 18.4$ ksi. This governs, because 75% of 20 ksi is 15 ksi < 18.4.

For A325 ⅞-in bolts with a capacity of 9.3 kips (slip-critical, Class A surface), the number of bolts needed is

$$n = \frac{18.4 \times 5}{9.3} = 10$$

Use 12 bolts.

Connection to Girder. The connection of each bracket to a girder must carry a shear of 59.4 kips. The number of bolts required is

$$n = \frac{59.4}{9.3} = 7$$

Use at least 8 bolts.

Bracket Stiffeners. Stiffener spacing on the brackets generally should not exceed the web depth. Locations of the stiffeners are shown in Fig. 11.29. Use pairs of 4 × ⅜-in plates, as in the main span, with ¼-in fillet welds.

Check of Bracket Section. Bending and shear stresses at an intermediate point on the bracket should be checked to ensure that, because of the reduction in depth, allowable stresses are not exceeded. For the purpose, a section midway between stringer S1 and the girder is selected. Depth of web there is $48 - 3.5(48 - 24)/9.5 = 39.15$ in. The dead load consists of 0.16 kip per ft from weight of bracket, 0.05 kip per ft from weight of concrete haunch, and 23 kips from the stringers. Thus, the dead-load moment is

$$M_{DL} = -\frac{0.16(6)^2}{2} - \frac{0.05(4.5)^2}{2} - 23 \times 3.5 = -83.9 \text{ ft-kips}$$

The dead-load shear is

$$V_{DL} = 0.16 \times 6 + 0.05 \times 4.5 + 23 = 24.2 \text{ kips}$$

Live load is 22 kips 1 ft from the section. Hence, the live-load and impact moments are

$$M_{LL} = -22 \times 1 = -22 \text{ ft-kips} \qquad M_I = -0.3 \times 22 = -6.6 \text{ ft-kips}$$

Live-load and impact shears are

$$V_{LL} = 22 \text{ kips} \qquad V_I = 0.3 \times 22 = 6.6 \text{ kips}$$

Moments and shears due to sidewalk live load are

$$M_{SLL} = -5.8 \times 3.5 = -20.3 \text{ ft-kips} \qquad V_{SLL} = 5.8 \text{ kips}$$

Hence, the total moments and shears at the section are

$$M = -132.8 \text{ ft-kips} \qquad V = 58.0 \text{ kips}$$

Shear stress in the web is

$$f_v = \frac{58.6}{39.15 \times 0.375} = 4.0 < 12 \text{ ksi}$$

The moment of inertia of the section is

$$I = 2 \times 6 \left(\frac{40.15}{2}\right)^2 + \frac{0.375(39.15)^3}{12} = 6,720 \text{ in}^4$$

and the section modulus is

$$S = \frac{6,720}{20.08} = 334 \text{ in}^3$$

So the maximum bending stress at the section is

$$f_b = \frac{132.8 \times 12}{334} = 4.8 < 20 \text{ ksi}$$

Therefore, the bracket section is satisfactory.

11.9.6 Design of a Girder Supporting Floorbeams

The girders will be made of Grade 50 steel. Simply supported, they span 137.5 ft, but have a loaded length of 140 ft. They will be made identical.

Loading. Most of the load carried by each girder is transmitted to it by the floorbeams as concentrated loads. Computations are simpler, however, if the floorbeams are ignored and the girder is treated as if it received loads only from the slab. Moments and shears computed with this assumption are sufficiently accurate for design purposes because of the relatively close spacing of the floorbeams. Thus, the dead load on the girders may be considered uniformly distributed (Table 11.49).

Sidewalk live load, because the span exceeds 100 ft, is determined from a formula, with loaded length of sidewalk $L = 140$ ft and sidewalk width $W = 3$ ft:

$$p = \frac{(0.03 + 3/L)(55 - W)}{50} = \frac{(0.03 + 3/140)(55 - 3)}{50}$$

$$= 0.0535 \text{ kip per ft}^2$$

Thus, the live load from the 3-ft sidewalk is

$$w_{SLL} = 0.0535 \times 3 = 0.160 \text{ kip per ft}$$

TABLE 11.49   Dead Load on Girder, kips per ft

Railing:	0.07
Sidewalk: $0.150 \times 1 \times 3$ =	0.45
Slab: $0.150 \times 27 \times 7.5/12$ =	2.53
Floorbeams and stringers:	0.40
Girder—assume:	0.60
Lateral bracing—assume:	0.10
Utilities and miscellaneous:	0.10
DL per girder:	4.25

Live load, for maximum effect on a girder, should be placed as indicated in Fig. 11.36. Because of load reductions permitted in accordance with number of lanes of traffic loaded, the number of lanes to be loaded is determined by trial. Let W = wheel load, kips. Then, if two lanes are loaded, with no reduction permitted, the load P, kips, distributed to the girder is

$$P_2 = \frac{(39.5 + 33.5 + 27.5 + 21.5)W}{35} = \frac{122W}{35} = 3.48W$$

If three lanes are loaded, with 10% reduction,

$$P_3 = \frac{0.9(122 + 15.5 + 9.5)W}{35} = \frac{132.3W}{359} = 3.78W > P_2$$

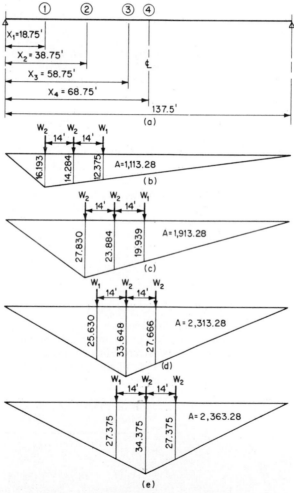

FIGURE 11.36 Moment influence lines for deck girder. (*a*) Location of four points on the girder for which influence diagrams are drawn. (*b*) Diagram for point 1. (*c*) Diagram for 2. (*d*) Diagram for 3. (*e*) Diagram for 4.

TABLE 11.50 Dead-Load Moments and Sidewalk Live-Load Moments, ft-kips

| | \multicolumn{4}{c}{Distance from support, ft} | | | |
	18.75	38.75	58.75	68.75
Influence area	1,113.3	1,913.3	2,313.3	2,363.3
$M_{DL} = Aw_{DL}$	4,732	8,132	9,832	10,044
$M_{SLL} = Aw_{SLL}$	178	306	370	378

And if all four lanes are loaded, with 25% reduction,

$$P_4 = \frac{0.75(147 + 3.5 - 2.5)W}{35} = \frac{111W}{35} = 3.17W < P_3$$

Therefore, loading in three lanes governs. The girder receives 3.78 wheel loads, or 1.89 axle loads.

Impact for loading over the whole span is taken as the following fraction of live-load stress:

$$I = \frac{50}{L + 125} = \frac{50}{137.5 + 125} = 0.19$$

Moments. Curves for maximum moments at points along the span will be drawn by plotting maximum moments at midspan and at each floorbeam (points 1 to 4 in Fig. 11.36a). These moments are calculated with the aid of influence lines drawn for moment at these points (Fig. 11.36b to e).

Dead-load moments are obtained by multiplying the uniform load $w_{DL} = 4.25$ kips per ft by the area A of the appropriate influence diagram. Moments due to sidewalk live loading are similarly calculated with uniform load $w_{SLL} = 0.16$ kip per ft. Dead-load moments are summarized in Table 11.50.

Maximum live-load moments are produced by truck loading on a 137.5-ft span. Since the girder receives 1.89 axle loads, it is subjected at 14-ft intervals to moving concentrated loads:

$$\text{Two } W_2 = 1.89 \times 32 = 60.48 \text{ kips} \qquad W_1 = 1.89 \times 8 = 15.12 \text{ kips}$$

For maximum moment at a point along the span, one load W_2 is placed at the point (Fig. 11.36b to e). The maximum moment then is the sum of the products of each load by the corresponding ordinate of the applicable influence diagram. Impact moments are 19% of the live-load moments. Table 11.51 summarizes maximum live-load and impact moments.

TABLE 11.51 Maximum Live-Load and Impact Moments, ft-kips

| | \multicolumn{4}{c}{Distance from support, ft} | | | |
	18.75	38.75	58.75	68.75
M_{LL}	2,030	3,384	4,096	4,149
M_I	386	644	778	788

Total maximum moments are given and the curve of maximum moments (moment envelope) is plotted in Fig. 11.37.

Reaction. Maximum reaction occurs with full load over the entire span. For dead load, with $w_{DL} = 4.25$ kips per ft,

$$R_{DL} = \frac{4.25 \times 140}{2} = 297.5 \text{ kips}$$

For sidewalk live load, with $w_{SLL} = 0.16$ kip per ft,

$$R_{SLL} = \frac{0.16 \times 140}{2} = 11.2 \text{ kips}$$

Lane loading governs for live load. For maximum reaction and shear, the uniform load of 0.64 kip per ft should cover the entire span and the 26-kip concentrated load should be placed at the support, in each design lane.

$$R_{LL} = \frac{1.89(26 + 0.64 \times 140)}{2} = 134 \text{ kips}$$

$$R_I = 0.19 \times 134 = 25.4 \text{ kips}$$

The total maximum reaction is $R = 468.1$ kips, say 470 kips.

Shears. Maximum live-load shears at floorbeam locations occur with truck loading between the beam and the far support. A heavy wheel should be at the beam in each design lane. The shears are readily computed with influence diagrams. For example, the influence line for shear at point 1 (Fig. 11.38*a*) is shown in Fig. 11.38*b*. Dead-load shear is obtained as the product of the uniform dead load $w_{DL} = 4.25$ kips per ft by the area of the complete influence diagram.

$$V_{DL} = 4.25(51.276 - 1.279) = 213 \text{ kips}$$

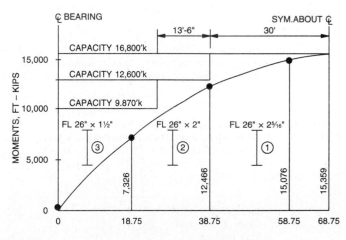

FIGURE 11.37 Moment diagram for deck girder and capacities of various sections.

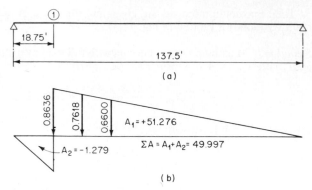

FIGURE 11.38 (a) Location of point 1 on girder. (b) Influence diagram for shear at point 1.

Sidewalk live-load shear is the product of the load $w_{SLL} = 0.16$ kip per ft and the larger of the positive or negative areas of the influence diagram.

$$V_{SLL} = 0.16 \times 51.276 = 8 \text{ kips}$$

Maximum live-load shear is the sum of the products of each load by the corresponding ordinate of the influence diagram (Fig. 11.38b).

$$V_{LL} = 60.48 \times 0.8636 + 60.48 \times 0.7618 + 15.12 \times 0.6600 = 108 \text{ kips}$$

The loaded length for impact is $137.5 - 18.75 = 118.75$ ft.

$$V_I = \frac{50}{118.75 + 125} 108 = 0.205 \times 108 = 22 \text{ kips}$$

Total maximum shear $V_1 = 351$ kips (Table 11.52).

Shears at other points are computed in the same way and are listed in Table 11.52.

Web Size.   Minimum depth-span ratio for a girder is 1:25. Greater economy and a stiffer member are obtained, however, with a deeper member when clearances permit. In this example, the web is made 110 in deep, $\frac{1}{15}$ of the span. With an

TABLE 11.52 Maximum Shear, kips

	Distance from support, ft			
	0	18.75	38.75	58.75
Dead load	293	213	127	42
Sidewalk live load	11	8	6	4
Live load	134	108	89	69
Impact	(At 0.19) 26	(At 0.205) 22	(At 0.223) 20	(At 0.245) 17
Total	464	351	242	132

allowable stress of 17 ksi for thin Grade 50 steel, the web thickness required for shear is

$$t = \frac{464}{17 \times 110} = 0.25 \text{ in}$$

Without a longitudinal stiffener, according to Table 10.15 thickness must be at least

$$t = \frac{110\sqrt{50}}{990} = 0.9 \text{ in, say } \tfrac{13}{16} \text{ in}$$

Even with a longitudinal stiffener, however, to prevent buckling, web thickness, from Table 10.15, must be at least

$$t = \frac{110\sqrt{50}}{1,980} = 0.393 \text{ in, say } \tfrac{7}{16} \text{ in}$$

though transverse stiffeners also are provided.

If the web were made $\tfrac{13}{16}$ in thick, it would weigh 304 lb per ft. If it were $\tfrac{7}{16}$ in thick, it would weigh 164 lb per ft, 140 lb per ft less. Since the longitudinal stiffener may weigh less than 10 lb per ft, economy favors the thinner web. Use a $110 \times \tfrac{7}{16}$-in web with a longitudinal stiffener.

Flange Size at Midspan.   For Grade 50 steel 4 in thick or less, $F_y = 50$ ksi and the allowable bending stress is 27 ksi. With a maximum moment at midspan, from Fig. 11.37, of 15,359 ft-kips, and distance between flange centroids of about 113 in, the required area of one flange is about

$$A_f = \frac{15,359 \times 12}{113 \times 27} = 60.4 \text{ in}^2$$

Assume a $26 \times 2\tfrac{5}{16}$-in plate for each flange. It provides an area of 60.1 in² and has a width-thickness ratio

$$b/t = 26/2.313 = 11.2$$

which is less than 20 permitted.

The trial section is shown in Fig. 11.39. Moment of inertia is calculated in Table 11.53. Distance from neutral axis to top or bottom of the girder is 57.31 in. Hence, the section modulus is

$$S = \frac{427,900}{57.31} = 7,466 \text{ in}^3$$

Maximum bending stress therefore is

$$f_b = \frac{15,359 \times 12}{7,466} = 24.7 < 27 \text{ ksi}$$

The section is satisfactory. Moment capacity supplied is

$$M_C = \frac{27 \times 7,466}{12} = 16,800 \text{ ft-kips}$$

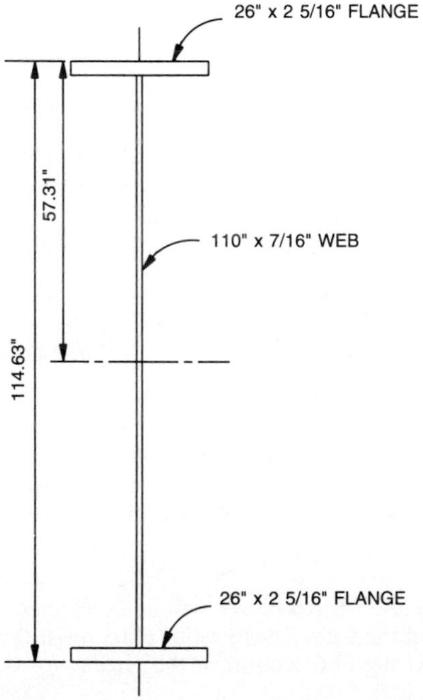

26" x 2 5/16" FLANGE

57.31"

110" x 7/16" WEB

114.63"

26" x 2 5/16" FLANGE

FIGURE 11.39   Cross section of deck girder at midspan.

TABLE 11.53 Moment of Inertia of Girder

Material	A	d	Ad^2 or I_o
2 flanges 26 × 2⁵⁄₁₆	120.3	56.16	379,400
Web 110 × ⁷⁄₁₆	48.1		48,500
			I = 427,900 in⁴

The flange thickness will be reduced between midspan and the supports, and the flange width will remain 26 in. Splices at the changes in thickness will be made with complete-penetration groove welds. For fatigue, the stress category at these splices is B. On the assumption that the structure supports a major highway with an ADTT less than 2500, the number of stress cycles of truck loading is 500,000. Since the bridge is supported by two simple-span girders and floorbeams, it is a non-redundant-path structure, and the allowable stress range is therefore 23 ksi at the splices.

Changes in Flange Size. At a sufficient distance from midspan, the bending moment decreases sufficiently to permit reducing the thickness of the flange plates to 2 in. The moment of inertia of the section reduces to 374,600, and the section modulus to 6,573. Thus, with 26 × 2-in flange plates, the section has a moment capacity of

$$M_C = \frac{6,573 \times 23}{12} = 12,600 \text{ ft-kips}$$

When this capacity is plotted in Fig. 11.37, the horizontal line representing it stays above the moment envelope until within 30 ft of midspan. Hence, flange size can be decreased at that point. Length of the 2⁵⁄₁₆-in plate then is 60 ft. (See Fig. 11.40.)

At a greater distance from midspan, thickness of the flange plates can be reduced to 1½ in. The moment of inertia drops to 290,930, and the section modulus to 5,150. Consequently, with 26 × 1½-in plates, the section has a moment capacity of

$$M_C = \frac{5,150 \times 23}{12} = 9,870 \text{ ft-kips}$$

When this capacity is plotted in Fig. 11.37, the horizontal line representing it stays above the moment envelope until within 49.5 ft of midspan. The 2-in plates can be cut off there. Length of the 26 × 2-in plates therefore is 39.5 − 30 = 13.5 ft and of the 26 × 1½-in plates, which extend to the ends of the girder, 26 ft (Fig. 11.40).

Flange-to-Web Welds. Each flange will be connected to the web by a fillet weld on opposite sides of the web. These welds must resist the horizontal shear between flange and web. At the end section of the girder, for determination of the shear, the static moment is

$$Q = 26 \times 1.5 \times 55.5 = 2,174 \text{ in}^3$$

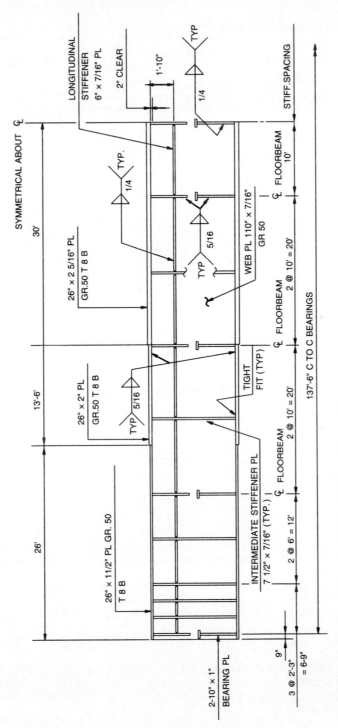

FIGURE 11.40 Details of deck plate girder for four-lane highway bridge.

The shear stress then is

$$v = \frac{VQ}{I} = \frac{464 \times 2174}{290,930} = 3.47 \text{ kips per in}$$

The minimum size of fillet weld permissible, governed by the thickest plate at the section, is $\frac{5}{16}$ in. With an allowable shear stress $F_v = 0.27F_u = 0.27 \times 65 = 17.6$ ksi, the allowable load per weld $17.55 \times 0.707 = 12.44$ kips per in, and for two welds, 24.89 kips per in. Hence, the capacity of two $\frac{5}{16}$-in fillet welds is $24.89 \times \frac{5}{16} = 7.78$ kips per in > 3.47 kips per in. Use two $\frac{5}{16}$-in welds. (See also the design of fillet welds in Art. 11.9.4.)

Intermediate Transverse Stiffeners. Where required, a pair of transverse stiffeners of Grade 36 steel will be welded to the girder web. Minimum width of stiffener is $26/4 = 6.5$ in $> (2 + 110/30 = 5.7$ in). Use a $7\frac{1}{2}$-in wide plate. Minimum thickness required is $\frac{7}{16}$ in. Try a pair of $7\frac{1}{2} \times \frac{7}{16}$-in stiffeners.

Maximum spacing of the transverse stiffeners can be computed from Eq. (10.31a). For the $110 \times \frac{7}{16}$-in girder web and a maximum shear at the support of 463 kips, the average shear stress is $463/48.1 = 9.62$ ksi. The web depth-thickness ratio $D/t_w = 110/(\frac{7}{16}) = 251$. Maximum spacing of stiffeners is limited to $110(270/251)^2 = 127$ in. Try a stiffener spacing $d_o = 80$ in. This provides a depth-spacing ratio $D/d_o = 110/80 = 1.375$. From Eq. (10.30d), for use in Eq. (10.31a), $k = 5[1 + (1.375)^2] = 14.45$ and $\sqrt{k/F_y} = \sqrt{14.45/50} = 0.537$. Since $D/t_w = 251$, C in Eq. (10.30a) is determined by the parameter $251/0.537 = 467 > 237$. Hence, C is given by Eq. (10.30c):

$$C = \frac{45,000k}{(D/t_w)^2 F_y} = \frac{45,000 \times 14.45}{251^2 \times 50} = 0.206$$

From Eq. (10.31a), the maximum allowable shear for $d_o = 80$ in is

$$F_v' = F_v \left[C + \frac{0.87(1 - C)}{\sqrt{1 + (d_o/D)^2}} \right]$$

$$= \frac{50}{3} \left[0.206 + \frac{0.87(1 - 0.206)}{\sqrt{1 + (80/110)^2}} \right] = 12.74 \text{ ksi} > 9.62 \text{ ksi}$$

Since the allowable stress is larger than the computed stress, the stiffeners may be spaced 80 in apart. The location of floorbeams, however, may make closer spacing preferable.

The AASHTO standard specifications limit the spacing of the first intermediate transverse stiffener to the smaller of $1.5D = 1.5 \times 110 = 165$ and the spacing for which the allowable shear stress in the end panel does not exceed

$$F_v = CF_y/3 = 0.206 \times 50/3 = 3.43 \text{ ksi} < 9.62 \text{ ksi}$$

Much closer spacing than 80 in is required near the supports. Try $d_o = 27$ in, for which $k = 88$ and $C > 1$. Hence, $F_v = 50/3 = 17$ ksi > 9.62 ksi. Spacing selected for intermediate transverse stiffeners between the supports and the first floorbeam is shown in Fig. 11.40.

At that beam, the shear stress is $f_v = 351/48.1 = 7.28$ ksi. Try a stiffener spacing $d_o = 10$ ft $= 120$ in, which is less than the 127-in limit. This provides $D/d_o = $

0.917, $k = 9.20$, and $C = 0.131$. The allowable shear for this spacing then is

$$F_v = \frac{50}{3} \left[0.131 + \frac{0.87(1 - 0.131)}{\sqrt{1 + (120/110)^2}} \right] = 10.69 \text{ ksi} > 7.28 \text{ ksi}$$

The 10-ft spacing is satisfactory. Actual spacing throughout the span is shown in Fig. 11.40.

The moment of inertia provided by each pair of stiffeners must satisfy Eq. (10.27), with J as given by Eq. (10.28).

$$J = 2.5 \left(\frac{110}{27} \right)^2 - 2 = 39.5$$

$$I = 120(\tfrac{7}{16})^3 \, 39.5 = 123 \text{ in}^4$$

The moment of inertia furnished by a pair of 7½-in-wide stiffeners is

$$I = \frac{(\tfrac{7}{16})(15,437)^3}{12} = 134 > 123 \text{ in}^4$$

Hence, the pair of 7½ × $\tfrac{7}{16}$-in stiffeners is satisfactory.

Longitudinal Stiffener.   One longitudinal stiffener of Grade 36 steel will be welded to the web. It should be placed with its centerline at a distance $110/5 = 22$ in below the bottom surface of the compression flange (Fig. 11.40).

Assume a 6-in wide stiffener. Then, by Eq. (10.34), the thickness required is

$$t = \frac{6\sqrt{23}}{71.2} = 0.40 \text{ in, say } \tfrac{7}{16} \text{ in}$$

Moment of inertia furnished with respect to the edge in contact with the web is

$$I = \frac{0.437(6)^3}{3} = 31.5 \text{ in}^4$$

With transverse stiffeners spaced 120 in apart, the moment of inertia required, by Eq. (10.33), is

$$I_{min} = 110(0.437)^3 \left[2.4 \left(\frac{120}{110} \right)^2 - 0.13 \right] = 25.1 < 31.5 \text{ in}^4$$

Therefore, use a 6 × $\tfrac{7}{16}$-in plate for the longitudinal stiffener.

Bearing Stiffeners.   A pair of bearing stiffeners of Grade 50 steel is provided at each support. They are designed to transmit the 464-kip end reaction between bearing and girder.

Try 10 × 1-in plates. With provision for clearing the flange-to-web fillet weld, the effective width of each plate is $10 - 0.75 = 9.25$ in. The effective bearing area is $2 \times 1 \times 9.25 = 18.5$ in^2. Allowable bearing stress is 40 ksi. Actual bearing stress is

$$f_p = \frac{464}{18.5} = 25.0 < 40 \text{ ksi}$$

The width-thickness ratio of the assumed plate $b/t = 10/1 = 10$ satisfies Eq. (10.26), with $F_y = 50$ ksi.

$$\frac{b}{t} = \frac{69}{\sqrt{50}} = 9.75 < 10$$

The pair of stiffeners is designed as a column acting with a length of web equal to 18 times the web thickness, or 7.88 in. Area of the column is

$$2 \times 10 \times 1 + 7.88 \times \frac{7}{16} = 23.44 \text{ in}^2$$

Buckling is prevented by the floorbeam connecting to the stiffeners. Hence, the stress in the stiffeners must be less than the allowable compressive stress of 27 ksi and need not satisfy the column formulas. For the 464-kip reaction, the compressive stress is

$$f_a = \frac{464}{23.44} = 19.8 < 27 \text{ ksi}$$

Therefore, the pair of 10 × 1-in bearing stiffeners is satisfactory.

Stiffener-web welds must be capable of developing the entire reaction. With fillet welds on opposite sides of each stiffener, four welds are used. They extend the length of the stiffeners, from the bottom of the 48-in-deep floorbeam to the girder tension flange. Thus, total length of the welds is $4(110 - 48) = 248$ in. Average shear on the welds is

$$v = \frac{464}{248} = 1.9 \text{ kips per in}$$

Weld size required to carry this shear is, with allowable stress $F_v = 0.27F_o = 17.6$ ksi,

$$\frac{1.9}{0.707 \times 17.6} = 0.152 \text{ in}$$

This, however, is less than the $\frac{5}{16}$-in minimum size of weld required for a 1-in-thick plate. Therefore, use $\frac{5}{16}$-in fillet welds.

11.9.7 Design of Horizontal Lateral Bracing

Each girder flange is subjected to half the transverse wind load. The top flange is assisted by the concrete deck in resisting the load and requires no lateral bracing. The following illustrates design of lateral bracing for the bottom.

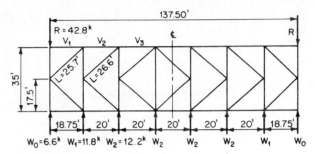

FIGURE 11.41 Lateral bracing system for deck-girder bridge.

Figure 11.41 shows the layout of the lateral truss system, which lies in a plane at the bottom of the floorbeams. The girders comprise the chords of the truss, and the floorbeams the transverse members, or posts. The truss must be designed to resist a wind load of 50 psf, but not less than 300 lb per lin ft, on the exposed area. The wind is considered a uniformly distributed, moving load acting perpendicular to the girders and reversible in direction.

The uniform load on the truss for an exposed depth of 12.14 ft (Table 11.54) is

$$w = 0.050 \times 12.14 = 0.61 \text{ kip per ft}$$

It is resolved into a concentrated load at each panel point (Fig. 11.41):

$$W_2 = 0.61 \times 20 = 12.2 \text{ kips}$$

$$W_1 = \frac{0.61(20 + 18.75)}{2} = 11.8 \text{ kips}$$

$$W_0 = \frac{0.61(18.75 + 1.5)}{2} = 6.6 \text{ kips}$$

The reaction at each support is

$$R = 2 \times 12.2 + 11.8 + 6.6 = 42.8 \text{ kips}$$

With the wind considered a moving load, maximum shear in each panel is:

$$V_1 = 42.8 - 6.6 = 36.2 \text{ kips}$$

$$V_2 = 36.2 - 11.8 \times \frac{118.75}{137.5} = 25.9 \text{ kips}$$

TABLE 11.54 Exposed Area, ft² per lin ft

Railing	0.91
Slab	1.83
Girder	9.40
Total	12.14

$$V_3 = 25.9 - 12.2 \times \frac{98.75}{137.5} = 17.1 \text{ kips}$$

$$V_4 = 17.1 - 12.2 \times \frac{78.75}{137.5} = 10.1 \text{ kips}$$

The shear is assumed to be shared equally by the two diagonals in each panel. Since the direction of the wind is reversible, the stress in each diagonal may be tension or compression. Design of the members is governed by compression.

The diagonals, being secondary compression members, are permitted a slenderness ratio L/r up to 140. (The effective length factor K is taken conservatively as unity.) For the end panel, the length c to c of connections is

$$L = 25.7 - 3 = 22.7 \text{ ft}$$

Hence, the radius of gyration should be at least $r = 22.7 \times 12/140 = 1.95$ in. Similarly, for interior panels, minimum $r = 23.6 \times 12/140 = 2.02$ in.

Assume for the diagonals a WT6 $\times$ 26.5 (Fig. 11.42). It has the following properties:

$$S_x = 3.54 \text{ in}^3 \qquad r_x = 1.51 \text{ in} \qquad r_y = 2.48 \text{ in} \qquad A = 7.80 \text{ in}^2 \qquad y = 1.02 \text{ in}$$

To permit the slenderness ratio about the vertical axis to govern the design, provide a vertical brace at midlength of each diagonal. The minimum slenderness ratio then is

$$\frac{L}{r_y} = \frac{22.7 \times 12}{2.48} = 110 < 140$$

Horizontal Buckling.   For a column of Grade 36 steel with this slenderness ratio and with bolted ends, the allowable compressive stress is

$$F_a = 16.98 - 0.00053 \left(\frac{L}{r}\right)^2 = 16.98 - 0.00053(110)^2 = 10.57 \text{ ksi}$$

Maximum stress occurs in the end panel where wind shear is a maximum, 36.2 kips. Each diagonal is assumed to carry half this, 18.1 kips. Thus, it is subjected to an axial force of

$$F = \frac{18.1 \times 25.7}{17.5} = 26.6 \text{ kips}$$

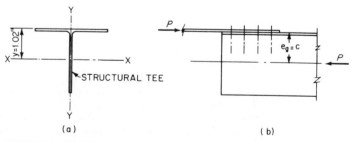

FIGURE 11.42   Diagonal brace. (*a*) Cross section. (*b*) Eccentric loading on the end connection of the diagonal.

This causes an average compressive stress in the diagonal of

$$f_a = \frac{26.6}{7.80} = 3.4 < 10.57 \text{ ksi}$$

Hence, the WT6 × 26.5 is adequate for resisting buckling in the horizontal direction.

Vertical Buckling.   Because of the T shape of the WT, its end connections load it eccentrically. Therefore, the diagonal should be checked for combined axial plus bending stresses and buckling in the vertical direction. The eccentricity and c distance from the neutral axis to the top of the compression flange is 1.02 in (Fig. 11.42). The slenderness ratio for buckling in the vertical direction, with a conservative value of $K = 1.0$ and provision for a midlength brace, is

$$\frac{L}{r_x} = \frac{12 \times 22.7/2}{1.51} = 90.2$$

Members subjected to combined axial compression and bending must satisfy

$$\frac{f_a}{F_a} + \frac{C_{mx} f_{bx}}{(1 - f_a/F'_{ex})F_{bx}} + \frac{C_{my} f_{by}}{(1 - f_a/F'_{ey})F_{by}} \leq 1.0$$

where $F'_e = \dfrac{\pi^2 E}{FS(K_b L_b/r_b)^2}$

$FS = 2.12$

C_m = coefficient defined in Art. 6.19.1 (1.0 is conservative)

The axial stress f_a is 3.4 ksi and the allowable stress is

$$F_a = 16.98 - 0.00053(KL/r_x)^2 = 16.98 - 0.00053(90.1)^2 = 12.7 \text{ ksi}$$

The bending stress f_b is 26.6 × 1.02/3.54 = 7.66 ksi. The allowable bending stress for Grade 36 steel in this case is $F_b = 20.0$ ksi.

$$F'_e = \frac{\pi^2(29,000)}{2.12(90.2)^2} = 16.6 \text{ ksi}$$

Substitution in the interaction equation gives

$$\frac{3.4}{12.7} + \frac{1.0 \times 7.66}{[1 - (3.4/16.6)]20.0} = 0.27 + 0.48 = 0.75 < 1.0\text{—OK}$$

Use the WT6 × 26.5 for all the diagonals.

Bracing Connections.   End connections of the laterals are to be made with A325 ⅞-in-dia. high-strength bolts. These have a capacity of 9.3 kips in slip-critical connections with Class A surfaces. The number of bolts required is determined by whichever is larger, 75% of the strength of the diagonal or the average of the calculated stress and strength of the diagonal.

TABLE 11.55 Net Area of Diagonal, in^2

Gross area:	7.80
Half web area: $-5.45 \times 0.345/2 =$	-0.94
Two holes: $-2 \times 1 \times 0.576 \quad =$	-1.15
Net area:	5.71

In the computation of the tensile strength of the T section, the effective area should be taken as the net area of the connected flange plus half the area of the outstanding web (Table 11.55). With an allowable stress of 20 ksi, the tensile capacity is

$$T = 5.71 \times 20 = 114 \text{ kips}$$

Compressive capacity with $F_a = 8.5$ ksi on the gross area is

$$C = 7.80 \times 8.5 = 66 < 114 \text{ kips}$$

Tensile capacity governs. Hence, the number of bolts required is determined by

$$0.75 \times 114 = 86 \text{ kips} > \left(\frac{26.6 + 114}{2} = 70 \text{ kips} \right)$$

and equals $86/9.3 = 9.2$. Use ten $\frac{7}{8}$-in high-strength bolts.

11.9.8 Expansion Shoes

Each girder transmits its end reactions to piers through one fixed and one expansion shoe. For a moderate climate, the latter should be designed to permit movements resulting from variations of temperature between $+50$ and $-70°F$, and to allow rotation of the girder ends under live loads. Both shoes are weldments, fabricated of Grade 36 steel. They must be capable of transmitting the maximum girder reaction of 464 kips and their own weight, assumed at 16 kips, or a total of 480 kips.

The expansion shoe incorporates a rocker, to permit the required movements, a sole plate attached to the girder bottom to transmit the reaction to the rocker, and a base plate, to distribute the load to the concrete pier (Fig. 11.43). AASHTO specifications require that the shoe permit a thermal movement of 1.25 in per 100 ft of span, or a total of $1.25 \times 140/100 = 1.75$ in. For the temperature range specified, the thermal movements expected are

$$\text{Expansion: } 0.0000065 \times 50 \times 140 \times 12 = 0.547 \text{ in}$$

$$\text{Shortening: } 0.0000065 \times 70 \times 140 \times 12 = 0.763 \text{ in}$$

$$\text{Total:} \qquad\qquad\qquad\qquad\qquad\qquad 1.310 \text{ in}$$

The maximum live-load deflection preferred by AASHTO specifications is $1/800$ of the span, or $140 \times 12/800 = 2.1$ in. If this deflection occurred, and if the elastic curve is assumed parabolic, the end rotation under live load of the girder would be

$$\theta = \frac{4 \times 2.1}{137.5 \times 12} = 0.005 \text{ radian}$$

The distance from the neutral axis to the center of curvature of the rocker is

$$R_r = 110/2 + 5 = 60 \text{ in}$$

With R_r as radius, live load causes an expansion of $60 \times 0.005 = 0.30$ in at each end of the girder. Thus, the rocker must be capable of accommodating a live-load expansion of $2 \times 0.3 = 0.60$ in.

Addition of this to the thermal expansion yields a total expansion of

$$e = 0.60 + 0.55 = 1.15 \text{ in}$$

Maximum shortening is 0.76 in. The 4-in web of the rocker (Fig. 11.43) permits movements up to $4/2 = 2$ in, expansion or contraction.

Rocker.   The rocker has a radius of 15 in, diameter of 30 in. By Eq. (10.20), the allowable bearing, if Grade 36 steel is used, is

$$P = \frac{3\sqrt{30}(36 - 13)}{20} = 19 \text{ kips per in}$$

To carry the 480-kip load, length of bearing should be at least $480/19 = 25.3$ in. Since the base of the rocker incorporates two $1\frac{5}{8}$-in-dia. holes for $1\frac{1}{2}$-in-dia. pintles, the computed length should be increased to $25.3 + 2 \times 1.625 = 28.6$ in. Use 30 in (Fig. 11.43*b*).

Web and Hinge.   The 4-in-thick rocker web rests on a curved steel plate with 4-in maximum thickness. Maximum eccentricity of loading on the web is 1.15 in, less

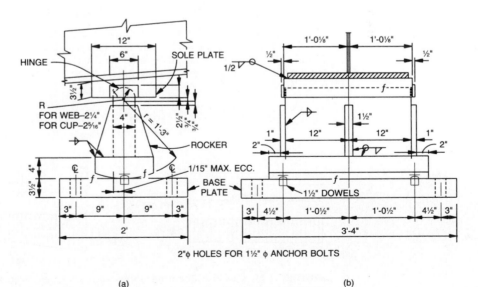

FIGURE 11.43   Expansion shoe. (*a*) Side view. (*b*) End view.

than half the web thickness. Therefore, the loading cannot cause bending in the curved plate.

The curved top of the web serves as a hinge (Fig. 11.43a), which bears against a cup in the sole plate. Thus, the compressive stress in the 24-in-long web equals the bearing stress in the hinge. The allowable bearing stress of 14 ksi for Grade 36 steel pins subject to rotation applies also to such hinges. The compressive stress in the web is

$$f_p = \frac{480}{4 \times 24} = 5.0 < 14 \text{ ksi}$$

Hence, the web and hinge are satisfactory.

Base Plate.   The rocker is seated on a base plate, which distributes the 480-kip load to the concrete pier. Allowable bearing stress on the concrete is $0.3f'_c = 0.3 \times 3 = 0.9$ ksi. Hence, the minimum net area of the plate is $480/0.9 = 533$ in². Since the plate incorporates four 2-in-dia. holes for 1½-in-dia. anchor bolts, the required area should be increased to $533 + 4 \times 3.14 = 596$ in². For a width of 24 in (Fig. 11.42a), length of plate should be at least $596/24 = 22.7$ in. Use 40 in to obtain adequate space beyond the rocker for the anchor bolts.

$$\text{Net area of base plate} = 40 \times 24 - 4 \times 3.14 = 947.4 > 533 \text{ in}^2$$

Thickness of plate must be large enough to keep bending stresses caused by the bearing pressure within the allowable. Under dead load, live load, and impact, the pressure may be considered substantially uniform (Fig. 11.44a). If thermal movements also occur, the pressure will be nonuniform and is usually assumed to vary linearly (Fig. 11.44b). The allowable bending stresses may be increased 25% when temperature stresses are included.

For a thermal movement of 1.15 in, the base pressures are

$$p = \frac{P}{A} \pm \frac{6P_e}{bd^2} = \frac{480}{947} \pm \frac{6 \times 480 \times 1.15}{40(24)^2} = 0.51 \pm 0.14$$

Therefore, the maximum pressure is 0.65 ksi and the minimum 0.37 ksi. Pressure directly under the load is $0.37 + (0.65 - 0.37)13.15/24 = 0.52$ ksi. Consider now a 1-in-wide strip of the base plate under this linearly varying pressure. The bending moment in the plate at the bearing point, 10.85 in from the nearest edge of the plate, is

$$M = \frac{0.52(10.85)^2}{2} + \frac{1}{2}(0.65 - 0.52)\frac{2}{3}(10.85)^2 = 35.7 \text{ in-kips}$$

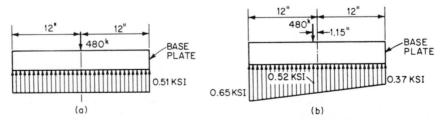

FIGURE 11.44   Base plate at expansion shoe. (a) Uniform pressure on the plate under dead load plus live load plus impact. (b) Linearly varying pressure on the plate with thermal loading added.

With the allowable stress increased 25%, the effective moment is

$$M_{\text{eff}} = \frac{35.7}{1.25} = 28.6 \text{ in-kips}$$

For dead load, live load, and impact, with the basic allowable stress $f_b = 20$ ksi, the base pressure is

$$p = \frac{480}{947} = 0.51 \text{ ksi}$$

The bending moment in a 1-in-wide strip of plate at the bearing point, 12 in from either plate edge, is

$$M = \frac{0.51(12)^2}{2} = 36.7 > 28.6 \text{ in-kips}$$

This moment governs. Thickness of base plate required then is

$$t = \sqrt{\frac{6M}{f_b}} = \sqrt{\frac{6 \times 36.7}{20}} = 3.3, \text{ say } 3.5 \text{ in}$$

Use a base plate 24 × 3½ in by 3 ft 4 in long.

11.9.9 Fixed Shoes

A fixed shoe is placed at the opposite end of each girder from the expansion shoe. The fixed shoe precludes translation at the girder end but permits rotation in the plane of the web. Like the expansion shoe, the fixed shoe is a weldment, fabricated of Grade 36 steel. Major components are a bearing bar with curved top to serve as a hinge, a sole plate attached to the girder bottom to transmit the reaction to the bearing bar, a base plate to distribute the load to the concrete pier, and three ribs attached to the bearing bar to improve stability and load distribution (Fig. 11.45).

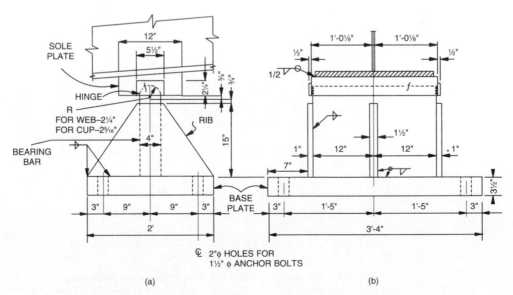

FIGURE 11.45 Fixed shoe. (*a*) Side view. (*b*) End view.

Loadings.   Several loading combinations must be investigated:

Group I. Dead load (*DL*) + live load (*LL*) + impact (*I*) at 100% of the basic allowable stress

Group II. *DL* + wind (*W*) at 125% of the basic allowable stress

Group III. *DL* + *LL* + *I* + 0.3*W* + longitudinal force (*LF*) + wind on moving live load (*WL*) at 125% of the basic allowable stress

The group I loading is a vertical load of 480 kips, as for the expansion shoe.

For the group II loading, the dead load is a vertical load from the girder of 297.5 kips plus the weight of the shoe, or a total of 307 kips. The wind imposes horizontal and vertical forces. The horizontal force is contributed by the 42.8-kip reaction of the horizontal lateral bracing system, which is assumed to be shared equally by the fixed shoes of the two girders. Each shoe therefore takes 42.8/2 = 21.4 kips horizontally, perpendicular to the girders.

The vertical wind loading is caused by overturning effects of the wind. The wind forces are the same as those used in the design of the lateral system.

$$\text{Railing force:} \qquad W_1 = 0.05 \times 0.91 \times {}^{140}\!/_2 = 3.2 \text{ kips}$$

$$\text{Slab and girder force: } W_2 = 0.05(1.83 + 9.40)^{140}\!/_2 = 39.5 \text{ kips}$$

W_1 acts 13.1 ft above the hinge of the shoe, and W_2, 6 ft above the hinge. Thus, these forces produce a moment about the hinge of

$$M_h = 3.2 \times 13.1 + 39.5 \times 6 = 279 \text{ ft-kips}$$

The moment is resisted by a couple consisting of vertical forces at each shoe, 35 ft apart. Hence, the moment induces an upward or downward force of 279/35 = 8 kips.

Total downward vertical force at the hinge then is 8 + 307 = 315 kips.

Similarly, the wind forces cause a moment about the bottom of the shoes, or top of pier, of

$$M_p = 279 + (3.2 + 39.5)1.3 = 335 \text{ ft-kips}$$

This induces vertical forces at top of pier of 335/35 ≈ 10 kips.

Total vertical downward force at top of pier then is 10 + 307 = 317 kips per shoe. The 21.4-kip horizontal force acts simultaneously with this.

For group III loading, the vertical load *DL* + *LL* + *I* = 480 kips. From the preceding calculation, 0.3*W* causes a horizontal force of 0.3 × 21.4 = 6.4 kips transversely and a vertical force of 0.3 × 8 = 2 kips at the hinge and 0.3 × 10 = 3 kips at the top of pier. *LF* is 5% of the live load (land loading for moment, including concentrated load but not impact) for two lanes of traffic heading in the same direction.

$$LL = 2(0.64 \times 140 + 18) = 216 \text{ kips}$$

The longitudinal force acting on one fixed shoe then is

$$LF = \frac{0.05 \times 216 \times 30.5}{35} = 10 \text{ kips}$$

WL is 0.1 kip per ft applied 6 ft above the roadway surface.

$$WL = 0.1 \times 140 = 14 \text{ kips}$$

Acting 17.1 ft above the hinge, it is shared equally by the four shoes. Thus, each shoe receives a transverse horizontal load of $14/4 = 4$ kips. Also, WL produces a moment about the hinge of $17.1 \times 14/2 = 120$ ft-kips and vertical forces on the shoes of $120/35 = 3$ kips.

Maximum downward vertical load on the hinge therefore is $480 + 2 + 3 = 485$ kips.

Similarly, WL causes a moment about the top of pier of $120 + 7 \times 1.3 = 129$ ft-kips and vertical forces at top of pier of $129/35 = 4$ kips. Total vertical downward force at top of pier then is $480 + 3 + 4 = 487$ kips per shoe. The 10-kip longitudinal force and a transverse wind force of $6.4 + 14/4 = 10$ kips act simultaneously with the vertical force.

Because the group II and III loadings are allowed a 25% increase in allowable stress, design of the shoe for vertical loading is governed by group I loading with basic allowable stresses.

Bearing Bar and Hinge.   Width and thickness of the bearing bar and radius of the hinge are made the same as the width and thickness of the rocker web and radius of the hinge, respectively, of the expansion shoe (Figs. 11.43 and 11.45).

Base Plate.   With a 40×24-in base plate, the same size as used for the expansion shoe, the maximum pressure on the concrete pier, 0.51 ksi, is the same as that under the base plate of the expansion shoe. Thickness of the plate for the fixed shoe, however, is governed by bending moment about the longitudinal axis at the outer face of the exterior rib, 7 in from the edge of the base plate. This moment is

$$M = \frac{0.51(7)^2}{2} = 12.5 \text{ in-kips}$$

Required thickness of base plate, with an allowable bending stress of 20 ksi, then is

$$t = \sqrt{\frac{6M}{f_b}} = \sqrt{\frac{6 \times 12.5}{20}} = 1.94 \text{ in}$$

Use the same size base plate as for the expansion shoe, $24 \times 3\frac{1}{2}$ in by 3 ft 4 in long.

11.9.10 Overturning Forces

The structure should be checked for resistance to overturning under group II or group III loading, plus an upward force. This force should be taken as 20 psf for group II and 6 psf for group III on deck and sidewalk. The force is assumed to act at the windward quarter point of the width of the structure, $54/4 = 13.5$ ft from the windward edge of the sidewalk, or $13.5 - 9.5 = 4.0$ ft from the windward girder.

With the larger uplift forces and lesser downward loads, group II loading governs the design. For this loading, the uplift force 4 ft from the windward girder is

$$P = 0.02 \times 54 \times 140 = 151 \text{ kips}$$

It causes an uplift reaction at each of the two windward shoes of

$$R_u = \frac{151}{2} \frac{35 - 4}{35} = 67 \text{ kips}$$

For group II loading, as calculated previously, wind produces an uplift of 10 kips at top of pier. Hence, the total uplift is $67 + 10 = 77$ kips. This is counteracted by the 307-kip downward dead-load reaction. Net force is $307 - 77 = 230$ kips downward.

Since there is no uplift at the shoes, tie-down bolts are not required. The horizontal forces, being small, will be resisted by friction. Minimum-size anchor bolts permitted by AASHTO specifications may be used. Use four 1½-in-dia. anchor bolts per shoe.

11.9.11   Seismic Evaluation of Bearings

Since this is a single-span bridge, only the connection between the bridge and the abutments must be designed to resist the longitudinal and transverse gravity reactions. If the bridge is assigned to AASHTO Seismic Performance Category A, the connection should be designed to resist a horizontal seismic force equal to 0.20 times the dead-load reaction. For allowable-stress design, a 50% increase in allowable stress is permitted for structural steel and a 33⅓% increase for reinforced concrete.

Expansion Shoe.   The total dead-load reaction is 297.5 kips. The transverse seismic (EQ) load $= 0.20 \times 297.5$ kips $= 59.5$ kips. For the sole-plate connection to the girder, try a ½-in fillet weld with a capacity of $½ \times 0.707 \times 0.27 \times 58 = 5.54$ kips per in. The required length of weld is $59.5/5.54 = 10.74$ in $\le (2 \times 12$ in $= 24$ in). For the cap-plate connection to the end of the sole plate, try a ½-in fillet weld. Length $= 10.74$ in $\le (6.0 + 2 \times 2.50 = 11$ in).

The rocker will be subjected to a transverse moment equal to $17.25 \times 59.5 = 1026$ in-kips and a vertical dead load of 297.5 kips. The eccentricity $e = 1,026/297.5 = 3.45$ in $\le 30/6$. There will be no uplift. Since the rocker base is 30 in long, it carries a load

$$P_{max} = \frac{297.5}{30} \left(1 + \frac{6 \times 3.45}{30} \right)$$

$$= 16.76 \text{ kips per in} < (1.50 \times 19 = 28.5 \text{ kips per in})$$

where 19 kips per in is the allowable bearing pressure (Art. 11.9.8).

The masonry plate will be subjected to a transverse moment of $20.75 \times 59.5 = 1235$ in-kips. The eccentricity $e = 1,235/297.5 = 4.15$ in $< (40/6 = 6.7$ in). There will be no uplift. The bearing pressure on the 24×40-in masonry plate is

$$P_{max} = \frac{297.5}{24 \times 40} \left(1 + \frac{6 \times 4.15}{40} \right) = 0.50 \text{ ksi} < (1.33 \times 0.90 = 1.20 \text{ ksi})$$

This is less critical than the service loads.

Two 1½-in dowels, each with an area of 1.76 in^2, secure the rocker to the masonry plate. The shear in these dowels is

$$f_v = \frac{59.5}{2 \times 1.76} = 16.9 \text{ ksi} \le (1.50 \times 12 = 18 \text{ ksi})$$

where 12 ksi is the allowable shear stress. Four 1½-in anchor bolts will provide sufficient connection to the abutment.

Fixed Shoe.   The total dead-load reaction is 297.5 kips. However, the fixed shoe must resist a longitudinal EQ component of 20% of the dead-load reaction, 59.5 kips, plus a transverse EQ component equal to 30% of that, 17.9 kips. Hence, the shoe will be subjected to a total force equal to

$$P = \sqrt{(59.5)^2 + (17.9)^2} = 62.1 \text{ kips}$$

The shoe will be anchored with four 1½-in anchor bolts. The shear on the bolts is

$$f_v = \frac{62.1}{4 \times 1.76} = 8.8 \text{ ksi} \le (1.50 \times 12 = 18 \text{ ksi})$$

The fixed-shoe masonry plate will be subjected to a longitudinal moment M_L and a transverse moment M_T.

$$M_L = 59.5 \times 20.75 = 1{,}235 \text{ in-kips}$$

which induces an eccentricity $e_L = 1{,}235/297.5 = 4.15$ in.

$$M_T = 17.9 \times 20.075 = 371 \text{ in-kips}$$

which induces an eccentricity $e_T = 371/297.5 = 1.25$ in. The bearing pressure on the plate is

$$P_{max} = \frac{297.5}{24 \times 40} \left(1 + \frac{6 \times 4.15}{24} + \frac{6 \times 1.25}{40} \right)$$

$$= 0.69 \text{ ksi} \le (1.33 \times 0.90 = 1.20 \text{ ksi})$$

The masonry-plate reaction is less than factored design load; therefore, the plate is satisfactory.

The AASHTO Seismic Performance Category A has lower seismic design requirements than Categories B to D. While rocker bearings may be adequate for Category A, other types of bearings, which are low profile, such as pot or elastomeric bearings, are generally preferred. In extreme cases, base-isolation bearings are used.

11.10   *THROUGH PLATE-GIRDER BRIDGES WITH FLOORBEAMS*

For long or heavily loaded bridge spans, restrictions on depth of structural system imposed by vertical clearances under a bridge generally favor use of through construction. Through girders support the deck near their bottom flange. Such spans preferably should contain only two main girders, with the railway or roadway between them (Fig. 11.46). In contrast, deck girders support the deck on the top flange (Art. 11.8).

The projection of the girders above the deck in through bridges may be objectionable for highway structures, because they obstruct the view from the bridge of pedestrians or drivers. But they may offer the advantage of eliminating the need for railings and parapets. For railroad bridges over highways, streets, or other facilities from which the bridges are highly visible to the general public, through girders provide a more attractive structure than through trusses.

The projection of the girders above the deck also has the disadvantage of requiring special provisions for bracing the compression flange of the girders. Deck girders usually require no special provision for this purpose, because when a rigid

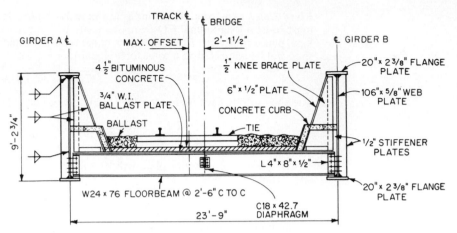

FIGURE 11.46 Cross section of through-girder railroad bridge.

deck is used, it provides the needed lateral support. Through girders should be laterally braced with gusset plates or knee braces with solid webs connected to the stiffeners.

In railroad bridges, spacing of the through girders should be at least ¹⁄₂₀ of the span, or should be adequate to ensure that the girders and other structural components provide required clearances for trains, whichever is greater.

Article 11.11 presents an example to indicate the design procedure for a through girder bridge with floorbeams. Because the example in Art. 11.9 dealt with highway loading, additional information is provided by designing a railroad bridge in the following example. Also, a curved alignment is selected, whereas the girders are kept straight, to illustrate the application of centrifugal forces to the structure. Note that because the girders are straight, the centerline of the track is offset from the centerline of the bridge. Design procedures not discussed in the example generally are the same as for deck girders (Art. 11.9) or plate-girder stringers (Art. 11.4).

11.11 EXAMPLE—ALLOWABLE-STRESS DESIGN OF A THROUGH PLATE-GIRDER BRIDGE

Two simply supported, welded, through plate girders carry the single track of a railroad bridge on an 86-ft span (Fig. 11.46). The girders are spaced 23.75 ft c to c. The track is on an 8° curve, for which the maximum design speed is 30 mph. Maximum offset of centerline of track for centerline of bridge is 2.12 ft. Live loads from the trains are distributed by ties, ballast, and a Grade 50W steel ballast plate to rolled-steel floorbeams spaced 2.5 ft c to c. These beams transmit the loads to the girders. Steel to be used is Grade 36. Loading is Cooper E65.

11.11.1 Design of Floorbeams

For convenience in computing maximum moment, the dead load on a floorbeam may be considered to consist of three parts: weight of track and load-distributing

TABLE 11.56 Dead Load on Floorbeam, kips per ft

Track: $0.200 \times 2.5/18.5$	$= 0.027$
Tie: $0.160/18.5$	$= 0.009$
Ballast: $0.120 \times 1 \times 2.5$	$= 0.300$
Bituminous concrete: $0.150 \times 2.5 \times 4.5/12$	$= 0.140$
¾-in ballast plate: 0.0306×2.5	$= \underline{0.077}$
Load over 18.5 ft:	0.553
Beam—assume:	0.080
Concrete curb: $0.150 \times 2.5 \times 2.5 \times 2 = 1.9$ kips	

material, spread over about 18.5 ft; weight of floorbeam and connections, distributed over the span, which is taken as 23.5 ft; and weight of concrete curb, which is treated as a concentrated load (Table 11.56). This loading produces a reaction

$$R_{DL} = \frac{0.553 \times 18.5}{2} + \frac{0.080 \times 23.5}{2} + 1.9 = 7.9 \text{ kips}$$

Maximum bending moment occurs at midspan and equals

$$M_{DL} = 7.9 \times 11.75 - 1.9 \times 10.50 - 0.553 \times 9.25 \times 4.63$$
$$- 0.080 \times 11.75 \times 5.88$$
$$= 44 \text{ ft-kips}$$

The live load P, kips, carried by the floorbeam can be computed from

$$P = 1.15AD/S \qquad S \geq d \tag{11.42}$$

where A = axle load, kips
$\quad S$ = axle spacing, ft
$\quad D$ = effective beam spacing, ft
$\quad d$ = actual beam spacing, ft

with D taken equal to d. The axle load $A = 65$ kips, and the axle spacing $S = 5$ ft.

$$P = \frac{1.15 \times 65 \times 2.5}{5} = 37.4 \text{ kips}$$

$P/2 = 18.7$ kips is applied as a concentrated load at each rail. Then loads cause a reaction

$$R_{LL} = \frac{18.7(11.98 + 7.27)}{23.5} = 15.3 \text{ kips}$$

Maximum moment occurs under a rail and equals

$$M_{LL} = 15.3 \times 11.52 = 176.5 \text{ ft-kips}$$

Impact is taken as 36% of wheel live-load stresses. Impact moment then is

$$M_I = 0.36 \times 176.5 = 63.5 \text{ ft-kips}$$

The total moment is 284 ft-kips. This requires a section modulus

$$S = \frac{284 \times 12}{20} = 170 \text{ in}^3$$

Use a W24 × 76, with $S = 176$ in³.

Maximum live-load floorbeam reaction is $37.4 - 15.3 = 22.1$ kips. The maximum floorbeam reaction is

$$R = 7.9 + 22.1 + 22.1 \times 0.36 = 38.0 \text{ kips}$$

11.11.2   Design of Girders

The girders will be made of Grade 36 steel. Simply supported, they span 86 ft. They will be made identical.

Dead Load.   Most of the load carried by each girder is transmitted to it by the floorbeams as concentrated loads. Computations are simpler, however, if the floorbeams are ignored and the girder is treated as if it received load from the ballast plate. Moments and shears computed with this assumption are sufficiently accurate for design purposes because of the relatively close spacing of the floorbeams. Thus, the dead load on the girder may be considered uniformly distributed. It is computed to be 3.765 kips per ft.

Maximum dead-load moment occurs at midspan and equals

$$M_{DL} = \frac{3.765(86)^2}{8} = 3{,}500 \text{ ft-kips}$$

Dead-load moments along the span are listed in Table 11.57. Maximum dead-load shear and the reaction is

$$R_{DL} = \frac{3.765 \times 86}{2} = 162 \text{ kips}$$

Live Load.   Computation of live-load moments, shears, and reactions is simplified with the aid of tables or charts. (See, for example, D. B. Steinman, "Locomotive Loadings for Railway Bridges," *ASCE Transactions*, vol. 86, pp. 606–723, and J. F. Leppmann and J. J. Kozak, "Bridge Engineering, Standard Handbook for Civil

TABLE 11.57   Moments, ft-kips, in Girders

Distance from supports, ft		Dead load, M_{DL}	Equivalent E10 loading q	$3.83q$	Live load M_{LL}, $3.83qL_1L_2/2$	Impact M_r, $0.175 \times 6.5qL_1L_2/2$	Centrifugal force C
L_1	L_2						
15	71	2,010	1.40	5.37	2,860	850	180
30	56	3,170	1.33	5.09	4,280	1,370	280
40	46	3,470	1.34	5.12	4,720	1,400	310
43	43	3,500	1.33	5.09	4,720	1,400	310

Engineers," 3d ed., McGraw-Hill Book Company, New York.) If figures are available for any magnitude of Cooper E loading, those for any other magnitude can be obtained by proportion.

Since the tracks are not centered between the girders, one girder will be more heavily loaded than the other. The amount of live load transmitted to the more heavily loaded girder may be obtained by taking moments about the other girder. Let q be the equivalent uniform live load for E10 loading, kips per ft. Then, $6.5q$ is the equivalent load for E65, and the girder receives $6.5q \times 14.0/23.75 = 3.83q$. Live-load moments along the span are listed in Table 11.57.

Maximum reaction and shear under E10 loading is 66.1 kips. Hence, the maximum reaction for E65 loading is

$$R_{LL} = 66.1 \times 3.83 = 254 \text{ kips}$$

Impact is taken as 17.5% of the axle live-load stresses. Therefore, the impact moment is

$$M_I = 0.175 \times 6.5 \times 1.33 \times 43^2/2 = 1,400 \text{ ft-kips}$$

and the impact reaction is

$$R_I = 0.175 \times 6.5 \times 1.33 \times 86/2 = 65 \text{ kips}$$

Centrifugal Force. This is computed as a percentage of live load, with speed $S = 30$ mph and degree of curve $D = 8°$.

$$C = 0.00117S^2D = 0.00117(30)^28 = 8.43\%$$

Application of this percentage to the equivalent load producing maximum moment yields the equivalent centrifugal force

$$C_e = 0.0843 \times 1.33 \times 6.5 = 0.73 \text{ kip per ft}$$

This force acts 6 ft above top of rail, or 10.8 ft above bottom of girder. It is resisted by a couple consisting of a vertical force at each girder equivalent to

$$q_C = \frac{0.73 \times 10.8}{23.75} = 0.332 \text{ kip per ft}$$

The maximum moment produced by these forces can be obtained by proportion from the maximum live-load moment:

$$M_C = \frac{4,720 \times 0.332}{5.09} = 310 \text{ ft-kips}$$

Similarly, the maximum shear and reaction equal

$$R_C = \frac{254 \times 0.336}{5.09} = 17 \text{ kips}$$

Longitudinal Force. The longitudinal force from trains should be taken as 15% of the live load, without impact, acting at base of rail. Application of this percentage to the equivalent load producing maximum reaction yields the equivalent longitudinal force

$$q_L = 0.15 \times 1.54 \times 6.5 = 1.50 \text{ kips per ft}$$

Since the rails will be continuous across the bridge, the longitudinal forces will be $1.50 \times 86/1,200 = 0.11$ kips per ft. This imposes on the girder a horizontal force of

$$H_L = \frac{0.11 \times 86 \times 14}{23.75} = 5.6 \text{ kips}$$

Because this force acts at top of rail, about 4 ft above the bottom of the girder, it causes a moment. This is resisted by a couple composed of a force at each support of the girder:

$$R_L = \frac{5.6 \times 4}{86} = 0.3 \text{ kips}$$

Wind Transverse to Bridge.   The wind may act on live load and structure in any horizontal direction. The wind load on the train should be taken as a moving load of 0.3 kip per ft, acting 8 ft above top of rail. Wind load on the structure should be taken as 0.03 ksf, acting on 1.5 times the vertical projection of the span.

Transverse to the bridge, wind on the live load, acting 12.8 ft above the bottom of the girder, imposes vertical forces on the girder of $0.3 \times 12.8/23.75 = 0.162$ kip per ft. This causes a midspan bending moment of

$$M_{WLL} = \frac{0.162(86)^2}{8} = 150 \text{ ft-kips}$$

Maximum shear and reaction equal

$$R_{WLL} = \frac{0.162 \times 86}{2} = 7 \text{ kips}$$

In addition, acting at each of the four girder supports is a transverse horizontal force

$$H_{WLL} = \frac{0.3 \times 86}{4} = 6.5 \text{ kips}$$

Transverse wind on a projection of 9.3 ft of structure imposes a load of

$$0.030 \times 9.3 \times 1.5 = 0.420 \text{ kip per ft}$$

It acts about 4.7 ft above the bottom of the girder. The resulting overturning moment causes vertical forces in the girders of $0.420 \times 4.7/23.75 = 0.083$ kip per ft. These forces produce a midspan bending moment

$$M_W = \frac{0.083(86)^2}{8} = 77 \text{ ft-kips}$$

The reaction is

$$R_W = \frac{0.083 \times 86}{2} = 3.6 \text{ kips}$$

Also, a transverse horizontal force acts at each of the four girder supports:

$$H_W = \frac{0.42 \times 86}{4} = 9 \text{ kips}$$

Longitudinal Wind.   Longitudinal wind on the live load transmitted to the girder equals $0.3 \times 14/23.75 = 0.18$ kip per ft. Acting 12.8 ft above the bottom of the girder, it imposes vertical and horizontal longitudinal forces at the supports:

$$R_{WL} = \frac{0.18 \times 86 \times 12.8}{86} = 2.3 \text{ kips}$$

$$H_{WL} = 0.18 \times 86 = 15.5 \text{ kips}$$

Similarly, the longitudinal wind on the structure is $0.420/2 = 0.210$ kip per ft per girder. It imposes vertical and horizontal longitudinal forces at the supports of

$$R_{WL} = \frac{0.21 \times 86 \times 4.7}{86} = 1.0 \text{ kip}$$

$$H_{WL} = 0.21 \times 86 = 18 \text{ kips}$$

Wind on Unloaded Bridge.   The structure also should be investigated for a transverse wind load of 50 psf on 1.5 times the vertical projection of the span. Moments and shears caused can be obtained by proportion from those previously computed.

$$M_W = \frac{77 \times 50}{30} = 128 \text{ ft-kips}$$

$$R_W = \frac{3.6 \times 50}{30} = 6 \text{ kips}$$

Loading Combinations.   Three loading combinations are investigated:

Case I. $DL + LL + I + C$ at full basic allowable stresses (Table 11.58)

Case II. Case I + wind on loaded bridge + longitudinal force at 125% of basic allowable stresses (Table 11.59)

Case III. Dead load + wind on unloaded bridge at 125% of basic allowable stresses (Table 11.60).

Moments and shears for case I are larger than those for case III and, when allowance is made for a 25% increase in allowable stresses for case II, also larger than those for case II. Hence, case I at basic allowable stresses governs the design.

The curve of maximum moments at various points of the span, or moment envelope (Fig. 11.47), can now be plotted for case I.

TABLE 11.58 Loading Case I—Maximum Moments and Shears

Type of loading	Moment, ft-kips	Shear, kips
DL	3,500	162
LL	4,720	254
I	1,400	65
C	310	17
Total	9,930	498

TABLE 11.59 Loading Case II—Maximum Moments and Shears

Type of loading	Moment, ft-kips	Shear, kips
Case I	9,930	498
Wind	230	11
LF	. . .	4
Total	10,160	513

TABLE 11.60 Loading Case III—Moments and Shears

Type of loading	Moment, ft-kips	Shear, kips
DL	3,500	162
Wind on bridge	128	6
Total	3,628	168

Web Size.   While depth of web has no effect on vertical clearances under through-girder bridges, it has several effects on economy. The deeper the girders, the less flange material required and the stiffer the members. But web thickness and number of stiffeners required usually increase. Also, girder spacing may have to be increased, because of wider gussets or knee braces needed for lateral bracing.

In this example, a web depth of 106 in is assumed. With an allowable stress of 12.5 ksi, the web thickness required for shear is

$$t = \frac{504}{12.5 \times 106} = 0.380 \text{ in}$$

To prevent buckling, however, even with transverse stiffeners, thickness should be at least $1/160$ of the clear distance between flanges.

$$t = \frac{106}{160} = 0.663 \text{ in, say } {}^{11}/_{16} \text{ in}$$

Use a $106 \times {}^{5}/_{8}$-in web.

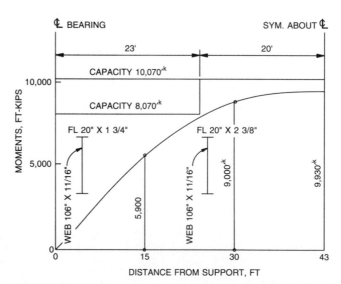

FIGURE 11.47   Moment envelope for through girder and capacities of cross sections.

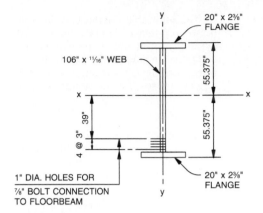

FIGURE 11.48 Cross section of through girder at midspan.

Flange Size at Midspan.   To select a trial size for the flange, assume an allowable bending stress $F_b = 20$ ksi and distance between flange centroids of about 110 in. Then, for a maximum moment of 9,930 ft-kips, the required area of one flange is about

$$A_f = \frac{9930 \times 12}{110 \times 20} = 54 \text{ in}^2$$

Assume a $20 \times 2\frac{3}{8}$-in plate for each flange, with an area of 47.5 in². Width-thickness ratio of outstanding portion is $10/2.38 = 4.2$, which is less than 12 permitted. The trial section is shown in Fig. 11.48. Moment of inertia of the section about the X axis is calculated in Table 11.61. Distance from neutral axis to top or bottom of the girder is 55.4 in. Hence, the gross and net section moduli are

$$S_g = \frac{347,100}{55.4} = 6,270 \text{ in}^3 \qquad S_{net} = \frac{340,700}{55.4} = 6.150 \text{ in}^3$$

The allowable tensile bending stress is 20 ksi. The actual tensile stress is

$$f_b = \frac{9,930 \times 12}{6,150} = 19.4 < 20 \text{ ksi}$$

TABLE 11.61   Moment of Inertia of Through Plate Girder at Midspan

Material	A	d	Ad² or I_o
2 flanges 20 × 2⅜	95.0	54.19	278,900
Web 106 × ¹¹⁄₁₆			68,200
			$I_g = 347,100 \text{ in}^4$
4 holes: $-1(\frac{5}{8})(39^2 + 42^2 + 45^2 + 48^2 + 51^2)$			$= -6,400$
			$I_{net} = 340,700 \text{ in}^4$

The allowable compressive bending stress is a function of l, the distance, in, between points of lateral support of the compression flange, and r_y, the radius of gyration, in, of the compression flange and that portion of the web area on the compression side of the axis of bending, about the axis in the plane of the web. For a rectangular section, $r = 0.289d$, where d = depth of section perpendicular to axis. Hence, for the compression flange,

$$r_y = 0.289 \times 20 = 5.78 \text{ in}$$

AREA specifications limit the spacing of lateral supports for the compression flange to a maximum of 12 ft for through girders. Since the knee braces are placed at floorbeam locations, which are 30 in apart, space the knee braces 10 ft = 120 in c to c. Then, the allowable compressive stress is the larger of the following:

$$F_b = 20 - 0.0004 \left(\frac{l}{r_y}\right)^2 = 20 - 0.0004 \left(\frac{120}{5.78}\right)^2 = 19.83 \text{ ksi}$$

$$F_b = \frac{10,500A_f}{ld} = \frac{10,500 \times 47.5}{120 \times 110.75} = 37.5 \text{ ksi}$$

but not to exceed 20 ksi. The actual compressive stress is

$$f_b = \frac{9,930 \times 12}{6,270} = 19.0 < 19.83 \text{ ksi}$$

The section is satisfactory. Moment capacity supplied is

$$M_C = \frac{20 \times 6,150}{12} = 10,070 \text{ ft-kips}$$

Intermediate Transverse Stiffeners.   For the web, the depth-thickness ratio $d/t = 106/(^{11}/_{16}) = 154$. This exceeds the AREA limit of 60 for an unstiffened web. Transverse stiffeners are required and spacing should not exceed $d = 332t/\sqrt{f_v} \leq 72$ in, where f_v is the shear stress, ksi.

$$f_v = \frac{498}{72.9} = 6.83 \text{ ksi}$$

For this shear, $d = 332(^{11}/_{16})/\sqrt{6.83} = 87 \text{ in} > 72$ in. Use a stiffener spacing of 60 in. Try a pair of plates at each location, with the width equal at least to $D/30 + 2 = {}^{106}/_{30} + 2 = 5.5 \text{ in} < 16$ in. Use $6 \times {}^3/_8$-in plates welded to the web for the intermediate stiffeners.

Change in Flange Size.   At a sufficient distance from midspan, the bending moment decreases sufficiently to permit reducing the thickness of the flange plates to $1^3/_4$ in. The net moment of inertia reduces to 265,000 in⁴ and the section modulus to 4,840 in³. Thus, with $20 \times 1^3/_4$-in flange plates, the section has a moment capacity of

$$M_C = \frac{20 \times 4,840}{12} = 8,070 \text{ ft-kips}$$

When this is plotted in Fig. 11.47, the horizontal line representing it stays above the moment envelope until within 20 ft of midspan. Hence, flange size can be decreased at that point. Length of the 2⅜-in plate then is 40 ft and of the 1¾-in plates, which extend to the end of the girder, 23 ft.

The flange plates will be spliced with complete-penetration groove welds. For calculation of fatigue stresses, the welded connection is Stress Category B and for this span of less than 100 ft should be designed for 2,000,000 cycles of loading. The allowable stress range is 14.5 ksi. The actual stress range for live loads plus impact is estimated to be

$$f_r = \frac{5{,}200 \times 12}{4{,}840} = 12.9 \text{ ksi} < 14.5 \text{ ksi—OK}$$

Flange-to-Web Welds.   AREA specifications require that the flange plates be connected to the web with continuous, full-penetration groove welds.

Knee Braces.   A knee brace with solid web (Fig. 11.49*a*) braces the compression flange of the girder at 10-ft intervals. Attached with bolts to the top of the floorbeam and welded to a girder stiffener, the brace extends from the floorbeam to the top flange of the girder, and from the web of the girder outward a maximum of 36 in. The outer edge is cut to a slope of 3 on 1. (Some railroads prefer a maximum slope of 2.8 on 1.) The length of this edge is 75 in.

Assume a ½-in-thick plate for the web. Since the 75-in length of the edge exceeds 60 × ½ = 30 in, the edge is stiffened with a 6 × ½-in plate. This plate is considered to act with 6 in of the web in transmitting the buckling force to the floorbeam (Fig. 11.49*b*). This force is assumed horizontal and equal to 2.5% of the force in the 20 × 2⅜-in compression flange. With a compressive stress in the flange of 19.0 ksi, the force to be resisted is

$$F = 0.025 \times 20 \times 2.375 \times 19.0 = 23 \text{ kips}$$

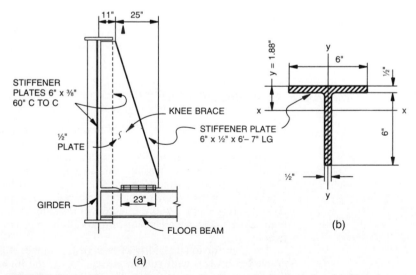

(a)

(b)

FIGURE 11.49   Knee brace for compression flange of through girder. (*a*) Elevation. (*b*) Cross section assumed effective.

The T section in Fig. 11.49*b* therefore is subjected, because of the 3 on 1 slope, to a force

$$P = 23 \times \tfrac{79}{25} = 72.7 \text{ kips}$$

Area of the T is $2 \times 6 \times \frac{1}{2} = 6$ in². Distance of the neutral axis from the outer surface of the flange is

$$y = \frac{3 \times 0.25 + 3 \times 3.5}{6} = 1.88 \text{ in}$$

Moments of inertia are computed to be $I_x = 24.83$ in⁴ and $I_y = 9.00$ in⁴. The latter governs. Thus, the least radius of gyration is

$$r_y = \sqrt{\frac{9.0}{6}} = 1,227 \text{ in}$$

The slenderness ratio then is $79/1.227 = 65$. Hence, treated as a column, the T section has an allowable compressive stress of

$$F_a = 21.5 - \frac{0.1kL}{r} = 21.5 - 0.1 \times 65 = 15.0 \text{ ksi}$$

And the brace has a capacity of

$$P = 15.0 \times 6 = 90.0 > 72.7 \text{ kips}$$

Therefore, the knee brace is satisfactory.

The number of $\frac{7}{8}$-in-dia. high-strength bolts required to transmit the 23-kip horizontal force to the floorbeam, with a capacity of 9.3 kips per bolt, is $23/(2 \times 9.3) = 2$. For sealing, however, the maximum bolt spacing is $4 + 4t = 4 + 4 \times \frac{1}{2} = 6$ in. Sealing controls. Use five $\frac{7}{8}$-in bolts 5 in c to c and two angles $4 \times 4 \times \frac{1}{2}$ in by 23 in long.

Other Details. Stiffeners are designed and located in the same way as for deck plate girders (Art. 11.9). They should be placed at floorbeams, but need not be at every beam. Other details also are treated in the same way as for plate girders.

11.12 COMPOSITE BOX-GIRDER BRIDGES

Box girders have several favorable characteristics that make their use desirable for spans of about 120 ft and up. Structural steel is employed at high efficiency, because a high percentage can be placed in wide flanges where the metal is very effective in resisting bending. Corrosion resistance is higher than in plate-girder and rolled-beam bridges. For, with more than half the steel surface inside the box, less steel, especially corners, which are highly susceptible, is exposed to corrosive influences. Also, the box shape is more effective in resisting torsion than the I shape used for plate girders and rolled beams. In addition, box girders offer an attractive appearance.

The high torsional rigidity of box girders makes this type of construction preferable for bridges with curved girders. Also, the high rigidity assists the deck in distributing loads transversely. This is illustrated in Fig. 11.50. A single load placed off center on a bridge with single-web girders is carried mainly by nearby girders.

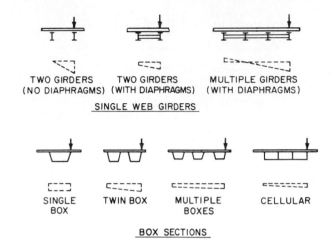

FIGURE 11.50 Comparison of lateral load distribution for single-web girders and box girders.

But similarly placed on a box-girder bridge, the load is supported nearly equally by all the girders. The effect of the deck is ignored in this illustration.

Depending on its width, a bridge may be supported on one or more box girders. Each girder may comprise one or more cells. For economy in long-span construction, the cells may be made wide and deep. Width, for example, may be 12 ft or more. Usual thickness of the concrete deck, however, generally limits spacing of the girder webs to about 10 ft and cantilevers to about 5 ft. Consequently, thicker slabs are justified to take advantage of the economies accruing from wider girder cells.

Some designers have found it advantageous to use an alternative scheme with narrow box girders. They place a pair of boxes near the roadway edges and distribute the loads to these girders through longitudinal stringers and transverse floorbeams, as is done in plate-girder construction (Art. 11.9). See, for example, Fig. 11.58.

Box girders may be simply supported or continuous. Since they generally are used principally in long spans, continuity is highly desirable for economy and increased stiffness. Also, use of high-strength steels is advantageous in the longer spans.

Box girders are adaptable to composite and orthotropic-plate construction. (The latter is discussed in Arts. 11.14 and 11.15.) With composite construction, only a narrow top flange is needed with each web. The flanges usually need be only wide enough for load distribution to the web and to provide required clearances and edge distances for welded shear connectors. Figures 11.51 and 11.52 show several types of box-girder bridges that have been constructed with and without composite construction.

Boxes may be rectangular or trapezoidal. (Triangular boxes with apex down have been used, but they have several disadvantages. They usually have to be deeper than rectangular or trapezoidal boxes. Also, because of smaller area, triangular boxes have less torsional resistance. Furthermore, the bottom flange often has to be a heavy built-up section, complicated by bent plates for connecting to the webs.) With trapezoidal boxes, fewer girders may be required, but a thicker bottom plate or thicker concrete slab may be needed than for rectangular boxes. Fabrication costs for either shape are about the same.

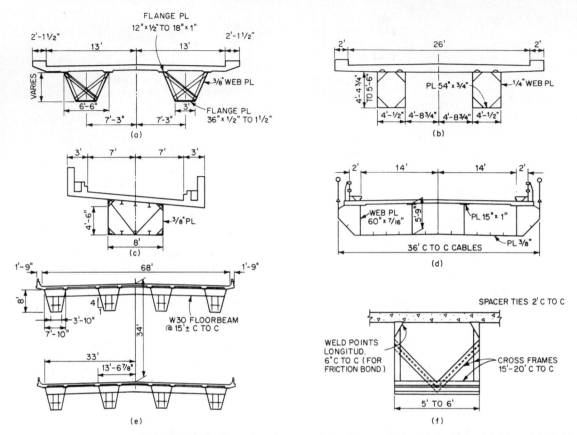

FIGURE 11.51 Examples of cross sections of composite box-girder highway bridges. (*a*) Rigid-frame construction with inclined legs, over Stillaguamish River. Spans are 50–160–85 ft and 200 ft c to c of leg pins. (*b*) Boxes with corners trussed for rigidity, in 110-ft span, King County, Wash. (*c*) Ramp with minimum horizontal radius of 67 ft and continuous spans of 58.5–52–73 ft, in Port Authority of New York and New Jersey Bus Terminal. (*d*) Suspension-bridge spans of 170–430–170 ft over Klamath River, Orleans, Calif. (*e*) Double-deck approach spans of 170–170 ft, Fremont Bridge, Portland, Ore. (*f*) Box UV girder proposed by Homer Hadley. V-shaped troughs formed by corner plates atop the webs are filled with concrete to secure composite action with concrete deck.

Construction costs for box-girder bridges often are kept down by shop fabrication of the boxes. Thus, designers should bear in mind the limitations placed by shipping clearances on the width of box girders as well as on length and depths. If the girders are to be transported by highway, and single box girders with widths exceeding about 12 ft are required, use of more but narrower girders may be more economical.

11.13 EXAMPLE—ALLOWABLE-STRESS DESIGN OF A COMPOSITE, BOX-GIRDER BRIDGE

Following is an example to indicate the design procedure for a bridge with box girders composite with a concrete deck. The procedure does not differ greatly from

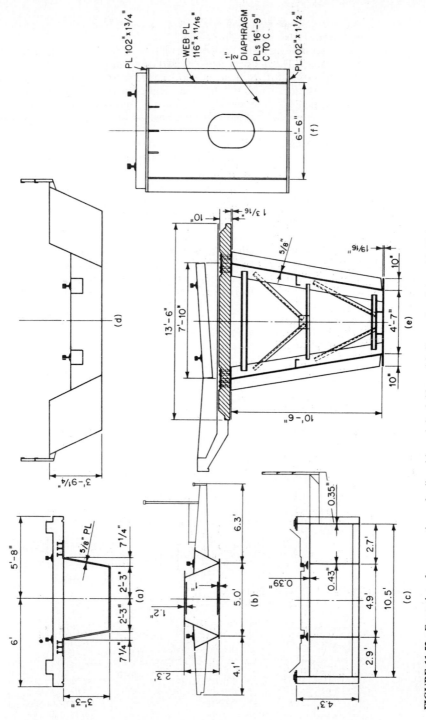

FIGURE 11.52 Examples of cross sections of railroad box-girder bridges. (*a*) Composite girder with 85-ft span for elevated rapid-transit system, San Francisco, Calif. (*b*) Rails bear directly on steel box of 56-ft span over Autobahn. Kirchwcyhe, Bremen, Germany. (*c*) Three-cell box over Alsstrasse, M-Gladbach, Germany, with 80-ft span. (*d*) Rigid-frame construction with 117-ft span carries track with radius of 1780 ft, near Frankfurt, Germany. (*e*) Precast-concrete deck bolted to girder for composite action in 152.5-ft span, Czechoslavakia. (*f*) Typical section of four 122-ft spans in Chester, Pa.

that for a single-web plate girder with composite deck (Art. 11.4). The example incorporates the major differences.

A two-lane highway bridge with simply supported, composite box girders will be designed. The deck is carried by two trapezoidal girders (Fig. 11.53). Top width of each box is 8 ft 6 in, as is the distance c to c of adjacent top flanges of the girders. Bottom width is 5 ft 10 in. Thus, the webs have a slope of 4 on 1. The girders span 120 ft. Structural steel to be used is Grade 36. Loading is HS20-44. Appropriate design criteria given in Sec. 10 will be used for this structure.

Concrete Slabs. The general design procedure outlined in Art. 11.2 for slabs on rolled beams also holds for slabs on box girders. A 7.5-in-thick concrete slab will be used with the box girders.

Design Criteria. AASHTO "Standard Specifications for Highway Bridges" apply to single-cell box girders where width c to c between top steel flanges is approximately equal to the distance c to c of adjacent top steel flanges of adjacent box girders. (The distance c to c of flanges of adjacent boxes should be between 0.8 and 1.2 times the distance c to c of the flanges of each box.) In this example, both the width and spacing equal 8 ft 6 in. Also, the deck overhang must not exceed 6 ft or 60% of the spacing. In this example, the overhang of 5.25 ft is nearly equal to $0.60 \times 8.5 = 5.1$ ft. Hence, AASHTO specifications for composite box girders may be used.

Loads, Moments, and Shears. Assume that the girders will not be shored during casting of the concrete slab. Hence, the dead load on each girder includes the

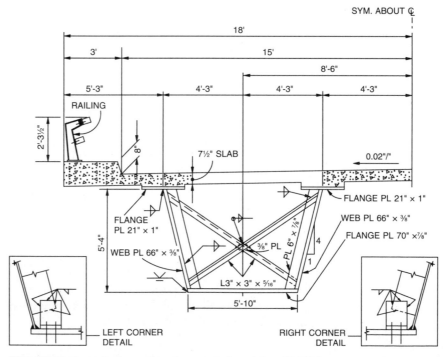

FIGURE 11.53 Half cross section of composite box girder with 120-ft span.

TABLE 11.62 Dead Loads, kips per ft, on Box Girder

(a) Dead load on steel box		(b) Dead load on composite section	
Slab: 0.150 × 18 × 7.5/12	= 1.69	Railing:	0.02
Haunches: 2 × 0.150 × 2 × 1/12	= 0.05	Safety walk: 0.150 × 3 × 8.38/12 =	0.31
Girder and framing details—assume:	0.60	SDL per girder:	0.33
DL per girder:	2.34		

weight of the 18-ft-wide half of the deck as well as weights of steel girders and framing details. This dead load will be referred to as DL (Table 11.62a). Maximum moment occurs at the center of the 120-ft span and equals

$$M_{DL} = \frac{2.34(120)^2}{8} = 4{,}210 \text{ ft-kips}$$

Maximum shear occurs at the supports and equals

$$V_{DL} = \frac{2.34 \times 120}{2} = 140.4 \text{ kips}$$

Railings and safety walks will be placed after the concrete slab has cured. This superimposed dead load will be designated SDL (Table 11.62b). Maximum moment occurs at midspan and equals

$$M_{SDL} = \frac{0.33(120)^2}{8} = 590 \text{ ft-kips}$$

Maximum shear occurs at supports and equals

$$V_{SDL} = \frac{0.33 \times 120}{2} = 19.8 \text{ kips}$$

The HS20-44 live load imposed may be a truck load or lane load. But for this span, truck loading governs. The center of gravity of the three axles lies between the two heavier loads and is 4.66 ft from the center load. Maximum moment occurs under the center axle load when its distance from midspan is the same as the distance of the center of gravity of the loads from midspan, or 4.66/2 = 2.33 ft. Thus, the center load should be placed $^{120}\!/_2 - 2.33 = 57.67$ ft from a support. Then, the maximum moment is

$$M_T = \frac{72(120/2 + 2.33)^2}{120} - 32 \times 14 = 1{,}880 \text{ ft-kips}$$

Under AASHTO specifications, the live-load bending moment for each girder is determined by applying to the girder the fraction W_L of a wheel load (both front and rear) as given by

$$W_L = 0.1 + 1.7R + \frac{0.85}{N_w} \tag{11.43}$$

where $R = N_w/N$, with $0.5 \leq R \leq 1.5$
$N_w = W_c/12$, reduced to nearest whole number
N = number of box girders
W_c = roadway width, ft, between curbs or between barriers if curbs are not used

In this example, $W_c = 30$, $N_w = {}^{30}\!/_{12} = 2$, $N = 2$, and $R = {}^{2}\!/_{2} = 1$. Therefore,

$$W_L = 0.1 + 1.7 \times 1 + \frac{0.85}{2} = 2.225 \text{ wheels} = 1.113 \text{ axles}$$

So the maximum live-load moment is

$$M_{LL} = 1.113 \times 1{,}880 = 2{,}110 \text{ ft-kips}$$

Though this moment does not occur at midspan as do the maximum dead-load moments, stresses due to M_{LL} may be combined with those from M_{DL} and M_{SDL} to produce the maximum stress, for all practical purposes.

For maximum shear with the truck load, the outer 32-kip load should be placed at the support. Then, the shear is

$$V_T = \frac{72(120 - 14 + 4.66)}{120} = 66.4 \text{ kips}$$

On the assumption that the live-load distribution is the same as for bending moment, the maximum live-load shear is

$$V_{LL} = 1.113 \times 66.4 = 73.8 \text{ kips}$$

Impact is taken as the following fraction of live-load stress:

$$I = \frac{50}{L + 125} = \frac{50}{120 + 125} = 0.204$$

Hence, the maximum moment due to impact is

$$M_I = 0.204 \times 2{,}100 = 430 \text{ ft-kips}$$

and the maximum shear due to impact is

$$V_I = 0.204 \times 73.8 = 15.1 \text{ kips}$$

MIDSPAN BENDING MOMENTS, FT-KIPS			END SHEAR, KIPS			
M_{DL}	M_{SDL}	$M_{LL} + M_I$	V_{DL}	V_{SDL}	$V_{LL} + M_I$	Total V
4,210	590	2530	140.4	19.8	88.9	249.1

Properties of Composite Section.   The 7.5-in thick roadway slab includes an allowance of 0.5 in for a wearing surface. Hence, the effective thickness of the concrete slab for composite action is 7 in. Half the width of the deck, 18 ft = 216 in, is considered to participate in the composite action with each box girder.

A trial section for a girder is assumed as shown in Fig. 11.54. Its neutral axis can be located by taking moments of web and flange areas about a horizontal axis at middepth of the web. This computation and those for the section moduli S_{st} and S_{sb} of the steel section alone are conveniently tabulated in Table 11.63. The moment of inertia of each inclined web I_x may be computed from

$$I_x = \frac{s^2}{s^2 + 1} I \qquad (11.44)$$

where s = slope of web with respect to horizontal axis
$\quad I$ = moment of inertia of web with respect to axis at middepth normal to the web = $ht^3/12$
$\quad h$ = depth of web in its plane
$\quad t$ = web thickness normal to its plane

In computation of the properties of the composite section, the concrete slab, ignoring the haunch area, is transformed into an equivalent steel area. For the purpose, for this bridge, the concrete area is divided by the modular ratio $n = 10$ for short-time loading, such as live loads and impact. For long-time loading, such as dead loads, the divisor is $3n = 30$, to account for the effects of creep. The computations of neutral-axis location and section moduli for the composite section are tabulated in Table 11.64. To locate the neutral axis, moments are taken about middepth of the girder webs.

Stresses in Composite Section. Since the girders will not be shored when the concrete is cast and cured, the stresses in the steel section for load *DL* are determined with the section moduli of the steel section alone (Table 11.63). Stresses for load *SDL* are computed with the section moduli of the composite section when $n = 30$ (Table 11.64*a*). And stresses in the steel for live loads and impact are calculated with section moduli of the composite section when $n = 10$ (Table 11.64*b*). Calculations for the stresses are given in Table 11.65.

The width-thickness ratio of the unstiffened compression flanges now can be checked, as for an I shape, by the general formula applicable for any stress level:

$$\frac{b}{t} = \frac{194}{\sqrt{F_y}} = \frac{194}{\sqrt{36}} = 32 > \frac{21}{1}$$

Hence, the trial section is satisfactory.

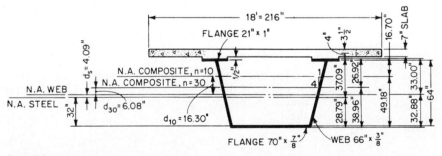

FIGURE 11.54 Locations of neutral axes of steel box alone and of composite box section.

TABLE 11.63 Steel Section for Maximum Moment in Box Girder

Material	A	d	Ad	Ad^2	I_o	I
Two top flanges 21 × 1	42.0	32.50	1,364	44,400		44,400
Two webs 66 × ⅜	49.5				16,900	16,900
Bottom flange 70 × ⅞	61.3	−32.44	−1,990	64,600		64,600
	152.8		−626			125,900

$d_s = -626/152.8 = -4.09$ in

$-4.09 \times 626 = -2,600$

$I_{NA} = 123,300$

Distance from neutral axis of steel section to:

$$\text{Top of steel} = 32 + 1.00 + 4.09 = 37.09 \text{ in}$$

$$\text{Bottom of steel} = 32 + 0.88 - 4.09 = 28.79 \text{ in}$$

Section moduli	
Top of steel	Bottom of steel
$S_{st} = 123,300/37.09 = 3,320$ in³	$S_{sb} = 123,300/28.79 = 4,260$ in³

Stresses in the concrete are determined with the section moduli of the composite section with $n = 30$ for SDL (Table 11.54a) and $n = 10$ for $LL + I$ (Table 11.54b). Therefore, the composite section is satisfactory. With the thickness specified for maximum moment, no changes in flange thicknesses are desirable. Use the section shown in Fig. 11.54 throughout the span.

Check of Web.   The 64-in vertical projection of the webs satisfies the requirements that the depth-span ratio for girder plus slab exceed 1:25 and for girder alone 1:30. The depth-thickness ratio of each web is $66/0.375 = 176$. This is close enough to the AASHTO specifications limiting requirement of $D/t \leq 170$ to be acceptable without longitudinal stiffeners. For the maximum compressive bending stress of 18.24 ksi, the maximum depth-thickness ratio permitted with transverse stiffeners but without longitudinal stiffeners is

$$\frac{D}{t} = \frac{727}{\sqrt{f_b}} = \frac{727}{\sqrt{18.24}} = 170$$

The design shear for the inclined web V_w equals the vertical shear V_v divided by the cosine of the angle of inclination θ of the web plate to the vertical. For a maximum shear of 249.1 kips and a slope of 4 on 1 (cos $\theta = 0.97$), the design shear is

$$V_w = \frac{V_v}{\cos \theta} = \frac{249.1}{0.97} = 257 \text{ kips}$$

TABLE 11.64   Composite Section for Maximum Moment in Box Girder

(a) For dead loads, $n = 30$						
Material	A	d	Ad	Ad^2	I_o	I
Steel section	152.8		-626			125,900
Concrete $216 \times 7/30$	50.4	37.0	1,863	68,900	200	69,100
	203.2		1,237			195,000

$d_{30} = 1,237/203.2 = 6.08$ in $\qquad\qquad\qquad -6.08 \times 1,237 = \quad -7,500$

$$I_{NA} = \quad 187,500$$

Distance from neutral axis of composite section to:

$$\text{Top of steel} = 33.00 - 6.08 = 26.92 \text{ in}$$

$$\text{Bottom of steel} = 32.88 + 6.08 = 38.96 \text{ in}$$

$$\text{Top of concrete} = 26.92 + 0.50 + 7 = 34.42 \text{ in}$$

Section moduli		
Top of steel	Bottom of steel	Top of concrete
$S_{st} = 187,500/26.92$	$S_{sb} = 187,500/38.96$	$S_c = 187,500/34.42$
$= 6,970$ in^3	$= 4,820$ in^3	$= 5,450$ in^3

(b) For live loads, $n = 10$						
Material	A	d	Ad	Ad^2	I_o	I
Steel section	152.8		-626			125,900
Concrete $216 \times 7/10$	151.2	37.0	5,588	206,700	600	207,300
	304.0		4,962			333,200

$d_{10} = 4,962/304.0 = 16.30$ in $\qquad\qquad\qquad -16.30 \times 4,962 = \quad -80,900$

$$I_{NA} = \quad 252,300$$

Distance from neutral axis of composite section to:

$$\text{Top of steel} = 33.00 - 16.30 = 16.70 \text{ in}$$

$$\text{Bottom of steel} = 32.88 + 16.30 = 49.18 \text{ in}$$

$$\text{Top of concrete} = 16.70 + 0.50 + 7 = 24.20 \text{ in}$$

Section moduli		
Top of steel	Bottom of steel	Top of concrete
$S_{st} = 252,300/16.70$	$S_{sb} = 252,300/49.18$	$S_c = 252,300/24.20$
$= 15,100$ in^3	$= 5,140$ in^3	$= 10,430$ in^3

TABLE 11.65 Stresses in Composite Box Girder, ksi

(a) Steel stresses

Top of steel (compression)	Bottom of steel (tension)
$DL: f_b = 4,210 \times 12/3,320 = 15.20$	$f_b = 4,210 \times 12/4,260 = 11.84$
$SDL: f_b = 590 \times 12/6,970 = 1.02$	$f_b = 590 \times 12/4,820 = 1.47$
$LL + I: f_b = 2,530 \times 12/15,100 = 2.02$	$f_b = 2,530 \times 12/5,080 = 5.92$
Total: 18.24 < 20	19.23 < 20

(b) Stresses at top of concrete

$$SDL: f_c = 590 \times 12/(5,450 \times 30) = 0.04$$
$$LL + I: f_c = 2,530 \times 12/(10,430 \times 10) = 0.29$$
$$\text{Total:} \qquad 0.33 < 1.0$$

With a cross-sectional area of 49.5 in², the web will be subjected to shearing stress considerably below the 12 kips permitted.

$$f_v = \frac{257}{49.5} = 5.2 < 12 \text{ ksi}$$

Maximum shears at sections along the span are given in Table 11.66.

Flange-to-Web Welds. Fillet welds placed on opposite sides of each girder web to connect it to each flange must resist the horizontal shear between flange and web. In this example, as is usually the case (see Art 11.4, for example), the minimum size of weld permissible for the thickest plate at the connection determines the size of weld. For both the ⅞-in bottom flange and the 1-in top flanges, the minimum size of weld permitted is ⁵⁄₁₆ in. Therefore, use a ⁵⁄₁₆-in fillet weld on opposite sides of each web at each flange.

TABLE 11.66 Maximum Shear in Composite Box Girder

	Distance from support, ft						
	0	10	20	30	40	50	60
DL, kips	160	134	108	81	54	27	0
LL + I, kips	89	81	73	65	57	49	41
Total, kips	249	215	181	146	111	76	41
f_v, ksi	5.03	4.35	3.66	2.95	2.24	1.54	0.83

Intermediate Transverse Stiffeners.   To determine if transverse stiffeners are required, the allowable shear stress F_v will be computed and compared with the average shear stress $f_v = 5.03$ ksi at the support.

$$F_v = [270(D/t)]^2 = (270/170)^2 = 2.52 \text{ ksi} < 5.03 \text{ ksi}$$

Therefore, transverse intermediate stiffeners are required.

Maximum spacing of stiffeners may not exceed $3 \times 64 = 192$ in or $D[260(D/t)]^2 = 64(260/170)^2 = 150$ in. Try a stiffener spacing $d_o = 90$ in. This provides a depth-spacing ratio $D/d_o = 64/90 = 0.711$. From Eq. (10.30d), for use in Eq. (10.31a), $k = 5[1 + (0.711)^2] = 7.53$ and $\sqrt{k/F_y} = \sqrt{7.53/36} = 0.457$. Since $D/t = 170$, C in Eq. (10.30a) is determined by the parameter $170/0.457 = 372 > 237$. Hence, C is given by

$$C = \frac{45,000k}{(D/t)^2 F_y} = \frac{45,000 \times 7.53}{170^2 \times 36} = 0.326$$

From Eq. (10.31a), the maximum allowable shear for $d_o = 90$ in is

$$F_v' = F_v \left[C + \frac{0.87(1 - C)}{\sqrt{1 + (d_o/D)^2}} \right]$$

$$= \frac{36}{3} \left[0.326 + \frac{0.87(1 - 0.326)}{\sqrt{1 + (90/64)^2}} \right] = 7.99 \text{ ksi} > 5.03 \text{ ksi}$$

Since the allowable stress is larger than the computed stress, the stiffeners may be spaced 90 in apart.

The AASHTO standard specifications limit the spacing of the first intermediate stiffener to the smaller of $1.5D = 1.5 \times 64 = 96$ in and the spacing for which the allowable shear stress in the end panel does not exceed

$$F_v = CF_y/3 = 0.326 \times 36/3 = 3.91 \text{ ksi} < 5.03 \text{ ksi}$$

Therefore, closer spacing is needed near the supports. Try $d_o = 45$ in, for which $k = 15.11$, $C = 0.654$, and $F_v = CF_y/3 = 0.654 \times {}^{36}\!/_3 = 7.85$ ksi > 5.03. Therefore, 45-in spacing will be used near the supports and 90-in spacing in the next 22.5 ft of girder, as shown in Fig. 11.55. Transverse stiffeners are omitted from the central 60 ft of girder, except at midspan.

Where required, a single plate stiffener of Grade 36 steel will be welded inside the box girder to each web. Minimum width of stiffeners is one-fourth the flange width, or $21/4 = 5.25 > 2 + 66/30 = 4.2$ in. Use a 6-in-wide plate. Minimum thickness required is $6/16 = \frac{3}{8}$ in. Try $6 \times \frac{3}{8}$-in stiffeners.

The moment of inertia provided by each stiffener must satisfy Eq. (10.27), with J as given by Eq. (10.28).

$$J = 2.5 \left(\frac{64}{90} \right)^2 - 2 = -0.73 \qquad \text{use } 0.5$$

$$I = 90(3/8)^3 0.5 = 2.37$$

The moment of inertia furnished is

$$I = \frac{(3/8)6^3}{3} = 27 > 2.37 \text{ in}^4$$

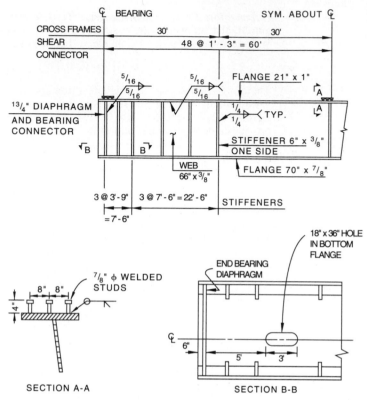

FIGURE 11.55 Locations of stiffeners, cross frames, and shear connectors for composite box girder.

Hence, the 6 × ⅜-in stiffeners are satisfactory. Weld them to the webs with a pair of ¼-in fillet welds.

Bearing Stiffeners.   Instead of narrow-plate stiffeners and a cross frame over the bearings, a plate diaphragm extending between the webs is specified. The plate diaphragm has superior resistance to rotation, displacement, and distortion of the box girder. Assume for the diaphragm a bearing length of 20 in at each web, or a total of 40 in. The allowable bearing stress is 29 ksi. Then, the thickness required for bearing is

$$t = \frac{249}{40 \times 29} = 0.22 \text{ in}$$

But the thickness of a bearing stiffener also is required to be at least

$$t = \frac{b'}{12} \sqrt{\frac{F_y}{33}} = \frac{20}{12} \sqrt{\frac{36}{33}} = 1.74 \text{ in}$$

Therefore, use a plate 64 × 1¾ in extending between the webs at the supports, with a 30-in-square access hole.

The welds to the webs must be capable of developing the entire 249-kip reaction. Minimum-size fillet weld for the 1¾-in diaphragm is 5⁄16 in. With two such welds at each web, their required length, with an allowable stress of 15.7 ksi, is

$$\frac{249}{4(5⁄16)0.707 \times 15.7} = 8 \text{ in}$$

Weld the full 66-in depth of web.

Shear Connectors.   To ensure composite action of concrete deck and box girders, shear connectors welded to the top flanges of the girders must be embedded in the concrete (Art 10.17). For this structure, ⅞-in-dia. welded studs are selected. They are to be installed in groups of three at specified locations to resist the horizontal shear between the steel section and the concrete slab (Fig. 11.55). With height $H = 4$ in, they satisfy the requirement $H/d \geq 4$, where d = stud diameter, in.

With $f'_c = 2,800$ psi for the concrete, the ultimate strength of a ⅞-in welded stud is, from Eq. (11.26),

$$S_v = 0.4d^2\sqrt{f'_c E_c} = 0.4(⅞)^2\sqrt{2.8 \times 2,900} = 27.6 \text{ kips}_,$$

This value is needed for determining the number of shear connectors required to develop the strength of the steel girder or the concrete slab, whichever is smaller. With an area $A_s = 152.8$ in², the strength of the girder is

$$P_1 = A_s F_y = 152.8 \times 36 = 5,500 \text{ kips}$$

The compressive strength of the concrete slab is

$$P_2 = 0.85f'_c bt = 0.85 \times 2.8 \times 216 \times 7 = 3,600 < 5,500 \text{ kips}$$

Concrete strength governs. Hence, from Eq. (11.25), the number of studs provided between midspan and each support must be at least

$$N_1 = \frac{P_1}{\phi S_v} = \frac{3,600}{0.85 \times 27.6} = 153$$

With the studs placed in groups of three, therefore, there should be at least 153⁄3 = 51 groups on each half of the girder.

Pitch is determined by fatigue requirements. The allowable load range, kips per stud, may be computed from Eq. (11.4). With $\alpha = 10.6$ for 500,000 cycles of load (AASHTO specifications),

$$Z_r = 10.6(0.875)^2 = 8.12 \text{ kips per stud}$$

At the supports, the shear range $V_r = 89$ kips, the shear produced by live load plus impact. Consequently, with $n = 10$ for the concrete, and the transformed concrete area equal to 151.2 in², and $I = 252,300$ in⁴ from Table 11.64b, the range of horizontal shear stress is

$$S_r = \frac{V_r Q}{I} = \frac{89 \times 151.2 \times 20.70}{252,300} = 1.107 \text{ kips per in}$$

Hence, the pitch required for stud groups near the supports is

$$p = \frac{3Z_r}{S_r} = \frac{3 \times 8.12}{1.107} = 22 \text{ in}$$

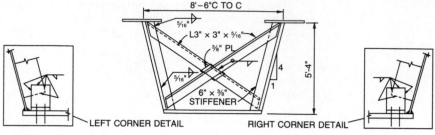

FIGURE 11.56   Intermediate cross frame.

Use a pitch of 14⅜ in to satisfy both this requirement and that for 51 groups of studs between midspan and each support (Fig. 11.55).

Intermediate Cross Frames.   Though intermediate cross frames or diaphragms are not required by standard specifications, it is considered good practice by many designers to specify such interior bracing in box girders to help maintain the shape under torsional loading. So in addition to the end bearing diaphragm, cross frames will be installed at 30-ft intervals. Minimum-size angles can be used (Fig. 11.56).

Camber.   The girders should be cambered to compensate for dead-load deflections under DL and SDL. For computation for DL, the moment of inertia I of the steel section alone should be used. For SDL, I should apply to the composite section with $n = 30$ (Table 11.64a). Both deflections can be computed from Eq. (11.5) with $w_{DL} = 2.34$ kips per ft and $w_{SDL} = 0.33$ kip per ft.

$$DL: \delta = 22.5 \times 2.34(120)^4/(29{,}000 \times 123{,}300) = 3.04 \text{ in}$$
$$SDL: \delta = 22.5 \times 0.33(120)^4/(29{,}000 \times 187{,}500) = \underline{0.29} \text{ in}$$
$$\text{Total:} \qquad\qquad\qquad\qquad\qquad\qquad\qquad\qquad\qquad 3.33 \text{ in}$$

Live-Load Deflection.   Maximum live-load deflection should be checked to ensure that it does not exceed $12L/800$. This deflection may be obtained with acceptable accuracy from Eq. (11.6), with

$$P_T = 8 \times 1.113 + 0.204 \times 8 \times 1.113 = 10.73 \text{ kips}$$

From Table 11.64b, for $n = 10$, $I = 252{,}300$ in⁴. Therefore,

$$\delta = \frac{324 \times 10.73}{29{,}000 \times 252{,}300}(120^3 - 555 \times 120 + 4{,}780) = 0.79 \text{ in}$$

And the deflection-span ratio is

$$\frac{0.79}{120 \times 12} = \frac{1}{1{,}800} < \frac{1}{800}$$

Thus, the live-load deflection is acceptable.

Other Details.   These may be treated in the same way as for I-shaped plate girders.

11.14   ORTHOTROPIC-PLATE GIRDER BRIDGES

In orthotropic-plate construction, a steel-plate deck is used instead of concrete. The plate is topped with a wearing surface that may or may not be concrete. The

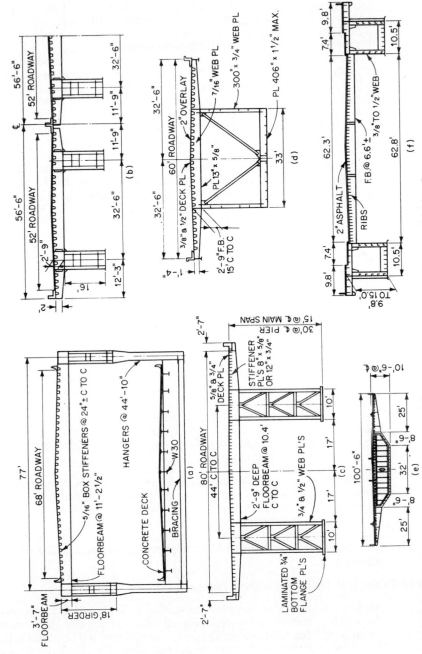

FIGURE 11.57 Examples of cross sections of orthotropic-plate highway bridges. (*a*) Fremont Bridge, Portland, Ore, incorporates continuous tied-arch spans of 448.3–255.3–448.3 ft. (*b*) Poplar St. Bridge over the Mississippi River at St. Louis has continuous spans of 300–500–600–500–265 ft. (*c*) San Mateo-Heyward Bridge over San Francisco Bay, Calif., contains three cantilever-type spans of 375–750–375 ft with a 375-ft suspended span and 187.5-ft cantilevers. (*d*) San Diego-Coronado Bridge over San Diego Bay, Calif., provides continuous spans of 600–600–500 ft. (*e*) Wye River Bridge, England, cable-stayed, spans 285–770–285 ft. (*f*) Severin River Bridge over the Rhine River, Cologne, Germany, also cable-stayed, has spans of 161–292–157–990–494–172 ft.

steel-plate deck serves the usual deck function of distributing loads to main carrying members, but it also acts as the top flange of those members (Art. 10.21). Because the deck provides a large area, orthotropic-plate construction is very efficient in resisting bending. With a lightweight wearing surface, furthermore, bridges of this type have relatively low dead load, a characteristic particularly important for keeping down the costs of long spans. Figure 11.57 shows some examples of cross sections that have been used for orthotropic-plate bridges.

These examples indicate that orthotropic plates often are used with box girders. In addition to low dead weight, this type of construction offers many of the advantages of composite box girders discussed in Art. 11.12. The examples, however, are all long-span bridges. It may also be economical for medium spans to use orthotropic plates with girders with inverted-T shapes.

For relatively simple types of orthotropic-plate bridges, such as those with the type of cross section shown in Fig. 11.58, approximate analyses by the Pelikan-Esslinger method (Art. 4.12) or similar methods give sufficiently accurate results for practical purposes. For more complex types, more accurate analyses may be desirable. These can be executed with the aid of computers. Special attention should be given to the stability of deep webs and wide flanges.

(T. G. Galambos, "Guide to Stability Design Criteria for Metal Structures," John Wiley & Sons, Inc., New York.)

11.15 EXAMPLE—DESIGN OF AN ORTHOTROPIC-PLATE BOX-GIRDER BRIDGE

Following is an example to indicate the design procedure for a bridge with box girders and an orthotropic-plate deck. The example incorporates the major differences in method from that for a single-web plate girder (Art. 11.4) and composite box girder (Art. 11.13).

A two-lane highway bridge with two simply supported box girders, inverted-T floorbeams, and trapezoidal, longitudinal ribs will be designed. The girders are rectangular in section, with webs 30 in c to c. They span 120 ft. The typical cross section in Fig. 11.58 shows a 30-ft roadway flanked by 3-ft-wide safety walks. The transverse floorbeams, spanning 30 ft between the girders, are spaced 15 ft c to c. The ribs, spaced 2 ft c to c, are continuous. Structural steel to be used for the box girders is Grade 36. Loading is H20-44. Appropriate design criteria in Sec. 10 will be used for this structure.

Design Procedure.   The Pelikan-Esslinger method of approximate analysis will be used. The bridge will be considered to comprise three systems, as defined in Art. 10.21. They contain four members (Art. 4.12): I—the plate; II—plate and rib; III—plate and floorbeam; and IV—girder and plate, including ribs. Member IV can be designed by conventional procedure and is treated first. The other members are designed in accordance with special theory applicable to orthotropic-plate construction.

11.15.1 Design of Girder with Orthotropic-Plate Flange

Member IV consists of one steel box section spanning 120 ft, seven ribs, and half the width of the steel-plate deck (Fig. 11.58). The ribs and this ⅜-in-thick plate act as part of the top flange of the girder. In addition, a 42 × ½ in plate atop the two webs of the box section also forms part of the top flange. Note that the ⅜-in plate, 180 in wide, slopes upward from the girder toward midspan at 0.02 in per in, or a total of 3.60 in. Consequently, because the ribs are welded to this plate, the center of gravity of the ribs rises an average of 2.30 in.

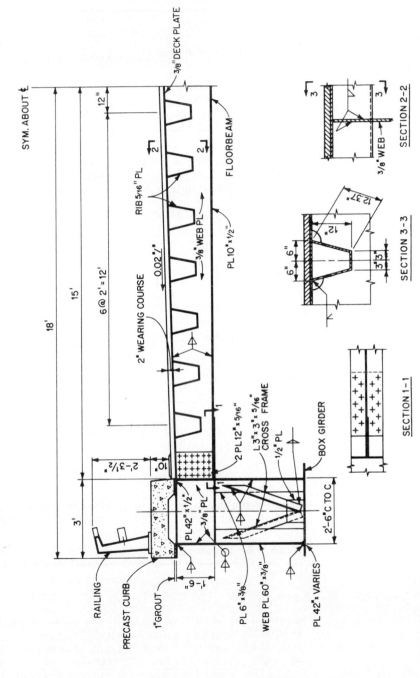

FIGURE 11.58 Half cross section of orthotropic-plate bridge with 120-ft span.

TABLE 11.67   Dead Load on Member IV, kips per ft

Railing:	0.025
Curb: $0.150 \times 3.0 \times 0.83$	$= 0.375$
Wearing course: $0.150 \times 15 \times \frac{2}{12}$	$= 0.375$
Deck plate: 0.153×15	$= 0.230$
Ribs: 7×0.0329	$= 0.230$
Floorbeams—assume:	0.050
Girder and details—assume:	0.500
DL per girder:	1.785, say 1.8

A $60 \times \frac{3}{8}$-in plate is assumed for each web, and a 42-in plate, $1\frac{1}{8}$ in thick, is selected for the bottom flange at midspan. Each rib is trapezoidal, a $\frac{5}{16}$-in bent plate, 12 in wide at the top, 6 in wide at the bottom (Figs. 11.58 and 11.61).

Loads, Moments, and Shears.   Member IV is subjected to a dead load consisting of its own weight and the weights of framing details, floorbeams, railing, curb, and wearing course (Table 11.67).

Dead-load moments and shears at 10-ft intervals along the span are listed in Tables 11.68 and 11.69.

The H20-44 live load imposed may be a truck load or a lane load. But for this span, lane loading governs. For bending moment, the 0.64-kip-per-ft uniform load should cover the entire span. For maximum moment at any point on the span, an 18-kip concentrated load should be placed at that point. For maximum shear at any point, a 26-kip concentrated load should be placed there, and the uniform load should extend from the point to the farthest support. For example, maximum moment occurs at midspan with the 18-kip load placed there and the uniform load over the entire span:

$$M_T = \frac{0.64(120)^2}{8} + \frac{18 \times 120}{4} = 1{,}152 + 540 = 1{,}692 \text{ ft-kips}$$

Maximum shear occurs at a support with the 26-kip load placed there and the uniform load over the entire span:

$$V_T = \frac{0.64 \times 120}{2} + 26 = 38.4 + 26 = 64.4 \text{ kips}$$

In distributing the two lanes of live load to the girders, the deck is restrained against rotation to some extent by the girders. It may be assumed to be somewhere between

TABLE 11.68   Moments in Orthotropic-Plate Box Girder, ft-kips

	Distance from support					
	10	20	30	40	50	60
M_{DL}	990	1,800	2,430	2,880	3,150	3,240
M_{LL}	660	1,200	1,610	1,910	2,090	2,150
M_I	130	240	330	390	430	440
Total	1,780	3,240	4,370	5,180	5,670	5,830

TABLE 11.69 Shears in Orthotropic-Plate Box Girder, kips

	Distance from support						
	0	10	20	30	40	50	60
V_{DL}	108	90	72	54	36	18	0
V_{LL}	82	71	60	99	38	27	17
V_I	17	15	12	10	8	6	3
Total	207	176	144	113	82	51	20
f_v, ksi	4.60	3.91	3.20	2.52	1.82	1.13	0.44

simply supported and fixed. In this example, either assumption gives about the same result. For the assumption of a simple support, with one truck wheel 2 ft from a girder, the load distributed by the deck to that girder is

$$W = \frac{28 + 22 + 16 + 10}{30} = 2.53 \text{ wheels}$$

$$= 1.27 \text{ axles}$$

Then, the maximum moment at midspan due to live load is

$$M_{LL} = 1.27 \times 1,692 = 2,150 \text{ ft-kips}$$

Similarly, maximum live-load shear at the support and maximum reaction is

$$V_{LL} = 1.27 \times 64.4 = 81.8 \text{ kips}$$

Maximum live-load moments and shears at 10-ft intervals along the span are listed in Tables 11.68 and 11.69.

Impact for maximum moments and maximum shear and reaction at the supports may be taken as the following fraction of live-load stress:

$$I = \frac{50}{L + 125} = \frac{50}{120 + 125} = 0.204$$

Impact moments and shears also are given in Tables 11.68 and 11.69.

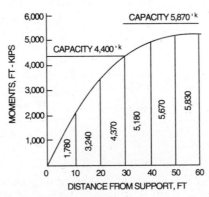

FIGURE 11.59 Moment envelope for girder (member IV).

These tables, in addition, give the total moments and shears at 10-ft intervals. The moments are used to plot the curve of maximum moments (moment envelope) in Fig. 11.59.

Web Size. Minimum depth-span ratio for a girder is 1:25. For a 120-ft span, therefore, the girders should be at least $120/25 = 4.8$ ft $= 58$ in deep. Use webs 60 in deep.

With an allowable shear stress of 12 ksi for Grade 36 steel, the thickness required for shear in each web is

$$t = \frac{207}{2 \times 12 \times 60} = 0.144 \text{ in}$$

With stiffeners, the web thickness, however, to prevent buckling, should be at least $\frac{1}{165}$ of the depth, or 0.364 in. Use two $60 \times \frac{3}{8}$-in webs, with an area of 45 in².

Check of Section at Midspan. The assumed girder section at midspan is shown in Fig. 11.58, with the bottom flange taken as $1\frac{1}{8}$ in thick. (See also Fig. 11.60.) The neutral axis is located by taking moments of areas of the components about middepth of the webs. This computation and those for moment of inertia and section moduli are given in Table 11.70. The neutral axis is found to be 12.36 in above middepth of web, based on the gross section. (Area of bolt holes is less than 15% of the flange area.)

Maximum bending stress occurs in the bottom flange and equals

$$f_b = \frac{5{,}830 \times 12}{3{,}520} = 19.9 < 20 \text{ ksi}$$

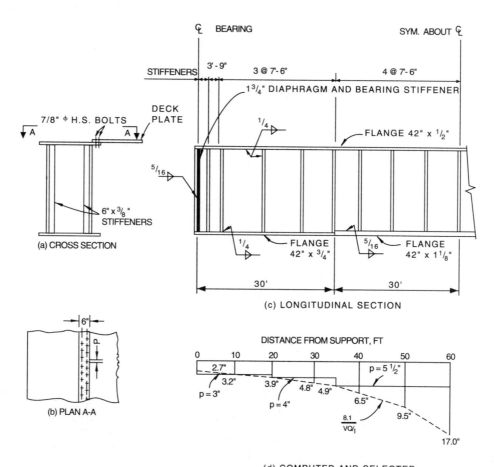

FIGURE 11.60 Details of box girder with orthotropic-plate top flange.

TABLE 11.70 Maximum Moment at Midspan of Box Girder with Orthotropic-Plate Flange

Material	A	d	Ad	Ad^2	I_o	I
Deck plate 180 × ⅜	67.50	32.49	2,192	71,300		71,300
Seven ribs	65.87	25.25	1,664	42,100		42,100
Top flange 42 × ½	21.00	30.25	635	19,200		19,200
Two webs 60 × ⅜	45.00				13,500	13,500
Bottom flange 42 × 1⅛	47.25	− 30.56	− 1,443	44,400		44,400
	246.62		3,048			190,500

$d = 3,048/246.62 = 12.36$ in

$$-12.36 \times 3,048 = -37,600$$
$$I_{NA} = 152,900$$

Distance from neutral axis to:

$$\text{Top of steel} = 30.50 + 0.38 + 0.02 \times 180 - 12.36 = 22.12 \text{ in}$$

$$\text{Bottom of steel} = 31.13 + 12.36 = 43.49 \text{ in}$$

$$\text{Bottom of rib} = 22.12 - 0.38 - 12 - 0.02 \times 12 = 9.50 \text{ in}$$

Section moduli		
Top of steel	Bottom of steel	Bottom of rib
$S_t = 152,900/22.12$	$S_b = 152,900/43.49$	$S_r = 152,900/9.50$
$= 6,910 \text{ in}^3$	$= 3,520 \text{ in}^3$	$= 16,100 \text{ in}^3$

The section is satisfactory. Moment capacity supplied is

$$M_C = \frac{20 \times 3,520}{12} = 5,870 \text{ ft-kips}$$

Change in Bottom Flange. At a sufficient distance from midspan, the bending moment decreases enough to permit reducing the thickness of the bottom flange to ¾ in. As indicated in Table 11.71, the moment of inertia reduces to 121,400 in⁴ and the section modulus of the bottom flange to 2,640 in³.

With the ¾-in bottom flange, the section has a moment capacity of

$$M_C = \frac{20 \times 2,640}{12} = 4,400 \text{ ft-kips}$$

When this is plotted on Fig. 11.59, the horizontal line representing it stays above the moment envelope until within 30 ft of midspan. Hence, flange size can be decreased at that point. Length of the 1⅛-in bottom-flange plate then is 60 ft and of the ¾-in plate, which extends to the end of the girder, 30 ft.

The flange plates will be spliced with complete-penetration groove welds. If these welds are ground smooth in the direction of stress and a transition slope of 1 to 2½ is provided for change in plate thickness, the connection falls in Stress Category B for fatigue calculations. The bridge may be treated as a redundant-load-path structure subjected to 500,000 cycles of loading. Hence, the allowable

TABLE 11.71 Section near Supports of Box Girder with Orthotropic-Plate Flange

Material	A	d	Ad	Ad^2	I
Midspan section	246.62		3,048		190,500
Flange decrease 42 × ⅜	− 15.75	− 30.94	487	− 15,100	− 15,100
	230.87		3,535		175,400

$d = 3,535/230.87 = 15.30$ in

$$-15.30 \times 3,535 = -54,000$$
$$I_{NA} = 121,400$$

Distance from neutral axis to:

Top of steel $= 30.50 + 0.38 + 0.02 \times 180 - 15.30 = 19.18$ in

Bottom of steel $= 30.75 + 15.30 = 46.05$ in

Bottom of rib $= 19.18 - 0.38 - 12 - 12 \times 0.02 = 6.56$ in

	Section moduli	
Top of steel	Bottom of steel	Bottom of rib
$S_t = 121,400/19.18$	$S_b = 121,400/46.05$	$S_r = 121,400/6.56$
$= 6,330$ in³	$= 2,640$ in³	$= 18,500$ in³

stress range $F_r = 29$ ksi. The stress range for live loads plus impact is

$$f_r = \frac{1,940 \times 12}{2,640} = 8.82 \text{ ksi} < 29 \text{ ksi—OK}$$

Flange-to-Web Welds. Each flange will be connected to each web by a fillet weld on opposite sides of the web. These welds must resist the horizontal shear between flange and web. Computations can be made, as for the plate-girder stringers in Art. 11.4, but minimum size of weld permissible for the thickest plate at the connection governs. Therefore, use ¼-in welds with the ½-in top flange, the ¾-in bottom flange, and ⁵⁄₁₆-in welds with the 1⅛-in bottom flange (Fig. 11.60c).

Bending Stresses in Deck. As part of member IV, the deck plate is subjected to the following maximum bending stresses:

At midspan	At flange change
$f_b = 5,830 \times 12/6,910 = 10.12$ ksi	$f_b = 4,370 \times 12/6,330 = 8.28$ ksi

At midspan, maximum stress at the bottom of a rib is

$$f_b = \frac{5,830 \times 12}{16,100} = 4.35 \text{ ksi}$$

Intermediate Transverse Stiffeners. To determine if transverse stiffeners are required, the allowable shear stress F_v will be compared with the average shear stress $f_v = 4.60$ ksi at the support. Web depth-thickness ratio $D/t = 60/(\⅜) = 160$.

$$F_v = [270/(D/t)]^2 = (270/160)^2 = 2.85 \text{ ksi} < 4.60 \text{ ksi}$$

Therefore, transverse intermediate stiffeners are required.

Maximum spacing of stiffeners may not exceed $3 \times 60 = 180$ in or $D[260/(D/t)]^2 = 60(260/160)^2 = 158$ in. Try a stiffener spacing $d_o = 90$ in. This provides a depth-spacing ratio $D/d_o = 60/90 = 0.667$. From Eq. (10.30d), for use in Eq. (10.31a), $k = 5[1 + (0.667)^2] = 7.22$ and $\sqrt{k/F_y} = \sqrt{7.22/36} = 0.445$. Since $D/t = 160$, C in Eq. (10.30a) is determined by the parameter $160/0.445 = 357 > 237$. Consequently, C is given by

$$C = \frac{45,000k}{(D/t)^2 F_y} = \frac{45,000 \times 7.22}{160^2 \times 36} = 0.352$$

From Eq. (10.31a), the maximum allowable shear for $d_o = 90$ in is

$$F'_v = F_v \left[C + \frac{0.87(1 - C)}{\sqrt{1 + (d_o/D)^2}} \right]$$

$$= \frac{36}{3} \left[0.352 + \frac{0.87(1 - 0.352)}{\sqrt{1 + (90/60)^2}} \right] = 7.98 \text{ ksi} > 4.60 \text{ ksi}$$

Since the allowable stress is larger than the computed stress, the stiffeners may be spaced 90 in apart.

The AASHTO standard specifications limit the spacing of the first intermediate stiffener to the smaller of $1.5D = 1.5 \times 60 = 90$ in and the spacing for which the allowable shear stress in the end panel does not exceed

$$F_v = CF_y/3 = 0.352 \times 36/3 = 4.22 \text{ ksi} < 4.60 \text{ ksi}$$

Therefore, closer spacing is needed near the supports. Try $d_o = 45$ in, for which $k = 13.89$, $C = 0.674$, and $F_v = CF_y/3 = 0.674 \times 36/3 = 8.09$ ksi > 4.60 ksi. Therefore, 45-in spacing will be used near the supports and 90-in spacing for the rest of the girder (Fig. 11.60c).

Where required, a single, vertical plate stiffener of Grade 36 steel will be welded inside each box girder to each web. (Longitudinal stiffeners are not required, since the ⅜-in web thickness exceeds $D\sqrt{f_b}/727 = 60\sqrt{19.1}/727 = 0.362$ in.) Width of transverse stiffeners should be at least

$$2 + \frac{D}{30} = 2 + \frac{60}{30} = 4 \text{ in}$$

Use a 6-in-wide plate. Minimum thickness required is ⁹⁄₁₆ = ⅜ in. Try $6 \times$ ⅜-in stiffeners.

The moment of inertia provided by each stiffener must satisfy Eq. (10.27), with J as given by Eq. (10.28).

$$J = 2.5 \left(\frac{60}{90} \right)^2 - 2 = -0.89 \text{—use } 0.5$$

$$I = 90(⅜)^3 0.5 = 2.37 \text{ in}^4$$

The moment of inertia furnished is

$$I = \frac{(⅜)6^3}{3} = 27 > 2.37 \text{ in}^4$$

Hence, the 6 × ⅜-in stiffeners are satisfactory. Weld them to the webs with a pair of ¼-in fillet welds.

Bearing Stiffeners.   Use a bearing diaphragm. See Art 11.13.

Intermediate Cross Frames.   Cross frames should be provided at the floorbeams (Fig. 11.58) to maintain the cross-sectional shape of the box girders.

Longitudinal Splice of Deck and Box Girder.   The ⅜-in deck plate is to be attached to the ½-in top flange of the box girder with A325 ⅞-in-dia, high-strength bolts in slip-critical connections with Class A surfaces (Fig. 11.60a). With an allowable stress of 15.5 ksi, each bolt has a capacity of 9.3 kips. The bolts must be capable of resisting the horizontal shear between the top flange and the deck plate. For determination of the pitch of the bolts along the longitudinal splice (Fig. 11.60b), the statical moment Q of the deck-plate area, including the ribs, about the neutral axis of the girder is needed.

For the girder section at midspan,

$$Q = 67.5 \times 20.13 + 65.87 \times 12.89 = 1{,}360 + 850 = 2{,}210 \text{ in}^3$$

Also, for this section, $Q/I = 2{,}210/152{,}900 = 0.0145$. For the section near the supports,

$$Q = 67.5 \times 17.19 + 65.87 \times 9.95 = 1{,}162 + 656 = 1{,}818 \text{ in}^3$$

And for this section, $Q/I = 1{,}818/121{,}400 = 0.0150$.

Multiplication of Q/I by the shear V, kips, yields the shear, kips per in, to be resisted by the bolts. The pitch required then is found by dividing VQ/I into the bolt capacity 9.3. The shears V can be obtained from Table 11.69. The computed pitch is shown by the dash line in Fig. 11.60d, while the pitch to be used is indicated by the solid line. For sealing, the maximum pitch, in, is

$$4 + 4t = 4 + 4 \times \tfrac{3}{8} = 5\tfrac{1}{2} \text{ in}$$

(See also "*Floorbeam Connections to Girders*" on p. 11.153.)

Camber.   The girders should be cambered to compensate for dead-load deflection. In computation of this deflection, an average moment of inertia may be used; in this case, 140,000 in⁴. The deflection can be computed from Eq. (11.5) with $w_{DL} = 1.8$ kips per ft.

$$\delta = \frac{22.5 \times 1.8(120)^4}{29{,}000 \times 140{,}000} = 2.1 \text{ in}$$

Live-Load Deflection.   Maximum live-load deflection should be checked to ensure that it does not exceed $12L/800$. This deflection may be computed with acceptable accuracy for lane loading from

$$\delta = \frac{144ML^2}{EI} \tag{11.45}$$

where $M = M_{LL} + M_I$ at midspan, ft-kips
$\quad\ \ L =$ span, ft
$\quad\ \ I =$ average moment of inertia, in⁴

For $M_{LL} + M_I = 2,380$ ft-kips, from Table 11.68, and an average $I = 140,000$ in⁴,

$$\delta = \frac{144 \times 2,380(120)^2}{29,000 \times 140,000} = 1.21 \text{ in}$$

And the deflection-span ratio is

$$\frac{1.21}{120 \times 12} = \frac{1}{1,190} < \frac{1}{800}$$

Thus, the live-load deflection is acceptable.

Other Details. These may be treated in the same way as for I-shaped plate girders.

11.15.2 Design of Ribs

Select Grade 50W steel for the ribs and deck plate for atmospheric corrosion resistance. This steel has a yield strength of 50 ksi in thicknesses up to 4 in. The trapezoidal section chosen for the ribs is shown in Fig. 11.61.

Stresses in the ribs, and in the deck plate as part of the ribs (member II), may be determined by orthotropic-plate theory (Art. 4.12). In the first stage of the calculations, the ribs are assumed to be continuous and supported by rigid floor-beams. In the second stage, midspan bending moments are increased, because the floorbeams actually are flexible. The decrease in rib moments at the supports, however, is ignored.

Rib Dead Load. Each rib supports its own weight and the weights of framing details, a 24-in-wide strip of deck plate, and a 2-in-thick wearing course (Table 11.72). Because ribs in adjoining spans are subjected to the same uniform loading, each rib may be treated as a fixed-end beam.

Dead-load moment at the support is

$$M_{DL} = -\frac{0.12(15)^2 12}{12} = -27 \text{ in-kips}$$

And at midspan,

$$M_{DL} = \frac{0.12(15)^2 12}{24} = 14 \text{ in-kips}$$

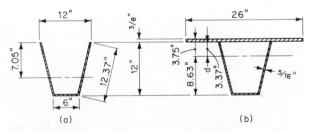

FIGURE 11.61 Sections through rib of orthotropic-plate deck. (*a*) Rib alone. (*b*) Rib with part of plate as top flange.

TABLE 11.72   Dead Load, kips per ft, on Rib

Rib: 30.74 × 0.00106	= 0.0327
Deck: 24 × 0.00128	= 0.0306
Wearing course: 0.150 × 2 × 2/12	= 0.0500
Details:	0.0032
DL per rib:	0.1165, say 0.12

Shear will not be computed because, with two webs per rib, shear stresses are negligible.

Effective Width of Rib Top Flange.   Before live-load moments can be determined for the ribs, certain properties of member II must be computed:

Effective span $s_e = 0.7s = 0.7 \times 15 \times 12 = 126$ in [Eq. (4.166)]

For determination of the effective width of the deck plate as the top flange of member II, take the effective width a_e at top of rib equal to the actual width a, and the effective rib spacing e_e equal to the actual spacing e between ribs. Then, $a_e/s_e = {}^{12}\!/_{126} = 0.1$ and $e_e/s_e = {}^{12}\!/_{126} = 0.1$. From Table 4.6, $a_o/a_e = 1.08$ and $e_o/e_e = 1.08$. Hence, the effective width of the top flange is

$$a_o' + e_o' = 1.08 \times 12 + 1.08 \times 12 = 26 \text{ in}$$

The resulting rib cross section is shown in Fig. 11.61b. The neutral axis can be located by taking moments of the areas of rib and deck plate about the top of rib. This computation and those for moment of inertia and section moduli are given in Table 11.73b.

The effective top-flange width for computing the relative rigidities of floorbeam and rib is 1.1 $(a + e) = 26.4$ in. The section properties given in Table 11.73b will be used, however, because the effect on stresses, in this case, is negligible.

Slenderness of Rib.   From Table 11.73b, the radius of gyration of the rib $r = \sqrt{I_{NA}/A} = \sqrt{376/19.16} = 4.43$ in, and the slenderness ratio is $L/r = 15 \times 12/4.43 = 41$. The yield strength F_y of the rib steel is 50 ksi. The maximum compressive stress in the deck plate F is 10.12 ksi (Art. 11.15.1). The maximum permissible slenderness ratio is

$$\frac{L}{r} = 1000 \sqrt{\frac{1.5}{F_y} - \frac{2.7F}{F_y^2}} = 1000 \sqrt{\frac{1.5}{50} - \frac{2.7 \times 10.12}{50^2}} = 138 > 41\text{---OK}$$

Torsional Rigidity.   The basic differential equation for an orthotropic plate with closed ribs [Eq. (4.178)] contains two parameters, the flexural rigidity D_y in the longitudinal direction and the torsional rigidity H. The latter may be computed from Eq. (4.181). For that computation, for the trapezoidal rib, by Eq. (4.183),

$$K = \frac{(12 + 6)^2 12^2}{(6 + 2 \times 12.37)/0.3125 + 12/0.375} = 357.5$$

With the shearing modulus $G = 11,200$ ksi,

$$GK = 11,200 \times 357.5 = 4.01 \times 10^6$$

TABLE 11.73 Properties of Ribs

(a) Rib without deck plate

Material	A	d	Ad	Ad^2	I_o	I
Two sides 12.37 × 5/16	7.73	6.00	46.4	278	79	357
Flange 5.38 × 5/16	1.68	11.84	19.9	236		236
	9.41		66.3			593
d = 66.3/9.41 = 7.05 in				$-7.05 \times 66.3 =$		-468
					$I_{NA} =$	125

(b) Rib with 26-in plate flange

Material	A	d	Ad	Ad^2	I
Rib alone	9.41		66.3		593
Top flange 26 × 3/8	9.75	−0.1875	−1.8	0.34	0
	19.16		64.5		593
d = 64.5/19.16 = 3.37 in				$-3.37 \times 64.5 =$	-217
				$I_{NA} =$	376

Distance from neutral axis to:

$$\text{Top of deck plate} = 0.375 + 3.37 = 3.75 \text{ in}$$

$$\text{Bottom of rib} = 12 - 3.37 = 8.63 \text{ in}$$

Section moduli

Top of deck plate	Bottom of rib
S_t = 376/3.75 = 100 in^3	S_b = 376/8.63 = 43.6 in^3

Also needed for computing H is the reduction factor v. It can be obtained approximately from Eq. (4.184), with

$$EI_p = \frac{29,000(0.375)^3}{10.92} = 140$$

Thus, the reciprocal of the reduction factor may be taken as

$$\frac{1}{v} = 1 + \frac{4.01 \times 10^6}{140} \frac{12^2}{12(12 + 12)^2} \left(\frac{\pi}{145.8}\right)^2$$

$$\times \left[\left(\frac{12}{12}\right)^3 + \left(\frac{12 - 6}{12 + 6} + \frac{6}{12}\right)^2\right] = 6.61$$

Then, by Eq. (4.181),

$$H = \frac{1}{2} \frac{4.01 \times 10^6}{6.61(12 + 12)} = 12,600$$

Flexural Rigidity.   The other parameter, the flexural rigidity in the longitudinal direction, for the basic differential equation for the orthotropic plate [Eq. (4.178)] can be obtained from Eq. (4.180):

$$D_y = \frac{29,000 \times 376}{12 + 12} = 454,000$$

Plate Parameter.   For use in determination of influence coefficients, the plate parameter is

$$\sqrt{\frac{H}{D_y}} = \sqrt{\frac{12,600}{454,000}} = 0.166$$

Relative Rigidity of Rib and Floorbeam.   For calculating the effects of floorbeam deflections on rib moments, the flexibility coefficient is needed. It can be obtained from Eq. (4.192). For the purpose, the moment of inertia of the rib I_r can be taken as I_{NA} in Table 11.73b, and the moment of inertia of the floorbeam I_f can be taken as the average of the moment of inertia at midspan and that at the ends, from Table 11.80.

$$I_f = \tfrac{1}{2}(2,730 + 1,950) = 2,340 \text{ in}^4$$

Thus, by Eq. (4.192), with $l = 0.7 \times 360 = 252$ in because the floorbeams will be considered fixed at the supports under live loads,

$$\gamma = \frac{(252)^4 376}{\pi^4 (180)^3 (12 + 12) 2,340} = 0.0477$$

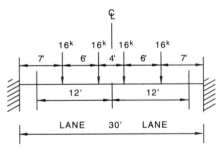

FIGURE 11.62   Positions of truck wheels for maximum moments in ribs and floorbeams.

Live Load as Fourier Series.   The roadway can be divided into two design lanes, centered in the 30-ft roadway. For maximum moments in the ribs, place a truck in each design lane close to the bridge centerline, as indicated in Fig 11.62. (Larger stresses result with the loads moved 1 ft off center, but the effect is small.) For use with the moment-influence coefficients to be computed, this loading has to be converted into a Fourier series. For the two axle loads in Fig. 11.62, Eq. (4.172) can be used for this purpose, with the wheel load $P = 16$ kips, spacing $c = 72$ in, and location of axle centers $x_1 = 120$ in and $x_2 = 240$ in. Because the loading is symmetrical, the integer n should be taken only as odd numbers. The load will be distributed, in this case, over the whole 30-ft roadway.

$$Q_n = \frac{4 \times 16}{360} \cos \frac{72n\pi}{2 \times 360} \left(\sin \frac{120n\pi}{360} + \sin \frac{240n\pi}{360} \right)$$

$$= 0.178 \cos 18°n(\sin 60°n + \sin 120°n) \qquad n = 1, 3, 5, \ldots$$

Substitution of $n = 1$ yields

$$Q_1 = 0.178 \times 0.951(0.866 + 0.866) = 0.293$$

For $n = 3$, the sines become zero, and for $n = 5$, the cosine is zero. Hence,

$$Q_3 = Q_5 = 0$$

For $n = 7$,

$$Q_7 = 0.178(-0.588)(0.866 + 0.866) = -0.181$$

For this example, only two terms of the Fourier series will be used. In general, however, additional terms are required for accuracy, because the computations require subtraction of nearly equal numbers.

Parameters for Influence Coefficients.   The ribs at this stage are considered continuous, with span $s = 180$ in, over rigid supports. For calculation of moment-influence coefficients, the parameter α_n given by Eq. (4.187) is needed.

$$\alpha_n = \frac{n\pi}{l}\sqrt{\frac{H}{D_y}}\sqrt{2} = \frac{n\pi}{360}\,0.166\sqrt{2} = 0.00205n$$

Values of functions of α_n for $n = 1$ and $n = 7$ needed for moment-influence coefficients are tabulated in Table 11.74. Also required is the carry-over factor κ_n given by Eq. (4.188). For this calculation, β_n and k_n are computed in Table 11.74a.

TABLE 11.74   Values of Functions for Computing Influence Coefficients

---

(*a*) Functions of rib span *s*

$\alpha_1 = 0.00205$		$\alpha_7 = 0.1435$	
$\alpha_1 s = 0.370$	$\alpha_1 s/2 = 0.185$	$\alpha_7 s = 2.59$	$\alpha_7 s/2 = 1.29$
$\sinh \alpha_1 s = 0.379$		$\sinh \alpha_7 s = 6.63$	
$\cosh \alpha_1 s = 1.069$	$\cosh \alpha_1 s/2 = 1.017$	$\cosh \alpha_7 s = 6.70$	$\cosh \alpha_7 s/2 = 1.954$
$\tanh \alpha_1 s = 0.354$	$\tanh \alpha_1 s/2 = 0.1730$	$\tanh \alpha_7 s = 0.989$	$\tanh \alpha_7 s/2 = 0.859$
$\coth \alpha_1 s = 2.82$		$\coth \alpha_7 s = 1.011$	

$$\beta_1 = \frac{0.379 - 0.370}{0.379} = 0.0238 \qquad \beta_7 = \frac{6.63 - 2.59}{6.63} = 0.608$$

$$k_1 = \frac{0.370 \times 2.82 - 1}{0.0238} = 1.823 \qquad k_7 = \frac{2.59 \times 1.011 - 1}{0.608} = 2.66$$

(*b*) Functions for *y* = 84

$0.00205 \times 84 = 0.172$	$0.01435 \times 84 = 1.206$
$\sinh 0.172 = 0.1728$	$\sinh 1.206 = 1.520$
$\cosh 0.172 = 1.014$	$\cosh 1.206 = 1.820$

(*c*) Functions for *y* = 90

$0.00205 \times 90 = 0.185$	$0.01435 \times 90 = 1.292$
$\sinh 0.185 = 0.1861$	$\sinh 1.292 = 1.683$
$\cosh 0.185 = 1.017$	$\cosh 1.292 = 1.957$

(*d*) Functions for *y* = 78

$0.00205 \times 78 = 0.160$	$0.01435 \times 78 = 1.12$
$\sinh 0.160 = 0.1607$	$\sinh 1.12 = 1.369$
$\cosh 0.160 = 1.013$	$\cosh 1.12 = 1.696$

$$\kappa_1 = \sqrt{1.823^2 - 1} - 1.823 = -0.30 \qquad \kappa_7 = \sqrt{2.66^2 - 1} - 2.66 = -0.20$$

Moment-Influence Coefficients. Substitution of the computed values in Eq. (4.189) yields the influence coefficients for bending moment at an unyielding support:

$$m_{O1} = -2.500(3.61 \sinh 0.00205y - \cosh 0.00205y - 0.00722y + 1)$$

$$m_{O7} = -61.5(1.04 \sinh 0.01435y - \cosh 0.01435y - 0.00667y + 1)$$

Substitution of the computed values in Eq. (4.190) gives the influence coefficients for bending moment at midspan when supports are rigid:

$$m_{C1} = 240 \sinh 0.00205y - 858(0.1730 \sinh 0.00205y - \cosh 0.00205y + 1)$$

$$m_{C7} = 17.8 \sinh 0.01435y - 12.6(0.859 \sinh 0.01435y - \cosh 0.01435y + 1)$$

Rib Live-Load Moments, Rigid Floorbeams. For maximum moment in a rib at a support with H20 loading, place the 16-kip truck wheels 7 ft from the support on one span ($y = 84$) and the 4-kip wheels 7 ft from the same support on the adjoining span ($y = 84$). With values tabulated in Table 11.74b, the moment-influence coefficients become, for $y = 84$, $m_{O1} = -8.25$ and $m_{O7} = -12.35$.

The bending moments at the support will be computed at the centerline of the bridge, $x = 180$ in. Then, by Eq. (4.169),

$$Q_{nx} = Q_1 - Q_7 + \cdots = 0.293 + 0.181 + \cdots$$

By Eq. (4.191), the moment at the support due to the 16-kip wheel at $y = 84$ is

$$M_O = -24(8.25 \times 0.293 + 12.35 \times 0.181) = -112 \text{ in-kips}$$

Because of the 4-kip wheel in the adjoining span, also at $y = 84$,

$$M_O = -112 \times 4/16 = -28 \text{ in-kips}$$

Thus, the live-load moment at the support is $M_{LL} = -112 - 28 = -140$ in-kips.

For maximum moment at midspan, place the 16-kip wheels there ($y = 90$). The 4-kip wheels will be on the adjoining span 6.5 ft from the support ($y = 78$). The midspan moments also will be computed for $x = 180$. The moment-influence coefficients become, with values from Table 11.74 for $y = 90$, $m_{C1} = 31.8$ and $m_{C7} = 23.9$. Then, for the 16-kip wheels, by Eq. (4.191),

$$M_C = 24(31.8 \times 0.293 + 23.9 \times 0.181) = 328 \text{ in-kips}$$

The effect of the 4-kip wheels on the midspan moment is found in several steps. First, the moment M_O at the support is computed. This requires determination of the moment-influence coefficients for $y = 78$. Then, the carry-over factors are used to calculate the midspan moment:

$$M_C = \frac{M_O(1 - \kappa_n)}{2} \tag{11.46}$$

With values from Table 11.74d, the moment-influence coefficients become $m_{O1} = -5.00$ and $m_{O7} = -12.6$. By Eq. (4.191), for the 4-kip wheels,

$$M_{O1} = -\tfrac{4}{16} \times 24 \times 5.00 \times 0.293 = -8.8 \text{ in-kips}$$

$$M_{O7} = -\tfrac{4}{16} \times 24 \times 12.6 \times 0.181 = -13.7 \text{ in-kips}$$

With Eq. (11.45), the midspan moment due to the 4-kip wheels is found to be

$$M_C = -8.8 \frac{1 + 0.300}{2} - 13.7 \frac{1 + 0.200}{2} = -14 \text{ in-kips}$$

Thus, the live-load moment at midspan with rigid supports is

$$M_{LL} = 328 - 14 = 314 \text{ in-kips}$$

Rib Live-Load Moments, Flexible Floorbeams.   Because the floorbeams actually are not rigid and deflect under live loading, end moments in ribs are less than they would be with rigid supports, and midspan moments are larger. The decrease in support moments in this case will be ignored. The increase in midspan moment, however, will be computed from Eq. (4.176).

From Table 4.4, the influence coefficient for reaction due to a load at midspan is 0.601. The flexibility coefficient has previously been computed to be $\gamma = 0.0477$. From Table 4.7, the influence coefficient for midspan moment in a continuous beam on elastic supports, with load at support 0, is 0.027. Taking into account only the effects of the two supports on either side of midspan, the influence coefficient for change in midspan moment is $2(0.601 \times 0.027) = 0.0324$. By Eq. (4.176), the change in midspan moment is

$$\Delta M_C = 0.293 \times 180 \times 24 \times 0.0324 = 41 \text{ in-kips}$$

Therefore, the live-load moment at midspan with flexible supports is

$$M_{LL} = 314 + 41 = 355 \text{ in-kips}$$

Impact.   For the 15-ft rib span, impact should be taken as 30% of live-load stresses.
At midspan, $M_I = 0.30 \times 355 = 106$ in-kips.
At supports, $M_I = 0.30(-140) = -42$ in-kips.

Maximum Rib Moments.   The design moments previously calculated for member II are summarized in Table 11.75.

In a similar way, stress reversals can be computed for investigation of fatigue stresses.

Rib Stresses.   Section properties for determination of member II stresses are given in Table 11.74*b*. Computations for maximum rib stresses at midspan and supports are given in Table 11.76. The compressive stress at the bottom of member II is augmented by the compressive stress induced when the rib acts as part of the top flange of member IV. This stress was previously computed to be 4.26 ksi. Hence, the total compressive stress at the bottom of the rib is

$$f_b = 4.80 + 4.26 = 9.06 < 1.25 \times 27 \text{ ksi}$$

TABLE 11.75   Rib Moments, in-kips

	M_{DL}	M_{LL}	M_I	Total M
Midspan	14	355	106	475
Supports	-27	-140	-42	-209

TABLE 11.76   Bending Stresses in Ribs

At midspan:	
Top of deck plate	$f_b = 475/100 = 4.75$ ksi (compression)
Bottom of rib	$f_b = 475/43.6 = 10.9 < 27$ ksi (tension)
At supports:	
Bottom of rib	$f_b = 209/43.6 = 4.80$ ksi (compression)

Rib Stability.   Closed ribs of the dimensions usually used have high resistance to buckling. In this case, therefore, there is no need to check the stability of the ribs, especially since compressive bending stresses are low.

11.15.3   Design of Floorbeam with Orthotropic-Plate Flange

Select Grade 50W steel for the floorbeams. This steel provides atmospheric corrosion resistance and has a yield strength $F_y = 50$ ksi in thicknesses up to 4 in.

The floorbeams are tapered. Web depth ranges from 21 in at midspan to 18 in at the box-girder supports (Fig. 11.63). The span is 30 ft. Spacing is 15 ft c to c. The floorbeams are considered simply supported for dead load, fixed end for live loads.

Floorbeam Dead Load.   Each beam supports its own weight and the weights of framing details, a 15-ft-wide strip of deck, including 14 ribs, and a 2-in-thick wearing course (Table 11.77).

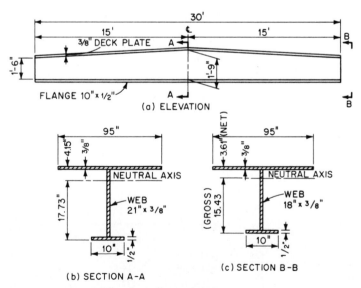

FIGURE 11.63   Floorbeam with orthotropic-plate top flange.

TABLE 11.77 Dead Load on Floorbeam, kips per ft, with Orthotropic-Plate Flange

Floorbeam	= 0.042
Deck plate: 0.153×15	= 0.230
Ribs: $14 \times 0.036 \times 15/30$	= 0.252
Wearing course: $0.150 \times 15 \times 2/12$	= 0.375
Details:	0.026
DL per beam:	0.925

Maximum dead-load moment occurs at midspan and equals

$$M_{DL} = \frac{0.925(30)^2}{8} = 104 \text{ ft-kips}$$

Maximum dead-load shear occurs at the supports and equals

$$V_{DL} = \frac{0.925 \times 30}{2} = 13.9 \text{ kips}$$

Floorbeam Live-Load Moments. Bending moments are computed in two stages. In the first, the floorbeams are assumed to act as rigid supports for the continuous ribs. In the second stage, the changes in moments due to flexibility of the floorbeams are calculated.

For maximum moments in a floorbeam in the first stage, the H20 truck live loads are placed in each of the two design lanes as indicated in Fig. 11.62. The 16-kip wheels are placed over the floorbeam. A 4-kip wheel is located 14 ft away on each of the adjoining rib spans. From Table 4.4, the reaction influence coefficient for this location is estimated to be 0.06. Hence, the wheel loads on the floorbeam equal

$$P = 16 \times 1.00 + 2 \times 4 \times 0.06 = 16.5 \text{ kips}$$

For this loading, the bending moment at the support is

$$M_{LL} = -\frac{16.5}{30^2} [7(23)^2 + 13(17)^2 + 17(13)^2 + 23(7)^2] = -210 \text{ ft-kips}$$

and the midspan moment is

$$M_{LL} = -210 + 2 \times 16.5 \times 13 - 16.5 \times 6 = 110 \text{ ft-kips}$$

The effects on these moments of floorbeam flexibility may be approximated in the following way, using the first term of the Fourier series for the loading in Fig. 11.62: For the 16-kip wheels, $Q_1 = 0.293$. Hence, $Q_1 = 0.293 \times 16.5/16 = 0.302$ for 16.5-kip wheels, and for 4-kip wheels, $Q_1 = 0.293 \times 4/16 = 0.073$. From Table 4.7, for $\gamma = 0.0477$, the reaction influence coefficient for unit load at the support is 0.75, and for unit load at an adjacent support, 0.14. From Table 4.4, the reaction influence coefficient for the 4-kip wheels 1 ft from the adjacent support is 0.99. Thus, for the first term of the Fourier series, the reaction of the loading on the floorbeam is

$$R = 0.302 \times 0.75 + 2 \times 0.073 \times 0.99 \times 0.14 = 0.247$$

Now, $Q_1 = 0.302$ corresponds to the loading that produced the live-load moments previously calculated on the assumption that the floorbeams were rigid. Also, $Q_1 = 0.247$ corresponds to loading with the same distribution on the floorbeam. Therefore, the bending moments can be found by proportion from those previously calculated. Thus, the moment at the support is

$$M_{LL} = -\frac{210 \times 0.247}{0.302} = -172 \text{ ft-kips}$$

and the moment at midspan is

$$M_{LL} = \frac{110 \times 0.247}{0.302} = 90 \text{ ft-kips}$$

Impact.   For a 30-ft span, impact is taken as 30%.
At midspan, $M_I = 0.30 \times 90 = 27$ ft-kips.
At supports, $M_I = 0.30(-172) = -52$ ft-kips.

Total Floorbeam Moments.   The design moments previously calculated are summarized in Table 11.78.

Properties of Floorbeam Sections.   For stress computations, an effective width s_o of the deck plate is assumed to act as the top flange of member III. For determination of s_o, the effective spacing of floorbeams s_f is taken equal to the actual spacing, 180 in. The effective span l_e, with the floorbeam ends considered fixed, is taken as $0.7 \times 30 \times 12 = 252$ in. Hence,

$$\frac{s_f}{l_e} = \frac{180}{252} = 0.715$$

From Table 4.6 for this ratio,

$$\frac{s_o}{s_f} = 0.53, \text{ and } s_o = 0.53 \times 180 = 95 \text{ in}$$

(Fig. 11.63b and c).
    The neutral axis of the floorbeam sections at midspan and supports can be located by taking moments of component areas about middepth of the web. This computation and those for moments of inertia and section moduli are given in Table 11.79.

Floorbeam Stresses.   These are determined for the total moments in Table 11.78 with the section properties given in Table 11.79. Calculations for the stresses at midspan and the supports are given in Table 11.80. Since the stresses are well within the allowable, the floorbeam sections are satisfactory.

TABLE 11.78 Moments, ft-kips, in Floorbeam with Orthotropic-Plate Flange

	M_{DL}	M_{LL}	M_I	Total M
Midspan	104	90	27	221
Supports	0	-172	-52	-224

TABLE 11.79 Floorbeam Moments of Inertia and Section Moduli

(a) At midspan

Material	A	d	Ad	Ad^2	I_o	I
Deck 95 × ⅜	35.6	10.69	381	4,070		4,070
Web 21 × ⅜	7.9				290	290
Bottom flange 10 × ½	5.0	−10.75	−54	580		580
	48.5		327			4,940

$d = 327/48.5 = 6.73$ in

$$-6.73 \times 327 = -2,210$$
$$I_{NA} = 2,730$$

Distance from neutral axis to:

$$\text{Top of deck plate} = 10.50 + 0.375 - 6.73 = 4.15 \text{ in}$$

$$\text{Bottom of rib} = 10.50 + 0.50 + 6.73 = 17.73 \text{ in}$$

Section moduli

Top of deck plate	Bottom of rib
$S_t = 2,730/4.15 = 658$ in^3	$S_b = 2,730/17.73 = 154$ in^3

(b) At supports, gross section

Material	A	d	Ad	Ad^2	I_o	I
Deck 95 × ⅜	35.6	9.19	327	3,010		3,010
Web 18 × ⅜	6.8				180	180
Bottom flange 10 × ½	5.0	−9.25	−46	430		430
	47.4		281			3,620

$d_g = 281/47.4 = 5.93$ in

$$-5.93 \times 281 = -1,670$$
$$\text{Gross } I_{NA} = 1,950$$

Distance from neutral axis to:

$$\text{Bottom of rib} = 9 + 0.50 + 5.93 = 15.43 \text{ in}$$

Section modulus, bottom of rib

$$S_b = 1,950/15.43 = 126 \text{ in}^3$$

(c) At supports, net section

Material	A	d	Ad	Ad^2	I_o	I
Gross section	47.4		281			3,620
Top-flange holes	−10.2	9.19	−94	−860		−860
Bottom-flange holes	−1.0	−9.25	9	−90		−90
Web holes	−2.3				−120	−120
	33.9		196			2,550

$d_{net} = 196/33.9 = 5.77$ in

$$-5.77 \times 196 = -1,130$$
$$\text{Net } I_{NA} = 1,420$$

(Continued on p. 11.152)

TABLE 11.79 Floorbeam Moments of Inertia and Section Moduli *(Continued)*

Distance from neutral axis to:

Top of deck plate $= 9 + 0.375 - 5.77 = 3.61$

Section modulus, top of deck plate

$S_t = 1{,}420/3.61 = 392 \text{ in}^3$

Floorbeam Shears. For maximum shear, the truck wheels are placed in each design lane as indicated in Fig. 11.64. The 16-kip wheels are placed over the floorbeam. A 4-kip wheel is located 14 ft away on each of the adjoining rib spans. Thus, with the floorbeams assumed acting as rigid supports for the ribs, the wheel load is 16.5 kips, as for maximum floorbeam moment. (The effects of floorbeam flexibility can be determined as for bending moments.)

This loading produces a simple-beam reaction of 41.8 kips. It also causes end moments of -202 and -86, which induce a reaction of $(-86 + 202)/30 = 3.9$ kips. Hence, the maximum live-load reaction and shear equal

$$V_{LL} = 41.8 + 3.9 = 45.7 \text{ kips}$$

Shear due to impact is

$$V_I = 0.30 \times 45.7 = 13.7 \text{ kips}$$

MAXIMUM FLOORBEAM SHEARS, KIPS

V_{DL}	V_{LL}	V_I	Total V
13.9	45.7	13.7	73.3

Allowable shear stress in the web for Grade 50W steel is 17 ksi. Average shear stress in the web is

$$f_v = \frac{73.3}{18 \times \text{⅜}} = 10.9 < 17 \text{ ksi}$$

Transverse stiffeners are not required.

Flange-to-Web Welds. The web will be connected to the deck plate and the bottom flange by a fillet weld on opposite sides of the web. These welds must resist

TABLE 11.80 Bending Stresses in Member III

At midspan:
  Top of deck plate $f_b = 221 \times 12/658 = 4.03$ ksi (compression)
  Bottom flange $f_b = 221 \times 12/154 = 17.2 < 27$ (tension)
At supports:
  Top of deck plate $f_b = 224 \times 12/392 = 6.9$ ksi (tension)
  Bottom flange $f_b = 224 \times 12/126 = 21.4 < 27$ ksi

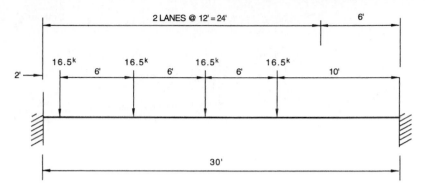

FIGURE 11.64 Positions of truck wheels for maximum shear in floorbeam.

the horizontal shear between flange and web. For the weld to the 10 × ½-in bottom flange, the minimum size fillet weld permissible with a ½-in plate, ¼ in, may be used. Shear, however, governs for the weld to the deck plate.

For computing the shear v, kips per in, between web and deck plate, the total maximum shear V is 73.3 kips and the moment of inertia of the floorbeam cross section I is 1,950 in.[4] The static moment of the deck plate is

$$Q = 35.6(4.15 - 0.19) = 141 \text{ in}^3$$

Hence, the shear to be carried by the welds is

$$v = \frac{VQ}{I} = \frac{73.3 \times 141}{1,950} = 5.30 \text{ kips per in}$$

The allowable stress on the weld is 18.9 ksi. So the allowable load per weld is 18.9 × 0.707 = 13.4 kips per in, and for two welds, 26.7 kips per in. Therefore, the weld size required is 5.30/26.7 = 0.20 in. Use ¼-in fillet welds.

Floorbeam Connections to Girders.   Since the bottom flange of the floorbeam is in compression, it can be connected to the inner web of each box girder with a splice plate of the same area. Use a 10 × ½-in plate, shop-welded to the girder and field-bolted to the floorbeam. With A325 ⅞-in-dia. high-strength bolts in slip-critical connections with Class A surfaces, the allowable load per bolt is 9.3 kips. If the capacity of the 10 × ½-in flange is developed at the allowable stress of 27 ksi, the number of bolts required in the connection is 27 × 5/9.3 = 15. Use 16.

The deck plate is spliced to the girder with ⅞-in-dia. high-strength bolts. To meet girder requirements, the pitch may vary from 3 to 5½ in (Fig. 11.60*d*). But the bolts also must transmit the tensile forces from the deck plate to the girder when the plate acts as the top flange of member III. The shear in the bolts from the girder compression is perpendicular to the shear from the floorbeam tension. Hence, the allowable load per bolt decreases from 9.3 to 9.3 × 0.707 = 6.6 kips. With an average tensile stress in the deck plate of 6.2 ksi, and a net area after deduction of holes of 35.6 − 10.2 = 25.4 in², the plate carries a tensile force of 25.4 × 6.2 = 158 kips. Thus, to transmit this force, 158/6.6 = 24 bolts are needed. If a pitch of 3 in is used in the 95-in effective width of the plate, 31 bolts are provided. Use a 3-in pitch for 4 ft on each side of every floorbeam.

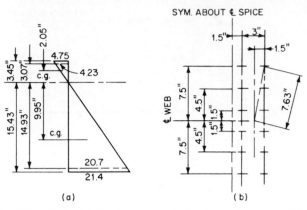

FIGURE 11.65 (*a*) Bending stresses in floorbeam at supports. (*b*) Bolted web connection of floorbeam to girder.

The web connection to the girder must transmit both vertical shear, $V = 73.3$ kips, and bending moment. The latter can be computed from the stress diagram for the cross section (Fig. 11.65*a*).

$$M = \tfrac{1}{2} \times \tfrac{3}{8}(4.23 \times 3.07 \times 2.05 + 20.7 \times 14.93 \times 9.95) = 581 \text{ in-kips}$$

Assume that the connection will be made with two rows of six bolts each, on each side of the connection centerline (Fig. 11.65*b*). The polar moment of inertia of these bolts can be computed as the sum of the moments of inertia about the x (horizontal) and y (vertical) axes.

$$
\begin{aligned}
I_x &= 4(1.5^2 + 4.5^2 + 7.5^2) = 315 \\
I_y &= 12(1.5)^2 \qquad\qquad = \underline{27} \\
&\qquad\qquad\qquad\qquad J = 342
\end{aligned}
$$

Load per bolt due to shear is

$$P_v = \frac{73.3}{12} = 6.1 \text{ kips}$$

Load on the outermost bolt due to moment is

$$P_m = \frac{581 \times 7.63}{342} = 12.95 \text{ kips}$$

The vertical component of this load is

$$P_v = \frac{12.95 \times 1.5}{7.63} = 2.5 \text{ kips}$$

and the horizontal component is

$$P_h = \frac{12.95 \times 7.5}{7.63} = 12.7 \text{ kips}$$

The total load on the outermost bolt is the resultant

$$P = \sqrt{(6.1 + 2.5)^2 + 12.7^2} = 15.3 < 2 \times 9.3$$

For the web connection plates, try two plates $17\frac{1}{2} \times \frac{5}{16}$ in. They have a net moment of inertia

$$I = 2\frac{(\frac{5}{16})17.5^3}{12} - 50 = 228 \text{ in}^4$$

To transmit the 581-in-kip moment in the web, they carry a bending stress of

$$f_b = \frac{581 \times 8.75}{228} = 22.3 < 27 \text{ ksi}$$

The assumed plates are therefore satisfactory if Grade 50W steel is used.

11.15.4 Design of Deck Plate

The deck plate is to be made of Grade 50W steel. This steel has a yield strength $F_y = 50$ ksi for the $\frac{3}{8}$-in deck thickness.

Stresses.   Bending stresses in the deck plate as the top flange of ribs (member II), floorbeams (member III), and girders (member IV) are relatively low. Combining the stresses of members II and IV yields $4.75 + 9.73 = 14.48 < 1.25 \times 27$ ksi.

Deflection.   The thickness of deck plate to limit deflection to $\frac{1}{300}$ of the rib spacing can be computed from Eq. (10.77). For a 16-kip wheel, assumed distributed over an area of $26 \times 12 = 312$ in^2, the pressure, including 30% impact, is

$$p = 1.3 \times 16/312 = 0.0667 \text{ ksi}$$

Required thickness with rib spacing $a = e = 12$ in is

$$t = 0.07 \times 12(0.0667)^{1/3} = 0.341 < 0.375 \text{ in}$$

The $\frac{3}{8}$-in deck plate is satisfactory.

11.16 *CONTINUOUS-BEAM BRIDGES*

Articles 11.1 and 11.3 recommended use of continuity for multispan bridges. Advantages over simply supported spans include less weight, greater stiffness, smaller deflections, and fewer bearings and expansion joints. Disadvantages include more complex fabrication and erection and often the costs of additional field splices.

Continuous structures also offer greater overload capacity. Failure does not necessarily occur if overloads cause yielding at one point in a span or at supports. Bending moments are redistributed to parts of the span that are not overstressed. This usually can take place in bridges because maximum positive moments and maximum negative moments occur with loads in different positions on the spans. Also, because of moment redistribution due to yielding, small settlements of supports have no significant effects on the ultimate strength of continuous spans. If, however, foundation conditions are such that large settlements could occur, simple-span construction is advisable.

While analysis of continuous structures is more complicated than that for simple spans, design differs in only a few respects. In simple spans, maximum dead-load moment occurs at midspan and is positive. In continuous spans, however, maximum dead-load moment occurs at the supports and is negative. Decreasing rapidly with distance from the support, the negative moment becomes zero at an inflection point near a quarter point of the span. Between the two dead-load inflection points in each interior span, the dead-load moment is positive, with a maximum about half the negative moment at the supports.

As for simple spans, live loads are placed on continuous spans to create maximum stresses at each section. Whereas in simple spans maximum moments at each section are always positive, maximum live-load moments at a section in continuous spans may be positive or negative. Because of the stress reversal, fatigue stresses should be investigated, especially in the region of dead-load inflection points. At supports, however, design usually is governed by the maximum negative moment, and in the midspan region, by maximum positive moment. The sum of the dead-load and live-load moments usually is greater at supports than at midspan. Usually also, this maximum is considerably less than the maximum moment in a simple beam with the same span. Furthermore, the maximum negative moment decreases rapidly with distance from the support.

The impact fraction for continuous spans depends on the length L, ft, of the portion of the span loaded to produce maximum stress. For positive moment, use the actual loaded length. For negative moment, use the average of two adjacent loaded spans.

Ends of continuous beams usually are simply supported. Consequently, moments in three-span and four-span continuous beams are significantly affected by the relative lengths of interior and exterior spans. Selection of a suitable span ratio can nearly equalize maximum positive moments in those spans and thus permit duplication of sections. The most advantageous ratio, however, depends on the ratio of dead load to live load, which, in turn, is a function of span length. Approximately, the most advantageous ratio for length of interior to exterior span is 1.33 for interior spans less than about 60 ft, 1.30 for interior spans between about 60 to 110 ft, and about 1.25 for longer spans.

When composite construction is advantageous (see Art. 11.1), it may be used either in the positive-moment regions or throughout a continuous span. Design of a section in the positive-moment region in either case is similar to that for a simple beam. Design of a section in the negative-moment regions differs in that the concrete slab, as part of the top flange, cannot resist tension. Consequently, steel reinforcement must be added to the slab to resist the tensile stresses imposed by composite action.

11.17 ALLOWABLE-STRESS DESIGN OF BRIDGE WITH CONTINUOUS, COMPOSITE STRINGERS

The structure is a two-lane highway bridge with overall length of 298 ft. Site conditions require a central span of 125 ft. End spans, therefore, are each 86.5 ft (Fig. 11.66a). The typical cross section in Fig. 11.66b shows a 30-ft roadway, flanked on one side by a 21-in-wide barrier curb and on the other by a 6-ft-wide sidewalk. The deck is supported by six rolled-beam, continuous stringers of Grade 36 steel. Concrete to be used for the deck is Class A, with 28-day strength $f'_c = 4,000$ psi and allowable compressive stress $f_c = 1,600$ psi. Loading is HS20-44. Appropriate design criteria given in Sec. 10 will be used for this structure.

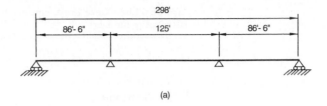

(a)

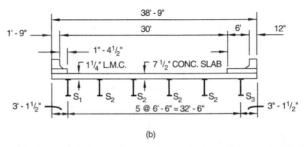

(b)

FIGURE 11.66 (*a*) Spans of a continuous highway bridge. (*b*) Typical cross section of bridge.

Concrete Slab. The slab is designed to span transversely between stringers, as in Art. 11.2. A 9-in-thick, two-course slab will be used. No provision will be made for a future 2-in wearing course.

Stringer Loads. Assume that the stringers will not be shored during casting of the concrete slab. Then, the dead load on each stringer includes the weight of a strip of concrete slab plus the weights of steel shape, cover plates, and framing details. This dead load will be referred to as *DL* and is summarized in Table 11.81.

Sidewalks, parapet, and barrier curbs will be placed after the concrete slab has cured. Their weights may be equally distributed to all stringers. Some designers, however, prefer to calculate the heavier load imposed on outer stringers by the cantilevers by taking moments of the cantilever loads about the edge of curb, as shown in Table 11.82. In addition, the six composite beams must carry the weight, 0.016 ksf, of the 30-ft-wide latex-modified concrete wearing course. The total superimposed dead load will be designated *SDL*.

The HS20-44 live load imposed may be a truck load or lane load. For these spans, truck loading governs. With stringer spacing $S = 6.5$ ft, the live load taken

TABLE 11.81   Dead Load, kips per ft, on Continuous Steel Beams

	Stringers S_1 and S_3	Stringers S_2
Slab	0.618	0.630
Haunch	0.102	0.047
Rolled beam and details—assume:	0.302	0.320
DL per stringer	1.040	0.997

TABLE 11.82 Dead Load, kips per ft, on Composite Stringers

	SDL	x	Moment
Barrier curb: 0.530/6 = 0.088		1.33	0.117
Sidewalk: 0.510/6	0.085	3.50	0.298
Parapet: 0.338/6	0.056	6.50	0.364
Railing: 0.015/6	0.002	6.50	0.013
	0.143		0.675
1¼-in LMC course	0.078		
SDL for S_2:	0.309		

Eccentricity for $S_1 = 0.117/0.088 + 6.5 + 1.38 = 9.21$ ft

Eccentricity for $S_3 = 0.675/0.143 + 6.5 - 3.88 = 7.34$ ft

SDL for $S_1 = 0.309 \times 9.21/6.5 = 0.438$

SDL for $S_3 = 0.309 \times 7.34/6.5 = 0.349$

by outer stringers S_1 and S_3 is

$$\frac{S}{4 + 0.25S} = \frac{6.5}{4 + 0.25 \times 6.5} = 1.155 \text{ wheels} = 0.578 \text{ axle}$$

The live load taken by S_2 is

$$\frac{S}{5.5} = \frac{6.5}{5.5} = 1.182 \text{ wheels} = 0.591 \text{ axle}$$

Sidewalk live load (*SLL*) on each stringer is

$$w_{SLL} = \frac{0.060 \times 6}{6} = 0.060 \text{ kip per ft}$$

The impact factor for positive moment in the 86.5-ft end spans is

$$I = \frac{50}{L + 125} = \frac{50}{86.5 + 125} = 0.237$$

For positive moment in the 125-ft center span,

$$I = \frac{50}{125 + 125} = 0.200$$

And for negative moments at the interior supports, with an average loaded span $L = (86.5 + 125)/2 = 105.8$ ft,

$$I = \frac{50}{105.8 + 125} = 0.217$$

Stringer Moments. The steel stringers will each consist of a single rolled beam of Grade 36 steel, composite with the concrete slab only in regions of positive moment. To resist negative moments, top and bottom cover plates will be attached in the region of the interior supports. To resist maximum positive moments in the center span, a cover plate will be added to the bottom flange of the composite section. In the end spans, the composite section with the rolled beam alone must carry the positive moments.

For a precise determination of bending moments and shears, these variations in moments of inertia of the stringer cross sections should be taken into account. But this requires that the cross sections be known in advance or assumed, and the analysis without a computer is tedious. Instead, for a preliminary analysis, to determine the cross sections at critical points, the moment of inertia may be assumed constant and the same in each span. This assumption considerably simplifies the analysis and permits use of tables of influence coefficients. (See, for example, "Moments, Shears, and Reactions for Continuous Highway Bridges," *American Institute of Steel Construction*.) The resulting design also often is sufficiently accurate to serve as the final design. In this example, dead-load negative moment at the supports, computed for constant moment of inertia, will be increased 10% to compensate for the variations in moment of inertia.

Curves of maximum moment (moment envelopes) are plotted in Figs. 11.67 and 11.68 for S_1 and S_2, respectively. Because total maximum moments at critical points are nearly equal for S_1, S_2, and S_3, the design selected for S_1 will be used for all stringers. (In some cases, there may be some cost savings in using shorter cover plates for the stringers with smaller moments.)

Properties of Negative-Moment Section.   The largest bending moment occurs at the interior supports, where the section consists of a rolled beam and top and bottom cover plates. With the dead load at the supports as indicated in Fig. 11.67 increased 10% to compensate for the variable moment of inertia, the moments in stringer S_1 at the supports are as follows:

S_1 Moments at Interior Supports, ft-kips

M_{DL}	M_{SDL}	$M_{LL} + M_I$	M_{SLL}	Total M
$-1,331$	-510	-821	-78	$-2,740$

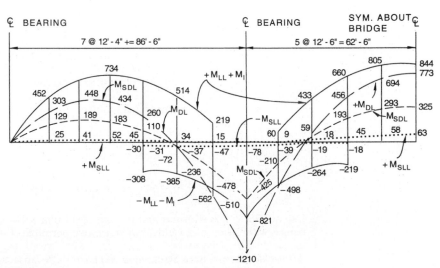

FIGURE 11.67   Maximum moments in outer stringer S_1.

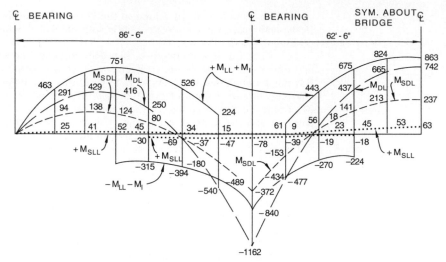

FIGURE 11.68 Maximum moments in interior stringer S_2.

For computing the minimum depth-span ratio, the distance between center-span inflection points can be taken approximately as $0.7 \times 125 = 87.5 > 86.5$ ft. In accordance with AASHTO specifications, the depth of the steel beam alone should be at least $87.5 \times 12/30 = 35$ in. Select a 36-in wide-flange beam. With an effective depth of 8.5 in for the concrete slab, allowing ½ in for wear, overall depth of the composite section is 44.5 in. Required depth is $87.5 \times 12/25 = 42 < 44.5$ in.

With an allowable bending stress of 20 ksi, the cover-plated beam must provide a section modulus of at least

$$S = \frac{2{,}740 \times 12}{20} = 1{,}644 \text{ in}^3$$

Try a W36 $\times$ 280. It provides a moment of inertia of 18,900 in^4 and a section modulus of 1,030 in^3, with a depth of 36.50 in. The cover plates must increase this section modulus by at least $1{,}644 - 1{,}030 = 614$ in^3. Hence, for an assumed distance between plates of 37 in, area of each plate should be about $614/37 = 16.6$ in^2. Try top and bottom plates 14 $\times$ 1⅜ in (area = 19.25 in^2). The 16.6-in flange width provides at least 1 in on both sides of the cover plates for fillet welding the plates to the flange.

The assumed section provides a moment of inertia of

$$I = 18{,}900 + 2 \times 19.25(18.94)^2 = 32{,}700 \text{ in}^4$$

Hence, the section modulus provided is

$$S = \frac{32{,}700}{19.63} = 1{,}666 > 1{,}644 \text{ in}^3$$

Use a W36 $\times$ 280 with top and bottom cover plates 14 $\times$ 1⅜ in. Weld plates to flanges with ⁵⁄₁₆-in fillet welds, minimum size permitted for the flange thickness.

Allowable Compressive Stress near Supports.   Because the bottom flange of the beam is in compression near the supports and is unbraced, the allowable compressive stress may have to be reduced to preclude buckling failure. AASHTO specifications, however, permit a 20% increase in the reduced stress for negative moments near interior supports. The unbraced length should be taken as the distance between diaphragms or the distance from interior support to the dead-load inflection point, whichever is smaller. In this example, if distance between diaphragms is assumed not to exceed about 22 ft, the allowable bending stress for a flange width of 16.6 in is computed as follows:

Allowable compressive stress F_b, ksi, on extreme fibers of rolled beams and built-up sections subject to bending, when the compression flange is partly supported, is determined from

$$F_b = \frac{50,000}{S_{xc}} C_b \left(\frac{I_{yc}}{l}\right) \sqrt{0.772\frac{J}{I_{yc}} + 9.87 \left(\frac{d}{l}\right)^2} \le 0.55 F_y \qquad (11.47)$$

where C_b = $1.75 + 1.05(M_1/M_2) + 0.3(M_1/M_2)^2 \le 2.3$
$\quad\;\; S_{xc}$ = section modulus with respect to the compression flange, in^3
$\qquad\;$ = 1,666 in^3
$\quad\;\; I_{yc}$ = moment of inertia of compression flange about vertical axis in plane of web, in^4 = $1.57 \times 16.6^3/12 = 598$ in^4
$\qquad l$ = length of unsupported flange between lateral = $22 \times 12 = 264$ in connections, knee braces, or other points of support, in
$\qquad J$ = torsional constant, in^4
$\qquad$ = $\frac{1}{3}(bt_{fc}^3 + bt_{ft}^3 + dt_w^3)$
$\qquad$ = $\frac{1}{3}[16.6(1.57)^3 + 16.6(1.57)^3 + 36.52(0.89)^3] = 51$ in^4
$\qquad d$ = depth of girder, in = 36.52 in
$\qquad M_1$ = smaller end moment in the unbraced length of the stringer
$\qquad$ = $-121 - 52 - 394 - 38 = -605$ ft-kip
$\qquad M_2$ = larger end moment in the unbraced length of the stringer
$\qquad$ = $-1331 - 510 - 821 - 78 = -2,740$ ft-kip
$\qquad C_b$ = $1.75 + 1.05(605/2,740) + 0.3(605/2,740)^2 = 2.00$

Substitution of the above values in Eq. (11.46) yields

$$F_b = \frac{50,000 \times 2.0}{1,666} \left(\frac{598}{264}\right) \sqrt{0.772(51/598) + 9.87(36.52/264)^2}$$

$$= 68.62 \text{ ksi} > (0.55 \times 36 = 19.8 \text{ ksi})$$

Use F_b = 19.8 ksi.

Cutoffs of Negative-Moment Cover Plates.   Because of the decrease in moments with distance from an interior support, the top and bottom cover plates can be terminated where the rolled beam alone has sufficient capacity to carry the bending moment. The actual cutoff points, however, may be determined by allowable fatigue stresses for the base metal adjacent to the fillet welds between flanges and ends of the cover plates. The number of cycles of load to be resisted for HS20-44 loading is 500,000 for a major highway. For Grade 36 steel and these conditions, the allowable fatigue stress range for this redundant-load-path structure and the Stress Category E' connection is F_r = 9.2 ksi.

Resisting moment of the W36 $\times$ 280 alone with F_r = 9.2 ksi is

$$M = \frac{9.2 \times 1,030}{12} = 790 \text{ ft-kips}$$

This equals the live-load bending-moment range in the end span about 12 ft from the interior support. Minimum terminal distance for the 14-in cover plate is $1.5 \times 14 = 21$ in. Try an actual cutoff point 12 ft 4 in from the support. At that point, the moment range is $219 - (-562) = 781$ ft-kips. Thus, the stress range is

$$F_r = \frac{781 \times 12}{1,030} = 9.1 \text{ ksi} < 9.2 \text{ ksi}$$

Fatigue does not govern. Use a cutoff 12 ft 4 in from the interior support in the end span.

In the center span, the resisting moment of the W36 equals the bending moment about 8 ft 4 in from the interior support. With allowance for the terminal distance, the plates may be cut off 10 ft 6 in from the support. Fatigue does not govern there.

Properties of End-Span Composite Section. The 9-in-thick roadway slab includes an allowance of 0.5 in for wear. Hence, the effective thickness of the concrete slab for composite action is 8.5 in.

The effective width of the slab as part of the top flange of the T beam is the smaller of the following:

$$\frac{1}{4} \text{ span} = \frac{1}{4} \times 86.5 \times 12 = 260 \text{ in}$$

$$\text{Overhang} + \text{half the spacing of stringers} = 37.5 + 78/2 = 76.5 \text{ in}$$

$$12 \times \text{slab thickness} = 12 \times 8.5 = 102 \text{ in}$$

Hence the effective width is 76.5 in (Fig. 11.69).

To resist maximum positive moments in the end span, the W36 × 280 will be made composite with the concrete slab. As in Art. 11.2, the properties of the end-span composite section are computed with the concrete slab, ignoring the haunch area, transformed into an equivalent steel area. The computations for neutral-axis locations and section moduli for the composite section are tabulated in Table 11.83.

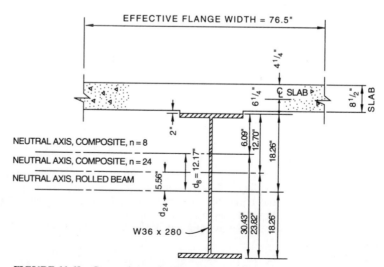

FIGURE 11.69 Composite section for end span of continuous girder.

To locate the neutral axes for $n = 24$ and $n = 8$ moments are taken about the neutral axis of the rolled beam.

Stresses in End-Span Composite Section.   Since the stringers will not be shored when the concrete is cast and cured, the stresses in the steel section for load DL are determined with the section moduli of the steel section alone. Stresses for load SDL are computed with section moduli of the composite section when $n = 24$. And stresses in the steel for live loads and impact are calculated with section moduli of the composite section when $n = 8$. See Table 11.68. Maximum positive bending moments in the end span are estimated from Fig. 11.67:

MAXIMUM POSITIVE MOMENTS IN END SPAN, FT-KIPS

M_{DL}	M_{SDL}	$M_{LL} + M_I$	M_{SLL}
434	183	734	52

Stresses in the concrete are determined with the section moduli of the composite section with $n = 24$ for SDL from Table 11.83a and $n = 8$ for $LL + I$ from Table 11.83b (Table 11.84).

Since the bending stresses in steel and concrete are less than the allowable, the assumed steel section is satisfactory for the end span.

Properties of Center-Span Section for Maximum Positive Moment.   For maximum positive moment in the middle portion of the center span, the rolled beam will be made composite with the concrete slab and a cover plate will be added to the bottom flange. Area of cover plate required A_{sb} will be estimated from Eq. (11.1) with $d_{cg} = 35$ in and $t = 8.5$ in.

MAXIMUM POSITIVE MOMENTS IN CENTER SPAN, FT-KIPS

M_{DL}	M_{SDL}	$M_{LL} + M_I$	M_{SLL}
773	325	844	63

$$A_{sb} = \frac{12}{20} \left(\frac{773}{35} + \frac{325 + 844 + 63}{35 + 8.5} \right) = 30.2 \text{ in}^2$$

The bottom flange of the W36 + 280 provides an area of 26.0 in². Hence, the cover plate should supply an area of about $30.2 - 26.0 = 4.2$ in². Try a $10 \times \frac{1}{2}$-in plate, area = 5.0 in².

The trial section is shown in Fig. 11.70. Properties of the cover-plated steel section alone are computed in Table 11.85. In determination of the properties of the composite section, use is made of the computations for the end-span composite section in Table 11.84. Calculations for the center-span section are given in Table 11.86. In all cases, the neutral axes are located by taking moments about the neutral axis of the rolled beam.

Midspan Stresses in Center Span.   Stresses caused by maximum positive moments in the center span are computed in the same way as for the end-span composite section (Table 11.87a). Stresses in the concrete are computed with the section moduli of the composite section with $n = 24$ for SDL and $n = 8$ for $LL + I$ (Table 11.87b).

TABLE 11.83   End-Span Composite Section

(a) For dead loads, $n = 24$

Material	A	d	Ad	Ad^2	I_o	I
Steel section	82.4				18,900	18,900
Concrete 76.5 × 7.75/24	24.7	24.14	596	14,400	120	14,520
	107.1		596			33,420
$d_{24} = 596/107.1 = 5.56$ in				$-5.56 \times 596 =$		$-3,320$
					$I_{NA} =$	30,100

Distance from neutral axis of composite section to:

Top of steel = 18.26 − 5.56 = 12.70 in

Bottom of steel = 18.26 + 5.56 = 23.82 in

Top of concrete = 12.70 + 2 + 7.75 = 22.45 in

Section moduli		
Top of steel	Bottom of steel	Top of concrete
$S_{st} = 30,100/12.70$ $= 2,370$ in^3	$S_{sb} = 30,100/23.82$ $= 1,264$ in^3	$S_c = 30,100/22.45$ $= 1,341$ in^3

(b) For live loads, $n = 8$

Material	A	d	Ad	Ad^2	I_o	I
Steel section	82.4				18,900	18,900
Concrete 76.5 × 8.5/8	81.3	24.51	1,993	48,840	490	49,330
	163.7		1,993			68,230
$d_8 = 1993/163.7 = 12.17$ in				$-12.17 \times 1,993 =$		$-24,260$
					$I_{NA} =$	43,970

Distance from neutral axis of composite section to:

Top of steel = 18.26 − 12.17 = 6.09 in

Bottom of steel = 18.26 + 12.17 = 30.43 in

Top of concrete = 6.09 + 2 + 8.5 = 16.59 in

Section moduli		
Top of steel	Bottom of steel	Top of concrete
$S_{st} = 43,970/6.09$ $= 7,220$ in^3	$S_{sb} = 43,970/30.43$ $= 1,445$ in^3	$S_c = 43,970/16.59$ $= 2,650$ in^3

TABLE 11.84 Stresses, ksi, in End Span for Maximum Positive Moment

(a) Steel stresses	
Top of steel (compression)	Bottom of steel (tension)
DL: $f_b = 434 \times 12/1,030 = 5.06$	$f_b = 434 \times 12/1,030 = 5.06$
SDL: $f_b = 183 \times 12/2,370 = 0.93$	$f_b = 183 \times 12/1,264 = 1.74$
$LL + I$: $f_b = 786 \times 12/7,220 = 1.31$	$f_b = 786 \times 12/1,445 = 6.53$
Total: $7.30 < 20$	$13.33 < 20$

(b) Stresses at top of concrete
SDL: $f_c = 183 \times 12/(1,341 \times 24) = 0.07$
$LL + I$: $f_c = 786 \times 12/(2,650 \times 8) = 0.44$
Total: $0.51 < 1.6$

Since the bending stresses in steel and concrete are less than the allowable, the assumed steel section is satisfactory. Use the W36 × 280 with 10 × ½-in cover plate on the bottom flange. Weld to flange with ⅜-in fillet welds, minimum size permitted for the flange thickness.

Cutoffs of Positive-Moment Cover Plate. Bending moments decrease almost parabolically with distance from midspan. At some point on either side of midspan, therefore, the bottom cover plate is not needed for carrying bending moment. After the plate is cut off, the remaining section of the stringer is the same as the composite section in the end span. Properties of this section can be obtained from Table 11.84. Try a theoretical cutoff point 12.5 ft on both sides of midspan.

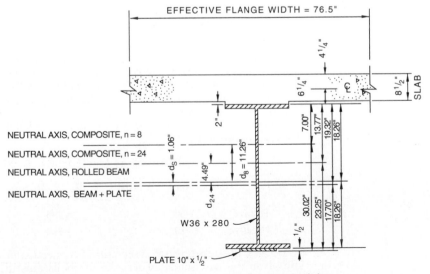

FIGURE 11.70 Composite section for center span of continuous girder.

TABLE 11.85 Rolled Beam with Cover Plate

Material	A	d	Ad	Ad^2	I_o	I
W36 × 280	82.4				18,900	18,900
Cover plate 10 × ½	5.0	−18.51	−93	1,710		1,710
	87.4		−93			20,610
$d_s = -93/87.4 = -1.06$ in					$-1.06 \times 93 =$	−100
					$I_{NA} =$	20,510

Distance from neutral axis of steel section to:

$$\text{Top of steel} = 18.26 + 1.06 = 19.32 \text{ in}$$

$$\text{Bottom of steel} = 18.26 + 0.50 - 1.06 = 17.70 \text{ in}$$

Section moduli	
Top of steel	**Bottom of steel**
$S_{st} = 20,510/19.32 = 1,062$ in^3	$S_{sb} = 20,510/17.70 = 1,159$ in^3

CENTER-SPAN MOMENTS, FT-KIPS, 12.5 FT FROM MIDSPAN

M_{DL}	M_{SDL}	$M_{LL} + M_I$	M_{SLL}
694	293	805	58

Calculations for the stresses at the theoretical cutoff point are given in Table 11.88. The composite section without cover plate is adequate at the theoretical cutoff point. With an allowance of $1.5 \times 10 = 15$ in for the terminal distance, actual cutoff would be about 14 ft from midspan. Since there is no stress reversal, fatigue does not govern there. Use a cover plate 10 × ½ in by 28 ft long.

Stringer design as determined so far is illustrated in Fig. 11.71.

Bolted Field Splice. The 298-ft overall length of the stringer is too large for shipment in one piece. Hence, field splices are necessary. They should be made where bending stresses are small. Suitable locations are in the center span near the dead-load inflection points. Provide a bolted field splice in the center span 20 ft from each support. Use A325 ⅞-in-dia high-strength bolts in slip-critical connections with Class A surfaces.

Bending moments at each splice location are identical because of symmetry. They are obtained from Fig. 11.67.

MOMENTS AT FIELD SPLICE, FT-KIPS

	M_{DL}	M_{SDL}	$M_{LL} + M_I$	M_{SLL}	Total M
Positive	−80	−50	280	10	160
Negative	−80	−50	−330	−20	−480

Because of stress reversal, a slip-critical connection must be used. Also, fatigue stresses in the base metal adjacent to the bolts must be taken into account for 500,000 cycles of loading. The allowable fatigue stress range ksi, in the base metal

TABLE 11.86 Center-Span Composite Section for Maximum Positive Moment

(a) For dead loads, $n = 24$

Material	A	d	Ad	Ad^2	I
End-span composite section	107.1		596		33,420
Cover plate $10 \times \frac{1}{2}$	5.0	-18.51	-93	1,710	1,710
	112.1		503		35,130
$d_{24} = 503/112.1 = 4.49$ in				$-4.49 \times 503 =$	$-2,260$
				$I_{NA} =$	32,870

Distance from neutral axis of composite section to:

$$\text{Top of steel} = 18.26 - 4.49 = 13.77 \text{ in}$$

$$\text{Bottom of steel} = 18.26 + 0.50 + 4.49 = 23.25 \text{ in}$$

$$\text{Top of concrete} = 13.77 + 2 + 7.75 = 23.52 \text{ in}$$

Section moduli		
Top of steel	Bottom of steel	Top of concrete
$S_{st} = 32,870/13.77$	$S_{sb} = 32,870/23.25$	$S_c = 32,870/23.52$
$= 2,387 \text{ in}^3$	$= 1,414 \text{ in}^3$	$= 1,398 \text{ in}^3$

(b) For live loads, $n = 8$

Material	A	d	Ad	Ad^2	I
End-span composite section	162.7		1,993		68,230
Cover plate $10 \times \frac{1}{2}$	5.0	-18.51	-93	1,710	1,710
	168.7		1,900		69,940
$d_8 = 1900/168.7 = 11.26$ in				$-11.26 \times 1,900 =$	$-21,390$
				$I_{NA} =$	48,550

Distance from neutral axis of composite section to:

$$\text{Top of steel} = 18.26 - 11.26 = 7.00 \text{ in}$$

$$\text{Bottom of steel} = 18.26 + 0.50 + 11.26 = 30.02 \text{ in}$$

$$\text{Top of concrete} = 7.00 + 2 + 8.5 = 17.50 \text{ in}$$

Section moduli		
Top of steel	Bottom of steel	Top of concrete
$S_{st} = 48,550/7.00$	$S_{sb} = 48,550/30.02$	$S_c = 48,550/17.50$
$= 6,936 \text{ in}^3$	$= 1,617 \text{ in}^3$	$= 2,774 \text{ in}^3$

TABLE 11.87   Stresses, ksi, in Center Span for Maximum Positive Moment

(a) Steel stresses	
Top of steel (compression)	Bottom of steel (tension)

DL: $f_b = 773 \times 12/1{,}062 = 8.73$	$f_b = 773 \times 12/1{,}159 = 8.00$
SDL: $f_b = 325 \times 12/2{,}387 = 1.63$	$f_b = 325 \times 12/1{,}414 = 2.76$
$LL + I$: $f_b = 903 \times 12/6{,}936 = \underline{1.56}$	$f_b = 903 \times 12/1{,}617 = \underline{6.70}$
Total: $\qquad\qquad\qquad\qquad 11.92 < 20$	$17.46 < 20$

(b) Stresses at top of concrete
SDL: $f_c = 325 \times 12/(1{,}398 \times 24) = 0.12$
$LL + I$: $f_c = 903 \times 12/(2{,}774 \times 8) = \underline{0.49}$
Total: $\qquad\qquad\qquad\qquad\qquad 0.61 < 1.6$

for tension or stress reversal for the Stress Category B connection and the redundant-load-path structure is 29 ksi. The allowable shear stress for bolts in a slip-critical connection is 15.5 ksi.

The web splice is designed to carry the shear on the section. Since the stresses are small, the splice capacity is made 75% of the web strength. For web strength $0.885 \times 36.5 = 32.3$ and $F_v = 12$ ksi,

$$V = 0.75 \times 32.3 \times 12 = 291 \text{ kips}$$

Each bolt has a capacity in double shear of $2 \times 0.601 \times 15.5 = 18.6$ kips. Hence, the number of bolts required is $291/18.6 = 16$. Use two rows of bolts on each side of the splice, each row with 10 bolts and 3-in pitch. Also, use on each side of the web a $30 \times \frac{9}{16}$-in splice plate, total area $= 33.7 > 32.3$ in^2.

The flange splice is designed to carry the moment on the section. With the allowable bending stress of 20 ksi, the W36 × 280 has a resisting moment of

$$M = \frac{1{,}030 \times 20}{12} = 1{,}720 > 480 \text{ ft-kips}$$

The average of the resisting and calculated moment is 1,100 ft-kips, which is less than $0.75 \times 1{,}720 = 1{,}290$ ft-kips. Therefore, the splice should be designed for a moment of 1,290 ft-kips. With a moment arm of 35 in, force in each flange is

$$P = \frac{1{,}290 \times 12}{35} = 442 \text{ kips}$$

TABLE 11.88   Tensile Stresses, ksi, 12.5 ft from Midspan

DL: $f_b = 694 \times 12/1{,}030 = 8.09$	
SDL: $f_b = 293 \times 12/1{,}264 = 2.78$	
$LL + I$: $f_b = 863 \times 12/1{,}445 = \underline{7.17}$	
Total: $\qquad\qquad\qquad\qquad 18.04 < 20$	

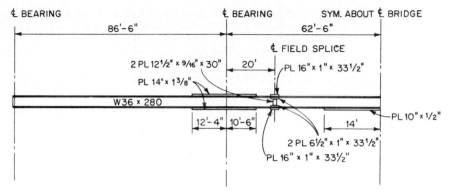

FIGURE 11.71   Cover plates and field splice for typical girder.

Then, the number of bolts in double shear required is 442/18.6 = 24. Use on each side of the splice four rows of bolts, each row with six bolts. But to increase the net section of the flange splice plates, the bolts in inner and outer rows should be staggered 1½ in.

The flange splice plates should provide a net area of 442/20 = 22.1 in². Try a 16 × 1-in plate on the outer face of each flange and a 6½-in plate on the inner face on both sides of the web. Table 11.89 presents the calculations for the net area of the splice plates. The plates can be considered satisfactory. See Fig. 11.71.

Shear in Web.   Maximum shear in the stringer totals 270 kips. Average shear stress in the web, which has an area of 32.3 in², is

$$f_v = \frac{270}{32.3} = 8.35 \text{ ksi}$$

With an allowable stress in shear of 12 ksi, the web has ample capacity. Furthermore, since the shear stress is less than 0.75 × 12 = 9 ksi, bearing stiffeners are not required.

Shear Connectors.   To ensure composite action of concrete slab and steel stringer, shear connectors welded to the top flange of the stringer must be embedded in the concrete (Art. 11.5.10). For this structure, ¾-in-dia. welded studs are selected. They are to be installed at specified locations in the positive-moment regions of the stringer in groups of three (Fig. 11.72b) to resist the horizontal shear at the top of the steel stringer. With height $H = 6$ in, they satisfy the requirement $H/d \geq 4$, where d = stud diameter, in.

TABLE 11.89   Net Area of Splice Plates, in²

Plate	Gross Area	Hole Area	$S^2/4g$	Net Area
16 × 1	16	−4	2(2.25/12 + 2.25/26.4)	12.55
2—6½ × 1	13	−4	2(2.25/12)	9.38
Total	29			21.93

Computation of number of welded studs required and pitch is similar to that for the simply supported stringer designed in Art. 11.2. With $f'_c = 4,000$ psi for the concrete, the ultimate strength of each stud is $S_u = 27.8$ kips. By Eq. (11.4), the allowable load range, kips per stud, for fatigue resistance is, for 500,000 cycles of load, $Z_r = 5.97$.

In the end-span positive-moment region, the strength of the rolled beam is

$$P_1 = A_s F_y = 82.4 \times 36 = 2,970 \text{ kips}$$

The compressive strength of the concrete slab is

$$P_2 = 0.85 f'_c bt = 0.85 \times 4.0 \times 75 \times 8.5 = 2,170 < 2,970 \text{ kips}$$

Concrete strength governs. Therefore, the number of studs to be provided between the point of maximum moment and both the support and the dead-load inflection point must be at least

$$N = \frac{P_2}{0.85 S_u} = \frac{2,170}{0.85 \times 27.8} = 92$$

Since the point of maximum moment is about 37 ft from the support, and the studs are placed in groups of three, there should be at least 32 groups within that distance. Similarly, there should be at least 32 groups in the 23 ft from the point of maximum moment and the dead-load inflection point.

Pitch is determined by fatigue requirements. The sloping lines in Fig. 11.72a represent the range of horizontal shear stress, kips per in, $S_r = V_r Q/I$, where V_r is the shear range, or change in shear caused by live loads, Q is the moment about the neutral axis of the transformed concrete area ($n = 8$), and I is the moment of

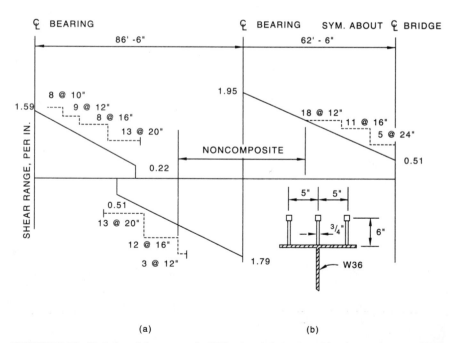

(a)   (b)

FIGURE 11.72   Variation of shear range (solid lines) and pitch selected for shear connectors (dash lines) for continuous girder.

inertia of the composition section. Shear resistance provided, kips per in, equals $3Z_r/p = 17.91/p$, where p is the pitch.

Spacing, in	10	12	14	16	20	24
Shear resistance, kips per in	1.79	1.49	1.28	1.12	0.90	0.75

The shear-connector spacings selected to meet the preceding requirements are indicated in Fig. 11.72a.

Additional connectors are required at the dead-load inflection point in the end span over a distance of one-third the effective slab width. The number depends on A_r the total area, in^2, of longitudinal slab reinforcement for the stringer over the interior support. With six No. 5 top bars and eight No. 5 bottom bars,

$$A_r = 14 \times 0.307 = 4.30 \text{ in}^2$$

The range of stress in the reinforcement may be taken as $f_r = 10$ ksi. Then, the additional connectors needed total

$$N_c = \frac{A_r f_r}{Z_r} = \frac{4.30 \times 10}{5.97} = 7.2, \text{ say } 9$$

These are indicated for the noncomposite region at the inflection point in Fig. 11.72a.

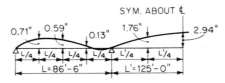

SYM. ABOUT ℄

0.71" 0.59" 0.13" 1.76" 2.94"

L/4 L/4 L/4 L/4 L'/4 L'/4

L = 86'-6"   L' = 125'-0"

FIGURE 11.73 Camber of girder to offset dead-load deflections.

In the center span also, concrete strength also determines the number of shear connectors required between mid-span and each dead-load inflection point. As in the end span, at least 18 groups of connectors should be provided. In addition, at least three groups are required at the inflection points within a distance of $^{75}\!/_3 = 25$ in. The pitch is determined by the shear range as for the end span. Figure 11.72a indicates the pitch selected.

Bearings. Fixed and expansion bearings for the continuous stringers are the same as for simply supported stringers.

Camber. Dead-load deflections can be computed by a method described in Sec. 3 for the actual moments of inertia along the stringer. The camber to offset these deflections is indicated in Fig. 11.73.

Live-Load Deflection. Maximum live-load deflection occurs at the middle of the center span and equals 1.39 in. The deflection-span ratio is $1.39/(125 \times 12) = 1/1,080$. This is less than 1/1,000, the maximum for bridges in urban areas and is satisfactory.

SECTION 12
TRUSS BRIDGES*

John M. Kulicki
Senior Vice President

Joseph E. Prickett
Associate

David H. LeRoy
Associate
Modjeski and Masters, Inc., Harrisburg, Pennsylvania

A truss is a structure that acts like a beam but with major components, or members, subjected primarily to axial stresses. The members are arranged in triangular patterns. Ideally, the end of each member at a joint is free to rotate independently of the other members at the joint. If this does not occur, secondary stresses are induced in the members. Also, if loads occur other than at panel points, or joints, bending stresses are produced in the members.

Though trusses were used by ancient Romans, the modern truss concept seems to have been originated by Andrea Palladio, a sixteenth century Italian architect. From his time to the present, truss bridges have taken many forms.

Early trusses might be considered variations of an arch. They applied horizontal thrusts at the abutments, as well as vertical reactions. In 1820, Ithiel Town patented a truss that can be considered the forerunner of the modern truss. Under vertical loading, the Town truss exerted only vertical forces at the abutments. But unlike modern trusses, the diagonals, or web systems, were of wood lattice construction and chords were composed of two or more timber planks.

In 1830, Colonel Long of the U.S. Corps of Engineers patented a wood truss with a simpler web system. In each panel, the diagonals formed an X. The next major step came in 1840, when William Howe patented a truss in which he used wrought-iron tie rods for vertical web members, with X wood diagonals. This was followed by the patenting in 1844 of the Pratt truss with wrought-iron X diagonals and timber verticals.

The Howe and Pratt trusses were the immediate forerunners of numerous iron bridges. In a book published in 1847, Squire Whipple pointed out the logic of using cast iron in compression and wrought iron in tension. He constructed bowstring trusses with cast-iron verticals and wrought-iron X diagonals.

These trusses were statically indeterminate. Stress analysis was difficult. Later, simpler web systems were adopted, thus eliminating the need for tedious and exacting design procedures.

To eliminate secondary stresses due to rigid joints, early American engineers constructed pin-connected trusses. European engineers primarily used rigid joints. Properly proportioned, the rigid trusses gave satisfactory service and eliminated

---

*Revised and updated from Sec. 12, "Truss Bridges," by Jack P. Shedd, in the first edition.

the possibility of frozen pins, which induce stresses not usually considered in design. Experience indicated that rigid and pin-connected trusses were nearly equal in cost, except for long spans. Hence, modern design favors rigid joints.

Many early truss designs were entirely functional, with little consideration given to appearance. Truss members and other components seemed to lie in all possible directions and to have a variety of sizes, thus giving the impression of complete disorder. Yet, appearance of a bridge often can be improved with very little increase in construction cost. By the 1970s, many speculated that the cable-stayed bridge would entirely supplant the truss, except on railroads. But improved design techniques, including load-factor design, and streamlined detailing have kept the truss viable. For example, some designs utilize Warren trusses without verticals. In some cases, sway frames are eliminated and truss-type portals are replaced with beam portals, resulting in an open appearance.

Because of the large number of older trusses still in the transportation system, some historical information in this section applies to those older bridges in an evaluation or rehabilitation context.

(H. J. Hopkins, "A Span of Bridges," Praeger Publishers, New York; S. P. Timoshenko, "History of Strength of Materials," McGraw-Hill Book Company, New York.)

12.1   TRUSS COMPONENTS

Principal parts of a highway truss bridge are indicated in Fig. 12.1; those of a railroad truss are shown in Fig. 12.2.

Joints are the intersections of truss members. Joints along upper and lower chords often are referred to as panel points. To minimize bending stresses in truss members, live loads generally are transmitted through floor framing to the panel points of either chord in older, shorter-span trusses. Bending stresses in members due to their own weight was often ignored in the past. In modern trusses, bending due to the weight of the members should be considered.

Chords are top and bottom members that act like the flanges of a beam. They resist the tensile and compressive forces induced by bending. In a constant-depth truss, chords are essentially parallel. They may, however, range in profile from nearly horizontal in a moderately variable-depth truss to nearly parabolic in a bowstring truss. Variable depth often improves economy by reducing stresses where chords are more highly loaded, around midspan in simple-span trusses and in the vicinity of the supports in continuous trusses.

Web members consist of diagonals and also often of verticals. Where the chords are essentially parallel, diagonals provide the required shear capacity. Verticals carry shear, provide additional panel points for introduction of loads, and reduce the span of the chords under dead-load bending. When subjected to compression, verticals often are called posts, and when subjected to tension, hangers. Usually, deck loads are transmitted to the trusses through end connections of floorbeams to the verticals.

Counters, which are found on many older truss bridges still in service, are a pair of diagonals placed in a truss panel, in the form of an X, where a single diagonal would be subjected to stress reversals. Counters were common in the past in short-span trusses. Such short-span trusses are no longer economical and have been virtually totally supplanted by beam and girder spans. X pairs are still used in lateral trusses, sway frames and portals, but are seldom designed to act as true counters, on the assumption that only one counter acts at a time and carries the maximum panel shear in tension. This implies that the companion counter takes

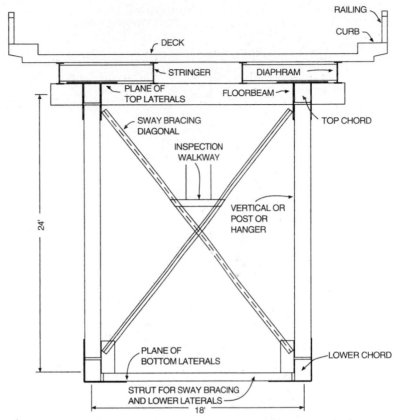

FIGURE 12.1   Cross section shows principal parts of a deck-truss highway bridge.

little load because it buckles. Current "Standard Specifications for Highway Bridges" of the American Association of State Highway and Transportation Officials (AASHTO), however, prefer that counters be made rigid. If adjustable counters are used, only one may be placed in each truss panel, and it should have open turnbuckles. Design of such members should include an allowance of 10 kips for initial stress. Sleeve nuts and loop bars should not be used.

End posts are compression members at supports of simple-span trusses. AASHTO specifications prefer that trusses have inclined end posts. Laterally unsupported hip joints should not be used.

Working lines are straight lines between intersections of truss members. To avoid bending stresses due to eccentricity, the gravity axes of truss members should lie on working lines. Some eccentricity may be permitted, however, to counteract dead-load bending stresses. Furthermore, at joints, gravity axes should intersect at a point. If an eccentric connection is unavoidable, the additional bending caused by the eccentricity should be included in the design of the members utilizing appropriate interaction equations.

AASHTO specifications require that members be symmetrical about the central plane of a truss. They should be proportioned so that the gravity axis of each section lies as nearly as practicable in its center.

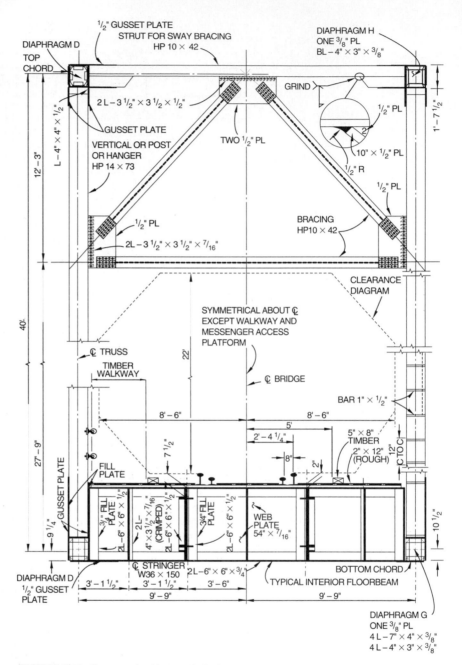

FIGURE 12.2   Cross section shows principal parts of a through-truss railway bridge.

Connections may be made with welds or high-strength bolts. American Railway Engineers Association (AREA) practice, however, excludes field welding, except for minor connections that do not support live load.

The deck is the structural element providing direct support for vehicular loads. Where the deck is located near the bottom chords (through spans), it should be supported by only two trusses.

Floorbeams should be set normal or transverse to the direction of traffic. They and their connections should be designed to transmit the deck loads to the trusses.

Stringers are longitudinal beams, set parallel to the direction of traffic. They are used to transmit the deck loads to the floorbeams. If stringers are not used, the deck must be designed to transmit vehicular loads to the floorbeams.

Lateral bracing should extend between top chords and between bottom chords of the two trusses. This bracing normally consists of trusses placed in the planes of the chords to provide stability and lateral resistance to wind. Trusses should be spaced sufficiently far apart to preclude overturning by design lateral forces.

Sway bracing may be inserted between truss verticals to provide lateral resistance in vertical planes. Where the deck is located near the bottom chords, such bracing, placed between truss tops, must be kept shallow enough to provide adequate clearance for passage of traffic below it. Where the deck is located near the top chords, sway bracing should extend the full-depth of the trusses.

Portal bracing is sway bracing placed in the plane of end posts. In addition to serving the normal function of sway bracing, portal bracing also transmits loads in the top lateral bracing to the end posts (Art. 12.5).

Skewed bridges are structures supported on piers that are not perpendicular to the planes of the trusses. The **skew angle** is the angle between the transverse centerline of bearings and a perpendicular to the longitudinal centerline of the bridge.

12.2   TYPES OF TRUSSES

Figure 12.3 shows some of the common trusses used for bridges. **Pratt trusses** have diagonals sloping downward toward the center and parallel chords (Fig. 12.3a). **Warren trusses,** with parallel chords and alternating diagonals, are generally, but not always, constructed with verticals (Fig. 12.3c) to reduce panel size. When rigid joints are used, such trusses are favored because they provide an efficient web system. Most modern bridges are of some type of Warren configuration.

Parker trusses (Fig. 12.3d) resemble Pratt trusses but have variable depth. As in other types of trusses, the chords provide a couple that resists bending moment. With long spans, economy is improved by creating the required couple with less force by spacing the chords farther apart. The Parker truss, when simply supported, is designed to have its greatest depth at midspan, where moment is a maximum. For greatest chord economy, the top-chord profile should approximate a parabola. Such a curve, however, provides too great a change in slope of diagonals, with some loss of economy in weights of diagonals. In practice, therefore, the top-chord profile should be set for the greatest change in truss depth commensurate with reasonable diagonal slopes; for example, between 40° and 60° with the horizontal.

K trusses (Fig. 12.3e) permit deep trusses with short panels to have diagonals with acceptable slopes. Two diagonals generally are placed in each panel to intersect at midheight of a vertical. Thus, for each diagonal, the slope is half as large as it would be if a single diagonal were used in the panel. The short panels keep down the cost of the floor system. This cost would rise rapidly if panel width were to increase considerably with increase in span. Thus, K trusses may be economical

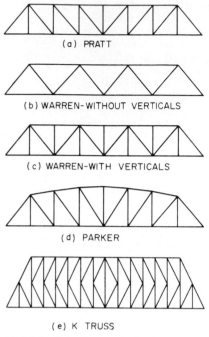

(a) PRATT

(b) WARREN-WITHOUT VERTICALS

(c) WARREN-WITH VERTICALS

(d) PARKER

(e) K TRUSS

FIGURE 12.3 Types of simple-span truss bridges.

for long spans, for which deep trusses and narrow panels are desirable. These trusses may have constant or variable depth.

Bridges also are classified as highway or railroad, depending on the type of loading the bridge is to carry. Because highway loading is much lighter than railroad, highway trusses generally are built of much lighter sections. Usually, highways are wider than railways, thus requiring wider spacing of trusses.

Trusses are also classified as to location of deck: deck, through, or half-through trusses. **Deck trusses** locate the deck near the top chord so that vehicles are carried above the chord. **Through trusses** place the deck near the bottom chord so that vehicles pass between the trusses. **Half-through trusses** carry the deck so high above the bottom chord that lateral and sway bracing cannot be placed between the top chords. The choice of deck or through construction normally is dictated by the economics of approach construction.

The absence of top bracing in half-through trusses calls for special provisions to resist lateral forces. AASHTO working-stress design specifications require that truss verticals, floorbeams, and their end connections be proportioned to resist a lateral force of at least 0.30 kip per lin ft, applied at the top chord panel points of each truss. The Guide Specification for Strength Design of Trusses contains similar requirements. The top chord of a half-through truss should be designed as a column with elastic lateral supports at panel points. The critical buckling force of the column, so determined, should be at least 50% larger than the maximum force induced in any panel of the top chord by dead and live loads plus impact. Thus, the verticals have to be designed as cantilevers, with a concentrated load at top-chord level and rigid connection to a floorbeam. This system offers elastic restraint to buckling of the top chord. The analysis of elastically restrained compression members is covered in T. V. Galambos, "Guide to Stability Design Criteria for Metal Structures," Structural Stability Research Council.

12.3 BRIDGE LAYOUT

Trusses, offering relatively large depth, open-web construction, and members subjected primarily to axial stress, provide large carrying capacity for comparatively small amounts of steel. For maximum economy in truss design, the area of metal furnished for members should be varied as often as required by the loads. To accomplish this, designers usually have to specify built-up sections that require considerable fabrication, which tend to offset some of the savings in steel.

Truss Spans. Truss bridges are generally comparatively easy to erect, because light equipment often can be used. Assembly of mechanically fastened joints in the field is relatively labor-intensive, which may also offset some of the savings in steel. Consequently, trusses seldom can be economical for highway bridges with spans less than about 450 ft.

Railroad bridges, however, involve different factors, because of the heavier loading. Trusses generally are economical for railroad bridges with spans greater than 150 ft.

The current practical limit for simple-span trusses is about 800 ft for highway bridges and about 750 ft for railroad bridges. Some extension of these limits should be possible with improvements in materials and analysis, but as span requirements increase, cantilever or continuous trusses are more efficient. The North American span record for cantilever construction in 1993 was 1,600 ft for highway bridges and 1,800 ft for railroad bridges.

For a bridge with several truss spans, the most economical pier spacing can be determined after preliminary designs have been completed for both substructure and superstructure. One guideline provides that the cost of one pier should equal the cost of one superstructure span, excluding the floor system. In trial calculations, the number of piers initially assumed may be increased or decreased by one, decreasing or increasing the truss spans. Cost of truss spans rises rapidly with increase in span. A few trial calculations should yield a satisfactory picture of the economics of the bridge layout. Such an analysis, however, is more suitable for approach spans than for main spans. In most cases, the navigation or hydraulic requirement is apt to unbalance costs in the direction of increased superstructure cost. Furthermore, girder construction is currently used for span lengths that would have required approach trusses in the past.

Panel Dimensions.   To start economic studies, it is necessary to arrive at economic proportions of trusses so that fair comparisons can be made among alternatives. Panel lengths will be influenced by type of truss being designed. They should permit slope of the diagonals between 40° and 60° with the horizontal for economic design. If panels become too long, the cost of the floor system substantially increases and heavier dead loads are transmitted to the trusses. A subdivided truss becomes more economical under these conditions.

For simple-span trusses, experience has shown that a depth-span ratio of 1:5 to 1:8 yields economical designs. (AASHTO "Standard Specifications for Highway Bridges" set a minimum of 1:10 for this ratio.) For continuous trusses with reasonable balance of spans, a depth-span ratio of 1:12 should be satisfactory. Because of the lighter live loads for highways, somewhat shallower depths of trusses may be used for highway bridges than for railway bridges.

Designers, however, do not have complete freedom in selection of truss depth. Certain physical limitations may dictate the depth to be used. For through-truss highway bridges, for example, it is impractical to provide a depth of less than 24 ft, because of the necessity of including suitable sway frames. Similarly, for through railway trusses, a depth of at least 30 ft is required. The trend toward double-stack cars encourages even greater minimum depths.

Once a starting depth and panel spacing have been determined, permutation of primary geometric variables can be studied efficiently by computer-aided design methods. In fact, preliminary studies have been carried out in which every primary truss member is designed for each choice of depth and panel spacing, resulting in a very accurate choice of those parameters.

Bridge Cross Sections.   Selection of a proper bridge cross section is an important determination by designers. In spite of the large number of varying cross-sections observed in truss bridges, actual selection of a cross section for a given site is not a large task. For instance, if a through highway truss were to be designed, the roadway width would determine the transverse spacing of trusses. The span and consequent economical depth of trusses would determine the floorbeam spacing, because the floorbeams are located at the panel points. Selection of the number of stringers and decisions as to whether to make the stringers simple spans between floorbeams or continuous over the floorbeams, and whether the stringers and floor-beams should be composite with the deck, complete the determination of the cross section.

Good design of framing of floor system members requires attention to details. In the past, many points of *stress relief* were provided in floor systems. Due to corrosion and wear resulting from use of these points of movement, however, experience with them has not always been good. Additionally, the relative movement that tends to occur between the deck and the trusses may lead to out-of-plane bending of floor system members and possible fatigue damage. Hence, modern detailing practice strives to eliminate small unconnected gaps between stiffeners and plates, rapid change in stiffness due to excessive flange coping, and other distortion fatigue sites. Ideally, the whole structure is made to act as a unit, thus eliminating distortion fatigue.

Deck trusses for highway bridges present a few more variables in selection of cross section. Decisions have to be made regarding the transverse spacing of trusses and whether the top chords of the trusses should provide direct support for the deck. Transverse spacing of the trusses has to be large enough to provide lateral stability for the structure. Narrower truss spacings, however, permit smaller piers, which will help the overall economy of the bridge.

Cross sections of railway bridges are similarly determined by physical requirements of the bridge site. Deck trusses are less common for railway bridges because of the extra length of approach grades often needed to reach the elevation of the deck. Also, use of through trusses offers an advantage if open-deck construction is to be used. With open decks, considerable corrosion of the lower chords of deck trusses may occur from drippings from freight cars.

After preliminary selection of truss type, depth, panel lengths, member sizes, lateral systems, and other bracing, designers should review the appearance of the entire bridge. Esthetics can often be improved with little economic penalty.

12.4   DECK DESIGN

For most truss members, the percentage of total stress attributable to dead load increases as span increases. Because trusses are normally used for long spans, and a sizable portion of the dead load comes from the weight of the deck, a lightweight deck is advantageous. It should be no thicker than actually required to support the design loading.

In the preliminary study of a truss, consideration should be given to the cost, durability, maintainability, inspectability, and replaceability of various deck systems, including transverse, longitudinal, and four-way reinforced concrete decks, orthotropic-plate decks, and concrete-filled or overlaid steel grids. Open-grid deck floors will seldom be acceptable for new fixed truss bridges but may be advantageous in rehabilitation of bridges and for movable bridges.

The design procedure for railroad bridge decks is almost entirely dictated by the proposed cross section. Designers usually have little leeway with the deck, because they are required to use standard railroad deck details wherever possible.

Deck design for a highway bridge is somewhat more flexible. Most highway bridges have a reinforced-concrete slab deck, with or without an asphalt wearing surface. Reinforced concrete decks may be transverse, longitudinal or four-way slabs.

- Transverse slabs are supported on stringers spaced close enough so that all the bending in the slabs is in a transverse direction.

- Longitudinal slabs are carried by floorbeams spaced close enough so that all the bending in the slabs is in a longitudinal direction. Longitudinal concrete slabs are practical for short-span trusses where floorbeam spacing does not exceed about 20 ft. For larger spacing, the slab thickness becomes so large that the resultant dead load leads to an uneconomic truss design. Hence, longitudinal slabs are seldom used for modern trusses.

- Four-way slabs are supported directly on longitudinal stringers and transverse floorbeams. Reinforcement is placed in both directions. The most economical design has a spacing of stringers about equal to the spacing of floorbeams. This restricts use of this type of floor system to trusses with floorbeam spacing of about 20 ft. As for floor systems with a longitudinal slab, four-way slabs are generally uneconomical for modern bridges.

12.5 LATERAL BRACING, PORTALS, AND SWAY FRAMES

Lateral bracing should be designed to resist the following: (1) Lateral forces due to wind pressure on the exposed surface of the truss and on the vertical projection of the live load. (2) Lateral forces due to centrifugal forces when the track or roadway is curved. (3) For railroad bridges, lateral forces due to the nosing action of locomotives caused by unbalanced conditions in the mechanism and also forces due to the lurching movement of cars against the rails because of the play between wheels and rails. Adequate bracing is one of the most important requirements for a good design.

Since the loadings given in American Association of State Highway and Transportation Officials and American Railway Engineering Association specifications only approximate actual loadings, it follows that refined assumptions are not warranted for calculation of panel loads on lateral trusses. The lateral forces may be applied to the windward truss only and divided between the top and bottom chords according to the area tributary to each. A lateral bracing truss is placed between the top chords or the bottom chords, or both, of a pair of trusses to carry these forces to the ends of the trusses.

Besides its use to resist lateral forces, another purpose of lateral bracing is to stiffen structures and prevent unwarranted lateral vibration. In deck-truss bridges, however, the floor system is much stiffer than the lateral bracing. Hence, the major purpose of lateral bracing is to true-up the bridges and to resist wind load during erection.

The portal usually is a sway frame extending between a pair of trusses whose purpose also is to transfer the reactions from a lateral-bracing truss to the end posts of the trusses, and, thus, to the foundation. This action depends on the ability of the frame to resist transverse forces.

The portal is normally a statically indeterminate frame. Because the design loadings are approximate, an exact analysis is seldom warranted. It is normally satisfactory to make simplifying assumptions. For example, a plane of contraflexure may be assumed halfway between the bottom of the portal knee brace and the bottom of the post. The shear on the plane may be assumed divided equally between the two end posts.

Sway frames are placed between trusses, usually in vertical planes, to stiffen the structure (Fig. 12.1 and 12.2). They should extend the full depth of deck trusses and should be made as deep as possible in through trusses. The AASHTO "Standard Specifications for Highway Bridges" require sway frames in every panel. But many bridges are serving successfully with sway frames in every other panel, even lift bridges whose alignment is critical. Some designs even eliminate sway frames entirely.

Diagonals of sway frames should be proportioned for slenderness ratio as compression members. With an X system of bracing, any shear load may be divided equally between the diagonals. An approximate check of possible loads in the sway frame should be made to ensure that stresses are within allowable limits.

12.6 RESISTANCE TO LONGITUDINAL FORCES

Acceleration and braking of vehicular loads, and longitudinal wind, apply longitudinal loads to bridges. In highway bridges, the magnitudes of these forces are generally small enough that the design of main truss members is not affected. In railroad bridges, however, chords that support the floor system might have to be increased in section to resist tractive forces. In all truss bridges, longitudinal forces are of importance in design of truss bearings and piers.

In railway bridges, longitudinal forces resulting from accelerating and braking may induce severe bending stresses in the flanges of floorbeams, at right angles to the plane of the web, unless such forces are diverted to the main trusses by traction frames. In single-track bridges, a transverse strut may be provided between the points where the main truss laterals cross the stringers and are connected to them (Fig. 12.4a). In double-track bridges, it may be necessary to add a traction truss (Fig. 12.4b).

When the floorbeams in a double-track bridge are so deep that the bottoms of the stringers are a considerable distance above the bottoms of the floorbeams, it may be necessary to raise the plane of the main truss laterals from the bottom of the floorbeams to the bottom of the stringers. If this cannot be done, a complete and separate traction frame may be provided either in the plane of the tops of the stringers or in the plane of their bottom flanges.

The forces for which the traction frames are designed are applied along the stringers. The magnitudes of these forces are determined by the number of panels of tractive or braking force that are resisted by the frames. When one frame is designed to provide for several panels, the forces may become large, resulting in uneconomical members and connections.

12.7 TRUSS DESIGN PROCEDURE

The following sequence may serve as a guide to the design of truss bridges:

• Select span and general proportions of the bridge, including a tentative cross section.
• Design the roadway or deck, including stringers and floorbeams.
• Design upper and lower lateral systems.

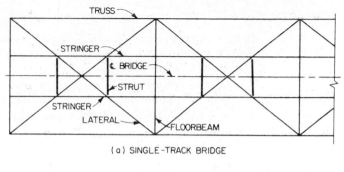

(a) SINGLE-TRACK BRIDGE

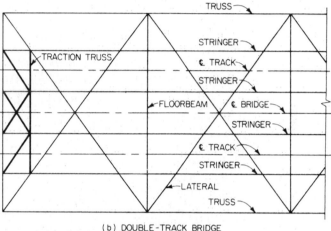

(b) DOUBLE-TRACK BRIDGE

FIGURE 12.4 Lateral bracing and traction trusses for resisting longitudinal forces on a truss bridge.

- Design portals and sway frames.
- Design posts and hangers that carry little stress or loads that can be computed without a complete stress analysis of the entire truss.
- Compute preliminary moments, shears, and stresses in the truss members.
- Design the upper-chord members, starting with the most heavily stressed member.
- Design the lower-chord members.
- Design the web members.
- Recalculate the dead load of the truss and compute final moments and stresses in truss members.
- Design joints, connections, and details.
- Compute dead-load and live-load deflections.
- Check secondary stresses in members carrying direct loads and loads due to wind.
- Review design for structural integrity, esthetics, erection, and future maintenance and inspection requirements.

12.7.1 Analysis for Vertical Loads

Determination of member forces using conventional analysis based on frictionless joints is often adequate when the following conditions are met:

1. The plane of each truss of a bridge, the planes through the top chords, and the planes through the bottom chords are fully triangulated.
2. The working lines of intersecting truss members meet at a point.
3. Cross frames and other bracing prevent significant distortions of the box shape formed by the planes of the truss described above.
4. Lateral and other bracing members are not cambered; i.e., their lengths are based on the final dead-load position of the truss.
5. Primary members are cambered by making them either short or long by amounts equal to, and opposite in sign to, the axial compression or extension, respectively, resulting from dead-load stress. Camber for trusses can be considered as a correction for dead-load deflection. (If the original design provided excess vertical clearance and the engineers did not object to the sag, then trusses could be constructed without camber. Most people, however, object to sag in bridges.) The cambering of the members results in the truss being out of vertical alignment until all the dead loads are applied to the structure (geometric condition).

When the preceding conditions are met and are rigorously modeled, three-dimensional computer analysis yields about the same dead-load axial forces in the members as the conventional pin-connected analogy and small secondary moments resulting from the self-weight bending of the member. Application of loads other than those constituting the geometric condition, such as live load and wind, will result in sag due to stressing of both primary and secondary members in the truss.

Rigorous three-dimensional analysis has shown that virtually all the bracing members participate in live-load stresses. As a result, total stresses in the primary members are reduced below those calculated by the conventional two-dimensional pin-connected truss analogy. Since trusses are usually used on relatively long-span structures, the dead-load stress constitutes a very large part of the total stress in many of the truss members. Hence, the savings from use of three-dimensional analysis of the live-load effects will usually be relatively small. This holds particularly for through trusses where the eccentricity of the live load, and, therefore, forces distributed in the truss by torsion are smaller than for deck trusses.

The largest secondary stresses are those due to moments produced in the members by the resistance of the joints to rotation. Thus, the secondary stresses in a pin-connected truss are theoretically less significant than those in a truss with mechanically fastened or welded joints. In practice, however, pinned joints always offer frictional resistance to rotation, even when new. If pin-connected joints freeze because of dirt, or rust, secondary stresses might become higher than those in a truss with rigid connections. Three-dimensional analysis will, however, quantify secondary stresses, if joints and framing of members are accurately modeled. If the secondary stress exceeds 4 ksi for tension members or 3 ksi for compression members, the AASHTO "Standard Specifications for Highway Bridges" require that excess be treated as a primary stress.

When the working lines through the centroids of intersecting members do not intersect at the joint, or where sway frames and portals are eliminated for economic or esthetic purposes, the state of bending in the truss members, as well as the rigidity of the entire system, should be evaluated by a more rigorous analysis than the conventional.

The attachment of floorbeams to truss verticals produces out-of-plane stresses, which should be investigated in highway bridges and must be accounted for in railroad bridges, due to the relatively heavier live load in that type of bridge. An analysis of a frame composed of a floorbeam and all the truss members present in the cross section containing the floorbeam is usually adequate to quantify this effect.

Deflection of trusses occurs whenever there are changes in length of the truss members. These changes may be due to strains resulting from loads on the truss, temperature variations, or fabrication effects or errors. Methods of computing deflections are similar in all three cases. Prior to the introduction of computers, calculation of deflections in trusses was a laborious procedure and was usually determined by energy or virtual work methods or by graphical or semigraphical methods, such as the Williot-Mohr diagram. With the widespread availability of matrix structural analysis packages, the calculation of deflections and analysis of indeterminant trusses are speedily executed.

(See also Arts. 3.30, 3.31, and 3.34 to 3.39.)

12.7.2 Computation of Wind Stresses in Trusses

The areas of trusses exposed to wind normal to their longitudinal axis are computed by multiplying widths of members as seen in elevation by the lengths center to center of intersections. The overlapping areas at intersections are assumed to provide enough surplus to allow for the added areas of gussets. The American Railway Engineering Association specifies that for railway bridges this truss area be multiplied by the number of trusses, on the assumption that the wind strikes each truss fully (except where the leeward trusses are shielded by the floor system). The American Association of State Highway and Transportation specifies that the area of the trusses and floor as seen in elevation be multiplied by a wind pressure that accounts for 1½ times this area being loaded by wind.

The area of the floor should be taken as that seen in elevation, including stringers, deck, railing, and railing pickets.

AREA specifies that when there is no live load on the structure, the wind pressure should be taken as at least 50 psf, which is equivalent to a wind velocity of about 125 mph. When live load is on the structure, reduced wind pressures are specified for the trusses plus full wind load on the live load: 30 psf on the bridge, which is equivalent to a 97-mph wind, and 300 lb per lin ft on the live load on one track applied 8 ft above the top of rail.

AASHTO specifies a wind pressure on the structure of 75 psf. Total force, lb per lin ft, in the plane of the windward chords should be taken as at least 300 and in the plane of the leeward chords, at least 150. When live load is on the structure, these wind pressures can be reduced 70% and combined with a wind force of 100 lb per lin ft on the live load applied 6 ft above the roadway.

Wind analysis is typically carried out with the aid of computers with a space truss and some frame members as a model. It is helpful, and instructive, to employ a simplified, noncomputer method of analysis to compare with the computer solution to expose major modeling errors that are possible with space models. Such a simplified method is presented in the following.

Idealized Wind-Stress Analysis of a through Truss with Inclined End Posts. The wind loads computed as indicated above are applied as concentrated loads at the panel points.

A through truss with parallel chords may be considered as having reactions to the top lateral bracing system only at the main portals. The effect of intermediate sway frames, therefore, is ignored. The analysis is applied to the bracing and to the truss members.

The lateral bracing members in each panel are designed for the maximum shear in the panel resulting from treating the wind load as a moving load; that is, as many panels are loaded as necessary to produce maximum shear in that panel. In design of the top-chord bracing members, the wind load, without live load, usually governs. The span for top-chord bracing is from hip joint to hip joint. For the bottom-chord members, the reduced wind pressure usually governs because of the considerable additional force that usually results from wind on the live load.

For large trusses, wind stress in the trusses should be computed for both the maximum wind pressure without live load and for the reduced wind pressure with live load and full wind on the live load. Because wind on the live load introduces an effect of "transfer," as described later, the following discussion is for the more general case of a truss with the reduced wind pressure on the structure and with wind on the live load applied 8 ft above the top of rail, or 6 ft above the deck.

The effect of wind on the trusses may be considered to consist of three additive parts:

- **Chord stresses** in the fully loaded top and bottom lateral trusses.

- **Horizontal component,** which is a uniform force of tension in one truss bottom chord and compression in the other bottom chord, resulting from transfer of the top lateral end reactions down the end portals. This may be taken as the top lateral end reaction times the horizontal distance from the hip joint to the point of contraflexure divided by the spacing between main trusses. It is often conservatively assumed that this point of contraflexure is at the end of span, and, thus, the top lateral end reaction is multiplied by the panel length, divided by the spacing between main trusses. Note that this convenient assumption does not apply to the design of portals themselves.

- **Transfer stresses** created by the moment of wind on the live load and wind on the floor. This moment is taken about the plane of the bottom lateral system. The wind force on live load and wind force on the floor in a panel length is multiplied by the height of application above the bracing plane and divided by the distance center to center of trusses to arrive at a total vertical panel load. This load is applied downward at each panel point of the leeward truss and upward at each panel point of the windward truss. The resulting stresses in the main vertical trusses are then computed.

The total wind stress in any main truss member is arrived at by adding all three effects: chord stresses in the lateral systems, horizontal component, and transfer stresses.

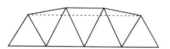

FIGURE 12.5 Top chord in a horizontal plane approximates a curved top chord.

Although this discussion applies to a parallel-chord truss, the same method may be applied with only slight error to a truss with curved top chord by considering the top chord to lie in a horizontal plane between hip joints, as shown in Fig. 12.5. The nature of this error will be described in the following.

Wind Stress Analysis of Curved-Chord Cantilever Truss.   The additional effects that should be considered in curved-chord trusses are those of the vertical components of the inclined bracing members. These effects may be illustrated by the behavior of a typical cantilever bridge, several panels of which are shown in Fig. 12.6.

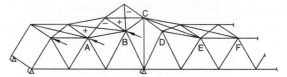

FIGURE 12.6 Wind on a cantilever truss with curved top chord is resisted by the top lateral system.

As transverse forces are applied to the curved top lateral system, the transverse shear creates stresses in the top lateral bracing members. The longitudinal and vertical components of these bracing stresses create wind stresses in the top chords and other members of the main trusses. The effects of these numerous components of the lateral members may be determined by the following simple method:

- Apply the lateral panel loads to the *horizontal projection* of the top-chord lateral system and compute all *horizontal components* of the chord stresses. The stresses in the inclined chords may readily be computed from these horizontal components.

- Determine at every point of slope change in the top chord all the vertical forces acting on the point from both bracing diagonals and bracing chords. Compute the truss stresses in the vertical main trusses from those forces.

- The final truss stresses are the sum of the two contributions above and also of any transfer stress, and of any horizontal component delivered by the portals to the bottom chords.

12.7.3   Computer Determination of Wind Stresses

For computer analysis, the structural model is a three-dimensional framework composed of all the load-carrying members. Floorbeams are included if they are part of the bracing system or are essential for the stability of the structural model.

All wind-load concentrations are applied to the framework at braced points. Because the wind loads on the floor system and on the live load do not lie in a plane of bracing, these loads must be "transferred" to a plane of bracing. The accompanying vertical loads required for equilibrium also should be applied to the framework.

Inasmuch as significant wind moments are produced in open-framed portal members of the truss, flexural rigidity of the main-truss members in the portal is essential for stability. Unless the other framework members are released for moment, the computer analysis will report small moments in most members of the truss.

With cantilever trusses, it is a common practice to analyze the suspended span by itself and then apply the reactions to a second analysis of the anchor and cantilever arms.

Some consideration of the rotational stiffness of piers about their vertical axis is warranted for those piers that support bearings that are fixed against longitudinal translation. Such piers will be subjected to a moment resulting from the longitudinal forces induced by lateral loads. If the stiffness (or flexibility) of the piers is not taken into account, the sense and magnitude of chord forces may be incorrectly determined.

12.7.4 Wind-Induced Vibration of Truss Members

When a steady wind passes by an obstruction, the pressure gradient along the obstruction causes eddies or vortices to form in the wind stream. These occur at stagnation points located on opposite sides of the obstruction. As a vortex grows, it eventually reaches a size that cannot be tolerated by the wind stream and is torn loose and carried along in the wind stream. The vortex at the opposite stagnation point then grows until it is shed. The result is a pattern of essentially equally spaced (for small distances downwind of the obstruction) and alternating vortices called the "Vortex Street" or "von Karman Trail." This vortex street is indicative of a pulsating periodic pressure change applied to the obstruction. The frequency of the vortex shedding and, hence, the frequency of the pulsating pressure, is given by

$$f = \frac{VS}{D} \tag{12.1}$$

where V is the wind speed, fps, D is a characteristic dimension, ft, and S is the Strouhal number, the ratio of velocity of vibration of the obstruction to the wind velocity (Table 12.1).

When the obstruction is a member of a truss, self-exciting oscillations of the member in the direction perpendicular to the wind stream may result when the frequency of vortex shedding coincides with a natural frequency of the member. Thus, determination of the torsional frequency and bending frequency in the plane perpendicular to the wind and substitution of those frequencies into Eq. (12.1) leads to an estimate of wind speeds at which resonance may occur. Such vibration has led to fatigue cracking of some truss and arch members, particularly cable hangers and I-shaped members. The preceding proposed use of Eq. (12.1) is oriented toward guiding designers in providing sufficient stiffness to reasonably preclude vibrations. It does not directly compute the amplitude of vibration and, hence, it does not directly lead to determination of vibratory stresses. Solutions for amplitude are available in the literature. See, for example, M. Paz, "Structural Dynamics Theory and Computation," Van Nostrand Reinhold, New York; R. J. Melosh and H. A. Smith, "New Formulation for Vibration Analysis," *ASCE Journal of Engineering Mechanics*, vol. 115, no. 3, March 1989.

C. C. Ulstrup, in "Natural Frequencies of Axially Loaded Bridge Members," *ASCE Journal of the Structural Division*, 1978, proposed the following approximate formula for estimating bending and torsional frequencies for members whose shear center and centroid coincide:

$$f_n = \frac{a}{2\pi} \left(\frac{k_n L}{I}\right)^2 \left[1 + \epsilon_p \left(\frac{KL}{\pi}\right)^2\right]^{1/2} \tag{12.2}$$

where f_n = natural frequency of member for each mode corresponding to $n = 1$, $2, 3, \ldots$
$k_n L$ = eigenvalue for each mode (see Table 12.2)
$\quad K$ = effective length factor (see Table 12.2)
$\quad L$ = length of the member, in
$\quad I$ = moment of inertia, in^4, of the member cross section
$\quad a$ = coefficient dependent on the physical properties of the member
$\quad\quad = \sqrt{EIg/\gamma A}$ for bending
$\quad\quad = \sqrt{EC_w g/\gamma I_p}$ for torsion
$\quad \epsilon_p$ = coefficient dependent on the physical properties of the member
$\quad\quad = P/EI$ for bending
$\quad\quad = \dfrac{GJA + PI_p}{AEC_w}$ for torsion

TABLE 12.1 Strouhal Number for Various Sections*

Wind direction	Profile	Strouhal number S	Profile	Strouhal number S
⇒		0.120		0.200
⇓		0.137		
⇓	d/2, d	0.144	b, d	
⇓	d/4, d	0.145	b/d 2.5 / 2.0 / 1.5 / 1.0 / 0.7 / 0.5	0.060 / 0.080 / 0.103 / 0.133 / 0.136 / 0.138
⇓	d/4, d/2, d/4, d	0.147		

*As given in "Wind Forces on Structures," *Transactions*, vol. 126, part II, p. 1180, American Society of Civil Engineers.

TABLE 12.2 Eigenvalue k_nL and Effective Length Factor K

Support condition	k_nL			K		
	$n = 1$	$n = 2$	$n = 3$	$n = 1$	$n = 2$	$n = 3$
(pin–pin)	π	2π	3π	1.000	0.500	0.333
(fixed–pin)	3.927	7.069	10.210	0.700	0.412	0.292
(fixed–fixed)	4.730	7.853	10.996	0.500	0.350	0.259
(fixed–free)	1.875	4.694	7.855	2.000	0.667	0.400

E = Young's modulus of elasticity, psi
G = shear modulus of elasticity, psi
γ = weight density of member, lb/in³
g = gravitational acceleration, in/s²
P = axial force (tension is positive), lb
A = area of member cross section, in²
C_w = warping constant
J = torsion constant
I_p = polar moment of inertia, in⁴

In design of a truss member, the frequency of vortex shedding for the section is set equal to the bending and torsional frequency and the resulting equation is solved for the wind speed V. This is the wind speed at which resonance occurs. The design should be such that V exceeds by a reasonable margin the velocity at which the wind is expected to occur uniformly.

12.8   TRUSS MEMBER DETAILS

The "Standard Specifications for Highway Bridges," American Association of State Highway and Transportation Officials (AASHTO) suggest the following shapes for truss members:

H sections, made with two side segments (composed of angles or plates) with solid web, perforated web, or web of stay plates and lacing. Modern bridges almost exclusively use H sections made of three plates welded together.

Channel sections, made with two angle segments, with solid web, perforated web, or web of stay plates and lacing. These are seldom used on modern bridges.

Single box sections, made with side channels, beams, angles and plates, or side segments of plates only. The side elements may be connected top and bottom with solid plates, perforated plates, or stay plates and lacing. Alternatively, they may be connected at the top with solid cover plates and at the bottom with perforated plates, or stay plates and lacing. Modern bridges use primarily four-plate welded box members. The cover plates are usually solid, except for access holes for bolting joints.

Double box sections, made with side channels, beams, angles and plates, or side segments of plates only. The side elements may be connected together with top and bottom perforated cover plates, or stay plates and lacing.

To obtain economy in member design, it is important to vary the area of steel in accordance with variations in total loads on the members. The variation in cross section plus the use of appropriate-strength grades of steel permit designers to use essentially the weight of steel actually required for the load on each panel, thus assuring an economical design.

With respect to shop fabrication of welded members, the H shape usually is the most economical section. It requires four fillet welds and no expensive edge preparation. Requirements for elimination of vortex shedding, however, may offset some of the inherent economy of this shape.

Box shapes generally offer greater resistance to vibration due to wind, to buckling in compression, and to torsion, but require greater care in selection of welding details. For example, various types of welded cover-plate details for boxes considered in design of the second Greater New Orleans Bridge and reviewed with several fabricators resulted in the observations in Table 12.3.

TABLE 12.3   Various Welded Cover-Plate Designs for Second Greater New Orleans Bridge

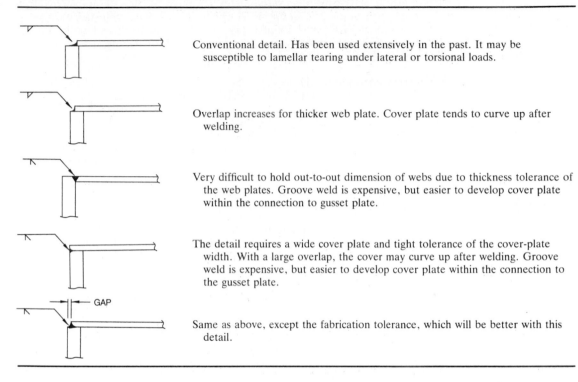

Conventional detail. Has been used extensively in the past. It may be susceptible to lamellar tearing under lateral or torsional loads.

Overlap increases for thicker web plate. Cover plate tends to curve up after welding.

Very difficult to hold out-to-out dimension of webs due to thickness tolerance of the web plates. Groove weld is expensive, but easier to develop cover plate within the connection to gusset plate.

The detail requires a wide cover plate and tight tolerance of the cover-plate width. With a large overlap, the cover may curve up after welding. Groove weld is expensive, but easier to develop cover plate within the connection to the gusset plate.

Same as above, except the fabrication tolerance, which will be better with this detail.

Additional welds placed inside a box member for development of the cover plate within the connection to the gusset plate are classified as AASHTO category E at the termination of the inside welds and should not be used. For development of the cover plate within the gusset-plate connection, groove welds, large fillet welds, larger gusset plates, or a combination of the last two should be used.

Tension Members.   These should be arranged so that there will be no bending in the members from eccentricity of the connections. If this is possible, then the total stress can be considered uniform across the entire net area of the member. At a joint, the greatest practical proportion of the member surface area should be connected to the gusset or other splice material.

Designers have a choice of a large variety of sections suitable for tension members, although box and H-shaped members are typically used. The choice will be influenced by the proposed type of fabrication and range of areas required for tension members. The design should be adjusted to take full advantage of the selected type. For example, welded plates are economical for tubular or box-shaped members. Structural tubing is available with almost 22 in^2 of cross-sectional area and might be advantageous in welded trusses of moderate spans. For longer spans, box-shape members can be shop-fabricated with almost unlimited areas.

Tension members for bolted trusses involve additional considerations. For example, as noted in the AASHTO Specifications, only 50% of the unconnected leg of an angle or tee can be considered effective, because of the eccentricity of the connection to the gusset plate at each end.

To minimize the loss of section for fastener holes and to connect into as large a proportion of the member surface area as practical, it is desirable to use a staggered fastener pattern. In Fig. 12.7, which shows a plate with staggered holes, the net width along Chain 1-1 equals plate width W, minus three hole diameters. The net width along Chain 2-2 equals W, minus five hole diameters, plus the quantity $S^2/4g$ for each of four gages, where S is the pitch and g the gage.

Compression Members.   These should be arranged so that there will be no bending in the member from eccentricity of connections. Though the members may contain fastener holes, the gross area may be used in design of such columns, on the assumption that the body of the fastener fills the hole. Welded box and H-shaped members are typically used for compression members in trusses.

Compression members should be so designed that the main elements of the section are connected directly to gusset plates, pins, or other members. It is desirable that member components be connected by solid webs. Care should be taken to ensure that the criteria for slenderness ratios, plate buckling, and fastener spacing are satisfied.

Posts and Hangers.   These are the vertical members in truss bridges. A post in a Warren deck truss delivers the load from the floorbeam to the lower chord. A hanger in a Warren through-truss delivers the floorbeam load to the upper chord.

Posts are designed as compression members. The posts in a single-truss span are generally made identical. At joints, overall dimensions of posts have to be compatible with those of the top and bottom chords to make a proper connection at the joint.

Hangers are designed as tension members. Although wire ropes or steel rods could be used, they would be objectionable for esthetic reasons. Furthermore, to provide a slenderness ratio small enough to maintain wind vibration within acceptable limits will generally require rope or rod area larger than that needed for strength.

Truss-Member Connections.   Main truss members should be connected with gusset plates and other splice material, although pinned joints may be used where the size of a bolted joint would be prohibitive. To avoid eccentricity, fasteners con-

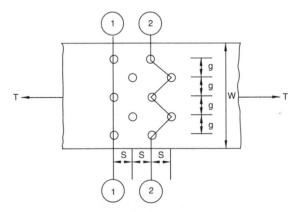

FIGURE 12.7   Chains of bolt holes used for determining the net section of a tension member.

necting each member should be symmetrical about the axis of the member. It is desirable that fasteners develop the full capacity of each element of the member. Thickness of a gusset plate should be adequate for resisting shear, direct stress, and flexure at critical sections where these stresses are maximum. Re-entrant cuts should be avoided; however, curves made for appearance are permissible.

12.9 MEMBER AND JOINT DESIGN—EXAMPLES

Design of a truss member by the working-strength method of the AASHTO "Standard Specifications for Highway Bridges"and the AASHTO "Guide Specification for Load-Factor Design of Truss Members" is illustrated in following examples. Also presented is the design of a connection in a Warren truss in which splicing of a truss chord occurs within a joint. Some designers prefer to have the chord run continuously through the joint and be spliced adjacent to the joint. Satisfactory designs can be produced using either approach. Chords of trusses that do not have a diagonal framing into each joint, such as a Warren truss, are usually continuous through joints with a post or hanger. Thus, many of the chord members are usually two panels long. Because of limitations on plate size and length for shipping, handling, or fabrication, it is sometimes necessary, however, to splice the plates within the length of a member. Where this is necessary, common practice is to offset the splices in the plates so that only one plate is spliced at any cross section.

12.9.1 Load-Factor Design of Truss Chord

A chord of a truss is to be designed to withstand a factored compression load of 7,878 kips and a factored tensile load of 1,748 kips. Corresponding service loads are 4,422 kips compression and 391 kips tension. The structural steel is to have a specified minimum yield stress of 36 ksi. The member is 46 ft long and the slenderness factor K is to be taken as unity. A preliminary design yields the cross section shown in Fig. 12.8. The section has the following properties:

$$A_g = \text{gross area} = 281 \text{ in}^2$$

$$I_{gx} = \text{gross moment of inertia with respect to } x \text{ axis}$$

$$= 97,770 \text{ in}^4$$

$$I_{gy} = \text{gross moment of inertia with respect to } y \text{ axis}$$

$$= 69,520 \text{ in}^4$$

$$w = \text{weight per linear foot} = 0.98 \text{ kips}$$

Ten 1¼-in.-dia. bolt holes are provided in each web at the section for the connections at joints. The welds joining the cover plates and webs are minimum size, ⅜ in, and are classified as AASHTO stress category B.

Although the AASHTO standard specification specifies a load factor for dead load of 1.30, the following computation uses 1.50 to allow for about 15% additional weight due to paint, diaphragms, weld metal and fasteners.

Compression in Chord from Factored Loads.   The uniform stress on the section is

$$f_c = 7878/281 = 28.04 \text{ ksi}$$

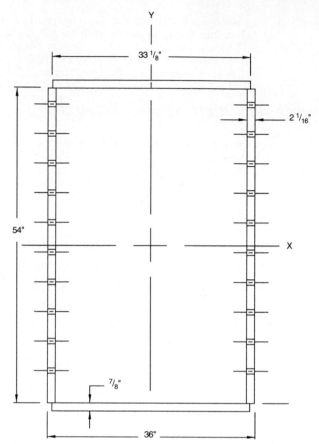

FIGURE 12.8   Cross section of a truss chord with a box section.

The radius of gyration with respect to the weak axis is

$$r_y = \sqrt{I_{gy}/A_g} = \sqrt{69,520/281} = 15.73 \text{ in}$$

and the slenderness ratio with respect to that axis is

$$\frac{KL}{r_y} = \frac{1 \times 46 \times 12}{15.73} = 35 < \left(\sqrt{\frac{2\pi^2 E}{F_y}} = 126\right)$$

where E = modulus of elasticity of the steel = 29,000 ksi. The critical buckling stress in compression is

$$F_{cr} = F_y\left[1 - \frac{F_y}{4\pi^2 E}\left(\frac{KL}{r_y}\right)^2\right] \tag{12.3}$$

$$= 36\left[1 - \frac{36}{4\pi^2 E}(35)^2\right] = 34.6 \text{ ksi}$$

As indicated in the AASHTO standard specification, the maximum strength of the concentrically loaded column is $P_u = A_g f_{cr}$ and

$$f_{cr} = 0.85 F_{cr} = 0.85 \times 34.6 = 29.42 \text{ ksi}$$

For computation of the bending strength, the sum of the depth-thickness ratios for the web and cover plates is

$$\Sigma \frac{s}{t} = 2 \times \frac{54}{2.0625} + 2 \times \frac{36 - 2.0625}{0.875} = 129.9$$

The area enclosed by the centerlines of the plates is

$$A = 54.875(36 - 2.0625) = 1{,}862 \text{ in}^2$$

Then, the design bending stress is given by

$$F_a = F_y \left[1 - \frac{0.0641 F_y S_g L \sqrt{\Sigma(s/t)}}{EA\sqrt{I_y}} \right] \qquad (12.4)$$

$$= 36 \left[1 - \frac{0.0641 \times 36 \times 3{,}507 \times 46 \times 12\sqrt{129.9}}{29{,}000 \times 1862\sqrt{69{,}520}} \right]$$

$$= 35.9 \text{ ksi}$$

For the dead load of 0.98 kips/ft, the dead-load factor of 1.50, the 46-ft span, and a factor of 1/10 for continuity in bending, the dead-load bending moment is

$$M_{DL} = 0.98(46)^2 \times 12 \times 1.50/10 = 3733 \text{ kip-in}$$

The section modulus is

$$S_g = I_{gx}/c = 97{,}770/(54/2 + 0.875) = 3507 \text{ in}^3$$

Hence, the maximum compressive bending stress is

$$f_b = M_{DL}/S_g = 3733/3507 = 1.06 \text{ ksi}$$

The plastic section modulus is

$$Z_g = 2(33.125 \times 0.875(54/2 + 0.875/2) + 2$$

$$\times 2 \times 2.0625 \times 54/2 \times 54/4 = 4598 \text{ in}^4$$

The ratio of the plastic section modulus to the elastic section modulus is $Z_g/S_g = 4{,}598/3{,}507 = 1.31$.

The AASHTO standard specifications require for combined axial load and bending that the axial force P and moment M satisfy the following equations:

$$\frac{P}{0.85 A_g F_{cr}} + \frac{MC}{M_u(1 - P/A_g F_e)} \leq 1.0 \qquad (12.5a)$$

$$\frac{P}{0.85 A_g F_y} + \frac{M}{M_p} \leq 1.0 \qquad (12.6a)$$

where M_u = maximum strength, kip-in, in bending alone
= $S_g f_a$
M_p = full plastic moment, kip-in, of the section
= ZF_y
Z = plastic modulus = $1.31 S_g$
C = equivalent moment factor, taken as 0.85 in this case
F_e = Euler buckling stress, ksi, with 0.85 factor = $0.85 E\pi^2/(KL/r_x)^2$

The effective length factor K is taken equal to unity and the radius of gyration r_x with respect to the x axis, the axis of bending, is

$$r_x = \sqrt{I_g/A_g} = \sqrt{97,770/281} = 18.65 \text{ in}$$

The slenderness ratio KL/r_x then is $46 \times 12/18.65 = 29.60$.

$$F_e = 0.85 \times 29,000\pi^2/29.60^2 = 278 \text{ ksi}$$

For convenience of calculation, Eq. (12.5a) can be rewritten, for $P = A_g F_c$, $0.85 F_{cr} = f_{cr}$, $M = S_g f_b$, and $M_u = S_g F_a$, as

$$\frac{f_c}{f_{cr}} + \frac{f_b}{F_a} \cdot \frac{C}{1 - P/A_g F_e} \leq 1.0 \qquad (12.5b)$$

Substitution of previously calculated stress values in Eq. (12.5b) yields

$$\frac{28.04}{29.42} + \frac{1.06}{35.9} \cdot \frac{0.85}{1 - 7878/(281 \times 278)} = 0.953 + 0.028$$

$$= 0.981 \leq 1.0$$

Similarly, Eq. (12.6a) can be rewritten as

$$\frac{f_c}{0.85 F_y} + \frac{f_b}{F_y Z/S_g} \leq 1.0 \qquad (12.6b)$$

Substitution of previously calculated stress values in Eq. (12.6b) yields

$$\frac{28.04}{0.85 \times 36} + \frac{1.06}{36 \times 1.31} = 0.916 + 0.022 = 0.938 \leq 1.0$$

The sum of the ratios, 0.981, governs (stability) and is satisfactory. The section is satisfactory for compression.

Local Buckling. The AASHTO specifications limit the depth-thickness ratio of the webs to a maximum of

$$d/t = 180/\sqrt{f_c} = 180/\sqrt{28.04} = 34.0$$

The actual d/t is $54/2.0625 = 26.2 < 34.0$—OK
Maximum permissible width-thickness ratio for the cover plates is

$$b/t = 213.4/\sqrt{f_c} = 213.4/\sqrt{28.04} = 40.3$$

The actual b/t is $33.125/0.875 = 37.9 < 40.3$—OK

Tension in Chord from Factored Loads. For determination of the design strength of the section, the effect of the bolt holes must be taken into account. According to the AASHTO specification, the gross area should not be used if the holes occupy more than 15% of the gross area. When they do, the excess above 15% should be deducted from the gross area to obtain the net area. The holes occupy $10 \times 1.25 = 12.50$ in of web-plate length, and 15% of the 54-in plate is 8.10 in. The excess is 4.40 in. Hence, the net area is $A_n = 281 - 2 \times 4.40 \times 2.0625 = 263$ in^2.

For computation of net moment of inertia, assume that the excess is due to 4 bolts, located 7 and 14 in on both sides of the neutral axis in bending about the x axis. Equivalent diameter of each hole is $4.40/4 = 1.10$ in. The deduction from the gross moment of inertia $I_g = 97,770$ in^4 then is

$$I_d = 2 \times 2 \times 1.10 \times 2.0625(7^2 + 14^2) = 2220 \text{ in}^4$$

Hence, the net moment of inertia I_n is $97,770 - 2,220 = 95,550$ in^4, and the net elastic section modulus is

$$S_n = \frac{95,550}{54/2 + 0.875} = 3428 \text{ in}^3$$

The stress on the net section for the axial tension load of 1,748 kips alone is

$$f_t = 1748/263 = 6.65 \text{ ksi}$$

The bending stress due to $M_{DL} = 3733$ kip-in, computed previously, is

$$f_b = 3733/3428 = 1.09 \text{ ksi}$$

For combined axial tension and bending, the sum of the ratios of required strength to design strength is

$$\frac{P}{P_u} + \frac{M}{M_p} = \frac{6.65}{36} + \frac{1.09}{36 \times 1.31} = 0.208 < 1\text{---OK}$$

The section is satisfactory for tension.

Fatigue at Welds. Fatigue is to be investigated for the truss as a nonredundant-path structure subjected to 500,000 cycles of loading. AASHTO standard specifications permit for these conditions and the category B welds between web plates and cover plates a stress range of 23 ksi. Maximum service loads on the chord are 391 kips tension and 4,422 kips compression. The stress range then is

$$f_{sr} = \frac{391 - (-4,422)}{281} = 17.1 \text{ ksi} < 23 \text{ ksi}$$

The section is satisfactory for fatigue.

12.9.2 Service-Load Design of Truss Chord

The truss chord designed in Art. 12.9.1 by load-factor design and with the cross section shown in Fig. 12.8 is designed for service loads in the following, for illustrative purposes. Properties of the section are given in Art 12.9.1.

Compression in Chord for Service Loads. The uniform stress in the section for the 4,422-kip load on the gross area $A_g = 281$ in^2 is

$$f_c = 4422/281 = 15.74 \text{ ksi}$$

The AASHTO standard specifications give the following formula for the allowable axial stress for $F_y = 36$ ksi:

$$F_a = 16.98 - 0.00053(KL/r_y)^2 \tag{12.7}$$

For the slenderness ratio $KL/r_y = 35$, determined in Art. 12.9.1, the allowable stress then is

$$F_a = 16.98 - 0.00053(35)^2 = 16.33 \text{ ksi} > 15.74 \text{ ksi—OK}$$

The allowable bending stress is $f_b = 20$ ksi. Due to the 0.98 kips/ft weight of the 46-ft-long chord, the dead-load bending moment with a continuity factor of $\frac{1}{10}$ is

$$M_{DL} = 0.98(46)^2 \times 12/10 = 2488 \text{ kip-in}$$

For the section modulus $S_{gx} = 97,770/27.875 = 3507 \text{ in}^3$, the dead-load bending stress is

$$f_b = 2488/3507 = 0.709 \text{ ksi}$$

For combined bending and compression, AASHTO specifications require that the following interaction formula be satisfied:

$$\frac{f_c}{F_a} + \frac{f_b}{F_b} \cdot \frac{C_m}{1 - f_c/F_e'} \tag{12.8}$$

The coefficient C_m is taken as 0.85 for the condition of transverse loading on a compression member with joint translation prevented. For bending about the x axis, with a slenderness ratio of $KL/r_x = 29.60$, as determined in Art. 12.9.1, the Euler buckling stress with a 2.12 safety factor is

$$F_e' = \frac{\pi^2 E}{2.12(KL/r_x)^2} = \frac{\pi^2 \times 29,000}{2.12(29.60)^2} = 154 \text{ ksi}$$

Substitution of the preceding stresses in Eq. (12.8) yields

$$\frac{15.74}{16.33} + \frac{0.709}{20} \cdot \frac{0.85}{1 - 15.74/154} = 0.964 + 0.034 = 0.998 < 1\text{—OK}$$

The section is satisfactory for compression.

Tension in Chord from Service Loads.   The section shown in Fig. 12.8 has to withstand a tension load of 391 kips on the net area of 263 in^2 computed in Art. 12.9.1. The allowable tensile stress F_t is 20 ksi. The uniform tension stress on the net section is

$$f_t = 391/263 = 1.49 \text{ ksi}$$

As computed in Art. 12.9.1, the moment of inertia of the net section is 95,550 in^4 and the corresponding section modulus is $S_n = 3,428 \text{ in}^3$. Also, as computed previously for compression in the chord, the dead-load bending moment $M_{DL} = 2,488$ kip-in. Hence, the maximum bending stress is

$$f_b = 2488/3428 = 0.726 \text{ ksi}$$

The allowable bending stress F_b is 20 ksi.

For combined axial tension and bending, the sum of the ratios of actual stress to allowable stress is

$$\frac{f_t}{F_t} + \frac{f_b}{F_b} = \frac{1.49}{20} + \frac{0.726}{20} = 0.075 + 0.036 = 0.111 < 1\text{—OK}$$

The section is satisfactory for tension.

Fatigue Design. See Art. 12.9.1.

12.10 *TRUSS JOINT DESIGN*

At every joint in a truss, working lines of the intersecting members preferably should meet at a point to avoid eccentric loading (Art. 12.1). While the members may be welded directly to each other, most frequently they are connected to each other by bolting to gusset plates. Angle members may be bolted to a single gusset plate, whereas box and H shapes may be bolted to a pair of gusset plates.

A gusset plate usually is a one-piece element. When necessary, it may be spliced with groove welds. When the free edges of the plate will be subjected to compression, they usually are stiffened with plates or angles. Consideration should be given in design to the possibility of the stresses in gusset plates during erection being opposite in sense to the stresses that will be imposed by service loads.

Gusset plates are sometimes designed by the *method of sections* based on conventional strength of materials theory. The method of sections involves investigation of stresses on various planes through a plate and truss members. Analysis of gusset plates by finite-element methods, however, may be advisable where unusual geometry exists.

Transfer of member forces into and out of a gusset plate invokes the potential for block shear around the connector groups and is assumed to have about a 30° angle of distribution with respect to the gage line, as illustrated in Fig. 12.9 (lines 1-5 and 4-6).

The following summarizes a procedure for load-factor design of a truss joint. Splices are assumed to occur within the truss joints. (See examples in Arts. 12.11 and 12.12.) The concept employed in the procedure can also be applied to working-stress design.

1. Lay out the centerlines of truss members to an appropriate scale and the members to a scale of $\frac{1}{2}$ in = 1 ft, with gage lines.

2. Detail the fixed parts, such as floorbeam, strut, and lateral connections.

3. Determine the grade and size of bolts to be used.

4. Detail the end connections of truss diagonals. The connections should be designed for the average of the design strength of the diagonals and the factored load they carry but not less than 75% of the design strength. The design strength should be taken as the smallest of the following: (*a*) member strength, (*b*) column capacity, and (*c*) strength based on the width-thickness ratio *b*/*t*. A diagonal should have at least the major portion of its ends normal to the working line (square) so that milling across the ends will permit placing of templates for bolt-hole alignment accurately. The corners of the diagonal should be as close as possible to the cover plates of the chord and verticals. Bolts for connection to a gusset plate should be centered about the neutral axis of the member.

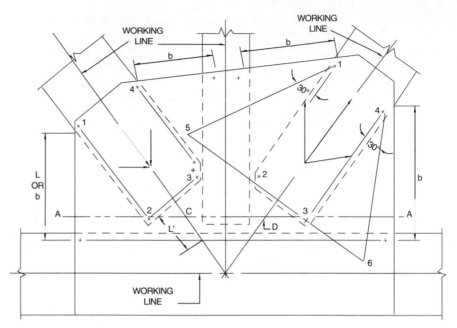

FIGURE 12.9 Typical design sections for a gusset plate.

5. Design fillet welds connecting a flange plate of a welded box member to the web plates, or the web plate of an H member to the flange plates, to transfer the connection load from the flange plate into the web plates over the length of the gusset connection. Weld lengths should be designed to satisfy fatigue requirements. The weld size should be shown on the plans if the size required for loads or fatigue is larger than the minimum size allowed.

6. Avoid the need for fills between gusset plates and welded-box truss members by keeping the out-to-out dimension of web plates and the in-to-in dimension of gusset plates constant.

7. Determine gusset-plate outlines. This step is influenced principally by the diagonal connections.

8. Select a gusset-plate thickness t to satisfy the following criteria, as illustrated in Fig. 12.9:

   a. The loads for which a diagonal is connected may be resolved into components parallel to and normal to line A-A in Fig. 12.9 (horizontal and vertical). A shearing stress is induced along the gross section of line A-A through the last line of bolts. Equal to the sum of the horizontal components of the diagonals (if they act in the same direction), this stress should not exceed $F_y/1.35\sqrt{3}$, where F_y is the yield stress of the steel, ksi.

   b. A compression stress is induced in the edge of the gusset plate along Section A-A (Fig. 12.9) by the vertical components of the diagonals (applied at C and D) and the connection load of the vertical or floorbeam, when compressive. The compression stress should not exceed the permissible column stress for the unsupported length of the gusset plate (L or b in Fig. 12.9). A stiffening angle should be provided if the slenderness ratio $L/r = L\sqrt{12}/t$ of the compression edge exceeds 120, or if the permissible column stress is exceeded. The L/r of the section formed by the angle plus a 12-in width of the gusset plate should be used to recheck that $L/r \leq 120$ and the permissible column stress is not exceeded. In addition to checking the L/r of the gusset in compression, the width-thickness ratio b/t of every free edge should be checked to ensure that it does not exceed $348/\sqrt{F_y}$.

   c. At a diagonal (Fig. 12.9),

$$V_1 + V_2 \geq P_d \tag{12.9}$$

where P_d = load from the diagonal, kips
   V_1 = shear strength, kips, along lines 1-2 and 3-4
      = $A_g F_y / \sqrt{3}$
   A_g = gross area, in^2, along those lines
   V_2 = strength, kips, along line 2-3 based on $A_n F_y$ for tension diagonals or $A_g F_a$ for compression diagonals
   A_n = net area, in^2, of the section
   F_a = allowable compressive stress, ksi

   The distance L' in Fig. 12.9 is used to compute F_a for sections 2-3 and 5-6.

   d. Assume that the connection stress transmitted to the gusset plate by a diagonal spreads over the plate within lines that diverge outward at 30° to the axis of the member from the first bolt in each exterior row of bolts, as indicated by path 1-5-6-4 (on the right in Fig. 12.9). Then, the stress on the section normal to the axis of the diagonal at the last row of bolts (along line 5-6) and included between these diverging lines should not exceed F_y on the net section for tension diagonals and F_a for compression diagonals.

9. Design the chord splice (at the joint) for the full capacity of the chords. Arrange the gusset plates and additional splice material to balance, as much as practical, the segment being spliced.

10. When the chord splice is to be made with a web splice plate on the inside of a box member (Fig. 12.10), provide extra bolts between the chords and the gusset on each side of the inner splice plate when the joint lies along the centerline of the floorbeam. This should be done because the diaphragm bolts at floorbeam connections deliver some floorbeam reaction across the chords. When a splice plate is installed on the outer side of the gusset, back of the floorbeam connection angles (Fig. 12.10), the entire group of floorbeam bolts will be stressed, both vertically and horizontally, and should not be counted as splice bolts.

11. Determine the size of standard perforations and the distances from the ends of the member.

12.11 EXAMPLE—LOAD-FACTOR DESIGN OF TRUSS JOINT

The joint shown in Fig. 12.10 is to be designed to satisfy the criteria listed in Table 12.4. Fasteners to be used are 1⅛-in-dia. A325 high-strength bolts in a slip-critical connection with Class A surfaces, with an allowable shear stress $F_v = 15.5$ ksi assume 16 ksi for this example. The bolts connecting a diagonal or vertical to a gusset plate then have a shear capacity, kips, for service loads

$$P_v = NA_v F_v = 16NA_v \tag{12.10}$$

where N = number of bolts and A_v = cross-sectional area of a bolt, in^2. For load-factor design, P_v is multiplied by a load factor. For example, for Group I loading,

$$1.5[D + (4/3)(L + I)] = 1.5(1 + R/3)P_v \tag{12.11}$$

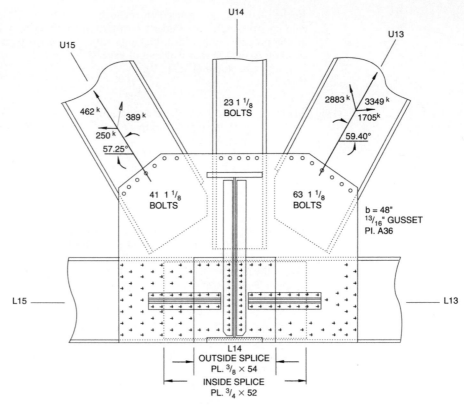

FIGURE 12.10   Truss joint for example of load-factor design.

where R = ratio of live load L to the total service load. Hence, for this loading, the load factor is $1.5(1 + R/3)$.

Diagonal U15-L14.   The diagonal is subjected to factored loads of 2,219 kips compression and 462 kips tension. It has a design strength of 2,379 kips. The AASHTO standard specifications require that the connection to the gusset plate

TABLE 12.4   Allowable Stresses for Truss Joint, ksi*

Design section	Yield stress of steel	
	36	50
Shear on line *A-A*	15.4	21.4
Shear on lines 1-2 and 3-4	20.8	28.9
Tension on lines 2-3 and 5-5	36.0	50.0

*See Figs. 12.9 and 12.10.

transmit 75% of the design strength or the average of the factored load and the design strength, whichever is larger. Thus, the design load for the connection is

$$P = (2219 + 2379)/2 = 2299 \text{ kips} > 0.75 \times 2379$$

The ratio of the service live load to the total service load for the diagonal is $R = 0.55$. Hence, for Group I loading on the bolts, the load factor is $1.5(1 + R/3) = 1.775$. For service loads, the 1⅛-in-dia. bolts have a capacity of 15.90 kips per shear plane. Therefore, since the member is connected to two gusset plates, the number of bolts required for diagonal U15-L14 is

$$N = \frac{2299}{2 \times 1.775 \times 15.90} = 41 \text{ per side}$$

Diagonal L14-U13. The diagonal is subjected to factored loads of a maximum of 3272 kips tension and a minimum of 650 kips tension. It has a design strength of 3425 kips. The design load for the connection is

$$P = (3272 + 3425)/2 = 3349 \text{ kips} > 0.75 \times 3425$$

The ratio of the service live load to the total service load is $R = 0.374$, and the load factor for the bolts is $1.5(1 + 0.374/3) = 1687$. Then, the number of 1⅛-in bolts required is

$$N = \frac{3349}{2 \times 1.687 \times 15.90} = 63 \text{ per side}$$

Vertical U14-L14. The vertical carries a factored compression load of 362 kips. It has a design strength of 1439 kips, limited by b/t at a perforation. The design load for the connection is

$$P = 0.75 \times 1439 = 1079 \text{ kips} > (362 + 1439)/2$$

Since the vertical does not carry any live load, the load factor for the bolts is 1.5. Hence, the number of 1⅛-in bolts required for the vertical is

$$N = \frac{1079}{2 \times 1.5 \times 15.90} = 23 \text{ per side}$$

Splice of Chord Cover Plates. Each cover plate of the box chord is to be spliced with a plate on the inner and outer face (Fig. 12.11). A36 steel will be used for the splice material, as for the chord. Fasteners are ⅞-in-dia. A325 bolts, with a capacity for service loads of 9.62 kips per shear plane. The bolt load factor is 1.791.

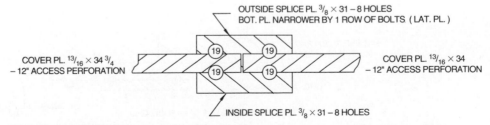

OUTSIDE SPLICE PL. ³/₈ × 31 − 8 HOLES
BOT. PL. NARROWER BY 1 ROW OF BOLTS (LAT. PL.)

COVER PL. ¹³/₁₆ × 34 ³/₄ − 12" ACCESS PERFORATION

COVER PL. ¹³/₁₆ × 34 − 12" ACCESS PERFORATION

INSIDE SPLICE PL. ³/₈ × 31 − 8 HOLES

FIGURE 12.11 Cross section of chord cover-plate splice for example of load-factor design.

The cover plate chord L14-L15 (Fig. 12.10) is $^{13}/_{16} \times 34^{3}/_{4}$ in but has 12-in-wide access perforations. Net area of the plate is 18.48 in². The cover plate for chord L13-L14 is $^{13}/_{16} \times 34$ in, also with 12-in-wide access perforations. Net area of this plate is 17.88 in². Design of the chord splice is based on the 17.88-in² area. The difference of 0.60 in² between this area and that of the larger cover plate will be made up on the L14-L15 side of the web-plate splice as "cover excess."

Where the net section of the joint elements is controlled by allowances for bolts, only the excess exceeding 15% of the gross section area is deducted from the gross area to obtain the net area. (This is the designer's interpretation of the applicable requirements for splices in the AASHTO "Standard Specifications for Highway Bridges." The current edition of the specifications should be consulted on this and other interpretations, inasmuch as the specifications are under constant reevaluation.)

The number of bolts needed for a cover-plate splice is

$$N = \frac{17.88 \times 36}{2 \times 1.791 \times 9.62} = 19 \text{ per side}$$

Try two splice plates, each $^{3}/_{8} \times 31$ in, with a gross area of 23.26 in². Assume eight 1-in-dia. bolt holes in the cross section. The area to be deducted for the holes then is

$$2 \times 0.375(8 \times 1 - 0.15 \times 31) = 2.51 \text{ in}^2$$

Consequently, the area of the net section is

$$A_n = 23.26 - 2.51 = 20.75 \text{ in}^2 > 17.88 \text{ in}^2\text{—OK}$$

Tension Splice of Chord Web Plate.   A splice is to be provided between the $1^{1}/_{4} \times 54$-in web of chord L14-L15 and the $1^{5}/_{8} \times 54$-in web of the L13-L14 chord. Because of the difference in web thickness, a $^{3}/_{8}$-in fill will be placed on the inner face of the $1^{1}/_{4}$-in web (Fig. 12.12). The gusset plate can serve as part of the needed splice material. The remainder is supplied by a plate on the inner face of the web and a plate on the outer face of the gusset. Fasteners are $1^{1}/_{8}$-in-dia. A325 bolts, with a capacity for service loads of 15.90 kips. Load factor is 1.791.

The web of the L13-L14 chord has a gross area of 87.75 in². After deduction of the 15% excess of seven $1^{1}/_{4}$-in-dia. bolt holes, the net area of this web is 86.69 in².

The web on the L14-L15 chord has a gross area of 67.50 in². After deduction of the 15% excess of seven bolt holes from the chord splice and addition of the "cover excess" of 0.60 in², the net area of this web is 67.29 in².

The gusset plate is $^{13}/_{16}$ in thick and 118 in high. Assume that only the portion that overlaps the chord web; that is, 54 in, is effective in the splice. To account

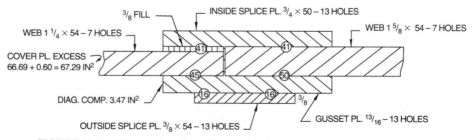

FIGURE 12.12   Cross section of chord web-plate slice for example of load-factor design.

for the eccentric application of the chord load to the gusset, an effectiveness factor may be applied to the overlap, with the assumption that only the overlapping portion of the gusset plate is stressed by the chord load.

The **effectiveness factor** E_f is defined as the ratio of the axial stress in the overlap due to the chord load to the sum of the axial stress on the full cross section of the gusset and the moment due to the eccentricity of the chord relative to the gusset centroid.

$$E_f = \frac{P/A_o}{P/A_g + Pey/I} \tag{12.12}$$

where P = chord load
  A_o = overlap area = $54t$
  A_g = full area of gusset plate = $118t$
   e = eccentricity of P = $118/2 - 54/2 = 32$ in
   y = $118/2 = 59$ in
   I = $118^3t/12 = 136,900t$ in^4

Substitution in Eq. (12.12) yields

$$E_f = \frac{P/54t}{P/118t + 32 \times 59P/136,900t} = 0.832$$

The gross area of the gusset overlap is $^{13}\!/_{16} \times 54 = 43.88$ in^2. After deduction of the 15% excess of thirteen $1\frac{1}{4}$-in-dia. bolt holes, the net area is 37.25 in^2. Then, the effective area of the gusset as a splice plate is $0.832 \times 37.25 = 30.99$ in^2.

In addition to the 67.29 in^2 of web area, the gusset has to supply an area for transmission of the 250-kip horizontal component from diagonal U15-L14 (Fig. 12.10). With $F_y = 36$ ksi, this area equals $250/(36 \times 2) = 3.47$ in^2. Hence, the equivalent web area from the L14-L15 side of the joint is $67.29 + 3.47 = 70.76$ in^2. The number of bolts required to transfer the load to the inside and outside of the web should be determined based on the effective areas of gusset that add up to 70.76 in^2 but that provide a net moment in the joint close to zero.

The sum of the moments of the web components about the centerline of the combination of outside splice plate and gusset plate is $3.47 \times 0.19 + 67.29 \times 1.22 = 0.66 + 82.09 = 82.75$ in^3. Dividing this by 2.59 in, the distance to the center of the inside splice plate, yields an effective area for the inside splice plate of 31.95 in^2. Hence, the effective area of the combination of the gusset and outside splice plates is $70.76 - 31.95 = 38.81$ in^2. This is then distributed to the plates in proportion to thickness: gusset, 24.96 in^2, and splice plate, 13.85 in^2.

The number of $1\frac{1}{8}$-in A325 bolts required to develop a plate with area A is given by

$$N = AF_y/(1.791 \times 15.90) = 36A/28.48 = 1.264A$$

Table 12.5 lists the number of bolts for the various plates.

TABLE 12.5 Number of Bolts for Plate Development

Plate	Area, in^2	Bolts
Inside splice plate	31.95	41
Outside splice plate	13.85	18
Gusset plate on L14-L15 side	$(13.85 + 24.96 - 3.47) = 35.34$	45
Gusset plate on L13-L14 side	$(13.85 + 24.96) = 38.81$	50

Check of Gusset Plates. At Section *A-A* (Fig. 12.10), each plate is 128 in wide and 118 in high, $\frac{13}{16}$ in thick. The design shear stress is 15.4 ksi (Table 12.4). The sum of the horizontal components of the loads on the truss diagonals is 1244 + 1.705 = 2949 kips. This produces a shear stress on section *A-A* of

$$f_v = \frac{2,949}{2 \times 128 \times \text{}^{13}\!/_{16}} = 14.18 \text{ ksi} < 15.4 \text{ ksi—OK}$$

The vertical component of diagonal U15-L14 produces a moment about the centroid of the gusset of 1,934 × 21 = 40,600 kip-in and the vertical component of U13-L14 produces a moment 2,883 × 20.5 = 59,100 kip-in. The sum of these moments is *M* = 99,700 kip-in. The stress at the edge of one gusset plate due to this moment is

$$f_b = \frac{6M}{td^2} = \frac{6 \times 99,700}{2(^{13}\!/_{16})128^2} = 22.47 \text{ ksi}$$

The vertical, carrying a 362-kip load, imposes a stress

$$f_c = \frac{P}{A} = \frac{362}{2 \times 128 \times \text{}^{13}\!/_{16}} = 1.74 \text{ ksi}$$

The total stress then is *f* = 22.47 + 1.74 = 24.21 ksi.

The width *b* of the gusset at the edge is 48 in. Hence, the width-thickness ratio is $b/t = 48/(^{13}\!/_{16}) = 59$. From step *b* in Art. 12.10, the maximum permissible *b/t* is $348/\sqrt{F_y} = 348/\sqrt{36} = 58 < 59$. The edge has to be stiffened. Use a stiffener angle 3 × 3 × ½ in.

For computation of the design compressive stress, assume the angle acts with a 12-in width of gusset plates. The slenderness ratio of the edge is 48/0.73 = 65.75. The maximum permissible slenderness ratio is

$$\sqrt{2\pi^2 E/F_y} = \sqrt{2\pi^2 \times 29,000/36} = 126 > 65.75$$

Hence, the design compressive stress is

$$f_a = 0.85F_y \left[1 - \frac{F_y}{4\pi^2 E} \left(\frac{L}{r} \right)^2 \right] \tag{12.13}$$

$$= 0.85 \times 36 \left[1 - \frac{36}{4\pi^2 \times 29,000} \left(\frac{48}{0.73} \right)^2 \right]$$

$$= 26.44 \text{ ksi} > 24.21 \text{ ksi—OK}$$

Next, the gusset plate is checked for shear and compression at the connection with diagonal U15-L14. The diagonal carries a factored compression load of 2,299 kips. Shear paths 1-2 and 3-4 (Fig. 12.9) have a gross length of 93 in. From Table 12.4, the design shear stress is 20.8 ksi. Hence, design shear on these paths is

$$V_d = 2 \times 20.8 \times 93 \times \text{}^{13}\!/_{16} = 3143 \text{ kips} > 2299 \text{ kips—OK}$$

Path 2-3 need not be investigated for compression. For compression on path 5-6, a 30° distribution from the first bolt in the exterior row is assumed (Art. 12.10, step 8*d*). The length of path 5-6 between the 30° lines is 82 in. The design stress, computed from Eq. (12.13) with a slenderness ratio of 52.9, is 27.9 ksi. The design strength of the gusset plate then is

$$P = 2 \times 27.9 \times 82 \times {}^{13}\!/_{16} = 3718 \text{ kips} > 2299 \text{ kips—OK}$$

Also, the gusset plate is checked for shear and tension at the connection with diagonal L14-U13. The diagonal carries a tension load of 3,272 kips. Shear paths 1-2 and 3-4 (Fig. 12.9) have a gross length of 98 in. From Table 12.4, the allowable shear stress is 20.8 ksi. Hence, the allowable shear on these paths is

$$V_d = 2 \times 20.8 \times 98 \times {}^{13}\!/_{16} = 3312 \text{ kips} > 3,272 \text{ kips—OK}$$

For path 2-3, capacity in tension with F_y = 36 ksi is

$$P_{23} = 2 \times 36 \times 27 \times {}^{13}\!/_{16} = 1580 \text{ kips}$$

For tension on path 5-6 (Fig. 12.9), a 30° distribution from the first bolt in the exterior row is assumed (Art. 12.10, step 8d). The length of path 5-6 between the 30° lines is a net of 83 in. The allowable tension then is

$$P_{56} = 2 \times 36 \times 83 \times {}^{13}\!/_{16} = 4856 \text{ kips} > 3272 \text{ kips—OK}$$

Welds to Develop Cover Plates.   The fillet weld sizes selected are listed in Table 12.6 with their capacities, for an allowable stress of 26.10 ksi. A $^5\!/_{16}$-in weld is selected for the diagonals. It has a capacity of 5.76 kips/in.

The allowable compressive stress for diagonal U15-L14 is 22.03 ksi. Then, length of fillet weld required is

$$\frac{22.03(\%)23\frac{1}{8}}{2 \times 5.76} = 38.7 \text{ in}$$

For F_y = 36 ksi, the length of fillet weld required for diagonal L14-U13 is

$$\frac{36(\frac{1}{2})23\frac{1}{8}}{2 \times 5.76} = 36.1 \text{ in}$$

12.12 EXAMPLE—SERVICE-LOAD DESIGN OF TRUSS JOINT

The joint shown in Fig. 12.13 is to be designed for connections with 1⅛-in-dia. A325 bolts with an allowable stress F_v = 16 ksi. Shear capacity of the bolts is 15.90 kips.

TABLE 12.6 Weld Capacities—Load-Factor Design

Weld size, in	Capacity of weld, kips per in
$^5\!/_{16}$	5.76
$\frac{3}{8}$	6.92
$^7\!/_{16}$	8.07
$\frac{1}{2}$	9.23

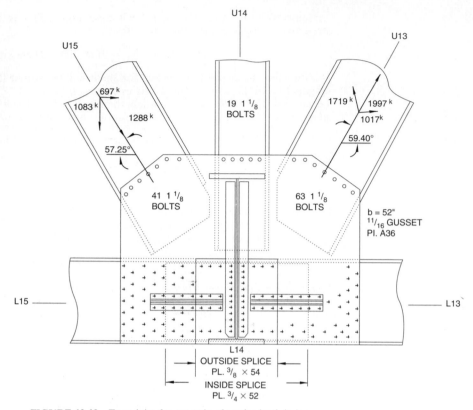

FIGURE 12.13 Truss joint for example of service-load design.

Diagonal U15-L14. The diagonal is subjected to loads of 1250 kips compression and 90 kips tension. The connection is designed for 1288 kips, 3% over design load. The number of bolts required for the connection to the $^{11}/_{16}$-in-thick gusset plate is

$$N = 1288/(2 \times 15.90) = 41 \text{ per side}$$

Diagonal L14-U13. The diagonal is subjected to a maximum tension of 1939 kips and a minimum tension of 628 kips. The connection is designed for 1997 kips, 3% over design load. The number of 1⅛-in-dia. A325 bolts required is

$$N = 1997/(2 \times 15.90) = 63 \text{ per side}$$

Vertical U14-L14. The vertical carries a compression load of 241 kips. The member is 74.53 ft long and has a cross-sectional area of 70.69 in². It has a radius of gyration $r = 10.52$ in and slenderness ratio of $KL/r = 74.53 \times 12/10.52 = 85.0$ with K taken as unity. The allowable compression stress then is

$$F_a = 16.98 - 0.00053(KL/r)^2 \tag{12.14}$$

$$= 16.98 - 0.00053 \times 85.0^2 = 13.15 \text{ ksi}$$

The allowable unit stress for width-thickness ratio b/t, however, is $11.10 < 13.15$ and governs. Hence, the allowable load is

$$P = 70.69 \times 11.10 = 785 \text{ kips}$$

The number of bolts required is determined for 75% of the allowable load:

$$N = 0.75 \times 785/(2 \times 15.90) = 19 \text{ bolts per side}$$

Splice of Chord Cover Plates.   Each cover plate of the box chord is to be spliced with a plate on the inner and outer face (Fig. 12.14). A36 steel will be used for the splice material, as for the chord. Fasteners are ⅞-in-dia. A325 bolts, with a capacity of 9.62 kips per shear plane.

The cover for L14-L15 (Fig. 12.13) is ¹³/₁₆ by 34¾ in but has 12-in-wide access perforations. Net area of the plate is 18.48 in². The cover plate for L13-L14 is ¹³/₁₆ × 34 in, also with 12-in-wide access perforations. Net area of this plate is 17.88 in². Design of the chord splice is based on the 17.88-in² area. The difference of 0.60 in² between this area and that of the larger cover plate will be made up on the L14-L15 side of the web-plate splice as "cover excess."

Where the net section of the joint elements is controlled by the allowance for bolts, only the excess exceeding 15% of the gross area is deducted from the gross area to obtain the net area, as in load-factor design (Art. 12.11).

For an allowable stress of 20 ksi in the cover plate, the number of bolts needed for the cover-plate splice is

$$N = \frac{17.88 \times 20}{2 \times 9.62} = 19 \text{ per side}$$

Try two splice plates, each ⅜ × 31 in, with a gross area of 23.26 in². Assume eight 1-in-dia. bolt holes in the cross section. The area to be deducted for the holes then is

$$2 \times 0.375(8 \times 1 - 0.15 \times 31) = 2.51 \text{ in}^2$$

Consequently, the area of the net section is

$$A_n = 23.26 - 2.51 = 20.75 \text{ in}^2 > 17.88 \text{ in}^2 \text{---OK}$$

Splice of Chord Web Plate.   A splice is to be provided between the 1¼ × 54-in web of chord L14-L15 and the 1⅝ × 54-in web of the L13-L14 chord. Because of the difference in web thickness, a ⅜-in fill will be placed on the inner face of the 1¼-in web (Fig. 12.15). The gusset plate can serve as part of the needed splice material. The remainder is supplied by a plate on the inner face of the web and a

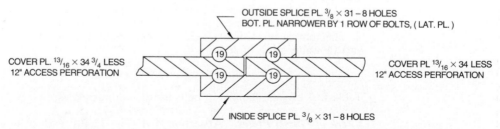

FIGURE 12.14   Cross section of chord cover-plate splice for example of service-load design.

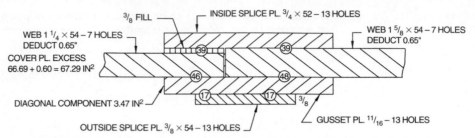

FIGURE 12.15   Cross section of chord web-plate splice for example of service load design.

plate on the outer face of the gusset. Fasteners are 1⅛-in-dia. A325 bolts, with a capacity of 15.90 kips.

The web of the L13-L14 chord has a gross area of 87.75 in². After deduction of the 15% excess of seven 1¼-in-dia. bolt holes, the net area of this web is 86.69 in².

The web of the L14-L15 chord has a gross area of 67.50 in². After deduction of the 15% excess of seven bolt holes from the chord splice and addition of the "cover excess" of 0.60 in², the net area of this web is 67.29 in².

The gusset plate is ¹¹/₁₆ in thick and 123 in high. Assume that only the portion that overlaps the chord web, that is, 54 in, is effective in the splice. To account for the eccentric application of the chord load to the gusset, an effectiveness factor E_f [Eq. (12.12)] may be applied to the overlap (Art. 12.11). The moment of inertia of the gusset is $123^3 t/12 = 155,100t$ in⁴.

$$E_f = \frac{P/54t}{P/123t + P(123/2 - 54/2)(123/2)/155,100t} = 0.849$$

The gross area of the gusset overlap is ¹¹/₁₆ × 54 = 37.13 in². After the deduction of the excess of thirteen 1¼-in-dia. bolt holes, the net area is 31.52 in². Then, the effective area of the gusset as a splice plate is 0.849 × 31.52 = 26.76 in².

In addition to the 67.29 in² of web area, the gusset has to supply an area for transmission of the 49-kip horizontal component from diagonal U15-L14. With an allowable stress of 20 ksi, the area is 49/(20 × 2) = 1.23 in². Hence, the equivalent web area from the L14-L15 side of the joint is 67.29 + 1.23 = 68.52 in². The number of bolts required to transfer the load to the inside and outside of the web should be based on the effective areas of gusset that add up to 68.52 in² but that provide a net moment in the joint close to zero.

The sum of the moments of the web components about the centerline of the combination of outside splice plate and gusset plate is 1.23 × 0.19 + 67.29 × 1.16 = 78.29 kip-in. Dividing this by 2.53, the distance to the center of the inside splice plate, yields an effective area for the inside splice plate of 30.94 in². Hence, the effective area of the combination of the gusset and outside splice plates is 68.52 − 30.94 = 37.58 in². This is then distributed to the plates as follows: gusset, 22.88 in², and outside splice plate, 14.70 in².

The number of 1⅛-in-dia. A325 bolts required to develop a plate with area A and allowable stress of 20 ksi is

$$N = 20A/15.90 = 1.258A$$

Table 12.7 lists the number of bolts for the various plates.

Check of Gusset Plates.   At section *A-A* (Fig. 12.10), each plate is 134 in wide and 123 in high, ¹¹/₁₆ in thick. The allowable shear stress is 10 ksi. The sum of the

TABLE 12.7   Number of Bolts for Plate Development

Plate	Area, in^2	Bolts
Inside splice plate	30.94	39
Outside splice plate	14.70	19
Gusset plate on L14-L15 side	$(14.70 + 22.88 - 1.16) = 36.42$	46
Gusset plate on L13-L14 side	$(14.70 + 22.88) = 37.58$	48

horizontal components of the loads on the truss diagonals is $697 + 1017 = 1714$ kips. This produces a shear stress on Section A-A of

$$f_v = \frac{1714}{2 \times 134 \times {}^{11}\!/_{16}} = 9.30 \text{ ksi} < 10 \text{ ksi—OK}$$

The vertical component of diagonal U15-L14 produces a moment about the centroid of the gusset of $1083 \times 21 = 22{,}740$ kip-in and the vertical component of U13-L14 produces a moment $1719 \times 20.5 = 35{,}240$ kip-in. The sum of these moments is 57,980 kip-in. The stress at the edge of one gusset plate due to this moment is

$$f_b = \frac{6M}{td^2} = \frac{6 \times 57{,}980}{2({}^{11}\!/_{16})134^2} = 14.09 \text{ ksi}$$

The vertical, carrying a 241-kip load, imposes a stress

$$f_c = \frac{P}{A} = \frac{241}{2 \times 134 \times {}^{11}\!/_{16}} = 1.31 \text{ ksi}$$

The total stress then is $14.09 + 1.31 = 15.40$ ksi

The width b of the gusset at the edge is 52 in. Hence, the width-thickness ratio is $b/t = 52/({}^{11}\!/_{16}) = 75.6$. From step $8b$ in Art. 12.10, the maximum permissible b/t is $348\sqrt{F_y} = 348/\sqrt{36} = 58 < 75.6$. The edge has to be stiffened. Use a stiffener angle $4 \times 3 \times \frac{1}{2}$ in.

For computation of the allowable compressive stress, assume the angle acts with a 12-in width of gusset plate. The slenderness ratio of the edge is $52/1.00 = 52.0$. The maximum permissible slenderness ratio is

$$\sqrt{2\pi^2 E / F_y} = \sqrt{2\pi^3 \times 29{,}000/36} = 126 > 52$$

Hence, the allowable stress from Eq. (12.14) is

$$F_a = 16.98 - 0.00053 \times 52^2 = 15.55 \text{ ksi} > 15.40 \text{ ksi—OK}$$

Next, the gusset plate is checked for shear and compression at the connection with diagonal U15-L14. The diagonal carries a load of 1,288 kips. Shear paths 1-2 and 3-4 (Fig. 12.9) have a gross length of 105 in. The allowable shear stress is 12 ksi. Hence, the allowable shear on these paths is

$$V_d = 2 \times 12 \times 105 \times {}^{11}\!/_{16} = 1733 \text{ kips} > 1288 \text{ kips—OK}$$

Path 2-3 need not be investigated for compression. For compression on path 5-6, a 30° distribution from the first bolt in the exterior row is assumed (Art. 12.10, step 8*d*). The length of path 5-6 between the 30° lines is 88 in. The allowable stress, computed from Eq. (12.14) with a slenderness ratio $KL/r = 0.5 \times 25/0.198 = 63$, is 14.88 ksi. This permits the gusset to withstand a load

$$P = 2 \times 14.88 \times 88 \times {}^{11}/_{16} = 1800 \text{ kips} > 1288 \text{ kips}$$

Also, the gusset plate is checked for shear and tension at the connection with diagonal L14-U13. The diagonal carries a tension load of 1,997 kips. Shear paths 1-2 and 3-4 (Fig. 12.9) have a gross length of 102 in. The allowable shear stress is 12 ksi. Hence, the allowable shear on these paths is

$$V_d = 2 \times 12 \times 102 \times {}^{11}/_{16} = 1683 \text{ kips}$$

For path 2-3, capacity in tension with an allowable stress of 20 ksi is

$$P_{23} = 2 \times 20 \times 21.6 \times {}^{11}/_{16} = 594 \text{ kips} > (1997 - 1683)\text{—OK}$$

For tension on path 5-6 (Fig. 12.9), a 30° distribution from the first bolt in the exterior row is assumed (Art. 12.10, step 8*d*). The length of path 5-6 between the 30° lines is a net of 88 in. The allowable tension then is

$$P = 2 \times 20 \times 88 \times {}^{11}/_{16} = 2420 \text{ kips} > 1997 \text{ kips—OK}$$

Welds to Develop Cover Plates.   The fillet weld sizes selected are listed in Table 12.8 with their capacities, for an allowable stress of 15.66 ksi. A $^{5}/_{16}$-in weld is selected for the diagonals. It has a capacity of 3.46 kips/in.

The allowable compressive stress for diagonal U15-L14 is 11.93 ksi. Then, length of fillet weld required is

$$\frac{11.93(\frac{7}{8})23\frac{1}{8}}{2 \times 3.46} = 34.9 \text{ in}$$

The allowable tensile stress for diagonal L14-U13 is 20.99 ksi. In this case, the required weld length is

$$\frac{20.99(\frac{1}{2})23\frac{1}{8}}{2 \times 3.46} = 35.1 \text{ in}$$

12.13   SKEWED BRIDGES

To reduce scour and to avoid impeding stream flow, it is generally desirable to orient piers with centerlines parallel to direction of flow; therefore skewed spans

TABLE 12.8   Weld Capacities—Service-Load Design

Weld size, in	Capacity of weld, kips per in
$^{5}/_{16}$	3.46
$^{3}/_{8}$	4.15
$^{7}/_{16}$	4.84
$^{1}/_{2}$	5.54

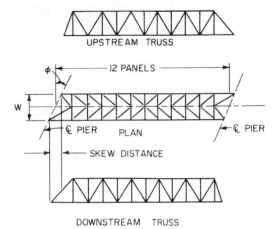

FIGURE 12.16 Skewed bridge with skew distance less than panel length.

may be required. Truss construction does not lend itself to bridges where piers are not at right angles to the superstructure (skewed crossings). Hence, these should be avoided and this can generally be done by using longer spans with normal piers. In economic comparisons, it is reasonable to assume some increased cost of steel fabrication if skewed trusses are to be used.

If a skewed crossing is a necessity, it is sometimes possible to establish a panel length equal to the skew distance $W \tan \phi$, where W is the distance between trusses and ϕ the skew angle. This aligns panels and maintains perpendicular connections of floorbeams to the trusses (Fig. 12.16). If such a layout is possible, there is little difference in cost of skewed spans and normal spans. Design principles are similar. If the skewed distance is less than the panel length, it might be possible to take up the difference in the angle of inclination of the end post, as shown in Fig. 12.16. This keeps the cost down, but results in trusses that are not symmetrical within themselves and, depending on the proportions, could be very unpleasing esthetically. If the skewed distance is greater than the panel length, it may be necessary to vary panel lengths along the bridge. One solution to such a skew is shown in Fig. 12.17, where a truss, similar to the truss in Fig. 12.16, is not symmetrical within itself and, again, might not be esthetically pleasing. The most desirable solution for skewed bridges is the alternative shown in Fig. 12.16.

Skewed bridges require considerably more analysis than normal ones, because the load distribution is nonuniform. Placement of loads for maximum effect, distribution through the floorbeams, and determination of panel point concentrations are all affected by the skew. Unequal deflections of the trusses require additional checking of sway frames and floor system connections to the trusses.

12.14 TRUSS BRIDGES ON CURVES

When it is necessary to locate a truss bridge on a curve, designers should give special consideration to truss spacing, location of bridge centerline, and stresses.

For highway bridges, location of bridge centerline and stresses due to centrifugal force are of special concern. For through trusses, the permissible degree of cur-

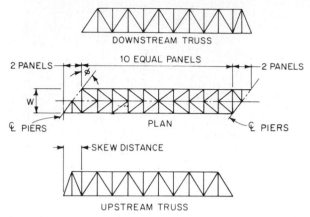

FIGURE 12.17 Skewed bridge with skew distance exceeding panel length.

vature is limited because the roadway has to be built on a curve, while trusses are planar, constructed on chords. Thus, only a small degree of throw, or offset from a tangent, can be tolerated. Regardless of the type of bridge, horizontal centrifugal forces have to be transmitted through the floor system to the lateral system and then to supports.

For railroad truss bridges, truss spacing usually provides less clearance than the spacing for highway bridges. Thus, designers must take into account tilting of cars due to superelevation and the swing of cars overhanging the track. The centerline of a through-truss bridge on a curve often is located so that the overhang at midspan equals the overhang at each span end. For bridges with more than one truss span, layout studies should be made to determine the best position for the trusses.

Train weight on a bridge on a curve is not centered on the centerline of track. Loads are greater on the outer truss than on the inner truss because the resultant of weight and centrifugal force is closer to the outer truss. Theoretically, the load on each panel point would be different and difficult to determine exactly. Because the difference in loading on inner and outer trusses is small compared with the total load, it is generally adequate to make a simple calculation for a percentage increase to be applied throughout a bridge.

Stress calculations for centrifugal forces are similar to those for any horizontal load. Floorbeams, as well as the lateral system, should be analyzed for these forces.

12.15 TRUSS SUPPORTS AND OTHER DETAILS

End bearings transmit the reactions from trusses to substructure elements, such as abutments or piers. Unless trusses are supported on tall slender piers that can deflect horizontally without exerting large forces on the trusses, it is customary to provide expansion bearings at one end of the span and fixed bearings at the other end.

Anchoring a truss to the support, a fixed bearing transmits the longitudinal loads from wind and live-load traction, as well as vertical loads and transverse wind. This bearing also must incorporate a hinge, curved bearing plate, pin arrangement, or elastomeric pads to permit end rotation of the truss in its plane.

An expansion bearing transmits only vertical and transverse loads to the support. It permits changes in length of trusses, as well as end rotation.

Many types of bearings are available. To ensure proper functioning of trusses in accordance with design principles, designers should make a thorough study of the bearings, including allowances for reactions, end rotations and horizontal movements. For short trusses, a rocker may be used for the expansion end of a truss. For long trusses, it generally is necessary to utilize some sort of roller support. See also Arts. 10.22 and 11.9.

Inspection Walkways.   An essential part of a truss design is provision of an inspection walkway. Such walkways permit thorough structural inspection and also are of use during erection and painting of bridges. The additional steel required to support a walkway is almost insignificant.

12.16 *CONTINUOUS TRUSSES*

Many river crossings do not require more than one truss span to meet navigational requirements. Nevertheless, continuous trusses have made possible economical bridge designs in many localities. Studies of alternative layouts are essential to ensure selection of the lowest-cost arrangement. The principles outlined in preceding articles of this section are just as applicable to continuous trusses as to simple spans. Analysis of the stresses in the members of continuous trusses, however, is more complex, unless computer-aided design is used. In this latter case, there is no practical difference in the calculation of member loads once the forces have been determined. However, if the truss is truly continuous, and, therefore, the truss in each span is statically indeterminant, the member forces are dependent on the stiffness of the truss members. This may make several iterations of member-force calculations necessary. But where sufficient points of articulation are provided to make each individual truss statically determinant, such as the case where a suspended span is inserted in a cantilever truss, the member forces are not a function of member stiffness. As a result, live-load forces need be computed only once, and dead-load member forces need to be updated only for the change in member weight as the design cycle proceeds. When the stresses have been computed, design proceeds much as for simple spans.

The preceding discussion implies that some simplification is possible by using cantilever design rather than continuous design. In fact, all other things being equal, the total weight of members will not be much different in the two designs if points of articulation are properly selected. More roadway joints will be required in the cantilever, but they, and the bearings, will be subject to less movement. However, use of continuity should be considered because elimination of the joints and devices necessary to provide for articulation will generally reduce maintenance, stiffen the bridge, increase redundancy and, therefore, improve the general robustness of the bridge.

SECTION 13
ARCH BRIDGES

Arthur W. Hedgren, Jr.*

Senior Vice President, HDR Engineering, Inc.,
Pittsburgh, Pennsylvania

Basic principles of arch construction have been known and used successfully for centuries. Magnificent stone arches constructed under the direction of engineers of the ancient Roman Empire are still in service after 2000 years, as supports for aqueducts or highways. One of the finest examples is the Pont du Gard, built as part of the water-supply system for the city of Nîmes, France.

Stone was the principal material for arches until about two centuries ago. In 1779, the first metal arch bridge was built. Constructed of cast iron, it carried vehicles over the valley of the Severn River at Coalbrookedale, England. The bridge is still in service but now is restricted to pedestrian traffic. Subsequently, many notable iron or steel arches were built. Included was Eads' Bridge, with three tubular steel arch spans, 502, 520, and 502 ft, over the Mississippi River at St. Louis. Though completed in 1874, it now carries large daily volumes of heavy highway traffic.

Until 1900, stone continued as a strong competitor of iron and steel. After 1900, concrete became the principal competitor of steel for shorter-span arch bridges.

Development of structural steels made it feasible to construct long-span arches economically. The 1675-ft Bayonne Bridge, between Bayonne, N.J., and Staten Island, N.Y., was completed in 1931. The 1000-ft Lewiston-Queenston Bridge over the Niagara River on the United States–Canadian border was put into service in 1962. Availability of more high-strength steels and improved fabrication techniques expanded the feasibility of steel arches for long spans. Examples include the 1255-ft-span Fremont Bridge in Portland, Ore., finished in 1973, and the 1700-ft-span New River Gorge Bridge near Fayetteville, W. Va., opened in 1977.

Nearly all the steel arches that have been built lie in vertical planes. Accordingly, this section discusses design principles for such arches. A few arch bridges, however, have been constructed with ribs inclined toward each other. This construction is effective in providing lateral stability and offers good appearance. Also, the decrease in average distance between the arch ribs of a bridge often makes possible the use of more economical Vierendeel-girder bracing instead of trussed bracing. Generally, though, inclined arches are not practicable for bridges with very wide

---

*Revised from Sec. 13, "Arch Bridges," by George S. Richardson (deceased), Richardson, Gordon and Associates, Pittsburgh, in *Structural Steel Designer's Handbook*, 1st ed., McGraw-Hill Book Company, New York.

roadways unless the span is very long, because of possible interference with traffic clearances.

13.1   TYPES OF ARCHES

In the most natural type of arch, the horizontal component of each reaction, or thrust, is carried into a buttress, which also carries the vertical reaction. This type will be referred to as the *true arch*. The application of arch construction, however, can be greatly expanded economically by carrying the thrust through a tie, a tension member between the ends of the span. This type will be referred to as a *tied arch*.

Either truss or girder may be used for the arch member. Accordingly, arch bridges are classified as *trussed* or *solid-ribbed*.

Arch bridges are also classified according to the degree of articulation. A *fixed arch*, in which the construction prevents rotation at the ends of the span, is statically indeterminate, so far as external reactions are concerned, to the third degree. If the span is articulated at the ends, it becomes *two-hinged* and statically indeterminate to the first degree. In recent years, most arch bridges have been constructed as either fixed or two-hinged. Sometimes a hinge is included at the crown in addition to the end hinges. The bridge then becomes *three-hinged* and statically determinate.

In addition, arch bridges are classified as *deck* construction when the arches are entirely below the deck. This is the most usual type for the true arch. Tied arches, however, normally are constructed with the arch entirely above the deck and the tie at deck level. This type will be referred to as a *through arch*. Both true and tied arches, however, may be constructed with the deck at some intermediate elevation between springing and crown. These types are classified as *half-through*.

The arch also may be used as one element combined with another type of structure. For example, many structures have been built with a three-span continuous truss as the basic structure and with the central span arched and tied. This section is limited to structures in which the arch type is used independently.

13.2   ARCH FORMS

A great variety of forms have been used for trussed or solid-ribbed arch bridges. The following are some of the principal forms used.

Lindenthal's Hell Gate Bridge over the East River in New York has trusses deep at the ends and shallow at the crown. The bottom chord is a regular arch form. The top chord follows a reversed curve transitioning from the deep truss at the end to the shallow truss at the center. Accordingly, it is customary to refer to arch trusses of this form as Hell-Gate-type trusses. In another form commonly used, top and bottom chords are parallel. For a two-hinged arch, a crescent-shaped truss is another logical form.

For solid-ribbed arches, single-web or box girders may be used. Solid-ribbed arches usually are built with girders of constant depth. Variable-depth girders, tapering from deep sections at the springing to shallower sections at the crown, however, have been used occasionally for longer spans. As with trussed construction, a crescent-shaped girder is another possible form for a two-hinged arch.

Tied arches permit many variations in form to meet specific site conditions. In a true arch (without ties), the truss or solid rib must carry both thrust and moment under variable loading conditions. These stresses determine the most effective depth of truss or girder. In a tied arch, the thrust is carried by the arch truss or solid rib, but the moment for variable loading conditions is divided between arch and tie, somewhat in proportion to the respective stiffnesses of these two members. For this reason, for example, if a deep girder is used for the arch and a very shallow member for the tie, most of the moment for variable loading is carried by the arch rib. The tie acts primarily as a tension member. But if a relatively deep member is used for the tie, it carries a high proportion of the moment, and a relatively shallow member may be used for the arch rib. In some cases, a truss has been used for the arch tie in combination with a shallow, solid rib for the arch. This combination may be particularly applicable for double-deck construction.

Rigid-framed bridges, sometimes used for grade-separation structures, are basically another form of two-hinged or fixed arch. The generally accepted arch form is a continuous, smooth-curve member or a segmental arch (straight between panel points) with breaks located on a smooth-curve axis. For a rigid frame, however, the arch axis becomes rectangular in form. Nevertheless, the same principles of stress analysis may be used as for the smooth-curve arch form.

The many different types and forms of arch construction make available to bridge engineers numerous combinations to meet variable site conditions.

13.3 SELECTION OF ARCH TYPE AND FORM

Some of the most important elements influencing selection of type and form of arch follow.

Foundation Conditions. If a bridge is required to carry a roadway or railroad across a deep valley with steep walls, an arch is probably a feasible and economical solution. (This assumes that the required span is within reasonable limits for arch construction.) The condition of steep walls indicates that foundation conditions should be suitable for the construction of small, economical abutments. Generally, it might be expected that under these conditions the solution would be a deck bridge. There may be other controls, however, that dictate otherwise. For example, the need for placing the arch bearings safely above high-water elevation, as related to the elevation of the deck, may indicate the advisability of a half-through structure to obtain a suitable ratio of rise to span. Also, variable foundation conditions on the walls of the valley may fix a particular elevation as much more preferable to others for the construction of the abutments. Balancing of such factors will determine the best layout to satisfy foundation conditions.

Tied-Arch Construction. At a bridge location where relatively deep foundations are required to carry heavy reactions, a true arch, transmitting reactions directly to buttresses, is not economical, except for short spans. There are two alternatives, however, that may make it feasible to use arch construction.

If a series of relatively short spans can be used, arch construction may be a good solution. In this case, the bridge would comprise a series of equal or nearly equal spans. Under these conditions, dead-load thrusts at interior supports would be balanced or nearly balanced. With the short spans, unbalanced live-load thrusts would not be large. Accordingly, even with fairly deep foundations, intermediate pier construction may be almost as economical as for some other layout with simple or continuous spans. There are many examples of stone, concrete, and steel arches in which this arrangement has been used.

The other alternative to meet deep foundation requirements is tied-arch construction. The tie relieves the foundation of the thrust. This places the arch in direct competition with other types of structures for which only vertical reactions would result from the application of dead and live loading.

There has been some concern over the safety of tied-arch bridges because the ties can be classified as fracture-critical members. A fracture-critical member is one that would cause collapse of the bridge if it fractured. Since the horizontal thrust of a tied-arch is resisted by its tie, most tied arches would collapse if the tie were lost. While some concern over fracture of welded tie girders is well-founded, methods are available for introducing redundancy in the construction of ties. These methods include using ties fabricated from multiple bolted-together components and multiple post-tensioning tendons. Tied arches often provide cost-effective and esthetically pleasing structures. This type of structure should not be dismissed over these concerns, because it can be easily designed to address them.

Length of Span.   Generally, determination of the best layout for a bridge starts with trial of the shortest feasible main span. Superstructure costs per foot increase rapidly with increase in span. Unless there are large offsetting factors that reduce substructure costs when spans are lengthened, the shortest feasible span will be the most economical.

Arch bridges are applicable over a wide range of span lengths. The examples in Art. 13.8 cover a range from a minimum of 193 ft to a maximum of 1700 ft. With present high-strength steels and under favorable conditions, spans on the order of 2000 ft are feasible for economical arch construction.

In addition to foundation conditions, many other factors may influence the length of span selected at a particular site. Over navigable waters, span is normally set by clearance requirements of regulatory agencies. For example, the U.S. Coast Guard has final jurisdiction over clearance requirements over navigable streams. In urban or other highly built-up areas, the span may be fixed by existing site conditions that cannot be altered.

Truss or Solid Rib.   Most highway arch bridges with spans up to 750 ft have been built with solid ribs for the arch member. There may, however, be particular conditions that would make it more economical to use trusses for considerably shorter spans. For example, for a remote site with difficult access, truss arches may be less expensive than solid-ribbed arches, because the trusses may be fabricated in small, lightweight sections, much more readily transported to the bridge site.

In the examples of Art. 13.8, solid ribs have been used in spans up to 1255 ft, as for the Fremont Bridge, Portland, Ore. For spans over 750 ft, however, truss arches should be considered. Also, for spans under this length for very heavy live loading, as for railroad bridges, truss arches may be preferable to solid-rib construction.

For spans over about 600 ft, control of deflection under live loading may dictate the use of trusses rather than solid ribs. This may apply to bridges designed for heavy highway loading or heavy transit loading as well as for railroad bridges. For spans above 1000 ft, truss arches, except in some very unusual case, should be used.

Articulation.   For true, solid-ribbed arches the choice between fixed and hinged ends will be a narrow one. In a true arch it is possible to carry a substantial moment at the springing line if the bearing details are arranged to provide for it. This probably will result in some economy, particularly for long spans. It is, however, common practice to use two-hinged construction.

An alternative is to let the arch act as two-hinged under partial or full dead load and then fix the end bearings against rotation under additional load.

Tied arches act substantially as two-hinged, regardless of the detail of the connection to the tie.

Some arches have been designed as three-hinged under full or partial dead load and then converted to the two-hinged condition. In this case, the crown hinge normally is located on the bottom chord of the truss. If the axis of the bottom chord follows the load thrust line for the three-hinged condition, there will be no stress in the top chord or web system of the truss. Top chord and web members will be stressed only under load applied after closure. These members will be relatively light and reasonably uniform in section. The bottom chord becomes the main load-bearing member.

If, however, the arch is designed as two-hinged, the thrust under all loading conditions will be nearly equally divided between top and bottom chords. For a given ratio of rise to span, the total horizontal thrust at the end will be less than that for the arrangement with part of the load carried as a three-hinged arch. Shifting from three to two hinges has the effect of increasing the rise of the arch over the rise measured from springing to centerline of bottom chord.

Esthetics.   For arch or suspension-type bridges, a functional layout meeting structural requirements normally results in simple, clean-cut, and graceful lines. For long spans, no other type offered serious competition so far as excellent appearance is concerned until about 1950. Since then, introduction of cable-stayed bridges and orthotropic-deck girder construction has made construction of good-looking girders feasible for spans of 2500 ft or more. Even with conventional deck construction but with the advantage of high-strength steels, very long girder spans are economically feasible and esthetically acceptable.

The arch then must compete with suspension, cable-stayed, and girder bridges so far as esthetic considerations are concerned. From about 1000 ft to the maximum practical span for aches, the only competitors are the cable-supported types.

Generally, architects and engineers prefer, when all other things are equal, that deck structures be used for arch bridges. If a through or half-through structure must be used, solid-ribbed arches are desirable when appearance is of major concern, because the overhead structure can be made very light and clean-cut (Figs. 13.5 to 13.8 and 13.15 to 13.18).

Arch Form as Related to Esthetics.   For solid-ribbed arches, designers are faced with the decision as to whether the rib should be curved or constructed on segmental chords (straight between panel points). A rib on a smooth curve presents the best appearance. Curved ribs, however, involve some increase in material and fabrication costs.

Another decision is whether to make the rib of constant depth or tapered.

One factor that has considerable bearing on both these decisions is the ratio of panel length to span. As panel length is reduced, the angular break between chord segments is reduced, and a segmental arch approaches a curved arch in appearance. An upper limit for panel length should be about $\frac{1}{15}$ of the span.

In a study of alternative arch configurations for a 750-ft span, four solid-ribbed forms were considered. An architectural consultant rated these in the following order:

Tapered rib, curved

Tapered-rib on chords

Constant-depth rib, curved

Constant-depth rib on chords

He concluded that the tapered rib, 7 ft deep at the springing line and 4 ft deep at the crown, added considerably to the esthetic quality of the design as compared with a constant-depth rib. He also concluded that the tapered rib would minimize the angular breaks at panel points with the segmental chord axis. The tapered rib on chords was used in the final design of the structure. The effect of some of these variables on economy is discussed in Art. 13.6.

13.4   COMPARISON OF ARCH WITH OTHER BRIDGE TYPES

Because of the wide range of span length within which arch construction may be used (Art. 13.3), it is competitive with almost all other types of structures.

Comparison with Simple Spans.   Simple-span girder or truss construction normally falls within the range of the shortest spans used up to a maximum of about 800 ft. Either true arches under favorable conditions or tied arches under all conditions are competitive within the range of 200 to 800 ft. (There will be small difference in cost between these two types within this span range.) With increasing emphasis on appearance of bridges, arches are generally selected rather than simple-span construction, except for short spans for which beams or girders may be used.

Comparison with Cantilever or Continuous Trusses.   The normal range for cantilever or continuous-truss construction is on the order of 500 to 1800 ft for main spans. More likely, a top limit is about 1500 ft. Tied arches are competitive for spans within the range of 500 to 1000 ft. True arches are competitive, if foundation conditions are favorable, for spans from 500 ft to the maximum for the other types. The relative economy of arches, however, is enhanced where site conditions make possible use of relatively short-span construction over the areas covered by the end spans of the continuous or cantilever trusses.

The economic situation is approximately this: For three-span continuous or cantilever layouts arranged for the greatest economy, the cost per foot will be nearly equal for end and central spans. If a tied or true arch is substituted for the central span, the cost per foot may be more than the average for the cantilever or continuous types. If, however, relatively short spans are substituted for the end spans of these types, the cost per foot over the length of those spans is materially reduced. Hence, for a combination of short spans and a long arch span, the overall cost between end piers may be less than for the other types. In any case, the cost differential should not be large.

Comparison with Cable-Stayed and Suspension Bridges.   Such structures normally are not used for spans of less than 500 ft. Above 3000 ft, suspension bridges are probably the most practical solution. In the shorter spans, self-anchored construction is likely to be more economical than independent anchorages. Arches are competitive in cost with the self-anchored suspension type or similar functional type with cable-stayed girders or trusses. There has been little use of suspension bridges for spans under 1000 ft, except for some self-anchored spans. For spans above 1000 ft, it is not possible to make any general statement of comparative costs. Each site requires a specific study of alternative designs.

13.5   ERECTION OF ARCH BRIDGES

Erection conditions vary so widely that it is not possible to cover many in a way that is generally applicable to a specific structure.

Cantilever Erection.   For arch bridges, except short spans, cantilever erection usually is used. This may require use of two or more temporary piers. Under some conditions, such as an arch over a deep valley where temporary piers are very costly, it may be more economical to use temporary tiebacks.

Particularly for long spans, erection of trussed arches often is simpler than erection of solid-ribbed arches. The weights of individual members are much smaller, and trusses are better adapted to cantilever erection. The Hell-Gate-type truss (Art. 13.2) is particularly suitable because it requires little if any additional material in the truss on account of erection stresses.

For many double-deck bridges, use of trusses for the arch ties simplifies erection when trusses are deep enough and the sections large enough to make cantilever erection possible and at the same time to maintain a clear opening to satisfy temporary nagivation or other clearance requirements.

Control of Stress Distribution.   For trussed arches designed to act as three-hinged, under partial or full dead load, closure procedures are simple and positive. Normally, the two halves of the arch are erected to ensure that the crown hinge is high and open. A top-chord member at the crown is temporarily omitted. The trusses are then closed by releasing the tiebacks or lowering temporary intermediate supports. After all dead load for the three-hinged condition is on the span, the top chord is closed by inserting the final member. During this operation, consideration must be given to temperature effects to ensure that closure conditions conform to temperature-stress assumptions.

If a trussed arch has been designed to act as two-hinged under all conditions of loading, the procedure may be first to close the arch as three-hinged. Then, jacks are used at the crown to attain the calculated stress condition for top and bottom chords under the closing erection load and temperature condition. This procedure, however, is not as positive and not as certain of attaining agreement between actual and calculated stresses as the other procedure described. (There is a difference of opinion among bridge engineers on this point.)

Another means of controlling stress distribution may be used for tied arches. Suspender lengths are adjusted to alter stresses in both the arch ribs and the ties.

Fixed Bases.   For solid-ribbed arches to be erected over deep valleys, there may be a considerable advantage in fixing the ends of the ribs. If this is not provided for in design, it may be necessary to provide temporary means for fixing bases for cantilever erection of the first sections of the ribs. If the structure is designed for fixed ends, it may be possible to erect several sections as cantilevers before it becomes necessary to install temporary tiebacks.

13.6 DESIGN OF ARCH RIBS AND TIES

Computers greatly facilitate preliminary and final design of all structures. They also make possible consideration of many alternative forms and layouts, with little additional effort, in preliminary design. Even without the aid of a computer, however, experienced designers can, with reasonable ease, investigate alternative layouts and arrive at sound decisions for final arrangements of structures.

Rise-Span Ratio.   The generally used ratios of rise to span cover a range of about 1:5 to 1:6. For all but two of the arch examples in Art. 13.8, the range is from a maximum of 1:4.7 to a minimum of 1:6.3. The flatter rise is more desirable for through arches, because appearance will be better. Cost will not vary materially within the rise limits of 1:5 to 1:6. These rise ratios apply both to solid ribs and to truss arches with rise measured to the bottom chord.

Panel Length.   For solid-ribbed arches fabricated with segmental chords, panel length should not exceed ⅟₁₅ of the span. This is recommended for esthetic reasons, to prevent too large angular breaks at panel points. Also, for continuously curved axes, bending stresses in solid-ribbed arches become fairly severe if long panels are used. Other than this limitation, the best panel length for an arch bridge will be determined by the usual considerations, such as economy of deck construction.

Ratio of Depth to Span.   In the examples in Art. 13.8, the true arches (without ties) with constant-depth solid ribs have depth-span ratios from 1:58 to 1:79. The larger ratio, however, is for a short span. A more normal range is 1:70 to 1:80. These ratios also are applicable to solid-ribbed tied arches with shallow ties. In such cases, since the ribs must carry substantial bending moments, depth requirements are little different from those for a true arch. For structures with variable-depth ribs, the depth-span ratio may be relatively small (Fig. 13.7).

For tied arches with solid ribs and deep ties, depth of rib may be small, because the ties carry substantial moments, thus reducing the moments in the ribs. For a number of such structures, the depth-span ratio ranges from 1:140 to 1:190, and for the Fremont Bridge, Portland, Ore., is as low as 1:314. Note that such shallow ribs can be used only with girder or trussed ties of considerable depth.

For truss arches, whether true or tied, the ratio of crown depth to span may range from 1:25 to 1:50. Depth of tie has little effect on depth of truss required. Except for some unusual arrangement, the moment of inertia of the arch truss is much larger than the moment of inertia of its tie, which primarily serves as a tension member to carry the thrust. Hence, an arch truss carries substantial bending moments whether or not it is tied, and required depth is not greatly influenced by presence or absence of a tie.

Single-Web or Box Girders.   For very short arch spans, single-web girders are more economical than box girders. For all the solid-ribbed arches in Art. 13.8, however, box girders were used for the arch ribs. These examples include a minimum span of 193 ft. Welded construction greatly facilitates use of box members in all types of structures.

For tied arches for which shallow ties are used, examples in Art. 13.8 show use of members made up of web plates with diaphragms and rolled shapes with post-tensioned strands. More normally, however, the ties, like solid ribs, would be box girders.

Truss Arches.   All the usual forms of bolted or welded members may be used in truss arches but usually, sealed, welded box members are preferred. These present a clean-cut appearance. There also is an advantage in the ease of maintenance.

Another variation of truss arches that can be considered is use of Vierendeel trusses (web system without diagonals). In the past, complexity of stress analysis for this type discouraged their use. With computers, this disadvantage is eliminated. Various forms of Vierendeel truss might well be used for both arch ribs and ties. There has been some use of Vierendeel trusses for arch bracing, as shown in the examples in Art. 13.8. This design provides an uncluttered, good-looking bracing system.

Dead-Load Distribution.   It is normal procedure for both true and tied solid-ribbed arches to use an arch axis conforming closely to the dead-load thrust line. In such cases, if the rib is cambered for dead load, there will be no bending in the rib under that load. The arch will be in pure compression. If a tied arch is used, the tie will be in pure tension. If trusses are used, the distribution of dead-load stress may be similarly controlled. Except for three-hinged arches, however, it will be necessary to use jacks at the crown or other stress-control procedures to attain the stress distribution that has been assumed.

Live-Load Distribution.   One of the advantages of arch construction is that fairly uniform live loading, even with maximum-weight vehicles, creates relatively low bending stresses in either the rib or the tie. Maximum bending stresses occur only under partial loading not likely to be realized under normal heavy traffic flow. Maximum live-load deflection occurs in the vicinity of the quarter point with live load over about half the span.

Wind Stresses.   These may control design of long-span arches carrying two-lane roadways or of other structures for which there is relatively small spacing of ribs compared with span length. For a spacing-span ratio larger than $1:20$, the effect of wind may not be severe. As this ratio becomes substantially smaller, wind may affect sections in many parts of the structure.

Thermal Stresses.   Temperature causes stress variation in arches. One effect sometimes neglected but which should be considered is that of variable temperature throughout a structure. In a through, tied arch during certain times of the day or night, there may be a large difference in temperature between rib and tie due to different conditions of exposure. This difference in temperature easily reaches 30°F and may be much larger.

Deflection.   For tied arches of reasonable rigidity, deflection under live load causes relatively minor changes in stress (secondary stresses). For a 750-ft span with solid-ribbed arches 7 ft deep at the springing line and 4 ft deep at the crown and designed for a maximum live-load deflection of $\frac{1}{800}$ of the span, the secondary effect of deflections was computed as less than 2% of maximum allowable unit stress. For a true arch, however, this effect may be considerably larger and must be considered, as required by design specifications.

Dead-Load to Total-Load Ratios.   For some 20 arch spans checked, the ratio of dead load to total load varied within the narrow range of 0.74 to 0.88. A common ratio is about 0.85. This does not mean that the ratio of dead-load stress to maximum total stress will be 0.85. This stress ratio may be fairly realistic for a fully loaded structure, at least for most of the members in the arch system. For partial live loading, however, which is the loading condition causing maximum live-load stress, the ratio of dead to total stress will be much lower, particularly as span decreases.

For most of the arches checked, the ratio of weight of arch ribs or, in the case of tied arches, weight of ribs and ties to, total load ranged from about 0.20 to 0.30. This is true despite the wide range of spans included and the great variety of steels used in their construction.

Use of high-strength steels helps to maintain a low ratio for the longer spans. For example, for the Fort Duquesne Bridge, Pittsburgh, a double-deck structure of 423-ft span with a deep truss as a tie, the ratio of weight of arch ribs plus truss ties to total load is about 0.22, or a normal factor within the range previously cited. For this bridge, arch ribs and trusses were designed with 77% of A440 steel and the remainder A36. These are suitable strength steels for this length of span.

For the Fort Pitt Bridge, Pittsburgh, with a 750-ft span and the same arrangement of structure with shallow girder ribs and a deep truss for the ties, the ratio of weight of steel in ribs plus trussed ties to total load is 0.33. The same types of steel in about the same percentages were used for this structure as for the Fort Duquesne Bridge. A higher-strength steel, such as A514, would have resulted in a much lower percentage for weight of arch ribs and trusses and undoubtedly in considerable economy. When the Fort Pitt arch was designed, however, the owner decided there had not been sufficient research and testing of the A514 steel to warrant its use in this structure.

For a corresponding span of 750 ft designed later for the Glenfield Bridge at Pittsburgh, a combination of A588 and A514 steels was used for the ribs and ties. The ratio of weight of ribs plus ties to total load is 0.19.

Incidentally, the factors for this structure, a single-deck bridge with six lanes of traffic plus full shoulders, are almost identical with the corresponding factors for the Sherman Minton Bridge at Louisville, Ky., an 800-ft double-deck structure with truss arches carrying three lanes of traffic on each deck. The factors for the Pittsburgh bridge are 0.88 for ratio of dead load to total load and 0.19 for ratio of weight of ribs plus ties to total load. The corresponding factors for the Sherman Minton arch are 0.85 and 0.19. Although these factors are almost identical, the total load for the Pittsburgh structure is considerably larger than that for the Louisville structure. The difference may be accounted for primarily by the double-deck structure for the latter, with correspondingly lighter deck construction.

For short spans, particularly those on the order of 250 ft or less, the ratio of weight of arch rib to total load may be much lower than the normal range of 0.20 to 0.30. For example, for a short span of 216 ft, this ratio is 0.07. On the other hand, for a span of only 279 ft, the ratio is 0.18, almost in the normal range.

A ratio of arch-rib weight to total load may be used by designers as one guide in selecting the most economical type of steel for a particular span. For a ratio exceeding 0.25, there is an indication that a higher-strength steel than has been considered might reduce costs and its use should be investigated, if available.

Effect of Form on Economy of Construction. For solid-ribbed arches, a smooth-curve axis is preferable to a segmental-chord axis (straight between panel points) so far as appearance is concerned. The curved axis, however, involves additional cost of fabrication. At the least, some additional material is required in fabrication of the arch because of the waste in cutting the webs to the curved shape. In addition to this waste, some material must be added to the ribs to provide for increased stresses due to bending. This occurs for the following reason: Since most of the load on the rib is applied at panel points, the thrust line is nearly straight between panel points. Curving the axis of the rib causes eccentricity of the thrust line with respect to the axis and thus induces increased bending moments, particularly for dead load. All these effects may cause an increase in the cost of the curved rib on the order of 5 to 10%.

For tied solid-ribbed arches for which it is necessary to use a very shallow tie, costs are larger than for shallow ribs and deep ties. (A shallow tie may be necessary to meet underclearance restrictions and vertical grades of the deck.) A check of a 750-ft span for two alternate designs, one with a 5-ft constant-depth rib and 12.5-ft-deep tie and the other with a 10-ft-deep rib and 4-ft-deep tie, showed that the latter arrangement, with shallow tie, required about 10% more material than the former, with deep tie. The actual increased construction cost might be more on the order of 5%, because of some constant costs for fabrication and erection that would not be affected by the variation in weight of material.

Comparison of a tapered rib with a constant-depth rib indicates a small-percentage saving in material in favor of the tapered rib. Thus, costs for these two alternatives would be nearly equal.

13.7 DESIGN OF OTHER ELEMENTS

A few special conditions relating to elements of arch bridges other than the ribs and ties should be considered in design of arch bridges.

Floor System. Tied arches, particularly those with high-strength steels, undergo relatively large changes in length of deck due to variation in length of tie under various load conditions. It therefore is normally necessary to provide deck joints at intermediate points to provide for erection conditions and to avoid high participation stresses.

Bracing. During design of the Bayonne Bridge arch (Art. 13.8), a study in depth explored the possibility of eliminating most of the sway bracing (bracing in a vertical plane between ribs). In addition to detailed analysis, studies were made on a scaled model to check the effect of various arrangements of this bracing. The investigators concluded that, except for a few end panels, the sway bracing could be eliminated. Though many engineers still adhere to an arbitrary specification requirement calling for sway bracing at every panel point of any truss, more consideration should be given to the real necessity for this. Furthermore, elimination of sway frames not only reduces costs but it also greatly improves the appearance of the structure. For several structures from which sway bracing has been omitted, there has been no adverse effect.

Various arrangements may be used for lateral bracing systems in arch bridges. For example, a diamond pattern, omitting cross struts at panel points, is often effective. Also, favorable results have been obtained with a Vierendeel truss.

In the design of arch bracing, consideration must be given to the necessity for the lateral system to prevent lateral buckling of the two ribs functioning as a single compression member. The lateral bracing thus is the lacing for the two chords of this member.

Hangers. These must be designed with sufficient rigidity to prevent adverse vibration under aerodynamic forces or as very slender members (wire rope or bridge strand). A number of long-span structures incorporate the latter device. Vibration problems have developed with some bridges for which rigid members with high slenderness ratios have been used. Corrosion resistance and provision for future replacement are other concerns which must be addressed in design of wire hangers.

13.8 EXAMPLES OF ARCH BRIDGES

Thanks to the cooperation of several engineers in private and public practice, detailed information on about 25 arch bridges has been made available. Sixteen have been selected from this group to illustrate the variety of arch types and forms in the wide range and span length for which steel arches have been used. Many of these bridges have been awarded prizes in the annual competition of the American Institute of Steel Construction.

The examples include only bridges constructed within the United States, though there are many notable arch bridges in other countries. A noteworthy omission is the imaginative and attractive Port Mann Bridge over the Fraser River in Canada. C.B.A. Engineering Ltd., consulting engineers, Vancouver, B.C., were the design engineers. By use of an orthotropic deck and stiffened, tied, solid-ribbed arch, an economical layout was developed with a central span of 1,200 ft, flanked by side spans of 360 ft each. A variety of steels were used, including A373, A242, and A7.

Following are data on arch bridges that may be useful in preliminary design. (*Text continues on page* 13.44.)

FIGURE 13.1

NEW RIVER GORGE BRIDGE

LOCATION: Fayetteville, West Virginia
TYPE: Trussed deck arch, 40 panels, 36 at 40± to 43± ft
SPAN: 1,700 ft   RISE: 353 ft   RISE/SPAN = 1:4.8
NO. OF LANES OF TRAFFIC: 4
HINGES: 2   CROWN DEPTH: 34 ft   DEPTH/SPAN = 1:50

AVERAGE DEAD LOAD:	LB PER FT
Deck slab and surfacing of roadway	8,600
Railings and parapets	1,480
Floor steel for roadway	3,560
Arch trusses	11,180
Arch bracing	1,010
Arch bents and bracing	2,870
TOTAL	28,700

SPECIFICATION FOR LIVE LOADING: HS20-44
EQUIVALENT LIVE + IMPACT LOADING PER ARCH FOR FULLY LOADED
    STRUCTURE: 1,126 lb per ft
TYPES OF STEEL IN STRUCTURE:

Arch	A588
Floor system	A588

OWNER: State of West Virginia
ENGINEER: Michael Baker, Jr., Inc.
FABRICATOR/ERECTOR: American Bridge Division, U.S. Steel Corporation
DATE OF COMPLETION: October, 1977

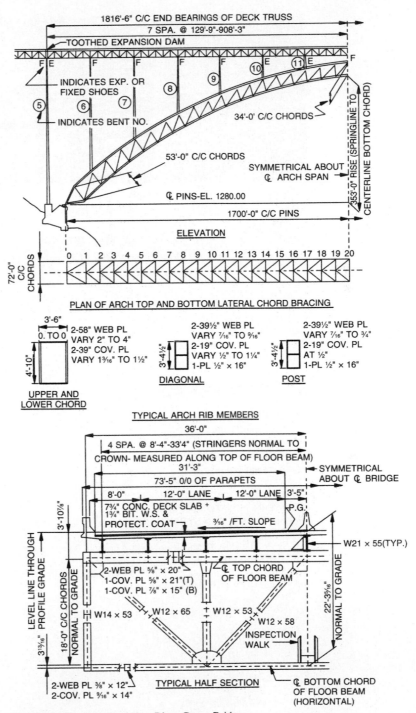

FIGURE 13.2 Details of New River Gorge Bridge.

FIGURE 13.3

BAYONNE BRIDGE

LOCATION: Between Bayonne, N.J., and Port Richmond, Staten Island, N.Y.
TYPE: Halt-through truss arch, 40 panels at 41.3 ft
SPAN: 1,675 ft   RISE: 266 ft   RISE/SPAN = 1:6.3
NO. OF LANES OF TRAFFIC: 4 plus 2 future rapid transit
HINGES: 2   CROWN DEPTH: 37.5 ft   DEPTH/SPAN = 1:45

AVERAGE DEAD LOAD:	LB PER FT
Track, paving	6,340
Floor steel and floor bracing	6,160
Arch truss and bracing	14,760
Arch hangers	540
Miscellaneous	200
TOTAL	28,000

SPECIFICATION LIVE LOADING:	LB PER FT
2 rapid-transit lines at 6,000 lb per ft	12,000
4 roadway lanes at 2,500 lb per ft	10,000
2 sidewalks at 600 lb	1,200
TOTAL (unreduced)	23,200

EQUIVALENT LIVE + IMPACT LOADING ON EACH ARCH FOR FULLY LOADED
    STRUCTURE WITH REDUCTION FOR MULTIPLE LANES AND LENGTH OF
    LOADING: 2,800 lb per lin ft
TYPES OF STEEL IN STRUCTURE: About 50% carbon steel, 30% silicon steel, and 20%
    high-alloy steel (carbon-manganese)
OWNER: The Port Authority of New York and New Jersey
ENGINEER: O. H. Ammann, Chief Engineer
FABRICATOR: American Bridge Co., U.S. Steel Corp. (also erector)
DATE OF COMPLETION: 1931

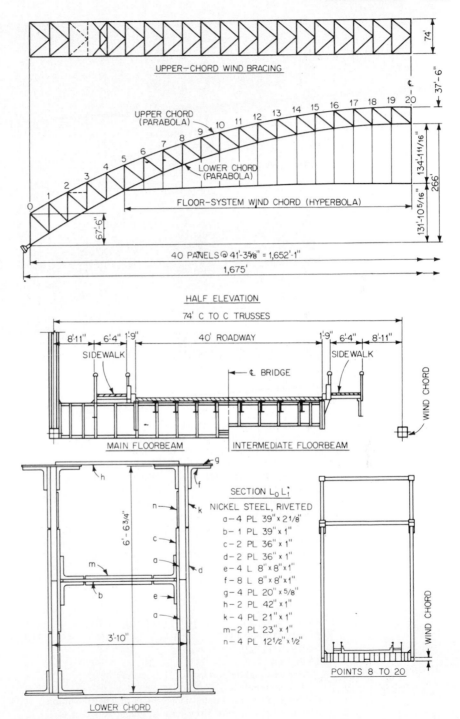

UPPER-CHORD WIND BRACING

UPPER CHORD
(PARABOLA)

LOWER CHORD
(PARABOLA)

FLOOR-SYSTEM WIND CHORD (HYPERBOLA)

74'

37'-6"

134'-11¹¹/₁₆"

266'

131'-10⁵/₁₆"

67'-6"

40 PANELS @ 41'-3⁵/₈" = 1,652'-1"

1,675'

HALF ELEVATION

74' C TO C TRUSSES

8'-11"   6'-4"   1'-9"   40' ROADWAY   1'-9"   6'-4"   8'-11"

SIDEWALK   SIDEWALK

₵ BRIDGE

WIND CHORD

MAIN FLOORBEAM   INTERMEDIATE FLOORBEAM

SECTION $L_0 L_1$

NICKEL STEEL, RIVETED

a– 4 PL 39" x 2¹/₈"
b– 1 PL 39" x 1"
c– 2 PL 36" x 1"
d– 2 PL 36" x 1"
e– 4 L 8" x 8" x 1"
f– 8 L 8" x 8" x 1"
g– 4 PL 20" x ⁵/₈"
h– 2 PL 42" x 1"
k– 4 PL 21" x 1"
m–2 PL 23" x 1"
n– 4 PL 12¹/₂" x ¹/₂"

6'-6³/₄"

3'-10"

WIND CHORD

POINTS 8 TO 20

LOWER CHORD

FIGURE 13.4 Details of Bayonne Bridge.

FIGURE 13.5

FREMONT BRIDGE

LOCATION: Portland, Oregon
TYPE: Half-through, tied, solid ribbed arch, 28 panels at 44.83 ft
SPAN: 1,255 ft   RISE: 341 ft   RISE/SPAN = 1:3.7
NO. OF LANES OF TRAFFIC: 4 each upper and lower roadways
HINGES: 2   DEPTH: 4 ft   DEPTH/SPAN = 1:314

AVERAGE DEAD LOAD:	LB PER FT
Decks and surfacing	10,970
Railings and Parapets	1,280
Floor steel for roadway	4,000
Floor bracing	765
Arch ribs	2,960
Arch bracing	1,410
Arch hangers or columns and bracing	1,250
Arch tie girders	4,200
TOTAL	26,835

SPECIFICATION FOR LIVE LOADING: AASHTO HS20-44
EQUIVALENT LIVE + IMPACT LOADING PER ARCH FOR FULLY LOADED
  STRUCTURE: 2,510 lb per ft
TYPES OF STEEL IN STRUCTURE:
  Arch ribs and tie girders A514, A588 A441, A36
  Floor system A588, A441, A36
OWNER: State of Oregon, Department of Transportation
ENGINEER: Parsons, Brinckerhoff, Quade & Douglas
FABRICATOR: American Bridge Division, U.S. Steel Corp.
ERECTOR: Murphy Pacific Corporation
DATE OF COMPLETION: 1973

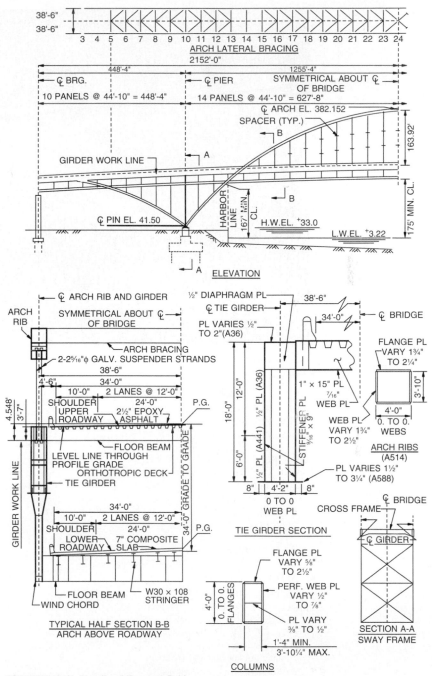

FIGURE 13.6 Details of Fremont Bridge.

FIGURE 13.7

ROOSEVELT LAKE BRIDGE

LOCATION: Roosevelt, Arizona, SR 188
TYPE: Half through, solid rib arch, 16 panels at 50 ft
SPAN: 1,080 ft   RISE: 230 ft   RISE/SPAN = 1:4.7
NO. OF LANES OF TRAFFIC: 2
HINGES: 0   CROWN DEPTH: 8 ft   DEPTH/SPAN = 1:135

AVERAGE DEAD LOAD:	LB PER FT
Deck slab, and surfacing of roadway	4,020
Railings and parapets	800
Floor steel for roadway	1,140
Floor bracing	190
Arch ribs	4,220
Arch bracing	790
Arch hangers	80
TOTAL	11,240

SPECIFICATION FOR LIVE LOADING: HS20-44
EQUIVALENT LIVE + IMPACT LOADING PER ARCH FOR FULLY LOADED
   STRUCTURE: 971 lb per ft
TYPES OF STEEL IN STRUCTURE:
   Arch ribs and ties A572
   Hanger floorbeams and stringers A572
   All others A36
OWNER: Arizona Department of Transportation
ENGINEER: Howard Needles Tammen and Bergendoff
CONTRACTOR: Edward Kraemer & Sons, Inc.
FABRICATOR: Pittsburgh DesMoines Steel Co./Schuff Steel
ERECTOR: John F. Beasley Construction Co.
DATE OF COMPLETION: October 23, 1991 Public Opening

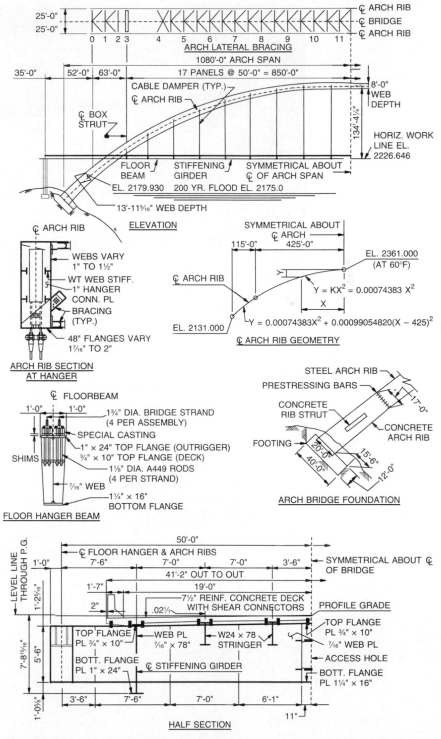

FIGURE 13.8 Details of Roosevelt Lake Bridge.

FIGURE 13.9

LEWISTON–QUEENSTON BRIDGE

LOCATION: Over the Niagara River between Lewiston, N.Y., and Queenston, Ontario
TYPE: Solid-ribbed deck arch, 23 panels at 41.6 ft
SPAN: 1,000 ft   RISE: 159 ft   RISE/SPAN = 1:6.3
NO. OF LANES OF TRAFFIC: 4
HINGES: 0   DEPTH: 13.54 ft   DEPTH/SPAN = 1:74
AVERAGE DEAD LOAD:

	LB PER FT
Deck slab and surfacing for roadway	5,700
Slabs for sidewalks	495
Railings and parapets	780
Floor steel for roadway and sidewalks	2,450
Floor bracing	110
Arch ribs	7,085
Arch bracing	1,060
Arch posts and bracing	1,390
Miscellaneous—utilities, excess, etc.	300
TOTAL	19,370

SPECIFICATION LIVE LOADING: HS20-S16-44
EQUIVALENT LIVE + IMPACT LOADING ON EACH ARCH FOR FULLY LOADED
    STRUCTURE: 1,357 lb per ft
TYPES OF STEEL IN STRUCTURE:

		%
Arch ribs	A440	100
Spandrel columns	A7	94
	A440	6
Rib bracing and end towers	A7	100
Floor system	A373 and A7	

OWNER: Niagara Falls Bridge Commission
ENGINEER: Hardesty & Hanover
FABRICATOR: Bethlehem Steel Co. and Dominion Steel and Coal Corp., Ltd., Subcontractor
DATE OF COMPLETION: Nov. 1, 1962

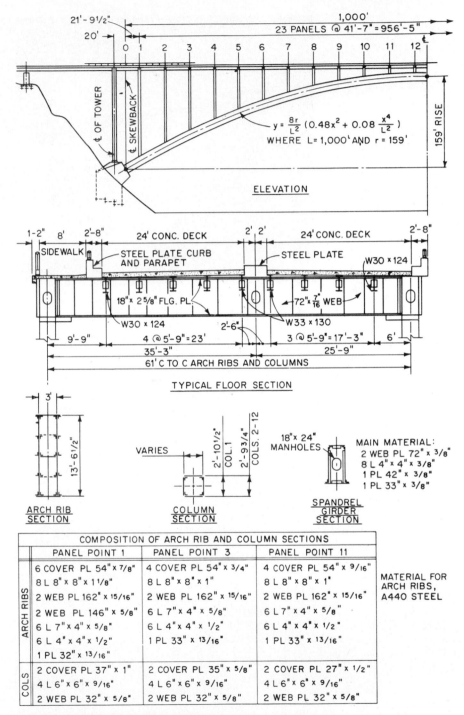

ELEVATION

$$y = \frac{8r}{L^2}\left(0.48x^2 + 0.08\frac{x^4}{L^2}\right)$$
WHERE L = 1,000' AND r = 159'

TYPICAL FLOOR SECTION

ARCH RIB SECTION

COLUMN SECTION

SPANDREL GIRDER SECTION

MAIN MATERIAL:
2 WEB PL 72" x 3/8"
8 L 4" x 4" x 3/8"
1 PL 42" x 3/8"
1 PL 33" x 3/8"

COMPOSITION OF ARCH RIB AND COLUMN SECTIONS			
	PANEL POINT 1	PANEL POINT 3	PANEL POINT 11
ARCH RIBS	6 COVER PL 54" x 7/8"	4 COVER PL 54" x 3/4"	4 COVER PL 54" x 9/16"
	8 L 8" x 8" x 1 1/8"	8 L 8" x 8" x 1"	8 L 8" x 8" x 1"
	2 WEB PL 162" x 15/16"	2 WEB PL 162" x 15/16"	2 WEB PL 162" x 15/16"
	2 WEB PL 146" x 5/8"	6 L 7" x 4" x 5/8"	6 L 7" x 4" x 5/8"
	6 L 7" x 4" x 5/8"	6 L 4" x 4" x 1/2"	6 L 4" x 4" x 1/2"
	6 L 4" x 4" x 1/2"	1 PL 33" x 13/16"	1 PL 33" x 13/16"
	1 PL 32" x 13/16"		
COLS	2 COVER PL 37" x 1"	2 COVER PL 35" x 5/8"	2 COVER PL 27" x 1/2"
	4 L 6" x 6" x 9/16"	4 L 6" x 6" x 9/16"	4 L 6" x 6" x 9/16"
	2 WEB PL 32" x 5/8"	2 WEB PL 32" x 5/8"	2 WEB PL 32" x 5/8"

MATERIAL FOR ARCH RIBS, A440 STEEL

FIGURE 13.10 Details of Lewiston-Queenston Bridge.

FIGURE 13.11

SHERMAN MINTON BRIDGE

LOCATION: On Interstate 64 over the Ohio River between Louisville, Ky., and New Albany, Ind.

TYPE: Tied, through, truss arch, 22 panels at 36.25 ft

SPAN: 800 ft   RISE: 140 ft   RISE/SPAN = 1:5.7

NO. OF LANES OF TRAFFIC: 6, double deck

HINGES: 2   CROWN DEPTH: 30 ft   DEPTH/SPAN = 1:27

AVERAGE DEAD LOAD:

	LB PER FT
Deck slab and surfacing for roadway	7,600
Slabs for sidewalks	1,656
Railings and parapets	804
Floor steel for roadway and sidewalks	2,380
Floor bracing	420
Arch trusses	3,400
Arch bracing	880
Arch hangers and bracing	160
Arch ties	1,040
Miscellaneous—utilities, excess, etc. (including future wearing surface)	1,680
TOTAL	20,020

SPECIFICATION LIVE LOADING: H20-S16

EQUIVALENT LIVE + IMPACT LOADING ON EACH ARCH FOR FULLY LOADED STRUCTURE: 1,755 lb per ft

TYPES OF STEEL IN STRUCTURE:

		%
Arch trusses	A514	69
	A242	18
	A373	13
Floor system	A242	36
	A7	62
	A373	2

OWNER: Indiana Department of Highways and Kentucky Transportation Cabinet

ENGINEER: Hazelet & Erdal, Louisville, Ky.

FABRICATOR: R. C. Mahon Co.

DATE OF COMPLETION: Dec. 22, 1961, opened to traffic

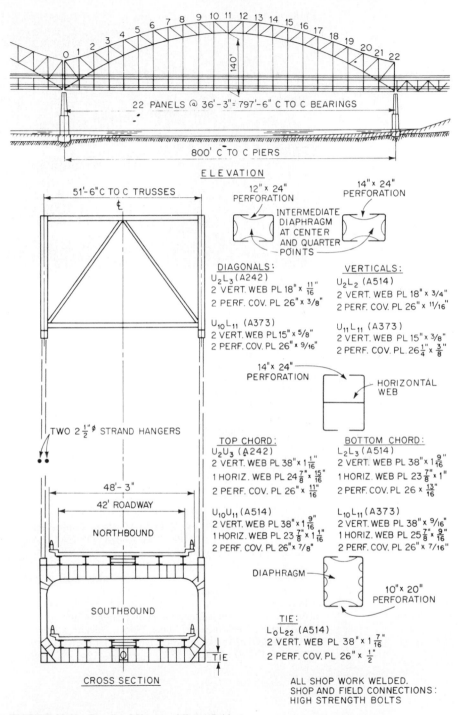

FIGURE 13.12 Details of Sherman Minton Bridge.

FIGURE 13.13

WEST END—NORTH SIDE BRIDGE

LOCATION: Pittsburgh, Pennsylvania, over Ohio River
TYPE: Tied, through, truss arch, 28 panels at 27.8 ft
SPAN: 778 ft   RISE: 151 ft   RISE/SPAN = 1:5.2
NO. OF LANES OF TRAFFIC: 4, including 2 street-railway tracks
HINGES: Two   CROWN DEPTH: 25 ft   DEPTH/SPAN = 1:31
AVERAGE DEAD LOAD: LB PER FT

Roadway, sidewalks, and railings	4,870
Floor steel and floor bracing	2,360
Arch trusses	4,300
Arch ties	2,100
Arch bracing	550
Hangers	360
Utilities and excess	600
TOTAL	15,140

SPECIFICATION LIVE LOADING: Allegheny County Truck & Street Car
EQUIVALENT LIVE + IMPACT LOADING ON EACH ARCH FOR FULLY LOADED
    STRUCTURE: 1,790 lb per ft
TYPES OF STEEL IN STRUCTURE:
    All main material in arch trusses and ties, including splice material—silicon steel.
    Floor system and bracing... A7
    Hangers... Wire rope
OWNER: Pennsylvania Department of Transportation
ENGINEER: Department of Public Works, Allegheny County
FABRICATOR: American Bridge Division, U.S. Steel Corp.
DATE OF COMPLETION: 1932

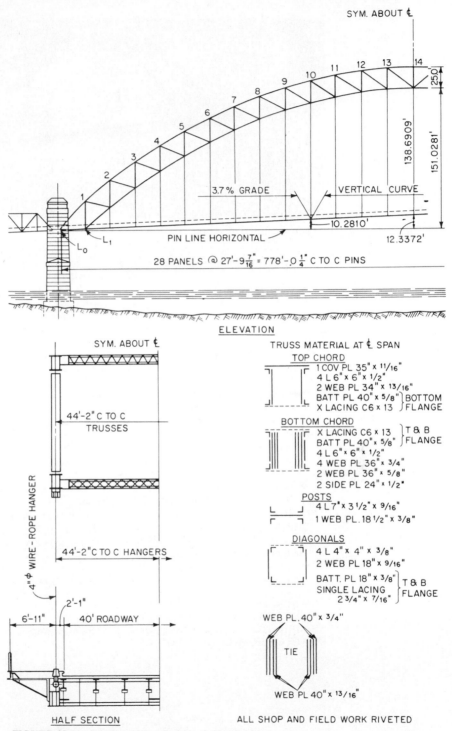

FIGURE 13.14 Details of West End-North Side Bridge.

FIGURE 13.15

FORT PITT BRIDGE

LOCATION: Pittsburgh, Pennsylvania, over the Monongahela River
TYPE: Solid-ribbed, tied, through arch, 30 panels at 25 ft
SPAN: 750 ft   RISE: 122.2 ft   RISE/SPAN = 1:6.2
NO. OF LANES OF TRAFFIC: 4, each level of double deck
HINGES: 2   DEPTH: 5.4 ft   DEPTH/SPAN = 1:139

AVERAGE DEAD LOAD:	LB PER FT
Deck slab and surfacing for roadway, slabs for sidewalks, railings and parapets, on both decks	16,100
Floor steel for roadway and sidewalks, on both decks	4,860
Floor bracing (truss bracing)	480
Arch ribs	5,480
Arch bracing	1,116
Arch hangers (included with rib and tie)	
Arch ties (trusses)	8,424
Miscellaneous—utilities, excess, etc.	400
TOTAL	36,860

SPECIFICATION LIVE LOADING: HS20-S16-44
EQUIVALENT LIVE + IMPACT LOADING ON EACH ARCH FOR FULLY LOADED
   STRUCTURE: 2,500 lb per ft

TYPES OF STEEL IN STRUCTURE:		%
Arch ribs and trussed ties	A242	64
	A7	36
Floor system	A242	90
	A7	10

OWNER: Pennsylvania Department of Highways
ENGINEER: Richardson, Gordon and Associates
FABRICATOR: American Bridge Division, U.S. Steel Corp.
DATE OF COMPLETION: 1957

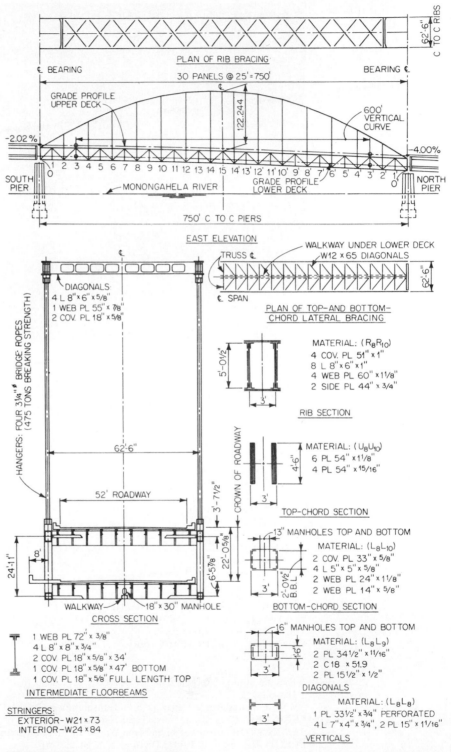

FIGURE 13.16 Details of Fort Pitt Bridge.

FIGURE 13.17

GLENFIELD BRIDGE

LOCATION: I-79 crossing of Ohio River at Neville Island, Pennsylvania
TYPE: Tied, through, solid-ribbed arch, 15 panels at 50 ft
SPAN: 750 ft   RISE: 124.4 ft   RISE/SPAN = 1:6
NO. OF LANES OF TRAFFIC: 6 plus 10-ft berms
HINGES: 0   CROWN DEPTH: 4 ft   DEPTH/SPAN = 1:187

AVERAGE DEAD LOAD:	LB PER FT
Deck slab and surfacing for roadway	13,980
Railings and parapets	1,090
Floor steel for roadway	3,397
Floor bracing	392
Arch ribs	2,563
Arch bracing	1,639
Arch hangers	94
Arch ties	3,400
Miscellaneous—utilities, excess, etc.	589
TOTAL	27,144

SPECIFICATION LIVE LOADING: H20-S16-44
EQUIVALENT LIVE + IMPACT LOADING ON EACH ARCH FOR FULLY LOADED
  STRUCTURE: 1,920 lb per ft

TYPES OF STEEL IN STRUCTURE:		%
Arch ribs and ties	A514	64
	A588	36
Rib and bottom-lateral bracing	A36	100
Hangers	Wire rope	

OWNER: Pennsylvania Department of Transportation
ENGINEER: Richardson, Gordon and Associates
FABRICATOR: Bristol Steel and Iron Works, Inc. and Pittsburgh DesMoines Steel Co.
ERECTOR: American Bridge Division, U.S. Steel Corp.
DATE OF COMPLETION: 1976

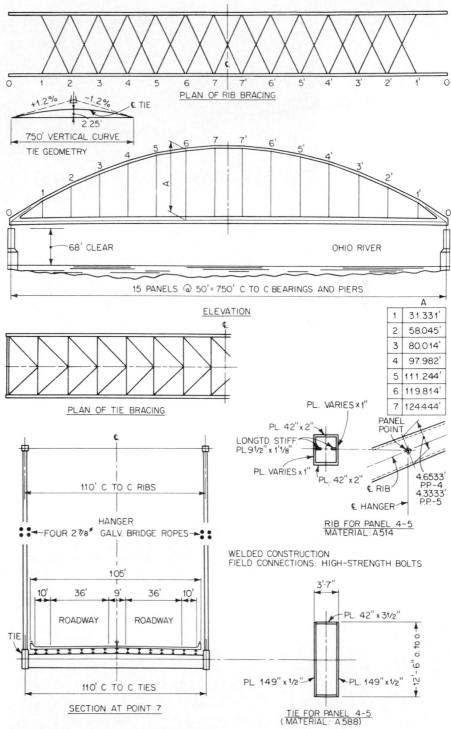

PLAN OF RIB BRACING

TIE GEOMETRY

+1.2%   −1.2%   ℄ TIE
2.25'
750' VERTICAL CURVE

ELEVATION

68' CLEAR

OHIO RIVER

15 PANELS @ 50' = 750' C TO C BEARINGS AND PIERS

	A
1	31.331'
2	58.045'
3	80.014'
4	97.982'
5	111.244'
6	119.814'
7	124.444'

PLAN OF TIE BRACING

PL. VARIES × 1"
PL. 42" × 2"
LONGTD. STIFF.
PL. 9½" × 1⅛"
PL. VARIES × 1"
PL. 42" × 2"

PANEL POINT
4.6533'
PP-4
4.3333'
PP-5
℄ RIB
℄ HANGER

RIB FOR PANEL 4-5
MATERIAL: A514

WELDED CONSTRUCTION
FIELD CONNECTIONS: HIGH-STRENGTH BOLTS

110' C TO C RIBS

HANGER
FOUR 2⅞" GALV. BRIDGE ROPES

105'

| 10' | 36' | 9' | 36' | 10' |

ROADWAY   ROADWAY

TIE

110' C TO C TIES

SECTION AT POINT 7

3'-7"
PL. 42" × 3½"
PL. 149" × ½"
PL. 149" × ½"
12'-6" o. to o.

TIE FOR PANEL 4-5
(MATERIAL: A588)

FIGURE 13.18 Details of Glenfield Bridge.

FIGURE 13.19

COLD SPRING CANYON BRIDGE

LOCATION: About 13.5 miles north of city limit of Santa Barbara, Calif.
TYPE: Solid-ribbed deck arch, 11 panels, 9 at 63.6 ft
SPAN: 700 ft   RISE: 119.2 ft   RISE/SPAN = 1:5.9
NO. OF LANES OF TRAFFIC: 2
HINGES: 2   DEPTH: 9 ft   DEPTH/SPAN = 1:78

AVERAGE DEAD LOAD:	LB PER FT
Deck slab and surfacing for roadway	3,520
Railings and parapets	1,120
Floor steel for roadway	620
Floor bracing	75
Arch ribs	3,400
Arch bracing	530
Arch posts and bracing	210
TOTAL	9,475

SPECIFICATION LIVE LOADING: H20-S16-44
EQUIVALENT LIVE + IMPACT LOADING ON EACH ARCH FOR FULLY LOADED
    STRUCTURE: 904 lb per ft
TYPES OF STEEL IN STRUCTURE:

Arch ribs	A373
Floor system	A373

OWNER: State of California
ENGINEER: California Department of Transportation
FABRICATOR: American Bridge Division, U.S. Steel Corp.
DATE OF COMPLETION: December, 1963

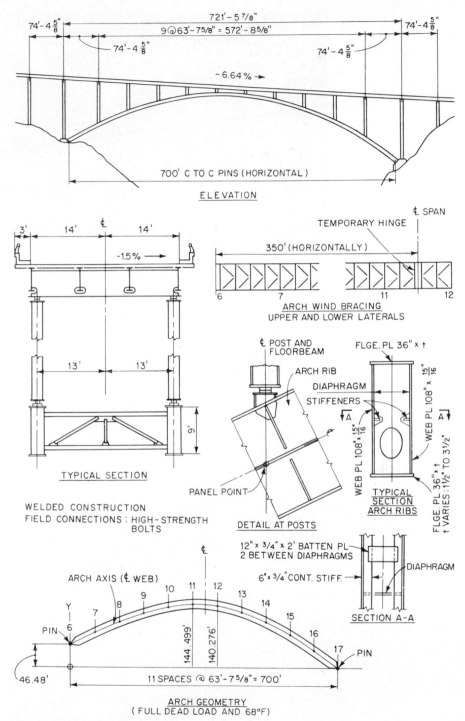

FIGURE 13.20 Details of Cold Spring Canyon Bridge.

FIGURE 13.21

BURRO CREEK BRIDGE

LOCATION: Arizona State Highway 93, about 75 miles southeast of Kingman, Arizona
TYPE: Trussed deck arch, 34 panels at 20 ft
SPAN: 680 ft   RISE: 135 ft   RISE/SPAN = 1:5.0
NO. OF LANES OF TRAFFIC: 2
HINGES: 2   CROWN DEPTH: 20 ft   DEPTH/SPAN = 1:34

AVERAGE DEAD LOAD:	LB PER FT
Deck slab and surfacing for roadway	3,140
Slab for sidewalks	704
Railings and parapets	470
Floor steel for roadway	800
Floor bracing	203
Arch trusses	2,082
Arch bracing	580
Arch posts and bracing	608
TOTAL	8,587

SPECIFICATION LIVE LOADING: H20-S16-44
EQUIVALENT LIVE + IMPACT LOADING ON EACH ARCH FOR FULLY LOADED
    STRUCTURE: 1,420 lb per ft

TYPES OF STEEL IN STRUCTURE:		%
Arch trusses	A441	61
	A36	39
Other components	A36	

OWNER: Arizona Department of Transportation
ENGINEER: Bridge Division
FABRICATOR: American Bridge Division, U.S. Steel Corp.
DATE OF COMPLETION: Mar. 23, 1966

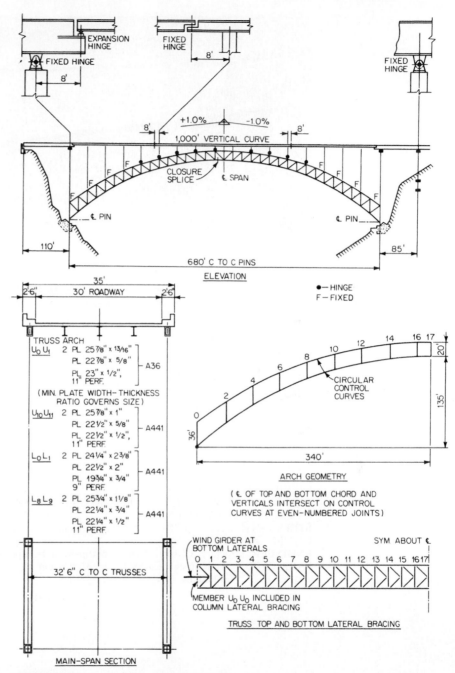

FIGURE 13.22 Details of Burro Creek Bridge.

FIGURE 13.23

COLORADO RIVER ARCH BRIDGE

LOCATION: Utah State Route 95 over Colorado River, near Garfield-San Juan county line
TYPE: Half-through, solid-ribbed arch, 21 panels, 19 at 27.5 ft
SPAN: 550 ft   RISE: 90 ft   RISE/SPAN = 1:6.1
NO. OF LANES OF TRAFFIC: 2
HINGES: 0   DEPTH: 7 ft   DEPTH/SPAN = 1:79
AVERAGE DEAD LOAD:

	LB PER FT
Deck slab and surfacing for roadway	2,804
Railings and parapets	605
Floor steel for roadway	615
Floor bracing	60
Arch ribs	2,200
Arch bracing	370
Arch hangers and bracing	61
TOTAL	6,715

SPECIFICATION LIVE LOADING: HS20-44
EQUIVALENT LIVE + IMPACT LOADING ON EACH ARCH FOR FULLY LOADED
   STRUCTURE: 952 lb per ft
STEEL IN THIS STRUCTURE: A36, except arch hangers, which are bridge strand.
OWNER: State of Utah
ENGINEER: Structures Division, Utah Department of Transportation
FABRICATOR: Western Steel Co., Salt Lake City, Utah
DATE OF COMPLETION: Nov. 18, 1966

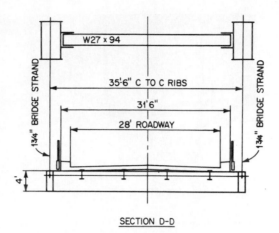

SECTION D-D

ALL SHOP WORK WELDED
ALL FIELD CONNECTIONS: 7/8"ϕ HIGH-STRENGTH BOLTS

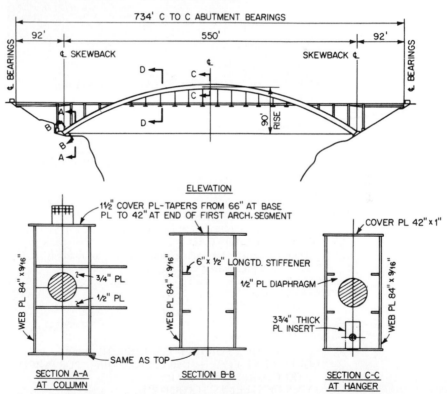

FIGURE 13.24 Details of Colorado River Arch Bridge.

FIGURE 13.25

SMITH AVENUE HIGH BRIDGE

LOCATION: Smith Avenue over Mississippi River in St. Paul, Minnesota
TYPE: Solid-ribbed, tied, deck arch, 26 panels at 40 ft
SPAN: 520 ft   RISE: 109.35 ft   RISE/SPAN = 1:4.8
NO. OF LANES OF TRAFFIC: 2
HINGES: 0   DEPTH: 8 ft   DEPTH/SPAN = 1:65

AVERAGE DEAD LOAD:	LB PER FT
Deck slab, sidewalks, railings and surfacing for roadway	9,370
Floor steel for roadway	920
Arch ribs	2,810
Arch bracing	360
Arch ties	200
Arch columns and bracing	300
TOTAL	13,960

SPECIFICATION FOR LIVE LOADING: HS20-44
EQUIVALENT LIVE + IMPACT LOADING PER ARCH FOR FULLY LOADED
   STRUCTURE: 2,250 lb per ft
TYPES OF STEEL IN STRUCTURE:
   Arch ribs and ties A588
   Floor system A588
OWNER: Minnesota Department of Transportation
ENGINEER: Strgar Roscoe Fausch/T. Y. Lin International
FABRICATOR: Lunda Construction
DATE OF COMPLETION: July 25, 1987

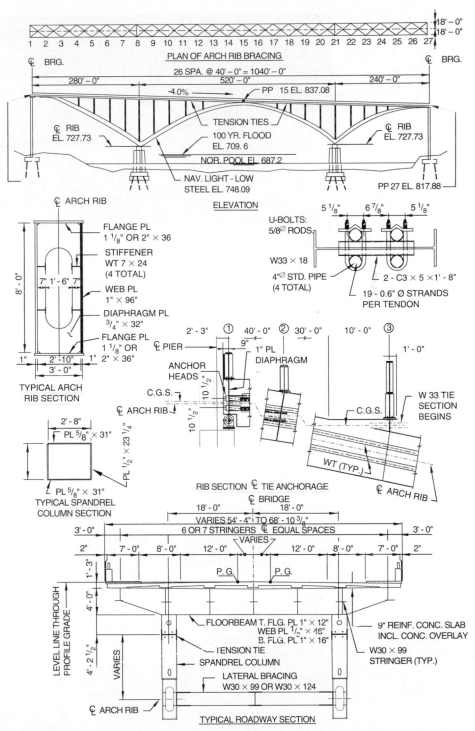

FIGURE 13.26 Details of Smith Avenue High Bridge.

FIGURE 13.27

LEAVENWORTH CENTENNIAL BRIDGE

LOCATION: Leavenworth, Kansas, over Missouri River
TYPE: Tied, through, solid-ribbed arch, 13 panels at 32.3± ft
SPAN: 420 ft   RISE: 80 ft   RISE/SPAN = 1:5.2
NO. OF LANES OF TRAFFIC: 2
HINGES: 0   DEPTH: 2.8 ft   DEPTH/SPAN = 1:150
AVERAGE DEAD LOAD:

	LB PER FT
Deck slab and surfacing for roadway	2,710
Railings and parapets (aluminum)	32
Floor steel for roadway	820
Floor steel for sidewalks	202
Floor bracing	116
Arch ribs	986
Arch bracing	420
Arch hangers and bracing	200
Arch ties	1,104
Miscellaneous—utilities, excess, etc.	110
TOTAL	6,700

SPECIFICATION LOADING: H20-S16-44
EQUIVALENT LIVE + IMPACT LOADING ON EACH ARCH FOR FULLY LOADED
  STRUCTURE: 885 lb per ft
TYPES OF STEEL IN STRUCTURE:

		%
Arch ribs	A7	25
	A242	75
Ties	A242	
Floor system and bracing	A7	
Hangers	A7	

OWNER: Kansas Department of Transportation and Missouri Highway and Transportation
  Department
ENGINEER: Howard, Needles, Tammen & Bergendoff
FABRICATOR: American Bridge Division, U.S. Steel Corp.
DATE OF COMPLETION: April, 1955

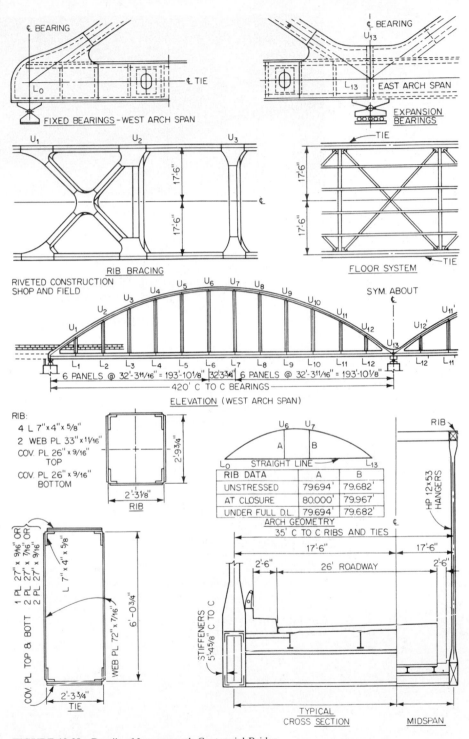

FIGURE 13.28 Details of Leavenworth Centennial Bridge.

FIGURE 13.29

NORTH FORK STILLAGUAMISH RIVER BRIDGE

LOCATION: Cicero, Snobomish County, Wash.
TYPE: Tied, through, solid-ribbed arch, 11 panels at 25.3 ft
SPAN: 278.6 ft   RISE: 51 ft   RISE/SPAN = 1:5.5
NO. OF LANES OF TRAFFIC: 2
HINGES: 0   DEPTH: 2 ft   DEPTH/SPAN = 1:139
AVERAGE DEAD LOAD:

	LB PER FT
Deck slab and surfacing for roadway	2,500
Railings and parapets	1,000
Floor steel for roadway	475
Floor bracing	59
Arch ribs	684
Arch bracing	400
Arch hangers or posts and bracing	83
Arch ties	799
TOTAL	6,000

SPECIFICATION LIVE LOADING: HS20
EQUIVALENT LIVE + IMPACT LOADING ON EACH ARCH FOR FULLY LOADED
    STRUCTURE: 1,055 lb per ft
TYPES OF STEEL IN STRUCTURE:

		%
Arch ribs and ties	A36	28
	A440 and A441	72
Floor system	A36	63
	A440 and A441	37
Hangers	A36	

OWNER: Washington Department of Transportation
ENGINEER: Bridges and Structures Division, Washington DOT
FABRICATOR: Northwest Steel Fabricators, Vancouver, Wash.
GENERAL CONTRACTOR: Dale M. Madden, Inc., Seattle, Wash.
DATE OF COMPLETION: 1966

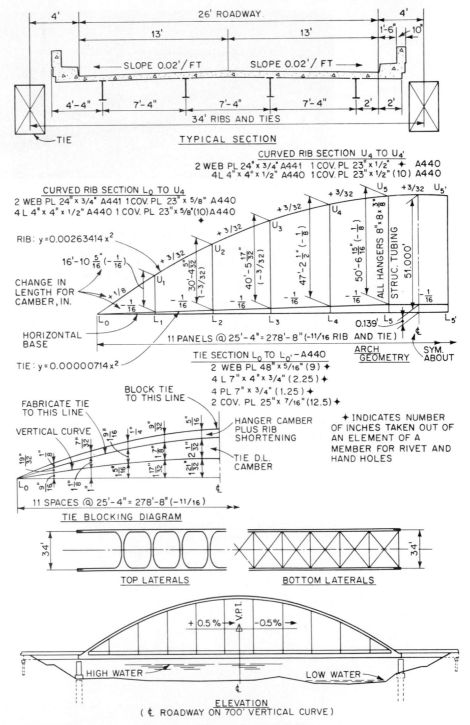

FIGURE 13.30 Details of North Fork Stillaguamish River Bridge.

FIGURE 13.31

SOUTH STREET BRIDGE OVER I-84

LOCATION: South Street over Route I-84, Middlebury, Conn.
TYPE: Solid-ribbed deck arch, 7 panels, 5 at 29 ft
SPAN: 193 ft   RISE: 29 ft   RISE/SPAN = 1:6.7
NO. OF LANES OF TRAFFIC: 2
HINGES: 2   DEPTH: 3.3 ft   DEPTH/SPAN = 1:58

AVERAGE DEAD LOAD:	LB PER FT
Deck slab and surfacing for roadway	4,000
Railings and parapets	500
Floor steel for roadway	560
Arch ribs	1,070
Arch bracing	230
Arch posts and bracing	20
Miscellaneous—utilities, excess, etc.	50
TOTAL	6,430

SPECIFICATION LIVE LOADING: H20-S16-44
EQUIVALENT LIVE + IMPACT LOADING ON EACH ARCH FOR FULLY LOADED
   STRUCTURE: 1,498 lb per ft

TYPES OF STEEL IN STRUCTURE:		%
Arch ribs	A373	98
	A242	2
Floor system	A373	

OWNER: Connecticut Department of Transportation
ENGINEER: Connecticut DOT
FABRICATOR: The Ingalls Iron Works Co.
DATE OF COMPLETION: 1964

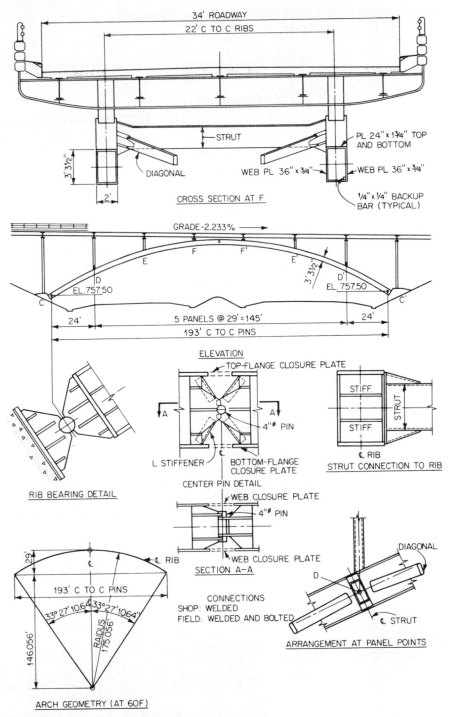

FIGURE 13.32 Details of South Street Bridge over I-84.

13.9 GUIDELINES FOR PRELIMINARY DESIGNS AND ESTIMATES

The usual procedure followed by most designers in preliminary designs of bridges involves the following steps:

1. Preliminary layout of structure
2. Preliminary design of floor system and calculation of weights and dead load
3. Preliminary layout of bracing systems and estimates of weights and loads
4. Preliminary estimate of weight of main load-bearing structure
5. Preliminary stress analysis
6. Check of initial assumptions for dead load

Preliminary Weight of Arch.   The ratios given in Art. 13.6 can guide designers in making a preliminary layout with nearly correct proportions. The 16 examples of arches in Art. 13.8 also can be helpful for that purpose and, in addition, valuable in making initial estimates of weights and dead loads.

Equations (13.1) and (13.2), shown graphically in Fig. 13.33, were developed to facilitate estimating weights of main arch members.

For a true arch of low-alloy, high-strength steel (without ties), the ratio R of weight of rib to total load on the arch may be estimated from

$$R = 0.032 + 0.000288S \qquad (13.1)$$

where S = span, ft.

For a tied arch of low-alloy, high-strength steel, the ratio R of weight of rib and tie to total load on the arch may be estimated from

$$R = 0.088 + 0.000321S \qquad (13.2)$$

This equation was derived from a study of seven of the structures in Art. 13.8 that are tied arches made of low-alloy, high-strength steels predominantly for ribs, trusses, and ties. Equation (13.1), however, is not supported by as many examples of actual designs and may give values on the high side for truss arches. Despite the small number of samples, both equations should give reasonably accurate estimates of weight for preliminary designs and cost estimates of solid-ribbed and truss arches and for comparative studies of different types of structures.

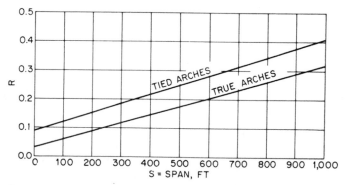

FIGURE 13.33   Chart gives ratio R of weight of rib, or rib and tie, to total load for arches fabricated predominantly with low-alloy steel.

With R known, the weight W, lb per ft, of arch, or arch plus tie, is given by

$$W = \frac{R(D + L)}{1 - R} \tag{13.3}$$

where D = dead load on arch, lb per ft, excluding weight of arch, or arch plus tie and L = equivalent live load plus impact, lb per ft, on arch when the structure is fully loaded. D is determined from preliminary design of bridge components other than arches and ties.

Effect of Type of Steel on Arch Weights.   The following approximate analysis may be used to determine the weight of arch rib or arch rib and tie based on the weight of arch for some initial design with one grade of steel and an alternative for some other grade with different physical properties.

Let F_b = basic unit stress for basic design, ksi
F_a = basic unit stress for alternative design, ksi
D = dead load, lb per ft, excluding weight of rib, or rib and tie
L = equivalent live load plus impact, lb per ft, for fully loaded structure
W_b = weight of rib, or rib and tie, lb per ft for basic design
W_a = weight of rib, or rib and tie, lb per ft, for alternate design
P_b = total load, lb, carried by 1 lb of rib, or 1 lb of rib and tie, for basic design
P_a = total load, lb, carried by 1 lb of rib, or 1 lb of rib and tie, for alternate design

The load supported per pound of member may be assumed proportional to the basic unit stress. Hence,

$$\frac{P_b}{P_a} = \frac{F_b}{F_a} \tag{13.4}$$

Also, the load per pound of member equals the ratio of the total load, lb per ft, on the arch to weight of member, lb per ft. Thus,

$$P_b = \frac{D + L + W_b}{W_b} = 1 + \frac{D + L}{W_b} \tag{13.5}$$

Similarly, and with use of Eq. (13.4),

$$P_a = 1 + \frac{D + L}{W_a} = \frac{P_b F_a}{F_b} \tag{13.6}$$

Solving for the weight of rib, or rib plus tie, gives

$$W_a = \frac{(D + L)F_b/F_a}{P_b - F_b/F_a} \tag{13.7}$$

Use of the preceding equations will be illustrated by application to the Sherman Minton Bridge (Figs. 13.11 and 13.12). Its arches were fabricated mostly of A514 steel. Assume that a preliminary design has been made for the floor system and bracing. A preliminary estimate of weight of truss arch and tie is required.

From the data given for this structure in Art. 13.8, the total load per arch, excluding truss arch and tie, is

$$D + L = \frac{15,580}{2} + 1755 = 9545 \text{ lb per ft}$$

From Eq. (13.2), or from Fig. 13.33, with span $S = 797.5$ ft, if the arch had been constructed of low-alloy steel, the ratio of weight of rib and tie to total load would be about

$$R = 0.088 + 0.000321 \times 797.5 = 0.34$$

By Eq. (13.3), the weight of rib and tie, if made of low-alloy steel, would have been

$$W_b = 9545 \times \frac{0.34}{1 - 0.34} = 4900 \text{ lb per ft}$$

For the A514 steel actually used for the arch, an estimate of weight of rib and tie may be obtained from Eqs. (13.5) and (13.7). When these formulas are applied, the following basic unit stresses may be used:

Normal grades of low-carbon steel—$F = 18$ ksi
Low-alloy, high-strength steels—$F = 24$ ksi
A514 high-strength steel—$F = 45$ ksi

These stresses make some allowance for reductions due to thickness, reductions due to compression, and other similar factors. A check against a number of actual designs indicates that these values give about the correct ratios for the above grades of steel. Accordingly, the calculations for estimating weight of rib plus tie of A514 steel are as follows:

$$\frac{F_b}{F_a} = \frac{24}{45} = 0.53$$

By Eq. (13.5),

$$P_b = 1 + \frac{9545}{4900} = 2.95$$

Then, by Eq. (13.7), the weight of rib and tie is estimated at

$$W_a = 9545 \times \frac{0.53}{2.95 - 0.53} = 2090 \text{ lb per ft}$$

Use 2100 in preliminary design calculations.

Weight of truss arch and tie as constructed was $\frac{1}{2}(3400 + 1040) = 2200$ lb per ft, checking the estimate within about 5%.

13.10 BUCKLING CONSIDERATIONS FOR ARCHES

Since all arches are subjected to large compressive stresses and also usually carry significant bending moments, stability considerations must be addressed. The American Association of State Highway and Transportation Officials (AASHTO) "Standard Specifications for Highway Bridges" contain provisions intended to ensure stability of structures.

For true arches, the design should provide stability in the vertical plane of the arch, with the associated effective buckling length, and also provide for moment amplification effects. For tied arches with the tie and roadway suspended from the arch, moment amplification in the arch rib need not be considered. For such arches, the effective length can be considered the distance along the arch between hangers. However, with the relatively small cross-sectional area of the cable hangers, the effective length may be slightly longer than the distance between hangers.

For prevention of buckling in the lateral direction, a lateral bracing system of adequate stiffness should be provided. Effective lengths equal to the distance between rib bracing points are usually assumed. Special consideration should be given to arch-end portal areas. Lateral torsional buckling for open I-section ribs is much more critical than for box ribs and must be prevented.

Local buckling of web and flange plates is avoided by designs conforming to the limiting plate width-to-thickness ratios in the AASHTO standard specifications. If longitudinal web stiffeners are provided, additional criteria for their design are available (Art. 10.13).

13.11 EXAMPLE—DESIGN OF TIED-ARCH BRIDGE

The typical calculations that follow are based on the design of the 750-ft, solid-ribbed arch for the Glenfield Bridge (Figs. 13.17 and 13.18). The original design was in accordance with AASHTO "Standard Specifications for Highway Bridges," 1965. The example has been revised in general in accordance with provisions in the 15th edition of the AASHTO specifications, inclusive of the 1991 "Interim Specifications," for the service-load-design method (ASD) (Sec. 10). Conditions that do not meet current code provisions are noted.

The structure is a tied, through arch with 50-ft-long panels. The tie has a constant depth of 12 ft 6 in. The arch rib is segmental (straight between panel points) and tapers in depth from 7 ft at the springing line to 4 ft at the crown.

The tied arches are assumed to be fabricated so that dead loads, except member dead loads between panel points, are carried by axial stresses. The floor system is assumed to act independently. Thus, it does not participate in the longitudinal behavior of the arches. The following illustrates the design of selected components and some typical structural details.

13.11.1 Design of Floor System

The floor system is designed for HS20-44 loading. See Sec. 11.

Slab Design.   Assumed cross sections of the roadway slab are shown in Fig. 13.34. Design is based on an allowable reinforcing steel stress $f_s = 20$ ksi and an allowable compressive concrete stress $f_c = 0.4f_c' = 1.2$ ksi, for 28-day strength $f_c' = 3$ ksi and modular ratio $n = 9$. The effective slab span $S = 7.08 - 0.96/2 = 6.60$ ft.

Slab Dead Load, psf

Concrete: $150 \times 0.667 = 100$
Future wearing surface $= \underline{30}$
Total $ = 130$

Calculations of bending moments in the slab take into account continuity. For dead load, maximum moment is taken to be $M = wS^2/10$, where w is the dead load, kips per ft². For live load, $M = 0.8P_{20}(S + 2)/32$, where 0.8 is the continuity

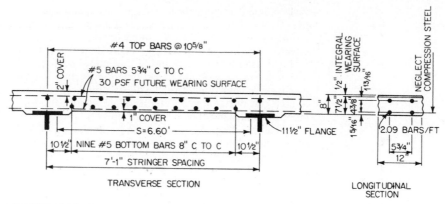

FIGURE 13.34 Concrete deck of tied-arch Glenfield Bridge (Fig. 13.18) is supported on steel stringers. (Current AASHTO standard specifications require that concrete cover over top reinforcing steel be at least 2½ in instead of the 2 in used for the Glenfield Bridge.)

factor, and P_{20} is a wheel load, 16 kips. The impact moment is obtained by applying a factor of 30% to the live-load moment.

Slab Moments, ft-kips

Dead load: $0.130(6.60)^2/10$	$= 0.566$
Live load: $0.8 \times 16(6.60 + 2)/32$	$= 3.440$
Impact: 0.30×3.440	$= 1.032$
Total	$= 5.038$

For the area of main reinforcement, $A_s = 0.65$ in²/ft, the distance from the center of gravity of compression in the slab to center of gravity of tension steel is computed to be $jd = 5.02$ in. For this moment arm, maximum stresses are computed to be $f_c = 1.017$ ksi for the concrete and $f_s = 18.8$ ksi in the steel, and are considered acceptable.

The percentage of the main reinforcement that must be supplied for distribution reinforcing in the longitudinal direction is computed from $220/\sqrt{S}$, but need not exceed 67%.

$$\frac{220}{\sqrt{6.60}} = 85.6 \qquad \text{Use } 67\%$$

Required area of distribution steel $= 0.67 \times 0.65 = 0.44$ in²/ft. No. 5 bars 8 in c to c supply an area of 0.47 in²/ft > 0.44.

Stringer Design. The stringers are designed as three-span continuous, noncomposite beams on rigid supports 50 ft apart (Fig. 13.35). (While floorbeams do not provide perfectly rigid supports, the effects of their flexibility have been studied and found to be small.) The 101-ft-wide roadway is assumed to be carried by 15 stringers, each a W33 × 130, made of A36 steel. Allowable stresses are taken to be $F_b = 20$ ksi in bending and $F_v = 12$ ksi in shear.

The dead load is considered to consist of three parts: the initial weight of stringer and concrete deck (Table 13.1*a*), the superimposed weight of median, railings, parapets, and future wearing surface (Table 13.1*b*), and a concentrated load at midspan from a diaphragm and connections (Fig. 13.35).

TABLE 13.1 Dead Load per Stringer, kips per ft

(a) Initial dead load

W33 × 130	= 0.130
8-in concrete slab: 0.150(⁸⁄₁₂)7.08	= 0.708
Concrete haunch	= 0.030
Total	0.868 kip per ft

(b) Superimposed dead load on 15 stringers

Bridge median	= 0.030
Two railings: 2 × 0.030	= 0.060
Two parapets	= 1.004
Future wearing surface: 0.03 × 101	= 3.030
Total	4.124

Superimposed dead load per stringer = 4.124/15 = 0.275 kip per ft
Total uniformly distributed dead load = 0.868 + 0.275 = 1.143 kips per ft
Concentrated dead load at midspan = 0.36 kips

For live load, the number of wheels to be carried by each stringer is determined from $S/5.5$, where S is the slab span, ft. Thus,

$$\text{Live-load distribution} = \frac{7.08}{5.5} = 1.288 \text{ wheels}$$

The impact fraction is computed from $I = 50/(L + 125) \le 0.3$, where L is the loaded length of member, ft.

$$I = \frac{50}{50 + 125} = 0.286$$

The typical stringer may be analyzed by moment distribution, with a computer or with appropriate tabulated influence ordinates. Design moment and reaction for the end spans are given in Table 13.2.

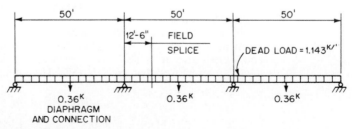

FIGURE 13.35 Stringers of Glenfield Bridge are continuous over three spans, with a splice in the center span.

TABLE 13.2   Design Moment and Reaction for a Stringer End Span

	Maximum bending moment, ft-kips, at 0.4 point of end span	Maximum reaction, kips, at first interior support
Dead load	230.5	64.6
Live load	317.7	48.5
Impact: 28.6%	90.8	13.9
Total	639.0	127.0

For a W33 × 130 with section modulus $S = 405$ in^3, the maximum bending stress is

$$f_b = \frac{M}{S} = 639 \times \frac{12}{405} = 18.9 \text{ ksi} < 20\text{—OK}$$

The maximum shear stress f_v in the web of the 33.10-in-deep beam is considerably less than the allowable. Thus, the W33 × 130 stringer is satisfactory. Furthermore, since $f_v < (0.75F_v = 9 \text{ ksi})$, bearing stiffeners are not required.

AASHTO specifications limit the deflection under live load plus impact to $L/800$, where L is the span. Computations give the deflection at the 0.4 point of the end span $\delta = 0.472$ in.

$$\frac{\delta}{L} = \frac{0.472}{50 \times 12} = \frac{1}{1,271} < \frac{1}{800}$$

The deflection is satisfactory.

Stringer Splice Design. Because of the 150-ft length of the three-span beam, it will be erected in two pieces. A field splice is located in the center span, 12 ft 6 in from a support (Fig. 13.35). The connection will be made with ⅞-in-dia A325 bolts. These are allowed 9.3 kips in single shear and 18.6 kips in double shear in a slip-critical connection with a Class A contact surface.

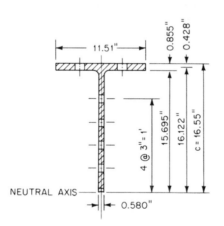

FIGURE 13.36   Stringer cross section at splice.

At the splice, maximum moments are +198 ft-kips and −196 ft-kips. Maximum shear is 48.8 kips.

Figure 13.36 gives important dimensions of a W33 × 130 and shows planned locations of bolt holes. This section has a gross moment of inertia $I = 6,699$ in^4 and a gross area $A = 38.26$ in^2, which has to be reduced by the percentage of flange area removed that exceeds 15%. Hole diameter = ⅞ + ⅛ = 1 in.

Hole Areas, in^2

Flange: $1 \times 0.855 \times 2 = 1.710$ (two holes)

Web: $1 \times 0.580 = 0.580$

The fraction of flange area removed in excess of 15% is:

$$1.710/(11.51 \times 0.855) - 0.15 = 0.024$$

The adjusted gross moment of inertia is then calculated as follows:

$$\text{Flange reduction} = 11.51 \times 0.855 \times 0.024 = 0.24 \text{ in}^2 \text{ per flange}$$

$$\text{Reduced } I = 6699 - (2 \times 0.24 \times 16.122^2) = 6574 \text{ in}^4$$

AASHTO specifications require that splices be designed for the average of the calculated stress, or moment, and the allowable capacity of the member, but not less than 75% of the capacity. In addition, the fatigue stress range must be within the allowable.

With an allowable bending stress of 20 ksi, the moment capacity of the reduced section of the stringer is

$$M_c = \frac{F_b I}{c} = 20 \times \frac{6574}{16.55 \times 12} = 662 \text{ ft-kips}$$

Thus, the section must be designed for at least $0.75 \times 662 = 497$ ft-kips.

The maximum moment at the splice is 198 ft-kips. Hence, the average moment for splice design is

$$M_{av} = \frac{1}{2}(198 + 662) = 430 \text{ ft-kips} < 497 \text{ ft-kips}$$

Consequently, the splice should be designed for 497 ft-kips.

The shear capacity of the web is

$$V_c = 12 \times 33.1 \times 0.580 = 230 \text{ kips}$$

Thus, the section could be designed for a shear of at least $0.75 \times 230 = 173$ kips.

The maximum shear at the splice is 48.8 kips. Hence, the average shear for splice design is

$$V_{av} = \frac{1}{2}(48.8 + 230) = 139.4 \text{ kips} < 173$$

Design for the average shear capacity, however, would be unduly conservative, since web thickness at the splice is not governed by shear stress. Since both shear and moment are directly related to the load, it appears reasonable to use a design shear equal to the maximum shear increased by the same proportion as the design moment.

$$V = 48.8 \times \frac{497}{198} = 122.5 \text{ kips}$$

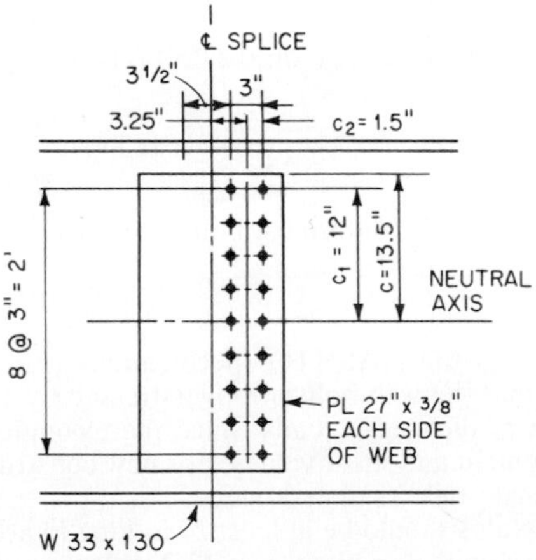

FIGURE 13.37 Web splice for stringer.

Stringer Web Splice. The size of splice plates required for the web (Fig. 13.37) is determined by the requirements for moment. The bolts are designed for shear and moment.

The moment to be carried by the bolts consists of two components. One is the moment due to the eccentricity of the shear.

$$M_v = 122.5 \times \frac{3.25}{12} = 33.2 \text{ ft-kips}$$

The other is the proportion of the 497-ft-kip design moment carried by the web. The web moment may be calculated by assuming that moments are proportional to moment of inertia. With a clear depth of 31.39 in between flanges, the web has a moment of inertia.

$$I_w = \frac{0.580(31.39)^3}{12} = 1495 \text{ in}^4$$

Consequently, the web moment is

$$M_w = 497 \times \frac{1495}{6574} = 113.0 \text{ ft-kips}$$

Thus, the web bolts must be designed for a moment of $113.0 + 33.2 = 146.2$ ft-kips.

For determining bolt loads, the polar moment of inertia is first computed, as the sum of the moments of inertia of the bolts about two perpendicular axes (Fig. 13.37).

Polar Moment of Inertia

$$2 \times 2 \times 270 \quad = 1080$$
$$9 \times 2(1.5)^2 \quad = \quad 41$$
$$I_B = 1121 \text{ in}^2$$

The maximum bolt loads are resolved into horizontal and vertical components P_H and P_V, respectively.

$$P_H = \frac{Mc_1}{I_B} = \frac{146.2 \times 12 \times 12}{1,121} = 18.78 \text{ kips}$$

$$P_V = \frac{Mc_2}{I_B} = \frac{146.2 \times 12 \times 1.5}{1,121} = 2.35 \text{ kips}$$

Also, shear imposes a vertical load on the 18 bolts of

$$P_s = \frac{122.5}{18} = 6.81 \text{ kips}$$

Hence, the total load on the outermost bolt is the resultant

$$P = \sqrt{(6.81 + 2.35)^2 + 18.78^2} = 20.89 \text{ kips} > 18.6 \text{ kips}$$

Because of changes in the AASHTO specifications regarding splice design, the example indicates that the web bolts are overstressed by 12%. Actually, the bolts are adequate to carry the design loads, since the example is based on the higher "75% capacity" moment and shear values. For new construction, the design should be altered to eliminate the calculated overstress.

The web splice plates should be at least 27 in long to accommodate the 18 bolts in two rows of 9 each, with 3-in pitch (Fig. 13.37). If ⅜-in-thick plates are selected, the area is more than adequate to resist the 122.5-kip shear. For resisting moment, the two plates supply a moment of inertia of

$$I = 2 \times 0.375 \left(\frac{27^3}{12}\right) = 1230 \text{ in}^4$$

Consequently, the maximum bending stress in the plates is

$$f_b = \frac{M_w c}{I} = 113.0 \times 12 \times \frac{13.5}{1230} = 14.88 \text{ ksi} < 20\text{—OK}$$

The shear stress for maximum design shear is

$$f_v = \frac{P}{A} = \frac{122.5}{2 \times 27 \times 0.375} = 6.05 \text{ ksi} < 12 \text{ ksi}$$

The two 27 × ⅜-in plates are satisfactory for strength.

For fatigue, the range of moment carried by the web is

$$M_w = (198 + 196)\frac{1495}{6574} = 89.6 \text{ ft-kips}$$

The maximum bending-stress range in the gross section of the web splice plates then is

$$f_b = \frac{89.6 \times 12 \times 13.5}{1230} = 11.80 \text{ ksi}$$

Fatigue in base-metal gross section at high-strength-bolted, slip-critical connections is classified by AASHTO as category B. For a redundant-load-path structure and 2,000,000 loading cycles, the allowable stress range is 18 ksi > 11.80 ksi. Thus, the two 27 × ⅜-in plates are satisfactory for fatigue.

Stringer Flange Splice. Each stringer flange is spliced with a ½-in-thick plate on the outer surface and two 9/16-in-thick plates on the inner surface, as indicated in Fig. 13.38. They are required to resist the design moment less the moment carried by the web splice. Thus, the splice moment is 497 − 113 = 384 ft-kips. With a stringer depth of 33.10 in and flange thickness of 0.86 in, the moment arm of the flange plates is about 33.10 − 0.86 = 32.24 in. Hence, the force, tension or compression, in each flange is

$$T = C = 384 \times \frac{12}{32.34} = 142.5 \text{ kips}$$

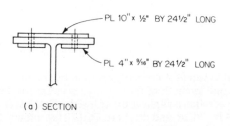

(a) SECTION

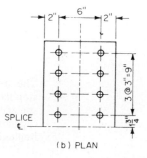

(b) PLAN

FIGURE 13.38 Flange splice for stringer.

TABLE 13.3   Net Area of Stringer Flange Plates, in²

$0.500(10 - 2 \times 1)$	$= 4.00$
$2 \times 0.5625(4 - 1)$	$= 3.38$
Total	$= 7.38 > 7.13$—OK

With an allowable stress of 20 ksi, the plate area required is

$$A = \frac{142.5}{20} = 7.13 \text{ in}^2$$

Table 13.3 indicates the net area supplied by the flange splice plates.

The flange splice plates must be checked for fatigue. The range of live-load moment at the splice is $198 + 196 = 394$ ft-kips. The stress range in the flanges then is

$$f = \frac{394 \times 12 \times 16.55}{6574} = 11.9 \text{ ksi} < 18 \text{ ksi}$$

The plates are satisfactory in fatigue.

The number of bolts required to transmit the flange force in double shear is $142.5/18.6 = 7.66$. The eight bolts supplied (Fig. 13.38) are satisfactory.

Design of Interior Floorbeam.   Floorbeams, spaced 50 ft c to c, are designed as hybrid girders over the center 60 ft of their 110-ft spans. The web is made of A588 steel ($F_y = 50$ ksi). Its depth varies from 109.1 in at centerline of ties to 116 in at centerline of bridge, to accommodate the cross slope of the roadway. For the flange, A514 steel ($F_y = 100$ ksi) as well as A588 is used, to keep flange thickness constant over the full width of bridge.

The dead and live loads on a typical interior floorbeam are indicated in Fig. 13.39. To the live load, an impact factor of $50/(110 + 125) = 0.213$ must be applied. For maximum moments and shears, however, live loads are reduced 25% because more than three lanes are loaded. Table 13.4 indicates the maximum moments and shears in the interior floorbeam.

The hybrid section is used in the region of maximum moment. Its properties required for bending analysis are given in Table 13.5.

Flange Check.   Based on these properties, the bending stress for maximum moment is

$$f_b = \frac{M}{S} = \frac{22,353 \times 12}{6789} = 39.51 \text{ ksi}$$

The allowable bending stress is governed by the buckling capacity of the compression flange and strength of the web. To account for the latter, AASHTO standard specifications require application of a reduction factor R to the allowable buckling capacity [Art. 10.20 and Eq. (10.76)]. The allowable compressive stress for a homogeneous beam is

$$F_b = 0.55F_y = 0.55 \times 100 = 55 \text{ ksi}$$

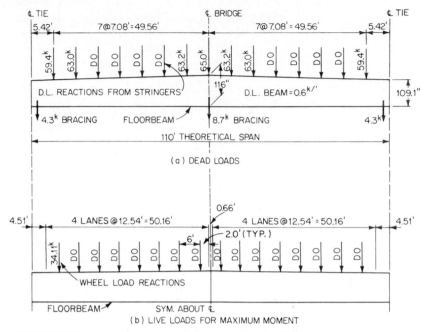

FIGURE 13.39 Loads on an interior floorbeam of Glenfield Bridge.

For use in Eq. (10.76),

$$\alpha = \frac{F_{yw}}{F_{yf}} = \frac{50}{100} = 0.5$$

$$\beta = \text{ratio of web area to tension-flange area} = \frac{65.25}{48} = 1.36$$

$$\psi = \frac{c}{2c} = \frac{60}{120} = 0.5$$

Substitution in Eq. (10.76) yields $R = 0.939$. Hence, the allowable stress is

$$F_b' = RF_b = 0.939 \times 55 = 51.6 \text{ ksi} > 39.51$$

The flange is satisfactory.

TABLE 13.4 Maximum Moments and Shears in Interior Floorbeam

	Moment, ft-kips, at bridge center	Shear, kips, at tie
Dead load	14,691	510.7
Live load	6,316	209.4
Impact	1,346	44.6
Total	22,353	764.7

TABLE 13.5 Properties of Hybrid Section of Floorbeam

Section	Steel	F_y	Area	d	I
Web: 116 × ⁹⁄₁₆	A588	50	65.25	...	73,167
Flanges: 2—24 × 2	A514	100	96.00	59.0	334,176
			161.25		407,343

Section modulus $S = 407,343/60 = 6,789$ in³

The width-thickness ratio b/t of the compression flange may not exceed 24 or

$$\frac{b}{t} = \frac{103}{\sqrt{f_b}} = \frac{103}{\sqrt{51.6}} = 14.3$$

Actual width-thickness ratio is $24/2 = 12 < 14.3$—OK

Longitudinal Stiffener for Floorbeam. Web design assumes longitudinal stiffening. The effective bending stress in the web is $f_b/R = 39.51/0.939 = 42.1$ ksi. For this stress, web thickness for web depth $D = 116$ in must be at least $D/340 = 116/340 = 0.34$ in and

$$t_w = \frac{D\sqrt{f_b}}{1450} = \frac{116\sqrt{42.1}}{1450} = 0.52 \text{ in}$$

Use a ⁹⁄₁₆-in web plate, $t_w \doteq 0.563$ in.

Location of the longitudinal stiffener is determined by $D/5 = 116/5 = 23.25$ in. Thus, it should be placed 1 ft 11 in from the bottom of the top flange of the floorbeam. Minimum moment of inertia required for the stiffener is

$$I = Dt^3 \left[2.4 \left(\frac{d_o}{D}\right)^2 - 0.13 \right] = 116(0.563)^3 \left[2.4 \left(\frac{85}{116}\right)^2 - 0.13 \right] = 23.9 \text{ in}^4$$

in which the spacing of transverse stiffeners at midspan is taken as $d_o = 85$ in. Bending stress in the longitudinal stiffener, which is 35 in from the neutral axis, is

$$f_b = 39.51 \times \frac{35}{60} = 23.04 \text{ ksi}$$

Use A588 steel for the stiffener, with allowable stress of 27 ksi. If a stiffener width $b' = 6$ in is assumed, the minimum thickness required is

$$t = \frac{b'\sqrt{f_b}}{71.2} = \frac{6\sqrt{39.51}}{71.2} = 0.53$$

Use a 6 × ⁹⁄₁₆-in plate for the longitudinal stiffener. Its moment of inertia exceeds the 23.9 in⁴ required.

Web Check. Allowable web shear is $F_v = 17$ ksi for A588 steel. Assume a thickness of ⁷⁄₁₆ in for the 109.3-in web at the support, as indicated by preliminary design.

Then, maximum shear stress at the support is

$$f_v = \frac{764.7}{109.3 \times 0.438} = 16.0 \text{ ksi} < 17$$

The $109.3 \times \frac{7}{16}$-in web is satisfactory.

Transverse Stiffeners for Floorbeam. In accordance with AASHTO standard specifications, the maximum spacing for transverse intermediate stiffeners should not exceed

$$D \left(\frac{260}{D/t_w} \right)^2 = 109.3 \left(\frac{260}{109.3/0.438} \right)^2 = 118.7 \text{ in} \leq 3D$$

Stiffener spacing is determined by the shear stress computed from Eq. (10.31a). For this computation, assume that the end spacing for the stiffeners is $d_o = 22$ in. Hence, $d_o/D = 22/109.3 = 0.201$. From Eq. (10.30d), for use in Eq. (10.31a), $k = 5[1 + (1/0.201)^2] = 128.8$ and $\sqrt{k/F_v} = \sqrt{128.8/50} = 1.60$. Since $D/t_w = 109.3/0.438 = 250$, C in Eq. (10.30a) is determined by the parameter $250/1.60 = 156 < 190$. Hence, $C = 1$, and the maximum allowable shear is $F'_v = F_y/3 = 50/3 = 16.67$ ksi > 16 ksi—OK.

The occurrence of simultaneous shear and bending in a panel when shear is larger than 60% of the allowable shear stress F_v requires that the allowable bending stress be limited to

$$F_s = F_y(0.754 - 0.34f_v/F_v) \tag{13.8}$$

A check of the stresses indicates that the member sizes are adequate for this condition.

Transverse-stiffener design requires computation of a factor J for determining required moment of inertia, where $J = 0.5$ or more:

$$J = 2.5 \left(\frac{D}{d_o} \right)^2 - 2 = 2.5 \left(\frac{109.3}{22} \right)^2 - 2 = 59.7$$

Hence, the moment of inertia required, with a stiffener spacing $d_o = 22$ in, is

$$I = d_o t_w^3 J = 22(0.438)^3 59.7 = 110 \text{ in}^4$$

The minimum gross cross-sectional area A for intermediate stiffeners is determined from

$$A = Y[0.15BDt_w(1 - C)(f_v/F_v) - 18t_w^2] \tag{13.9}$$

where $B = 1.0$ for stiffeners in pairs
$Y = F_{yw}/F_{ys} = \frac{50}{36} = 1.39$
$C = 1.0$ from a previous calculation

When $C = 1.0$, Eq (13.9) yields a negative number, indicating that the moment of inertia of the stiffener controls the stiffener size. The stiffener width should be at least $2 + D/30 = 2 + 116/30 = 5.86$ in and at least one-fourth the flange width, or $24/4 = 6$ in.

For the first transverse stiffeners, a pair of stiffeners $7 \times \frac{1}{2}$ in is specified. These have a moment of inertia

$$I = 0.5(14)^3/12 = 114.3 \text{ in}^4 > 110.0 \text{ in}^4$$

For the remaining intermediate transverse stiffeners, $d_o = 42.5$ in is used. For computation of required I,

$$J = 2.5(109.3/42.5)^2 - 2 = 14.53 > 0.50$$

For this value of J,

$$I = 42.5(0.438)^3 14.53 = 51.9 \text{ in}^4$$

Use a 6-in-wide stiffener. Thickness must be at least $\frac{1}{16}$ of this, or $\frac{3}{8}$ in. Moment of inertia furnished by a pair of $6 \times \frac{3}{8}$ in stiffeners is

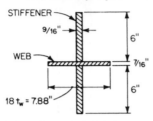

$$I = \frac{0.375(12)^3}{12} = 54.0 \text{ in}^4 > 51.9 \text{ in}^4$$

Hence, a pair of $6 \times \frac{3}{8}$-in stiffeners is satisfactory. Use A36 steel.

A check of stiffeners between the first and second stringers indicates required $I = 53.5 \text{ in}^4 < 54.0$.

FIGURE 13.40 Bearing stiffeners for floor-beam.

Bearing Stiffeners for Floorbeam.   Stiffeners must be provided under the stringers. Use a pair of A36 stiffener plates. They act, with a strip of web $18t_w = 7.88$ in long, as a column transmitting the stringer reactions to the floorbeam (Fig. 13.40). Minimum stiffener thickness permitted, assuming a 6-in wide plate, is

$$t = \frac{b'}{12} \sqrt{\frac{F_y}{33}} = \frac{6}{12} \sqrt{\frac{36}{33}} = 0.523 \text{ in}$$

Use $6 \times \frac{9}{16}$-in plates. Moment of inertia furnished is

$$I = \frac{0.563(12.44)^3}{12} = 90.2 \text{ in}^4$$

Area of the equivalent column is

$$A = 7.88 \times 0.438 + 2 \times 6 \times 0.563 = 10.20 \text{ in}^2$$

The radius of gyration thus is

$$r = \sqrt{\frac{I}{A}} = \sqrt{\frac{90.2}{10.20}} = 2.97 \text{ in}$$

For a length of about 115 in (maximum stiffener depth at stringers), the column then has a slenderness ratio $KL/r = 115/2.97 = 38.7$. The allowable compressive stress for this column is

$$F_a = 16.98 - 0.00053(KL/r)^2 = 16.98 - 0.00053(38.7)^2 = 16.19 \text{ ksi}$$

Actual compressive stress under a maximum stringer reaction of 127 kips is

$$f_a = \frac{127}{10.20} = 12.4 \text{ ksi} < 16.19 \text{ ksi}$$

The bearing stiffeners provide an effective bearing area, outside the flange-to-web weld, of

$$A_b = 2 \times 0.563(6 - 0.5) = 6.19 \text{ in}^2$$

The stringer reaction imposes a bearing stress of $127/6.19 = 20.5$ ksi. Since this is less than the 29 ksi permitted, the $6 \times \frac{9}{16}$ in stiffeners are satisfactory.

Flange-to-Web Welds for Floorbeam. These are made the minimum size permitted for a 2-in flange, $\frac{5}{16}$-in fillet welds. To check these welds, the properties of the floorbeam at the support given in Table 13.6 are needed. Shear at the top of the web is

$$v = \frac{VQ}{I} = \frac{764.7 \times 2{,}675}{345{,}381} = 5.91 \text{ kips per in}$$

Thus, each of the two fillet welds connecting the web to the flange is subjected to a unit shear

$$f_v = \frac{5.91}{2 \times 0.707 \times 0.3125} = 13.4 \text{ ksi}$$

This is less than the 15.7 ksi allowed by AASHTO specifications for welds made with E70XX welding electrodes. For fatigue the maximum shear range is $209.4 + 44.6 = 254.0$ kips. The shear stress range at the top of the web is

$$v = 254.0 \times 2675/345{,}381 = 1.97 \text{ kips per in}$$

The shear stress on the throat of the fillet weld is

$$f_v = 1.97/(2 \times 0.707 \times 0.3125) = 4.46 \text{ ksi}$$

This detail falls into AASHTO fatigue stress category F and, for a redundant-load-path structure and 2,000,000 loading cycles, the allowable stress range is 9 ksi > 4.46 ksi. Thus, the $\frac{5}{16}$-in welds are satisfactory.

13.11.2 Design of Arch Rib

Arch, tie, and hangers were analyzed by computer with the system assumed acting as an indeterminate plane frame. The live load was taken as a moving load of 1.92 kips per ft, without a concentrated load or impact.

The design procedure for the arch rib will be illustrated by the calculations for a rib section 54.78 ft long at panel point U_3 (Fig. 13.18). The assumed cross section,

TABLE 13.6 Properties of Floorbeam at Support

Section	Area	d	I
Web: $109\frac{3}{8} \times \frac{7}{16}$	47.8	...	47,649
Flanges: 2—24 × 2	96.0	55.69	297,732
			345,381

At top of web, $Q = 24 \times 2 \times 55.69 = 2675 \text{ in}^3$

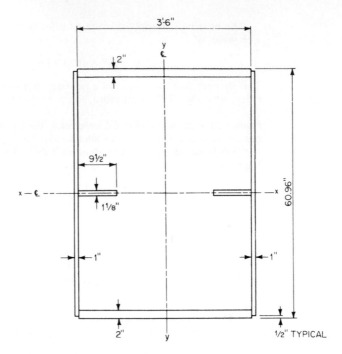

FIGURE 13.41   Arch-rib cross section.

of A514 steel, is shown in Fig. 13.41. The section properties given in Table 13.7 are needed. From the computer analysis, the loads on the arch section are:

	Thrust, kips	Moments, ft-kips
Dead load	−8430	367
Live load	−686	1285
Total	−9116	1652

TABLE 13.7   Properties of Arch Section

Section	Area	Axis x-x			Axis y-y		
		d_y	I_o	I_t	d_x	I_o	I_t
2—59.96 × 1	119.92	...	35,928	35,928	21.5	...	55,433
2—42 × 2	168.00	29.48	...	146,004	...	24,696	24,696
2—9½ × 1⅛	21.38				16.25	160	5,806
	309.30			181,932			85,935

$$\text{Radius of gyration } r_y = \sqrt{85{,}935/309.30} = 16.67 \text{ in}$$

$$r_x = \sqrt{181{,}932/309.30} = 24.24 \text{ in}$$

$$\text{Slenderness ratio } L/r_y = 54.78 \times 12/16.67 = 39.4$$

$$L/r_x = 54.78 \times 12/24.24 = 27.1$$

$$\text{Section modulus } S_x = 181{,}932/(60.96/2) = 5969 \text{ in}^3$$

The arch rib is subjected to both axial compressive stresses and bending stresses. For buckling in the vertical plane, the allowable compressive stress is

$$F_a = \frac{F_y}{2.12}\left[1 - \frac{(KL/r)^2 F_y}{4\pi^2 E}\right] = \frac{100}{2.12}\left[1 - \frac{(39.4)^2 100}{(4\pi^2)29,000}\right] = 40.77 \text{ ksi}$$

The allowable bending stress for A514 steel is $F_b = 55$ ksi. The axial compressive stress in the rib is

$$f_a = P/A = 9,116.0/309.30 = 29.47 \text{ ksi}$$

The bending stress in the rib is

$$f_b = M/S_x = 1,652 \times 12/5,969 = 3.32 \text{ ksi}$$

The combined axial and bending stresses are required to satisfy the equation

$$f_a/F_a + f_b/F_b \leq 1.0 \tag{13.10}$$

Substitution of the preceding stresses in Eq. (13.10) yields

$$29.47/40.77 + 3.32/55.0 = 0.72 + 0.06 = 0.78\text{---OK}$$

[Equation (13.11), p. 13.64, should be used instead of the simpler Eq. (13.10) but the difference in result is trivial for this example.]

Plate Buckling in Arch Rib.   Compression plates are checked to ensure that width-thickness ratios b/t meet AASHTO specifications. However, the arch rib does not fall into the limits of applicability of the AASHTO equations because

$$f_b/(f_a + f_b) = 3.32/(29.47 + 3.32) = 0.10 < 0.20$$

Hence, the check is performed as follows: Because stiff longitudinal stiffeners will be attached to the web (Fig. 13.42), assume that a node will occur at the web middepth. Thus, the D/t_w ratio is checked for a solid web based on the clear distance between the longitudinal stiffener and flange: $D/t_w = 27.92/1 = 27.92$. Since the stress gradient is small for this case, the D/t_w limit will be based on the total stress $(f_a + f_b)$ rather than the axial stress alone. Therefore, the limit is

$$D/t_w \leq 158/\sqrt{f_a + f_b} \leq 158/\sqrt{29.47 + 3.32} = 27.61 \approx 27.92$$

Since the limit is exceeded by only about 1%, the depth-thickness ratio is acceptable.

The width-thickness ratio for the flange plates between webs is limited to a maximum of

$$b/t_f = 134/\sqrt{f_a + f_b} \leq 47$$

$$= 134/\sqrt{29.47 + 3.32} = 23.5 < 47$$

The width-thickness ratio of the flange plates is

$$b/t_f = 42/2 = 21 < 23.5\text{---OK}$$

Longitudinal Stiffener in Arch Rib.   The required moment of inertia of the longitudinal stiffener (Fig. 13.42) about an axis at its base, parallel to the web, is

$$I_s = 0.75Dt_w^3 = 0.75\ 56.96(1)^3$$

$$= 42.7 \text{ in}^4$$

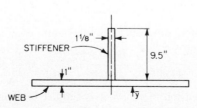

FIGURE 13.42 Stiffener on arch-rib web.

The stiffener provides a moment of inertia of

$$I = bh^3/3 = 1.125(9.5)^3/3 = 321.5 \text{ in}^4 > 42.7 \text{—OK}$$

The width-thickness ratio of the stiffener is governed by

$$b'/t_s = 51.4/\sqrt{f_a + f_b/3} \le 12$$

$$= 51.4/\sqrt{29.47 + (3.32/3)} = 9.3 < 12$$

The width-thickness ratio of the stiffener is

$$b'/t_s = 9.5/1.125 = 8.4 < 9.3 \text{—OK}$$

It is not necessary to check the arch rib for fatigue, since it is not subject to tensile stresses.

Although the design procedure presented in the preceding is reasonable, new designs should conform to current AASHTO provisions.

13.11.3   Design of Tie

Design procedure for the tie will be illustrated for a section at panel point L_3 (Fig. 13.18). The assumed cross section, of A588 steel, is shown in Fig. 13.43. The tie is subjected to combined axial tension and bending. In this case, the axial stress is so large that no compression occurs on the section due to bending. The allowable stress for A588 steel in axial tension or bending is 27 ksi.

The section properties given in Table 13.8 are needed.

From the computer analysis of the arch-tie system, the loads on the tie section are

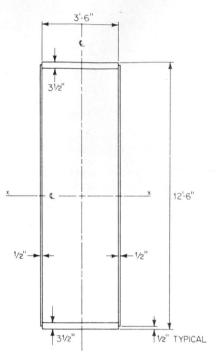

FIGURE 13.43   Tie cross section.

	Axial tension, kips	Moment, ft-kips
Dead load	7,696	471
Live load	418	16,436
Total	8,114	16,907

Based on the preceding properties, the maximum combined stress is

$$f = \frac{P}{A} + \frac{Mc}{I} = \frac{8,114}{443} + \frac{16,907 \times 12}{24,710} = 18.32 + 8.21 = 26.53 \text{ ksi} < 27$$

Since the tie is a tension member, it is subject to fatigue. It was checked as a nonredundant-load-path structure and found to be acceptable. Thus, the tie section is satisfactory.

TABLE 13.8 Properties of Tie Section

Section	Area A	d	Ad^2	I_o	I_t
PL 2—149 × ½	149.0			276,000	276,000
PL 2—42 × 3½	294.0	73.25	1,577,000		1,577,000
	443.0				1,853,000

Section modulus $S_x = 1,853,000/75 = 24,710$ in^3

13.11.4 Design of Hangers

All hangers consist of four 2⅞-in dia bridge ropes (breaking strength 758 kips per rope). With a safety factor of 4, the allowable load per rope is 189 kips.

From the computer analysis of the arch-tie system, the most highly stressed hanger is L_4U_4 (Fig. 13.18). It carries a 630-kip dead load and 99.5-kip live load, for a total of 729.5 kips. This is carried by four ropes. Thus, the load per rope is

$$P = \frac{729.5}{4} = 182.4 \text{ kips} < 189\text{—OK}$$

The live-load stress range for the hangers is small and considered acceptable. If a lower safety factor is used, or a larger live-load stress range is present, a more detailed fatigue investigation should be made. Also, provisions must be made to eliminate any possible aerodynamic vibrations of the hangers.

13.11.5 Bottom Lateral Bracing

The plan of the bracing used in the plane of the tie is shown in Fig. 13.18. Figure 13.44 shows the section used for the diagonal in the panel between L_0 and L_1. Steel is A36.

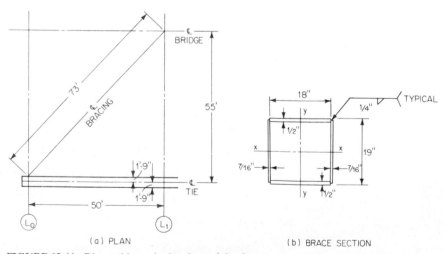

(a) PLAN   (b) BRACE SECTION

FIGURE 13.44 Diagonal brace in the plane of the ties.

Because of lateral wind, the axial load on the 73-ft-long diagonal is 295 kips. The member also is subjected to bending due to its own weight. The section properties given in Table 13.9 are needed.

Weight of the member is

$$w = 34.18 \times 0.0034 = 0.12 \text{ kip per ft}$$

This produces a maximum bending moment, at midspan, of

$$M = \frac{wL^2}{8} = \frac{0.12(73)^2}{8} = 79.9 \text{ ft-kips}$$

Maximum dead-load stress produced then is

$$f_b = \frac{79.9 \times 12}{211} = 4.54 \text{ ksi}$$

Wind will induce an axial compressive stress in the diagonal of

$$f_a = \frac{295}{34.18} = 8.63 \text{ ksi}$$

Members subjected to combined axial compression and bending must satisfy

$$\frac{f_a}{F_a} + \frac{C_{mx}f_{bx}}{(1 - f_a/F'_{ex})F_{bx}} + \frac{C_{my}f_{by}}{(1 - f_a/F'_{ey})F_{by}} \leq 1.0 \qquad (13.11)$$

where $F'_e = \dfrac{\pi^2 E}{FS(K_b L_b/r_b)^2}$

FS = safety factor = 2.12

TABLE 13.9 Properties of Diagonal in Bottom Lateral Bracing

Section	Area A	Axis x-x				Axis y-y			
		d_x	I_o	Ad_x^2	I_x	d_y	I_o	Ad_y^2	I_y
PL 2—18½ × ⁷⁄₁₆	16.18	...	462	...	462	9.22	...	1375	1375
PL 2—18 × ½	18.00	9.25	...	1540	1540	...	486	...	486
	34.18				2002				1861

Distance c to c connections = 73 − 2 = 71 ft

Radius of gyration $r_x = \sqrt{2002/34.18} = 7.65$ in

$r_y = \sqrt{1861/34.18} = 7.38$ in

Effective length factor K = 0.75 (truss-type member connections)

Slenderness ratio KL/r_y = 0.75(71 × 12)/7.38 = 86.6 < 140

Slenderness ratio KL/r_x = 0.75(71 × 12)/7.65 = 83.5 < 140

Section modulus S_x = 2002/9.5 = 211 in³

C_m = coefficient defined for Eq. (6.67) (may be taken conservatively equal to unity)

The allowable axial stress is

$$F_a = 16.98 - 0.00053(KL/r_y)^2 = 16.98 - 0.00053(86.6)^2 = 13.0 \text{ ksi}$$

The allowable bending stress for A36 steel is $F_b = 20$ ksi. For dead-load bending about the strong axis,

$$F'_e = \frac{\pi^2(29,000)}{2.12(83.5)^2} = 19.36 \text{ ksi}$$

When wind stresses are combined with dead-load stresses (Group II loading), the allowable stresses may be increased 25%. Substitution of wind-load and dead-load stresses in Eq. (13.11) gives

$$\frac{8.63}{1.25 \times 13.0} + \frac{1.0 \times 4.54}{(1 - 8.63/19.36)20 \times 1.25} = 0.53 + 0.33 = 0.86 < 1.0\text{—OK}$$

Plate Buckling in Lateral Brace.   Compression plates are checked to ensure that width-thickness ratios b/t meet AASHTO specifications. The compressive stress is taken as $f_a = (8.63 + 4.54)/1.25 = 10.54$ ksi.

Width-Thickness Ratios

	½-in top plate	⁷⁄₁₆-in vertical plate
Actual b/t	$18/0.5 = 36$	$18.5/0.438 = 42.3$
Allowable b/t	$126.5/\sqrt{f_a} = 39 < 45$ max	$158/\sqrt{f_a} = 48.7 < 50$ max

The brace section is satisfactory.

13.11.6   Rib Bracing

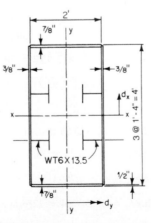

FIGURE 13.45 Section for brace between arch ribs.

The plan of the bracing used for the arch rib is shown in Fig. 13.18. Figure 13.45 shows the section used for a brace in the first panel of bracing. Steel is A36.

Rib bracing is designed to carry its own weight, wind on ribs and rib bracing, and an assumed buckling shear from compression of the ribs. Loads on the first-panel brace are given in Table 13.10, and section properties are computed in Table 13.11. The maximum bending stress produced by total load is

$$f_b = \frac{1120 \times 12}{1285} + \frac{134.5 \times 12}{657}$$

$$= 10.5 + 2.5 = 13.0 \text{ ksi}$$

The total axial stress is

$$f_a = 324.7/93.2 = 3.5 \text{ ksi}$$

TABLE 13.10   Loads on Brace Between Arch Ribs

	P, kips	M_x, ft-kips	M_y, ft-kips
Dead load	...	1120	67.5
Wind	58.7	...	67.0
Buckling	266.0		
Total	324.7	1120	134.5

For combined stresses with wind, allowable stresses may be increased 25%. Axial and bending loads are evaluated for combined stresses with Eq. (13.11) with $C_m = 1$.

$$F'_{ex} = \frac{\pi^2(29,000)}{2.12(29.0)^2} = 160.5 \text{ ksi}$$

$$F'_{ey} = \frac{\pi^2(29,000)}{2.12(56.5)^2} = 42.3 \text{ ksi}$$

$$F_a = 16.98 - 0.00053(56.5)^2 = 15.3 \text{ ksi}$$

$$F_b = 20.0 \text{ ksi (A36 steel)}$$

TABLE 13.11   Properties of Rib Brace

			Axis x-x				Axis y-y		
Section	A	d_x	I_o	Ad_x^2	I_x	d_y	I_o	Ad_y^2	I_y
PL 2—47 × ⅜	35.2	...	6490	...	6,490	12.19	...	5230	5230
PL 2—24 × ⅞	42.0	23.56	...	23,300	23,300	...	2020	...	2020
WT 6 × 13.5	16.0	8.00	33	1,024	1,057	7.23	46	836	882
	93.2				30,847				8132

Radius of gyration $r_x = \sqrt{30,847/93.2} = 18.2$ in

$$r_y = \sqrt{8,132/93.2} = 9.35 \text{ in}$$

Unsupported length = 58.7 ft

Effective length factor $K = 0.75$ (truss-type member connections)

Slenderness ratio $KL/r_x = 0.75 \times 58.7 \times 12/18.2 = 29.0$

Slenderness ratio $KL/r_y = 0.75 \times 58.7 \times 12/9.35 = 56.5$

Section modulus $S_x = 30,847/24 = 1285$ in^3

$$S_y = 8132/12.38 = 657 \text{ in}^3$$

$$\frac{3.5}{1.25 \times 15.3} + \frac{1.0 \times 10.5}{(1 - 3.5/160.5)20.0 \times 1.25} + \frac{1.0 \times 2.5}{(1 - 3.5/42.3)20.0 \times 1.25}$$

$$= 0.18 + 0.43 + 0.11 = 0.72 < 1.0\text{—OK}$$

Plate Buckling in Brace.   Compression plates are checked to ensure that width-thickness ratios b/t meet AASHTO specifications. Compressive stress is taken as $f_a = (3.5 + 13.0)/1.25 = 13.2$ ksi.

Width-Thickness Ratios

	⅜-in Web	⅞-in Flange
Actual b/t	$16/0.375 = 42.7$	$24/0.875 = 27.4$
Allowable b/t	$158/\sqrt{f_a} = 45.3 < 50$ max	$126.5/\sqrt{f_a} = 36.3 < 45$ max

The brace section is satisfactory.

SECTION 14
CABLE-SUSPENDED BRIDGES

Walter Podolny, Jr.
Bridge Division, Office of Engineering,
Federal Highway Administration,
U.S. Department of Transportation, Washington, D.C.

Few structures are as universally appealing as cable-supported bridges. The origin of the concept of bridging large spans with cables, exerting their strength in tension, is lost in antiquity and undoubtedly dates back to a time before recorded history. Perhaps primitive humans, wanting to cross natural obstructions such as deep gorges and large streams, observed a spider spinning a web or monkeys traveling along hanging vines.

14.1 EVOLUTION OF CABLE-SUSPENDED BRIDGES

Early cable-suspended bridges were footbridges consisting of cables formed from twisted vines or hide drawn tightly to reduce sag. The cable ends were attached to trees or other permanent objects located on the banks of rivers or at the edges of gorges or other natural obstructions to travel. The deck, probably of rough-hewn plank, was laid directly on the cable. This type of construction was used in remote ages in China, Japan, India, and Tibet. It was used by the Aztecs of Mexico, the Incas of Peru, and by natives in other parts of South America. It can still be found in remote areas of the world.

From the sixteenth to nineteenth centuries, military engineers made effective use of rope suspension bridges. In 1734, the Saxon army built an iron-chain bridge over the Oder River at Glorywitz, reportedly the first use in Europe of a bridge with a metal suspension system. However, iron chains were used much earlier in China. The first metal suspension bridge in North America was the Jacob's Creek Bridge in Pennsylvania, designed and erected by James Finley in 1801. Supported by two suspended chains of wrought-iron links, its 70-ft span was stiffened by substantial trussed railing and timber planks.

Chains and flat wrought-iron bars dominated suspension-bridge construction for some time after that. Construction of this type was used by Thomas Telford in 1826 for the noted Menai Straits Bridge, with a main span of 580 ft. But 10 years before, in 1816, the first wire suspension bridges were built, one at Galashiels, Scotland, and a second over the Schuylkill River in Philadelphia.

A major milestone in progress with wire cable was passed with erection of the 1,010-ft suspended span of the Ohio River Bridge at Wheeling, Va. (later W.Va.), by Charles Ellet, Jr., in 1849. A second important milestone was the opening in 1883 of the 1,595.5-ft wire-cable-supported span of the Brooklyn Bridge, built by the Roeblings.

In 1607, a Venetian engineer named Faustus Verantius published a description of a suspended bridge partly supported with several diagonal chain stays (Fig. 14.1a). The stays in that case were used in combination with a main supporting suspension (catenary) cable. The first use of a pure stayed bridge is credited to Löscher, who built a timber-stayed bridge in 1784 with a span of 105 ft, (Fig. 14.2a). The pure-stayed-bridge concept was apparently not used again until 1817 when two British engineers, Redpath and Brown, constructed the King's Meadow Foot-bridge (Fig. 14.1b) with a span of about 110 ft. This structure utilized sloping wire cable stays attached to cast-iron towers. In 1821, the French architect, Poyet, suggested a pure cable-stayed bridge (Fig. 14.2b) using bar stays suspended from high towers.

The pure cable-stayed bridge might have become a conventional form of bridge construction had it not been for an unfortunate series of circumstances. In 1818, a composite suspension and stayed pedestrian bridge crossing the Tweed River near Dryburgh-Abbey, England (Fig. 14.1c) collapsed as a result of wind action. In 1824, a cable-stayed bridge crossing the Saale River near Nienburg, Germany (Fig. 14.1d) collapsed, presumably from overloading. The famous French engineer C. L. M. H. Navier published in 1823 a prestigious work wherein his adverse comments on the failures of several cable-stayed bridges virtually condemned the use of cable stays to obscurity.

Despite Navier's adverse criticism of stayed bridges, a few more were built shortly after the fatal collapses of the bridges in England and Germany, for example, the Gischlard-Arnodin cable bridge (Fig. 14.2c) with multiple sloping cables hung from two masonry towers. In 1840, Hatley, an Englishman, used chain stays in a parallel configuration resembling harp strings (Fig. 14.2d). He maintained the parallel spacing of the main stays by using a closely spaced subsystem anchored to the deck and perpendicular to the principal load-carrying cables.

The principle of using stays to support a bridge superstructure did not die completely in the minds of engineers. John Roebling incorporated the concept in his suspension bridges, such as his Niagara Falls Bridge (Fig. 14.3); the Old St. Clair Bridge in Pittsburgh (Fig. 14.4); the Cincinnati Bridge across the Ohio River, and the Brooklyn Bridge in New York. The stays were used in addition to vertical suspenders to support the bridge superstructure. Observations of performance indicated that the stays and suspenders were not efficient partners. Consequently, although the stays were comforting safety measures in the early bridges, in the later development of conventional catenary suspension bridges the stays were omitted. The conventional suspension bridge was dominant until the latter half of the twentieth century.

The virtual banishment of stayed bridges during the nineteenth and early twentieth centuries can be attributed to the lack of sound theoretical analyses for determination of the internal forces of the total system. The failure to understand the behavior of the stayed system and the lack of methods for controlling the equilibrium and compatibility of the various highly indeterminate structural components appear to have been the major drawback to further development of the concept. Furthermore, the materials of the period were not suitable for stayed bridges.

Rebirth of stayed bridges appears to have begun in 1938 with the work of the German engineer Franz Dischinger. While designing a suspension bridge to cross the Elbe River near Hamburg (Fig. 14.5), Dischinger determined that the vertical deflection of the bridge under railroad loading could be reduced considerably by incorporating cable stays in the suspension system. From these studies and his later

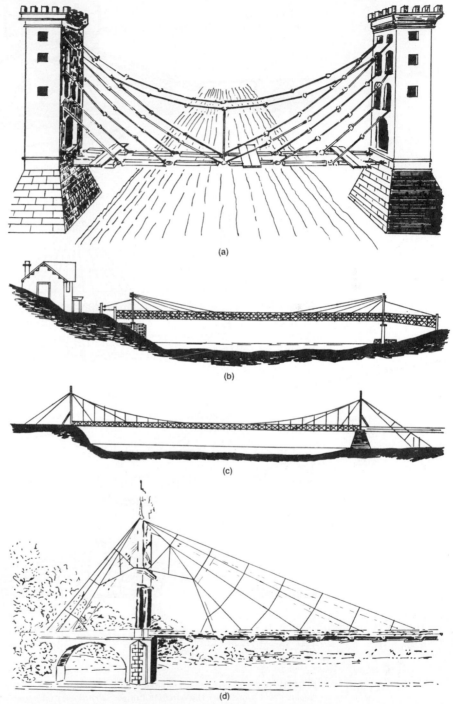

FIGURE 14.1 (*a*) Chain bridge by Faustus Verantius, 1607. (*b*) King's Meadow Footbridge. (*c*) Dryburgh-Abbey Bridge. (*d*) Nienburg Bridge. (*Reprinted with permission from K. Roik et al, "Schrägseilbrüchen," Wilhelm Ernst & Sohn, Berlin.*)

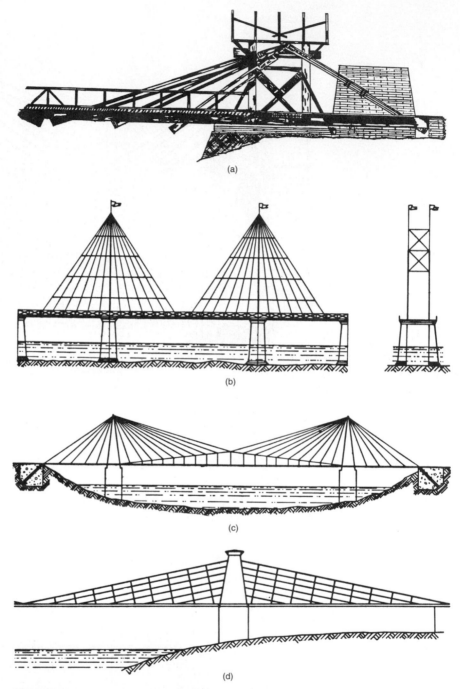

FIGURE 14.2 (*a*) Löscher-type timber bridge. (*b*) Poyet-type bridge. (*c*) Gischlard-Arnodin–type sloping-cable bridge. (*d*) Hatley chain bridge. (*Reprinted with permission from H. Thul, "Cable-Stayed Bridges in Germany," Proceedings of the Conference on Structural Steelwork, 1966, The British Constructional Steelwork Association, Ltd., London.)*

FIGURE 14.3   Niagara Falls Bridge.

design of the Strömsund Bridge in Sweden (1955) evolved the modern cable-stayed bridge. However, the biggest impetus for cable-stayed bridges came in Germany after World War II with the design and construction of bridges to replace those that had been destroyed in the conflict.

(W. Podolny, Jr., and J. B. Scalzi, "Construction and Design of Cable-Stayed Bridges," 2d ed., John Wiley & Sons, Inc., New York; R. Walther et al., "Cable-Stayed Bridges," Thomas Telford, London; D. P. Billington and A. Nazmy, "His-

FIGURE 14.4   Old St. Clair Bridge, Pittsburgh.

FIGURE 14.5   Bridge system proposed by Dischinger. *(Reprinted with permission from F. Dischinger, "Hangebrüchen for Schwerste Verkehrslasten," Der Bauingenieur, Heft 3 and 4, 1949.)*

tory and Aesthetics of Cable-Stayed Bridges," *Journal of Structural Engineering*, vol. 117, no. 10, October 1990, American Society of Civil Engineers.)

14.2   CLASSIFICATION OF CABLE-SUSPENDED BRIDGES

Cable-suspended bridges that rely on very high strength steel cables as major structural elements may be classified as suspension bridges or cable-stayed bridges. The fundamental difference between these two classes is the manner in which the bridge deck is supported by the cables. In suspension bridges, the deck is supported at relatively short intervals by vertical suspenders, which, in turn, are supported from a main cable (Fig. 14.6a). The main cables are relatively flexible and thus take a profile shape that is a function of the magnitude and position of loading. Inclined cables of the cable-stayed bridge (Fig. 14.6b), support the bridge deck directly with relatively taut cables, which, compared to the classical suspension bridge, provide relatively inflexible supports at several points along the span. The nearly linear geometry of the cables produces a bridge with greater stiffness than the corresponding suspension bridge.

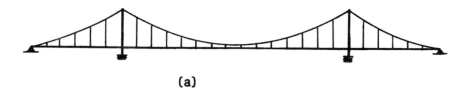

(a)

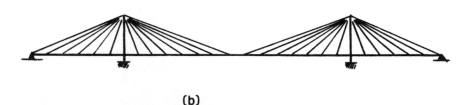

(b)

FIGURE 14.6   Cable-suspended bridge systems: (*a*) suspension and (*b*) cable-stayed. *(Reprinted with permission from W. Podolny, Jr. and J. B. Scalzi, "Construction and Design of Cable-Stayed Bridges," 2d ed., John Wiley & Sons, Inc., New York.)*

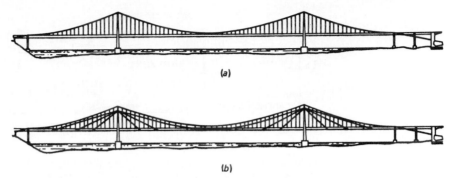

FIGURE 14.7 The Salazar Bridge. (*a*) elevation of the bridge in 1993; (*b*) elevation of future bridge. *(Reprinted with permission from W. Podolny, Jr., and J. B. Scalzi, "Construction and Design of Cable-Stayed Bridges," John Wiley & Sons, Inc., New York.)*

Cable-suspended bridges are generally characterized by economy, lightness, and clarity of structural action. These types of structures illustrate the concept of form following function and present graceful and esthetically pleasing appearance. Each of these types of cable-suspended bridges may be further subclassified; those subclassifications are presented in articles that follow.

Many early cable-suspended bridges were a combination of the suspension and cable-stayed systems (Art. 14.1). Such combinations can offer even greater resistance to dynamic loadings and may be more efficient for very long spans than either type alone. The only contemporary bridge of this type is Steinman's design for the Salazar Bridge across the Tagus River in Portugal. The present structure, a conventional suspension bridge, is indicated in Fig. 14.7*a*. In the future cable stays are to be installed to accommodate additional rail traffic (Fig. 14.7*b*).

(W. Podolny, Jr., and J. B. Scalzi, "Construction and Design of Cable-Stayed Bridges," 2d ed., John Wiley & Sons, Inc., New York.)

14.3 CLASSIFICATION AND CHARACTERISTICS OF SUSPENSION BRIDGES

Suspension bridges with cables made of high-strength, zinc-coated, steel wires are suitable for the longest of spans. Such bridges usually become economical for spans in excess of 1000 ft, depending on specific site constraints. Nevertheless, many suspension bridges with spans as short as 300 or 400 ft have been built, to take advantage of their esthetic features.

The basic economic characteristic of suspension bridges, resulting from use of high-strength materials in tension, is lightness, due to relatively low dead load. But this, in turn, carries with it the structural penalty of flexibility, which can lead to large deflections under live load and susceptibility to vibrations. As a result, suspension bridges are more suitable for highway bridges than for the more heavily loaded railroad bridges. Nevertheless, for either highway or railroad bridges, care must be taken in design to provide resistance to wind- or seismic-induced oscillations, such as those that caused collapse of the first Tacoma Narrows Bridge in 1940.

14.3.1 Main Components of Suspension Bridges

A pure suspension bridge is one without supplementary stay cables and in which the main cables are anchored externally to anchorages on the ground. The main

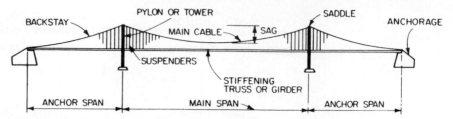

FIGURE 14.8 Main components of a suspension bridge.

components of a suspension bridge are illustrated in Fig. 14.8. Most suspension bridges are stiffened; that is, as shown in Fig. 14.8, they utilize horizontal stiffening trusses or girders. Their function is to equalize deflections due to concentrated live loads and distribute these loads to one or more main cables. The stiffer these trusses or girders are, relative to the stiffness of the cables, the better this function is achieved. (Cables derive stiffness not only from their cross-sectional dimensions but also from their shape between supports, which depends on both cable tension and loading.)

For heavy, very long suspension spans, live-load deflections may be small enough that stiffening trusses would not be needed. When such members are omitted, the structure is an unstiffened suspension bridge. Thus, if the ratio of live load to dead load were, say, 1:4, the midspan deflection would be of the order of $\frac{1}{100}$ of the sag, or 1/1,000 of the span, and the use of stiffening trusses would ordinarily be unnecessary. (For the George Washington Bridge as initially constructed, the ratio of live load to dead load was approximately 1:6. Therefore, it did not need a stiffening truss.)

14.3.2 Types of Suspension Bridges

Several arrangements of suspension bridges are illustrated in Fig. 14.9. The main cable is continuous, over saddles at the pylons, or towers, from anchorage to

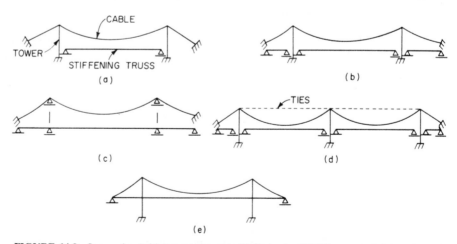

FIGURE 14.9 Suspension-bridge arrangements. (*a*) One suspended span, with pin-ended stiffening truss. (*b*) Three suspended spans, with pin-ended stiffening trusses. (*c*) Three suspended spans, with continuous stiffening truss. (*d*) Multispan bridge, with pin-ended stiffening trusses. (*e*) Self-anchored suspension bridge.

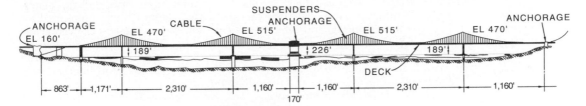

FIGURE 14.10   San Francisco-Oakland Bay Bridge.

anchorage. When the main cable in the side spans does not support the bridge deck (side spans independently supported by piers), that portion of the cable from the saddle to the anchorage is virtually straight and is referred to as a straight backstay. This is also true in the case shown in Fig. 14.9a where there are no side spans.

Figure 14.9d represents a multispan bridge. This type is not considered efficient, because its flexibility distributes an undesirable portion of the load onto the stiffening trusses and may make horizontal ties necessary at the tops of the pylons. Ties were used on several French multispan suspension bridges of the nineteenth century. However, it is doubtful whether tied towers would be esthetically acceptable to the general public. Another approach to multispan suspension bridges is that used for the San Francisco–Oakland Bay Bridge (Fig. 14.10). It is essentially composed of two three-span suspension bridges placed end to end. This system has the disadvantage of requiring three piers in the central portion of the structure where water depths are likely to be a maximum.

Suspension bridges may also be classified by type of cable anchorage, external or internal. Most suspension bridges are *externally anchored* (earth-anchored) to a massive external anchorage (Fig. 14.9a to d). In some bridges, however, the ends of the main cables of a suspension bridge are attached to the stiffening trusses, as a result of which the structure becomes *self-anchored* (Fig. 14.9e). It does not require external anchorages.

The stiffening trusses of a self-anchored bridge must be designed to take the compression induced by the cables. The cables are attached to the stiffening trusses over a support that resists the vertical component of cable tension. The vertical upward component may relieve or even exceed the dead-load reaction at the end support. If a net uplift occurs, a pendulum-link tie-down should be provided at the end support.

Self-anchored suspension bridges are suitable for short to moderate spans (400 to 1,000 ft) where foundation conditions do not permit external anchorages. Such conditions include poor foundation-bearing strata and loss of weight due to anchorage submergence. Typical examples of self-anchored suspension bridges are the Paseo Bridge at Kansas City, with a main span of 616 ft, and the former Cologne-Mülheim Bridge (1929) with a 1,033-ft span.

Another type of suspension bridge is referred to as a *bridle-chord bridge*. Called by Germans *Zügelgurtbrücke*, these structures are typified by the bridge over the

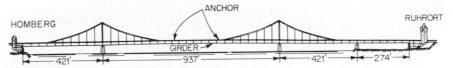

FIGURE 14.11   Bridge over the Rhine at Ruhrort-Homberg, Germany, a bridle-chord type.

Rhine River at Ruhrort-Homberg (Fig. 14.11), erected in 1953, and the one at Krefeld-Urdingen, erected in 1950. It is a special class of bridge, intermediate between the suspension and cable-stayed types and having some of the characteristics of both. The main cables are curved but not continuous between towers. Each cable extends from the tower to a span, as in a cable-stayed bridge. The span, however, also is suspended from the cables at relatively short intervals over the length of the cables, as in suspension bridges.

A distinction to be made between some early suspension bridges and modern suspension bridges involves the position of the main cables in profile at midspan with respect to the stiffening trusses. In early suspension bridges, the bottom of the main cables at maximum sag penetrated the top chord of the stiffening trusses and continued down to the bottom chord (Fig. 14.4, for example). Because of the design theory available at the time, the depth of the stiffening trusses was relatively large, as much as 1/40 of the span. Inasmuch as the height of the pylons is determined by the sag of the cables and clearance required under the stiffening trusses, moving the midspan location of the cables from the bottom chord to the top chord increases the pylon height by the depth of the stiffening trusses. In modern suspension bridges, stiffening trusses are much shallower than those used in earlier bridges and the increase in pylon height due to midspan location of the cables is not substantial (as compared with the effect in the Williamsburg Bridge in New York City where the depth of the stiffening trusses is 25% of the main-cable sag).

Although most suspension bridges employ vertical suspender cables to support the stiffening trusses or the deck structural framing directly (Fig. 14.8), a few suspension bridges, for example, the Severn Bridge in England and the Bosporus Bridge in Turkey, have inclined or diagonal suspenders (Fig. 14.12). In the vertical-suspender system, the main cables are incapable of resisting shears resulting from external loading. Instead, the shears are resisted by the stiffening girders or by displacement of the main cables. In bridges with inclined suspenders, however, a truss action is developed, enabling the suspenders to resist shear. (Since the cables can support loads only in tension, design of such bridges should ensure that there always is a residual tension in the suspenders; that is, the magnitude of the compression generated by live-load shears should be less than the dead-load tension.) A further advantage of the inclined suspenders is the damping properties of the system with respect to aerodynamic oscillations.

(N. J. Gimsing, "Cable-Supported Bridges—Concept and Design," John Wiley & Sons, Inc., New York.)

14.3.3 Suspension Bridge Cross Sections

Figure 14.13 shows typical cross sections of suspension bridges. The bridges illustrated in Fig. 14.13*a*, *b*, and *c* have stiffening trusses, and the bridge in Fig. 14.13*d*

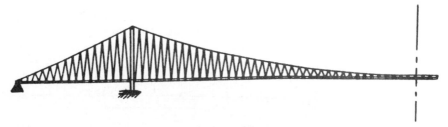

FIGURE 14.12 Suspension system with inclined suspenders.

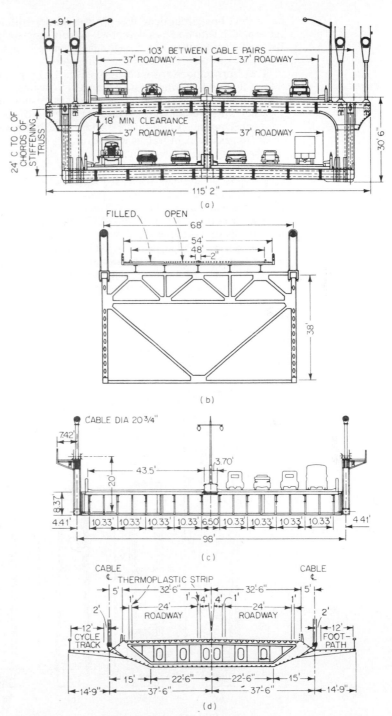

FIGURE 14.13 Typical cross sections of suspension bridges: (*a*) Verrazano Narrows, (*b*) Mackinac, (*c*) Triborough, (*d*) Severn.

has a steel box-girder deck. Use of plate-girder stiffening systems, forming an H-section deck with horizontal web, was largely superseded after the Tacoma Narrows Bridge failure by truss and box-girder stiffening systems for long-span bridges. The H deck, however, is suitable for short spans.

The Verrazano Narrows Bridge (Fig. 14.13*a*), employs 6-in-deep, concrete-filled, steel-grid flooring on steel stringers to achieve strength, stiffness, durability, and lightness. The double-deck structure has top and bottom lateral trusses. These, together with the transverse beams, stringers, cross frames, and stiffening trusses, are conceived to act as a tube resisting vertical, lateral, and torsional forces. The cross frames are rigid frames with a vertical member in the center.

The Mackinac Bridge (Fig. 14.13*b*) employs a 4¼-in, steel-grid flooring. The outer two lanes were filled with lightweight concrete and topped with bituminous-concrete surfacing. The inner two lanes were left open for aerodynamic venting and to reduce weight. The single deck is supported by stiffening trusses with top and bottom lateral bracing as well as ample cross bracing.

The Triborough Bridge (Fig. 14.13*c*) has a reinforced-concrete deck carried by floorbeams supported at the lower panel points of through stiffening trusses.

The Severn Bridge (Fig. 14.13*d*) employs a 10-ft-deep torsion-resisting box girder to support an orthotropic-plate deck. The deck plate is stiffened by steel trough shapes, and the remaining plates, by flat-bulb stiffeners. The box was faired to achieve the best aerodynamic characteristics.

14.3.4 Suspension Bridge Pylons

Typical pylon configurations, shown in Fig. 14.14, are portal frames. For economy, pylons should have the minimum width in the direction of the span consistent with stability but sufficient width at the top to take the cable saddle.

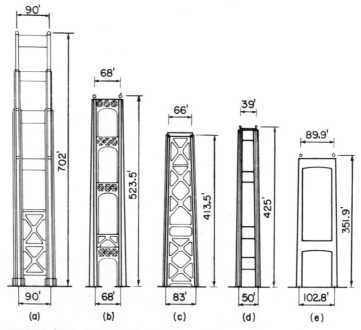

FIGURE 14.14 Suspension-bridge pylons: (*a*) Golden Gate, (*b*) Mackinac, (*c*) San Francisco-Oakland Bay, (*d*) First Tacoma Narrows, (*e*) Walt Whitman.

Most suspension bridges have cables fixed at the top of the pylons. With this arrangement, because of the comparative slenderness of pylons, top deflections do not produce large stresses. It is possible to use rocker pylons, pinned at the base and top, but they are restricted to use with short spans. Also, pylons fixed at the base and with roller saddles at the top are possible, but limited to use with medium spans. The pylon legs may, in any event, be tapered to allow for the decrease in area required toward the top.

The statical action of the pylon and the design of details depend on the end conditions.

Simply supported, main-span stiffening trusses are frequently suspended from the pylons on short pendulum hangers. Dependence is placed primarily on the short center-span suspenders to keep the trusses centered. In this way, temperature effects on the pylon can be reduced by half.

A list of major modern suspension bridges is provided in Table 14.1.

14.4 CLASSIFICATION AND CHARACTERISTICS OF CABLE-STAYED BRIDGES

The cable-stayed bridge has come into wide use since the 1950s for medium- and long-span bridges, because of its economy, stiffness, esthetic qualities, and ease of erection without falsework. Cable-stayed bridges utilize taut cables connecting pylons to a span to provide intermediate support for the span. This principle has been understood by bridge engineers for at least two centuries, as indicated in Art. 14.1.

Cable-stayed bridges are economical for bridge spans intermediate between those suited for deck girders (usually up to 600 to 800 ft but requiring extreme depths, up to 33 ft) and the longer-span suspension bridges (over 1,000 ft). The cable-stayed bridge, thus, finds application in the general range of 600- to 1,600-ft spans, but spans as long as 2,600 ft may be economically feasible.

A cable-stayed bridge has the advantage of greater stiffness over a suspension bridge. Cable-stayed single or multiple box girders possess large torsional and lateral rigidity. These factors make the structure stable against wind and aerodynamic effects.

14.4.1 Structural Characteristics of Cable-Stayed Bridges

The true action of a cable-stayed bridge is considerably different from that of a suspension bridge. As contrasted with the relatively flexible main cables of the latter, the inclined, taut cables of the cable-stayed structure furnish relatively stable point supports in the main span. Deflections are thus reduced. The structure, in effect, becomes a continuous girder over the piers, with additional intermediate, elastic (yet relatively stiff) supports in the span. As a result, the stayed girder may be shallow. Depths usually range from $\frac{1}{60}$ to $\frac{1}{80}$ of the main span, sometimes even as small as $\frac{1}{100}$ of the span.

Cable forces are usually balanced between the main and flanking spans, and the structure is internally anchored; that is, it requires no massive masonry anchorages. Second-order effects of the type requiring analysis by a deflection theory are of relatively minor importance for the common, self-anchored type of cable-stayed bridge, characterized by compression in the main bridge girders.

TABLE 14.1   Major Suspension Bridges

Name	Location	Length of main span		Year completed
		ft	m	
Akashi Kaikyo[1]	Japan	6066	1990	1998
Great Belt Link[1]	Zealand-Sprago, Denmark	5328	1624	1996
Humber River	Hull, England	4626	1410	1981
Verrazano Narrows	New York City, U.S.A.	4260	1298	1964
Golden Gate	San Francisco, U.S.A.	4200	1280	1937
Mackinac Straits	Michigan, U.S.A.	3800	1158	1957
Minami Bisan-Seto	Japan	3668	1118	1988
Second Bosphorus[1]	Istanbul, Turkey	3576	1090	
First Bosphorus	Istanbul, Turkey	3524	1074	1973
George Washington	New York City, U.S.A.	3500	1067	1931
Tagus River[2]	Lisbon, Portugal	3323	1013	1966
Forth Road	Queensferry, Scotland	3300	1006	1964
Kita Bisan-Seto	Japan	3300	1006	1988
Severn	Beachley, England	3240	988	1966
Shimotsui Straits	Japan	3136	956	1988
Ohnaruto	Japan	2874	876	1985
Tacoma Narrows I[3]	Tacoma, Wash., U.S.A.	2800	853	1940
Tacoma Narrows II	Tacoma, Wash., U.S.A.	2800	853	1950
Innoshima	Japan	2526	770	1983
Kanmon Straits	Kyushu-Honshu, Japan	2336	712	1973
Angostura	Ciudad Bolivar, Venezuela	2336	712	1967
San Francisco-Oakland Bay[4]	San Francisco, Calif., U.S.A.	2310	704	1936
Bronx-Whitestone	New York City, U.S.A.	2300	701	1939
Pierre Laporte	Quebec, Canada	2190	668	1970
Delaware Memorial[5]	Wilmington, Del., U.S.A.	2150	655	1951 1968
Seaway Skyway	Ogdensburg, N.Y., U.S.A.	2150	655	1960
Walt Whitman	Philadelphia, Pa., U.S.A.	2000	610	1957
Tancarville	Tancarville, France	1995	608	1959
Lillebaelt	Lillebaelt Strait, Denmark	1969	600	1970
Ambassador	Detroit, Mich., U.S.A.–Canada	1850	564	1929
Skyway[3]	(Chicago World's Fair), U.S.A.	1850	564	1933
Throgs Neck	New York City, U.S.A.	1800	549	1961
Benjamin Franklin[2]	Philadelphia, Pa., U.S.A.	1750	533	1926
Skjomen	Narvik, Norway	1722	525	1972
Kvalsund	Hammerfest	1722	525	1977
Kleve-Emmerich	Emmerich, Germany	1640	500	1965

TABLE 14.1 Major Suspension Bridges (*Continued*)

| Name | Location | Length of main span | | Year completed |
		ft	m	
Bear Mountain	Peckskill, N.Y., U.S.A.	1632	497	1924
Wm. Preston Lane, Jr.[5]	Near Annapolis, Md., U.S.A.	1600	488	1952 1973
Williamsburg[2]	New York City, U.S.A.	1600	488	1903
Newport	Newport, R.I., U.S.A.	1600	488	1969
Chesapeake Bay	Sandy Point, Md., U.S.A.	1600	488	1952
Brooklyn[2,7]	New York City, U.S.A.	1595	486	1883
Lions Gate	Vancouver, B.C., Canada	1550	472	1939
Hirato Ohashi	Hirato, Japan	1536	468	1977
Sotra	Bergen, Norway	1535	468	1971
Vincent Thomas	San Pedro–Terminal Is., Calif., U.S.A.	1500	457	1964
Mid-Hudson	Poughkeepsie, N.Y., U.S.A.	1500	457	1930
Manhattan[2]	New York City, U.S.A.	1470	448	1909
MacDonald Bridge	Halifax, Nova Scotia, Canada	1447	441	1955
A. Murray Mackay	Halifax, Nova Scotia, Canada	1400	426	1970
Triborough	New York City, U.S.A.	1380	421	1936
Alvsborg	Goteborg, Sweden	1370	418	1967
Hadong-Namhae	Pusan, South Korea	1325	404	1973
Baclan	Garrone R., Bordeaux, France	1292	394	1967
Amu-Darja R.	Buhara-Ural, Russia	1280	390	1964
Clifton[3]	Niagara Falls, N.Y., U.S.A.	1268	386	1869
Cologne-Rodenkirchen I[3]	Cologne, Germany	1240	378	1941
Cologne-Rodenkirchen II	Cologne, Germany	1240	378	1955
St. Johns	Portland, Ore., U.S.A.	1207	368	1931
Wakato	Wakamatsu-Tobata, Japan	1205	367	1962
Mount Hope	Bristol, R.I., U.S.A.	1200	366	1929
St. Lawrence R.	Ogdensburg, N.Y.–Prescot, Ont.	1150	351	1960
Ponte Hercilio[2,6]	Florianapolis, Brazil	1114	340	1926
Bidwell Bar Bridge	Oroville, Calif., U.S.A.	1108	338	1971
Middle Fork Feather River	California, U.S.A.	1105	337	1964
Varodd, Topdalsfjord	Kristiansand, Norway	1105	337	1956
Tamar Road	Plymouth, Great Britain	1100	335	1961
Deer Isle	Deer Isle, Me., U.S.A.	1080	329	1939
Rombaks	Narvik, Nordland, Norway	1066	325	1964
Maysville	Maysville, Ky., U.S.A.	1060	323	1931
Ile d'Orleans	Paris, France	1059	323	1935
Ohio River	Cincinnati, Ohio, U.S.A.	1057	322	1867

TABLE 14.1 Major Suspension Bridges (*Continued*)

Name	Location	Length of main span		Year completed
		ft	m	
Otto Beit	Zambezi R., Zimbabwe	1050	320	1939
Dent	N. Fork, Clearwater R., Id., U.S.A.	1050	320	1971
Niagara[3]	Lewiston, N.Y., U.S.A.	1040	317	1850
Cologne-Mulheim I[3]	Cologne, Germany	1033	315	1929
Cologne-Mulheim II	Cologne, Germany	1033	315	1951
Miampimi	Mexico	1030	314	1900
Wheeling	West Virginia, U.S.A.	1010	308	1848 1856
Konohana[8,9]	Osaka, Japan	984	300	1990
Elizabeth[6]	Budapest, Hungary	951	290	1903 1964
Tjeldsund	Harstad, Norway	951	290	1967
Grand' Mère	Quebec, Canada	948	289	1929
Cauca River	Columbia	940	287	1894
Peace River	British Columbia, Canada	932	284	1950
Aramon	France	902	275	1901
Cornwall-Masena	St. Lawrence R., New York–Ontario	900	274	1958
Brevik	Telemark, Norway	892	272	1962
Royal Gorge	Arkansas R., Canon City, Colo., U.S.A.	880	268	1929
Freiburg (Saane)[3]	Switzerland	870	265	1834
Kjerringstraumen	Nordland, Norway	853	260	1975
Ambassador	Detroit, Mich., U.S.A.	850	259	1929
Railway Bridge[3]	Niagara River, N.Y., U.S.A.	821	250	1854
Dome, Grand Canyon	Dome, Ariz., U.S.A.	800	244	1929
Point[3,6]	Pittsburgh, Pa., U.S.A.	800	244	1877
Rochester	Rochester, Pa., U.S.A.	800	244	1896
Niagara River	Lewiston, N.Y., U.S.A.	800	244	1899
Thousand Is. International	St. Lawrence R., U.S.A.–Canada	800	244	1938
Waldo Hancock	Penobscot R., Bucksport, Me., U.S.A.	800	244	1931
Anthony Wayne	Maumee R., Toledo, Ohio, U.S.A.	785	239	1931
Parkersburg	Parkersburg, W.Va., U.S.A.	775	236	1916
Footbridge[3]	Niagara R., N.Y., U.S.A.	770	235	1847
Vernaison	France	764	233	1902
Cannes Ecluse	France	760	232	1900
Ohio River	E. Liverpool, Ohio, U.S.A.	750	229	1905
Gotteron	Freiburg, Switzerland	746	227	1840
Iowa-Illinois Mem. I[3]	Moline, Ill., U.S.A.	740	226	1934

TABLE 14.1   Major Suspension Bridges (*Continued*)

| Name | Location | Length of main span | | Year completed |
		ft	m	
Iowa-Illinois Mem. II	Moline, Ill., U.S.A.	740	226	1959
Monogahela R.	S. 10th St., Pittsburgh, Pa., U.S.A.	725	221	1933
Rondout	Kingston, N.Y., U.S.A.	705	215	1922
Ohio River	E. Liverpool, Ohio, U.S.A.	705	215	1896
Clifton[3,6]	Bristol, England	702	214	1864
Ohio River[6]	St. Marys, Ohio, U.S.A.	700	213	1929
Ohio River[3,6]	Point Pleasant, Ohio, U.S.A.	700	213	1928
General U. S. Grant	Ohio R., Portsmouth, Ohio, U.S.A.	700	213	1927
Airline	St. Jo, Texas, U.S.A.	700	213	1927
Red River	Nacona, Texas, U.S.A.	700	213	1924
Ohio River	Steubenville, Ohio, U.S.A.	700	213	1904
Ohio River	Steubenville, Ohio, U.S.A.	689	210	1928
Isere	Veurey, France	688	210	1934
Hungerford[3,6]	London, England	676	206	1845
Mississippi R.[3]	Minneapolis, Minn., U.S.A.	675	206	1877
Lancz[6]	Budapest, Hungary	663	202	1845
White River	Des Arc, Arkansas, U.S.A.	650	198	1928
Roche Bernard[3]	Vilaine, France	650	198	1836
Caille[3]	Annecy, France	635	194	1839
Columbia R.	Beebee, Wash., U.S.A.	632	193	1919

[1]Under Construction 1992. [2]Railroad and highway. [3]Not standing. [4]Twin spans. [5]Twin bridges. [6]Eyebar chain. [7]Includes cable stays. [8]Self-anchored. [9]Monocable.

14.4.2   Types of Cable-Stayed Bridges

Cable-stayed bridges may be classified by the type of material they are constructed of, by the number of spans stay-supported, by transverse arrangement of cable-stay planes, and by the longitudinal stay geometry.

A concrete cable-stayed bridge has both the superstructure girder and the pylons constructed of concrete. Generally, the pylons are cast-in-place, although in some cases, the pylons may be precast-concrete segments above the deck level to facilitate the erection sequence. The girder may consist of either precast or cast-in-place concrete segments. Examples are the Talmadge Bridge in Georgia and the Sunshine Skyway Bridge in Florida.

All-steel cable-stayed bridges consist of structural steel pylons and one or more stayed steel box girders with an orthotropic deck (Fig. 14.15). Examples are the Luling Bridge in Louisiana and the Meridian Bridge in California (also constructed as a swing span).

Other so-called steel cable-stayed bridges are, in reality, composite structures with concrete pylons, structural-steel edge girders and floorbeams (and possibly

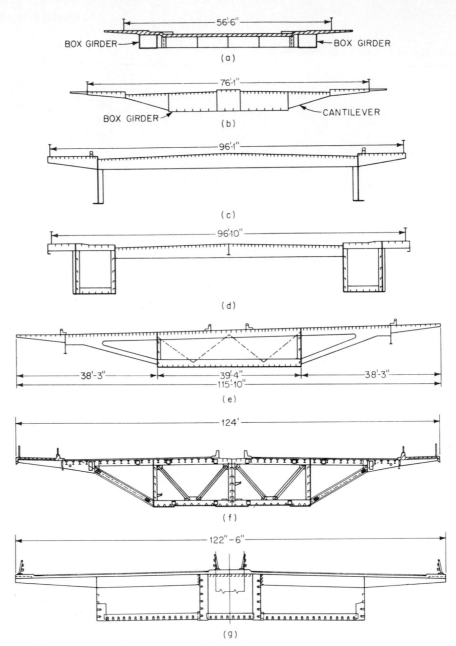

FIGURE 14.15 Typical cross sections of cable-stayed bridges: (*a*) Büchenauer Bridge with composite concrete deck and two steel box girders, (*b*) Julicherstrasse crossing with orthotropic-plate deck, box girder, and side cantilevers. (*c*) Kniebrucke with orthotropic-plate deck and two solid-web girders. (*d*) Severn Bridge with orthotropic-plate deck and two box girders. (*e*) Bridge near Maxau with orthotropic-plate deck, box girder, and side cantilevers. (*f*) Leverkusen Bridge with orthotropic-plate deck, box girder, and side cantilevers. (*g*) Lower Yarra Bridge with composite concrete deck, two box girders, and side cantilevers. (*Adapted from A. Feige, "The Evolution of German Cable-Stayed Bridges—An Overall Survey," Acier-Stahl-Steel (English version), no. 12, December 1966 (reprinted in the AISC Journal, July 1967.)*

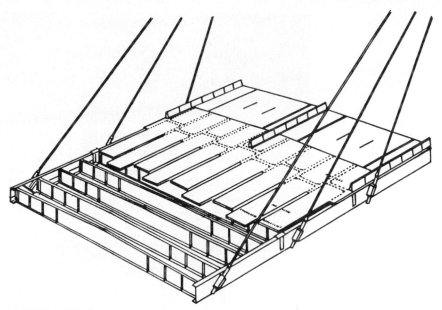

FIGURE 14.16 Composite steel-concrete superstructure girder of a cable-stayed bridge.

stringers), and a composite cast-in-place or precast plank deck. The precast deck concept is illustrated in Fig. 14.16.

In general, span arrangements are single span; two spans, symmetrical or asymmetrical; three spans; or multiple spans. Single-span cable-stayed bridges are a rarity, usually dictated by unusual site conditions. An example is the Ebro River Bridge at Navarra, Spain (Fig. 14.17). Generally, back stays are anchored to deadman anchorage blocks, analogous to the simple-span suspension bridge (Fig. 14.9a).

14.4.3 Span Arrangements in Cable-Stayed Bridges

A few examples of two-span cable-stayed bridges are illustrated in Fig. 14.18. In two-span, asymmetrical, cable-stayed bridges, the major spans are generally in the range of 60 to 70% of the total length of stayed spans. Exceptions are the Batman Bridge (Fig. 14.18g) and Bratislava Bridge (Fig. 14.18h), where the major spans are 80% of the total length of stayed spans. The reason for the longer major span is that these bridges have a single back stay anchored to the abutment rather than several back stays distributed along the side span.

Three-span cable-stayed bridges (Fig. 14.19) generally have a center span with a length about 55% of the total length of stayed spans. The remainder is usually equally divided between the two anchor spans.

Multiple-span cable-stayed bridges (Fig. 14.20) normally have equal length spans with the exception of the two end spans, which are adjusted to connect with approach spans or the abutment. The cable-stay arrangement is symmetrical on each side of the pylons. For convenience of fabrication and erection, the girder has "drop-in" sections at the center of the span between the two leading stays. The ratio of drop-in span length to length between pylons varies from 20%, when a single stay emanates from each side of the pylon, to 8% when multiple stays emanate from each side of the pylon.

FIGURE 14.17   Ebro River Bridge, Navarra, Spain. *(Reprinted with permission from Stronghold International, Ltd.)*

14.4.4 Cable-Stay Configurations

Transverse to the longitudinal axis of the bridge, the cable stays may be arranged in a single or double plane with respect to the longitudinal centerline of the bridge and may be positioned in vertical or inclined planes (Fig. 14.21). Single-plane systems, located along the longitudinal center line of the structure (Fig. 14.21*a*) generally require a torsionally stiff stayed box girder to resist the torsional forces developed by unbalanced loading. The laterally displaced vertical system (Fig. 14.21*b*) has been used for a pedestrian bridge. The V-shaped arrangement (Fig. 14.21*e*), has been used for cable-stayed bridges supporting pipelines. This variety of transverse-stay geometry leads to numerous choices of pylon arrangements (Fig. 14.22).

There are four basic stay configurations in elevation (Fig. 14.23): radiating, harp, fan, and star. In the radiating system, all stays converge at the top of the pylon. In the harp system, the stays are parallel to each other and distributed over the height of the pylon. The fan configuration is a hybrid of the radiating and the harp system. The star system was used for the Norderelbe Bridge in Germany primarily for its esthetic appearance. The variety of configurations in elevation leads to a wide variation of geometric arrangements, as indicated by Fig. 14.23.

The number of stays used for support of the deck ranges from a single stay on each side of the pylon to a multistay arrangement, as illustrated in Figs. 14.18 to 14.20. Use of a few stays leads to large spacing between attachment points along the girder. This necessitates a relatively deep stayed girder and large concentrations of stay force to the girder, with attendant complicated connection details. A large number of stays has the advantage of reduction in girder depth, smaller diameter stays, simpler connection details, and relative ease of erection by the cantilever method. However, the number of terminal stay anchorages is increased and there are more stays to install.

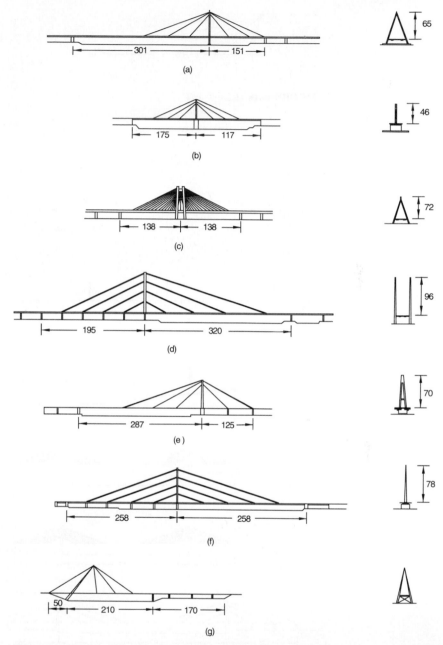

FIGURE 14.18 Examples of two-span cable-stayed bridges (dimensions in meters): (*a*) Cologne, Germany; (*b*) Karlsruhe, Germany; (*c*) Ludwigshafen, Germany; (*d*) Kniebrucke-Dusseldorf, Germany; (*e*) Manheim, Germany; (*f*) Dusseldorf-Oberkassel, Germany; (*g*) Batman, Australia; (*h*) Bratislava, Czechoslovakia.

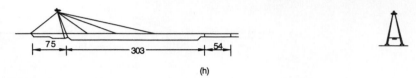

(h)

FIGURE 14.18   (*Continued*)

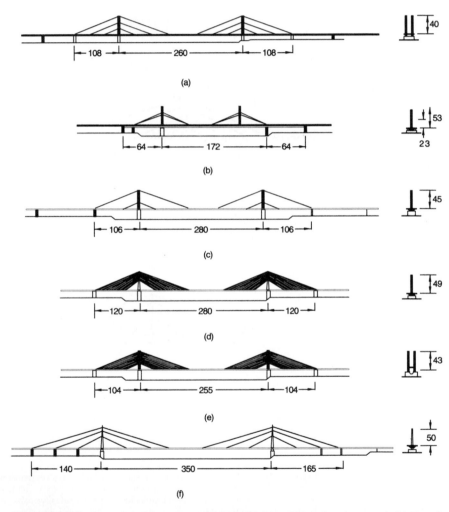

FIGURE 14.19   Examples of three-span cable-stayed bridges (dimensions in meters): (*a*) Dussel-dorf-North, Germany; (*b*) Norderelbe, Germany; (*c*) Leverkusen, Germany; (*d*) Bonn, Germany; (*e*) Rees, Germany; (*f*) Duisburg, Germany; (*g*) Stromsund, Sweden; (*h*) Papineau, Canada; (*i*) Onomichi, Japan.

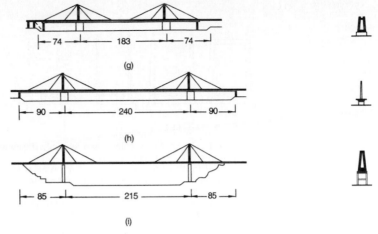

(g)

(h)

(i)

FIGURE 14.19 (*Continued*)

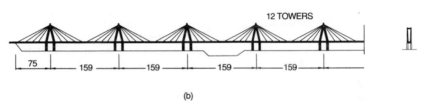

6 TOWERS

(a)

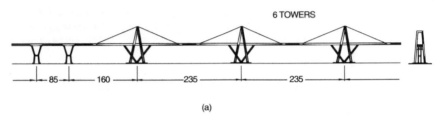

12 TOWERS

(b)

FIGURE 14.20 Examples of multispan cable-stayed bridges (dimensions in meters): (*a*) Maracaibo, Venezuela, and (*b*) Ganga Bridge, India.

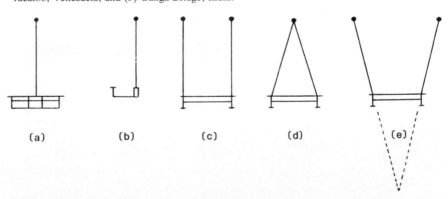

(a) (b) (c) (d) (e)

FIGURE 14.21 Cross sections of cable-stayed bridges showing variations in arrangements of cable stays. (*a*) Single-plane vertical. (*b*) Laterally displaced vertical. (*c*) Double-plane vertical. (*d*) Double-plane inclined. (*e*) Double-plane V-shaped. (*Reprinted with permission from W. Podolny, Jr., and J. B. Scalzi, "Construction and Design of Cable-Stayed Bridges," 2d ed., John Wiley & Sons, Inc., New York.*)

14.23

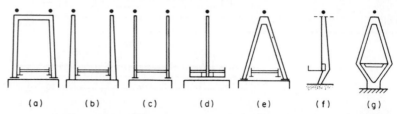

FIGURE 14.22 Shapes of pylons used for cable-stayed bridges. (*a*) Portal frame with top cross member. (*b*) Pylon fixed to pier and without top cross member. (*c*) Pylon fixed to girders and without top cross member. (*d*) Axial pylon fixed to superstructure. (*e*) A shaped pylon. (*f*) Laterally displaced pylon fixed to pier. (*g*) Diamond-shaped pylon. *(Reprinted with permission from A. Feige, "The Evolution of German Cable-Stayed Bridges—An Overall Survey," Acier-Stahl-Steel (English version), no. 12, December 1966 (reprinted in the AISC Journal, July 1967.)*

A list of major modern cable-stayed bridges is provided in Table 14.2.
(W. Podolny, Jr., and J. B. Scalzi, "Construction and Design of Cable-Stayed Bridges," 2d ed., John Wiley & Sons, Inc., New York.)

14.5 CABLE-SUSPENDED BRIDGES FOR RAIL LOADING

Because of flexibility and susceptibility to vibration under dynamic loads, pure suspension bridges are rarely constructed for railway spans. They are sometimes used, however, where dead load constitutes a relatively large proportion of the total load. Where provisions for both railway and highway traffic is necessary, as for the future extension of the Salazar Bridge (Fig. 14.7), the addition of inclined cable stays from the pylon to the stiffening girder is advantageous or a cable-stayed bridge may be used, for increased stiffness.

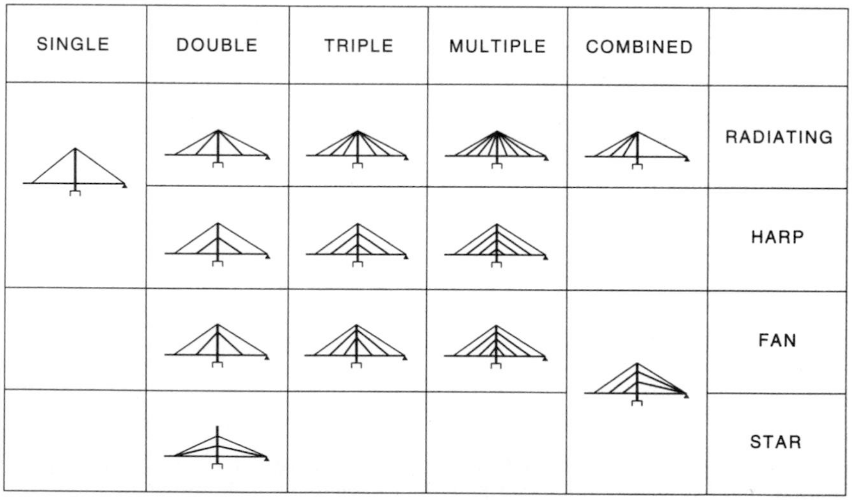

SINGLE	DOUBLE	TRIPLE	MULTIPLE	COMBINED	
◿	◿	◿	◿	◿	RADIATING
	◿	◿	◿		HARP
	◿	◿	◿		FAN
				◿	
	◿				STAR

FIGURE 14.23 Stay configurations for cable-stayed bridges.

TABLE 14.2   Major Cable-Stayed Bridges

Name	Location	Length of main or major span		Year completed
		ft	m	
Tatara[1]	Ehime, Japan	2920	890	
Normandie[1]	Le Havre, France	2808	856	
Meiko Higashi[1]	Aichi, Japan	1936	590	
Tsurumi[1]	Kanagawa, Japan	1673	510	
Iguchi[1]	Ehime, Japan	1608	490	
Higashi Kobe[1]	Hyogo, Japan	1591	485	
Annacis	Vancouver, B.C., Canada	1526	465	1983
Yokohama Bay	Kanagawa, Japan	1509	460	1989
Second Hooghly R.[1]	Calcutta-Howrah, India	1484	452	
Dao Kanong, Chao Phraya R.	Bangkok, Thailand	1476	450	1987
Barrios De Luna	Spain	1444	440	1983
Helgeland	Sandnessjoen, Nordland, Norway	1394	425	1991
Hitsuishijima	Kagawa, Japan	1378	420	1988
Iwagurojima	Kagawa, Japan	1378	420	1988
Meiko Higashi[1]	Aichi, Japan	1345	410	
Meiko Nishi	Aichi, Japan	1329	405	1985
Bridge over the Waal River	Ewijck, Holland	1325	404	1976
Saint Nazaire	Saint Nazaire, France	1325	404	1975
Rande	Vigo, Spain	1313	400	1978
Dame Point	Jacksonville, Fla., U.S.A.	1300	396	1988
Houston Ship Channel[1]	Baytown, Texax, U.S.A.	1250	381	
Hale Boggs Memorial	Luling, La., U.S.A.	1222	372	1983
Dusseldorf-Flehe	Rhine R., Germany	1205	367	1979
Tjorn Bridge, Askerofjord	Near Gothenburg, Sweden	1201	366	1981
Sunshine Skyway	Tampa, Fla., U.S.A.	1200	366	1987
Yamatogawa	Osaka, Japan	1165	355	1982
Tenpozan	Osaka, Japan	1148	350	1989
Duisburg-Newenkamp	Germany	1148	350	1970
Novi Sad	Yugoslavia	1135	346	1981
West Gate	Melbourne, Australia	1102	336	1974
Talmadge Memorial Bridge	Savannah, Ga., U.S.A.	1100	335	1990
Puente Brazo Largo[2]	Rio Parana, Argentina	1083	330	1976
Zarate[2]	Rio Parana, Argentina	1083	330	1975
Glebe Island Bridge[1]	Sydney, Australia	1083	330	
Kohlbrand	Hamburg, Germany	1066	325	1974
Int. Guadiana Bridge	Portugal-Spain	1063	324	1991

TABLE 14.2 Major Cable-Stayed Bridges (*Continued*)

Name	Location	Length of main or major span ft	m	Year completed
Pont de Brotonne, Seine R.	Rouen, France	1050	320	1977
Kniebrucke	Rhine R., Dusseldorf, Germany	1050	320	1969
Daugava R.	Riga, Latvia	1024	312	1981
Dartford-Thurrock Bridge	Thames R., Great Britain	1001	305	1991
Erskine, River Clyde	Glasgow, Scotland	1000	305	1971
Bratislava	Danube R., Czechoslovakia	994	303	1972
Severin	Cologne, Germany	990	302	1959
Moscovsky, Dnieper R.	Kiev, Ukraine	984	300	1976
Pasco-Kennewick	Washington, U.S.A.	981	299	1978
Donaubrucke	Deggenau, Germany	951	290	1974
Coatzacoalcos	Mexico	945	288	1979
Kurt-Schumacher	Mannheim-Ludwigshafen, Germany	941	287	1972
Wadi Kuf	Sipac, Libya	925	282	1971
Leverkusen	Germany	919	280	1965
Friedrich-Ebert	Bonn, Germany	919	280	1967
Rheinbrucke	Speyer, Germany	902	275	1974
East Huntington	East Huntington, W.Va., U.S.A.	900	274	1985
Bayview Bridge	Quincy, Ill., U.S.A.	900	274	1987
Ewijk, Wall R.	Near Ewijk, Netherlands	886	270	1976
River Waal	Tiel, Holland	876	267	1974
Theodor Heuss	Dusseldorf, Germany	853	260	1957
Burton	New Brunswick, Canada	850	259	1970
Oberkassel	Dusseldorf, Germany	846	258	1973
Arade Bridge	Portimão, Portugal	840	256	1991
Rees	Rees-Kalkar, Germany	837	255	1967
Savh River Railroad	Belgrade, Yugoslavia	833	254	1977
Weirton-Steubenville	West Virginia, U.S.A.	820	250	1990
Tokachi Chuo	Hokkaido, Japan	820	250	1989
Yobuko	Saga, Japan	820	250	1988
Suehiro	Tokushima, Japan	820	250	1975
Chaco/Corrientes	Parana River, Argentina	804	245	1973
Papineau-Leblanc	Montreal, Canada	790	241	1969
Aomori	Aomon, Japan	787	240	1991
Kessock	Inverness, Scotland	787	240	1981
Kamone	Osaka, Japan	787	240	1975
Kemi[1]	Wakayama, Japan	784	239	

TABLE 14.2   Major Cable-Stayed Bridges (*Continued*)

| Name | Location | Length of main or major span | | Year completed |
		ft	m	
Sugawara	Osaka, Japan	780	238	1989
Cochrane	Mobile, Ala., U.S.A.	780	238	1991
Lake Maracaibo	Venezuela	771	235	1962
Neuwied	Rhine R., Germany	770	235	1977
Wye River Bridge	England	770	235	1966
Clark Bridge Replacement[1]	Alton, Ill., U.S.A.	756	230	
Chesapeake and Delaware Canal Bridge[1]	Delaware, U.S.A.	750	229	
Donaubrucke	Hainburg, Austria	748	228	1972
Penang	Malaysia	738	225	1980
Luangawa	Zambia	730	223	1968
Katsushika	Tokyo, Japan	722	220	1987
Rokko	Hyogo, Japan	722	220	1976
Long's Creek	New Brunswick, Canada	713	217	1967
Toyosato	Osaka, Japan	709	216	1970
Onomichi	Hiroshima, Japan	705	215	1968
Donaubrucke	Linz, Austria	705	215	1972
Godsheide	Hasselet, Belgium	690	210	1978
Polcevera Viaduct	Genoa, Italy	689	210	1967
Chalkis	Greece	689	210	1984
Arno	Firenze, Italy	676	206	1976
Batman	Tasmania, Australia	675	206	1968
Burlington Bridge[1]	Burlington, Iowa, U.S.A.	660	201	
Shin Inagawa[1]	Osaka, Japan	656	200	
Yodogawa	Osaka, Japan	656	200	1987
Chung Yang	Taiwan	656	200	1984
Neches River	Texas, U.S.A.	640	195	1991
James River Bridge	Near Richmond, Va., U.S.A.	630	192	1989
Sakitama	Saitama, Japan	623	190	1991
Ashigara	Kanagawa, Japan	607	185	1991
Stromsund	Sweden	600	183	1955
Admiyak	Baghdad, Iraq	599	183	1983
Carpineto	Province Potenza, Italy	594	181	1977
Galekopperbrug	Utrecht, Holland	590	180	1974
Suigo	Chiba, Japan	587	179	1977
Lanaye	Belgium	581	177	1987
Maxau	Maxau, Germany	575	175	1966

TABLE 14.2 Major Cable-Stayed Bridges (*Continued*)

Name	Location	Length of main or major span ft	m	Year completed
Karlsruhe	Germany	574	175	1965
Norderelbe	Hamburg, Germany	564	172	1963
Wandre, Meuse, R.	Belgium	551	168	1990
Ben-Ahin, Meuse R.	Belgium	551	168	1989
Daikoku	Kanagawa, Japan	541	165	1974
Massena	Paris, France	530	162	1969
Steyregger	Danube R., Austria	529	161	1979
Ishikari Kako	Hokkaido, Japan	525	160	1972
Arakawa	Tokyo, Japan	525	160	1971
George Street	Newport, Gwent, Wales	500	152	1964
Chichibu	Saitama, Japan	499	152	1985
Yelcho River Bridge	Chile	492	150	1989
Olympic Bridge	Korea	492	150	1984
Isere Viaduct	Valence-Grenoble, France	486	148	1991
Mainbrucke[2]	Hoechst, Germany	486	148	1972
Ebro River	Navarra, Spain	480	146	1979
Matsukawa[1]	Fukushima, Japan	476	145	
Magliana	Rome, Italy	476	145	1967
Hattabara[1]	Hiroshima, Japan	472	144	
Gassho	Toyama, Japan	472	144	1979
Eisai	Kanagawa, Japan	472	144	1977
Dnieper River	Kiev, Ukraine	472	144	1965
Toko[1]	Hokkaido, Japan	459	140	
Bannaguro	Hokkaido, Japan	459	140	1990

[1]Under construction 1992. [2]Railroad and highway. [3]Twin bridges. [4]Swing span.

An important consideration in the design for rail loading (including rapid-transit trains) is the positioning of the tracks with respect to the transverse centerline of the deck structure. In the Williamsburg Bridge (Fig. 14.24*a*), the railway is positioned adjacent to the centerline, greatly minimizing torsional forces. In the Manhattan Bridge (Fig. 14.24*b*), the railway is positioned outboard of the centerline, resulting in large torsional forces. As a result of this positioning, the Manhattan Bridge, over the years, has suffered damage and had to be retrofitted with a torsion tube to increase its resistance to torsional forces.

The Zarate-Brazo Largo Bridges in Argentina (two identical structures) are unique cable-stayed bridges not only from the standpoint of supporting highway and railroad traffic, but also in that the rail line is on one side of the structures. This positioning necessitated an increased stiffness of the stays on the railroad side

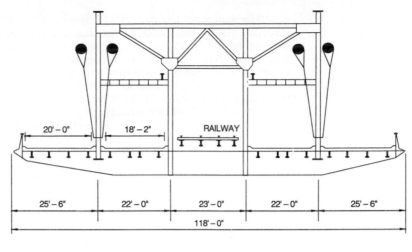

(a)

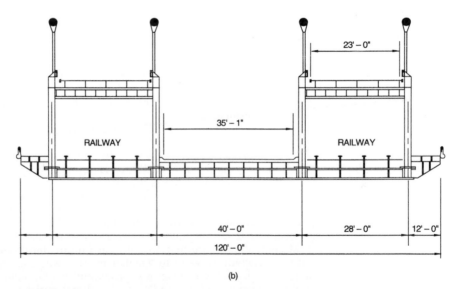

(b)

FIGURE 14.24 Position of rail loading on two suspension bridges: (*a*) Williamsburg Bridge and (*b*) Manhattan Bridge.

(see W. Podolny, Jr., and J. B. Scalzi, "Construction and Design of Cable-Stayed Bridges," 2d ed., John Wiley & Sons, Inc., New York.)

14.6 SPECIFICATIONS AND LOADINGS FOR CABLE-SUSPENDED BRIDGES

"Standard Specifications for Highway Bridges," American Association of State Highway and Transportation Engineers (AASHTO), covers ordinary steel bridges, generally with spans less than 500 ft. Specifications of the American Railway Engineering Association (AREA) for steel railway bridges apply to spans not exceeding 400 ft. There are no standard American specifications for longer spans than these. AASHTO and AREA specifications, however, are appropriate for design of local areas of a long-span structure, such as the floor system. A basically new set of specifications must be written for each long-span bridge to incorporate the special features brought about by site conditions, long spans, sometimes large traffic capacities, flexibility, aerodynamic and seismic conditions, special framing, and sophisticated materials and construction processes.

Structural analysis is usually applied to the following loading conditions: dead load, live load, impact, traction and braking, temperature changes, displacement of supports (including settlement), wind (both static and dynamic effects), seismic effects, and combinations of these. Guidelines for loadings on long-span bridges are given in P. G. Buckland, "North American and British Long-Span Bridge Loads," *Journal of Structural Engineering*, vol. 117, no. 10, October 1991, American Society of Civil Engineers (ASCE). Recommendations for stay cables are presented in "Recommendations for Stay-Cable Design, Testing and Installation," Committee on Cable-Stayed Bridges, Post-Tensioning Institute. See also "Guide for the Design of Cable-Stayed Bridges," ASCE Committee on Cable-Stayed Bridges.

14.7 CABLES

The concept of bridging long spans with cables, flexible tension members, antecedes recorded history (Art. 14.1). Known ancient uses of metal cables include the following: A short length of copper cable discovered in the ruins of Ninevah, near Babylon, is estimated to have been made in about 685 B.C. in the Kingdom of Assyria. A piece of bronze rope was discovered in the ruins of Pompeii, which was destroyed by the eruption of Mt. Vesuvius in 79 A.D. The Romans made cables of wires and rope; on display in the Museo Barbonico at Naples, Italy, is a 1-in-dia., 15-in-long specimen of their lang-lay bronze rope, in which the direction of lay of both wires and strands is the same.

These early specimens of rope consisted of hand-made wires. In succeeding centuries, the craftsmanship reached such a state of the art that only a very close inspection reveals that wires were hand-made. Viking craftsmanship produced such uniform wire that some authorities believe that mechanical drawing was used.

Machine-drawn wire first appeared in Europe during the fourteenth century, but there is controversy as to whether the first wire rope resembling the current uniform, high-quality product was produced by a German, A. Albert (1834), or an Englishman named Wilson (1832). The first American machine-made wire rope was placed in service in 1846. Since then, with technological improvements, such as advances in manufacturing processes and introduction of high-strength steels, the quality of strand and rope has advanced to that currently available.

In structural applications, cable is generally used in a generic sense to indicate a flexible tension member. Several types of cables are available for use in cable-supported bridges. The form or configuration of a cable depends on its makeup; it can be composed of parallel bars, parallel wires, parallel strands or ropes, or locked-coil strands (Fig. 14.25). Parallel bars are not used for suspension bridges because of the curvature requirements at the pylon saddles. Nor are they used in cable-stayed bridges where a saddle is employed at the pylon, but they have been utilized in a stay where it terminates and is anchored at the pylon.

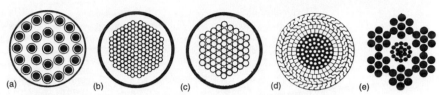

FIGURE 14.25 Various types of cables used for stays: (*a*) parallel bars, (*b*) parallel wires, (*c*) parallel strands, (*d*) helical lock-coil strands (*e*) ropes. *(Courtesy of VSL International, Ltd.)*

14.7.1 Definition of Terms

Cable. Any flexible tension member, consisting of one or more groups of wires, strands, ropes or bars.

Wire. A single, continuous length of metal drawn from a cold rod.

Prestressing wire. A type of wire usually used in posttensioned concrete applications. As normally used for cable stays, it consists of 0.25-in-dia. wire produced in the United States in accordance with ASTM A421 Type BA.

Structural strand (with the exception of parallel-wire strand). Wires helically coiled about a center wire to produce a symmetrical section (Fig. 14.26), produced in the United States in accordance with ASTM A586.

Lay. Pitch length of a wire helix.

Parallel-wire strand. Individual wires arranged in a parallel configuration without the helical twist (Fig. 14.26).

Locked-coil strand. An arrangement of wires resembling structural strands except that the wires in some layers are shaped to lock together when in place around the core (Fig. 14.26).

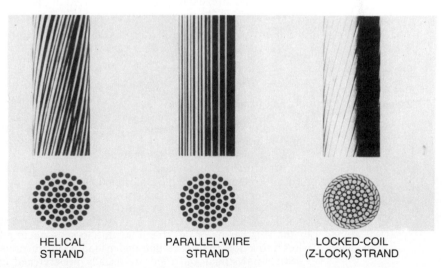

HELICAL STRAND	PARALLEL-WIRE STRAND	LOCKED-COIL (Z-LOCK) STRAND

FIGURE 14.26 Types of strands. *(Courtesy of Bethlehem Steel Corporation.)*

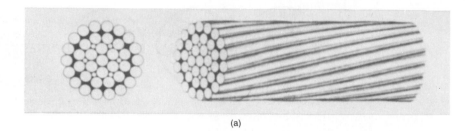

(a)

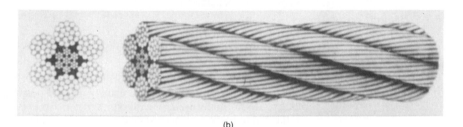

(b)

FIGURE 14.27 Configuration of (*a*) structural strand and (*b*) structural rope. *(Reprinted with permission from J. B. Scalzi et al., "Design Fundamentals of Cable Roof Structures," ADUSS 55-3580-01, U.S. Steel Corporation.)*

Structural rope. Several strands helically wound around a core that is composed of a strand or another rope (Fig. 14.27), produced in the United States in accordance with ASTM A603.

Prestressing strands. A 0.6-in-dia. seven-wire, low-relaxation strand generally used for prestressed concrete and produced in the United States in accordance with ASTM A416 (used for stay cables).

Bar. A solid, hot-rolled bar produced in the United States in accordance with ASTM A722 Type II (used for cable stays).

14.7.2 Structural Properties of Cables

A comparison of nominal ultimate and allowable tensile stress for various types of cables is presented in Table 14.3.

Structural strand has a higher modulus of elasticity, is less flexible, and is stronger than structural rope of equal size. The wires of structural strand are larger than those of structural rope of the same nominal diameter and, therefore, have a thicker zinc coating and better resistance to corrosion (Art. 14.9).

The total elongation or stretch of a structural strand is the result of several component deformations. One of these, termed constructional stretch, is caused by the lengthening of the strand lay due to subsequent adjustment of the strand wires into a denser cross section under load. Constructional stretch is permanent.

Structural strand and rope are usually prestretched by the manufacturer to approach a condition of true elasticity. Prestretching removes the constructional stretch inherent in the product as it comes from the stranding or closing machines.

TABLE 14.3 Comparison of Nominal Ultimate and Allowable Tensile Stress for Various Types of Cables, ksi

Type	Nominal tensile strength, F_{pu}	Allowable tensile strength, F_t
Bars, ASTM A722 Type II	150	$0.45F_{pu} = 67.5$
Locked-coil strand	210	$0.33F_{pu} = 70$
Structural strand, ASTM A586*	220	$0.33F_{pu} = 73.3$
Structural rope, ASTM A603*	220	$0.33F_{pu} = 73.3$
Parallel wire	225	$0.40F_{pu} = 90$
Parallel wire, ASTM A421	240	$0.45F_{pu} = 108$
Parallel strand, ASTM A416	270	$0.45F_{pu} = 121.5$

*Class A zinc coating (see Art. 14.9).

Prestretching also permits, under prescribed loads, the accurate measuring of lengths and marking of special points on the strand or rope to close tolerances. Prestretching is accomplished by the manufacturer by subjecting the strand to a predetermined load for a sufficient length of time to permit adjustment of the component parts to that load. The prestretch load does not normally exceed 55% of the nominal ultimate strength of the strand.

In bridge design, careful attention should be paid to correct determination of the cable modulus of elasticity, which varies with type of manufacture. The modulus of elasticity is determined from a gage length of at least 100 in and the gross metallic area of the strand or rope, including zinc coating, if present. The elongation readings used for computing the modulus of elasticity are taken when the strand or rope is stressed to at least 10% of the rated ultimate stress or more than 90% of the prestretching stress. The minimum modulus of elasticity of prestretched structural strand and rope are presented in Table 14.4. The values in the table are for normal prestretched, structural, helical-type strands and ropes; for parallel wire strands, the modulus of elasticity is in the range of 28,000 to 28,500 ksi.

For cable-stayed bridges, it is also necessary to use an equivalent reduced modulus of elasticity E_{eq} to account for the reduced stiffness of a long, taut cable due to sag under its own weight, especially during erection when there is less tension.

TABLE 14.4 Minimum Modulus of Elasticity of Prestretched Structural Strand and Rope*

Type	Diameter, in	Modulus of elasticity, ksi
Strand	½ to 2⁹⁄₁₆	24,000
	2⅝ and larger	23,000
Rope	⅜ to 4	20,000

*For Class B or Class C weight of zinc-coated outer wires, reduce modulus 1,000 ksi.

The formula for this equivalent modulus was developed by J. H. Ernst:

$$E_{eq} = \frac{E}{1 + \dfrac{E(\gamma l)^2}{12\sigma_m{}^3}\left[\dfrac{(1 + \mu)^4}{16\mu^2}\right]} \tag{14.1}$$

where E = modulus of elasticity of the steel from test
  $\sigma_m = (\sigma_u + \sigma_o)/2 = \sigma_o(1 + \mu)/2$ = average stress
  σ_u and σ_o = upper and lower stress limits, respectively
  $\mu = \sigma_u/\sigma_o$
  γ = weight of cable per unit of length per unit of cross-sectional area
  l = horizontal projected length of cable

The bracketed term in the denominator becomes unity when $\sigma_o = \sigma_u$, that is, when the stress is constant. The reduction in modulus of elasticity of the cable due to sag is a major factor in limiting the maximum spans of cable-stayed bridges.

The effects of creep of cables of cable-supported bridges should be taken into account in design. Creep is the elongation of cables under large, constant stress, for instance, from dead loads, over a period of time. The effects can be evaluated by modification of the cable equation in the deflection theory. As an indication of potential magnitude, an investigation of the Cologne-Mulheim Suspension Bridge indicated that, in a 100-year period, the effects of cable creep would be the equivalent of about one-fourth the temperature drop for which the bridge was designed.

14.7.3 Erection of Cables

Until the 1960s, parallel-wire, suspension-bridge main cables were formed with a spinning wheel carrying one wire at a time (and more recently two or four wires) over the pylons from anchorage to anchorage (Fig. 14.28). Not only were the wires

FIGURE 14.28 Transfer of wire from a spinning wheel to an eyebar-and-shoe arrangement at an anchorage.

FIGURE 14.29 Parallel wire strand (*a*) before compaction from an hexagonal arrangement into a round cross section, and (*b*) after compaction.

spun aerially individually, but each wire had to be removed from the spinning wheel at the anchorages, looped over a circular or semicircular strand shoe, then looped again over the spinning wheel for a return trip (Art. 14.18). Furthermore, wires had to be adjusted individually, then banded into strands and readjusted (Fig. 14.29*a*), and finally compacted into a circular cross section (Fig. 14.29*b*). This process is time-consuming, costly, and hazardous.

Prefabricated parallel-wire strands are an economical alternative. Large main cables of suspension bridges may be made up of many such strands, laid parallel to each other in a selected geometric pattern. In the commonly used hexagonal, there may be 19, 37, 61, 91, or 127 large strands. In a rectangular pattern, there may be 6 or more strands in each horizontal row and 6 or more vertical rows, with suitable spacers. The strands may have up to 233 wires each, all shop-fabricated, socketed, tested, and packaged on reels. Their use can yield a tremendous saving in erection time over the older process of aerial spinning of cables on the site.

For the Newport Bridge, which was completed in 1969, shop-fabricated, parallel-wire strands form the cables. Each cable is made up of 4,636 wires, each 0.202 in in diameter, shop-fabricated into 76 parallel wire strands of 61 wires each. Thus, in place of thousands of spinning-wheel trips previously necessary, only 152 trips of a hauling rope were needed to form the two cables. Furthermore, thousands of sag adjustments of individual wires were eliminated from the field operation.

From a design point of view, parallel-wire cables are superior to cables made of helical-wire strands. Straight, parallel-laid wires deliver the full strength and modulus of elasticity of the steel, whereas strength and modulus of elasticity are both reduced (by about one-eighth) with helical placement. On the other hand, from the bridge-erection standpoint, standard helical-strand-type cables are superior to field-assembled parallel-wire type. Strands are readily erected and adjusted, with a minimum of equipment and manpower. Therefore, they have been used on many small- to moderate-sized suspension bridges. Prefabricated parallel-wire strands, however, combine the erection advantages of strand-type cables with the superior in-place characteristics of parallel-wire cables.

For smaller cable bridges, cables with few strands may be arranged in an open form with strands separated. But for longer bridges, the strands are arranged in a closed form (Fig. 14.29a) in either a hexagonal or other geometrical pattern. They then may be compacted by machine (Fig. 14.29b) and wrapped for protection. Note that a group of helical-type strands cannot be compacted into as dense a mass as a group of parallel-wire strands.

Cable-stayed bridges formerly used traditional structural strands or locked-coil strands for the stays. Since then, stays composed of prestressing steels are generally used. Cable stays for cable-stayed bridges are similar to posttensioning tendons in that they consist of the following primary elements:

• Prestressing steel (parallel wires, strands, or bars)

• Sheathing (duct), which encapsulates the prestressing steel and may be a steel pipe or a high-density polyethylene pipe (HDPE)

• A material that fills the void between the prestressing steel and the sheathing and may be a cementitious grout, petroleum wax, or other appropriate material

• Anchorages

There are two basic methods of manufacture and installation of stays: (1) assembly on site in final position and (2) prefabricated installation. Both methods have been successfully employed. Given various constraints for a specific project or site, it is generally a question of economics as to which method is employed. Prefabrication may be accomplished either at a factory remote from the construction site or, if feasible, at the project site (possibly on the bridge deck). Normally, factory-prefabricated stays are delivered to the site reeled on drums, complete with the bundle of wires or strands, the HDPE sheathing, and anchorages. (This method cannot be used with prestressing bars or steel pipe sheathing.) Usually, one or both anchorages are fitted to the stay.

At the site, prefabricated stays usually are erected into final inclined position either by crane or by a temporary guying system that is erected between anchorage points from which the stays are suspended. The stays are brought into final position by means of a winch or other suitable hydraulic equipment.

When a guying system is used, site assembly of stays in final position begins with installation of the guying system. The sheathing, either steel or HDPE, is then positioned in the final inclined position. Next, the strand or wire-bundle stays are pulled through the sheathing by winches. When strands are used, the push-through method may be employed. In that case, by use of specialized equipment, individual strands are pushed into the sheathing. Parallel prestressing bars are somewhat more difficult to install inasmuch as bar couplers must also be installed at intervals along the stay length.

14.8 CABLE SADDLES, ANCHORAGES, AND CONNECTIONS

Saddles atop towers of suspension bridges may be large steel castings in one piece (Fig. 14.30) or, to reduce weight, partly of weldment (Fig. 14.31). The size of the saddle may be determined by the permissible lateral pressures on the cables, which are a function of the radius of curvature of the saddle. Other saddles of special design may be required at side piers to deflect the anchor-span cables to the anchorages. Also, splay saddles may be needed at the anchorages.

In cable-stayed bridges, where the cable stays converge to the top of a pylon (radiating configuration) and are continuous over the pylon, massive saddles, similar to those for suspension-bridge towers, are used. For the types of cable-stay configurations where the stays are distributed along a cellular-type pylon, similar (but smaller) saddles may be used at the pylon. If the pylon is solid concrete, the saddles are generally steel pipe, bent to the appropriate degree of curvature and embedded in the concrete.

Suspension-bridge anchorages for the main cables are usually massive concrete blocks designed to resist, with mass and friction, the overturning and sliding effects of the main-cable pull. (Where local conditions permit, as with the Forth Road Bridge, the cables may be anchored in tunnels in rock.) The anchorages contain embedded steel eyebar chains to which the main wire cables are connected. A typical arrangement, as used for the Verrazano Narrows Bridge is shown in Fig. 14.32. A saddle is installed where the strands diverge to attach to the eyebars. Strand wires loop over a strand shoe and are attached to an eyebar (Fig. 14.28— see discussion of spinning in Art. 14.18).

A slightly different concept was used for the Newport Bridge (Fig. 14.33). In this case, the prefabricated strands of the main cable diverge and pass through 78 pipes held in position by a structural steel framework and transfer their loads to the anchorage through a bearing-type anchorage socket. The whole supporting framework is eventually encased in concrete. In this anchorage-block arrangement, the strand sockets bear on the back of the anchorage block instead of connecting with a tension linkage at the front of the block.

In suspension bridges, the suspender cables are attached to the main cables by cable bands. These are usually made of paired, semicylindrical steel castings with clamping bolts. There are basically two arrangements for attaching the suspenders. The first is typified by the detail used for the Forth Road Bridge (Fig. 14.34). In this arrangement, the cable band has grooves to accommodate looping of the structural rope over the main cable. Because of the bending of the suspender over the main cable, structural rope is used for the suspender, to take advantage of its flexibility. The second basic arrangement for attaching a suspender to a cable band was used for the Hennepin Avenue Bridge (Fig. 14.35). In this case, the suspender is attached to the cable band by standard zinc-poured sockets. Since bending of

FIGURE 14.30 Pylon saddle.

the suspender is not required, the suspender generally is a structural strand. Properly attached, zinc-poured sockets can develop 100% of the strength of strands and wire rope.

The end fittings or sockets of structural strand or rope are standardized by manufacturers and may be swaged or zinc-poured. These fittings include open or

FIGURE 14.31 Pylon saddle used for the Hennepin Avenue Bridge.

closed sockets of drop-forged or cast steel. Some are illustrated in Fig. 14.36. Fatigue must be considered in designing bridge cables that depend on zinc-poured socketing, particularly if they are subject to a wide range of stress.

The attachments of suspenders to girders depend on the type of girder detail. Generally, the end fitting of a suspender is a swaged or zinc-poured type. Where there are multiple strands or ropes in a suspender, the fitting may be specially made.

Early cable-stayed bridges had stays consisting of parallel structural strands or locked-coil strands. These strands had conventional zinc-poured sockets. Because of concern with the low fatigue strength of structural strand with zinc-poured sockets, a new type of socket, called a HiAm (high amplitude) socket, was developed in 1968 by Prof. Fritz Leonhardt in conjunction with Bureau BBR Ltd., Zurich. It was intended for use with stays consisting of parallel ¼-in.-dia. prestressing wires that terminate with button heads (ASTM A421 Type BA) in an anchor plate in the socket. The anchor socket is filled with steel balls and an epoxy-and-zinc dust binder. This type of anchorage increases the fatigue resistance to about twice that for zinc-poured sockets. The HiAm sockets were used in the United States for the Pasco-Kennewick, Luling, and East Huntington cable-stayed bridges. After those bridges were constructed, seven-wire prestressing strand came into general use, and several types of anchorages were developed to accommodate parallel prestressing strands in cable stays.

14.9 CORROSION PROTECTION OF CABLES

In the past, the method of protecting the main cables of suspension bridges against corrosion was by coating the steel with a red lead paste, wrapping the cables with

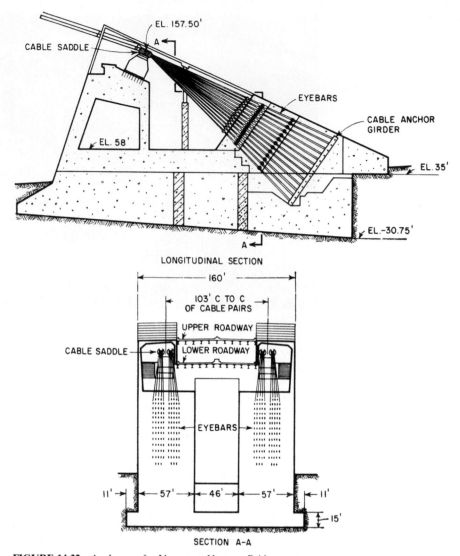

FIGURE 14.32   Anchorage for Verrazano Narrows Bridge.

galvanized, annealed wires, and applying a red lead paint. This method has met with a varying degree of success from excellent for the Brooklyn Bridge to poor for the General U. S. Grant Bridge at Portsmouth, Ohio. A potential defect in this system is that, as the cable stretches and shortens under live loads and temperature changes, some separation of adjacent turns of wire wrapping may occur. Depending on the degree of separation, the paint may crack and permit leakage of water and contaminants into the cable.

Alernative protection systems that have been used for some suspension bridges are as follows:

Bidwell Bar Bridge. This 1108 ft-span bridge has 11-in-dia., parallel, structural strand cables (Fig. 14.37). The protective system consists of the following components: plastic filler pieces extruded from black polyethylene; a covering of

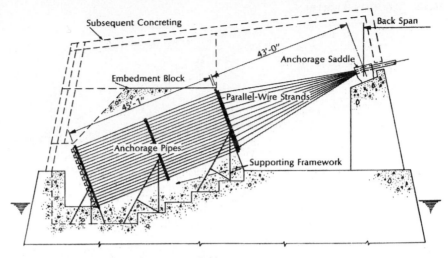

FIGURE 14.33 Anchorage for Newport Bridge.

nylon film; a "first pass" glass-reinforced acrylic-resin covering consisting of one layer of glass-fiber mat, two layers of glass cloth, and several coats of acrylic resin; a weather coat of acrylic resin; and a finished coat of acrylic resin containing a sand additive to give the surface a rough texture. This type of covering was developed by Bethlehem Steel in conjunction with DuPont.

Newport, R.I., Bridge. The protective system for the cables of this bridge is the same as that described for the Bidwell Bar Bridge. However, since the cables consist of parallel wires, the black polyethylene filler pieces were not required.

General U. S. Grant Bridge, Ohio. The protective system comprises spiral-wrapped neoprene sheet and hypalon paint, a proprietary system developed by U. S. Steel.

Second Chesapeake Bay Bridge (William Preston Lane, Jr., Memorial). This has the same protective covering as applied to the General U. S. Grant Bridge.

Hennepin Avenue Bridge. The protective system consists of a wrapped neoprene sheet and hypalon paint system.

The Bidwell Bar Bridge was constructed in 1964 for the California Department of Water Resources. The protective cable covering has been performing satisfactorily. In the early 1970s, some corrosion was discovered at the cable bands, presumably resulting from shrinkage of the covering. Bethlehem Steel corrected the condition by rewrapping a short portion at the cable bands and recalking. A 1991 inspection indicated no distress in the cable covering.

The similar system applied to the Newport Bridge (installed in 1969) is still performing satisfactorily. A 1980 inspection indicated that some crazing of the top surface had occurred in some areas, but these were superficial and did not extend through the thickness. These areas were patched. There also were some signs of distress at the cable bands, in the calking groove. As a result of thermal contraction of the covering, the calking had worked loose (presumably the same condition as that in the Bidwell Bar Bridge). Repairs were made with a more resilient type calk that accommodates thermal movement.

The system developed by U. S. Steel and applied to the Second Chesapeake Bay Bridge in 1973, the General U. S. Grant Bridge in 1980, and the Hennepin Avenue Bridge in 1990, has been performing satisfactorily. This type of system was also used for rewrapping the Brooklyn Bridge cables in 1986.

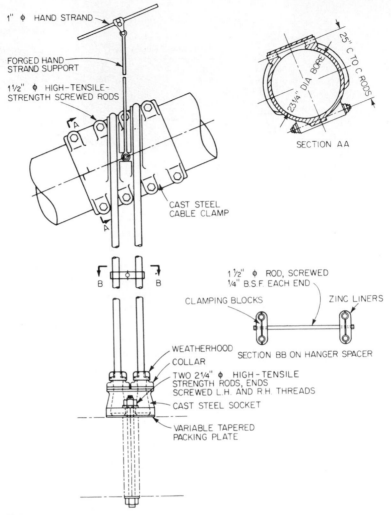

FIGURE 14.34 Cable band and suspender detail used for the Forth Road Bridge. *(Reprinted with permission from Sir Gilbert Roberts, "Forth Road Bridge," Institution of Civil Engineers, London.)*

Table 14.5 presents a partial listing of suspension bridges with appropriate statistics and the corrosion protection used for the main cables.

An area where the main cable is particularly vulnerable extends from the splay saddle to the eyebars in the anchorage blocks. The only corrosion protection available is the zinc coating of the wires. Depending on environmental conditions in the anchorage blocks, the galvanizing may have a life expectancy on the order of 20 years. Serious corrosion in this area occurred in the Brooklyn, Williamsburg, and Manhattan Bridges, requiring corrective measures. In the rehabilitation of one anchorage of the Manhattan Bridge, dehumidification equipment was included to control humidity in the anchorage block.

TABLE 14.5 Cable Construction and Corrosion Protection for Some Suspension Bridges

Name	Location	Year	No. of cables	Cable dia., in	No. of strands	No. of wires or strands	Wire dia., in	Cable construction[1]	Corrosion protection[2]
Brooklyn Bridge	Brooklyn, N.Y.	1883	4	15 5/8	19	282	0.184[3]	AS	CWR
Williamsburg	New York, N.Y.	1903	4	18 3/4	37	208	0.192[4]	AS	Note[5]
Manhattan	New York, N.Y.	1910	4	20 3/4	37	256	0.195	AS	CWR
George Washington	New York, N.Y.	1931	4	36	61	434	0.196	AS	CWR
San Francisco–Oakland Bay	California	1936	2	28 3/4	37	472	0.195	AS	CWR
Bronx-Whitestone	New York, N.Y.	1939	2	21 1/2	37	266		AS	CWR
Mackinac Straits	Michigan	1957	2	24 1/4	37	340	0.196	AS	CWR
Walt Whitman	Philadelphia, Pa.	1957	2	23 1/8	37	308	0.196	AS	CWR
Throgs Neck	New York, N.Y.	1961	2	23	37	296	0.1875	AS	CWR
Bidwell Bar	State Rt. 62 Feather R., Calif.	1964	2	11	37[6]			PHSS	GRAR
Verrazano Narrows	New York, N.Y.	1964	4	35 7/8	61	428		AS	CWR
Forth Road	Queensferry, Scotland	1964	2	24	37	314	0.196	AS	CWR
Tagus (Salazar)	Lisbon, Portugal	1966	2	23 1/16	37	304	0.196	AS	CWR
Severn River	Beachley, England	1966	2	20	19	440	0.196	AS	CWR
Newport	Newport, R.I.	1969	2	15 1/4	76	61	0.202	FPWS	GRAR
Bosporus	Istanbul, Turkey	1973	2	23	19	548	0.2	AS	CWR
Humber	England	1980	2	27 1/2	37	404	0.2	AS	CWR
Gen U. S. Grant	Portsmouth, Ohio	1927	2	7 1/8	3	486	0.162[4]	SFPW	CWR
		1940	2	7 13/16	19[7]			PHSS	CWR
		1980	2		8			PHSS	NSHP
Hennepin Ave.	Minneapolis, Minn.	1990	4	15 3/8	19[9]			PHSS	NSHP

[1]Cable construction: AS = aerial spinning, PHSS = parallel helical structural strand, SFPW = site fabricated parallel wire strand, FPWS = factory fabricated prefabricated parallel wire strand.

[2]Corrosion protection: CWR = conventional wire wrapping and red lead, GRAR = glass reinforced acrylic resin, NSHP = neoprene sheet and hypalon paint.

[3]Deduced average diameter of the galvanized wire, average bare-wire diameter 0.181 in.

[4]Ungalvanized wires.

[5]Between 1916 and 1922, the original canvas wrapping and steel sheet protection of the cables was removed and replaced by galvanized wrapping wire.

[6]31 helical structural strands, 1 11/16-in dia., and 6 helical strands, 1 1/8-in dia.

[7]13 helical structural strands, 1 3/4-in dia., and 6 helical strands, 1 1/4-in dia., Class A coating.

[8]Same configuration as under note 7; i.e., same basic steel area, but changed coating from Class A to Class C.

[9]13 helical structural strands, 3 3/8-in dia., and 6 helical strands, 2 5/8-in dia.

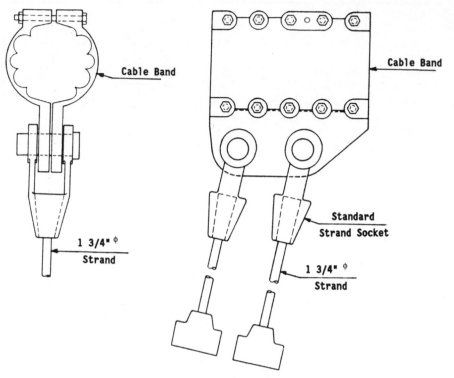

FIGURE 14.35 Cable band and suspender detail used for the Hennepin Avenue Bridge.

Suspender Corrosion. Generally, corrosion of suspenders is likely to occur at the anchorage sockets at the stiffening trusses and at retainer castings on top of those trusses. This may be attributed to two possible sources: salt spray from roadway deicing salts, or moisture that enters the interstices of the strand or rope at an upper level and trickles down to the socket or casting.

A 1974 report on the condition of the suspenders of the Golden Gate Bridge revealed that there was considerable reduction in suspender area due to corrosion that occurred as high as 150 ft above the roadway. This could be attributed to saltwater mist or fog.

For corrosion protection, U. S. Steel developed a procedure for extruding high-density black polyethylene over strands and rope. In many applications, this jacket also reduces vibration fatigue. For this purpose, particular attention is given to sealing the ends and minimizing the bending of wires at the nose of the socket.

Galvanizing. Wires can be protected against corrosion by galvanizing, a sacrificial coating of zinc that prevents corrosion of the steel so long as the coating is unbroken. Corrosion protection of the individual wires in a structural strand or rope is provided by various thicknesses of zinc coating, depending on the location of the wire in the strand or rope and the degree of corrosive environment expected. The effectiveness of the zinc coating is proportional to its thickness, measured in ounces per square foot of surface area of the uncoated wire. Class A zinc coating varies from 0.40 to 1.00 oz/ft^2 depending on the nominal diameter of the coated wire. A Class B or C coating is, respectively, 2 or 3 times as heavy as the Class A coating.

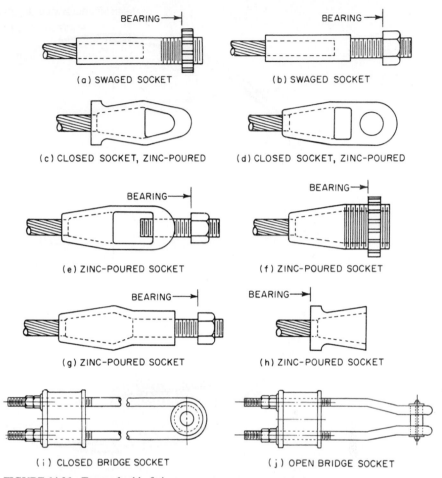

FIGURE 14.36 Types of cable fittings.

Generally, there are three basic combinations of coating: Class A coating throughout all wires; Class A coating for the inner wires and Class B for the outer wires; and Class A coating for the inner wires and Class C for the outer wires, depending on the degree of protection desired. Other coating thicknesses and arrangements are possible.

The heavier zinc coatings displace more of the steel area. This necessitates a reduction in rated breaking strength of strand or rope. ASTM A586 and A603 specify minimum breaking strengths required for various sizes of strand or rope in accordance with the three combinations of coating previously described. For other combinations of coating, the manufacturer should be consulted as to minimum breaking strength and modulus of elasticity.

Galvanizing has some disadvantages. Depending on environmental conditions, for example, galvanizing may be expected to last only about 20 years. Also, the possibility that hot-dip galvanizing may cause hydrogen embrittlement is of concern. (There is some indication, however, that, with current technology, the hot-dip galvanizing method is not as likely to cause hydrogen embrittlement as previously.) In addition, it may be difficult to meet specifications for a Class C coating with the

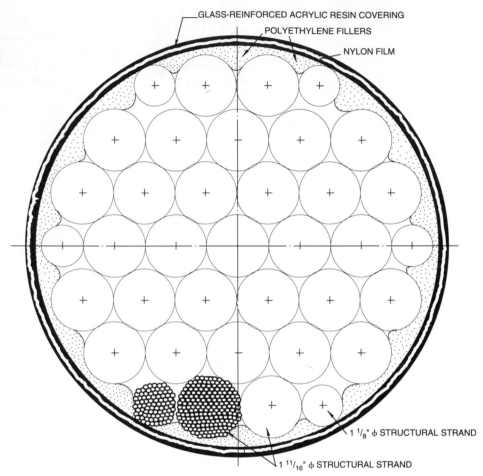

FIGURE 14.37 Cable corrosion protection system used for the Bidwell Bar Bridge. *(Reprinted with permission from J. L. Durkee and M. IABSE, "Advancements in Suspension Bridge Cable Construction," Paper No. 27, Symposium on Suspension Bridges, Lisbon, November 1966.)*

hot-dip method. Furthermore, wire with hot-dip galvanizing may not have the fatigue resistance that wire coated by electrolytic galvanizing has.

Protection of Stays.   In early cable-stayed bridges, stays, consisting of locked-coil or structural strands, were protected against corrosion by galvanizing and paint. Nevertheless, extensive corrosion occurred (S. C. Watson and D. G. Srafford, "Cables in Trouble," *Civil Engineering*, vol. 58, no. 4, April 1988, American Society of Civil Engineers). Contemporary stays, in contrast, are similar to external tendons generally used for posttensioned concrete. They consist of prestressing steel, sheathing, corrosion-protection materials, and anchorages. There is no evidence of trouble from corrosion with such stays.

The Schillerstrasse footbridge in Stuttgart, Germany, completed in 1961, was the first cable-stayed bridge to employ a sheathed and cement-grout-injected stay system. The stays consist of a bundle of parallel prestressing wire encapsulated in a polyethylene (PE) pipe and injected with cement grout. The purpose of the PE sheathing is twofold: to provide a form for the cement grout and to serve as a corrosion barrier. The stays have been inspected on numerous occasions and have shown no signs of corrosion. The first use of this system in the United States was for the Pasco-Kennewick Bridge, completed in 1978. The stays of the bridge were inspected in 1990. After 12 years in service, the exposed wire was as bright and as good as the day it was installed, indicating that with proper care and procedures for installation, cementitious grout can be an effective corrosion inhibitor.

A sheathing of high-density polyethylene pipe is airtight. A ¼-in thickness provides the same vapor barrier as a 35-ft-thick concrete wall. However, the HDPE pipe must be handled with care. If abused, as in the case of the Luling Bridge (related to excessive grout pressure), the pipe may, in time, develop longitudinal cracks. In addition, the cement-grout column may develop transverse cracks from cyclic tension in the stays, among other reasons. Thus, there is need to prevent direct access to the bare prestressing steel by corrosive agents.

Alternative materials to cement grout or means of providing additional corrosion barriers have been sought to increase corrosion protection of steel stays. Such materials as grease, wax, polymer-cement grout, and polybutadiene polyurethane have been tried with varying degrees of success.

The generally preferred method is to provide a multiple-barrier system, in which one or more materials take over the protective function for a material that has failed or been compromised. Currently in the United States, two methods are favored, for use in conjunction with an HDPE outer sheathing and cementitious grout. One is epoxy coating of the individual seven-wire prestressing strands in the cable. The other is use of individual greased and sheathed strands, the so-called monostrand method, an adaptation of the monostrand tendons used for posttensioned floor slabs and parking garages. Future possibilities include a ceramic coating developed for corrosion protection of automobile brake cables, which may be adapted for protection of cable stays.

14.10 STATICS OF CABLES

The following summary of elementary statics of cables applies to completely flexible and inextensible cables but includes correction for elastic stretch. The formulas derive from the fundamental differential equation of a cable shape

$$y'' = -\frac{w}{H} \tag{14.2}$$

where y'' = second derivative of the cable ordinate with respect to x

x = distance, measured normal to the cable ordinate, from origin of coordinates to point where y'' is taken

H = horizontal component of cable tension produced by w

w = distributed load, which may vary with x

Two cases are treated: catenary, the shape taken by a cable when the load is uniformly distributed over its length, and parabola, the shape taken by a cable when the load is uniformly distributed over the projection of the span normal to the load.

Table 14.6 lists equations for symmetrical cable. These equations, however, may be extended to asymmetrical cables, as noted later.

The derivation of the equations considered the cable as inextensible. Actually, the tension in the cable stretches it. The stretch, in, of half the cable length may be estimated from

$$\Delta s = \frac{(T + H)s}{2AE} \tag{14.3}$$

TABLE 14.6 Equations for Catenary and Parabolic Cables*

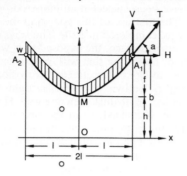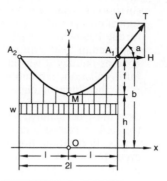

	Catenary	Parabola
Cable ordinate y†	$y = (e^{x/h} + e^{-x/h})h/2$ $= h \cosh x/h$	$y = h + x^2/2h$
Coordinate b of attachment points A_1 and A_2	$b = f + h$ $= h \cosh l/h$	$b = f + h$ $= h + l^2/2h$ $= (l^2 + 4f^2)/2f$
Sag-span ratio a	$a = f/2l$	$a = f/2l$
Slope y' of cable	$y' = \sinh x/h$	$y' = x/h$
Ordinate h of cable low point	$h = H/w$	$h = H/w$ $= l^2/2f$ $= l/4a$ $= f/8a^2$
Sag f	$f = h(\cosh l/h - 1)$	$f = l^2/2h$
Half length s of cable	$s = \sqrt{b^2 - h^2}$ $= \sqrt{f^2 + 2hf}$ $= \sqrt{2fb - f^2}$	$s = \dfrac{l}{2h} \sqrt{h^2 + l^2} + \dfrac{h}{2} \times [\log_e (l + \sqrt{h^2 + l^2}) - \log_e h]$ $= \dfrac{l}{2h} \sqrt{h^2 + l^2} + \dfrac{h}{2} \sinh^{-1} \dfrac{l}{h}$ $\approx l \left(1 + \dfrac{8}{3} a^2 - \dfrac{32}{5} a^4 + \dfrac{256}{7} a^6 - \cdots \right)$
Angle α of cable at A_1 and A_2	$\tan \alpha = \sin l/h$ $= \dfrac{1}{h} \sqrt{f^2 + 2fh}$ $= \dfrac{1}{b - f} \sqrt{2fb - f^2}$ $= s/h$ $= \dfrac{1}{h} \sqrt{b^2 - h^2}$ $\cos \alpha = h/b$ $= h/(h + f)$ $= (b - f)/b$ $\sin \alpha = \dfrac{1}{b} \sqrt{2fb - f^2}$ $= s/b$ $= \dfrac{1}{b} \sqrt{b^2 - h^2}$ $= \dfrac{1}{h + f} \sqrt{f^2 + 2fh}$	$\tan \alpha = l/h$ $= \sqrt{2f/h}$ $= \sqrt{2f/(b - f)}$ $= 2f/l$ $= 4a$ $= 1/\sqrt{1 + 16a^2}$ $\cos \alpha = h/\sqrt{h^2 + l^2}$ $= h/\sqrt{h^2 + 2fh}$ $= (b - f)/\sqrt{b^2 - f^2}$ $= l/\sqrt{l^2 + 4f^2}$ $\sin \alpha = \sqrt{2f/(b + f)}$ $= l/\sqrt{h^2 + l^2}$ $= \sqrt{2f/(h + 2f)}$ $= 2f/\sqrt{l^2 + 4f^2}$ $= 4a/\sqrt{1 + 16a^2}$

TABLE 14.6 Equations for Catenary and Parabolic Cables* (*Continued*)

	Catenary	Parabola
Vertical component V of cable tension	$V = w\sqrt{b^2 - h^2}$ $ = w\sqrt{f^2 + 2fh}$ $ = w\sqrt{2fb - f^2}$ $ = ws$	$V = w\sqrt{2fh}$ $ = wl$ $ = 4wah$
Horizontal component H of cable tension	$H = wh$ $ = w(b - f)$	$H = wh$ $ = wl^2/2f$ $ = wl/4a$ $ = wf/8a^2$
Cable tension T	$T = wb$	$T = w\sqrt{h^2 + l^2}$ $ = w\sqrt{2fh + h^2}$ $ = wh\sqrt{1 + 16a^2}$

*Adapted from H. Odenhausen, "Statical Principles of the Application of Steel Wire Ropes in Structural Engineering," *Acier-Stahl-Steel*, no. 2, pp. 51–65, 1965.

†Since

$$\cosh \frac{x}{h} = 1 + \frac{(x/h)^2}{2!} + \frac{(x/h)^4}{4!} + \frac{(x/h)^6}{6!} + \cdots$$

the parabolic profile (obtained by dropping the third and subsequent terms) is an approximation for the catenary. The accuracy of this approximation improves as sag f becomes smaller.

> where s = half the length of cable, in
> T = cable tension, kips, at point of attachment
> H = horizontal component of cable tension, kips
> A = cross-sectional area of cable, in²
> E = modulus of elasticity of cable steel, ksi

Properties of asymmetrical cables may be obtained by determining first the properties of their component symmetrical elements.

For a parabolic cable (Fig. 14.38), determine point C on the cable, which lies on a horizontal line through a point of attachment. The horizontal distance of C

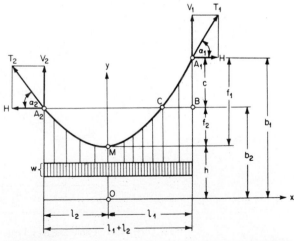

FIGURE 14.38 Cable assumes parabolic shape when subjected to a uniform load acting over its horizontal projection.

from the support at the cable high point may be computed from $2l_1 - l$, where the cable span $l = l_1 + l_2$, after l_1 has been found from

$$l_1 = \frac{f_1 l}{c}\left(1 \pm \sqrt{1 - \frac{c}{f_1}}\right) \tag{14.4}$$

where l_1, l_2 = horizontal distances from M, the cable low point, to the high and low supports A and B, respectively
  f_1 = cable sag measured from high point
  c = vertical distance between points of support

The portion of the cable between C and the lower support is symmetrical. Its ordinates, slope, length, and cable tension may be computed from the equations in Table 14.6.

For a catenary cable (Fig. 14.39), point C on a horizontal line through the lower support may be located by stepwise solution of the equation $y = h \cos x/h$ for a symmetrical catenary. An initial solution may be obtained by use of a parabola. Substitution in the exact equation then yields more accurate values. When distances l_1 and l_2 of C from the supports have been determined, the ordinates, slope, length, and cable tension of the symmetrical portion of the cable may be computed.

The portion of the cable from C to the high point is an oblique cable (Fig. 14.40), a special case of the asymmetrical cable. Its properties can be obtained with the equations in Table 14.6 and Eq. (14.4) by treating the oblique cable as part of a symmetrical one.

(See also Art. 4.8.)

14.11 SUSPENSION-BRIDGE ANALYSIS

Structural analysis of a suspension bridge is that step in the design process whereby, for given structural geometry, materials, and sizes, the moments and shears in

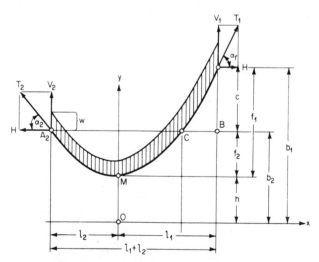

FIGURE 14.39 Cable assumes a catenary shape when subjected to a uniform load acting over its length.

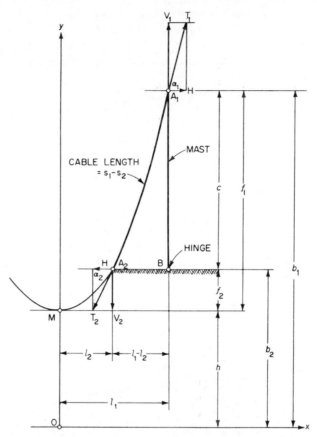

FIGURE 14.40   Part of catenary between low point and a support.

stiffening trusses, axial loads in cables and suspenders, and deflections of all elements are determined for given loads and temperature changes. The stress analysis usually is carried out in two broad categories: static and dynamic.

14.11.1   Static Analysis—Elastic Theory

Before the Manhattan Bridge was designed about 1907, suspension bridges were analyzed by the classical theory of structures, the so-called elastic, or first-order, theory of indeterminate analysis. This neglects the deformations of the structural geometry under load in formulation of the equations of equilibrium. The earliest theory was developed by Rankine, who assumed that a stiffening truss distributes the loads uniformly to the cable from which it is suspended. The elastic theory is advantageous because the resulting equations are linear in the loads and internal forces, and linear superposition applies for internal forces caused by different loads. Distortions of the structural geometry under live load, however, can cause a gross overstatement of moments, shears, and deflections calculated by the elastic theory. This theory, therefore, is seldom used, except as a basis for preliminary design or for design of bridges with short spans or rigid stiffening trusses for which large distortions are not possible. (See also Art. 14.11.2.)

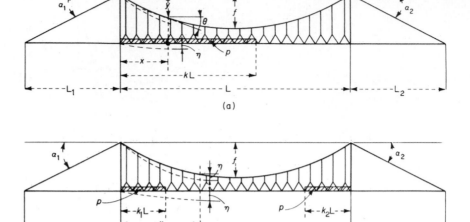

FIGURE 14.41 Suspension bridge with unloaded side spans and pin-ended main-span truss. (*a*) Single uniform load extending from a pylon into the main span. (*b*) Uniform load extending from both pylons into the main span.

The elastic-theory equations following apply to the structure in Fig. 14.41 with unloaded side spans and pin-ended, main-span stiffening truss. This structure has one redundant, the horizontal component H of cable reaction. An equation for determining H is obtained by making the structure statically determinate by cutting the cable at its low point and applying H there. The gap that is opened at the cut by loads on the stiffening truss must equal the oppositely directed movement at that point produced by H. These deflections can be calculated by the virtual work or dummy-unit-load method, and the equation can readily be solved for H.

For loads,

$$H = -\frac{\delta_{ao}}{\delta_{aa}} \tag{14.5}$$

where $\delta_{aa} = \displaystyle\int \frac{M_a{}^2\,ds}{EI} + \int \frac{N_a{}^2\,ds}{AE}$

$\qquad \delta_{ao} = \displaystyle\int \frac{M_a M_o\,ds}{EI} + \int \frac{N_a N_o\,ds}{AE}$

$\quad M_a$ = statically determinate moment due to unit horizontal force applied at cut end of cable

$\quad M_o$ = statically determinate moment due to loads

$\quad N_a$ = statically determinate axial forces due to unit horizontal force applied at cut end of cable

$\quad N_o$ = statically determinate axial forces due to loads

$\quad E$ = modulus of elasticity of stiffening-truss steel

$\quad I$ = moment of inertia of stiffening truss

$\quad A$ = cross-sectional area of member subjected to axial force

For temperature change,

$$H_t = -\frac{\delta_{at}}{\delta_{aa}}$$ (14.6)

where $\delta_{at} = \displaystyle\int \frac{\epsilon_t t \sec \theta}{A_c E_c} ds = \int \frac{ds}{dx} \frac{\epsilon_t t}{A_c E_c} ds$

ϵ_t = coefficient of thermal expansion
t = temperature change
A_c = cross-sectional area of cable
E_c = modulus of elasticity of cable steel

Assumptions. To evaluate Eqs. (14.5) and (14.6), the following conditions are assumed for the structure in Fig. 14.41:

1. The cable takes the shape of a parabola under dead load w.
2. Elongation of suspenders and shortening of pylons are so small that they can be neglected.
3. Spacing of suspenders is so small relative to span that the suspenders can be considered a continuous sheet.
4. The horizontal component of cable tension in side spans equals the horizontal component of cable tension in main span. This holds if the cable is fixed to the top of flexible pylons or to a movable saddle atop the pylons.
5. The stiffening truss acts as a beam of constant moment of inertia simply supported at the ends. Under dead load, it is straight and horizontal. Usually erected so that it carries none of the dead load, the stiffening truss therefore is stressed only by live load and temperature changes.

Thus, the horizontal component of cable tension due to temperature rise t may be computed from

$$H_t = -\frac{3EI\epsilon_t t L_t}{f^2 NL}$$ (14.7)

where f = cable sag
L = length of main span
$L_t = \displaystyle\int_0^S (ds/dx) \, ds + L_1 \sec^2 \alpha_1 + L_2 \sec^2 \alpha_2$
$\approx L + {}^{16}\!/_3 a^2 L + L_1 \sec^2 \alpha_1 + L_2 \sec^2 \alpha_2$
$a = f/L$
S = length of main-span cable
$N = \dfrac{8}{5} + \dfrac{3EI}{A_c E_c f^2} (1 + 8a^2) + \dfrac{3EIL_1}{A_c E_c L f^2} \sec^3 \alpha_1 + \dfrac{3EIL_2}{A_c E_c L f^2} \sec^3 \alpha_2$
L_1, L_2 = lengths of side spans
α_1, α_2 = angle with respect to horizontal of side-span cables

The horizontal component of cable tension due to a uniform live load p extending a distance kL from either end of the main span may be computed from

$$H_p = \frac{pL}{5Na} \left(\frac{5}{2} k^2 - \frac{5}{2} k^4 + k^5 \right)$$ (14.8)

For maximum cable stress ($k = 1$), the horizontal component due to dead load is

$$H_w = \frac{wL}{8a} \tag{14.9}$$

For live load over the whole span,

$$H_p = \frac{pL}{5Na} \tag{14.10}$$

The sum yields the maximum horizontal component of cable tension:

$$H_{max} = \frac{L}{a}\left(\frac{w}{8} + \frac{p}{5N}\right) \tag{14.11}$$

For maximum moment at distance x *from pylon:*
A. When $0 \leqq x \leqq NL/4$, solve for k (Fig. 14.41a):

$$k + k^2 - k^3 = \frac{NL}{4(L - x)} \tag{14.12}$$

Maximum positive moment with loaded length kL then may be obtained from

$$M_{max} = \frac{px}{2}(L - x)\left\{1 - \frac{8}{5N}[1 - \tfrac{1}{2}(1 - k)^4(2 + 3k)]\right\} \tag{14.13a}$$

Maximum negative moment with loaded length $L - kL$ may be computed from

$$M'_{max} = -\frac{2}{5}\frac{px(L - x)}{N}(1 - k)^4(2 + 3k) \tag{14.13b}$$

B. When $NL/4 \leqq x \leqq L/2$, solve for k_1 and k_2 (Fig. 14.41b):

$$1 - 2k_1^2 + k_1^3 = \frac{NL}{4x} \tag{14.14}$$

$$1 - 2k_2^2 + k_2^3 = \frac{NL}{4(L - x)} \tag{14.15}$$

Maximum negative moment with loaded lengths k_1L and k_2L may be obtained from

$$M = -\frac{2}{5}\frac{px(L - x)}{N}[k_1^4(5 - 3k_1) + k_2^4(5 - 3k_2)] \tag{14.16}$$

Maximum positive moment with loaded length $L - L(k_1 + k_2)$ may be calculated from

$$M_{max} = \frac{1}{2}px(L - x)\left\{1 - \frac{4}{5N}[2 - k_1^4(5 - 3k_1) - k_2^4(5 - 3k_2)]\right\} \tag{14.17}$$

For maximum shear at distance x *from pylon:*
A. When $0 \leqq x \leqq (1 - N/4)L/2$, solve for k_o:

$$k_o + k_o^2 - k_o^3 = \frac{NL}{4(L - 2x)} \tag{14.18}$$

Maximum positive shear with load between distances x and k_oL from a pylon may be obtained from

$$V_{max} = V_1 - V_2 \tag{14.19}$$

$$V_1 = \frac{pL}{2}\left(1 - \frac{x}{L}\right)^2\left[1 - \frac{8}{5N}\left(\frac{1}{2} - \frac{x}{L}\right)\left(2 + \frac{4x}{L} + \frac{x^2}{L^2} - 2\frac{x^3}{L^3}\right)\right] \tag{14.20a}$$

$$V_2 = \frac{pL}{2}(1 - k_o)^2\left[1 - \frac{8}{5N}\left(\frac{1}{2} - \frac{x}{L}\right)(2 + 4k_o + k_o^2 - 2k_o^3)\right] \tag{14.20b}$$

B. When $(1 - N/4)L/2 \leqq x \leqq L/2$,

$$V = \frac{pL}{2}\left(1 - \frac{x}{L}\right)^2\left[1 - \frac{8}{5N}\left(\frac{1}{2} - \frac{x}{L}\right)\left(2 + \frac{4x}{L} + \frac{x^2}{L^2} - 2\frac{x^3}{L^3}\right)\right] \tag{14.21}$$

(S. P. Timoshenko and D. H. Young, "Theory of Structures," 2d ed., McGraw-Hill Book Co., Inc., New York; A. G. Pugsley, "Theory of Suspension Bridges," Edward Arnold, Ltd., London.)

14.11.2   Static Analysis—Deflection Theory

Distortions of structural geometry of long suspension spans under live load may be very large. As a consequence, the elastic theory (Art. 14.11.1) gives unduly conservative moments, shears, and deflections. For economy, therefore, a deflection theory, also referred to as an exact or second-order theory, that accounts for effects of deformations should be used.

With the notation and assumptions given for the elastic theory in Art. 14.11.1, a differential equation can be written for the structure in Fig. 14.41 to include the vertical deflection η of the cable (and stiffening truss) at any point x. This equation expresses the flexural relationship between the horizontal component of cable tension H under dead and live loads and the stiffening-truss deflection under uniform live load p:

$$EI\eta'''' = p + H_p y'' + H\eta'' \tag{14.22}$$

where each prime represents a differentiation with respect to x.

Equation (14.22) by itself is not sufficient for solution for the two unknowns, η and the horizontal component of cable tension H_p due to live load. (*Note:* H can be expressed in terms of H_p.) Therefore, an additional compatibility equation is necessary. It expresses the cable condition that the total horizontal projection of cable length between anchorages remains unchanged.

The differential equations are not linear, and linear superposition is technically not applicable. This would imply that the use of influence lines for handling moving live loads is not permissible.

In the conventional deflection theory, however, the differential equations are linearized over a small range by assuming that the exponential terms containing H in the solution of the equations are constant during integration (even though that assumption is not valid for a particular loading case). This assumption may be made because, for example, the loading length for maximum moment at a point is not greatly affected by the magnitude of H. With this quasi-linear theory, an average value of H, or two values, H_{min} and H_{max}, may be used as a basis for drawing linearized influence lines as in first-order theory. With two influence lines (maximum and minimum) thus available, the results can be interpolated for more accurate values of H.

H. Bleich and S. P. Timoshenko suggested that the zero points of the influence lines be determined in this quasi-linear theory to establish the most unfavorable live-loading position. Then, the final results may be calculated by the classical theory with the live load in this position.

Besides the preceding classical differential-equation approach, a trigonometric-series method also is useful. Other advantageous procedures include successive approximation by relaxation theory, simultaneous-linear-equation approach of the flexibility-coefficient methods, elastic-foundation analogy, and analogy of an axially loaded beam.

Much of the literature on classical suspension-bridge theory deals with the effects of minor terms neglected in the assumptions of the deflection theory. S. P. Timoshenko gave an excellent account of the effect of horizontal displacements of the cable, elongation of suspenders, shear deflections, and temperature changes in the cable. (S. P. Timoshenko and D. H. Young, "Theory of Structures," 2d ed., McGraw-Hill, Inc., New York.) Other investigators have extended the theory to stiffening trusses that are continuous (such as in the Salazar Bridge) or have variable moments of inertia, widely spaced or inclined suspenders, or multiple main spans.

In general, inclined suspenders can have an important effect on results. Continuous spans are of advantage primarily in short bridges, but the advantage diminishes with long spans. Simple supports are preferred because they avoid settlement problems.

The following treatment of the classical approach is based on A. A. Jakkula's generalization of the work of many investigators. It is restricted here to the case of a suspension bridge with unloaded backstays and a two-hinged stiffening truss (Figs. 14.41 and 14.42). This presentation is useful because it has been extended to configurations with loaded backstays, or other variations of the suspension system, and has been programmed for computers.

Advances in computational methods have prompted several new approaches to analysis of suspension-bridges that differ from the classical deflection theory in that they adapt discrete mathematical models to computer programming. For example, if the suspenders are treated as finitely spaced elements (instead of an assumed continuous sheet as in classical methods), the analysis becomes that of an open-panel truss. Solution is required of a set of simultaneous transcendental equations,

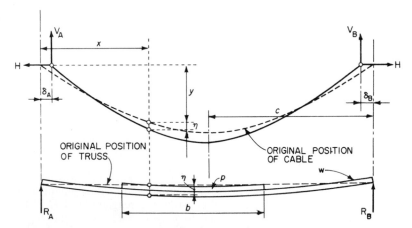

FIGURE 14.42　Original and deflected positions of the main-span cable and stiffening truss for the bridge in Fig. 14.41. Cable and truss have equal deflections at distance x from a pylon.

which are nonlinear because of the effects of distortion. Such solutions involve iterative techniques, the use of the Newton-Raphson method, and other sophisticated mathematical operations. An analogous continuous formulation adapted to computer use has also been proposed.

The solution of the fundamental differential equation [Eq. (14.22)] can be expressed in terms of hyperbolic functions or exponential functions. In the latter form, the solution is

$$\eta = C_1 e^{ax} + C_2 e^{-ax} + \frac{1}{H_w + H_s}\left(M_1 - \frac{p}{a^2}\right) - \frac{H_s}{H_w + H_s}\left(y - \frac{8f}{a^2 L^2}\right) \quad (14.23)$$

where x = horizontal distance from cable support to point where deflection η is measured

y = vertical distance from cable support to cable at point where deflection η is measured

e = 2.71828

a = sag-span ratio f/L

f = cable sag

L = length of main span

H_w = horizontal component of cable tension produced by uniform dead load w

H_s = horizontal component of cable tension produced by all causes other than dead load

M_1 = bending moment in stiffening truss calculated as if the truss were a simple beam independent of the cable

p = total uniform live load per cable

Constants C_1 and C_2 are integration constants to be evaluated, for each load position, from the end conditions and conditions of continuity.

Another method for finding the equation of the deflected truss is to represent deflection by a trigonometric series. If the stiffening truss is considered a free body, it will be in equilibrium under the force system indicated in Fig. 14.43. The truss is acted on by dead load w over the entire span, live load p over any length of span $k_2L - k_1L$, and suspender pull $w + q$, where q is the portion of the uniform live load carried by the cable.

The deflection of the truss can be given by the trigonometric series:

$$\eta = \sum_{n=1}^{\infty} a_n \sin\frac{n\pi x}{L} = a_1 \sin\frac{\pi x}{L} + a_2 \sin\frac{2\pi x}{L} + a_3 \sin\frac{3\pi x}{L} + \cdots \quad (14.24)$$

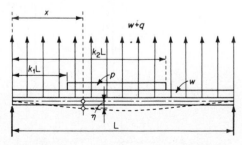

FIGURE 14.43 Forces acting on the truss of Fig. 14.41.

The Fourier coefficients a_n are determined by energy methods.

$$a_n = \frac{(\cos n\pi k_1 - \cos n\pi k_2)qL/n\pi - (1 - \cos n\pi)wL\beta/n\pi}{n^4 EI\pi^4/2L^3 + (1 + \beta)H_w\pi^2 n^2/2L} \qquad (14.25)$$

where $\beta = H_s/H_w$
E = modulus of elasticity of stiffening-truss steel
I = moment of inertia of stiffening truss

In the following development of the deflection theory, the trigonometric-series solution is used. It converges rapidly and avoids difficulties with the integration constants C_1 and C_2, which hold only for values of x for which M_1 has the same algebraic form. (Thus, the loading in Fig. 14.43 requires evaluation of six constants of integration.)

Equations (14.23) to (14.25) contain the unknown H_s. It must be determined before deflections can be numerically evaluated for any particular case. The energy method may be used for this purpose: the work done by the suspender forces moving through the deflection undergone by the cable is equated to the internal work done by the internal stress in the cable moving through the deformation suffered by the cable.

The force system acting on the cable treated as a free body is shown in Fig. 14.44. Application of the energy equation yields the so-called cable condition:

$$\frac{H_s L_c}{A_c E_c} + \epsilon_t t L_t = \frac{8f}{L^2} \int_0^L \eta \, dx - \frac{1 + \beta}{2 + \beta} \int_0^L \eta'' \eta \, dx \qquad (14.26)$$

where ϵ_t = coefficient of thermal expansion
t = temperature change
$\beta = H_s/H_w$
A_c = cross-section area of cable
E_c = modulus of elasticity of cable steel
$L_t = \displaystyle\int_0^S (ds/dx) \, ds + L_1 \sec^2 \alpha_1 + L_2 \sec^2 \alpha_2$
$\qquad \approx L + {}^{16}\!/_3 a^2 L + L_1 \sec^2 \alpha_1 + L_2 \sec^2 \alpha_2$
L_1, L_2 = lengths of side spans

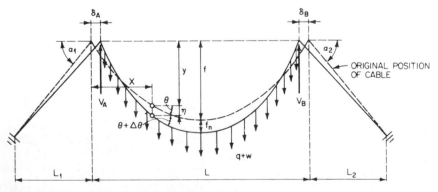

FIGURE 14.44   Forces acting on the cable of Fig. 14.41.

$\alpha_1, \alpha_2 =$ angle with respect to horizontal of side-span cables

$$L_c = \int_0^S (ds/dx)^2 \, ds + L_1 \sec^3 \alpha_1 + L_2 \sec^3 \alpha_2$$
$$\approx L + 8a^2 L + L_1 \sec^3 \alpha_1 + L_2 \sec^3 \alpha_2$$

$S =$ length of main-span cable

$a = f/L$

Substitution of Eq. (14.24) in Eq. (14.26) yields the Timoshenko "exact" form of the equation for H_s.

$$\frac{H_s L_c}{A_c E_c} \pm \epsilon_t L_t - \frac{16f}{L\pi}\left(a_1 + \frac{a_3}{3} + \frac{a_5}{5} + \cdots\right)$$
$$- \frac{1+\beta}{2+\beta}\left(\frac{\pi^2}{2L}\right)[a_1{}^2 + (2a_2)^2 + (3a_3)^2 + \cdots] = 0 \quad (14.27)$$

The last term on the left side of the equation accounts for the actual distribution of live load to the cable. If this term is neglected, the simpler Timoshenko approximate solution is obtained. Direct solution of Eq. (14.27) is possible only by successive approximations.

Successive differentiation of Eq. (14.27) with respect to x yields:

Stiffening-truss angular deflection

$$\phi_x = \eta' = \frac{\pi}{L}\sum_{n=1}^{\infty} n a_n \cos\frac{n\pi x}{L} \quad (14.28)$$

Moment

$$M_x = -EI\eta'' = \frac{EI\pi^2}{L^2}\sum_{n=1}^{\infty} n^2 a_n \sin\frac{n\pi x}{L} \quad (14.29)$$

Shear

$$V_x = -EI\eta''' = \frac{EI\pi^3}{L^3}\sum_{n=1}^{\infty} n^3 a_n \cos\frac{n\pi x}{L} \quad (14.30)$$

An alternative form of the equation for H_s, known as the Melan equation, derived from Eqs. (14.23) and (14.26).

$$H_s{}^2 \frac{L_c}{A_c E_c} + H_s\left(\frac{H_w L_c}{A_c E_c} \pm \epsilon_t L_t + \frac{16f^2}{3L} - \frac{64f^2 K_2}{L^4}\right)$$
$$+ \left(\frac{pfLK_1}{3} \pm \epsilon_t L_t + H_w + \frac{8fpK_3}{L^2}\right) = 0 \quad (14.31)$$

where $K_1 = k_2{}^2(4k_2 - 6) - k_1{}^2(4k_1 - 6)$

$$K_2 = \frac{4 + aL(e^{aL} - e^{-aL}) - 2(e^{aL} - e^{-aL})}{a^3(e^{aL} - e^{-aL})}$$

$$K_3 = \frac{(e^{aL} - 1)(e^{-ak_2 L} - e^{-ak_1 L}) + (e^{-aL} - 1)(e^{ak_2 L} - e^{ak_1 L})}{a^3(e^{aL} - e^{-aL})}$$
$$+ \frac{(e^{aL} - e^{-aL})(ak_2 L - ak_1 L)}{a^3(e^{aL} - e^{-aL})}$$

This form of the equation frequently is useful for determining H_p or H_s directly. Once either has been evaluated, however, the deflections are more readily determined by the series method. Moments are then calculated from

$$M = M_1 - (H_w + H_s)\eta - H_{sy} \tag{14.32}$$

Example. The Ambassador Bridge (Table 14.1), with unloaded backstays and a two-hinged stiffening truss, has the following properties:

$$f = 205.6 \text{ ft} \qquad A_c = 240.89 \text{ in}^2$$
$$L = 1850 \text{ ft} \qquad E_c = 27,000 \text{ ksi}$$
$$L_1 = 984.2 \text{ ft} \qquad E = 30,000 \text{ ksi}$$
$$L_2 = 833.9 \text{ ft} \qquad I = 113.71 \text{ ft}^4$$
$$\alpha_1 = 20°32' \qquad \alpha_2 = 24°$$

From the bridge data:

$$w = 6.2 \text{ kips per ft for the east cable}$$

$$H_w = 12,920 \text{ kips for the east cable}$$

The structure is analyzed for live loads of 0.2 to 2.0 kips per ft in increments of 0.2. These live loads are placed in various positions: over the entire span, the end half, the center half, the end quarter, the quarter nearest the center, and the center quarter. Analysis is made by both the Timoshenko approximate and exact forms of Eq. (14.27).

Approximate Method. Equation (14.27) becomes, when the proper values of the given data are used:

$$H_p = 848.60269a_1 + 282.86755a_3 + 169.72053a_5 \tag{14.33}$$

The coefficients a_1, a_3, and a_5 contain $\beta = H_p/H_w$. So a method of successive approximations must be used. First, a value of H_p or β is assumed. Then, this value is used in Eq. (14.25) to get values of a_1, a_3, and a_5. These, in turn, are substituted in Eq. (14.33) to obtain H_p. This computed value of H_p will not agree with the assumed value unless by accident the correct value of H_p had been guessed.

This procedure is repeated again. Thus, for two assumed values of β, β_1, and β_2, two calculated values of H_p, H_{p1}, and H_{p2} are obtained. On a graph, the straight line $H_p = \beta H_w = 12,920\beta$ is drawn (Fig. 14.45), and the points β_1,H_{p1} and β_2,H_{p2} plotted. A straight line between these points intersects the line $H_p = 12,920\beta$ at the correct value of H and β. (As many as six points were plotted, and always the calculated value of H_p lay on a straight line.)

The preceding procedure was used in calculating 60 values of H_p. Each was checked by finding correct values of a_1, a_3, and a_5 with Eq. (14.31). Values of H_p yielded by both methods are given in Table 14.7. The values of H_p from Fig. 14.45 are the values most nearly correct, since the check values of H_p

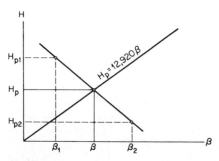

FIGURE 14.45 Chart for linear interpolation of horizontal component H_p of the cable tension.

TABLE 14.7 Cable Tension Component H_p, kips

H_p = values from Fig. 14.45, H'_p = check value from Eq. (14.31)

Live load, kips per ft	$k_1 = 0$ $k_2 = 1$		$k_1 = 0$ $k_2 = 0.5$		$k_1 = 0.25$ $k_2 = 0.75$		$k_1 = 0$ $k_2 = 0.25$		$k_1 = 0.25$ $k_2 = 0.50$		$k_1 = 0.375$ $k_2 = 0.625$	
	H_p	H'_p	H_p	H'_p	H_p	H'_p	H_p	H'_p	H_p	H'_p	H_p	H'_p
0.2	385	386	193	193	269	268	59	59	134	135	144	146*
0.4	770	765	385	383	536	538	117	117	269	268	289	287
0.6	1151	1160*	578	575	803	811*	176	174*	403	403	432	433
0.8	1534	1524	769	770	1069	1079	234	236	537	534	576	576
1.0	1912	1915	961	956	1336	1331	292	293	671	666*	719	720
1.2	2291	2281	1152	1149	1600	1601	351	352	804	804	863	857
1.4	2668	2658	1343	1341	1863	1879	409	413	937	938	1005	1006
1.6	3041	3054	1534	1524*	2127	2139	468	465	1070	1070	1148	1145
1.8	3416	3419	1723	1722	2391	2383	526	526	1203	1203	1291	1286
2.0	3789	3786	1913	1906	2651	2661	584	587	1335	1337	1433	1422
		0.78*		0.65*		1.00*		1.15*		0.75*		1.39*

*Maximum difference, % of H_p.

often changed several kips when H_p from Fig. 14.45 was changed a few tenths of a kip.

In calculating the deflections, it was found necessary, especially for unsymmetrical loading, to use five terms of Eq. (14.24). Table 14.8 gives deflections, ft, for a typical point.

TABLE 14.8 Deflections, ft, at $x = 0.2L$

Live load, kips per ft	$k_1 = 0$ $k_2 = 1$	$k_1 = 0$ $k_2 = 0.5$	$k_1 = 0.25$ $k_2 = 0.75$	$k_1 = 0$ $k_2 = 0.25$	$k_1 = 0.25$ $k_2 = 0.50$	$k_1 = 0.375$ $k_2 = 0.625$
0.2	0.2744	0.6962	0.0594	0.4122	0.2879	−0.0214
0.4	0.5440	1.3267	0.1229	0.7951	0.5445	−0.0448
0.6	0.8249	2.0758	0.1908	1.1888	0.8622	−0.0646
0.8	1.0842	2.7167	0.2582	1.5838	1.1277	−0.0829
1.0	1.3448	3.2418	0.3170	2.0394	1.3385	−0.1000
1.2	1.6241	3.8637	0.3906	2.3625	1.6768	−0.1211
1.4	1.8926	4.4757	0.4723	2.7514	1.8599	−0.1322
1.6	2.1746	5.0835	0.5435	3.1326	2.1137	−0.1499
1.8	2.4355	5.6719	0.6056	3.5173	2.3668	−0.1634
2.0	2.6870	6.2535	0.6937	4.0276	2.7363	−0.1801

TABLE 14.9 H_p, kips, Obtained by Exact
Method [Eq. (14.27)]

Live load, kips per ft	$k_1 = 0$ $k_2 = 0.5$	$k_1 = 0$ $k_2 = 0.25$
0.2	193	59
0.4	386	117
0.6	579	176
0.8	773	235
1.0	966	294
1.2	1,159	353
1.4	1,352	412
1.6	1,545	472
1.8	1,738	531
2.0	1,931	591

Timoshenko Exact Method.   For the full Eq. (14.27), Table 14.9 gives the values
of H_p, kips, for two of the load distributions. The results show that the actual
distribution of live load did not increase the value of H_p more than 1%.

(S. O. Asplund, "Structural Mechanics: Classical and Matrix Methods," Prentice-Hall, Inc., Englewood Cliffs, N.J.

E. Egervary, "Bases of a General Theory of Suspension Bridges Using a Matrical
Method of Calculation," *International Association of Bridge and Structural Engineers* (*IABSE*) *Publication*, vol. 16, pp. 149–184, 1956.

M. Esslinger, "Suspension Bridge Design Calculations by Electronic Computer," *Acier-Stahl-Steel*, no. 5, pp. 223–230, 1962.

A. A. Jakkula, "Theory of the Suspension Bridge," *IABSE Publication*, vol.
4, pp. 333–358, 1936.

C. P. Kuntz, J. P. Avery, and J. L. Durkee, "Suspension-bridge Truss Analysis
by Electronic Computer," *ASCE Conference on Electronic Computation*, Nov. 20–
21, 1958.

D. J. Peery, "An Influence-line Analysis for Suspension Bridges," *ASCE Transactions*, vol. 121, pp. 463–510, 1956.

T. J. Poskitt, "Structural Analysis of Suspension Bridges," *ASCE Proceedings*,
ST1, February, 1966, pp. 49–73.

G. C. Priester, "Application of Trigonometric Series to Cable Stress Analysis
in Suspension Bridges," *Engineering Research Bulletin* 12, University of Michigan,
1929.

A. G. Pugsley, "Theory of Suspension Bridges," Edward Arnold (Publishers),
Ltd., London.

A. G. Pugsley, "A Flexibility Coefficient Approach to Suspension Bridge Theory," *Institute of Civil Engineering Proceedings*, vol. 32, 1949.

S. A. Saafan, "Theoretical Analysis of Suspension Bridges," *ASCE Proceedings*,
ST4, August, 1966, pp. 1–12.

S. P. Timoshenko, "The Stiffness of Suspension Bridges," *ASCE Transactions*,
vol. 94, pp. 377–405, 1930.

S. P. Timoshenko and D. H. Young, "Theory of Structures," 2d ed., McGraw-Hill Book Company, New York.)

14.12   PRELIMINARY SUSPENSION-BRIDGE DESIGN

Since suspension bridges are major structures, it is desirable even in preliminary design to proceed into rather detailed refinement of the involved mathematical computations. Often, complete deflection-theory analysis is advisable at that stage. Such refinement is economically feasible with computers. Two procedures for preliminary design are described in the following.

Preliminary design may be started by examining pertinent site factors (clearance requirements, roadway width, foundation materials, etc.) and studying the details of existing structures of similar proportions and conditions. Table 14.10 gives typical data. Such data should be used with discretion, however, because of major differences in codes with regard to live loads, safety factors, allowable working stresses, and deflections. There also may be significant differences in details, such as roadway structure, which has a major effect on dead loads; as well as different underclearances, lengths of side spans, wind conditions, and other site conditions that influence the weight of steel required. Many published weights per unit area may be misleading because of inclusion of sidewalks, bicycle paths, and other elements in the widths of continental bridges.

Span Ratios.   With straight backstays, the ratio of side to main spans may be about 1:4 for economy. For suspended side spans, this ratio may be about 1:2. Physical conditions at the site may, however, dictate the selected span proportions.

Sag.   The sag-span ratio is important. It determines the horizontal component of cable force. Also, this ratio affects height of towers, pull on anchorages, and total stiffness of the bridge. For minimum stresses, the ratio should be as large as possible for economy, say 1:8 with suspended side spans, or 1:9 with straight backstays. But the towers may then become high. Several comparative trials should be made. For the Forth Road Bridge, the correct sag-span ratio of 1:11 was thus determined. The general range of this ratio in practice is 1:8 to 1:12, with an average around 1:10.

Truss Depth.   Stiffening-truss depths vary from $\frac{1}{60}$ to $\frac{1}{170}$ the span. Aerodynamic conditions, however, play a major role in shaping the preliminary design, and some of the criteria given in Art. 14.16 should be studied at this stage.

Other Criteria.   Allowable stresses in main cables may vary from 80 to 86 ksi. Permissible live-load deflections in practice are seldom specified but usually do not exceed $\frac{1}{300}$ the span. In Europe, greater reliance is placed on limiting the radius of curvature of the roadway (thus, to 600 or 1,000 meters); or to limiting the cross slope of the roadway under eccentric load (thus, to about 1%); or to limiting the vertical acceleration under live load (thus, to 0.31 meter per sec^2).

14.12.1   Preliminary Design by Steinman-Baker Procedure

Analysis by elastic theory is sufficiently accurate for short spans and designs with deep, rigid stiffening systems that limit deflections to small amounts. The simple calculations of elastic theory are also useful, however, for preliminary designs and

TABLE 14.10 Details of Major Suspension Bridges*

Item	Verrazano Narrows Bridge	Golden Gate Bridge	Mackinac Bridge	George Washington Bridge	Salazar Bridge (Portugal)	Forth Bridge (Scotland)	Severn Bridge (England–Wales)	Tacoma Narrows Bridge II	San Francisco–Oakland Bay Bridge†	Bronx–Whitestone Bridge	Delaware Memorial Bridge I	Walt Whitman Bridge
Length of main span, ft	4,260	4,200	3,800	3,500	3,323	3,300	3,240	2,800	2,310	2,300	2,150	2,000
Length of each side span, ft	1,215	1,125	1,800	610/650	1,586	1,340	1,000	1,100	1,160	735	750	770
Length of suspended structure, ft	6,690	6,450	7,400	4,760	6,495	5,980	5,240	5,000	10,450	3,770	3,650	3,540
Length including approach structure, ft	13,700	8,981	19,205	5,800	10,575	8,244	7,640	5,979	43,500	7,995	10,750	11,687
Width of bridge (c to c cables), ft	103	90	68	106	77	78	75	60	66	74	57	79
Number of traffic lanes	12	6	4	14‡	4	4	4	4	9	6	4	7
Height of towers above MHW, ft	690	746	552	595	625	512	470	500	447	377	440	378
Clearance at center above MHW, ft	228	215	148	220‡	246	150	120	187	203	150	175	150
Deepest foundation below MHW, ft	170	115	210	75	260	106	75	224	235	165	115	107
Diameter of cable, in	35⅞	36	24½	36	23 1/16	24	20	20¼	28¾	22	19¾	23⅜
Length of one cable, ft	7,205	7,650	8,683	5,235	7,899	7,000	5,600	5,500	5,080	4,166	4,015	3,845
Number of wires per cable	26,108	27,572	12,580	26,474	11,248	11,618	8,300	8,702	17,464	9,842	8,284	11,396
Total length of wire used, miles	142,500	80,000	41,000	105,000	33,600	30,800	18,000	20,000	70,800	14,800	12,600	16,600
Year of completion	1964	1937	1957	1931	1966	1964	1966	1950	1936	1939	1951	1957

* Courtesy of *Engineering News-Record.*
† Twin spans.
‡ With new lower deck.

estimates if tabular percentage corrections are applied, based on experience with the deflection theory.

The corrections depend principally on the magnitude of the dead load and on the flexibility of the structure. The magnitude of the corrections increase with the deflection η and with the horizontal component of cable tension H_w under dead load. They therefore increase with span L and dead load w, while decreasing with truss stiffness EI and cable sag f. D. B. Steinman expressed this in a simple parameter S, the stiffness factor, such that

$$S = \frac{1}{L}\sqrt{\frac{EI}{H_w}} = \frac{1}{L^2}\sqrt{\frac{8fEI}{w}} \tag{14.34}$$

This value is used in the Steinman-Baker charts, Fig. 14.46, to obtain the percentage C to be applied to elastic-theory shears and moments to get deflection-theory shears and moments.

Roughly, the percentage reductions from the approximate theory are proportional to $1/S$, which might be called a flexibility factor; that is, the magnitude of the reduction increases considerably with long spans and heavy dead load, and diminishes with stiffness and sag.

The Steinman-Baker charts are based on the following proportions: side span, one-half the main span; sag-span ratio, 0.1; moment of inertia I, constant; cable design stress, 80 ksi; modulus of elasticity E, 29,000 ksi; ratio of dead to live load, 3. Further refinement of C for other proportions is suggested as follows:

For unloaded side spans, increase C by 2½% of its value.

For sag-span ratio = 0.12 (or 0.08) instead of 0.10, decrease (or increase) C by 2% of its value.

For cable stress = 120 (or 40) instead of 80 ksi, increase (or decrease) C by ½% of its value.

For $I/I_1 = 0.75$ instead of 1.00, increase C by 1½% of its value.

For $L_1/L = 0.25$ instead of 0.50, increase C by 2% of its value.

For load ratio $w/p = 5$ instead of 3, add 1% to C; for $w/p = 2$, subtract 1% from C; for $w/p = 1½$, subtract 2% from C.

In elastic-theory analysis for preliminary design of a bridge with two cable planes, the bridge may be treated as plane frameworks loaded in each plane; that is, the action as a space structure may be disregarded. The alleviating effects of torsion of the stiffening girders, of the unloading action of the cross frames or diaphragms between the girders, and of the participation of the connection of the pylons may all be left for more refined later analysis.

14.12.2 Preliminary Design by Hardesty-Wessman Procedure

Hardesty and Wessman presented an approximate, partly empirical, preliminary design method based on the distortion of an unstiffened cable. The maximum moments at the quarter point and at the center of the main span at constant mean temperature are computed in two major steps:

1. The deflections η' of an unstiffened cable under partial live load, for various ratios of live to dead load p/w, are obtained from Fig. 14.47. These charts were developed with average live-load lengths from a study of bridges in service and

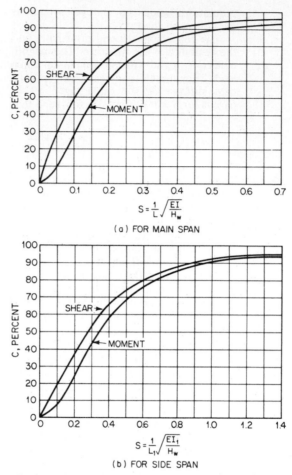

FIGURE 14.46   Steinman-Baker correction curves for stresses obtained by elastic theory for suspension bridges. *(Reprinted with permission from D. B. Steinman, "A Practical Treatise on Suspension Bridges," 2d ed., John Wiley & Sons, Inc., New York.)*

also based on the assumptions that the cable length is unchanged and the pylon tops do not move.

2. Corrections then are made for the effect of adding the stiffening truss (which reduces deflection η'). A trial moment of inertia is used and corrected later if necessary. Equations (14.35) are used for a first estimate of the maximum horizontal components of cable tension $H_w + H_p$. Then, Eqs. (14.36) are used to determine the bending moment M_t induced in the stiffening truss when it is bent to the deflections η' of the unstiffened cable.

Maximum positive moment at a quarter point, with live load at the same end over a length of $0.4L$, where L = main span, ft, is

$$H_w + H_p = \frac{1}{f + \eta'_c} (0.125wL^2 + 0.040pL^2) \qquad (14.35a)$$

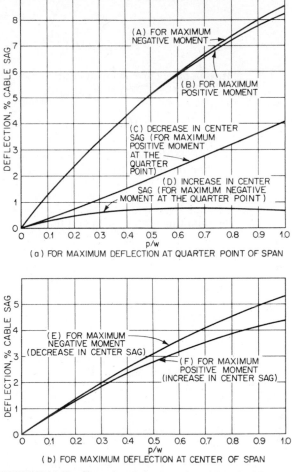

FIGURE 14.47 Chart gives deflections of unstiffened cables under partial load. *(Reprinted with permission from S. Hardesty and H. E. Wessman, "Preliminary Design of Suspension Bridges," ASCE Transactions, vol. 104, 1939.)*

where f = cable sag, ft
 η_c' = deflection at center of unstiffened cable, ft
 w = uniform dead load
 p = uniform live load

Maximum negative moment at a quarter point, with live load at the opposite end over a length of $0.6L$, is

$$H_w + H_p = \frac{1}{f + \eta_c'} (0.125wL^2 + 0.085pL^2) \qquad (14.35b)$$

Maximum positive moment at the center, with live load at the center over a length of $0.3L$, is

$$H_w + H_p = \frac{1}{f + \eta_c'} (0.125wL^2 + 0.0638pL^2) \qquad (14.35c)$$

Maximum negative moment at the center, with live load over a length of $0.35L$ at each end, is

$$H_w + H_p = \frac{1}{f + \eta_c'} (0.125wL^2 + 0.0613pL^2) \qquad (14.35d)$$

In each of the preceding equations, the quantity within the parentheses is the bending moment at the center of a simple span due to dead load over the entire span and live load over a part of the span. The deflection η_c' is positive when downward and negative when upward.

Since $H_w = wL^2/8f$ is known, Eqs. (14.35) yield a trial value of H_p, with which the following bending moments in the truss can be computed:

Maximum positive moment at quarter points:

$$M_t = 47.0 \frac{EI\eta'}{L^2} \qquad (14.36a)$$

where I = moment of inertia of stiffening truss and E = modulus of elasticity of truss steel.

Maximum negative moment at quarter points:

$$M_t = 43.0 \frac{EI\eta'}{L^2} \qquad (14.36b)$$

Maximum positive moment at center:

$$M_t = 65.8 \frac{EI\eta'}{L^2} \qquad (14.36c)$$

Maximum negative moment at center:

$$M_t = 59.2 \frac{EI\eta'}{L^2} \qquad (14.36d)$$

Since the truss is neither infinitely flexible nor infinitely stiff, the cable will be forced back only a part of the distance η'. The moment in the truss is reduced from M_t to

$$M = M_t \frac{(H_w + H_p)\eta'}{M_t + (H_w + H_p)\eta'} \qquad (14.37)$$

And the deflection is reduced to

$$\eta = \eta' \frac{(H_w + H_p)\eta'}{M_t + (H_w + H_p)\eta'} \qquad (14.38)$$

Finally, changes in length of cable due to live load or temperature, and the sag changes caused by movement of pylon tops, cable stress, and temperature, are combined in one change in the center sag. These effects are

Change in length of cable due to stress:

Main span:
$$\Delta L_s = \frac{H_p L}{A_c E_c}(1 + {}^{16}\!/_3 a^2) \tag{14.39}$$

where $a = f/L$

A_c = cross-sectional area of cable

E_c = modulus of elasticity of cable steel

Unloaded side span:
$$\Delta L_{1s} = \frac{H_p L_1}{A_c E_c}\sec^2 \alpha_1 \tag{14.40}$$

where α_1 = angle side-span cable makes with horizontal.

Change in length of cable due to temperature:

Main span:
$$\Delta L_t = \epsilon_t \Delta t L(1 + {}^{8}\!/_3 a^2) \tag{14.41}$$

where ϵ = coefficient of expansion and Δt = temperature change.

Unloaded side span:
$$\Delta L_{1t} = \epsilon_t \Delta t L_1 \sec \alpha_1 \tag{14.42}$$

Change in sag in main span due to temperature:

$$\Delta f = \frac{15}{16a(5 - 24a^2)}\Delta L \tag{14.43}$$

Change in sag due to movement of pylon top:

$$\Delta f = \frac{15 - 40a^2 + 288a^4}{16a(5 - 24a^2)}(2\Delta L_1 \sec \alpha_1) \tag{14.44}$$

Moment caused by change in sag:

Main span at center:
$$M = 9.4\frac{EI}{L^2}\Delta f \tag{14.45}$$

Main span at quarter-point:
$$M = 7.6\frac{EI}{L^2}\Delta f \tag{14.46}$$

The corrected value of the sag allows a second trial value of H_p to be obtained. Then, the process is repeated.

In applying this method to preliminary design, an arbitrary moment of inertia is selected, based on a tentative chord section and truss depth. The procedure is repeated with other values of I, say one-half and double those in the first analysis. Final selection may be based on limiting values of desired deflection and grade change due to load.

(D. B. Steinman, "A Practical Treatise on Suspension Bridges," 2d ed., John Wiley & Sons, Inc., New York; S. Hardesty and H. E. Wessman, "Preliminary Design of Suspension Bridges," *Transactions of the American Society of Civil Engineers*, vol. 104, 1939.)

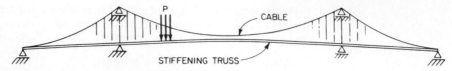

FIGURE 14.48   Self-anchored suspension bridge.

14.13   *SELF-ANCHORED SUSPENSION BRIDGES*

Self-anchored suspension bridges differ from the type discussed in Arts. 14.11 and 14.12 only in that external anchorages are dispensed with (see Art. 14.3). Unlike the externally anchored type, self-anchored suspension bridges may properly be analyzed by the elastic theory, since the effect of distortions of the structural geometry under live load is practically eliminated. The structure is also not stressed by uniform temperature change of cables and stiffening girders. The analysis is thus simpler. But the favorable reductions of bending moments that occur with externally anchored suspension bridges are lost. Furthermore, the effect of axial load in the stiffening girder must be considered, as well as the effect of girder camber.

For a symmetrical three-span structure with continuous stiffening girders (Fig. 14.48), a plane system (cable, suspenders, and girder) has three redundants. C. H. Gronquist derived in simple form the elastic-theory equations for determining the redundants for a continuous stiffening-truss system. He took into account camber and its action in reducing cable and truss stress by archlike action. He also demonstrated that the equations for the horizontal component of cable tension H_p under live load for the self-anchored bridge, with girder camber and shortening eliminated, are the same as the elastic-theory equations for an externally anchored suspension bridge.

(C. H. Gronquist, "Simplified Theory of the Self-Anchored Suspension Bridge," *Transactions of the American Society of Civil Engineers*, vol. 107, 1942.)

14.14   *CABLE-STAYED BRIDGE ANALYSIS*

The static behavior of a cable-stayed girder can best be gaged from the simple, two-span example of Fig. 14.49. The girder is supported by one stay cable in each span, at E and F, and the pylon is fixed to the girder at the center support B. The static system has two internal cable redundants and one external support redundant.

If the cable and pylon were infinitely rigid, the structure would behave as a continuous four-span beam AC on five rigid supports A, E, B, F, and C. The cables

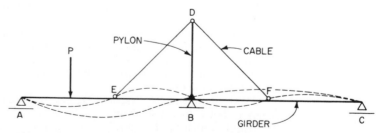

FIGURE 14.49   Dash lines indicate deflected positions of a cable-stayed girder.

are elastic, however, and correspond to springs. The pylon also is elastic, but much stiffer because of its large cross section. If cable stiffness is reduced to zero, the girder assumes the shape of a deflected two-span beam ABC.

Cable-stayed bridges of the nineteenth century differed from those of the 1960s in that their stays constituted relatively soft spring supports. Heavy and long, the stays could not be highly stressed. Usually, the cables were installed with significant slack, or sag. Consequently, large deflections occurred under live load as the sag decreased. In contrast, modern cables are made of high-strength steel, are relatively short and taut, and have low weight. Their elastic action may therefore be considered linear, and an equivalent modulus of elasticity may be used [Eq. (14.1)]. The action of such cables then produces something more nearly like a four-span beam for a structure such as the one in Fig. 14.49.

If the pylon were hinged at its base connection with a stayed girder at B, rather than fixed, the pylon would act as a pendulum column. This would have an important effect on the stiffness of the system, for the spring support at E would become more flexible. The resulting girder deflection might exceed that due to the elastic stretch of the cables. In contrast, the elastic shortening of the pylon has no appreciable effect.

Relative girder stiffness plays a dominant role in the structural action. The stayed girder tends to approach a beam on rigid supports A, E, B, F, C as girder stiffness decreases toward zero. With increasing girder stiffness, however, the support of the cables diminishes, and the bridge approaches a girder supported on its piers and abutments A, B, C.

In a three-span bridge, a side-span cable connected to the abutment furnishes more rigid support to the main span than does a cable attached to some point in the side span. In Fig. 14.49, for example, the support of the load P in the position shown would be improved if the cable attachment at F were shifted to C. This explains why cables from the pylon top to the abutment are structurally more efficient, though not as esthetically pleasing as other arrangements.

The stiffness of the system also depends on whether the cables are fixed at the towers (at D, for example, in Fig. 14.49) or whether they run continuously over (or through) the pylons. Some early designs with more than one cable to a pylon from the main span required one of the cables to be fixed to the pylon and the others to be on movable saddle supports. Most contemporary designs fix all the stays to the pylon.

The curves of maximum-minimum girder moments for all load variations usually show a large range of stress. Designs providing for the corresponding normal forces in the girder may require large variations in cross sections. By prestressing the cables or by raising or lowering the support points, it is possible to achieve a more uniform and economical moment capacity. The amount of prestressing to use for this purpose may be calculated by successively applying a unit force in each of the cables and drawing the respective moment diagrams. Then, by trial, the proper multiples of each force are determined so that, when their moments are superimposed on the maximum-minimum moment diagrams, an optimum balance results.

("Guidelines for the Design of Cable-Stayed Bridges," Committee on Cable-Stayed Bridges, American Society of Civil Engineers.)

14.14.1 Static Analysis—Elastic Theory

Cable-stayed bridges may be analyzed by the general method of indeterminate analysis with the equations of virtual work.

The degree of internal redundancy of the system depends on the number of cables, types of connections (fixed or movable) of cables with the pylons, and the

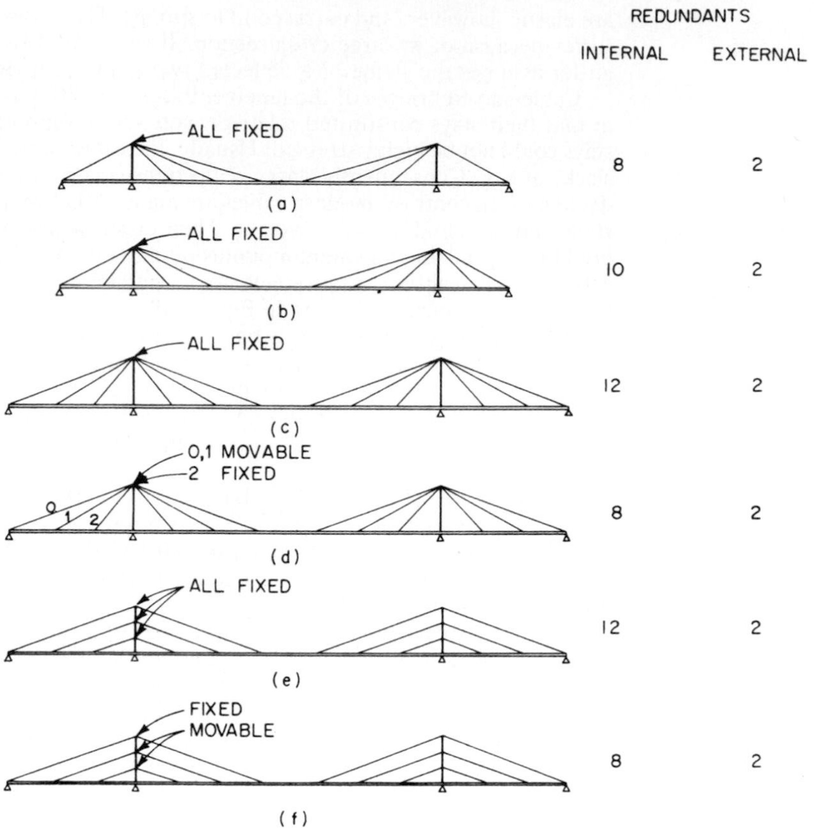

FIGURE 14.50 Number of internal and external redundants for various types of cable-stayed bridges.

nature of the pylon connection at its base with the stayed girder or pier. The girder is usually made continuous over three spans. Figure 14.50 shows the order of redundancy for various single-plane systems of cables.

If the bridge has two planes of cables, two stayed girders, and double pylons, it usually also must be provided with a number of intermediate cross diaphragms in the floor system, each of which is capable of transmitting moment and shear. The bridge may also have cross girders across the top of the pylons. Each of these cross members adds two redundants, to which must be added twice the internal redundancy of the single-plane structure, and any additional reactions in excess of those needed for external equilibrium as a space structure. The redundancy of the space structure is very high, usually of the order of 40 to 60. Therefore, the methods of plane statics are normally used, except for large structures.

For a cable-stayed structure such as that illustrated in Fig. 14.51a, it is convenient to select as redundants the bending moments in the stayed girder at those points where the cables and pylons support the girder. When these redundants are set equal to zero, an articulated, statically determinate base system is obtained, Fig. 14.51b. When the loads are applied to this choice of base system, the stresses in the cables do not differ greatly from their final values; so the cables may be dimensioned in a preliminary way.

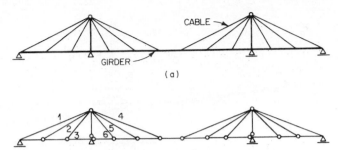

FIGURE 14.51 Cable-stayed bridge with three spans. (*a*) Girder is continuous over the three spans. (*b*) Insertion of hinges in the girder at cable attachments makes system statically determinate.

Other approaches are also possible. One is to use the continuous girder itself as a statically indeterminate base system, with the cable forces as redundants. But computation is generally increased.

A third method involves imposition of hinges, for example at *a* and *b* (Fig. 14.52), so placed as to form two coupled symmetrical base systems, each statically indeterminate to the fourth degree. The influence lines for the four indeterminate cable forces of each partial base system are at the same time also the influence lines of the cable forces in the real system. The two redundant moments X_a and X_b are treated as symmetrical and antisymmetrical group loads, $Y = X_a + X_b$ and $Z = X_a - X_b$, to calculate influence lines for the 10-degree indeterminate structure shown. Kern moments are plotted to determine maximum effects of combined bending and axial forces.

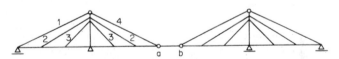

FIGURE 14.52 Hinges at *a* and *b* reduce the number of redundants for a cable-stayed girder continuous over three spans.

A similar concept is illustrated in Fig. 14.53, which shows the application of independent symmetric and antisymmetric group stress relationships to simplify calculations for an 8-degree indeterminate system. Thus, the first redundant group X_1 is the self-stressing of the lowest cables in tension to produce $M_1 = +1$ at supports.

The above procedures also apply to influence-line determinations. Typical influence lines for two bridge types are shown in Fig. 14.54. These demonstrate that the fixed cables have a favorable effect on the girders but induce sizable bending moments in the pylons, as well as differential forces on the saddle bearings.

Note also that the radiating system in Fig. 14.50*c* and *d* generally has more favorable bending moments for long spans than does the harp system of Fig. 14.54. Cable stresses also are somewhat lower for the radiating system, because the steeper cables are more effective. But the concentration of cable forces at the top of the pylon introduces detailing and construction difficulties. When viewed at an angle, the radiating system presents esthetic problems, because of the different intersection angles when the cables are in two planes. Furthermore, fixity of the cables at

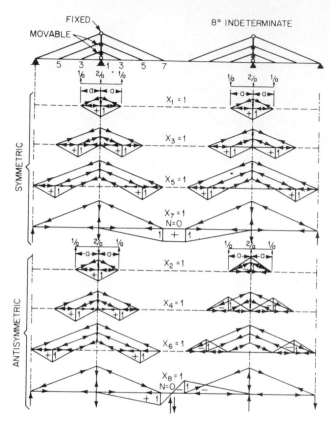

FIGURE 14.53 Forces induced in a cable-stayed bridge by independent symmetric and antisymmetric group loadings. *(Reprinted with permission from O. Braun, "Neues zur Berchnung Statisch Unbestimmter Tragwerke, "Stahlbau, vol. 25, 1956.)*

pylons with the radiating system in Fig. 14.50*c* and *d* produces a wider range of stress than does a movable arrangement. This can adversely influence design for fatigue.

A typical maximum-minimum moment and axial-force diagram for a harp bridge is shown in Fig. 14.55.

The secondary effect of creep of cables (Art. 14.7) can be incorporated into the analysis. The analogy of a beam on elastic supports is changed thereby to that of a beam on linear viscoelastic supports. Better stiffness against creep for cable-stayed bridges than for comparable suspension bridges has been reported. (K. Moser, "Time-Dependent Response of Suspension and Cable-Stayed Bridges," International Association of Bridge and Structural Engineers, 8th Congress Final Report, 1968, pp. 119–129.)

(W. Podolny, Jr., and J. B. Scalzi, "Construction and Design of Cable-Stayed Bridges," 2d ed., John Wiley & Sons, Inc., New York.)

14.14.2 Static Analysis—Deflection Theory

Distortion of the structural geometry of a cable-stayed bridge under action of loads is considerably less than in comparable suspension bridges. The influence on stresses

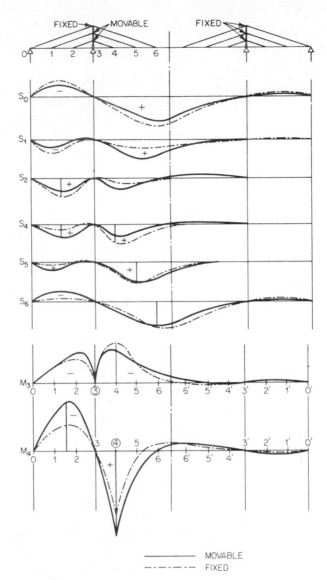

FIGURE 14.54 Typical influence lines for a three-span cable-stayed bridge showing the effects of fixity of cables at the pylons. *(Reprinted with permission from H. Homberg, "Einflusslinien von Schrägseilbruchen," Stahlbau, vol. 24, no. 2, 1955.)*

of distortion of stayed girders is relatively small. In any case, the effect of distortion is to increase stresses, as in arches, rather than the reverse, as in suspension bridges. This effect for the Severin Bridge is 6% for the stayed girder and less than 1% for the cables. Similarly, for the Düsseldorf North Bridge, stress increase due to distortion amounts to 12% for the girders.

The calculations, therefore, most expeditiously take the form of a series of successive corrections to results from first-order theory (Art. 14.14.1). The mag-

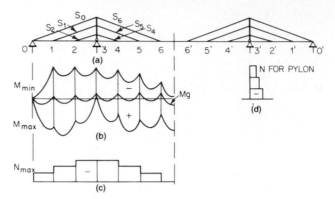

FIGURE 14.55 Typical moment and force diagrams for a cable-stayed bridge. (*a*) Girder is continuous over three spans. (*b*) Maximum and minimum bending moments in the girder. (*c*) Compressive axial forces in the girder. (*d*) Compressive axial forces in a pylon.

nitude of vertical and horizontal displacements of the girder and pylons can be calculated from the first-order theory results. If the cable stress is assumed constant, the vertical and horizontal cable components V and H change by magnitudes ΔV and ΔH by virtue of the new deformed geometry. The first approximate correction determines the effects of these ΔV and ΔH forces on the deformed system, as well as the effects of V and H due to the changed geometry. This process is repeated until convergence, which is fairly rapid.

14.15 PRELIMINARY DESIGN OF CABLE-STAYED BRIDGES

In general, the height of a pylon in a cable-stayed bridge is about $\frac{1}{6}$ to $\frac{1}{8}$ the main span. Depth of stayed girder ranges from $\frac{1}{60}$ to $\frac{1}{80}$ the main span and is usually 8 to 14 ft, averaging 11 ft. Live-load deflections usually range from $\frac{1}{400}$ to $\frac{1}{500}$ the span.

To achieve symmetry of cables at pylons, the ratio of side to main spans should be about 3:7 where three cables are used on each side of the pylons, and about 2:5 where two cables are used. A proper balance of side-span length to main-span length must be established if uplift at the abutments is to be avoided. Otherwise, movable (pendulum-type) tiedowns must be provided at the abutments.

Wide box girders are mandatory as stayed girders for single-plane systems, to resist the torsion of eccentric loads. Box girders, even narrow ones, are also desirable for double-plane systems to enable cable connections to be made without eccentricity. Single-web girders, however, if properly braced, may be used.

Since elastic-theory calculations are relatively simple to program for a computer, a formal set may be made for preliminary design after the general structure and components have been sized.

Manual Preliminary Calculations for Cable Stays.   Following is a description of a method of manual calculation of reasonable initial values for use as input data for design of a cable-stayed bridge by computer. The manual procedure is not precise but does provide first-trial cable-stay areas. With the analogy of a continuous, elastically supported beam, influence lines for stay forces and bending mo-

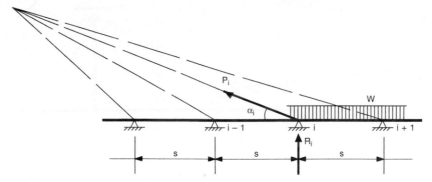

FIGURE 14.56 Cable-stayed girder is supported by cable force P_i at ith point of cable attachment. R_i is the vertical component of P_i.

ments in the stayed girder can be readily determined. From the results, stress variations in the stays and the girder resulting from concentrated loads can be estimated.

If the dead-load cable forces reduce deformations in the girder and pylon at supports to zero, the girder acts as a beam continuous over rigid supports, and the reactions can be computed for the continuous beam. Inasmuch as the reactions at those supports equal the vertical components of the stays, the dead-load forces in the stays can be readily calculated. If, in a first-trial approximation, live load is applied to the same system, the forces in the stays (Fig. 14.56) under the total load can be computed from

$$P_i = \frac{R_i}{\sin \alpha_i} \tag{14.47}$$

where R_i = sum of dead-load and live-load reactions at i and α_i = angle between girder and stay i. Since stay cables usually are designed for service loads, the cross-sectional area of stay i may be determined from

$$A_i = \frac{R_i}{\sigma_a \sin \alpha_i} \tag{14.48}$$

where σ_a = allowable unit stress for the cable steel.

The allowable unit stress for service loads equals $0.45 f_{pu}$, where f_{pu} = the specified minimum tensile strength, ksi, of the steel. For 0.6-in-dia., seven-wire prestressing strand (ASTM A416), f_{pu} = 270 ksi and for ¼-in-dia. ASTM A421 wire, f_{pu} = 240 ksi. Therefore, the allowable stress is 121.5 ksi for strand and 108 ksi for wire.

The reactions may be taken as $R_i = ws$, where w is the uniform load, kips per ft, and s, the distance between stays. At the ends of the girder, however, R_i may have to be determined by other means.

Determination of the force P_o acting on the back-stay cable connected to the abutment (Fig. 14.57) requires that the horizontal force F_h at the top of the pylon be computed first. Maximum force on that cable occurs with dead plus live loads on the center span and dead load only on the side span. If the pylon top is assumed immovable, F_h can be determined from the sum of the forces from all the stays, except the back stay:

$$F_h = \sum \frac{R_i}{\tan \alpha_i} - \sum \frac{R_i'}{\tan \alpha_i'} \tag{14.49}$$

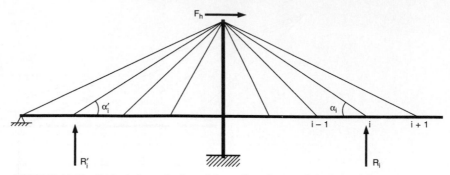

FIGURE 14.57 Cables induce a horizontal force F_h at the top of a pylon.

where R_i, R_i' = vertical component of force in the ith stay in the main span and side span, respectively

α_i, α_i' = angle between girder and the ith stay in the main span and side span, respectively

Figure 14.58 shows only the pylon and back-stay cable to the abutment. If, in Fig. 14.58, the change in the angle α_o is assumed to be negligible as F_h deflects the pylon top, the load in the back stay can be determined from

$$P_o = \frac{F_h h_t^3 \cos \alpha_o}{3l_o(E_cI/E_sA_s) + h_t^3 \cos^2 \alpha_o} \tag{14.50}$$

If the bending stiffness E_cI of the pylon is neglected, then the back-stay force is given by

$$P_o = F_h/\cos \alpha_o \tag{14.51}$$

where h_t = height of pylon
l_o = length of back stay
E_c = modulus of elasticity of pylon material
I = moment of inertia of pylon cross section

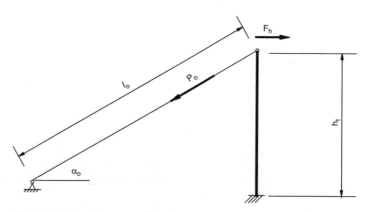

FIGURE 14.58 Cable force P_o in backstay to anchorage and bending stresses in the pylon resist horizontal force F_h at the top of the pylon.

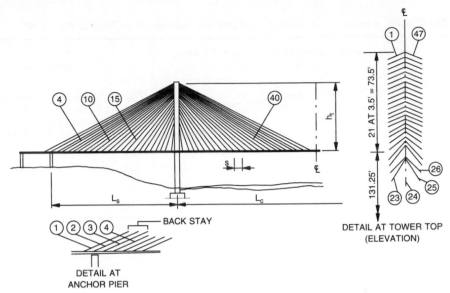

FIGURE 14.59   Half of a three-span cable-stayed bridge. Properties of components are as follows:

Girder		Tower	
Main span L_c	940 ft	Height h_d	204.75 ft
Side span L_b	440 ft	Area A	120 ft^2
Stay spacing s	20 ft	Moment of inertia I	3620 ft^4
Area A	101.4 ft	Elastic modulus E_t	45,000 ksi
Moment of inertia I	48.3 ft^4		
Elastic modulus E_g	47,000 ksi	Stays	
		Elastic modulus E_s	28,000 ksi

(Reprinted with permission from W. Podolny, Jr., and J. B. Scalzi, "Construction and Design of Cable-Stayed Bridges," 2d ed., John Wiley & Sons, Inc. New York.)

E_s = modulus of elasticity of cable steel
A_s = cross-sectional area of back-stay cable

For the structure illustrated in Fig. 14.59, values were computed for a few stays from Eqs. (14.47), (14.48), (14.49), and (14.51) and tabulated in Table 14.11*a*. Values for the final design, obtained by computer, are tabulated in Table 14.11*b*.

Inasmuch as cable stays 1, 2, and 3 in Fig. 14.59 are anchored at either side of the anchor pier, they are combined into a single back-stay for purposes of manual calculations. The edge girders of the deck at the anchor pier were deepened in the actual design, but this increase in dead weight was ignored in the manual solution. Further, the simplified manual solution does not take into account other load cases, such as temperature, shrinkage, and creep.

Influence lines for stay forces and girder moments are determined by treating the girder as a continuous, elastically supported beam. From Fig. 14.60, the following relationships are obtained for a unit force at the connection of girder and stay:

$$P_i = \frac{1}{\sin \alpha_i} \qquad \Delta l_{si} = \frac{P_i l_{si}}{A_{si} E_s} = \delta_i \sin \alpha_i$$

TABLE 14.11   Comparison of Manual and Computer Solution for the Stays in Fig. 14.59*

Stay number	(a) According to Eqs. (14.47), (14.48), (14.49), and (14.51)					(b) Computer solution			
	R_{DL}, kips	P_{DL}, kips	R_{DL+LL}, kips	P_{DL+LL}, kips	A, in²	P_{DL}, kips	P_{DL+LL}, kips‡	Number of 0.6-in strands§	Strand area, in² §
Back stay‡	—	2596	—	3969	32.667	2775	3579	136	29.512
4	360	824	400	916	7.539	851	1049	40	8.680
10	360	684	400	760	6.255	695	797	31	6.727
15	360	550	400	611	5.029	558	654	25	5.425
40	360	734	400	815	6.708	756	878	34	7.378

*Reprinted with permission from W. Podolny, Jr., and J. B. Scalzi, "Construction and Design of Cable-Stayed Bridges," 2d ed., John Wiley & Sons, Inc., New York.
†Stays No. 1, 2, and 3 combined into one back stay.
‡Maximum live load.
§Per plane of a two-plane structure.

which lead to

$$\delta_i = \frac{l_{si}}{A_{si}E_s \sin^2 \alpha_i}$$

With Eq. (14.48) and $l_{si} = h_t \sin \alpha_i$, the deflection at point i is given by

$$\delta_i = \frac{h_t \sigma_a}{R_i E_s \sin^2 \alpha_i} \qquad (14.52)$$

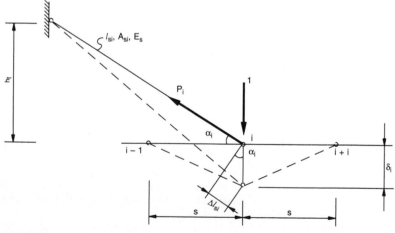

FIGURE 14.60   Unit force applied at point of attachment of ith cable stay to girder for determination of spring stiffness.

With R_i taken as $s(w_{DL} + w_{LL})$, the product of the uniform dead and live loads and the stay spacing s, the spring stiffness of cable stay i is obtained as

$$k_i = \frac{1}{\delta_i s} = \frac{(w_{DL} + w_{LL})E_s \sin^2 \alpha_i}{h_t \sigma_a} \qquad (14.53)$$

For a vertical unit force applied on the girder at a distance x from the girder-stay connection, the equation for the cable force P_i becomes

$$P_i = \frac{\xi W s}{2 \sin \alpha_i} \, \eta_p \qquad (14.54)$$

where $\eta_p = e^{-\xi x}(\cos \xi x + \sin \xi x)$

$$\xi = \sqrt[4]{\frac{k_i}{4 E_c I}} \qquad (14.55)$$

The bending moment M_i at point i may be computed from

$$M_i = \frac{W}{4\xi} e^{-\xi x}(\cos \xi x - \sin \xi x) = \frac{W}{4\xi} \eta_m \qquad (14.56)$$

where $\eta_m = e^{-\xi x}(\cos \xi x - \sin \xi x)$.

(W. Podolny, Jr., and J. B. Scalzi, "Construction and Design of Cable-Stayed Bridges," 2d ed., John Wiley & Sons, Inc., New York.)

14.16 AERODYNAMIC ANALYSIS OF CABLE-SUSPENDED BRIDGES

The wind-induced failure on November 7, 1940, of the Tacoma Narrows Bridge in the state of Washington shocked the engineering profession. Many were surprised to learn that failure of bridges as a result of wind action was not unprecedented. During the slightly more than 12 decades prior to the Tacoma Narrows failure, 10 other bridges were severely damaged or destroyed by wind action (Table 14.12). As can be seen from Table 14.12a, wind-induced failures have occurred in bridges with spans as short as 245 ft up to 2800 ft. Other "modern" cable-suspended bridges have been observed to have undesirable oscillations due to wind (Table 14.12b).

14.16.1 Required Information on Wind at Bridge Site

Prior to undertaking any studies of wind instability for a bridge, engineers should investigate the wind environment at the site of the structure. Required information includes the character of strong wind activity at the site over a period of years. Data are generally obtainable from local weather records and from meteorological records of the U.S. Weather Bureau. However, caution should be used, because these records may have been attained at a point some distance from the site, such as the local airport or federal building. Engineers should also be aware of differences in terrain features between the wind instrumentation site and the structure site that may have an important bearing on data interpretation. Data required are wind

TABLE 14.12 Long-Span Bridges Adversely Affected by Wind*

(a) Severely damaged or destroyed

Bridge	Location	Designer	Span, ft	Failure date
Dryburgh Abbey	Scotland	John and William Smith	260	1818
Union	England	Sir Samuel Brown	449	1821
Nassau	Germany	Lossen and Wolf	245	1834
Brighton Chain Pier	England	Sir Samuel Brown	255	1836
Montrose	Scotland	Sir Samuel Brown	432	1838
Menai Straits	Wales	Thomas Telford	580	1839
Roche-Bernard	France	Le Blanc	641	1852
Wheeling	U.S.A.	Charles Ellet	1010	1854
Niagara-Lewiston	U.S.A.	Edward Serrell	1041	1864
Niagara-Clifton	U.S.A.	Samuel Keefer	1260	1889
Tacoma Narrows I	U.S.A.	Leon Moisseiff	2800	1940

(b) Oscillated violently in wind

Bridge	Location	Year built	Span, ft	Type of stiffening
Fyksesund	Norway	1937	750	Rolled I beam
Golden Gate	U.S.A.	1937	4200	Truss
Thousand Island	U.S.A.	1938	800	Plate girder
Deer Isle	U.S.A.	1939	1080	Plate girder
Bronx-Whitestone	U.S.A.	1939	2300	Plate girder
Long's Creek	Canada	1967	713	Plate girder

*After F. B. Farquharson et al., "Aerodynamic Stability of Suspension Bridges," University of Washington Bulletin 116, parts I through V, 1949–1954.

velocity, direction, and frequency. From these data, it is possible to predict high wind speeds, expected wind direction and probability of occurrence.

The aerodynamic forces that wind applies to a bridge depend on the velocity and direction of the wind and on the size, shape, and motion of the bridge. Whether resonance will occur under wind forces depends on the same factors. The amplitude of oscillation that may build up depends on the strength of the wind forces (including their variation with amplitude of bridge oscillation), the energy-storage capacity of the structure, the structural damping, and the duration of a wind capable of exciting motion.

The wind velocity and direction, including vertical angle, can be determined by extended observations at the site. They can be approximated with reasonable conservatism on the basis of a few local observations and extended study of more general data. The choice of the wind conditions for which a given bridge should be designed may always be largely a matter of judgment.

At the start of aerodynamic analysis, the size and shape of the bridge are known. Its energy-storage capacity and its motion, consisting essentially of natural modes of vibration, are determined completely by its mass, mass distribution, and elastic properties and can be computed by reliable methods.

The only unknown element is that factor relating the wind to the bridge section and its motion. This factor cannot, at present, be generalized but is subject to reliable determination in each case. Properties of the bridge, including its elastic forces and its mass and motions (determining its inertial forces), can be computed and reduced to model scale. Then, wind conditions bracketing all probable conditions at the site can be imposed on a section model. The motions of such a dynamic section model in the properly scaled wind should duplicate reliably the motions of a convenient unit length of the bridge. The wind forces and the rate at which they can build up energy of oscillation respond to the changing amplitude of the motion. The rate of energy change can be measured and plotted against amplitude. Thus, the section-model test measures the one unknown factor, which can then be applied by calculation to the variable amplitude of motion along the bridge to predict the full behavior of the structure under the specific wind conditions of the test. These predictions are not precise but are about as accurate as some other features of the structural analysis.

14.16.2   Criteria for Aerodynamic Design

Because the factor relating bridge movement to wind conditions depends on specific site and bridge conditions, detailed criteria for the design of favorable bridge sections cannot be written until a large mass of data applicable to the structure being designed has been accumulated. But, in general, the following criteria for suspension bridges may be used:

- A truss-stiffened section is more favorable than a girder-stiffened section.
- Deck slots and other devices that tend to break up the uniformity of wind action are likely to be favorable.
- The use of two planes of lateral system to form a four-sided stiffening truss is desirable because it can favorably affect torsional motion. Such a design strongly inhibits flutter and also raises the critical velocity of a pure torsional motion.
- For a given bridge section, a high natural frequency of vibration is usually favorable:

    For short to moderate spans, a useful increase in frequency, if needed, can be attained by increased truss stiffness. (Although not closely defined, moderate spans may be regarded as including lengths from about 1,000 to about 1,800 ft.)

    For long spans, it is not economically feasible to obtain any material increase in natural frequency of vertical modes above that inherent in the span and sag of the cable.

    The possibility should be considered that for longer spans in the future, with their unavoidably low natural frequencies, oscillations due to unfavorable aerodynamic characteristics of the cross section may be more prevalent than for bridges of moderate span.

- At most bridge sites, the wind may be broken up; that is, it may be nonuniform across the site, unsteady, and turbulent. So a condition that could cause serious oscillation does not continue long enough to build up an objectionable amplitude. However, bear in mind:

    There are undoubtedly sites where the winds from some directions are unusually steady and uniform.

There are bridge sections on which any wind, over a wide range of velocity, will continue to build up some mode of oscillation.

- An increase in stiffness arising from increased weight increases the energy-storage capacity of the structure without increasing the rate at which the wind can contribute energy. The effect is an increase in the time required to build up an objectionable amplitude. This may have a beneficial effect much greater than is suggested by the percentage increase in weight, because of the sharply reduced probability that the wind will continue unchanged for the greater length of time. Increased stiffness may give added structural damping and other favorable results.

Although more specific design criteria than the above cannot be given, it is possible to design a suspension bridge with a high degree of security against aerodynamic forces. This involves calculation of natural modes of motion of the proposed structure, performance of dynamic-section-model tests to determine the factors affecting behavior, and application of these factors to the prototype by suitable analysis.

Most long-span bridges built since the Tacoma bridge failure have followed the above procedures and incorporated special provisions in the design for aerodynamic effects. Designers of these bridges usually have favored stiffening trusses over girders. The second Tacoma Narrows, Forth Road, and Mackinac Straits Bridges, for example, incorporate deep stiffening trusses with both top and bottom bracing, constituting a torsion space truss. The Forth Road and Mackinac Straits Bridges have slotted decks. The Severn Bridge, however, has a streamlined, closed-box stiffening girder and inclined suspenders. Some designs incorporate longitudinal cable stays, tower stays, or even transverse diagonal stays (Deer Isle Bridge). Some have unloaded backstays. Others endeavor to increase structural damping by frictional or viscous means. *All have included dynamic-model studies as part of the design.*

14.16.3   Wind-Induced Oscillation Theories

Several theories have been advanced as models for mathematical analysis to develop an understanding of the process of wind excitation. Among these are the following.

Negative-Slope Theory.   When a bridge is moving downward while a horizontal wind is blowing (Fig. 14.61a), the resultant wind is angled upward (positive angle of attack) relative to the bridge. If the lift coefficient C_L, as measured in static tests, shows a variation with wind angle α such as that illustrated by curve A in Fig. 14.61b, then, for moderate amplitudes, there is a wind force acting downward

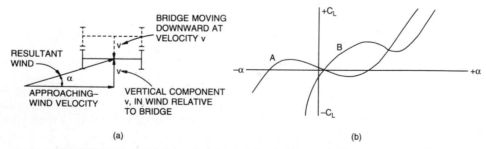

FIGURE 14.61   Wind action on a cable-stayed bridge. (*a*) Downward bridge motion develops upward wind component. (*b*) Lift coefficient C_L depends on angle of attack α of the wind.

on the bridge while the bridge is moving downward. The bridge will therefore move to a greater amplitude than it would without this wind force. The motion will, however, be halted and reversed by the action of the elastic forces. Then, the vertical component of the wind also reverses. The angle of attack becomes negative, and the lift becomes positive, tending to increase the amplitude of the rebound. With increasing velocity, the amplitude will increase indefinitely or until the bridge is destroyed. A similar, though more complicated situation, would apply for torsional or twisting motion of the bridge.

Vortex Theory.   This attributes aerodynamic excitation to the action of periodic forces having a certain degree of resonance with a natural mode of vibration of the bridge. Vortices, which form around the trailing edge of the airfoil (bridge deck), are shed on alternating sides, giving rise to periodic forces and oscillations transverse to the deck.

Flutter Theory.   The phenomenon of flutter, as developed for airfoils of aircraft and applied to suspension-bridge decks, relates to the fact that the airfoil (bridge deck) is supported so that it can move elastically in a vertical direction and in torsion, about a longitudinal axis. Wind causes a lift that acts eccentrically. This causes a twisting moment, which, in turn, alters the angle of attack and increases the lift. The chain reaction becomes catastrophic if the vertical and torsional motions can take place at the same coupled frequency and in appropriate phase relation.

F. Bleich presented tables for calculation of flutter speed v_F for a given bridge, based on flat-plate airfoil flutter theory. These tables are applicable principally to trusses. But the tables are difficult to apply, and there is some uncertainty as to their range of validity.

A. Selberg has presented the following formula for flutter speed:

$$v_F = 0.88\omega_2 b \sqrt{\left[1 - \left(\frac{\omega_1}{\omega_2}\right)^2\right] \frac{\sqrt{\nu}}{\mu}} \qquad (14.57)$$

where ν = mass distribution factor for specific section = $2r^2/b^2$ (varies between 0.6 and 1.5, averaging about 1)

μ = $2\pi\rho b^2/m$ (ranges between 0.01 and 0.12)

m = mass per unit length

b = half width of bridge

ρ = mass density of air

ω_1 = circular vertical frequency

ω_2 = circular torsional frequency

r = mass radius of gyration

Selberg has also published charts, based on tests, from which it is possible to approximate the critical wind speed for any type of cross section in terms of the flutter speed.

Applicability of Theories.   The vortex and flutter theories apply to the behavior of suspension bridges under wind action. Flutter appears dominant for truss-stiffened bridges, whereas vortex action seems to prevail for girder-stiffened bridges. There are mounting indications, however, these are, at best, estimates of aerodynamic behavior. Much work has been done and is being done, particularly in the spectrum approach and the effects of nonuniform, turbulent winds.

14.16.4 Design Indices

Bridge engineers have suggested several criteria for practical design purposes. O. H. Ammann, for example, developed two analytical-empirical indices that were applied in the design of the Verrazano Narrows Bridge, a vertical-stiffness index and a torsional-stiffness index.

Vertical-Stiffness Index S_v. This is based on the magnitude of the vertical deflection of the suspension system under a static downward load covering one-half the center span. The index includes a correction to allow for the effect of structural damping of the suspended structure and for the effect of different ratios of side span to center span.

$$S_v = \left(8.2 \frac{W}{f} + 0.14 \frac{I}{L^4}\right)\left(1 - 0.6 \frac{L_1}{L}\right) \tag{14.58}$$

where W = weight of bridge, lb per lin ft
$\quad f$ = cable sag, ft
$\quad I$ = moment of inertia of stiffening trusses and continuous stringers, in^2 by ft^2
$\quad L$ = length of center span, in thousands of feet
$\quad L_1$ = length of side span, in thousands of feet

Torsional-Stiffness Index S_t. This is defined as the maximum intensity of sinusoidal loads, of opposite sign in opposite planes of cables, on the center span and producing 1-ft deflections at quarter points of the main span. This motion simulates deformations similar to those in the first asymmetric mode of torsional oscillations.

$$S_t = \left(\frac{B}{A} + 1\right) \frac{\pi^2}{4} \frac{W}{f} \tag{14.59}$$

where $A = \dfrac{b^2}{2} \dfrac{H_w}{E}$

$\quad B = \dfrac{2bd(A_vU_vA_hU_h)}{(b/d)A_vU_v + (d/b)A_hU_h}$

$\quad W$ = weight of bridge, lb per lin ft
$\quad f$ = cable sag, ft
$\quad H_w$ = horizontal component of cable load due to dead load (half bridge), kips
$\quad b$ = distance between centerlines of cables, or centerlines of pairs of cables, ft
$\quad d$ = vertical distance between top and bottom planes of lateral bracing, ft
$\quad E$ = modulus of elasticity of truss steel, ksf
$\quad A_v$ = area of the diagonals in one panel of vertical truss, ft^2
$\quad A_h$ = area of diagonals in one panel of horizontal lateral bracing (two members for X or K bracing), ft^2
$\quad U_v$ = sin^2 γ_v cos γ_v
$\quad U_h$ = sin^2 γ_h cos γ_h
$\quad \gamma_h$ = angle between diagonals and chord of horizontal truss
$\quad \gamma_v$ = angle between diagonals and chord of vertical truss

Typical values of these indices are listed in Table 14.13 for several bridges.

Other indices and criteria have been published by D. B. Steinman. In connection with these, Steinman also proposed that, unless aerodynamic stability is otherwise

TABLE 14.13 Stiffness Indices and First Asymmetric Mode Natural Frequencies*

Bridge	Structural parameters									Vertical motions		Torsional motions	
	L, ft	L_1, ft	f, ft	W, lb per ft	I, in² ft²	b, ft	d, ft	A, ft⁴	B, ft⁴	Stiffness index	Frequency, cycles per min	Stiffness index	Frequency, cycles per min
Verrazano Narrows Bridge	4,260	1,215	390	36,650	180,000	101.25	24	130.8	144.5	702	6.2	448	11.9
George Washington Bridge, 8-lane single deck complete	3,500	650	319	28,570	168	106	...	...	...	654	6.7	221	8.2
George Washington Bridge, 14-lane double deck complete	3,500	650	326	40,000	66,000	106	30	126.5	163.7	950	6.7	694	13.2
Golden Gate Bridge with upper lateral system only	4,200	1,125	475	21,300	88,000	90	...	...	...	342	5.6	111	7.0
Golden Gate Bridge with double lateral system	4,200	1,125	475	22,800+	88,000	90	25	51.3	75.5	364	5.6	292	11.0
Tacoma Narrows original with 2-lane single deck (very unfavorable aerodynamic characteristics)	2,800	1,100	232	5,700	2,567	39	...	...	...	158	8.0	61	10.0

*From M. Brunner, H. Rothman, M. Fiegen, and B. Forsyth, "Verrazano-Narrows Bridge: Design of Superstructure," *Journal of the Construction Division*, vol. 92, no. CO2, March 1966, American Society of Civil Engineers.

assured, the depth, ft, of stiffening girders and stiffening trusses should be at least $L/120 + (L/1,000)^2$, where L is the span, ft. Furthermore, EI of the stiffening system should be at least $bL^4/120\sqrt{f}$, where b is the width, ft, of the bridge and f the cable sag, ft.

14.16.5 Natural Frequencies of Suspension Bridges

Dynamic analyses require knowledge of the natural frequencies of free vibration, modes of motion, energy-storage relationships, magnitude and effects of damping, and other factors.

Two types of vibration must be considered: bending and torsion.

Bending. The fundamental differential equation [Eq. (14.22)] and cable condition [Eq. (14.26)] of the suspension bridge in Fig. 14.41 can be transformed into

$$EI\eta'''' - H\eta'' = \omega^2 m\eta + H_p y'' \qquad (14.60)$$

$$\frac{H_p L_c}{E_c A_c} + y'' \int_0^L \eta \, dx = 0 \qquad (14.61)$$

where ω = circular natural frequency of the bridge
η = deflection of stiffening truss or girder
m = bridge mass = w/g
y = vertical distance from cable to the line through the pylon supports
w = dead load, lb per lin ft
g = acceleration due to gravity = 32.2 ft/s²

From these equations, the basic Rayleigh energy equation for bending vibrations can be derived:

$$\int EI\eta''^2 \, dx + H \int \eta'^2 \, dx + \frac{E_c A_c}{L_c} \left(y'' \int \eta \, dx \right)^2 = \omega^2 \int m\eta^2 \, dx \qquad (14.62)$$

Symbols are defined in "Torsion," following. After ω has been determined from this, the natural frequency of the bridge $\omega/2\pi$, Hz, can be computed.

Torsion. The Rayleigh energy equations for torsion are

$$EC_s \int \phi''^2 \, dx + \left(GI_T + \frac{b^2 H}{2} \right) \int \phi'^2 \, dx + \frac{E_c A_c}{2L_c} \left(y''b \int \phi \, dx \right)^2$$

$$+ EI_y y_M \int \eta''\phi'' \, dx = \omega^2 I_p \int \phi^2 \, dx \qquad (14.63)$$

$$EI_y y_M \int \phi''\eta'' \, dx + EI_y \int \eta''^2 \, dx = \omega^2 m \int \eta^2 \, dx \qquad (14.64)$$

where ϕ = angle of twist, radians
E = modulus of elasticity of stiffening girder, ksf
G = modulus of rigidity of stiffening girder, ksf
I_T = polar moment of inertia of stiffening girder cross section, ft⁴
I_p = mass moment of inertia of stiffening girder per unit of length, kips-sec²
I_y = moment of inertia of stiffening girder about its vertical axis, ft⁴
C_s = warping resistance of stiffening girder relative to its center of gravity, ft⁶

> b = horizontal distance between cables, ft
> H = horizontal component of cable tension, kips
> A_c = cross-sectional area of cable, ft^2
> E_c = modulus of elasticity of cable, ksf
> L_c = $\int \sec^3 \alpha \, dx$
> α = angle cable makes with horizontal, radians
> y_M = ordinate of center of twist relative to the center of gravity of stiffening girder cross section, ft
> ω = circular frequency, radians per sec
> m = $m(x)$ = mass of stiffening girder per unit of length, kips-sec^2/ft^2

Solution of these equations for the natural frequencies and modes of motion is dependent on the various possible static forms of suspension bridges involved (see Fig. 14.9). Numerous lengthy tabulations of solutions have been published.

14.16.6 Damping

Damping is of great importance in lessening of wind effects. It is responsible for dissipation of energy imparted to a vibrating structure by exciting forces. When damping occurs, one part of the external energy is transformed into molecular energy, and another part is transmitted to surrounding objects or the atmosphere. Damping may be internal, due to elastic hysteresis of the material or plastic yielding and friction in joints, or Coulomb (dry friction), or atmospheric, due to air resistance.

14.16.7 Aerodynamics of Cable-Stayed Bridges

The aerodynamic action of cable-stayed bridges is less severe than that of suspension bridges, because of increased stiffness due to the taut cables and the widespread use of torsion box decks.

14.16.8 Stability Investigations

It is most important to note that the validation of stability of the completed structure for expected wind speeds at the site is mandatory. However, this does not necessarily imply that the most critical stability condition of the structure occurs when the structure is fully completed. A more dangerous condition may occur during erection, when the joints have not been fully connected and, therefore, full stiffness of the structure has not yet been realized. In the erection stage, the frequencies are lower than in the final condition and the ratio of torsional frequency to flexural frequency may approach unity. Various stages of the partly erected structure may be more critical than the completed bridge. The use of welded components in pylons has contributed to their susceptibility to vibration during erection.

Because no exact analytical procedures are yet available, wind-tunnel tests should be used to evaluate the aerodynamic characteristics of the cross section of a proposed deck girder, pylon, or total bridge. More importantly, the wind-tunnel tests should be used during the design process to evaluate the performance of a number of proposed cross sections for a particular project. In this manner, the wind-tunnel investigations become a part of the design decision process and not a postconstruction corrective action. If the wind-tunnel evaluations are used as an after-the-fact verification and they indicate an instability, there is the distinct risk that a redesign of a retrofit design will be required that will have undesirable ramifications on schedules and availability of funding.

(F. Bleich and L. W. Teller, "Structural Damping in Suspension Bridges," *ASCE Transactions*, vol. 117, pp. 165–203, 1952.

F. Bleich, C. B. McCullough, R. Rosecrans, and G. S. Vincent, "The Mathematical Theory of Vibration of Suspension Bridges," Bureau of Public Roads, Government Printing Office, Washington, D.C.

F. B. Farquharson, "Wind Forces on Structures Subject to Oscillation," *ASCE Proceedings*, ST4, July, 1958.

A Selberg, "Oscillation and Aerodynamic Stability of Suspension Bridges," *Acta Polytechnia Scandinavica*, Civil Engineering and Construction Series 13, Trondheim, 1961.

D. B. Steinman, "Modes and Natural Frequencies of Suspension Bridge Oscillations," *Transactions Engineering Institute of Canada*, vol. 3, no. 2, pp. 74–83, 1959.

D. B. Steinman, "Aerodynamic Theory of Bridge Oscillations," *ASCE Transactions*, vol. 115, pp. 1180–1260, 1950.

D. B. Steinman, "Rigidity and Aerodynamic Stability of Suspension Bridges," *ASCE Transactions*, vol. 110, pp. 439–580, 1945.

"Aerodynamic Stability of Suspension Bridges," 1952 Report of the Advisory Board on the Investigation of Suspension Bridges, *ASCE Transactions*, vol. 120, pp. 721–781, 1955.)

R. L. Wardlaw, "A Review of the Aerodynamics of Bridge Road Decks and the Role of Wind Tunnel Investigation," U. S. Department of Transportation, Federal Highway Administration, Report No. FHWA-RD-72-76.

A. G. Davenport, "Buffeting of a Suspension Bridge by Storm Winds," *ASCE Journal of the Structural Division*, vol. 115, ST3, June, 1962.

"Guidelines for Design of Cable-Stayed Bridges," ASCE Committee on Cable-Stayed Bridges.

W. Podolony, Jr., and J. B. Scalzi, "Construction and Design of Cable-Stayed Bridges," 2d ed., John Wiley & Sons, Inc., New York.

E. Murakami and T. Okubu, "Wind-Resistant Design of a Cable-stayed Bridge," International Association for Bridge and Structural Engineering, Final Report, 8th Congress, New York, September 9–14, 1968.)

14.17 *SEISMIC ANALYSIS OF CABLE-SUSPENDED STRUCTURES*

For short-span structures (under about 500 ft) it is commonly assumed in seismic analysis that the same ground motion acts simultaneously throughout the length of the structure. In other words, the wavelength of the ground waves are long in comparison to the length of the structure. In long-span structures, such as suspension or cable-stayed bridges, however, the structure could be subjected to different motions at each of its foundations. Hence, in assessment of the dynamic response of long structures, the effects of traveling seismic waves should be considered. Seismic disturbances of piers and anchorages may be different at one end of a long bridge than at the other. The character or quality of two or more inputs into the total structure, their similarities, differences, and phasings, should be evaluated in dynamic studies of the bridge response.

Vibrations of cable-stayed bridges, unlike those of suspension bridges, are susceptible to a unique class of vibration problems. Cable-stayed bridge vibrations cannot be categorized as vertical (bending), lateral (sway), and torsional; almost every mode of vibration is instead a three-dimensional motion. Vertical vibrations, for example, are introduced by both longitudinal and lateral shaking in addition to vertical excitation. In addition, an understanding is needed of the multimodal contribution to the final response of the structure and in providing representative values of the response quantities. Also, because of the long spans of such structures, it is necessary to formulate a dynamic response analysis resulting from the multi-support excitation. A three-dimensional analysis of the whole structure and substructure to obtain the natural frequencies and seismic response is advisable. A qualified specialist should be consulted to evaluate the earthquake response of the structure.

("Guide Specifications for Seismic Design of Highway Bridges," American Association of State Highway and Transportation Officials; "Guidelines for the Design of Cable-Stayed Bridges," ASCE Committee on Cable-Stayed Bridges.

A. M. Abdel-Ghaffar, and L. I. Rubin, "Multiple-Support Excitations of Suspension Bridges," *Journal of the Engineering Mechanics Division*, ASCE, vol. 108, no. EM2, April, 1982.

A. M. Abdel-Ghaffar, and L. I. Rubin, "Vertical Seismic Behavior of Suspension Bridges," *The International Journal of Earthquake Engineering and Structural Dynamics*, vol. 11, January–February, 1983.

A. M. Abdel-Ghaffar, and L. I. Rubin, *"Lateral Earthquake Response of Suspension Bridges," Journal of the Structural Division*, ASCE, vol. 109, no. ST3, March, 1983.

A. M. Abdel-Ghaffar, and J. D. Rood, "Simplified Earthquake Analysis of Suspension Bridge Towers," *Journal of the Engineering Mechanics Division*, ASCE, vol. 108, no. EM2, April, 1982.)

14.18 *ERECTION OF CABLE-SUSPENDED BRIDGES*

The ease of erection of suspension bridges is a major factor in their use for long spans. Once the main cables are in position, they furnish a stable working base or platform from which the deck and stiffening truss sections can be raised from floating barges or other equipment below, without the need for auxiliary falsework. For the Severn Bridge, for example, 60-ft box-girder deck sections were floated to the site and lifted by equipment supported on the cables.

Until the 1960s, the field process of laying the main cables had been by spinning (Art. 14.7.3). (This term is actually a misnomer, for the wires are neither twisted nor braided, but are laid parallel to and against each other.) The procedure (Fig. 14.62) starts with the hanging of a catwalk at each cable location for use in construction of the bridge. An overhead cableway is then installed above each catwalk. Loops of wire (two or four at a time) are carried over the span on a set of grooved spinning wheels. These are hung from an endless hauling rope of the cableway until arrival at the far anchorage. There, the loops are pulled off the spinning wheels manually and placed around a semicircular strand shoe, which connects them by an eyebar or bolt linkage to the anchorage (Fig. 14.28). The wheels then start back to the originating anchorage. At the same time, another set of wheels carrying wires starts out from that anchorage. The loops of wire on the latter set of wheels are also placed manually around a strand shoe at their anchorage destination. Spinning proceeds as the wheels shuttle back and forth across the span. A system of counterweights keeps the wires under continuous tension as they are spun.

The wires that come off the bottom of the wheels (called dead wires) and that are held back by the originating anchorage are laid on the catwalk in the spinning

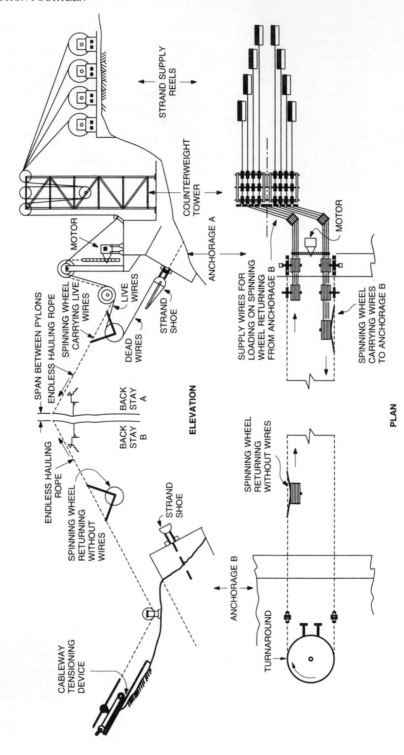

FIGURE 14.62 Scheme for spinning four wires at a time for the cables of the Forth Road Bridge.

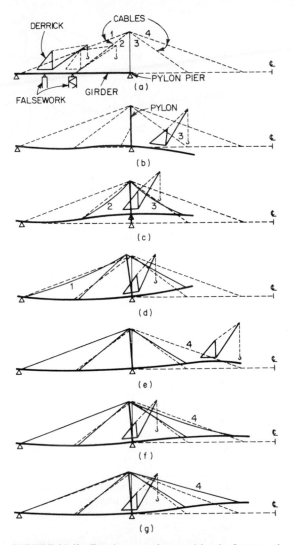

FIGURE 14.63 Erection procedure used for the Strömsund Bridge. (*a*) Girder, supported on falsework, is extended to the pylon pier. (*b*) Girder is cantilevered to the connection of cable 3. (*c*) Derrick is retracted to the pier and the girder is raised, to permit attachment of cables 2 and 3 to the girder. (*d*) Girder is reseated on the pier and cable 1 is attached. (*e*) Girder is cantilevered to the connection of cable 4. (*f*) Derrick is retracted to the pier and cable 4 is connected. (*g*) Preliminary stress is applied to cable 4. (*h*) Girder is cantilevered to midspan and spliced to its other half. (*i*) Cable 4 is given its final stress. (*j*) The roadway is paved, and the bridge takes its final position. (*Reprinted with permission from H. J. Ernst, "Montage Eines Seilverspannten Balkens im Grossbrucken-bau," Stahlbau, vol. 25, no. 5, May 1956.*)

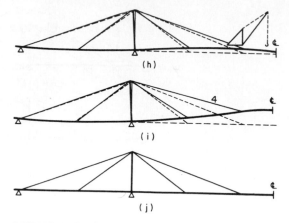

FIGURE 14.63   (*Continued*)

process. The wires passing over the wheels from the unreelers and moving at twice the speed of the wheels, are called *live wires*.

As the wheels pass each group of wire handlers on the catwalks, the dead wires are temporarily clipped down. The live wires pass through small sheaves to keep them in correct order. Each wire is adjusted for level in the main and side spans with come-along winches, to ensure that all wires will have the same sag.

The cable is made of many strands, usually with hundreds of wires per strand (Art. 14.7). All wires from one strand are connected to the same shoe at each anchorage. Thus, there are as many anchorage shoes as strands. At saddles and anchorages, the strands maintain their identity, but throughout the rest of their length, the wires are compacted together by special machines. The cable usually is forced into a circular cross section of tightly bunched parallel wires.

The usual order of erection of suspension bridges is substructure, pylons and anchorages, catwalks, cables, suspenders, stiffening trusses, floor system, cable wrapping, and paving.

Cables are usually coated with a protective compound. The main cables are wrapped with wire by special machines, which apply tension, pack the turns tightly against one another, and at the same time advance along the cable. Several coats of protective material, such as paint, are then applied. For alternative wrapping, see Art. 14.9.

Typical cable bands are illustrated in Figs. 14.34 and 14.35. These are usually made of paired, semicylindrical steel castings with clamping bolts, over which the wire-rope or strand suspenders are looped or attached by socket fittings.

Cable-stayed structures are ideally suited for erection by cantilevering into the main span from the piers. Theoretically, erection could be simplified by having temporary erection hinges at the points of cable attachment to the girder, rendering the system statically determinate, then making these hinges continuous after dead load has been applied. The practical implementation of this is difficult, however, because the axial forces in the girder are large and would have to be concentrated in the hinges. Therefore, construction usually follows conventional tactics of cantilevering the girder continuously and adjusting the cables as necessary to meet the required geometrical and statical constraints. A typical erection sequence is illustrated in Fig. 14.63.

Erection should meet the requirements that, on completion, the girder should follow a prescribed gradient; the cables and pylons should have their true system lengths; the pylons should be vertical, and all movable bearings should be in a neutral position. To accomplish this, all members, before erection, must have a deformed shape the same as, but opposite in direction to, that which they would have under dead load. The girder is accordingly cambered, and also lengthened by the amount of its axial shortening under dead load. The pylons and cable are treated in similar manner.

Erection operations are aided by raising or lowering supports or saddles, to introduce prestress as required. All erection operations should be so planned that the stresses during the erection operations do not exceed those due to dead and live load when the structure is completed; otherwise loss of economy will result.

INDEX